AF327326

Handbook of Self Assembled Semiconductor Nanostructures for Novel Devices in Photonics and Electronics

Handbook of Self Assembled Semiconductor Nanostructures for Novel Devices in Photonics and Electronics

Edited by

Mohamed Henini

ELSEVIER

Amsterdam • Boston • Heidelberg • London • New York • Oxford • Paris
San Diego • San Francisco • Singapore • Sydney • Tokyo

Elsevier
The Boulevard, Langford Lane, Kidlington, Oxford OX5 1GB, UH
Radarweg 29, PO Box 211, 1000 AE Amsterdam, The Netherlands

First edition 2008

British Library Cataloguing in Publication Data
A catalogue record for this book is available from the British Library

Library of Congress Cataloging-in-Publication Data
A catalog record for this book is available from the Library of Congress

ISBN: 978-0-08-046325-4

For information on all Elsevier publications
visit our website at books.elsevier.com

Typeset by Charon Tec Ltd., A Macmillan Company. (www.macmillansolutions.com)

Printed and bound in Great Britain

08 09 10 11   10 9 8 7 6 5 4 3 2 1

Working together to grow
libraries in developing countries

www.elsevier.com | www.bookaid.org | www.sabre.org

ELSEVIER   BOOK AID International   Sabre Foundation

Contents

Chapter 26   Semiconductor Quantum Dots for Biological Applications   773

Chapter 27   Quantum Dot Modification and Cytotoxicity   799

Preface

In 1969, Leo Esaki (1973 Nobel Laureate) and Ray Tsu from IBM, USA, proposed research on "man-made crystals" using a semiconductor superlattice (a semiconductor structure comprising several alternating ultra-thin layers of semiconductor materials with different properties). This invention was perhaps the first proposal to advocate the engineering of a new semiconductor material, and triggered a wide spectrum of experimental and theoretical investigations. However, the study of what are now called low dimensional structures (LDS) began in the late 1970s when sufficiently thin epitaxial layers were first produced following developments in the technology of epitaxial growth of semiconductors, mainly pioneered in industrial laboratories for device purposes.

The LDS are materials structures whose dimensions are comparable with interatomic distances in solids (i.e. nanometre, nm). Their electronic properties are significantly different from the same material in bulk form. These properties are changed by quantum effects. At the inception of their investigation it was already clear that such structures were of great scientific interest and excitement and their novel properties caused by quantum effects offered potential for application in new devices. Moreover, these complex LDS offer device engineers new design opportunities for tailor-made new generation electronic devices. The LDS could be considered as a new branch of condensed matter physics because of the large variety of possible structures and the changes in the physical processes.

One of the promising fabrication methods to produce and study structures with a dimension less than one such as quantum dots, in order to realize novel devices that make use of low-dimensional confinement effects, is self-organization. The quantum dots are often referred to as *artificial atoms* due to their small size (typically $\sim 10\,\text{nm}$). The self-assembling mechanism allows the formation of tens of billions of dots per cm^2 with a high degree of uniformity in a single-step process. Self-assembled nanostructured materials offer a number of advantages over conventional material technologies in a widerange of sectors. The research on self-assembled nanostructures involves a strong interdisciplinary combination, for example material science, physics, chemistry, biology, electronics and optoelectronics.

Twenty-eight chapters is not a lot in which to convey the current successes of modern self-organization research. However, the topics covered will give the reader a comprehensive overview of the field of self-assembled nanostructures and a better evaluation of future trends.

The two parts of this book deal mainly with two fabrication methods, namely epitaxial and colloidal techniques that make feasible the design of artificial nanostructures. Each part introduces subjects on properties, characterization and applications of self-assembled nanostructures. We trust that this book will provide an excellent introduction to the self-organization of nanostructures to a large number of researchers and scientists active in the nanostructures science and technology area, and will fill an important gap in the market.

I would like to record my thanks to all the authors who have done so much hard work to achieve this successful book and acknowledge the invaluable help of the staff at Elsevier. Their efforts are greatly appreciated.

Finally, I would like to thank my family for their tolerance and understanding during the long hours and many mood swings accompanying my efforts as editor of this book.

Mohamed Henini

School of Physics and Astronomy,
University of Nottingham, Nottingham NG7 2RD, UK

CHAPTER 1

Self-organized Quantum Dot Multilayer Structures

Gunther Springholz

*Institut für Halbleiter- und Festkörperphysik, Johannes Kepler
Universität Linz, A-4040 Linz, Austria*

1.1 Introduction

Strained-layer heteroepitaxy is a powerful tool for the fabrication of self-assembled semiconductor nanostructures [1–4]. It is based on the tendency of strained layers to spontaneously form self-organized corrugated surface structures during growth. This process can be driven by energetic as well as by kinetic effects such as strain-induced coherent islanding, faceting, kinetic step-bunching or step-meandering. One particularly effective process is the formation of three-dimensional nanoislands in the Stranski–Krastanow (SK) growth mode [4–7]. This islanding is driven by the highly efficient elastic strain relaxation within the islands arising from their lateral elastic expansion or compression in the directions of their free side faces [4–11]. For islands larger than a critical size, the relaxed elastic energy outweighs the corresponding increase in surface energy, leading to an effective lowering of the total energy of the system [4–10]. Therefore, this island formation occurs in a large number of high-misfit heteroepitaxial material systems. When the islands are subsequently embedded in a higher energy band gap matrix material, the electronic carriers are spatially confined and thus, self-assembled quantum dots with atomic-like optical and electronic properties are formed [1–3, 12–14]. Because these epitaxial quantum dots are essentially defect free, their electronic properties are often superior to those fabricated by conventional lithographic processing and etching techniques.

Owing to the statistical nature of growth, ensembles of self-assembled dots exhibit considerable variations in size and sometimes even shape. This results in a large inhomogeneous broadening of the energy levels and optical transitions [1–3, 12]. In addition, there is little control over the lateral arrangement and position of the nanoislands. Both factors pose considerable limitations for practical device applications. Three-dimensional stacking of self-assembled quantum dots in multilayer or superlattice structures has turned out to be one of the most effective means for controlling the vertical and lateral arrangement of the dots. Apart from the possible improvements in uniformity [15–21], this also increases the total amount of active material in quantum dot devices and allows the tuning of electronic wave functions by quantum mechanical coupling across the spacer layers [22–24]. In this way, quantum dot molecules [23–26] have been obtained, which are of particular interest for solid-state quantum computation devices [26, 27].

Multilayering has been found to yield significant improvements in the uniformity of dots [17, 18, 28–31] and may even result in the formation of ordered quantum dot superstructures [18, 28]. Experimentally, different types of vertical and lateral dot correlations have been observed in multilayer structures, ranging from vertically aligned dot columns in Ge/Si [22, 32–41] or InAs/GaAs [15, 16, 31, 42–45] dot superlattices, to vertical anti-correlations for

II–VI [46–48] and III–V superlattices [60–65, 87, 88], oblique replication on high-indexed surfaces [142] as well as *fcc*-like dot stackings for IV–VI semiconductor multilayers [18, 30, 49–57]. The actual type of correlation has been found to depend on a large number of parameters such as the spacer thickness [15, 35, 36, 45, 47, 49–57, 87, 88], dot size [45, 55, 56], elastic material properties [18, 47, 58, 59], surface orientation [58], growth conditions [45, 56] as well as chemical composition of dots and spacer layers [60–66]. In addition, it has been found that lateral dot ordering is particularly effective for multilayers with staggered interlayer dot correlations [18, 30, 50, 57, 58, 60].

In this chapter, the mechanisms for vertical and lateral ordering in quantum dot multilayer structures and the resulting different dot stackings are described. The particular emphasis is on the prototype Si/Ge, InAs/GaAs and PbSe/PbEuTe material systems in which extensive studies have been carried out. The chapter is organized as follows: first, a brief general overview of the different possible interaction mechanisms and various dot stackings for different material systems is given in Section 1.2. In Section 1.3, the elastic interactions between buried and surface dots are treated in detail, introducing the far-field and near-field approximations and analysing the effect of the elastic anisotropy and growth orientation. It is shown that the elastic anisotropy plays a crucial role in the ordering process and is responsible for the formation of staggered dot stackings. In Section 1.4, the derived theoretical predictions are compared with experimental results, and growth simulations of the stacking and ordering processes are described in Section 1.5. Sections 1.6 to 1.8 treat various aspects for the three well-studied InAs/GaAs, Si/Ge and PbSe/PbEuTe multilayer systems, including the correlation lengths, as well as stacking and ordering transitions as a function of spacer thickness and growth conditions. In section 1.9, other interaction mechanisms such as surface morphology, surface segregation and alloy decomposition are discussed, and a brief summary and outlook are given in the final section.

1.2   Mechanisms for interlayer correlation formation

The formation of interlayer correlations in multilayer structures obviously requires some kind of mechanism through which the dots in the buried layers influence the dot growth in the subsequent layers. By this mechanism, the information on the dot positions is conveyed from one layer to another and thus vertical and lateral correlations are formed. Conceptually, one can think of three kinds of such mechanisms, namely, (i) chemical processes such as surface segregation, (ii) non-planarized corrugated surface morphologies, or (iii) long-range elastic interactions due to the strain fields emerging from the buried dots. These mechanisms are illustrated schematically in Fig. 1.1 and may produce different kinds of correlations in dependence of the interaction process.

Nucleation of Stranski–Krastanow dots is a rather complex process that sensitively depends on parameters such as surface stress, lattice mismatch, thickness and composition of the wetting layer, free energies and local curvature of the growth surface, surface step structure as well as surface kinetics during growth. On a planar and chemically uniform substrate, these parameters are invariant across the surface and thus homogeneous dot nucleation at random surface sites occurs. In multilayer structures, however, the buried dots beneath the surface induce significant variations of strain, topography and/or chemical composition of the surface, which may lead to preferential nucleation at particular surface sites that are linked to the position of the buried dots. This preferential nucleation can be induced, e.g. by a local enhancement of the growth rate, local changes in surface diffusivity, as well as by local decreases in the critical wetting layer thickness or energy barrier for island nucleation and results in spatial correlations between surface and subsurface dots. Apart from this, the existence of interlayer interactions is also manifested by the significant changes in dot size [19, 28, 29, 31, 36, 40, 43, 51, 67, 68, 69, 70, 71, 72, 73], density [17, 19, 28, 31, 43, 51, 67], lateral arrangement [18, 19, 28, 30, 31, 51, 74], shape [34, 51, 67–72, 73], and critical thickness for dot nucleation [39, 40, 75] observed in many experiments. Moreover, different interlayer correlations may be formed, depending on the details of the interaction mechanisms and growth conditions, as is illustrated schematically in Fig. 1.1.

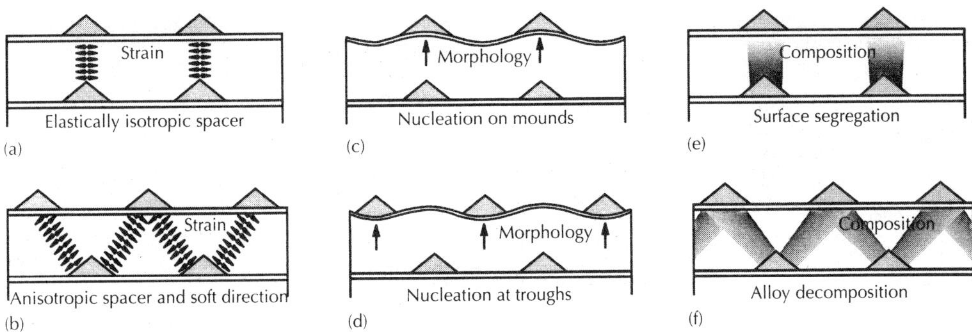

Figure 1.1   Possible mechanisms for formation of interlayer correlations in self-assembled quantum dot multilayers. Left-hand side: Interlayer correlations caused by the elastic strain fields emerging from subsurface dots and subsequent dot nucleation at the minima of strain on the epilayer surface. Depending on the elastic properties of the spacer layer as well as surface orientation, these minima may be localized (a) above or (b) between the buried dots, which will give rise to either a vertical dot alignment or a staggered dot stacking, respectively. Centre: Interlayer correlations caused by non-planar surface morphologies resulting from incomplete surface planarization during dot overgrowth. Depending on the dominant mechanism of capillary or stress-driven surface mass transport, subsequent dots may nucleate either on top of the mounds (c) or in the troughs in between (d), again giving rise to different types of interlayer correlations. Right-hand side: Correlated nucleation induced by non-uniformities in the chemical composition of the spacer layer due to (e) surface segregation or (f) alloy decomposition. Clearly, all three mechanisms may induce vertical dot alignments or staggered dot stackings in multilayer structures.

As already mentioned above, interlayer correlation may be caused by three main mechanisms, namely, (i) elastic lattice deformations around the buried dots due to the dot/substrate lattice mismatch [15, 17, 18, 29, 58], (ii) corrugations in surface topography due to incomplete surface planarization [76], and (iii) surface segregation or alloy decomposition within the spacer layer that often occurs in highly strained material systems [62–66]. In the case of the interaction via the elastic strain fields (Fig. 1.1a and b), preferential dot nucleation is generally induced at the strain minima on the wetting layer surface. Depending on the elastic properties of the materials, these minima may be on top (a) or *between* the buried dots (b), resulting in a vertical or staggered dot stacking, respectively. This will be treated in detail in sections 1.3 to 1.5. Similar dot stackings may also result from surface corrugations of the spacer layer [77] when planarization is incomplete during spacer layer growth (Fig. 1.1c and d). Depending on whether subsequent dots nucleate preferentially at convex or concave surface areas, again vertical or staggered dot stackings may be formed as shown schematically in Fig. 1.1c and d, respectively. Finally, surface segregation of dot material (Fig. 1.1e) or alloy decomposition of the spacer layer (Fig. 1.1f) [62–66] may produce a non-uniform chemical composition of the surface above the dots. This can affect the growth of subsequent dots due to the resulting variations in chemical potential, strain and effective wetting layer thicknesses. As will be discussed in section 1.9, this again may cause a vertical alignment or staggered dot stackings in dependence of the details of the experimental conditions. Generally, all three mechanisms may act simultaneously and thus amplify or counteract each other. Therefore, it is often not easy to determine which is the main driving force for correlation formation for a given material system and growth conditions.

Experimentally, indeed different dot stackings have been found for different multilayer structures. Figure 1.2 shows representative examples as revealed by cross-sectional transmission electron microscopy (TEM) studies. The most common case of vertical dot alignment is illustrated in Fig. 1.2a for a self-assembled InAs/GaAs quantum dot superlattice [44], in which case the vertical dot alignment was found to persist up to 50 nm GaAs spacer layer thicknesses [15, 16, 42–45, 78]. A similar vertical alignment is also found for SiGe/Si dot superlattices up to 70 nm thick spacers [32–40], as well as for InP/GaInP [70] and GaN/AlN [79–86] multilayers as shown in detail in Sections 1.4–1.7. Staggered dot stackings have been found for IV–VI [18, 49, 50, 57], III–V [87, 88] as well as II–VI semiconductors [46–48] and examples are depicted in Fig. 1.2b and e. Staggered stackings were also observed for self-assembled InAs/AlInAs quantum *wire* superlattices [60–65] as exemplified by Fig. 1.2d. For a given material system even transitions

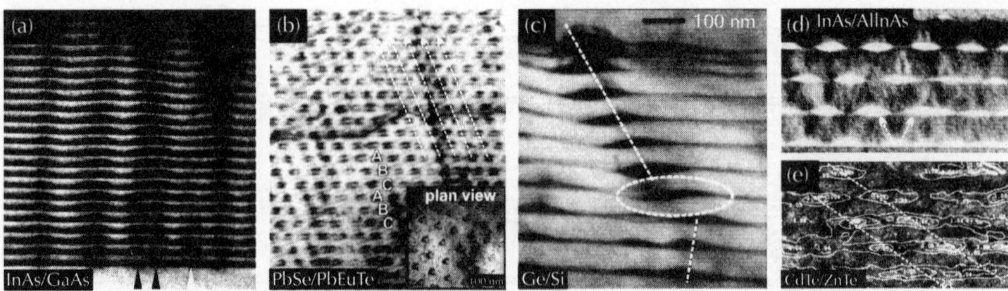

Figure 1.2 Examples of different types of interlayer dot stackings in self-assembled quantum dot multilayers observed by cross-sectional transmission electron microscopy: (a) vertically aligned (001) InAs quantum dot superlattice with 20 nm GaAs spacer layers. Adapted from Darhuber et al. [38]. (b) Fcc-like ABCABC... dot stacking in a (111)-oriented superlattice of 5 ML PbSe dots alternating with 45 nm PbEuTe spacers. Inset: Plan-view transmission electron microscopy micrograph showing the 2D hexagonal dot ordering within the growth plane. After Springholz et al. [18, 49, 51]. (c) Inclined dot correlations in a 1.2 nm Ge/40 nm Si (001) dot superlattice. Adapted from Sutter et al. [69]. (d) Vertically anticorrelated InAs/AlInAs quantum wire superlattice on InP (001) with 3ML/10 nm thicknesses, respectively. Adapted from Brault et al. [64, 65]. (e) Anticorrelated 2 ML CdTe islands intercalated with 15 ML ZnTe spacers. The white contour lines indicate iso lattice parameters, i.e. the chemical composition extracted from the atomically resolved transmission electron microscopy image. Adapted from Mackowski et al. [48].

between different interlayer correlations have been observed as exemplified by Fig. 1.2c which shows a Ge/Si dot superlattice where the initial vertical alignment switches to an oblique dot replication due to an instability in the planarization process [76]. Transitions between different dot stackings as a function of spacer layer thickness were also found for PbSe/PbEuTe [50, 57], CdSe/ZnSe [47] and InGaAs/GaAs dot superlattices [87, 88]. This will be discussed in detail in Sections 1.4, 1.6 and 1.8.

1.3   Strain-field interactions in multilayer structures

Strain is the major driving force for self-assembled Stranski–Krastanow quantum dot formation [4–10]. Therefore, the elastic strain fields produced by buried quantum dots play a crucial role for the formation of interlayer correlations in multilayer structures. The strain fields of the buried dots are caused by the strong elastic lattice deformation induced by the large lattice mismatch between the dots and the surrounding matrix material. These strain fields extend up to the epilayer surface and, during subsequent dot layer growth, impose a bias on the diffusion current of deposited adatoms due to the concomitant gradients in the surface chemical potential. At the surface strain minima therefore a local enhancement of growth and preferential island nucleation may occur. This may also be enforced by a strain-induced local reduction of the island nucleation barrier.

The strain fields created by the buried dots depend on a large variety of parameters such as size, shape, depth and chemical composition of the buried dots, the spacer material as well as the elastic properties and the crystallographic orientation of the growth surface. In order to understand the different interlayer correlations and stacking types, obviously, all details of the elastic strain fields must be taken into account. For this purpose, it is useful to distinguish between two limiting cases, namely, (i) the *far-field limit*, where the dot depth is large as compared to the dot dimensions, and (ii) the *near-field limit*, where the buried dots are very near to the growth surface. In the far-field limit, the internal structure as well as the actual size and shape of the dots can be ignored, i.e. the dots can be treated as simple point stress sources. This drastically simplifies the calculations and it turns out that the surface strain distribution produced by each buried dot is then solely determined by the elastic properties of the matrix material and the surface orientation. Thus, the far-field limit is particularly instructive to reveal the general trends of the elastic dot interactions. According to calculations and experimental observations, the far-field limit applies well when the dot depth, i.e. the spacer thickness, exceeds about two times the size of the dots [18, 48]. For the near-field case, the actual geometry of the dots as well as their depth strongly influences the strain distributions, which therefore have to be evaluated separately for each material system.

1.3.1   The isotropic point-source model

1.3.1.1   Surface strain distribution

For the simplest case of an elastically isotropic matrix material, the far-field stresses created by an individual point-like buried dot can be derived analytically. As shown by several works [89, 90], the buried dot in this case induces a radially symmetric strain distribution $\varepsilon_\parallel(r) = 1/2(\varepsilon_{xx} + \varepsilon_{yy})$ on the surface that is independent of surface orientation. This strain distribution can be written as:

$$\varepsilon_\parallel(r) = -\frac{P}{d^3} \cdot \left[\frac{2 - r^2/d^2}{(1 + r^2/d^2)^{5/2}} \right]. \tag{1.1}$$

In this expression, d is the dot depth underneath the surface, $r = (x^2 + y^2)^{1/2}$ is the radial distance from the centre above the dot, P is the strength of the point stress source given by $P = \varepsilon_0 V_0(1 + \nu)/\pi$ [90, 90], where ε_0 is the dot/matrix lattice mismatch, V_0 is the dot volume and ν the Poisson's ratio. If we consider the case of a surface wetting layer of the same material as the dots as is illustrated schematically in Fig. 1.3a, then Eq. 1.1 represents the amount by which the global layer–substrate lattice mismatch strain ε_0 is locally reduced on the wetting layer due to the presence of the buried dot. This is accounted for by the negative prefactor of Eq. 1.1.

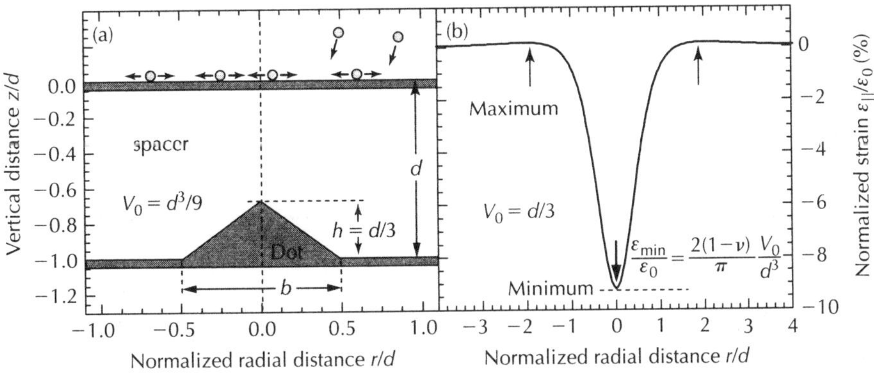

Figure 1.3   (a) Schematic illustration of a buried pyramidal Stranski–Krastanow dot located at a depth d *below the surface with a height of* h = d/3, *a square base of* b = d *and a volume of* $V_0 = d^3/9$. *(b) Corresponding calculated normalized surface strain energy distribution* $\varepsilon_\parallel/\varepsilon_0$ *plotted as a function of normalized radial distance* r/d *from the centre above the dot calculated within the point-source approximation (Eq. 1.1) for an elastically isotropic medium with* $\nu = 0.3$. *The arrows indicate the minima and maxima of the strain distribution.*

The relative strain variation $\varepsilon_\parallel/\varepsilon_0$ given by Eq. 1.1 is plotted in Fig. 1.3b as a function of normalized surface coordinates r/d. Obviously, directly above the buried dot, the strain is significantly reduced by a value of $\Delta\varepsilon = \varepsilon_\parallel(r = 0) = -2(1 + \nu)/\pi (V_0/d^3) \cdot \varepsilon_0$. Considering the case of a buried dot pyramid with square base $b = d$ and a height of $h = d/3$, the dot volume is $V_0 = d^3/9$ and thus the original misfit strain ε_0 is reduced by about 10% above the buried dot (see Fig. 1.3b). Also, there exist two weak side maxima at a lateral distance of $r/d = 2$ at which the strain is slightly increased by about 0.2%. Since the shape of the normalized strain distribution is invariant as a function of normalized surface coordinates r/d, the width (FWHM) of the central strain minimum is inversely proportional to the dot depth, i.e. FWHM $= 1/d$, and the strain reduction above the dot decreases inversely proportional to the cube of the dot depth. This is a general property of the far-field point-source solution that also applies for elastically anisotropic materials.

For subsequent dot layer deposition, the variation in the strain energy $E_s(x, y)$ on the wetting layer surface results in a gradient in the surface chemical potential $\Delta\mu \sim \Delta E_s$ across the surface. This produces a diffusional bias of surface adatoms towards the strain minima on the surface, where preferred dot nucleation will occur. The strain energy distribution on the surface is calculated using $E(x, y) = 1/2\ c_{ijkl}\varepsilon_{ij}(x, y)\varepsilon_{kl}(x, y)$ where c_{ijkl} are the elastic constants of the wetting layer, and $\varepsilon_{ij}(x, y)$ are the tensor components of the surface strain distribution. The latter is given by the sum of the misfit strain ε_0 and the additional inhomogeneous strain fields $\varepsilon_{\parallel}(r)$ induced by the buried dots. In the far-field limit $\varepsilon_{\parallel}(r) \ll \varepsilon_0$ and thus the isotropic surface strain energy variation $\Delta E_s(r) = E_s(r) - E_s^{2D}$ can be approximated by [18, 19]:

$$\Delta E_s(r) = -(1 + \nu)\frac{E_s^{2D} \cdot V_0}{\pi \cdot d^3} \cdot \left[\frac{2 - r^2/d^2}{(1 + r^2/d^2)^{5/2}}\right] \tag{1.2a}$$

with:

$$E_s^{2D} = 2\mu(1 + \nu)/(1 - \nu) \cdot \varepsilon_0^2 \tag{1.2b}$$

where E_s^{2D} is the constant background strain energy within the biaxially strained 2D wetting layer arising from the homogeneous dot/matrix lattice mismatch. The strain energy variation (Eq. 1.2) shows exactly the same form of the strain energy distribution of Eq. 1.1 with only a different prefactor. Therefore, the strain energy distribution $\Delta E_s(r)$ is radially symmetric as well with a minimum directly above the buried dot with an amplitude of $\Delta E_{s,\mathrm{min}} = -2(1 + \nu)/\pi \cdot E_s^{2D}V_0/d^3$.

1.3.1.2   Lateral ordering in 1D growth simulations

According to the isotropic strain calculations of Fig. 1.3, a single buried dot will always produce one central strain minimum on the surface directly along the growth direction. This would lead to a vertical dot alignment in multilayer structures. For a whole array of buried dots, however, there is also a lateral overlap of the strain fields of neighbouring dots which modifies the surface strain distribution. Based on a simple one-dimensional growth model, Tersoff *et al.* [17] have shown that this overlap may cause a lateral ordering of the dots as well. The effect of overlapping strain fields is illustrated in Fig. 1.4a [17], where the surface strain distribution produced by four

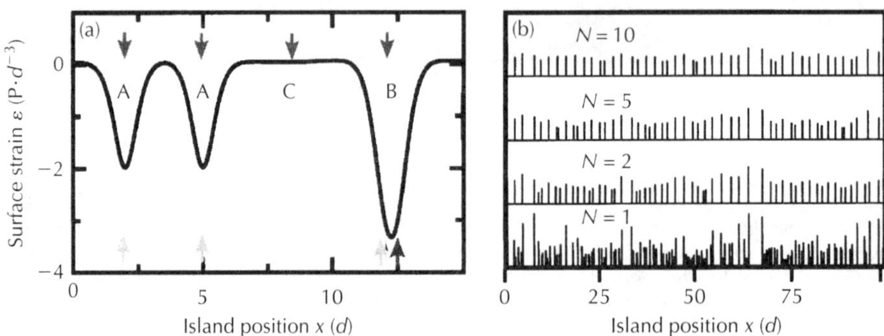

Figure 1.4   Self-organized lateral ordering of strained Stranski–Krastanow dots in a multilayer structure predicted by an isotropic 1 + 1 dimensional growth model by Tersoff et al. [17]. (a) Strain distribution produced by four subsurface dots located at different positions x below the surface as indicated by the upward pointing arrows, calculated using the isotropic point-source approximation. The downward arrows at the top indicate local strain minima on the surface above the dots, representing favoured positions for dot nucleation. Properly spaced dots (labelled A) replicate vertically along the growth direction, whereas for closely spaced dots only one new strain minimum and thus one new dot in position B will be formed in the subsequent layer. For widely spaced buried dots, a weak local strain minimum is produced at point C, where an additional new dot is assumed to grow. (b) Deterministic growth simulation assuming dot nucleation at each local surface strain minimum induced by the buried dots. The vertical lines indicate the lateral position of the dots in the bilayer N = 1, 2, 5 and 10 of the superlattice structure, starting from a random distribution in the initial dot layer N = 1. Clearly, a rapid equalization of the lateral dot separations due to the lateral strain interactions occurs. The height of each vertical line represents the size of each dot, which becomes more uniform with increasing N. Adapted from Teichert et al. [21].

buried dots located at different positions x within an interface below the surface is plotted (the position of the buried dots is indicated by the upward pointing arrows). Obviously, three different cases can be distinguished: (i) for widely spaced buried dots (labelled A in Fig. 1.4a), the surface strain minima are exactly on top of the dots. Thus, in the subsequent layers, these dots will be replicated along the growth direction. (ii) For closely spaced dots (labelled B in Fig. 1.4a), the surface strain minima are merged to one and thus only one new dot will nucleate on the surface in the middle between the buried dots. (iii) When the spacing of the buried dots is very wide, an additional weak local strain minimum is produced at point C in between the buried dots, where subsequently an additional new dot may start to grow. As a result of this strain field overlap, the dot positions gradually change from one layer to another as multilayer growth proceeds.

To demonstrate the possible lateral ordering induced by this effect, Tersoff *et al.* [17] have performed a deterministic one-dimensional growth simulation based on the assumption that at every local surface strain minimum a new dot is created with a volume that is proportional to the area of the Voronoi polygons around each surface dot that represents the average capture area of adatoms for each dot. After each dot layer growth, the dots are covered by a spacer layer with thickness d and the subsequent surface strain distribution is recalculated. Starting from an initial random dot layer, this sequence is repeated N times to produce the desired multilayer structure.

The results of this deterministic model are shown in Fig. 1.4 (b), where the lateral dot positions in the $N = $ 1st, 2nd, 5th and 10th dot layers are plotted as vertical lines with a length proportional the volume of the individual dots. Starting with a randomly distributed dot layer with high density, one can see that the lateral dot spacing as well as dot size becomes progressively more uniform as the number of periods increases. This demonstrates that in a multilayer structure a lateral dot ordering may be induced. It is noted, however, that for elastically isotropic materials this conclusion applies only for a one-dimensional system (see Section 1.5 for details). Therefore, for more realistic growth models, the elastic anisotropy, random surface diffusion as well as the size and shape of the dots must be taken into account. The 1D simulation also yields a significant narrowing of the dot size distribution of up to a factor of 5 and the preferred final lateral dot separation is found to be equal to 3.5 times the spacer thickness [17]. Thus, the spacer layer acts like a band pass filter for a certain lateral dot separation.

1.3.2   The effect of elastic anisotropy

Most semiconductors exhibit a rather high elastic anisotropy. As a result, the surface strain distribution produced by a buried dot is significantly altered compared to the isotropic model. In particular, the strain distributions show a strong dependence on surface orientation. To obtain the resulting strain distributions, the equilibrium stress equations must be solved by taking the true elastic properties of the matrix material around the dot as well as the boundary condition of a free surface with vanishing normal surface stresses into account. This can be done using a Fourier or Green's function method [18, 58, 59, 91–93]. Alternatively, finite element methods [50, 94–100] or atomistic calculations using semi-empirical atom potentials [100, 103] have been used, which all yield very similar results for the strain fields well outside of the buried dots [100].

In cubic materials, the main elastic anisotropy axes are the crystallographic $\langle 100 \rangle$ and $\langle 111 \rangle$ directions, in which Young's modulus E_{hkl} has its extremal values. This is illustrated in Fig. 1.5, where the variation of E_{hkl} as a function of crystallographic direction $[hkl]$ is plotted for several semiconductors such as PbTe and PbSe (a) and Si, Ge, GaAs and ZnSe (b). Correspondingly, the $\langle 100 \rangle$ or $\langle 111 \rangle$ directions represent the elastically *hard* or *soft* directions of cubic materials. The degree of deviation from the isotropic case is characterized by the dimensionless anisotropy ratio:

$$A = 2c_{44}/(c_{11} - c_{12}) \approx E_{111}/E_{100} \tag{1.3}$$

which is essentially equal to the ratio of the elastic moduli E_{hkl} along the main $\langle 111 \rangle$ and $\langle 100 \rangle$ anisotropy axes. For isotropic materials, E_{hkl} does not depend on the direction of the applied stresses and thus $A = 1$. For anisotropic materials, one has to consider two opposite cases, namely, that A is either larger or smaller than one. In the first case of $A > 1$, the $\langle 111 \rangle$ directions

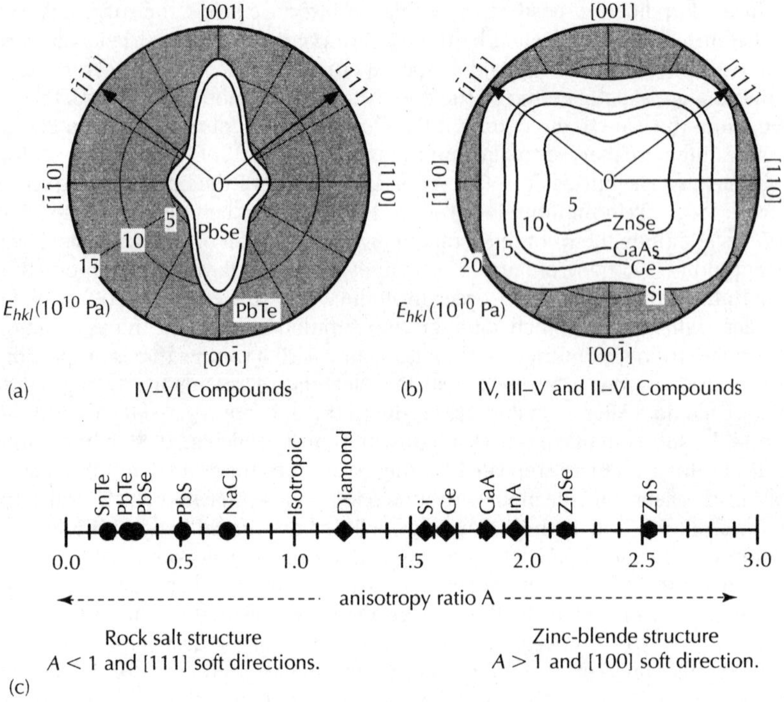

Figure 1.5   *Polar plots of the Young's modulus E_{hkl} as a function of [hkl] direction within the $(\bar{1}10)$ plane. (a) IV–VI compounds PbTe and PbSe with rock salt structure and elastic anisotropy ratio A < 1, i.e. $\langle 001 \rangle$ as elastically hard directions with maximum E_{hkl} value. (b) Group IV, III–V and II–VI semiconductors (Si, Ge, GaAs and ZnSe) with zinc-blende structure and $\langle 111 \rangle$ as elastically hard axes (A > 1). Bottom: (c) Elastic anisotropy ratio $A = 2c_{44}/(c_{11} - c_{12})$ plotted for various cubic semiconductors.*

are the elastically hard directions and the $\langle 001 \rangle$ directions the soft directions, i.e. $E_{111}/E_{100} > 1$. As shown in Fig. 1.5b, this essentially applies to all group IV, III–V and II–VI semiconductors with diamond or zinc-blende crystal structures because the chemical bonds are along the $\langle 111 \rangle$ directions. The anisotropy is largest for the II–VI compounds, with $A = 2.04$ for ZnTe and 2.53 for ZnS (Fig. 1.5c). For C, Si and Ge, A increases from 1.21, 1.56 to 1.64, respectively, and for the III–V compounds A ranges from 1.83 for GaAs to 2.08 for InAs. In the opposite case of $A < 1$, now $\langle 100 \rangle$ are the elastically hard directions and $\langle 111 \rangle$ the soft directions ($E_{111}/E_{100} < 1$, see Fig. 1.5a). This applies, e.g., for materials with rock salt crystal structure in which the nearest neighbours are along the $\langle 100 \rangle$ directions. In particular, for the narrow gap IV–VI semiconductors the elastic anisotropy is particularly large, with $A = 0.18$, 0.27 and 0.51 for SnTe, PbTe and PbS, respectively.

1.3.2.1   *Far-field strain distributions*

With respect to the elastic strain fields, obviously, the strongest changes will occur for materials with high elastic anisotropy. Thus, when A strongly deviates from one, the strain distributions will become strongly dependent on surface orientation. Selecting two materials with large anisotropy but opposite directions of the anisotropy axes, the normalized surface strain energy distributions $\Delta E_s/E_s^{2D}$ induced by a buried point-like stress source are shown in Fig. 1.6 for PbTe (top row) and ZnSe (centre row) for different surface orientations and scaled surface coordinates r/d. It is evident that not only the energy distributions strongly differ as a function of surface orientation, but that they also show the opposite trend when A is larger (ZnSe) or smaller than one (PbTe). In particular, it is found that only when the surface orientation is parallel to the elastically *hard* direction (i.e. (100) for ZnSe and (111) for PbTe) the strain energy minimum resides exactly above the buried dot, whereas in all other cases the energy minima are laterally displaced by a value r_{min}. Even more, when the surface is close to an elastically soft direction, the central energy minimum splits into several minima, as shown in panels (c) to (f) of Fig. 1.6 for PbTe and ZnSe.

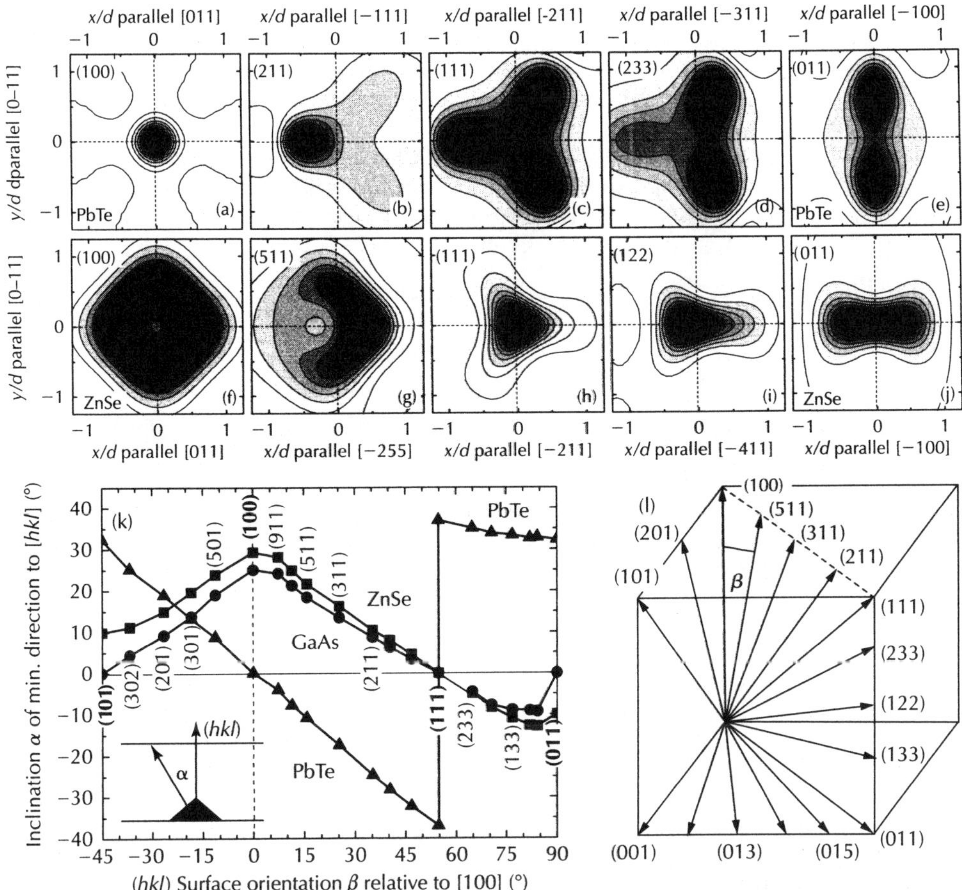

Figure 1.6   Top: *Calculated surface strain energy distributions* $\Delta E_s(x, y)$ *above a point-like strained buried quantum dot for different surface orientations of PbTe (top row,* A = 0.28*) and ZnSe (second row,* A = 2.52*). The energy distributions are shown as iso-strain contour plots as a function of reduced surface coordinates* x/d *and* y/d, *where* d *is the dot depth below the surface. Darker colours indicate areas of lower strain energy. For most surface orientations the strain energy minimum is displaced from the centre above the buried dot. When the surface normal is an elastically soft direction, a splitting into several side minima occurs. Bottom: (k) Inclination angle* α *at which the energy minima appear on the surface relative to the (hkl) surface normal plotted as a function of the angle* β *between the (hkl) surface and the (100) plane. The different (hkl) orientations are indicated in (l).*

In the far-field or point-source limit, the directions where the surface minima are formed are unique for each surface orientation and elastic anisotropy ratio. This is because in this approximation the elastic strain fields do not depend on the structure of the dots but only on the elastic properties of the matrix material. In addition, the surface strain distributions scale strictly linearly with the dot depth. This means that for each surface orientation there exists a well-defined characteristic correlation angle $\alpha = \arctan(r_{min}/d)$ along which the energy minima appear on the surface relative to the surface normal direction. The directions within the surface along which the surface energy minima are laterally displaced are also well defined and are given by the projection of the elastically hard direction closest to the surface normal onto the surface plane. For example, for $(n11)$ surface orientations, the lateral energy minima displacement within the surface plane is along $[\bar{2}nn]$ for $A > 1$ and along $[2\bar{n}\bar{n}]$ for $A < 1$ and $n \geq 1$. Likewise, for $(1nn)$ surfaces, the lateral displacement within the surface is along $[\bar{2}n,1,1]$ for $A > 1$ and along $[n,n^2,1-n^2]$ for $A < 1$ and $n > 1$. Finally, for $(01n)$ surfaces, the displacement is along $[n+1,n,1]$ for $A > 1$ and along $[011]$ for $A < 1$.

Figure 1.6k shows the systematic variation of the correlation angle α as a function of the (hkl) surface orientation for three materials with different elastic anisotropy (GaAs, ZnSe, and PbTe). In this plot, the (hkl) surface orientation is parameterized in terms of the angle β between $[hkl]$ and the [100] direction. Clearly, there is a systematic variation of α as the surface orientation is tilted from (100) through (111) and (101) (see Fig. 1l). In particular, the largest interlayer correlation angle appears when the surface is parallel to the elastically soft direction (i.e. (100) for $A > 1$ and (111) for $A < 1$), whereas the energy minima are almost vertically aligned when the surface is close to the elastically hard direction. Again it is evident that the behaviour is opposite for materials with anisotropy ratio larger or smaller than one.

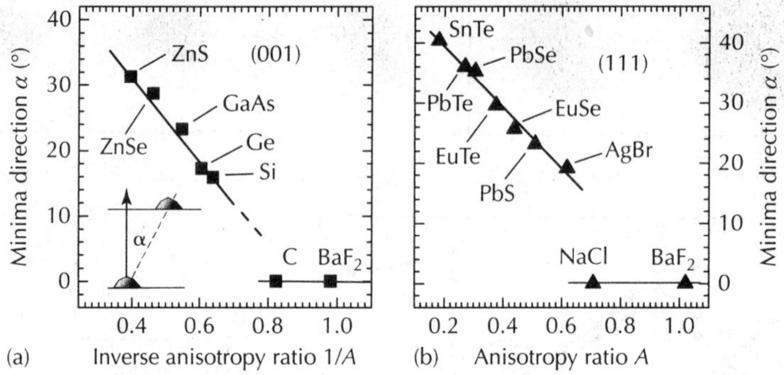

Figure 1.7   Direction α of the surface strain energy minima relative to a buried force nucleus plotted versus elastic anisotropy of the matrix material for (a) the (001) surface and the group IV and the zinc-blende III–V and II–VI semiconductors and (b) for materials with rock salt structure (IV–VI semiconductors, etc.) and (111) surface orientation. In both cases, the surface normal direction is parallel to the elastically soft direction. The elastic anisotropy is determined by the anisotropy ratio $A = 2c_{44}/(c_{11} - c_{12})$. Adapted from Holy et al. [56].

If we now compare materials with the same hard axis (i.e. ZnSe and GaAs in Fig. 1.6k), one can see that the larger the elastic anisotropy ($A_{ZnSe} > A_{GaAs}$), the larger the correlation angle and thus the larger the lateral displacement of the energy minima. A systematic analysis of this trend shows that the displacements and thus the correlation angles α depend in a linear way on the anisotropy ratio for $A > 1$, respectively, on its reciprocal value for $A < 1$ [56]. This is shown in Fig. 1.7, where the energy minima directions α are plotted for the most relevant high symmetry (111) and (100) surface orientations as a function of elastic anisotropy. For the (100) direction and the III–V and II–VI semiconductors (Fig. 1.7a), a splitting of the energy minima occurs if the anisotropy exceeds the critical value of $A_c > 1.4$, and beyond this value α varies linearly according to [58]:

$$\alpha_{100} = 56° \times (1 - 1.1 \cdot A^{-1}) \quad \text{for (100)} \quad \text{and} \quad A > 1.4. \tag{1.4a}$$

Thus, α_{100} increases from 16° to 23°, 28° and 32° for Si, GaAs, ZnSe and ZnS, respectively. For the (111) surface direction and materials with rock salt structure (Fig. 1.7b), a splitting occurs for $A_c < 0.6$, and below this value, α varies according to [58]:

$$\alpha_{111} = 50° \times (1 - A) \quad \text{for (111)} \quad \text{and} \quad A < 0.6. \tag{1.4b}$$

Therefore, α_{111} increases from 19° to 36° and 41° for PbS, PbTe and SnTe, respectively. A similar behaviour (increasing correlation angle with increasing elastic anisotropy) also applies for other surface orientations. Apart from the changes in α direction, with increasing anisotropy the depths of the energy minima also change. For the (100) surface the depth of the energy minima increases with increasing anisotropy ratio, whereas for the (111) surface the depth decreases with increasing A [58]. This is due to the fact that the anisotropy ratio represents the ratio of Young's modulus between the [111] and [100] directions.

1.3.2.2　Dot stackings in the far-field limit

For multilayers with large spacer thickness compared to the dot size and little lateral overlap of neighbouring dot strain fields, the far-field strain calculations can be directly used to predict the interlayer dot correlations in multilayer structures based on the assumption that the surface dots are small enough to occupy just one strain minimum on the surface. From the calculations shown in Fig. 1.6, it then follows that for most materials and growth orientations *inclined* interlayer dot correlations should be formed, with dot correlation angles corresponding to those plotted in Fig. 1.6k. This applies, e.g., to multilayer growth on high-indexed surfaces such as (211), (311), (511) in which case the dots should replicate along oblique directions. This has indeed been found by recent experiments by Schmidbauer *et al.* [142] for InAs/GaAs dot superlattice growth on high-indexed (n11) substrate orientations. In fact, from the strain calculations for elastically anisotropic materials an *exact* vertical dot alignment along the growth direction is expected only when the surface orientation is parallel to the elastically hard axis, i.e. either for the (111) or (100) surface of cubic materials. This is because in the far-field limit only for these surface orientations the strain minimum resides directly above the buried dots (see Fig. 1.6a and h). As shown in the next section, this also applies to multilayers with small spacer thicknesses.

A particular situation arises when the surface orientation is aligned or is close to the elastically soft crystal axis. If the elastic anisotropy is sufficiently large, then the surface strain distribution splits into several side minima above each buried dot (see Figs. 1.6 and 1.7). When one new dot nucleates at each of these minima, *staggered* dot stackings will be formed. For (001) growth and an elastic anisotropy A larger than 1.5, four side energy minima occur above each buried dot (Fig. 1.6f). These minima define a preferred square arrangement of dots in the subsequent growth plane, with the previous dot located in the centre of the square below the surface. This dot arrangement is replicated again in the subsequent dot layers, which in total results in an *ABAB...* stacking sequence of dots, as shown schematically in Fig. 1.8a. In an ideal case, this will yield an overall *body-centred tetragonal* 3D dot arrangement in the structures with a lateral alignment of the dots along the in-plane ⟨100⟩ surface directions.

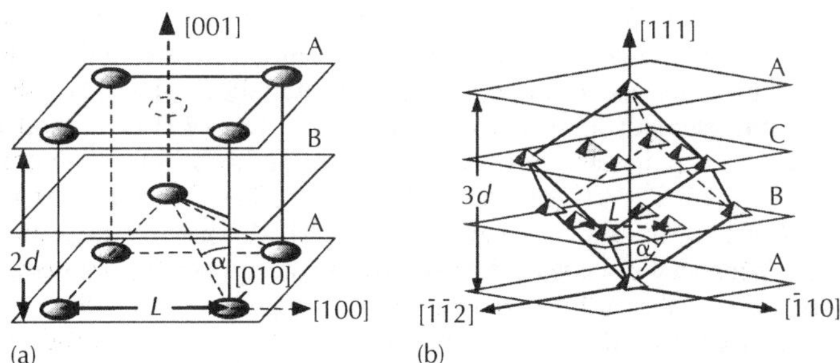

Figure 1.8　Staggered three-dimensional dot stackings expected from the point-source strain calculations for materials with high elastic anisotropy and growth along the elastically soft direction. (a) ABAB... stacking and resulting body-centred tetragonal dot lattice for (001) growth and anisotropy ratio A ≫ 1.5, (b) fcc-like ABCABC... dot stacking and resulting trigonal 3D dot lattice for the (111) growth orientation and materials with A < 0.6, as observed for PbSe/PbEuTe superlattices (see Fig. 1.2b).

For the (111) growth orientation and A smaller than 0.6, three side minima appear in the energy distributions (see Fig. 1.6c), which create a triangle with equal sides along the ⟨2 1̄ 1̄⟩ surface directions. This induces a triangular or hexagonal dot arrangement in the subsequent layer centred above the buried dots underneath the surface. Each of these new dots will induce the same triangular arrangement in the subsequent layer, and thus it takes in total three dot layers until the dots appear again at the same position along the growth direction. This replication thus produces an *ABCABC...* dot stacking sequence that is shown schematically in Fig. 1.8b. The resulting dot arrangement is similar to the atom stacking in face-centred cubic lattices but in general the ratio between the lateral dot spacing within the 2D sheet of hexagonally ordered dots and the vertical spacer thickness will not be equal to 1.155 as in *fcc* lattices. Therefore, the dot crystal lattice is expanded or compressed in the (111) direction, i.e. the 3D dot arrangement represents an overall *trigonal* lattice of dots. As shown in Fig. 3.4e, also for the (110) growth orientation two well-defined side energy minima may occur on the surface above the dots. Accordingly, this may result in the formation of vertical sheets of 2D rhombohedrally ordered dots in a multilayer structure.

1.3.3 Near-field strain interactions

For multilayers with small spacer layer thicknesses, the buried dots can no longer be approximated as simple point stress sources but their actual size, shape and gradients in compositions must be taken into account. These parameters obviously differ strongly from one material to another and moreover depend on the growth and overgrowth conditions, the dot layer thickness and utilized growth technique (see, e.g., Stangl *et al.* [92] for a review). For (001) SiGe/Si dots, e.g., hut cluster islands with {105} facets are formed at low growth temperatures and small coverages [21, 124], whereas for thicker dot layers and higher temperatures dome-shaped islands with {113} facets are formed [125–127], with several transitional shapes in between [126, 127]. Even steeper dots with {111} side facets have been observed for SiGe dots grown by liquid phase epitaxy [104]. For (001) InAs/GaAs islands, on the other hand, multifaceted islands composed of {317}, {011} and {111} facets have been reported [128, 129], whereas for (111) PbSe islands pyramids with triangular base and {100} side facets were observed [18, 130]. On high-indexed surfaces, asymmetric island shapes have been found, such as for InAs on GaAs (113)A [105] and Ge on (113) Si surfaces [106]. A further complication arises from the fact that during overgrowth significant changes in dot shape and composition occur due to intermixing with the surrounding matrix material [92, 107–113]. This intermixing strongly depends on the growth conditions [109, 110] as well as the chemical composition of the spacer layer [112, 113] and changes the chemical composition of the dots as well [91, 110–113]. As a consequence, no general solution of the near-field strain interactions can be given but each particular experimental situation and material system must be considered separately.

To calculate the strain fields of near-surface dots several methods have been used [92]. If the elastic constants of dots and matrix material do not differ much, the strain fields can be obtained by convoluting the point-source solution with the given dot shape [18, 92]. Alternatively, one can also apply the finite element method [50, 110, 114, 118, 142] or atomistic calculations using semi-empirical atom potentials [119–122]. These methods have been employed extensively for InAs/GaAs [93, 117, 119] and Si/Ge [110, 118–121] dots but also for other materials such as InN/AlN [84, 123]. As shown by Pryor *et al.* [119], all three methods give quite similar results for the strain fields well outside of the buried dots, as applies for the surface strain fields well above the dots relevant for multilayer structures. As a general trend, for near-surface dots the strain fields are focused towards the surface normal direction, i.e. the surface energy minima are confined more closely to the region directly above the buried dots. This arises from the fact that the free surface allows a very efficient strain relaxation due to the outward or inward relaxation of the surface lattice planes.

1.3.3.1 (100) surfaces

For (100) surfaces, the changes of the strain energy distributions in dependence of the dot depth are demonstrated in Fig. 1.9 for InAs dots embedded in GaAs. In this example, the dots were assumed as truncated square base InAs pyramids with a fixed base width of 20 nm and a height of 7 nm (see insert of Fig. 1.9), similar to what has been found in cross-sectional scanning tunnelling microscopy studies [72, 73, 111]. For simplicity, variations in the chemical composition within the dots due to intermixing [92, 111] were neglected, i.e. pure InAs dots were

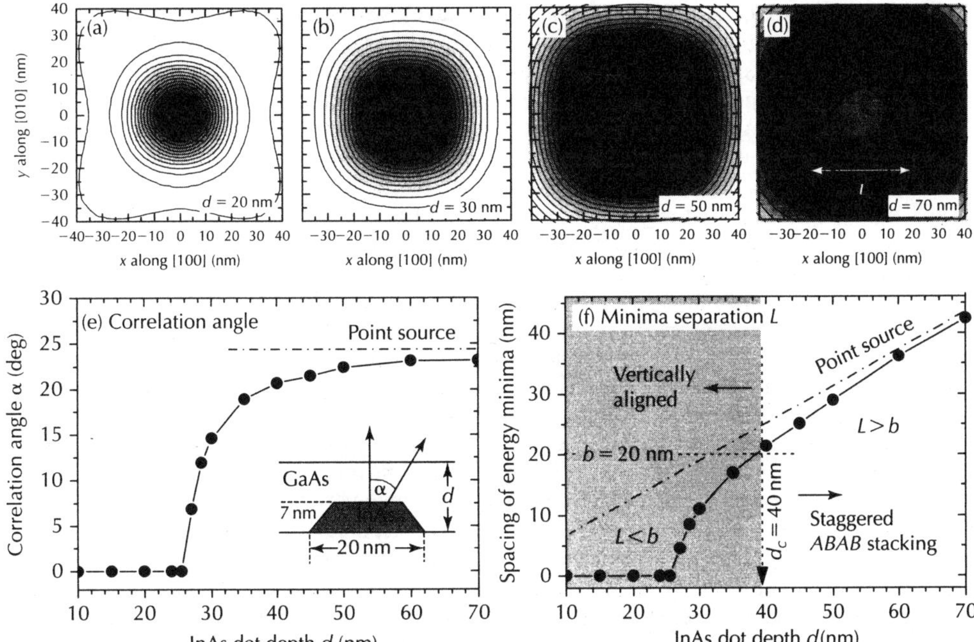

Figure 1.9   Top: *Iso-strain energy contour plots of the surface strain energy distribution above a buried InAs quantum dot with truncated pyramidal shape, 20 nm base and 7 nm height located in a GaAs matrix at different dot depths of* d = 20, 30, 50 *and* 70 nm *from (a) to (d), respectively. The darker colour corresponds to lower strain energy on the wetting layer, and the separation of the contour lines decreases each by a factor of four from 0.2 meV to 0.003 meV per atom pair from (a) to (d), respectively. The dashed squares indicated the base of the InAs islands. Bottom: (e) Direction* α *of the energy minima relative to the surface normal as well as (f) lateral separation* L *of the energy minima plotted as a function of the InAs dot depth. The dash-dotted lines indicate the results of the point-source model for comparison.*

assumed. The strain energy distributions were calculated using the semi-analytical method [92] and the GaAs spacer thickness was varied from 10 to 70 nm. Figure 1.9a–d shows representative energy distributions for dot depths of $d = 20$, 30, 50 and 70 nm, respectively. Whereas for large dot depths of $d \geq 50$ nm (i.e. d larger than two times the dot width), the surface strain distributions exhibit four side minima along $\langle 110 \rangle$ in good agreement with the point-source model (compare Fig. 1.9c and d with Fig. 1.6f), with decreasing dot depth the lateral spacing of the energy minima rapidly shrinks and eventually they merge into one single minimum located directly above the dot (*cf.* Fig. 1.9a). This clearly demonstrates the focusing of the strain fields in the vertical growth direction when the dots reside very near to the growth surface.

Figure 1.9e shows the directional angle α of the surface energy minima relative to the surface normal as a function of the GaAs spacer thickness. For large spacer thicknesses, α converges to the value of 24° obtained from the far-field (point-source) approximation. At spacer thicknesses lower than 40 nm, however, α rapidly drops to zero such that at $d \leq 25$ nm only one central minimum is formed. Thus, a spacer thickness of more than four times the dot height is required in order to get any splitting of the strain energy minima on the surface. Similar results were also obtained by other works for slightly different geometries for the buried dots [93, 94, 96]. The dependence of the lateral energy minima separation L on dot depth is shown in Fig. 1.9f. For large d, α is constant and L increases essentially linearly with increasing spacer thickness as expected from the point source approximation (dashed line in Fig. 1.9f). Below $d = 40$ nm, L rapidly decreases and drops to zero when d reaches 26 nm. As a consequence, for superlattices with small spacer thicknesses the InAs dots will always be vertically aligned. To obtain a staggered *ABAB...* dot stacking, the spacing of the energy minima L must be larger than the base width b of the dots which is indicated by the horizontal dashed line in Fig. 1.9f. Accordingly, the GaAs spacer thickness must be larger than a critical thickness of $d_c \sim 40\,\mathrm{nm}$ (see Fig. 1.9f). This is in agreement with recent experiments by Wang *et al.* [87], where a transition between vertically aligned and staggered InGaAs/GaAs dots was observed at spacer thicknesses around 40 to 50 nm. On the other hand, the depth of the energy minima becomes very shallow at large dot depths. Therefore, there is only a limited range of parameters where a staggered dot stacking can be achieved. This will be discussed in more detail in Section. 1.3.4.

The transition in InAs dot stacking clearly depends on the given dot size, i.e. for larger dots with wider base b, the transition to a staggered dot stacking will occur at a larger critical spacer thicknesses d_c. If the aspect ratio, i.e. shape of the dots, is constant, the length scale of the strain distributions scales linearly with the system size, whereas the directional angles α of the surface strain minima are constant for a given b/d ratio. The minima separation L can then be easily recalculated from Fig. 1.9 by applying a linear scaling factor of b/b_0 where b_0 is the 20 nm base width used in the present calculations. Thus, for two times larger dots ($b/b_0 = 2$), the minima angle α of Fig. 1.9 appear at two times larger spacer thicknesses at which the minima separation L is twice as large. Therefore, in this case the expected transition from vertically aligned to staggered dots is shifted to a two times larger spacer thickness of $d_c \sim 80\,\mathrm{nm}$.

The conclusions from the InAs/GaAs strain calculations apply for all zinc-blende III–V and II–VI semiconductors because the elastic anisotropy is quite similar or even larger than for the InAs/GaAs system (*cf.* Fig. 1.5). For CdSe dots in ZnSe, e.g., surface strain distributions have been calculated by Quek and Liu [94] for pyramidal dots with 20 nm base and 10 nm height using the finite element method. As a result, a vertically aligned surface strain minimum was found up to 20 nm ZnSe spacers, but a lateral minima splitting of 16 nm was found already for 30 nm spacers [94]. This means that due to the higher elastic anisotropy of ZnSe compared to GaAs ($A = 2.3$ versus 1.83, respectively), the splitting occurs already at a spacer thickness of three times the island height. At 40 nm ZnSe thickness, the minima spitting is already as large as 33 nm, indicating that under these conditions the strain field is already close to the point-source approximation (splitting of 42 nm).

We have also performed strain calculations for the Ge/Si (100) case assuming pyramidal dots with 50 nm base and 10 nm height. As a result, we find that even up to 100 nm Si spacers only one vertically aligned surface strain energy minimum is formed above each dot. This is due to the smaller elastic anisotropy. Even in the far-field limit the splitting of the surface energy minima is about 40% smaller than for the InAs/GaAs case. Since Ge dots also exhibit a much wider island shape, spacer thicknesses much larger than 100 nm would be required to obtain a sufficiently large energy minima splitting to induce a staggered *ABAB*... dot stacking in the SiGe/Si system. This is well beyond the spacer layer thicknesses up to which interlayer correlations have been observed in this material system [19, 32, 34–36] (see Section 1.7). As a consequence, there seems to be little chance for obtaining a staggered dot stacking in the Si/Ge system.

1.3.3.2   (111) surfaces

For materials with elastically hard axis along [111] (i.e. with $A > 1$ such as for SiGe, III–V and II–VI materials), the point-source model already yields only a single energy minimum directly above each dot. Therefore, for extended dots and thin spacer layers no change in this respect occurs in the near-field limit but only the overall shape of the energy distributions around the central minimum is somewhat changed in dependence on the lateral extend and shape of the buried dots. The situation is different for IV–VI materials where [111] is the elastically soft direction and thus a three-fold minima splitting is found in the far-field limit. We have evaluated in detail the changes of the (111) near-field strain distributions as a function of dot size [55] and spacer thickness [50] using the finite element method. Figure 1.10a to c shows the calculated strain energy distributions for pyramidal PbSe dots within a PbTe matrix at a fixed dot depth of 42 nm but variable dot height h (see inset of Fig. 1.10e). The shape of the dots was kept constant with a fixed aspect ratio of $h/b = 2.45$. Although the spacer thickness is constant in this case, clearly the separation of the three surface strain minima L_{min} indicated by the arrows in Fig. 1.10a–c shrinks as the size of the buried dots relative to the spacer thickness increases, i.e. as the top of the buried dot pyramids

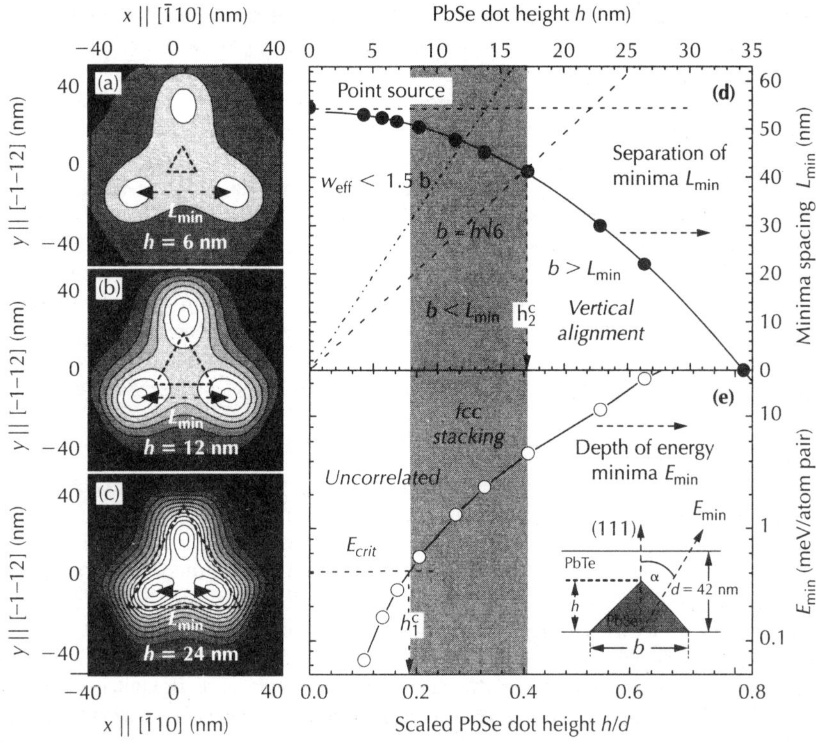

Figure 1.10 Left-hand side *(a) to (c): Iso-strain energy contour plots of the surface strain energy distribution above pyramidal PbSe quantum dots with different heights of 6, 12 and 24 nm, respectively, buried below a 42 nm PbTe (111) spacer layer. The dot base is indicated by the dashed triangles. The contour line separation is 0.046, 0.13 and 0.65 meV/atom pair from (a) to (c), respectively, and the brighter colour corresponds to lower strain energies. Right-hand side: Plot of the dependence of the separation* L_{min} *(d) and depth* E_{min} *(e) of the energy minima as a function of PbSe dot height. In (d), the change of the dot base width* $b = h\sqrt{6}$ *as a function of dot height is plotted as a dashed line as well as the effective width (dash-dotted line) of the denuded zone* $w_{eff} = 1.5 \times b$ *around each dot where further dot nucleation is suppressed. In the left region, the depth of the energy minima becomes insufficient to cause a correlated dot nucleation. From experiments this is deduced about 0.3 meV/atom pair [50]. The central region corresponds to the conditions where an fcc-like dot stacking is predicted. In the right region the minima separation is smaller than the dot base width and therefore the dots are expected to be vertically aligned. These predictions are in quite good agreement with experimental observations [55].*

moves closer to the epilayer surface. This is shown in Fig. 1.10d, where the separation L_{min} of the energy minima is plotted as a function of PbSe dot height.

For small dots, the minima separation is equal to that of the point-source model (horizontal dashed line). As the dot height increases above 10 nm, the minima separation gradually decreases and reaches zero for a dot height larger than about 30 nm. This is again caused by the focusing of the strain fields in the surface normal direction. Similar results were obtained for PbSe dots with fixed 12 nm height but varying dot depths [50] (see Section 1.8 for details), in which case a splitting of the surface strain minima occurs only when the spacer thickness is larger than about 16 nm, whereas for thinner spacers again only one vertically aligned central strain minimum is formed. Thus, for this material system, due to the substantially higher elastic anisotropy compared to III-V materials, a splitting of the strain minima occurs when the spacer thickness is two times larger than the dot height. In addition, the overall strain distributions are found to agree well with those of the point-source model at a d/h ratio above 3, where the correlation direction of the strain minima from the finite element calculations of $\alpha = 31°$ are already close to the far-field value of $36°$. This means that the point-source approximation can be readily applied for spacer thickness three times larger than the dot height.

Concerning the dot stacking in multilayers, for formation of an *fcc*-like dot stacking as shown in Fig. 1.8b, the minima separation must be larger than the dot base width $b = h\sqrt{6}$ indicated by the dashed triangles in Fig. 1.10a–c. From the crossing of the diagonal dashed $b = h\sqrt{6}$ line in Fig. 1.10d with the $L_{min}(h)$ curve, one can infer that this condition is met only when the ratio between the dot height and spacer thickness is smaller than about 0.4, i.e. for dots smaller than 15 nm for $d = 42$ nm, which agrees well with our experimental observations [55]. Figure 1.10e also shows the dependence of the energy minima depth on dot size. Clearly, the depth decreases very rapidly with decreasing dot height, since the elastic strength of the buried dots is proportional to their volume that scales as h^3 (*cf.* Eq. 1.2). From experimental studies on PbSe dot superlattices, no interlayer correlations were observed when the depth of the energy minima is smaller than about 0.3 meV/atom pair [50]. According to Fig. 1.10e, this point is reached when the dot height drops below 7 nm, in good agreement with experimental observations [55]. Thus, *fcc* stacked PbSe/PbTe dot multilayers can only be obtained for certain dot sizes and spacer thicknesses. This general conclusion holds for all material systems and is discussed in Section 1.3.4.

1.3.3.3   High-indexed surfaces

For high-indexed surfaces, only recently finite element calculations of the strain fields of buried dots have been performed by Schmidbauer *et al.* [142] in order to model the interlayer correlations of InGaAs/GaAs dot superlattices. For a constant 33 nm GaAs spacer thickness, it was found that for the $(n11)$B surfaces the inclined energy minima directions above lens-shaped InGaAs dots with 5 nm height and 30 nm width agree well with those predicted by the point-source model (see Fig. 1.6k). The results of these calculations are presented in more detail in Section 1.4.3 and are displayed in Fig. 1.17c. Although, the finite element calculations were not extended to thinner spacer layers, from the results of the previous sections it is evident that for near-surface dots the surface minima will be confined closer to the surface normal direction. In this case, the interlayer dot replication angles will become smaller than predicted by the far-field approximation of Fig. 1.6k. The fact that the experimentally measured interlayer correlation directions in InGaAs/Gas dot superlattices on high-indexed surfaces [142] were found to agree well with the angles predicted by the point-source model again shows that this approximation works well already for spacer thicknesses of 33 nm as used in these experiments.

1.3.4   Stacking conditions and replication angles

According to the near-field strain calculations, in multilayers with small spacer thicknesses the dots should be always vertically aligned, whereas staggered or inclined dot correlations can occur only for spacers exceeding a certain critical thickness value d_c^1. With increasing spacer thickness, on the other hand, the strain fields of buried dots rapidly decay such that above a certain critical thickness d_c^2 no interlayer correlations will be formed any more. Up to now, we have also not considered the influence of overlapping strain fields of neighbouring buried dots that may modify the surface strain distributions and thus the interlayer correlations as well. These three aspects are analysed in more detail in the following to derive some generic rules for the stacking and correlations in multilayer structures. These conditions will be compared with experimental results in Sections 1.5–1.8.

1.3.4.1   Correlated–uncorrelated transition

As revealed by various growth studies, interlayer dot correlations in multilayer structures typically persist to spacer thicknesses up to around 40 to 70 nm for different material systems [15, 35, 36, 45, 50] as will be shown in detail in Sections 1.6–1.8 for the InAs/GaAs, Si/Ge and PbSe/PbEuTe systems. Although the limiting thickness for interlayer correlations d_c^2 actually will depend on parameters such as growth conditions [45, 56] and size of dots [55], it is remarkable that the experimental values are quite similar for the different material systems. The critical thickness for interlayer correlations is mainly determined by the magnitude of the surface strain energy minima induced by the buried dots. These can be easily calculated since for large spacer thicknesses the point-source approximation holds. For a given (hkl) surface orientation, the depth of the energy minima $\Delta E_{s,\min}$ produced by buried dots then depends only on the dot volume V_0 and dot depth d, as well as the lattice mismatch ε_0 and the elastic constants and orientation of the materials. Similar to the isotropic case (Eq. 1.2), in the far-field limit the depth of the energy minima $\Delta E_{s,\min}$ can be calculated by the simple expression of

$$\Delta E_{s,\min} = -C_{hkl} \cdot E_s^{2D} V_0/d^3. \tag{1.5}$$

In this expression, $E_s^{2D} = K_{hkl} \cdot \varepsilon_0^2$ is the strain energy density of the biaxially strained 2D wetting layer, where $K_{hkl} = 2\mu(1 + \nu)/(1 - \nu)$ is given by the elastic constants and growth orientation. For the three material systems, e.g. Si/Ge, InAs/GaAs, PbSe/PbTe, the 2D misfit strain energy E_s^{2D} turns out to be of the same order of magnitude of $\sim 100\,\mathrm{meV}$/atom pair. C_{hkl} is a dimensionless constant that takes into account the elastic anisotropy and the dependence of the surface strain distributions on the (hkl) surface orientation. From the far-field strain calculations of Section 1.3.2, $C_{100} = 0.5$ for (100) Si and GaAs and 0.56 for ZnSe, and it is $C_{111} = 0.69$ for (111) PbTe or PbSe (see also [58]). For the isotropic case, $C_{iso} = 2(1 + \nu)/\pi = 0.83$ according to Eq. 1.2.

To assess the remaining parameter of the island volume, it is noted that self-assembled quantum dots usually exhibit a well-defined shape that is defined by low-energy side facets. For Ge hut cluster islands (which we consider here), these are {105} facets [19, 124], or higher-indexed facets for the larger dome-shaped islands [19, 125–127]. For InAs islands, several different facets have been reported [128, 129], but for simplicity we approximate them by pyramids with {113} facets. (111) PbSe islands exhibit a triangular pyramidal shape with {100} facets [130]. Using typical dot base widths of 60 nm for Ge and 30 nm for InAs and PbSe islands, one can estimate the corresponding island volumes to be 7200, 4270 and 1600 nm^3, respectively, which is assumed to be preserved during overgrowth. Using these values for Eq. 1.5, it turns out that the experimental cut-off spacer thickness d_c^2 of about 50 nm for the depths of the surface strain energy minima $\Delta E_{s,\min}$ is only 1–2 meV per atom pair for all three material systems. This shows that surprisingly low surface strain energy variations are sufficient in order to induce a correlated dot nucleation in multilayer structures. In fact, taking into account the significant intermixing of Stranski–Krastanow dots with the surrounding matrix material [92, 107–113] the magnitude of the energy minima will be further reduced. Although the interaction energies represent only rough estimates, it is evident that the obtained critical magnitude of surface strain energy variations is not only a factor of more than 20 smaller than the homogeneous misfit strain energies E_s^{2D} of the 2D wetting layer, but also more than one order of magnitude lower than the typical thermal energies k_BT during growth. It is therefore an open question to explain how such small energy variations can influence diffusion of surface adatoms enough in order to trigger correlated dot nucleation in multilayer structures. As a result, the exact microscopic mechanism of correlated dot nucleation still remains to be clarified.

1.3.4.2 *Vertically aligned–staggered stacking transitions*

According to the near-field strain calculations, for near-surface dots the strain fields are strongly focused along the surface normal direction. Therefore, in multilayers with small spacer thicknesses the dots should be always vertically aligned. In the far-field limit, on the other hand, several side minima appear on the surface when the growth is along the elastically soft direction and the anisotropy is sufficiently large. Thus, for large spacer thicknesses staggered *ABAB...* or *ABCABC...* dot stackings should be formed. As discussed in Section 1.3.3, generally, the lower the elastic anisotropy, the smaller the far-field correlation angles α of Eq. 1.4 and, therefore, the larger the spacer thickness where the minima spitting and stacking transition occurs. For InAs/GaAs (100), e.g., the splitting occurs when the dot depth over the dot height d/h is about ≈ 4, for CdSe/ZnSe (100) this occurs at a $d/h \approx 3$ and for PbSe/PbTe (111) at a $d/h \approx 2$. However, a splitting alone is not sufficient in order to get a staggered dot stacking. For this the splitting of the energy minima $L_{\min}$ must be larger than the given base width b of the surface dots, i.e.:

$$L_{\min} > b \tag{1.6}$$

Otherwise, the dots cannot occupy each individual strain minimum on the surface and thus follow the correlation defined by the strain energy distributions. This is the first condition for the formation of a staggered dot stacking.

According to Figs. 1.8–1.10, the separation of the energy minima L_{min} as a function of spacer thickness d can be written as:

$$L_{min.100} = \sqrt{2} \cdot d \cdot \tan \alpha_{100} \quad \text{for (100)} \tag{1.7a}$$

$$L_{min.111} = \sqrt{3} \cdot d \cdot \tan \alpha_{111} \quad \text{for (111)} \tag{1.7b}$$

Where α_{hkl} is the correlation direction of the surface strain minima with respect to the buried dot derived from the strain calculations. Combined with Eq. 1.6, the critical thickness d_c^1 for the transition to a staggered dot stacking is given by:

$$d_{c.100}^1 = b/(\sqrt{2} \tan \alpha_{100}) \quad \text{for (100)} \tag{1.8a}$$

$$d_{c.111}^1 = b/(\sqrt{3} \tan \alpha_{111}) \quad \text{for (111)} \tag{1.8b}$$

When the spacer thickness is sufficiently large, α_{hkl} approaches the constant far-field values given explicitly by Eq. 1.4. Using the corresponding value of $\alpha_{100} = 23°$ for InAs/GaAs (100), a transition from vertically aligned dots to a staggered $ABAB...$ stacking is expected at a critical spacer layer thickness of $d_c^1 = 30\,nm$ for InAs dots with a base width of $b = 20\,nm$. As shown in Fig. 1.9e, at this spacer thickness, however, the minima direction α is significant lower than $23°$ and significantly varies with spacer thickness due to the near-field effects. Thus, Eq. 1.8 cannot be solved explicitly in this case. From the plot of L_{min} over d of Fig. 1.9f, evidently the critical spacer thickness where the staggered stacking condition Eq. 1.9 is fulfilled (horizontal dashed line) is at a somewhat larger value of $d_c^1 = 40\,nm$, which agrees well with recent experimental observations [87] (see Section 1.4.3). Therefore, by using the far-field approximation the critical thickness for the staggered stacking transition is underestimated by about 30%.

For the (111) case, similar arguments apply. Using the far-field value of $\alpha_{111} = 36°$ and a typical lateral dot size of $b = 30\,nm$ for PbSe/PbTe dots, the condition of Eq. 1.8b yields a critical thickness of $d_c = 24\,nm$ for the formation of an fcc-type $ABCABC...$ dot stacking. As shown in Section 1.8, the far-field approximation again underestimates the critical thickness values by about 25% which according to near-field calculations is rather expected at a critical thickness value of $30\,nm$ as observed by experiments [50].

1.3.4.3   Replication angles and superposition of strain fields

For densely spaced buried dots, the overlapping strain fields of neighbouring buried dots must be taken into account in the surface strain calculations. Since for the initial dot layer of a multi-layer structure the dots are usually randomly arranged, the evolution of surface strain fields and resulting dot arrangements as a function of number of superlattice periods can be realistically modelled only by growth simulation as described in Section 1.5. For a simplified treatment, however, one can consider instead a periodic array of buried quantum dots, as has been used, e.g., by Shchukin *et al.* [59]. The situation is illustrated for the case of an array of buried InAs dots with a lateral spacing of $l_{dots} = 70\,nm$ along the [011] direction embedded in GaAs (100) as shown in Fig. 1.11. In this case, the InAs dots were assumed as truncated pyramids with 7 nm height and 20 nm square base (as in Fig. 1.9) and the normalized surface strain energy distributions $\rho_{strain} = E_s/E_s^{2D}$ are plotted versus lateral position x along [011] for three different spacer thicknesses of $d_s = 46$, 39 and 26 nm from (a) to (c), respectively. For the small spacer thickness, one strain energy minimum is formed above each dot, i.e. subsequent dots will be vertically aligned. At $d_s = 39\,nm$, the energy minima are split up into two side minima, as indicated by the arrows and dashed lines in Fig. 1.11b. When the spacer thickness further increases, the side minima of neighbouring dots move closer and closer such that at a spacer thickness of $d_s = 46\,nm$ one single energy minimum is formed *in between* the subsurface dots (see Fig. 1.11a). Therefore, a staggered $ABAB...$ dot stacking will be formed beyond this spacer thickness.

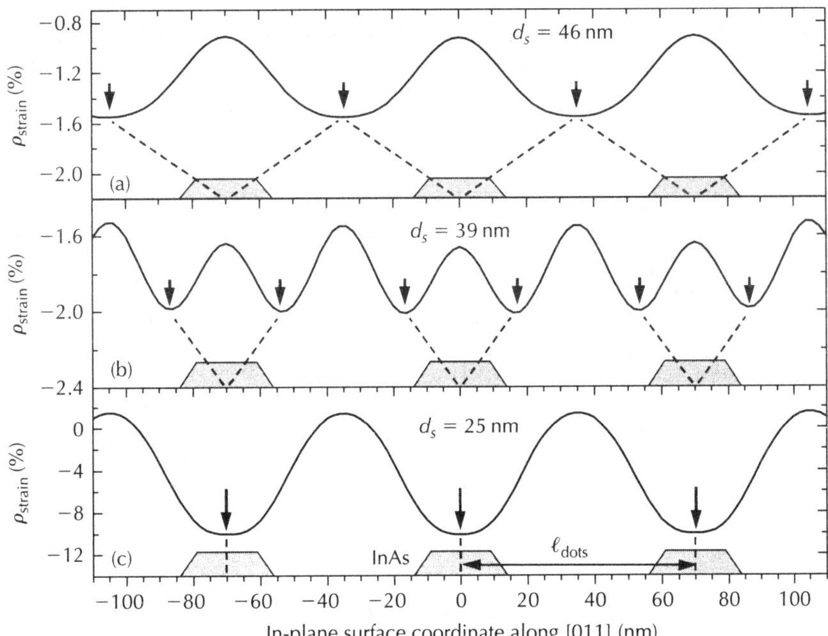

Figure 1.11 Normalized surface strain energy distribution $\rho_{strain} = E_s/E_s^{2D}$ above periodic arrays of InAs dots embedded in GaAs (100) at a fixed lateral spacing of l_{dots} of 70 nm along $\langle 011 \rangle$ but three different GaAs spacer thicknesses d_s of (a) 46 nm, (b) 39 nm and (c) 25 nm. The InAs dots are assumed as square-shaped truncated pyramids with 7 nm height and 20 nm base along the $\langle 010 \rangle$ directions, identical to those assumed in Fig. 1.9, corresponding to a 28 nm cross-section along $\langle 011 \rangle$. The position of the energy minima are indicated by the arrows.

Figure 1.12 shows the position of the surface strain energy minima x_{min} on top of the InAs dot array as obtained from these calculations as a function of spacer thickness d_s (solid line). Also shown for comparison are the surface strain minima positions that would arise from isolated dots alone (dashed line). Obviously, due to the overlap of the strain fields of the subsurface dots, a single surface strain minimum is formed in between the dots already at a spacer thickness of $d_s = 46$ nm instead of $d_s = 80$ nm (crossing of the dashed lines in Fig. 1.12) expected when the superposition

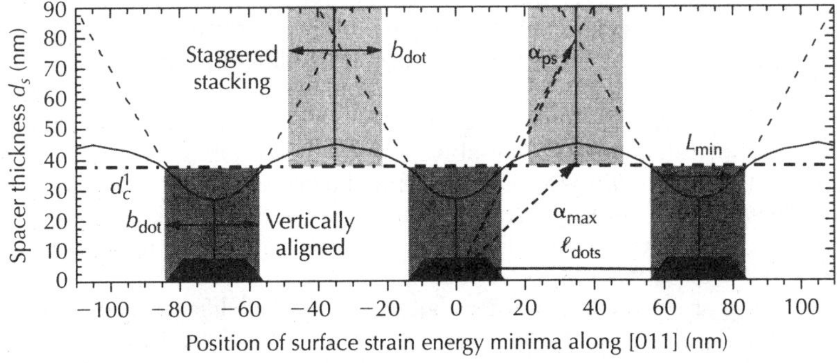

Figure 1.12 Position of the surface strain energy minima (solid line) above a periodic array of InAs dots in GaAs (100) as a function of GaAs spacer thickness d_s obtained from the calculations of Fig. 1.11. The InAs dot spacing is $l_{dots} = 70$ nm along the $\langle 011 \rangle$ direction and the dots assumed as square-shaped, truncated pyramids with 7 nm height and 20 nm or 28 nm base width b_{dots} along $\langle 010 \rangle$ or $\langle 011 \rangle$, respectively. The energy minima positions for isolated non-interacting dots are represented as a dashed line for comparison. The horizontal dash-dotted line represents the critical spacer thickness d_c^1 at which the transition from vertical to staggered dot stacking is expected. L_{min} is the lateral spacing of the energy minima, and α the resulting angle of interlayer dot replication, ranging from $\alpha_{max} \cong 50°$ to $\alpha_{ps} \cong 23°$ (point source).

of the strain fields is neglected. With respect to the transition between vertically aligned and staggered dots, it is evident that this should take place when the lateral splitting L_{min} of the side minima above each dot becomes larger than *half* of the lateral dot spacing l_{dots}, i.e.

$$L_{min} > l_{dots}/2 \tag{1.9}$$

Under this condition, indicated by the horizontal dash-dotted line in Fig. 1.12, the minima produced by neighbouring dots, i.e. $(l_{dots} - L_{min})$ are closer to each other than the minima produced by the same dot. According to Fig. 1.12, for the chosen InAs dot array with $l_{dots} = 70$ nm, this will occur at a critical spacer thickness $d_c^1 = 38$ nm, but evidently this critical thickness will be shifted to larger (or lower) values when the spacing within the dot array increases (or decreases).

For a simple estimate of the critical thickness d_c^1 for the stacking transition, it is noted that at the transition point, the lateral energy minima splitting above the dot array is nearly the same as that for the individual dots (compare solid and dashed lines in Fig. 1.12). If we further take into account that for single InAs/GaAs (100) dots the energy minima splitting along $\langle 011 \rangle$ is given by $L_{min} = 2d \tan \alpha$ (note the $\sqrt{2}$ difference to Eq. 1.7a), the critical thickness d_c^1 condition for the stacking transition can be recast in the form:

$$d_c^1 \cong l_{dots}/(4\tan \alpha_{100}). \tag{1.10}$$

If we further neglect the dependence of α_{100} as a function of spacer thickness (*cf.* Fig. 1.9) and simply use the far-field value of $\alpha_{ps} = 24°$ for InAs/GaAs (100) according to Eq. (1.4a), an approximate critical thickness of $d_c^1 = 0.56 \cdot l_{dots}$ is obtained, which for the given set of parameters yields a critical thickness $d_c^1 = 39$ nm, in reasonable good agreement with the value $d_c^1 = 38$ nm $= 0.56 \cdot l_{dots}$ obtained from the array calculations (horizontal dash-dotted line in Fig. 1.12). Almost identical results were obtained by Kunert and Schöll [93] from strain calculations for square arrays of InAs dot pyramids with 20 nm width and 5 nm height. The critical thickness values obtained for arrays of extended dots is almost identical to that calculated using the far-field α value, because the smaller correlation angle α_{100} of the near-field calculation is compensated by the energy minima shift caused by the lateral overlap of the strain fields. Thus, the overall agreement with the simple point-source approximation is remarkably good.

In real multilayer structures, obviously, the average lateral spacing of the disordered dots in the first dot layer strongly depends on the growth condition. Therefore, the growth conditions *and* spacer thickness must be precisely tuned in order to meet the condition for a staggered dot stacking. For example, for InAs dots spaced at either 50 or 100 nm, the critical thickness d_c^1 changes from 26 to 52 nm. Also, it is noted that critical thickness derived from the periodic array calculations (Eq. 1.10) gives very similar results compared to thickness predicted by Eq. (1.8) based on the comparison of the lateral dot size b with the energy minimum spacing L_{min}. Practically, all three conditions of Eq. 1.5–1.10 must be met in order to obtain a staggered dot stacking in multilayer structures ($d_c^1 < d_s < d_c^2$), which may be difficult or even impossible to achieve for a given material system. Shchukin *et al.* [59] has also applied the model of interacting dot arrays to II–VI semiconductor multilayers. It was found that due to the higher elastic anisotropy and larger α_{100} for II–VI compounds (see Fig. 1.7), a transition to a staggered dot stacking should occur already at a critical spacer thickness of $d_c^1 \cong 0.35 \cdot l_{dots}$, which also seems to fit with the experimental observations [47, 48].

While the model of interacting dots does not yield a marked difference in the stacking condition compared to the single dot model, the resulting interlayer dot correlation angle in a multilayer structure changes significantly when the spacer thickness approaches d_c^1 due to the lateral shift of the surface energy minima positions (see Fig. 1.12). For a periodic dot array, i.e. for $d \geq d_c^1$, the dots in each subsequent layer will nucleate in the middle between neighbouring buried dots. Thus, the lateral shift Δ of the dot position with respect to the buried dots is fixed at a constant value $\Delta = l_{dots}/2$ independent of spacer thickness. As a result, the interlayer replication angle α_{corr} is no longer constant but varies with changing spacer thickness according to

$$\alpha_{corr} = \tan^{-1}(l_{dots}/2d) \tag{1.11}$$

At the critical thickness $d \approx l_{dots}/2$, and thus α_{corr} is as large as $\alpha_{max} = 50°$ for the given InAs dot array (see arrow in Fig. 1.12), compared to $\alpha_{ps} = 23°$ expected for a single buried dot. This increase in interlayer correlation angle is in good agreement with the experimental values reported by Wang *et al.* for InGaAs/GaAs dots at a comparable spacer thickness [87]. According to Eq. 1.11, for thicker spacer layers, the dot replication angle should decrease and approach the value of 23° of the single point-source model. In fact, such difference in correlation angle has been recently reported by Gutierrez *et al.* [88] for InGaAs/GaAs multilayers, where for well-separated isolated dots the interlayer correlation angle was found to be around 23° close to the single dot model, whereas for densely spaced dots with large strain field overlap a much larger value of around 50° was observed. Finally, it is noted that according to strain calculations, when L becomes as large as $\sim 3 \cdot l_{dots}/2$, the staggered dot stacking of periodic arrays should switch back to a vertical dot alignment (see Shchukin *et al.* [59]). However, since at such large spacer thicknesses the strain interaction is very weak, there is little chance that this transition can be actually obtained by experiments.

1.4   Comparison with experimental results

Experimentally, interlayer correlations in multilayer structures have been studied mainly for low-indexed surface orientations of material systems such as InAs/GaAs (100) [15, 16, 42–45], Si/Ge (100) [19, 22, 32, 35–38] and PbSe/PbEuTe (111) [18, 30, 49–57]. Since in these cases, the growth direction is parallel to the elastically soft direction, staggered dot stackings are expected in the far-field limit, whereas the dots should be vertically aligned for near-surface dots. For high-indexed surfaces, recent experimental studies have been reported for InGaAs/GaAs multilayers grown on $(n11)$B surfaces, and the results will be compared to theoretical predictions as well.

1.4.1   Vertically aligned dots

Vertically aligned dots are generally expected under three conditions, namely, (i) the materials exhibit a small elastic anisotropy, (ii) the growth axis is along the elastically hard direction, or (iii) the spacer layer thickness is small, i.e. less than about 2–3 times the dot height. In all three cases, the surface strain minima produced by the subsurface dots will be directly above the buried dots, i.e. the dots will replicate in the vertical growth direction. Experimentally, for multilayers with small spacer thicknesses, a vertical dot alignment has been found in experiments. This applies not only to (100) InAs/GaAs [15, 16, 31, 42–45, 78], InGaP/InP [77], SiGe/Si [19, 22, 32, 34–38] or GaN/AlN [80–85] multilayers, but also for (111) PbSe/PbTe superlattices [50–57]. Various examples are compiled in Fig. 1.13 for (a) InAs dot-in-a-well superlattices with 30 nm GaAs spacers (Gutierez *et al.* [78]), (b) to (d) InP/Ga$_{0.52}$In$_{0.48}$P dot multilayers on GaAs (100) with three different spacer layer thicknesses of 16, 4, and 2 nm, respectively (see Zundel *et al.* [77]), as well as (e) an 80 period GaN/AlN dot superlattice on a 6H–SiC substrate consisting of six monolayers GaN quantum dots alternating with 10 nm thick AlN barriers (see Sarigiannidou *et al.* [84]). Another example was also shown in Fig. 1.2a for an InAs/GaAs dot superlattice with 20 nm spacers (Darhuber *et al.* [44]). Clearly, in all cases a well-defined vertical dot alignment is formed and additional results are described in Sections 1.6 and 1.7 for the InAs/GaAs (100) and SiGe/Si (100) systems. The predominant vertical dot alignment observed for the large majority of experimental studies is due to the fact that in most works spacer thicknesses smaller than 40 nm were used, in which case the elastic strain fields and thus the strain minima are vertically aligned due to free surface relaxation. Moreover, for the SiGe/Si (100) system as well as the wurzite GaN/AlN system, the elastic anisotropy is not very large and, therefore, vertically aligned dots are observed for all spacer thicknesses up to which interlayer correlations persist. To our knowledge, other materials with low elastic anisotropy have not been studied yet.

Concerning the growth along the elastically hard direction, i.e. (111) for SiGe, III–V and II–VI compounds or (100) for IV–VI compounds, it is noted that these growth orientations are less commonly used in these material systems and therefore little work has been carried out in this respect. It also turns out, that for these surface orientations, under the usual growth conditions

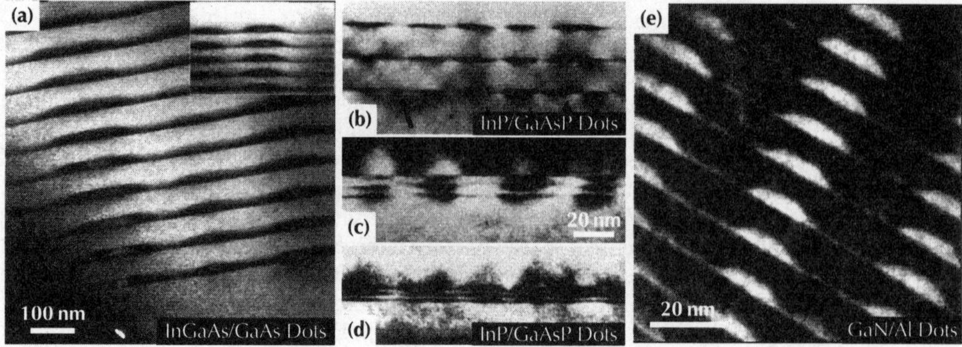

Figure 1.13   *Cross-sectional transmission electron micrographs of various vertically aligned quantum dot multilayer structures. (a) 10 period superlattice of 2.9 ML InAs dots in 8 nm $In_{0.15}Ga_{0.85}As$ wells separated by 30 nm GaAs spacer layers grown on GaAs (100). Inset: Superlattice of $In_{0.5}Ga_{0.5}As$ quantum dots separated by 12 nm GaAs spacer layers (see Gutierez et al. [78]). Centre: 3 ML InP/$Ga_{0.52}In_{0.48}P$ quantum dot multilayers on GaAs (100) with three different $Ga_{0.52}In_{0.48}P$ spacer thicknesses of 16, 4, and 2 nm from (b) to (d), respectively (see Zundel et al. [77]). (e) 80 period GaN/AlN quantum dot superlattice consisting of six monolayers GaN alternating with 10 nm AlN spacers, grown by plasma-assisted molecular beam epitaxy on 6H–SiC substrate. The image represents the Ga elemental map from energy filtered transmission electron microscopy (see Sarigiannidou et al. [84]).**

the lattice mismatch strain tends to be relaxed plastically by misfit dislocations rather than by coherent 3D islanding. This has been found, e.g., for InAs on GaAs (111) [131–133] as well as for PbTe/PbSe (100) [134, 135] by scanning tunnelling microscopy studies. Thus, self-assembled Stranski–Krastanow dots seem to be difficult to obtain. For the SiGe/Si (111) system, 3D island formation has been observed (see, e.g., [136, 137]), but to our knowledge, no multilayer structures have been investigated so far. From the strain calculations described above, a vertical dot alignment should occur for this growth orientation.

1.4.2   Fcc-like dot stacking

For materials with $\langle 111 \rangle$ as elastically soft directions ($A < 1$), a three-fold splitting of the surface strain minima is predicted in the far-field limit for (111) surfaces. Therefore, an *fcc*-like *ABCABC...* dot stacking should take place if the elastic anisotropy as well as spacer thickness is sufficiently large. This is met by the IV–VI semiconductors with their particularly large elastic anisotropy. For self-assembled (111) PbSe/PbEuTe dot superlattices with PbEuTe spacer thicknesses between 40 and 55 nm, indeed an *fcc*-type dot stacking has been found experimentally [18–20, 49, 50, 53–57]. Figure 1.14 shows the (222) reciprocal space maps of a 60 period PbSe/PbEuTe dot superlattice with 47 nm PbEuTe spacers recorded along two different azimuth directions of (a) $[\bar{2}11]$ and (b) $[\bar{1}\bar{1}2]$ that are rotated by 60° with respect to each other. Clearly, a large number of satellite peaks are observed in the q_z as well as q_x directions perpendicular, respectively, parallel to the layer surface. This proves that the dots are highly ordered both laterally and vertically. A striking feature in the reciprocal space maps is that the lateral satellite peaks are not aligned along the horizontal q_x direction but rather along *inclined* directions as indicated by the dashed lines in Fig. 1.14a and b. This proves that the PbSe dots are correlated along directions oblique to the growth direction. Moreover, the mirror symmetric arrangement of the satellite peaks in the two reciprocal space maps indicates the existence of a 3 m symmetry of the 3D lattice of dots, in contrast to a hexagonal 6 mm symmetry that would be expected if the dots were vertically aligned. This indicates the formation of a trigonal 3D lattice of dots. The corresponding dot arrangement is shown schematically in Fig. 1.8 and corresponds to hexagonally ordered 2D dot layers that are *fcc*-like stacked in the vertical direction [18, 49].

Within each dot layer, the dots are highly ordered in the lateral direction as is proven by the atomic force microscopy image [18, 30, 49] of the final layer of the multilayer structure shown in Fig. 1.14d. From the satellite spacings in the reciprocal space maps, the lateral dot separation

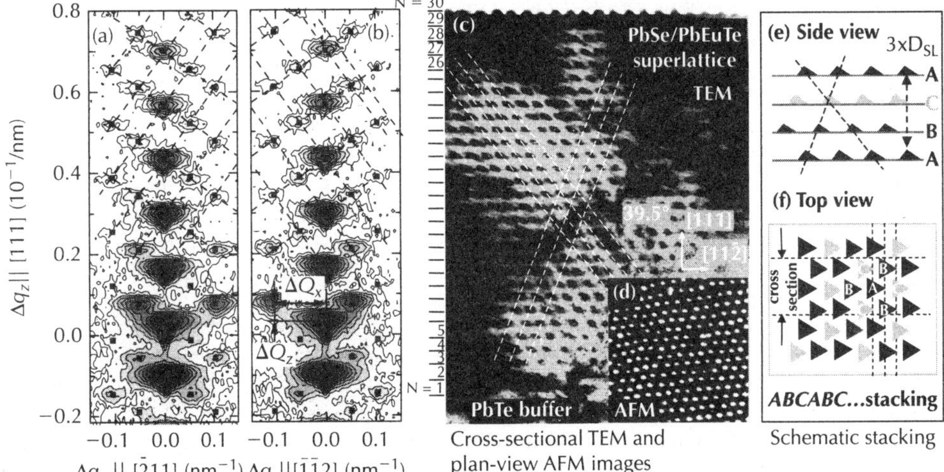

$\Delta q_z \parallel [111]$ $(10^{-1}/nm)$

$\Delta q_x \parallel [\bar{2}11]$ (nm^{-1}) $\Delta q_x \parallel [\bar{1}\bar{1}2]$ (nm^{-1})

Cross-sectional TEM and plan-view AFM images

Schematic stacking

Figure 1.14 Left-hand side: *X-ray reciprocal space maps of a 60 period PbSe/PbEuTe quantum dot superlattice with 47 nm period recorded around the (222) Bragg reflection along the (a) [$\bar{2}$11] and (b) [$\bar{1}\bar{1}$2] azimuth directions which are mirror symmetric with respect to each other due to the 3 mm trigonal symmetry of the dot arrangement. The squares indicate the expected intensity maxima for an fcc-like ABCABC... dot stacking sequence.* Centre: *Cross-sectional transmission electron microscopy image (c) and plan-view atomic force microscopy image (1 × 1 μm²) of the final dot layer (d) of a corresponding 30 period dot superlattice. The dashed lines in (c) indicate the oblique interlayer correlation direction of the dots.* Right-hand side: *Schematic illustration of the ABCABC... dot stacking as seen in cross-section (e) and plan view (f). See Springholz et al. [18–20, 49].*

within the dot planes is determined as 68 nm, which exactly fits with the strain energy minima separation calculated by the far-field strain interaction model [18]. With the given 47 nm vertical separation of the dot layers, a trigonal lattice constant of $a_0 = 61$ nm and a trigonal angle of $\alpha = 39.5°$ are obtained for this sample. The latter is again in good agreement with the correlation angle predicted by the far-field model (Section 1.3.2). From the analysis of dot superlattices with different spacer thicknesses it is found that the in-plane dot separation within the dot layers scales linearly with spacer thickness [18, 50]. Thus, the trigonal lattice constant of the 3D dot arrangement can be tuned continuously by changing the spacer thicknesses.

Figure 1.14c shows the cross-sectional transmission electron micrograph of an *fcc*-stacked PbSe dot superlattice with 30 periods, demonstrating that the dots are aligned in directions inclined to the growth axis as indicated by the dashed lines. According to the schematic drawings of Fig. 1.14e and f, the dot arrangement in the transmission electron microscopy cross-section agrees well with the expected *ABCABC...* stacking sequence. It is also obvious that a highly regular periodic lattice of dots is formed in all directions and that the dot ordering sets are already within the first superlattice layers, in agreement with atomic force microscopy studies [18, 30, 49]. This underlines the highly efficient lateral ordering process that is described in detail in Sections 1.5 and 1.8. Further growth experiments show that the *fcc*-type dot stacking persists up to spacer thicknesses of about 55 nm [50], after which the strain fields become too weak to enforce a correlated dot nucleation. For spacer thicknesses lower than about 40 nm, a transition to a vertical dot alignment is found [50–57], in agreement with the near-field strain calculations presented in the previous section. Thus, this material system has proven to be an excellent model system for investigation of ordering and stacking types in multilayer structures.

1.4.3 Anticorrelated and staggered dot stackings

For materials with anisotropy ratio larger than about 1.5, a strain minima splitting occurs for (100) surfaces in the far-field limit (see Fig. 1.9). Therefore, a body centred–tetragonal dot stacking as shown in Fig. 1.8a should be formed under appropriate growth conditions. Although this applies, e.g., to the Ge/Si and InAs/GaAs (100) systems, experimentally, mostly a vertical dot alignment has been observed in these cases. This results from the fact that the elastic anisotropy and thus the lateral splitting of the surface strain minima is not as large as for the IV–VI compounds. From the near-field strain calculations (Section 1.3.3.1), in addition, only one central vertically aligned strain minimum is expected for spacer thicknesses smaller than ~ 30 nm for InAs/GaAs and ~ 100 nm for the SiGe/Ge. Even for larger spacer thicknesses, the energy minima splitting of $L_{min} = \sqrt{2}d \tan \alpha_{100}$ is still limited due to the small α_{100} values below $23°$, and for large spacer thicknesses, the elastic strain fields of the buried dots are rather small. Therefore, for Si/Ge or III–V structures it is difficult to obtain a staggered dot stacking for the possible range of spacer thicknesses.

An exception represents the work of Wang *et al.* [87], in which a staggered or anticorrelated dot stacking was found in InGaAs/GaAs multilayers with spacer thicknesses larger than 40 nm. In their cross-sectional scanning tunnelling microscopy study of interlayer correlations as a function of spacer thickness it was found that for a GaAs spacer thickness of $d_s = 25$ nm the dots are vertically aligned, for 34 nm spacers the stacking is not very well defined, but for larger spacer thicknesses of 54 nm a clear staggered or anticorrelated dot stacking in consecutive layers is formed. This is demonstrated by the corresponding cross-sectional scanning tunnelling microscopy images displayed in Fig. 1.15a to c, and this stacking transition is revealed even more clearly by the enlarged STM images shown in Fig. 1.15d and e. These results agrees well with the predictions from the near-field strain calculations discussed in Sections 1.3.3–1.3.4. A similar staggered dot stacking was also found for low density InAs/GaAs dot superlattices by Gutierrez *et al.* [88], where for 25 nm GaAs spacers the InAs dots were vertically aligned, whereas for 35 nm spacers the dots were correlated along oblique directions inclined by $23°$ to the surface normal, as expected from the far-field model. On the other hand, in the same study for InGaAs/GaAs dots with high dot density, a much higher correlation angle of $50°$ was found, similar to the angle deduced by Wang *et al.* [87] in Fig. 1.15. As was already discussed in Section 1.3.4 this difference can be explained by considering the effect of overlapping strain fields of closely spaced buried dots [93].

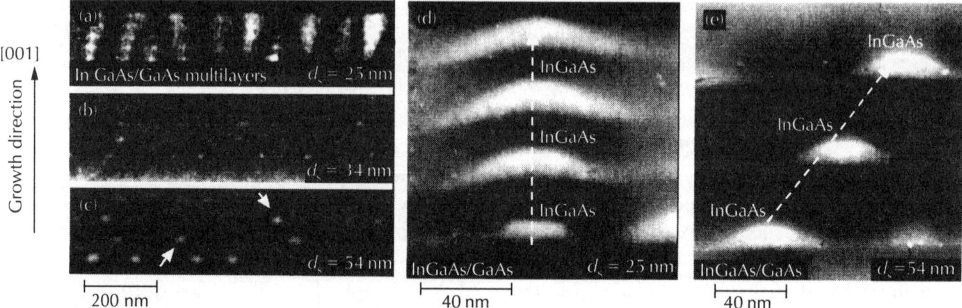

Figure 1.15 Cross-sectional scanning tunnelling microscopy images of $In_{0.5}Ga_{0.5}As$/GaAs quantum dot multilayers with different GaAs spacer thicknesses d_s of (a) 25 nm, (b) 32 nm and (c) 54 nm. For multilayer with thin spacers, the dots are vertically aligned, whereas for the sample with thick spacers an anticorrelated staggered dot stacking is observed with an interlayer correlation angle of about 50° to the surface normal. Enlarged cross-sectional STM images of the sample with 25 and 54 nm spacers are shown in (d) and (e), respectively. Adapted from Wang et al. [87].

Oblique correlations and staggered stackings were also observed for self-assembled InAs/InAlAs [60–65, 138, 139], InAs/InP [140, 188] and InAs/InGaP [187] quantum *dash* or quantum *wire* multilayers on (100)-oriented InP substrates, as is exemplified by Fig. 1.16a. In these structures, due to anisotropy of surface kinetics or kinetic step bunching, the usual InAs quantum

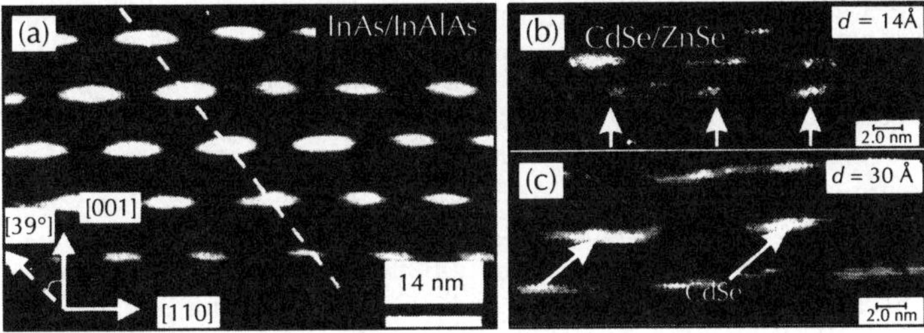

Figure 1.16   Left-hand side: *(a) Cross-sectional transmission electron microscopy images of a vertically anticorrelated InAs/InAlAs quantum wire superlattice with 6 ML InAs wires alternating with 10 nm In$_{0.52}$Al$_{0.48}$As spacer layers. Adapted from Hanxuan Li et al. [138].* Right-hand side: *Processed TEM image of multi-sheet arrays of 2D CdSe islands separated by ZnSe layers with (b) 1.5 and (c) 3 nm with a contrast proportional to the local lattice parameter, i.e. Cd composition obtained by lattice analysis. The multilayer with smaller spacers exhibits a preferred vertical correlation of the CdSe islands, whereas for the thicker spacers the islands are anticorrelated. Adapted from Shchukin et al. [4] and Krestnikov et al. [47].*

dots are strongly elongated along the [−110] surface direction so that 1D quantum wires rather than 0D quantum dots are formed. As illustrated in Figs 1.16a and 1.2d, in several studies an anticorrelated stacking of these wires was found, with the interlayer correlation angles varying over a rather wide range between 20° and 50° in dependence of the growth conditions, material composition and spacer thicknesses. For one-dimensional buried wire structures aligned along the [−110] directions, the strain fields on (100) surfaces have the same basic properties as for buried quantum dots, except that instead of four only two strain minima will appear on the surface laterally displaced with respect to the buried wire, but the correlation angles in the far-field limit are close to the 23° derived for InAs/GaAs quantum dots. Although the observed anticorrelation complies with the basic trend derived from the anisotropic strain field calculations, due to the fact that an anticorrelated stacking is found also for small spacer thicknesses additional interlayer interaction mechanisms have been invoked as the origin for the anticorrelated stacking in this case [62–65]. This will be further discussed in section 1.9. A particularly interesting mixed interlayer stacking was also found for two-fold stacked InAs quantum dots on lithographically prepatterned (001) GaAs substrate templates [141]. In this case, two different dot types are found in the second layer, one in the "on top" positions and one in the "staggered" positions. As shown by Heidemeyer *et al.* [141], this special dot stacking is caused by elastic anisotropy of the spacer layers and the overlapping strain fields.

For II–VI semiconductors, the elastic anisotropy is significantly larger than for III–V compounds. Thus, there is a stronger tendency for the formation of staggered dot stackings. Indeed, anticorrelated dot stackings have been found for (100) CdSe/ZnSe [46, 47] as well as CdTe/ZnTe [48] multilayers. This is exemplified by the cross-sectional transmission electron microscopy images of Fig. 1.16c and 1.14e, respectively. The anticorrelation resembles the theoretically predicted *ABAB...* stacking shown in Fig. 1.8a that arises from the four-fold splitting of the surface strain distributions. In addition, a local energy *maximum* is formed directly above each buried dot (see Fig. 1.6f), which means that this point is unfavourable for subsequent island nucleation. Since the four energy minima are separated only by a weak saddle point (see Fig. 1.6f), there is no strong lateral ordering tendency during growth, i.e. the dots remain rather disordered in the lateral direction. As a consequence, no well-ordered centred-tetragonal dot lattice but only an anticorrelated dot arrangement is formed. The experimentally derived interlayer correlation angles of these structures [46–48] are also found to be significantly larger than those expected from the point-source model. For example, α is deduced as 40° from the transmission electron microscopy image of Fig. 1.2e as compared to the expected far-field value of 28°. This was explained by Shchukin *et al.* [59] by considering the elastic interaction energy between sheets of periodic 2D arrays of strained nanoislands. Calculating the interaction energy as a function of the relative vertical and lateral displacements of the island arrays in successive layers it was found that for a certain spacer thickness range the energy is minimized when the islands are *anticorrelated* in successive layers. For II–VI multilayer, thus, a transition from a vertical dot alignment to an anticorrelated stacking was predicted for spacer thicknesses exceeding more than one third of the lateral array period. As shown in Fig. 1.16b and c, this is in reasonable agreement with the experimental observations [47, 48].

1.4.4   Oblique replication on high-indexed surfaces

Quantum dot multilayers on vicinal or high-indexed surfaces represent a particularly interesting case for interlayer correlation formation. This is due to the highly anisotropic surface strain distributions induced by buried dots as well as the resulting oblique interlayer dot correlations predicted by the strain calculations (Fig. 1.6). Vicinal and high-indexed surfaces also exhibit a high anisotropy of almost all surface properties such as surface diffusion and surface and step edge energies which result in anisotropic dot shapes and a higher tendency of lateral ordering (see, e.g. Refs. [105, 106, 142, 144, 146, 195]). However, only recently, systematic studies by Schmidbauer *et al.* [142, 143] have deduced the interlayer correlations in high-indexed multilayer structures. The investigated structures consisted of 16.5 period InGaAs/GaAs quantum dot superlattices with 10 monolayer InGaAs dots alternating with 120 monolayer GaAs spacer layers grown on $(n11)B$ GaAs substrates with $n = 3, 4, 5, 7$ and 9. Apart from the remarkable lateral ordering attributed to anisotropic surface diffusion [142, 144], the direction of the interlayer dot correlations was found to be inclined to the surface normal towards the in-plane $[2nn]$ directions as predicted by the far-field strain model.

The interlayer dot correlation angles α in these samples were determined by diffuse X-ray reciprocal space mapping [142] as exemplified in Fig. 1.17a and b for the $(511)B$ superlattice. The such determined correlation angles of 8, 17, 19, 36 and 21° for the $(311)B$, $(411)B$, $(511)B$, $(711)B$ and $(911)B$ superlattices, respectively, are plotted as full diamonds ($\blacklozenge$) in Fig. 1.17c as

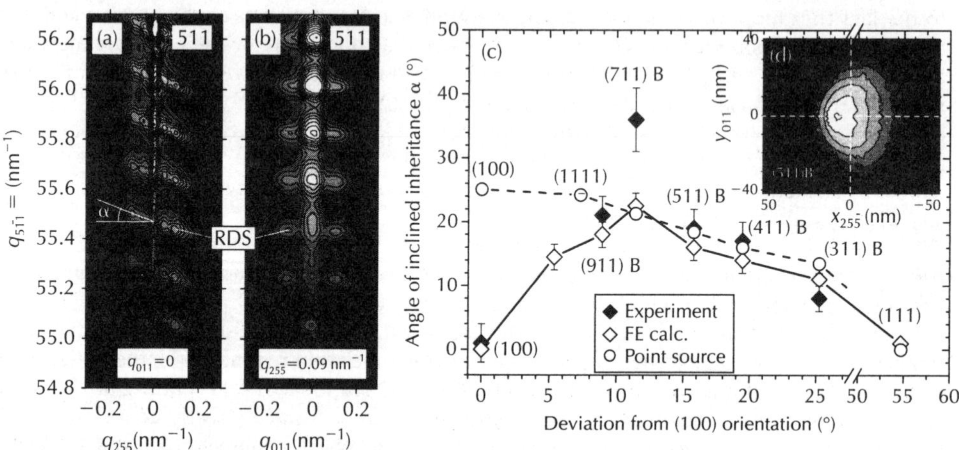

Figure 1.17   Interlayer correlations for 16.5 period $In_{0.4}Ga_{0.6}As/GaAs$ *superlattices with 120 monolayer GaAs spacers grown on high-indexed* $(n11)B$ *substrates. The diffuse X-ray reciprocal space maps of the* $(511)B$ *sample around the* (511) *reciprocal lattice point are shown in (a) and (b) for two different azimuth directions of the X-ray beam. For the map recorded along the* $[225]$ *direction, the diffuse scattering around the lateral satellite peaks are concentrated along lines inclined by the angle* α *to the horizontal direction, whereas for the other map along* $[011]$ *the side maxima are horizontally aligned. This indicates an inclined interlayer correlation of the InGaAs quantum dots towards the* $[255]$ *direction. The experimentally determined correlation angle* α *(angle of inclined inheritance) for the different samples is plotted in (c) as* ($\blacklozenge$) *versus surface orientation, together with the surface strain energy minima direction calculated by the finite element method* ($\diamond$) *as well as by the point source approximation* ($\circ$)*. For the finite element calculations, the buried dots at a depth of 30 nm are assumed to be lens shaped with 5 nm height and 30 nm with, and the example of the resulting surface strain energy distribution is shown in (d) for the* $(511)B$ *surface. Adapted from Schmidbauer* et al. *[142].*

a function of the tilt of the $(n11)$ surface to the (100) direction. The experimental correlation angels are remarkably close to the respective angles of 13.5, 16.3, 18.4, 21.5 and 22.5° predicted by the point-source model that are represented by the open circles (○) in Fig. 1.17c. The elastic strain fields of the buried dots were also calculated by Schmidbauer *et al.* using the finite element method [142], assuming lens shaped InGaAs dots with 5 nm height and 30 nm width at a depth of 30 nm GaAs below the surface. The example of such a surface strain energy distribution is shown in Fig. 1.17d for the (511)B surface, showing the clear shift of the strain energy minimum along the in-plane [255] surface direction arising from the elastic anisotropy of the GaAs matrix. The correlation angles deduced from the finite element calculations ((◇) in Fig. 1.17c) are found to agree well with the measured correlation directions (◆). In addition, these angles are quite close to the point-source approximation (o), except for the surface orientations close to (100) for which only one surface minimum is formed in the finite element calculations. The lateral ordering of the dots in the superlattices on high-indexed substrates was found to be rather good [142, 144, 145] and is discussed in more detail in Section 1.6.1. For the Ge/Si system, up to now only (113) and (311) dot superlattices have been investigated with Si spacer thicknesses in the range of 30 to 60 monolayers [146]. From cross-sectional transmission electron microscopy, some indication for inclined dot replication can be inferred but this was not clarified in more detail.

1.5   Monte Carlo growth simulations

In multilayer structures, the dots in the first layer are usually completely disordered just as in single dot layers grown under the same conditions. In the subsequent dot layers, the lateral arrangement of dots changes due to the interlayer strain interactions. Therefore, the questions arise (i) how this rearrangement evolves as a function of the number of deposited superlattice periods, (ii) what final structure is eventually formed, and (iii) which conditions are required such that a self-organized lateral ordering occurs. Lateral dot ordering obviously requires a significant overlap of the strain fields of neighbouring subsurface dots and/or a splitting of the surface strain energy minima. Otherwise, each dot will merely be replicated along the strain minima direction, i.e. the disordered dot arrangement of the first layer will be just replicated from one layer to another, which is expected for low-density dilute dot systems with lateral dot separations that are large compared to the spacer layer thickness.

When the dot density is high, the strain fields of subsurface dots strongly overlap. Thus, the locations of the local surface strain minima on the spacer layer sensitively depend on the relative position of the buried dots. Depending on the arrangement of the buried dots, the strain minima will be laterally shifted with respect to the buried dot positions. In particular, for closely spaced subsurface dots the strain minima may merge into one minimum and new minima may be introduced between widely spaced buried dots [17]. As a result, the dots in each new layer will be rearranged until a quasi steady state configuration is reached, which under favourable conditions exhibits some short-range lateral order of dots [17, 58, 147]. A much more pronounced effect is expected when the surface energy minima are split into several side minima due to the elastic anisotropy [18, 19, 58]. The strain distributions then significantly change in each layer and thus a much faster dot rearrangement and much more efficient lateral ordering will take place.

To assess the lateral ordering tendency in multilayer structures, it is instructive to simulate the dot rearrangements taking place during multilayer growth. This can be done by introducing a few assumptions, namely, (i) that the dot nucleation sites are determined by the local minima of the surface strain fields of the buried dots due to the attraction of the deposited adatoms and (ii) that the size of the dots created at each energy minimum is proportional to the surface area from which the adatoms are collected [17]. Applying these assumptions to an isotropic and deterministic continuum growth model, Tersoff *et al.* [17] have shown that in 1 + 1 dimensions a rapid homogenization of the size and lateral dot spacings in a superlattice structure may occur. In our work, we have extended this model to 2 + 1 dimensions and included explicitly the elastic anisotropy and growth orientation of the materials [58] in order to determine the 2D arrangement of the dots within multilayer structures. In addition, the Monte Carlo method was used to simulate the growth of the dots in each layer. The total simulation procedure thus consists of the following steps [58, 147]:

1. Random deposition of atoms.
2. Surface diffusion of the adatoms along the gradients in the surface chemical potential determined by the strain fields of the buried dots until the nearest strain energy minimum is reached. The atoms gathered in a particular minimum create a quantum dot with a size proportional to the number of collected adatoms. The latter is proportional to the Voronoi polygonal area around each island.
3. After deposition of 10^4 adatoms, the surface is covered by a spacer layer with a given thickness and the new surface energy distribution due to the buried islands is calculated by summation of the strain fields of all islands in the first subsurface dot layer. The small contribution of the lower dot layers in the superlattice stack is neglected.

This sequence is repeated N times for the desired number of superlattice periods. The finite lateral extent of the dots is taken into account by excluding nucleation of new dots within a certain radius R_d around already existing dots. This denuded zone also fixes the dot density in the first layer and it can be associated with an effective surface diffusion length. As further simplification, we consider the buried islands as point-like stress sources, corresponding to the far-field limit of strain interactions, and we also ignore the strain fields of the surface dots during the growth of the dot layers. This is because these strain fields are very small at the stage of dot nucleation and thus are not expected to influence much the nucleation sites of the dots. To speed up the simulations and thus allow treating very large surface areas ($>1 \, \mu m^2$) and large dot ensembles, the hopping of deposited adatoms is performed on a surface mesh of 2.5×2.5 nm and whole adatom clusters of this size are moved at the same time.

Figure 1.18 shows the simulation results for three representative cases, namely, (a) an elastically isotropic multilayer with arbitrary growth orientation, (b) a (100)-oriented SiGe dot superlattice where the Ge dots are separated by 20 nm Si spacer layers, and (c) a (111)-oriented PbSe dot superlattice with 50 nm PbEuTe spacer layers. In Fig. 1.18, for each case, the dot positions within 9th and 10th, respectively, 10th, 11th, and 12th superlattice layers are represented by

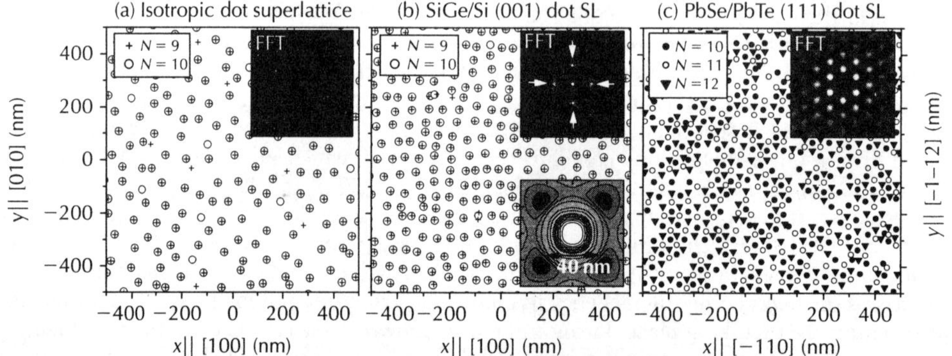

Figure 1.18   Monte Carlo growth simulations of self-organized quantum dot superlattices showing the island positions in the last superlattice dot layers. (a) 9th (+) and 10th (○) layer of an elastically isotropic superlattice, (b) 9th (+) and 10th (○) layer of a (001) SiGe/Si dot superlattice with 20 nm Si spacer layers, (c) 10th (•), 11th (○) and 12th (▼) layer for a (111) PbSe/PbEuTe superlattice with 50 nm PbEuTe spacers. The insets show the 2D Fourier (FFT) power spectrum of the island positions in the last superlattice layer. Lower insert in (b): elastic energy distributions above a buried SiGe island shown as iso-strain energy plot with the darker areas indicating regions of higher strain energy. Thus, there exist four repulsive strain energy maxima along the $\langle 011 \rangle$ surface directions, which induce a preferential in-plane dot alignment along the $\langle 010 \rangle$ surface directions. See Holy et al. [58] and Springholz et al. [19] for details.

different symbols and plotted on top of each other. For the elastically isotropic case (Fig. 1.18a), the dots are always vertically aligned on top of each other but there is obviously not any lateral ordering tendency. This is revealed by the Fourier transform power spectrum of the final dot positions shown in the insert. The clear discrepancy to the isotropic 1 + 1D growth simulations by Tersoff *et al.* [17] is due to the fact that a uniform dot spacing in one dimension is not sufficient to produce an ordered structure in two dimensions if there is no additional in-plane anisotropy.

For the SiGe/Si (100) growth simulation (Fig. 1.18b), because the assumed denuded zone R_d of 50 nm is significantly larger than the lateral splitting of the energy minima induced by the elastic anisotropy, the dots are nearly perfectly aligned in the growth direction. In addition, however, there is also a weak lateral ordering tendency along the in-plane $\langle 010 \rangle$ surface directions as is indicated by the arrows in the 2D FFT power spectrum of the dot positions shown in the upper inset of Fig. 1.18b. A detailed analysis shows that this preferred dot alignment is due to the fact that above each dot four *maxima* in strain energy appear on the surface along the $\langle 011 \rangle$ directions away from the energy minima. This is shown by the lower inset of Fig. 1.18b, where the strain energy distribution on the 20 nm Si spacer above a buried Ge dot is depicted over an extended surface area and where the strain *maxima* are indicated by the dark colour. These maxima cause a *repulsive* interlayer dot interaction and, therefore, the subsequent surface dots tend to nucleate along the $\langle 010 \rangle$ surface directions relative to each other. The same result is also found in simulations for (100) InAs/GaAs dot multilayers because the symmetry and basic shape of the strain distribution are similar to those for the SiGe system. Due to the higher elastic anisotropy, however, the observed lateral ordering tendency is even enhanced.

For the third case of (111)-oriented PbSe/PbEuTe dot superlattice growth simulations shown in Fig. 1.18c, clearly, the lateral ordering is not only much more efficient but also the expected *ABCABC...* dot stacking is well reproduced (compare relative dot positions in the 10th, 11th and 12th layers). Even more, the lateral dot spacing in the final layer of 63 nm is exactly equal to the spacing of the surface energy minima $L = \sqrt{3}d \tan \alpha$ (with $d_s = 50$ nm and $\alpha = 36°$) calculated from the point-source model and thus reproduces very well the experimental results [18, 30, 50] (see also section 1.6.1 for details). Figure 1.19a–d shows the evolution of the PbSe dot arrangement in the simulations as a function of the superlattice period at $N = 1, 5, 10$ and 15, respectively, with the FFT power spectra of the dot positions depicted as inserts. Whereas the dots in the initial layer are completely disordered, already after five superlattice periods a clear hexagonal lateral ordering has taken place as signified by the appearance of satellite peaks in the FFT power

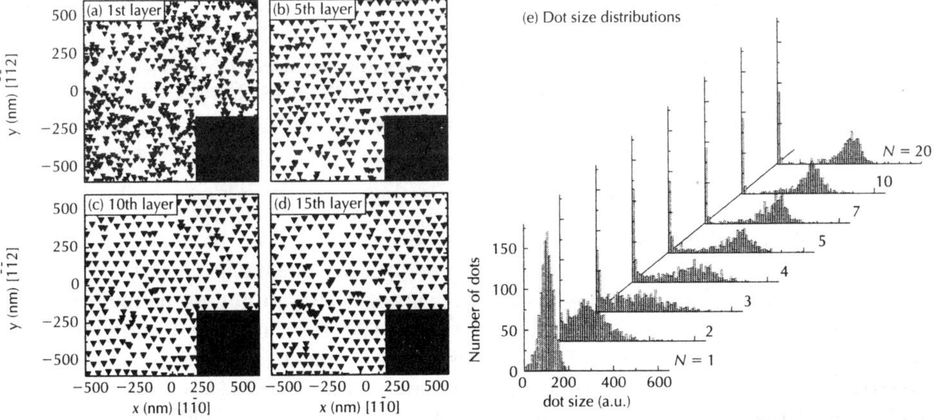

Figure 1.19   *Monte Carlo simulation of self-organized (111) PbSe/PbEuTe dot superlattice growth, showing the island positions (triangles) on the surface after the N = 1st, 5th, 10th, and 15th superlattice period from (a) to (d), respectively. The insets show the 2D Fourier power spectra of the island positions indicating the progressing 2D hexagonal in-plane ordering. Panel (e) shows the evolution of the dot size histograms as a function of number of superlattice periods N, showing a significant reduction of the relative width of the distributions after about four superlattice periods. Adapted from Springholz et al. [49].*

spectra (see inserts). With increasing number of periods, these satellite peaks become increasingly more well defined and additional higher-order satellites appear. Thus, the hexagonal ordering rapidly improves.

Figure 1.19e shows the corresponding evolution of the size distribution as a function of the number of superlattice periods. For the initial dot layer, the density and size are determined merely by the chosen capture radius and deposited number of adatoms. With increasing number of periods, obviously, a transition from the initially rather small dots to about a factor of two larger dots in the multilayer stack occurs, which is caused by the reduced dot density induced by the interlayer correlations. During the concomitant lateral ordering process, the size distribution at first significantly broadens for the first few superlattice layers but then rapidly narrows, indicating a progressive size homogenization. After about ten layers, the relative width of the size distribution normalized to the average dot size saturates at a value of about half of that of the initial layer. This indicates that the lateral ordering in multilayers can improve the uniformity of self-assembled quantum dots. The observed transient broadening of the size distribution is caused by the extensive rearrangement of the dot positions in the first superlattice layers, which gives rise to a larger variation in the dot size (see Fig. 1.19e). This transient behaviour is in remarkable agreement with our experimental observations described in section 1.6. Other growth simulations focusing of various other aspects of ordering and correlation formation such as the influence of the growth temperature can be found in [148–151].

An important conclusion from the growth simulations is that there actually exist two different ordering mechanisms in multilayer structures. The first one is based on *repulsive* strain fields from the subsurface dots, which results in an alignment of the surface dots along certain surface directions in which the mutual dot repulsion is locally reduced. The second mechanism is caused by the *attractive* strain fields of buried dots and is operative when the surface strain energy distribution above each dot splits into several side minima. In the first case, the dots are still aligned along the vertical growth direction and the lateral ordering tendency is rather weak and essentially only of short-range type. In the second case, the interlayer dot correlations are of staggered type and the lateral ordering is more efficient and of longer range. In both cases, the anisotropy of the strain fields on the surface is a crucial prerequisite for the ordering process, without this anisotropy no lateral ordering is found even for idealized growth simulations. This conclusion also holds for the near-field strain interactions when the dots can no longer be approximated by point sources. Lateral ordering is counteracted by thermal disorder induced by random surface diffusion and nucleation at other surface sites. Thus, in general the ordering process in actual growth experiments will be more gradual and less efficient than in the above described growth simulations and it will also depend significantly on the growth conditions. In addition, it is noted that in the near-field limit, anisotropic surface strain fields may also be introduced by anisotropies in the island shapes. Thus, for higher-indexed growth orientations and growth conditions that produce highly anisotropic dot shapes, a more efficient lateral dot ordering can be expected. This conclusion is also supported by recent growth experiments [60–65, 142, 144].

1.6 InGaAs/GaAs multilayers

Self-assembled InGaAs quantum dots have attracted great interest due to the high photoluminescence efficiency achieved in this material system [1–3]. This allows investigation of the fundamental electronic properties of single dots or single stacks of dots using high-resolution optical spectroscopy [12–14]. In addition, efficient lasers (see, e.g. Refs. [1, 168–170]) and detectors have been realized with this system and there are interesting possibilities for single photon sources and quantum information processing devices. Therefore, this material system has been investigated in great detail. In fact, the first interlayer correlations in self-assembled quantum dot multilayers were observed in InAs/GaAs dot superlattices [42, 43], which has stimulated detailed further investigations on this self-organization process.

1.6.1   Pairing probability as a function of spacer thickness

For InAs/GaAs quantum dot superlattices, it was found that the InAs dots are vertically aligned along the (100) growth direction forming one-dimensional dot columns up to GaAs spacer thicknesses of around 40 nm [15, 16, 45] as shown in Fig. 1.20. For larger spacer thicknesses, this interlayer correlation is lost, i.e. the dots nucleate independently from those in the previous layers. This transition has been studied by Xie *et al.* [15] using cross-sectional transmission electron microscopy (Fig. 1.20a–c) from which the pairing probability of the dots in subsequent layers was determined as a function of spacer thickness, as shown in Fig. 1.20d. For spacer thicknesses up to about 15 nm, the interlayer pairing probability was found to be equal to one, but then gradually decreases until at about 55 nm spacer thickness no interlayer correlations were found any more. Similar results were reported in a cross-sectional scanning tunnelling microscopy study of Legrand *et al.* [45], showing that the pairing probability depends also on the InAs dot size, i.e. for larger dots the interlayer correlations persist to larger spacer thicknesses. In both studies, the spacer thickness $d_{1/2}$ at which the interlayer pairing probability drops to one half was found to be in the range of 20 to 30 nm for a growth temperature around 500°C and InAs thicknesses of 2–2.3 monolayers. This yields InAs dots with a diameter of 10 to 20 nm. For samples with different dot size, Legrand *et al.* [45] found that the crossover spacer thickness $d_{1/2}$ scales linearly with the dot size due to the corresponding increasing strength of the strain interactions (see also Section 1.3.4).

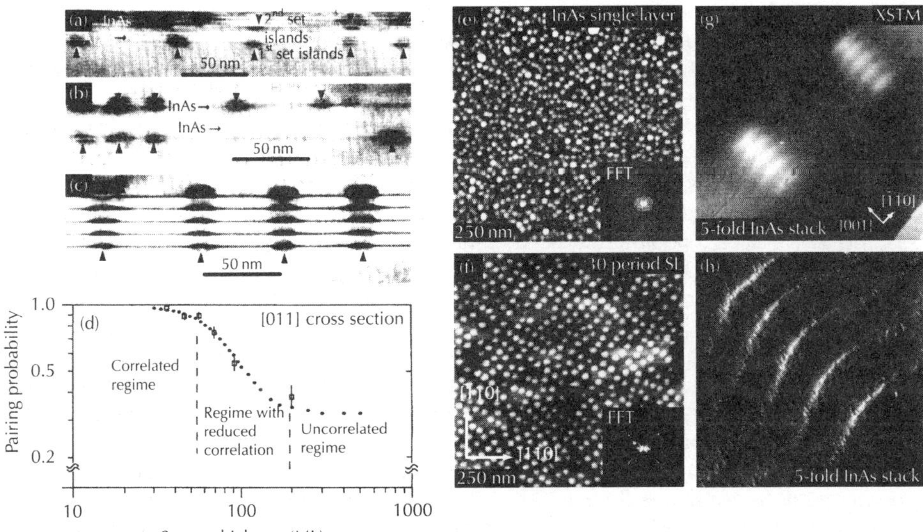

Figure 1.20   Left-hand side: *Cross-sectional transmission electron microscopy images of InAs/GaAs multilayers with GaAs spacer thicknesses of (a) 42, (b) 92 and (c) 36 monolayers (ML), where for the latter the InAs dots are perfectly aligned along the growth direction. Adapted from Xie* et al. *[15]. (d) Pairing probability (open squares) of these InAs dots deduced from the transmission electron microscopy images as a function of GaAs spacer thickness. The filled circles show the fit to the experimental data using a model described in [15]. Centre: Atomic force microscopy images of a single 3 ML InAs dot layer (e) and of a 30 period InAs/GaAs (001) dot superlattice (f) with 6.5 nm GaAs spacers. The horizontal axis is along the [110] direction. The 2D Fourier transform (FFT) power spectra of the topography images, shown as insets, indicate a lateral ordering of the dots along the ⟨100⟩ directions. Adapted from Solomon* et al. *[74]. Right-hand side: Cross-sectional scanning tunnelling microscopy images of five-fold stacks of self-assembled 2.4 ML InAs dots separated by 10 nm GaAs spacer layers showing the strong changes in dot shape and size as a function of the number of bilayers: (g) 150 × 150 nm² topography image and (f) enlarged 55 × 55 nm² current image. Adapted from Bruls* et al. *[73].*

The pairing probability in the vertical dot columns was explained by Xie *et al.* [15] using a model in which the lowering of the strain energy above the buried dots leads to an attraction of mobile surface adatoms [15, 45]. The local nucleation probability was then assumed to be proportional to the amount of accumulated InAs atoms at the strain minima. For small spacer thickness and correspondingly deeper energy minima, all deposited adatoms within the surface diffusion length are attracted, leading to a unity dot pairing probability. With increasing spacer thickness, the depth of the energy minima and thus the pairing probability diminishes until the minima become too weak to induce a correlated dot nucleation. The predictions of this model are represented as full circles in Fig. 1.20d and were found to agree well with the experimental data (open squares) [15, 45]. Also, a gradual transition from correlated to uncorrelated multilayers was found in Monte Carlo superlattice growth simulations [149–151].

1.6.2   Lateral ordering

With respect to the lateral dot ordering in (100) InAs/GaAs superlattices, Solomon *et al.* [74] have shown a clear short-range dot ordering for superlattices with small GaAs spacer thicknesses $d < 10$ nm. As shown in Fig. 1.20f the dots are then preferentially aligned along the lateral $\langle 010 \rangle$ directions, whereas for the single dot reference layer no lateral order was found (Fig. 1.20e). This is evidenced by the corresponding Fourier transform power spectra of the atomic force microscopy images shown in the inset. For the superlattice sample, four satellite peaks are seen, indicating a preferred square arrangement of the dots. This is explained by Monte Carlo growth simulations shown in Fig. 1.18 and arises from strain-induced repulsion between the buried dots and the second nearest surface layer dots (see Section 1.5). The same type of lateral ordering was also found by Darhuber *et al.* [38] for InAs dot superlattices with 20 nm GaAs spacers using high-resolution X-ray diffraction reciprocal space mapping. Solomon *et al.* [31] have also reported a significant size homogenization of the dots, with a decrease of the FWHM of the dot height distribution from $\pm 20\%$ for single InAs dot reference layers to $\pm 8\%$ in the 20 period dot superlattice. This was also accompanied by a 25% narrowing of the photoluminescence line widths in the superlattice to 54 meV [16, 31]. A similar narrowing was reported by Nakata *et al.* [75] for superlattices with 3 nm spacers, observing a line width decrease from 90 meV to 27 meV, and comparable results were obtained by He *et al.* [152]. Also, for vertically aligned InP/GaInP quantum dot superlattices a significant photoluminescence narrowing from 41 to 16 meV was found [77].

For vertically aligned (100) InGaAs/GaAs dot superlattices with In concentration of around 50% and 120 monolayer GaAs spacer thickness, another type of ordering has been recently reported by Wang *et al.* [153–156]. In this case, the dots were found to be aligned in one-dimensional dot chains along the $[01\bar{1}]$ surface direction, as is exemplified by the atomic force microscopy image shown in Fig. 1.21a [145]. This has been attributed to the anisotropy of the surface diffusion and the dot shapes [153–156]. The same group has also found a hexagonal ordering when the superlattices are grown under As_2 instead of As_4 flux [157].

Concerning the ordering of (100) InGaAs/GaAs multilayers with staggered dot stacking [87, 88] occurring at larger spacer thicknesses as described in Section 1.4.3, no detailed studies on the lateral dot ordering have been reported yet. From the growth simulations and strain calculations described in the previous sections, however, it is expected that the ordering is much more efficient than for superlattices with vertically aligned dots. First experimental evidence that this is really the case has been recently reported by Gutierrez *et al.* [88], where a square ordering of InGaAs dots within the growth plane was found for a staggered InGaAs/GaAs (100) multilayer using plan-view transmission electron microscopy. A particular difference of the lateral ordering between vertically aligned and staggered InGaAs dot superlattices is that, according to the strain calculations and Monte Carlo growth simulations, the ordering is of short range type for the former and of long range for the latter, as is also indicated by the experimentals [74, 88]. This results from the fact that for vertically aligned superlattices lateral ordering is due to repulsion of next-nearest surface dots due to the surface strain *maxima* induced by the buried dots along the $\langle 010 \rangle$ surface directions (see, e.g., inset of Fig. 1.18b), whereas for the staggered dot stacking, lateral ordering is caused by the square arrangement of the surface strain *minima* aligned along the $\langle 010 \rangle$ surface directions (see, e.g., Fig. 1.9c and d).

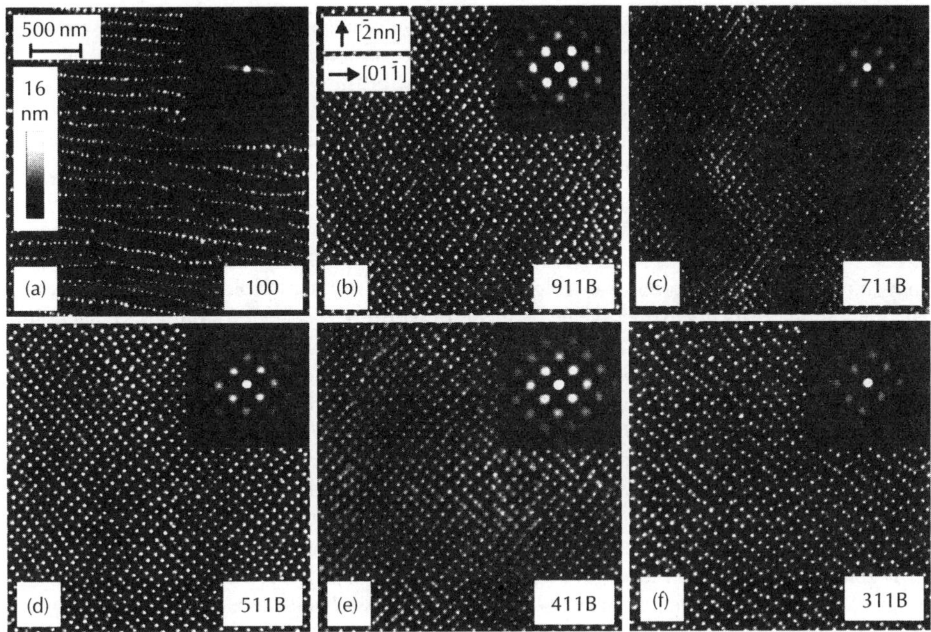

Figure 1.21   Atomic force microscopy images of 16.5 period $In_{0.4}Ga_{0.6}As/GaAs$ superlattices with ten monolayer InGaAs alternating with 120 monolayer GaAs grown on (100), (911)B, (711)B, (511)B, (411)B and (311)B substrates from (a) to (f), respectively. For the different samples on (n11) substrates, the horizontal direction of the atomic force microscopy images is along $[01\bar{1}]$ and the vertical direction along $[\bar{2}nn]$. The insets show the 2D autocorrelation images calculated from the atomic force microscopy images, showing the good nearly square lateral ordering of the dots along the in-plane $\langle\bar{1}n0\rangle$ surface directions rotated by 45° to the horizontal $[01\bar{1}]$ direction. Adapted from Lytvyn et al. [145]. The oblique interlayer correlation of dots in these samples is shown in Fig. 1.17.

An even more efficient 2D lateral ordering as well as an inclined interlayer dot correlation was found by the group of Salamo *et al.* for InGaAs/GaAs superlattices grown on high-indexed (*n*11)B substrates [142, 144, 145]. This is evidenced by the corresponding series of atomic force microscopy images for 16.5 period superlattices on (*n*11)B substrates with $n = 9, 7, 5, 4$ and 3 shown in Fig. 1.21b to f, respectively, reported by Lytvyn *et al.* [145]. Clearly, the dots are preferentially aligned along the in-plane $\langle\bar{1}n0\rangle$ directions, resulting in an efficient square ordering in the superlattice stack, although a faint ordering was also found for single dot layers on the high indexed substrates [145]. The ordering was found to be best for the superlattice grown on the (411)B orientation and was attributed to the high in-plane anisotropy of the Young's modulus within the growth plane with a strong minimum along the $\langle\bar{1}n0\rangle$ direction and pronounced maxima along the $\langle01\bar{1}\rangle$ and $\langle\bar{2}nn\rangle$ directions [145]. Lateral ordering of InGaAs/GaAs dots on (311)B surfaces was also studied by Lee *et al.* [158], who found only the formation of one-dimensional dot chains on the surface. In this case, however, metal organic vapour deposition was used and the GaAs spacer thickness was only 12 nm. On (311)B InGaAs/GaAs dot superlattices a peculiar short-range ordering was also recently reported by Lippen *et al.* [159, 160]. As a consequence, the exact ordering mechanisms are not yet resolved and remain to be investigated in detail.

1.6.3   Sizes, shapes and critical wetting layer thickness

One general feature of vertically aligned InAs dot superlattices is the significant increase of the dot size and broadening of shape along the vertical dot columns. This is illustrated in Fig. 1.20g and h by cross-sectional scanning tunnelling images of five-fold stacked 2.4 ML InAs dots separated by 10 nm GaAs spacer layers by Bruls *et al.* [72, 73]. Similar changes were also found in numerous cross-sectional scanning tunnelling [67–72] and transmission electron microscopy [15, 16, 44] studies. From a statistical evaluation of 20 period InAs/GaAs dot superlattices,

Solomon *et al.* [31] have found a 50% increase in dot height as well as 25% increase in dot width as compared to that of a single InAs dot reference layer. Similar changes were reported by Nakata *et al.* [75], who found in addition a significant decrease of the critical wetting layer thickness for dot nucleation from $h_c = 1.6$ monolayers for the first to 1.1 monolayers for the subsequent layers in the multilayer stack. The latter seems to be the main reason for the substantial increase in dot size because correspondingly more material is available for dot formation.

1.6.4 Photoluminescence

As a result of the increase in dot size, a significant red shift of the photoluminescence emission of vertically aligned InAs dots superlattice by around 100 meV has been found with respect to the photoluminescence of single InAs dot layers [16, 31, 74, 75, 152]. The size increase should also broaden the overall photoluminescence emission in spite of the narrowing of the size distribution in the final dot layer. Therefore, contradictory results have been reported by various groups. While on the one hand some groups have reported a significant narrowing of the photoluminescence emission of InAs multilayers, as mentioned above [16, 31, 75, 152, 161], others have found little changes [162, 163] or even a broadening of the photoluminescence emission [164–166]. This difference may be explained also by taking the electronic coupling between the dots into account. For very small spacer thicknesses (as applies to most experimental studies), there is a strong overlap of the electronic wave functions of the dots in adjacent layers and thus the carriers are actually delocalized over several stacked dots. As a result, a certain averaging over the vertical inhomogeneities in the dot sizes occurs. By tunnelling, the excited carriers may be transferred to the lowest energy eigenstates along the dot columns and thus a narrower photoluminescence is obtained. For intermediate spacer thicknesses around 20 nm, the electronic coupling becomes small and thus the differently red-shifted photoluminescence emission of successive dot layers is superimposed. As a result, the emission may be significantly broader as compared to that of a single dot layer [167], an effect that should be also strongly temperature dependent. For even larger spacer thicknesses, the structural interaction between the dots also becomes negligible and thus the multilayer emission will eventually approach that of a several times repeated single dot layer [162, 163]. In spite of these complications, in most studies a notable increase in photoluminescence efficiency was achieved with stacked InAs dots [16, 31, 74, 75, 152, 162, 163]. Therefore, InAs multilayer dot structures have been successfully utilized to improve the performance of quantum dot lasers (see, e.g., [1, 4, 168–170] for reviews).

A special technique has been developed by Fafard *et al.* [164] to reduce the variations in dot sizes in vertically aligned InAs multilayers. It is based on an indium flush step after each InAs layer after partial capping of the dots with a thin GaAs layer. This indium flush step consists of a short growth interruption during which the excess InAs that sticks out of the GaAs layer is desorbed by a short annealing step at 610°C before continuing with superlattice growth at the 520°C growth temperature. This reduces the height of the InAs dots to the thickness of the GaAs capping layer and the layer-to-layer dot height uniformity is significantly improved [164]. As a result, very narrow photoluminescence emission was obtained from these multilayers such that even the s, p, and d shell structure of the electronic transitions could be observed [164].

1.7 Ordering in SiGe/Si dot superlattices

For SiGe/Si (100) quantum dot multilayers, the general trends are quite similar as for InAs/GaAs superlattices. From cross-sectional transmission electron microscopy [19, 32–40] as well as high resolution X-ray diffraction [36, 38] studies, a predominant vertical alignment of SiGe dots was found for Si spacer layers up to 50–70 nm. This is demonstrated in Fig. 1.22a by the transmission electron microscopy micrograph of a vertically aligned SiGe/Si dot superlattice with 20 nm Si spacers grown by MBE at a temperature of 620°C [147]. The degree of interlayer correlation, i.e. the interlayer pairing probability of Ge/Si multilayers as a function of spacer thickness, has been studied by Kienzle *et al.* [35] and Stangl *et al.* [36] for a constant Ge thickness of 6.5 ML.

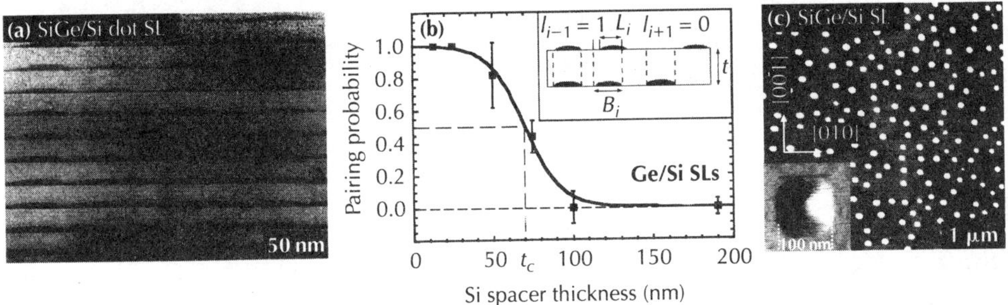

Figure 1.22   (a) *Cross-sectional transmission electron micrograph of a self-assembled (100)* $Si_{0.5}Ge_{0.5}/Si$ *dot superlattice showing the typical vertical alignment of the dots along the growth direction. The SiGe and Si layer thicknesses are 2.1 and 20 nm, respectively. Adapted from Holy et al. [147]. (b) Pairing probability of 6.5 ML Ge dots in five-stack Ge/Si multilayers grown at 620°C determined by cross-sectional transmission electron microscopy as a function of Si spacer thickness. For spacer thicknesses smaller than 40 nm, a nearly perfect dot alignment is obtained. For larger thicknesses the pairing probability decreases such that above 100 nm, no interlayer correlations are found. Adapted from Kienzle et al. [35]. (c) Atomic force microscopy image of the last SiGe dot layer of a 20 period SiGe/Si dot superlattice with 5.5 ML SiGe dots alternating with 30 nm Si spacers. Inset: Image of a single SiGe dot, illustrating the typical pyramidal island shape with {105} side facets.*

The results are shown in Fig. 1.22b, where the probability of vertical dot alignment, defined as the degree of base area overlap of subsequent dots in the vertical directions as defined by the inset in Fig. 1.22b, is plotted versus spacer layer thickness. For Si spacers up to 30 nm, a perfect vertical alignment with 100% pairing probability is found. For thicker spacers the pairing probability gradually decreases until at 100 nm no interlayer correlations are formed. At 70 nm spacer thickness, a pairing probability of 50% was found. Thus, interlayer correlations persist up to significantly larger spacer thickness than for the InAs/GaAs case, which is attributed to the fact that the Ge islands in these studies were about a factor of 5–10 times larger in diameter (85 nm) [34] than the typical InAs dots of InAs/GaAs structures [15, 45]. In the same study, it was also found that along the vertical Ge dot columns the dot size increases linearly with layer index number [35], which is most pronounced for superlattices with very thin spacers.

Apart from the changes in dot size, several studies have shown that in Si/Ge multilayers the critical thickness for island formation decreases with increasing number of superlattice periods. The results of a systematic UHV-CVD growth study of Le Thanh *et al.* [40] using *in situ* reflection high-energy electron diffraction is shown in Fig. 1.23 where the critical thickness of the 2D–3D transition is plotted as a function of the number N of deposited Si/Ge bilayers for Si spacer thicknesses of 22 and 2.5 nm for a constant 4 ML thickness of the Ge layers. For the thicker Si spacers, the critical thickness decreases by a factor of two within the first four superlattice periods and saturates thereafter at a value of about two monolayers compared to four monolayers for the first Ge layer. For the 2.5 nm thin Si spacers, the reduction is even larger, saturating at a critical thickness of about 1.4 monolayers within the superlattice stack (see Fig. 1.23a). The dependence of the critical thickness of the second Ge layer on the Si spacer thickness is depicted in Fig. 1.23c [40], showing that for spacer thicknesses lower than 90 nm, the critical thickness of the second Ge layer decreases linearly with decreasing spacer thickness, whereas for thicker spacer the critical thickness for the second Ge dot layer approaches the value of the first Ge layer. This indicates that at this point, no interlayer interactions occur any more, which agrees qualitatively with the results of Kienzle *et al.* [35] displayed in Fig. 1.22b. A similar behaviour was found by Schmidt *et al.* [39] for MBE grown five-period Si/Ge multilayers with 25 nm thick Si spacers, where the Ge wetting layer thickness was found to decrease from 3.8 ML for the first layer to 2.5 ML in the fifth layer. These changes in the critical wetting layer thickness have also been shown to significantly influence the photoluminescence spectra from multilayer samples as is discussed in more detail, e.g., in [22, 40].

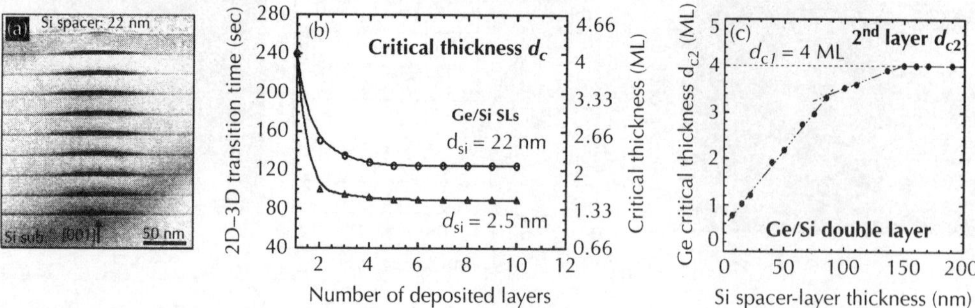

Figure 1.23 (a) *Change of the critical thickness for Ge islanding as a function of the number of deposited layers N in a multilayer consisting of ten periods of 4 ML Ge alternating with 22 nm (○) or 2.5 nm (△) Si spacer layers. The 2D–3D transition was determined using* in situ *reflection high-energy electron diffraction and the growth was carried out using UHV-CVD at a temperature of 550°C. (b) Change of the critical thickness d_{c2} of the second Ge layer as a function of the Si spacer thickness at a growth temperature of 600°C. The critical thickness d_{c1} = 4 ML of the first Ge layer is indicated by the dashed horizontal line. Adapted from Le Thanh et al. [40].*

Concerning the lateral dot ordering, one has to distinguish between two different cases according to the density and spacing of the dots in the first superlattice layer that strongly depend on the composition and thickness of the SiGe dot layers and the chosen growth conditions. For the case of high-density SiGe hut cluster islands deposited at temperatures around 550°C, Teichert *et al.* [17, 21, 28] have demonstrated a quite effective lateral ordering of SiGe hut islands along the ⟨100⟩ directions, with the formation of a preferred square nearest neighbour dot arrangement. This is illustrated by the atomic force microscopy images and corresponding Fourier transform power spectra of Teichert *et al.* [21, 28] obtained for $Si_{0.25}Ge_{0.75}$/Si multilayers with N = 1, 2 and 20 periods shown in Fig. 1.23a to c, respectively. Clearly, the lateral ordering progresses rapidly with increasing number of deposited periods. This is accompanied by an almost three-fold increase of the dot size but the uniformity of the dots in the final dot layer significantly improves. This is demonstrated by the data displayed in Fig. 1.23d–f, where the evolution of the lateral dot spacings, the lateral aspect ratio as well as root mean square roughness that is proportional to the dot height are plotted versus N, respectively.

In the second case of low-density dots formed by growth of pure Ge at higher temperatures as used in most other studies [33–36, 38–40], the dots of single layers are already much more uniform and widely spaced. As a result, the Ge dots in these multilayers show only a very faint lateral ordering tendency (see Fig. 1.22c) even at small spacer thicknesses, with only a very weak tendency of a lateral alignment of the dots along the ⟨100⟩ directions, as has been found, e.g., by X-ray diffraction studies [32, 36, 38]. These findings are attributed to the weak lateral overlap of the strain field of the dots and the weak ordering tendency is in good agreement with our Monte Carlo growth simulations [58, 147] (see section 1.5 and Fig. 1.18). In these simulations, the lateral dot alignment along ⟨100⟩ as found in the experiments was explained to result from the elastic anisotropy of the Si spacer material. The apparent discrepancy between the work of Teichert *et al.* [21, 28] and those of the other groups [33–36, 38–40] is ascribed to the different growth conditions, which for the former result in closely spaced dots with a large lateral strain field overlap, whereas in the other works the lateral dot interaction is rather weak.

1.8 PbSe/PbEuTe dot superlattices

Self-assembled PbSe/PbEuTe quantum dot superlattices represent a particularly interesting system for investigation of interlayer correlations in multilayer structures. On the one hand, the elastic anisotropy is particularly large and, therefore, an exceedingly efficient vertical and lateral ordering takes place [18, 30]. On the other hand, different dot stacking types occur in dependence of spacer thicknesses as well as growth conditions [50–56, 57]. Because in the IV–VI compounds

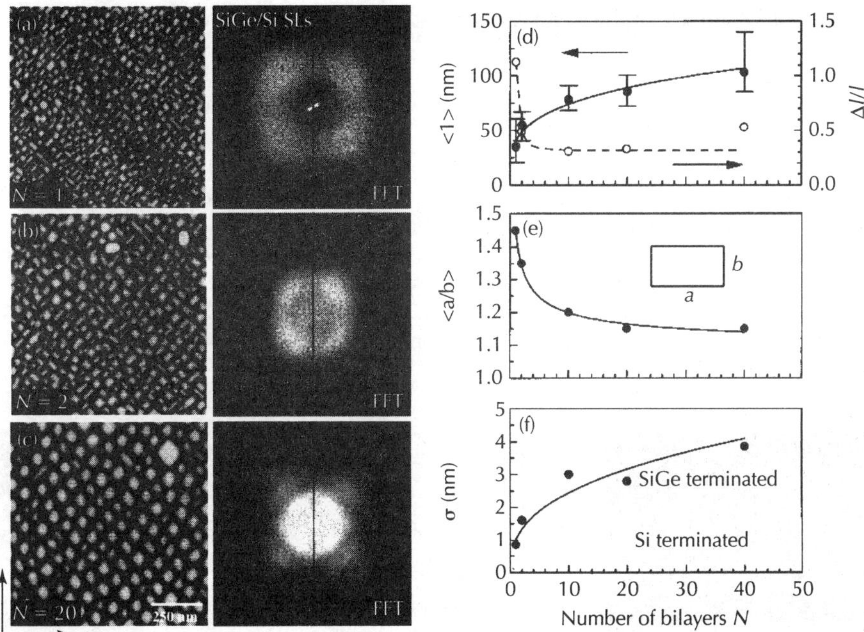

Figure 1.24   Atomic force microscopy images (left column) and corresponding 2D power spectra (right column) of a SiGe surface layer of $Si_{0.25}Ge_{0.75}/Si$ multilayers with increasing number of multilayer periods N = 1, 2 and 20 from (a) to (c), respectively (see schematic drawings on the left). The SiGe layer thickness is 2.5 nm and the Si spacer thickness 10 nm. Panels (d) to (f) show the evolution of characteristic island parameters as a function of number of superlattice periods N. (d) Mean island spacing <l> along [100] (solid line) with the error bars indicating the width of the size distribution, and relative width $\Delta l/<l>$ of the distribution of island spacings (dashed line) w. (e) Average aspect ratio of the SiGe island bases <a/b> (see inset). (f) Overall root mean square roughness of the surface topography. Adapted from Teichert et al. [21, 28].

the (111) direction is the elastically soft direction, *fcc*-like dot stackings are formed in (111) multilayer structures. Therefore, this system is particularly well suited for testing the theoretical predictions of superlattice growth models. In addition, extensive systematic studies on interlayer correlation formation have been carried out in this material system.

Self-assembled PbSe dots are produced by heteroepitaxial growth on PbEuTe (111) at a critical coverage of 2.5 monolayers [130]. For the (100) growth orientation, this islanding transition is suppressed because misfit dislocation formation sets in already at a smaller thickness of one monolayer [134, 135]. Therefore, self-assembled quantum dots cannot be obtained in this growth orientation. PbSe dots are under tensile strain because their lattice constant is 5.4% smaller than that of the underlying PbEuTe buffer layers. This is in contrast to the compressively strained dots present in most other studied material systems, but from an energetic point of view this does not make any difference because the sign of strain is removed in the elastic energy calculations. The growth properties of single PbSe dot layers have been studied in detail by our group [130, 171], showing that the basic behaviour is quite similar to that of other self-assembled quantum dot systems. In particular, PbSe dots show a well-defined pyramidal shape with {100} side facets and very narrow size distributions [130]. The size and dot density can be readily controlled by adjusting the growth temperature or dot layer thickness [171], with the dot height being typically in the range of 60 to 200 Å, and the density between $1 \times 10^{10} - 20 \times 10^{10} \, cm^{-2}$. Because of the narrow energy band gaps, PbSe dots are also of interest for fabrication of mid-infrared quantum lasers as demonstrated in our previous work [172]. Multilayer structures are always grown on thick strain relaxed PbTe buffer layers as virtual substrates. In the superlattice structures, the $Pb_{1-x}Eu_xTe$ spacer layer composition is typically in the range of $x = 4$ to 10%. By proper adjustment of the ternary composition, strain symmetrized superlattices can be produced [30] in which the number of deposited bilayers is not limited by misfit dislocation formation.

1.8.1 Stackings as a function of spacer thickness

For PbSe/PbEuTe dot superlattices, the dependence of the vertical and lateral dot correlations on spacer thickness was studied for spacer thicknesses varying from 100 to 800 Å, while the PbSe dot layer thickness of five monolayers as well as the growth conditions were kept constant. The lateral dot ordering is illustrated in Fig. 1.25a–d, where representative atomic force microscopy images of the final dot layer after 30 superlattice periods are depicted for spacer thicknesses of 160, 320, 465 and 660 Å, respectively. In addition, the average lateral dot spacing formed in the multilayer structure is plotted in Fig. 1.25e for a large number of samples as a function of vertical superlattice period d_{SL}.

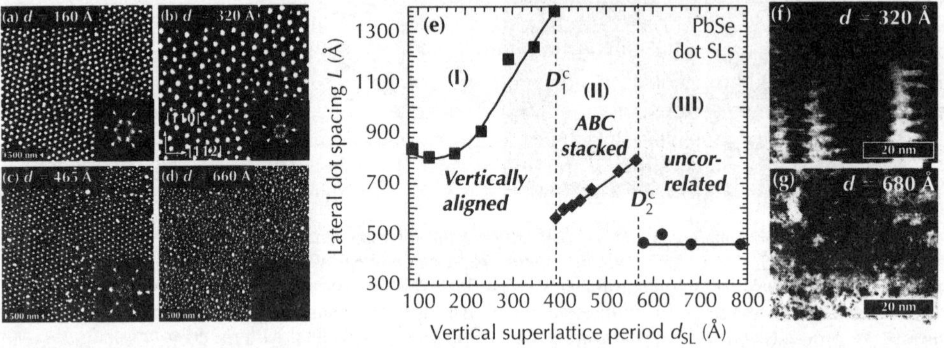

Figure 1.25   Left-hand side: Atomic force microscopy surface images of 30 period PbSe/PbEuTe dot superlattices with different spacer layer thickness of (a) 160, (b) 320, (c) 465 and (d) 660 Å, demonstrating the different dot spacings and ordering tendency in the structures as evidenced by the FFT power spectra shown as insets. (e) Preferred lateral dot spacing L plotted as a function of vertical superlattice period: squares indicate superlattices with vertically aligned dots, diamonds those with fcc-like ABCABC... stacking and full circles those of uncorrelated superlattices. Right-hand side: Cross-sectional transmission electron microscopy images of a vertically aligned (f) and an uncorrelated (g) PbSe/PbEuTe dot superlattice with spacer thicknesses of 320 and 680 Å, respectively. See Springholz et al. [50, 51].

For spacer thicknesses lower than about 400 Å (Fig. 1.25a and b), there is a clear hexagonal ordering tendency of the dots in the lateral directions [51] that is also evidenced by six peaks in the Fourier transform power spectra of the atomic force microscopy images shown as inserts. The lateral spacing of the dots is larger than 800 Å and rapidly increases with increasing spacer thickness (see solid line in region I of Fig. 1.25e). As demonstrated by the cross-sectional transmission electron microscopy image of a sample with 320 Å spacer thickness shown in Fig. 1.25f, the dots in these samples are vertically aligned along the growth direction.

For samples with intermediate spacer thicknesses between 400 and 550 Å, the average lateral dot spacing abruptly drops by more than a factor of two from 1300 Å to 580 Å and a very efficient 2D hexagonal lateral ordering takes place (see Fig. 1.25c) [18, 30, 49]. From detailed high resolution X-ray diffraction [18–20, 49, 50, 53, 55–57] as well as cross-sectional transmission electron microscopy [19, 20, 50, 52, 54] studies already presented in Section 1.4.2, for these samples, a well-defined *fcc*-like interlayer dot stacking (see Fig. 1.14) is formed, corresponding to an overall trigonal three-dimensional lattice of dots as shown in Fig. 1.8a. In addition, the lateral dot spacing within the superlattice stacks increases linearly with increasing spacer thickness or superlattice period d_{SL}, as indicated by the data and solid line depicted in region II in Fig. 1.25e. In this range of spacer thicknesses, the interlayer correlation angle is also essentially constant at about 39° [18], which means that the lattice constant of the resulting three-dimensional trigonal dot lattice can be tuned over a range from 500 to 700 Å just by changes in the spacer thickness [18].

With further increasing spacer thickness, the lateral ordering of the dots becomes weaker and weaker until at about 600 Å no lateral ordering (see Fig. 1.25d) and not interlayer correlations are observed any more. The latter is illustrated by the cross-sectional transmission electron microscopy image of a superlattice with 680 Å spacers depicted in Fig. 1.25g, where obviously the dots in each layer nucleate at random sites independent of the dots in the previous layers. Thus, these superlattices represent merely a repetition of uncorrelated single dot layers and therefore, no dot rearrangements or lateral ordering occurs. As a consequence, the average lateral dot spacing on top of the superlattice stack (full circles in region III of Fig. 1.25e) is always equal to the constant value of about 500 Å as found for single dot reference layers grown under the same conditions.

1.8.2   Lateral ordering

1.8.2.1   Fcc-stacked superlattices

For PbSe/PbEuTe dot superlattices with intermediate spacer thicknesses between 420 and 520 Å the remarkably efficient lateral ordering process is illustrated in Fig. 1.26, where the atomic force microscopy images of the last PbSe dot layer of series of superlattices consisting of $N = 1, 10, 30$ and 100 periods are depicted [30]. In theses samples, the PbSe and $Pb_{1-x}Eu_xTe$ layer thicknesses were kept constant at 5 ML and 470 Å, respectively, and identical growth conditions were used. As shown in Fig. 1.26a, for the single layer $N = 1$ the islands are distributed randomly on the surface without any preferred lateral correlation direction. With an increasing number N of periods, a rapidly progressing ordering sets in. Already after ten periods, the dots are aligned in single and double rows along the $\langle \bar{1}10 \rangle$ directions (Fig. 1.26b). Measurements on samples with fewer than ten bilayers show that this ordering commences first with the formation of small patches of hexagonally ordered regions, which subsequently enlarge and join to form row-type structures. With further increasing period number (Fig. 1.26c and d), larger and larger ordered regions are formed, such that for samples with large N the perfect hexagonal 2D arrangement is disrupted only by single point defects, such as missing dots, dots at interstitial positions, or occasionally by additionally inserted dot rows ("dislocations") (see Fig. 1.26d for $N = 100$ periods).

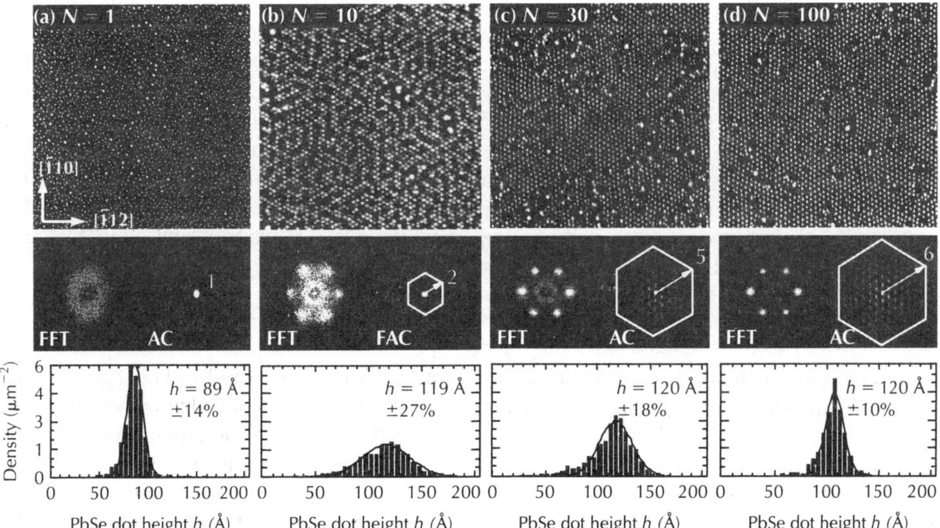

Figure 1.26   Top row: *Atomic force microscopy images of PbSe/PbEuTe superlattices with increasing number of superlattice periods of* N = 1, 10, 30 *and 100 from (a) to (d), respectively. Each bilayer consists of five monolayers PbSe dots alternating with 480 Å PbEuTe spacers (image size: 3 × 3 μm²). Centre row: Fourier (FFT) power spectra (left) and 1 × 1 μm² auto correlation (AC) spectra (right) of the atomic force microscopy images characterizing the hexagonal lateral ordering process. Bottom row: Corresponding dot height histograms deduced from the atomic force microscopy images. The full lines represent Gaussian fits of the histograms. See Pinczolits et al. [30] for details.*

The development of the lateral ordering was characterized by Fourier transformation (FFT) as well as auto correlation (AC) analysis, as shown in the middle panel of Fig. 1.26. For the single dot layer, the FFT power spectrum exhibits only a broad and diffuse ring that indicates an average in-plane dot distance of $800\,\text{Å}$ without any preferred nearest neighbour direction. The relative width (FWHM) of this ring is $\pm 47\%$, which indicates a substantial variation of the lateral dot separations. Also, the autocorrelation spectrum does not exhibit any structure outside the central maximum, indicating the lack of any correlations in the lateral dot positions. In contrast, for the ten bilayer sample the FFT spectrum (Fig. 1.26b) already shows six well-separated side maxima. This corresponds to a preferred spacing between the dot *rows* of $488\,\text{Å}$. Six side maxima also appear in the autocorrelation spectrum, indicating that the next nearest neighbour dots are along the $\langle \bar{1}10 \rangle$ directions, with a preferred distance of $680\,\text{Å}$.

For the 30 and 100 period superlattices, the peaks in the FFT spectra drastically sharpen and many higher-order satellite peaks appear (see Fig. 1.26c and d). The FWHM of the satellite peaks narrows from $\pm 47\%$ for $N = 1$ to $\pm 6\%$ for the 100 period superlattice, i.e. the dot spacings become extremely well defined. Many higher-order peaks also appear in the autocorrelation spectra, which shows that large perfectly ordered dot domains are formed, with a nearly perfect registry of the dot position up to ten nearest neighbouring dots. As indicated in the autocorrelation images of Fig. 1.26, average domain radii of 1, 2, 5 and 6 hexagonal unit cells are obtained for $N = 1$, 10, 30 and 100 periods, respectively. This underlines the efficiency of the lateral ordering process.

A particular feature of the superlattice structures is that neither the lateral dot spacing nor the average dot size changes with increasing number of periods – in contrast to the observations for InAs or SiGe dot superlattices (see, e.g., Sections 1.6 and 1.7 and Fig. 1.20). This is proven by the island height histograms of different samples depicted in the lower panel of Fig. 1.26. After an initial transient from the disordered to the ordered dot state, a constant average dot height of 120 Å is found for all layers (Fig. 1.26). In addition, no change in the critical wetting layer thickness [30] and no changes in the faceted pyramidal island shapes were observed. To determine how the lateral ordering of the dots affects the size homogeneity, the width of the size distributions was determined by Gaussian fits as indicated in Fig. 1.26 by the solid lines. For the single PbSe dot layer the width of the histogram is $\pm 14\%$. Although the lateral ordering starts already in the first few superlattice layers, the height distribution at first actually broadens to $\pm 27\%$ after ten superlattice periods, and only thereafter decreases to reach a final value of $\pm 10\%$ for $N = 100$. This transient broadening can be understood in terms of a *mismatch* of the average lateral dot spacing in the first PbSe layer with respect to the preferred spacing in the superlattice stack that is determined by the elastic interlayer interactions. In the first superlattice layers, this mismatch is accommodated by the formation of many defects and missing rows in the hexagonally ordered dot regions, and because the size of the dots near these defects deviates from those within the ordered regions, a broadening of the size distribution is induced. Once, however, the ordered dot domains have a larger size, a notable size homogenization occurs. This transient effect has also been found in the Monte Carlo growth simulations (Fig. 1.18).

1.8.2.2   *Vertically aligned superlattices*

The properties of vertically aligned PbSe dot superlattices are summarized in Fig. 1.27, where on the left-hand side the atomic force microscopy images of three vertically aligned dot superlattices with spacer thickness of 105–$330\,\text{Å}$ are shown, together with the corresponding evolution of the dot size distributions depicted in the centre panel. Clearly, there is a very strong tendency for a lateral dot alignment along the in-plane $\langle \bar{1}10 \rangle$ surface directions. As a result, 2D hexagonally ordered dot arrays are formed, which is evidenced by the six-fold symmetric satellite peaks in the FFT power spectra of the atomic force microscopy images shown as inserts. In addition, the dot sizes and spacings are not only much larger as compared to that of the single dot reference layer, but also systematically increase with increasing spacer thickness. This is proven by the corresponding decrease in the FFT satellite peak separations in Fig. 1.27. The lateral ordering is best for spacer thicknesses between 100 and 220 Å, for which the FFT satellite peaks are most narrow in width and for which even weaker second-order FFT satellites are visible. For larger

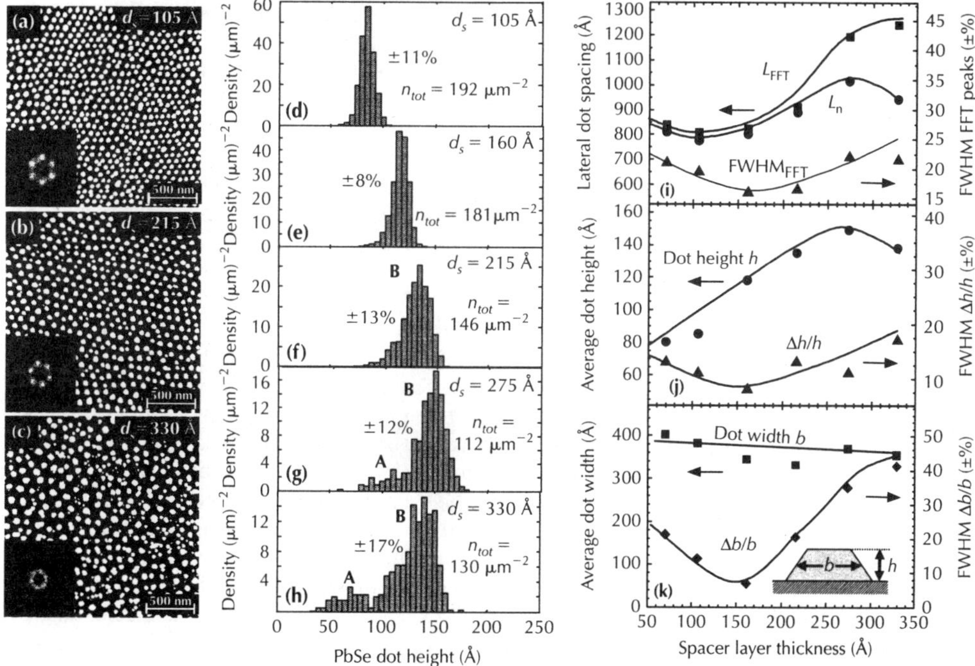

Figure 1.27   Left-hand side: *Atomic force microscopy surface images of vertically aligned PbSe/PbEuTe quantum dot superlattices with PbEuTe spacer layer thicknesses of 105, 215 and 330 Å, from (a) to (c), respectively. The number of 50 superlattice periods and the 5 ML thickness of the PbSe dot layer is constant for all samples. The insets show the 2D FFT power spectra of the atomic force microscopy images. Centre: Dot height histograms obtained for superlattices with spacer thicknesses of 105, 160, 215, 275 and 330 Å, from (d) to (h), respectively. Right-hand side: Dot parameters plotted as a function of the spacer layer thickness: (i) lateral dot spacing L, obtained from the FFT satellite peaks (■) as well as the PbSe dot density (●). Also plotted is the FWHM (▲) of the FFT satellite peaks. (j) Average PbSe dot height h (●) and corresponding FWHM ($\Delta h/h$) of the height histogram peaks (▲). (k) Average dot width b (■) and corresponding FWHM ($\Delta b/b$) of the histograms of the dot widths (◆) plotted as a function of spacer thickness. Adapted from Raab et al. [51].*

spacer thicknesses, the satellites become increasingly smeared out and they almost disappear for the superlattice with 330 Å spacers (Fig. 1.27c). This is due to the increasing disorder in the dot arrangement and the appearance of a second type of smaller PbSe dots nucleated between the larger dots. Because the 330 Å spacer thickness of this sample is already close to the transition to the *fcc*-like dot stacking, these interstitial dots obviously represent local *fcc*-stacked regions in which the lateral dot separation is a factor of two smaller than those for the vertically aligned dots (see Fig. 1.25e).

For a quantitative analysis of the lateral ordering process in the vertically aligned structures, the FFT satellite peak widths and separations Δk were deduced. The preferred lateral dot spacing L is then obtained from $L_{FFT} = 1/(\Delta k \cdot \sin 60°)$ or, alternatively, from the dot density n by using $L_n = 1/\sqrt{n \cdot \sin 60°}$. The results are plotted in Fig. 1.27f as a function of spacer thickness. Clearly, the preferred lateral dot spacing increases strongly with increasing spacer thickness, but does not follow a strict linear dependence as observed for the *fcc*-stacked superlattices. The relative width of the FFT satellite peaks (triangles in Fig. 1.27f) shows a clear minimum of $\pm 16\%$ at a spacer thickness of $d_s = 160$ Å, but for thinner as well as thicker spacer layers, the FWHM increases to about $\pm 21\%$. Thus, the best lateral ordering occurs for 160 Å spacers. For larger spacer thicknesses, also the mean dot separation obtained from the dot densities does not agree any more with the value deduced from the FFT peak spacing. This is due to the appearance of the smaller dots in the atomic force microscopy images, which increases the dot density but does not affect much the separation of the larger dots.

The dependence of the dot size and shape as a function of spacer thickness was deduced from a statistical analysis that also yields the width of the size distributions. Fig. 1.27d–h shows the height histograms for the samples with spacer thicknesses of 105 to 330 Å. Clearly, for spacer thicknesses increasing from 85 to 275 Å, the dot height rapidly increases from $h = 85$ Å to $h = 149$ Å, respectively. For thicker spacers, however, a small left-hand shoulder (A) starts to emerge at smaller dot heights, and this shoulder becomes even more pronounced for the sample with 330 Å spacers. Also, the average dot size does not increase anymore but rather slightly decreases to 138 Å. Both effects are caused by the formation of additional smaller dots on the surface, which reduces the overall amount of PbSe available for the larger dots.

Figure 1.27j and k summarize the dependence of the average dot height h and dot width b as a function of spacer thickness. As indicated in Fig. 1.27j, up to $d_s = 275$ Å, the average dot height increases linearly with increasing spacer thickness, whereas the dot width of about 350 Å remains essentially constant (Fig. 1.27k). This translates into a flattening of the dot shape for small spacers, indicating that the dot growth is enhanced in the lateral direction. Atomic force microscopy images recorded with selected sharp atomic force microscopy tips show that these dots assume a truncated pyramidal shape with triangular base and {100} side facets, as shown schematically in the inset of Fig. 1.27k. For very thin spacer layers, the PbSe dots are rather flat with an aspect ratio of only about 1:5, whereas for the thicker spacers the aspect ratio increases to about 1:3. This is still below the value of 1:2.2 of the pyramidal dots of single dot layers that do not show any flattening of the island apex. The modifications of dot shape are obviously induced by the elastic strain fields of the buried dots, which are strongest for the thinnest spacer layers but decay rapidly as the spacer thickness increases. Therefore, for spacer thicknesses larger than 400 Å, the dots exhibit the same pyramidal shape known for unperturbed single dot layers.

Perhaps the most interesting feature is the pronounced narrowing of the size distribution for the well-ordered vertically aligned samples. From the dependence of the width of the size distributions plotted in Fig. 1.27j and k as a function of spacer thickness it is found that the FWHM decreases from $\pm 13\%$ to $\pm 8\%$ when d_s increases from 80 to 160 Å, after which it increases again to above $\pm 15\%$ for $d_s = 330$ Å. A similar, but even more pronounced, trend is observed for the variation of the lateral dot widths (Fig. 1.27k), which again shows a pronounced minimum for 160 Å spacers. Thus, the highest uniformity is obtained for the superlattice with the best hexagonal ordering, demonstrating that the lateral ordering produces a higher uniformity of the dot ensembles.

1.8.2.3   Order parameters derived by X-ray diffraction studies

The vertical and lateral ordering was also characterized by anomalous X-ray diffraction performed at the ESRF synchrotron light source in Grenoble with an X-ray photon energy tuned to the M shell absorption edge of the Pb atoms. In this way, the structure scattering factor of the PbEuTe matrix is drastically reduced to about 50 times below that of PbSe and therefore the scattering contrast between the dots and matrix is drastically enhanced. The resulting anomalous reciprocal space maps are depicted in Fig. 1.28a–d for the superlattices with 105, 165, 215 and 465 Å spacer thicknesses, respectively. Clearly, for all samples a large number of satellite peaks is observed in the vertical q_z as well as lateral q_x direction. However, whereas for the samples with thin spacers the lateral satellite peaks are aligned parallel to the q_x direction, for the sample with thick spacers they are aligned along inclined directions (dashed lines in Fig. 1.28). In addition, reciprocal space maps recorded along different azimuth directions indicate a six-fold symmetry for the samples with small spacers as compared to a three-fold symmetry for the other sample. This is additional evidence of the different interlayer dot stackings in the samples, forming a trigonal *fcc*-like dot lattice for the latter and a vertically aligned 3D hexagonal dot lattice for the former.

The quality of the dot ordering process was assessed from cross-sectional line scans of the reciprocal space maps shown in Fig. 1.28e and f. For the sample with *fcc* stacking (Fig. 1.28f) a much larger number of lateral satellites can be resolved as compared to those of the vertically aligned dot samples (Fig. 1.28e). In addition, for the latter a significant increase in the peak widths with increasing q_x scattering vector occurs. For a quantitative analysis, the widths of the satellites were derived by fitting the cross-sectional profiles by Gaussians, with the fits represented as solid lines in Fig. 1.28e and f. The resulting FWHM of the lateral satellite peaks are plotted

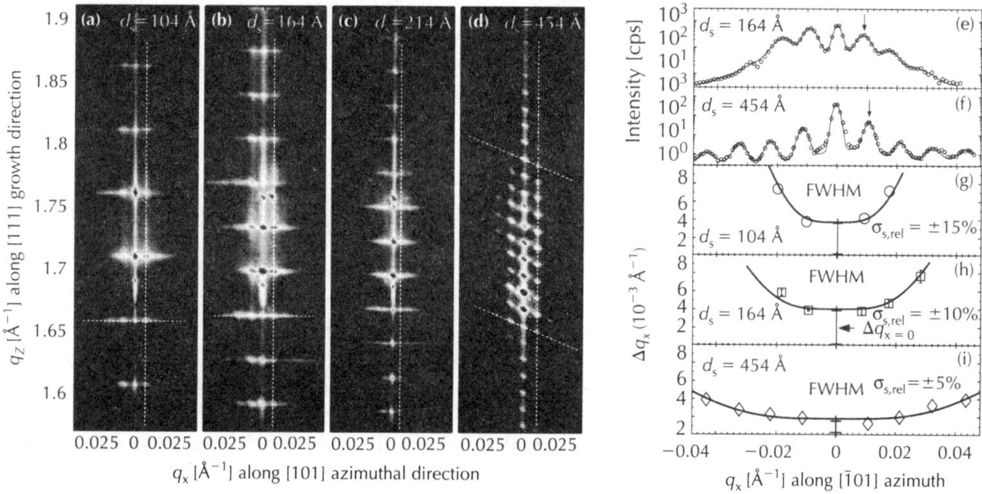

Figure 1.28 Left-hand side: *(111) Reciprocal space maps recorded by anomalous synchrotron X-ray diffraction at 5.1 Å wavelength of four PbSe quantum dot superlattices with different PbEuTe spacer thicknesses d_s = 104, 164, 214 and 454 Å, from (a) to (d), respectively. The arrangement of the lateral satellite peaks along q_x indicates a 3D hexagonal dot lattice structure and vertical dot alignment for the superlattices with small spacer thickness (a) to (c), as compared to the 3D trigonal dot lattice with fcc dot stacking for the sample with 454 Å spacers (d). Right-hand side: Cross-sectional q_x line scans for the superlattice samples with (e) 164 Å and (f) 454 Å spacers recorded along the horizontal, respectively, inclined dashed lines in the reciprocal space maps (b) and (d). (g) to (i) Full width at half maximum (FWHM) Δq_x of the lateral satellite peaks (open symbols) along q_x plotted as a function of q_x scattering vector for the samples with d_s = 104, 164 and 454 Å. The solid lines represent the fit of the data with a short-range order scattering model from which the dot order parameters such as variance of the dot spacings σ_L as well as the average domain size is obtained. Adapted from Lechner et al. [57].*

in Fig. 1.28g–j versus in-plane scattering vector q_x. For the vertically aligned dot superlattices, the lateral peak width rapidly increases, whereas for the superlattice with *fcc* stacking the FWHM are almost constant. To deduce the corresponding order parameters of the dot structures, this dependence was fitted using a modified short-range ordering scattering model, in which the dependence of the satellite peak width Δq_x as a function of scattering vector q_x is described as [57, 91, 173]:

$$\Delta q_x \approx \sqrt{q_x^4 \sigma_L^4 / L^2 + M^2}. \tag{1.12}$$

Here, L is the average lateral dot distance obtained from the satellite peak spacing, σ_L is the variance in the nearest neighbour dot distance within the growth plane and M is the relative size of the ordered dot domain in units of L. By fitting the experimental data with this relation (solid lines in Fig. 1.28g–j), σ_L and M can be deduced. M basically corresponds to the offset of Δq_x at $q_x = 0$, and σ_L is determined by the parabolic increase of Δq_x at higher q_x values. For the well-ordered vertically aligned samples (Fig. 1.28g and h), $\sigma_L = 117$ and 86 Å, which corresponds to a relative variance or dispersion of the lateral dot spacings of ± 15 and $\pm 10\%$, respectively. For the *fcc*-stacked sample σ_L is only 36 Å, corresponding to a much smaller dispersion of only $\pm 5\%$. This indicates that the lateral dot ordering is significantly better than that of the vertically aligned samples. The same trend also applies for the average domain sizes, with $M = 2$ for the vertically aligned samples as compared to $M = 5$ for the superlattice with *fcc* stacking. This indicates that the ordering process for superlattices with staggered dot stacking is much more efficient compared to that for vertically aligned dots, which agrees well with the results from the Monte Carlo growth simulations (Section 1.5). From an analogous measurement of the satellite peak widths in the vertical q_z direction, it is found that this width is quite small for all four samples and that the widths are nearly independent of q_z. This indicates that the layer-to-layer dot correlation along the growth axis is nearly perfect, with a correlation length as large as 25 superlattice periods for the vertically aligned samples and of seven periods for the superlattice with *fcc* stacking. This difference is caused by the decreasing strain field interactions with increasing spacer thickness, eventually approaching zero in the far-field limit.

1.8.3   Interlayer correlations as a function of dot size

Apart from the spacer thickness dependence, we have also investigated the influence of the dot layer thickness and the growth temperature on the interlayer correlation formed in self-assembled PbSe dot superlattices [55, 56]. As shown by studies of single dot layers [130, 171], both parameters strongly affect the size and density of the dots and thus the configuration of the starting layer which is expected to alter the evolution of the subsequent growth process. Two series of samples were investigated. In the first series, the PbSe dot layer thickness was varied from 1 to 8 ML, while keeping the substrate temperature of 360°C, the spacer thickness of 410 Å and the number of 100 superlattice periods constant [54]. In the second series, the substrate temperature was varied from 320 to 400°C whereas the spacer and dot layer thicknesses were kept constant at 420 Å and 5 monolayers and the number of superlattice periods was 30 [56].

Figure 1.29a–d shows the atomic force microscopy surface images of the superlattices with PbSe thicknesses of 3, 4, 5 and 8 monolayers. At 3 ML PbSe coverage (Fig. 1.29a), the PbSe dots are obviously randomly distributed over the surface without a preferred nearest neighbor direction (see featureless diffuse FFT power spectrum of the atomic force microscopy image depicted as inset). When the PbSe layer thickness increases to 4 ML, this disordered dot arrangement gives way to a well-ordered hexagonal dot arrangement (see Fig. 1.29b) where the dots are aligned along the $\langle \bar{1}10 \rangle$ surface directions with a well-defined lateral dot spacing of 580 Å. The good lateral ordering is evidenced by the appearance of sharp, six-fold symmetric satellite peaks in the FFT power spectrum. The same ordering is also observed for PbSe thicknesses of 5 and 6 ML, as shown in Fig. 1.29c. For thicker PbSe layers, however, the dots start to cluster and large "super" dots are formed. This is illustrated in Fig. 1.29d by the atomic force microscopy image of the 8 ML superlattice sample. Although between the large dots a few smaller dots remain that still show

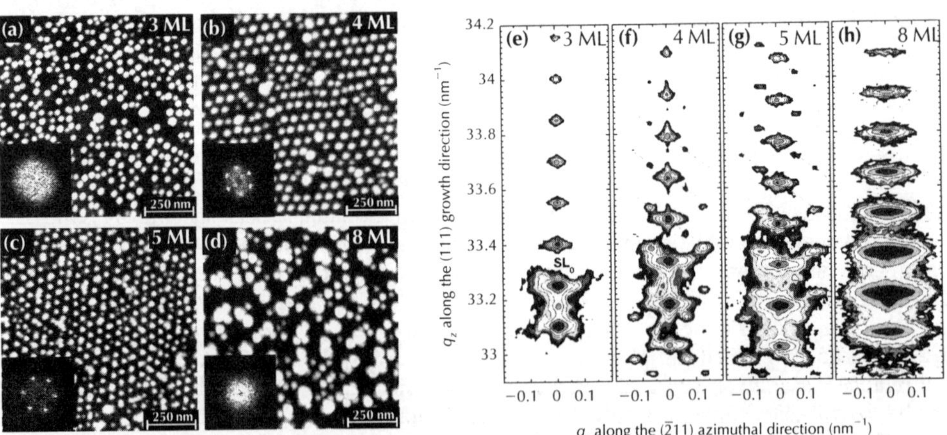

Figure 1.29   Left-hand side: *Atomic force microscopy surface images of 100 period PbSe quantum dot superlattices with constant 410 Å PbEuTe spacers but varying PbSe dot layer thickness of 3, 4, 5 and 8 monolayers for (a) to (d), respectively. The FFT power spectra shown as insets indicate a different lateral ordering for the samples. Right-hand side: (222) X-ray reciprocal space maps for the same samples. For the 3 ML superlattice (e), no lateral satellites are observed due to the lack of any lateral or vertical dot correlations. For the 4 and 5 ML superlattices (f) and (g), the lateral satellite indicates an fcc-like dot stacking, whereas for the 8 ML superlattice (h) the large peak broadening along q_x indicates the lack of lateral ordering but a vertical dot alignment along the growth direction. Adopted from Springholz et al. [55].*

some signs of a preferred hexagonal dot coordination, no distinct satellite peaks appear in the FFT power spectrum. This indicates the existence of a rather disordered overall dot arrangement.

To determine the corresponding interlayer dot correlations, high-resolution X-ray diffraction reciprocal space maps were recorded along the $[\bar{2}11]$ azimuth direction. The results are shown in Fig. 1.29e–h for superlattices with 3, 4, 5 and 8 ML PbSe thicknesses. For the 3 ML dot sample (Fig. 1.29e), only vertically aligned satellite peaks along q_z appear. Around the 0th order SL0 superlattice peak there is a strongly enhanced diffuse scattering caused by the presence of self-assembled quantum dots at the heterointerfaces. Because this scattering component is absent for the higher-order satellites, there is no spatial correlation of the dots from one layer to the next layer, i.e. the superlattice represents an uncorrelated repetition of disordered single dot layers, similar to PbSe dot superlattices with thick spacer layers.

For the superlattices with intermediate PbSe thicknesses (see Fig. 1.29f and g for 4 and 5 ML, respectively), satellite peaks are observed in both the vertical q_z and lateral q_x direction. Thus, the dots are highly correlated both vertically *and* laterally, creating a three-dimensional ordered lattice of dots. As discussed in detail in the previous section, because the lateral satellites are aligned along lines inclined by 38° to the vertical q_z direction an *fcc*-like dot stacking is formed. For the samples with larger PbSe thicknesses, again a striking change occurs, in which the well-defined lateral satellites are replaced by a strong correlated diffuse broadening of *all* satellites along the lateral q_x direction. This indicates that although the dots are disordered in the lateral direction, the interface corrugation is replicated from layer to layer along the growth axis. Thus, the dots are vertically aligned in columns as was found for PbSe superlattices with small spacer thickness.

From a statistical analysis of the dot size distributions [55, 56], the average dot height of the final layer was found to increase from 48 Å for the 3 ML superlattice to 125 Å for the 8 ML sample, and the narrowest size distribution is formed for the well-ordered samples with 4 and 5 ML PbSe thicknesses, for which the FWHM of the histograms is $\pm14\%$, whereas it is more than $\pm25\%$ for the disordered 3 and 8 ML superlattices [55, 56]. Thus, an *fcc*-like dot stacking is formed only for the superlattices with an average dot height between 80 and 120 Å. This demonstrates that the kind of interlayer correlation and dots stacking formed in the superlattice structures does not depend only on the spacer thickness but also on the dot size.

For the sample series with varied growth temperature, a similar behaviour was found [54, 56]. At growth temperatures below 340°C, the dot density is very high and the average dot size smaller than 70 Å. As a result, no interlayer correlations are formed and no lateral ordering occurs [56]. At temperatures between 360 and 380°C, in contrast, atomic force microscopy and X-ray diffraction measurements reveal a good hexagonal lateral ordering as well as a well-defined *fcc*-like vertical dot stacking. Again in this case, the average dot height is around a value of 90 Å with a very low $\pm14\%$ size dispersion. At higher temperatures, very large and disordered dots are formed with a large average height of 160 Å and the X-ray diffraction data indicates a preferential vertical dot alignment in the samples [56]. This indicates again that an *fcc*-type dot stacking is formed only for a very limited range of PbSe dot sizes and that the dot size is an important parameter in the interlayer correlation formation.

1.8.4 *Phase diagram for vertical and lateral dot ordering*

To explain the changes in the interlayer dot correlations as a function of spacer thickness and dot size, the dependence of the elastic strain fields and surface energy distributions on these parameters must be taken into account. For this purpose, we have performed a series of finite element calculations for pyramidal PbSe dots of different size but invariant shape located in a (111)-oriented PbEuTe matrix at various depths d below the surface [50]. In accordance with our atomic force microscopy studies [130], the shape of the dots was modelled as a triangular pyramid with $\{100\}$ side facets and constant aspect ratio of $b/h = \sqrt{6}$, where b is the island base and h the island height. A general conclusion of these calculations is that the surface strain energy distributions plotted in terms of dimensionless scaled surface coordinates x/d depend only on the *ratio* d/b of the island depth to the island size. As a consequence, the directions α of the surface energy minima are determined exclusively by the d/b ratio.

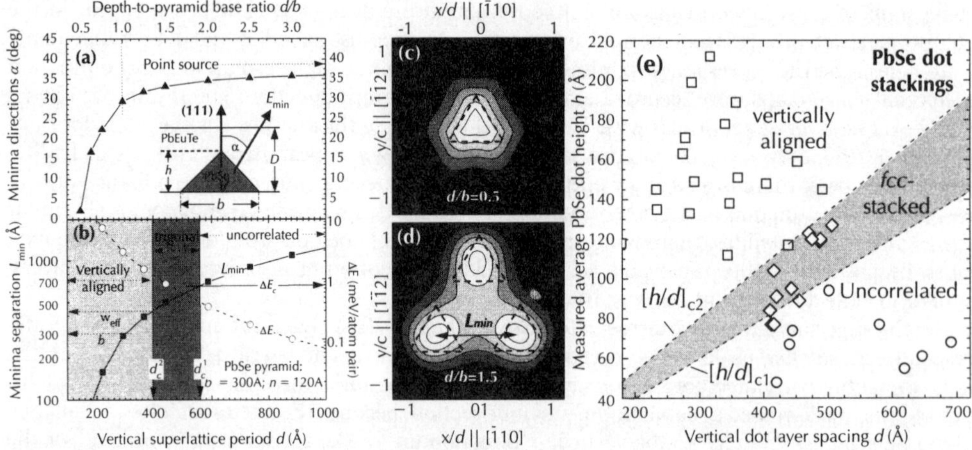

Figure 1.30   The influence of the spacer layer thickness, i.e. vertical superlattice period d *as well as the PbSe dot size on the strain energy distributions and interlayer correlations in (111) PbSe/PbEuTe quantum dot superlattices. (a) Direction* α *of the surface strain energy minima above pyramidal PbSe islands plotted as a function of the renormalized vertical dot layer separation* d/b, *where* b *is the lateral dot base width and* d *is the superlattice period. The surface strain energy distributions for* d/b = 0.5 *and* 1.5 *are plotted in (c) and (d) as a function of the reduced surface coordinates* x/d. *The energy separation between the contour lines is* 6 meV *and* 0.16 meV, *respectively. (b) Calculated lateral separation* L_min *of the energy minima (■) and depth of the minima* ΔE *(○) plotted versus superlattice period for PbSe pyramids with a fixed base width of* b = 300 Å *and a dot height of* h = 120 Å. (e) *Phase diagram of different dot stackings in PbSe dot superlattices as a function of vertical dot layer spacing and PbSe dot size: (□) vertically aligned dots, (◇) fcc-stacked and (○) uncorrelated superlattices. Data obtained from X-ray diffraction, transmission electron and atomic force microscopy measurements. The phase boundary lines are plotted according to Eqs (1.13) and (1.14). See also Springholz* et al. *[50].*

Figure 1.30c and d shows two representative scaled strain energy distributions for $d/b = 0.5$ and 1.5, respectively. Clearly, for spacer thicknesses larger than the island base, i.e. $d/b > 1$, the strain energy distributions closely resemble that obtained by the point-source model (compare Fig. 1.30d with Fig. 1.6c), with three well-separated energy minima along the $\langle \bar{1}\,\bar{1}\,2\rangle$ directions. Thus, for large d/b values the results converge to that of the far-field point-source model. This is also evidenced by Fig. 1.30a, where the minima direction α is plotted versus d/b which converges to 36° when d/b is larger than one. Thus, the experimentally observed *fcc* dot stacking as well as the measured and calculated interlayer correlation directions are in good agreement with each other for $d/b > 1$. However, when d/b decreases below 1, α rapidly decreases, reaching zero already for $d/b \leq 0.5$ (see Fig. 1.30a). This means that the three side energy minima are joined to one central minimum exactly above the buried dot, as is illustrated by the strain energy distribution depicted in Fig. 1.30c for $d/b = 0.5$. At small spacer thicknesses and/or large dot sizes, the *fcc* dot stacking is therefore replaced by a vertical dot alignment, in good agreement with the experimental observations.

Figure 1.30b shows the dependence of the energy minima separation L_{min} as well as their depth ΔE_{min} as a function of the vertical superlattice period d for PbSe dots with 120 Å height and 300 Å base, as measured for the sample series described in Section 1.8.1. The differently shaded regions indicate the different interlayer correlations observed by the experiments. For large spacers, L increases linearly with d, but rapidly drops to zero at d below 200 Å. On the other hand, the depth of the energy minima continuously decreases as the spacer thickness increases due to the decay of the elastic strain fields. Similar calculations for a constant spacer thickness but varying dot size are depicted in Fig. 1.10 [55].

To deduce the conditions at which the transition between the differently correlated dot structures occur, one has to compare the minima separation L_{min} and the dot base width b as indicated by the dashed triangles in Fig. 1.30c and d. For the case of small dots or thick spacers, the minima separation is much larger than the dot base width ($b < L_{min}$). Therefore, the dots on the surface can easily occupy just one single energy minimum and as growth proceeds, an *fcc*-like

ABCABC... stacking sequence is formed. For *very* small dots and large spacer thicknesses, however, the energy minima become very shallow because the strain fields decrease linearly with decreasing dot volume in the far-field limit (see Eq. 1.3). Experimentally, no interlayer dot correlations were found for vertical superlattice periods beyond the critical value of $d_c^1 = 560$ Å (see Fig. 1.25e). From Fig. 1.30b, at this point, the depth of the strain energy minima induced by one subsurface dot is as small as $\Delta E = 0.4$ meV/atom pair. Because in the multilayers each surface energy minimum is produced by the superposition of the strain fields of three subsurface dots, the corresponding minimal interaction energy required for interlayer correlation formation is actually three times this value, i.e. $\Delta E_{crit} = 1.2$ meV for the given growth conditions. Surprisingly, this value is more than one order of magnitude smaller than the thermal energies at the given growth temperature. This may be taken as an indication that the nucleation of Stranski–Krastanov islands is mostly governed by the critical nucleus size and nucleation barrier rather than by the hopping and diffusion process of the single surface adatoms.

For large dots and small spacer thicknesses, the energy minima separation successively decreases (Fig. 1.30b) and they eventually merge into one single minimum. Experimentally, the transition from the *fcc* dot stacking to a vertical alignment already occurs before this point, namely, at a spacer thickness of $d_c^2 = 400$ Å (see Fig. 1.30e), i.e. at a d/b value of about 1.3. This is because for *fcc*-stacked dots, not only must the energy minima separation be larger than the dot base widths, but additionally a certain denuded zone around each dot must be taken into account where further dot nucleation is kinetically suppressed. According to our experimental data, the size of this denuded zone must be about 1.6 times the base width, which is the effective width of the minimal surface area required for each dot.

For a given material system, the two phase boundary conditions can be written in a rather generalized form because under the condition of an invariant island shape there is a fixed relationship between the dot base and dot height and the dot volume is proportional to h^3. The first phase boundary corresponds to the cut-off length of the interlayer correlations that is determined by the minimal interaction energy E_{crit} required for correlated dot nucleation. This value is specific for each material system and growth condition, but according to Eq. 1.3, it is determined only by ratio of the volume over the dot depth V_0/h. As $V_0 \sim h^3$, this cut-off condition can be reformulated as:

$$[h/d]_{c1} = \sqrt[3]{\frac{E_{crit}}{E_s^{2D} \cdot C_{hkl} \cdot \delta}} \tag{1.13}$$

where $\delta = V_0/h^3$ is constant for a given island shape, and E_0 is the surface strain energy density of the uniformly strained 2D reference layer. Using for PbSe the values $E_s^{2D} = 142$ meV, $C_{hkl} = 0.69$ (see Section 1.3.4), $\delta = \sqrt{3}/2$ for {100} faceted pyramids and the experimental value of $E_{crit} = 1.4$ meV from above, one obtains a critical ratio of $[h/d]_{c1} = 0.22$ below which no interlayer correlations should be formed in the superlattice structures.

Concerning the second boundary condition, for a fixed island shape the correlation angle α of the energy minima is constant for a given d/b (see Fig. 1.30a) and likewise for a fixed h/d ratio. Since the diameter of the denuded zone w_{eff} is proportional to b and thus $w_{eff}/h = \kappa = $ constant, the condition of $w_{eff} = L_{min}$ for the transition between the vertical dot alignment and *fcc* stacking can be written for the (111) case as:

$$[h/d]_{c2} = \frac{\sqrt{3} \cdot \tan \alpha_{111}}{\kappa} \tag{1.14}$$

For the (100) growth orientation, the $\sqrt{3}$ factor simply has to be replaced by $\sqrt{2}$ because of the different arrangement of the energy minima (Section 1.3). Applying the appropriate parameters for the PbSe case ($\kappa = 3.9$ and $\alpha_{111} = 38°$), one obtains a critical ratio of $[h/d]_{c2} = 0.28$ above which the PbSe dots should be always vertically aligned. Thus, an *fcc*-like dot stacking should occur only in the range of $0.22 < [h/d] < 0.28$.

Compiling in Fig. 1.30 the large body of experimental data obtained by X-ray diffraction, transmission electron and atomic force microscopy for more than 50 different PbSe quantum dot superlattices with various spacer thicknesses and dot sizes, a complete phase diagram of dot stackings is obtained for this material system in which the phase boundaries follow exactly the dashed boundary lines defined by the stacking conditions of Eqs (1.13) and (1.14). Moreover, this phase diagram clearly demonstrates that staggered dot stackings can be generally obtained only in a narrow window of parameters confined by the phase boundaries represented by Eqs (1.13) and (1.14). Its width is determined not only by the elastic properties, the surface orientation and the characteristic island shapes but also by the epitaxial growth conditions that will determine the effective cut-off energy value E_{crit}. For materials with small elastic anisotropy (e.g. for Si/Ge (100)) $[h/d]_{c2}$ will become smaller than $[h/d]_{c1}$. This means that no set of parameters exist in which a staggered dot stacking may be formed, which poses a strong limitation on the conditions and materials in which a staggered stacking can be obtained.

1.9   Other mechanisms for interlayer correlation formation

Elastic interactions are certainly the most important factor for interlayer correlations formed in quantum dot multilayer structures. However, as already mentioned in Section 1.2, there exist at least two other alternative but less obvious mechanisms that may contribute to interlayer correlation formation. These mechanisms are correlated dot nucleation mediated by (i) non-planarized surface topographies (Fig. 1.1c and d) or (ii) lateral compositional modulations of the spacer layer (Fig. 1.1e and f). Although up to now little work has been carried out to clarify these mechanisms, each of them may in principle give rise to different interlayer dot stacking types (see Fig. 1.1), depending on the intricate details of the interaction and nucleation process.

1.9.1   Morphologic correlations

With respect to the correlations mediated by the growth morphology, one first has to realize that each quantum dot layer itself represents a highly corrugated surface structure. Although during overgrowth, this 3D surface will be eventually planarized to minimize the surface energy, a corrugated non-planar surface structure can be retained when the capping process is incomplete. This can be expected, e.g., for thin spacer layers or when planarization is hindered by sluggish surface kinetics. Obviously, a corrugated surface will influence the subsequent island nucleation and, because the surface corrugations are linked to dots below the surface, in this way interlayer dot correlations can be produced as well. That the surface morphology plays a profound role in self-assembled dot nucleation has been recently established by investigation of self-assembled quantum dot growth on prepatterned substrates. In these studies it has been demonstrated that dot nucleation can be triggered by lithographically patterned surface sites [174–178]. In fact, in this way a near-perfect position control of self-assembled quantum dots has been achieved for the Ge on Si [174] as well as InAs on GaAs material systems [176].

For overgrowth of Stranski–Krastanow islands, actually two different growth scenarios may occur with quite different resulting surface morphologies. On the one hand, when the surface diffusivity of spacer adatoms and surface capillarity forces are small, surface planarization is rather slow and mound-like structures are retained above the buried islands when the spacer thickness is not very large. This situation is illustrated schematically in Fig. 1.1c and d. On the contrary, when surface mass transport is dominated by the stress fields of the buried dots, the mobile adatoms during spacer deposition are repelled from the surface above the buried dots due to the opposite sign of strain [15]. As a result, the growth of the spacer layer can be locally suppressed and surface depressions or pits are formed above the dots [15]. This has been reported, e.g., for InAs islands overgrown by GaAs [179] or InP [180], as well as for PbSe dots overgrown with PbEuTe [112]. This means that the actual type of spacer morphology strongly depends on the chosen growth conditions.

A second complication arises from the fact that the reaction of dot nucleation to the presence of non-planar corrugated surface morphologies itself depends on the mechanism that dominates the dot layer surface mass transport. If lateral mass transport during wetting layer growth is dominated by capillary forces, then dot material will accumulate at the concave surface areas and as a result preferred dot nucleation at the troughs of the surface morphology will occur as illustrated in panel (d) of Fig. 1.1. Experimentally, this behaviour has been found for Ge growth on prepatterned Si, where Ge islands were found to nucleate preferentially at the bottom of pits etched into the Si substrate [174, 181]. A similar behaviour was also reported for InAs islands grown over GaAs pit patterns [176]. On the other hand, when mass transport is dominated by stress-driven surface diffusion, the opposite behaviour may take place because adatoms then diffuse preferentially towards the convex areas of the surface where part of the misfit strain can be elastically relaxed. As a result, the dot material will accumulate at the tops or edges of patterned surface structures where, as a result, preferential dot nucleation will occur. This behaviour has been reported, e.g., for InAs quantum dots deposited on GaAs surface ridge or mesa structures [182–184], as well as for Ge grown over Si mesas [185, 186]. Both effects can be further altered when large differences in the free surface energies or diffusivities exist on differently oriented areas of the surface morphology. In the light of these complexities, the outcome of an actual growth experiment is difficult to predict but it is clear that this correlation mechanism can give rise to both a vertical dot alignment as well as a staggered stacking. An example for the profound effect of the spacer morphology on the interlayer dot correlation is shown in Fig. 1.2c for a Ge/Si dot superlattice investigated by Sutter *et al.* [76]. In this case, due to the strong increase of the Ge dot size in the superlattice stack, at a higher layer index no complete planarization of the thin Si spacer layers was achieved. As a result, the vertical dot alignment switches to an oblique dot replication at a certain point of growth (see Fig. 1.2c) [76]. Other examples of non-vertical dot stackings possibly related to non-planarized spacer layer morphologies include self-assembled InP/GaInP quantum dot stacks [187] as well as InAs/InP [188] and InAs/InAlAs [62, 65, 189] *quantum wire* multilayers, in which not only oblique interlayer dot alignments with varying correlation angles but also staggered wire stackings have been observed. Corrugated surface morphologies often appear in epitaxial growth on vicinal or high-index surfaces due to kinetic [190, 191] or strain-induced [21, 192, 193] step bunching or due to spontaneous surface facetation [4]. As shown by previous works, this can also lead to the formation of correlated interface corrugations in superlattice structures [21, 194–196], which combined with the growth of self-assembled Stranski–Krastanow dots may lead to morphology driven correlation formation in quantum dot multilayers [197–201].

1.9.2 Correlations induced by composition

The third mechanism for interlayer correlations is based on the formation of lateral chemical composition variations within the spacer layer induced by the buried dots. As is indicated schematically in Fig. 1.1e and f, these variations may originate from two processes, namely, from preferential surface segregation of dot material above the buried islands (Fig. 1.1e), or from strain- or morphology-driven alloy decomposition of the spacer material (Fig. 1.1f). The first mechanism will be operative for strongly segregating heteroepitaxial materials. This applies to many self-assembled quantum dot systems because they are usually composed of materials with large differences in lattice constants and binding energies, which are the two major driving forces for surface segregation. As surface segregation will tend to cause an enhanced accumulation of dot material directly above the buried dots, the subsequent wetting layer growth may be locally enhanced and subsequent dots will therefore tend to nucleate on top of the previous islands. Practically, this effect is superimposed by the simultaneous action of the elastic strain fields of the buried dots that for thin spacer layers produce a vertical dot alignment as well. As a result, it is unclear up to now how much surface segregation actually contributes to interlayer correlation formation in multilayer structures. Several studies have indicated that the critical wetting layer for island nucleation in multilayers is significantly reduced with increasing number of deposited layers (see Sections 1.6 and 1.7 and [39, 40, 75], which may be taken as an indication that surface segregation could be an important parameter in multilayer growth. Surface segregation strongly depends on the growth conditions and can be altered by the use of surfactants [202]. This may provide a tool for controlling and studying its effect in multilayer structures in more detail.

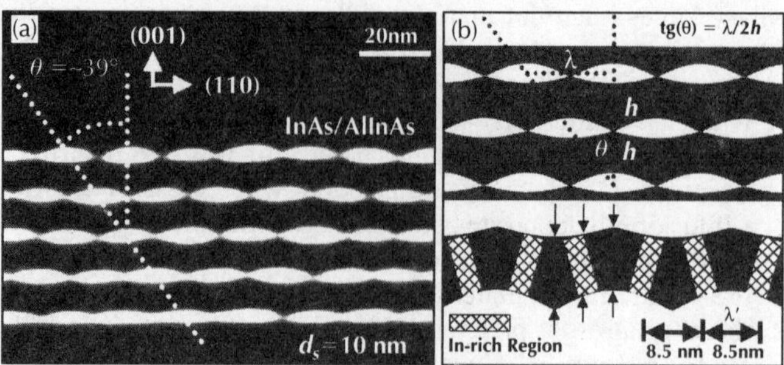

Figure 1.31 Staggered stacking in InAs/AlInAs quantum wire superlattices: (a) Cross-sectional transmission electron micrograph of five periods of 4.6 monolayers InAs alternating with 10 nm $Al_{0.48}In_{0.52}As$ grown on a (100) InP substrate, with an interlayer correlation angle of $\sim39°$ between the quantum wires. (b) Schematic illustration of the interlayer correlation formation process that is proposed to be caused by alloy decomposition of the AlInAs spacer layers into V-shaped In-rich regions. Adapted from Sun et al. [62].

For multilayers with multi-component ternary or quaternary alloys as spacer layers, lateral compositional variation can be formed when there exists a tendency for the alloys to spinodally decompose into regions of different chemical composition. This effect is often driven by strain or surface corrugations during epitaxial growth. Alloy phase separation is quite common in III–V ternary or quaternary alloys for certain chemical compositions and growth conditions (see, e.g., [203] for a review). The resulting lateral variations of the chemical composition in the spacer layer may not only cause a composition modulation within the wetting layer, but will also induce additional strain variations due to the concomitant variations in alloy lattice constant. This may amplify or counteract the strain fields from the buried dots, but in any case will modify the dot nucleation process and thus contribute to interlayer correlation formation.

The most prominent example of this mechanism is the staggered stacking found in self-assembled InAs/AlInAs (001) quantum wire superlattices [60–65, 139], as is illustrated by the transmission electron microscopy images shown in Figs 1.2d, 1.16a and 1.31a. In this case, the ternary AlInAs spacer layers show a strong tendency for alloy decomposition due to an immiscibility gap [66, 204]. Under the presence of surface roughness or strain variations, lateral phase separation in In-rich and Al-rich regions occurs [204–208]. As a result, V-shaped In-rich regions emerge from the sides faces of the buried InAs wires as is illustrated schematically in Fig. 1.31b and has been reported by Brault *et al.* [64, 65]. A corresponding cross-sectional transmission electron micrograph with well-visible In-rich regions is shown in Fig. 1.2d. Subsequently, InAs quantum wires nucleate preferentially at the intersections of In-rich V arms of neighbouring buried wires, which gives rise to a staggered *ABAB...* interlayer stacking as shown schematically in Fig. 1.31b and is clearly visible by transmission electron microscopy (Figs 1.31a and 1.16a). This stacking type has been observed consistently by several groups for varying AlInAs spacer layer thicknesses from 3 to 30 nm [60–65, 138, 139], with interlayer correlation angles around 40° (see Fig. 1.31a). Supporting evidence for this mechanism comes from the fact that no such staggered stacking has been found when the AlInAs spacer layers were replaced by GaInAs, GaInP or InP spacers, for which a chemical decomposition does not take place and for which therefore, only the usual vertical dot alignment was found [61, 77, 139, 209]. This underlines that the staggered correlations could be induced by chemical effects. A model for this process has been recently developed by Priester and Grenet [66] that correctly predicts the oblique correlations in such structures. On the other hand, since the AlInAs spacer layers on top of the InAs quantum wires are not always completely planarized prior to subsequent InAs growth [62, 65, 210] the surface morphology could also play a significant role in the formation of the staggered correlations in these material systems. In fact, an inclined alignment of InAs quantum wires was also found recently for pure InP spacer layers [188] in which alloy decomposition cannot occur.

1.10   Summary and outlook

In summary, a large variety of different types of correlations and stackings can be formed in self-assembled quantum dot multilayer structures. This can be caused by different mechanisms, of which the elastic interactions mediated by the strain fields of the buried dots play the most important role. The strain interactions have been modelled in great detail, and the derived theoretical predictions of stacking and ordering are in quite good agreement with experimental observations. Apart from the elastic strain fields, interlayer correlations may be also induced by corrugated surface topographies or by interactions based on chemical composition variations within the spacer material. Depending on the material properties, growth orientation, layer thicknesses and growth conditions all of these processes may induce a vertical dot alignment as well as staggered dot stackings and even for a single material system the interlayer correlation may change from one stacking type to another. Interlayer stackings have also a profound effect on the lateral arrangement of the dots. In particular, a lateral ordering of the dots can be induced with a resulting homogenization of dot sizes as desired for device applications. Staggered dot stackings are particularly effective in this respect, because the dot nucleation sites are determined by the interactions with several neighbouring dots below the surface. For vertically aligned dots, the initial lateral arrangement is mainly replicated from layer to layer with only a weak lateral ordering tendency.

While the stackings due to elastic interactions are already quite well understood, there are still ample uncertainties and open issues to be resolved for the other interaction mechanisms. This is due to the fact that they strongly depend on the growth conditions and are always superimposed by the simultaneous action of the strain field interactions. Therefore, more work is still needed to clarify their actual role in multilayer growth. On the other hand, the interplay between the different mechanisms may be utilized to create new and more complex interlayer stacking types and novel quantum dot superstructures. This may be achieved, e.g., by alternating the material compositions, spacer thicknesses as well as growth conditions during multilayer growth, or by combining compressively and tensily strained dots. Another promising approach is to combine interlayer stacking with site-controlled dot nucleation on lithographically pre-patterned substrate templates. The latter will allow the tailoring of the initial dot arrangement and thus may ultimately lead to the synthesis of fully controlled three-dimensional nanostructures.

Acknowledgements

The author would like to thank V. Holy, M. Pinczolits, P. Mayer, A. Raab, R.T. Lechner, J. Stangl, P. Simicek, D. Lugovyy, S. Zerlauth, F. Schäffler, L. Salamanca-Riba, H.H. Kang, and G. Bauer for long-term collaborations. This work was supported by the Fonds zur Förderung der Wissenschaftlichen Forschung (IRON and Project P17166-TPH) and the Gesellschaft für Mikro- und Nanoelektronik of Austria.

References

1. D. Bimberg, M. Grundmann, and N.N. Ledentsov. *Quantum Dot Heterostructures* (Wiley, Chichester, 1998).
2. D. Leonard, M. Krishnamurty, C.M. Reaves, S.P. Denbaar, and P. Petroff, Direct formation of quantum-sized dots from uniform coherent islands of InGaAs on GaAs surfaces, *Appl. Phys. Lett.* **63**, 3203 (1993).
3. J.M. Moison, F. Houzay, F. Barthe, L. Leprince, E. Andre, and O. Vatel, Self-organized growth of regular nanometer-scale InAs dots on GaAs, *Appl. Phys. Lett.* **64**, 196 (1994).
4. V.A. Shchukin, N.N. Ledentsov, and D. Bimberg. *Epitaxy of Nanostructures* (Springer Verlag, Berlin, 2004). V.A. Shchukin and D. Bimberg, Spontaneous ordering of nanostructures on crystal surfaces, *Rev. Mod. Phys.* **71**, 1125 (1999).

5. D.J. Srolovitz, On the stability of surfaces of stressed solids, *Acta Metall.* **37**, 621 (1989).
6. C.W. Snyder, B.G. Orr, D. Kessler, and L.M. Sander, Effect of strain on surface morphology in highly strained InGaAs films, *Phys. Rev. Lett.* **66**, 3032 (1991).
7. D.J. Eaglesham and M. Cerullo, Dislocation-free Stranski-Krastanow growth of Ge on Si(100), *Phys. Rev. Lett.* **64**, 1943 (1990).
8. H. Gao, Stress concentration at slightly undulating surfaces, *J. Mech. Solids* **39**, 443 (1991).
9. C. Ratsch and A. Zangwill, Equilibrium theory of the Stranski–Krastanov epitaxial morphology, *Surf. Sci.* **293**, 123 (1993).
10. J. Tersoff and F.K. LeGoues, Competing relaxation mechanisms in strained layers, *Phys. Rev. Lett.* **72**, 3570 (1994).
11. I. Daruka and A.-L. Barabási, Dislocation-free island formation in heteroepitaxial growth: a study at equilibrium, *Phys. Rev. Lett.* **79**, 3708 (1997).
12. J.Y. Marzin, J.M. Gerard, A. Izrael, D. Barrier, and G. Bastard, Photoluminescence of single InAs quantum dots obtained by self-organized growth on GaAs, *Phys. Rev. Lett.* **73**, 716 (1994).
13. R. Leon, P.M. Petroff, D. Leonhard, and S. Fafard, Spatially resolved visible luminescence of self-assembled semiconductor quantum dots, *Science* **267**, 1966 (1995). E. Dekel, D. Gershoni, E. Ehrenfreund, D. Spektor, J.M. Garcia, and P.M. Petroff, Multiexciton spectroscopy of a single self-assembled quantum dot, *Phys. Rev. Lett.* **80**, 4991 (1998),
14. A. Zrenner, A close look on single quantum dots, *J. Chem. Phys.* **112**, 7790 (2000).
15. Q. Xie, A. Madhukar, P. Chen, and N. Kobayashi, Vertically self-organized InAs quantum box islands on GaAs (100), *Phys. Rev. Lett.* **75**, 2542 (1995).
16. G.S. Solomon, J.A. Trezza, A.F. Marshall, and J.S. Harris, Jr, Vertically aligned and electronically coupled growth induced InAs islands in GaAs, *Phys. Rev. Lett.* **76**, 952 (1996).
17. J. Tersoff, C. Teichert, and M.G. Lagally, Self-organization in growth of quantum dot superlattices, *Phys. Rev. Lett.* **76**, 1675 (1996).
18. G. Springholz, V. Holy, M. Pinczolits, and G. Bauer, Self-organized growth of three dimensional quantum-dot crystals with *fcc*-like stacking and a tunable lattice constant, *Science* **282**, 734 (1998).
19. G. Springholz, M. Pinczolits, V. Holy, S. Zerlauth, I. Vavra, and G. Bauer, Vertical and lateral ordering in self-organized quantum dot superlattices, *Physica E.* **9**, 149 (2001).
20. G. Springholz, V. Holy, P. Mayer, M. Pinczolits, A. Raab, R.T. Lechner, G. Bauer, H. Kang, and L. Salamanca-Riba, Self-organized ordering in self-assembled quantum dot superlattices, *Mat. Sci. Eng. B.* **88**, 143 (2002).
21. C. Teichert, Self-organisation of nanostructures in semiconductor heteroepitaxy, *Physics Reports* **365**, 335 (2002).
22. O.G. Schmidt and K. Eberl, Multiple layers of self-asssembled Ge/Si islands: photoluminescence, strain fields, material interdiffusion, and island formation, *Phys. Rev. B.* **61**, 13721 (2000).
23. O.G. Schmidt, K. Eberl, and A. Rau, Strain and band-edge alignment in single and multiple layers of self-assembled Ge/Si and GeSi/Si islands, *Phys. Rev. B.* **62**, 16715 (2000).
24. G. Ortner, M. Bayer, A. Larionov, V.B. Timofeev, A. Forchel, Y.B. Lyanda-Geller, T.L. Reinecke, P. Hawrylak, S. Fafard, and Z. Wasilewski, Fine structure of excitons in InAs/GaAs coupled quantum dots: a sensitive test of electronic coupling, *Phys. Rev. Lett.* **90**, 086404 (2003).
25. I. Shtrichman, C. Metzner, B.D. Gerardot, W.V. Schoenfeld, and P.M. Petroff, Photoluminescence of a single InAs quantum dot molecule under applied electric field, *Phys. Rev. B.* **65**, 081303 (2002).
26. M. Bayer, P. Hawrylak, K. Hinzer, S. Fafard, M. Korkusinski, Z.R. Wasilewski, O. Stern, and A. Forchel, Coupling and entangling of quantum states in quantum dot molecules, *Science* **291**, 451 (2001).
27. See, e.g., P.M. Petroff, A. Lorke, and A. Imamoglu, Epitaxially self-assembled quantum dots, *Physics Today* **54/4**, 46 (2001). and references therein.
28. C. Teichert, L.J. Peticolas, J.C. Bean, J. Tersoff, and M.G. Lagally, Stress-induced self-organization of nanoscale structures in SiGe/Si multilayer films, *Phys. Rev. B.* **53**, 16334 (1996).
29. F. Liu, S.E. Davenport, H.M. Evans, and M.G. Lagally, Self-organized replication of 3D coherent island size and shape in multilayer heteroepitaxial films, *Phys. Rev. Lett.* **82**, 2528 (1999).
30. M. Pinczolits, G. Springholz, and G. Bauer, Evolution of hexagonal lateral ordering in strain-symmetrized PbSe/PbEuTe quantum-dot superlattices, *Phys. Rev. B.* **60**, 11524 (1999).

31. G.S. Solomon, S. Komarov, J.S. Harris, and Y. Yamamoto, Increased size uniformity through vertical quantum dot columns, *J. Crys. Growth* **175/176**, 707 (1997).

32. L. Vescan, W. Jäger, C. Dieker, K. Schmidt, A. Hartmann, and H. Lüth, Formation of heterogeneous thickness modulations during epitaxial growth of LPCVD-$Si_{1-x}Ge_x$ Si quantum well structures, *Mat. Res. Soc. Symp. Proc.* **263**, 23 (1992).

33. P. Schittenhelm, G. Abstreiter, A. Darhuber, G. Bauer, P. Werner, and A. Kosogov, Growth of self-assembled homogeneous SiGe-dots on Si (100), *Thin Solid Films* **294**, 291 (1997).

34. P. Schittenhelm, C. Engel, F. Findeis, G. Abstreiter, A.A. Darhuber, G. Bauer, A.O. Kosogov, and P. Werner, Self-assembled Ge dots: growth, characterization, ordering, and applications, *J. Vac. Sci. Technol. B.* **16**, 1575 (1998).

35. O. Kienzle, F. Ernst, M. Rühle, O.G. Schmidt, and K. Eberl, Modified Stranski–Krastanov growth in stacked layers of self-assembled islands, *Appl. Phys. Lett.* **74**, 269 (1999).

36. J. Stangl, T. Roch, G. Bauer, I. Kegel, T.H. Metzger, O.G. Schmidt, K. Eberl, O. Kienzle, and F. Ernst, Vertical correlation of SiGe islands in SiGe/Si superlattices: X-ray diffraction versus transmission electron microscopy, *Appl. Phys. Lett.* **77**, 3953 (1999).

37. E. Mateeva, P. Sutter, J.C. Bean, and M.G. Lagally, Mechanism of organization of three-dimensional islands in SiGe/Si multilayers, *Appl. Phys. Lett.* **71**, 3233 (1997).

38. A. Darhuber, P. Schittenhelm, V. Holy, J. Stangl, G. Bauer, and G. Abstreiter, High-resolution x-ray diffraction from multilayered self-assembled Ge dots, *Phys. Rev. B.* **55**, 15652 (1997).

39. O.G. Schmidt, O. Kienzle, Y. Hao, K. Eberl, and F. Ernst, Modified Stranski–Krastanov growth in stacked layers of self-assembled islands, *Appl. Phys. Lett.* **74**, 1272 (1999).

40. V. Le Thanh, V. Yam, P. Boucaud, F. Fortune, C. Ulysse, D. Bouchier, L. Vervoort, and J.-M. Lourtioz, Vertically self-organized Ge/Si(001) quantum dots in multilayer structures, *Phys. Rev. B.* **60**, 5851 (1999). V. Le Thanh, V. Yam, L.H. Nguyen, Y. Zenhg, P. Boucaud, D. Debarre, and D. Bouchier, Vertical ordering in multilayers of self-assembled Ge/Si(001) quantum dots, *J. Vac. Sci. Technol. B.* **20**, 1259 (2002).

41. M. Herbst, C. Schramm, K. Brunner, T. Asperger, H. Riedl, G. Abstreiter, A. Vörckel, H. Kurz, and E. Müller, Structural and optical properties of vertically correlated Ge island layers grown at low temperatures, *Mat. Sci. Engineering B.* **89**, 5 (2002).

42. L. Goldstein, F. Glas, J.Y. Marzin, M.N. Charasse, and G. Le Roux, Growth by molecular beam epitaxy and characterization of InAs/GaAs strained-layer superlattices, *Appl. Phys. Lett.* **47**, 1099 (1985).

43. S. Guha, A. Madhukar, and K.C. Rajkumar, Onset of incoherency and defect introduction in the initial stages of molecular beam epitaxial growth of highly strained In_xGa_{1-x} As on GaAs(100), *Appl. Phys. Lett.* **57**, 2110 (1990).

44. A.A. Darhuber, V. Holy, J. Stangl, G. Bauer, A. Krost, F. Heinrichsdorff, M. Grundmann, D. Bimberg, V.M. Ustinov, P.S. Kop'ev, A.O. Kosogov, and P. Werner, Lateral and vertical ordering in multilayered self-organized InGaAs quantum dots studied by high resolution x-ray diffraction, *Appl. Phys. Lett.* **70**, 955 (1997).

45. B. Legrand, J.P. Nys, B. Grandidier, D. Stievenard, A. Lemaitre, J.M. Gerard, and V. Thierry-Mieg, Quantum box size effect on vertical self-alignment studied using cross-sectional scanning tunneling microscopy, *Appl. Phys. Lett.* **74**, 2608 (1999).

46. M. Strassburg, V. Kutzer, U.W. Pohl, A. Hoffmann, I. Broser, N.N. Ledentsov, D. Bimberg, A. Rosenauer, U. Fischer, D. Gerthsen, I.L. Krestnikov, M.V. Maximov, P.S. Kop'ev, and Zh.I. Alverov, Gain studies of (Cd, Zn) Se quantum islands in a ZnSe matrix, *Appl. Phys. Lett.* **72**, 942 (1998).

47. L. Krestnikov, M. Straßburg, M. Caesar, A. Hoffmann, U.W. Pohl, D. Bimberg, N.N. Ledentsov, P.S. Kop'ev, Zh.I. Alferov, D. Litvinov, A. Rosenauer, and D. Gerthsen, Control of the electronic properties of CdSe submonolayer superlattices via vertical correlation of quantum dots, *Phys. Rev. B.* **60**, 8695 (1999).

48. G. Mackowski, T. Karczewski, J. Wojtowicz, J. Kossut, S. Kret, A. Szczepanska, P. Duczewski, G. Prechtl, and W. Heiss, Structural and optical evidence of island correlation in CdTe/ZnTe superlattices, *Appl. Phys. Lett.* **78**, 3884 (2001).

49. G. Springholz, M. Pinczolits, V. Holy, P. Mayer, K. Wiesauer, T. Roch, and G. Bauer, Self-organized growth of three-dimensional IV VI semiconductor quantum dot crystals with *fcc*-like vertical stacking and tunable lattice constant, *Surf. Sci.* **454–456**, 657 (2000).

50. G. Springholz, M. Pinczolits, P. Mayer, V. Holy, G. Bauer, H.H. Kang, and L. Salamanca-Riba, Tuning of vertical and lateral correlations in self-organized $PbSe/Pb_{1-x}Eu_xTe$ quantum-dot superlattices, *Phys. Rev. Lett.* **84**, 4669 (2000).

51. A. Raab, R.T. Lechner, and G. Springholz, Self-organized lateral ordering for vertically aligned PbSe/PbEuTe quantum dot superlattices, *Appl. Phys. Lett.* **80**, 1273 (2002).

52. A. Raab, G. Springholz, R.T. Lechner, I. Vavra, H.H. Kang, and L. Salamanca-Riba, Atomic force microscopy and transmission electron microscopy study of self-organized ordering in vertically aligned PbSe quantum dot superlattices, *Mat. Res. Soc. Symp. Proc.* **696**, 209 (2002).

53. V. Holy, J. Stangl, G. Springholz, M. Pinczolits, G. Bauer, I. Kegel, and T.H. Metzger, Lateral and vertical ordering of self-assembled PbSe quantum dots studied by high-resolution X-ray diffraction, *Physica B.* **283**, 65 (2000).

54. H.H. Kang, L. Salamanca-Riba, M. Pinczolits, G. Springholz, V. Holy, and G. Bauer, TEM investigation of self-organized PbSe quantum dots as a function of spacer layer thickness and growth temperature, *Mat. Sci. Eng. B.* **80**, 104 (2001).

55. G. Springholz, A. Raab, R.T. Lechner, and V. Holy, The dot size dependence of vertical and lateral ordering in self-organized PbSe/PbEuTe quantum dot superlattices, *Appl. Phys. Lett.* **82**, 799 (2003).

56. A. Raab, G. Springholz, and R.T. Lechner, The growth temperature and coverage dependence of vertical and lateral ordering in self-assembled PbSe quantum dot superlattices, *Phys. Rev. B.* **67**, 165321 (2003).

57. R.T. Lechner, T. Schülli, V. Holy, G. Springholz, J. Stangl, A. Raab, T.H. Metzger, and G. Bauer, Ordering parameters of self-organized 3D quantum dot lattices determined by anomalous x-ray diffraction, *Appl. Phys. Lett.* **84**, 885 (2004).

58. V. Holy, G. Springholz, M. Pinczolits, and G. Bauer, Strain induced vertical and lateral correlations in quantum dot superlattices, *Phys. Rev. Lett.* **83**, 356 (1999).

59. V.A. Shchukin, D. Bimberg, V.G. Malyshkin, and N.N. Ledentsov, Vertical correlations and anticorrelations in multisheet arrays of two-dimensional islands, *Phys. Rev. B.* **57**, 12262 (1998).

60. H. Li, J. Wu, Z. Wang, and T. Daniels-Race, High-density InAs nanowires realized in situ on (100) InP, *Appl. Phys. Lett.* **75**, 1173 (1999).

61. J. Wu, Y.P. Zeng, Z.Z. Sun, F. Lin, B. Xu, and Z.G. Wang, Self-assembled InAs quantum wires on InP(001), *J. Crys. Growth* **21**, 1803 (2000).

62. Z. Sun, S.F. Yoon, J. Wu, and Z. Wang, Origin of the vertical-anticorrelation arrays of InAs/InAlAs nanowires with a fixed layer-ordering orientation, *J. Appl. Phys.* **91**, 6021 (2004).

63. Z.Z. Sun, S.F. Yoon, J. Wu, and Z.G. Wang, Size self-scaling in stacked InAs/InAlAs nanowire multilayers, *Appl. Phys. Lett.* **85**, 5061 (2004).

64. J. Brault, M. Gendry, O. Marty, M. Pitaval, J. Olivares, G. Grenet, and G. Hollinger, Staggered vertical self-organization of stacked InAs/InAlAs quantum wires on InP(001), *Appl. Surface Sci.* **162–163**, 584 (2000).

65. B. Salem, G. Brèmond, M. Hjiri, F. Hassen, H. Maaref, O. Marty, J. Brault, and M. Gendry, Effect of spacer layer thickness on the optical properties of stacked InAs/InAlAs quantum wires grown on InP (001), *Mat. Sci. Engineering B.* **101**, 259 (2003).

66. C. Priester and G. Grenet, Role of alloy spacer layers in non top-on-top vertical correlation in multistacked systems, *Phys. Rev. B.* **64**, 125312 (2001).

67. W. Wu, J.R. Tucker, G.S. Solomon, and J.S. Harris, Jr, Atomically resolved STM of vertically ordered InAs quantum dots, *Appl. Phys. Lett.* **71**, 1083 (1997).

68. B. Lita, R.S. Goldman, J.D. Phillips, and P.K. Bhattacharya, Nanometer-scale studies of vertical organization and evolution of stacked self-assembled InAs/GaAs quantum dots, *Appl. Phys. Lett.* **74**, 2824 (1999).

69. B. Lita, R.S. Goldman, J.D. Phillips, and P.K. Bhattacharya, Interdiffusion and surface segregation in stacked self-assembled InAs/GaAs quantum dots, *Appl. Phys. Lett.* **75**, 2797 (1999).

70. H. Eisele, O. Flebbe, T. Kalka, C. Preinesberger, F. Heinrichsdorff, A. Krost, D. Bimberg, and M. Dähne-Prietsch, Cross-sectional scanning-tunneling microscopy of stacked InAs quantum dots, *Appl. Phys. Lett.* **75**, 106 (1999).

71. O. Flebbe, H. Eisele, T. Kalka, F. Heinrichsdorff, A. Krost, D. Bimberg, and M. Dähne-Prietsch, Atomic structure of stacked InAs quantum dots grown by metal-organic chemical vapor deposition, *J. Vac. Sci. Technol. B.* **17**, 1639 (1999).

72. D.M. Bruls, J.W.A.M. Vugs, P.M. Koenraad, M. Hopkinson, M.S. Skolnick, F. Long, and S.P.A. Gill, Determination of the shape and indium distribution of low-growth-rate InAs quantum dots by cross-sectional scanning tunneling microscopy, *Appl. Phys. Lett.* **81**, 1708 (2002).

73. D.M. Bruls, P.M. Koenraad, H.W.M. Salemink, J.H. Wolter, M. Hopkinson, and M.S. Skolnik, Stacked low-growth-rate InAs quantum dots studied at the atomic level by cross-sectional scanning tunneling microscopy, *Appl. Phys. Lett.* **82**, 3758 (2003).

74. G.S. Solomon, S. Komarov, and J.S. Harris Jr, Nearest-neighbor spatial ordering of strain-induced islands using a subsurface island superlattice, *J. Cryst. Growth* **201–202**, 1190 (1999).

75. Y. Nakata, Y. Sugiyama, T. Fuatsugi, and N. Yokoyama, Self-assembled structures of closely stacked InAs islands grown on GaAs by molecular beam epitaxy, *J. Cryst. Growth* **175/176**, 713 (1997).

76. P. Sutter, E. Mateeva-Sutter, and L. Vescan, Oblique stacking of three-dimensional dome islands in Ge/Si multilayers, *Appl. Phys. Lett.* **78**, 1736 (2001).

77. M.K. Zundel, P. Specht, K. Eberl, N.Y. Jin-Philipp, and F. Philipp, Structural and optical properties of vertically aligned InP quantum dots, *Appl. Phys. Lett.* **71**, 2972 (1997).

78. M. Gutierrez, M. Hopkinson, H.Y. Liu, J.S. Ng, M. Herrera, D. Gonzalez, R. Garcia, and R. Beanland, Strain interactions and defect formation in stacked InGaAs quantum dot and dot-in-well structures, *Physica E.* **26**, 245 (2005).

79. J.L. Rouviere, J. Simon, N. Pelekanos, B. Daudin, and G. Feullet, Preferential nucleation of GaN quantum dots at the edge of AlN threading dislocations, *Appl. Phys. Lett.* **75**, 2632 (1999).

80. K. Hoshino, S. Kako, and Y. Arakawa, Formation and optical properties of stacked GaN/AlN self-assembled quantum dots grown by metalorganic chemical vapor deposition, *Appl. Phys. Lett.* **85**, 1262 (2004).

81. N. Gogneau, F. Fossard, E. Monroy, S. Monnoye, H. Mank, and B. Daudin, Effects of stacking on the structural and optical properties of self-organized GaN/AlN quantum dots, *Appl. Phys. Lett.* **84**, 4224 (2004).

82. V. Chamard, T.H. Metzger, E. Bellet-Amalric, B. Daudin, C. Adelmann, H. Mariette, and G. Mula, Structure and ordering of GaN quantum dot multilayers, *Appl. Phys. Lett.* **79**, 1971 (2001).

83. V. Chamard, T. Schülli, M. Sztucki, T.H. Metzger, E. Sarigiannidou, J.L. Rouvière, M. Tolan, C. Adelmann, and B. Daudin, Strain distribution in nitride quantum dot multilayers, *Phys. Rev. B.* **69**, 125327 (2004).

84. E. Sarigiannidou, E. Monroy, B. Daudin, J.L. Rouvière, and A.D. Andreev, Strain distribution in GaN/AlN quantum-dot superlattices, *Appl. Phys. Lett.* **87**, 203112 (2005).

85. S. Founta, F. Rol, E. Bellet-Amalric, E. Sarigiannidou, B. Gayral, C. Moisson, H. Mariette, and B. Daudin, Growth of GaN quantum dots on nonpolar A-plane SiC by molecular-beam epitaxy, *Phys. Stat. Sol. (b)* **243**, 3968 (2006).

86. F. Guillot, E. Bellet-Amalric, E. Monroy, M. Tchernycheva, L. Nevou, L. Doyennette, F.H. Julien, L.S. Dang, T. Remmele, M. Albrecht, T. Shibata, and M. Tanaka, Si-doped GaN/AlN quantum dot superlattices for optoelectronics at telecommunication wavelengths, *J. Appl. Phys.* **100**, 044326 (2006).

87. X.-D. Wang, N. Liu, C.K. Shih, S. Govindaraju, and A.L. Holmes Jr, Spatial correlation–anticorrelation in strain-driven self-assembled InGaAs quantum dots, *Appl. Phys. Lett.* **85**, 1356 (2004).

88. M. Gutiérrez, M. Herrera, D. González, R. García, and M. Hopkinson, Role of elastic anisotropy in the vertical alignment of In(Ga)As quantum dot superlattices, *Appl. Phys. Lett.* **88**, 193118 (2006).

89. A.A. Maradudin and R.F. Wallis, Elastic interactions of point defects in a semi-infinite medium, *Surf. Sci.* **91**, 423 (1980).

90. S.M. Hu, Stress from a parallelepipedic thermal inclusion in a semispace, *J. Appl. Phys.* **66**, 2741 (1989).

91. M.A. Krivoglaz, X-ray and Neutron Scattering in Nonideal Crystals Vol. I (Springer Verlag, Berlin, 1996).

92. J. Stangl, V. Holy, and G. Bauer, Structural properties of self-organized semiconductor nanostructures, *Rev. Modern Physics* **76**, 725 (2004).

93. R. Kunert and E. Schöll, Strain-controlled correlation effects in self-assembled quantum dot stacks, *Appl. Phys. Lett.* **89**, 153103 (2006).

94. S.S. Quek and G.R. Liu, Effects of elastic anisotropy on the self-organized ordering of quantum dots, *Nanotechnology* **14**, 752 (2003).

95. Q.X. Pei, C. Lu, and Y.Y. Wang, Effect of elastic anisotropy on the elastic fields and vertical alignment of quantum dots, *J. Appl. Phys.* **93**, 1487 (2003).

96. A.E. Romanov, G.E. Beltz, W.T. Fischer, P.M. Petroff, and J.S. Speck, Elastic fields of quantum dots in subsurface layers, *J. Appl. Phys.* **89**, 4523 (2001).

97. P. Liu, Y.W. Zhang, and C. Lu, Three-dimensional finite-element simulations of the self-organized growth of quantum dot superlattices, *Phys. Rev. B.* **68**, 195314 (2003).

98. T. Benabbas, Y. Androussi, and A. Lefebvre, A finite-element study of strain fields in vertically aligned InAs islands in GaAs, *J. Appl. Phys.* **86**, 1945 (1999).

99. C. Priester, Modified two-dimensional to three-dimensional growth transition process in multistacked selforganized quantum dots, *Phys. Rev. B.* **63**, 153303 (2001).

100. C. Pryor, J. Kim, L.W. Wang, A.J. Williamson, and A. Zunger, Comparison of two methods for describing the strain profiles in quantum dots, *J. Appl. Phys.* **83**, 2548 (1998).

101. M.A. Makeev and A. Madhukar, Simulations of atomic level stresses in systems of buried Ge/Si islands, *Phys. Rev. Lett.* **86**, 5542 (2001). M.A. Makeev, W. Yu, and A. Madhukar, Stress distributions and energetics in the laterally ordered systems of buried pyramidal Ge/Si (001) islands: an atomistic simulation study, *Phys. Rev. B.* **68**, 195301 (2003).

102. I. Daruka, A.-L. Barabási, S.J. Zhou, T.C. Germann, P.S. Lomdahl, and A.R. Bishop, Moleculardynamics investigation of the surface stress distribution in a Ge/Si quantum dot superlattice, *Phys. Rev. B.* **60**, R2150 (1999).

103. Y. Kikuchi, H. Sugii, and K. Shintani, Strain profiles in pyramidal quantum dots by means of atomistic simulation. *J. Appl. Phys.* **89**, 1191 (2001).

104. M. Hanke, M. Schmidbauer, R. Köhler, F. Syrowatka, A.-K. Gerlitzke, and T. Boeck, Equilibrium shape of SiGe Stranski–Krastanow islands on silicon grown by liquid phase epitaxy, *Appl. Phys. Lett.* **84**, 5228 (2004).

105. Y. Temko, T. Suzuki, and K. Jacobi, Shape and growth of InAs quantum dots on GaAs (113)A, *Appl. Phys. Lett.* **82**, 2142 (2003).

106. J.-H. Zhu, C. Miesner, K. Brunner, and G. Abstreiter, Strain relaxation of faceted Ge islands on Si(113), *Appl. Phys. Lett.* **75**, 2395 (1999).

107. J.M. García, G. Medeiros-Ribeiro, K. Schmidt, T. Ngo, J.L. Feng, A. Lorke, J. Kotthaus, and P.M. Petroff, Intermixing and shape changes during the formation of InAs self-assembled quantum dots, *Appl. Phys. Lett.* **71**, 2014 (1997).

108. A. Rastelli, M. Kummer, and H. von Känel, Reversible shape evolution of Ge islands on Si (001), *Phys. Rev. Lett.* **87**, 256101 (2001).

109. A. Rastelli, E. Müller, and H. von Känel, Shape preservation of Ge/Si (001) islands during Si capping, *Appl. Phys. Lett.* **80**, 1438 (2002).

110. A. Hesse, J. Stangl, V. Holy, T. Roch, G. Bauer, O.G. Schmidt, U. Denker, and B. Struth, Effect of overgrowth on shape, composition, and strain of SiGe islands on Si (001), *Phys. Rev. B.* **66**, 085321 (2002).

111. N. Liu, J. Tersoff, O. Baklenov, A.L. Holmes Jr, and C.K. Shih, Nonuniform composition profile in $In_{0.5}Ga_{0.5}As$ alloy quantum dots, *Phys. Rev. Lett.* **84**, 334 (2000).

112. L. Abtin, G. Springholz, and V. Holy, Surface exchange and shape transitions of PbSe quantum dots during overgrowth, *Phys. Rev. Lett.* **97**, 266103 (2006).

113. G. Springholz, L. Abtin, and V. Holy, Determination of shape and composition of buried PbSe quantum dots using scanning tunneling microscopy, *Appl. Phys. Lett.* **90**, 113119 (2007).

114. Q.X. Pei, C. Lu, and Y.Y. Wang, Effect of elastic anisotropy on the elastic fields and vertical alignment of quantum dots, *J. Appl. Phys.* **93**, 1487 (2003).

115. A.E. Romanov, G.E. Beltz, W.T. Fischer, P.M. Petroff, and J.S. Speck, Elastic fields of quantum dots in subsurface layers, *J. Appl. Phys.* **89**, 4523 (2001).

116. P. Liu, Y.W. Zhang, and C. Lu, Three-dimensional finite-element simulations of the self-organized growth of quantum dot superlattices, *Phys. Rev. B.* **68**, 195314 (2003).

117. T. Benabbas, Y. Androussi, and A. Lefebvre, A finite-element study of strain fields in vertically aligned InAs islands in GaAs, *J. Appl. Phys.* **86**, 1945 (1999).

118. C. Priester, Modified two-dimensional to three-dimensional growth transition process in multistacked selforganized quantum dots, *Phys. Rev. B.* **63**, 153303 (2001).

119. C. Pryor, J. Kim, L.W. Wang, A.J. Williamson, and A. Zunger, Comparison of two methods for describing the strain profiles in quantum dots, *J. Appl. Phys.* **83**, 2548 (1998).

120. M.A. Makeev and A. Madhukar, Simulations of atomic level stresses in systems of buried Ge/Si islands, *Phys. Rev. Lett.* **86**, 5542 (2001). M.A. Makeev, W. Yu, and A. Madhukar, Stress distributions and energetics in the laterally ordered systems of buried pyramidal Ge/Si (001) islands: an atomistic simulation study, *Phys. Rev. B.* **68**, 195301 (2003).

121. I. Daruka, A.-L. Barabási, S.J. Zhou, T.C. Germann, P.S. Lomdahl, and A.R. Bishop, Molecular-dynamics investigation of the surface stress distribution in a Ge/Si quantum dot superlattice, *Phys. Rev. B.* **60**, R2150 (1999).

122. Y. Kikuchi, H. Sugii, and K. Shintani, Strain profiles in pyramidal quantum dots by means of atomistic simulation, *J. Appl. Phys.* **89**, 1191 (2001).

123. B. Jogai, Three-dimensional strain field calculations in multiple InN/AlN wurtzite quantum dots, *J. Appl. Phys.* **90**, 699 (2001).

124. Y.W. Mo, D.E. Savage, B.S. Schwartzentruber, and M.G. Lagally, Kinetic pathway in Stranski–Krastanov growth of Ge on Si (001), *Phys. Rev. Lett.* **65**, 1020 (1990).

125. M.A. Lutz, R.M. Feenstra, P.M. Mooney, J. Tersoff, and O.J. Chu, Facet formation in strained $Si_{1-x}Ge_x$ films, *Surf. Sci.* **316**, L1075 (1993).

126. G. Medeiros-Ribeiro, A.M. Bratkowski, T.I. Kamins, D.A.A. Ohlberg, and R.S. Williams, Shape transition of germanium nanocrystals on a silicon (001) surface from pyramids to domes, *Science* **279**, 353 (1998).

127. F.M. Ross, R.M. Tromp, and M.C. Reuter, Transition states between pyramids and domes during Ge/Si island growth, *Science* **286**, 1931 (1999).

128. J. Marquez, L. Geelhaar, and K. Jacobi, Atomically resolved structure of InAs quantum dots, *Appl. Phys. Lett.* **78**, 2309 (2001).

129. G. Costantini, C. Manzano, R. Songmuang, O.G. Schmidt, and K. Kern, InAs/GaAs (001) quantum dots close to thermodynamic equilibrium, *Appl. Phys. Lett.* **82**, 3194 (2003).

130. M. Pinczolits, G. Springholz, and G. Bauer, Direct formation of self-assembled quantum dots under tensile strain by heteroepitaxy of PbSe on PbTe (111), *Appl. Phys. Lett.* **73**, 250 (1998). M. Pinczolits, G. Springholz, and G. Bauer, Molecular beam epitaxy of highly faceted self-assembled IV–VI quantum dots with bimodal size distribution, *J. Crystal Growth* **201/202**, 1126 (1999).

131. H. Yamaguchi, M.R. Fahy, and B.A. Joyce, Inhibitions of three dimensional island formation in InAs films grown on GaAs(111)A surface by molecular beam epitaxy, *Appl. Phys. Lett.* **69**, 776 (1996).

132. B.Z. Nosho, L.A. Zepeda-Ruiz, R.I. Pelzel, W.H. Weinberg, and D. Maroudas, Surface morphology in InAs/GaAs(111)A heteroepitaxy: Experimental measurements and computer simulations, *Appl. Phys. Lett.* **75**, 829 (1999).

133. L.A. Zepeda-Ruiz, R.I. Pelzel, W.H. Weinberg, and D. Maroudas, Interfacial stability and structure in InAs/GaAs(111)A heteroepitaxy: Effects of buffer layer thickness and film compositional grading, *Appl. Phys. Lett.* **77**, 3352 (2000).

134. G. Springholz and K. Wiesauer, Nanoscale dislocation patterning in PbTe/PbSe (001) lattice-mismatched heteroepitaxy, *Phys. Rev. Lett.* **88**, 015507 (2002).

135. K. Wiesauer and G. Springholz, Near-equilibrium strain relaxation and misfit dislocation interactions in PbTe on PbSe (001) heteroepitaxy, *Appl. Phys. Lett.* (2003).

136. B. Voigtländer, Fundamental processes in Si/Si and Ge/Si epitaxy studied by scanning tunneling microscopy during growth, *Surf. Sci. Rep.* **43**, 127 (2001).

137. N. Motta, Self-assembling and ordering of Ge/Si(111) quantum dots: scanning microscopy probe studies, *J. Phys.: Condens. Matter* **14**, 8353–8378 (2002).

138. H. Li, T. Daniels-Race, and M.-A. Hasan, Lateral correlation of InAs/AlInAs nanowire superlattices on InP (001), *J. Vac. Sci. Technol. B.* **19**, 1471 (2001).

139. H. Li, T. Daniels-Race, and M.-A. Hasan, Effects of the matrix on self-organization of InAs quantum nanostructures grown on InP substrates, *Appl. Phys. Lett.* **80**, 1367 (2002).

140. N. Chauvin, B. Salem, G. Bremond, C. Bru-Chevallier, G. Guillot, C. Monat, P. Regreny, M. Gendry, B. Lalmic, and O. Marty, Photoluminescence studies of stacked InAs/InP quantum sticks, *J. Crys. Growth* **275**, 2327 (2005).

141. H. Heidemeyer, U. Denker, C. Müller, and O.G. Schmidt, Morphology response to strain field interferences in stacks of highly ordered quantum dot arrays, *Phys. Rev. Lett.* **91**, 196103 (2003).

142. M. Schmidbauer, Sh. Seydmohamadi, D. Grigoriev, Zh.M. Wang, Yu.I. Mazur, P. Schäfer, M. Hanke, R. Köhler, and G.J. Salamo, Controlling planar and vertical ordering in three-dimensional (In,Ga)As quantum dot lattices by GaAs surface orientation, *Phys. Rev. Lett.* **96**, 066108 (2006).

143. D. Grigoriev, M. Schmidbauer, P. Schäfer, S. Besedin, Yu.I. Mazur, Zh.M. Wang, G.J. Salamo, and R. Köhler, Three-dimensional self-ordering in an InGaAs/GaAs multilayered quantum dot structure investigated by x-ray diffuse scattering, *J. Phys. D.* **38**, A154 (2005).

144. Zh.M. Wang, Sh. Seydmohamadi, J.H. Lee, and G.J. Salamo, Surface ordering of (In,Ga)As quantum dots controlled by GaAs substrate indexes, *Appl. Phys. Lett.* **85**, 5031 (2004).

145. P.M. Lytvyn, V.V. Strelchuk, O.F. Kolomys, I.V. Prokopenko, M.Ya. Valakh, Yu.I. Mazur, Zh.M. Wang, G.J. Salamo, and M. Hanke, Two-dimensional ordering of (In,Ga)As quantum dots in vertical multilayers grown on GaAs (100) and (n11), *Appl. Phys. Lett.* **91**, 173118 (2007).

146. H. Omi and T. Ogino, Self-organization of Ge islands on high-index Si substrates, *Phys. Rev. B.* **59**, 7521 (1999).

147. V. Holy, J. Stangl, Z. Zerlauth, F. Schäffler, G. Bauer, N. Darowski, D. Lübbert, and U. Pietsch, Lateral arrangement of self-assembled quantum dots in an SiGe/Si superlattice, *J. Phys. D.* **32**, A234 (1999).
V. Holý, A. A. Darhuber, J. Stangl, S. Zerlauth, F. Schäffler, G. Bauer, N. Darowski, D. Lübbert, U. Pietsch, and I. Vávra, Coplanar and grazing incidence x-ray-diffraction investigation of self-organized SiGe quantum dot multilayers, *Phys. Rev. B*, **58**, 7934 (1998).

148. F. Liu, S.E. Davenport, H.M. Evans, and M.G. Lagally, Self-organized replication of 3D coherent island size and shape in multilayer heteroepitaxial films, *Phys. Rev. Lett.* **82**, 2528 (1998).

149. C.S. Lee, B. Kahang, and A.L. Barabasi, Spatial ordering of stacked quantum dots, *Appl. Phys. Lett.* **78**, 984 (2001).

150. P.M. Lam and S. Tan, Kinetic Monte Carlo model of self-organized quantum dot superlattices, *Phys. Rev. B.* **64**, 035321 (2001).

151. M. Meixner and E. Schöll, Kinetically enhanced correlation and anticorrelation effects in self-organized quantum dot stacks, *Phys. Rev. B.* **67**, 121202 (2003).

152. J. He, R. Nötzel, P. Offermans, P.M. Koenraad, Q. Gong, G.J. Hamhuis, T.J. Eijkemans, and J.H. Wolter, Formation of columnar (In,Ga)As quantum dots on GaAs (100), *Appl. Phys. Lett.* **85**, 2771 (2004).

153. Z.M. Wang, K. Holmes, Yu.I. Mazur, and G.J. Salamo, Fabrication of (In,Ga)As quantum-dot chains on GaAs(100), *Appl. Phys. Lett.* **84**, 1931 (2004). Z.M. Wang, Yu.I. Mazur, G.J. Salamo, P.M. Lytvin, V.V. Strelchuk, and M.Ya. Valakh, Persistence of (In,Ga)As quantum-dot chains under index deviation from GaAs(100), *Appl. Phys. Lett.* **84**, 4681 (2004).

154. Yu.I. Mazur, W.Q. Ma, X. Wang, Z.M. Wang, G.J. Salamo, M. Xiao, T.D. Mishima, and M.B. Johnson, InGaAs/GaAs three-dimensionally-ordered array of quantum dots, *Appl. Phys. Lett.* **83**, 987 (2003).

155. Zh.M. Wang, H. Churchill, C.E. George, and G.J. Salamo, High anisotropy of lateral alignment in multilayered (In,Ga)As/GaAs(100) quantum dot structures, *J. Appl. Phys.* **96**, 6908 (2004). Zh.M. Wang, Yu.I. Mazur, K. Holmes, and G.J. Salamo, Control on self-organization of InGaAs/GaAs(100) quantum-dot chains, *J. Vac. Sci. Technol. B.* **23**, 1732 (2005).

156. Zh.M. Wang, Yu.I. Mazur, J.L. Shultz, G.J. Salamo, T.D. Mishima, and M.B. Johnson, One-dimensional postwetting layer in InGaAs/GaAs(100) quantum-dot chains, *J. Appl. Phys.* **99**, 033705 (2006).

157. M.L. Hussein, E. Marega, and G. Salamo, Isotropic lateral ordering of III-V quantum dots over GaAs (001) by self-assembly, *Mat. Res. Soc. Symp. Proc.* **959**, 17 (2007).

158. J.-S. Lee, M. Sugisaki, H.-W. Ren, S. Sugou, and Y. Masumoto, Spontaneous lateral alignment of multistacked In$_{0.45}$Ga$_{0.55}$As quantum dots on GaAs (311)B substrate, *J. Crys. Growth* **200**, 77 (1999).

159. T. van Lippen, R. Nötzel, G.J. Hamhuis, and J.H. Wolter, Self-organized lattice of ordered quantum dot molecules, *Appl. Phys. Lett.* **85**, 118 (2004).

160. T. van Lippen, R. Nötzel, and J.H. Wolter, Ordering of quantum dot molecules by self-organization, *J. Vac. Sci. Technol. B.* **23**, 1693 (2005). T. van Lippen, R. Nötzel, G. J. Hamhuis, and J. H. Wolter, Ordered quantum dot molecules and single quantum dots formed by self-organized anisotropic strain engineering, *J. Appl. Phys.* **97**, 044301 (2005).

161. Y. Sugiyama, Y. Nakata, T. Futatsugi, M. Sugawara, Y. Awano, and N. Yokoyama, Narrow photoluminescence line width of closely stacked InAs self-assembled quantum dot structures, *Jpn. J. Appl. Phys.* **36**, L158 (1997).

162. L. Chu, M. Arzberger, G. Böhm, and G. Abstreiter, Influence of growth conditions on the photoluminescenece of self-assembled InAs/GaAs quantum dots, *J. Appl. Phys.* **85**, 2355 (1999).

163. V. Celibert, E. Tranvouez, G. Guillot, C. Bru-Chevallier, L. Grenouillet, P. Duvaut, P. Gilet, P. Ballet, and A. Million, MBE growth optimization and optical spectroscopy of InAs/GaAs quantum dots emitting at 1.3 µm in single and stacked layers, *J. Crys. Growth* **275**, 2313 (2005).

164. Z.R. Wasilewski, S. Fafard, and J.P. McCaffrey, Size and shape engineering of vertically stacked self-assembled quantum dots, *J. Cryst. Growth* **201–202**, 1131 (1999).

165. S. Fafard, M. Spanner, J.P. McCaffrey, and Z.R. Wasilewski, Coupled InAs/GaAs quantum dots with well-defined electronic shells, *Appl. Phys. Lett.* **76**, 2268 (2000).

166. S. Fafard, Z.R. Wasilewski, C.Ni. Allen, D. Picard, M. Spanner, J.P. McCaffrey, and P.G. Piva, Manipulating the energy levels of semiconductor quantum dots, *Phys. Rev. B.* **59**, 15368 (1999).

167. T. Nakaoka, J. Tatebayashi, and Y. Arakawa, Spectroscopy on single columns of vertically aligned InAs quantum dots, *Physica E.* **21**, 409 (2004).

168. N.N. Ledentsov, N. Kirstaedter, M. Grundmann, D. Bimberg, V.M. Ustinov, I.V. Kochnev, P.S. Kop'ev, and Zh.I. Alferov, Three-dimensional arrays of self-ordered quantum dots for laser applications, *Microelectron. J.* **28**, 915 (1997).

169. N.N. Ledentsov, V.M. Ustinov, V.A. Shchukin, P.S. Kop'ev, Zh.I. Alferov, and D. Bimberg, Quantum dot heterostructures: fabrication, properties, lasers (Review), *Semiconductors* **32**, 343 (1998).

170. D. Bimberg, M. Grundmann, F. Heinrichsdorff, N.N. Ledentsov, V.M. Ustinov, A.E. Zhukov, A.R. Kovsh, M.V. Maximov, Y.M. Shernyakov, B.V. Volovik, A.F. Tsatsul'nikov, P.S. Kop'ev, and Zh.I. Alferov, Quantum dot lasers: breakthrough in optoelectronics, *Thin Solid Films* **367**, 235 (2000).

171. A. Raab and G. Springholz, Controlling the size and density of self-assembled PbSe quantum dots by adjusting the substrate temperature and layer thickness, *Appl. Phys. Lett.* **81**, 2457 (2002).

172. G. Springholz, T. Schwarzl, W. Heiss, G. Bauer, M. Aigle, H. Pascher, and I. Vavra, Mid-infrared surface-emitting PbSe/PbEuTe quantum dot lasers, *Appl. Phys. Lett.* **79**, 1225 (2001).

173. V. Holy, J. Stangl, G. Springholz, M. Pinczolits, G. Bauer, I. Kegel, and T.H. Metzger, Lateral and vertical ordering of self-assembled PbSe quantum dots studied by high-resolution X-ray diffraction, *Physica B.* **283**, 65 (2000).

174. Z. Zhong, A. Halilovic, T. Fromherz, F. Schäffler, and G. Bauer, 2D periodic positioning of self-assembled Ge islands on prepatterned Si (001) substrates, *Appl. Phys. Lett.* **82**, 4779 (2003).

175. Z. Zhong and G. Bauer, Site controlled and size-homogeneous Ge islands on prepatterned Si (001) substrates, *Appl. Phys. Lett.* **84**, 1922 (2004). Z. Zhong, A. Halilovic, M. Mühlberger, F. Schäffler, and G. Bauer, Ge island formation on stripe patterned Si (001) substrates, *Appl. Phys. Lett.* **82**, 445 (2003); and Positioning of self-assembled Ge islands on stripe-patterned Si (001) substrates, *J. Appl. Phys.* **93**, 6258 (2003).

176. R. Songmuang, S. Kiravittaya, and O.G. Schmidt, Formation of lateral quantum dot molecules around self-assembled nanoholes, *Appl. Phys. Lett.* **82**, 2892 (2003).

177. Y. Nakamura, O.G. Schmidt, N.Y. Jin-Phillipp, S. Kiravittaya, C. Müller, K. Eberl, H. Gräbeldinger, and H. Schweizer, Vertical alignment of laterally ordered InAs and InGaAs quantum dot arrays on patterned (001) GaAs substrates, *J. Crys. Growth* **242**, 339 (2002).

178. Y. Nakamura, N. Ikeda, S. Ohkouchi, Y. Sugimoto, H. Nakamura, and K. Asakawa, Regular array of InGaAs quantum dots with 100-nm-periodicity formed on patterned GaAs substrates, *Physica E.* **21**, 551 (2004).

179. J.M. García, G. Medeiros-Ribeiro, K. Schmidt, T. Ngo, J.L. Feng, A. Lorke, J. Kotthaus, and P.M. Petroff, Intermixing and shape changes during the formation of InAs self-assembled quantum dots, *Appl. Phys. Lett.* **71**, 2014 (1997).

180. T. Raz, D. Ritter, and G. Bahir, Formation of InAs self-assembled quantum rings on InP, *Appl. Phys. Lett.* **82**, 1706 (2003).

181. O.G. Schmidt, N.Y. Jin-Phillipp, C. Lange, U. Denker, K. Eberl, R. Schreiner, H. Gräbeldinger, and H. Schweizer, Long-range ordered lines of self-assembled Ge islands on a flat Si (001) surface, *Appl. Phys. Lett.* **77**, 4139 (2000).

182. D.S.L. Mui, D. Leonard, L.A. Coldren, and P.M. Petroff, Surface migration induced self-aligned InAs islands grown by molecular beam epitaxy, *Appl. Phys. Lett.* **66**, 1620 (1995).

183. H. Lee, J.A. Johnson, M.Y. He, J.S. Speck, and P.M. Petroff, Strain-engineered self-assembled semiconductor quantum dot lattices (buried stressors), *Appl. Phys. Lett.* **78**, 105 (2001).

184. R.L. Williams, G.C. Aers, P.J. Poole, J. Lefebvre, D. Chithrani, and B. Lamontagne, Controlling the self-assembly of InAs/InP quantum dots, *J. Crys. Growth* **223**, 321 (2001).

185. T. Kitajima, B. Liu, and S.R. Leone, Two-dimensional periodic alignment of self-assembled Ge islands on patterned Si(001) surfaces, *Appl. Phys. Lett.* **80**, 497 (2002).

186. E. Tevaarwerk, P. Rugheimer, O.M. Castellini, D.G. Keppel, S.T. Utley, D.E. Savage, M.G. Lagally, and M.A. Eriksson, Electrically isolated SiGe quantum dots on mesa structures, *Appl. Phys. Lett.* **80**, 4626 (2002).

187. A. Fantini, F. Phillipp, C. Kohler, J. Porsche, and F. Scholz, Investigation of self-assembled InP-GaInP quantum dot stacks by TEM, *J. Crys. Growth* **244**, 129 (2002).

188. H.R. Gutierrez, M.A. Cotta, and M.M.G. de Carvalho, Vertical stacks of InAs quantum wires in a InP matrix, *J. Crys. Growth* **254**, 1 (2003).

189. Z.G. Wang and J. Wu, Controllable growth of semiconductor nanometer structures, *Microelectron. J.* **34**, 379 (2003).

190. C. Schelling, G. Springholz, and F. Schäffler, Kinetic growth instabilities on vicinal Si (001) surfaces, *Phys. Rev. Lett.* **83**, 995 (1999).

191. C. Schelling, M. Mühlberger, G. Springholz, and F. Schäffler, $Si_{1-x}Ge_x$ growth instabilities on vicinal Si(001) substrates: kinetic vs. strain-induced effects, *Phys. Rev. B.* **64**, 041301 (2001).

192. Y.H. Phang, C. Teichert, M.G. Lagally, L.J. Peticolos, J.C. Bean, and E. Kasper, Correlated-interface roughness anisotropy in SiGe/Si superlattices, *Phys. Rev. B*, **50**, 14435 (1994).

193. J. Tersoff, Y.H. Phang, Z. Zhang, and M.G. Lagally, Step-bunching instability of vicinal substrates under stress, *Phys. Rev. Lett.* **75**, 2730 (1995).

194. V. Holy, A.A. Darhuber, J. Stangl, G. Bauer, J. Nützel, and G. Abstreiter, Oblique roughness replication in strained SiGe/Si multilayers, *Phys. Rev. B.* **57**, 12435 (1998).

195. A.A. Darhuber, J. Zhu, V. Holy, J. Stangl, P. Mikulik, K. Brunner, G. Abstreiter, and G. Bauer, Highly regular self-organization of step bunches during growth of SiGe on Si(113), *Appl. Phys. Lett.* **73**, 1535 (1998).

196. M. Hanke, M. Schmidbauer, R. Köhler, H. Kirmse, and M. Pristovsek, Lateral short range ordering of step bunches in InGaAs/GaAs superlattices, *Appl. Phys.* **95**, 1736 (2004).

197. J.-H. Zhu, K. Brunner, and G. Abstreiter, Two-dimensional ordering of self-assembled Ge islands on vicinal Si(001) surfaces with regular ripples, *Appl. Phys. Lett.* **73**, 620 (1998).

198. K. Brunner, J. Zhu, C. Miesner, G. Abstreiter, O. Kienzle, and F. Ernst, Self-organized periodic arrays of SiGe wires and Ge islands on vicinal Si substrates, *Physica E.* **7**, 881 (2000).

199. J. Stangl, T. Roch, V. Holy, M. Pinczolits, G. Springholz, G. Bauer, I. Kegel, T.H. Metzger, J. Zhu, K. Brunner, G. Abstreiter, and D. Smilgies, Strain-induced self-organized growth of nanostructures: from step bunching to ordering in quantum dot superlattices, *J. Vac. Sci. Technol. B.* **18**, 2187 (2000).

200. T. Mano, R. Nötzel, D. Zhou, G.J. Hamhuis, T.J. Eijkemans, and J.H. Wolter, Complex quantum dot arrays formed by combination of self-organized anisotropic strain engineering and step engineering on shallow patterned substrates, *J. Appl. Phys.* **97**, 014304 (2005).

201. L. Fitting, M.E. Ware, J.R. Haywood, J.H. Walter, and R.J. Nemanich, Self-organized nanoscale Ge dots and dashes on SiGe/Si superlattices, *J. Appl. Phys.* **98**, 024317 (2005).

202. See, e.g., M. Horn-von-Hoegen, Surfactants: perfect heteroepitaxy of Ge on Si (111), *Appl. Phys. A.* **59**, 503 (1994). E. Tournie, N. Grandjean, A. Trampert, J. Massies, and K. Ploog, Surfactant-mediated molecular-beam epitaxy of III–V strained-layer heterostructures, *J. Cryst. Growth* **150**, 460 (1995).

203. A. Zunger and S. Mahajan, in: *Handbook on Semiconductors*, eds. T.S. Moss, and S. Mahajan. Vol. **3**, (Elsevier Science, Amsterdam, 1994), 1399 pp.

204. C. Priester and G. Grenet, Surface roughness and alloy stability interdependence in lattice-matched and lattice-mismatched heteroepitaxy, *Phys. Rev. B.* **61**, 91602 (2000).

205. J.E. Guyer and P.W. Voorhees, Morphological stability of alloy thin films, *Phys. Rev. Lett.* **74**, 4031 (1995); and Morphological stability of alloy thin films, *Phys. Rev. B.* **54**, 11710 (1996),

206. J. Tersoff, Stress-driven alloy decomposition during step-flow growth, *Phys. Rev. Lett.* **77**, 2017 (1996).

207. B.J. Spencer, P.W. Voorhees, and J. Tersoff, Enhanced instability of strained alloy films due to compositional stresses, *Phys. Rev. Lett.* **84**, 2449 (2000).

208. M.V. Maximov, A.F. Tsatsul'nikov, B.V. Volovik, D.S. Sizov, Yu.M. Shernyakov, I.N. Kaiander, A.E. Zhukov, A.R. Kovsh, S.S. Mikhrin, V.M. Ustinov, Zh.I. Alferov, R. Heitz, V.A. Shchukin, N.N. Ledentsov, D. Bimberg, Yu.G. Musikhin, and W. Neumann, Tuning quantum dot properties by activated phase separation of an InGa(Al)As alloy grown on InAs stressors, *Phys. Rev. B.* **62**, 16671 (2000).

209. D. Fuster, M. Ujué González, L. González, Y. González, T. Ben, A. Ponce, and S.I. Molina, Stacking of InAs/InP (001) quantum wires studied by in situ stress measurements: role of inhomogeneous stress fields, *Appl. Phys. Lett.* **84**, 4723 (2004).

210. Z.G. Wang and J. Wu, Controllable growth of semiconductor nanometer structures, *Microelectron. J.* **34**, 379 (2003).

CHAPTER 2

InAs Quantum Dots on Al$_x$Ga$_{1-x}$As Surfaces and in an Al$_x$Ga$_{1-x}$As Matrix

Aaron Maxwell Andrews, Matthias Schramböck, and Gottfried Strasser

Technical University of Vienna, Floragasse 7, 1040 Wien, Austria

2.1 Introduction

Self-assembled semiconductor quantum dots (QDs) have opened the door for new and interesting physical phenomena based on reduced dimensionality [1–5], where the properties of the QDs are highly dependent on the principal materials. Although the dot size, dot shape, and dot density fine-tune the dot ensemble properties, the matrix and dot material have an enormous impact on the possible applications of the QD devices. This is relevant whether incorporating the QDs into existing technologies, including GaAs-based lasers [6–13] and detectors [14, 15], or as in the case of core-shell nanoparticles [16, 17], changing the rules altogether and forcing a new approach to realize functionality. This chapter will focus on the growth and devices of InAs QDs in an Al$_x$Ga$_{1-x}$As matrix and show how to independently control the QD density and QD size on various Al$_x$Ga$_{1-x}$As surfaces. The QD density control is critical for device doping and design, while the size is essential for energy level control, including those above the GaAs band edge where unipolar intraband devices operate. This will help bridge the gap between Al$_x$Ga$_{1-x}$As/GaAs-based devices and the extensive GaAs-based QD research.

2.2 Quantum dot formation

2.2.1 Strained heteroepitaxial growth

In lattice-mismatched heteroepitaxy, the film is uniformly stressed when it grows coherently on a thick substrate, as illustrated in Fig. 2.1. In the case of a one-dimensional (1D) lattice mismatch, the non-zero component of stresses in the film is given as [18]:

$$\sigma_{xx}^{1D} = \frac{2G}{1-\nu}\varepsilon_m \tag{2.1}$$

where the x-axis is chosen along the direction of lattice mismatch, which is characterized by the misfit parameter (misfit strain) ε_m, and the y-axis is perpendicular to the film/substrate interface. The film material is assumed to be elastically isotropic with shear modulus G and Poisson ratio ν. In the case of equi-biaxial mismatch, the stress state in the film is given as:

$$\sigma_{xx}^{2D} = \sigma_{zz}^{2D} = 2G\frac{1+\nu}{1-\nu}\varepsilon_m \tag{2.2}$$

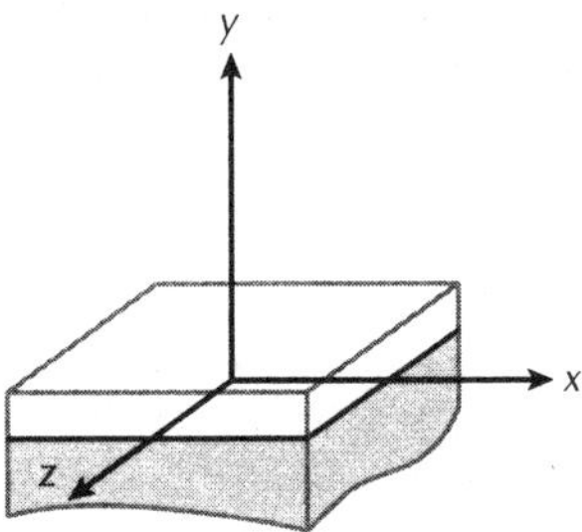

Figure 2.1   Schematic for the coordinate system used for strained thin film equations. The dark grey substrate is assumed to be infinitely thick.

where the misfit strain ε_m is defined by the crystal lattice parameters of the film a_f and substrate a_s materials:

$$\varepsilon_m = \frac{a_s - a_f}{a_f}. \tag{2.3}$$

The strain in the perpendicular direction for biaxially strained cubic (100) crystals can be written as:

$$\varepsilon_\perp = -2\frac{C_{12}}{C_{11}}\varepsilon_m \tag{2.4}$$

where the C_{ij} are the elastic stiffness constants. In the case of InAs on GaAs, where $\varepsilon_m = -0.06687$, C_{11} and C_{12} are 8.329×10^{11} and 4.526×10^{11} dynes/cm^2 (GaAs 11.88×10^{11} and 5.38×10^{11} dynes/cm^2), then $\varepsilon_\perp = 0.07267$ and a strained monolayer (ML) height of 3.2493 Å, but for a more general solution Eq. 2.4 can be rewritten:

$$\varepsilon_\perp = -2\frac{\nu}{1-\nu}\varepsilon_m. \tag{2.5}$$

The stored elastic energy w per unit area of the film/substrate interface associated with the stresses is proportional to the film thickness h:

$$w^{2D} = 2(1+\nu)w^{1D} = 2G\frac{1+\nu}{1-\nu}\varepsilon_m^2 h. \tag{2.6}$$

Increasing the stored energy with increasing film thickness h will eventually lead to the onset of a variety of relaxation processes in the elastically stressed film.

For growing films, misfit dislocation (MD) generation at the interface between the film and substrate has been shown to be the most common mechanism for the relaxation of elastic stress [19–23]. In the majority of cases, the MDs are associated with threading dislocations (TDs), which are concomitant to MDs but have their lines going through the film to the free surface [24, 25]. The Matthews–Blakeslee [26] critical thickness h_c for MD generation may be derived [22] by considering the energetics of a combined MD–TD configuration in a stressed film:

$$h_c = \frac{b}{\varepsilon_m(1+\nu)8\pi\cos\lambda}(1-\cos^2\beta)\ln\left(\frac{\alpha_o h_c}{b}\right) \approx \frac{b}{\varepsilon_m} \tag{2.7}$$

where $b = |\vec{b}|$ is the magnitude of the dislocation Burgers vector $\vec{b}$, λ is the angle between the Burgers vector and a line that lies in the film/substrate interface normal to the MD line, β is the angle between the MD line and $\vec{b}$, and α_o is the dislocation core cut-off parameter. This is an equilibrium critical thickness, meaning there is still no relaxation through MD formation. As film growth continues past h_c, the film can begin to relax to release this additional energy, but the metastable conditions can continue without an active source of dislocations.

2.2.2   Quantum dot nucleation on $Al_xGa_{1-x}As$ surfaces

There are three main growth modes for thin film epitaxy. The lattice mismatch, as well as chemical and structural similarity (or differences), determine the growth mode. Figure 2.2 illustrates the three possible growth modes: (a) Frank–van der Merwe or layer-by-layer, (b) Stranski–Krastanov (SK) or layer-by-layer with islands, and (c) Volmer–Weber or island growth.

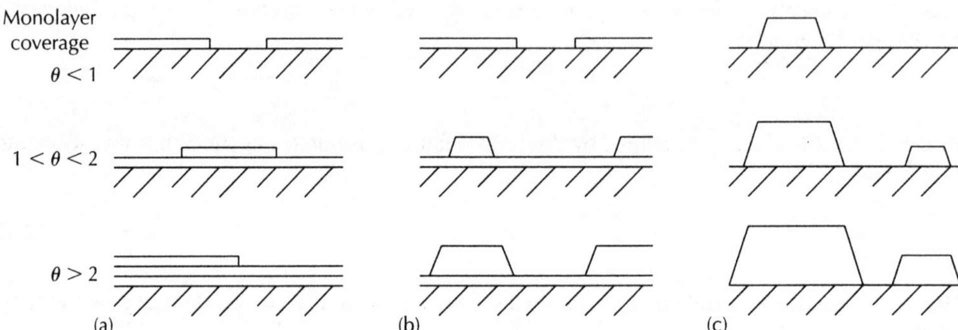

Figure 2.2   Schematic representation of the three growth modes in epitaxial films. (a) Frank–van der Merwe or layer-by-layer growth. (b) Stranski–Krastanov or layer-by-layer plus island growth. (c) Volmer–Weber or island growth.

Growth of strained films with $\varepsilon \geq 2\%$ results in an intermediate growth mode of layer-by-layer plus islands. The initial growth is layer-by-layer for a few monolayers before stable islands on these strained monolayers are formed. This is due to a decrease in the binding energy characteristics for layer-by-layer growth [27]. As film growth continues, the coherent islanding is followed by strain relaxation. For $In_xGa_{1-x}As/GaAs$ or Ge_xSi_{1-x}/Si, the stability and shape of the islands are typically strain dependent because they share the same crystal structure as the substrate, but the substrate growth orientation still plays a critical roll. Growth on a high surface free energy crystal plane can result in faceting and islanding [28]. Thus, this intermediate growth mode can be controlled by changing surface energy, growth temperature, and growth rate.

In SK growth, the roughening of the surface is a way to relieve the local strain at the surface [29]. Changing the surface morphology is a balance between the strain energy per unit volume and the interfacial/surface energy per unit area. The negative effect of surface roughening is that the troughs and cusps formed are regions of stress concentration that have the potential for defect nucleation [30]. The islands formed are highly strained because they are still coherent with the substrate and coalescence of the islands does not result in additional defects because they are still oriented to the initial wetting layer. This growth mode is limited to the formation of quantum dots [31, 32]. If growth is interrupted in Stranski–Krastanov mode and the sample is allowed to equilibrate, the coherent wetting layer will ripen into islands, consuming the wetting layer. Submonolayer InAs depositions can also form QDs, without the SK grown mode, when a three-dimensionally patterned substrate, vicinal substrate, buried stressor, or interdiffusion facilitates the QDs [2]. Since the islands are still coherent, growth can continue and the islands can be buried in the structure forming the 3D quantum dots or layers of coupled quantum dots.

Utilizing the SK growth mode and molecular-beam epitaxy (MBE) [33], dislocation-free self-assembled semiconductor quantum dots (QDs) were first reported in 1985 in the InAs/GaAs material system [1]. Later, characterization of the InAs QD nucleation and growth mechanism was achieved through the use of a shadow mask and a non-rotated substrate during the InAs deposition, where the critical thickness for InAs QDs on a 530°C GaAs surface was determined to be 1.5 ML [34]. The earliest measurable QD density begins in the 10^6 cm^{-2} range and after an additional 0.021 ML is in the 10^8 cm^{-2} range (1 μm^{-2}). This dependence of QD density on ML deposition continues until 1.9 ML and a density of 4×10^{10} cm^{-2}, beyond which the QDs begin to ripen or grow in size. However, the growth rate and temperature has a great impact on the QD nucleation, growth, and final uniformity [35]. Reducing the growth rate to less than 0.01 ML/s ($\approx$0.01 μm/h), what has been classified as a (*very*) low growth rate (LGR), greatly reduces the QD density and inhomogeneous size distribution, while increasing the average QD size. Additional

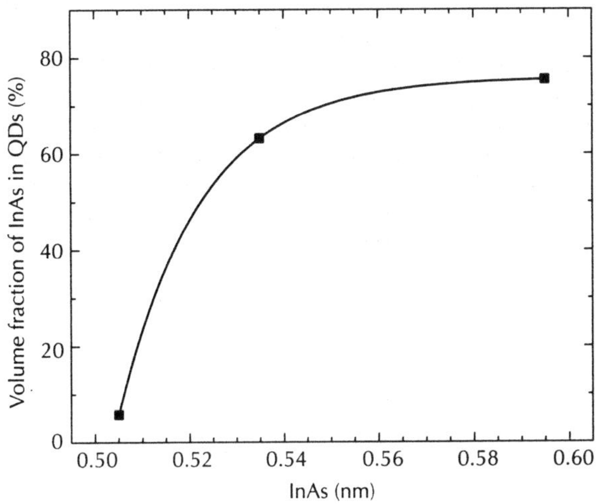

Figure 2.3   InAs volume fraction incorporated into quantum dots as a percentage of the total InAs deposited on a GaAs surface at 490°C [36].

techniques (e.g. repeated growth and desorption) can improve the size distribution further, but they will not be discussed here [4].

Figure 2.3 shows the results of a different study into the incorporation of the InAs into the QDs [36]. In this case, the volume of the QDs was measured by AFM and then compared to the total InAs thickness deposited on a GaAs surface at 490°C. The polynomial curve fit for the volume fraction indicates a critical thickness of 0.502 nm (1.545 ML), while a modified Belehradek curve fit, typically of the form:

$$\rho = \rho_0 (\Theta - \Theta_c)^\alpha \tag{2.8}$$

where ρ is volume fraction in this case, Θ is the InAs thickness, Θ_c is the critical thickness, and α is the exponent, shows a critical thickness of 0.504 nm (1.551 ML). The results from both studies are highly dependent on growth conditions: As$_2$ vs As$_4$, pressure of arsenic, growth rate, substrate temperature, etc.

Based on strain energy, Eqs 2.1–2.6, and mismatch strain ε_m one would not expect a change in strain relaxation for InAs QDs on AlAs surfaces and, in fact, this has been studied and the critical thickness for QD nucleation was found to be almost identical to that on GaAs surfaces [37, 38].

However, epitaxial growth on AlAs proceeds differently than for GaAs, because the migration of surface adatoms is reduced and higher growth temperatures are typically required to reach optimal growth conditions. The surface diffusion length of Al for similar temperatures on a (100) substrate is almost 30 times lower than Ga, due to the increased activation energy of 1.0 eV with respect to 0.8 eV [39]. For comparison, the In surface diffusion lengths on GaAs are calculated to be 3–10 times that of Ga [40]. On AlAs surfaces the nucleation thickness does not change, but the time and/or deposition thickness required for a complete 3D RHEED pattern increases 20–30% due to the reduced surface diffusion [38]. Increasing the arsenic overpressure has a minimal effect on the QDs on a GaAs surface, but it improves the In surface diffusion on the AlAs surfaces and enables the formation of larger QDs. Surface STM measurements also show a reduced intermixing of the wetting layer with the AlAs substrate [41]. This reduced diffusion leads to 2D InAs island formation, instead of a complete wetting layer, which is the precursor to the 3D QDs [42].

2.2.3   Calibrating InAs growth rate

Since it has been shown that the QD growth rate determines the QD nucleation density [35], the first step to controlling QD density is to control the growth rate [43]. Figure 2.4 shows the QD density measured by AFM for an InAs growth rate of 0.05 μm/h at 500 and 510°C and for the growth rate of 0.01 μm/h at 500°C. For the faster InAs deposition rate at 500°C the QD density increases with increasing InAs deposition, as observed in [34]. Raising the substrate temperature by 10°C reduces the QD density, which still has a strong dependence on the InAs deposition thickness. It appears that the QD size and density are not independent factors under these growth conditions. By lowering the growth rate to 0.01 μm/h the QD density is relatively independent of the InAs deposition thickness. The reduced growth rate makes QD size a tunable factor by adjusting the InAs thickness, but it does not provide a method for controlling the QD density.

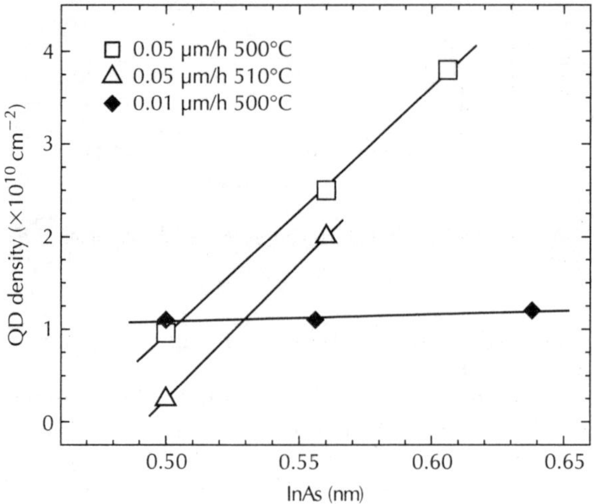

Figure 2.4   InAs QD density dependence on growth rate and temperature for a GaAs surface. Open markers denote an InAs growth rate of 0.05 μm/h, while solid markers denote a growth rate of 0.01 μm/h.

The growth rates of InAs used for QD growth are typically 0.001–0.1 μm/h. These low growth rates are preferred to improve QD size uniformity [35] and can be calibrated with multiple techniques: reflection high energy electron diffraction (RHEED) oscillations, high-resolution X-ray diffraction (XRD), and photoluminescence (PL). An additional growth and pause technique has been used to reduce the average InAs growth rate, but an InAs deposition rate of ≤0.05 μm/h produced results that were identical to the reduced growth rates without the pauses.

The first method utilizes the oscillations in RHEED intensity during layer-by-layer growth [44]. The time to grow one monolayer can be determined by the time between intensity oscillations. The InAs growth rate is determined by growing In$_x$Ga$_{1-x}$As and then subtracting the growth rate of GaAs. This reduces the lattice mismatch and allows for multiple intensity oscillations and repeat measurements for more accurate peak fitting, making this a relatively rapid process, but the monolithic growth of In$_x$Ga$_{1-x}$As quantum wells below subsequent quantum dots showed up to a 6% error in the growth rate measurement [45]. This method complicates the growth of ternary or quaternary QDs even more.

The other techniques for growth rate calibration are more time consuming, but more accurate. The growth of a strained In$_x$Ga$_{1-x}$As/GaAs superlattice at temperatures below 530°C is measured by high-resolution XRD [46] or PL. With XRD, the thickness and therefore growth rate of the GaAs and InAs can be independently measured, but the results are usually described in micrometres per hour (µm/h) instead of the typical monolayers per second (ML/s) and deposition thickness is in nanometres instead of ML. For this technique, all thickness values measured are for strained InAs on GaAs. It is important to keep in mind the critical thickness h_c when designing an InAs growth rate calibration superlattice. Growing GaAs on top of In$_x$Ga$_{1-x}$As does not relieve the strain energy in the film and subsequent layers can result in relaxation.

2.3   Control of quantum dot size and density

The majority of InAs-based QDs are grown by molecular-beam epitaxy (MBE) on nominally on-axis (100) GaAs substrates. The oxide is thermally desorbed under an arsenic overpressure at a substrate temperature of ~630°C. This can lead to a roughening of the growth surface and the creation of additional surface monolayer steps. Although this roughing is on the order of 2 Å RMS, the QDs readily nucleate at a step edge as shown in Fig. 2.5. There are two common methods to improve the surface roughness resulting from the oxide desorption: atomic hydrogen cleaning and the growth of a short period superlattice. The atomic hydrogen cleaning utilizes thermally cracked H$_2$ molecules to produce H* that bonds to the oxide on the surface. The oxygen leaves the sample surface as H$_2$O at temperatures well below the onset of thermal desorption producing a potentially smoother substrate surface. The second method utilizes growth to smooth the substrate surface, with alternating AlAs and GaAs layers. This smoothing superlattice can easily be

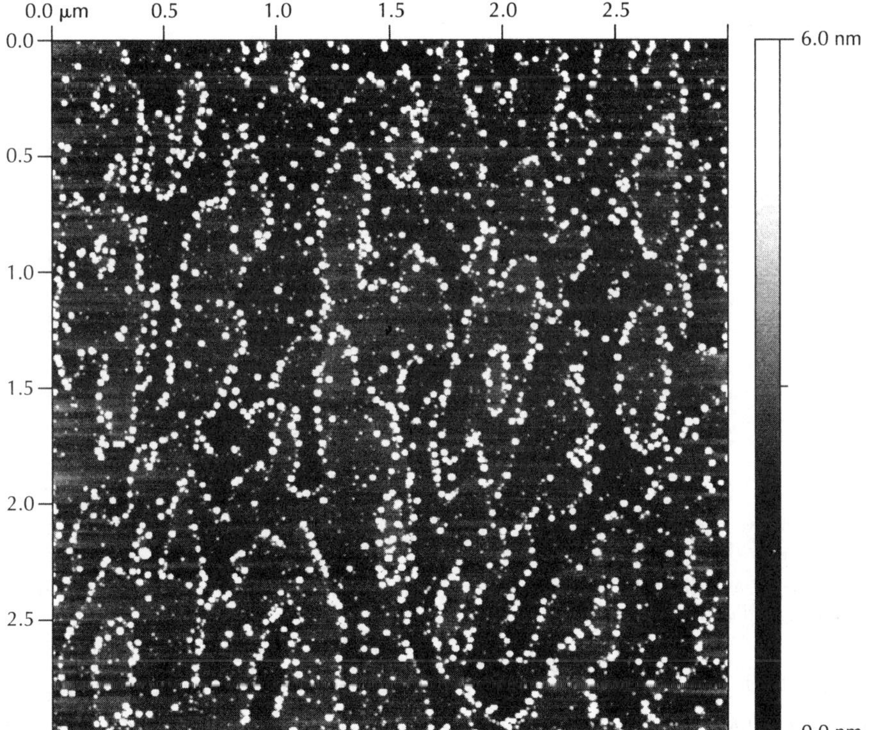

Figure 2.5   AFM micrograph of 0.500 nm InAs (1.54 ML) on a GaAs surface at 490°C. Note the dot locations coinciding with the surface step edges [36].

grown after the oxide has been thermally desorbed in any buffer layer, but the AlAs/GaAs super-lattice produces barriers, wells, and subbands below the subsequent structure that may interfere with electrical, optical, and/or the physical properties of the desired QD structure.

Growth of high-quality GaAs and AlAs films have been reported over a wide range of tempera-tures (200–620 and 580–650°C, respectively), but that is not true for InAs films. Typically, GaAs is grown at 580–600°C and the $Al_xGa_{1-x}As$ layers are grown 10–50°C hotter to optimize the carrier mobility, surface roughness, and crystalline defects, but the In incorporation coefficient, the fraction of In that incorporates into the growing crystal, begins to fall below 1.0 between 520 and 530°C and desorption of the In proceeds quickly above 570°C [47–49].

2.3.1 QD nucleation and growth

A set of samples were grown, with an As_4 beam equivalent pressure of 8e-6 torr, to determine the effect of temperature on the InAs QD density on four different $Al_xGa_{1-x}As$ surfaces ($x = 0, 0.15$, 0.3, and 0.45) [43]. Figure 2.6 is a plot of the QD density as a function of substrate temperature for 0.556 nm of InAs at 0.01 μm/h deposited on the four $Al_xGa_{1-x}As$ surfaces. The GaAs surface shows a change in QD density of almost two orders of magnitude from $3 \times 10^8 cm^{-2}$ at 520°C to $1.1 \times 10^{10} cm^{-2}$ at 500°C, greatly illustrating the importance of stable and accurate substrate temperature measurements. As the Al content of the surface increases, so does the QD density. The plot shows the similar QD density behaviour for the $Al_xGa_{1-x}As$ surfaces with increasing tem-perature, where the dependence on temperature is greatest for the highest Al content surface. The reduction of QD density with increasing temperature reaches an apparent zero point QD density for all $Al_xGa_{1-x}As$ surfaces at 523°C, under these conditions. The higher temperature and thus sur-face mobility favours the step flow growth mode over the 2D layer-by-layer growth mode used as a precursor to the SK growth mode [50]. The InAs wetting layers on AlAs surfaces form high-density 2D islanding that potentially acts as a precursor for 3D island growth [42].

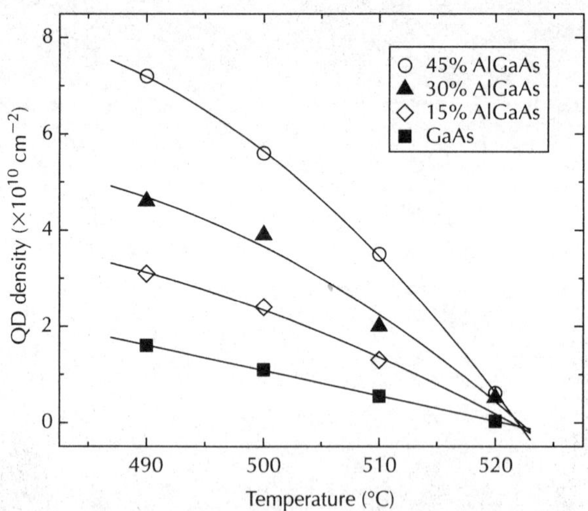

Figure 2.6   QD density dependence on temperature and $Al_xGa_{1-x}As$ surface composition for 0.556 nm InAs at 0.01 μm/h. Curve fits are to aid the eye.

The QD energies are determined by the QD shape and are dominated by the smallest dimension: height. The height of the QD samples in Fig. 2.6 are plotted in Fig. 2.7. The $Al_xGa_{1-x}As$ surface has a large influence on the In surface diffusion and thus the QD height behaviour. For the GaAs sur-face, there is an increase in QD height with increasing surface temperature. Compared to Fig. 2.6, the QD density is falling while the QDs are becoming larger. For the 15% Al surface, the QD height

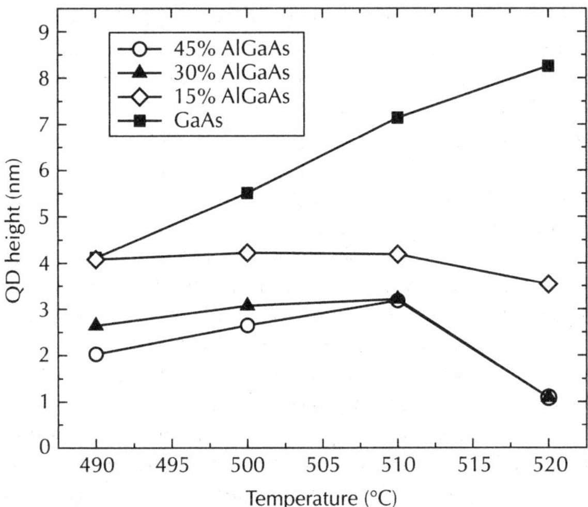

Figure 2.7   QD height dependence on temperature and Al$_x$Ga$_{1-x}$As surface composition for 0.556 nm InAs.

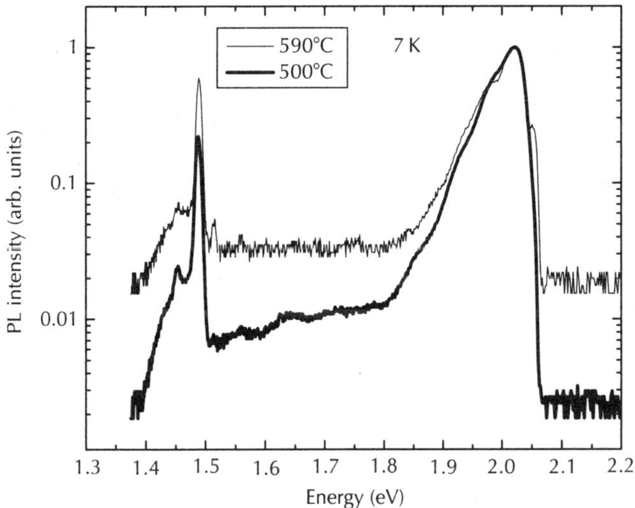

Figure 2.8   Photoluminescence of 2 μm Al$_{0.45}$Ga$_{0.55}$As layers grown on a GaAs substrate. The first sample, thick line, was grown at 500°C and the second, thin line, was grown at 590°C [36].

appears to be mostly independent of the surface temperature during growth. For the 30 and 45% Al surfaces, they show similar behaviour of QD height increasing with temperature, but there is a maximum at 510°C and by 520°C there is a significant reduction in QD height. Perhaps due to the reduced surface diffusion on the Al$_x$Ga$_{1-x}$As surface, the wetting layer is more stable and the critical thickness for QD nucleation is achieved locally, but the In on the surface cannot diffuse for QD ripening, thus producing smaller dots.

2.4   Changing the confining matrix

To achieve the 3D confinement, self-assembled semiconductor QDs are produced with low band gap material in a higher band gap matrix. A list of useful semiconductor properties and references can be found at the Ioffe Physico-Technical Institute webpage [51] and more detailed information on III–V compound semiconductor band parameters can be found in Vurgaftman *et al.* [52].

For the GaAs-based material system, these are InAs-based QDs in an $Al_xGa_{1-x}As$ matrix where AlAs and GaAs can be alloyed with the InAs to control the QD composition and size [37, 38, 41], but before incorporating InAs QDs into an $Al_xGa_{1-x}As$ matrix at temperatures $\sim100°C$ below convention, the material quality and surface roughness had to be characterized. $Al_{0.33}Ga_{0.67}As$ and $Al_{0.45}Ga_{0.55}As$ test structures, $2\,\mu m$ thick, grown under As_4 at 8×10^{-6} torr, and temperatures of 500–510°C and 590–600°C were analysed by PL and AFM. Both high and low temperature samples showed strong PL intensity peaks for the $Al_xGa_{1-x}As$ layers, while the sample grown at 590–600°C shows approximately three times the GaAs substrate peak intensity. This is attributed to the higher electron mobility observed in higher temperature $Al_xGa_{1-x}As$ growth. The surface roughness was remarkably similar for both growth temperatures, due to the relatively low Al surface diffusion. These results suggest that an $Al_xGa_{1-x}As$ matrix grown around 500°C will be of sufficient quality for strong QD luminescence.

Band gaps of $Al_xGa_{1-x}As$ for the direct gap Γ-valley and indirect X- and L-valleys are plotted in Fig. 2.9. Since the band gap is temperature dependent, the solid lines are 300 K values and dotted lines for 4 K. For practical purposes, thin films with an aluminium concentration of up to 0.45 can be used for direct gap $Al_xGa_{1-x}As$ without a significant problem, but higher Al fractions can result in the X-valley energy level being lower than the upper QD states, quenching the desired higher energy and intraband–intradot photoluminescence [53].

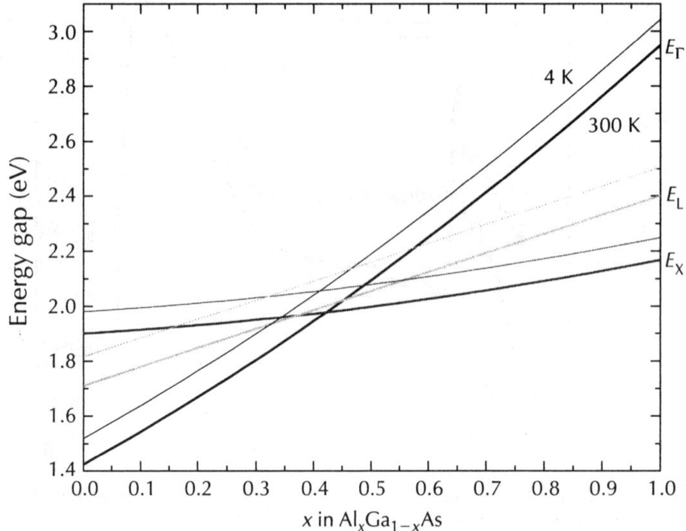

Figure 2.9   Energy gap for $Al_xGa_{1-x}As$ for the three conduction band valleys: Γ, X, L. Solid lines are at 300 K and vertically offset dotted lines are at 4 K. Calculated from [51, 52].

2.5   Overgrowth of quantum dots

Just to review, growth conditions for both MBE and MOCVD are typically chosen to optimize several growth parameters: maximizing adatom surface mobility, while minimizing surface roughness and heterostructure interdiffusion. This window between thermal roughening and statistical roughening must encompass all of the adatoms in the desired layer. The growth of self-assembled QDs uses deviations from the "optimal" growth conditions to engineer three-dimensional structures. In this case, a large mismatch ($\varepsilon_m \geq 0.02$) layer is grown under conditions to maximize the adatom surface mobility, while not minimizing the surface roughness.

The layer forms QDs to relieve the strain energy, but what happens when the QD is then overgrown with an oppositely mismatched material? The overgrowth of self-assembled semiconductor QDs is still a very active field of research [54, 55]. The evolution of the surface atoms with the addition of the matrix or other capping material raises many questions. With lattice-matched heteroepitaxial crystal growth, the interdiffusion of the constituent layers can easily be measured by various destructive and non-destructive techniques. Strained heteroepitaxial layers can be more difficult to characterize, but the same techniques still apply. For a non-planar strained structure, the analysis and characterization of the crystal growth becomes very complicated and *in situ* AFM or STM is required. InAs QDs overgrown with GaAs follow a multi-step process. Initially the impinging surface atoms erode the QD height and produce a region of In$_x$Ga$_{1-x}$As at the QD base. The substrate temperature plays a great roll in the degree of erosion and intermixing. The growth rate of the GaAs capping layer influences the final surface geometry above the QD peak. A slow growth rate allows the top of the QD to overgrow in a smooth fashion, while a faster growth rate does not preferentially grow over the area with the largest strain and thus forms an asymmetric pit above the QD peak. Regretfully, overgrowth of InAs QDs with Al$_x$Ga$_{1-x}$As has not been studied in such detail but one can assume that due to the lower diffusion rate of Al$_x$Ga$_{1-x}$As one can expect one of three results shown in Fig. 2.10. The reduced surface diffusion for Al could produce two opposite results – Fig. 2.10a shows the growth of an Al$_x$Ga$_{1-x}$As layer where the new material does not grow over the QD, Fig. 2.10b shows the growth of an Al$_x$Ga$_{1-x}$As layer where the new material does not migrate on the surface, instead it mimics the surface topography, while Fig. 2.10c shows a representation that is a mix between (a) and (b) that is closer to the overgrowth observed for GaAs. A detailed study has not been published, but TEM pictures reveal the overgrowth to be closest to schematics (a) and (c) [56].

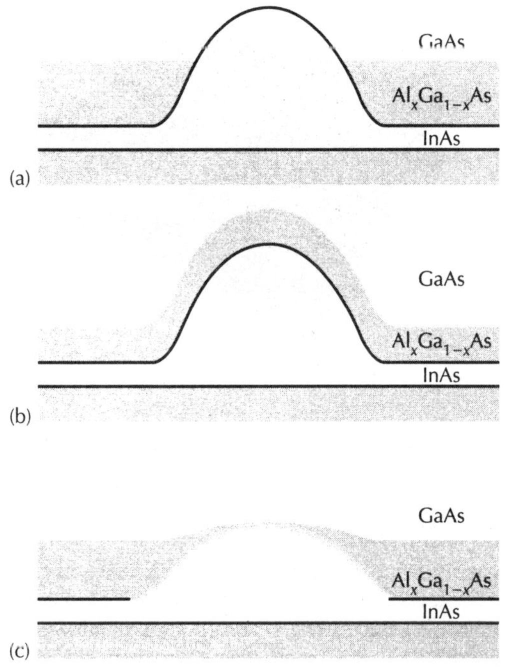

Figure 2.10   Schematic for the overgrowth of InAs quantum dots with thin Al$_x$Ga$_{1-x}$As layers. Growth possibilities (a) Al$_x$Ga$_{1-x}$As does not grow at the top of the QD, (b) Al$_x$Ga$_{1-x}$As does not diffuse across the surface of the QD, (c) Al$_x$Ga$_{1-x}$As diffuses, erodes, and intermixes as seen in GaAs overgrowth. As interpreted from a TEM micrograph in [56].

Creating a 30 period buried QD structure, for PL studies, with 50 nm $Al_xGa_{1-x}As$ spacer layers is not straightforward, as shown in Fig. 2.11. If the QD height or density is too high or the growth temperature of the $Al_xGa_{1-x}As$ layer is too low, the smooth starting surface will not recover in the 50 nm and the new QDs will preferentially nucleate on the surface steps. This effect will repeat every period until the sample surface roughness is greater than the period itself. Often by changing just one growth parameter, e.g. a higher temperature in Fig. 2.12, the growth of the QD periods is as desired, due to the dependence of QD density and size on temperature.

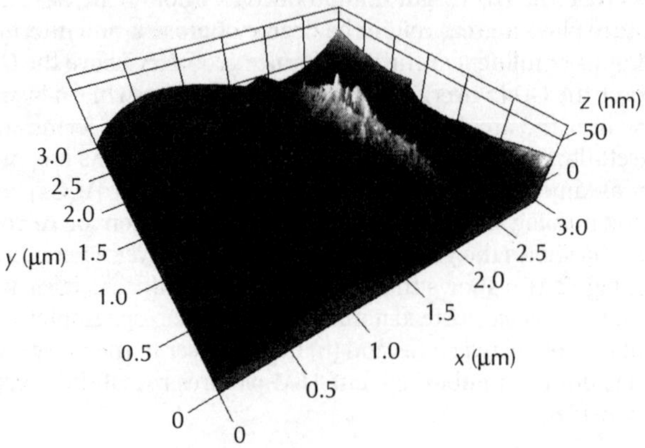

Figure 2.11   *Surface of a 30 period QD stack in $Al_{0.45}Ga_{0.55}As$ for T_g 510°C and 0.556 nm InAs [36].*

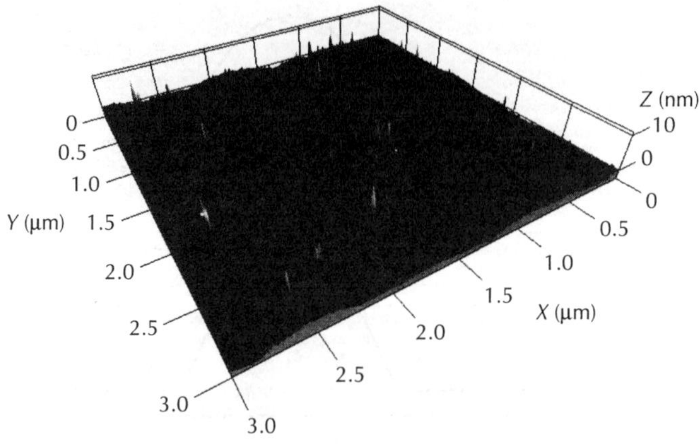

Figure 2.12   *Surface of a 30 period QD stack in $Al_{0.45}Ga_{0.55}As$ for T_g 520°C and 0.556 nm InAs [36].*

2.5.1   QD characterization

Photoluminescence measurements were performed to measure the band gap of the $Al_xGa_{1-x}As$ matrix material, energy level of the InAs wetting layer, and the energy levels of the QDs in the $Al_xGa_{1-x}As$ matrix. For the PL measurements a pulsed diode pumped solid-state laser operating at a wavelength of 532 nm is used. The samples were measured at low (2 W cm^{-2}) and high (500 W cm^{-2}) excitation densities to probe the ground state and excitated state transitions in the QDs. The emission of radiation from the photo-excited carriers is measured with a cooled photomultiplier tube and a lock-in amplifier. The measurements were performed at low temperature (4 K).

Samples prepared for PL were a stack of 30 times a layer of InAs QDs with a 50 nm Al$_x$Ga$_{1-x}$As spacer layer, so that the QDs are not preferentially nucleated from the strain field of the underlying QD layer [57, 58]. The temperature was kept constant from the start of the first QD layer through each spacer layer (GaAs or Al$_x$Ga$_{1-x}$As), with a three- minute pause for temperature stabilization before each new QD layer. The final QD layer was left uncapped for AFM analysis.

2.5.2   Inhomogeneous broadening of QD size

The growth of InAs QDs in an Al$_x$Ga$_{1-x}$As matrix produces increasing PL line widths with increasing Al concentration [59]. If it is assumed that the QD height is the dominant factor in the PL energy from a single QD then we can naïvely assume the PL FWHM from an ensemble of QDs is proportional to:

$$FWHM_{PL} \propto \left(\frac{1}{h - \Delta h/2}\right)^2 - \left(\frac{1}{h + \Delta h/2}\right)^2 \tag{2.9}$$

where h is the average QD height and Δh is the FWHM of the QD height distribution. Figure 2.13 shows the proportional PL FWHM calculated from the data measured by AFM as it depends on the Al content in the Al$_x$Ga$_{1-x}$As surface. This value is directly related to the inhomogeneous broadening of the PL peaks that would be observed from such QDs after they were buried. From 490 to 510°C, an increased substrate surface temperature results in a more uniform height distribution, but as the Al content increases, so does the calculated PL FWHM expected from the QDs, thus explaining the observed PL peak broadening for QDs in an Al$_x$Ga$_{1-x}$As matrix. This broadening is the result of a combination of a reduced QD height with a similar or broader height distribution. At 520°C there is a large increase in the calculated proportional PL FWHM for the 15, 30, and 45% Al$_x$Ga$_{1-x}$As samples, because of the greatly reduced average QD height without a reduction in the QD height distribution. Previously, the inhomogeneous broadening of QD PL peaks for Al$_x$Ga$_{1-x}$As has been attributed to the surface morphology created by using a ternary in the QD or matrix material [59, 60]. The AFM results used in Fig. 2.13 show this PL observation to additionally be the result of the inhomogeneous size distribution. While not a contradictory finding, since the surface morphology most certainly influences the QD nucleation, surface morphology alone does not address QD size or size distribution.

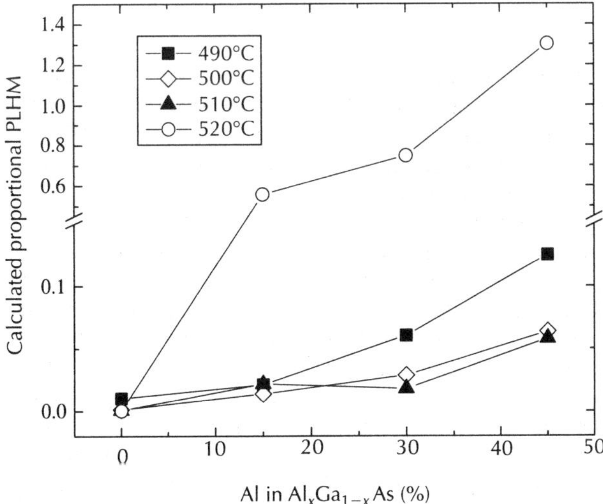

Figure 2.13   Calculated proportional PL FWHM dependence on temperature and Al$_x$Ga$_{1-x}$As surface composition for 0.556 nm InAs. This value is calculated from AFM height distribution data that directly leads to inhomogeneous PL peak broadening.

To further understand the behaviour of the In on the $Al_xGa_{1-x}As$ surfaces in these samples, the volume fraction of InAs incorporated into the QDs is plotted in Fig. 2.14. This is based on the volume of InAs incorporated into the QDs, as measured by the AFM software, divided by the InAs deposition amount of 0.556 nm. For all sample surfaces, the InAs volume incorporation is almost zero at 520°C, indicating that strain-driven QDs are unstable despite the Al content. At temperatures of 500–510°C the InAs volume incorporation into QDs increases significantly, increasing more as the Al content of the surfaces increases. The scaling with the Al content may be interrelated to the increased QD density. By 490°C there appears to be the onset of a "freezing out" for the 45% $Al_xGa_{1-x}As$ surface, most likely due to the drastically reduced surface diffusion, discussed earlier. It is assumed that the remaining InAs volume that is not incorporated into the QDs remains in the wetting layer. This would be slightly less than one monolayer for these samples.

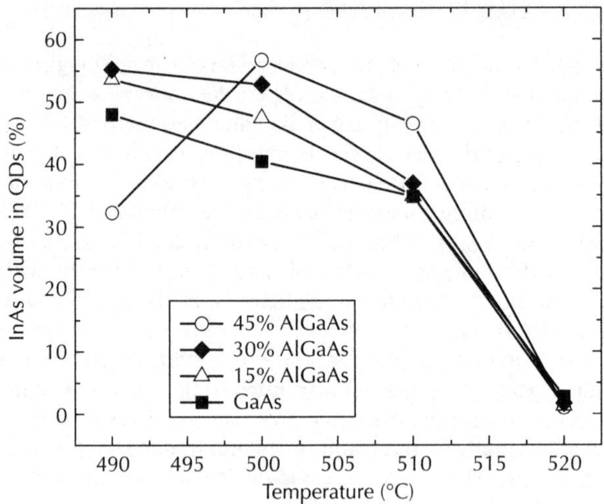

Figure 2.14   InAs volume fraction incorporated into QDs as determined by substrate temperature and $Al_xGa_{1-x}As$ surface composition for 0.556 nm InAs. The collective volume of the QDs is determined by the AFM software.

The QD density and QD height for $Al_xGa_{1-x}As$ surfaces can be decoupled through temperature and growth rate, but the incorporation of QDs into a repeating planar-stacked structure requires some attention. Smooth, high-quality, $Al_xGa_{1-x}As$ films can be difficult to achieve at low temperatures. This is further complicated when growing an InAs QD/AlGaAs superlattice. If the QDs are large, the capping layer will recover the planar surface more slowly, which can also be growth rate dependent [61], and the reduced surface diffusion of adatoms on $Al_xGa_{1-x}As$ surfaces compounds the problem. While QD heights of 2–10 nm or more are achievable on a single surface, only a planar $Al_xGa_{1-x}As$ surface is achieved for our PL stacks with QD densities in the high 10^{10} cm^{-2} with QD heights below 4 nm. Larger QDs required a reduced QD density, i.e. higher InAs deposition temperature. The results of these PL studies at ~4 K are plotted in Fig. 2.15. The PL peak energies for the bulk GaAs, $Al_{0.3}Ga_{0.7}As$, $Al_{0.33}Ga_{0.67}As$, and $Al_{0.45}Ga_{0.55}As$ are included for reference. The wetting layer and QD ground state energy are plotted for 12 samples of various $Al_xGa_{1-x}As$ matrix compositions. All samples are the result of 30 periods of planar InAs QDs grown on the $Al_xGa_{1-x}As$ surfaces and then buried by a 50 nm spacer layer. The variations in peak QD energy levels for data points at the same Al composition are due to variations in the QD size, achieved through the controlled ripening by InAs deposition. Increasing the Al composition in the matrix results in a blue shift of the QD ground state energy, which can be attributed to both the matrix material and the smaller QDs the $Al_xGa_{1-x}As$ surface produces. At a matrix composition of $Al_{0.45}Ga_{0.55}As$ three samples were produced with a peak QD energy level above

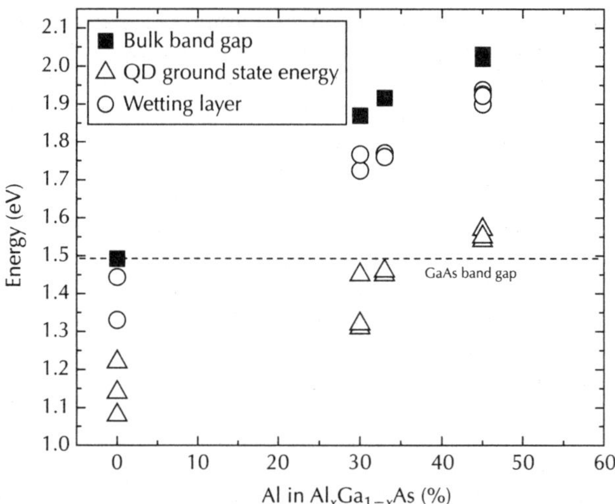

Figure 2.15   *Summary of PL peaks for 12 buried InAs QDs in Al$_x$Ga$_{1-x}$As samples at ~4 K. The solid squares denote the peak energies for bulk GaAs, Al$_{0.30}$Ga$_{0.70}$As, Al$_{0.33}$Ga$_{0.67}$As, and Al$_{0.45}$Ga$_{0.55}$As. The open circles denote the peak energies for the wetting layer in the Al$_x$Ga$_{1-x}$As matrix. The open triangles denote the peak ground state energies of the QDs in the Al$_x$Ga$_{1-x}$As matrix.*

the GaAs band edge, although the peak spread of 0.1 eV actually represents a very small change in InAs deposition. The greatest limitation now for InAs QDs in Al$_x$Ga$_{1-x}$As is the inhomogeneous size distribution that is directly related to the Al content in the Al$_x$Ga$_{1-x}$As surfaces.

2.6   Applications

2.6.1   Quantum dot detectors

Detection of mid- and far-infrared radiation using self-assembled QDs was reported for the first time in 1997 and 1998, respectively [62, 63], as an alternative to quantum well infrared photodetectors (QWIPs), where the quantum wells are employed for detection of radiation. QD infrared photodetectors (QDIPs) exploit intraband photoexcitation of electrons from confined states in the dots into levels at the continuum, Fig. 2.16. Compared to QWIPs, QDIPs are expected to show an improved performance due to the 3D confinement of the QDs, regarding the detection of normal incident light, as QDIPs can only detect light polarized in the growth direction due

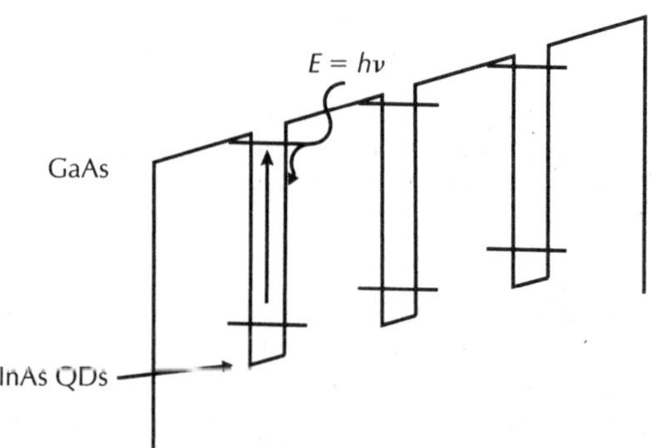

Figure 2.16   *Schematic of the potential profile for QDIPs. The detection mechanism is by intraband photoexcitation and then tunnelling into the continuum.*

to intersubband selection rules [64, 65] originating from the confinement in the growth direction only. Experiments showed that these selection rules, although based on the effective mass approximation, are followed quite closely [66]. QDIPs are also expected to have higher operation temperatures, lower dark current and higher photoconductor responsivity than QWIPs because of the potentially long excited electron lifetime due to reduced phonon emission.

Effort has been focused on improving the performance of QDIPs. To increase the responsivity, structures were designed where the carrier transport is shifted to a neighbouring channel with high electron mobility. As transport channels, InGaAs QWs or modulation doped AlGaAs/GaAs channels where used [67–69]. To lower the dark currents, which are caused by thermal excitation of the carriers in the dots, AlGaAs blocking layers were introduced. Due to the higher band gap of AlGaAs compared to GaAs these barriers can effectively suppress the flow of dark current [70–72].

However, there are limitations to exploiting the inherent advantages of QDs for detectors. It seems that a strong and dominant normal incidence photoresponse has not been achieved so far. A reason for this is thought to be the weak in-plane confinement, which is responsible for the normal incidence detection. Due to this weak in-plane confinement there exist several states in the dot, this is depicted schematically in Fig. 2.17. The transitions within the dots consume most of the in-plane oscillator strength and therefore dominate, but these transitions do not contribute to the resulting photocurrent as the electrons cannot escape from the dot [73]. In the growth direction, the confinement is strong and the dominant transition is from the confined state to the continuum, which does give rise to an intense photocurrent.

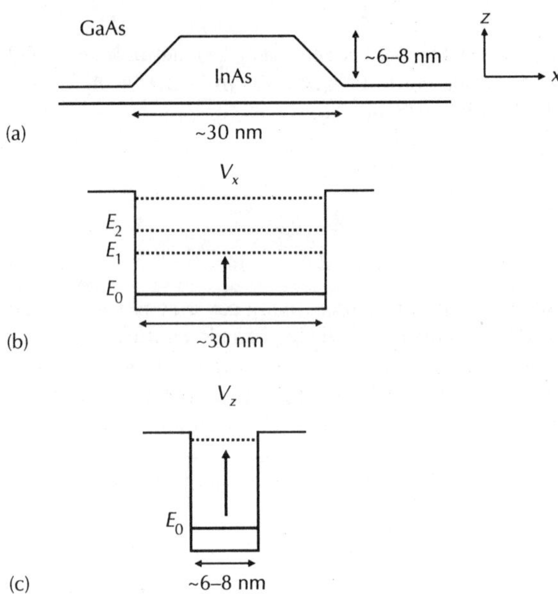

Figure 2.17   Illustration of an (a) InAs QD in a GaAs matrix with typical dot dimensions where x and y directions are in-plane. Also shown are transitions in the (b) in-plane direction or (c) growth direction. The confinement is illustrated by narrow (z) or wide (x or y) quantum wells. The upward arrows indicate the strongest transitions for z and x polarized light [73].

The in-plane confinement is determined by the QD diameter, which for InAs QDs on GaAs is typically ~ 30 nm [5]. In order to overcome this limitation, the dot dimensions have to be decreased. For efficient detectors the QD density should be high. To have a comparable absorption to QWIPs dot densities should be in the range of $(3–50) \times 10^{10}\,\mathrm{cm}^{-2}$ depending on the number of electrons per dot. The densities for InAs QDs on GaAs are in the range of $(0.03 - 1.6) \times 10^{10}\,\mathrm{cm}^{-2}$, depending on the growth conditions, Fig. 2.6. $\mathrm{Al}_x\mathrm{Ga}_{1-x}\mathrm{As}$ surfaces offer both – high dot densities with small dot sizes. On $\mathrm{Al}_{0.45}\mathrm{Ga}_{0.55}\mathrm{As}$ surfaces densities as high as

$7.2 \times 10^{10}\,\text{cm}^{-2}$ can readily be achieved. As was shown in Fig. 2.15 the excitonic ground state energy in the dot can be tuned by varying the Al content of the confining matrix. Depending on the Al content, it was shown to increase about $\sim 450\,\text{meV}$ and can be adjusted from well below the GaAs band edge to above the GaAs band gap; tuning of the wavelength dependent detector response over a wide range is possible.

2.6.2 Quantum dot quantum-cascade emitters

Since 1997 it has been a goal to incorporate QDs into a complete quantum cascade emitter structure [74] to improve the efficiency and temperature performance. A significant reduction in the efficiency of quantum cascade lasers (QCLs) is caused by longitudinal optical (LO) phonon depletion of the upper laser level. By embedding QDs in the active region of a QCL this scattering mechanism is expected to be reduced, due to the phonons' small energy dispersion and the discrete energy levels [75–77].

An estimation of the dimensions of the QDs necessary to efficiently suppress phonon emission is given by Eq. 2.10:

$$(3\pi^2\hbar^2)/(2\,m^*\,L^2) > \hbar\omega_{\text{LO}} \tag{2.10}$$

To avoid LO-phonon emission the energy level spacing between the excited and ground states in the dot has to exceed the optical phonon energy $\hbar\omega_{\text{LO}}$, which amounts to $E_{\text{QD, LO}} \sim 32\,\text{meV}$ in an InAs QD with strain taken into account [78]. The maximum dot size L can be estimated from the energy spacing in a square well of size L. For GaAs, with an effective mass $m^* = 0.067\,\text{m}$, this implies dots smaller than $L \approx 20\,\text{nm}$ in all three dimensions. Typical QD heights are well above this limit as can be seen in Fig. 2.7. Concerning the in-plane dot dimensions, InAs dots grown on GaAs surfaces show typical diameters of $\sim 30\,\text{nm}$, as mentioned previously. InAs QDs grown on AlGaAs or AlAs surfaces have typical diameters at or below the calculated maximum QD size L [79].

For the design of devices which exploit intraband transitions in QDs it is necessary to investigate the intraband relaxation mechanisms in quasi-zero-dimensional QDs. Because of the discrete density of electronic states in QDs and weak energy dispersion of the longitudinal optical (LO)-phonons it was suggested that efficient electron non-radiative relaxation is only possible for energies close to the LO-phonon energy, also known as the phonon bottleneck effect [80]. It has been shown that the phonon bottleneck effects can be significantly suppressed in QDs, leading to decay times from excited states to ground states in the range of 10–50 ps [81]. This is due to the discrete nature of the confined states in QDs, which causes a strong coupling between electrons and phonons, so-called polarons. Efficient relaxation of excited polarons even for quite large detunings ($\sim 20\,\text{meV}$) from the LO-phonon energy has been demonstrated [82]. Recently, the effect of strong electron–phonon coupling on the ground state to first excited state transition energy has been observed in annealed, self-assembled InAs/GaAs QDs [83].

Mid-infrared luminescence from a resonant tunnelling QD unipolar device was first demonstrated [84] with a single layer of self-assembled InAs QDs in an AlAs barrier. By using composition grading, electrons were resonantly injected into the upper QD states. The electrons were resonantly extracted from the lower QD states through a superlattice miniband. The superlattice additionally forms an electron filter that prevents the carriers from directly tunnelling out of the dot before relaxing to a lower energy level and thus the emission of a mid-IR photon is encouraged. The graded barrier for QD injection limits this to half of a quantum cascade with no engineered population inversion. Emission from this structure also showed anisotropic polarization due to the anisotropy in QD shape in the growth plane [9].

Mid-infrared electroluminescence from a quantum-dot-quantum-well cascade structure, consisting of ten periods, with AlInAs QDs in the active region, has also been demonstrated [85]. In this design the QDs were embedded in the AlGaAs barriers of the structure. The electroluminescence only showed a distinct peak at low currents, where a resonant subensemble of quantum dots with a small inhomogeneous broadening contributes to the signal. For higher currents a spectral broadening of the electroluminescence was observed.

Figure 2.18 shows a schematic of a grown structure, where the QDs are incorporated into a complete quantum cascade emitter structure [9]. The QDs embedded in the $Al_{0.45}Ga_{0.55}As$ barriers were combined with an injector and extractor. The electrons are resonantly injected into the excited state of the dot and after relaxation to the dot ground state, the carriers are then extracted resonantly from the QD ground state by the lower miniband. Figure 2.19 shows the conduction band diagram and calculated moduli squared wave functions of such a QD quantum cascade (QC) emitter structure under a bias of 40 kV/cm. The energy separation between the QD ground state and excited state (100 meV) was taken from interband photoluminescence measurements, shown earlier, performed on comparable dots and matched with the separation between the upper miniband and the lower miniband. At a reverse bias of $\sim 20\,kVcm^{-1}$ the upper miniband should be in alignment with an excited QD state and resonantly inject carriers

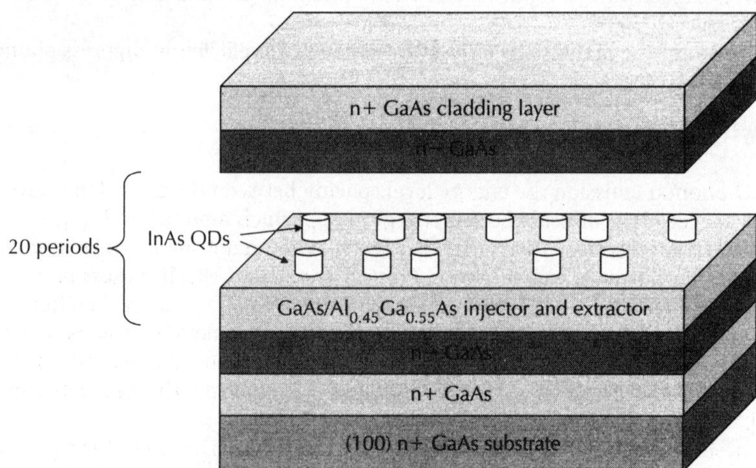

Figure 2.18   Schematic of the QD quantum cascade structure. One period consists of one layer of self-assembled InAs QDs embedded in the thickest AlGaAs barrier of a 42.5 nm thick GaAs/AlGaAs superlattice. The whole structure is composed of 20 periods.

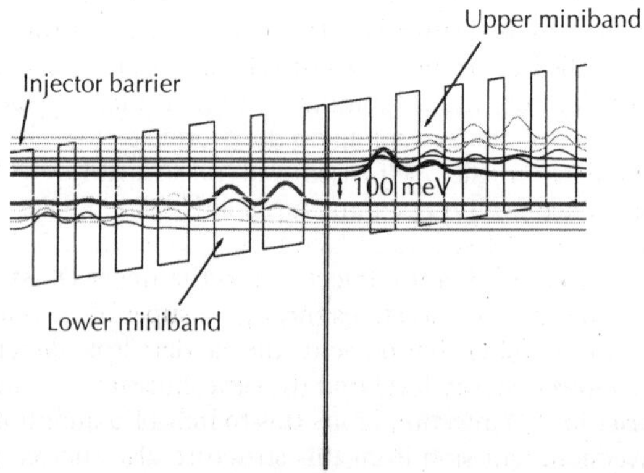

*Figure 2.19   Conduction band diagram and moduli squared wavefunctions for a QD quantum cascade emitter structure at 40 kV/cm bias. The QDs are depicted schematically as a narrow quantum well. The layer sequence (nm) of one cascade starting from the injector barrier is **2.6**, 2.8, **2.0**, 2.8, **1.8**, 3.0, **1.7**, 3.4, **2.8**, 4.0, **1.4**, 4.5, **2.3**, InAs QDs, **4.6**, 2.8 where bold numbers refer to $Al_{0.45}Ga_{0.55}As$ barriers and regular type to GaAs wells. Underlined numbers refer to doped layers ($n_{Si} = 2.5 \times 10^{17}\,cm^{-2}$). The separation between injecting and extracting miniband state is designed to be 100 meV. The QDs are embedded asymmetrically in an AlGaAs barrier. So that with a QD height of 2.3 nm this is meant to centre the QDs within the barrier [36].*

into the structure. From the comparable QD calibration growth the QD density in the structure was estimated to be $4 \times 10^{10}\,\mathrm{cm}^{-2}$ and the doping of $1.4 \times 10^{11}\,\mathrm{cm}^{-2}$ for each period was chosen to have a sufficient number of carriers to fill the lower energy levels of the QDs.

Figure 2.20 shows high-resolution transmission electron microscopy (HR-TEM) images of the grown GaAs/AlGaAs superlattice with the embedded QDs [86]. Also shown is a reference structure without QDs (Figure 2.20a). The bright regions visible in the images correspond to the AlGaAs barriers whereas the darker areas are the GaAs wells. In Figure 2.20a the InAs wetting layer (WL) can be seen as a thin black line between the asymmetric AlGaAs barriers. In Figure 2.20b the QDs appear as dark regions. The size of the dots is estimated to be $\sim$10 nm in diameter and $\sim$2 nm in height.

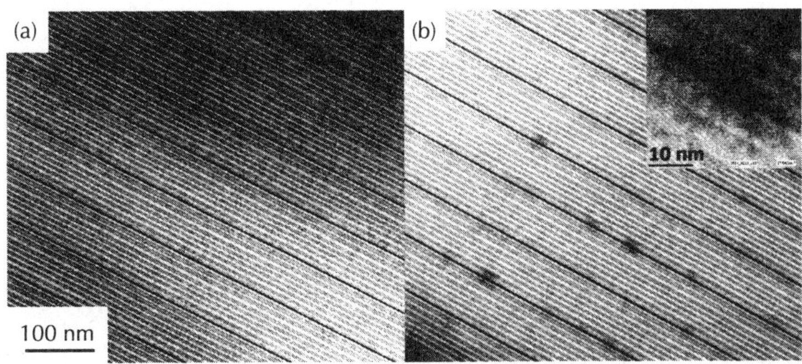

Figure 2.20   High-resolution transmission electron microscopy images of (a) the quantum cascade structure without QDs and (b) the quantum cascade structure with the QDs embedded in the AlGaAs barriers. The inset shows a close-up of a single QD [86].

I–V curves for continuous wave operation at $T = 4.2$ K for the structures with and without QDs are shown in Fig. 2.21. Under positive bias both structures show a similar I–V characteristic. Under negative bias, both I–V characteristics show an onset of the current at around $10\,\mathrm{keVcm}^{-1}$. As this onset is present in both structures (with and without QDs) apparently the QDs are not involved in this carrier transport. This parasitic current flow is common below threshold in QCL structures. At an electric field of about $20\,\mathrm{keVcm}^{-1}$, the bias voltage for which resonant injection and extraction of carriers from the QD was designed, only the structure with the QDs shows a constantly increasing current flow. The device without QDs shows a significantly smaller almost constant current flow, which only increases again at considerably higher electric fields.

To gain insight into the transport properties of the structures it is interesting to look at the current density and its relation to carrier transfer time for one period. Under the assumption of operating in the strong coupling regime the current density J can be estimated:

$$J = Ne/\tau \tag{2.11}$$

where N is the sheet carrier density, e is the electronic charge and τ is the carrier transfer time for one period [87, 88]. In a QCL, the current density is determined by the lifetime of the upper laser level and the doping level. For the structure with the QDs a lower limit can be estimated assuming all QDs contribute to carrier transport and only one electron at a time is resonantly injected and extracted from the QD levels. In this case, $N = 4 \times 10^{10}\,\mathrm{cm}^{-2}$, the estimated QD density. For the current density of 1000 A cm^{-2} the transfer time τ is calculated to be 6.4 ps, which is in reasonable agreement with the 4.7 ps measured by time-resolved photoinduced IR absorption measurements [89]. For comparison, typical values for QCLs are $\sim$1 ps [90]. For the device without QDs the current density is 90 A cm^{-2}, $N = 1.4 \times 10^{11}\,\mathrm{cm}^{-2}$, which is the doping per period, and therefore $\tau = 0.25$ ns.

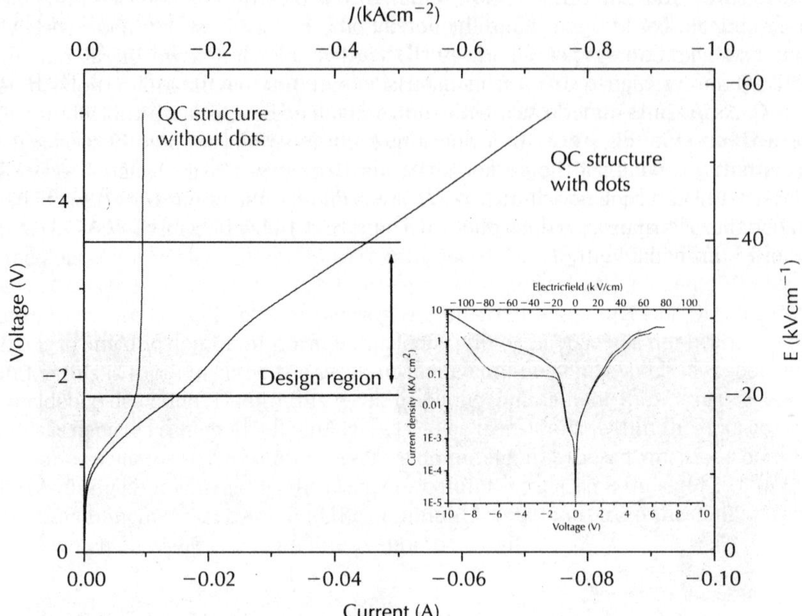

Figure 2.21   I–V characteristic measured at T = 4 K for the two quantum cascade structures with and without QDs in the active region. The large image shows the I–V for negative bias voltages only. The inset shows the absolute value of the current density for positive and negative bias voltages. The region for which resonant injection into and extraction from the QDs was designed is labelled as design region.

Extraction of the carriers from the QD ground state is most efficiently if the ground state aligns with the extracting miniband states. Interband photoluminescence studies of comparable QDs grown on $Al_{0.45}Ga_{0.65}As$ surfaces showed that the lowest energy level in the QDs are situated about 45 meV above the GaAs bandedge. Efficient carrier extraction from the QD ground state should thus be possible. This is an important prerequisite to achieve population inversion in such a device.

Despite these results, the discussed QD QC structure has not shown electroluminescence so far. Which means the desired optical transition is not occurring and other relaxation mechanisms inside the dot are more efficient. The first possibility is a phonon scattering process, which is supposed to be slow due to the proposed phonon bottleneck [75]. The second mechanism is an Auger relaxation process. Because energy has to be conserved, relaxation due to an Auger process is almost impossible if the initial and the final state of the electron are discrete, as the electronic states are to be assumed in a QD. A sequential intradot Auger process is possible, where two carriers, initially captured in dot excited states, undergo an irreversible Auger scattering to final states where one particle remains bound in the dot with a lower energy while the other has been ejected from the dot into a wetting layer continuum state [90]. This process is more likely to occur at high current densities as the dot has to be occupied with more than one electron in the excited states. As both processes are non-radiative relaxation processes they can prevent emission of mid-infrared photons from the QD QC structure.

References

1. L. Goldstein, F. Glas, J.Y. Marzin, M.N. Charasse, and G. Le Roux, *Appl. Phys. Lett.* **47**(10), 1099 (1985).
2. N.N. Ledenstov, D. Bimberg, F. Hopfer, A. Mutig, V.A. Schukin, A.V. Savel'ev, G. Fiol, E. Stock, H. Eisele, M. Dähne, D. Gerthsen, U. Fischer, D. Litvinov, A. Resenauer, S.S. Mikhrin, A.R. Kovsh, N.D. Zakharov, and P. Werner, *Nanoscale Res. Lett.* **2**(9), (2007). Published Online,

3. M. Henini, *Nanoscale Res. Lett.* **1**(1), 32 (2007).

4. O.G. Schmidt, S. Kiravittaya, Y. Nakamura, H. Heidemeyer, R. Songmuang, C. Müller, N.Y. Jin-Phillipp, K. Eberl, H. Wawra, S. Christiansen, H. Gräbeldinger, and H. Schweizer, *Surf. Sci.* **514**, 10 (2002).

5. J. Stangl, V. Holy, and G. Bauer, *Reviews of Modern Physics* **76**, 725 (2004).

6. N. Kirstaedter, N.N. Ledenstov, M. Grundmann, D. Bimberg, V.M. Ustinov, S.S. Ruvimov, M.V. Maximov, P.S. Kopev, Zh.I. Alferov, U. Richter, P. Werner, U. Gosele, and J. Heydenreich, *Electron. Lett.* **30**, 1416 (1994).

7. H. Ishikawa, H. Shoji, Y. Nakata, K. Mukai, M. Sugawara, M. Egawa, N. Otsuka, Y. Sugiyama, T. Futatsugi, and N. Yokoyama, *J. Vac. Sci. Tech. A.* **16**(2), 794 (1998).

8. Q. Cao, S.F. Yoon, C.Y. Liu, and C.Y. Ngo, *Nanoscale Res. Lett.* **2**(6), 303 (2007).

9. D. Wasserman, C. Gmachl, S.A. Lyon, and E.A. Shaner, *Appl. Phys. Lett.* **88**, 191118 (2006).

10. B. Aslan, H.C. Liu, J.A. Gupta, Z.R. Wasilewski, G.C. Aers, S. Raymond, and M. Buchanan, *Appl. Phys. Lett.* **88**, 043103 (2006).

11. Y. Arakawa and H. Sakaki, *Appl. Phys. Lett.* **40**(11), 939 (1982).

12. R.P. Mirin, J.P. Ibbetson, K. Nishi, A.C. Gossard, and J.E. Bowers, *Appl. Phys. Lett.* **67**(25), 3795 (1995).

13. M. Asada, Y. Miyamoto, and Y. Suematsu, *IEEE J. Quantum Electron.* **22**(9), 1915 (1986).

14. D.P. Nguyen, N. Regnault, R. Ferreira, and G. Bastard, *Phys. Rev. B.* **71**(24), 245329 (2005).

15. F. Schrey, L. Rebohle, R. Müller, G. Strasser, K. Unterrainer, D.P. Nguyen, N. Regnault, R. Ferreira, and G. Bastard, *Phys. Rev. B.* **72**(15), 155310 (2005).

16. R.E. Bailey, A.M. Smith, and S. Nie, *Physica E.* **25**, 1 (2004).

17. M. Achermann, M.A. Petruska, S. Kos, D.L. Smith, D.D. Koleske, and V.I. Klimov, *Nature* **429**, 624 (2004).

18. H. Gao, *J. Mech. Phys. Solids* **42**, 741 (1994).

19. J.W. Matthews, F.R.N. Nabarro. *Dislocations in Solids*, Vol. 2 (Amsterdam, North-Holland, 1979), 461 pp.

20. B.W. Dodson and J.Y. Tsao, *Appl. Phys. Lett.* **51**, 1325 (1987).

21. E.A. Fitzerald, *Mater. Sci. Rep.* **7**, 87 (1991).

22. L.B. Freund, *MRS Bulletin* **17**(7), 52 (1992).

23. R. Beanland, D.J. Dunstan, and P.J. Goodhew, *Adv. Phys.* **45**, 87 (1996).

24. S. Kishino, M. Ogirima, and K. Kurata, *J. Electrochem. Soc.* **119**, 618 (1972).

25. J.S. Speck, M.A. Brewer, G.E. Beltz, A.E. Romanov, and W. Pompe, *J. App. Phys.* **80**, 3808 (1996).

26. J.W. Matthews and A.E. Blakeslee, *J. Cryst. Growth* **27**, 118 (1974).

27. M.A. Herman and J. Sitter. *Molecular Beam Epitaxy: Fundamentals and Current Status* (Springer-Verlag, Berlin, 1989, 215.

28. G. Wulff, *Z. Kristallogr. Mineral.* **34**, 449 (1901).

29. H. Gao and W.D. Nix, *Annu. Rev. Mater. Sci.* **29**, 173 (1999).

30. B.J. Spencer and D.I. Meiron, *Acta. Metall. Mater.* **42**, 3629 (1994).

31. B.D. Gerardot, I. Shtrichman, D. Hebert, and P.M. Petroff, *J. Cryst. Growth* **252**, 44 (2003).

32. A.E. Romanov, P.M. Petroff, and J.S. Speck, *Appl. Phys. Lett.* **74**, 2280 (1999).

33. I.N. Stranski and Von.L. Krastanow, *Akad. Wiss. Lit. Mainz Math. Naturwiss. Kl. IIb* **146**, 797 (1939).

34. D. Leonard, K. Pond, and P.M. Petroff, *Phys. Rev. B.* **50**(16), 11687 (1994).

35. R. Murray, D. Childs, S. Malik, P. Siverns, C. Roberts, J.-M. Hartmann, and P. Stavrinou, *Jpn. J. Appl. Phys. Part 1* **38**, 528 (1999).

36. Unpublished data, Courtesy of A.M. Andrews *et al.*, Technische Universität Wien.

37. R. Leon, S. Fafard, D. Leonard, J.L. Merz, and P.M. Petroff, *Appl. Phys. Lett.* **67**(4), 521 (1995).

38. K. Pierz, Z. Ma, U.F. Keyser, and R.J. Haug, *J. Cryst. Growth* **249**, 477 (2003).

39. S. Koshiba, Y. Nakamura, M. Tsuchiya, H. Noge, H. Kano, Y. Nagamune, T. Noda, and H. Sakaki, *J. Appl. Phys.* **76**(7), 4138 (1994).

40. E. Penev, S. Stojkovic', P. Kratzer, and M. Scheffler, *Phys. Rev. B.* **69**, 115335 (2004).

41. P. Offermans, P.M. Koenraad, J.H. Wolter, K. Pierz, M. Roy, and P.A. Maksym, *Physica E.* **26**, 236 (2005).

42. P. Ballet, J.B. Smathers, and G.J. Salamo, *Appl. Phys. Lett.* **75**(3), 337 (1999).

43. A.M. Andrews, M. Schramböck, T. Roch, W. Schrenk, E. Gornik, and G. Strasser, *J. Mater. Sci: Mater. Electron.* Online September, (2007).

44. M.A. Herman and J. Sitter. *Molecular Beam Epitaxy: Fundamentals and Current Status* (Springer-Verlag, Berlin, 1989). 141,

45. D. Leonard, M. Krishnamurthy, C.M. Reaves, S.P. Denbaars, and P.M. Petroff, *Appl. Phys. Lett.* **63**(23), 3202 (1993).

46. A. Krost, G. Bauer, and J. Woitok, in: *Optical Characterization of Epitaxial Semiconductor Layers*, edited byG. Bauer and , W. Richter. (Springer, Berlin, 1996, Ch. 6.

47. G. Colayni and R. Venkat, *J. Cryst. Growth* **211**, 21–26 (2000).

48. J.M. Moison, C. Guille, F. Houzay, F. Barthe, and M. Van Rompay, *Phys. Rev B.* **40**(9), 6149 (1989).

49. K.R. Evans, R. Kaspi, J.E. Ehret, M. Skowronski, and C.R. Jones, *J. Vac. Sci. Technol. B,* **13**(4), 1820 (1995).

50. *Molecular Beam Epitaxy*, edited by, R.F.C. Farrow. (Noyes Publications, New Jersey, 1995) 703 pp.

51. http://www.ioffe.ru/SVA/NSM/.

52. I. Vurgaftman, J.R. Meyer, and L.R. Ram-Mohan, *J. Appl. Phys.* **89**(11), 5815 (2001).

53. P. Dawson, Z. Ma, K. Pierz, and E.O. Göbel, *Appl. Phys. Lett.* **81**(13), 2349 (2002).

54. Q. Gong, P. Offermans, R. Nötzel, P.M. Koenraad, and J.H. Wolter, *Appl. Phys. Lett.* **85**(23), 5697 (2004).

55. G. Costantini, A. Rastelli, C. Manzano, P. Acosta-Diaz, R. Songmuang, G. Katsaros, O.G. Schmidt, and K. Kern, *Phys. Rev. Lett.* **96**, 226106 (2006).

56. I. Hapke-Wurst, *Resonanter Magnetotransport durch selbstorganisierte InAs Quantenpunkte*, Ph.D. Thesis (2002), p. 47.

57. P. Howe, E.C. Le Ru, E. Clarke, B. Abbey, R. Murray, and T.S. Jones, *J. Appl. Phys.* **95**(6), 2998 (2004).

58. V. Holy, G. Springholz, M. Pin czolits, and G. Bauer, *Phys. Rev. Lett.* **83**(2), 356 (1999).

59. A. Polimeni, A. Patane, M. Henini, L. Eaves, and P.C. Main, *Phys. Rev. B.* **59**(7), 5064 (1999).

60. C. Ni` Allen, P. Finnie, S. Raymond, Z.R. Wasilewski, and S. Fafard, *Appl. Phys. Lett.* **79**(17), 2701 (2001).

61. G. Costantini, A. Rastelli, C. Manzano, P. Acosta-Diaz, R. Songmuang, G. Katsaros, O.G. Schmidt, and K. Kern, *Phys. Rev. Lett.* **96**, 226106 (2006).

62. K.W. Berryman, S.A. Lyon, and M. Segev, *Appl. Phys. Lett.* **70**(14), 1861 (1997).

63. J. Phillips, K. Kamath, and P. Bhattacharya, *Appl. Phys. Lett.* **72**(16), 2020 (1998).

64. D.D. Coon and R.P.G. Karunasiri, *Appl. Phys. Lett.* **45**, 649 (1984).

65. L.C. West and S.J. Eglash, *Appl. Phys. Lett.* **46**, 1156 (1985).

66. H.C. Liu, M. Buchanan, and Z.R. Wasilewski, *Appl. Phys. Lett.* **72**(14), 1682 (1998).

67. L. Chu, A. Zrenner, M. Bichler, and G. Abstreiter, *Appl. Phys. Lett.* **79**(14), 2249 (2001).

68. M.D. Kim, A.G. Choo, T.I. Kim, S.S. Ko, D.H. Baek, and S.C. Hong, *J. Cryst. Growth.* **227**, 1162 (2001).

69. S.-W. Lee, K. Hirakawa, and Y. Shimada, *Appl. Phys. Lett.* **75**(10), 1428 (1999).

70. O. Baklenov, Z.H. Chen, E.T. Kim, I. Mukhametzhanov, A. Madhukar, F. Ma, Z. Ye, B. Yang, and J. Campbell, *Proc. 58th IEEE Device Research Conf.* Denver, CO, June 2000, p. 171.

71. Z. Ye, J.C. Campbell, Z.H. Chen, E.-T. Kim, and A. Madhukar, *IEEE J. Quant. Electron.* **38**(9), 1234 (2002).

72. Z.H. Chen, O. Baklenov, E.T. Kim, I. Mukhametzhanov, J. Tie, A. Madhukar, Z. Ye, and J.C. Campbell, *Infrared Phys. Tech.* **42**, 479 (2001).

73. H.C. Liu, *Opto-Electron. Rev.* **11**, 1 (2003).

74. N.S. Wingreen and C.A. Stafford, *IEEE J. Quant Electron.* **33**(7), 1170 (1997).

75. H. Benisty, C.M. Sotomayor-Torrès, and C. Weisbuch, *Phys. Rev. B.* **44**(19), 10945 (1991).

76. M.J. Steer, D.J. Mowbray, W.R. Tribe, M.S. Skolnick, and M.D. Sturge, *Phys. Rev. B.* **54**(24), 17738 (1996).

77. Y. Toda, O. Moriwaki, M. Nishioka, and Y. Arakawa, *Phys. Rev. Lett.* **82**(22), 4114 (1999).

78. M. Grundmann, O. Stier, and D. Bimberg, *Phys. Rev. B.* **52**(16), 11969 (1995).

79. I. Hapke-Wurst, U. Zeitler, U. Fl. Keyser, R.J. Haug, K. Pierz, and Z. Ma, *Appl. Phys. Lett.* **82**(8), 1209 (2003).

80. R. Heitz, H. Born, F. Guffarth, O. Stier, A. Schliwa, A. Hoffmann, and D. Bimberg, *Phys. Rev. B.* **64**(24), 241304(R) (2007).
81. E.A. Zibik, L.R. Wilson, R.P. Green, G. Bastard, R. Ferreira, P.J. Phillips, D.A. Carder, J.P.R. Wells, J.W. Cockburn, M.S. Skolnick, M.J. Steer, and M. Hopkinson, *Phys. Rev. B.* **70**, 161305 (2004).
82. X.-Q. Li, H. Nakayama, and Y. Arakawa, *Phys. Rev. B.* **59**, 5069 (1999).
83. E.A. Zibik, W.H. Ng, L.R. Wilson, M.S. Skolnick, J.W. Cockburn, M. Gutierrez, M.J. Steer, and M. Hopkinson, *Appl. Phys. Lett.* **90** (2007).
84. D. Wasserman and S.A. Lyon, *Appl. Phys. Lett.* **81**(15), 2848 (2002).
85. N. Ulbrich, J. Bauer, G. Scarpa, R. Boy, D. Schuh, G. Abstreiter, S. Schmult, and W. Wegscheider, *Appl. Phys. Lett.* **83**(8), 1530 (2003).
86. Unpublished data, Courtesy of F. Schäffler, JKU Linz.
87. C. Sirtori, F. Capasso, J. Faist, A.L. Hutchinson, D.L. Sivco, and A.Y. Cho, *IEEE J. Quant. Electron.* **34**(9), 1722 (1998).
88. J. Faist, *Appl. Phys. Lett.* **90**, 253512 (2007).
89. T. Müller, F. F. Schrey, G. Strasser, and K. Unterrainer, *Appl. Phys. Lett.* **83**(17), 3572 (2003).
90. H. Page, C. Becker, A. Robertson, G. Glastre, V. Ortiz, and C. Sirtori, *Appl. Phys. Lett.* **78**, 3529 (2001).

CHAPTER 3

Optical Properties of In(Ga)As/GaAs Quantum Dots
for Optoelectronic Devices

E.M. Clarke and R. Murray

Imperial College London

3.1 Introduction

The emergence of quantum well (QW) physics in the 1980s following developments in epitaxial growth illustrated the important benefits that can be obtained through quantum confinement of carriers along one direction in a semiconductor device. QW lasers have threshold currents an order of magnitude smaller than bulk double heterostructure devices and also exhibit significantly improved temperature sensitivity of the threshold current. Using different materials combinations it is possible to produce QW lasers over a wide wavelength range 400–3000 nm, many of which have found their way into consumer products. This success led, quite naturally, to an investigation of the effects that further confinement might bring. Although quantum wires (QWs) have generated interest they have failed to match the technological potential of self-assembled quantum dots (QDs) discovered, almost by accident, in 1985 [1]. It was not until the early 1990s that the potential of this method for confining carriers was recognized and the last decade has seen the publication of over ten thousand QD papers. In this chapter we will survey the growth, structure and optical properties of self-assembled QDs, choosing InAs/GaAs as the prototypical material system, and outline the current and future status of QD optoelectronic devices, particularly lasers. The emphasis will be to link the device characteristics to the optical properties of a confined system which are, in turn, limited by the epitaxial growth. In addition to their incorporation in conventional optoelectronic devices such as lasers and LEDs, QDs can be incorporated into novel device structures which utilize some of their unique optical properties. In particular, the extremely narrow spectral linewidth of emission from single InAs QDs and the ability to isolate single photon emission from QDs enable the realization of efficient on-demand single photon sources and also the possibility of triggered entangled photon sources. Also, the spatial confinement of carriers and the modified electronic states in QDs suggest that QD structures may be used as a vehicle for manipulation of the spin state of carriers, using concepts from the emerging field of spintronics. It is hoped that this will lead to new devices suitable for quantum information applications.

A brief overview of the growth of In(Ga)As/GaAs QD structures is now presented, demonstrating the numerous considerations that must be taken into account for successful realization of QD-based devices.

3.2 Growth of In(Ga)As/GaAs QDs

QDs have been fabricated by various methods, including lithography, growth of nanocrystals, natural monolayer fluctuations in the thickness of ultra-thin QWs and using surface voltage gates in a 2D electron gas (2DEG) structure to electrostatically define lateral confinement regions. The most widespread methods for QD fabrication involve molecular beam epitaxy (MBE) or metal-organic chemical vapour deposition (MOCVD) of lattice-mismatched semiconductors. Initially, growth proceeds in a layer-by-layer fashion (as for QWs) but after a critical thickness the increasing strain between the layers makes it energetically favourable for the strain to be relieved by formation of 3D islands. This is the Stranski–Krastanov growth mode. Depending upon the amount of material deposited, the islands are free of dislocations [2, 3] and following overgrowth (capping) they form nanoscale inclusions within the bulk matrix, with dimensions comparable with the de Broglie wavelength of electrons in the crystal.

Growth of InAs on GaAs follows the Stranski–Krastanov model although precise details are still the subject of debate. There is a mismatch of around 7% between the lattice constants of InAs and GaAs and the initial 2D pseudomorphic layer known as the wetting layer persists until a critical InAs thickness θ_{crit} of $\sim 1.7\,\mathrm{ML}$ (depending on growth conditions) is reached. At the critical thickness, 3D islands form and the island density increases very rapidly over a small coverage interval ($<0.1\,\mathrm{ML}$) [4, 5]. The transition from 2D to 3D growth can be clearly seen during MBE growth of InAs/GaAs layers using reflection high-energy electron diffraction (RHEED), as shown in Fig. 3.1. RHEED is a very useful tool for *in situ* monitoring of MBE growth, particularly for growth of QD structures. A high energy (15 keV) electron beam is aimed at a glancing angle to the growth surface and a diffraction pattern is obtained that is characteristic of the surface. Under growth conditions suitable for In(Ga)As/GaAs growth the surface is always As terminated and the surface reconstruction observed depends on both the As overpressure and the substrate temperature. Under conditions typical for InAs QD growth, the surface reconstruction of the GaAs(001) surface prior to InAs deposition is likely to be $c(4 \times 4)$, as seen in Fig. 3.1a. Once InAs deposition commences the surface reconstruction rapidly assumes a (1×3) character typical of InAs surfaces, but the RHEED pattern is still streaky (Fig. 3.1b), as a consequence of the Laue diffraction condition. Once θ_{crit} is reached the RHEED pattern changes abruptly from the streaky pattern to a spotty pattern seen in Fig. 3.1c, indicating the increased surface roughness due to 3D island formation.

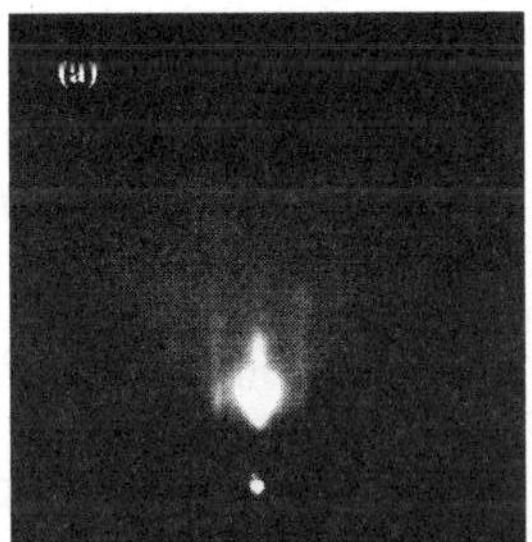

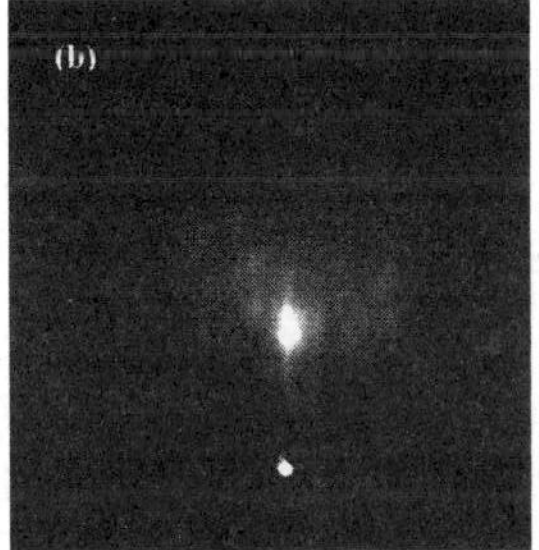

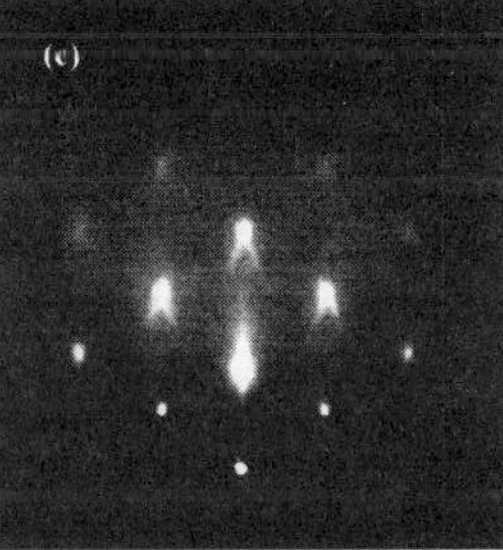

Figure 3.1   RHEED images obtained along the $[-110]$ direction prior to and during deposition of InAs on GaAs: (a) GaAs(001) $c(4 \times 4)$ surface reconstruction before InAs deposition, (b) RHEED pattern immediately after deposition of InAs commences, (c) RHEED pattern once 2D–3D transition is reached. Note the chevrons present on the central spots. RHEED images courtesy of Dr Tomasz Krzyzewski, Imperial College London.

The 3D island density quickly reaches saturation (after $\sim 1.8\,\mathrm{ML}$ [6]) and further deposition causes existing islands to ripen to an equilibrium size, and the initially broad island size distribution narrows significantly [7, 8]. As InAs deposition continues beyond θ_{crit}, the spots in the RHEED pattern become increasingly clear and chevrons may be observed (Fig. 3.1c), indicating well-defined facets on the 3D islands. From the RHEED patterns, the structure of the uncapped islands can be inferred, as discussed elsewhere in this book.

A $1\,\mu\mathrm{m} \times 1\,\mu\mathrm{m}$ atomic force microscopy (AFM) image of an uncapped InAs/GaAs QD ensemble is shown in Fig. 3.2. Some variation in the size of the islands can be observed, while their distribution on the surface is apparently random. The island density in Fig. 3.2 is $1.2 \times 10^{10}\,\mathrm{cm}^{-2}$, which is typical for InAs/GaAs QDs grown at a low InAs deposition rate [9]. For MBE growth, the QD density may be varied between 10^8 and $10^{11}\,\mathrm{cm}^{-2}$ ($1\,\mu\mathrm{m}^{-2}$ to $1000\,\mu\mathrm{m}^{-2}$), depending on conditions such as temperature, InAs growth rate, As pressure and even As species ($\mathrm{As_2}$ or $\mathrm{As_4}$).

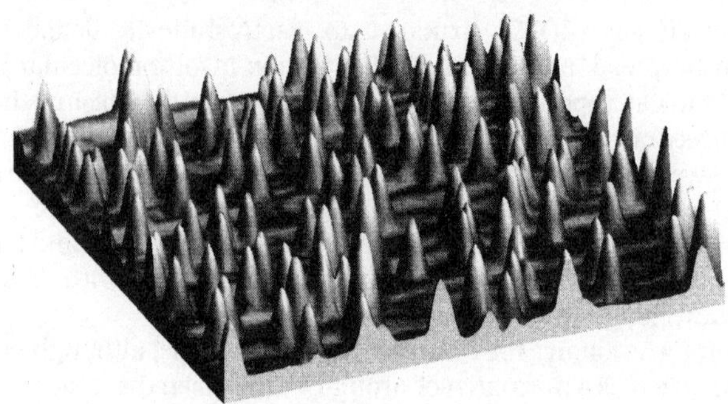

Figure 3.2   1 μm × 1 μm atomic force microscope (AFM) image of an uncapped InAs/GaAs QD ensemble, grown at low InAs deposition rate (0.014 MLs^{-1}). Note the variation in QD size and placement on the surface.

The size of fully mature (coherent) islands also depends on growth conditions but will typically be 3–10 nm in height and have a base diameter of a few tens of nm. Under certain growth conditions a bimodal size distribution with two distinct families of QDs may be obtained [10–12]. This bimodal QD distribution can be exploited for some applications in order to select a subset of QDs, as will be discussed later in this chapter. As more InAs is deposited, QDs will eventually coalesce and dislocate. Once dislocation has occurred, elastic strain is relieved and the dislocated islands will grow rapidly in comparison with coherent islands [4, 13].

In order for the QDs to be incorporated into useful device structures, subsequent overgrowth of material (capping) is required. As the InAs islands are capped, the spotty features in the RHEED pattern shown in Fig. 3.1c become less distinct and eventually the pattern recovers a streaky appearance associated with a nominally flat surface. At this point a GaAs-type surface reconstruction is recovered. Capping has a profound effect on the structural as well as optical properties of the QDs, and is discussed more fully in the chapter by Ulloa *et al*. The emission wavelength of the QDs is blueshifted due to increased compressive stress but the intensity of emission from the capped QDs is significantly enhanced as fewer carriers undergo non-radiative recombination at surface states [14]. The height of the InAs islands collapses during the capping process: uncapped islands of ~10 nm height reduce to only ~2 nm height after overgrowth of just 2 nm GaAs. Indium atoms are transferred from the QDs into the capping material to form an InGaAs alloy [15]. GaAs growth is initially more favourable on the wetting layer, which is still lattice matched to the underlying substrate, rather than on the tops of the InAs islands that are partially relaxed. This is particularly evident for larger QDs grown using low growth rates, where complete relaxation is observed [16]. These QDs do not collapse and retain a much higher In content once capped [17]. At lower growth temperatures (<480°C), Ga/In intermixing between the QDs and the capping layer is suppressed, but at the expense of increased defect formation due to poorer GaAs quality [18]. Scanning transmission electron microscopy (STEM) and energy dispersive X-ray analysis have questioned the traditional picture of Stranski–Krastanov growth resulting in 3D islands on top of the 2D wetting layer. These measurements indicate that capped QDs are inclusions of high In-fraction alloy located within a dilute InGaAs confining layer formed by intermixing between the wetting layer, QDs and the GaAs cap [19, 20]. Changes in the QD shape are observed during GaAs capping, with an elongation of the QD in the [1 $\bar{1}$0] direction [17].

With their smaller band gap, QDs capture and confine electrons and holes. Their small size means that the carrier wavefunctions strongly overlap, leading to efficient radiative recombination. Just as for QWs, several electron and hole states exist within each QD, each one giving rise to an optical transition depending on the parity of the states.

A low temperature (10 K) photoluminescence (PL) spectrum obtained from an InAs/GaAs QD ensemble is shown in Fig. 3.3. The ground state (GS), first excited state (X1) and second excited state (X2) of the QD ensemble are resolved in this PL spectrum, with an approximately equal spacing of 60 meV. The laser spot excites ~10^6 QDs, with each QD contributing a narrow emission line (small homogeneous linewidth) but due to the variation in size and composition of the

QDs as indicated in Fig. 3.2, the PL feature for each optical transition is inhomogeneously broadened to $\sim$28 meV. This inhomogeneous broadening can vary between 15 and 100 meV depending on growth conditions. The inset to Fig. 3.3 shows a spectrum obtained from the same sample using low excitation power. In this case only GS emission is observed, suggesting that carrier relaxation through the QD states is rapid compared with the radiative lifetime of X1 and X2.

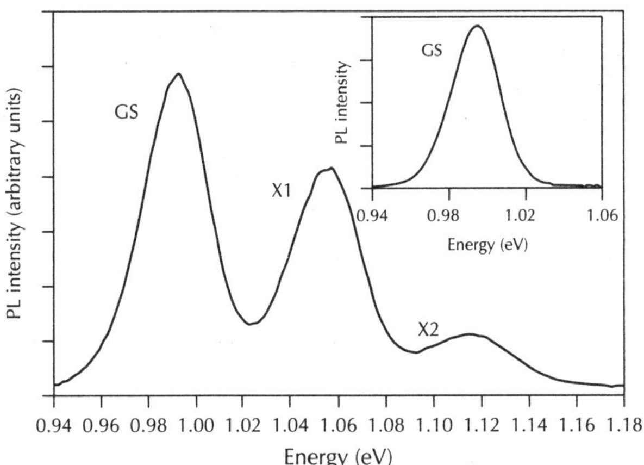

Figure 3.3   PL spectrum obtained at 10 K using high power excitation ($\sim$500 Wcm^{-2}) from an Ar$^+$ laser. Inset: 10 K PL spectrum from the same QD sample using low excitation power ($\sim$0.1 Wcm^{-2})

The 10 K GS emission wavelength of 1250 nm for the QD ensemble shown in Fig. 3.3 will shift beyond 1300 nm at room temperature. InAs/GaAs QDs have attracted attention because of the long emission wavelength that can be achieved by QDs in comparison to QWs. InGaAs/GaAs QW devices operate to wavelengths up to 1000 nm at room temperature, but extension of QW emission to long wavelengths is difficult due to the build-up of strain as the In content is increased. This explains the technological interest in InAs QDs where the formation of 3D islands relieves strain and the QDs can remain coherent and In rich. There are several growth recipes that can extend room temperature emission of InAs/GaAs QD devices to 1300 nm: submonolayer deposition with interrupts [21, 22], low growth rates [23], capping QDs with InGaAs rather than GaAs [24] or enclosing the QDs within an InGaAs QW (a dots-in-a-well or DWELL structure) [25]. More complicated growth techniques have extended emission to 1500 nm, by growth on metamorphic InGaAs buffers [26] or by growth of closely stacked layers (QD bilayers) [27]. Control of strain in QD structures is discussed in more depth in the chapter by Seravalli *et al.* Figure 3.4 shows room temperature PL spectra obtained from samples grown using two of these techniques, giving 1300 nm emission from a single InGaAs-capped InAs/GaAs QD layer and emission >1500 nm from an InGaAs-capped InAs/GaAs QD bilayer.

Extension of the emission wavelength allows GaAs-based devices to compete with the InP-based devices used for applications in these wavelength ranges, and also allows more simple growth and fabrication of long-wavelength vertical cavity structures such as resonant-cavity light-emitting diodes (RCLEDs) [28] and vertical-cavity surface-emitting lasers (VCSELs) [29], where monolithic growth of high-quality GaAs/Al(Ga)As distributed Bragg reflector (DBR) mirrors is possible. Figure 3.5 shows the room temperature reflectivity spectrum obtained from a cavity LED structure incorporating a layer of low-density QDs, suitable for a QD-based single photon source operating at low temperature at 1200 nm. The measured reflectivity (circles) agrees very well with the expected reflectivity from the cavity design (solid line), with a high reflectivity corresponding to the DBR stop bands and a sharp cavity resonance (the discrepancy in the depth of the cavity resonance between the calculated and experimental values is due to the resolution of the measurement), demonstrating the high control over growth conditions that is possible with MBE growth of GaAs-based vertical cavity structures. Room temperature GS emission from the QDs incorporated into the structure is at $\sim$1300 nm, which accounts for the discrepancy

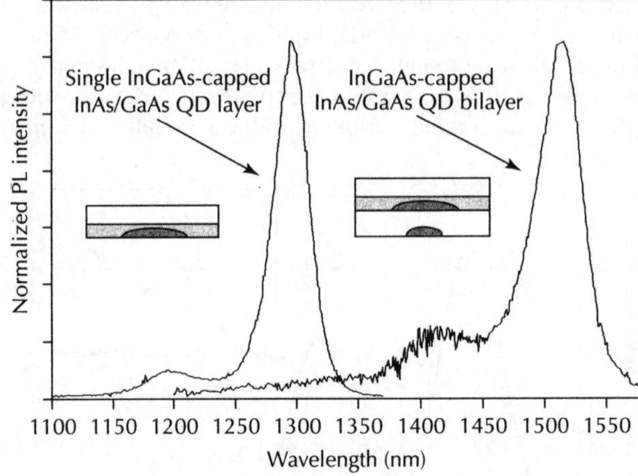

Figure 3.4   Room temperature PL spectra obtained from a single InGaAs-capped InAs/GaAs QD layer emitting at 1300 nm and an InGaAs-capped InAs/GaAs QD bilayer emitting at 1515 nm. Noise around 1400 nm (at the X1 transition) is due to water absorption.

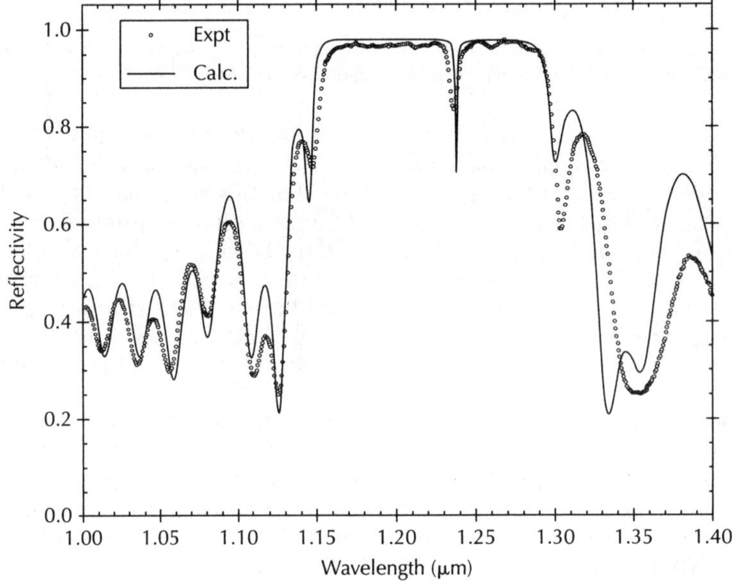

Figure 3.5   Room temperature reflectivity spectrum from a GaAs/AlAs cavity LED structure incorporating InAs/ GaAs QDs (circles), with the expected reflectivity from the cavity design (solid line). The structure is designed for low-temperature operation; the room temperature emission of the QDs is around 1300 nm. Structure design and calculated reflectivity spectrum courtesy of Dr Paul Stavrinou, Imperial College London.

between the experimental and calculated spectra at this point (the calculated spectrum considers a uniform GaAs cavity without QDs).

3.3   Stacked QD layers

Although high values for material gain have been reported for QDs [30, 31], their low surface coverage compared with bulk or QWs significantly limits the maximum gain or absorption available from a single QD layer, with obvious implications for many device structures. For example, in QD laser structures where lasing from the ground state is desired, the low gain from a single QD layer can be overcome in part through long cavities and high reflection coatings to reduce mirror losses, but the most common method is to incorporate several QD layers. However, for this strategy to be effective these layers must provide coincident gain. This is not straightforward since growth of subsequent QD layers may be affected by the underlying layers, either by strain fields or by changes in surface morphology resulting from the buried QDs. These effects place a limitation on the minimum separation between layers and therefore the maximum number of QD layers that can be accommodated within the active region of a laser. Transmission electron microscopy (TEM) studies [32] have shown that a strong vertical correlation of QD layers occurs for separations of 20 nm or less. It was clear from this investigation that the QDs in the upper layers are larger than those in the lower layers and this was confirmed by scanning tunnelling microscopy (STM) [33]. The increased size of QDs in subsequent layers of QD stacks has been attributed to strain-induced In migration causing preferential QD nucleation above buried QDs. This reduces the time taken to reach the critical thickness θ_{crit} for the 2D–3D transition [34] and deposition of the same amount of In will lead to larger QDs. In addition, In segregation from lower layers [35] and enhanced In/Ga intermixing during capping [36, 37] is also possible. All these factors can have a significant effect on the emission energy and details of both redshifts [38–40] and blueshifts [41–43] of emission compared with single layer structures have been reported.

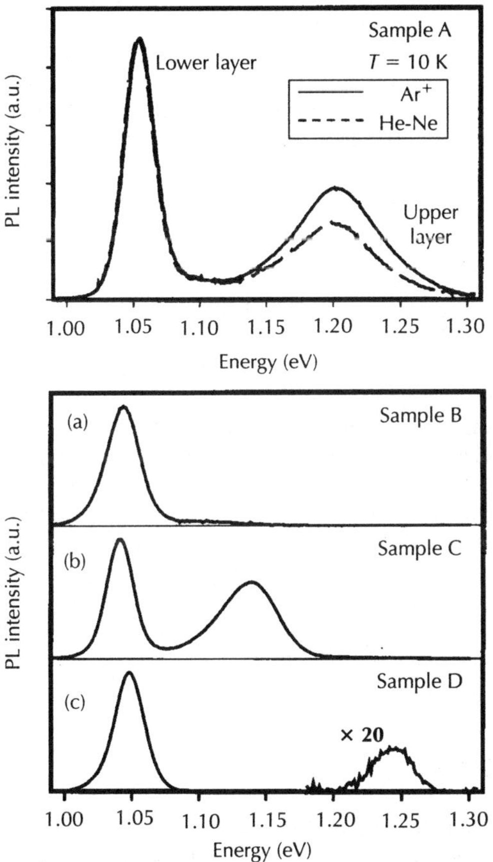

*Figure 3.6   Low temperature (10 K) PL spectra obtained from samples containing two InAs/GaAs QD layers, separated by GaAs spacer layers of 40 nm (samples A and B), 20 nm (sample C) and 10 nm (sample D). Reprinted with permission from Le Ru et al., J. Appl. Phys. **91**, 1365 (2002) [44]. Copyright 2002, American Institute of Physics.*

Figure 3.6 shows low temperature PL spectra obtained from samples containing two QD layers separated by GaAs spacer layers of different thickness [44]. The PL excitation intensity is low ($0.1\,\mathrm{Wcm^{-2}}$) so emission from excited states is negligible. Both QD layers were grown under the same conditions (by deposition of 2.3 ML InAs at a substrate temperature of 495°C) and are separated by 40 nm (samples A and B), 20 nm (sample C) or 10 nm GaAs (sample D). For samples A and D, the GaAs spacer layer was grown at 495°C, the same temperature used for QD growth, whereas the growth temperature for samples B and C was increased to 580°C after the first 10 nm GaAs. After the required GaAs thickness had been deposited the growth surface was annealed under an As flux for 10 minutes. This annealing step desorbs segregated In from the surface [45] and reduces surface undulations inferred from *in situ* reflection high-energy electron diffraction (RHEED) during growth. This was confirmed by subsequent AFM studies of annealed and unannealed GaAs surfaces [44]. The importance of the annealing step is illustrated in the comparison of PL from samples A and B. For QDs grown on the unannealed surface, emission is blueshifted, despite the large separation between layers. The higher energy peak at 1.20 eV can be unambiguously assigned to emission from the second QD layer by excitation with either an $\mathrm{Ar^+}$ laser or an HeNe laser. The latter source penetrates deeper into the sample and generates a greater proportion of emission from the lower QD layer. In contrast, only a single peak is observed from sample B, where the annealing step has successfully recovered a flat GaAs surface. As the separation between the QD layers is reduced to 20 nm (sample C), strain from the first QD layer affects growth of the second layer and emission is blueshifted, despite desorption of surface In and smoothing of the GaAs surface by annealing. By reducing the separation between layers to 10 nm (sample D), the two QD layers become electronically coupled and strong emission is detected only from the QDs in the first layer, due to carrier redistribution to the lowest energy QD states [38, 39]. For multiple QD-layer laser structures it is crucial to have the gain peak of all layers coincident. A conclusion gained from the annealing study is that a spacer separation of at least 40 nm is required for coincident emission. In order to determine whether the integrated emission scales with the number of QD layers, two samples were compared; one consisting of a single QD layer, the other containing three QD layers separated by 40 nm (annealed) GaAs spacers. In order to avoid problems associated with carrier diffusion, a metal mask with apertures 60–210 μm in diameter was deposited on the surface of each sample. A PL study then showed unequivocally that the PL intensity obtained from the multiple QD layer sample was indeed three times that observed from the single layer [44]. Thus, by following this recipe, coincident gain from all layers is assured.

Electronically coupled QDs have attracted attention for structures designed to manipulate either carrier charge or spin as qubits for quantum information applications and this is covered more fully elsewhere in this book. For these applications it is desirable for the coupled QDs to be identical; this is a considerable challenge for the reasons described above. In order to improve the uniformity of QDs in a multiple layer structure, the so-called "indium flush" growth technique can be used [45, 46]. QDs in each layer are partially capped by a thin layer of a few nm GaAs before growth is interrupted and the substrate temperature is raised to desorb surface indium before growth of the remaining GaAs spacer layer. This has the effect of truncating the QDs in each layer so that they have the same height (although there may still be compositional variation between the QDs). In conclusion, self-assembled QDs require considerable growth expertise to extract desirable optical properties for use in devices.

3.4 Energy states in QDs

The optical properties of a solid are dictated by the electronic structure; specifically the density of states (DOS). As the dimensionality is reduced, the resulting modification to the density of states (DOS) is responsible for many of the improvements in the optical properties of QDs, including higher material and differential gain, lower threshold current densities (J_{th}) in laser structures [30] and even temperature insensitive operation (corresponding to a characteristic temperature $T_0 = \infty$) [47].

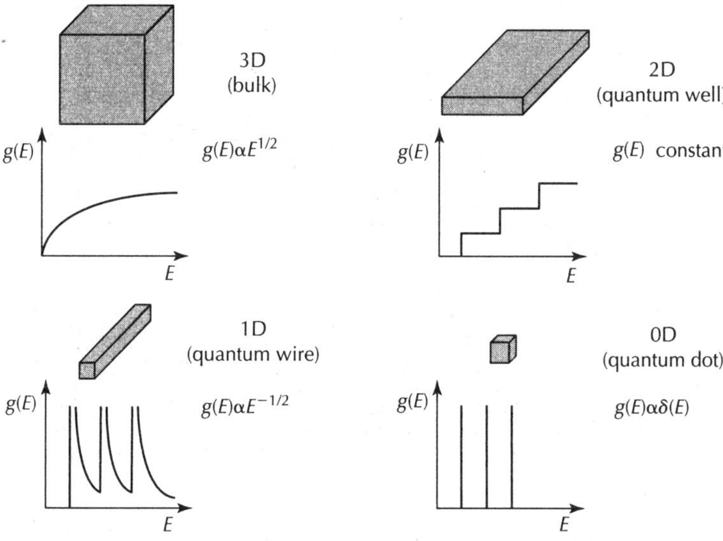

Figure 3.7   Density of states (DOS) functions g(E) *in quantum-confined systems.*

Figure 3.7 shows the conduction band DOS functions for systems with increasing confinement. For bulk systems, the familiar $g(E) \alpha E^{1/2}$ relationship is observed but for QWs, $g(E)$ is constant for each subband; for quantum wires with 2D confinement $g(E)$ is proportional to $E^{-1/2}$ and for 3D confinement a δ-function DOS is obtained. The most significant aspect of this DOS modification is that an increased number of states are available near the band edge as confinement is increased. The influence of the DOS is further illustrated by considering the probability of an optical transition between two states as given by Fermi's golden rule:

$$W_{i \to f} = \frac{2\pi}{\hbar} \left| M_{if} \right|^2 g(E)$$

where $W_{i \to f}$ is the transition rate between initial and final states and M_{if} is the optical matrix element. Modification of the DOS due to confinement will increase the transition rate at the band edge and the increased electron–hole overlap due to the spatial localization of carriers will increase M_{if}. The increased overlap will enhance the Coulomb interactions between the electron and hole, forming excitons.

There have been many attempts at modelling the electronic states of QDs in the last ten years, using a variety of methods. These include effective mass approximations [48], 8-band $\mathbf{k} \cdot \mathbf{p}$ theory [49–51] and atomistic pseudopotential calculations [52]. These approaches rely on detailed knowledge of the size, shape, composition and strain state of the QD. Numerous experimental approaches have been used to determine these parameters, including STM [53, 54], TEM [55–57] and X-ray diffraction [58], but it remains exceedingly difficult to completely determine the QD structure.

Early effective mass models for QDs with a shallow confining potential predicted a single bound electron state and multiple hole levels [48, 59, 60]. These models were applied to optical emission spectra obtained from QD ensembles that exhibited short wavelength emission and a large inhomogeneous broadening which masked the presence of excited state transitions. However, with improvements in growth methods, QD samples with reduced inhomogeneous broadening were realized and excited state emission was clearly observed [61] (see Figs 3.3 and 3.4). Itskevich *et al.* showed that excited state transitions occur between electron and hole levels with the same quantum number (E_0–H_0, E_1–H_1, etc.) [62], demonstrating that parity is a good quantum number. Figure 3.8 shows the results of applying pressure to a high-quality QD sample. With increasing pressure there is a crossover between the conduction band Γ and X valleys and the optical transition becomes indirect. If only one electron state (E_0) is involved, PL emission from all the states

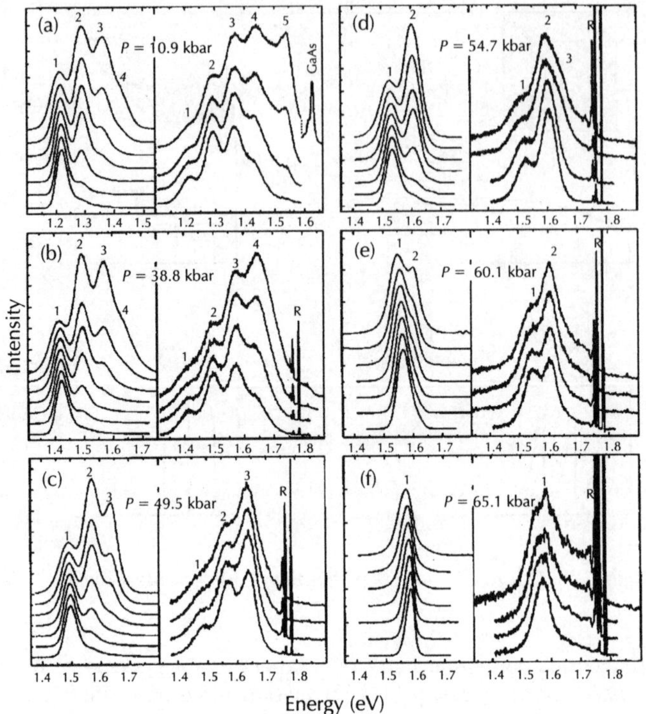

*Figure 3.8   PL spectra obtained from an InAs/GaAs QD sample at a range of excitation conditions (left panels: 10–300 Wcm^{-2} cw, right panels: 20–200 Wcm^{-2} pulsed) for increasing pressure, showing the effect of pressure on QD states, labelled 1–5. Reprinted with permission from Itskevich et al., Phys. Rev. B **60**, R2185 (1999) [62], Copyright 1999 by the American Physical Society.*

would quench simultaneously. However, the results showed that increasing pressure quenched the optical transitions sequentially, showing that more than one electron state is present.

Use of multiple band **k · p** or pseudopotential models allows a more rigorous simulation of the QD energy levels, particularly regarding the strain state of the QD [49–52]. These approaches predict the existence of multiple confined electron levels, consistent with excited state emission resulting from transitions between states of the same quantum number. They also lead to a reduction of the error in the estimated excitonic band gap, which is significantly overestimated in some effective mass models [63]. Confinement lifts the heavy-hole/light-hole degeneracy with the hole ground state having predominantly heavy-hole character, although excited states may be more mixed [64] and experimental evidence for the presence of pure light hole states in QDs is inconclusive [65, 66].

For these calculations structural (compositional) parameters must be assumed. Most models consider a pyramidal QD of uniform In(Ga)As composition. Due to their increased effective mass, hole wavefunctions are then expected to be confined towards the apex of the pyramidal QD, with the electron wavefunction located towards the base. This would reduce the electron–hole overlap in the QD and so diminish the optical transition strength but would also result in a permanent dipole in the QD. Measurements of the photocurrent with increasing applied bias for QDs in p-i-n or n-i-p structures, shown in Fig. 3.9, did indeed demonstrate the existence of a dipole, but this was in the opposite direction to that expected for QDs of uniform composition, suggesting instead that the hole is confined towards the base of the QD and the electron towards its apex [67]. This suggests of a non-uniform InGaAs composition within the QD, with increasing In content towards the top of the QD. This is borne out by STM [53] and X-ray diffraction measurements [58]. The composition gradient in QDs may be due to both indiffusion of Ga from the substrate during QD growth [68] and In segregation [69]. Eight-band **k · p** and pseudopotential calculations of InAs/GaAs QDs

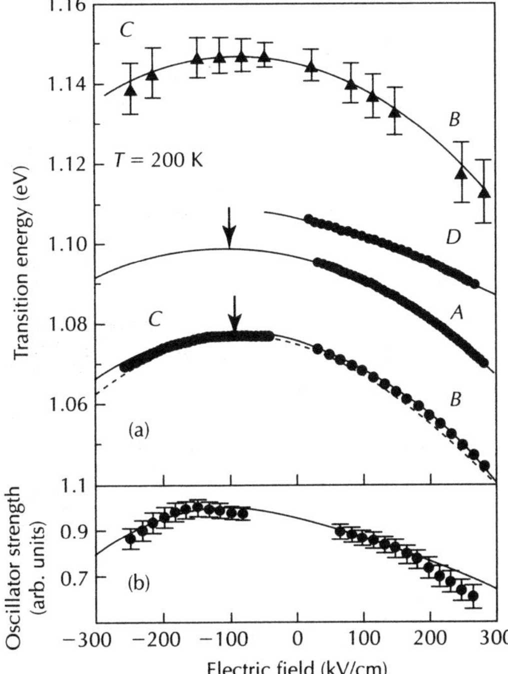

Figure 3.9   (a) Dependence of the peak photocurrent from the ground state (circles) and 1st excited state (triangles) of QDs incorporated into p-i-n or n-i-p structures on applied electric field, demonstrating the Stark shift of QD emission. The peak emission shift occurs for a negative applied field, indicating the orientation of the dipole moment. (b) Oscillator strength for the ground state transition of the QDs shown in curve BC in (a). Reprinted with permission from Fry et al., Phys. Rev. Lett. 84, 733 (2000) [67]. Copyright 2000 by the American Physical Society.

[70, 71] and InGaAs/GaAs QDs [72] which takes into account an increasing In content from the bottom to the top of the QD do replicate the observed dipole orientation, with the hole wavefunction confined at the base of the QD and the electron wavefunction confined towards the apex. These calculations also demonstrate the sensitivity of the exciton emission energy to changes in In content, with an increase in average In content of the QD from 50 to 60% resulting in a redshift of around 70 meV, while changes in the profile of the In distribution of the QD redshift the emission by 20–40 meV [72]. From studies of the Stark shift on emission from single InGaAs QDs, the magnitude of the In:Ga composition gradient was found to be dependent on the In content of the QD [73].

However, to a good first approximation, QDs can be represented by a parabolic potential [74], regardless of the details of the QD size, shape, composition and strain state. This is consistent with the equal energy spacing between states seen in QD PL spectra (for example, in Fig. 3.3 and in [75]) and by capacitance–voltage and far-infrared absorption measurements of InAs/GaAs QDs determining the charging energies of the QD states [76]. For In(Ga)As/GaAs QDs, the QD height is significantly smaller than the QD diameter (especially once capped). By comparison with atomic systems, we would expect the electron ground state in a QD to be s-like with a degeneracy of 2 (due to spin), and the first excited state to be p-like with a degeneracy of 4. This has been demonstrated in transmission experiments on InAs/GaAs QDs in a field-effect structure [77], in which additional electrons were controllably added to the QDs (as shown in Fig. 3.10). Absorption from the ground state of the QDs is quenched for an electron occupancy $N_e = 2$ electrons per QD, corresponding to a filled ground state. Absorption from the first excited state is considerably reduced at $N_e \sim 4$ and completely quenched by $N_e \sim 6$.

For PL emission resulting from interband recombination of electrons and holes, the energy separation between the ground state (GS) and first excited state (X1) PL emission represents the sum of the electron level separation and the hole level separation. Typical GS–X1 separations

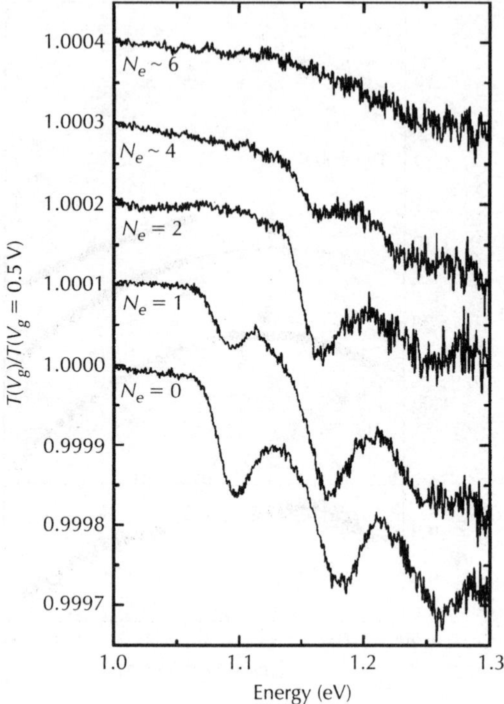

Figure 3.10   Transmission spectra obtained from InAs/GaAs QDs in a field-effect structure with increasing electron occupancy (N_e) in the QDs, showing quenching of absorption as the QD states are filled. Reprinted with permission from Warburton et al., Phys. Rev. Lett. **79**, *5282 (1997) [77], Copyright 1997 by the American Physical Society.*

in PL spectra are $\sim$50–70 meV, although this may be increased by variation of the QD growth conditions [78–80]. The intraband separation between states in either the conduction band or the valence band is not straightforward to determine experimentally, but experiments using far-infrared (intraband) absorption to modulate near-infrared (interband) emission suggest typical electron level separations of 45–55 meV and hole level separations of 10–15 meV [81, 82], as expected considering the greater effective mass for holes.

The traditional picture of carrier capture and relaxation processes in QDs is shown schematically in Fig. 3.11. For most device applications, carriers are either electrically injected or optically excited into the GaAs matrix surrounding the QDs (1). Carriers are then captured via the wetting layer (WL) (2) into the QDs (3) and then rapidly relax to the ground state (4). Carrier capture and relaxation occurs on the order of ps. Once carriers reach the ground state (E_0), radiative recombination may take place (5). Typical radiative lifetimes for carriers in QDs are around 1 ns. If excitation or carrier injection is sufficient for multiple carriers to be captured by each QD, state filling can occur and radiative recombination may be observed from excited states. The discrete energy levels restrict the number of energy states available for carriers in a QD, and this restriction was predicted to severely inhibit carrier relaxation in QDs, since LO-phonon energies do not match the energy separation between QD states (the so-called phonon bottleneck) [83, 84]. This would have a major impact on device characteristics, for example gain recovery in lasers and semiconductor optical amplifiers. However, as mentioned above, carrier capture and relaxation in QDs is on the ps timescale [85, 86]. In order to circumvent the phonon bottleneck, efficient carrier relaxation mechanisms via multiphonon processes [87] and by carrier–carrier scattering [88, 89] have been proposed. Also the existence of 2D-like continuum states extending through the ladder of QD states to the ground state has been suggested [90] following photoluminescence excitation (PLE) measurements. As shown in Fig. 3.12, PLE spectra obtained from both QD ensembles (Fig. 3.12a) and from a single InGaAs/GaAs QD (Fig. 3.12b) show a rising

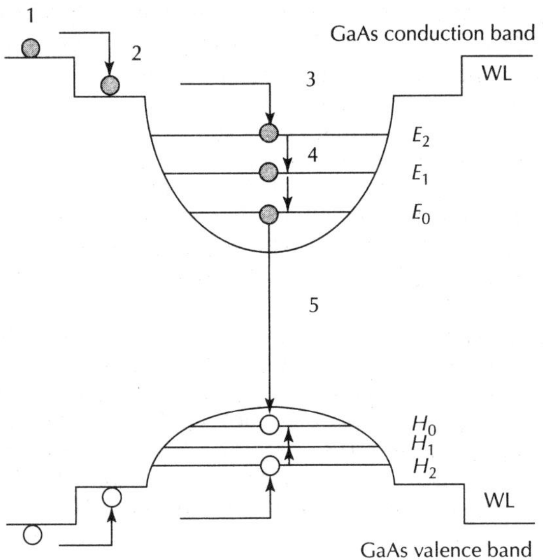

Figure 3.11   *Schematic diagram illustrating carrier capture from GaAs barrier into a QD via the wetting layer and carrier relaxation within the QD.*

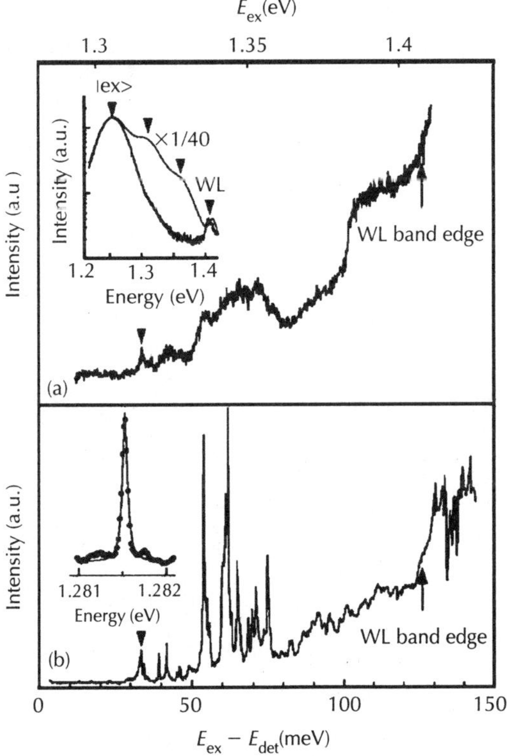

Figure 3.12   *(a) Far-field PLE spectrum obtained from an InGaAs/GaAs QD ensemble with a detection energy $E_{det} = 1.282\,eV$. The inset shows PL spectra obtained at low and high excitation powers from the same sample with an excitation energy $E_{ex} = 1.959\,eV$. (b) Near-field PLE spectrum obtained from a single InGaAs/GaAs QD; the inset shows a PL spectrum obtained from the QD. Emission peaks due to an excitation energy $E_{ex} - E_{det} \sim 30$ and $65\,meV$ are attributed to Raman scattering. Reprinted with permission from Toda et al., Phys. Rev. Lett. **82**, 4114 (1999) [90], Copyright 1999 by the American Physical Society.*

background extending from an excitation energy ~50 meV above the detection energy up to the WL band edge due to these continuum states. Coupling between the continuum states and QD states may provide an efficient carrier relaxation mechanism, as carriers can easily relax through the continuum states and then transfer to the QD ground state.

Consensus has not been reached regarding the origin of the continuum and its possible role in QD carrier relaxation. It has been attributed to a band tail of deep states due to carrier localization at the rough interfaces of the WL [91] or to be a result of crossed transitions between WL valence band states and bound electron states in the QD [92, 93]. Coupling of the QD ground state to the continuum states via LO-phonon emission provides a mechanism for carrier relaxation directly to the QD ground state rather than through the excited states, thus circumventing the phonon bottleneck [94, 95].

Alternatively, carrier relaxation in QDs may be influenced by the formation of polarons by strong coupling of electrons [96–98] (or excitons [99]) to LO-phonons. The polaron lifetime (and thus the carrier relaxation) is expected to be determined by the lifetime of the LO-phonon component [98, 100, 101]. Due to phonon anharmonicity, the LO-phonon component of the polaron can relax by emission of two phonons [102]. The two-phonon emission can be of sufficient energy to match the separation between the electron states, allowing the electron to relax between the states and circumventing the phonon bottleneck. However, polaron effects have only been seen in samples where the electron–hole (Auger) relaxation method is not present, for example in intraband measurements on *n*-doped QDs. If electrons and holes are present, Auger scattering provides an efficient mechanism for electron relaxation and the polaron mechanism is not required [103]. For most devices, with high electron and hole densities, Auger scattering is by far the most likely process for carrier relaxation.

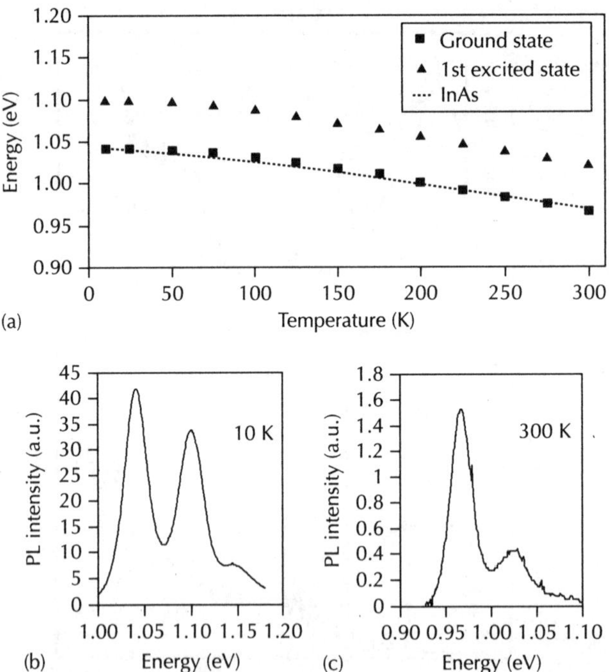

Figure 3.13   (a) Temperature dependence of the GS and X1 peak emission from a sample containing a low density (1.5 × 10⁹ cm⁻²) InAs/GaAs QD layer, with 10 K (b) and 300 K (c) spectra obtained from the sample. Also shown in (a) is the temperature dependence of the InAs band gap energy, calculated from Varshni relations and then shifted to match the QD GS transition at low temperature [104].

The operating temperature of QD-based devices can vary significantly depending on the application. Conventional optical telecommunication devices such as lasers and SOAs are expected to operate around room temperature or above (typically 20–80°C). Infrared photodetectors may be expected to work with liquid nitrogen cooling at 77 K and novel devices such as QD single photon sources have been demonstrated at low temperatures. However, the optical properties of QDs are very different in these regimes, due to temperature dependent carrier capture and escape mechanisms and the interplay between radiative and non-radiative processes.

Figure 3.13a shows the temperature variation of the peak PL emission energy of the ground state and first excited state of an InAs/GaAs QD sample. The sample contains a single QD layer with a very low QD density of $1.5 \times 10^9 \, \mathrm{cm}^{-2}$. Thus significant excited state emission is observed even for moderate PL excitation power over the temperature range 10–300 K. As the temperature is increased, the peak emission energy of both states redshifts, following a similar trend to that of the InAs band gap [105–107]. In some cases, especially for samples exhibiting a large inhomogeneous broadening or a bimodal distribution of QDs, an increased redshift is observed for the QD ensemble peak in the mid-temperature range (100–200 K), following a sigmoidal variation with temperature [108, 109]. Figure 3.13 also shows 10 K (panel b) and 300 K (panel c) PL spectra obtained from the sample. At 300 K, the PL intensity is reduced with respect to the low temperature value by a factor of 20 due to the low QD density where carrier recapture is less likely. A reduction in the PL intensity by a factor of ~6 is seen in samples with higher QD density. Note that the degree of state filling, as shown by the ratio of the ground state to first excited state emission intensity, is also reduced due to thermal escape of carriers from the QDs.

The temperature dependence of PL intensity from In(Ga)As/GaAs QDs is best described using an Arrhenius plot, as shown in Fig. 3.14, showing the integrated PL intensity (plotted on a logarithmic scale) at various temperatures between 70 and 335 K (plotted as $1/T$ or $1000/T$ for reasons outlined below) for two InAs/GaAs QD samples, A and B, obtained from the same epitaxial wafer. Sample A contains as-grown InAs/GaAs QDs whereas sample B has been subjected to post-growth rapid thermal annealing at 750°C for 10 s, which causes interdiffusion of In and Ga in the QDs, blueshifting their emission energy and thus reducing the potential barrier height for carrier escape to the WL or GaAs matrix [110]. Note that the QD density in both samples is the same. The PL spectra were obtained using a low excitation density of $\sim 4 \, \mathrm{Wcm}^{-2}$, and the contribution of excited state emission is small.

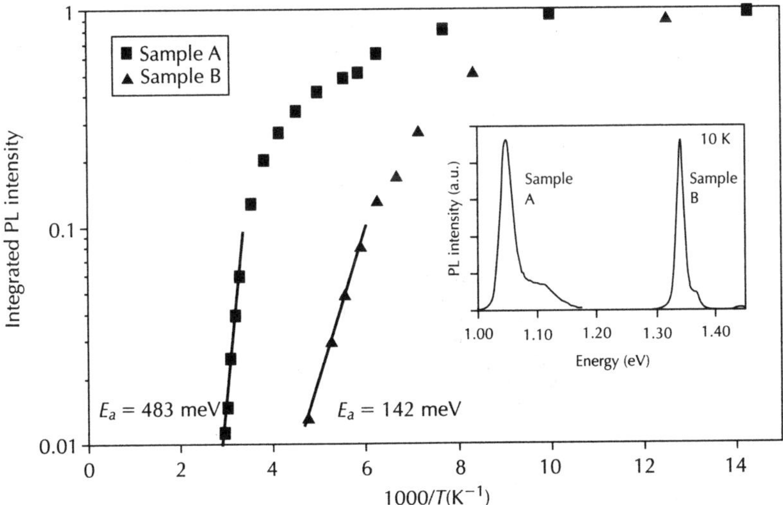

Figure 3.14   Arrhenius plot of the temperature dependence of PL intensity from InAs/GaAs QD samples with different confining potentials. Inset: 10 K PL spectra obtained from the two samples.

Three regimes can be identified in the Arrhenius plots: (1) at low temperatures (<100 K for the QDs presented in Fig. 3.14) the integrated PL intensity from the QDs remains constant, (2) at high temperatures the PL intensity is strongly quenched with increasing temperature, following a dependence of the form:

$$I = C \exp\left(E_a / k_B T\right)$$

where E_a is an activation energy for carrier escape from the QDs, and (3) at intermediate temperatures the PL intensity starts to drop but does not immediately follow the exponential quenching behaviour. This mid-temperature range coincides with the sigmoidal variation of the PL emission energy mentioned above and also with a narrowing of the PL linewidth that is frequently observed [109, 111–113]. These effects are attributed to a redistribution of carriers from small QDs on the high-energy side of the QD spectral distribution to larger, low-energy QDs, due to preferential thermal escape from the small QDs with lower potential barrier height.

From the gradient of the Arrhenius plot at high temperatures, the activation energy E_a can be determined, which is in good agreement with the difference between the GaAs band gap and the QD ground state energy ($E_a = E_{GaAs} - E_{QD}$) [114]. The quenching of the PL is attributed to non-radiative recombination of carriers in the GaAs barrier [115].

The temperature dependence of PL intensity (normalized to have the same integrated intensity at 10 K) from an InAs/GaAs QD sample at different excitation powers is shown in Fig. 3.14 [114]. The excitation density is varied between $25\,\mathrm{mWcm}^{-2}$ and $4\,\mathrm{Wcm}^{-2}$, and is low enough such that excited state emission can be neglected. From the gradients of the fit lines at high temperatures, a common activation energy $E_a = 270\,\mathrm{meV}$ is obtained for all excitation powers, as expected. However, as the excitation density is reduced, quenching of the PL intensity occurs at lower temperatures. This suggests that at higher temperatures ($>150\,\mathrm{K}$), the regime where carrier escape is observed, the excitation dependence of the PL intensity becomes superlinear. The inset to Fig. 3.15 shows that C follows a quadratic dependence with the PL intensity $I(P)$. If the excitation density is further increased so that the average carrier occupancy of the QDs is greater than one electron–hole pair per QD, the dependence changes to a linear one: $I(P) \alpha P$ [114]. This behaviour is attributed to the independent capture and escape of electrons and holes in QDs: for low excitation, with an average QD occupancy of less than one electron–hole pair per QD, the probability of capture and one electron *and* one hole by a QD (necessary for PL emission) will depend quadratically on the excitation power. This illustrates the importance of the statistical nature of carrier capture and escape in QD ensembles: the carrier distribution across the ensemble

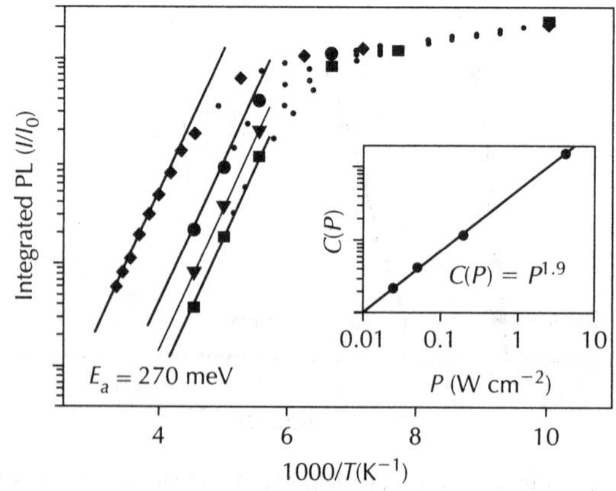

*Figure 3.15   Arrhenius plots of the temperature dependence of PL intensity (normalized at 10 K) from an InAs/ GaAs QD sample for a range of excitation levels: $25\,mWcm^{-2}$ (squares), $50\,mWcm^{-2}$ (triangles), $0.2\,Wcm^{-2}$ (circles) and $4\,Wcm^{-2}$ (diamonds). Fit lines at high temperature are used to extract the activation energy, E_a, and coefficient of the exponential, C. Inset: Dependence of C on excitation power. Reprinted with permission from Le Ru et al., Phys. Rev. B **67**, 245318 (2003) [114]. Copyright 2003 by the American Physical Society.*

will not be uniform and some QDs may contain multiple carriers while others may be unoccupied. It is important to consider the random population of QDs when formulating models of QD carrier dynamics, rather than assuming a uniform population of carriers across the ensemble as in conventional rate equation approaches, particularly when excited state population is important [116].

3.5 Single QD spectroscopy

Emission from QD ensembles is dominated by inhomogeneous broadening due to variations in QD size and composition. Under typical excitation conditions (e.g. laser excitation with a $100\,\mu m$ spot size on a QD sample with a QD density of $10^{10}\,cm^{-2}$), PL experiments will probe around 10^6 QDs. In order to isolate emission from single QDs, control of growth conditions to reduce QD density, processing of the QD sample for example by defining small apertures in a metal mask or etching narrow-diameter mesas in the sample, and specialized (usually low temperature) measurement is required. Far-field imaging, using a confocal microscope, typically gives a spatial resolution of $\sim 1\,\mu m$, which is suitable for spectroscopy of apertures or mesa structures. For greater resolution near-field scanning optical microscopy (NSOM) can be used, to obtain a spatial resolution as low as $50\,nm$ [117]. The resulting spectra show a series of very narrow lines associated with each QD [118, 119], with linewidths of only a few μeV. By using power-dependent PL and excitation that is either resonant or non-resonant with the QD energy levels, spectral lines can be attributed to single exciton, charged exciton and biexciton complexes in the QD [120, 121]. Due to Coulomb interactions, the emission energy of the biexciton may be shifted with respect to the exciton emission by the biexciton binding energy. Values reported for the biexciton binding energy in In(Ga)As/GaAs QDs vary but are on the order of a few meV. The biexciton binding energy may even be negative, suggesting a repulsive Coulomb interaction between the two excitons (though still stable due to the confinement provided by the QD), such that the biexciton emission energy is greater than the exciton emission energy [122, 123].

Many models assume a high degree of symmetry of QDs, for example D_{2d} or C_{4v} group symmetry (for example, pyramidal QDs). In practice many QDs are of lower symmetry, often due to elongation in the $[1\bar{1}0]$ direction. This has implications for the electron and hole wavefunctions within the QD. One manifestation of this is in the fine structure splitting of exciton emission that is observed from many QDs [124]. The fine structure results from a lifting of the degeneracy of the spin-up and spin-down bright exciton states ($m = \pm 1$) due to the electron–hole exchange interaction, which mixes these radiative states into linear combinations:

$$|-1\rangle, |+1\rangle \rightarrow \frac{1}{\sqrt{2}}\left(|+1\rangle + |-1\rangle\right), \frac{1}{\sqrt{2}}\left(|+1\rangle - |-1\rangle\right).$$

This can be observed by polarization-resolved PL obtained from single In(Ga)As QDs, as shown in Fig. 3.16b [124]. Due to radiative selection rules that are discussed in more detail later in this chapter, recombination involving the $m = \pm 1$ excitons results in emission of circularly polarized light. When the states are split into the linear combinations of the $m = \pm 1$ states by the exchange interaction as above, a linearly polarized doublet aligned along the in-plane crystallographic axes of the QD will be observed, as seen in the lower panel of Fig. 3.16b (for zero applied magnetic field), separated in energy by the fine structure splitting δ_1. The exchange interaction is enhanced by asymmetry [125] and also by carrier confinement [126]. In QDs this is particularly significant due to the elongation of the QDs and the 3D confinement leading to a large electron–hole wavefunction overlap [124, 127].

The fine structure splitting of the exciton emission from InGaAs/GaAs QDs is typically on the order of tens of μeV [128]. If a magnetic field is applied in the growth direction of the QDs, then the exciton states undergo Zeeman splitting and for sufficiently large applied fields the Zeeman splitting will exceed the exchange splitting, de-mixing the states and circularly polarized emission is then observed, as shown in Fig. 3.16b. As will be discussed later in this chapter, the fine

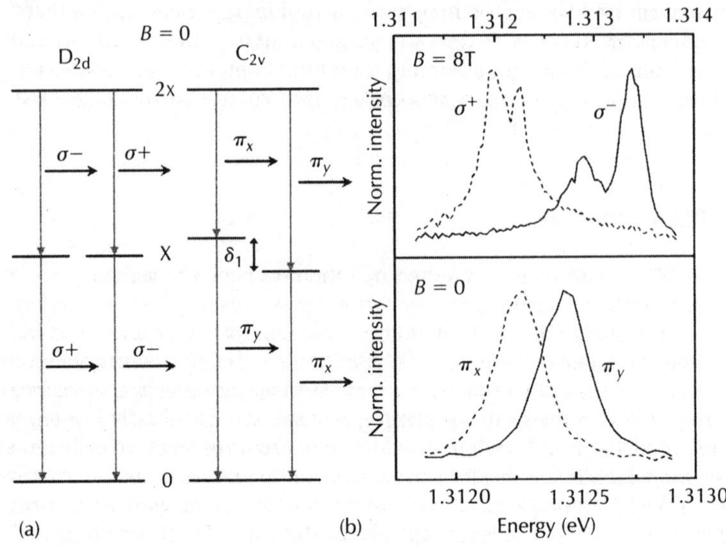

*Figure 3.16   (a) Schematic diagram of exciton transitions in symmetric QD (group symmetry D_{2d}) with no fine structure splitting or an elongated QD with reduced symmetry (C_{2v}) with fine structure splitting δ_1. (b) Polarization-resolved spectra obtained from a single InGaAs/GaAs QD with reduced symmetry. At zero applied magnetic field the exciton emission is split into two linear polarizations; with an 8 T applied magnetic field, circular polarization is observed. (b) Reprinted with permission from Bayer et al., Phys. Rev. Lett. **82**, 1748 (1999) [124], Copyright 1999 by the American Physical Society.*

structure and exchange interaction are particularly important for entangled photon sources and as a spin dephasing mechanism.

The homogeneous broadening of QD emission is important not only in its relation to carrier dynamics within the QD but may also directly affect the performance of QD-based devices, since many of their predicted advantages assume a delta-function like DOS with negligible homogeneous broadening of the states. For many devices it is the inhomogeneous broadening of emission due to variation in QD size, strain and composition that most influences device characteristics but as growth methods are improved, highly uniform QD ensembles can be realized with significantly reduced emission linewidth, so the homogeneous broadening may become a substantial component of the overall linewidth. Single QD layers with a room temperature PL linewidth of 19 meV have been demonstrated (with an associated variation in QD height of $\sim$5% from AFM measurements) [129] and bilayer QD ensembles with exceptionally narrow PL linewidths of $\sim$10 meV at 10 K (17.5 meV at 300 K) [130] have been reported.

The homogeneous linewidth (γ) of the GS transition from In(Ga)As/GaAs QDs has been determined by a variety of experimental techniques and values of γ can vary significantly according to temperature and excitation levels. The measured GS radiative lifetime of $\sim$1 ns implies a lifetime broadening in the μeV range, but early studies directly measuring single QD PL emission reported low temperature ($<$10 K) PL linewidths of tens of μeV [117, 119, 131]. Subsequent high-resolution single QD spectroscopy indicated a very narrow low temperature (2 K) PL linewidth of 3.4 μeV, corresponding to $\gamma = 1.8\,\mu$eV (equivalent to a dephasing time $T_2 \sim 730$ ps) once the experimental spectral resolution was deconvolved [132]. As the temperature is increased, the single QD PL linewidth significantly increases, with NSOM experiments on single InGaAs QDs reporting a homogeneous linewidth of 12 meV at 300 K. Figure 3.17 shows PL spectra obtained from single QDs over a temperature range from 2 to 300 K for two different classes of QDs; those with a shallow potential (type 1) with two confined electron shells (six electron states) or a deeper potential (type 2) with three confined electron shells (12 electron states) [132]. The behaviour of both types is similar: below 100 K, sharp emission lines are observed ($<$100 μeV) but significant broadening to 3–5 meV occurs at higher temperatures. As the temperature is increased, the single QD emission lineshape also changes. At low temperature, the lineshape is well approximated

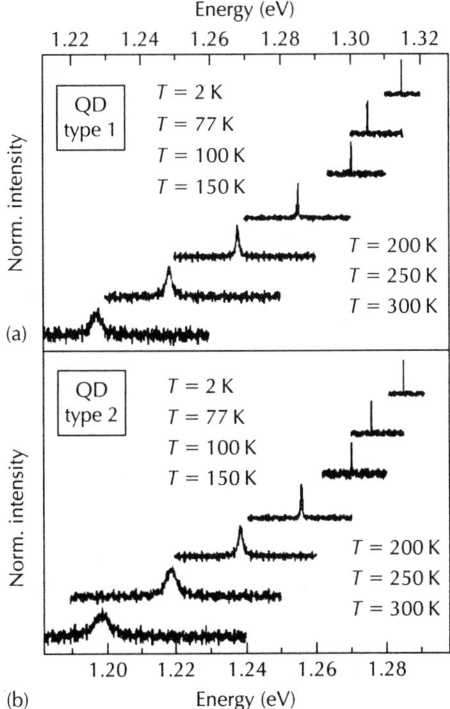

*Figure 3.17   PL spectra obtained from single InGaAs/GaAs QDs over a temperature range of 2–300 K for QDs with a shallow confinement potential (type 1) and deeper confinement potential (type 2). Reprinted with permission from Bayer and Forchel, Phys. Rev. B **65**, 041308 (2002) [132]. Copyright 2002 by the American Physical Society.*

by a Lorentzian, but at higher temperatures (>50 K) the lineshape has two components: a central Lorentzian known as the zero-phonon line (ZPL) and sidebands associated with acoustic phonon interactions. The sidebands rapidly dominate the overall linewidth of the emission as the temperature is further increased [133].

The homogeneous broadening can also be determined indirectly from an analysis of the longitudinal mode grouping in QD laser spectra, spectral hole burning (SHB) or four-wave mixing (FWM) in SOA structures. The latter two techniques can be used to determine the polarization decay characterized by the dephasing time T_2, which is related to the homogeneous broadening according to:

$$\gamma = \frac{2\hbar}{T_2}.$$

These methods for determining the homogeneous broadening involve the QD ensemble rather than individual QDs and are performed under much higher excitation conditions than that used for single QD spectroscopy, so they more closely match the operating conditions of many devices such as QD lasers, SOAs or saturable absorbers. SHB may be observed during pump–probe experiments, where the pump pulse saturation of the QD absorption in a spectral region around the pump energy is detected in the absorption or transmission spectrum of the probe. For a narrow pump pulse of spectral width $W \ll \gamma/\hbar$, the spectral hole will have a width of 2γ. Using this method, Borri *et al.* [134] have observed spectral hole burning corresponding to $\gamma \sim 3$ meV ($T_2 = 290$ fs) at room temperature and for low excitation densities. Similar values for the exciton dephasing time (and thus the homogeneous broadening) under the same conditions were also obtained by this study using transient FWM. This approach has been extended to determine the dephasing time over a range of temperatures [135, 136], which show significantly increased dephasing times at low (5–7 K) temperatures of several hundred picoseconds, corresponding to a homogeneous broadening of as low as 2 μeV [135], which is limited by the exciton lifetime. Such long dephasing times indicate that QD exciton coherence is maintained over sufficiently long timescales, which is essential for quantum information applications.

Although their emission is usually dominated by inhomogeneous broadening, the homogeneous linewidth can influence the operation of devices such as QD lasers, for example in the mode profile of laser emission spectra [137]. During operation of such devices, the carrier density in the active region of the device will be greater than that seen in single QD spectroscopy and the operating temperatures are typically around room temperature or above. Under these conditions, there may be a large increase in the homogeneous broadening, for example due to carrier–carrier interactions. As can be seen in Fig. 3.21 later in this chapter, at low temperatures wide lasing spectra involving many longitudinal modes may be observed from QD lasers, whereas at higher temperatures narrow lasing lines of mode groups separated by several meV are seen. Sugawara *et al.* [137] have attributed this to a collective lasing action as QDs are coupled due to homogeneous broadening, and from laser spectra obtained from InGaAs/GaAs QD lasers have determined a value for the homogeneous broadening of 16–19 meV.

As has been outlined so far, aspects of the growth and resulting structure of In(Ga)As/GaAs QDs significantly influence their optical properties. The remaining sections of the chapter use examples of QD devices to illustrate how the optical properties impact on device performance. Other types of QD devices, including photodetectors and superluminescent LEDs, are described elsewhere in this book but we shall concentrate on conventional devices: QD lasers, VCSELs and semiconductor optical amplifiers (SOAs) and on the development of novel devices (single photon sources and spin-LEDs) that exploit some of the unique properties of QDs.

3.6 Quantum dot lasers

One of the main driving forces in QD research has been the promise of improved optoelectronic devices. Early predictions have largely been successfully demonstrated and it is likely that commercial products will appear within the next few years. However, as discussed in the previous sections, the growth of QD devices is more complicated than QWs and questions remain concerning repeatability for large-scale production. It should also be remembered that QW laser diodes, present in CD, DVD players, printers and many sensing applications, required a decade of research and development before they became commercially viable.

The majority of research has concentrated on development of QD lasers, for which performance improvements including low and temperature-insensitive threshold current density [30, 47], zero linewidth enhancement factor [138, 139] and increased modulation speed [140] had been predicted. For QD-based SOAs, fast gain recovery [141, 142] and pattern-effect-free amplification [143, 144] is expected and has since been demonstrated. The broad emission and gain that can be obtained from QDs due to the variation in size and composition of the many QDs within an ensemble has also been exploited for high-power, broadband emission from superluminescent light-emitting diodes (SLEDs), which may be used for eye-safe biomedical imaging applications. Output powers of several hundred mW have been demonstrated from QD-based SLEDs [145] and by including multiple layers of QDs within an SLED structure (for which the optical properties of each layer have been controllably varied), emission can be further broadened, to ~120 nm for a 1300 nm emitting device [146, 147]. The inhomogeneously broadened optical response of a QD ensemble can also be exploited in saturable absorber structures, either as semiconductor saturable absorber mirrors (SESAMs), used to mode lock solid state lasers for ultrashort (fs) pulse generation [148], or as absorber sections integrated into QD lasers for direct modulation [149]. The broad absorption spectrum, ultrafast absorption recovery, long-wavelength operation and easy incorporation into structures containing high-quality distributed Bragg reflector (DBR) mirrors all make In(Ga)As/GaAs QDs particularly attractive for SESAM structures, and mode-locking of solid state lasers with QD-SESAMs has been demonstrated [150, 151]. Mode locking of QD lasers using saturable absorber sections has realized transform-limited 7 ps pulse generation with

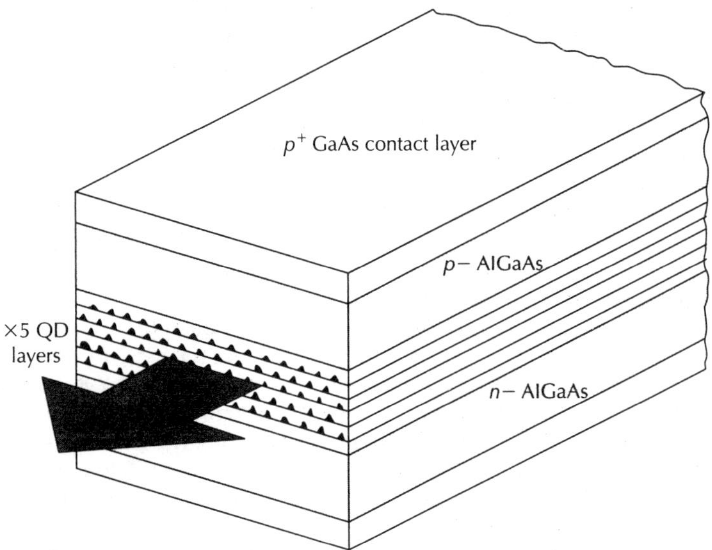

Figure 3.18   Schematic of a broad area QD laser. Five layers of QDs embedded in GaAs form the active region which is typically 300–500 nm thick. N-type (lower) and p-type (upper) AlGaAs cladding layers (shown in grey) provide current injection and also optical confinement. The topmost layer is a heavily doped p+ GaAs contact layer. Ridge waveguide (or gain guided stripe) geometries limit the number of lateral modes.

modulation speeds of 20 GHz [152] and two-section lasers have also been developed for switching lasing between different QD energy states [153, 154].

A typical broad area QD laser (see Fig. 3.18) will have a cavity width of 50 μm and length $L = 300$–500 μm. The design follows that developed for QW lasers. Carriers are injected through the doped AlGaAs layers into the active region. The lower refractive index of these cladding regions confines the optical modes within the active region. The (uncoated) end facets reflect $R \sim 30\%$ of the light which is sufficient to obtain lasing provided the gain, g, is high enough. To obtain better beam quality, ridge waveguide structures limit the lateral extent of the optical mode and lower the current requirements. The condition for lasing based on a photon round trip within the cavity is:

$$\Gamma g = \frac{1}{L} \ln\left(\frac{1}{R}\right) + \alpha_i$$

where Γ is the optical confinement factor (the ratio of the total QD volume to the total waveguide volume) and α_i is the intrinsic cavity loss (scattering, reabsorption). In principle, an ideal QD laser structure should achieve population inversion with occupancy of only one electron–hole pair per QD which would correspond to an exceptionally low threshold current density, J_{th}. However, the previous sections have shown that a QD ensemble consists of QDs having a range of confining potentials and lateral/vertical dimensions which results in a large inhomogeneous linewidth that broadens the gain spectrum. Lasing will then also be determined by the number of QDs contributing to the gain which will depend on the homogeneous linewidth of the QDs. In addition, internal strains may result in the generation of point defects which increase α_i, thereby increasing J_{th}. Nevertheless, the maximum gain available from a QD layer depends primarily on the dot density which can be controlled (within limits) by the growth conditions. High densities $(5$–$10 \times 10^{10}\,\mathrm{cm}^{-2})$ can be obtained at relatively high InAs deposition rates but the emission wavelength is invariably too short $(\sim 1\,\mu\mathrm{m})$ for telecommunications applications. Longer wavelengths require different growth conditions, usually involving slow InAs or submonolayer deposition to produce larger islands but this comes at a cost of a lower QD density $(1$–$2 \times 10^{10}\,\mathrm{cm}^{-2})$ and consequently gain. Such structures require long cavities, high reflection coatings or multiple

QD layers, otherwise lasing occurs in the first excited state which has twice the degeneracy and therefore twice the gain of the ground state. The latter problem has been solved largely through the development of DWELL structures which have QD densities around $5 \times 10^{10}\,\mathrm{cm}^{-2}$ but with emission wavelengths close to 1300 nm.

The first demonstration of lasing from self-assembled InAs/GaAs QDs was reported in 1994 [155]. The device lased at a wavelength corresponding to the first excited state but with a rather high threshold current of $950\,\mathrm{Acm}^{-2}$. A $T_0 = 330\,\mathrm{K}$ (see later discussion) was measured at low temperatures but much smaller values (corresponding to increased variation of threshold current with temperature) were measured at room temperature. These low T_0 values have been attributed to thermal escape of carriers and are a common feature of most QD lasers. Similar results were reported by other groups [156, 157]. Submonolayer and low growth rate techniques generate larger QDs with deeper confining potentials which push the emission to longer wavelengths [22], although no significant improvements in T_0 were reported. In addition, long cavities were required in order to achieve ground state lasing (a cavity length of 7.8 mm was used in similar devices [158]). Although material gain, g, from QDs is higher than that achieved by bulk and QW active regions (around $10^4\,\mathrm{cm}^{-1}$ for QDs compared with a few hundred cm^{-1} for QWs) [30, 31], the optical confinement factor Γ is typically 10^{-4}, so the modal gain, given by $g_{mod} = \Gamma g$, is relatively small. Reported values of the modal gain from the ground state of a QD laser are typically only 5–$10\,\mathrm{cm}^{-1}$ per layer [31, 159–164].

In order to increase ground state gain, multiple QD layers must be incorporated into the laser structure [165]. As discussed previously this is not trivial but growth of GaAs spacer layers (of appropriate thickness) at high temperature ($580°\mathrm{C}$ for MBE growth) also has the advantage of suppressing defect formation in GaAs, which further improves QD laser performance [166, 167]. Indeed, growth of QD laser structures incorporating three InAs QD-DWELL layers where each spacer was annealed resulted in a record low threshold current density of $J_{th} = 17\,\mathrm{Acm}^{-2}$ for devices with high-reflectivity facet coatings and $J_{th} = 32.5\,\mathrm{Acm}^{-2}$ for devices with as-cleaved facets [168].

The active region of directly modulated semiconductor lasers is subject to variations in refractive index, n, due to variations in carrier density, N, which in turn leads to undesirable frequency chirp. This phenomenon is usually characterized by the linewidth enhancement factor, α, given by:

$$\alpha = -\frac{4\pi}{\lambda}\left(\frac{dn}{dN}\right)\left(\frac{dg}{dN}\right)^{-1}.$$

Typical values for InP-based QW lasers are in the range 2–3. A symmetric gain spectrum implies $dn/dN = 0$ which should be the case for a QD laser lasing on the GS. Therefore QD lasers should exhibit a very small α factor. Measurements of α are rather difficult but initial results based on subthreshold amplified spontaneous emission (ASE) oscillations suggested that $\alpha \sim 0.1$ for a QD laser operating at 1220 nm [139] and similar results were reported by other groups [169, 170]. More recent measurements [171] have challenged these earlier findings and a comparison of 980 nm QW and QD lasers yielded relatively high values between 2.5 and 4.5 for both types. This was attributed to modifications to the gain spectrum by the presence of the QD excited states. This was followed by a report of a giant α-factor of 60 [172] in a QD laser operated at high bias (just prior to dual state lasing–see later discussion). Subsequent theoretical predictions are in agreement with these findings; ASE measurements performed under weak injection will result in a low α-factor whereas other methods, FM/AM response and pump–probe, performed at much higher current injection result in very large α-factors. In fact large α-factors may have advantages for semiconductor optical amplifier wavelength conversion where non-linear properties such as cross-phase modulation are increased with α.

The threshold current of a semiconductor laser is found to vary with temperature according to:

$$J_{th}(T) = J_0 \exp\left(T/T_0\right)$$

where T_0 is a characteristic temperature which increases with decreasing dimensionality [47]. Consequently, an ideal QD, with a δ-function DOS with no excited states (or at least excited states that are far enough removed from the lasing state so that carrier redistribution between states and carrier escape from the QDs does not occur), will have a high T_0. High T_0 values are desirable since they alleviate the need for temperature stabilization circuitry for the laser which can significantly add to the cost. As a guide, InGaAsP/InP QW lasers designed for telecommunications applications at 1300 nm have typical T_0 values of 50–60 K. This relatively small value is attributed to thermally induced carrier escape arising from the small conduction band offset. The first QD lasers did exhibit large T_0 values but only at low temperatures; for example, a T_0 of 350 K was measured between 50 and 120 K [86]. At higher temperatures T_0 values comparable with QWs were deduced. For QDs emitting at longer wavelengths, the temperature insensitivity persists to higher temperatures (200–250 K) which can be attributed to a deeper confinement potential for carriers but this value reduced to 35 K at room temperature [22]. Again the reason cited was carrier escape although in this case it was proposed that it was escape from the WL to the GaAs matrix and subsequent non-radiative recombination. Subsequent work has revealed rather complicated temperature behaviour resulting from carrier escape and the lack of a universal Fermi level in the QD ensemble.

Figure 3.19a shows results obtained from lasers grown in our laboratory which are typical of those reported by other groups. The device is a 5 mm × 20 μm InAs/GaAs QD laser containing five QD layers grown at a low growth rate of $0.013\,\mathrm{MLs}^{-1}$ giving a QD density of $\sim 1.7 \times 10^{10}\,\mathrm{cm}^{-2}$ per layer. The temperature dependence of J_{th} indicates a relatively large T_0 ($=192\,\mathrm{K}$) for the range 120–200 K but interestingly between 200 and 240 K the threshold current decreases corresponding to a *negative T_0* ($=-53\,\mathrm{K}$) which has been reported by several other groups [173, 174]. The threshold current then rapidly increases above 250 K, with $T_0 = 30\,\mathrm{K}$ between 280 and 310 K. At 320 K the gain is no longer sufficient for lasing in the ground state and lasing then switches to the first excited state. Figure 3.19b and c compares the temperature dependence of the PL intensity and linewidth with the variation of T_0. It can be seen that the onset of the negative T_0 regime for this laser occurs at a similar temperature to the onset

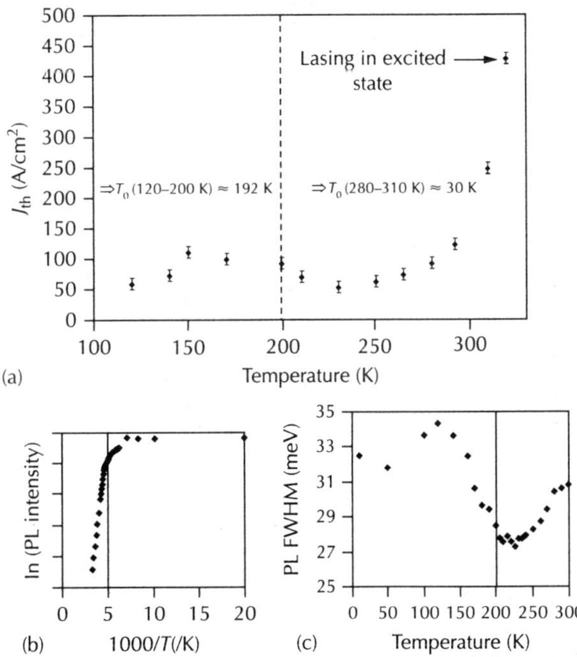

Figure 3.19   (a) Temperature dependence of J_{th} of a 5 mm × 20 μm ridge-waveguide InAs/GaAs QD laser, showing temperature insensitive operation at low temperature but a rapid increase of J_{th} with temperature around room temperature. At 320 K, lasing switches to the first excited state. (b) Arrhenius plot of the temperature dependence of PL intensity from the unprocessed laser structure. (c) Variation of the PL ground state FWHM with temperature from the unprocessed laser structure. The dashed lines marking 200 K are guides for the eye.

of PL intensity quenching and the minimum PL linewidth. The consensus is that a negative T_0 is due to thermal escape and redistribution of carriers within the QD ensemble [111, 113, 114]. This effect is more apparent in structures exhibiting a large inhomogeneous linewidth arising from a bimodal distribution (carrier diffusion within a QD layer) or stacked layers (diffusion of carriers between layers).

In order to extend the temperature-insensitive (high T_0) operation of QD lasers to higher temperatures, a method of inhibiting carrier escape is required. This can, in principle, be achieved by increasing the separation of the confined states or confining the carriers in a deeper potential. Tuning the QD diameter has resulted in an increased separation of the GS and X1 PL peaks of up to 104 meV [78], resulting in a high T_0 of ~180 K between 210 and 295 K for a laser containing a single QD layer and a T_0 of ~250 K over the same temperature range for a similar laser containing two QD layers. Unfortunately, the operating wavelength of 1230 nm at room temperature is too short for telecoms applications. An alternative strategy which has been applied successfully to QW lasers is to p-dope the wells. Modulation doping near the QDs to provide an excess of holes in the dots [175, 176] inhibits carrier escape and thermal spreading across QD states. This is expected to be more effective for holes than for electrons due to their reduced confinement and more closely spaced energy levels. By p-doping of the QD active region, high T_0 values (>200 K) at room temperature and above have been achieved [176–179], although frequently this is at the expense of a higher value of J_{th} [176, 180].

Figure 3.20 compares the variation in threshold current with temperature for undoped and p-doped QD lasers. The structures show quite different behaviour. The origin of the shift of the low T_0 regime to higher temperatures in p-doped QD lasers is currently the subject of much debate. The original motivation for p-doping the QDs was to saturate the hole levels in the QD. An excess hole concentration ensures that the ground state hole energy levels are always filled so that the gain will be limited only by the electron level occupancy. Since the electron energy states are more widely separated, temperature insensitive operation of the QD laser is expected to persist to higher temperatures. Auger recombination (which increases with p-doping and reduces with temperature) has also been cited as a reason for the very high T_0 values [180] although this has recently been disputed [181].

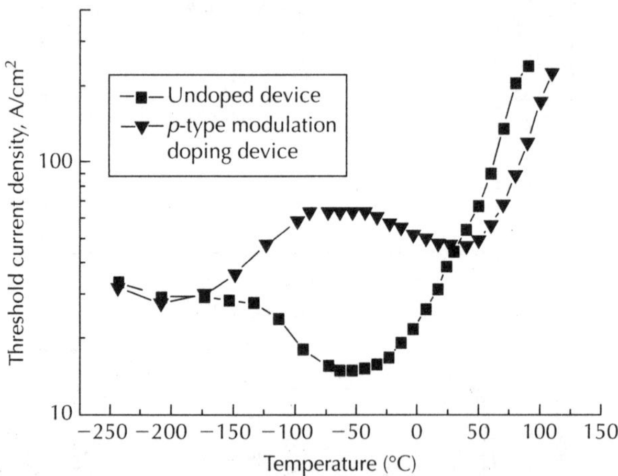

Figure 3.20   Temperature dependence of J_{th} *for undoped and p-doped 3 mm × 15 μm InAs/GaAs QD lasers operating near 1300 nm at room temperature. From Badcock et al.,* Electron. Lett. **42**, 922 (2006) [177].

Early reports of lasing in QD structures revealed significant spectral structure at low temperatures [182, 183] which was attributed to independent lasing of groups of dots. Lasing within a group depends on the homogeneous linewidth of dots which are linked via the photon interaction within the cavity. Thus, at low temperature many sharp lasing features are present across the gain spectrum. With increasing temperature phonon scattering increases the homogeneous linewidth and groups merge to leave one or only a few laser lines at room temperature. Figure 3.21

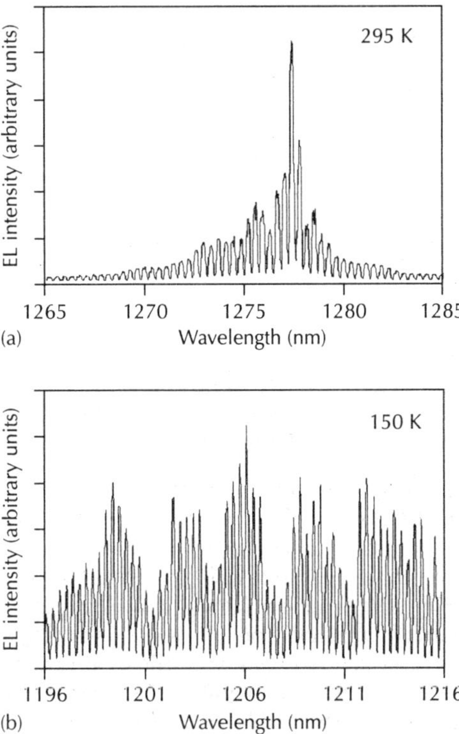

Figure 3.21 EL spectra around emission peak obtained from InAs/GaAs QD laser operating just above threshold ($J = 1.05\,J_{th}$) in pulsed mode (1% duty cycle) at (a) 295 K, and (b) 150 K. A broad longitudinal mode distribution is observed at low temperature but this narrows at higher temperatures.

exemplifies this effect. It shows EL spectra obtained just above threshold from a $470 \times 10\,\mu m$ ridge waveguide QD laser containing InAs QDs in a DWELL structure. Longitudinal modes are clearly visible because of the short cavity length and the device requires high-reflectivity facet coatings to achieve GS lasing at room temperature. At 150 K groups of longitudinal modes are evident due to low carrier escape rates, but by 295 K thermal equilibrium has been established across the QD ensemble and a single envelope containing three or four longitudinal modes is seen. The temperature sensitivity of the threshold current of this laser follows a similar trend to the device presented in Fig. 3.19 above, with a high T_0 at low temperatures ($T_0 = 330$ K at 150 K) but a significantly lower T_0 at room temperature ($T_0 = 75$ K at 295 K).

Narrowing of the laser mode distribution at high temperature has been attributed to preferential carrier redistribution to QDs participating in lasing (since carrier escape is less likely from these QDs due to the increase in the radiative recombination rate by stimulated emission) and a reduction in gain at higher temperatures resulting in a narrower spectral region around the gain peak at which cavity losses can be overcome and lasing can occur [184]. At low temperature, where carrier escape is not significant, QDs will contribute to lasing in all spectral regions with sufficient gain. Alternatively, as proposed by Sugawara *et al.* [137], homogeneous broadening may lead to collective lasing action in a narrow lasing line.

Figure 3.22 shows electroluminescence (EL) spectra obtained around room temperature from an InAs/GaAs QD laser operating just above threshold. The device has a large cavity length (5 mm) so individual longitudinal modes are not resolved in these spectra. As the current is increased more modes participate in lasing and the spectral width of the emission becomes broader [182], demonstrating the inhomogeneous nature of the gain in QD lasers; this is quite unlike bulk and QW lasers in which the gain is clamped at the laser threshold. The carrier occupancy in the QD ensemble continues to increase even after lasing begins and QDs with emission energies away from the gain

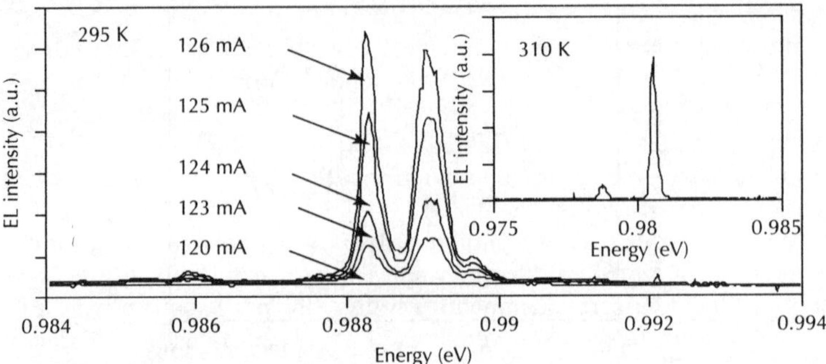

Figure 3.22   Room temperature electroluminescence (EL) spectra obtained for a range of pulsed currents (pulse duration 10 μs, duty cycle 1%) around the laser threshold from a 5 mm × 20 μm ridge-waveguide InAs/GaAs QD laser. Longitudinal modes for the 5 mm cavity are too closely spaced to be resolved, but a splitting of modes into two groups is observed. Inset: 310 K EL spectrum obtained near threshold (256 mA). One mode grouping now dominates emission.

peak can eventually provide enough gain for lasing at higher currents. Also evident from Fig. 3.22 are grouped longitudinal modes, an effect that has been observed by numerous groups [182, 184–187]. Such behaviour is not seen in QW lasers and is attributed to the inhomogeneous gain.

There have been a few reports of dual state lasing involving the ground state and first excited state of the QDs [188–190]. An example of this is shown in Fig. 3.23, where lasing spectra obtained over a range of current injection from a 5 mm × 5 μm ridge-waveguide InAs/GaAs QD laser demonstrates that the laser first switches on in the ground state at a threshold current of 55 mA and then at a current of 95 mA additional laser lines corresponding to lasing in the QD X1 state appear. Dual state lasing is observed between 95 mA and 220 mA, with additional laser modes appearing with increasing current before the GS lasing is quenched. The quenching of the GS lasing coincides with the appearance of an additional mode group in X1.

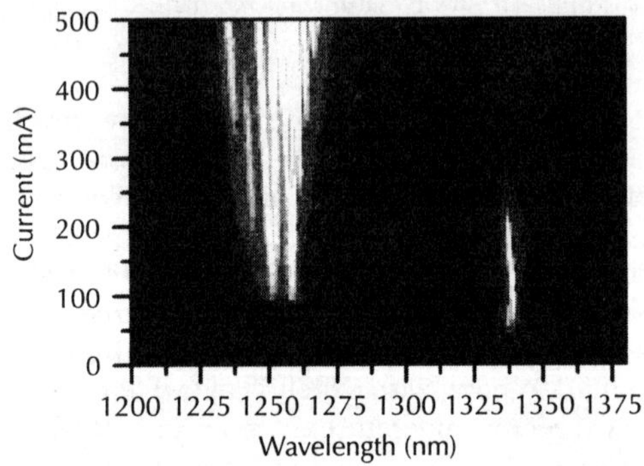

Figure 3.23   Electroluminescence spectra as a function of current for an InAs/GaAs QD laser exhibiting dual state lasing.

Although 1300 nm InAs/GaAs QD lasers are now well established further extension of the emission towards 1550 nm has been more problematic. Despite isolated reports of 1500 nm emission from MOCVD-grown InGaAs-capped QDs [191], it is difficult to achieve room temperature emission beyond ~1400 nm using InGaAs capping or DWELL structures [192]. Longer wavelength lasers have been produced using QD bilayers [193] or by growing QDs on a metamorphic InGaAs buffer [26, 194, 195]. In the former case room temperature ground state lasing was achieved at 1430 nm with a relatively low $J_{th} = 134\,\text{Acm}^{-2}$; in the latter case room temperature ground state lasing at 1515 nm has been achieved, although with rather high J_{th} due to defects in the structure [194]. For this technique [26], a thick (1.2 μm) $In_{0.2}Ga_{0.8}As$ buffer is deposited on the GaAs substrate and the laser structure is grown using InGaAs and InAlGaAs rather than GaAs and AlGaAs for the active and cladding regions, respectively, thereby reducing the strain experienced by the QDs in the structure. A similar approach has also been reported in achieving long wavelength (1450 nm) lasing at room temperature from a tunnel injection laser [196], in which QD layers are electronically coupled to a QW. This broad area (1 mm × 80 μm) device, containing six coupled QW–QD layers and with high-reflectivity facet coatings, exhibits a remarkably low room temperature $J_{th} = 63$ Acm^{-2}, despite use of the metamorphic buffer.

3.7   Vertical and resonant cavity structures

One of the advantages of QD growth on GaAs substrates is the ability to include distributed Bragg reflectors (see Fig. 3.5) into structures to limit the number of photon modes that can couple to the QDs. Al(Ga)As/GaAs DBRs have a large refraction index contrast such that high reflectivities can be achieved with a modest number of periods. Electrical injection is complicated by the high diffusivity of Be (the most common p-dopant in MBE-grown GaAs) towards the GaAs/AlAs interfaces which results in large series resistances. This problem can be overcome through the use of C doping although the high operating temperature of C-cells make them more difficult to control than Be cells.

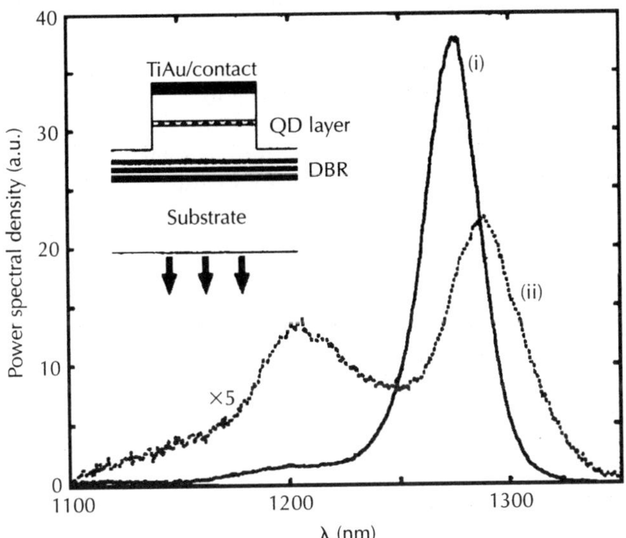

*Figure 3.24   Power spectral density of a QD RCLED. The structure is designed for substrate emission. Excited state emission is inhibited by the cavity and the useful light coupled to a fibre below the substrate is increased by a factor of ~10 compared with the control (no DBR). From Gray et al., Electron. Lett. **35**, 242 (1999) [28].*

A schematic of a QD resonant cavity–LED (RC–LED) structure is shown in Fig. 3.24. A single layer of QDs is incorporated in a λ-cavity consisting of a lower AlGaAs/GaAs DBR and an upper Ti/Au contact which acts as the top mirror (avoiding the problem of growing a p-doped DBR) [28]. The cavity resonance is broad since the DBR mirror reflectivity is constrained to allow substrate emission, but excited state emission is arranged to lie outside the stop band. Although there is imperfect matching of the cavity resonance and the QD GS emission, a comparison with the control sample (no DBR) demonstrates a factor of ten improvement in the coupling to an optical fibre butted against the substrate. QD resonant cavity photodiodes for detection at 1300 nm [197] or 1060 nm [198] have also been reported where the top mirror was an MgF/ZnSc DBR. These devices had a reported detection efficiency of 49% and 65%, respectively, with extremely small spectral bandwidths <2 nm.

Vertical cavity surface emitting lasers (VCSELs) incorporate one or more layers of QDs between two DBRs. Their low excitation volume means they are low-power devices with high directionality

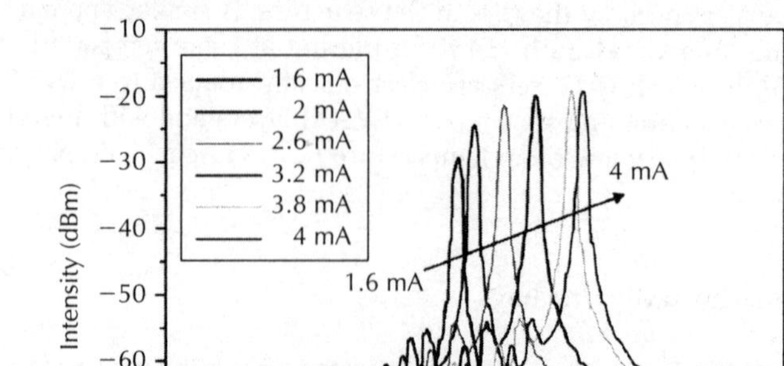

Figure 3.25   Lasing from a QD VCSEL. Note the low operating current (threshold = 7.6 kAcm^{-2}). A T_0 = 450 K was measured for this device. From Chang et al., IEEE Photon. Technol. Lett. **18**, *847 (2006) [204], © 2006 IEEE.*

and high-temperature emission wavelength stability. Relatively few studies have been made on QD VCSELs and most devices operate at wavelengths of 1000 nm or less [29, 199–203], although 980 nm is a target wavelength for short distance optical interconnects. Recently, 1300 nm versions have been demonstrated [204–206] as shown in Fig. 3.23. This device uses C-doping and contains five QD layers in a $3\lambda/2$ cavity. In order to confine the injected current to the centre of the structure (in this case a 26 μm mesa) the AlGaAs layers are deliberately oxidized to allow current only within a 5 μm central region. This technique is commonly used to lower the threshold current of VCSELs by producing a layer of Al_2O_3 between the cavity layer and one of the Bragg reflectors using a steam environment at 350 to 500°C. The rate of formation of the oxide layer is proportional to the content of Al in the material; thus the oxide forms first in those layers with the highest Al content. The refractive index of the oxide layer is lower than that of the semiconductor and also acts as a waveguide for the emitted light. Although this is an impressive result it is not clear that such devices can become commercially viable.

3.8   Semiconductor optical amplifiers

As the speed of data communications increases, regeneration (amplification) of optical pulses by conversion to electrical signals becomes increasingly difficult and optical amplifiers are seen as the key to "all optical" networks. The erbium doped fibre amplifier has been one of the major innovations in optical communications but is confined to the 1550 nm wavelength range where attenuation is already low. Developments of fibre amplifiers (doped with praseodymium) for the 1300 nm range have been slow and considerable effort has been put into semiconductor optical amplifiers (SOAs) having bulk or QW active regions. SOAs are essentially laser diodes without end mirrors and are frequently called travelling wave amplifiers. To prevent laser oscillation (which results in ripples in the gain spectrum) the facets can be antireflection coated ($R < 0.001\%$) or a tilted waveguide geometry can be used. At present QW SOAs are available from 850 to 1600 nm utilizing GaAs- and InP-based materials systems and can provide up to 30 dB of gain. Figure 3.26a shows how the gain of an SOA decreases as the input signal power increases. Carrier depletion leads to gain saturation and this can cause significant signal distortion, as shown in Fig. 3.26b. This is known as the pattern effect. If the gain change is too slow to follow the signal level transient, which is the case for conventional SOAs, or if the bit rate is high, the output waveform is distorted as shown on the right part of the diagram.

   The gain recovery in conventional bulk and QW SOAs exhibits two characteristic time constants; a short one (subps) which arises from strong non-linear effects such as spectral hole burning

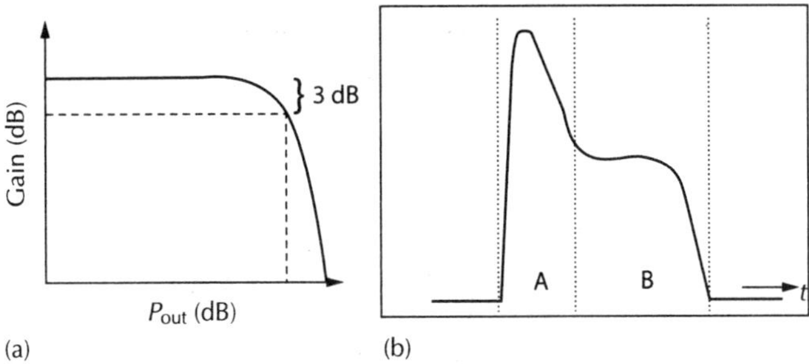

Figure 3.26   *Gain saturation due to carrier depletion (a) leads to distortion (pattern effect) in the output pulse. If the bit rate is lower than the gain recovery the leading part of the pulse (part A) is amplified with unsaturated gain, but the trailing part (B) is distorted.*

and subsequent fast intraband scattering, and a longer one of order 100 ps which is attributed to slow carrier refilling by current injection and diffusion. QD SOAs offer ultrafast (zero pattern effect) non-linear gain recovery nearly an order of magnitude faster than QWs or bulk material (see Fig. 3.27) with amplification at bit rates of 40–160 Gbits [207] and a spectrally broad gain (fast temporal response) giving them distinct advantages over their bulk and QW counterparts. In addition, 1300 nm operation on a GaAs substrate makes these devices commercially very attractive. A problem for single pass travelling wave SOAs is the relatively low gain. Although many QD layers can be incorporated into the active region a 20 dB gain will require devices longer than 5 mm. Bulk and QW SOAs are polarization sensitive due in part to the waveguide geometry but also to the valence band symmetry. Recently, polarization insensitivity has been demonstrated in a QD SOA through close stacking of many dot layers to form a columnar dot structure [208].

Finally, it should be noted that SOAs exhibit significant non-linear effects due to a changing carrier population during operation. These can cause problems such as frequency chirping and cross-modulation of signals. However, these can be put to advantage in the generation of new frequencies due to four wave mixing (FWM). This is a coherent non-linear process that mixes a pump beam of frequency ω_0 with that of a signal beam of frequency $\omega_0 - \Omega$ to generate a conjugate signal at $\omega_0 + \Omega$ as shown in Fig. 3.28 [141, 143]. Such optical wavelength converters will play a crucial role in all optical networks. Thus a combination of large bandwidth gain and

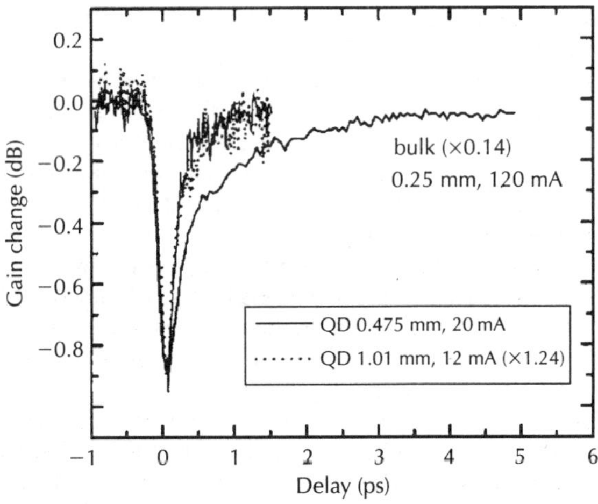

Figure 3.27   *Comparison of the gain compression dynamics of a QD (at two current inputs) and a bulk InP-based SOA. The gain recovery is some seven times faster for the QD device. From Bimberg and Ribbat,* Microelectron. J. **34**, *323 (2003) [206].*

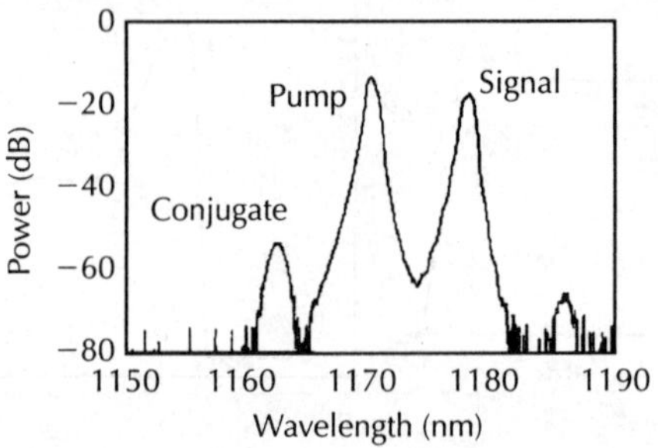

*Figure 3.28　Wavelength conversion by four-wave mixing of a pump and signal beam in a QD SOA. Reprinted with permission from Akiyama et al., Appl. Phys. Lett. **77**, 1753 (2000) [141], Copyright 2000, American Institute of Physics.*

wavelength range, fast recovery and polarization insensitivity should provide an alternative to QW SOAs in future fibre networks.

3.9　Single photon sources

Single photon sources are attractive for a variety of applications, including metrology [209], quantum key distribution [210] and linear optical quantum computing [211, 212]. Single photon emission has been demonstrated from a variety of sources, including single molecules [213, 214] and colour centres in crystals, for example nitrogen vacancies in diamond [215–217], and also from QDs [218–220]. QDs are particularly attractive as single photon sources because they can be incorporated into standard optoelectronic device structures allowing electrical operation of the source and integration into optical fibre communication systems. As mentioned earlier, typical growth conditions yield QD ensembles with QD densities of $\sim 10^{10} \, \mathrm{cm}^{-2}$ ($100 \, \mu\mathrm{m}^{-2}$). However, in order to isolate single QD emission in practical device structures a QD density approaching $10^8 \, \mathrm{cm}^{-2}$ ($1 \, \mu\mathrm{m}^{-2}$) is required – two orders of magnitude less than for normal growth conditions. Growth on patterned substrates has been shown to result in low QD densities, by preferential QD nucleation at nanoholes [221] or growth of InGaAs QDs embedded in GaAs/AlGaAs pyramidal structures [222], and single photon emission has been observed from QDs grown by both of these methods [222, 223]. Alternatively, it is possible to significantly reduce the QD density on conventional substrates through accurate control of growth conditions, for example by reduction of the InAs QD growth rate.

The standard method for evaluating a single photon source is to perform a correlation measurement following the Hanbury–Brown and Twiss (HBT) experiment [224], as shown in Fig. 3.29a. Emission from the source is incident on a beam-splitter and directed to one of two photodetectors and the time delay (τ) between detection events is evaluated. From the correlation between detection events, a second-order correlation function $g^{(2)}(\tau)$ is obtained which establishes the nature of the light source [209]. Attenuated laser sources (such as those currently used in commercial quantum cryptography systems) are Poissonian sources, i.e. the probability of a light pulse from the source containing n photons follows a Poissonian distribution $P(n)$. This results in a constant $g^{(2)}(\tau) = 1$. However, for an ideal single photon source $g^{(2)}(0) = 0$, since the single photon from the source must be directed by the beam-splitter to one or other of the detectors and simultaneous detection events are impossible.

Single photon emission from QDs relies on the cascade emission process [225, 226]. Due to Coulomb interactions, the biexciton emission energy is spectrally shifted with respect to the single exciton. As carriers are captured, relax and recombine in a single QD, emission will be characterized by multiple exciton complexes until the final electron–hole pair recombines at the exciton energy (which can be spectrally resolved). HBT correlation measurements of the exciton emission demonstrate photon antibunching ($g^{(2)}(0) < 1$) [218, 225] and cross-correlation measurements using a modified HBT set-up (in which monochromators are used so that one

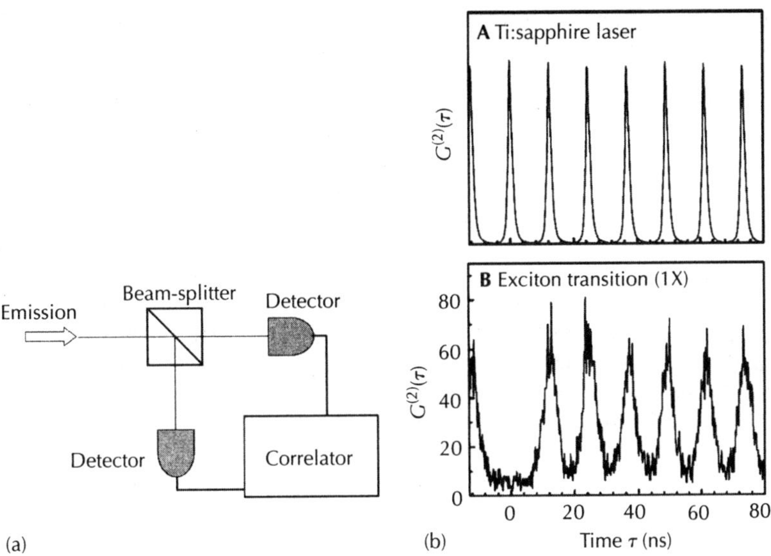

Figure 3.29   *(a) Schematic diagram of the Hanbury–Brown and Twiss experiment. (b) Correlation functions* $g^{(2)}(\tau)$ *obtained from a pulsed Ti:sapphire laser (A), representing a Poissonian source, or from the exciton transition of a QD (B), for which there is suppression of the correlation signal at zero delay* $\tau = 0$. *Part (b) from Michler et al., Science **290**, 2282 (2000) [218].*

photodetector will detect emission from the biexciton and the other detects emission from exciton recombination) demonstrate conclusively that sequential emission of the biexciton then the exciton occurs [226].

For a practical single photon source it is advantageous to place the QDs within a microcavity in order to increase the photon extraction efficiency. Although a QD single photon source that does not utilize a microcavity structure has been demonstrated [220], the high refractive index of GaAs ($\sim$3.5) results in a low critical angle for total internal reflection at the GaAs/air interface and so only a small fraction of photons emitted from the QD will escape the device. Placing the QDs within a planar microcavity structure, with a DBR mirror below the QD layer, has been shown to improve the photon collection by a factor of ten [227]. Further modification of the microcavity design by introducing lateral confinement of the optical modes, for example by using micropillars [228, 229], microdisks [230] or photonic crystal cavities [231, 232], can result in considerable alteration in emission from QDs, with significant enhancement of spontaneous emission into a cavity mode due to the Purcell effect [233]. The spontaneous emission rate from InAs/GaAs QDs has been enhanced by five times when the QDs were incorporated into a micropillar and by 15 times when the QDs were incorporated in a microdisk [230]. Considering micropillars, the spontaneous emission coupling coefficient β, the fraction of spontaneous emission in a single cavity mode, increases with decreasing pillar diameter, and values as high as $\beta = 0.78$ have been reported for a 0.5 μm diameter micropillar [234]. Efficient coupling of the spontaneous emission into a cavity mode in this manner should improve the overall efficiency of single photon sources. However, the extraction efficiency may be reduced for small micropillar diameters (less than 3 μm) as scattering due to the roughness of the etched sidewalls becomes significant, causing deterioration in the cavity Q-factor [235]. The overall efficiency of the single photon source is also dependent on the collection optics of the system; if the numerical aperture of the collection optics is less than 1 then overall extraction efficiencies from small cavity diameters may be reduced. This places a limitation on the optimum micropillar cavity diameter to around 1–2 μm. Single photon emission from a single InAs/GaAs QD in a 1.9 μm diameter pillar microcavity with a photon collection efficiency of 10% has been demonstrated [236]. This further emphasizes the requirement for a very low QD density ($\sim$$10^8$ cm^{-2}) in order to isolate a single QD in a QD single photon source device structure.

Currently the majority of QD single photon sources that have been demonstrated operate at wavelengths <1000 nm, mainly for compatibility with Si-based detectors. For integration with optical fibre-based telecommunications systems, emission at 1300 or 1550 nm is desirable. Single photon emission at 1300 nm has been demonstrated from single InAs/InP QDs [237] but this material system is not suitable for incorporation into micropillar structures due to the lack of high-quality DBR mirrors. In order to extend low temperature emission from low density InAs/GaAs QDs to 1300 nm, two approaches have been reported. The first method uses bimodal QD ensembles grown under conditions so that there is a very low density of large, long-wavelength QDs, along with a higher density of smaller QDs (emission from these small QDs can be spectrally filtered). These QDs have been incorporated into micropillar structures and single photon emission from the devices has been observed [238]. Alternatively, low density InAs/GaAs QDs can be grown using very low InAs growth rates. As the growth rate is reduced the resulting QD density is also reduced, with a concomitant increase in QD size [239]. Figure 3.30 shows the variation of InAs/GaAs QD density as the InAs growth rate is reduced. If the growth rate is reduced to these very low levels, a QD density of $\sim 10^8 \, \text{cm}^{-2}$ can be achieved [240].

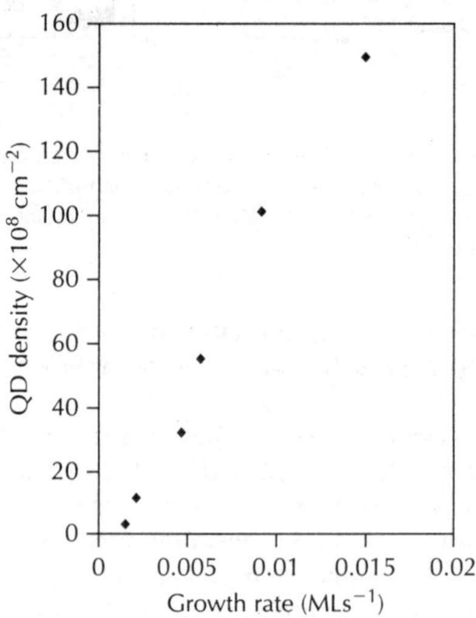

Figure 3.30   *Variation of QD density in InAs/GaAs QD samples grown at a substrate temperature of 500°C at low growth rates.*

As the growth rate is reduced and the resulting QD size is increased, the QD emission wavelength is extended but saturates at around 1200 nm at 10 K for GaAs-capped QDs [9]. However, by capping the QDs with InGaAs, extension of the low-temperature emission of the QDs to beyond 1300 nm can be achieved [240]. These QDs have been incorporated into planar microcavities and antibunching of the exciton emission from single QDs has been observed [241]. Very recently the first electrically-driven QD single photon source operating near 1300 nm was demonstrated [242].

3.10   Entangled photon sources

There has been growing interest in extending the concept of the QD single photon source to realize polarization-entangled photon sources, for quantum information applications including alternative key distribution methods in quantum cryptography [243]. This requires photons from the source to be indistinguishable [244, 245]. By a photon interference experiment, Santori *et al.* have shown that consecutive photons from an InAs/GaAs QD micropillar single photon source were indeed indistinguishable [246] and this has been used to generate polarization-entangled photon pairs using post-selected photons from the source [247]. This method of entanglement utilizes just the exciton emission from the QD single photon source, but the cascade emission process in QDs can provide sequential emission from the biexciton and then the exciton [226] and it has been proposed that this can produce pairs of entangled photons [248].

Following this proposal, we assume the polarization of the emitted photons is due to selection rules discussed in more depth in the next section on spintronics. The biexciton comprises two electrons with spin $J_z = \pm 1/2$ and two heavy holes with spin $J_z = \pm 3/2$. For bright transitions where $\Delta J = 1$ then according to the selection rules the emitted photon will have positive or negative circular polarization σ^+ or σ^-. Assuming there is no spin-flip during the cascade process then the first photon emitted will have a polarization of σ^+ or σ^- and the photon resulting from the second exciton recombination will have the opposite polarization; so the two-photon state should be entangled, as described by:

$$|\psi\rangle = \frac{1}{\sqrt{2}}\left(|\sigma^+\rangle_1|\sigma^-\rangle_2 + |\sigma^-\rangle_1|\sigma^+\rangle_2\right)$$

To achieve entanglement the two decay paths, as shown in Fig. 3.16 earlier, must be indistinguishable. First reports showed emission of classically correlated photon pairs from the biexciton cascade, illustrating a clear relationship between the polarization of the two photons, but not entanglement [249, 250]. This is attributed to the lifting of the spin degeneracy due to asymmetry of the QD resulting in fine structure splitting, as discussed previously. The fine structure splitting presents a major obstacle to the realization of entangled photon sources based on the biexciton–exciton cascade process because the two photons can now be distinguished according to the sequence of the exciton recombination (which decay path was taken). Entanglement has been achieved using photons from the biexciton cascade in QDs with fine structure splitting, but only by spectral filtering to post-select photons from a spectral region where emission from the two exciton polarizations overlap [251]. In order for the photons to be indistinguishable, the fine structure splitting must be less than the homogeneous linewidth [252]. Numerous approaches have been taken in order to control the fine structure splitting. One is to compensate for the asymmetry in the QD by applying an external lateral electric [253, 254] or magnetic field [255], or by applying a uniaxial stress along the [110] crystal direction [256]. Alternatively, post-growth annealing of the QDs can be used to tune the fine structure splitting [128, 252, 257, 258]. The splitting follows a broadly linear trend with the emission energy, as shown in Fig. 3.31, passing through zero fine structure splitting at around 1.4 eV, which also matches a similar trend

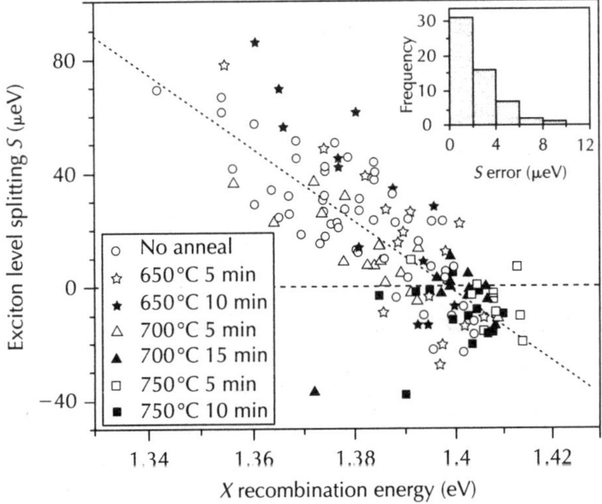

Figure 3.31   Fine structure splitting from single InGaAs/GaAs QDs under various annealing conditions. Inset: *Histogram of errors in the fine structure measurements. Reprinted with permission from Young et al., Phys., Rev. B* **72**, *113305 (2005) [128]. Copyright 2005 by the American Physical Society.*

from unannealed QDs of different sizes [259]. The reduced (and zero) fine structure splitting in these shallow QDs is attributed to a reduction in the piezoelectric effect that causes the electron and hole wavefunctions to point in orthogonal directions.

By selecting QDs with emission around 1.4 eV and negligible fine structure splitting, recent experiments have now demonstrated entanglement of photons from the biexciton cascade [260, 261]. Prospects for a QD-based entangled photon source at longer wavelengths are currently low, although very few studies have considered QDs emitting at these wavelengths [262, 263]. Indirect measurements of InAs/GaAs QDs emitting at 1260 nm at 7 K suggest a fine structure splitting of 20–90 μeV [262], although STM studies of InGaAs-capped QDs suggest that these QDs may be much more symmetric than similar GaAs-capped, because preferential In migration out of the QD along the [1$\bar{1}$0] direction during capping is inhibited by the InGaAs cap [264].

3.11 Spin-LEDs and the potential for QDs in spintronic devices

Exploitation of the spin of carriers in semiconductor devices is currently the focus of considerable research interest. The use of carrier spin in addition to charge may result in improvement in device performance [265] and development of new types of devices including non-volatile memory devices and the use of the spin of nuclei or carriers as information bits in quantum computing ("qubits") [266, 267].

Research concerning spin in semiconductors focuses on development of magnetic materials, spin injection into semiconductors and subsequent spin transport, manipulation and detection. One type of device that is useful to evaluate spin injection from magnetic materials into semiconductor device structures is the spin-light-emitting diode (spin-LED) [268–270], where electrical injection of carriers from a magnetic contact or through a layer of magnetic semiconductor gives a spin-polarized population of carriers, which then radiatively recombine in the active region of the LED. The polarization of the emitted light can then be related to the spin of the carriers from radiative selection rules. For a useful spin-LED, we require efficient transfer of injected spin to the LED active region and subsequent carrier relaxation and radiative recombination with the minimum of spin relaxation.

The radiative selection rules apply to both absorption and emission of light. The selection rules for absorption of circularly polarized light in bulk GaAs are illustrated in Fig. 3.32a. Excitation by circularly polarized light introduces a net spin polarization of the photo-excited electrons, with a maximum of 50% net spin polarization, since both heavy and light hole states are excited but heavy hole transitions are favoured by a ratio of 3:1 [271]. In strained semiconductors or for low-dimensional systems, the degeneracy of the valence band may be lifted and the spin polarization thus increased, since excitation will now involve one type of hole state, as in Fig. 3.32b.

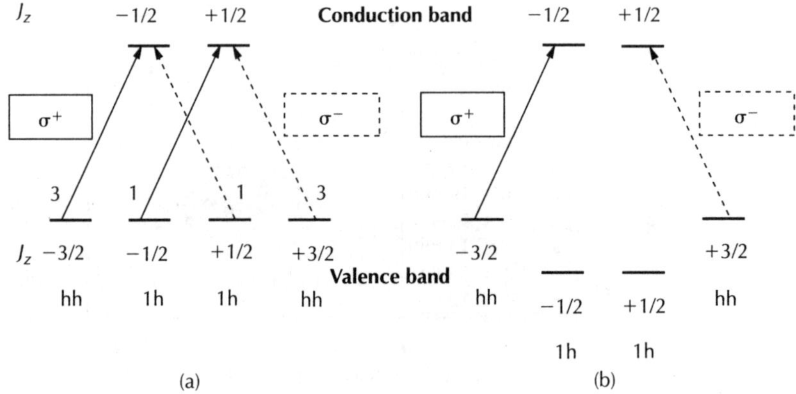

Figure 3.32   *Polarization selection rules for absorption of circularly polarized light into (a) bulk, unstrained GaAs (numbers by the arrows represent the relative transition strengths). (b) A strained and/or low dimensional system where the degeneracy of the heavy and light hole states has been lifted.*

As shown above, in bulk GaAs around the Γ-point the heavy and light holes are mixed and this leads to very rapid spin relaxation of holes, so in the majority of cases it is the electron spin that is exploited. The dominant spin relaxation mechanism in undoped bulk III–V semiconductors, especially at low temperature, is the D'yakanov–Perel mechanism [271]. This arises because in III–V semiconductors there is no centre of symmetry in the unit cell (so-called bulk inversion asymmetry), which leads to a lifting of the degeneracy of the conduction band spin states, equivalent to applying a small magnetic field. If the direction of motion of the electron differs from the direction of this effective magnetic field, precession of the electron will lead to spin relaxation. However, in QDs, since there is 3D confinement of the electron, the D'yakonov–Perel mechanism is expected to be suppressed so long spin relaxation times are predicted. Indeed, in early studies long spin relaxation times of excitons in QDs, exceeding the exciton radiative lifetime, were reported [272, 273]. These long spin relaxation times and the lifting of the heavy hole/light hole degeneracy makes QDs attractive for the active region of spin-LEDs.

The first QD-based spin-LEDs were demonstrated by Chye *et al.* [274], using structures incorporating GaMnAs spin aligner layers [275, 276]. Since GaMnAs is natively p-doped, injection of spin polarized holes is straightforward; to inject spin polarized electrons a reverse-biased Zener diode junction formed at a GaMnAs:(n-doped)GaAs interface was used, in which the spin polarised electrons in the valence band tunnel across the junction into the n-GaAs conduction band. The circular polarization of light emitted from the QDs was comparable to that emitted from QWs in similar structures.

For magnetic semiconductor layers such as GaMnAs only low-temperature operation is possible since the Curie temperature of these materials is low (up to $\sim 110\,\mathrm{K}$ for GaMnAs). For operation closer to room temperature an alternative is needed, and this is commonly achieved by using ferromagnetic contacts such as Fe. The maximum spin injection from an Fe contact is around 45%; however, early attempts at spin injection into semiconductors from Fe contacts produced little or no detected circular polarization of emitted light. This is because ohmic contacts are not suitable for efficient spin injection since the conductance mismatch between the layers results in significant spin dephasing due to carrier scattering at the interface [277]. This may be overcome using a tunnel junction, for example with a Schottky barrier in reverse bias [278, 279].

Two examples of InAs/GaAs QD spin-LEDs with Fe contacts have been reported recently [280, 281], using similar spin-LED structures but different measurement techniques. For successful measurement of circular polarization of emission due to spin polarization of carriers, the geometry of the system must be correct: the validity of the selection rules for optical transitions in QWs and QDs depends on the direction of absorption/emission such that normal to the QW or QD plane the rules outlined above are reliable. However, for in-plane emission the selection rules are ill-defined and optical polarization is not detected [282, 283], thus the carriers must have a component of spin in the direction normal to the QD plane (the growth direction). The thin Fe contact layer has its magnetization aligned along its plane, so for normal injection conditions the injected carrier spin will be in the plane and undetectable. To overcome this, measurements are either made using the Faraday geometry [280] or using the oblique Hanle effect [281]. For the Faraday geometry, a large magnetic field (several tesla) is applied in the growth direction in order to align the magnetization of the Fe contact out of the plane. Injected spins will then be orientated in the growth direction and the selection rules will be valid. Li *et al.* [280] observed a noticeable increase in the intensity of positive circularly polarized light (σ^+) with respect to negative circularly polarized light (σ^-) emitted from the LED for an applied field of 3T (enough to saturate the Fe magnetization in the growth direction), with a circular polarization (defined as $P_{circ} = [I(\sigma^+) - I(\sigma^-)]/[I(\sigma^+) + I(\sigma^-)]$) of around 5%. Interestingly, this circular polarization remained roughly constant throughout the temperature range 80–300K, in contrast to reports for QW spin-LEDs, which show a reduction in circular polarization with increased temperature [284].

The present authors have demonstrated electrical spin injection into an InAs/GaAs QD spin-LED using the oblique Hanle effect [281]. For the Hanle effect measurements, a small magnetic field ($<100\,\mathrm{mT}$) is applied at an angle $\theta = 45°$ to the growth direction. The magnetic field is not sufficient to rotate the Fe magnetization out of the plane, so the injected spin is in the direction of the plane. However, once the spin is injected into the semiconductor, it will precess around the direction of the applied field, with a Larmor frequency $\Omega = g^* \mu_B \mathbf{B}/\hbar$, where g^* is the effective Landé

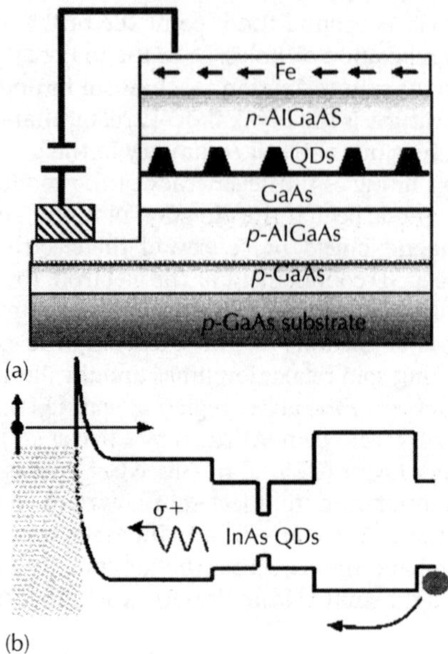

*Figure 3.33   (a) Schematic of spin-LED structure, (b) flatband diagram of the spin-LED. Reprinted with permission from Itskos et al., Appl. Phys. Lett. **88**, 022113 (2006) [281], Copyright 2006, American Institute of Physics.*

factor for the electrons and μ_B is the Bohr magneton. This gives a component of the spin in the growth direction that can be detected in the circular polarization of the light emitted from the LED.

The device consists of an n-i-p light-emitting diode containing a QD active region back to back with a Schottky diode formed by the Fe/n–AlGaAs barrier, such that when the LED is in forward bias then the Schottky diode is in reverse bias. Spin polarized electrons tunnel across the Schottky barrier from the Fe contact and unpolarized holes are injected from the p-contact. Emission in the growth direction is collected through the semi-transparent Fe contact. A schematic of the spin-LED structure and a flatband diagram of the device are shown in Fig. 3.33.

The circular polarization, P_{circ}, is analysed as a function of applied magnetic field. Once effects not related to spin injection (such as magneto-optical circular dichroism induced by the Fe contact and Zeeman splitting of the QD exciton) are subtracted, a dependence of P_{circ} on applied field is obtained, as shown in Fig. 3.34, which can be fitted by a Lorentzian function given by [271, 285]:

$$S_z = S_{\text{init}} \frac{T_S}{\tau} \frac{\left(\Omega T_S\right)^2 \cos\theta \sin\theta}{1+\left(\Omega T_S\right)^2}$$

where S_z is the spin component in the growth direction, which is equal to P_{circ} from the selection rules, T_S is the QD ground state exciton spin lifetime, defined by:

$$\frac{1}{T_S} = \frac{1}{\tau} + \frac{1}{\tau_S}$$

where τ is the exciton radiative lifetime and τ_S is the exciton spin relaxation time. The half-width $B_{1/2}$ of the Lorentzian is given by:

$$B_{1/2} = \frac{\hbar}{g^* \mu_B T_S}$$

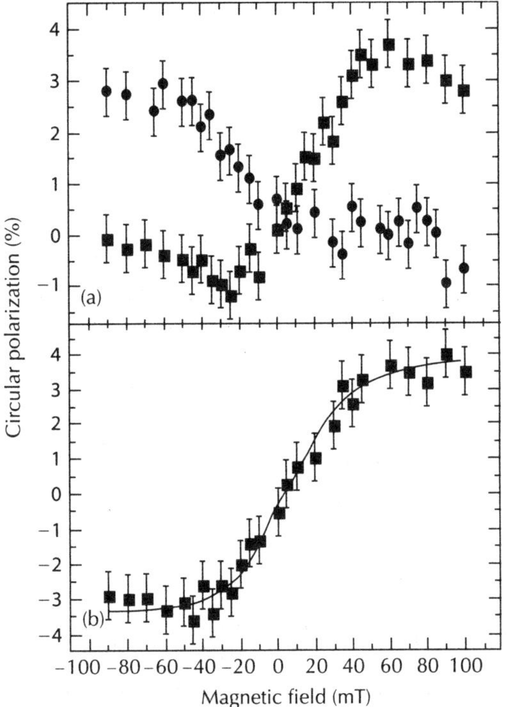

Figure 3.34   Dependence of the circular polarization of the ground state EL (black squares) and corresponding PL obtained using unpolarized excitation from a spin-LED device on applied oblique magnetic field. (b) Circular polarization after correction for non-spin injection related effects (determined by the PL response). Reprinted with permission from Itskos et al., Appl. Phys. Lett. **88**, *022113 (2006) [281], Copyright 2006, American Institute of Physics.*

Fitting the Hanle curve to the data yields a value of the scaled QD ground state exciton lifetime $g^*T_S = \hbar/\mu_B B_{1/2} = 510 \pm 70\,\mathrm{ps}$. From the saturation maximum of the Hanle curve we can obtain a spin polarization of $7.5 \pm 0.7\%$ in the ground state of the QDs. Using values for the ground state radiative lifetime $\tau = 800\,\mathrm{ps}$ [286] and a g^* factor of -1.7 [287] reported for QDs of similar size and emission energy, an initial spin injection from the Fe contact of $20 \pm 3.25\%$ can be estimated. Although the value of $P_{circ} = 7.5\%$ is the highest currently reported for (undoped) QD-based spin-LEDs, by comparison of the detected circular polarization with the injected spin it is clear that there is significant spin relaxation in the structure somewhere between injection into the bulk and subsequent capture, relaxation and radiative recombination of carriers in the QDs.

There are two further spin relaxation mechanisms which may have a significant effect on carrier spin in QDs: electron–hole exchange interaction and hyperfine interaction between electrons and nuclei. As mentioned above, electron–hole exchange interaction causes mixing of the pure spin-up and spin-down exciton states into linear combinations, permitting efficient spin relaxation. Rapid spin relaxation of $\tau_s \sim 100\,\mathrm{ps}$ in InAs/GaAs QDs has been attributed to the exchange interaction [288]. This may appear to be in contradiction to the early reports of long spin lifetimes in QDs but is in fact consistent. Gotoh *et al.* [272] report a low temperature (4 K) spin lifetime of $\tau_s = 0.9\,\mathrm{ns}$ but used quasi-resonant excitation (100 meV difference between the excitation energy and the PL detection energy) of highly symmetric InGaAs/AlGaAs quantum disks. Paillard *et al.* [273] examined polarization- and time-resolved PL obtained from InAs/GaAs QDs under strictly resonant excitation, with a variable magnetic field in the growth direction. For zero applied field, long-lived linear polarization following linearly polarized excitation was observed, which confirms that the exciton states are mixed by the exchange interaction (excitation by circularly polarized light at zero applied field results in an absence of circularly polarized PL). For a 2.5 T applied magnetic field, the spin-up and spin-down states are Zeeman-split and strictly resonant

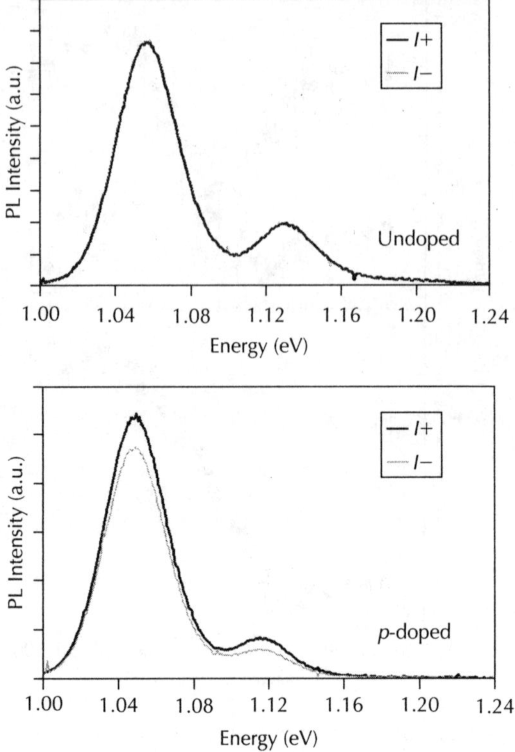

Figure 3.35   10 K cw polarization-resolved PL spectra obtained from an undoped sample containing a single QD layer and a p-doped sample containing similar QDs with a doping level of 10 acceptors per QD, with excitation into GaAs barrier by σ+ circularly polarized light. Emission is resolved for σ+ and σ− polarized light, denoted by I+ and I−, respectively. A clear circular polarization of emission from the p-doped sample is observed. Data courtesy Edmund Harbord, Imperial College London.

excitation by circularly polarized light of the unmixed spin-up and spin-down states results in circularly polarized PL that remains unchanged over the duration of the PL decay. These results thus show that the exchange interaction results in long-lived mixed states but short-lived pure spin-up or spin-down states. Subsequent studies have demonstrated quantum beating between the mixed states [289–291].

The spin dephasing effects of the exchange interaction can be circumvented by doping the QDs. Considering the X^+ trion state consisting of one electron and two holes, the coupling terms in the Hamiltonian introduced by the exchange interaction between the electron and spin-up hole are cancelled out by the terms introduced by the exchange interaction between the electron and the spin-down hole. When exciton recombination occurs, it involves a pure spin state ($|+1>$ or $|-1>$) resulting in circularly polarized emission. *P*-doping is also expected to enhance the circular polarization of the QD emission due to increased carrier capture rate [292], and a decrease in the exciton lifetime in the QD [293].

These factors result in a significant enhancement of circularly polarized emission from QDs, such that it can even be detected in cw PL using excitation into the bulk GaAs, as shown in Fig. 3.35. For an undoped InAs/GaAs QD sample, shown in Fig. 3.35a, there is no discernible difference in the PL spectra analysed for σ+ or σ− polarized light following σ+ polarized excitation, whereas for a similar QD sample that is modulation doped with a doping level of 10 acceptors/ QD, a clear enhancement of the σ+ polarized emission is observed (Fig. 3.35b), corresponding to a circular polarization $P_{circ} = 5\%$ at the GS emission peak and $P_{circ} = 15\%$ for the X1 emission peak. The increased P_{circ} observed for the excited state emission is due to Pauli spin blocking of the carriers [294]. Incorporation of *p*-doped QDs into a spin-LED structure with a Co ferromagnetic contact has been reported recently [295], with a maximum $P_{circ} = 15\%$ (with a 2 T magnetic field applied in the Faraday geometry), which persists up to 70 K (in comparison to $P_{circ} = 7.5\%$ for an undoped spin-LED [281]). It should be pointed out that for measurements on spin-LEDs using the Faraday configuration, the large applied magnetic field may introduce Zeeman splitting of the spin states that will overcome the exchange interaction effects. Interestingly, if the QDs are n-doped, negative circular polarization is observed: for example, for $\sigma+$ excitation, $\sigma-$ emission is obtained [296–298].

Another spin relaxation mechanism arises from the hyperfine interaction between electrons and nuclei [299, 300]. The electron spin in a QD interacts with a large number of nuclear spins (from up to 10^5–10^6 per dot). These are randomly orientated, resulting in an effective magnetic field, $\mathbf{B_N}$, about which the electron precesses. The electron precession is fast in comparison to either nuclear spin precession due to the electron or nuclear spin relaxation (for a GaAs QD, the electron precession period about $\mathbf{B_N}$ is $\sim 1\,\text{ns}$, the nuclear spin precession period is $\sim 1\,\mu\text{s}$ and nuclear spin relaxation time is $\sim 100\,\mu\text{s}$ [299]; for an InAs QD, the timescales would be similar), so the electron sees an effectively constant $\mathbf{B_N}$. For each QD in an ensemble, the effective field $\mathbf{B_N}$ has a different magnitude and direction. For an isotropic distribution of nuclear spin, a third of the effective field will be orientated along the same direction as the optically orientated electron spin (in the growth direction), and the other two-thirds will contribute to dephasing of the electron spin. Dephasing via the hyperfine interaction is masked in undoped due to the dominant dephasing effect of the exchange interaction, but has been observed by Braun *et al.* [301] in polarization- and time-resolved PL obtained from p-doped InAs/GaAs QDs: a decay seen in the circular polarization of PL from the QDs to one third of the original value takes place within 800 ps, after which the circular polarization remained unchanged over the radiative lifetime of the exciton. The decay time of the electron spin due to the hyperfine interaction was thus estimated to be 500 ps. However, the hyperfine interaction can be suppressed by applying a small magnetic field (100 mT) in the growth direction, which aligns the nuclear spins [301]. Alternatively, aligning the nuclear spins by exploiting the hyperfine interaction between photo-injected, spin-polarized electrons and the nuclei has been proposed [302] and has been recently demonstrated by a number of groups [303–305].

By control of the spin dephasing mechanisms in QDs, for example by applying a magnetic field, significant enhancement of the electron spin lifetimes can be achieved. Recently, an electron spin lifetime of over 20 ms at low temperature (1 K) was reported from InGaAs/GaAs QDs in a 4 T external magnetic field [306]. However, this is an emerging field of research and many aspects of spin dynamics in QDs have yet to be clarified before significant advances towards device applications can be made.

3.12　Conclusions

There have been significant advances in our understanding of the fundamental properties of QDs over the last decade. Controlling growth is crucial and underpins all the developments in QD optoelectronic devices. Although the emphasis worldwide has been on fabricating lasers for telecommunications applications at 1300 nm to take advantage of the increased confinement, research has already moved on to novel devices such as single photon sources which are beginning to challenge the performance of systems offering quantum key distribution based on attenuated lasers. Investigations of spin injection, manipulation and detection represent the pinnacle of semiconductor research today which has come full circle from collective properties of electrons in bands to artificial atoms.

Acknowledgements

The authors would like to thank Paul Stavrinou, Grigorios Itskos, Edmund Harbord and Peter Spencer for supplying some of the data presented here and Edik Refailov of the university of Dundee, U.K. for supplying the QD laser used in figure 3.21.

References

1. L. Goldstein, F. Glas, J.Y. Marzin, M.N. Charasse, and G. Le Roux, *Appl. Phys. Lett.* **47**, 1099 (1985).
2. D.J. Eaglesham and M. Cerullo, *Phys. Rev. Lett.* **64**, 1943 (1990).
3. S. Guha, A. Madhukar, and K.C. Rajkumar, *Appl. Phys. Lett.* **57**, 2110 (1990).
4. D. Leonard, K. Pond, and P.M. Petroff, *Phys. Rev. B.* **50**, 11687 (1994).
5. J.M. Gérard, I. Cabrol, J.Y. Marzin, N. Lebouché, and J.M. Moison, *Mat. Sci. Eng. B.* **37**, 8 (1996).
6. T.J. Krzyzewski, P.B. Joyce, G.R. Bell, and T.S. Jones, *Phys. Rev. B.* **66**, 201302 (2002).
7. J.M. Moison, F. Houzay, F. Barthe, L. Leprince, E. André, and O. Vatel, *Appl. Phys. Lett.* **64**, 196 (1994).
8. N.P. Kobayashi, T.R. Ramachandran, P. Chen, and A. Madhukar, *Appl. Phys. Lett.* **68**, 3299 (1996).
9. P.B. Joyce, T.J. Krzyzewski, G.R. Bell, T.S. Jones, E.C. Le Ru, and R. Murray, *Phys. Rev. B.* **64**, 235317 (2001).
10. H. Kissel, U. Müller, C. Walther, W.T. Masselink, Yu.I. Mazur, G.G. Tarasov, and M.P. Lisitsa, *Phys. Rev. B.* **62**, 7213 (2000).
11. G. Saint-Girons, G. Patriarche, L. Largeau, J. Coelho, A. Mereuta, J.M. Moison, J.M. Gérard, and I. Sagnes, *Appl. Phys. Lett.* **79**, 2157 (2001).
12. U.W. Pohl, K. Pötschke, A. Schliwa, F. Guffarth, D. Bimberg, N.D. Zakharov, P. Werner, M.B. Lifshits, V.A. Shchukin, and D.E. Jesson, *Phys. Rev. B.* **72**, 245332 (2005).
13. J. Drucker, *Phys. Rev. B.* **48**, 18203 (1993).
14. H. Saito, K. Nishi, and S. Sugou, *Appl. Phys. Lett.* **73**, 2742 (1998).
15. J. Garcia, G. Medeiros-Ribeiro, K. Schmidt, T. Ngo, J. Feng, A. Lorke, J. Kotthaus, and P. Petroff, *Appl. Phys. Lett.* **71**, 2014 (1997).
16. I. Kegel, T.H. Metzger, A. Lorke, J. Piesl, J. Stangl, G. Bauer, K. Nordland, W.V. Schoenfeld, and P.M. Petroff, *Phys. Rev. B.* **63**, 035318 (2001).
17. P.B. Joyce, T.J. Krzyzewski, G.R. Bell, and T.S. Jones, *Appl. Phys. Lett.* **79**, 3615 (2001).
18. Q. Xie, P. Chen, A. Kalburge, T. Ramachandran, A. Nayfonov, A. Konkar, and A. Madhukar, *J. Crys. Growth* **150**, 357 (1995).
19. U. Woggon, W. Langbein, J.M. Hvam, A. Rosenauer, T. Remmele, and D. Gerthsen, *Appl. Phys. Lett.* **71**, 377 (1997).
20. P.D. Siverns, S. Malik, G. McPherson, D. Childs, C. Roberts, R. Murray, B.A. Joyce, and H. Davock, *Phys. Rev. B.* **58**, R10127 (1998).
21. R.P. Mirin, J.P. Ibbetson, K. Nishi, A.C. Gossard, and J.E. Bowers, *Appl. Phys. Lett.* **67**, 3795 (1995).
22. D.L. Huffaker, G. Park, Z. Zou, O.B. Shchekin, and D.G. Deppe, *Appl. Phys. Lett.* **73**, 2564 (1998).
23. R. Murray, D. Childs, S. Malik, P. Siverns, C. Roberts, J.-M. Hartmann, and P. Stavrinou, *Jpn. J. Appl. Phys* **38**, 528 (1999).
24. K. Nishi, H. Saito, S. Sugou, and J.-S. Lee, *Appl. Phys. Lett.* **74**, 1111 (1999).
25. V.M. Ustinov, N.A. Maleev, A.E. Zhukov, A.R. Kovsh, A.Yu. Egorov, A.V. Lunev, B.V. Volovik, I.L. Krestnikov, Yu.O. Musikhin, N.A. Bert, P.S. Kopev, and Zh.I. Alferov, *Appl. Phys. Lett.* **74**, 2815 (1999).
26. N.N. Ledentsov, A.R. Kovsh, A.E. Zhukov, N.A. Maleev, S.S. Mikhrin, A.P. Vasil'ev, E.S. Semenova, M.V. Maximov, Yu.M. Shernyakov, N.V. Kryzhanovskaya, V.M. Ustinov, and D. Bimberg, *Electron. Lett.* **39**, 1126 (2003).
27. E.C. Le Ru, P. Howe, T.S. Jones, and R. Murray, *Phys. Rev. B.* **67**, 165303 (2003).
28. J.W. Gray, D. Childs, S. Malik, P. Siverns, C. Roberts, P.N. Stavrinou, M. Whitehead, R. Murray, and G. Parry, *Electron. Lett.* **35**, 242 (1999).
29. J.A. Lott, N.N. Ledentsov, V.M. Ustinov, N.A. Maleev, A.E. Zhukov, A.R. Kovsh, M.V. Maximov, B.V. Volovik, Zh.I. Alferov, and D. Bimberg, *Electron. Lett.* **36**, 1384 (2000).
30. M. Asada, Y. Miyamoto, and Y. Suematsu, *IEEE J. Quantum Electron.* **22**, 1915 (1986).
31. N. Kirstaedter, O.G. Schmidt, N.N. Ledentsov, D. Bimberg, V.M. Ustinov, A.Yu. Egorov, A.E. Zhukov, M.V. Maximov, P.S. Kop'ev, and Zh.I. Alferov, *Appl. Phys. Lett.* **69**, 1226 (1996).
32. Q. Xie, A. Madhukar, P. Chen, and N.P. Kobayashi, *Phys. Rev. Lett.* **75**, 2542 (1995).

33. W. Wu, J.R. Tucker, G.S. Solomon, and J.S. Harris, Jr, *Appl. Phys. Lett.* **71**, 1083 (1997).
34. O.G. Schmidt, O. Kienzle, Y. Hao, K. Eberl, and F. Ernst, *Appl. Phys. Lett.* **74**, 1272 (1999).
35. B. Lita, R.S. Goldman, J.D. Phillips, and P.K. Bhattacharya, *Appl. Phys. Lett.* **75**, 2797 (1999).
36. O.G. Schmidt and K. Eberl, *Phys. Rev. B.* **61**, 13721 (2000).
37. K. Eberl, M. Lipinski, Y.M. Manz, N.Y. Jin-Phillipp, W. Winter, C. Lange, and O.G. Schmidt, *Thin Solid Films* **380**, 183 (2000).
38. G.S. Solomon, J.A. Trezza, A.F. Marshall, and J.S. Harris, Jr, *Phys. Rev. Lett.* **76**, 952 (1996).
39. S. Fafard, M. Spanner, J.P. McCaffrey, and Z.R. Wasilewski, *Appl. Phys. Lett.* **76**, 2268 (2000).
40. N.N. Ledentsov, V.A. Shchukin, M. Grundmann, N. Kirstaedter, J. Böhrer, O. Schmidt, D. Bimberg, V.M. Ustinov, A.Yu. Egorov, A.E. Zhukov, P.S. Kop'ev, S.V. Zaitsev, N.Yu. Gordeev, Zh.I. Alferov, A.I. Borokov, A.O. Kosogov, S.S. Ruminov, P. Werner, U. Gösele, and J. Heydenreich, *Phys. Rev. B.* **54**, 8743 (1996).
41. M. Colocci, A. Vinattieri, L. Lippi, F. Bogani, M. Rosa-Clot, S. Taddei, A. Bosacchi, S. Franchi, and P. Frigeri, *Appl. Phys. Lett.* **74**, 564 (1999).
42. M.O. Lipinski, H. Schuler, O.G. Schmidt, K. Eberl, and N.Y. Jin-Phillipp, *Appl. Phys. Lett.* **77**, 1789 (2000).
43. H. Heidemeyer, S. Kiravittaya, C. Müller, N.Y. Jin-Phillipp, and O.G. Schmidt, *Appl. Phys. Lett.* **80**, 1544 (2002).
44. E.C. Le Ru, A.J. Bennett, C. Roberts, and R. Murray, *J. Appl. Phys.* **91**, 1365 (2002).
45. Z.R. Wasilewski, S. Fafard, and J.P. McCaffrey, *J. Crys. Growth* **201/202**, 1131 (1999).
46. S. Fafard, Z.R. Wasilewski, C.Ni. Allen, D. Picard, M. Spanner, J.P. McCaffrey, and P.G. Piva, *Phys. Rev. B.* **59**, 15368 (1999).
47. Y. Arakawa and H. Sakaki, *Appl. Phys. Lett.* **40**, 939 (1982).
48. M. Grundmann, O. Stier, and D. Bimberg, *Phys. Rev. B.* **52**, 11969 (1995).
49. H. Jiang and J. Singh, *Phys. Rev. B.* **56**, 4696 (1997).
50. C. Pryor, *Phys. Rev. B.* **57**, 7190 (1998).
51. O. Stier, M. Grundmann, and D. Bimberg, *Phys. Rev. B.* **59**, 5688 (1999).
52. L.-W. Wang, J. Kim, and A. Zunger, *Phys. Rev. B.* **59**, 5678 (1999).
53. N. Liu, J. Tersoff, O. Baklenov, A.L. Holmes, Jr, and C.K. Shih, *Phys. Rev. Lett.* **84**, 334 (2000).
54. D.M. Bruls, J.W.A.M. Vugs, P.M. Koenraad, H.W.M. Salemink, J.H. Wolter, M. Hopkinson, M.S. Skolnick, F. Long, and S.P.A. Gill, *Appl. Phys. Lett.* **81**, 1708 (2002).
55. A. Rosenauer, U. Fischer, D. Gerthsen, and A. Förster, *Appl. Phys. Lett.* **71**, 3868 (1997).
56. A. Lemaître, G. Patriarche, and F. Glas, *Appl. Phys. Lett.* **85**, 3717 (2004).
57. P. Wang, A.L. Bleloch, M. Falke, P.J. Goodhew, J. Ng, and M. Missous, *Appl. Phys. Lett.* **89**, 072111 (2006).
58. I. Kegel, T.H. Metzger, A. Lorke, J. Peisl, J. Stangl, G. Bauer, J.M. García, and P.M. Petroff, *Phys. Rev. Lett.* **85**, 1694 (2000).
59. M. Grundmann, N.N. Ledentsov, O. Stier, J. Böhrer, D. Bimberg, V.M. Ustinov, P.S. Kop'ev, and Zh.I. Alferov, *Phys. Rev. B.* **53**, R10509 (1996).
60. P.N. Brounkov, A. Polimeni, S.T. Stoddart, M. Henini, L. Eaves, P.C. Main, A.R. Kovsh, Yu.G. Musikhin, and S.G. Konnikov, *Appl. Phys. Lett.* **73**, 1092 (1998).
61. M.J. Steer, D.J. Mowbray, W.R. Tribe, M.S. Skolnick, M.D. Sturge, M. Hopkinson, A.G. Cullis, C.R. Whitehouse, and R. Murray, *Phys. Rev. B.* **54**, 17738 (1996).
62. I.E. Itskevich, M.S. Skolnick, D.J. Mowbray, I.A. Trojan, S.G. Lyapin, L.R. Wilson, M.J. Steer, M. Hopkinson, L. Eaves, and P.C. Main, *Phys. Rev. B.* **60**, R2185 (1999).
63. L.W. Wang, A.J. Williamson, A. Zunger, H. Jiang, and J. Singh, *Appl. Phys. Lett.* **76**, 339 (2000).
64. O. Stier, in: M. Grundmann (ed.) *Nano-optoelectronics: Concepts, Physics and Devices* (Springer-Verlag, Berlin, 2002).
65. A.W.E. Minnaert, A.Yu. Silov, W. van der Vleuten, J.E.M. Haverkort, and J.H. Wolter, *Phys. Rev. B.* **63**, 075303 (2001).
66. S. Cortez, O. Krebs, P. Voisin, and J.M. Gérard, *Phys. Rev. B.* **63**, 233306 (2001).
67. P.W. Fry, I.E. Itskevich, D.J. Mowbray, M.S. Skolnick, J.J. Finley, J.A. Barker, E.P. O'Reilly, L.R. Wilson, I.A. Larkin, P.A. Maksym, M. Hopkinson, M. Al-Khafaji, J.P.R. David, A.G. Cullis, G. Hill, and J.C. Clark, *Phys. Rev. Lett.* **84**, 733 (2000).
68. P.B. Joyce, T.J. Krzyzewski, G.R. Bell, B.A. Joyce, and T.S. Jones, *Phys. Rev. B.* **58**, R15981 (1998).

69. T. Walther, A.G. Cullis, D.J. Norris, and M. Hopkinson, *Phys. Rev. Lett.* **86**, 2381 (2001).

70. A.J. Williamson, L.W. Wang, and A. Zunger, *Phys. Rev. B.* **62**, 12963 (2000).

71. W. Sheng and J.-P. Leburton, *Phys. Rev. B.* **63**, 161301 (2001).

72. J. Shumway, A.J. Williamson, A. Zunger, A. Passaseo, M. DeGiorgi, R. Cingolani, M. Catalano, and P. Crozier, *Phys. Rev. B.* **64**, 125302 (2001).

73. J.J. Finley, M. Sabathil, P. Vogl, G. Abstreiter, R. Oulton, A.I. Tartakovskii, D.J. Mowbray, M.S. Skolnick, S.L. Liew, A.G. Cullis, and M. Hopkinson, *Phys. Rev. B.* **70**, 201308 (2004).

74. A. Wojs, P. Hawrylak, S. Fafard, and L. Jacak, *Phys. Rev. B.* **54**, 5604 (1996).

75. S. Raymond, S. Fafard, P.J. Poole, A. Wojs, P. Hawrylak, S. Charbonneau, D. Leonard, R. Leon, P.M. Petroff, and J.L. Merz, *Phys. Rev. B.* **54**, 11548 (1996).

76. M. Fricke, A. Lorke, J.P. Kotthaus, G. Medeiros-Ribeiro, and P.M. Petroff, *Europhys. Lett.* **36**, 197 (1996).

77. R.J. Warburton, C.S. Dürr, K. Karrai, J.P. Kotthaus, G. Medeiros-Ribeiro, and P.M. Petroff, *Phys. Rev. Lett.* **79**, 5282 (1997).

78. O.B. Shchekin, G. Park, D.L. Huffaker, and D.G. Deppe, *Appl. Phys. Lett.* **77**, 466 (2000).

79. Y.Q. Wei, S.M. Wang, F. Ferdos, J. Vukusic, A. Larsson, Q.X. Zhao, and M. Sadeghi, *Appl. Phys. Lett.* **81**, 1621 (2002).

80. H.Y. Liu, C.M. Tey, I.R. Sellers, T.J. Badcock, D.J. Mowbray, M.S. Skolnick, R. Beanland, M. Hopkinson, and A.G. Cullis, *J. Appl. Phys.* **98**, 083516 (2005).

81. B.N. Murdin, A.R. Hollingworth, J.A. Barker, D.G. Clarke, P.C. Findlay, C.R. Pidgeon, J.-P.R. Wells, I.V. Bradley, S. Malik, and R. Murray, *Phys. Rev. B.* **62**, R7755 (2000).

82. I.R. Sellers, D.J. Mowbray, T.J. Badcock, J.-P.R. Wells, P.J. Phillips, D.A. Carder, H.Y. Liu, K.M. Groom, and M. Hopkinson, *Appl. Phys. Lett.* **88**, 081108 (2006).

83. U. Bockelmann and G. Bastard, *Phys. Rev. B.* **42**, 8947 (1990).

84. H. Benisty, C.M. Sotomayor-Torrès, and C. Weisbuch, *Phys. Rev. B.* **44**, 10945 (1991).

85. S. Marcinkevičius and R. Leon, *Appl. Phys. Lett.* **76**, 2406 (2000).

86. K. Mukai, N. Ohtsuka, H. Shoji, and M. Sugawara, *Appl. Phys. Lett.* **68**, 3013 (1996).

87. R. Heitz, M. Veit, N.N. Ledentsov, A. Hoffmann, D. Bimberg, V.M. Ustinov, P.S. Kop'ev, and Zh.I. Alferov, *Phys. Rev. B.* **56**, 10435 (1997).

88. B. Ohnesorge, M. Albrecht, J. Oshinowo, A. Forchel, and Y. Arakawa, *Phys. Rev. B.* **54**, 11532 (1996).

89. T.S. Sosnowski, T.B. Norris, H. Jiang, J. Singh, K. Kamath, and P. Bhattacharya, *Phys. Rev. B.* **57**, R9423 (1998).

90. Y. Toda, O. Moriwaki, M. Nishioka, and Y. Arakawa, *Phys. Rev. Lett.* **82**, 4114 (1999).

91. C. Kammerer, G. Cassabois, C. Voisin, C. Delelande, Ph. Roussignol, and J.M. Gérard, *Phys. Rev. Lett.* **87**, 207401 (2001).

92. A. Vasanelli, R. Ferreira, and G. Bastard, *Phys. Rev. Lett.* **89**, 216804 (2002).

93. R. Oulton, J.J. Finley, A.I. Tartakovskii, D.J. Mowbray, M.S. Skolnick, M. Hopkinson, A. Vasanelli, R. Ferreira, and G. Bastard, *Phys. Rev. B.* **68**, 235301 (2003).

94. E.W. Bogaart, J.E.M. Haverkort, T. Mano, T. van Lippen, R. Nötzel, and J.H. Wolter, *Phys. Rev. B.* **72**, 195301 (2005).

95. G. Rainò, G. Visimberga, A. Salhi, M. De Vittorio, A. Passaseo, R. Cingolani, and M. De Giorgi, *Appl. Phys. Lett.* **90**, 111907 (2007).

96. T. Inoshita and H. Sakaki, *Phys. Rev. B.* **56**, R4355 (1997).

97. S. Hameau, Y. Guldner, O. Verzelen, R. Ferreira, G. Bastard, J. Zeman, A. Lemaître, and J.M. Gérard, *Phys. Rev. Lett.* **83**, 4152 (1999).

98. S. Sauvage, P. Boucaud, R.P.S.M. Lobo, F. Bras, G. Fishman, R. Prazeres, F. Glotin, J.M. Ortega, and J.-M. Gérard, *Phys. Rev. Lett.* **88**, 177402 (2002).

99. O. Verzelen, R. Ferreira, and G. Bastard, *Phys. Rev. Lett.* **88**, 146803 (2002).

100. X.-Q. Li, H. Nakayama, and Y. Arakawa, *Phys. Rev. B.* **59**, 5069 (1999).

101. E.A. Zibik, L.R. Wilson, R.P. Green, G. Bastard, R. Ferreira, P.J. Phillips, D.A. Carder, J.-P.R. Wells, J.W. Cockburn, M.S. Skolnick, M.J. Steer, and M. Hopkinson, *Phys. Rev. B.* **70**, 161305 (2004).

102. O. Verzelen, R. Ferreira, and G. Bastard, *Phys. Rev. B.* **62**, R4809 (2000).

103. G.A. Narvaez, G. Bester, and A. Zunger, *Phys. Rev. B.* **74**, 075403 (2006).

104. S. Adachi, ed. *Handbook on Physical Properties of Semiconductors Volume 2: III–V Compound Semiconductors* (Kluwer Academic, Boston, 2004).

105. S. Fafard, S. Raymond, G. Wang, R. Leon, D. Leonard, S. Charbonneau, J.L. Merz, P.M. Petroff, and J.E. Bowers, *Surf. Sci.* **361/362**, 778 (1996).

106. P.G. Eliseev, H. Li, G.T. Liu, A. Stintz, T.C. Newell, L.F. Lester, and K.J. Malloy, *IEEE J. Sel. Topic. Quantum Electron.* **7**, 135 (2001).

107. G. Ortner, M. Schwab, M. Bayer, R. Pässler, S. Fafard, Z. Wasilewski, P. Hawrylak, and A. Forchel, *Phys. Rev. B.* **72**, 085328 (2005).

108. L. Brusaferri, S. Sanguinetti, E. Grilli, M. Guzzi, A. Bignazzi, F. Bogani, L. Carraresi, M. Colocci, A. Bosacchi, P. Frigeri, and S. Franchi, *Appl. Phys. Lett.* **69**, 3354 (1996).

109. Z.Y. Xu, Z.D. Lu, X.P. Yang, Z.L. Yuan, B.Z. Zheng, J.Z. Xu, W.K. Ge, Y. Wang, J. Wang, and L.L. Chang, *Phys. Rev. B.* **54**, 11528 (1996).

110. S. Malik, C. Roberts, R. Murray, and M. Pate, *Appl. Phys. Lett.* **71**, 1987 (1997).

111. D.I. Lubyshev, P.P. González-Borrero, E. Marega, Jr, E. Petitprez, N. La Scala, Jr, and P. Basmaji, *Appl. Phys. Lett.* **68**, 205 (1996).

112. H. Lee, W. Yang, and P.C. Sercel, *Phys. Rev. B.* **55**, 9757 (1997).

113. S. Sanguinetti, M. Henini, M. Grassi Alessi, M. Capizzi, P. Frigeri, and S. Franchi, *Phys. Rev. B.* **60**, 8276 (1999).

114. E.C. Le Ru, J. Fack, and R. Murray, *Phys. Rev. B.* **67**, 245318 (2003).

115. K. Mukai, N. Ohtsuka, and M. Sugawara, *Appl. Phys. Lett.* **70**, 2416 (1997).

116. M. Grundmann and D. Bimberg, *Phys. Rev. B.* **55**, 9740 (1997).

117. A. Zrenner, *J. Chem. Phys.* **112**, 7790 (2000).

118. J.-Y. Marzin, J.-M. Gérard, A. Izraël, D. Barrier, and G. Bastard, *Phys. Rev. Lett.* **73**, 716, (1994).

119. M. Grundmann, J. Christen, N.N. Ledentsov, J. Böhrer, D. Bimberg, S.S. Ruvimov, P. Werner, U. Richter, U. Gösele, J. Heydenreich, V.M. Ustinov, A.Yu. Egorov, A.E. Zhukov, P.S. Kop'ev, and Zh.I. Alferov, *Phys. Rev. Lett.* **74**, 4043 (1995).

120. E. Dekel, D. Gershoni, E. Ehrenfreund, J.M. Garcia, and P.M. Petroff, *Phys. Rev. B.* **61**, 11009 (2000).

121. J.J. Finley, A.D. Ashmore, A. Lemaître, D.J. Mowbray, M.S. Skolnick, I.E. Itskevich, P.A. Maksym, M. Hopkinson, and T.F. Krauss, *Phys. Rev. B.* **63**, 073307 (2001).

122. S. Rodt, R. Heitz, A. Schliwa, R.L. Sellin, F. Guffarth, and D. Bimberg, *Phys. Rev. B.* **68**, 035331 (2003).

123. A.S. Lenihan, M.V. Gurudev Dutt, D.G. Steel, S. Ghosh, and P. Bhattacharya, *Phys. Rev. B.* **69**, 045306 (2004).

124. M. Bayer, A. Kuther, A. Forchel, A. Gorbunov, V.B. Timofeev, F. Schäfer, J.P. Reithmaier, T.L. Reinecke, and S.N. Walck, *Phys. Rev. Lett.* **82**, 1748 (1999).

125. H.W. van Kesteren, E.C. Cosman, W.A.J.A. van der Poel, and C.T. Foxon, *Phys. Rev. B.* **41**, 5283 (1990).

126. E. Blackwood, M.J. Snelling, R.T. Harley, S.R. Andrews, and C.T.B. Foxon, *Phys. Rev. B.* **50**, 14246 (1994).

127. D. Gammon, E.S. Snow, B.V. Shanabrook, D.S. Katzer, and D. Park, *Phys. Rev. Lett.* **76**, 3005 (1996).

128. R.J. Young, R.M. Stevenson, A.J. Shields, P. Atkinson, K. Cooper, D.A. Ritchie, K.M. Groom, M.J. Steer, H.Y. Liu, and M. Hopkinson, *Phys. Rev. B.* **72**, 113305 (2005).

129. O.G. Schmidt, S. Kiravittaya, Y. Nakamura, H. Heidemeyer, R. Songmuang, C. Müller, N.Y. Jin-Phillipp, K. Eberl, H. Wawra, S. Christiansen, H. Gräbeldinger, and H. Schweizer, *Surf. Sci.* **514**, 10 (2002).

130. Z. Mi and P. Bhattacharya, *J. Appl. Phys.* **98**, 023510 (2005).

131. K. Ota, N. Usami, and Y. Shiraki, *Physica E.* **2**, 573 (1998).

132. M. Bayer and A. Forchel, *Phys. Rev. B.* **65**, 041308 (2002).

133. I. Favero, G. Cassabois, R. Ferreira, D. Darson, C. Voisin, J. Tignon, C. Delalande, G. Bastard, Ph. Roussignol, and J.M. Gérard, *Phys. Rev. B.* **68**, 233301 (2003).

134. P. Borri, W. Langbein, J. Mørk, J.M. Hvam, F. Heinrichsdorff, M.-H. Mao, and D. Bimberg, *Phys. Rev. B.* **60**, 7784 (1999).

135. P. Borri, W. Langbein, S. Schneider, U. Woggon, R.L. Sellin, D. Ouyang, and D. Bimberg, *Phys. Rev. Lett.* **87**, 157401 (2001).
136. D. Birkedal, K. Leosson, and J.M. Hvam, *Phys. Rev. Lett.* **87**, 227401 (2001).
137. M. Sugawara, K. Mukai, Y. Nakata, H. Ishikawa, and A. Sakamoto, *Phys. Rev. B.* **61**, 7595 (2000).
138. M. Willatzen, T. Tanaka, Y. Arakawa, and J. Singh, *IEEE J. Quantum Electron.* **30**, 640 (1994).
139. T.C. Newell, D.J. Bossert, A. Stintz, B. Fuchs, K.J. Malloy, and L.F. Lester, *IEEE Photon. Technol. Lett.* **11**, 1527 (1999).
140. A.V. Uskov, Y. Boucher, J. Le Bihan, and J. McInerney, *Appl. Phys. Lett.* **73**, 1499 (1998).
141. T. Akiyama, O. Wada, H. Kuwatsuka, T. Simoyama, Y. Nakata, K. Mukai, M. Sugawara, and H. Ishikawa, *Appl. Phys. Lett.* **77**, 1753 (2000).
142. T.W. Berg, S. Bischoff, L. Magnusdottir, and J. Mørk, *IEEE Photon. Technol. Lett.* **13**, 541 (2001).
143. M. Sugawara, N. Hatori, T. Akiyama, Y. Nakata, and H. Ishikawa, *Jpn. J. Appl. Phys.* **40**, L488 (2001).
144. A.V. Uskov, T.W. Berg, and J. Mørk, *IEEE J. Quantum Electron.* **40**, 306 (2004).
145. Z.Y. Zhang, Z.G. Wang, B. Xu, P. Jin, Z.Z. Sun, and F.Q. Liu, *IEEE Photon. Technol. Lett.* **16**, 27 (2004).
146. L.H. Li, M. Rossetti, A. Fiore, L. Occhi, and C. Velez, *Electron. Lett.* **41**, 41 (2005).
147. S.K. Ray, K.M. Groom, R. Alexander, K. Kennedy, H.Y. Liu, M. Hopkinson, and R.A. Hogg, *J. Appl. Phys.* **100**, 103105 (2006).
148. U. Keller, K.J. Weingarten, F.X. Kartner, D. Kopf, B. Braun, I.D. Jung, R. Fluck, C. Honninger, N. Matuschek, and J. Aus der Au, *IEEE J. Sel. Topic. Quantum Electron.* **2**, 435 (1996).
149. M.G. Thompson, C. Marinelli, K.T. Tan, K.A. Williams, R.V. Penty, I.H. White, I.N. Kaiander, R.L. Sellin, D. Bimberg, D.-J. Kang, M.G. Blamire, F. Visinka, S. Jochum, and S. Hansmann, *Electron. Lett.* **39**, 1121 (2003).
150. E.U. Rafailov, S.J. White, A.A. Lagatsky, A. Miller, W. Sibbett, D.A. Livshits, A.E. Zhukov, and V.M. Ustinov, *IEEE Photon. Technol. Lett.* **16**, 2439 (2004).
151. H.C. Lai, A. Li, K.W. Su, M.L. Ku, Y.F. Chen, and K.F. Huang, *Opt. Lett.* **30**, 480 (2005).
152. M.G. Thompson, C. Marinelli, X. Zhao, R.L. Sellin, R.V. Penty, I.H. White, I.N. Kaiander, D. Bimberg, D.-J. Kang, and M.G. Blamire, *Electron. Lett.* **41**, 248 (2005).
153. M.A. Cataluna, W. Sibbett, D.A. Livshits, J. Weimert, A.R. Kovsh, and E.U. Rafailov, *Appl. Phys. Lett.* **89**, 081124 (2006).
154. H.-Y. Wang, H.-C. Cheng, S.-D. Lin, and C.-P. Lee, *Appl. Phys. Lett.* **90**, 081112 (2007).
155. N. Kirstaedter, N.N. Ledentsov, M. Grundmann, D. Bimberg, V.M. Ustinov, S.S. Ruvimov, M.V. Maximov, P.S. Kop'ev, Zh.I. Alferov, U. Richter, P. Werner, U. Gösele, and J. Heydenreich, *Electron. Lett.* **30**, 1416 (1994).
156. R. Mirin, A. Gossard, and J. Bowers, *Electron. Lett.* **32**, 1732 (1996).
157. K. Kamath, P. Bhattacharya, T. Sosnowski, T. Norris, and J. Phillips, *Electron. Lett.* **32**, 1374 (1996).
158. G.T. Liu, A. Stintz, H. Li, K.J. Malloy, and L.F. Lester, *Electron. Lett.* **35**, 1163 (2000).
159. G. Park, O.B. Shcheckin, S. Csutak, D.L. Huffaker, and D.G. Deppe, *Appl. Phys. Lett.* **75**, 3267 (1999).
160. E. Herrmann, P.M. Smowton, H.D. Summers, J.D. Thomson, and M. Hopkinson, *Appl. Phys. Lett.* **77**, 163 (2000).
161. P.W. Fry, L. Harris, S.R. Parnell, J.J. Finley, A.D. Ashmore, D.J. Mowbray, M.S. Skolnick, M. Hopkinson, G. Hill, and J.C. Clark, *J. Appl. Phys.* **87**, 615 (2000).
162. P.M. Smowton, E. Herrmann, Y. Ning, H.D. Summers, P. Blood, and M. Hopkinson, *Appl. Phys. Lett.* **78**, 2629 (2001).
163. T. Amano, T. Sugaya, and K. Komori, *IEEE Photon. Technol. Lett.* **18**, 619 (2006).
164. A. Salhi, L. Martiradonna, G. Visimberga, V. Tasco, L. Fortunato, M.T. Todaro, R. Cingolani, A. Passaseo, and M. De Vittorio, *IEEE Photon. Technol. Lett.* **18**, 1735 (2006).
165. H. Shoji, Y. Nakata, K. Mukai, Y. Sugiyama, M. Sugawara, N. Yokoyama, and H. Ishikawa, *Jpn. J. Appl. Phys.* **35**, L903 (1996).
166. H.Y. Liu, I.R. Sellers, T.J. Badcock, D.J. Mowbray, M.S. Skolnick, K.M. Groom, M. Gutiérrez, M. Hopkinson, J.S. Ng, J.P.R. David, and R. Beanland, *Appl. Phys. Lett.* **85**, 704 (2004).

167. H.Y. Liu, I.R. Sellers, M. Gutiérrez, K.M. Groom, W.M. Soong, M. Hopkinson, J.P.R. David, R. Beanland, T.J. Badcock, D.J. Mowbray, and M.S. Skolnick, *J. Appl. Phys.* **96**, 1988 (2004).

168. I.R. Sellers, H.Y. Liu, K.M. Groom, D.T. Childs, D. Robbins, T.J. Badcock, M. Hopkinson, D.J. Mowbray, and M.S. Skolnick, *Electron. Lett.* **40**, 1412 (2004).

169. P.K. Kondratko, S.-L. Chuang, G. Walter, T. Chung, and N. Holonyak, *Appl. Phys. Lett.* **83**, 4818 (2003).

170. S. Fathpour, P. Bhattacharya, S. Pradhan, and S. Ghosh, *Electron. Lett.* **39**, 1443 (2003).

171. D. Rodriguez, I. Esquivias, S. Deubert, J.P. Reighmaier, A. Forchel, M. Krakowski, M. Calligaro, and O. Parillaud, *IEEE J. Quantum Electron.* **41**, 117 (2005).

172. B. Dagens, A. Markus, J.X. Chen, J.-G. Provost, D. Make, O. Le Gouezigou, J. Landreau, A. Fiore, and B. Thedrez, *Electron. Lett.* **41**, 323 (2005).

173. A.E. Zhukov, V.M. Ustinov, A. Yu. Egorov, A.R. Kovsh, A.F. Tsatsul'nikov, N.N. Ledentsov, S.V. Zaitsev, N. Yu. Gordeev, P.S. Kop'ev, and Z.I. Alferov, *Jpn. J. Appl. Phys.* **36**, 4216 (1997).

174. Z. Zou, O.B. Shchekin, G. Park, D.L. Huffaker, and D.G. Deppe, *IEEE Photon. Technol. Lett.* **10**, 1673 (1998).

175. O.B. Shchekin and D.G. Deppe, *Appl. Phys. Lett.* **80**, 2758 (2002).

176. O.B. Shchekin and D.G. Deppe, *Appl. Phys. Lett.* **80**, 3277 (2002).

177. T.J. Badcock, H.Y. Liu, K.M. Groom, C.Y. Jin, M. Gutiérrez, M. Hopkinson, D.J. Mowbray, and M.S. Skolnick, *Electron. Lett.* **42**, 922 (2006).

178. S.S. Mikhrin, A.R. Kovsh, I.L. Krestnikov, A.V. Kozhukhov, D.A. Livshits, N.N. Ledentsov, Yu.M. Shernyakov, I.I. Novikov, M.V. Maximov, V.M. Ustinov, and Zh.I. Alferov, *Semicond. Sci. Technol.* **20**, 340 (2005).

179. I.P. Marko, N.F. Massé, S.J. Sweeney, A.D. Andreev, A.R. Adams, N. Hatori, and M. Sugawara, *Appl. Phys. Lett.* **87**, 211114 (2005).

180. S. Fathpour, Z. Mi, P. Bhattacharya, A.R. Kovsh, S.S. Mikhrin, I.L. Krestnikov, A.V. Kozhukhov, and N.N. Ledentsov, *Appl. Phys. Lett.* **85**, 5164 (2004).

181. T.J. Badcock, R.J. Royce, D.J. Mowbray, M.S. Skolnick, H.Y. Liu, M. Hopkinson, K.M. Groom, and Q. Jiang, *Appl. Phys. Lett.* **90**, 111102 (2007).

182. L. Harris, D.J. Mowbray, M.S. Skolnick, M. Hopkinson, and G. Hill, *Appl. Phys. Lett.* **73**, 969 (1998).

183. M. Sugawara, K. Mukai, and Y. Nakata, *Appl. Phys. Lett.* **74**, 1561 (1999).

184. A. Patanè, A. Polimeni, M. Henini, L. Eaves, P.C. Main, and G. Hill, *J. Appl. Phys.* **85**, 625 (1999).

185. E.P. O'Reilly, A.I. Onischenko, E.A. Avrutin, D. Bhattacharya, and J.H. Marsh, *Electron. Lett.* **34**, 2035 (1998).

186. L. Harris, A.D. Ashmore, D.J. Mowbray, M.S. Skolnick, M. Hopkinson, G. Hill, and J. Clark, *Appl. Phys. Lett.* **75**, 3512 (1999).

187. A. Patanè, A. Polimeni, L. Eaves, M. Henini, P.C. Main, P.M. Smowton, E.J. Johnston, P.J. Hulyer, E. Herrmann, G.M. Lewis, and G. Hill, *J. Appl. Phys.* **87**, 1943 (2000).

188. A.E. Zhukov, A.R. Kovsh, D.A. Livshits, V.M. Ustinov, and Z.I. Alferov, *Semicond. Sci. Technol.* **18**, 774 (2003).

189. A. Markus, J.X. Chen, C. Paranthoën, A. Fiore, C. Platz, and O. Gauthier-Lafaye, *Appl. Phys. Lett.* **82**, 1818 (2003).

190. A. Markus, M. Rossetti, V. Calligari, J.X. Chen, and A. Fiore, *J. Appl. Phys.* **98**, 104506 (2005).

191. J. Tatebayashi, M. Nishioka, and Y. Arakawa, *Appl. Phys. Lett.* **78**, 3469 (2001).

192. L. Seravalli, M. Minelli, P. Frigeri, S. Franchi, G. Guizzetti, M. Patrini, T. Ciabattoni, and M. Geddo, *J. Appl. Phys.* **101**, 024313 (2007).

193. L.H. Li, M. Rossetti, A. Fiore, and G. Patriarche, *Electron. Lett.* **42**, 638 (2006).

194. I.I. Novikov, N.Yu. Gordeev, M.V. Maximov, Y.M. Shernyakov, A.E. Zhukov, A.P. Vasil'ev, E.S. Semenova, V.M. Ustinov, N.N. Ledentsov, D. Bimberg, N.D. Zakharov, and P. Werner, *Semicond. Sci. Technol.* **20**, 33 (2005).

195. L.Ya. Karachinsky, T. Kettler, N.Yu. Gordeev, I.I. Novikov, M.V. Maximov, Y.M. Shernyakov, N.V. Kryzhanovskaya, A.E. Zhukov, E.S. Semenova, A.P. Vasil'ev, V.M. Ustinov, N.N. Ledentsov, A.R. Kovsh, V.A. Shchukin, S.S. Mikhrin, A. Lochmann, O. Schulz, L. Reissmann, and D. Bimberg, *Electron. Lett.* **41**, 478 (2005).

196. Z. Mi, P. Bhattacharya, and J. Yang, *Appl. Phys. Lett.* **89**, 153109 (2006).

197. J.C. Campbell, D.L. Huffaker, H. Deng, and D.G. Deppe, *Electron. Lett.* **33**, 1337 (1997).

198. O. Baklenov, H. Nie, K.A. Anselm, J.C. Campbell, and B.G. Streetman, *Electron. Lett.* **34**, 694 (1998).
199. H. Saito, K. Nishi, I. Ogura, S. Sugou, and Y. Sugimoto, *Appl. Phys. Lett.* **69**, 3140 (1996).
200. D.L. Huffaker, O. Baklenov, L.A. Graham, B.G. Streetman, and D.G. Deppe, *Appl. Phys. Lett.* **70**, 2356 (1997).
201. D.L. Huffaker, L.A. Graham, and D.G. Deppe, *Electron. Lett.* **33**, 1225 (1997).
202. S. Reitzenstein, A. Bazhenov, A. Gorbunov, C. Hofmann, S. Münch, A. Löffler, M. Kamp, J.P. Reithmaier, V.D. Kulakovskii, and A. Forchel, *Appl. Phys. Lett.* **89**, 051107 (2006).
203. F. Hopfer, A. Mutig, M. Kuntz, G. Fiol, D. Bimberg, N.N. Ledentsov, V.A. Shchukin, S.S. Mikhrin, D.L. Livshits, I.L. Krestnikov, A.R. Kovsh, N.D. Zakharov, and P. Werner, *Appl. Phys. Lett.* **89**, 141106 (2006).
204. Y.H. Chang, P.C. Peng, W.K. Tsai, G. Lin, F.I. Lai, R.S. Hsiao, H.P. Yang, H.C. Yu, K.F. Lin, J.Y. Chi, S.C. Wang, and H.C. Kuo, *IEEE Photon. Technol. Lett.* **18**, 847 (2006).
205. V.M. Ustinov, A.E. Zhukov, N.A. Maleev, A.R. Kovsh, S.S. Mikhrin, B.V. Volovik, Yu.G. Musikhin, Yu.M. Shernyakov, M.V. Maximov, A.F. Tsatsul'nikov, N.N. Ledentsov, Z.I. Alferov, J.A. Lott, and D. Bimberg, *J. Crys. Growth* **227/228**, 1155 (2001).
206. D. Bimberg and C. Ribbat, *Microelectron. J.* **34**, 323 (2003).
207. T. Akiyama, N. Hatori, Y. Nakata, H. Ebe, and M. Sugawara, *Electron. Lett.* **38**, 1139 (2002).
208. T. Kita, N. Tamura, O. Wada, M. Sugawara, Y. Nakata, H. Ebe, and Y. Arakawa, *Appl. Phys. Lett.* **88**, 211106 (2006).
209. M. Oxborrow and A.G. Sinclair, *Contemp. Phys.* **46**, 173 (2005).
210. N. Gisin, G. Ribordy, W. Tittel, and H. Zbinden, *Rev. Mod. Phys.* **74**, 145 (2002).
211. E. Knill, R. Laflamme, and G.J. Milburn, *Nature* **409**, 46 (2001).
212. P. Kok, W.J. Munro, K. Nemoto, T.C. Ralph, J.P. Dowling, and G.J. Milburn, *Rev. Mod. Phys.* **79**, 135 (2007).
213. Th. Basché, W.E. Moerner, M. Orrit, and H. Talon, *Phys. Rev. Lett.* **69**, 1516 (1992).
214. B. Lounis and W.E. Moerner, *Nature* **407**, 491 (2000).
215. C. Kurtsiefer, S. Mayer, P. Zarda, and H. Weinfurter, *Phys. Rev. Lett.* **85**, 290 (2000).
216. A. Beveratos, R. Brouri, T. Gacoin, A. Villing, J.-P. Poizat, and P. Grangier, *Phys. Rev. Lett.* **89**, 187901 (2002).
217. R. Hubbard, Yu.B. Ovchinnikov, J. Cheung, N. Fletcher, R. Murray, and A.G. Sinclair, *J. Mod. Opt.* **54**, 441 (2007).
218. P. Michler, A. Kiraz, C. Becher, W.V. Schoenfeld, P.M. Petroff, L. Zhang, E. Hu, and A. Imamoğlu, *Science* **290**, 2282 (2000).
219. R.M. Thompson, R.M. Stevenson, A.J. Shields, I. Farrer, C.J. Lobo, D.A. Ritchie, M.L. Leadbeater, and M. Pepper, *Phys. Rev. B.* **64**, 201302 (2001).
220. Z. Yuan, B. Kardynal, R.M. Stevenson, A.J. Shields, C.J. Lobo, K. Cooper, N.S. Beattie, D.A. Ritchie, and M. Pepper, *Science* **295**, 102 (2002).
221. P. Atkinson, S.P. Bremner, D. Anderson, G.A.C. Jones, and D.A. Ritchie, *J Vac Sci. Technol. B.* **24**, 1523 (2006).
222. M.H. Baier, E. Pelucchi, E. Kapon, S. Varoutsis, M. Gallart, I. Robert-Philip, and I. Abram, *Appl. Phys. Lett.* **84**, 648 (2004).
223. P. Atkinson, M.B. Ward, S.P. Bremner, D. Anderson, T. Farrow, G.A.C. Jones, A.J. Shields, and D.A. Ritchie, *Jpn. J. Appl. Phys.* **45**, 2519 (2006).
224. R. Hanbury-Brown and R.Q. Twiss, *Nature* **177**(4497) p. 27 (1956).
225. C. Santori, M. Pelton, G. Solomon, Y. Dale, and Y. Yamamoto, *Phys. Rev. Lett.* **86**, 1502 (2001).
226. E. Moreau, I. Robert, L. Manin, V. Theirry-Mieg, J.M. Gérard, and I. Abram, *Phys. Rev. Lett.* **87**, 183601 (2001).
227. A.J. Bennett, D.C. Unitt, P. See, A.J. Shields, P. Atkinson, K. Cooper, and D.A. Ritchie, *Appl. Phys. Lett.* **86**, 181102 (2005).
228. J.M. Gérard, B. Sermage, B. Gayral, B. Legrand, E. Costard, and V. Thierry-Mieg, *Phys. Rev. Lett.* **81**, 1110 (1998).
229. M. Pelton, C. Santori, J. Vučković, B. Zhang, G.S. Solomon, J. Plant, and Y. Yamamoto, *Phys. Rev. Lett.* **89**, 233602 (2002).
230. J.M. Gérard and B. Gayral, *IEEE J. Lightwave Technol.* **17**, 2089 (1999).

231. T.D. Happ, I.I. Tartakovskii, V.D. Kulakovskii, J.-P. Reithmaier, M. Kamp, and A. Forchel, *Phys. Rev. B.* **66**, 041303 (2002).
232. D.G. Gevaux, A.J. Bennett, R.M. Stevenson, A.J. Shields, P. Atkinson, J. Griffiths, D. Anderson, G.A.C. Jones, and D.A. Ritchie, *Appl. Phys. Lett.* **88**, 131101 (2006).
233. E.M. Purcell, *Phys. Rev.* **69**, 681 (1946).
234. G.S. Solomon, M. Pelton, and Y. Yamamoto, *Phys. Rev. Lett.* **86**, 3903 (2001).
235. W.L. Barnes, G. Björk, J.M. Gérard, P. Jonsson, J.A.E. Wasey, P.T. Worthing, and V. Zwiller, *Eur. Phys. J. D.* **18**, 197 (2002).
236. A.J. Bennett, D.C. Unitt, P. Atkinson, D.A. Ritchie, and A.J. Shields, *Opt. Express.* **13**, 50 (2005).
237. K. Takemoto, Y. Sakuma, S. Hirose, T. Usuki, N. Yokoyama, T. Miyazawa, M. Takatsu, and Y. Arakawa, *Jpn. J. Appl. Phys.* **43**, L993 (2004).
238. M.B. Ward, O.Z. Karimov, D.C. Unitt, Z.L. Yuan, P. See, D.G. Gevaux, A.J. Shields, P. Atkinson, and D.A. Ritchie, *Appl. Phys. Lett.* **86**, 201111 (2005).
239. P.B. Joyce, T.J. Krzyzewski, G.R. Bell, T.S. Jones, S. Malik, D. Childs, and R. Murray, *Phys. Rev. B.* **62**, 10891 (2000).
240. B. Alloing, C. Zinoni, V. Zwiller, L.H. Li, C. Monat, M. Gobet, G. Buchs, A. Fiore, E. Pelucchi, and E. Kapon, *Appl. Phys. Lett.* **86**, 101908 (2005).
241. C. Zinoni, B. Alloing, C. Monat, V. Zwiller, L.H. Li, A. Fiore, L. Lunghi, A. Gerardino, H. de Riedmatten, H. Zbinden, and N. Gisin, *Appl. Phys. Lett.* **88**, 131102 (2006).
242. M.B. Ward, T. Farrow, P. See, Z.L. Yuan, O.Z. Karimov, A.J. Bennett, A.J. Shields, P. Atkinson, K. Cooper, and D.A. Ritchie, *Appl. Phys. Lett.* **90**, 063512 (2007).
243. A.K. Ekert, *Phys. Rev. Lett.* **67**, 661 (1991).
244. C.K. Hong, Z.Y. Ou, and L. Mandel, *Phys. Rev. Lett.* **59**, 2044 (1987).
245. Y.H. Shih and C.O. Alley, *Phys. Rev. Lett.* **61**, 2921 (1988).
246. C. Santori, D. Fattal, J. Vučkovič, G.S. Solomon, and Y. Yamamoto, *Nature* **419**, 594 (2002).
247. D. Fattal, K. Inoue, J. Vučkovič, C. Santori, G.S. Solomon, and Y. Yamamoto, *Phys. Rev. Lett.* **92**, 037903 (2004).
248. O. Benson, C. Santori, M. Pelton, and Y. Yamamoto, *Phys. Rev. Lett.* **84**, 2513 (2000).
249. C. Santori, D. Fattal, M. Pelton, G.S. Solomon, and Y. Yamamoto, *Phys. Rev. B.* **66**, 045308 (2002).
250. R.M. Stevenson, R.M. Thompson, A.J. Shields, I. Farrer, B.E. Kardynal, D.A. Ritchie, and M. Pepper, *Phys. Rev. B.* **66**, 081302 (2002).
251. N. Akopian, N.H. Lindner, E. Poem, Y. Berlatzky, J. Avron, D. Gershoni, B.D. Gerardot, and P.M. Petroff, *Phys. Rev. Lett.* **96**, 130501 (2006).
252. A. Greilich, M. Schwab, T. Berstermann, T. Auer, R. Oulton, D.R. Yakovlev, M. Bayer, V. Staverache, D. Reuter, and A. Wieck, *Phys. Rev. B.* **73**, 045323 (2006).
253. K. Kowalik, O. Krebs, A. Lemaître, S. Laurent, P. Senellart, P. Voisin, and J.A. Gaj, *Appl. Phys. Lett.* **86**, 041907 (2005).
254. B.D. Gerardot, S. Seidl, P.A. Dalgarno, R.J. Warburton, D. Granados, J.M. Garcia, K. Kowalik, O. Krebs, K. Karrai, A. Badolato, and P.M. Petroff, *Appl. Phys. Lett.* **90**, 041101 (2007).
255. R.M. Stevenson, R.J. Young, P. See, D.G. Gevaux, K. Cooper, P. Atkinson, I. Farrer, D.A. Ritchie, and A.J. Shields, *Phys. Rev. B.* **73**, 033306 (2006).
256. S. Seidl, M. Kroner, A. Högele, K. Karrai, R.J. Warburton, A. Badolato, and P.M. Petroff, *Appl. Phys. Lett.* **88**, 203113 (2006).
257. W. Langbein, P. Borri, U. Woggon, V. Stavarache, D. Reuter, and A.D. Wieck, *Phys. Rev. B.* **69**, 161301 (2004).
258. A.I. Tartakovskii, M.N. Makhonin, I.R. Sellers, J. Cahill, A.D. Andreev, D.M. Whittaker, J.-P.R. Wells, A.M. Fox, D.J. Mowbray, M.S. Skolnick, K.M. Groom, M.J. Steer, H.Y. Liu, and M. Hopkinson, *Phys. Rev. B.* **70**, 193303 (2004).
259. R. Seguin, A. Schliwa, S. Rodt, K. Pötschke, U.W. Pohl, and D. Bimberg, *Phys. Rev. Lett.* **95**, 257402 (2005).
260. R.M. Stevenson, R.J. Young, P. Atkinson, K. Cooper, D.A. Ritchie, and A.J. Shields, *Nature* **439**, 179 (2006).
261. R.J. Young, R.M. Stevenson, P. Atkinson, K. Cooper, D.A. Ritchie, and A.J. Shields, *New J. Phys.* **8**, 29 (2006).

262. A.I. Tartakovskii, R.S. Kolodka, H.Y. Liu, M.A. Migliorato, M. Hopkinson, M.N. Makhonin, D.J. Mowbray, and M.S. Skolnick, *Appl. Phys. Lett.* **88**, 131115 (2006).
263. N.I. Cade, H. Gotoh, H. Kamada, H. Nakano, and H. Okamoto, *Phys. Rev. B.* **73**, 115322 (2006).
264. W.M. McGee, T.J. Krzyzewski, and T.S. Jones, *J. Appl. Phys.* **99**, 043505 (2006).
265. J. Rudolph, D Hägele, H.M. Gibbs, G. Khitrova, M. Oestreich, *Appl. Phys. Lett.* **82**, 4516 (2003).
266. D. Loss and D.P. DiVincenzo, *Phys. Rev. A.* **57**, 120 (1998).
267. B.E. Kane, *Nature* **393**, 133 (1998).
268. R. Fiederling, M. Keim, G. Reuscher, W. Ossau, G. Schmidt, A. Waag, and L.W. Molenkamp, *Nature* **402**, 787 (1999).
269. B.T. Jonker, Y.D. Park, B.R. Bennett, H.D. Cheong, G. Kioseoglou, and A. Petrou, *Phys. Rev. B.* **62**, 8180 (2000).
270. W. Van Roy, P. Van Dorpe, J. De Boeck, and G. Borghs, *Mat. Sci. Eng. B.* **126**, 155 (2006).
271. F. Meier and B.P. Zachachrenya, eds, *Optical Orientation*, (North-Holland, Amsterdam, 1984).
272. H. Gotoh, H. Ando, H. Kamada, A. Chavez-Pirson, and J. Temmyo, *Appl. Phys. Lett.* **72**, 1341 (1998).
273. M. Paillard, X. Marie, P. Renucci, T. Amand, A. Jbeli, and J.M. Gérard, *Phys. Rev. Lett.* **86**, 1634 (2001).
274. Y. Chye, M.E. White, E. Johnston-Halperin, B.D. Gerardot, D.D. Awschalom, and P.M. Petroff, *Phys. Rev. B.* **66**, 201301 (2002).
275. H. Ohno, *Science* **281**, 951 (1998).
276. Y. Ohno, D.K. Young, B. Beschoten, F. Matsukura, H. Ohno, and D.D. Awschalom, *Nature* **402**, 790 (1999).
277. G. Schmidt, D. Ferrand, L.W. Molenkamp, A.T. Filip, and B.J. van Wees, *Phys. Rev. B.* **62**, R4790 (2000).
278. E.I. Rashba, *Phys. Rev. B.* **62**, R16267 (2000).
279. A.T. Hanbicki, O.M.J. van't Erve, R. Magno, G. Kioseoglou, C.H. Li, B.T. Jonker, G. Itskos, R. Mallory, M. Yasar, and A. Petrou, *Appl. Phys. Lett.* **82**, 4092 (2003).
280. C.H. Li, G. Kioseoglou, O.M. van't Erve, M.E. Ware, D. Gammon, R.M. Stroud, B.T. Jonker, R. Mallory, M. Yasar, and A. Petrou, *Appl. Phys. Lett.* **86**, 132503 (2005).
281. G. Itskos, E. Harbord, S.K. Clowes, E. Clarke, L.F. Cohen, R. Murray, P. Van Dorpe, and W. Van Roy, *Appl. Phys. Lett.* **88**, 022113 (2006).
282. R. Fiederling, P. Grabs, W. Ossau, G. Schmidt, and L.W. Molenkamp, *Appl. Phys. Lett.* **82**, 2160 (2003).
283. C.E. Pryor and M.E. Flatté, *Phys. Rev. Lett.* **91**, 257901 (2003).
284. C.H. Li, G. Kioseoglou, O.M. van't Erve, A.T. Hanbicki, B.T. Jonker, R. Mallory, M. Yasar, and A. Petrou, *Appl. Phys. Lett.* **85**, 1544 (2004).
285. V.F. Motsnyi, P. Van Dorpe, W. Van Roy, E. Goovaerts, V.I. Safarov, G. Borghs, and J. De Boeck, *Phys. Rev. B.* **68**, 245319 (2003).
286. S. Malik, E.C. Le Ru, D. Childs, and R. Murray, *Phys. Rev. B.* **63**, 155313 (2001).
287. T. Nakaoka, T. Saito, J. Tatebayashi, and Y. Arakawa, *Phys. Rev. B.* **70**, 235337 (2004).
288. I. Favero, G. Cassabois, C. Voisin, C. Delalande, Ph. Roussignol, R. Ferreira, C. Couteau, J.P. Poizat, and J.M. Gérard, *Phys. Rev. B.* **71**, 233304 (2005).
289. A.S. Lenihan, M.V. Gurudev Dutt, D.G. Steel, S. Ghosh, and P.K. Bhattacharya, *Phys. Rev. Lett.* **88**, 223601 (2002).
290. A.I. Tartakovskii, J. Cahill, M.N. Makhonin, D.M. Whittaker, J.-P.R. Wells, A.M. Fox, D.J. Mowbray, M.S. Skolnick, K.M. Groom, M.J. Steer, and M. Hopkinson, *Phys. Rev. Lett.* **93**, 057401 (2004).
291. M. Sénès, B. Urbaszek, X. Marie, T. Amand, J. Tribollet, F. Bernardot, C. Testelin, M. Chamarro, and J.-M. Gérard, *Phys. Rev. B.* **71**, 115334 (2005).
292. K. Gündoğdu, K.C. Hall, T.F. Boggess, D.G. Deppe, and O.B. Shchekin, *Appl. Phys. Lett.* **84**, 2793 (2004).
293. J. Siegert, S. Marcinkevičius, and Q.X. Zhao, *Phys. Rev. B.* **72**, 085316 (2005).
294. V.K. Kalevich, M. Paillard, K.V. Kavokin, X. Marie, A.R. Kovsh, T. Amand, A.E. Zhukov, Yu.G. Musikhin, V.M. Ustinov, E. Vanelle, and B.P. Zakharchenya, *Phys. Rev. B.* **64**, 045309 (2001).

295. L. Lombez, P. Renucci, P.F. Braun, H. Carrère, X. Marie, T. Amand, B. Urbaszek, J.L. Gauffier, P. Gallo, T. Camps, A. Arnoult, C. Fontaine, C. Deranlot, R. Mattana, H. Jaffrès, J.-M. George, and P.H. Binh, *Appl. Phys. Lett.* **90**, 081111 (2007).
296. S. Cortez, O. Krebs, S. Laurent, M. Sénès, X. Marie, P. Voisin, R. Ferreira, G. Bastard, J.-M. Gérard, and T. Amand, *Phys. Rev. Lett.* **89**, 207401 (2002).
297. V.K. Kalevich, I.A. Merkulov, A. Yu. Shiryaev, K.V. Kavokin, M. Ikezawa, T. Okuno, P.N. Brunkov, A.E. Zhukov, V.M. Ustinov, and Y. Masumoto, *Phys. Rev. B.* **72**, 045325 (2005).
298. S. Laurent, M. Senes, O. Krebs, V.K. Kalevich, B. Urbaszek, X. Marie, T. Amand, and P. Voisin, *Phys. Rev. B.* **73**, 235302 (2006).
299. I.A. Merkulov, Al.L. Efros, M. Rosen, *Phys. Rev. B.* **65**, 205309 (2002).
300. A.V. Khaetskii, D. Loss, and L. Glazman, *Phys. Rev. Lett.* **88**, 86802 (2002).
301. P.-F. Braun, X. Marie, L. Lombez, B. Urbaszek, T. Amand, P. Renucci, V.K. Kalevich, K.V. Kavokin, O. Krebs, P. Voisin, and Y. Masumoto, *Phys. Rev. Lett.* **94**, 116601 (2005).
302. A. Imamočlu, E. Knill, L. Tian, and P. Zoller, *Phys. Rev. Lett.* **91**, 017402 (2003).
303. P.-F. Braun, B. Urbaszek, T. Amand, X. Marie, O. Krebs, B. Eble, A. Lemaitre, and P. Voisin, *Phys. Rev. B.* **74**, 245306 (2006).
304. A.I. Tartakovskii, T. Wright, A. Russell, V.I. Fal'ko, A.B. Van'kov, J. Skiba-Szymanska, I. Drouzas, R.S. Kolodka, M.S. Skolnick, P.W. Fry, A. Tahraoui, H.-Y. Liu, and M. Hopkinson, *Phys. Rev. Lett.* **98**, 026806 (2007).
305. P. Maletinsky, C.W. Lai, A. Badolato, and A. Imamoğlu, *Phys. Rev. B.* **75**, 035409 (2007).
306. M. Kroutvar, Y. Ducommun, D. Heiss, M. Bichler, D. Schuh, G. Abstreiter, and J.J. Finley, *Nature* **432**, 81 (2004).

CHAPTER 4

Cavity Quantum Electrodynamics with Semiconductor Quantum Dots

P. Senellart and I. Robert-Philip

*CNRS – Laboratoire de Photonique et Nanostructures,
Route de Nozay, F-91460 Marcoussis, France*

4.1 Introduction

The fundamental transition of single quantum dots is well described in terms of a two-level system, very much like an atom. Its spontaneous emission, which arises from the relaxation of the emitting system placed on its excited state, is not an intrinsic property of the emitter itself but is rather a property of the two-level system coupled to its electromagnetic vacuum environment. Since the spatial and spectral properties of the electromagnetic field are determined by all the boundary conditions present in the local surroundings (for instance, mirrors, interfaces, etc.), matter–field interactions in general and in particular spontaneous emission decay are expected to exhibit significant changes within the environment. For example, the presence of a microcavity around the emitting two-level system can accelerate or inhibit its spontaneous emission or even make it reversible. This research field, aimed at controlling spontaneous emission with the use of wavelength-sized cavities, is called cavity quantum electrodynamics (CQED) [1]. Two regimes can occur. First, the spontaneous emission rate can be enhanced or reduced compared with its vacuum level by the presence of the microresonator (this is called the weak coupling regime); second, the usual irreversible spontaneous emission can be changed to a reversible exchange of energy between the emitter and the cavity mode with a frequency given by the Rabi frequency (the system, composed of the single emitter and the cavity modes, is then in the strong coupling regime).

First observed in atom physics [2–5], the transposition of these effects to solid-state systems was inaugurated in 1992 with the demonstration of exciton-photon strong coupling in a planar microcavity embedding quantum wells [6]. Another conceptual and technological breakthrough was the observation of spontaneous emission exaltation that occurs when dissipation overwhelms the fundamental Rabi dynamic [7]. In this context, the ability to create semiconductor quantum dots was a turning point in the recent history of cavity effects. These nanoemitters can efficiently capture and confine electrons and holes in the three directions of space, making them less sensitive to surrounding defects (for example, those induced by semiconductor surfaces whose impact becomes significant when cavity dimension decreases). This three-dimensional confinement of carriers in individual quantum dots also makes them valuable candidates for the study of cavity quantum electrodynamics effects applied to semiconductor materials, because, unlike bulk or quantum-well materials, these nanoemitters exhibit sharper optical features in the spectrum as determined by the density of states.

In their pioneering work [7], J.M. Gérard *et al.* used micropillar cavities loaded with an ensemble of quantum dots. Since then, weak coupling and strong coupling from a single quantum dot have been experimentally demonstrated using a wide variety of cavity geometries. In this chapter, we describe these recent results on single quantum dots, stressing the cavity and quantum dot requirements for the implementation of such CQED effects. We consider some representative applications and outline the challenges for the future.

4.2 Basics of cavity quantum electrodynamics

4.2.1 Optical confinement and light–matter interaction

Electromagnetic field properties are strongly dependent on the presence of boundaries in the local environment, such as the presence of a microcavity. For example, the presence of the resonator alters the optical density of electromagnetic states: like for electron motion quantization in quantum dots, the density of optical states is modified with respect to its value in free space, when reducing the optical cavity dimensionality to two, one or zero dimensions. In the ideal zero-dimensional limit, the density of optical modes, which is related to the capacity of the cavity to confine light in time, is represented by a sequence of infinitely narrow impulse functions at precise, discrete, allowed energies: the resonator sustains discrete optical modes and confines light at these fixed resonant frequency values indefinitely (i.e. without loss). Moreover, wavelength-size spatial confinement ensures that these precise resonant frequencies are well separated and sparsely distributed throughout the spectrum.

Deviation from these ideal conditions is described by the quality factor Q and the modal volume V of each mode sustained by the cavity. The quality factor describes the coupling of the cavity mode to the outside continuum of optical modes (i.e. losses). It is proportional to the light confinement time: the decay of the energy stored in the optical mode at pulsation ω is proportional to $e^{-\omega t/Q}$. The decay rate $1/Q$ is thus the bandwidth at half-maximum of the cavity mode and Q can be expressed as a function of the mode wavelength λ and its spectral linewidth $\Delta\lambda$ as follows:

$$Q = \frac{\lambda}{\Delta\lambda}. \tag{4.1}$$

If the quality factor reflects the ability of the cavity to store light in time, the modal volume V represents its ability to store light in space. It is usually defined in terms of the electrical field energy in the cavity mode as follows:

$$V = \frac{\int_{\text{SPACE}} \varepsilon(\vec{r})\,|\,E(\vec{r})\,|^2\,d^3r}{\max(\varepsilon(\vec{r})\,|\,E(\vec{r})\,|^2)}. \tag{4.2}$$

Usually, this modal volume is expressed in units of effective optical wavelengh $(\lambda/n)^3$, where n is the refractive index of the cavity medium.

Consider now the system formed by a single quantum dot inserted in an optical microcavity. The energy of the first excited state of the quantum dot is supposed to be resonant to the cavity mode energy. Two states of the cavity–quantum dot system are coupled through ligth–matter interaction: the state with the quantum dot in its excited state with no occupation of the photon mode $|e,0\rangle$, and the state with the quantum dot in its ground state and one photon occupancy in the cavity mode $|g,1\rangle$. In the dipolar approximation, the coupling constant (referred to as ω_1 in Fig. 4.1) is proportional to the $\langle e,0|\vec{E}(\vec{r}_{em}) \cdot \vec{d}\,|\,g,1\rangle$ where the electron-hole pair trapped in the quantum dot is described by a point dipole $\vec{d}$ and $\vec{E}(\vec{r}_{em})$ is the elecetric field evaluated at the quantum dot location. Because the quantum dot is inserted in a semiconductor matrix, the electron–hole pair trapped in the quantum dot experiences phonon coupling as well Coulomb interaction with carriers in the surrounding quantum dot. Moreover, because the cavity is not perfect,

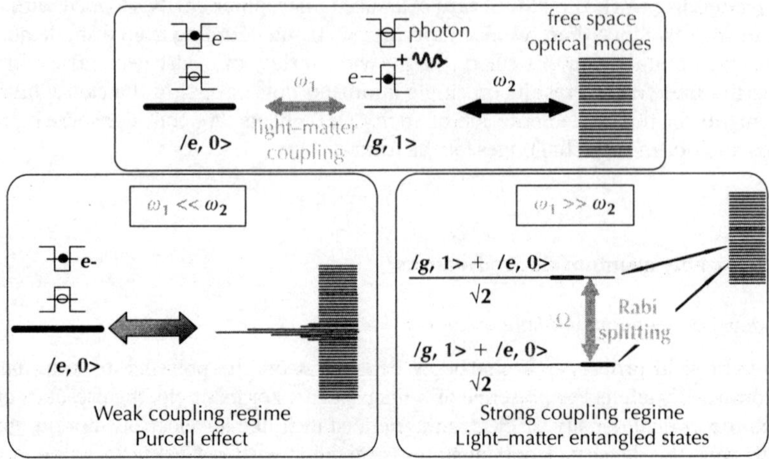

Figure 4.1   Schematics of cavity quantum electrodynamics for a single quantum dot in an optical microcavity. Top: General case. Left: Weak coupling regime. Right: Strong coupling regime.

the optical mode of the cavity is coupled to the optical mode of the free space through a constant ω_2 proportional to Q^{-1}. For the sake of simplicity, we only consider the coupling to the free space optical mode and neglect Coulomb and phonon interaction.

If $\omega_1 << \omega_2$, when the electron–hole pair radiatively recombines, the emitted photon will rapidly escape from the cavity. To describe this regime correctly, one needs to consider that the quantum dot in its excited state is coupled to a continuum of optical mode. The effect of the cavity is to change the density of the final state for the photon emission and to enhance or decrease the intensity of electric field at the quantum dot position. The quantum dot spontaneous emission remains irreversible but it can be accelerated or inhibited. The system is in the weak coupling regime, in the so-called Purcell effect [8].

If $\omega_1 >> \omega_2$, i.e. if the cavity losses are minimal, when the electron–hole pair radiatively recombines, the emitted photon will oscillate inside the cavity before escaping outside. During these oscillations, it will be periodically reabsorbed and re-emitted by the quantum dot with a period given by the Rabi frequency. In this so-called strong coupling regime, one can consider that the cavity–quantum dot system is isolated from any other interaction, so that $|e,0>$ and $|g,1>$ are no longer the eigenstates of the system. At exact resonance, the eigenstates are entangled light–matter states:

$$\frac{|e,0\rangle \pm |g,1\rangle}{\sqrt{2}},$$

split by the Rabi frequency $\Omega/\hbar$ (Fig. 4.1, right). In the following, we detail the description of these two regimes and discuss the key parameters that control the establishment of exaltation, inhibition of spontaneous emission as well as strong coupling regime.

4.2.2   *Spontaneous emission control – Purcell effect*

Hereafter, we shall focus on the weak coupling regime, which occurs when dissipation due to cavity losses overwhelms the fundamental Rabi dynamic.

In the electric dipolar approximation, the spontaneous emission decay Γ of a monochromatic two-level system, coupled to a continuum of optical modes, is given by Fermi's golden rule [1]:

$$\Gamma = \frac{2\pi}{\hbar} \left| \left\langle e,0 \left| \vec{E}(\vec{r}_{em}) \cdot \vec{d} \right| g,1 \right\rangle \right|^2 \rho(E = \hbar\omega_0) \tag{4.3}$$

where the emitter is described by a point dipole $\vec{d}$ undergoing harmonic oscillation at at angular frequency ω_0 around a point $\vec{r}_{em}$. $\rho(E = \hbar\omega)$ is the density of states of the electromagnetic field at energy E. For a single cavity mode oscillating at ω_c and exhibiting a quality factor Q, this density of optical modes can be expressed as follows:

$$\rho\left(E = \hbar\omega\right) = \frac{1}{\pi} \frac{\dfrac{2Q}{\hbar\omega_c}}{1 + \left(\dfrac{2Q}{\hbar\omega_c}\right)^2 \left(\hbar\omega - \hbar\omega_c\right)^2}. \tag{4.4}$$

The field operator for the cavity eigenmode reads:

$$\bar{E}(\vec{r}) = -i\sqrt{\frac{\hbar\omega_c}{2\varepsilon V}} \left(\vec{u}(\vec{r})a - \vec{u}^*(\vec{r})a^\dagger\right) \tag{4.5}$$

where $\vec{u}(\vec{r})$ is the normalized mode spatial distribution and describes the polarization and the relative amplitude of the electric field at location $\vec{r}$ ($\vec{u}(\vec{r})$ is normalized to 1 at its maximum). a and $a^\dagger$ are the photon annihilation and creation operators in the mode and V is its effective modal volume. Taking equations 4.4 and 4.5 into account, the spontaneous emission decay given by 4.3 can be expressed as:

$$\Gamma = \frac{2Q}{\hbar\varepsilon V}\left|\vec{u}(\vec{r}_{em}) \cdot \vec{d}\right|^2 \frac{1}{1 + \left(\dfrac{2Q}{\hbar\omega_c}\right)^2 \left(\hbar\omega_0 - \hbar\omega_c\right)^2}. \tag{4.6}$$

The spontaneous emission rate of the same emitter in bulk material of permittivity $\varepsilon = \varepsilon_0 n^2$ is:

$$\Gamma_0 = \frac{n\omega_0^3 d^2}{3\pi\varepsilon_0 \hbar c^3}. \tag{4.7}$$

We can therefore deduce the modification of the spontaneous emission rate induced by the interaction of the dipole with a single cavity mode as follows:

$$F = \frac{\Gamma}{\Gamma_0} = F_p \frac{\left|\vec{u}(\vec{r}_{em}) \cdot \vec{d}\right|^2}{\left|\vec{d}\right|^2} \frac{1}{1 + \left(\dfrac{2Q}{\hbar\omega_c}\right)^2 \left(\hbar\omega_0 - \hbar\omega_c\right)^2} \tag{4.8}$$

where F_p, the so-called Purcell factor, is [8]:

$$F_p = \frac{3}{4\pi^2} \frac{Q}{V/\left(\lambda_0/n\right)^3} \tag{4.9}$$

where λ_0 is the dipole wavelength in free space.

The modification of the spontaneous emission rate by the presence of the cavity, given by equation 4.8, is therefore the product of three contributions. The first one F_p is the largest spontaneous emission enhancement induced by the cavity mode that can be obtained; it depends only on optical characteristics of the optical mode, more precisely its quality factor and its effective volume. In other words, the Purcell factor is a figure of merit for the cavity alone and describes its ability to increase the coupling of an ideal emitter with the vacuum field, via a local enhancement of its intensity (small Vs) and/or of the effective mode density (high Qs). Notice that the amplitude of this acceleration (or Purcell effect) depends on the ratio Q/V whereas the visibility of the strong coupling regime between a single quantum dot and the cavity field depends on the ratio Q^2/V, as we shall see in the next paragraph. The second and third terms in equation 4.8 highlight the resonant character of the Purcell effect. The first one, proportional to $|\vec{u}(\vec{r}_{em}) \cdot \vec{d}|$, takes into account the relative position of the emitter to the spatial maximum of the mode electric field and the relative orientation of the emitting dipole to the electric field. If the emitting dipole is located at a node of the electric field, the spontaneous emission will be inhibited by the presence of the cavity. The last term describes the spectral detuning between the emitter and the cavity mode. For a dipole off spectral resonance with the cavity mode, the spontaneous emission will be suppressed. In conclusion, the interaction of a single dipole with a single cavity mode can give rise to two main effects:

- The inhibition of spontaneous emission, which occurs when the emitting dipole is located at a node of the electric field and/or is far off resonance with the cavity mode. In the latter case, the spontaneous emission suppression will be proportional to $1/(Q/V)$.
- The acceleration of spontaneous emission, whose largest amplitude is reached for a punctual monochromatic dipole on spectral and spatial resonance with a high-Q/V cavity mode and parallel to the mode electric field. In this context, the spontaneous emission acceleration is proportional to Q/V.

In practice, the emitting dipole may be randomly orientated, so that the modification of the spontaneous emission rate should be averaged over the dipole orientation, which reduces the amplitude of the modification by a factor of 3 (this factor would be equal to 2 if the dipole were randomly orientated in the plane of the electric field). Furthermore, the cavity confined mode might be g-fold degenerated (for instance $g = 2$ for the fundamental mode of pillars or disks with circular cross-section whereas $g = 1$ for the fundamental mode of a post with elliptical cross-section), which increases the density of states of the cavity field by a factor of g. Eventually, for a single emitter with an in-plane dipole randomly orientated in the cavity electric field plane and coupled to a single cavity mode with degeneracy g, the modification of the spontaneous emission rate should be expressed as:

$$F = \frac{\Gamma}{\Gamma_0} = \frac{g}{2} F_p \, |\vec{u}(\vec{r}_{em})|^2 \, \frac{1}{1 + \left(\dfrac{2Q}{\hbar\omega_c}\right)^2 \left(\hbar\omega_0 - \hbar\omega_c\right)^2} . \tag{4.10}$$

More generally, the radiative spontaneous recombination can occur either in the single cavity mode or in other "leaky" optical modes. In Fermi's golden rule formalism, the probabilities of emitting in the various final states add on, so that the global modification of the spontaneous emission rate reads:

$$F = \frac{\Gamma}{\Gamma_0} = \frac{g}{2} F_p \, |\vec{u}(\vec{r}_{em})|^2 \, \frac{1}{1 + \left(\dfrac{2Q}{\hbar\omega_c}\right)^2 \left(\hbar\omega_0 - \hbar\omega_c\right)^2} + \frac{\Gamma_{leak}}{\Gamma_0} . \tag{4.11}$$

where Γ_{leak} is the spontaneous emission rate in the leaky modes.

Let's now come back to the validity of this approach. We have supposed here that the the photon lifetime in the cavity mode (related to the incoherent field damping from the cavity into the continuum of external modes) is much shorter than the period of the Rabi oscillation (related to the coherent coupling of the dipole with the cavity field): in other words, the emitted photon is dissipated before being reabsorbed by the emitter. This condition reads:

$$\Omega \frac{Q}{\omega_c} \ll 1. \tag{4.12}$$

Furthermore, we have considered up to this point a punctual monochromatic emitter. This comes into play in the use of Fermi's golden rule, which holds only if a discrete state is coupled to a continuum of states. In practice, emitters are not perfectly monochromatic and the validity of such approach will depend on the relative amplitude between the natural homogeneous width of the optical transition and the spectral width of the cavity mode. Fermi's golden rule can only apply if:

$$\Gamma_0 \frac{Q}{\omega_c} \ll 1 \tag{4.13}$$

The large spectral width of commonly used solid-state emitters, for instance semiconductor quantum wells ($\sim 1\,\mathrm{nm}$, even at low temperature), is a major hindrance in this respect, leading to spontaneous modification factors of the order of a few units in the weak coupling regime [64, 65]. From this point of view, semiconductor quantum dots appear as valuable candidates for the implementation of CQED effects in the solid state, since they exhibit a very small spectral broadening at low temperature. For instance, the homogeneous linewidth of the fundamental transition in InAs/GaAs quantum dots at low temperature ($<10\,\mathrm{K}$) ranges between a few $\mu\mathrm{eV}$ (for resonant pumping [66, 67] and a few tens of $\mu\mathrm{eV}$ when increasing the energy detuning between the optical excitation and the optical fundamental transition [68].

4.2.3 Strong coupling regime

When the interaction between the electron–hole pair trapped in the quantum dot and the optical mode is large compared to any dephasing mechanism, the eigenstates of the system are mixed exciton–photon states split at resonance by the Rabi splitting $\Omega = 2\hbar g$ with $\hbar g = |\langle \vec{d} \cdot \vec{E} \rangle|$. Assuming that the dipole is parallel to the electric field, the coupling constant g can be expressed in terms of the oscillator strength of the optical transition of the quantum dot

$$f = 2m\omega_0 d^2 / (e^2 \hbar) \tag{4.14}$$

using:

$$g = \sqrt{\frac{1}{4\pi\varepsilon_0\varepsilon_R} \frac{\pi e^2 f}{mV}}. \tag{4.15}$$

In this equation, m is the free electron mass.

As discussed below, dephasing mechanisms play a key role in the implementation of CQED in solid-state systems: cavity losses as well as interaction between the carriers trapped in the quantum dot and their surroundings compete with the light–matter interaction. To include the dephasing mechanism in the system description, one can consider the coupling to these reservoirs by introducing the linewidth of both the photon (γ_C) and the exciton transition (γ_X). Considering the density matrix for the quantum dot–cavity system, the luminescence spectrum can be calculated in the weak excitation limit [69, 10]. For $g > |\gamma_X - \gamma_C|/4$ and at resonance between the cavity mode and the exciton energy, the spontaneous emission spectrum consists of two lines split by $\Omega_R = 2\hbar\sqrt{g^2 - (\gamma_C - \gamma_X)^2/16}$ with the average linewidth $(\gamma_X + \gamma_C)/2$.

For an InAs self-assembled quantum dot, the linewidth of the exciton line stands around a few tens of $\mu\mathrm{eV}$ whereas the typical linewidth for the mode resonance is around a few hundreds of $\mu\mathrm{eV}$. In a first approximation, the exciton linewidth can be neglected so that the strong coupling regime is reached for $g > \gamma_C/4$. Since the photon mode linewidth is given by $\gamma_C = \omega_0/Q$ (see Eq. 4.1), the strong coupling regime occurs for:

$$\sqrt{\frac{4}{\varepsilon_0\varepsilon_R} \frac{e^2 f}{mV} \frac{Q}{\omega_0}} > 1. \tag{4.16}$$

This condition is necessary but obsviously not sufficient. First, without any preferential orientation of the quantum dot the orientation of the quantum dot emission dipole is random: the strength of the cavity–quantum dot coupling must be corrected by the cosinus of the angle between the quantum dot exciton dipole and the electric field of the optical mode. However, this calculation stands only if the quantum dot is located at the exact maximum of the electromagnetic field. Finally, when increasing temperature or reducing the radiative lifetime of the excitonic transition, the broadening of the exciton line can become comparable to the cavity mode linewidth.

4.3   Implementation of cavity quantum electrodynamics in the solid state

4.3.1   *The resonator: a semiconductor microcavity*

In solid-state physics, two main mechanisms for light confinement are used. The first one exploits the total internal reflection at the interface between two media with different refractive index: total internal reflection bounces light to the region of high refractive index when light strikes the interface at a steep angle. The second one exploits the destructive interference between optical waves impinging on a medium with a periodically modulated index of refraction, so-called photonic crystals: for waves with a wavelength close to the period of modulation, the scattering of light on the photonic crystal combines with destructive interference. This interference cancels out light of certain wavelengths and the photonic crystal acts as a high-quality reflector: Analogous to the electronic band gap in semiconductors, such periodic structures exhibit a certain frequency range where light cannot propagate through the structure. This frequency range is known as the photonic band gap. Photonic crystals can be divided into three categories, namely one-dimensional (1D), two-dimensional (2D) and three-dimensional (3D) crystals according to the dimension of the periodicity. A common one-dimensional photonic crystal is the Bragg mirror composed of multiple layers of dielectric films. 2D and 3D photonic crystals employ the Bragg reflection in more than one spatial direction.

One can use total internal reflection to make three-dimensionally confined resonators, provided one can force wavefronts inside the cavity to interfere with themselves. This is achieved with the "whispering gallery" resonators. They are essentially circular disks [45], toroids [46] or spheres [47] in which the light circulates around close to the dielectric interface. Such modes are especially low in losses and can reach quality factors far exceeding what can be obtained in other solid-state microcavities: experimental quality factors of the whispering gallery modes sustained by silica microspheres [48–50] or toroids [51] reach values up to 10^8–10^9, while keeping a relatively small modal volume: $V \sim 10^2$–10^4 $(\lambda/n)^3$. Microdisk resonators with diameters as small as 1–$10\,\mu$m have also been processed in III–V semiconductor materials (see Fig. 4.2 left) and support modes with a high-quality factor (up to 10^4–10^5[52, 53]) with modal volume as low as a few $(\lambda/n)^3$.

Another strategy to confine light in the three directions of space would be to isolate a point defect, such as one or more missing periods, into a three-dimensional photonic crystal [54, 55]. A photonic crystal with a complete band gap in the microwave range was created in 1991 [56]. However, it appeared to be far from trivial to reach a full photonic band gap in the near-infrared and visible spectral range. This was achieved only recently [57–60]. Until then, control over light emission in 3D photonic crystals was demonstrated experimentally [61–63].

While various techniques for the fabrication of photonic crystals with full three-dimensional band gaps have been proposed and demonstrated with different levels of success, the incorporation of microcavities in a controllable way is a critical step. In view of the difficulties in fabricating deterministic photonic defects into 3D photonic crystal structures, a lot of effort has been devoted to less demanding structures involving a hybrid confinement: in this context, light confinement combines total internal reflection in one or two spatial directions and interference effects in the other directions of space. Such cavities are more amenable to fabrication than three-dimensional photonic crystal resonators, but retain or approximate many of the latter's desirable properties, such as 3D confinement of light.

One example is the micropillar cavity (see Fig. 4.2 centre). In micropost resonators, the transverse mode confinement results from total internal reflection at the semiconductor–air

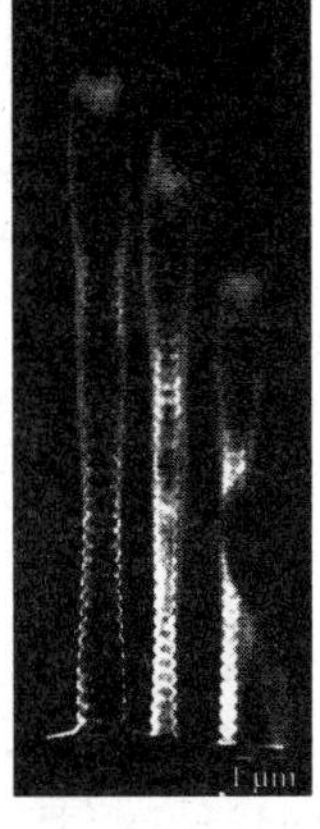

Figure 4.2   Images of different semiconductor cavity geometries, such as (left) micron-diameter GaAs microdisks, (centre) AlGaAs/GaAs micropillars and (right) photonic crystal cavities etched on a 180 nm thick GaAs membrane (Source: CNRS-LPN).

interface, while confinement in the vertical direction is provided by Bragg reflection. These cavities, obtained by etching a planar microcavity bounded by two Bragg reflectors, offer small cavity volume and relatively high Q. Quality factors as high as 165 000 have been reported for 4 μm diameter III–V semiconductor pillars, which sustain modes with modal volume of the order of a few tens of $(\lambda/n)^3$ [31]. The main advantage of micropillars compared with other types of microcavities is that light escapes very efficiently in the direction parallel to the pillar axis in a single-lobe Gaussian-like pattern, which is well suited for light collection and coupling to optical fibres, for example [32].

Another example is the 2D photonic crystal cavity in a slab waveguide, consisting of a 2D array of holes perforated through a thin membrane (see Fig. 4.2 right) [33]. In these cavities, the photonic-band gap effect is used for strong light confinement in the transverse directions, and total internal reflection at the air–slab interface ensures light confinement in the longitudinal direction. The main advantage of such resonators is their relative ease of fabrication (since only a thin slab of a few hundred nm thickness has to be pierced) associated with the wide variety of possible 2D photonic crystal patterns (due to the inherent flexibility in hole shape, size, and pattern). While many 2D photonic patterns (square lattice of holes, quasi-crystal-based arrangement of holes, etc.) have been proposed and sometimes demonstrated, by far the most widely studied arrangement is the hexagonal one. It has reasonably good band gaps for accessible values of the index contrast and filling factor. In these structures, the cavity is formed by the introduction of a point defect in the 2D photonic crystal: the in-plane photonic band gap, added to the vertical waveguiding effect, leads to efficient 3D photon confinement even though no omnidirectional band gap for photon propagation exists.

Different cavity geometries have been investigated, aimed at increasing Q factors while keeping the modal volume as small as possible. The simplest geometry consists of removing a finite number of holes in a perfect array of holes. For instance, the Ln cavities are formed by a line of n missing holes in a triangular array of holes. The so-called Hn cavities are formed by unperforated hexagons in a regular triangular photonic crystal with ΓM-type boundaries; these resonators are described by the size n of their side in units of crystal period. Recently, several numerical and experimental studies have shown that these microcavities are highly valuable candidates for achieving high-quality factors with wavelength-sized modal volumes. By finely tuning the holes' position and/or radius at the cavity termination, it is possible to enhance the Q factor of these resonators. This stronger light confinement results from the progressive increase of the reflectivity at the cavity boundaries and the progressive decrease of the mode group velocity while finely tuning the arrangement of holes [34]. For instance, Q factors of 45 000 with ultra-small volumes close to $(\lambda/n)^3$ have been measured in an L3 microcavity etched in a Si membrane, by slightly shifting the position of the holes surrounding the defect region [35].

Another approach is the so-called double heterostructure cavity. The cavity is composed of a line defect structure whose geometry (width, longitudinal periodicity, etc.) is locally changed, so that a mode gap is created at the location of the geometry modulation: the band-edge frequency of the fundamental line defect mode is lowered at the location of the geometry modulation, which forms a well that can trap photons at appropriate frequencies. For instance, by changing locally the width of a line defect, cavities with Q factors reaching 8×10^5 and a modal volume of the order of $(\lambda/n)^3$ have been fabricated on Si-slab membranes [36]. Similarly, by changing locally the longitudinal period of a line defect, cavities with Q factors reaching 10^6 and a modal volume of the order of $(\lambda/n)^3$ have been demonstrated on silicon slab waveguides [37, 38]. Such resonators made of III–V semiconductor materials exhibit Q factors exceeding 10^5 [39, 40].

4.3.2   The emitter: a single semiconductor quantum dot

In 1994, the emission of electron–hole pairs trapped in single quantum dots was experimentally evidenced [41, 42]. These observations were almost simultaneously reported in two different systems. Marzin and co-workers reported on the micro-photoluminescence on single InAs quantum dots. InAs quantum dots are formed during the epitaxial growth of an InAs quantum well in GaAs barriers: because of the lattice parameter mismatch between the two materials, after the growth of 1.7 monolayers of InAs, strength relaxation leads to formation of 3D InAs islands with lateral size around 20 nm and height around 3 nm [43]. After encapsulating the quantum dots with GaAs, the InAs island forms an energy trap for electron–hole pairs in all directions. The gap difference between InAs (0.41 eV) and GaAs (1.51 eV) is so large that the Coulomb interaction between the electron and the hole is a perturbation as compared to the quantum confinement of the carrier (around 10 meV as compared to a few 100 meV). To describe the first excited state of the quantum dot, the term "electron–hole pair" is therefore more appropriate than the term "exciton".

The other first demonstration of single quantum dot emission is observed in a very different confinement regime [42]. The sample consists of a thin (10 monolayers) GaAs quantum well in AlGaAs barriers. Even though molecular beam epitaxy is known to produce atomically smooth interfaces, atomic roughness persists at the quantum–well interface. With standard growth conditions, the roughness takes place on a small length scale, around 1–10 nm: an inhomogeneous broadening of the 2D exciton line is observed both in micro- and macro-photoluminescence measurements [44]. However, when performing growth interruption at the quantum–well interface during the growth process, the roughness can take place on a larger length scale: between 10 and 200 nm [26, 42]. The lateral size of the defect is then larger than the Bohr radius of the 2D exciton confined in the quantum well so that a quasi 2D-exciton can be weakly localized in the interface defect. These interface fluctuation quantum dots, often referred to as natural quantum dots, present 0D density of states for the excitons, despite a small binding energy, around 15 meV [26].

Both quantum dot types are promising emitters for solid-state quantum electrodynamics. The discrete density of states, commonly leading to the "macro-atom" image, has been demonstrated by observing single photon emission from both InAs [27, 28] and GaAs quantum dots [29]. To demonstrate the Purcell effect, both InAs and GaAs quantum dots are equally interesting: the typical emission linewidth (around a few tens of μeV) and their small size as compared to the emission wavelength make them ideal candidates to demonstrate the Purcell effect in a small effective volume microcavity.

However, to reach an exciton–photon strong coupling regime, one needs to maximize the exciton oscillator strength f. As discussed now, the oscillator strength of the exciton confined in the quantum dot can be much larger for natural quantum dots than for self-assembled single quantum dots. For an interband transition, omitting numerical coefficients, the oscillator strenth is given by [30]:

$$f \propto \left| \langle u_c | \vec{r} | u_v \rangle \right|^2 \quad \left| \int \Psi_X(\vec{r}_e, \vec{r}_h)\, d\vec{r}_e\, d\vec{r}_h \right|^2 \tag{4.17}$$

where u_c and u_v are the rapidly varying part of the Bloch function for the electron and the hole and $\Psi_X(\vec{r}_e, \vec{r}_h)$ is the excitonic envelope function.

For self-assembled InAs single quantum dots, the lateral confinement resulting from the quantum confinement is of the order of or smaller than the 2D exciton Bohr radius of the exciton in the wetting layer. In this case, one should consider independent quantization of electron and hole in the plane rather than exciton confinement as a whole particle. The electron–hole pair wave function then reads [12]:

$$\Psi_X(\vec{r}_e,\vec{r}_h) \propto \varphi_e(\vec{\rho}_e)\varphi_h(\vec{\rho}_h)\xi_e(z_e)\xi_h(z_h) \tag{4.18}$$

where $\vec{\rho}_e$ (resp. $\vec{\rho}_h$) describes the in-plane wave function for the electron (resp. hole). The oscillator strength of the optical transition is then given by the overlap between the electron and hole wave functions in all three direction of space:

$$f \propto \iint \varphi_e(\vec{\rho})\varphi_h(\vec{\rho})d\vec{\rho} \quad \int \xi_e(z)\xi_h(z)\,dz \tag{4.19}$$

Equation 4.19 indicates that the oscillator strength of self-assembled InAs quantum dots hardly depends on the precise geometry of the quantum dot. Indeed, the size dispersion among self-assembled quantum dots is small as compared to the exciton Bohr radius, so that the electron–hole overlap is very similar for various quantum dots. The typical calculated oscillator strength for self-assembled InAs quantum dots emitting around 1.3 eV is around $f = 10$.

Consider now natural GaAs quantum dots. For quantum dots whose lateral size exceeds the 2D exciton Borh radius, one can conveniently separate the exciton centre of mass motion and the electron–hole relative motion. The exciton wave function reads:

$$\Psi_X(\vec{r}_e,\vec{r}_h) \propto g_{\text{loc}}(\vec{R}_{CM})f_{1S}(\vec{\rho})\xi_e(z_e)\xi_h(z_h)$$

where $f_{1S}(\vec{\rho})$ represents the in-plane relative motion and $g_{\text{loc}}(\vec{R}_{CM})$ the localized centre of mass movement. In GaAs natural quantum dots, the monolayer fluctuation at the quantum–well interface localizes the exciton centre of mass on a scale roughly propotional to the lateral quantum dot size (an experimental evidence of this phenomenon can be found in [9]). As a result, for large GaAs quantum dots, the oscillator strength is proportional to the area of the interface defect.

Different theoretical models demonstrated this dependence of the oscillator strength on the quantum dot size in 1994 [12]. In 1999, Andreani and co-workers revisited this theoretical prediction in the context of a cavity–quantum dot strong coupling regime. Figure 4.3 presents the calculated oscillator strength for natural quantum dots formed at the interface of a 5 nm wide GaAs quantum well with AlAs barriers. The calculated oscillator strength presents a minimum value around the exciton 2D Bohr radius. This minimum is around $f = 50$, five times large than for self-assembled InAs quantum dots. Most interesting is the calculated oscillator strength for

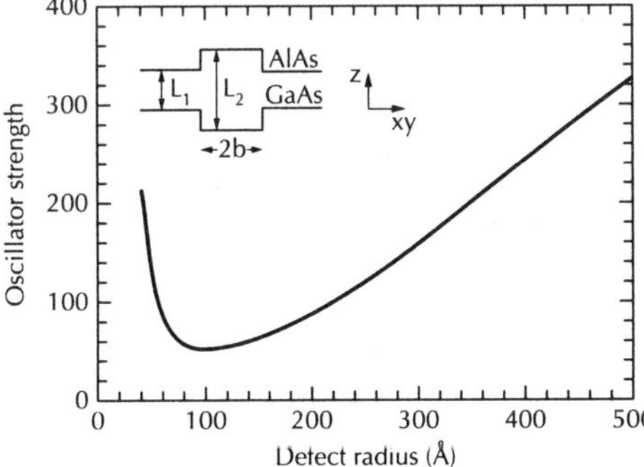

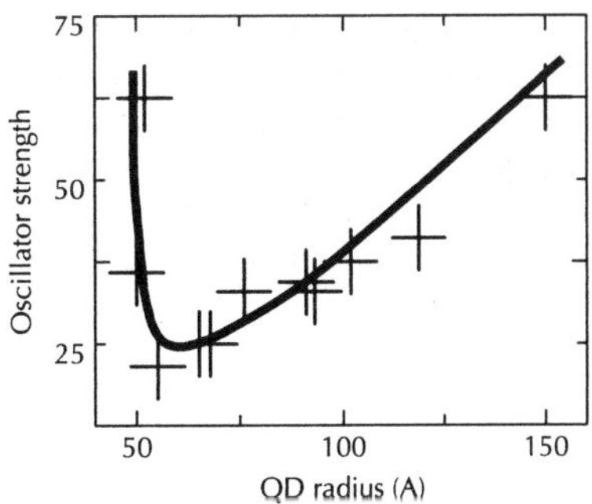

Figure 4.3   Left: *Calculated oscillator strength for an exciton trapped in a monolayer fluctuation in a 4 nm wide GaAs/AlAs quantum well as a function of the quantum dot radius from [10]. Right: Experimental measurement of the oscillator strength for an exciton trapped in a monolayer fluctuation in a 3 nm wide GaAs/Al*$_{0.3}$*Ga*$_{0.7}$*As quantum well [11].*

large quantum dot radius: it can reach values as large as $f = 300$. These theoretical predictions are quite promising: if one can find appropriate growth conditions, one can control the oscillator strength of the quantum dot excitonic transition.

Experimental estimation of the strength of the optical transition of quantum dots has first been obtained through time resolved photoluminescence measurements on quantum dot ensembles. Provided that the capture time into the quantum dot is short [13] and non-radiative processes are negligible, the decay time of the photoluminescence under non-resonant excitation gives a measure of the exciton radiative lifetime. Measurements on InAs quantum dots demonstrate a radiative decay time around 0.8–1.5 ns [14–16] corresponding to an oscillator strength between 9 and 15. Time resolved PL on single InGaAs quantum dots demonstrated radiative decay time around 600–1300 ps [17, 18]. More recently, the absorption of a single quantum dot exciton line has been measured, giving access to both the homogeneous linewidth of the exciton transition and its oscillator strength. To extract the small absorption contribution to the overall signal, Karrai *et al.* introduced modulation of the excitonic transition through the Stark effect by applying a periodic electric field. An oscillator strength of $f = 12$ was also deduced from this technique. Measurement of GaAs natural quantum dots oscillator strength has also been obtained through differential absorption techniques [19]. Oscillator strengths between 45 and 180 were reported using this technique. To prove the dependence of the oscillator strength with the quantum dot lateral size, time resolved photoluminescence on single GaAs quantum dots was performed. The radiative decay time appeared to vary from dot to dot between 100 ps and 230 ps corresponding to $f = 35$–75. For each measurement, the quantum dot lateral size was estimated through excitation of the photoluminescence technique [20]. Doing so, the non-monotonous dependence of the oscillator strength with the quantum dot radius was experimentally demonstrated (right part of Fig. 4.3) for quantum dot sizes below 30 nm. Detailed analysis of the exciton and bi-exciton power dependence gives an estimation of the capture time of carriers into the quantum dot (around 100 ps) [21]. For larger quantum dot size, the expected radiative lifetime gets shorter than the capture time into the quantum dot so that time resolved photoluminescence no longer gives access to the radiative decay of the transition. However, when optimizing growth conditions, GaAs quantum dots as large as 100–200 nm can be obtained [22], for which oscillator strength around a few hundred are expected.

As we will discuss in Section 4.5, natural InGaAs quantum dots have also been used to demonstrate the strong coupling regime. [23, 24] show that the quantum dot size strongly depends on the In content of the quantum dot layer. For In content around 30%, large elongated quantum dots are formed, with typical sizes of 30×100 nm. The oscillator strength for these quantum dots is estimated to be around 50. Finally, another option could be explored to develop fast radiative quantum dots as demonstrated in [25]. Large InAs self-assembled quantum dots emitting around 0.95 eV ($1.3 \mu m$) are grown by molecular beam epitaxy. A large quantum dot size leads to $f = 30$. To increase further the oscillator strengh, rapid thermal annealing at high temperature is used: In/Ga interdiffusion blue shifts the quantum dot emission and increases the quantum dot size in all directions. Oscillator strengths around $f = 100$ could be obtained with this technique.

In the present section, we have discussed the need to insert large oscillator strength quantum dots in ultimate nanocavities to demonstrate a strong coupling regime. Note that fast radiative quantum dots are also of great interest for the generation of indistinguishable photons as well as entangled photon pairs. Section 4.7.1 reports on the demonstration of indistinguishable photons using the Purcell effect to limit the dephasing processes during the quantum dot emission. Natural, fast, radiative quantum dots could also be promising candidates for such applications.

4.4   The weak coupling regime

4.4.1   Spontaneous emission inhibition

The conditions for observing spontaneous emission suppression are quite stringent. In order to determine the requirements for implementing such an effect, let's turn again to equation 4.11. The first term, related to emission into the cavity mode, vanishes as the quantum dot fundamental transition line is far detuned from the cavity resonance or if the quantum dot is located at a node of the cavity field. The second term Γ_{leak}/Γ_0 describes the possible decay due to emission into residual modes. Generally, this latter contribution to spontaneous emission decay precludes the implementation of spontaneous emission suppression: the confined optical modes in most resonators are superimposed on a quasi-continuum of "leaky" photon modes along the cavity, so that the off-resonance emission is little changed from its value in a homogeneous medium. However, it becomes possible to reduce or even suppress the contribution of this residual decay channel by use of 3D photonic crystal [54]. In these structures, the local density of optical states in the photonic crystal for emission inside the photonic band gap is reduced or even vanishes, removing also any additional decay channel within this spectral range. Therefore, for quantum dots emitting within the photonic band gap, suppression of spontaneous emission is made possible. However, as mentioned above, the fabrication of photonic crystals with full three-dimensional band gap is quite severe, even if the recent progress in the field is spectacular.

Hopefully, partial inhibition of spontaneous emission can be observed by use of 2D hexagonal photonic crystals etched on a suspended membrane, due to their "partial" TE band gap. For structures with adjusted geometries (whose parameters are the membrane thickness, holes radius and lattice spacing), such photonic crystal membranes exhibit a wide band gap for TE guided modes (i.e. modes with an electric field in the central plane of the slab pointing along the membrane), but no gap for TM modes (i.e. modes with an electric field pointing along the holes axis) [33]. Consider now a single dipole located vertically in the middle of the membrane. For dipoles orientated perpendicular to the membrane and therefore only coupled to TM modes, no gap in the emission rate can be observed but only a weak dependence on frequency, as expected from the absence of a 2D band gap for the TM-guided modes of the membrane. Conversely, for dipole orientations in the plane of the membrane and therefore only coupled to TE modes, a deep inhibition in the spontaneous emission rate should occur for emission in the photonic band gap. This latter effect has been demonstrated by incorporating single self-assembled InAs/GaAs quantum dots in H1 and L3 photonic crystal slab cavities etched on a GaAs suspended membrane [70–73]. The dipole orientation of the fundamental transition of these emitters is perpendicular to the holes' axis [74], so that the emitter couples only to the TE modes of the photonic crystal: The exciton "feels" the TE band gap but does not "see" the absence of TM band gap, enabling therefore the implementation of spontaneous emission suppression. Indeed, time resolved photoluminescence experiments on off-resonant quantum dots in such cavities indicate up to five-fold rate quenching due to the reduction of the local photon density of states in the photonic band gap [70].

4.4.2 Spontaneous emission acceleration

The quantum analysis of spontaneous emission based on Fermi's golden rule asserts that the largest enhancement of spontaneous emission rate of an emitting dipole is achieved if the dipole is on resonance spatially and spectrally with a high-Q/V single cavity mode and is pointing along the cavity electric field (see Eq. 4.11). The amplitude of this acceleration will be strongly dependent on the dipole orientation and the spatial and spectral matching of the emitter dipole with respect to the cavity field.

As mentioned above, progress in micro fabrication techniques has allowed a three-dimensional engineering of the refractive index on the wavelength scale and a rich diversification of the microcavity designs. For instance, micropost resonators, photonic disks and 2D photonic band gap slab microcavities sustain a discrete set of resonant modes with high-Q/V factors and have the potential to display a significant Purcell effect, provided a convenient emitter is used. Self-assembled InAs quantum dots constitute an appealing class of light emitters in this respect. Owing to their discrete density of electronic states, individual quantum dots exhibit a single, very narrow emission line under weak excitation conditions, which allows implementing significant CQED effects in high-F_P microcavities. However, the random nucleation position and size (and consequently emission wavelength) of such nano-emitters present a challenge to the efficient spectral and spatial coupling of the dot to the cavity mode. It should be repeated that the spontaneous emission rate of a given quantum dot into a cavity mode depends on its location and emission wavelength, which govern, respectively, the amplitude of the electric field it feels and the density of modes to which it is coupled.

In their pioneering work [7], J.-M. Gérard *et al.* used ensembles of InAs/GaAs self-assembled quantum dots embedded in GaAs/AlGaAs micropost cavities. A spontaneous emission rate enhancement by a factor of up to 5 was selectively observed for the quantum dots which were on resonance with one-cavity mode. The discrepancy between the expected Purcell factor ($\sim$30) and the observed enhancement results from a statistical averaging of the effect over all resonant dots of random spatial and spectral distributions with respect to the cavity mode. More recently, similar experiments have been carried out using only one self-assembled InAs/GaAs quantum dot. In this context, it should be possible to fully benefit from the Purcell effect, providing exact resonance energy and location of the dot with respect to the cavity mode.

The condition of spatial resonance is a real challenge, even if recent progress is quite promising in this field (through localization of the dot nucleation site or localization of the cavity around the randomly located dot, for instance – see paragraph 4.6.1). Therefore, most experiments performed up to now rely on a random spatial overlap between the individual quantum dot and the cavity mode and the spatial resonance condition is partially statistically fulfilled while scanning a large number of processed cavities embedding dilute arrays of dots.

The spectral resonance condition can be more easily satisfied. Two strategies can be used: tuning the cavity mode resonance to the energy of the fundamental transition of the dot or tuning the energy of the fundamental transition of the dot to the cavity mode resonance. This first approach can easily be implemented in some cavity geometries, mainly photonic crystal slab cavities whose resonance is tuned through successive cycles of oxidation and wet etching of the slab membrane [75, 76]. Another technique consists of varying, on the same sample, the cavities' geometry (for example, the diameters for micropost resonators) in order to shift spectrally the resonance conditions from one cavity to another on the same sample and cover a large spectral range. The second approach, consisting of spectral tuning of the dot transition to the cavity mode, exploits either the spectral dispersion of the dots by using dots of different sizes or the spectral tuning of the fundamental transition wavelength of the quantum dot as a function of temperature: when increasing the temperature from 5 to 50 K, the dot exciton wavelength red shifts by about 2 nm while the resonant wavelength of the mode red shifts by only less than 1 nm. The two can then be brought into resonance at a precise temperature (see Fig. 4.4 left). This effect has been successfully implemented in many different cavity geometries: microdisk [77], micropillar [78], and 2D photonic crystal slab cavity [73].

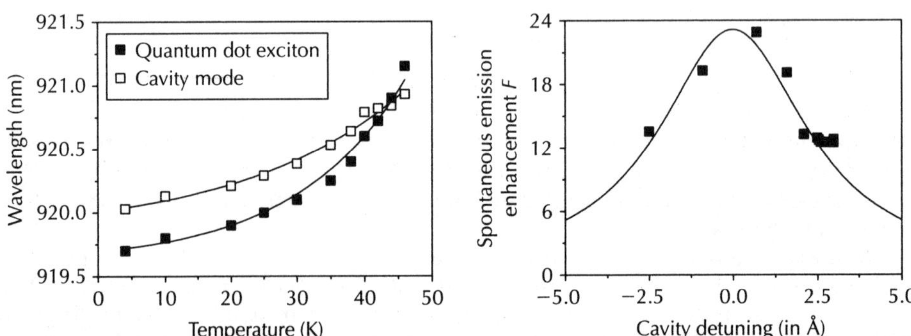

Figure 4.4   Left: *Spectral resonance matching of a single InAs quantum dot exciton to a micropillar cavity mode via temperature tuning, for an InAs/GaAs dot embedded in a 400 nm diameter GaAs/AlGaAs micropost with Q factor of 1600. Right: Amplitude of the spontaneous emission enhancement F as a function of spectral detuning, fitted by a Lorentzian.*

While combining these different strategies to couple a single dot to a cavity mode, significant Purcell effects on single dot have been recently observed at low temperature ($<$50 K). Direct measurements of spontaneous emission decay on resonant single quantum dots indicate an enhancement of the spontaneous emission dynamics of the order of a few units either in micro-disks ($F \sim 6$ [77]), micropillars ($F \sim 10$ [79]) or photonic crystal slab resonators ($F \sim 10$ [70,

73]). The resonant character of this effect can be highlighted when tuning the dot wavelength with respect to the cavity resonance through sample temperature increase, for example: when the exciton wavelength is slightly detuned from the cavity spectral resonance, the amplitude of the spontaneous emission rate enhancement decreases (see Fig. 4.4 right).

4.5 The strong coupling regime

To reach a strong coupling regime, one needs not only to develop cavities with high quality factor and small effective volume: one needs to maximize the quantity $Q\sqrt{f/V}$ (see Eq. 4.16). In 2001, Gérard estimated the possibility of reaching a strong coupling regime using one of the state of the art existing microcavities using self-assembled InAs quantum dots [80]. The authors calculated that the limit between weak and strong coupling regime could be reached using either microdisk microcavities or photonic crystal microcavities. To achieve light–matter entanglement, one needed either to improve the cavity performances or to use larger oscillator strength quantum dots. Three years later, the strong coupling regime was demonstrated in three groups, with different strategies both concerning the quantum dot and the cavity system [81, 82, 22]. In these three demonstrations, the same experimental techniques were used to demonstrate the strong coupling regime. In the following, we detail the demonstration of the strong coupling regime we obtained using GaAs quantum dots inserted in microdisk microcavities, which sustain whispering gallery modes (WGM). We then present the other demonstrations and discuss the figure of merit for the three systems.

4.5.1 First demonstrations of strong coupling regime

A way to demonstrate a strong coupling regime is to bring an exciton transition in resonance with the cavity mode transition. At resonance, anticrossing between the two lines is a signature of light–matter entanglement. Micro-photoluminescence is an efficient tool to monitor an excitonic transition energy. However, one could wonder if it is also an efficient tool to monitor the energy of an optical resonance. Indeed, optical mode energies are usually identified by inserting a large density of quantum dots providing a broadband emission source inside the microcavity. The quantum dot emission is then filtered by the cavity so that optical mode energies appear as well as identified lines on the photoluminescence spectrum [83, 84]. When only a few dots are embedded inside the microcavity, one might wonder how to identify the mode resonance. For high-quality factor microcavities ($Q = 10$–$20\ 000$), micro-photoluminescence experiments show that most of the time an emission line is observed at the energy of the optical mode. The origin of this line has not been fully understood up to now. The emission is referred to as a "background signal" filtered by the cavity mode. Yet, some recent experimental studies indicate that the emission filtered by the cavity mode is correlated to emission from single dots even though cavity mode exciton detuning is large [85, 86].

For a typical quality factor of $15\ 000$, the photon mode linewidth is around $100\ \mu eV$. When studying the emission from a single microcavity, one needs to find a way to identify the exciton lines and the "photon" lines. Figure 4.5 shows typical spectra observed on a single microdisk embedding GaAs natural quantum dots. Several lines are observed on the spectrum with similar linewidths. To identify the origin of the lines, temperature is increased. Figure 4.5a presents the absolute spectral shifts for a quantum dot exciton (quantum dot X) and a whispering gallery mode (WGM) resonance. When increasing temperature, both the exciton and the WGM energy decrease. The first one varies like the gap of bulk GaAs. The second decreases due to increase of the AlGaAs refractive index with temperature. These characteristic spectral behaviours are observed for every optical mode and quantum dot exciton on every microdisk so that varying temperature allows identifying exciton or optical mode origin of the emission lines. Finally, temperature tuning can be used to bring an exciton resonance into the cavity mode resonance.

Considering the spectral shift with temperature for each line observed in Fig. 4.5b, one can attribute the two highest energy lines to the emission of a small background within a whispering gallery mode and to the radiative recombination of a quantum dot exciton (quantum dot X).

When increasing temperature, a crossing of the two lines is observed at 37 K: the system is in the weak coupling regime. However, for a 2 μm diameter microdisk the mode effective volume is around $V = 0.07 \mu m^3$. Considering a quantum dot with an excitonic oscillator strength of $f = 100$ placed at an antinode, the system is expected to be in the strong coupling regime, with a Rabi splitting of 400 μeV. The observation of weak coupling regime is the signature that this quantum dot is most probably not positioned at a maximum of the electromagnetic field, which is the last condition required to observe a strong coupling regime.

To find a quantum dot exciton in both spectral and spatial resonance with the optical mode, many individual microcavities are studied. Figure 4.6a presents similar measurements performed on another 2 μm microdisk. Mainly, two lines are observed at each temperature. When increasing temperature from 4 K up to 30 K, the upper line shifts more than the lower line. This shows that the upper line can be assigned to a quantum dot exciton emission, and the lower line to an emission within a WGM. Yet, when increasing further temperature, the observed spectral shifts differ both from the one of an exciton and an optical mode: we observe an anti-crossing between the two lines, spectral signature of the strong coupling regime. The eigenstates of the system are not the exciton and photon states any more, but two exciton–photon mixed states, whose mixing depends on temperature. Figure 4.6 presents the emission energy of the upper and lower lines as a function of temperature. The dashed line indicates the spectral shift of weakly coupled exciton and WGM lines. The anti-crossing takes place at 30 K, where the energy separation between the upper and lower line reaches the minimum value of 400 μeV. The solid lines present the calculated coupled mode energy, considering the energy of the uncoupled WGM and quantum dot exciton mode and a Rabi splitting of 400 μeV. From the experimental Rabi splitting, with an effective volume of $V = 0.07 \mu m^3$, an exciton oscillator strength of $f = 100$ is deduced. This value is a minimum value since there is probably a spatial mismatch between the quantum dot and the optical mode antinode.

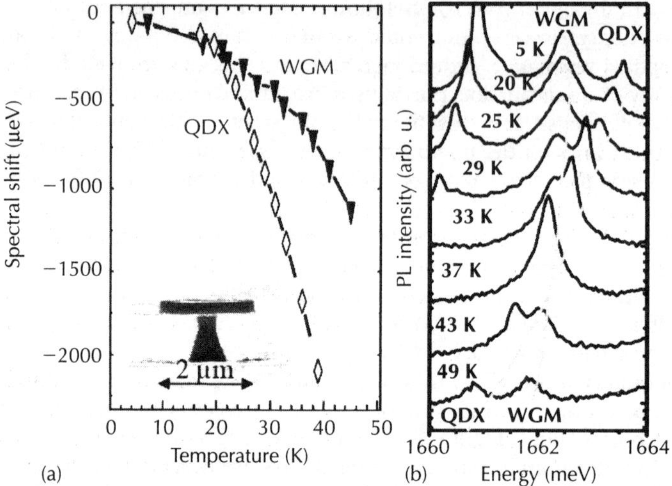

Figure 4.5   (a) Absolute spectral shift of the quantum dot exciton emission energy (diamonds) and of the WGM energy (triangles) as a function of temperature. Inset: *SEM image of a 2 μm microdisk. (b) Photoluminescence spectra under non-resonant continuous excitation for a particular microdisk of sample 2, for various temperatures.*

Figure 4.6c summarizes the measured linewidths of the upper and lower lines as a function of temperature. At low temperature (5–25 K), the minimum value of the photon-like (lower line) linewidth is 140 μeV ($Q = 12000$). The emission linewidth of the exciton-like upper line below resonance is around 280 μeV. At resonance, because of the half-exciton/half-photon nature of the eigenstates, their emission lines present the same linewidth (180 μeV), close to the expected average of 210 μeV.

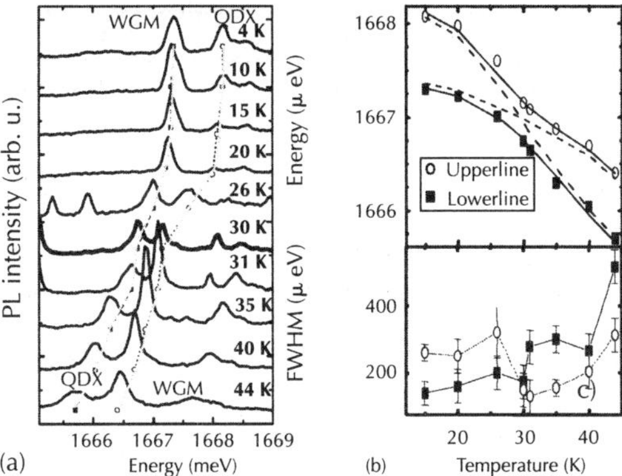

Figure 4.6   (a) Photoluminescence spectra under non-resonant continuous excitation for a particular microdisk for various temperatures. (b): Symbols: emission energy of the upper and lower lines as a function of temperature. Line: spectral shift of a quantum dot X and WGM. Continuous: calculated energy of the two coupled states. (c) Emission linewidth of the lower (circles) and upper (squares) lines as a function of temperature.

The same procedure based on spectral tuning through temperature has been used for the first demonstrations of the strong coupling regime (Fig. 4.7). Reithmaier and co-workers have chosen to insert "natural" InGaAs quantum dots in micropillar microcavities [81]. The success of this demonstration relies on two remarkable achievements. On the one hand, whereas the quality factor of micropillar microcavities is known to decrease strongly for diameters below $2\,\mu$m [23], Reithmaier has succeeded in increasing significantly this quality factor both by achieving quality factors over 100 000 for the planar cavity and improving the etching process [31]. On the other hand, a new type of quantum dot was developed leading to oscillator strength around 50. Poloczek and co-workers [24] have studied the influence of In content and strength on the growth of InGaAs quantum dots. For an indium content around 36%, large highly anisotropic quantum dots appear with a typical size of $100\,$nm $\times$ $20\,$nm presenting $f = 50$. The strong coupling regime was first reported in this system using a $1.5\,\mu$m diameter with $Q = 8000$ and effective volume of $V = 0.3\,\mu$m^3. This relatively large effective volume is compensated by the large oscillator strength of the quantum dot so that a Rabi splitting at a resonance of $170\,\mu$eV was observed.

Finally, the strong coupling regime was demonstrated using photonic crystal microcavities. The strong coupling regime has been evidenced twice by inserting InAs self-assembled quantum dots in modified L3-type optical microcavities [82, 87]. Despite the small oscillator strength of InAs quantum dots, the strong coupling regime is reached due to the small effective volume of these cavities $V = 0.04\,\mu$m^3.

Note that in the demonstration reported by Henessy and co-workers [87], a third line is observed together with the Rabi doublet. The appearance of this line is most probably due to spectral wandering of the quantum dot emission. The quantum dot emission can exhibit both exciton and charged exciton line, indicating that the capture of carrier into the quantum dot can be independent of electrons and holes. If the exciton–photon system is in the strong coupling regime, the charged exciton is detuned so that when charged excitons are trapped in the quantum dot, the energy of the optical mode is hardly modified.

Finally, we would like to mention another demonstration of strong coupling regime using a solid-state emitter, namely, CdSe nanorods inserted at the surface of a silicon microsphere. Despite the large effective volume of the cavity, a strong coupling regime is reached using large oscillator strength nanorods. The tuning between the cavity mode and the exciton emission is observed through spectral wandering of the line, as dominant effect for nanocrystals [88].

The figure of merit for the strong coupling regime is the Rabi splitting over linewidth ratio at resonance $4\hbar g/(\gamma_X + \gamma_C)$. Indeed, this ratio is a measure of the number of Rabi oscillations the system can undergo before the photon escapes outside the cavity or the exciton is scattered to a higher energy state. A summary of the four strong coupling realizations is presented below. A maximum figure of merit of 2 has been demonstrated up to now as compared to 5 for single atoms trapped in an optical microcavity [89].

	[81]	[82]	[22]	[87]
Cavity	Micropillar	Photonic crystal slab	Microdisk	Photonic crystal slab
Energy (eV)	1.32	1.05	1.66	1.32
Q	7350	6000	12 000	13 300
V (μm^3)	0.3	0.04	0.07	≈ 0.04
Quantum dot	In$_{0.3}$Ga$_{0.7}$As	InAs	GaAs	InAs
Size	30×100 nm^2	<25 nm	>44 nm	–
f	50	10	100	≈ 10
γ_X	140 μeV	140 μeV	180 μeV	35 μeV
Rabi splitting				
$2\hbar g$	140 μeV	170 μeV	410 μeV	150 μeV
$2\hbar g/\gamma$	1	1.3	2.2	2

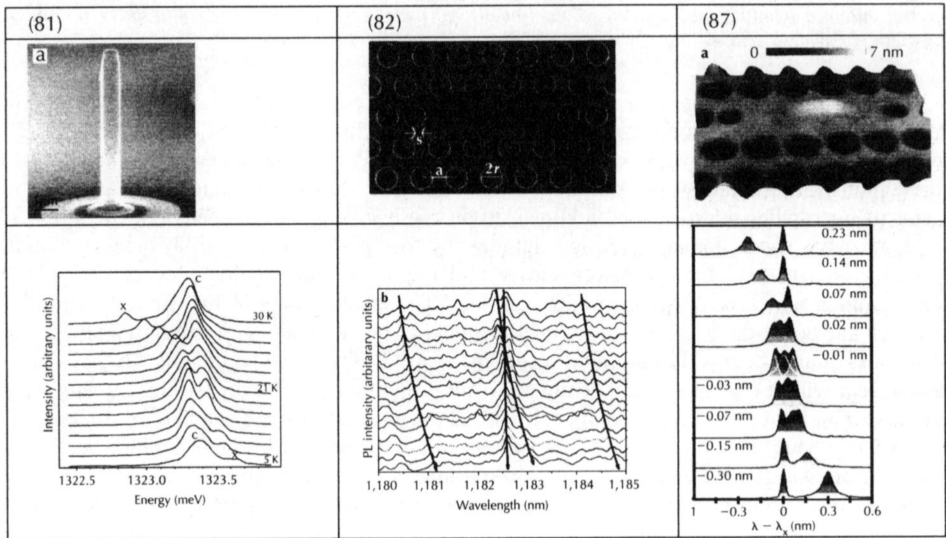

Figure 4.7   Three demonstrations of strong coupling regime. Top: SEM images (left and middle) and AFM image of the cavity system. Bottom: experimentally measured anti-crossing between the exciton and photon lines.

4.5.2   Some perspectives

We shall discuss below some quantum information processes based on the strongly coupled dot–cavity system (section 4.7.5) as well as the technological challenges that need to be addressed (section 4.6). Naturally, the physics of a strong coupling regime in a solid-state system has yet to be explored, as it has been investigated in atom-based systems.

One signature of the light–matter entanglement should be observed by performing crossed correlation measurement on the Rabi doublet. To perform such an experiment, the Rabi doublet over linewidth ratio will have to be significantly increased. However, recently, photon cross-correlation measurement on an exciton line and a photon line far from resonance show clear antibunching [87]. This first experimental study of the physics of a strong coupling regime reports interesting yet little understood correlation measurements.

Because of the fermionic nature of the electron–hole pairs trapped in a quantum dot, the Jaynes–Cumming ladder should be observed when more than one photon is introduced in the cavity. Such observations have been reported for a small number of photons in an atom–cavity system [90] and could be implemented with quantum dots.

Interesting features are also expected on the bi-exciton line for an exciton experiencing strong coupling regime with an optical mode analogue to the Autler–townes doublet [91]. Finally, the strong optical non-linearity resulting from the strong coupling regime could be used to realize photon blockade [92] as it has been recently reported for an atom-based system [93].

4.6   Towards a deterministic cavity–dot coupling

As presented in section 4.7, controlling the spontaneous emission rate for an atomic-like emitter in a solid-state system is very promising for applications in quantum cryptography, and future quantum computation protocols. The successful demonstrations of Purcell for single quantum dots as well as a strong coupling regime are important milestones in the new fields of solid-state quantum optics. However, to be able to investigate this new system, technological challenges need to be addressed.

To experience cavity quantum electrodynamics effects, the quantum dot has to be located precisely with respect to the electric field: either at an antinode (spontaneous emission exaltation, strong coupling regime) or at a node (spontaneous emission inhibition). Up to now, in all the experimental works except one [76], no control of the quantum dot position within the cavity is demonstrated. Many cavities are investigated to find a well-positioned quantum dot.

Finally, temperature tuning which has been used to reach the spectral resonance between the exciton and the photon mode has its limitations: first, spectral tuning is only possible for an initial detuning around 1–2 meV. Moreover, when increasing the temperature, the exciton line gets thermally broadened, which can prevent the establishment of a strong coupling regime. As a result, a more flexible way to tune the spectral resonance between the cavity mode and the excitonic transition is needed. In the present section, we report on various techniques that could allow these spatial and spectral tunings.

4.6.1   Spatial tuning

Most cavity effects on single quantum dots have been reported either on self-assembled quantum dots or natural quantum dots. In both cases, the quantum dot position is not controlled by the growth techniques.

However, techniques to control the growth position of single quantum dots have been investigated. One technique consists of patterning GaAs substrates using wet etching yielding to the formation of tetrahedral pyramids. The sample is then re-introduced in the growth chamber to grow AlGaAs barriers and a thin GaAs layer. The pyramidal geometry leads a confinement of the carriers in all space directions at the top of the pyramid [94, 95]. Single photon emission from these quantum dots' has been evidenced [96] showing the good optical quality of these dots' emission. However, up to now, these quantum dots have not been inserted in an optical microcavity.

More recently, InAs site controlled growth has been demonstrated. After substrate patterning through electron-beam lithography and reactive ion etching, GaAs/InAs re-growth was performed [97]. Photoluminescence on such a quantum dot ensemble shows non-radiative processes for the first quantum dot layer. However, using quantum dot vertical stacking, the optical properties of the dot layer improve when growing several layers of quantum dots. Power dependent measurements show that non-radiative processes are negligible after the sixth stack, with an exciton radiative lifetime around 1 ns [97]. Another work from the same team shows that depending on the growth conditions, self-aligned quantum dot pairs could be realized with the same technique [98]. These quantum dot pairs could be interesting in the context on quantum information processing as we explain in section 4.7.5. Finally, the same technique was used to realize organized GaAs quantum dots: good results were achieved on single quantum dot photoluminescence and single photon emission [99].

Controlled strain has also been used to grow quantum dot lattices. However, up to now, such a technique yields the formation of high-density quantum dot arrays, useful for laser devices but not for cavity quantum electrodynamics effects [100–102].

Last but not least, we would like to mention the work of Xie and Solomon on spatial order-ing of quantum dots in microdisks [103]. In this work, an Al-rich layer is grown below an GaAs layer, in a first growth step. Then microdisks are realized through optical lithography and wet etching. Surface deoxidation is performed before introducing the microdisk sample into the MBE chamber for quantum dot growth. The authors show that the microdisk geometry leads to pref-erential In adatom surface diffusion so that InAs quantum dots are grown within 100 nm from the microdisk edges. The edge linear quantum dot density is shown to depend strongly on the disk diameter, so that a 3 μm microdisk with one or two dots located at the edges are realized. Even though cavity effects have not yet been reported in this system, single spectroscopy and high-quality factor indicate that this system could be very promising.

Another approach has been developped by the group of Professor Imamoğlu (Zurich, Switzerland) to control the quantum dot position within the cavity. In a first demonstration [76], vertically aligned quantum dot stacks were grown up to the sample surface. The surface quantum dot acted as a tracer for the buried quantum dot, whose optical transition is at a higher energy than the upper dots. A matrix of gold markers and scanning electron microscopy was used to map the isolated quantum dot position. Controlled Purcell effect was first demonstrated with this technique. However, the lower energy absorption of the upper dots resulted in quality factor limitation for the optical mode. More recently, the same technique was realized without the tracer quantum dots. The buried quantum dot position was measured using AFM techniques. In this last report a strong coupling regime was observed [87].

4.6.2   Spectral tuning

Up to now, no real control of the quantum dot emission energy has been reached with standard growth techniques. To tune the exciton–photon mode resonance, several techniques to control the mode energy after the cavity realization have been investigated.

The first technique relies on chemical etching [85]. After the realization of a photonic crys-tal microcavity, a few monolayers of material are removed from the crystal surface by etching a self-formed native oxide. Doing so, the mode resonance wavelength can be tuned in a range of 80 nm, by 2–3 nm steps. This technique has been successfully used to demonstrate a strong cou-pling regime [87].

This first technique reveals the extreme dependence on the hole size of the mode energy in photonics crystal cavities. Thin film deposition in a helium flow cryostat has also been observed to tune the cavity mode energy [104]. This extreme dependence of the photon mode energy can be an advantage but also a critical drawback, the energy of the optical mode being determined by the quality of the chamber vacuum. Note that this effect is much more sensitive for photonic crystal cavities than for microdisk cavities and presumably micropillar cavities. However, depos-ing a thin glass slide seems to prevent this unwanted effect [104].

Nonetheless, when introducing a controlled flow of Xe in the cryostat, forming an epilayer on the photonic crystal cavity, the energy of the optical mode has been tuned in resonance with two quantum dot exciton lines, showing a strong coupling regime [105].

Finally, *in situ* laser microprocessing has been used to tune either the exciton or the photon res-onance [106]. Using high excitation power, the temperature of a microdisk microcavity embed-ding InAs quantum dots can be raised during a low-temperature photoluminescence experiment. Increasing the disk temperature, controlled desorption is reached, showing red shift of the optical mode. Increasing the disk temperature further, In and Ga atom interdiffusion leads to blue shift of the exciton transition. This technique is used to demonstrate exciton–photon resonance.

4.7   Applications of solid-state cavity quantum electrodynamics

Cavity quantum electrodynamics effects can find various potential applications in implement-ing novel photonic devices. Hereafter, we shall discuss several applications of CQED effects in the solid state: highly efficient light sources [27, 107, 108], generation of indistinguishable single photons, generation of polarization entangled photons, engineering of ultra-low threshold lasers and realization of quantum optical gates. We will shortly address the delicate question of quantum computing based on strongly coupled quantum dots in a cavity, stressing our point on practical first steps one could foresee in a mid-term project. Other interesting topics, such as ultra-fast nanolasers [109], will not be reviewed because of space limitations.

4.7.1   Efficient single photon source: possibilities and limitations

Purcell effect has been proposed to increase the extraction of quantum dot emission [27, 107]. Because of the high index in III–V compounds (around 3.5), most of the light emitted by a single quantum dot is reflected at the air–semiconductor interface so that only a small fraction (around 4%) of the emitted light can be extracted. When a single quantum dot placed in an optical microcavity experiences the Purcell effect, its emission rate into the cavity mode is increased by a factor F_P as compared to the leaky optical modes. As a result, the fraction of the spontaneous emission that is emitted into the mode is given by $\beta = F_p/(F_p + \Gamma_{\mathrm{leak}}/\Gamma_0)$, considering a perfectly resonant optical mode, and a perfectly positioned quantum dot. Without inhibition effect (such as reported in [7]), $\beta \approx F_p/(F_p + 1)$ so that 75% of the quantum dot emission is extracted in the optical mode for a Purcell effect of only 3.

Another possible application of the Purcell effect could be to increase the repetition rate of the single photon source. Actually, contrary to atoms or molecules, semiconductor quantum dots present spontaneous emission rates in the nanosecond range. Together with the Purcell effect, one could expect to reach a 10–100 GHz repetition rate single photon source. Altogether, semiconductor quantum dots embedded in optical microcavities appear promising to realize a high repetition rate efficient single photon source.

However, we would like to stress some limitations for such an application. First of all, the relevant parameter for single photon source efficiency is not only the β factor: the radiation pattern of the optical mode is also crucial for optical fibre coupling, for instance. In that matter, the far-field radiation pattern of micropillar microcavities has been experimentally measured [32]: for $2\,\mu\mathrm{m}$ diameter micropillars, the radiation pattern of the fundamental mode spreads over 20° and broadens when decreasing the pillar diameter. Obviously, since the Purcell effect scales-as Q/V, a high-quality factor should be preferred to a small effective volume microcavity.

Moreover, the scheme for realistic single photon source using quantum dots [110] relies on a non-resonant pulsed excitation creating high-energy electron–hole pairs in the wetting layer. Whenever capture processes into quantum dots are much faster that radiative recombination, one can consider that excitons are instantaneously trapped in the lowest energy states of the quantum dots. The system then returns to its ground state by emitting several photons. Only a single photon is emitted at the lowest energy for each excitation pulse. However, if the exciton radiative lifetime is shortened through the Purcell effect, excitons may remain in the wetting layer after the recombination of the low energy exciton. Another exciton can then be captured into the quantum dot.

Considering capture processes, one can estimate the limitation of the Purcell effect for efficient single photon source. A model for calculating the second-order autocorrelation function $g^2(0)$ is developed in [21] to account for photon correlation measurements on fast radiative natural GaAs quantum dots. This simple model considers a non-resonant excitation and includes capture processes. The same model is used to calculate $g^2(0)$ for various Purcell effects on single InAs quantum dots. Figure 4.8 presents the expected $g^2(0)$ for various Purcell factors. A strong increase in the probability that the quantum dot emits more than one photon as compared to a weak coherent pulse increases drastically with increasing F_p. Photon correlation measurements have been performed on single quantum dots experiencing the Purcell effect. Most of the time the amplitude of the zero delay correlation peaks is larger for quantum on resonance with the optical mode than off resonance [28, 72]. These observations are well accounted for by the present model. From these simple considerations, one can conclude that the most efficient single photon source will be easily obtained with moderate quality factor, 2–$3\,\mu\mathrm{m}$ diameter micropillar microcavities. Neither a smaller effective volume nor a larger quality factor is needed for such an application.

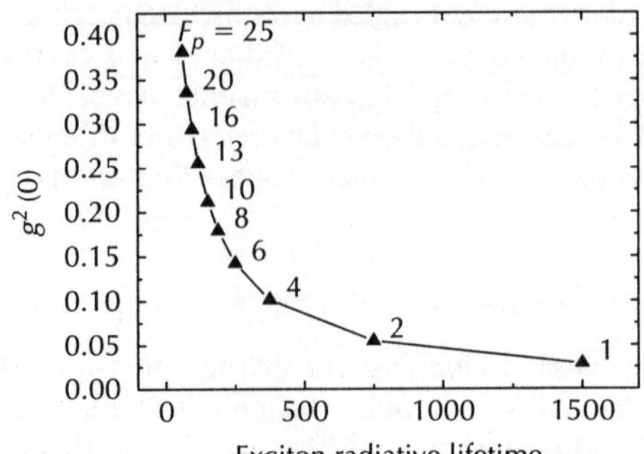

Figure 4.8   Calculated second-order correlation function as a function of the exciton radiative lifetime, corresponding to a Purcell factor indicated as a label on the plot. The exciton radiative lifetime without Purcell effect is 1.5 ns, the bi-exciton radiative lifetime is 750 ps, the capture time into the quantum dot is 50 ps and the radiative decay time of the wetting layer exciton population is 150 ps. A mean population of one exciton is created at the origin of time.

4.7.2   Indistinguishable single photon sources

In quantum mechanics, when two identical photons collide, they cannot be distinguished from one another, even in principle, either through their relevant intrinsic physical properties (polarization, energy, temporal profile, etc.) or through their respective trajectories. In this situation, the description of the two-particle system becomes ambiguous and the so-called exchange degeneracy appears. The problem of finding the correct and unambiguous description for such systems requires the introduction of a new postulate for quantum mechanics: the symmetrization postulate. This postulate distinguishes two classes of particles according to their collective behaviour in indistinguishable situations. These are bosons and fermions. In some experimental situations, the results of the experiment can only be deduced from this postulate. This is the case when two identical particles are incident on a 50/50 beam splitter. In that case, bosons emerge together along the same output port of the beam splitter (see Fig. 4.9), while fermions always follow two different output pathways. These coalescence and anti-coalescence phenomena result from the quantum interference between the amplitude probabilities of the four different output configurations (either both particles are transmitted or reflected; or one particle is reflected and the other one transmitted) [111]: these amplitude probabilities add coherently and an interference term appears, between the two paths in which the particles leave by different output ports. In other

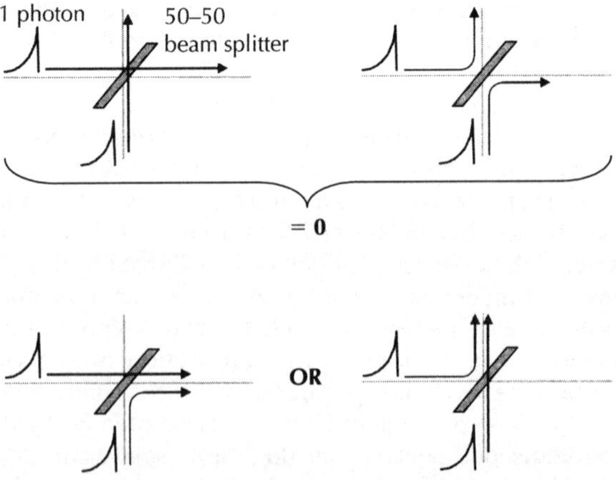

Figure 4.9   Output configurations when two identical photons are incident on a 50–50 beam splitter and collision characteristics: the two photons follow the same output port.

words, this is the indistinguishability of the "alternative amplitudes" that leads to the "final event". The interference is destructive for bosons and constructive for fermions.

More generally, two identical single photons can be defined as bosonic individual particles for which the symmetrization postulate has to be applied. In the phenomenon of two-photon interference, these two single photons arriving simultaneously by two different input ports on a beam splitter, both leave by the same output port, so that a two-photon state is constructed out of two distinct one-photon Fock states. This coalescence effect was first observed in parametric down conversion (in which a non-linear crystal splits a single photon into two photons of lower energy) [112]. It also occurs with single independently generated photons, for instance single photons emitted by single molecules [113], single atoms [114, 115] or single quantum dots. However, the usual spontaneous emission lifetimes of excitonic transitions in self-assembled quantum dots ranges from 500 ps to 1.5 ns, whereas the decoherence time of the excitonic dipole caused by phonon scattering is lower than a few hundred ps for non-resonant pumping: the natural width of spontaneous emission is of about the same order as the homogeneous broadening caused by phonon scattering. This precludes the generation of indistinguishable single photons from self-assembled quantum dots buried in bulk material, since successive single photons emitted through the radiative recombination of the exciton will be "marked" in energy by the random exciton–phonon interaction which causes the dephasing of the emitting exciton state (with a characteristic dephasing time T_2^*). In order to reduce the impact of dephasing on the emission process and thus restore the indistinguishability of the emitted photons, the radiative lifetime of the emitter (denoted by T_1) must dominate over the dephasing time T_2^* in determining the overall coherence time of the photon wave train, T_2, defined by:

$$\frac{1}{T_2} = \frac{1}{T_2^*} + \frac{1}{2T_1}. \tag{4.20}$$

While it is extremely difficult to eliminate the dephasing processes and lengthen T_2^*, it is possible to reduce the impact of decoherence phenomena by shortening the radiative lifetime of the emitter. This can be achieved by embedding the quantum dot in a microcavity, thus taking advantage of cavity quantum electrodynamics effects, in particular the Purcell effect. This idea of exploiting the Purcell effect has been successfully implemented by use of micropillar cavities [78, 116, 117] or photonic crystal slab resonators [118].

In these experiments, the microcavity source, cooled below 10 K, is excited twice by a pair of sequential laser pulses and resonant with an absorption of the quantum dot (typically 30–40 meV above the exciton transition energy, thus corresponding to an excited state of the exciton). This quasi-resonant excitation on a trapped excited state of the quantum dot exciton allows the reduction of the impact of dephasing processes through a relative lengthening of the pure dephasing time (of the order of a few hundred ps). Upon each excitation cycle, the dot emits at most two single photon pulses in the microcavity at the exciton wavelength. These emitted photons exit the microcavity and are combined on a 50–50 beam splitter with a relative time delay τ (see Fig. 4.10 left). They are then detected by single-photon detectors placed along the two output ports of the beam splitter and a correlation measurement is carried out, in order to infer the probability for the two photons to follow two different output pathways as a function of the time delay τ between the two single photon pulses. This probability is related to the visibility of the two-photon quantum interference. For distinguishable single particles, this probability is equal to 50% whatever the value of the time delay, while for ideal indistinguishable single photons impinging simultaneously on the beam splitter ($\tau = 0$), it would cancel out. In this experiment (Fig. 4.10 right), the visibility of the quantum interference presents a dip at null delay, which indicates that the two photons in the majority both go to the same output when they arrive simultaneously on the beam splitter. This strong reduction of simultaneous photocounting probabilities is strongly dependent on the overlapping of the two pulses and reaches its maximum for "perfect" wavefront matching ($\tau = 0$). The interference disappears when the delay time τ between the two particles exceeds the pulse width $2T_1$ (of about a few hundred ps).

This high visibility of the quantum interference in these two photon experiments ($>70\%$) results from the high Purcell effect the single quantum dot is submitted to: the spontaneous emission

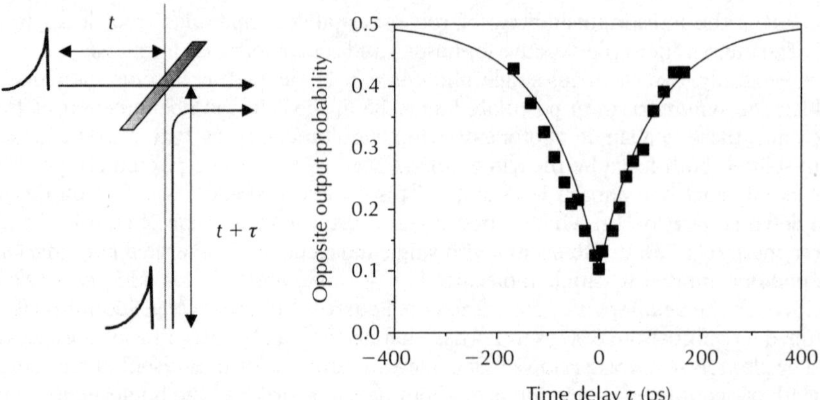

Figure 4.10   Collision experiment (left) used to observe the quantum interference between two identical single photons and (right) influence of the wavefront matching on the quantum interference visibility as a function of the time delay τ between both particles, emerging from a single InAs/GaAs quantum dot embedded in a 400 nm diameter GaAs/AlGaAs Pillar with a Q factor of ∼1500. The spontaneous emission lifetime T_1 of the quantum dot exciton is decreased to 105 ps by the presence of the cavity, while the pure dephasing time is of the order of 660 ps.

lifetime shortening due to the presence of the microcavity restores partially the single photon's indistinguishability. The remaining imperfection is likely to arise from several decoherence mechanisms. When the quantum dot is excited by a laser pulse, the photogenerated electron–hole pair is initially in an excited state and relaxed to its lowest excitonic state through phonon emission before emitting a single photon. The ratio of this relaxation time, which can reach tens of picoseconds, to the exciton radiative lifetime T_1, may limit the visibility of this quantum interference. Decoherence by phonon scattering is another possible mechanism, since even at low temperature ($T < 10$ K), the pure dephasing time in these experiments is long (of the order of a few hundred ps) but not infinite, thus contributing to the overall coherence time of the photon wave train (see Eq. 4.20).

4.7.3   Proposal for entangled photons sources

Recent proposals for quantum communication and quantum information protocols provide a significant incentive to develop practical entangled two-photon sources. One requirement for such sources is that the emission time of the photons is periodic with a precisely defined clock frequency. Bi-exciton (denoted XX) radiative cascade in quantum dots is a candidate system for such sources [119] (see Fig. 4.11a). In the first step of the cascade, a photon is emitted with random polarization. Conservation of angular momentum requires that the polarization of the photon emitted in the second step of the cascade is fixed relative to the first photon: the pair of photons is in a polarization entangled state. Entanglement requires two decay paths with different polarizations which are otherwise indistinguishable. However, in typical quantum dots, the excitonic relay level (denoted X) is split by the anisotropic exchange interaction, caused by in-plane anisotropy of the exciton wave function [120] (see Fig. 4.11b). Therefore, the energy splitting of the excitonic relay level provides information about which pathways the two photons were released along, via the energy of the emitted photons, and entanglement is destroyed.

In order to erase the "which path" information and recover polarization entanglement with a high fidelity, it is crucial to reduce the excitonic energy splitting within the radiative linewidth of the relay levels. First proofs of principle have been recently established, where entanglement between the frequency and polarization degrees of freedom has been measured in photon pairs generated by the cascade emission from single quantum dots. These experiments use either strong spectral filtering discarding unentangled photons pairs [121], lateral magnetic field [122] or thermal annealing [123], to tune the exciton fine splitting close to zero. However, the entangled photon collection is kept rather low, which precludes the practical use of such sources in quantum information processing platforms. Moreover, the degree of entanglement is still limited

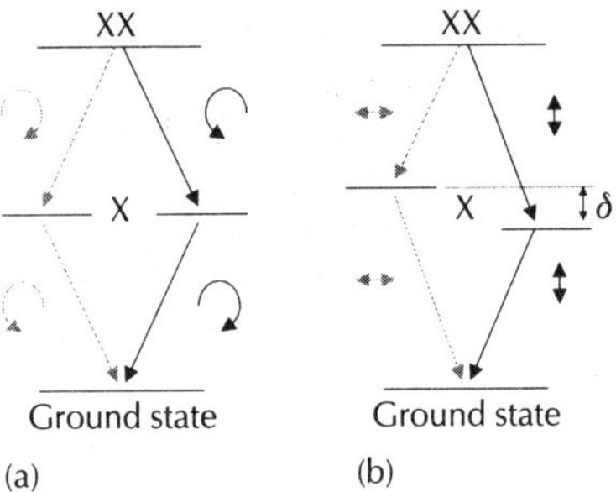

Figure 4.11   Schematic description of the bi-excitonic cascade in a typical quantum dot (a) with zero energy splitting on the relay excitonic level leading to polarization of maximally entangled photons and (b) with an energy splitting δ leading to two collinearly polarized photons in a single preferred basis.

by dot dephasing and residual excitonic fine-structure splitting. It is believed that these limitations can be to a large extent overcome by increasing the photon emission rate, through the embedding of the quantum dot in a semiconductor microcavity [124, 125].

Indeed, application of an optical microcavity resonant with the excitonic transition of the cascade increases the exciton decay rate and therefore is another means to enhance the entanglement visibility. First, a fast emission of the excitonic photon reduces the time during which dephasing degrades the intermediate relay state in the bi-excitonic cascade. Second, an increased excitonic decay rate results in an enhanced homogeneous linewidth of the excitonic transitions, and therefore increases the overlap between the wave packets corresponding to horizontally and vertically polarized photons released through exciton recombination: the excitonic photon is released before quantum beats between the two relay levels occur. Third, the reduction of the exciton transition dipole lifetime decreases the probability of re-exciting the system before emission of the second photon, when increasing the excitation rate. For typical quantum dots with excitonic energy splitting lower than 30 μeV, Purcell factors as high as 30 should allow for restoration of polarization entanglement. However, the requirements for the implementation of such an effect are stringent. In the first place, the exciton energy splitting must be kept low (<30 μeV) regarding achievable Purcell effects with state of the art fabrication techniques. Moreover, the cavity has to support two degenerate polarization modes in the transition dipoles plane, in order to preserve the single photon pair's polarization correlations. If not, the transition dipoles will have modified coupling constants, which will favour one recombination path and thus corrupt entanglement.

4.7.4   Proposal for ultra-low threshold lasers

Whether useful or detrimental, spontaneous emission plays a key role in laser devices. Spontaneous emission into the so-called lasing mode is desirable since it provides photons that are subsequently amplified by stimulated emission. By contrast, spontaneous emission into other non-lasing modes is prejudicial, because it tends to increase the threshold power required to initiate lasing. In order to reduce by several orders of magnitude the minimum threshold power of semiconductor lasers, one strategy would be to implement the Purcell effect. As mentioned above, the Purcell effect is the enhancement of the spontaneous emission rate of a single emitting dipole coupled to a high-Q/V cavity mode. The concomitant effect is the spatial redistribution of spontaneous emission and the preferential funnelling of light in the cavity mode. Indeed, the ratio of spontaneous emission funnelled into the lasing mode depends on the spontaneous emission decay rate into the lasing modes Γ and on the spontaneous emission decay rate into the non-lasing modes Γ_{leak} as follows:

$$\beta = \frac{\Gamma}{\Gamma + \Gamma_{leak}}. \tag{4.21}$$

This spontaneous emission coupling factor β can approach unity in nanocavity lasers with a large Purcell effect, with serious implications on the nature of light emission. In conventional semiconductor lasers, this β factor is very low, of the order of 10^{-5} and the laser threshold is marked by a sudden increase of output power and an abrupt quenching of the laser's quantum noise as the laser switches from spontaneous emission to coherent laser-like output. The switch takes place within a very well-defined threshold region with a relative width of $\beta^{1/2}$ [126]. With increasing β, the step-like "threshold" marked by a kink in the output intensity gradually changes to an s-shaped smooth intensity transition and disappears up to the so-called "threshold-less laser" in the limit $\beta \sim 1$ [127]: in this limit, the input–output curve is linear with no abrupt transition from a non-lasing to a lasing state. Therefore, for these microlasers with high β factor, the threshold is poorly defined both in terms of intensity and fluctuations. This is not surprising considering that the relative threshold width $\beta^{1/2}$ becomes of order unity for $\beta \sim 1$ and such lasers are hence commonly referred to as thresholdless lasers.

In the regime of efficient spontaneous emission funnelling (β close to one), a more accurate interpretation of the lasing onset and the nature of the emitted light is given by a intra-cavity photon statistics analysis. One criterion is the Fano factor as suggested by Rice and Carmichael [126]. It is defined by:

$$K = \frac{\langle \delta n^2 \rangle}{\langle n \rangle} = \langle n \rangle \left(g^2(0) - 1 \right) \tag{4.22}$$

where $g^2(0)$ is the second-order correlation function, n is the intra-cavity photon number, and $\delta n = n - n_0$, with n_0 the steady-state photon number. In conventional lasers, the Fano factor exhibits a sharp peak with a height of roughly $2\beta^{1/2}$ on threshold and the peak has a relative width of $\beta^{1/2}$. It is equal to one both below and above threshold. The transition to lasing can thus be determined through second-order correlation measurements, where bunching is observed below threshold ($g^2(0) > 1$) and Poissonian photon statistics are observed well above threshold ($g^2(0) = 1$). The different photon statistics between conventional lasers and high-β factor lasers are revealed by the evolution of $g^2(0)$ below threshold. In a conventional laser with $\beta << 1$ and below threshold, $g^2(0)$ reaches the value of 2 for thermal light and decreases abruptly to unity in the threshold region. For large values of β, this transition becomes broader, while $g^2(0)$ deviates from 2 well below threshold and decreases gradually to the value of 1 for ideal 0D laser cavities with infinite Q factor.

The implementation of thresholdless lasers requires therefore the use of nanocavities with large β factors. As mentioned above, one strategy uses the Purcell effect. However, the bottleneck blocking the demonstration of thresholdless operation arises from the resonant character of the Purcell effect. The wavelengths and positions of the quantum dots in the nanocavity are random. If the cavity mode is resonant with respect to the wavelength and position of the quantum dot, the dot emission is mainly coupled to the single cavity mode, allowing thresholdless operation to occur. On the contrary, if the wavelengths and positions of quantum dots are not perfectly matched with the cavity mode, the amplitude of the Purcell effect is reduced and the ratio of dot emission coupled to non-lasing modes increases, preventing thresholdless lasing operation. Nevertheless, nearly thresholdless laser operation in such a system is feasible even in the off-resonant case and was recently achieved by use of photonic crystal–slab cavities [86, 128] or micropillars [129] containing an array of self-assembled quantum dots. A smooth transition from spontaneous emission with significant photon bunching to stimulated emission with Poissonian photon statistics has been observed and β factors as high as 0.85 have been measured.

The ultimate thresholdless laser would be the single dot laser. The scaling of lasers to a small number of emitters has been investigated for two decades using atomic emitters, such that recently a one-atom laser has been demonstrated [130]. Single quantum dots isolated in high-β factor cavities may be a promising route to solid-state, single emitter lasing. In this regime, a fraction β of spontaneously emitted photons will be captured by the laser mode of the optical cavity. If the cavity mode is on resonance with emission, the captured photons will remain in the cavity for an average storage time Q/ω_c before leaking out. If the quantum dot is excited again and emits a second photon in the cavity mode before the first photon leaks out, lasing occurs. In order to reach threshold, then, it is necessary to implement high-oscillator strength quantum dots and a cavity with very high Q as well as a very high β that captures a reasonable fraction of the spontaneous emission [131]. These single emitter lasers would allow the generation of intensity-squeezed (or number-squeezed) states [132].

4.7.5   Possible applications for quantum information processing

With the development of optical studies of single quantum dots, theoretical works have investigated the possibility of implementing quantum information processing based on single quantum dots. Two different types of qbits are explored: one based on spin, the other based on orbital degree of freedom. The latter approach uses exciton states with typically 1 ns limited lifetime, so that ultra-fast quantum computing operations are proposed based on optical pulses. The former approach relies on the spin states of a resident electron in the quantum dot. This approach benefits from much longer coherence times [133, 134] but the implementation of quantum gates is slower than with excitonic states [135, 136]. Both approaches meet a common difficulty which is that only neighbouring quantum dots are coupled through dipole–dipole interaction [137, 138]. Recently, several theoretical approaches have proposed to mediate the coupling between distant quantum dots using a high-quality factor cavity mode [139–142]. The qbits are based on the spin state of an excess electron using either an in-plane [139] or off-plane [140] magnetic field. In these protocols, each quantum dot is individually addressed using near-field optics. Moreover, the distant quantum dots need to present the same transition energy as well as maximum coupling to the optical mode. The experimental conditions to implement these protocols are far from being gathered and will not be gathered in the near future either.

However, on the way to gathering these experimental conditions, important milestones have to be realized. One is the control of the charge state of a single quantum dot while being inserted in an optical microcavity. The control of the charge state of a single quantum dot has been successfully demonstrated using Schottky gates [143]. A first demonstration of such a charge control has very recently been reported in a VECSEL-like structure [144] with a structure confining photons only in the growth direction. To implement both cavity effects and charge states control, the same technique could be used, and the optical field confinement could be obtained through lateral oxidation [145]. Another promising approach could be inspired by the work of Muller and co-workers, using high-quality factors and small effective volume on a planar cavity presenting a local variation of the cavity thickness [146]. This technique, based on regrowth on a pattern 2D microcavity, could reasonably be modified for contact implementation.

Another key ingredient for these protocols is a non-destructive measurement of the spin state of the electron trapped in the quantum dot. First demonstrations of the spin state of a single electron have very recently been reported [144, 147]. The information on the spin state is obtained through the conditional Faraday rotation of a spectrally detuned laser. Sensitivity is a crucial matter for these experiments, which have already been shown to be greatly enhanced using cavity effect [143] and will be even more improved in high-quality factor small effective volume cavities.

Finally, cavity effects with several quantum dots have to be demonstrated. To realize such a demonstration precise control of the spectral and spatial tuning between the cavity and the quantum dot is necessary.

4.8   Summary and conclusions

Engineering the interaction of light with matter allows one to tune important properties of solids like the spontaneous emission rate or the spontaneous emission coupling factor into a laser mode. Light–matter interaction in semiconductors using quantum dots has been investigated in various microcavities. Inhibition and exaltation of a single quantum dot spontaneous emission has been demonstrated in various systems, and has already led to several applications in quantum cryptography. Strong coupling regime has recently been demonstrated reaching the ultimate limit for light–matter coupling in a solid-state system. Technological challenges now need to be addressed to reach either Purcell effect or strong coupling regime on demand. Indeed, controlling solid-state quantum electrodynamics will eventually open new fields of investigation, such as ultra-low threshold laser or quantum computation based on the coupling between a single spin and a cavity mode.

Acknowledgements

We would like to thank I. Abram, J. Bloch, A. Beveratos, R. Braive, A. Cavanna, S. Guilet, J. Hours, S. Laurent, L. Le Gratiet, A. Lemaitre, D. Martrou, A. Miard, G. Patriarche, E. Peter, I. Sagnes, and S. Varoutsis who largely contributed to this work. We thank P. Lalanne, J.M. Gérard and P. Voisin for fruitful discussions. This work was partly supported by ANR JCJC Micados, SANDIE European Network, ACN NanoQub, ACN Polqua, by the "Région Ile de France" and the "Conseil Général de l'Essonne".

References

1. S. Haroche. Cavity quantum electrodynamics, Course 13 in: *Fundamental Systems in Quantum Optics*, (Elesvier Science Publishers, 1992).
2. P. Goy, J.M. Raymond, M. Gross, and S. Haroche, Observation of cavity-enhanced single-atom spontaneous emission, *Phys. Rev. Lett.* **50**, (1903/1983).
3. R.G. Hulet, E.S. Hilfer, and D. Kleppner, Inhibited spontaneous emission by a Rydberg atom, *Phys. Rev. Lett.* **55**, 2137 (1985).
4. F. Bernardot, P. Nussenzweig, M. Brune, J.M. Raimond, and S. Haroche, Vacuum Rabi splitting observed on a microscopic atomic sample in a microwave cavity, *Euro. Phys. Lett.* **17**, 33 (1991).
5. R.J. Thompson, G. Rempe, and H.J. Kimble, Observation of normal-mode splitting for an atom in an optical microcavity, *Phys. Rev. Lett.* **68**, 1132 (1992).
6. C. Weisbuch, M. Nishioka, A. Ishikawa, and Y. Arakawa, Observation of the coupled exciton–photon mode splitting in a semiconductor quantum microcavity, *Phys. Rev. Lett.* **69**, 3314–3317 (1992).
7. J.-M. Gérard, B. Sermage, B. Gayral, B. Legrand, E. Costard, and V. Thierry-Mieg, Enhanced spontaneous emission by quantum boxes in a monolithic optical microcavity, *Phys. Rev. Lett.* **81**, 1110–1113 (1998).
8. E. Purcell, Spontaneous emission probabilites at radio frequencies, *Phys. Rev.* **69**, 681 (1946).
9. K. Matsuda, T. Saiki, S. Nomura, M. Mihara, Y. Aoyagi, S. Nair, and T. Takagahara, Near-field optical mapping of exciton wave functions in a GaAs quantum dot, *Phys. Rev. Lett.* **91**, 177401 (2003).
10. L.C. Andreani, G. Panzarini, and J.-M. Gérard, Strong-coupling regime for quantum boxes in pillar microcavities: Theory, *Phys. Rev. B.* **60**, 13276 (1999).
11. J. Hours, P. Senellart, E. Peter, A. Cavanna, and J. Bloch, Exciton radiative lifetime controlled by the lateral confinement energy in a single quantum dot, *Phys. Rev. B.* **71**, 161306 (2005).
12. A.V. Kavokin, Exciton oscillator strength in quantum wells: from localized to free resonant states, Phys. Rev. B 50 8000.
13. J.-M. Gérard, *Confined Electrons and Photons, NATO Advanced Studies Institute, Series B: Physics*, edited by E. Burstein and C. Weisbuch Vol. 357, (Plenum, New York, 1995), 19955 pp.
14. G. Wang, S. Fafard, D. Leonard, J.E. Bowers, J.L. Merz, and P.M. Petroff, Time-resolved optical characterization of InGaAs/GaAs quantum dots, *Appl. Phys. Lett.* **64**, 2815 (1994).
15. H. Yu, S. Lycett, C. Roberts, and R. Murray, Time resolved study of self-assembled InAs quantum dots, *Appl. Phys. Lett.* **69**, 4087 (1996).
16. M. Paillard, X. Marie, E. Vanelle, T. Amand, V.K. Kalevich, A.R. Kovsh, and A.E. Zhukovand V. M. Ustinov, Time-resolved photoluminescence in self-assembled InAs/GaAs quantum dots under strictly resonant excitation, *Appl. Phys. Lett.* **76**, 76 (2000).
17. R.M. Thompson, R.M. Stevenson, A.J. Shields, I. Farrer, C.J. Lobo, D.A. Ritchie, M.L. Leadbeater, and M. Pepper, Single-photon emission from exciton complexes in individual quantum dots, *Phys. Rev. B.* **64**, 201302 (2001).
18. C. Santori, G.S. Solomon, M. Pelton, and Y. Yamamoto, Time-resolved spectroscopy of multiexcitonic decay in an InAs quantum dot, *Phys. Rev. B.* **65**, 073310 (2002).

19. J.R. Guest *et al.*, Measurement of optical absorption by a single quantum dot exciton, *Phys. Rev. B.* **65**, 241310 (2002).

20. E. Peter, J. Hours, P. Senellart, A. Vasanelli, A. Cavanna, J. Bloch, and J.-M. Gérard, Phonon sidebands in exciton and biexciton emission from single GaAs quantum dots, *Phys. Rev. B.* **69**, 041307 (2004).

21. E. Peter, S. Laurent, J. Bloch, J. Hours, S. Varoutsis, D. Martrou, I. Robert-Philip, A. Beveratos, A. Lemaître, A. Cavanna, G. Patriarche, and P. Senellart, Fast radiative quantum dots: from single to multiple photon emission, *Appl. Phys. Lett.* **90**, 223118 (2007).

22. E. Peter, P. Senellart, D. Martrou, A. Lemaître, J. Hours, J.-M. Gérard, and J. Bloch, Exciton–photon strong-coupling regime for a single quantum dot embedded in a microcavity, *Phys. Rev. Lett.* **95**, 067401 (2005).

23. A. Löffler, J.P. Reithmaier, G. Sek, C. Hofmann, S. Reitzenstein, M. Kamp, and A. Forchel, Semiconductor quantum dot microcavity pillars with high-quality factors and enlarged dot dimensions, *Appl. Phys. Lett.* **86**, 111105 (2005).

24. P. Poloczek, G. Sek, J. Misiewicz, A. Löffler, J.P. Reithmaier, and A. Forchel, Optical properties of low-strained InxGa1–xAs/GaAs quantum dot structures at the two-dimensional–three-dimensional growth transition, *J. Appl. Phys.* **100**, 013503 (2006).

25. W. Langbein, P. Borri, U. Woggon, V. Stavarache, D. Reuter, and A.D. Wieck, Radiatively limited dephasing in InAs quantum dots, *Phys. Rev. B.* **70**, 033301 (2004).

26. D. Gammon, E.S. Snow, B.V. Shanabrook, D.S. Katzer, and D. Park, Homogeneous linewidths in the optical spectrum of a single gallium arsenide quantum dot, *Science* **273**, 87 (1996).

27. M. Pelton, C. Santori, J. Vuckovic, B. Zhang, G.S. Solomon, J. Plant, and Y. Yamamoto, Efficient source of single photons: a single quantum dot in a micropost microcavity, *Phys. Rev. Lett.* **89**, 233602 (2002).

28. P. Michler, A. Kiraz, C. Becher, W.V. Schoenfeld, P.M. Petroff, L. Zhang, E. Hu, and A. Imamoglu, A quantum dot single-photon turnstile device, *Science* **290**, 2282 (2000).

29. J. Hours, S. Varoutsis, M. Gallart, J. Bloch, I. Robert-Philip, A. Cavanna, I. Abram, F. Laruelle, and J.-M. Gérard, Single photon emission from individual GaAs quantum dots, *Appl. Phys. Lett.* **82**, 2206 (2003).

30. G. Bastard, Wave Mechanics Applied to Semiconductor Heterostructures, *Editions de Physique* (1996).

31. S. Reitzenstein, C. Hofmann, A. Gorbunov, M. Strauß, S.H. Kwon, C. Schneider, A. Löffler, S. Höfling, M. Kamp, and A. Forchel, AlAs/GaAs micropillar cavities with quality factors exceeding 150.000, *Appl. Phys. Lett.* **90**, 251109 (2007).

32. H. Rigneault, J. Broudic, B. Gayral, and J.-M. Gérard, Far-field radiation from quantum boxes located in pillar microcavities, *Opt. Lett.* **26**, 1595 (2001).

33. S. Johnson, S. Fan, P.R. Villeneuve, J.D. Joannopoulos, and L.A. Kolodziejski, Guided modes in photonic crystal slabs, *Phys. Rev. B.* **60**, 5751 (1999).

34. C. Sauvan, P. Lalanne, and J.P. Hugonin, Slow-wave effect and mode-profile matching in photonic crystal microcavities, *Phys. Rev. B.* **71**, 165118 (2005).

35. Y. Akahane, T. Asano, B.S. Song, and S. Noda, High-Q photonic nanocavity in a two-dimensional photonic crystal, *Nature (London)* **425**, 944 (2003).

36. E. Kuramochi, M. Notomi, S. Mitsugi, A. Shinya, T. Tanabe, and T. Watanabe, Ultrahigh-Q photonic crystal nanocavities realized by the local width modulation of a line defect, *Appl. Phys. Lett.* **88**, 041112 (2006).

37. B.-S. Song, S. Noda, T. Asano, and Y. Akahane, Ultra-high-Q photonic double heterostructure nanocavity, *Nature Mat.* **4**, 207 (2005).

38. T. Asano, B.-S. Song, and S. Noda, Analysis of the experimental Q factors (∼1 million) of photonic crystal nanocavities, *Opt. Express* **14**, 1996 (2006).

39. E. Weidner, S. Combrié, N.-V.-Q. Tran, A. De Rossi, J. Nagle, S. Cassette, A. Talneau, and H. Benisty, Achievement of ultrahigh quality factors in GaAs photonic crystal membrane nanocavity, *Appl. Phys. Lett.* **89**, 221104 (2006).

40. R. Herrmann, T. Sünner, T. Hein, A. Löffler, M. Kamp, and A. Forchel, Ultrahigh-quality photonic crystal cavity in GaAs, *Opt. Lett.* **31**, 1229 (2006).

41. J.-Y. Marzin, J.-M. Gérard, A. Izraël, D. Barrier, and G. Bastard, Photoluminescence of single InAs quantum dots obtained by self-organized growth on GaAs, *Phys. Rev. Lett.* **73**, 716 (1994).

42. K. Brunner, G. Abstreiter, G. Böhm, G. Tränkle, and G. Weimann, Sharp-line photoluminescence and two-photon absorption of zero-dimensional biexcitons in a GaAs/AlGaAs structure, *Phys. Rev. Lett.* **73**, 1138 (1994).

43. A. Lemaître, G. Patriarche, and F. Glas, Composition profiling of InAs/GaAs quantum dots, *Appl. Phys. Lett.* **85**, 3717 (2004).

44. K. Leosson, J.R. Jensen, W. Langbein, and J.M. Hvam, Exciton localization and interface roughness in growth-interrupted GaAs/AlAs quantum wells, *Phys. Rev. B.* **61**, 10322 (2000).

45. S.L. McCall, A.F.J. Levi, R.E. Slusher, S.J. Pearton, and R.A. Logan, Whispering-gallery mode microdisk lasers, *Appl. Phys. Lett.* **60**, 289 (1992).

46. V.S. Ilchenko, M.L. Gorodetsky, X.S. Yao, and L. Maleki, Microtorus: a high-finesse microcavity with whispering-gallery modes, *Opt. Lett.* **26**, 256 (2001).

47. V.B. Braginsky, M.L. Gorodetsky, and V.S. Ilchenko, Quality-factor and nonlinear properties of optical whispering-gallery modes, *Phys. Lett. A.* **137**, 393–397 (1989).

48. L. Collot, V. Lefèvre-Seguin, M. Brune, J.-M. Raimond, and S. Haroche, Very high Q whispering gallery mode resonances observed on fused silica microspheres, *Eur. Phys. Lett.* **23**, 327 (1993).

49. M.L. Gorodetsky, A.A. Savchenkov, and V.S. Ilchenko, Ultimate Q of optical microsphere resonators, *Opt. Lett.* **21**, 453 (1996).

50. D.W. Vernooy, V.S. Ilchenko, H. Mabuchi, E.W. Streed, and H.J. Kimble, High-Q measurements of fused-silica microspheres in the near infrared, *Opt. Lett.* **23**, 247–249 (1998).

51. D.K. Armani, T.J. Kippenberg, S.M. Spillane, and K.J. Vahala, Ultra-high-Q toroid microcavity on a chip, *Nature* **421**, 925–928 (2003).

52. B. Gayral, J.-M. Gérard, A. Lemaître, C. Dupuis, L. Manin, and J.-L. Pelouard, High-Q wet-etched GaAs microdisks containing InAs quantum boxes, *Appl. Phys. Lett.* **75**, 1908–1910 (1999).

53. K. Srinivasan, M. Borselli, T.J. Johnson, P.E. Barclay, O. Painter, A. Stintz, and S. Krishna, Optical loss and lasing characteristics of high-quality-factor AlGaAs microdisk resonators with embedded quantum dots, *Appl. Phys. Lett.* **86**, 151106 (2005).

54. E. Yablonovitch, Inhibited spontaneous emission in solid-state physics and electronics, *Phys. Rev. Lett.* **58**, 2059 (1987).

55. S. John, Strong localisation of photons in certain disordered dielectric superlattices, *Phys. Rev. Lett.* **58**, 2486 (1987).

56. E. Yablonovitch, T.J. Mitter, and K.M. Leung, Photonic band structure: the face-centered-cubic case employing nonspherical atoms, *Phys. Rev. Lett.* **67**, 2295 (1991).

57. S.Y. Lin, J.G. Fleming, D.L. Hetherington, B.K. Smith, R. Biswas, K.M. Ho, M.M. Sigalas, W. Zubrzycki, S.R. Kurtz, and J. Bur, A three-dimensional photonic crystal operating at infrared wavelengths, *Nature* **394**, 251–254 (1998).

58. S. Noda, K. Tomoda, N. Yamamoto, and A. Chutinan, Full three-dimensional photonic bandgap crystals at near-infrared wavelengths, *Science* **289**, 604–606 (2000).

59. A. Blanco, E. Chomski, S. Grabtchak, M. Ibisate, S. John, S.W. Leonard, C. Lopez, F. Meseguer, H. Miguez, J.P. Mondia, G.A. Ozin, O. Toader, and H.M. van Driel, Large-scale synthesis of a silicon photonic crystal with a complete three-dimensional bandgap near 1.5 micrometres,, *Nature (London)* **405**, 437 (2000).

60. Y.A. Vlasov, X.Z. Bo, J.C. Sturm, and D.J. Norris, On-chip natural assembly of silicon photonic bandgap crystals, *Nature (London)* **414**, 289 (2001).

61. M.J.A. de Dood, A. Polman, and J.G. Fleming, Modified spontaneous emission from erbium-doped photonic layer-by-layer crystals, *Phys. Rev. B.* **67**, 115106 (2003).

62. S. Ogawa, M. Imada, S. Yoshimoto, M. Okano, and S. Noda, Control of light emission by 3D photonic crystals, *Science* **305**, 227 (2004).

63. P. Lodahl, A.F. Van Driel, I.S. Nikolaev, A. Irman, K. Overgaag, D. Vanmaekelbergh, and W.L. Vos, Controlling the dynamics of spontaneous emission from quantum dots by photonic crystals, *Nature* **430**, 654–657 (2004).

64. G. Bourdon, I. Robert, R. Adams, K. Nelep, I. Sagnes, J.M. Moison, and I. Abram, Room temperature enhancement and inhibition of spontaneous emission in semiconductor microcavities, *Appl. Phys. Lett.* **77**, 1345 (2000).

65. M. Fujita, S. Takahashi, Y. Tanaka, T. Asano, and S. Noda, Simultaneous inhibition and redistribution of spontaneous light emission in photonic crystal, *Science* **308**, 1296 (2005).
66. P. Borri, W. Langbein, S. Schneider, U. Woggon, R.L. Sellin, D. Ouyang, and D. Bimberg, Ultralong dephasing time in InGaAs quantum dots, *Phys. Rev. Lett.* **87**, 157401 (2001).
67. D. Birkedal, K. Leosson, and J.M. Hvam, Long lived coherence in self-assembled quantum dots, *Phys. Rev. Lett.* **87**, 227401 (2001).
68. C. Kammerer, C. Voisin, G. Cassabois, C. Delalande, Ph. Roussignol, F. Klopf, J.P. Reithmaier, A. Forchel, and J.M. Gérard, Line narrowing in single semiconductor quantum dots: toward the control of environment effects, *Phys. Rev. B.* **66**, 041306(R) (2002).
69. H.J. Carmichael, R.J. Brecha, M.G. Raizen, H.J. Kimble, and P.R. Rice, Subnatural linewidth averaging for coupled atomic and cavity-mode oscillators, *Phys. Rev. A.* **40**, 5516 (1989).
70. D. Englund, D. Fattal, E. Waks, G. Solomon, B. Zhang, T. Nakaoka, Y. Arakawa, Y. Yamamoto, and J. Vuckovic, Controlling the spontaneous emission rate of single quantum dots in a two-dimensional photonic crystal, *Phys. Rev. Lett.* **95**, 013904 (2005).
71. A. Kress, F. Hofbauer, N. Reinelt, M. Kaniber, H.J. Krenner, R. Meyer, G. Böhm, and J.J. Finley, Manipulation of the spontaneous emission dynamics of quantum dots in two-dimensional photonic crystals, *Phys. Rev. B.* **71**, 241304(R) (2005).
72. W.-H. Chang, W.-Y. Chen, H.-S. Chang, T.-P. Hsieh, J.-I. Chyi, and T.-M. Hsu, Efficient single-photon sources based on low-density quantum dots in photonic-crystal nanocavities, *Phys. Rev. Lett.* **96**, 117401 (2006).
73. D.G. Gevaux, A.J. Bennett, R.M. Stevenson, A.J. Shields, P. Atkinson, J. Griffiths, D. Anderson, G.A.C. Jones, and D.A. Ritchie, Enhancement and suppression of spontaneous emission by temperature tuning InAs quantum dots to photonic crystal cavities, *Appl. Phys. Lett.* **88**, 131101 (2006).
74. S. Cortez, O. Krebs, P. Voisin, and J.M. Gérard, Polarization of the interband optical dipole in InAs/GaAs self-organized quantum dots, *Phys. Rev. B.* **63**, 233306 (2001).
75. G.C. DeSalvo, C.A. Bozada, J.L. Ebel, D.C. Look, J.P. Barrette, C.L.A. Cerny, R.W. Dettmer, J.K. Gillespie, C.K. Havasy, T.J. Jenkins, K. Nakano, C.I. Pettiford, T.K. Quach, J.S. Sewell, and G.D. Via, Wet chemical digital etching of GaAs at room temperature, *J. Electrochem. Soc.* **143**, 3652 (1996).
76. A. Badolato, K. Hennessy, M. Atatüre, J. Dreiser, E. Hu, P.M. Petroff, and A. Imamoglu, Deterministic Coupling of Single Quantum dots to Single Nanocavity Modes, *Science* **308**, 1158 (2005).
77. A. Kiraz, P. Michler, C. Becher, B. Gayral, A. Imamoglu, L. Zhang, E. Hu, W.V. Schoenfeld, and P.M. Petroff, Cavity-quantum electrodynamics using a single InAs quantum dot in a microdisk structure, *Appl. Phys. Lett.* **78**, 3932 (2001).
78. S. Varoutsis, S. Laurent, P. Kramper, A. Lemaître, I. Sagnes, I. Robert-Philip, and I. Abram, Restoration of photon indistinguishability in the emission of a semiconductor quantum dot, *Phys. Rev. B.* **72**, 041303(R) (2005).
79. M. Schwab, H. Kurtze, T. Auer, T. Berstermann, M. Bayer, J. Wiersig, N. Baer, C. Gies, F. Jahnke, J.P. Reithmaier, A. Forchel, M. Benyoucef, and P. Michler, Radiative emission dynamics of quantum dots in a single cavity micropillar, *Phys. Rev. B.* **74**, 045323 (2006).
80. J.-M. Gérard and P. Michler ed., single quantum dots, Topics *Appl. Phys.* **90**, 283–327 (2003), Springer-Verlag Berlin Heidelberg 2003.
81. J.P. Reithmaier, G. Sek, A. Löffler, C. Hofmann, S. Kuhn, S. Reitzenstein, L.V. Keldysh, V.D. Kulakovskii, T.L. Reinecke, and A. Forchel, Strong coupling in a single quantum dot–semiconductor microcavity system, *Nature* **432**, 197–200 (2004).
82. T. Yoshie, A. Scherer, J. Hendrickson, G. Khitrova, H.M. Gibbs, G. Rupper, C. Ell, O.B. Shchekin, and D.G. Deppe, Vacuum Rabi splitting with a single quantum dot in a photonic crystal nanocavity, *Nature* **432**, 200–203 (2004).
83. J.-M. Gérard, D. Barrier, J.Y. Marzin, R. Kuszelewicz, L. Manin, E. Costard, V. Thierry-Mieg, and T. Rivera, Quantum boxes as active probes for photonic microstructures: the pillar microcavity case, *Appl. Phys. Lett.* **69**, 449 (1996).
84. S.L. McCall, A.F.J. Levi, R.E. Slusher, S.J. Pearton, and R.A. Logan, Whispering-gallery mode microdisk lasers, *Appl. Phys. Lett.* **60**, 289 (1992).

85. K. Hennessy, A. Badolato, A. Tamboli, P.M. Petroff, E. Hu, M. Atatüre, J. Dreiser, and A. Imamoglu, Tuning photonic crystal nanocavity modes by wet chemical digital etching, *Appl. Phys. Lett.* **87**, 021108 (2005).

86. S. Strauf, K. Hennessy, M.T. Rakher, Y.-S. Choi, A. Badolato, L.C. Andreani, E.L. Hu, P.M. Petroff, and D. Bouwmeester, Self-tuned quantum dot gain in photonic crystal lasers, *Phys. Rev. Lett.* **96**, 127404 (2006).

87. K. Hennessy, A. Badolato, M. Winger, D. Gerace, M. Atatüre, S. Gulde, S. Fält, E.L. Hu, and A. Imamog lu, Quantum nature of a strongly coupled single quantum dot–cavity system, *Nature* **445**, 896–899 (2007).

88. N. LeThomas, U. Woggon, O. Schöps, M.V. Artemyev, M. Kazes, and U. Banin, Cavity QED with semiconductor nanocrystals, *Nano Lett.* **6**, 557 (2006).

89. A. Boca, R. Miller, K.M. Birnbaum, A.D. Boozer, J. McKeever, and H.J. Kimble, Observation of the vacuum Rabi spectrum for one trapped atom, *Phys. Rev. Lett.* **93**, 233603 (2004).

90. M. Brune, F. Schmidt-Kaler, A. Maali, J. Dreyer, E. Hagley, J.M. Raimond, and S. Haroche Quantum Rabi oscillation: a direct test of field quantization in a cavity, *Phys. Rev. Lett.* **76**, 1800 (1996).

91. S.H. Autler and C.H. Townes, Stark effects in rapidly varying fields, *Phys. Rev.* **100**, 703 (1955).

92. A. Auffèves-Garnier, C. Simon, J.-M. Gérard, and J.-P. Poizat, Giant optical nonlinearity induced by a single two-level system interacting with a cavity in the Purcell regime, *Phys. Rev. A.* **75**, 053823 (2007).

93. K.M. Birnbaum, A. Boca, R. Miller, A.D. Boozer, T.E. Northup, and H.J. Kimble, Photon blockade in an optical cavity with one trapped atom, *Nature* **436**, 87–90 (2005).

94. Y. Sugiyama, Y. Sakuma, S. Muto, and N. Yokoyama, Novel InGaAs/GaAs quantum dot structures formed in tetrahedral-shaped recesses on (111)B GaAs substrate using metalorganic vapor phase epitaxy, *Appl. Phys. Lett.* **67**, 256 (1995).

95. A. Hartmann, L. Loubies, F. Reinhardt, and E. Kapon, Self-limiting growth of quantum dot heterostructures on nonplanar {111}B substrates, *Appl. Phys. Lett.* **71**, 1314 (1997).

96. M. H. Baier, E. Pelucchi, E. Kapon, S. Varoutsis, M. Gallart, I. Robert-Philip, and T. Abram, Single photon emission from site-controlled pyramidal quantum dots, *Appl. Phys. Lett.* **84** 648 (2004).

97. S. Kiravittaya, A. Rastelli, and O.G. Schmidt, Photoluminescence from seeded three-dimensional InAs/GaAs quantum-dot crystals, *Appl. Phys. Lett.* **88**, 043112 (2006).

98. S. Kiravittaya, H. Heidemeyer, and O.G. Schmidt, Lateral quantum-dot replication in three-dimensional quantum-dot crystals, *Appl. Phys. Lett.* **86**, 263113 (2005).

99. S. Kiravittaya, M. Benyoucef, R. Zapf-Gottwick, A. Rastelli, and O.G. Schmidt, Ordered GaAs quantum dot arrays on GaAs(001): single photon emission and fine structure splitting, *Appl. Phys. Lett.* **89**, 233102 (2006).

100. J. Coelho, G. Patriarche, F. Glas, G. Saint-Girons, I. Sagnes, and L. Largeau, Buried dislocation networks designed to organize the growth of III–V semiconductor nanostructures, *Phys. Rev. B.* **70**, 155329 (2004).

101. H. Lee, J.A. Johnson, M.Y. He, J.S. Speck, and P.M. Petroff, Strain-engineered self-assembled semiconductor quantum dot lattices, *Appl. Phys. Lett.* **78**, 105 (2001).

102. K.M. Kim, Y.J. Park, Y.M. Park, C.K. Hyon, E.K. Kim, and J.H. Park, Alignment of InAs quantum dots on a controllable strain-relaxed substrate using an InAs/GaAs superlattice, *J. Appl. Phys.* **92**, 5453 (2002).

103. Z.G. Xie and G.S. Solomon, Spatial ordering of quantum dots in microdisks, *Appl. Phys. Lett.* **87**, 093106 (2005).

104. S. Strauf, M.T. Rakher, I. Carmeli, K. Hennessy, C. Meier, A. Badolato, M.J.A. DeDood, P.M. Petroff, E.L. Hu, E.G. Gwinn, and D. Bouwmeester, Frequency control of photonic crystal membrane resonators by monolayer deposition, *Appl. Phys. Lett.* **88**, 043116 (2006).

105. S. Mosor, J. Hendrickson, B.C. Richards, J. Sweet, G. Khitrova, H.M. Gibbs, T. Yoshie, A. Scherer, O.B. Shchekin, and D.G. Deppe, Scanning a photonic crystal slab nanocavity by condensation of xenon, *Appl. Phys. Lett.* **87**, 141105 (2005).

106. A. Rastelli, A. Ulhaq, S. Kiravittaya, L. Wang, A. Zrenner, and O.G. Schmidt, In situ laser microprocessing of single self-assembled quantum dots and optical microcavities, *Appl. Phys. Lett.* **90**, 073120 (2007).

107. E. Moreau, I. Robert, J.-M. Gérard, I. Abram, L. Manin, and V. Thierry-Mieg, Single-mode solid-state single photon source based on isolated quantum dots in pillar microcavities, *Appl. Phys. Lett.* **79**, 2865 (2001).

108. W.L. Barnes, G. Björk, J.M. Gérard, P. Jonsson, J.A.E. Wasey, P.T. Worthing, and V. Zwiller, Solid-state single photon sources: light collection strategies, *Eur. Phys. J. D.* **18**, 197 (2002).

109. H. Altug, D. Englund, and J. Vuckovic, Ultrafast photonic crystal nanocavity laser, *Nat. Phys.* **2**, 484 (2006).

110. J.-M. Gerard and B. Gayral, Strong Purcell effect for InAs quantum boxes in three-dimensional solid-state microcavities, *J. Lightwave Technol.* **17**, 2089 (1999).

111. H. Fearn and R. Loudon, Quantum theory of the lossless beam splitter, *Opt. Comm.* **64**, 485 (1987).

112. A. Kiraz, M. Ehrl, Th. Hellerer, O.E. Mustecaploglu, C. Bräuchle, and A. Zumbusch, Indistinguishable Photons from a Single Molecule, *Phys. Rev. Lett.* **94**, 223602 (2005).

113. C.K. Hong, Z.Y. Ou, and L. Mandel, Measurement of subpicosecond time intervals between two photons by interference, *Phys. Rev. Lett.* **59**, 2044 (1987).

114. Th. Legero, T. Wilk, M. Hennrich, G. Rempe, and A. Kuhn, Quantum beat of two single photons, *Phys. Rev. Lett.* **93**, 070503 (2004).

115. J. Beugnon, M.P.A. Jones, J. Dingjan, B. Darquié, G. Messin, A. Browaeys, and P. Grangier Quantum interference between two single photons emitted by independently trapped atoms, *Nature* **440**, 779 (2006).

116. C. Santori, D. Fattal, J. Vuckovic, G.S. Salomon, and Y. Yamamoto, Indistinguishable photons from a single-photon device, *Nature* **419**, 594 (2002).

117. A.J. Bennett, D.C. Unitt, A.J. Shields, P. Atkinson, and D.A. Ritchie, Influence of exciton dynamics on the interference of two photons from a microcavity single-photon source, *Opt. Express* **20**, 7772 (2005).

118. S. Laurent, S. Varoutsis, L. Le Gratiet, A. Lemaître, I. Sagnes, F. Raineri, A. Levenson, I. Robert-Philip, and I. Abram, Indistinguishable single photons from a single-quantum dot in a two-dimensional photonic crystal cavity, *Appl. Phys. Lett.* **87**, 163107 (2005).

119. O. Benson, C. Santori, M. Pelton, and Y. Yamamoto, Regulated and entangled photons from a single quantum dot, *Phys. Rev. Lett.* **84**, 2513 (2000).

120. M. Bayer, G. Ortner, O. Stern, A. Kuther, A.A. Gorbunov, A. Forchel, P. Hawrylak, S. Fafard, K. Hinzer, T.L. Reinecke, S.N. Walck, J.P. Reithmaier, F. Klopf, and F. Schäfer, Fine structure of neutral and charged excitons in self-assembled In(Ga)As/(Al)GaAs quantum dots, *Phys. Rev. B.* **65**, 195315 (2002).

121. N. Akopian, N.H. Lindner, E. Poem, Y. Berlatzky, J. Avron, D. Gershoni, B.D. Gerardot, and P.M. Petroff, Entangled photon pairs from semiconductor quantum dots, *Phys. Rev. Lett.* **96**, 130501 (2006).

122. R.M. Stevenson, R.J. Young, P. Atkinson, K. Cooper, D.A. Ritchie, and A.J. Shields, A semiconductor source of triggered entangled photon pairs, *Nature* **439**, 179 (2006).

123. R.J. Young, R.M. Stevenson, P. Atkinson, K. Cooper, D.A. Ritchie, and A.J. ShieldsImproved fidelity of triggered entangled photons from single quantum dots, *New J. Phys.* **8**, 29 (2006).

124. T.M. Stace, G.J. Milburn, and C.H.W. Barnes, Entangled two-photon source using biexciton emission of an asymmetric quantum dot in a cavity, *Phys. Rev. B.* **67**, 085317 (2003).

125. F. Troiani, J.I. Perea, and C. Tejedor, Cavity-assisted generation of entangled photon pairs by a quantum-dot cascade decay, *Phys. Rev. B.* **74**, 235310 (2006).

126. F. De Martini and G.R. Jacobovitz, Anomalous spontaneous–stimulated-decay phase transition and zero-threshold laser action in a microscopic cavity, *Phys. Rev. Lett.* **60**, 1711 (1988).

127. P.R. Rice and H.J. Carmichael, Photon statistics of a cavity-QED laser: a comment on the laser-phase-transition analogy, *Phys. Rev. A.* **50**, 4318 (1994).

128. Z.G. Xie, S. Götzinger, W. Fang, H. Cao, and G.S. Solomon, Influence of a single quantum dot state on the characteristics of a microdisk laser, *Phys. Rev. Lett.* **98**, 117401 (2007).

129. S.M. Ulrich, C. Gies, S. Ates, J. Wiersig, S. Reitzenstein, C. Hofmann, A. Löffler, A. Forchel, F. Jahnke, and P. Michler, Photon statistics of semiconductor microcavity lasers, *Phys. Rev. Lett.* **98**, 043906 (2007).

130. J. McKeever, A. Boca, A.D. Boozer, J.R. Buck, and H.J. Kimble, Experimental realization of a one-atom laser in the regime of strong coupling, *Nature* **425**, 268 (2003).
131. M. Pelton and Y. Yamamoto, Ultralow threshold laser using a single quantum dot and a microsphere cavity, *Phys. Rev. A.* **59**, 2418 (1999).
132. B. Jones, S. Ghose, J.P. Clemens, P.R. Rice, and L.M. Pedrotti, Photon statistics of a single atom laser, *Phys. Rev. A.* **60**, 3267 (1999).
133. J.M. Kikkawa, I.P. Smorchkova, N. Samarth, and D.D. Awschalom, *Room-Temperature Spin Memory in Two-Dimensional Electron Gases* **277**, 1284–1287 (1997).
134. S. Cortez, O. Krebs, S. Laurent, M. Senes, X. Marie, P. Voisin, R. Ferreira, G. Bastard, J.-M. Gérard, and T. Amand, Optically driven spin memory in n-doped InAs-GaAs quantum dots, *Phys. Rev. Lett.* **89**, 207401 (2002).
135. D. Loss and D.P. DiVincenzo, Quantum computation with quantum dots, *Phys. Rev. A.* **57**, 120 (1998).
136. D.P. DiVincenzo, D. Bacon, J. Kempe, G. Burkard, and K.B. Whaley, Universal quantum computation with the exchange interaction, *Nature* **408**, 339–342 (2000).
137. E. Biolatti, R.C. Iotti, P. Zanardi, and F. Rossi, Quantum information processing with semiconductor macroatoms, *Phys. Rev. Lett.* **85**, 5647 (2000).
138. X.-Q. Li and Y. Arakawa, Single qubit from two coupled quantum dots: an approach to semiconductor quantum computations, *Phys. Rev. A.* **63**, 012302 (2001).
139. A. Imamoglu, D.D. Awschalom, G. Burkard, D.P. DiVincenzo, D. Loss, M. Sherwin, and A. Small Quantum information processing using quantum dot spins and cavity QED, *Phys. Rev. Lett.* **83**, 4204 (1999).
140. M. Feng, I. D'Amico, P. Zanardi, and F. Rossi, Spin-based quantum-information processing with semiconductor quantum dots and cavity QED, *Phys. Rev. A.* **67**, 014306 (2003).
141. M.N. Leuenberger, Fault-tolerant quantum computing with coded spins using the conditional Faraday rotation in quantum dots, *Phys. Rev. B.* **73**, 075312 (2006).
142. F. Meier and D.D. Awschalom, Spin-photon dynamics of quantum dots in two-mode cavities, *Phys. Rev. B.* **70**, 205329 (2004).
143. R.J. Warburton, C.S. Dürr, K. Karrai, J.P. Kotthaus, G. Medeiros-Ribeiro, and P.M. Petroff, Charged Excitons in Self-Assembled Semiconductor Quantum Dots, *Phys. Rev. Lett.* **79**, 5282 (1997).
144. J. Berezovsky, M.H. Mikkelsen, O. Gywat, N.G. Stoltz, L.A. Coldren, and D.D. Awschalom, Nondestructive optical measurements of a single electron spin in a quantum dot, *Science* **314**, 1916 (2006).
145. D.J.P. Ellis, A.J. Bennett, A.J. Shields, P. Atkinson, and D.A. Ritchie, Oxide-apertured microcavity single-photon emitting diode, *Appl. Phys. Lett.* **90**, 233514 (2007).
146. A. Muller, D. Lu, L. Ahn, D. Gazula, S. Quadery, S. Freisem, D.G. Deppe, and C.K. Shih, Self-aligned all-epitaxial microcavity for cavity QED with quantum dots, *Nano Lett.* **6**, 2920–2924 (2006).
147. M. Atatüre, J. Dreiser, A. Badolato, and A. Imamoglu, Observation of Faraday rotation from a single confined spin, *Nat. Physics* **3**(101), (2007).

CHAPTER 5

InAs Quantum Dot Formation Studied at the Atomic Scale
by Cross-sectional Scanning Tunnelling Microscopy

J.M. Ulloa, P. Offermans and P.M. Koenraad

Department of Applied Physics, Eindhoven University of Technology, The Netherlands

5.1   Introduction

5.1.1   Quantum dot formation

Self-assembled quantum dots (QDs) have attracted much attention in the last years [1, 2]. These nanostructures are very interesting from a scientific point of view because they form nearly ideal zero-dimensional systems in which quantum confinement effects become very important. These unique properties also make them very interesting from a technological point of view. For example, InAs QDs are employed in QD lasers [3, 4], single electron transistors [5], mid-infrared detectors [6, 7], single-photon sources [8, 9], etc. InAs QDs are commonly created by the Stranski–Krastanov growth mode when InAs is deposited on a substrate with a bigger lattice constant, like GaAs or InP [10]. Above a certain critical thickness of InAs, three-dimensional islands are spontaneously formed on top of a wetting layer (WL) to reduce the strain energy. Once created, the QDs are subsequently capped, a step which is required for any device application.

For any application based on InAs QDs, an accurate control of their electronic properties is required. The electron and hole states in the dot are very sensitive to the QD size, shape and composition (as well as to the surrounding material), and therefore an accurate control of the QD structural properties is necessary for applications. This explains the strong effort dedicated in the last years to the structural characterization of QDs [11], which is essential in order to have a deep understanding of the QD formation process. Although a lot of effort has been dedicated to study surface QDs, there are relatively few studies focused on the effect of the capping process [12–20]. Some of these studies have already shown significant differences in size, shape and composition between uncapped and capped QDs. For example, an important collapse of the QD height has been reported for InAs/GaAs QDs capped with GaAs [14, 15, 17–19], revealing the big influence of the capping process on the structural properties of the QDs. Indeed, critical issues affecting the dot take place during capping like dot decomposition, intermixing, segregation, As/P exchange and composition modulation in the capping layer. This means that the real buried QDs must be studied in order to understand the complete QD formation process. Cross-sectional scanning tunnelling microscopy is an especially useful technique for this purpose, because it allows the structure of the capped dots at the atomic scale to be assessed.

5.1.2   Cross-sectional scanning tunnelling microscopy (X-STM)

The scanning tunnelling microscope is part of the family of scanning probe microscopes, which are able to provide direct real space information of a physical property at the atomic scale. This is done by scanning a probe across the surface of a sample while recording the measured signal, which results in an image of the surface. In the case of scanning tunnelling microscopy (STM), invented in 1981 by Binnig and Rohrer [21], the probe consists of a metallic tip which is brought into tunnelling contact with a (semi-)conducting surface. The quantum mechanical tunnelling effect allows the electron to tunnel through the vacuum barrier even when the energy of the electrons is lower than the potential barrier (Fig. 5.1a). Since the tunnelling current depends exponentially on the distance between tip and sample, an atomic-scale height map of the surface density of states and topography can be acquired by adjusting the tip sample distance during scanning, using a feedback loop that keeps the tunnelling current constant (Fig. 5.1b).

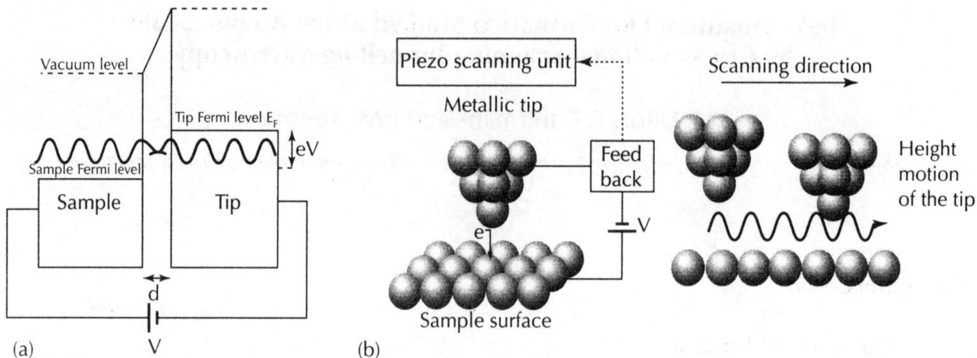

Figure 5.1   (a) Schematic showing the overlap and decay of tip and sample wave functions in the vacuum barrier. (b) Schematic of a scanning tunnelling microscopy with feedback loop.

In cross-sectional scanning tunnelling microscopy (X-STM), the direct visualization of nanostructures, which are generally embedded in several layers of semiconducting material, is possible by cleaving the sample and thereby exposing the cross-sectional surface of the buried nanostructures. The semiconductor wafers are cleaved along a natural cleavage plane to expose a cross-section of the epitaxial layers grown on the wafer. In the case of III/V semiconductors, the (110) and $(1\bar{1}0)$ cleavage planes of the zinc-blende crystal show a 1×1 surface unit cell reconstruction where the group III elements move into the surface while the group V elements move outward (Fig. 5.2). This buckling behaviour causes the energetic position of the dangling bond surface states of the type III and type V elements to move above and below the semiconductor band gap in the bulk, respectively. Therefore, depending on the polarity of the applied bias between tip and sample, either the empty states of the type III elements (Vsample > 0) or the filled states of the type V elements (Vsample < 0) are imaged (Fig. 5.3). Since the (110) and $(1\bar{1}0)$ surface planes contain only half of the (001) layers within the zinc-blende crystal, the atomic rows observed in the cross-sectional STM image are separated by a bilayer distance in the growth [001] direction, which is equal to the lattice constant a_0. Because there are no surface states present in the band gap of most cleaved III/V semiconductors, the Fermi level at the cleaved surface is unpinned, i.e. it is not forced to assume a specific value in the band gap independent of its value in the bulk material. This allows the electronic properties of the bulk to be probed at energies near the band gap. The bulk states generally contribute with an electronic contrast to STM images that is not related to the actual physical topography of the surface, but governed by factors such as variations in the band gap, doping level and electron affinity.

The electronic contribution to the apparent height in the STM image can be suppressed to a large extent by scanning at high voltages [22, 23]. Since the tunnelling probability decays exponentially with the height of the effective tunnelling barrier, the states with the highest energy contribute most to the tunnelling current. If a sufficiently large voltage is applied during empty state imaging of a heterostructure, the states contributing most to the tunnelling current have

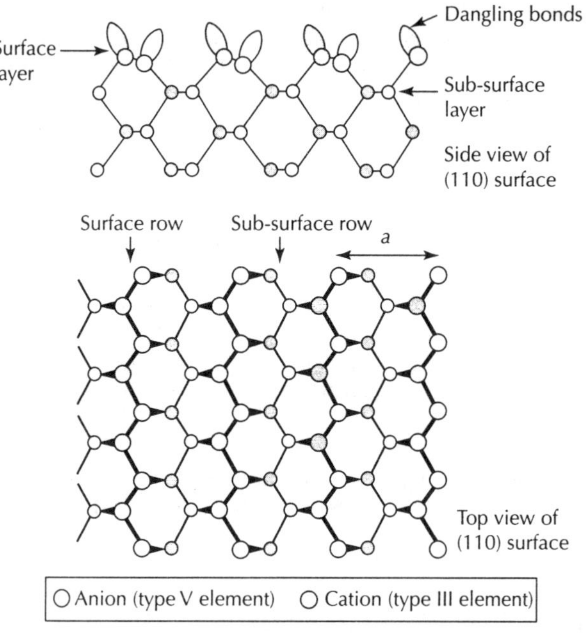

Figure 5.2   Schematic top and side view of III/V semiconductor (110) surface.

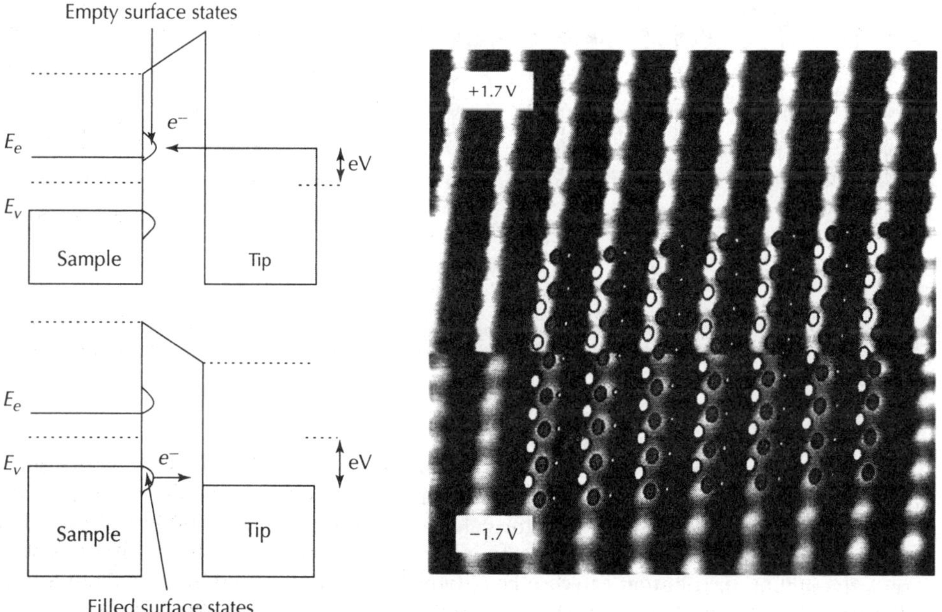

Figure 5.3   Schematic of tunnelling into empty and filled surface states, and STM image showing the atomic corrugation at both polarities.

energies far above the conduction band edge and the contribution of the lower lying states is negligible. Similarly, the voltage dependent sensitivity of the tunnelling current on doping level can be overcome by injecting electrons above the depletion region. In the case of filled states tunnelling (electron tunnelling from sample to tip), the valence band offset in a heterostructure is smaller with respect to the total barrier height than the conduction band offset, since the effective barrier height for tunnelling is increased by the band gap. Therefore, the resulting electronic contrast is smaller than in the case of empty states tunnelling (electron tunnelling from tip to sample). Thus, as demonstrated in [22, 23], at high (preferably negative) sample voltages, the electronic contribution to the height contrast is minimized, enabling the acquisition of the true (filled states) topography of the structure, such as the outward relaxation of the cleaved surface.

5.1.2.1 X-STM topography and strain analysis

When two materials with a different lattice constant are used in a heterostructure, it will accumulate strain. Stranski–Krastanov grown InAs quantum dots are the ultimate example of such strained heterostructures, but also (In,Ga) As quantum wells have built-in strain. When such a structure is cleaved, it reduces its built-in tensile or compressive strain by deforming the cleaved surface. Regions under compressive strain bulge outward while tensile strain depresses the cleaved surface (Fig. 5.4). The strain can be measured in two ways with cross-sectional scanning tunnelling microscopy:

- Strain in the plane of the surface can be deduced from the lattice spacing.
- Distortion normal to the surface can be measured using (filled states) topography imaging, as described before.

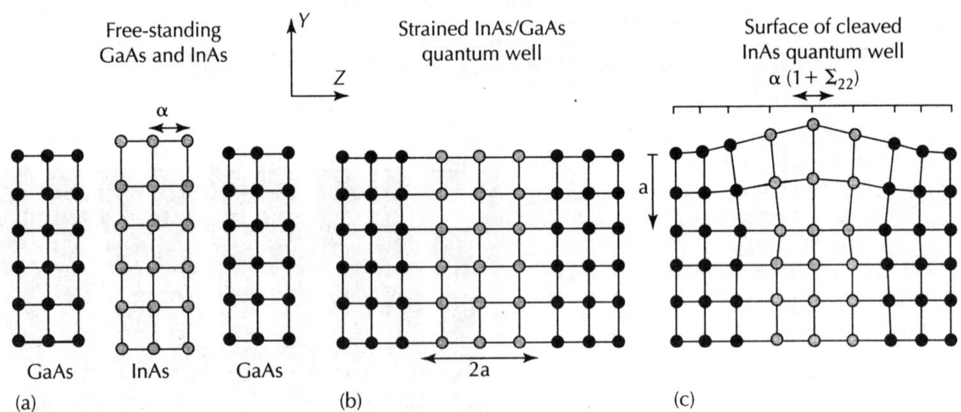

Figure 5.4   Strain relaxation at the cleaved surface of a strained quantum well. (a) Natural lattices of freestanding GaAs and InAs. (b) Due to lattice mismatch the InAs well is strained in GaAs. (c) The strain is released at the cleaved surface by outward relaxation.

The measured outward displacement and strain (measured by the change in lattice spacing) can be used to determine the indium composition of a strained (In,Ga) As nanostructure, by comparing the experimental data with the calculated relaxation and strain using elasticity theory [24]. In the case of quantum wells with a low concentration of indium (<30%), or wetting layers, the indium distribution can also be obtained directly by counting the indium atoms, as will be shown in the next sections. However, in the case of InAs quantum dots the indium concentration is too high to distinguish individual indium atoms, and the only way to determine its composition is by fitting to the measured outward relaxation or change in lattice constant. To calculate the outward relaxation of a cleaved QD we use the package ABAQUS. It is based on continuum elasticity theory and performs a finite element calculation to solve the 3D problem, in which an isotropic material is considered. The measured size and shape of the QDs is introduced in the model and the composition is varied until the calculated profile fits to the experimental one. The real size and shape of the QDs can only be known after scanning a large number of dots, since they will be cleaved at a random position. Figure 5.5 shows the calculated outward relaxation

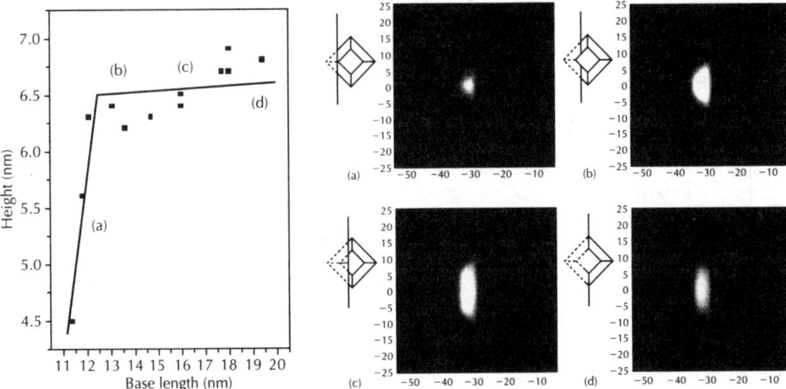

Figure 5.5　Expected (solid line) and measured by X-STM (squares) size distribution of cleaved InAs/GaAs QDs. The different sizes correspond to dots cleaved in different positions, as shown by the calculated outward relaxation for different cleavage planes.

of the surface of a cleaved InAs/GaAs QD and WL as a function of the position of the cleavage plane. The dots are considered to be truncated pyramids. The corresponding expected size distribution is shown by the solid line in the graph, in which the results of a real X-STM measurement are also shown (squares). Only after scanning a large number of cleaved dots can the maximum base length of the cross-section of the dot be determined, which is indicative for a cleavage near the dot centre. Only the outward relaxation profile of the dots cleaved close to the centre (dots in the region (d)) are used to compare with the model.

5.1.2.2　Experimental set-up

5.1.2.2.1　Sample and tip preparation

Small rectangular pieces (about $3.5 \times 10\,\text{mm}^2$) are cleaved from a wafer containing the semiconductor heterostructure of interest. These wafers are normally $350–550\,\mu\text{m}$ thick and are polished down with aluminum oxide powder to a thickness of about $100\,\mu\text{m}$, for easy cleavage. For an STM measurement it is essential to have good electric contact with the sample. Therefore, metallic contacts are evaporated at the top surface of the sample after treating it with an N2/H2 plasma. A small scratch of about $0.5–1.5\,\text{mm}$, which extends to a small notch at the side of the sample, is made at the top surface of the sample using a diamond pen. This scratch facilitates the cleavage of the sample and provides a fixed starting point for the propagation of the cleavage plane. The sample is clamped on a sample holder between two metal bars that can be screwed together, as shown in Fig. 5.6. Only one corner of the sample is clamped for the unconstrained propagation of the cleavage plane. Between the metal bars and the two surfaces of the sample, at the position where the sample is to be clamped, thin slices of indium are placed. The sample holder is heated in order to melt the indium before tightening the screws. The indium provides an even pressure distribution on the sample, preventing it from cleaving during tightening of the screws and slipping out of the holder when the sample is degassed subsequently in the UHV system. The sample is cleaved in the STM chamber using a gentle touch by a manipulator (the so-called "wobble stick") just before the measurement as shown in Fig. 5.6. A characteristic crack pattern is also shown in Fig. 5.6 [25]. After cleavage, the sample is placed into the sample stage and moved towards a mounted tip, with the help of a CCD camera equipped with a tele-lens. First, the sample is brought close to the tip and then further approached until tunnelling contact is done automatically by the system. After the first tunnelling contact, the tip is retracted and moved several $100\,\text{nm}$ towards the edge of the cleavage surface where the grown layers (epilayers) are located. The process of making tunnelling contact and moving towards the epilayers is repeated until the tip moves over the edge of the sample. At this point tunnelling contact is not achieved, which indicates that the location of the epilayers is within several $100\,\text{nm}$.

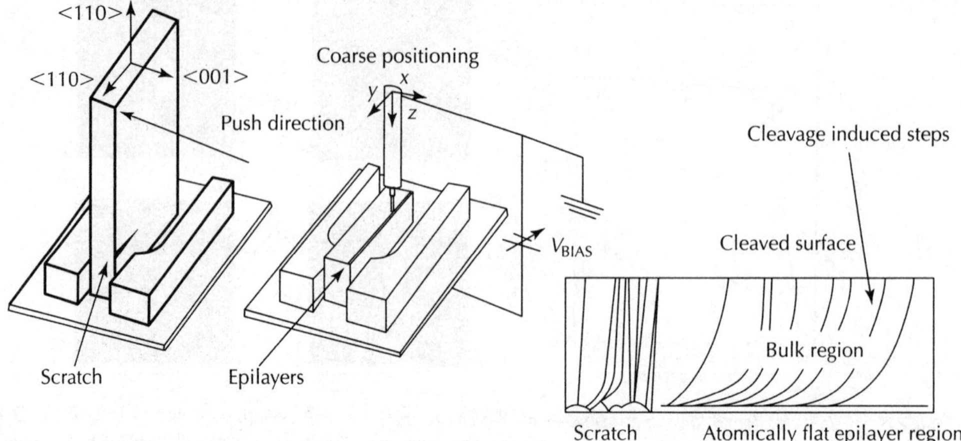

Figure 5.6   Schematic view of the sample cleavage process.

The tips are made of 99.97% pure polycrystalline tungsten wire with a diameter of 0.25 mm. A short piece of this wire ($\sim$5 mm) is spot welded onto an Omicron tip holder and cleaned for use in UHV. The tips are then electrochemically etched with a 2.0 molar potassium hydroxide solution. The top 1–1.5 mm of the tip is put into the solution and a positive voltage (4–5 V) is applied to the tip wire. A platinum–iridium (90%/10%) spiral serves as a counter-electrode. As the reaction products dissolve in the etchant, they sink down along the wire, which is visible from the local change of the diffractive index of the solution. Because of the geometry, the reaction velocity is the highest at the point where the tungsten wire penetrates the surface of the solution. This causes necking of the wire at the surface of the etching solution, as the reaction products owing down the tip shield the rest of the wire. Eventually, the wire will break at the neck leaving a very sharp tip. A current limiter is used to interrupt the etching process immediately ($<$1 μs) after the breaking of the wire. The tips are degassed after loading into the UHV system and treated with a 700 eV argon ion bombardment in order to mechanically stabilize the tip.

5.1.2.2.2   The STM unit and the UHV system

A commercially available room-temperature STM unit (Omicron STM-1 TS2) is used. The STM set-up is very sensitive to external vibrations, as the tip–sample distance during tunnelling is only a few angstrom. Moreover, the apparent height of the atomic corrugation visible in the STM images is only 20 pm. Therefore, the set-up is vibration isolated with different damping systems. The scanner unit is suspended on a set of springs and stabilized by an eddy current damping system. The eddy current damping system consists of copper fins that surround the scanner and are placed between permanent magnets when the scanner unit is in its suspended position. The STM unit, together with the rest of the set-up, is mounted on a heavy metal tabletop. Rubber dampers between this tabletop and the supporting frame filter out high-frequency vibrations. Low-frequency vibrations are suppressed by four active damping elements that are mounted between the floor and the frame. In this active damping system several motion detectors are present and several actuator coils counteract the detected motion of the system. Finally, the entire set-up is standing on a heavy concrete platform that is decoupled from the building.

A factor of critical importance in cross-sectional scanning tunnelling microscopy on III/V semiconductors is the production of an atomically flat cross-sectional surface that is free of contamination/oxidation, in order to obtain an electronically unpinned surface. Therefore, cleavage of the samples and the subsequent measurement are performed in a home-built ultra-high vacuum set-up (Fig. 5.7). The central vacuum chamber (STM), in which the STM unit is positioned, is pumped down by an ion-getter pump (IGP) with a titanium sublimation element (TSE) to a pressure lower than 5×10^{-11} torr (Varian VacIon Plus, 300 litre/sec). During X-STM

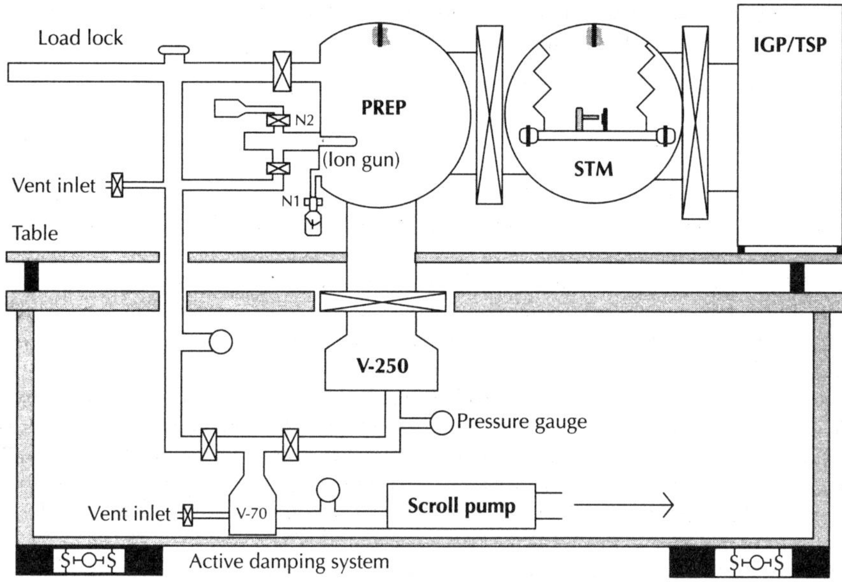

Figure 5.7   Schematic drawing of the UHV system containing the STM and damping system.

measurements the inner panel of the IGP/TSE can be cooled down with liquid nitrogen, which enhances the pump speed by a factor of two. In the preparation chamber (PREP), which is separated by a valve from the STM chamber, the tips and samples can be degassed with a baking unit and tips can be treated with an argon ion bombardment. Two oil-free turbo-molecular pumps (Varian V-250 and V-70), installed in series, keep the preparation chamber at a pressure of approximately 6×10^{-10} torr. The preparation chamber is connected to a load lock, for loading and unloading the tips and samples without seriously affecting the pressure in the preparation and STM chamber. The load lock is pumped with one of the turbo-molecular pumps (V-70) to a pressure of about 10^{-6} torr. An XDS5 dry scroll pump provides the necessary pre-vacuum of about 10^{-1} torr. During the X-STM measurements all pumps are switched off to prevent vibrations, except the IGP/TSE as this pump contains no moving parts. Two wobble sticks in the STM and preparation chamber are used for manipulation and transportation of the tips and samples.

5.2   Formation of the wetting layer

The formation of InAs wetting layers (WLs) has attracted relatively little attention compared to quantum dot (QD) formation [26–28]. In the simple picture of Stranski–Krastanov growth, after the build-up of a critical amount of strain, 2D layer growth is followed by QD formation. It has become increasingly clear, however, that such a simple picture is far from reality. Recently, In incorporation during pseudomorphic InAs/GaAs growth and QD formation was observed by *in situ* stress measurements [29].

In this section we analyse the composition and segregation of InAs WLs by either directly counting the indium atoms or by analysis of the outward displacement of the cleaved surface as measured by X-STM.

As mentioned in the introduction, the measured outward relaxation profile at high negative voltage can be compared to the calculated one to deduce the In concentration in the WL. By modelling the indium segregation, the outward displacement of the segregated WL can be calculated by integration of the analytical expression derived by Davies for the outward displacement of a cleaved quantum well [24]. Several models for indium segregation have been proposed

[30–33]. We use the phenomenological model of Muraki *et al.* [31], which has been shown to describe well the indium composition $x(n)$ of InAs WLs [28]:

$$x(n) = \begin{cases} 0, & n < 1 \\ (1 - R^n), & 1 \leq n \leq N \\ (1 - R^N)R^{n-N}, & n > N \end{cases} \tag{5.1}$$

where n is the monolayer (ML) index, N is the total amount of deposited indium and R is the indium segregation coefficient. N and R are determined by fitting the calculated relaxation profile to the measured relaxation profile.

The WLs were grown by molecular-beam epitaxy (MBE) on doped GaAs (100) wafers. In sample A, three different sets of WLs were grown at 495°C by deposition of 1.5 ML, 2.0 ML and 2.5 ML of InAs, respectively, at a growth rate of 0.1 ML/s. Each layer was repeated two times, separated by a 50 nm GaAs buffer layer, also grown at 495°C. A growth interruption of 10 seconds has been applied after the growth of each layer. No dot formation was observed for the layers with 1.5 ML indium deposition. In sample B, two sets of WLs were grown at 480°C by deposition of 2.0 ML of InAs at a high and a low growth rate of 0.1 ML/s and 0.01 ML/s, respectively. Each layer was repeated two times and capped by a 20 nm GaAs layer grown at 480°C, followed by a 30 nm GaAs layer grown at 580°C. A growth interruption of 10 seconds has been applied after the growth of each layer. In sample C, one set of InAs layers was grown in GaAs while a second set was grown in AlAs barriers. The InAs layers were grown at 500°C by deposition of 1.9 ML of InAs in a cycled way, i.e. with a 3 second pause after each deposition of 0.25 ML, at a growth rate of 0.043 ML/s. The following layer sequence was used: 20 nm GaAs/1.9 ML InAs/40 nm GaAs/1.9 ML InAs/40 nm GaAs/50 nm GaAs (doped 1×10^{18} cm^{-2})/20 nm GaAs/4 × (20 nm AlAs/1.9 ML InAs/20 nm AlAs/40 nm GaAs). To reduce interface roughness, the bottom AlAs barriers were grown at 600°C followed by a growth interruption prior to InAs deposition.

Sample A was used to study the effect of the amount of indium deposition on the WL formation. We measured the relaxation profiles of the WLs and fitted these with calculated relaxation profiles, by adjusting the fit parameters N and R. The resulting segregation profiles were verified by counting directly the number of indium atoms in the WL as a function of distance in growth direction. For the counting procedure, we selected four high-quality images of each layer, such as the one shown in Fig. 5.8. The relaxation and segregation profiles are shown in Fig. 5.9.

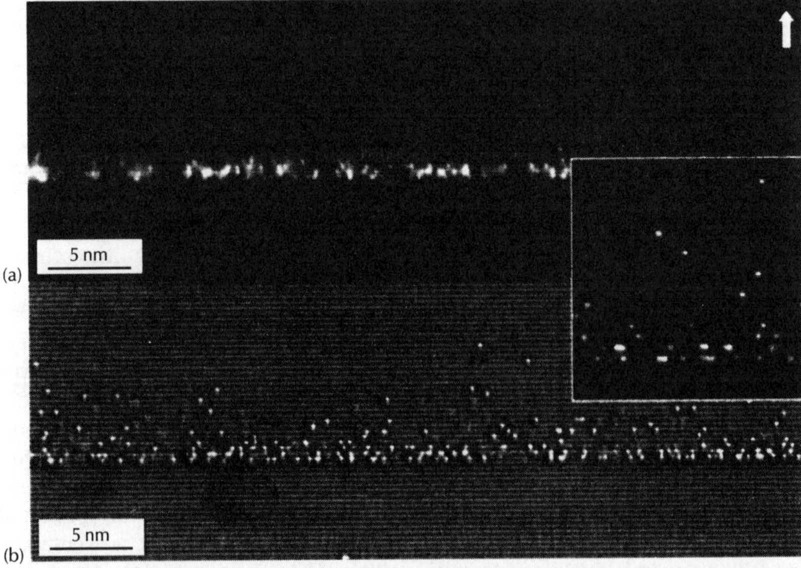

Figure 5.8   (a) *Empty states X-STM image of a segregated 2.0 ML InAs WL. The arrow indicates the growth direction.* (b) *The same image treated with a high-pass Fourier filter. The inset shows an enlarged view of part of the image.*

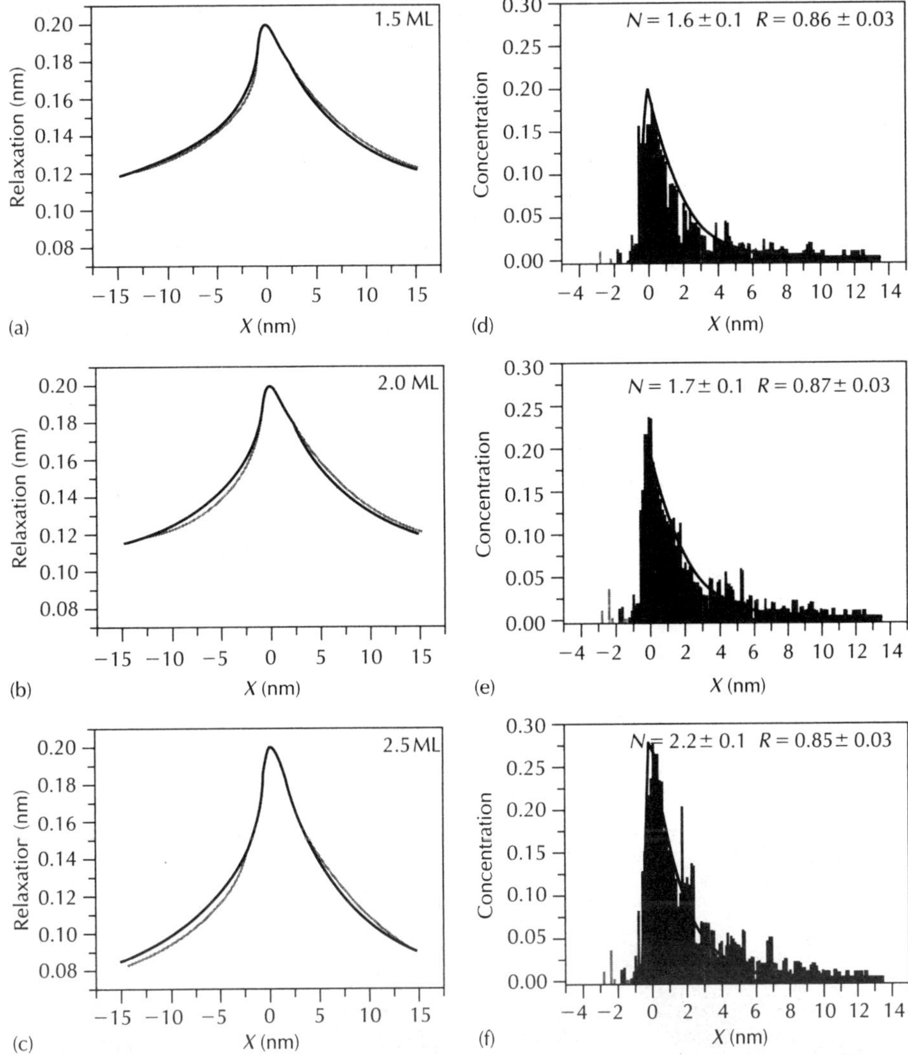

Figure 5.9   Measured and fitted relaxation profiles of the 1.5 ML (a), 2.0 ML (b) and 2.5 ML (c) InAs WLs of sample A. The black curves in (d), (e) and (f) show the segregation profiles corresponding to the fitted relaxation profiles in (a), (b) and (c). The columns indicate the counted indium concentration in the WL as a function of distance X in the growth direction.

In all three cases, we find an excellent agreement between the indium profile determined from the outward relaxation of the surface, and the direct counting procedure. For the 1.5 ML WL, the measured amount of indium N corresponds to the deposited amount, since no indium has gone into dot formation. For the 2.0 ML and 2.5 ML WLs, however, we find a clear indium enrichment of the WL, despite dot formation.

In Stranski–Krastanov growth mode, strain builds up until the critical amount of indium for dot formation is deposited [34]. It has been shown that only part of the deposited amount of indium contributes to the strain, by incorporation into the lattice, while the remaining indium forms a floating layer on the surface [29]. During dot formation, part of the floating indium is transferred by lateral mass transport to the dots. The amount of indium that remains in the dots, however, is strongly reduced by the capping process, which dissolves the top of the dots back into the WL, as we will show in sections 5.2 and 5.4. The dissolved indium adds to the remaining floating indium, and is eventually incorporated into the lattice, during continued capping. That is the reason for the observed indium enrichment of the WL after dot formation.

We used sample B to study the effect of a reduced growth rate on the WL formation. It is known that a reduced growth rate leads to an increased QD size and a reduced QD density. However, it is not *a priori* clear how this will affect the formation of the segregated WL in the buried structure. In Fig. 5.10 we show the average measured and calculated relaxation profiles of the InAs WLs of sample B. The dashed lines indicate the relaxation profiles calculated directly from indium atom counting using different images. We find that the high (0.1 ML/s) and low (0.01 ML/s) growth rate InAs WLs can be described by the same parameters within errors. However, there is, as expected, a marked difference in the size of the QDs, shown in Fig. 5.11. Whereas the QD grown at the high growth rate appears as a rather flat, disk-like shape with a height of 3 nm (Fig. 5.11a), the QDs grown at low rate show an indium distribution with a reversed truncated cone shape [35] with a height of 5.4 nm (Fig. 5.11b).

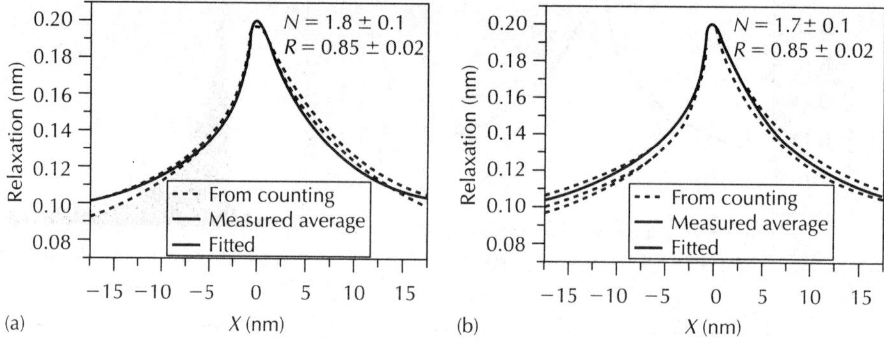

Figure 5.10   Measured and fitted relaxation profiles of the high (a) and low (b) growth rate InAs WLs of sample B. The dashed lines indicate the relaxation profiles calculated directly from counted segregation profiles.

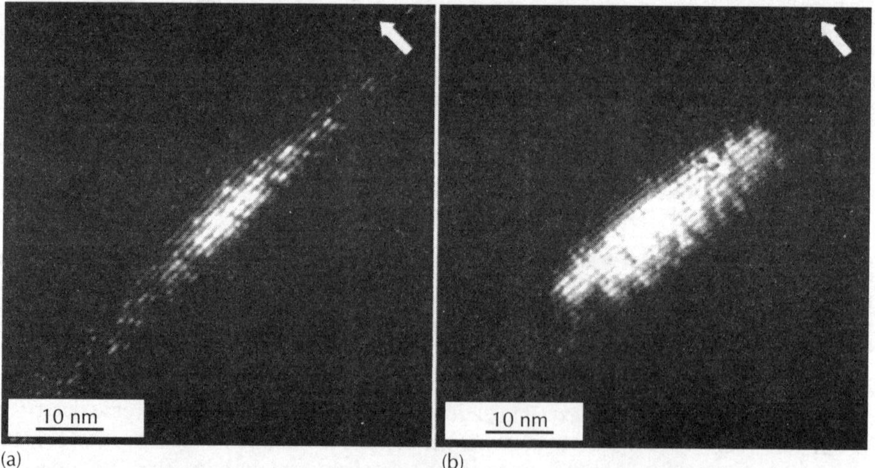

Figure 5.11   X-STM images of typical quantum dots found in the high (a) and low (b) growth rate WLs. The arrow indicates the growth direction.

We studied the effect of the host material by analysing the segregation of InAs WLs grown in the AlAs barriers of sample C, and comparing this to the segregation of InAs WLs grown, under the same growth conditions, in a GaAs matrix. It has been shown that QD formation in the InAs/AlAs system is kinetically limited due to a reduced lateral In migration on the AlAs surface, because of the larger Al–In bond strength [36]. In the next section the marked differences in the structural properties of the dots grown in GaAs and AlAs will be discussed [37]. Figure 5.12

shows the averaged measured and fitted relaxation profiles. We find that the vertical indium segregation in AlAs and GaAs can be described by almost the same parameters, in agreement with [27]. This indicates that, in contrast to the lateral In migration, the vertical indium segregation is strain driven rather than determined by the chemical bond strength.

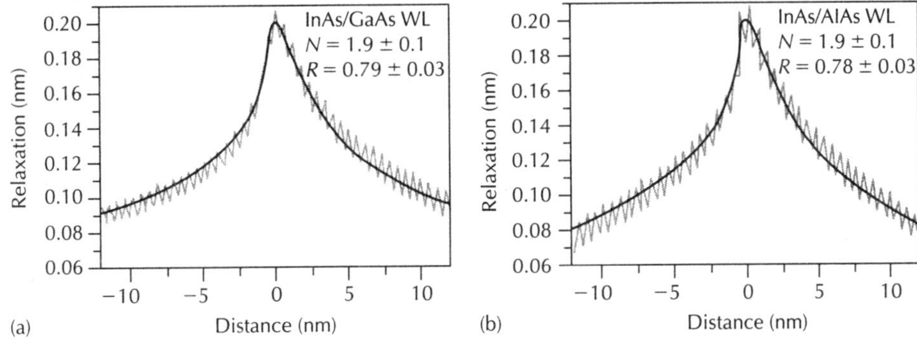

Figure 5.12   Measured and fitted relaxation profiles of the segregated InAs WLs in GaAs (a) and AlAs (b).

It is known that vertical indium segregation is reduced at lower growth temperatures [38]. As will be shown in section 5.4.1, we have reported that reducing the growth temperature to 300°C after capping of the WLs with three monolayers of GaAs leads to a dramatic reduction of the indium segregation [19]. However, such a capping procedure also leads to the almost complete dissolution of the QDs into the capping layer due to their partial coverage.

Another drastic example of the effects of partial capping is the formation of quantum rings (QRs). QRs can be grown by the partial capping of QDs with 2 nm of GaAs and subsequent annealing [39]. Recently, we observed that during this process, a second layer of indium accumulates on the surface of the capping layer, which is due to vertical segregation of indium from the WL and to lateral migration on the surface of indium atoms that have been expelled from the QDs during QR formation [40–43]. After continued capping, the second layer of indium itself forms a segregated indium distribution.

Finally, we show in Fig. 5.13 an overview image of the WLs of sample B. Surprisingly, one of the layers showed a shallow V-groove in which a QD was formed. The V-groove was unintentionally created on the GaAs substrate. It can clearly be seen that a large amount of indium atoms have accumulated in the V-groove. By comparing the extent of the indium segregation inside and

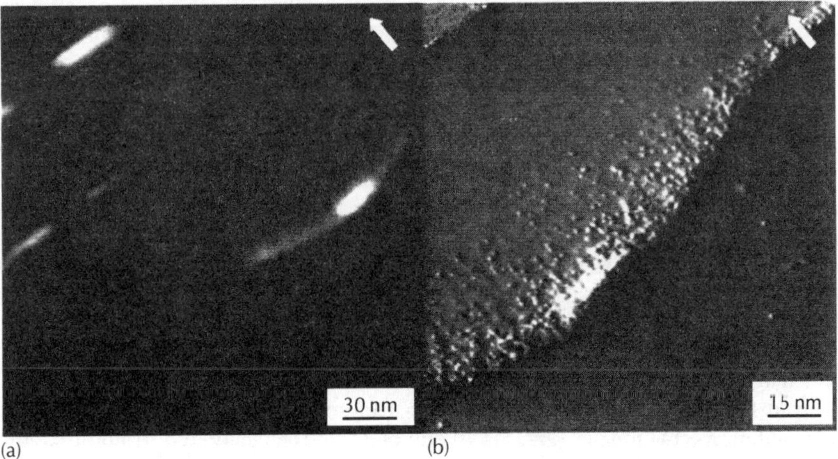

Figure 5.13   (a) X-STM overview image showing a shallow V-groove in one of the layers of sample B. (b) X-STM current image showing an enlarged view of the V-groove. The arrow indicates the growth direction.

outside the V-groove, it can be seen that during GaAs overgrowth, the indium segregation and migration facilitates a rapid planarization of the growth front, in the presence of indium atom accumulation in the V-groove.

To summarize, we have shown that the composition of (segregated) InAs WLs can be determined by either directly counting the indium atoms or by analysis of the outward displacement of the cleaved surface as measured by X-STM. We used this approach to study the effects of the deposited amount of indium, the InAs growth rate, and the host material on the formation of the WLs. We conclude that the formation of (segregated) WLs is a delicate interplay between surface migration, strain-driven segregation and the dissolution of quantum dots during overgrowth.

5.3   Dependence of the QD structural properties on the substrate material (GaAs vs AlAs)

Self-assembled InAs quantum dots embedded in an AlAs matrix have been of recent interest because of their larger confinement potential compared to InAs QDs in a GaAs matrix and their use in resonant tunnelling devices [44]. In order to understand the formation and the electronic properties of InAs QDs in AlAs, detailed information about the composition of the QDs is needed. Several studies have been reported on the effect of indium supply and growth temperature on the structural and optical properties of InAs QDs grown on AlAs [45, 46]. It has been shown that InAs QDs grown on AlAs exhibit smaller sizes and larger densities compared to InAs QDs that were grown on GaAs under similar growth conditions. This has been attributed to a reduced diffusion of In adatoms on the AlAs surface due to a higher surface roughness and the larger Al-In bond strength [45].

In this section we use X-STM measurements to determine the composition of InAs QDs and WLs grown on AlAs, and compare this to InAs QDs that were grown on GaAs under identical growth conditions.

The QDs were grown by MBE on doped GaAs (100) wafers. During growth of the QDs the substrate temperature was maintained at 500°C. A 1.9 ± 0.1 monolayer (ML) of InAs was grown in a cycled way, i.e. with a 3 second pause after each deposition of 0.25 ML, at a slow growth rate of 0.043 ML/s. After a 500 nm doped (1×10^{18} cm^{-2}) GaAs buffer layer the following sequence was grown: 20 nm GaAs/1.9 ML InAs/40 nm GaAs/1.9 ML InAs/40 nm GaAs/50 nm GaAs (doped 1×10^{18} cm^{-2})/20 nm GaAs/4 × (20 nm AlAs/1.9 ML InAs/20 nm AlAs/40 nm GaAs). To reduce interface roughness, the bottom AlAs barriers were grown at 600°C followed by a growth interruption prior to InAs deposition.

Figure 5.14a shows a large-scale filled-states topography X-STM image of the structure. Three layers of QDs grown on AlAs and two layers of QDs grown on GaAs are visible in the image. Compared to GaAs, the QDs grown on AlAs show a smaller size and have a significantly larger density of about 3×10^{11} cm^{-2}. Many QDs were imaged and the largest ones selected for analysis. It can then be assumed that these QDs are cleaved near their middle.

In Fig. 5.14b and c we compare high-voltage filled-states topography images of individual QDs grown on GaAs and AlAs, respectively. These images show the surface relaxation which varies with the local indium distribution in the QDs. From the contrast in the images it can be seen that the InAs/AlAs QD has a more homogeneous surface relaxation than the InAs/GaAs QD. We calculated the outward relaxation and the strain distribution of the QDs with the finite element calculation package ABAQUS. The QD shape was modelled by truncated pyramids with sizes determined by the X-STM measurements while the indium distribution was varied in order to get the optimal fit to the measured outward relaxation. The best results were obtained by allowing the modelled QDs to be cleaved at a plane 1 nm above their diagonal. The diagonal base length and the height of the InAs/GaAs QD are 28.4 nm and 6 nm, respectively. For the InAs/AlAs QD the diagonal base length is 19 nm and the height is 4.2 nm. The calculated relaxation of the cleaved surface of the QDs is shown in Fig. 5.14d and e using the same colour scale as in the corresponding X-STM images (Fig. 5.14b and c). From the calculated strain distribution, lattice constant profiles were derived which were used to verify the fitting results.

Figure 5.15 shows the measured and calculated outward relaxation profiles (a) and (b) and lattice constant profiles (c) and (d) taken in the growth direction through the centre of the QDs.

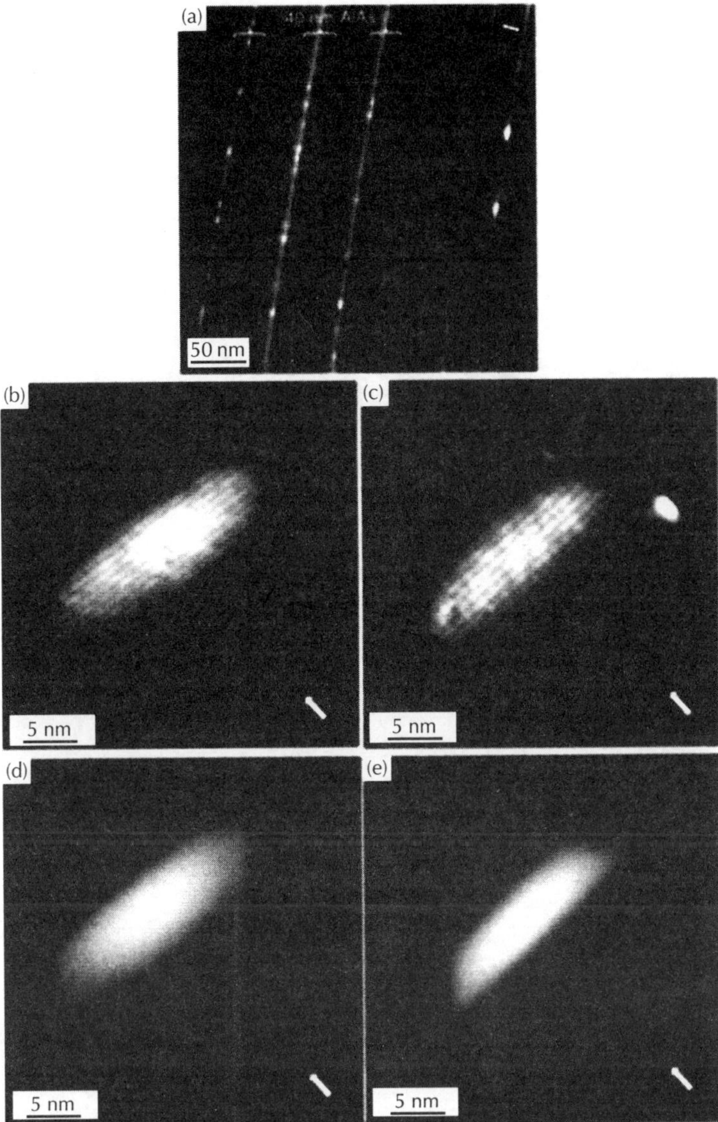

Figure 5.14   Filled states topography X-STM image of (a) three QD layers embedded in 40 nm thick AlAs barriers and two QD layers grown in GaAs, (b) InAs/GaAs QD, (c) InAs/AlAs QD. (d) and (e) show the calculated outward relaxation corresponding to (b) and (c). The colour scale for (b) and (d) is 0 (dark) to 600 pm (bright). The colour scale for (c) and (e) is 0 (dark) to 450 pm (bright). $V_{sample} = -3V$.

From the change in lattice constant, which is determined by the strain distribution in and around the QDs, it can be seen that there is compressive strain above and below the QDs.

For the InAs/GaAs QD, there is a clear increase in lattice constant towards the top of the QD, which indicates an increasing indium concentration. This can also be seen by the slight asymmetry in the relaxation profile of the InAs/GaAs QD. From X-STM and photocurrent experiments, it has been shown that low-growth rate InAs/GaAs QDs have an increasing indium concentration in the growth direction [47, 48]. However, other groups have reported InGaAs QDs with laterally non-uniform indium compositions showing an inverted-triangle, trumpet or truncated reversed-cone shape [26, 49–51]. We find that the indium distribution of our low-growth rate

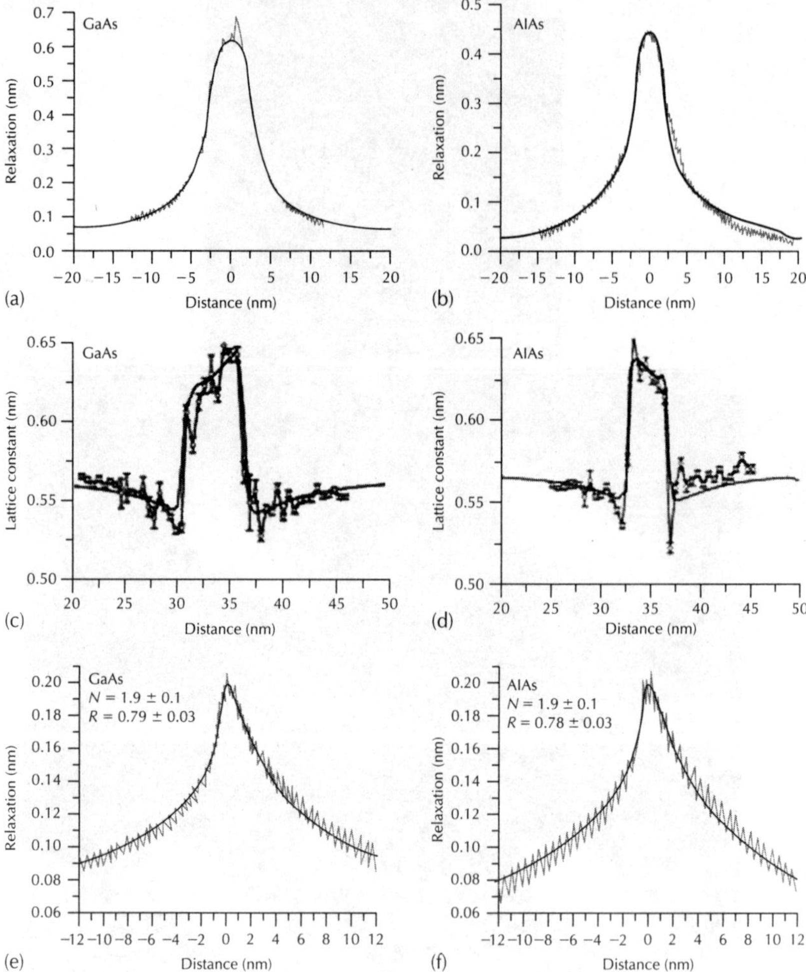

Figure 5.15 Calculated and measured relaxation profiles through the centre of the QD in the growth direction for an InAs/GaAs QD (a) and an InAs/AlAs QD (b). (c) and (d) are the corresponding calculated and measured lattice constant profiles. (e) and (f) show the measured and calculated relaxation profiles of the WLs in GaAs and AlAs.

InAs/GaAs QDs mostly resembles the trumpet shape proposed by [26] which we describe with a linear gradient in both the growth direction and lateral direction. Along the centre of the QD, the indium concentration increases with 80% at the base to 100% at the top of the QD. The gradient in the lateral direction is dependent on position along the growth direction. At the base of the QD, it varies linearly from 80 to 40% from the core to the perimeter, while at the top of the QD it remains constant. For the InAs/AlAs QD best fit results were obtained by an indium gradient of 85 to 70% decreasing from base to top of the QD. In this case there is no indication for a lateral gradient in the indium composition.

We attribute the observed differences in indium composition and size of the QDs in GaAs and AlAs to the combined effects of (a) the reduced diffusion of In on AlAs compared to GaAs, (b) the reduced intermixing for the InAs WL on AlAs, and (c) the capping process. It is known that the indium accumulation in QDs is determined by strain minimization during growth. However, the preferential indium aggregation at the In-rich region of the dot is limited by the lateral diffusion of indium in the case of growth on the AlAs substrate, which results in a reduced QD size and increased QD density.

Furthermore, top-view STM experiments have shown that the surface reconstruction of the InAs wetting layer on AlAs indicates less intermixing for the WL [52], which together with the reduced mobility of In on AlAs, explains the homogeneous indium distribution in the base of the InAs/AlAs QDs. It has been proposed that the growth of dots on AlAs is initiated by 2D islands which develop into small 3D islands for higher InAs depositions [52]. This is in contrast with the growth of dots on GaAs where dot formation is initiated by small indium-rich nucleation centres which develop into trumpet-shaped indium distributions by the preferential diffusion of In to the apex of the dot [26]. These growth mechanisms are supported by our observation of the indium distribution inside the dots.

The observed decrease of the indium concentration towards the top of the InAs/AlAs QD might be caused by the residual incorporation of AlAs in the top of the dot during the capping process. In order to reduce the total strain field in the QD some capping material is incorporated in the top of the dot. In the case of GaAs capping, the diluted part of the dot is very mobile and therefore disappears very quickly during capping, causing levelling of the QDs. The InAs diluted by Al, however, is much less mobile due to the stronger Al–In bond strength and therefore more diluted material at the top remains in place.

In Fig. 5.15e and f we show measured and calculated relaxation profiles of segregated WLs in GaAs and AlAs that were grown at a higher temperature of 530°C. These WLs did not show dot formation. The indium segregation profile of the WLs can be simply described by the phenomenological model of Muraki *et al.* described in the previous section [27, 31]. Based on this model, we calculated the outward relaxation of the WLs using the analytical expressions by Davies *et al.* [24]. The optimal fit to the experimental relaxation profiles was obtained with $N = (1.9 \pm 0.1)ML$, $R = 0.79 \pm 0.03$ and $N = (1.9 \pm 0.1)ML$, $R = 0.78 \pm 0.03$ for the InAs WLs in GaAs and AlAs, respectively, where N is the total amount of deposited In and R the segregation coefficient. The obtained values for N are in agreement with the nominal deposited amount of indium. We do not find a significant difference in the segregation coefficients of indium in AlAs and GaAs as was concluded by Schowalter *et al.* [27]. The small apparent difference between the relaxation profiles of the WLs can be attributed to the difference in lattice constants of GaAs and AlAs. We conclude that InAs segregation does not play a role in the difference in the formation of QDs in AlAs and GaAs.

In summary, InAs QDs grown on GaAs are larger than those grown on AlAs, and they show both a lateral and a normal gradient in the In composition, while InAs/AlAs QDs show only a vertical gradient and with an opposite sign (In concentration decreasing from bottom to top). The WLs on GaAs and AlAs do not show significant differences, so we suggest that the segregation of the WL is mainly strain driven, whereas the formation of the QDs is also determined by growth kinetics. In particular, the observed differences between QDs are due to the reduced mobility of In in AlAs due to the higher Al–In bond strength, which modifies both the QD formation and capping processes.

5.4 Capping process of InAs quantum dots

In previous sections we have studied relevant aspects of the WL and QD formation process. Nevertheless, once created, the QDs are subsequently capped, a step which is required for any device application. Although a lot of effort has been dedicated to understand the QD growth mechanism, there are relatively few studies focused on the effect of the capping process [12–20]. Some of these studies have already shown significant differences in size, shape and composition between uncapped and capped QDs. For example, an important collapse of the QD height has been reported for InAs/GaAs QDs capped with GaAs [14, 15, 17–19], revealing the big influence of the capping process on the structural properties of the QDs.

In this section we use X-STM to analyse the capping process of InAs QDs. The effect of the capping temperature and growth interruptions is studied, as well as the impact of using different materials in the capping layer. The possibility of controlling the QD height with a double capping method is also discussed.

5.4.1 Capping temperature and growth interruptions

Levelling of InAs/GaAs QDs after deposition of thin GaAs cap layers has been clearly revealed by top-view STM [18, 58] and atomic force microscopy (AFM) [54]. The levelling process has been attributed to the additional strain build-up between the cap layer and the partially relaxed InAs QDs [18, 54] to destabilize the QDs. These experiments, however, lack information about the shape and, in particular, the residual height of the QDs which is the most important parameter determining the electronic properties. In this section, we study the GaAs-capping process of InAs QDs on GaAs(100) by X-STM. The role of the capping temperature and growth interruptions is discussed. Detailed and accurate results of the QD levelling are presented, which are essential for understanding the capping process and the control of the structural and electronic properties of InAs QDs.

The samples were grown by solid source MBE on Si-doped n-type GaAs(100) substrates. After oxide desorption at 580°C, a 150 nm thick GaAs buffer layer was grown. Then the substrate temperature was lowered to 500°C for deposition of 2.1 ML InAs. Formation of InAs QDs was verified by the sharp transition from streaky to spotty of the reflection high-energy electron diffraction pattern. Thereafter, for samples A–D different capping procedures were applied: (A) deposition of 10 nm GaAs at 500°C followed by 150 nm GaAs growth at 580°C; (B) cooling down the sample to 300°C before capping the InAs QDs by 10 nm GaAs at the same temperature and growth of 150 nm GaAs at 580°C; (C) capping the InAs QDs by 3 ML GaAs at 500°C, then cooling down the sample to 300°C for deposition of 150 nm GaAs; and (D) capping the InAs QDs by 3 ML GaAs at 500°C followed by a growth interruption (GI) of time t, deposition of 10 nm GaAs at the same temperature, and growth of 50 or 150 nm GaAs at 580°C. In sample D five such InAs QD layers were inserted with GI times t of 0, 20, 40, 60, and 90 seconds, separated by 60 nm GaAs. An extra 30 nm GaAs spacer was grown as a marker between the QD layers with GI times t of 40 and 60 seconds. The time for cooling down samples B and C from 500 to 300°C was 4 minutes. The growth rates were 0.58 and 0.06 ML/s for GaAs and InAs, respectively, and the As_4 beam equivalent pressure was 1×10^{-5} torr.

Figure 5.16a shows the filled state's topography X-STM image of the InAs QDs in sample A which are capped in the conventional way by 10 nm GaAs at 500°C and 150 nm GaAs at 580°C. The marked part of the image is shown in Fig. 5.16b after treatment with a local mean equalization filter [55] to enhance atomic details by removing the large scale background contrast. The bright horizontal lines are the top zig-zag rows of the (110) surface, which are separated by one bilayer (BL), i.e. two MLs. The topographical contrast in Fig. 5.16a is due to the outward relaxation of the cleaved surface of the compressively strained InAs QDs, revealing their cross-sectional shape. The bright spots in Fig. 5.16b correspond to In atoms in the top layer of the cleaved surface. The height of the InAs QDs in sample A is measured as 8 BLs by counting the number of atomic rows.

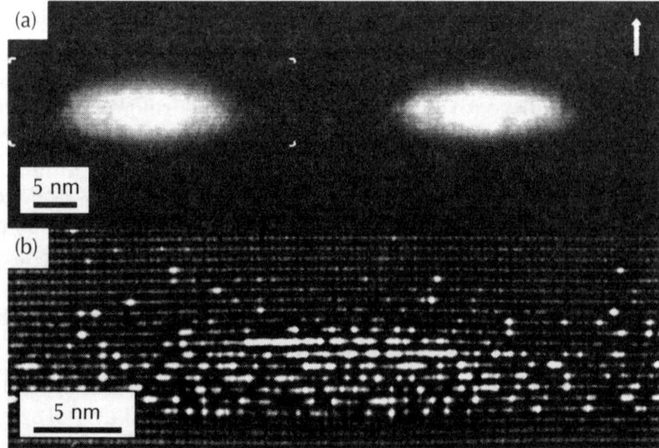

Figure 5.16   (a) Filled states topography X-STM image of the InAs QDs in sample A with conventional capping by 10 nm GaAs at 500°C and 150 nm GaAs at 580°C. $V_{sample} = -3.0V$. A part of the image (a) marked by four corners is treated by a local mean equalization filter and shown in (b). In (a) the arrow indicates the growth direction. The black-to-white height contrast in (a) is 0 to 0.5 nm.

It is well established that InAs QDs buried in the conventional way of sample A exhibit a reduced height compared to unburied ones due to QD levelling [18, 53, 54]. In order to determine the QD height reduction, the shape change during overgrowth of the InAs QDs in sample B is strongly suppressed by capping them at 300°C. The filled state's topography X-STM image of an InAs QD in sample B is shown in Fig. 5.17a with the filtered image in Fig. 5.17b. The InAs QD exhibits very sharp and well-defined interfaces confirming the suppressed QD levelling, atom diffusion, and segregation [29] and, thus, the preservation of the QD shape [56, 57]. The height of the InAs QD is 12 BLs which is 4 BLs larger than that of the QDs in sample A. This indicates that during conventional capping at 500°C the QD height is reduced by about one third of the original one.

Figure 5.17   (a) Filled states topography X-STM image of the InAs QD in sample B with the GaAs cap grown at 300°C. V$_{sample}$ = − 3.0 V. *A part of the image (a) marked by four corners is treated by a local mean equalization filter and shown in (b). In (a) the arrow indicates the growth direction. The black-to-white height contrast in (a) is 0 to 0.5 nm.*

When InAs QDs are capped only by a very thin GaAs layer, strong QD levelling or QD collapse occurs. Figure 5.18 shows the filled state's topography and filtered images of such InAs QDs in sample C. The QDs are capped at 500°C by 3 ML GaAs and subsequently overgrown at 300°C to further maintain the shape. During the thin GaAs capping and cooling down, the levelling of the InAs QDs leads to a rather homogeneous (In,Ga)As layer in between the QDs due to In detachment from the QD tops, Ga/In intermixing, and In segregation with a thickness of about 4 BL, which is much thicker than the original InAs wetting layer. Intermixing with the GaAs substrate [58] expected during the growth of InAs at 500°C additionally contributes approximately 3 ML of GaAs to the (In,Ga)As layer, which is derived by subtracting the thicknesses of deposited InAs (2.1 ML) and GaAs (3 ML) from the total (In,Ga)As thickness of 4 BL. After levelling of the QDs, unincorporated In floating on the surface is pinned there by the low-temperature GaAs capping and forms an In-rich layer marked by the arrow in Fig. 5.17b. The sharp interface between this layer and the low-temperature GaAs cap confirms that In segregation and diffusion in growth direction are strongly suppressed for GaAs overgrowth at 300°C. Most interestingly, the InAs QDs are completely levelled to the thickness of the (In,Ga)As layer in between them, which is much smaller than the height of the QDs observed in sample A. This suggests that the conventional, continuous GaAs capping at 500°C in sample A not only drives QD levelling during the initial stage, like the thin GaAs capping in sample C, but also quenches the levelling process when the QDs become buried.

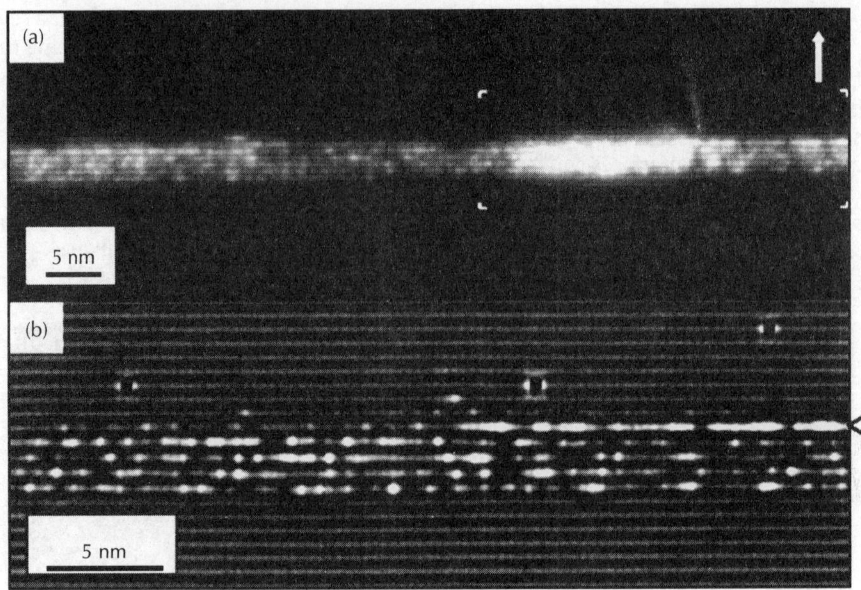

Figure 5.18   (a) Filled states topography X-STM image of the InAs QD in sample C capped by 3 ML GaAs at 500°C, followed by 150 nm GaAs grown at 300°C. $V_{sample} = -3.0$ V. A part of the image (a) marked by four corners is treated by a local mean equalization filter and shown in (b). In (a) the arrow indicates the growth direction. The black-to-white height contrast in (a) is 0 to 0.4 nm. The arrow in (b) points to the In-rich layer.

To assess the time scale of QD levelling which is, thus, crucial for the final size and shape of the buried QDs, varying GI times are inserted in sample D after deposition of 3 ML GaAs at 500°C on the InAs QDs prior to GaAs overgrowth. Figure 5.19 shows the filled state's topography image of the InAs QDs in sample D with GI times of 0, 20, 40, 60, and 90 seconds in subsequent layers. Clearly, the height of the InAs QDs capped without GI (first layer) is significantly larger than that of the QDs with insertion of 20 seconds GI. No significant further decrease in QD height is observed when the GI time is increased to 90 seconds. Hence, QD levelling during thin GaAs capping and GI takes place on a time scale of less than 20 seconds. For the present GaAs growth rate, this is comparable to the time required for growing several nanometres of GaAs to fully bury the QDs by continuous overgrowth. This indicates that both the driving and quenching of the QD levelling in conventional capping take place on a similar timescale. The size and shape of the buried QDs are therefore determined by a delicate interplay between driving and quenching of the QD levelling which is controlled by the GaAs growth rate and growth temperature.

A model based on the above experimental results is proposed for the growth of InAs QDs embedded in GaAs. The growth of InAs commences in the two-dimensional (2D) layer-by-layer mode until the InAs thickness reaches the critical value of 1.7 ML for InAs QD nucleation to reduce the accumulated strain. The InAs QDs are formed by In atoms transported massively from the 2D InAs layer, leaving a thin wetting layer on the surface. The whole system is stable at the minimum of the total energy composed of the surface energy, the strain energy, and the interface energy. Subsequent capping of the InAs QDs by GaAs, on the other hand, introduces extra strain energy between the GaAs cap and the InAs QD layer, resulting in an unstable system and the consequent QD levelling process. In atoms are redistributed from the InAs QD tops to the areas in between them during the QD levelling. They contribute to a several nanometres thick (In,Ga)As layer with an exponential In composition decay due to In segregation and Ga/In intermixing during overgrowth [28], reducing the lattice mismatch and, hence, the total energy of the system. Thus, the thickness and the In composition profile of the (In,Ga)As layer in between the InAs QDs strongly depends on the QD levelling and In segregation. It is important to note that the QD levelling is very sensitive to the substrate temperature and is strongly suppressed at low growth temperatures, where it becomes more and more difficult to thermally break the In-As bonds. In addition to

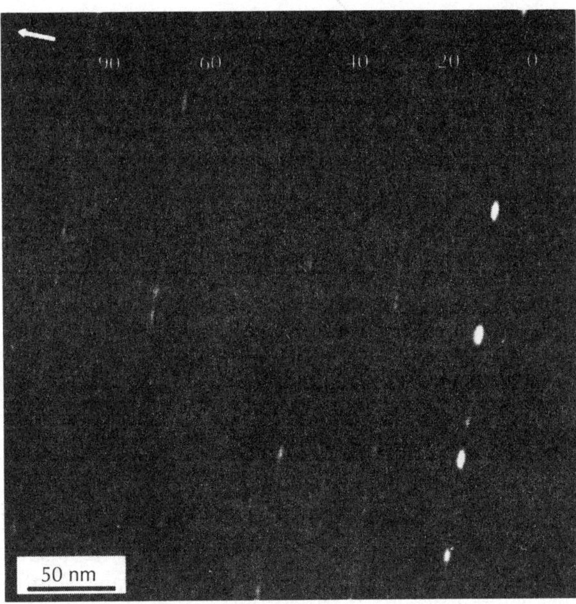

Figure 5.19   Filled states topography X-STM image of the five InAs QD layers in sample D. The InAs QDs are capped by 3 ML GaAs at 500°C, followed by a GI and a 10 nm GaAs cap grown at 500°C plus a GaAs separation layer at 580°C. $V_{sample} = -3.0\,V$. The increasing GI times of 0, 20, 40, 60, and 90 seconds are noted in the image. The arrow indicates the growth direction. The black-to-white height contrast is 0 to 0.5 nm.

inducing QD levelling, the GaAs cap buries the InAs QDs, thereby quenching the levelling process during continued overgrowth. Therefore, the size and shape of the embedded InAs QDs are determined by a delicate interplay between driving and quenching of the QD levelling during capping, which depends strongly on the growth rate and growth temperature of the GaAs cap.

In summary, we have investigated the GaAs-capping process of InAs QDs grown by molecular-beam epitaxy on GaAs(100) substrates. In its initial stage, GaAs capping induces levelling of the QDs to drastically decrease their height. During continuous capping the QD levelling is quenched when the QDs become buried. Both phenomena, driving and quenching of the QD levelling, take place on a similar time scale and are very sensitive to the GaAs growth rate and growth temperature. This understanding opens up an efficient route for controlling the size and shape of buried QDs.

5.4.2   Capping with different materials

We have seen how the QD structure is strongly affected by capping with GaAs and how this process is very sensitive to the growth temperature and growth interruptions. Since GaAs is the most commonly used capping material for InAs/GaAs QDs, the existing structural studies of buried InAs/GaAs QDs have been mainly devoted to GaAs-capped QDs. Nevertheless, different materials such as InGaAs and GaAsSb are nowadays used to cap InAs/GaAs QDs in an effort to extend its emission wavelength to the technologically interesting 1.3–155 μm region [59–64]. For InAs/InP QDs, capping materials other than InP, like InGaAsP, have also successfully been used for laser applications [65–67].

The use of different capping materials strongly affects the emission wavelength and therefore should strongly affect the QD electronic and/or structural properties, such as size, shape, composition, strain, band offsets, etc. Indeed, critical issues occurring during capping like dot decomposition, intermixing, segregation, As/P exchange and phase separation in the capping layer depend on the capping material. To understand the impact of the capping material on the structural properties of the QDs is consequently of crucial importance.

In this section, we have used X-STM to analyse at the atomic scale how capping with different materials influences the structural properties of InAs QDs in GaAs and InP. The role of the different effects occurring during capping (intermixing, segregation, As/P exchange, compositional modulation in the capping layer, etc.) is determined. First, we study the capping with strained layers to reveal the role of strain. For that we used InAs QDs grown on (100) GaAs capped with InGaAs and GaAsSb strain reducing layers. Then we study the capping with lattice matched layers (with respect to the substrate), where the chemical effects could dominate the process. For that we study InAs dots grown on (311)B InP that were capped by either InP, InGaAs or InGaAsP (the last two materials being nominally lattice matched to InP). While strong morphological differences are found when strained layers are used, the difference between various lattice-matched capping materials is more subtle, although also relevant for device applications.

5.4.2.1   Capping with strained layers

When InAs/GaAs QDs are capped with a material which has a larger lattice constant than GaAs (and consequently closer to the one of InAs), the strain induced in the dot during capping could be smaller, which would red shift the emission wavelength. In addition, the modified strain difference could induce differences in dot size, shape and composition since dot decomposition during capping could be influenced by the strain. This is what we analyse in this section.

5.4.2.1.1   InGaAs capping of InAs/GaAs QDs

Capping with InGaAs has become popular in the last few years because it allows the emission wavelength of InAs/GaAs QDs [59–63] to be increased considerably. It is therefore very interesting to study how the structural properties of InAs QDs are affected by an InGaAs strain reducing layer.

The sample used in this study was grown by solid source MBE on an n^+ Si-doped (100) GaAs substrate. 2.7 MLs of InAs were deposited at 500°C and 0.1 ML/s on an intrinsic GaAs buffer layer. The QD layer was capped with a nominally 6 nm thick $In_{0.15}Ga_{0.85}As$ layer grown at 500°C at a growth rate of 0.5 ML/s.

Figure 5.20 shows a high voltage filled states image of an InAs/GaAs QD capped with InGaAs. The InGaAs layer is not well appreciated in this image because the contrast is dominated by the cleavage induced defects close to the dot. The dot is 7.0 ± 0.5 nm high, considerably higher than the typical GaAs-capped QDs, which are 3–5 nm high, as shown in previous sections [15, 18, 19]. Nevertheless, the top facet is not as well defined as in the GaAs case and the In content decreases gradually in the growth direction (that is why the error in determining the height is larger). The In composition in the capping layer was deduced by the analysis of the outward relaxation. The fit shown in Fig. 5.21 was found when an 8.5 nm thick layer (as measured from the X-STM images) with 17% In content is considered in the calculation. The In content is close to the nominal value (15%), although the layer is considerably thicker than the nominal 6 nm. The lattice

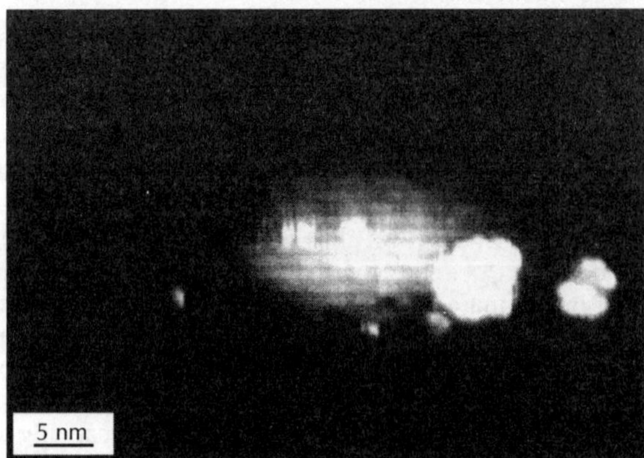

Figure 5.20   Topography image of an InAs/GaAs QD capped with InGaAs (V = −3 V, I$_{tunnel}$ = 55 pA). The big white regions are cleavage induced defects. Growth direction: [001], lateral direction: [110].

mismatch between InAs and the capping layer is now 0.056, about 17% smaller than in the case in which GaAs capping is used. We think that the reduced dot decomposition during capping is related to this smaller lattice mismatch; the result is that the dot height is increased by factor of ~2 compared to GaAs capping. This interpretation is in agreement with previously reported results in which QDs capped with InGaAs instead of GaAs are shown to retain their shape during the initial stages of capping [71, 72]. It could be argued that the increase in dot height is due to strain induced compositional modulation in the capping layer [68]. Although part of the increase in dot height could be due to that effect, we think that its contribution should be very small, since no traces of compositional modulation are observed in the capping layer. The role of the strain can be clarified by using GaAsSb as the capping material, because in this case the InAs dot height cannot increase due to phase separation in the capping layer.

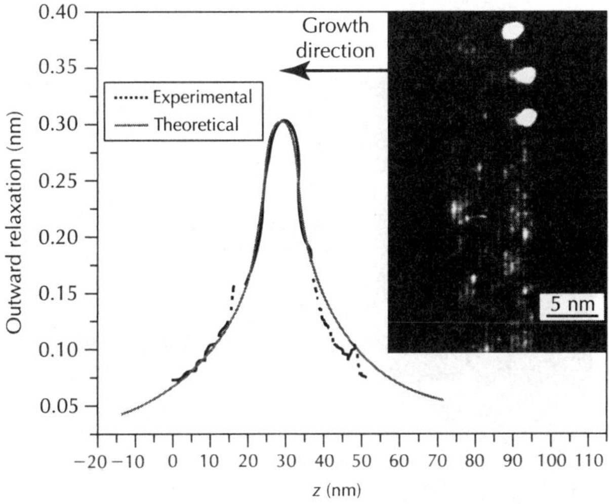

Figure 5.21   Measured (dotted line) and calculated (solid line) outward relaxation profiles of the InGaAs capping layer. A 8.5 nm thick layer (as measured from the images) with a 17% In content was considered in the calculation. The inset shows a high-pass filtered topography image of the InGaAs capping layer ($V = -3V$, $I_{tunnel} = 55 pA$). The big white regions are cleavage induced defects.

5.4.2.1.2   GaAsSb capping of InAs/GaAs QDs

In the last few years, GaAsSb capping layers have also been used to increase the emission wavelength of InAs/GaAs QDs [63, 69] and room temperature photoluminescence at $1.6\,\mu m$ has recently been reported from GaAsSb-capped In(Ga)As/GaAs QDs [64, 70]. The strong red shift observed by using GaAsSb instead of GaAs capping layers has been attributed to a type II band alignment [64, 70]. However, the structural properties of these QDs have not been studied, despite the fact that they could be significantly different from those of GaAs-capped QDs. Due to the larger lattice constant of GaAsSb compared to GaAs, the strain induced in the dot during capping could be smaller, which would further red shift the emission wavelength and could affect the dot size, shape and composition, as was the case with InGaAs capping.

The studied samples were grown by solid source MBE on n^+ Si-doped (100) GaAs substrates. In the first sample (sample A), 2.7 monolayers (ML) of InAs were deposited at 500°C and 0.1 ML/s on an intrinsic GaAs buffer layer. The QDs were capped with a nominally 6 nm thick $GaAs_{0.75}Sb_{0.25}$ layer grown at 475°C. In the second sample (sample B), two QD layers were grown under the same conditions (2.7 ML of InAs deposited at 500°C and 0.1 ML/s). The first layer was capped with GaAs, and the second with 6 nm $GaAs_{0.75}Sb_{0.25}$. The capping temperature was 500°C in both cases. A layer of surface QDs was also grown for atomic force microscopy measurements.

A number of individual QDs were analysed in sample A in order to extract information concerning their size, shape and composition. Figure 5.22 shows a high voltage filled states topography image of a QD in this sample. The group V elements are imaged in this measurement so that the bright spots are Sb atoms in the As matrix. Sb segregation into the GaAs layer is clearly observed. Contrary to what happens when capping with GaAs at similar temperatures [12], the capping layer covers all of the dot. This behaviour cannot be attributed to a larger bond strength preventing Ga migration on the growth surface, since the Ga–Sb bond is weaker than the Ga–As one (45.9 and 50.1 Kcal/mol, respectively). The reason is likely to be the smaller strain that exists between the partially relaxed InAs QDs and the GaAsSb capping layer. This smaller strain can be accommodated by the system at the present growth temperature (475°C) without inducing the migration of capping material away from the top of the dots. The dot is a full pyramid with a diagonal base length of 32 ± 2 nm. The height is 9.5 ± 0.2 nm, much larger than that of typical GaAs-capped QDs (3–5 nm high) [15,18,19]. This result indicates that the QDs are not dissolved during capping with GaAsSb.

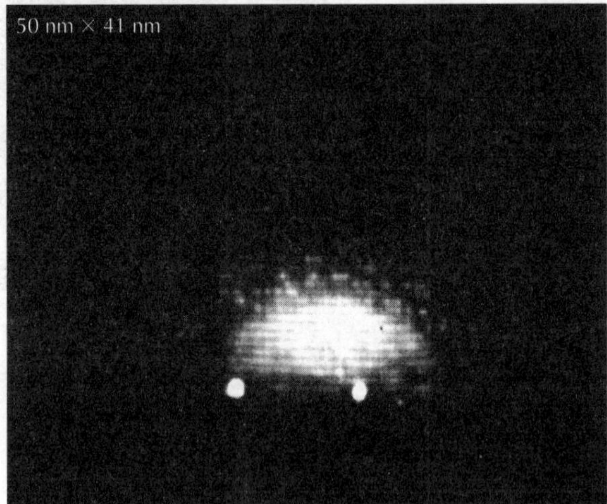

Figure 5.22   Filled states topography image of an InAs/GaAs QD capped with GaAsSb (V = − 3 V). The bright spots are Sb atoms in the As matrix. The white circles and the dark feature are cleavage induced defects.

The Sb composition in the capping layer can be obtained directly by counting the individual Sb atoms in the filled states images, for example Fig. 5.23a. The Sb concentration profile as a function of distance along the growth direction is plotted in Fig. 5.23b. The average Sb content is 22%. This result can be checked by also analysing the outward relaxation of the cleaved surface. The relaxation of the capping layer was compared to calculations from continuum elasticity theory. The fit shown in Fig. 5.23c was obtained for a 5 nm thick layer (as measured from the X-STM images) with a 25% Sb composition. The agreement between theory and experiment is quite good in the region of the layer itself, but deviates above the layer. This deviation is due to the effect of Sb segregation, which is not included in the model but is clearly present in the structure (see Fig. 5.23a and the asymmetry in the Sb profile obtained by atom counting). The obtained Sb concentration of 22–25% agrees well with the nominal growth value. The lattice mismatch between this GaAsSb composition and relaxed InAs is 0.048, about 28% smaller than in the case in which GaAs capping is used. We propose that the fact that dot decomposition during capping is suppressed is related to the reduced strain between the dot and the capping layer, although chemical effects due to the presence of Sb could also be relevant. The fact that the lattice mismatch is now considerably smaller than in the InGaAs-capped sample studied before could explain the fact that the dots are now completely preserved while they are still partially dissolved when capping with $In_{0.15}Ga_{0.85}As$, although the different chemistry in the two cases must be taken into account.

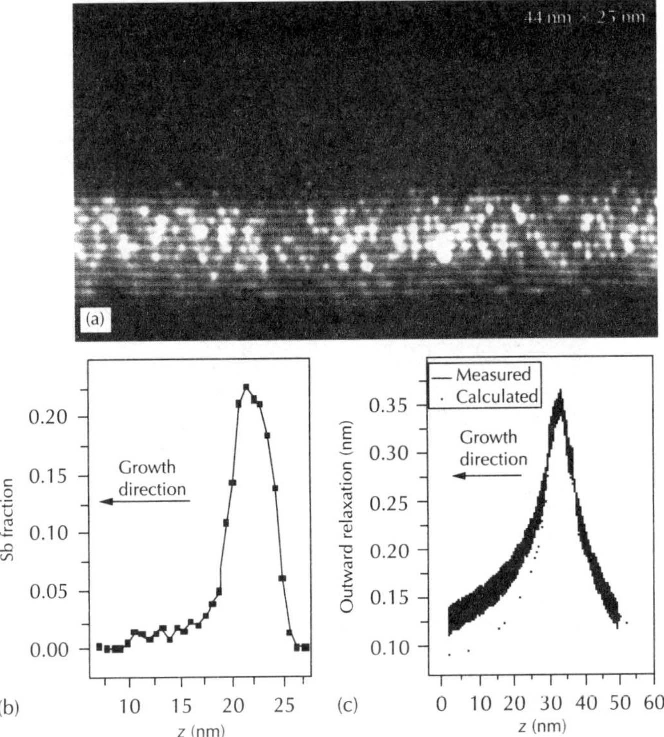

Figure 5.23   (a) Topography image of the GaAsSb capping layer (V = −3V). The bright spots are Sb atoms in the As matrix. Sb segregation into the GaAs layer is clearly observed. (b) Sb concentration profile in the capping layer is obtained by counting individual Sb atoms. (c) Measured (solid line) and calculated (dotted line) outward relaxation profiles of the GaAsSb capping layer. A 5 nm thick layer (as measured from (a)) with a 25% Sb content was used in the calculation.

The composition of the QDs can also be estimated from an analysis of the outward relaxation. The size and shape of the dots obtained from the images are used as input parameters for the model and the composition is changed until the calculated outward relaxation profile of the dot fits the experimentally determined one. We assume that the dot with the largest observed base length is cleaved through the middle and that the dots have a square base with the edges along the <100> directions. The largest measured baselength was 32 nm (see Fig. 5.22) which corresponds to a square with 23 nm. Figure 5.24 shows the fit to the outward relaxation profile through the middle of the dot, obtained when a trumpet-like shape In composition inside the QD is included in the model. The fit deviates in the region above the dot, probably due to the fact that Sb segregation is not included in the calculation. The In content increases from 80% at the bottom to 100% at the top of the QD and from 40% in the corners to 80% in the centre of the base (see inset of Fig. 5.24). This is the same non-uniform In composition that we found in section 5.3 in GaAs-capped InAs/GaAs QDs [73] and similar to what has been previously proposed in [26]. It is attributed to In–Ga intermixing. The presence of In-deficient corners in the present GaAsSb-capped QDs indicates that they are not necessarily created by a redistribution of material from the top of the island to the bottom during capping, as has been recently reported [74]. In the present case, the top of the dot is not dissolved during capping but the In-poor corners are still present, hence they must originate during an earlier stage of the QD formation process.

In order to confirm the observed differences between QDs capped with GaAs and GaAsSb, sample B was studied. This sample contains QDs capped with both GaAs and GaAsSb, with both capping layers grown at the same temperature as the InAs QDs. From an analysis of the measured outward relaxation, the Sb concentration on the GaAsSb layer was found to be 24%. However, the properties are different to those of sample A because the GaAsSb capping layer does not

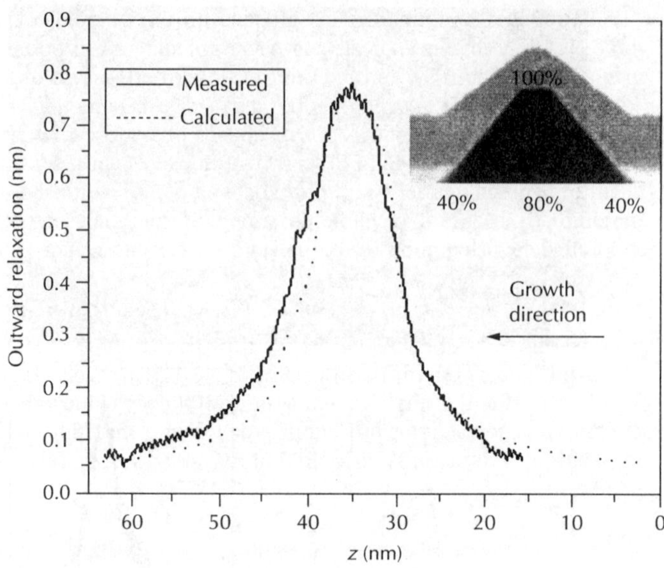

Figure 5.24   *Measured (solid line) and calculated (dotted line) outward relaxation profiles through the centre of a QD. The inset shows the In distribution inside the QD used in the calculation.*

completely cover the dots, with their apex remaining uncovered (see Fig. 5.25a). As the only difference between samples B and A is that the capping temperature was 25°C higher in sample B, it can be concluded that this increase in temperature is sufficient to allow the capping material to migrate from the relaxed apex of the dot, minimizing the strain [12]. This observation reveals that the final dot/GaAsSb capping layer configuration is very sensitive to the capping temperature. Despite this fact, differences in the QD structure for GaAs and GaAsSb capping are clearly revealed in Fig. 5.25. The GaAsSb-capped QDs (Fig. 5.25a) are again full pyramids, with a height of 8.3 ± 0.2 nm, while the GaAs-capped QDs (Fig. 5.25b) are truncated pyramids, with a height of only 3.8 ± 0.2 nm. QDs capped with GaAs are significantly dissolved during capping, while the shape of QDs capped with GaAsSb ($\sim 25\%$ Sb) at the same temperature is preserved. This result is confirmed by atomic force microscopy measurements on uncapped surface QDs, which are found to have a height of 8 ± 1 nm, in good agreement with the height of the GaAsSb-capped QDs. The base length of the strongly dissolved GaAs-capped QDs is smaller than that of the preserved GaAsSb-capped dots (20 vs 26 nm, respectively), indicating that the material redistribution during capping is not from the apex to the base of the dot [74], but occurs instead to the wetting layer (WL), as was also shown in section 5.2. This is in agreement with the observation of a significantly higher In content in the WL of GaAs-capped QDs (1.8 MLs vs 0.8 MLs in GaAsSb-capped QDs), obtained by counting individual In atoms in the X-STM empty states images.

The present results indicate that strain could be playing an important role in inducing dot decomposition during capping, and that decreasing the dot/capping layer lattice mismatch could strongly reduce dot decomposition. The fact that InAs/GaAs QDs capped with GaAsSb are much bigger than those capped with GaAs should also be taken into account when explaining the observed red shift of the emission wavelength for GaAsSb-capped QDs.

In conclusion, X-STM has been used to study at the atomic scale the effect of InGaAs and GaAsSb capping layers on the structural properties of self-assembled InAs/GaAs QDs. QDs capped with a GaAsSb layer with 22–25% Sb are much larger than typical GaAs-capped QDs. GaAsSb-capped QDs exhibit a full pyramidal shape of 8.3–9.5 nm height, while the same dots capped with GaAs are truncated pyramids with a height of only 3.8 nm. QDs capped with $In_{0.15}Ga_{0.85}As$ are between those two extreme cases. This finding indicates that dot decomposition during capping is suppressed by using $GaAs_{0.75}Sb_{0.25}$, this is likely related to the reduced dot/capping layer lattice mismatch, although chemical effects due to the presence of Sb could also be relevant.

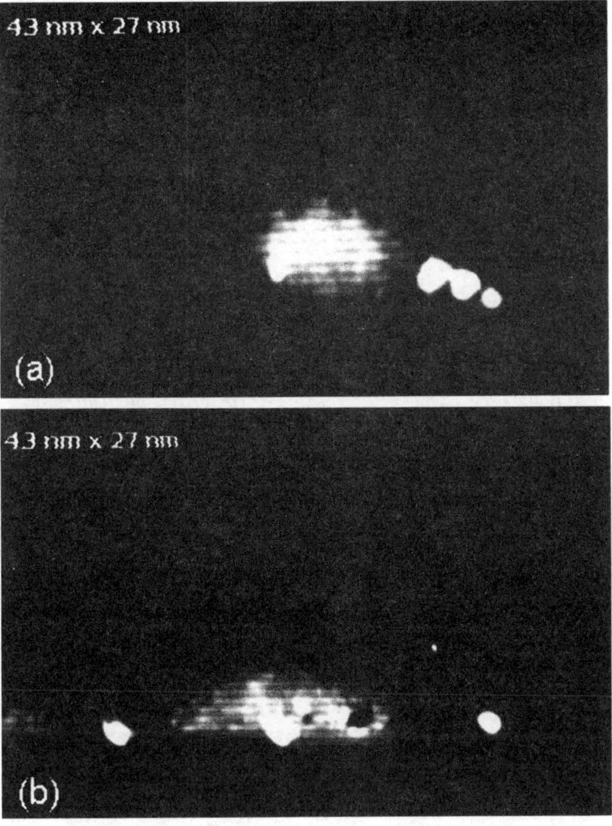

Figure 5.25   Filled states topography images of two QDs capped with (a) GaAsSb and (b) GaAs (V = −3V). The white circles and the dark features are cleavage induced defects.

5.4.2.2   Capping with lattice-matched layers

The results shown in the previous section seem to indicate that strain could be playing a mayor role in inducing dot decomposition during the capping process. Nevertheless, chemical effects should also be relevant, specially in the case of InAs/InP QDs, in which the As/P exchange reaction is known to be present. In order to study the chemical effects, different capping materials were used, all of them with the same lattice parameter, so the differences due to strain are eliminated.

The samples used in this study were grown by gas source MBE on an Si-doped (311)B-oriented InP substrate. This substrate orientation is very attractive for laser applications because, in comparison with InAs dot formation on conventional (100) InP substrates, a higher density of dots having a smaller size dispersion has been achieved on (311)B InP substrates [65, 66]. Indeed, room temperature lasers with low threshold current density emitting at $1.5\,\mu m$ were recently demonstrated [67]. In the analysed sample the growth temperature was set to 480°C. The QDs were formed by depositing 2.1 (100) equivalent monolayers (ML) of InAs at $0.33\,ML/s$ on InP buffer layers. A low As flux was supplied to the surface during the InAs deposition to enhance the formation of small QDs [21]. After island formation, a 30 s growth interruption under As flux was performed before the growth of the capping layer. Three QD layers, separated by 40 nm, were grown under the same conditions but capped with different materials: 40 nm of InP in the first layer, 20 nm of lattice-matched $In_{0.53}Ga_{0.47}As$ in the second layer (followed by 20 nm of InP) and lattice-matched $In_{0.87}Ga_{0.13}As_{0.285}P_{0.715}$ in the third one.

A large-scale filled states X-STM image of the studied structure is shown in Fig. 5.26, in which the three QD and capping layers can be observed. The QD layers capped with InP, InGaAs and InGaAsP are labelled A, B and C, respectively.

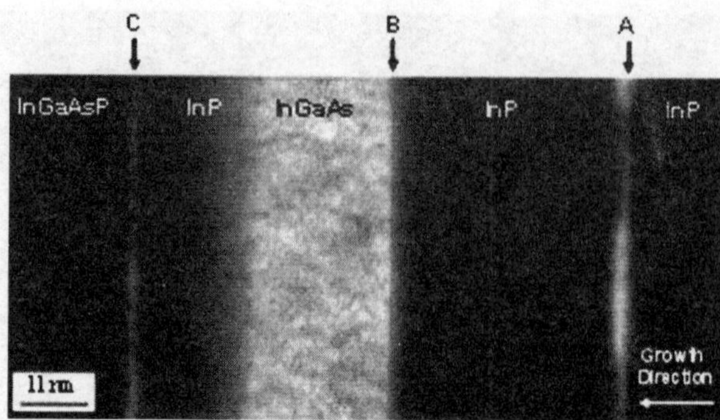

Figure 5.26   Large scale X-STM image of the structure showing the entire layer stack which comprises the InAs dot layers labelled with A, B and C ($V_{sample} = -2.5\,V$, $I_{tunnel} = 100\,pA$). Growth direction: [113], lateral direction: [1$\bar{1}$0].

A number of individual QDs were analysed within each layer in order to extract information relative to its composition, size and shape. Figure 5.27 shows the high-voltage filled states image of a single dot in layer A. Atomic details are resolved in this image, in which the group V elements, i.e. As and P, are imaged. From the homogeneity of the contrast it can be deduced that the QD composition is quite uniform and close to 100% InAs. It should be noticed that these dots are very different from the InAs/GaAs dots studied before. They resemble the shape of a quantum disc, since they have a very small aspect ratio.

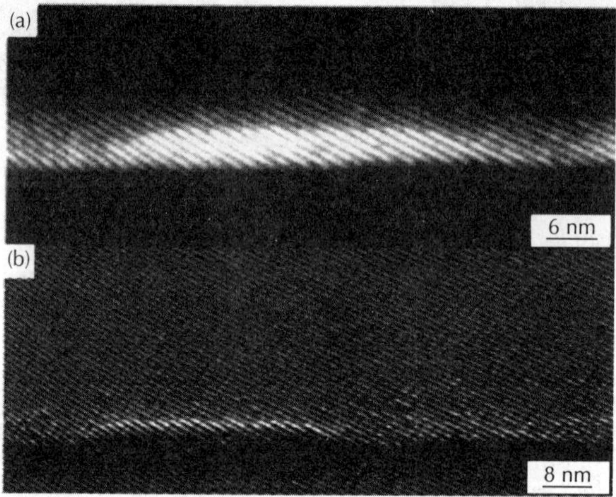

Figure 5.27   (a) X-STM topography image and (b) current image of a single InAs dot and the wetting layer (layer A) capped with InP ($V_{sample} = -3.0\,V$, $I_{tunnel} = 100\,pA$). Both of the images show the low aspect ratio of the dot having a flat top facet. In the current image As atoms can be clearly seen within the P rows of the InP capping layer. Growth direction: [113], lateral direction: [1$\bar{1}$0].

All the observed dots in this layer have a similar truncated pyramidal shape, with a flat top facet. The height and base length distribution of a number of dots showed an average height of $2.9 \pm 0.2\,$nm, and a maximum base length of $32 \pm 2\,$nm. The difference in the outward relaxation of the dots and the WL, as indicated by the brightness in the STM image, is quite small (see Fig. 5.27a). This indicates a high As concentration in the WL. Moreover, the measured magnitude of the outward relaxation of the WL is much higher than that observed in InAs/GaAs WLs [75] (see section 5.2). This is surprising as the compressive strain in the InAs/GaAs case is more than twice that of the InAs/InP system and thus for equally thick wetting layers in both systems a reduction by a factor of more than two in the outward relaxation for the case of InAs/InP should be expected. We explain this by the presence of a very thick WL which contains much more InAs material than the nominal 2.1 ML that were deposited during the dot formation.

The outward relaxation of the WL was again calculated by means of the analytical expression derived in [24], which assumes that the elastic response is linear and isotropic. The WL was modelled including the effect of the asymmetric As profile, which is likely created during the switching between phosphorus to arsenic flux or by As carryover [76]. This can be seen in Fig. 5.27b in which the bright spots in the capping layer correspond to As atoms in the InP matrix. Just as in the case of an InAs WL in GaAs, in which there is an asymmetric In profile due to segregation, we used the phenomenological model of Muraki *et al.* [31] to model the As profile. The total amount of deposited As is determined by fitting the calculated relaxation profile to the measured one. The result of this fitting and the corresponding As profile are shown in Fig. 5.28. We obtained a total amount of InAs of 4 ML. The latter is almost twice the nominal value of deposited InAs. The origin of this extra InAs is the As/P exchange reaction at the InAs/InP interface during the dot formation process. During the growth interrupts used before and after the dot formation process, the structure is kept under a As flux promoting the exchange of P by As and thus increasing the amount of deposited InAs. Such exchange has also been reported to give rise to the formation of InAs QDs and InAs quantum wires on InP surfaces where only a growth interrupt was used under an As flux without the additional In deposition [77, 78]. The large amount of InAs in the wetting layer is supported by the distribution of As in the WL obtained by directly counting the As atoms in the X-STM images. In Fig. 5.28 we plot the number of counted As atoms as a function of the distance in the growth direction. This method is accurate for As concentrations lower than about 25%, because above those values it becomes complicated to distinguish individual As atoms in InP. The profile based on the counting is shown in Fig. 5.28 together with the profile that was used to fit the outward relaxation. The agreement is quite good in the range of validity of the counting method.

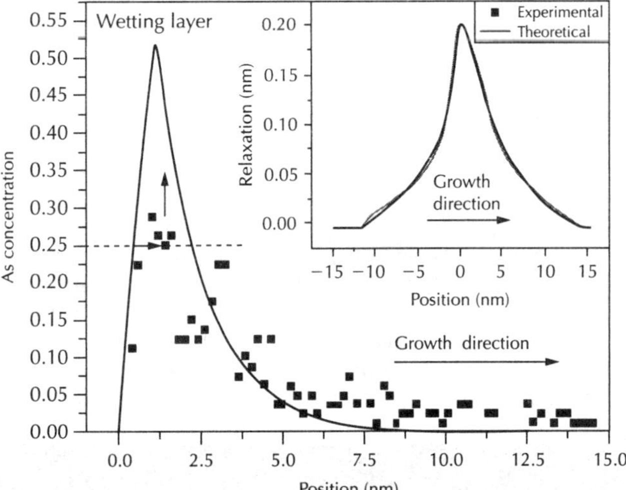

Figure 5.28   Distribution profile of As atoms from the InAs WL (layer A). Above a concentration of 25%, atom counting is not reliable and only a lower limit of the real value is obtained. The inset depicts the measured (dotted line) and the calculated (solid line) outward relaxation of the InAs wetting layer based on the As profile that is indicated by the solid line in the main figure.

The small height of the dots capped with InP suggests As/P exchange during the capping process. This effect should be eliminated by using InGaAs as the capping material and therefore lattice-matched $In_{0.53}Ga_{0.47}As$ was used in the second QD layer. Two dots in this layer are shown in Fig. 5.29. The height and base length distribution of a number of single dots was again investigated, giving an average height of 3.5 ± 0.2 nm and a maximum length of 29 ± 2 nm. These dots are on average 0.6 nm higher than those capped with InP, which corresponds to ~ 2 ML of InAs. This indicates that the height of the dots was reduced by ~ 2 ML due to dot decomposition induced by As/P exchange during InP capping. This value is in good agreement with the 2 ML reduction of the In(As,P)/InP quantum well width reported in [79]. The top facet of the dots capped with InGaAs is less well defined and is more curved than that of the dots capped with InP (compare Figs. 5.27 and 5.29). This is likely due to composition modulation in the capping layer, which we analyse next.

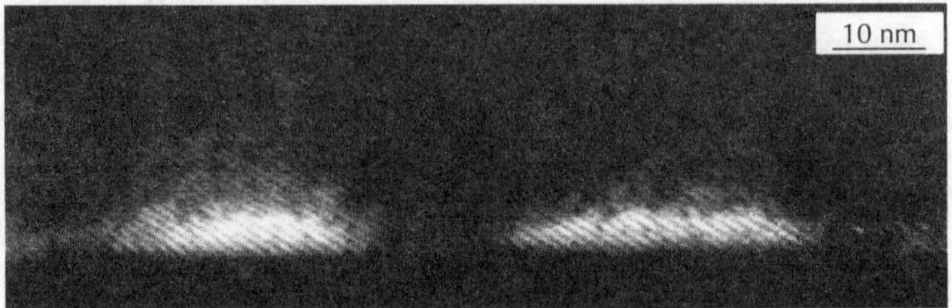

Figure 5.29　X-STM topography image of two adjacent InAs dots (in layer B) capped with InGaAs, ($V_{sample} = -3.0$ V, $I_{tunnel} = 100$ pA). The image reveals the presence of a strain induced lateral phase separation in the InGaAs capping layer. Growth direction: [113], lateral direction: [1$\bar{1}$0].

The inhomogeneous topographic contrast in the InGaAs layer (Fig. 5.29) reveals the presence of an inhomogeneous strain distribution, which must be due to the presence of In-rich (brighter) and Ga-rich (darker) regions. This phase separation is a strain-driven process in which the In adatoms on the growth surface migrate towards the regions on top of the dots to minimize the strain, creating a columnar-like In-rich region above the dots. This process has been observed in columnar InGaAs QDs grown on GaAs [81], as well as in InAs/GaAs QDs, where capping with InGaAs has been shown to induce an increase in the dot size [68]. In our case, the phase separation directly affects the QDs by creating a rough top interface in which the In content decreases gradually.

In the third layer, a lattice-matched InGaAsP alloy was used as the capping material (Fig. 5.30). The average dot height was 3.4 ± 0.2 nm, and the maximum measured base length 27 ± 2 nm. As in the case of InGaAs, the height of the dots is higher as those capped with InP where the dot is partially dissolved due to the As/P exchange. This means that, although there is phosphorous in the capping layer, there is no dot height reduction due to As/P exchange. The shape resembles that of a truncated pyramid, with a flat top interface. More remarkable is the fact that the phase separation in the capping layer is much weaker than in the InGaAs. We think that the Ga–P bond strength (54.9 kcal/mol) which is stronger than the In–As bond strength (48.0 kcal/mol) and Ga–As bond strength (50.1 kcal/mol) limits the phase segregation in InGaAsP as compared to InGaAs. The resulting weak composition modulation does not affect the dot shape or size, as is evidenced by the well-defined dot facets. Indeed, this material has been successfully used as capping layer in InAs/InP (311)B QD lasers emitting near 1.55 μm with low threshold current density [67].

Figure 5.30　X-STM image of two InAs dots (in layer C) capped with InGaAsP alloy showing a flat base and top ($V_{sample} = -2.5$ V, $I_{tunnel} = 100$ pA). Growth direction: [113], lateral direction: [1$\bar{1}$0].

In summary, X-STM has been used to analyse at the atomic scale the effect of different lattice-matched capping materials on the structural properties of self-assembled InAs/InP (311)B QDs. The As/P exchange on the InAs/InP interface during the growth interrupts is shown to increase the amount of InAs in the wetting layers by ~ 2 ML. The As/P exchange takes place also on the dot surface when the QDs are capped with InP, reducing the dot height by about 2 ML. This phenomenon can be avoided by using InGaAs as the capping material, but in that case a strong strain-driven phase separation appears, giving rise to In-rich regions above the dots and a degradation of the dot interface. If the quaternary alloy InGaAsP is used instead of InGaAs, the phase separation is much weaker and well-defined interfaces are obtained.

Nevertheless, the structural differences between InAs/InP QDs capped with different lattice-matched materials are much smaller than those found in the InAs/GaAs system when differences in strain are involved. This indicates that the dot/capping layer strain plays an important role during the capping process.

5.4.3   Double capping process

We have shown in the previous section how the As/P exchange reaction can modify the InAs/InP QDs. Indeed, this effect can be intentionally used to control the height of the QDs by using a modified capping procedure which we analyse in this section. One of the main potential applications of self-assembled InAs/InP QDs is as the active region of high-performance lasers emitting at $1.55\,\mu m$ [65, 67, 82, 83], the long-range optical fibre communication wavelength. For that purpose, an accurate control of the emission wavelength and of the QD size distribution is of crucial importance. One method to reduce the QD height dispersion is based on a discontinuous capping process [84], where the capping is performed in two sequences. First, a capping layer is deposited, with a thickness smaller than the QD height. Then, a growth interruption under phosphorous flux is performed, during which the uncovered part of the dots disappears due to As/P exchange [65, 85], leading, in principle, to a uniform QD height equal to the thickness of the first thin capping layer. Finally, the capping layer growth is completed. This modified capping procedure, named double cap (DC), was shown to allow wavelength emission control and to drastically decrease the photoluminescence (PL) linewidth [84]. Indeed, room temperature lasers based on double-capped InAs/InP QDs with low threshold current density emitting close to $1.55\,\mu m$ were recently demonstrated on (311)B substrates [67]. On (100) substrates, the DC process was also used for device elaboration by metal organic vapour phase epitaxy (MOVPE) [86]. However, it has been recently observed that the preferential growth of the first cap layer occurs on the edge of the QDs. This should induce an inaccurate control of the QD heights by the DC process in MOVPE [91].

The initial stage of the capping process of InAs QDs is a very complex phenomenon [12, 14, 87, 88], sensitive to many growth parameters [19, 71]. It has mainly been studied for InAs/GaAs QDs, and very little is known in the case of InAs/InP QDs [80, 89–91], in which the As/P exchange reaction complicates the description. It is therefore difficult to know what happens during and after the deposition of the first capping layer in the double capping method, and a detailed structural characterization of the process is required.

We have used X-STM to study the double capping process of InAs/InP QDs grown by MBE at the atomic scale. The effect of first capping layer with different thicknesses and composition has been analysed.

The samples were grown by gas source MBE on an Si-doped (311)B-oriented InP substrate. The growth temperature was set to 480°C. The QDs were formed by depositing 2.1 (100) equivalent MLs of InAs on both InP and $In_{0.87}Ga_{0.13}As_{0.285}P_{0.715}$ (lattice matched to InP) buffer layers. In order to make the analysis easier, a high As beam equivalent pressure was supplied to the surface during the InAs deposition to favour the formation of large QDs [66]. After island formation, a 30 s growth interruption under As flux was performed before the growth of the capping layer. Then, the first capping layer (CL1) was deposited and dot growth was interrupted for 60 s maintaining a group V overpressure. Finally, a thick (50 or 100 nm) second capping layer (CL2) was grown. Three different samples were studied. In sample A, three QD layers were grown with different thicknesses of CL1: 1.5, 2.5 and 3.5 nm (from now on called layer 1, layer 2 and layer 3, respectively). The buffer and the capping layers consisted of $In_{0.87}Ga_{0.13}As_{0.285}P_{0.715}$. In sample B, the buffer, CL1 and CL2 were made of InP. Finally, in sample C, the buffer and CL2 were made of InP, while CL1 consisted of $In_{0.87}Ga_{0.13}As_{0.285}P_{0.715}$. In samples B and C four QD layers were grown with different thicknesses of CL1: 2, 3, 4 and 5 nm (layers 1, 2, 3 and 4, respectively).

First, we present results on sample A, with InGaAsP buffer and capping layers. Figure 5.31 shows a large-scale filled states image of this sample. The three QD layers separated by ~50 nm are observed. The inhomogeneous contrast in the buffer and capping layers is due to the random alloy fluctuations in the InGaAsP. Remarkably, the QDs have a very large base length. The maximum measured base length was 60 nm, which gives aspect ratios as low as 0.04. It is clear from Fig. 5.31 that the height of the QDs increases when the thickness of CL1 increases. Moreover, the height of the QDs in each layer fits very well with the nominal thickness of CL1. After analysing 37 dots in this sample, the measured average height in layer 1 (CL1 = 1.5 nm) was 1.6 nm, in layer 2 (CL1 = 2.5 nm) 2.5 nm, and in layer 3 (CL1 = 3.5 nm) 3.4 nm (see the triangles in Fig. 5.34). This indicates that the double capping procedure is controlling the QD height with high accuracy.

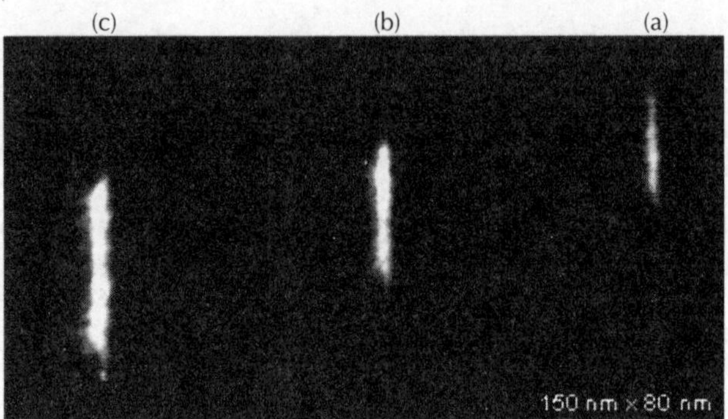

Figure 5.31   Large scale filled states image of sample A showing the three QD layers in which CL1 is 1.5, 2.5 and 3.5 nm thick in (a), (b) and (c), respectively. The growth direction is from right to left. $V_{sample} = -2.5 V$, $I_{tunnel} = 50 pA$.

In sample B, buffer and both capping layers consist of InP and the thickness of CL1 was increased up to 5 nm. Figure 5.32 shows the current image of one dot in layer 2 of this sample. On such specific QDs, shallow trenches in the InP buffer close to the QDs are observed. We assume that the trenches are mainly formed just after the dot formation during the growth interrupt under As pressure. Mass transport from wetting layer to quantum dots leads to the formation of a dewetted InP surface on which As/P exchange reaction occurs. The extra InAs formed migrate to the dots and a trench is formed around the QDs. The fact that these trenches were not present in sample A (see Fig. 5.1), where the dots were grown on InGaAsP, supports this explanation (As/P exchange is reduced on an InGaAsP surface) [90]. Similar observations have been previously reported for InAs QDs grown on (001) InP, and explained in a similar way [92].

Figure 5.32 is representative of what was observed when CL1 was 2 or 3 nm thick. The dots are similar to the ones in sample A: they have a flat top facet and their height fits very well with the thickness of CL1 (the average height was 2.2 and 2.9 nm in layers 1 and 2, respectively). Nevertheless, the situation is different when CL1 is 4 and 5 nm thick. In this case, the measured average heights of the QDs are 3.1 and 4 nm, respectively. Thus the QD heights are not accurately controlled by the first cap layer thickness in this range. These results confirm the PL measurement as a function of the thickness of CL1 previously published [84]. For CL1 thicker than 4 nm, the PL energy control by the DC process is lost.

On sample C, CL1 and CL2 consist of InGaAsP and InP, respectively. The thicknesses of CL1 were set at 2, 3, 4 and 5 nm. Figure 5.33 shows a filled states topography image of layers 2, 3 and 4. Atomic details are resolved in this image, in which the group V elements, i.e. As and P, are imaged. As the QDs are grown on InP, the trenches are present again, but it is clear from image 5.33c that it is a global effect affecting the whole WL. CL1 appears as a dark region with inhomogeneous contrast between the WL and CL2. A bright region in CL1 on top of the QDs is observed in layers 3 and 4, indicating an As- and/or In-rich region. This is likely due to strain induced composition modulation, similar to what was found in the previous section for InGaAs capping [90]. The thickness of CL1 in layers 2, 3 and 4, measured from the top of the WL, is 2.9, 3.8, 4.8 nm, respectively.

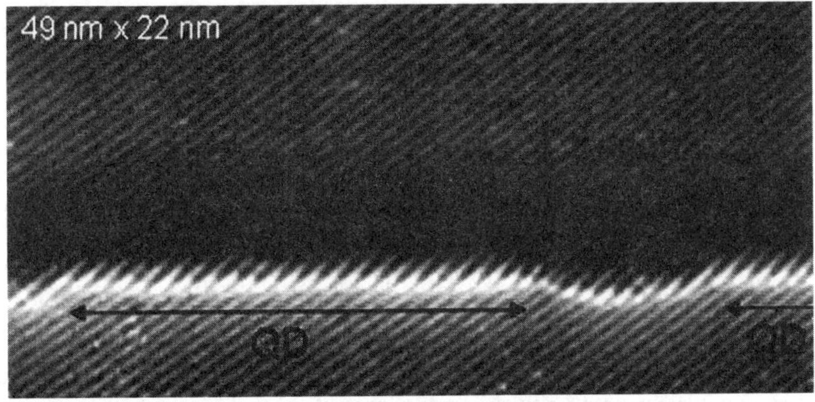

Figure 5.32 Current image of a QD in sample B corresponding to layer 2 (CL1 = 3 nm). Part of a second QD appears on the right and a shallow trench is clearly observed between the two QDs. $V_{sample} = -2.5\,V$, $I_{tunnel} = 52\,pA$.

This is in good agreement with the nominal values. Figure 5.33a shows how in layer 2 the height of the QD is levelled to the thickness of CL1. This is also the case in layer 1 (not shown). As in the previous sample, when the thickness of CL1 is increased to 4 nm, the height of the QDs is slightly smaller than CL1 (see Fig. 5.33b). This difference is much bigger when CL1 is 5 nm thick. In this case the height of the QD is more or less half of the thickness of CL1, as shown in Fig. 5.33c. Moreover, a strong decrease of the QD density is observed in this layer. Complete island dissolution, as previously observed for large InGaAs QDs on GaAs, may be effective in this case [93]. Complementary experiments to clarify this question are still in progress.

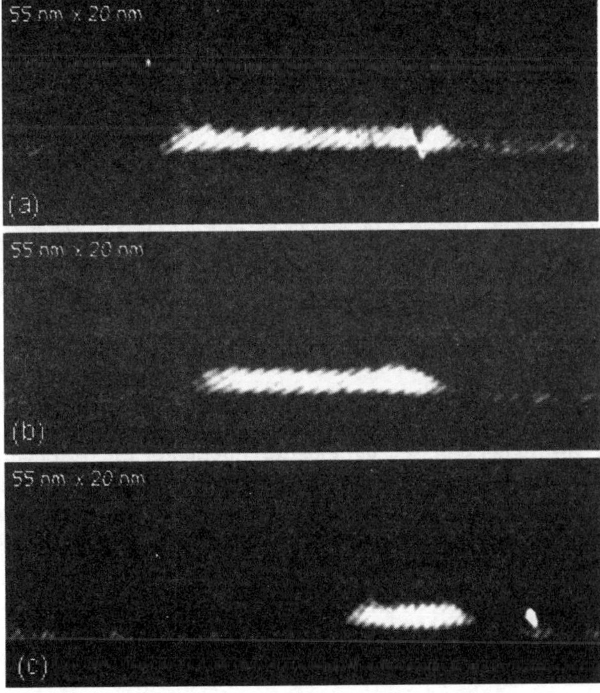

Figure 5.33 Topography image of three QDs in sample C corresponding to (a) layer 1 (CL1 = 2 nm), (b) layer 3 (CL1 = 4 nm) and (c) layer 4 (CL1 = 5 nm). CL1 appears as a dark region between the WL and CL2. The bright spots are As atoms in the P matrix. $V_{sample} = -3.0\,V$, $I_{tunnel} = 53\,pA$.

All the results are summarized in Fig. 5.34, in which the average height of the QDs in each layer is plotted as a function of the nominal thickness of CL1 for the three analysed samples. The solid curve represents the ideal situation where the dot height corresponds exactly to the thickness of CL1. The double capping method works well for first capping layers up to ∼3.5 nm, independently of the materials used. This is very important because the desired QD height for 1.55 μm applications (∼2.7 nm) lies in that range. For thicker layers, a reduction in the dot height compared to the thickness of the first capping layer is observed, indicating that the control over the QD height with the DC process is lost for capping layers thicker than ∼3.5 nm. This can be understood in terms of a transition from a double capping to a classical (one-step) capping process when the first capping layer is thick enough to completely cover the dots. Once the dots are covered, the growth interrupt has no effect on them. Since the height of similar classically capped QDs has been measured in the previous section to be around 3.5 nm, that should be the maximum possible height obtained and the DC method should not work for CL1 thicker than that. Indeed, 3.5 nm is approximately the saturation value observed for thick capping layers if Fig. 5.34 (except for the special case of 5 nm in sample C, in which a strong decrease of the QD density is also observed).

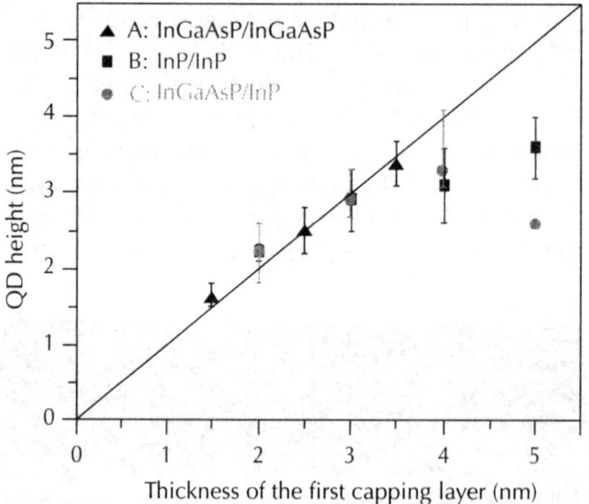

Figure 5.34   Average QD height in each layer as a function of the nominal thickness of CL1 for the three analysed samples. The error is taken as the distance to the maximum and minimum values measured. The solid curve represents the ideal situation where the dot height corresponds exactly to the thickness of CL1. A significant deviation is observed for thick first capping layers.

In summary, our X-STM measurements show that the double capping method applied to InAs/ InP (311)B QDs grown by MBE allows a very accurate control of the QD height when the thickness of the first capping layer is not higher than ∼3.5 nm, independently of the material used in the capping layer. For higher values a significant reduction of the QD height is observed. This is explained by the fact that 3.5 nm is the height of classically capped QDs and therefore for thicker capping layers the double capping process becomes a classical capping process since the QDs would be completely covered by the capping layer before the growth interrupt.

5.5   Conclusions

X-STM has been used to study at the atomic scale the formation of self-assembled QDs. This technique allows determining the size, shape and composition of the buried nanostructures and provides vital information to understand the QD formation process.

X-STM measurements show that the size and composition profile of InAs self-assembled quantum dots depend strongly on the substrate on which the dots are formed. In particular, InAs QDs grown on GaAs are larger than those grown on AlAs, and the In gradients inside the dot from bottom to top have opposite signs (increasing in InAs/GaAs QDs and decreasing in InAs/AlAs QDs) [73]. These differences are mainly due to the reduced mobility of In in AlAs due to the stronger Al-In bond strength.

The capping process is found to be a critical step in QD formation because it strongly modifies the QD structure. In its initial stage, GaAs capping induces leveling of the QDs to drastically decrease their height. During continuous capping the QD leveling is quenched when the QDs become buried. Both phenomena, driving and quenching of the QD leveling take place on a similar time scale and are very sensitive to the GaAs growth rate and growth temperature (decreasing the capping temperature can strongly reduce dot decomposition) [19]. This dot decomposition during capping can also be modified by using different capping materials [94]. It is strongly reduced when the dots are capped with $In_{0.15}Ga_{0.85}As$ and completely suppressed when $GaAs_{0.75}Sb_{0.25}$ is used [95]. The result is that QDs capped with these strain reducing layers are much taller than GaAs-capped QDs. This suggests that the dot/capping layer strain plays a mayor role in inducing dot decomposition during capping. When the different capping materials are lattice matched to the substrate, the differences in the QD properties are more subtle and can be dominated by chemical effects: InAs/InP QDs capped with InP have a 2 MLs smaller height than those capped with InGaAs or InGaAsP due to As/P exchange induced decomposition [90]. The effect of the As/P exchange can be intentionally used to control the QD height by using a double capping method. We found that the double capping method allows a very accurate control of the QD height when the thickness of the first capping layer is not higher than ~3.5 nm. For higher values a significant reduction of the QD height is observed. These results are interpreted in terms of a transition from a double capping to a classical capping process when the first capping layer is thick enough to completely cover the dots [96].

Acknowledgements

The authors would like to specially thank the growers of the analysed samples: Q. Gong and R. Nötzel, COBRA-Department of Applied Physics, Eindhoven University of Technology, The Netherlands; A. Letoublon and N. Bertru, Laboratoire d'études des nanostructures semiconductrices (LENS), INSA-Rennes, France; M.J. Steer, H.Y. Liu, and M. Hopkinson, Department of Electronic and Electrical Engineering, University of Sheffield, UK; K. Pierz, Physikalisch-Technische Bundesanstalt, Germany. The authors would also like to thank John H. Davies for help with the strain calculations and Peter Maxime and Marvin Roy for help with the ABAQUS software. Thanks also to C. Çelebi, I. Drouzas, A. Simon, and E. Gapihan for helping with the measurements and the interpretation of the data and to D.J. Mowbray for helpful discussion. This work has been supported by the European Union through the SANDiE Network of Excellence (contract No. NMP4-CT-2004-500101).

References

1. D. Bimberg, M. Grundmann, and N.N. Ledentsov, *Quantum Dot Heterostructures* (Wiley, Chichester, UK, 1999).
2. M.S. Skolnick and D.J. Mowbray, *Annu. Rev. Mater. Res.* **34**, 181 (2004).
3. S. Fafard, K. Hinzer, S. Raymond, M. Dion, J. McCaffrey, Y. Feng, and S. Charbonneau, *Science* **274**, 1350 (1996).
4. F. Heinrichsdorff, M.-H. Mao, N. Kirstaedter, A. Krost, D. Bimberg, A.O. Kosogov, and P. Werner, *Appl. Phys. Lett.* **71**, 22 (1997).
5. K. Yano, T. Ishii, T. Hashimoto, T. Kobayashi, F. Murai, and K. Seki, *IEEE Trans. Electron. Devices* **41**, 1628 (1994).
6. S.Y. Wang, S.D. Lin, H.W. Wu, and C.P. Lee, *Appl. Phys. Lett.* **78**, 1023 (2001).
7. A.D. Stiff, S. Krishna, P. Bhattacharya, and S.P. Kennerly, *Appl. Phys. Lett.* **79**, 421 (2001).

8. P. Michler, A. Kiraz, C. Becher, W.V. Schoenfeld, P.M. Petroff, L. Zhang, E. Hu, and A. Imamoğlu, *Science* **290**, 2282 (2000).
9. Z. Yuan, B.E. Kardynal, R.M. Stevenson, A.J. Shields, C.J. Lobo, K. Cooper, N.S. Beattie, D.A. Ritchie, and M. Pepper, *Science* **295**, 102 (2002).
10. D.J. Eaglesham and M. Cerullo, *Phys. Rev. Lett.* **64**, 1943 (1990).
11. J. Stangl, V. Holý, and G. Bauer, *Rev. Mod. Physics* **76**, 725 (2004).
12. Q. Xie, P. Chen, and A. Madhukar, *Appl. Phys. Lett.* **65**, 2051 (1994).
13. X.W. Lin, J. Washburn, Z. Liliental-Weber, E.R. Weber, A. Sasaki, A. Wakahara, and Y. Nabetani, *Appl. Phys. Lett.* **65**, 1677 (1994).
14. J.M. Garcia, G. Medeiros-Ribeiro, K. Schmidt, T. Ngo, J.L. Feng, A. Lorke, J. Kotthaus, and P.M. Petroff, *Appl. Phys. Lett.* **71**, 2014 (1997).
15. G.D. Lian, J. Yuan, L.M. Brown, G.H. Kim, and D.A. Ritchie, *Appl. Phys. Lett.* **73**, 49 (1998).
16. P.D. Siverns, S. Malik, G. McPherson, D. Childs, C. Roberts, R. Murray, and B.A. Joyce, *Phys. Rev. B.* **58**, R10127 (1998).
17. R. Leon, J. Wellman, X.Z. Liao, J. Zou, and D.J.H. Cockayne, *Appl. Phys. Lett.* **76**, 1558 (2000).
18. P.B. Joyce, T.J. Krzyzewski, P.H. Steans, G.R. Bell, J.H. Neave, and T.S. Jones, *Surf. Sci.* **492**, 345 (2001).
19. Q. Gong, P. Offermans, R. Nötzel, P.M. Koenraad, and J.H. Wolter, *Appl. Phys. Lett.* **85**, 5697 (2004).
20. G. Costantini, A. Rastelli, C. Manzano, P. Acosta-Diaz, R. Songmuang, G. Katsaros, O.G. Schmidt, and K. Kern, *Phys. Rev. Lett.* **96**, 226106-1 (2006).
21. G. Binning, H. Rohrer, Ch. Gerber, and E. Weibel, *Appl. Phys. Lett.* **40**, 178 (1982).
22. R.M. Feenstra, *Physica B.* **273**, 796 (1999).
23. D.M. Bruls, P.M. Koenraad, M. Hopkinson, J.H. Wolter, and H.W.M. Salemink, *Appl. Surf. Sci.* **190**, 258 (2002).
24. J.H. Davies, D.M. Bruls, J.W.A.M. Vugs, and P.M. Koenraad, *J. Appl. Phys.* **91**, 4171 (2002).
25. K. Sautho, M. Wenderoth, A.J. Heinrich, M.A. Rosentreter, K.J. Engel, T.C.G. Reusch, and R.G. Ulbrich, *Phys. Rev. B.* **60**, 4789 (1999).
26. M.A. Migliorato, A.G. Cullis, M. Fearn, and J.H. Jefferson, *Phys. Rev. B.* **65**, 115316 (2002).
27. M. Schowalter, A. Rosenauer, D. Gerthsen, M. Arzberger, M. Bichler, and G. Abstreiter, *App. Phys. Lett.* **79**, 4426 (2001).
28. A. Rosenauer, D. Gerthsen, D.V. Dyck, M. Arzberger, G. Böhm, and G. Abstreiter, *Phys. Rev. B.* **64**, 245334 (2001).
29. J.P. Silveira, J.M. García, and F. Briones, *Appl. Surf. Sci.* **188**, 75 (2002).
30. J.M. Moison, C. Guille, F. Houzay, F. Barthe, and N.V. Rompay, *Phys. Rev. B.* **40**, 6149 (1989).
31. K. Muraki, S. Fukatsu, Y. Shiraki, and R. Ito, *Appl. Phys. Lett.* **61**, 557 (1992).
32. N. Grandjean, J. Massies, and M. Leroux, *Phys. Rev. B.* **53**, 998 (1996).
33. O. Dehaese, X. Wallart, and F. Mollot, *Appl. Phys. Lett.* **66**, 52 (1995).
34. A.G. Cullis, D.J. Norris, T. Walther, M.A. Migliorato, and M. Hopkinson, *Phys. Rev. B.* **66**, 081305(R) (2002).
35. A. Lenz, R. Timm, H. Eisele, C. Hennig, S.K. Becker, R.L. Sellin, U.W. Pohl, D. Bimberg, and M. Dähne, *Appl. Phys. Lett.* **81**, 5150 (2002).
36. K. Pierz, Z. Ma, U.F. Keyser, and R.J. Haug, *J. Cryst. Growth* **249**, 477 (2003).
37. P. Offermans, P. Koenraad, J. Wolter, K. Pierz, M. Roy, and P. Maksym, *Physica E.* **26**, 236 (2005).
38. Y.C. Kao, F.G. Celli, and H.Y. Liu, *J. Vac. Sci. Technol. B.* **11**, 1023 (1993).
39. D. Granados and J.M. García, *Appl. Phys. Lett.* **82**, 2401 (2003).
40. L.G. Wang, P. Kratzer, M. Scheffler, and Q.K.K. Liu, *Appl. Phys. A.* **73**, 161 (2001).
41. Z.R. Wasilewski, S. Fafard, and J.P. McCaffrey, *J. Cryst. Grouth* **201/202**, 1131 (1999).
42. E. Steimetz, T. Wehnert, H. Kirmse, F. Poser, J.-T. Zettler, W. Neumann, and W. Richter, *J. Cryst. Grouth* **221**, 592 (2000).
43. A. Lenz, H. Eisele, R. Timm, S.K. Becker, R.L. Sellin, U.W. Pohl, D. Bimberg, and M. Dähne, *Appl. Phys. Lett.* **85**, 3848 (2004).
44. I. Hapke-Wurst *et al.*, *Semicond. Sci. Technol.* **14**, L41–L43 (1999).
45. K. Pierz, Z. Ma, U.F. Keyser, and R.J. Haug, *J. Cryst. Growth* **249**, 477–482 (2003).
46. P. Dawson, Z. Ma, K. Pierz, and E.O. Göbel, *Appl. Phys. Lett.* **81**, 2349 (2002).

47. P.W. Fry *et al.*, *Phys. Rev. Lett.* **84**, 733 (2000).

48. D.M. Bruls, J.W.A.M. Vugs, P.M. Koenraad, H.W.M. Salemink, J.H. Wolter, M. Hopkinson, M.S. Skolnick, F. Long, and S.P.A. Gill, *Appl. Phys. Lett.* **81**, 1708 (2002).

49. N. Liu, J. Tersoff, O. Baklenov, A.L. holmes Jr, and C.K. Shih, *Phys. Rev. Lett.* **84**, 334 (2000).

50. O. Wolst, M. Kahl, M. Schardt, S. Malzer, and G.H. Dohler, *Phys. E.* **17**, 554 (2003).

51. A. Lenz *et al.*, *Appl. Phys. Lett.* **81**, 5150 (2002).

52. P. Ballet, J.B. Smathers, and G.J. Salamo, *Appl. Phys. Lett.* **75**, 337 (1999).

53. P.B. Joyce, J. Krzyzewski, R. Bell, and S. Jones, *Appl. Phys. Lett.* **79**, 3615 (2001).

54. Q. Gong, R. Nötzel, J. Hamhuis, J. Eijkemans, and H. Wolter, *Appl. Phys. Lett.* **81**, 1887 (2002).

55. This method subtracts the mean value of the nearest neighbour pixels within the filter window:
$$O(x, y) = I(x, y) - \frac{1}{M \times N} \sum_{i=-N/2}^{N/2} \sum_{j=-M/2}^{M/2} I(x + i, y + j).$$ By choosing a size of about $1 \times 1\,\mathrm{nm}^2$ for the filter window, long range fluctuations in the image are suppressed while atomic details are preserved.

56. A. Rastelli, E. Müller, and H. von Känel, *Appl. Phys. Lett.* **80**, 1438 (2002).

57. M. Stoffel, U. Denker, S. Kar, H. Sigg, and G. Schmidt, *Appl. Phys. Lett.* **83**, 2910 (2003).

58. P.B. Joyce, J. Krzyzewski, R. Bell, A. Joyce, and S. Jones, *Phys. Rev. B.* **58**, R15981 (1998).

59. V.M. Ustinov and A.E. Zhukov, *Semicond. Sci. Technol.* **15**, R41 (2000).

60. K. Nishi, H. Saito, and S. Sugou, *Appl. Phys. Lett.* **74**, 1111 (1999).

61. E.C. Le Ru, P. Howe, T.S. Jones, and R. Murray, *Phys. Status Solidi C.* **0**, 1221 (2003).

62. H.Y. Liu, M. Hopkinson, C.N. Harrison, M.J. Steer, R. Frith, I.R. Sellers, D.J. Mowbray, and M.S. Skolnick, *J. Appl. Phys.* **93**, 2931 (2003).

63. H.Y. Liu, M.J. Steer, T.J. Badcock, D.J. Mowbray, M.S. Skolnick, P. Navaretti, K.M. Groom, M. Hopkinson, and R.A. Hogg, *Appl. Phys. Lett.* **86**, 143108 (2005).

64. J.M. Ripalda, D. Granados, Y. Gonzalez, A.M. Sanchez, S.I. Molina, and J.M. Garcia, *Appl. Phys. Lett.* **87**, 202108 (2005).

65. S. Frechengues, N. Bertru, V. Drouot, B. Lambert, S. Robinet, S. Loualiche, D. Lacombe, and A. Ponchet, *Appl. Phys. Lett.* **74**, 3356 (1999).

66. P. Caroff, N. Bertru, A. Le Corre, O. Dehaese, T. Rohel, I. Alghoraibi, H. Folliot, and S. Loualiche, *Jpn. J. Appl. Phys.* **44**, L1069 (2005).

67. P. Caroff, C. Paranthoën, C. Platz, O. Dehaese, H. Folliot, N. Bertru, C. Labbé, R. Piron, E. Homeyer, A. Le Corre, and S. Loualiche, *Appl. Phys. Lett.* **87**, 243107 (2005).

68. M.V. Maximov, A.F. Tsatsul'nikov, B.V. Volovik, D.S. Sizov, Yu.M. Shernyakov, I.N. Kaiander, A.E. Zhukov, A.R. Kovsh, S.S. Mikhrin, V.M. Ustinov, Zh.I. Alferov, R. Heitz, V.A. Shchukin, N.N. Ledentsov, D. Bimberg, Yu.G. Musikhin, and W. Neumann, *Phys. Rev. B.* **62**, 16671 (2000).

69. K. Akahane, N. Yamamoto, and N. Ohtani, *Physica E Amsterdam* **21**, 295 (2004).

70. H.Y. Liu, M.J. Steer, T.J. Badcock, D.J. Mowbray, M.S. Skolnick, F. Suarez, J.S. Ng, M. Hopkinson, and J.P.R. David, *J. Appl. Phys.* **99**, 046104 (2006).

71. R. Songmuang, S. Kiravittaya, and O.G. Schmidt, *J. Cryst. Growth* **249**, 416 (2003).

72. W.M. McGee, T.J. Krzyzewski, and T.S. Jones, *J. Appl. Phys.* **99**, 043505 (2006).

73. P. Offermans, P.M. Koenraad, J.H. Wolter, K. Pierz, M. Roy, and P.A. Maksym, *Phys. Rev. B.* **72**, 165332 (2005).

74. G. Costantini, A. Rastelli, C. Manzano, P. Acosta-Diaz, R. Songmuang, G. Katsaros, O.G. Schmidt, and K. Kern, *Phys. Rev. Lett.* **96**, 226106 (2006).

75. P. Offermans, P.M. Koenraad, R. Nötzel, J.H. Wolter, and K. Pierz, *Appl. Phys. Lett.* **87**, 111903 (2005).

76. W. Seifert, D. Hessman, X. Liu, and L. Samuelson, *J. Appl. Phys.* **75**, 1501 (1994).

77. H. Yang, P. Ballet, and G.J. Salamo, *J. Appl. Phys.* **89**, 7871 (2001).

78. B. Wang, F. Zhao, Y. Peng, Z. Jin, Y. Li, and S. Liu, *Appl. Phys. Lett.* **72**, 2433 (1998).

79. P. Disseix, C. Payen, J. Leymarie, and A. Vasson, *J. Appl. Phys.* **88**, 4612 (2000).

80. P.J. Poole, R.L. Williams, J. Lefebvre, and S. Moisa, *J. Cryst. Growth* **257**, 89 (2003).

81. J. He, R. Nötzel, P. Offermans, P.M. Koenraad, Q. Gong, G.J. Hamhuis, T.J. Eijkemans, and J.H. Wolter, *Appl. Phys. Lett.* **85**, 2771 (2004).

82. S. Fafard, Z. Wasilewski, J. McCaffrey, S. Raymond, and S. Charbonneau, *Appl. Phys. Lett.* **68**, 991 (1996).

83. H. Marchand, P. Desjardins, S. Guillon, J.-E. Paultre, Z. Bougrioua, R.Y.-F. Yip, and R.A. Masut, *Appl. Phys. Lett.* **71**, 527 (1997).
84. C. Paranthoën, N. Bertru, O. Dehaese, A. Le Corre, S. Loualiche, B. Lambert, and G. Patriarche, *Appl. Phys. Lett.* **78**, 1751 (2001).
85. K. Ozasa, Y. Aoyagi, Y.J. Park, and L. Samuelson, *Appl. Phys. Lett.* **71**, 797 (1997).
86. Y. Sakuma, M. Takeguchi, K. Takemoto, S. Hirose, T. Usuki, and N. Yokoyama, *J. Vac. Sci. Technol. B.* **23**, 1741 (2005).
87. N.N. Ledenstov, V.A. Shchukin, M. Grundmann, N. Kirstaedter, J. Böhrer, O. Schmidt, D. Bimberg, V.M. Ustinov, A.Yu. Egorov, A.E. Zhukov, P.S. Kop'ev, S.V. Zaitsev, N.Yu. Gordeev, Zh.I. Alferov, A.I. Borovkov, A.O. Kosogov, S.S. Ruvimov, P. Werner, U. Gösele, and J. Heydenreich, *Phys. Rev. B.* **54**, 8743 (1996).
88. Z.R. Wasilewski, S. Farad, and J.P. McCaffrey, *J. Cryst. Growth* **201**, 1131 (1999).
89. T. Raz, D. Ritter, and G. Bahir, *Appl. Phys. Lett.* **82**, 1706 (2003).
90. C. Çelebi, J.M. Ulloa, P.M. Koenraad, A. Simon, A. Letoublon, and N. Bertru, *Appl. Phys. Lett.* **89**, 023119 (2006).
91. G. Saint-Girons, G. Patriarche, A. Michon, G. Beaudoin, I. Sagnes, K. Smaali, and M. Troyon, *Appl. Phys. Lett.* **89**, 031923 (2006).
92. H.R. Gutiérrez, M.A. Cotta, J.R.R. Bortoleto, and M.M.G. de Carvalho, *J. Appl. Phys.* **92**, 7523 (2002).
93. J.-S. Lee, H.-W. Ren, S. Sugou, and Y. Masumoto, *J. Appl. Phys.* **84**, 6686 (1998).
94. J.M. Ulloa, C. Çelebi, P.M. Koenraad, A. Simon, E. Gapihan, A. Letoublon, N. Bertru, J. Drouzas, D.J. Mowbray, M.J. Steer and M. Hopkinson, *J. Appl. Phys*, **101**, 081707 (2007).
95. J.M. Ulloa, I.W.D. Drouzas, P.M. Koenraad, D.J. Mowbray, M.J. Steer, H.Y. Liu and M. Hopkinson. *Appl. Phys. Lett.* **90**, 213105 (2007).
96 J.M. Ulloa, P.M. Koenraad, E. Gapihan, A. Létoublon and N. Bertru, *Appl. Phys. Lett.* **91**, 073106 (2007).

CHAPTER 6

Growth and Characterization of Structural and Optical Properties of Polar and Non-polar GaN Quantum Dots

B. Gayral and B. Daudin

CEA-CNRS group "Nanophysique et Semiconducteurs", CEA-Grenoble, INAC/SP2M, 17 rue des Martyrs, 38054 Grenoble Cedex 9, France

6.1 Introduction

GaN has become a recent member of the quantum dot-forming semiconductor materials, as a colateral consequence of the huge effort devoted during the past ten years to III-N semiconductor materials, in relation to their potentialities for light emission/detection in a wavelength range spanning from IR to UV.

More generally, the interest in GaN and GaInN quantum dots (QDs) is motivated by several features specific to the nitride family. First of all, and although the situation is rapidly changing due to a continuous flux of newcomers in the business of bulk, self-standing, GaN and AlN wafers production, the high cost of these products means that most nitride devices are presently grown on sapphire or SiC. Due to the lattice mismatch, nitride thick layers and heterostructures grown on these substrates usually contain about 10^7 to 10^{10} dislocations/cm^2, depending on growth technique. Despite such a large density of defects, it will be shown that GaN QDs, due to their reduced size, are free of structural defects, contrary to quantum wells (QWs), and consequently exhibit remarkable optical properties. At this stage, it should be noted that the high efficiency, demonstrated since the mid-1990s, of light-emitting diodes (LEDs) and laser diodes (LDs) with InGaN QWs in the active layer is far from being fully understood. The origin of such an intense luminescence is currently assigned to the presence of localization centres in InGaN QWs, the nature of which is still controversial [1]. Whatever the nature of this localization is, the experimental evidence is that quantum dot-like centres are present in the InGaN QWs and that they are responsible for both a reduced diffusion of carriers to non-radiative centres and an enhanced radiative recombination, further motivating an interest in *intentional* growth of InGaN QDs for a better control of LED optical properties.

One specificity of nitrides is the large band gap difference between InN (0.7 eV), GaN (3.47 eV) and AlN (6.2 eV). As a consequence, carrier confinement in GaN (GaInN) QDs embedded in AlN or in AlGaN (GaN) is expected to be efficient at room temperature, with no significant thermal escape, which potentially opens the way to QD-based practical devices.

As extensively reviewed in Chapter 7, an additional specificity of nitride semiconductors with respect to other families is the presence of an internal electric field along the [0001] polar direction, as a result of the non-centrosymmetric wurtzite structure. This electric field, which is a combination of spontaneous and piezoelectric polarization, may be as high as 9–10 MV/cm in the case of GaN/AlN heterostructures [2]. As a consequence, both a red shift of the photoluminescence (PL) and a very long PL decay time resulting from field-induced carrier separation limit the efficiency of practical devices. Non-polar nitride heterostructures are a current option to overcome these difficulties: for this purpose, m-plane or r-plane sapphire as well as $[11\bar{2}0]$ and $[1\bar{1}00]$ SiC are used to grow non-polar $[11\bar{2}0]$ and $[1\bar{1}00]$ GaN and GaN/AlGaN heterostructures.

It is our aim to review here the growth and structural characterization of $[0001]$ and $[11\bar{2}0]$ GaN QDs. As a general overview of nitride QDs optical properties is given in Chapter 7, the present one will be rather focused on single dot spectroscopy, with the final prospect of providing the interested reader with a review of the state of the art on GaN QDs, by combining Chapters 6 and 7.

6.2   Epitaxial growth of nitrides

6.2.1   Generalities

Up to very recently, the lack of bulk nitride substrates, which has often been invoked as a serious limitation for the structural quality of the c-plane epilayers, is a particularly relevant issue for the case of the a-plane ones. To date, very few publications report on successful growth of bulk a-plane nitride layers exhibiting reasonably good structural quality. In a recent article, however, Paskova *et al.* managed to grow low-defect-density homoepitaxial a-plane GaN on bulk a-plane GaN substrates, exhibiting a dislocation density in the $10^5\,\mathrm{cm}^{-2}$ range. These substrates, originating from Kyma technologies, were obtained by slicing a c-plane HVPE-grown GaN boule parallel to the $[11\bar{2}0]$ plane [3]. However, this type of substrate, although already available to some extent, is not yet widely used by the GaN community, and the vast majority of groups are still using pseudosubstrates, e.g. substrates with a nature different from that of the epitaxial layer, to perform growth.

Two types of pseudosubstrates are currently used for the heteroepitaxial growth of a-plane GaN or AlN layers: $[1\bar{1}02]$ r-plane sapphire [4–10] and $[11\bar{2}0]$ a-plane SiC [11–14]. This type of growth results in a high density of structural defects such as stacking faults ($\sim 10^5\,\mathrm{cm}^{-1}$) and dislocations ($\sim 10^9\,\mathrm{cm}^{-2}$) [3, 4, 14], an issue that has been partially solved by lateral epitaxial growth techniques employing SiO_2 stripe patterns [15]. To date, a-plane GaN QDs have been achieved only on a-plane SiC; this is why we will focus our attention on the works carried out on this type of substrate.

Although MOCVD is generally used for the commercial growth of nitrides, most of the results reviewed in this chapter refer to MBE. The reason for this is familiar to growers: MBE being a low pressure technique allows one to use reflection high-energy electron diffraction (RHEED) in order to *in situ* control the growth process at a fraction of the monolayer scale. Such a tool is of course of particular interest for the growth of QDs and has also been extensively used for studying the specific GaN growth kinetics.

As a matter of fact, nitride growth mechanisms are strongly dependent on experimental conditions. Regarding the c-plane, two different GaN growth regimes have been identified as a function of the Ga flux (with the N flux kept constant), namely the N-rich regime (with an excess of N with respect to Ga) and the Ga-rich regime (with an excess of Ga with respect to N) [16]. In the latter, an auto-surfactant effect of Ga (a segregation of Ga on the growing GaN surface) has been put in evidence [17], and a coverage diagram, describing the Ga surface coverage during growth as a function of Ga flux and growth temperature, has been established [18].

Importantly, it has to be emphasized that the nature of the active N itself, i.e. ammonia or N plasma, drastically influences growth kinetics. Actually, the above considerations on the growth diagram as well as most experimental results reported in this chapter refer to plasma-assisted molecular beam epitaxy (MBE), active N being obtained by radio-frequency cracking of N_2 in a plasma cell.

It should be noted that the situation is different when active nitrogen results from cracking of ammonia. To our knowledge, no systematic study of Ga coverage has been performed in such a case but it is expected that adatom kinetics be drastically influenced by residual H resulting from ammonia decomposition.

Regarding the *a*-plane, it has been established that growth also strongly depends on Ga/N flux ratio value. More precisely, it has been found that GaN growth in Ga-rich conditions results in a rough surface whereas growth in N-rich conditions leads to a smooth surface, as supported by the AFM images of [11$\bar{2}$0] GaN grown on [11$\bar{2}$0] SiC shown in Fig. 6.1. It is worth emphasizing that this situation is opposite to what is observed for polar nitrides. It suggests that surface diffusion barriers are lower in N-rich conditions than in Ga-rich conditions for [11$\bar{2}$0] GaN, which again is the opposite of what is theoretically predicted [19] and experimentally observed [20] in the case of [0001] GaN growth.

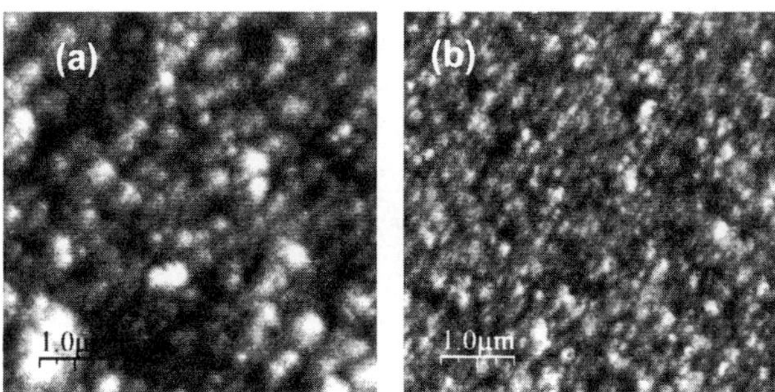

Figure 6.1   (a) (11–20) GaN grown in N-rich conditions. Rms roughness: 1.6 nm; (b) in Ga-rich conditions. Rms roughness: 2.7 nm. The thickness is 100 nm.

Figure 6.2 shows the Ga coverage diagram experimentally established for homoepitaxy of [0001] and [11–20] GaN. Whereas a self-regulated 2 MLs Ga coverage may be found for a proper Ga/N ratio value for an [0001] GaN surface [18, 21], no self-regulated regime is observed on an [11$\bar{2}$0] surface [22]. The Ga coverage diagram in Fig. 6.2 allows one to define several regimes, i.e. different Ga/N ratio values for a fixed active N flux.

Regarding Fig. 6.2a which corresponds to [0001] GaN, four regimes limited by experimentally determined lines can be identified: regime A corresponds to under-stoichiometry (N-rich) conditions which result in a rough surface, consistent with the reduced adatom diffusion path [19]. Regime B corresponds to the dynamic equilibrium on the surface of a Ga excess comprised between 0 and 1 ML. Regime C corresponds to a self-regulated Ga coverage of exactly 2 MLs [18, 21]. For higher Ga flux, Ga accumulation and Ga droplet formation are observed. Note that for a given substrate temperature, progressively increasing Ga flux allows one to successively experience the four regimes under consideration. Also note that the extension of each regime along the Ga flux axis is strongly dependent on growth temperature. In particular, for a substrate temperature below 680°C, only two regimes are observed, namely the N-rich one and the Ga-accumulation one.

The case of [11$\bar{2}$0] GaN is illustrated in Fig. 6.2b. Note that this diagram has been established in static conditions, by measuring the desorption time of Ga from an [11$\bar{2}$0] surface, following its exposure to increasing Ga flux. The full line in the Figure corresponds to the GaN growth rate at about 730°C which was simply added to static data in order to determine the Ga coverage diagram during growth [22]. In this diagram, A′ regime corresponds to nitrogen-rich conditions which lead to a *smooth* surface, as shown in Fig. 6.1 and *contrary* to the case of [0001] GaN. Regime B′ corresponds to the dynamical equilibrium on the surface of a Ga excess comprised between 0 and 1 ML, followed by Ga accumulation for higher Ga fluxes. Note that only three regimes are observed in the case of an [11$\bar{2}$0] GaN surface. In particular, no regime corresponding to a self-regulated coverage of Ga has been found for this cristallographic orientation.

Figure 6.2 proved to be very useful to identify the suitable growth conditions to achieve the Stranski–Krastanov growth mode of GaN on AlN giving rise to the formation of QDs, and to understand how the growth mode can be controlled by playing with the excess of Ga.

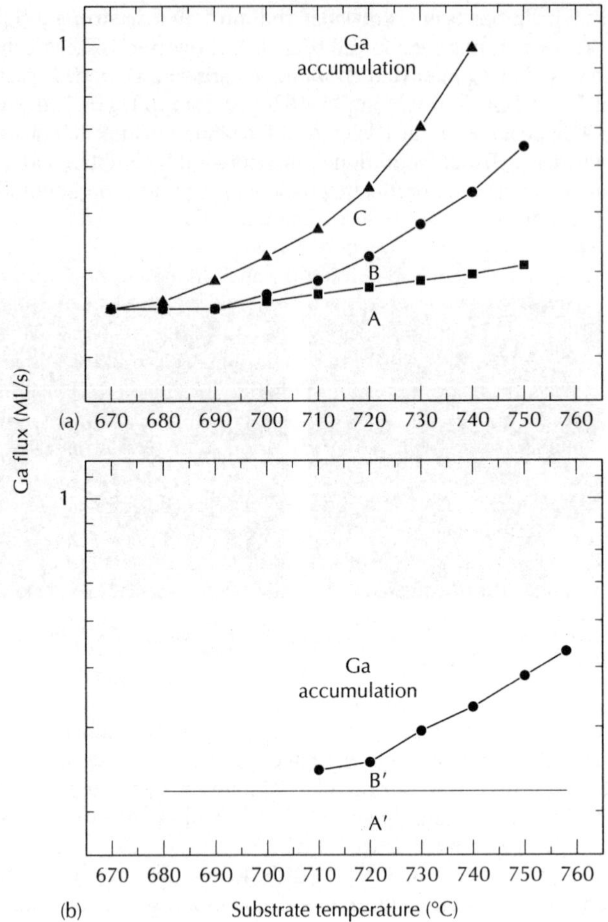

Figure 6.2　Ga flux dependence of the growth diagram for homoepitaxy of GaN, as a function of substrate temperature. (a) for [0001] GaN. The N flux is 0.35 ML/s. Four regions are identified: A: nitrogen-rich regime; B: dynamically stable Ga coverage between 0 and 1 ML; C: self-regulated Ga bilayer regime; Ga accumulation regime (b) for [11$\bar{2}$0] GaN. The N flux is 0.28 ML/s. Three regions are identified: A′: nitrogen-rich regime; B′: dynamically stable Ga coverage between 0 and 1 ML; Ga accumulation regime. The frontier line between A′ and B′ corresponds to the GaN growth rate at about 730°C.

In the case of heteroepitaxial growth of [0001] GaN on [0001] AlN, it has been established that regimes A and B lead to the formation of GaN QDs, according to the Stranski–Krastanov growth mode. By contrast, in regime C, the presence of a Ga bilayer results in stabilization of a 2D GaN layer and completely inhibits QD formation. However, it has been shown that by stopping GaN growth after deposition of a limited number of MLs and by letting the surface evolve under vacuum, desorption of the Ga bilayer leads to an instability of 2D GaN and to its spontaneous reorganization into 3D islands on a wetting layer: this is the *modified* SK growth mode, which results in the formation of GaN QDs similar to those obtained in regimes A and B [23, 24].

In the case of [11$\bar{2}$0] GaN grown on [11$\bar{2}$0] AlN, regime A′ leads to the formation of a 2D layer and can be used to grow quantum wells. By contrast, regime B′ leads to the formation of QDs according to the SK growth mode [25].

It should again be emphasized that QW and QD growth conditions are reversed for [0001] and [11$\bar{2}$0] orientations, possibly due to a change in adatom mobility on the [11$\bar{2}$0] surface as a function of Ga excess. As an alternative explanation of the observed features, it is also possible that surface energy of (11$\bar{2}$0) surfaces in Ga-rich conditions be sufficiently weak to make dot formation an energetically favourable process, whereas a supposed high surface energy in N-rich conditions would favour the stabilization of the 2D GaN layer.

6.2.2   Growth of GaN QDs

Due to the 2.5% lattice mismatch between GaN and AlN, growth of GaN deposited on AlN obeys the Stranski–Krastanov (SK) mode, with formation of three-dimensional islands for a GaN amount larger than a critical thickness of about 2.3 monolayers (MLs) [26]. However, occurrence of this growth mode strongly depends on the used Ga/N ratio value, i.e. on surface kinetics. It has been discussed in a previous section that SK growth mode is observed when growing GaN in N-rich conditions, that is in conditions of low adatom mobility [25]. Above stoichiometry, for Ga/N > 1, QD formation is rapidly followed by coalescence, which is assigned to an increase in Ga diffusion length on a Ga-rich surface. Next, in the extreme case of Ga/N $\gg 1$, a Ga bilayer is rapidly formed on growing GaN. The existence of this self-regulated Ga bilayer has been demonstrated both theoretically [21] and experimentally [18]. It results in the opening of a new diffusion path for N between the two Ga layers and a deep change in surface kinetics. When the Ga flux is high enough so that this Ga bilayer is completed before reaching the critical thickness, the 2D/3D transition is found to be completely inhibited and no dot formation is observed.

Actually, when stopping the GaN growth after a few MLs and letting the sample evolve under vacuum, it has been demonstrated that Ga bilayer desorption is followed by a reorganization of the GaN bidimensional layer and the formation of GaN QDs similar in shape and morphology to those grown in the N-rich regime, which suggests that the inhibition of the transition observed in the Ga/N $\gg 1$ regime is not of a kinetic nature but rather results from the competition between elastic and surface energy [24]. It was proposed that this peculiar growth mode, namely the growth of a 2D, Ga bilayer-stabilized, GaN followed by dot formation under vacuum be named the *modified* SK growth mode to differentiate it from the former one.

As mentionned above, the nature of active N source in MBE is a determinant parameter. Along this view, it should be noted that the above considerations about SK and *modified* SK growth mode of GaN QDs on AlN are only valid in the case of plasma-assisted MBE. When using ammonia as an N source, it has been demonstrated that QD formation only occurs on growth interruption [27], suggesting that GaN 2D growth is stabilized in the presence of ammonia, in a way phenomenologically similar to what is found to occur in the presence of a Ga bilayer.

As a remarkable consequence of the growth mode dependence as a function of the Ga/N ratio value, Fig. 6.3 illustrates the possiblity of growing either GaN QDs or QWs, at a given substrate temperature, by simply tuning the Ga flux.

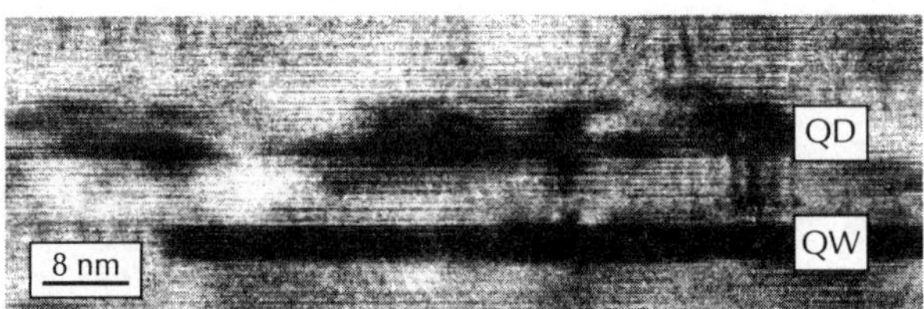

Figure 6.3   Quantum wells and quantum dots of [0001] GaN grown at 750°C and embedded in AlN. The QW was grown in regime C, the QDs in regime B (see Fig. 6.2a).

Consistent with the speculations about adatom diffusion as a function of Ga/N in the case of [11$\bar{2}$0] GaN (see Fig. 6.2), the conditions for growing [11$\bar{2}$0] GaN QDs and QWs have been found to be Ga-rich and N-rich, respectively, which is contrary to what is observed for [0001] GaN. As expected from the Ga coverage diagram shown in Fig. 6.2, no self-regulated Ga coverage regime is present for [11$\bar{2}$0] GaN and no *modified* SK growth mode was found.

Figure 6.4 shows high-resolution transmission electron microscopy (HRTEM) pictures of [11$\bar{2}$0] GaN QDs and QWs.

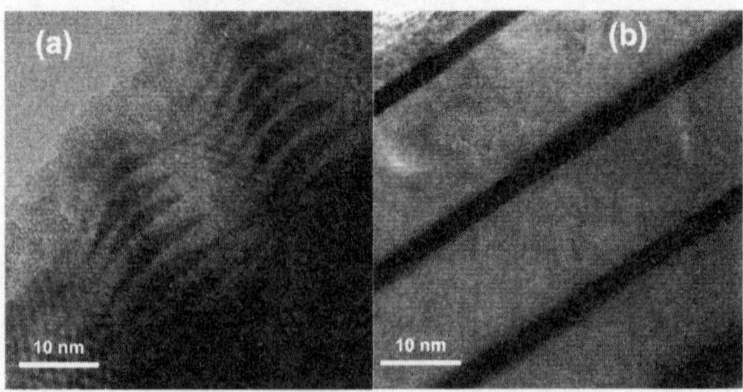

Figure 6.4   (a) [11$\bar{2}$0] GaN QDs embedded in AlN. Growth in regime B'. (b) QWs grown in regime A' (see Fig. 6.2b).

As a conclusion of this section, it appears that GaN QD formation strongly depends on kinetic conditions. It should be noted at this stage that the inhibition of the 2D/3D transition due to the self-surfactant effect of Ga is not limited to Ga but has also been observed in the case of [0001] GaN surface saturation by Eu. Furthermore, in the case of ammonia MBE, it has been found that [0001] GaN QD formation is only observed when growth is stopped and the material is allowed to evolve under vacuum. From all these results it is inferred that the occurrence of 2D/3D transition in the case of a GaN/AlN system is governed to a large extent by surface energy, in addition to the effect of lattice mismatch.

6.2.2.1   Growth of [0001] GaN QDs

Growth of GaN QDs on AlGaN by MOCVD was first reported by Tanaka *et al.* [28]. In that work, the reduced lattice mismatch due to low Al content of the AlGaN buffer was insufficient to promote the SK growth mode. Then, the surface was seeded with Si in order to achieve a micro-masking which resulted in localized GaN growth.

The SK growth mode of GaN on AlN by MBE was first observed in 1997 by Daudin *et al.* [26] and next by Damilano *et al.* [29]. More recently, it was demonstrated that the SK growth mode of GaN on AlN could also be observed in MOCVD [30] and for MBE growth of GaN on AlGaN with an Al content larger than about 50% [31].

Combining *in situ* RHEED observations and HRTEM images such as the one shown in Fig. 6.5, it has been established that [0001] GaN QDs are truncated hexagonal pyramids with {1$\bar{1}$03} facets.

Figure 6.5   HRTEM observation along [2,$\bar{1}$,$\bar{1}$,0] of a GaN QD embedded in AlN. The 2 H hexagonal stacking is clearly visible. The distance between each plane is c/2.

A major issue in GaN QD nucleation, related to its optical properties, is the role of structural defects. In particular, one may wonder whether threading edge dislocations as well as screw and mixed ones, the density of which is typically 10^7 to 10^{10}cm^{-2} in AlN or GaN templates, play a role.

To adress this issue, plan view HRTEM observations have been made. As shown in Fig. 6.6, it appears that most QDs nucleate close to a threading edge dislocation, due to the local distortion of the lattice. However, the dots themselves are always free of dislocations and can be viewed as a nanoisland of perfect material embedded in a very defective matrix [32].

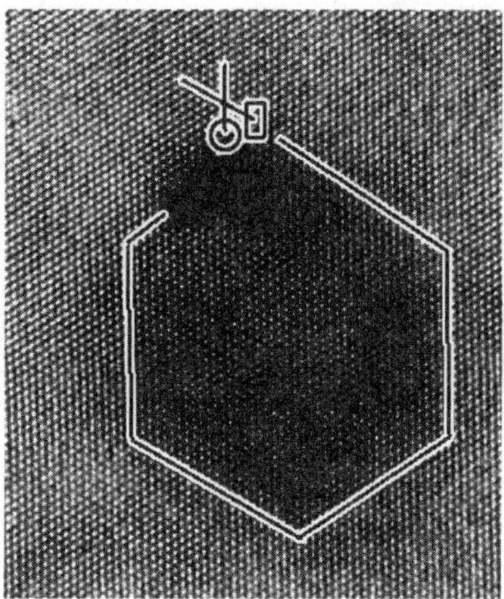

Figure 6.6   HRTEM plan view of a GaN QD embedded in AlN. Note the threading edge dislocation by the side of the dot, acting as a nucleation centre (after Fig. 6.7[32]).

Further atomic force microscopy experiments shown in Fig. 6.7 reveal an inhomogeneous dot distribution at the microscopic scale. Consistent with results of Fig. 6.6, this loop-like arrangement has been assigned to the inhomogeneous distribution of TDs which are shown (see Fig. 6.7b) to decorate the edges of steps on GaN and AlN surfaces.

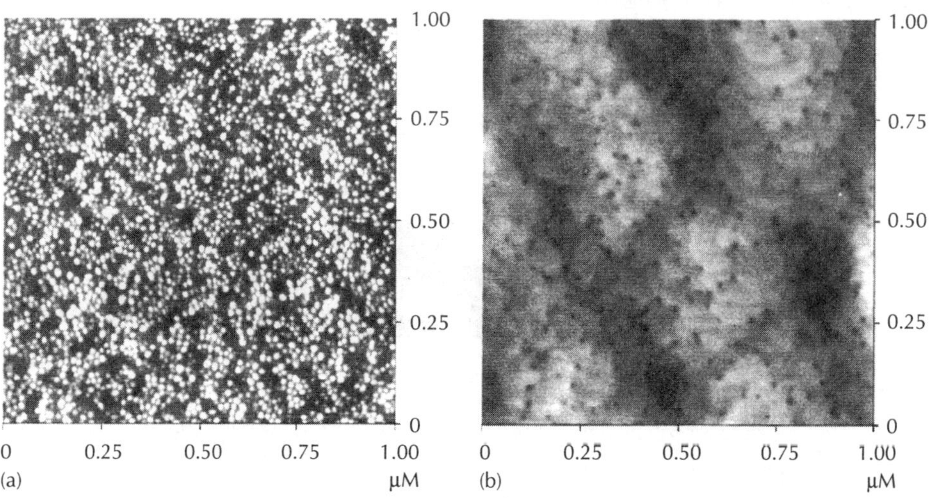

Figure 6.7   (a) AFM images of GaN QDs with a characteristic loop-like planar distribution associated with the distribution of TDs. (b) AFM image of 1 mm thick GaN grown on [0001] SiC showing threading edge dislocations at the edge of atomic steps.

6.2.2.2 Capping of [0001] GaN QDs

It has been found that the capping of [0001] GaN QDs by AlN is associated with a homogeneous decrease in size, of about one monolayer (see Fig. 6.8). Further Rutherford backscattering spectrometry (RBS) experiments have demonstrated that this size reduction is not due to dot–barrier interdiffusion but is rather assigned to a vertical exchange between Ga and Al, due to the higher stability of AlN at temperatures used for QD growth [33].

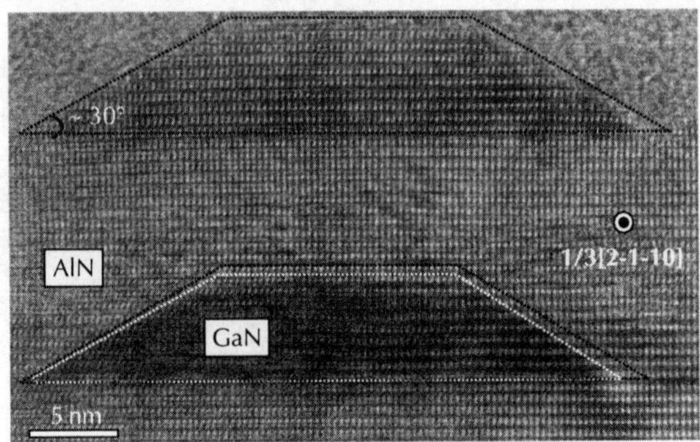

Figure 6.8   HRTEM image of [0001] GaN QDs on a surface and embedded in AlN. Note the homogeneous size reduction of the embedded dot with respect to the one on the surface (after [33]).

More precisely, a detailed study of GaN QD capping by AlN has shown that, following an initial wetting of dots with AlN up to 4 ML, further AlN growth proceeds in between dots up to 14 MLs which corresponds to AlN smoothing. As a whole, the AlN capping mechanism is schematized in Fig. 6.9 [34]. Starting from uncapped GaN QDs (a), their wetting by AlN is first observed (b). This stage is associated with the Ga/Al vertical exchange illustrated in Fig. 6.8. Next, homogeneous wetting of dots proceeds up to 4 MLs of AlN deposited (c). Further AlN growth preferentially occurs in between dots up to AlN smoothing (d). In a next stage, 2D homogeneous AlN growth is observed (e). No change in dot shape and no noticeable interdiffusion are associated with capping, making the GaN/AlN system a model one and facilitating the study of GaN QD structural properties, as extensively discussed in next sections.

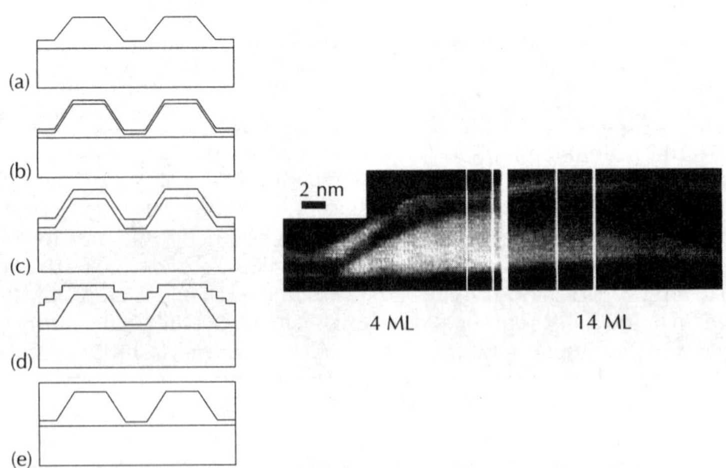

Figure 6.9   HRTEM image of [0001] GaN QDs (bright) covered with 4 and 14 ML of AlN (dark), respectively. Note the 2 ML thick GaN marker on AlN capping, allowing one to visualize AlN morphology (after [34]).

6.2.2.3   [000$\bar{1}$] GaN QDs

As a result of the non-centrosymmetry of wurtzite structure, [0001] and [000$\bar{1}$] directions are unequivalent, the one being metal terminated and the other one N terminated [35]. As a consequence, depending on the polarity of AlN or AlGaN templates, growth of [0001] or [000$\bar{1}$] GaN can be achieved. Despite of differences in surface energy and surface kinetics between [0001] and [000$\bar{1}$] surfaces [36], it has been demonstrated that GaN QDs can equally be grown on N-terminated AlN. The vertical correlation of stacked planes of dots has been demonstrated, for AlN spacer thickness similar to the thickness leading to vertical correlation of [0001] dots. As a whole, the structural and optical properties of [000$\bar{1}$] GaN QDs have been found to be rather similar to those of their [0001] counterpart [37].

6.2.2.4   Growth of [11$\bar{2}$0] GaN QDs

Morphology of [11$\bar{2}$0] GaN QDs is more complex: scanning electron microscopy images shown in Fig. 6.10 illustrate an asymmetry of dots along the [0001] polar direction. As a whole, various shapes of dots have been identified, from triangular to trapezoidal. Combined with HRTEM experiments shown in Fig. 6.11 and analysis of the facets shown in RHEED pattern, the shape shown as an inset in Fig. 6.11 has been proposed for [11$\bar{2}$0] GaN dots.

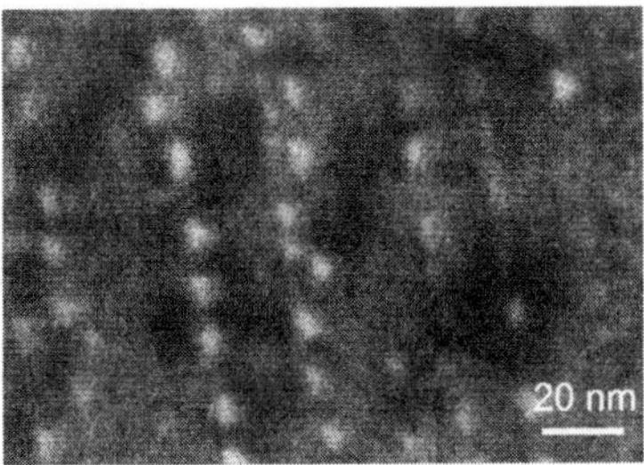

Figure 6.10   Scanning electron microscopy image of [11$\bar{2}$0] GaN QDs. Note the coexistence of several shapes (after [25]).

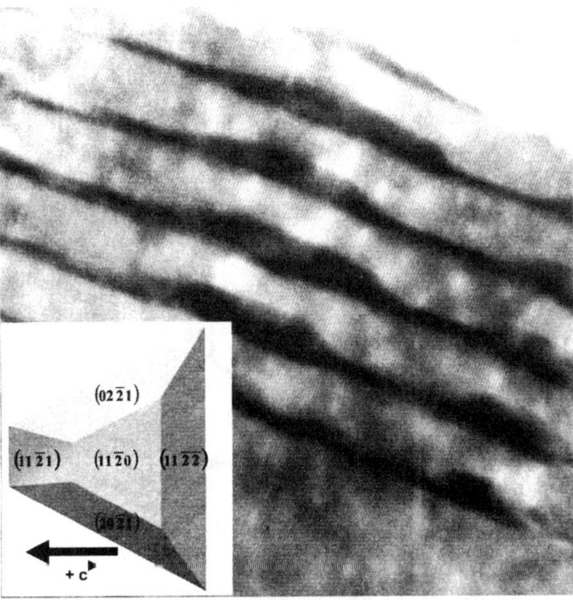

Figure 6.11   HRTEM images of [11$\bar{2}$0] GaN QDs viewed along the [1$\bar{1}$00] zone axis. An asymmetry of the dots is shown, i.e. two types of facets are seen along [0001] inclined by about 16° and 32° with respect to the basal plane (after [25]).

The asymmetry of dots could be related to the direction of the [0001] polar axis by performing convergent electron beam diffraction (CBED) experiments. As illustrated in the inset in Fig. 6.11, it was demonstrated that triangularly shaped [11$\bar{2}$0] GaN QDs point towards the $+c$ direction. The dot asymmetry in the [0001] direction has been tentatively assigned to a faster growth rate along $+c$ than along $-c$ (see [25]).

6.3   Structural properties of GaN QDs

The structural properties of QDs are delicate to carry out: either dots are let uncapped on the substrate but may be modified by oxidation or they are capped and the process of capping itself may modify their chemical composition. It has been shown in section 6.2.2.2 that GaN QDs can be considered as a model system as *no interdiffusion* between dots and barrier has been seen to date. This makes GaN QDs particularly suitable for study by HRTEM, medium energy ion scattering, X-ray diffraction and Raman spectroscopy, which are the techniques usually used for obtaining insight into dot structural properties and their variation as a function of capping and/or stacking. Note that Raman spectroscopy is extensively presented in Chapter 7 and will not be considered further here.

6.3.1   [0001] GaN QDs

Among the techniques allowing one to characterize the structural properties of QDs, namely their strain state with respect to the surrounding barrier, medium energy ion scattering (MEIS) is a relatively new technique and has been used in a few cases to determine the strain profile of nanostructures at the monolayer scale [38, 39]. The principle of the technique is shown schematically in Fig. 6.12. It consists of measuring the angular and energetic distribution of medium energy protons (typically 100 keV) backscattered from a target. As shown in Fig. 6.12, the angular distribution exhibits dips aligned with crystallographic directions, which result from the occurrence of a shadow cone for backscattered ions, due to the presence of atomic columns. The angular position of this dip depends on the strain and may be easily related to the c/a ratio, c and a being the out-of plane and in-plane lattice parameters, respectively. Due to the strong energy loss of 100 keV protons, the depth resolution is as good as the monolayer, allowing one to extract the strain profile of buried/unburied QDs as a function of their height.

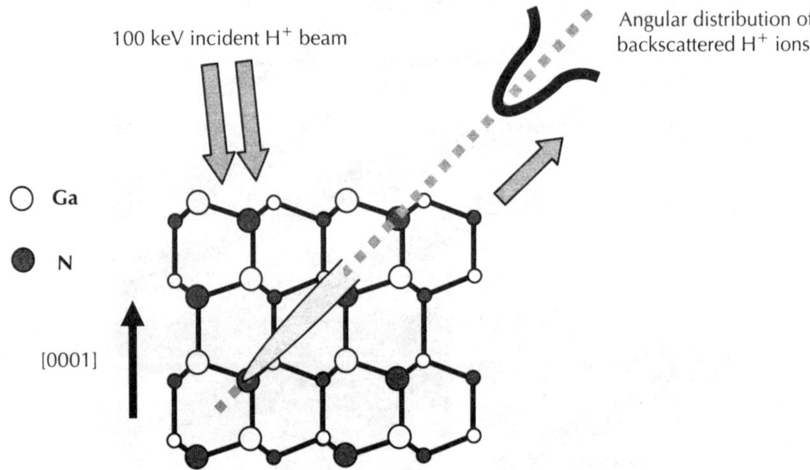

Figure 6.12   Schematics of an MEIS experiment.

The case of GaN QDs covered by 20 ML of AlN, sufficient to smooth AlN surfaces and consistent with section 6.2.2.2, is shown in Fig. 6.13 and compared to theoretical calculations [40].

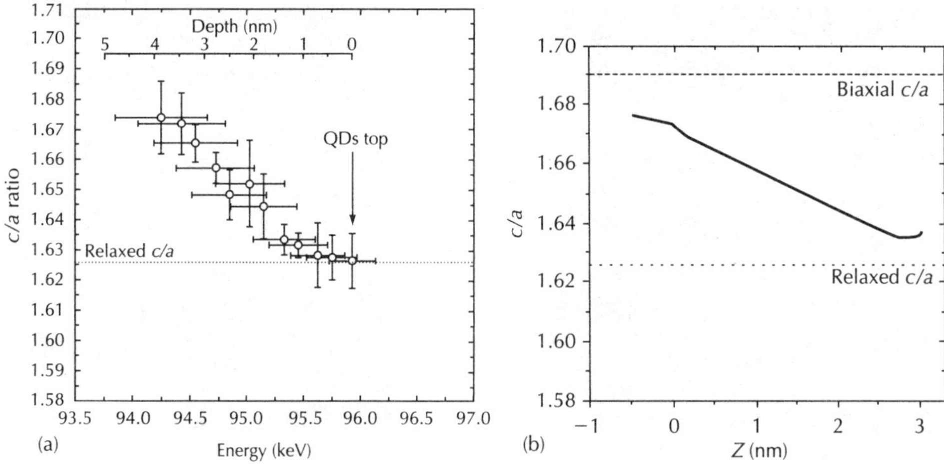

Figure 6.13   (a) c/a profile as a function of depth for GaN QDs covered by 20 MLs of AlN. (b) Theoretical calculations (after [40]).

Although the agreement is satisfactory, it has to be noted that the upper part of QDs is experimentally found to be more relaxed than theoretically predicted, indicating that the effective QD strain state likely depends partially on extrinsic parameters such as the presence of threading edge dislocations in their close vicinity.

Alternatively, quantitative information on QD strain state may be obtained by performing quantitative analysis of HRTEM images using the geometrical phase analysis technique [41, 42]. The result for [0001] GaN QDs is displayed in Fig. 6.14 which shows maps of the local c and a lattice parameters, respectively. From these images, it is inferred that compressed GaN QDs induce strain in the surrounding AlN matrix. In particular, it is found that AlN above the dots exhibits an expanded a lattice parameter, i.e. AlN is in tension and constitutes a privileged nucleation centre for the upper GaN dots, which leads to the build-up of vertical correlation of GaN QDs discussed in section 6.4.

Combining the results in Fig. 6.14, it is possible to extract a map of the c/a ratio value which is also shown. An average value of 1.64 is found in the GaN QDs, allowing one to conclude that GaN QDs are relaxed to a large extent, as c/a would be 1.69 in the hypothesis of a biaxial strain and equal to 1.625 in a bulk GaN crystal.

If we now compare these HRTEM results with MEIS experiments previously discussed, a rather good agreement is found. However, it should be noted that the sample preparation requested for HRTEM experiments may lead to additional strain relaxation assigned to a thin foil effect.

6.3.2   [11$\bar{2}$0] GaN QDs

It has been shown in section 6.2.2.3 that nucleation of [11$\bar{2}$0] GaN QDs is highly anisotropic and that dots are found to be aligned along the [1$\bar{1}$00] direction. Their strain profile was studied by MEIS [43]. In that case, two perpendicular directions were used to measure the strain profile in the {0001} and {1$\bar{1}$00} planes, respectively (see Fig. 6.15).

The evolution of the $\sqrt{3} \times a_{11\bar{2}0}/a_{1\bar{1}00}$ and $a_{11\bar{2}0}/c$ ratios as a function of depth is shown in Fig. 6.16. The data are compared to the case of a GaN layer completely strained in both in-plane directions on a relaxed AlN [37] (labelled GaN/AlN in Fig. 6.16), and to the case of relaxed GaN (corresponding to $\sqrt{3} \times a_{11\bar{2}0}/a_{1\bar{1}00} = 1$ in Fig. 6.16a and $a_{11\bar{2}0}/c = 0.615$ in Fig. 6.16b). However, it is worth noting that the AN buffer layer which as been deposited on SiC prior to GaN QD growth is probably far from being completely relaxed. In such a case, the strain state of GaN should lie between that of GaN/relaxed AlN and that of GaN/SiC, in good agreement with experimental results (see Fig. 6.16).

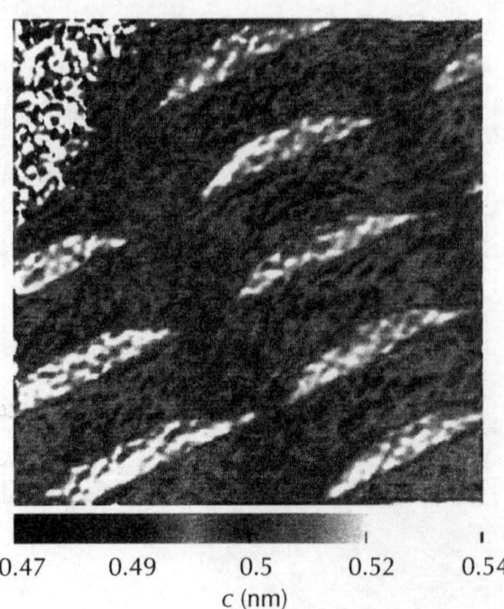

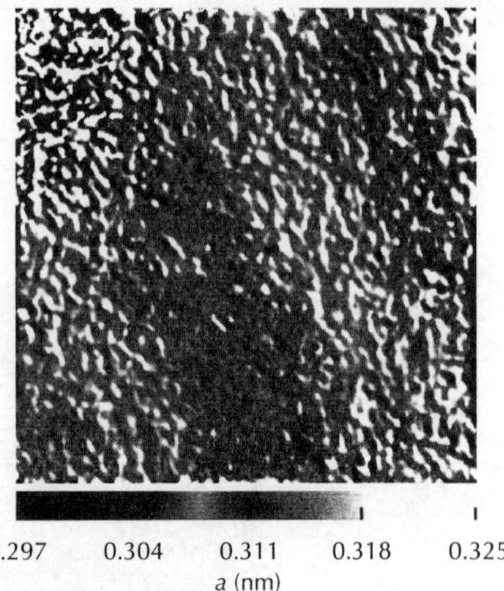

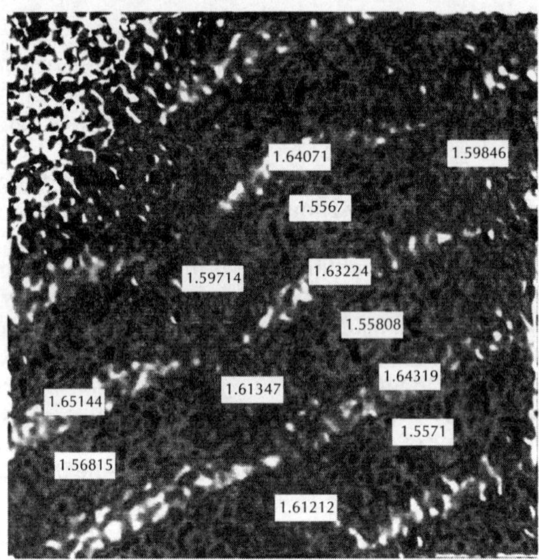

Figure 6.14   Maps of the whole $2\,K \times 2\,K$ CCD image shown in Fig. 6.12. (a) Map of the c lattice parameter; (b) map of the a lattice parameter; (c) c/a mapping extracted from results displayed in Fig. 6.13a and b. Note that the relaxed values of c and a are 0.5185 nm and 0.3189 nm, respectively, leading to a c/a ratio of 1.625.

The evolution of the $\sqrt{3} \times a_{11\bar{2}0}/a_{1\bar{1}00}$ ratio, corresponding to the strain profile along $[1\bar{1}00]$ (Fig. 6.16a), puts in evidence a decrease of the strain from the GaN/AlN interface to the top of the dots and wetting layer: while the initial value at the GaN/AlN interface is found to be larger than the value of a completely strained GaN, the strain at the top of the dots is intermediate between the completely strained and the relaxed values. We now turn to the evolution of the $a_{11\bar{2}0}/c$ ratio, corresponding to the strain profile along [0001] (Fig. 6.16b). It appears that the initial strain value at the GaN/AlN interface is lower than the corresponding initial value in the $[11\bar{2}0]$ direction.

As a whole, the strain profiles along the two non-equivalent $[1\bar{1}00]$ and [0001] directions are remarkably different. The systematic enhanced strain relaxation at the base of the dots along the [0001] direction has been interpreted as possibly due to a plastic process [43]. Indeed, in the case of a GaN layer deposited on a relaxed AlN, the in-plane GaN strain at the GaN/AlN interface amounts to -2.4% along $[1\bar{1}00]$ and -3.9% along [0001]. The greater in-plane strain along

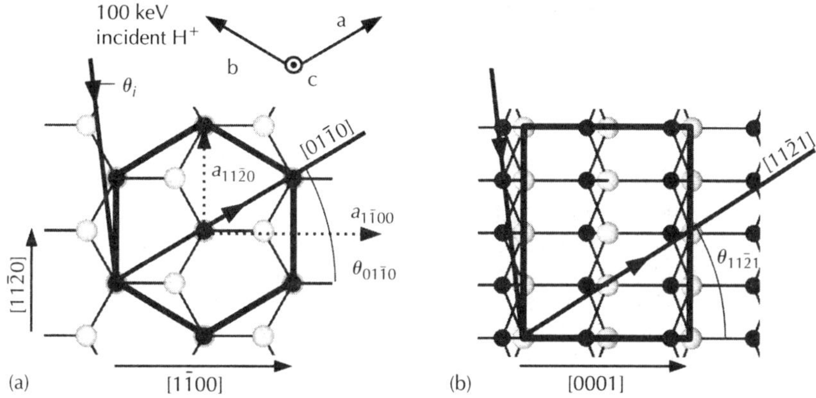

Figure 6.15   Scheme of the scattering geometry in the [0001] (a) and [11̄00] (b) planes for medium energy ion scattering (after [43]).

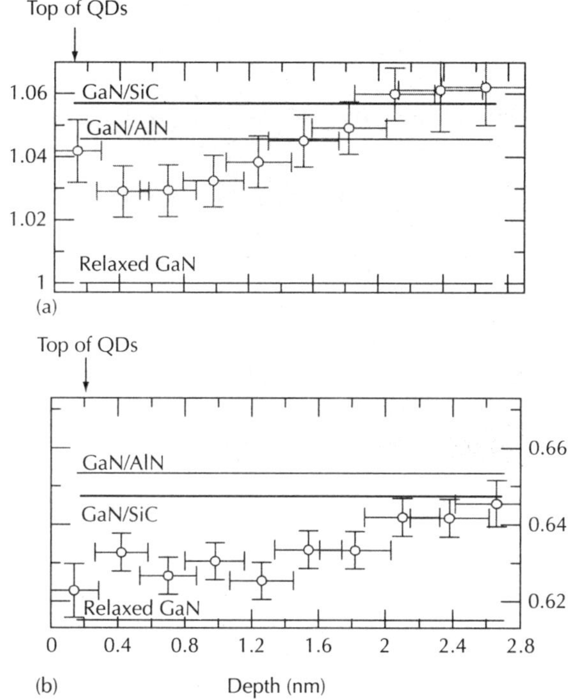

Figure 6.16   Depth profile of the $\sqrt{3} \times a_{112̄0}/a_{11̄00}$ (a) and $a_{112̄0}/c$ (b) ratios for [112̄0] GaN QDs. The GaN/ AlN label refers to a GaN layer completely strained in both in-plane directions on a relaxed AlN.

[0001] could favour plastic strain relaxation through the formation of misfit dislocations at the GaN/AlN interface along the [0001] direction. These defects would partly account for the 3.9% total strain, while additional elastic relaxation would accommodate part of the residual lattice mismatch. The combination of these elastic and plastic relaxation mechanisms could account for the smaller residual strain measured along [0001].

Interestingly, misfit dislocations associated with this hypothetical plastic relaxation have been observed by HRTEM, as evidence that misfit dislocations are generated in order to accommodate the stress in the [0001] direction. Very roughly, the lattice mismatch accommodated through the formation of these dislocations can be estimated at 1 to 2%. Note that this value is consistent with the negative offset of the strain profile displayed in Fig. 6.16b with respect to the one displayed in Fig. 6.16a.

6.4   Vertical correlation of stacked QDs

Vertical correlation of stacked QD planes is observed when the spacer between adjacent planes is sufficiently thin to ensure a modulation of the surface elastic potential induced by the presence of buried dots. As a consequence, nucleation of the upper QD plane preferentially occurs in the regions of the spacer locally extended by the presence of dots below. This mechanism has been described by Tersoff *et al.* [44]. It also works for GaN QDs and results in an homogenization of their size distribution associated with an average size increase [45]. Figure 6.17 shows an HRTEM image of vertically correlated GaN QDs planes embedded in AlN. The AlN spacer between adjacent planes is about 6 nm. As discussed in section 6.3.1, such a spacer is thin enough to mediate the deformation induced by the presence of the lower QD plane up to the surface. Moreover, it has been consistently shown by Raman spectroscopy (*cf.* Chapter 7) and by X-ray diffraction (see section 6.5) that vertical correlation can be experimentally observed for an AlN spacer thinner than about 8 nm.

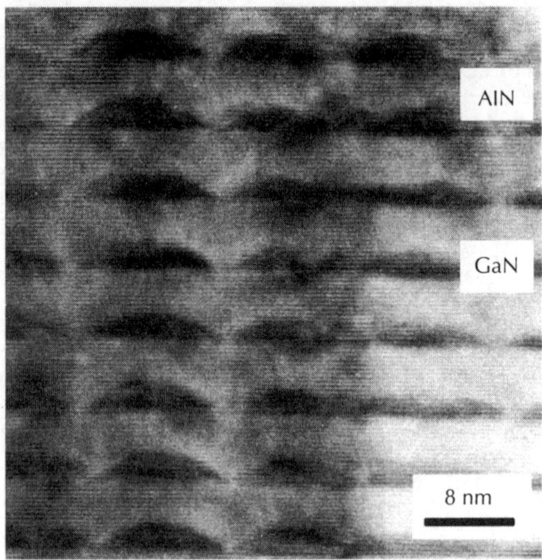

Figure 6.17   HRTEM image of a superlattice of vertically correlated GaN QD planes embedded in AlN.

It is worth noting that according to the model of Tersoff [44], vertical correlation, far of only consisiting of a vertical alignment of dots, implies an elastic interaction of the dot/barrier system which eventually results in a size filtering of dots and an homogenization of their population. Such a behaviour is illustrated in Fig. 6.18 which displays photoluminescence (PL) results of vertically correlated [0001] GaN dots as a function of the number of stacked planes. To properly interpret the results in Fig. 6.18, one must recall that optical properties of [0001] GaN dots are dominated by the presence of an internal electric field which results in a marked red shift of PL energy (see Chapter 7). One consequence of the electric field, which may be as strong as 10 MV/cm, is to emphasize the dot size dispersion effect. This is seen in Fig. 6.18 where the peak corresponding to three stacked planes of dots is as wide as 750 meV. For an increasing number of planes, which are correlated to the build-up of vertical correlation, one observes both (i) a drastic reduction in PL line width and (ii) an increased red shift whereas the left side of the PL peak is almost unchanged. These variations are consistent with an increase in the average dot size and an homogenization of dot population, as a signature of vertical correlation of successively stacked planes of dots. The driving force of the correlation has been identified in Fig. 6.14, i.e. the elastic deformation of the surrounding AlN barrier by GaN dots, acting as a nucleation filter.

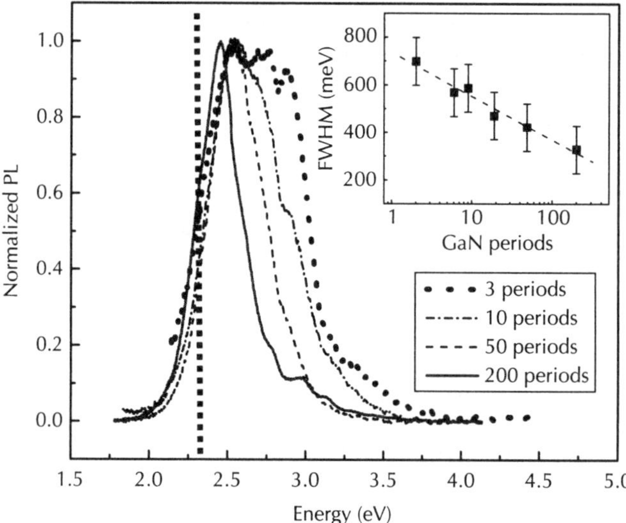

Figure 6.18   Photoluminescence of stacked GaN QDs planes as a function of the number of planes. Inset: Variation of FWHM of photoluminescence peaks as a function of the number of stacked planes.

6.5   X-ray diffraction analysis of GaN QDs

Structural characterization of GaN QDs by X-ray diffraction deserves special attention for several reasons. First of all, development of Synchrotron facilities has made feasible the characterization of single QD planes, despite the small amount of material under consideration, typically a few MLs. However, the epitaxial relationship between nanostructures and their substrate is a source of additional difficulties as both QDs and the substrate are diffracting in almost the same region of reciprocal space, making difficult – if not impossible – the deconvolution of each contribution. Furthermore, the situation is even more complicated due to the small size of QDs which leads to a marked diffraction peak widening.

Most of these difficulties can be overcome by the recent development of anomalous diffraction techniques under grazing incidence. First of all, multiwavelength anomalous diffraction (MAD) allows one to extract the structure factor of a specific element and, thus, the average strain, composition, and size of the corresponding region [46]. Second, the fine structure oscillations measured above the absorption edge in diffraction condition, known as extended diffraction anomalous fine structure (EDAFS), provide the local environment of the absorbing atoms in the diffraction-selected region [47]. Then, a proper combination of MAD and EDAFS makes possible a comprehensive study of the strain state of nanostructures, allowing one to determine both in-plane and out-of-plane strain.

These techniques have been recently applied to the case of GaN QDs [48, 49] – Fig. 6.19a shows the diffracted intensity as a function of X-ray energy. In the case of uncapped GaN QDs deposited on a thin AlN buffer on [0001] SiC, the Ga K edge is at 10.367 keV. Clearly, the intensity of the shoulder on the edge of the peak at $h = 3$ corresponding to SiC is strongly energy dependent, as a chemical signature of the presence of Ga in diffracting material. Following an extraction procedure described in [48] and [49], the contribution of GaN is shown in Fig. 6.19b. The position of the Ga signal maximum along [10$\bar{1}$0] is directly related to the average in-plane strain state in the QDs.

By performing MAD experiments at the maximum of the Ga partial structure factor, as a function of the AlN deposit on top of the QD planes, it was possible to extract the x Al/Ga proportion of the isostrain volume selected by diffraction [49]. The result is shown in Fig. 6.20. Up to 4–5 ML, the Al proportion, which is near to zero for free-standing QDs, increases linearly, and stabilizes

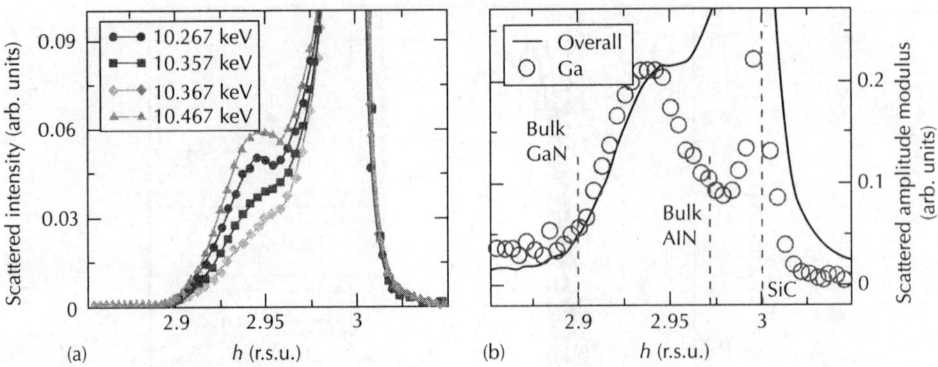

Figure 6.19   *(a) Diffracted intensity measured at four energies across the Ga K edge at 10.367 keV, along the [10$\bar{1}$0] reciprocal space direction, labelled in reciprocal space units (r.s.u.) by h. h = 3 is the position of the SiC substrate, used as a scale reference. (b) Extracted modulus of the Ga region's cattered amplitude, as deduced by MAD, compared to the square root of the overall scattered intensity measured at 10.267 keV. Grazing incidence angle, α, =0.3° (after [49]).*

above 4–5 ML, with a trend to saturate for further AlN coverages. Provided that AlN on top of the QDs is pseudomorphic to GaN for low coverages [50], and as no appreciable intermixing occurs inside the GaN QDs [49, 33], the variation of the Al proportion x in the isostrain region up to 4–5 ML indicates a uniform increase of the amount of AlN on top of the GaN QDs, followed by a change in the AlN growth process, which leads to AlN with an in-plane strain state different from that in the QDs. These results are consistent with the capping model described in section 6.2.2.2, as an illustration of the capabilities of anomalous diffraction techniques.

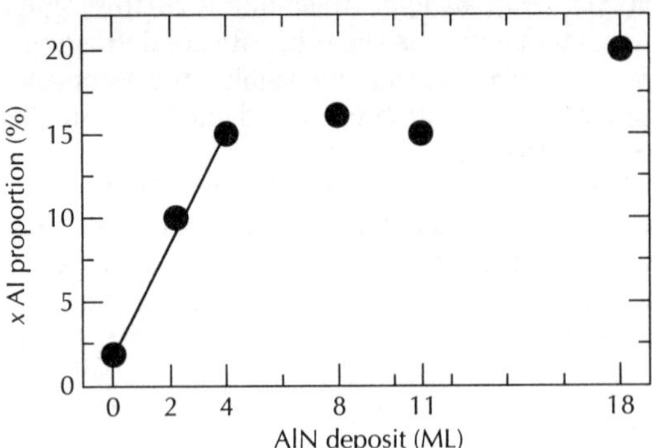

Figure 6.20   *Al proportion x of the isostrain region, as a function of the AlN deposit on top of the QD plane (after [49]).*

Anomalous X-ray diffraction in grazing incidence has also been used *in situ* to determine the progressive strain changes in both GaN QDs and the AlN barrier as a function of AlN thickness [50]. Results are shown in Fig. 6.21. They illustrate the combined tension/compression in the interacting AlN/GaN system. Interestingly, almost no strain variation is observed beyond an AlN thickness of about 8 nm (1 ML is about 0.25 nm), as evidence that the "memory" of the subjacent GaN QD plane is lost for thicker AlN layers. This result is consistent with the maximum AlN thickness which was found to make possible vertical correlation of successive planes of dots (see section 6.4) and also to Raman spectroscopy results leading to a similar conclusion (see Chapter 7).

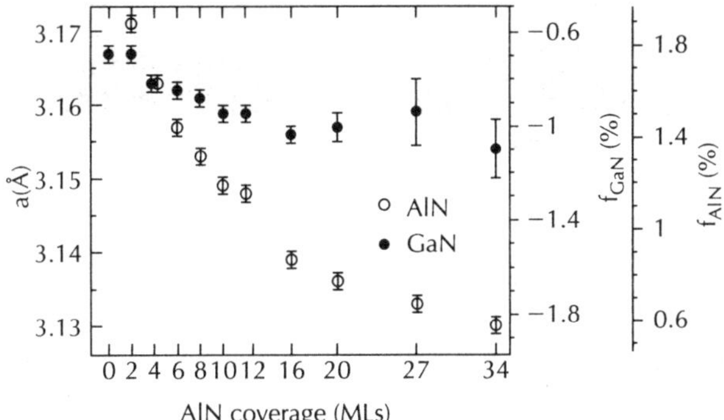

Figure 6.21 *In-plane lattice parameter and mismatches – relative to bulk AlN or GaN – for AlN and GaN deduced from* F_{Al+N} *and* F_{Ga} *extractions. The error bars in the last three points for GaN are large due to weak extracted* F_{Ga} *(after [50]).*

6.6 Optical properties of single GaN QDs

After the physics of the growth and the study of the structural properties of GaN/AlN QDs, this section discusses the optical properties of these heterostructures. The previous chapter described in detail the optical properties of polar QDs. The following section focuses more particularly on the optical properties of non-polar [11$\bar{2}$0] QDs, with an emphasis on the photoluminescence experiments on single quantum dots.

6.6.1 *Photoluminescence of ensembles of non-polar GaN QDs*

The overall optical properties of [11$\bar{2}$0] quantum dots are not obvious to forecast. Indeed, in that case, the polar axis (c axis) is in the plane of the layers. For a [1120] quantum well, the situation is quite simple: the polar axis is parallel to the GaN/AlN interfaces. There is thus no discontinuity of the polarization perpendicular to the interfaces so that there is no internal electric field. Despite the wurtzite cell, a non-polar quantum well thus behaves like a "standard" quantum well for which only confinement effects are present. This was confirmed experimentally in several publications [51, 52].

For quantum dots, the situation is more complicated. Indeed, the polarization vector does have a component along the normal to the quantum dot lateral facets. Therefore, a sheet charge density due to this polarization discontinuity appears on the QDs lateral facets. This charge density will thus generate an internal electric field that is likely to induce a sizeable quantum confined Stark effect. The geometry of the situation is, however, very different from the situation of polar QDs and QWs.

For polar heterostructures, the sheet charge densities created by the polarization discontinuities form a "condensator-like" structure in which two charged planes face each other giving rise to a quasi-uniform electric field inside the confined structure (even for quantum dots). For non-polar quantum dots, the geometry is quite different (see Fig. 6.22): only the two side facets of the QDs that are crossed by the c-axis carry a charge. This is thus a three-dimensional electrostatic problem, and the precise determination of the internal electric field requires the knowledge of the dot geometry, the strain distribution and the residual doping of the samples. Preliminary calculations by A. Cros and co-workers (see Chapter 7) have shown that the electric field should be much reduced (by at least one order of magnitude) compared to polar quantum dots [53]. Let us note that the issue of the residual doping of the quantum dots (usually MBE-grown non-intentionally doped III-N structures show a n-type character) is crucial here as it can be shown that each lateral facet of a non-polar dot carries the equivalent of between one and two elementary charges, so that the presence of a few electrons in each dot can drastically screen the internal electric field.

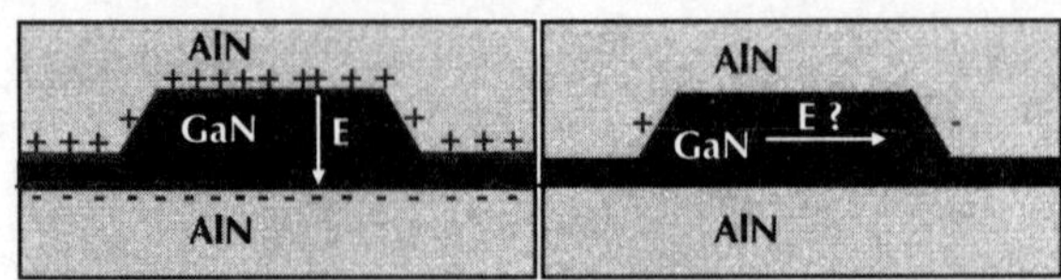

Figure 6.22   Geometry difference between a polar QD and a non-polar QD. On the left, a polar QD (c-axis pointing upwards), with planar capacitor-like geometry. On the right, a [11$\bar{2}$0] QD (c-axis in the plane of the page, pointing leftward) with few charges deposited on two side facets. The typical QD size is 20 nm diameter and 2 nm height.

The *a priori* determination of the optical properties of these non-polar QDs being a complex task requiring a precise knowledge of the structural and electronic properties of the grown structures, our approach was to probe experimentally the effect of the internal electric field on the optical properties of the [11$\bar{2}$0] quantum dots.

The first experiment (Fig. 6.23) was to perform a low-temperature (5 K) photoluminescence experiment on an ensemble of QDs, as a function of excitation power density. The sample is excited with a cw doubled argon laser, emitting at 244 nm (5.1 eV). The excitation power is varied over more than five orders of magnitude; the curves are vertically offset for clarity. The first observation is that the photoluminescence occurs above the GaN band gap, so that the confinement effects dominate over the quantum confined Stark effect. This is quite unlike polar quantum dots of similar sizes for which the quantum confined Stark effect leads to a spectacular red shift of the luminescence of the QDs, which can be up to 500 meV below the GaN band gap [54, 27]. Moreover, there is no indication of screening over the range of excitation densities that was probed. Again this is unlike polar GaN QDs for which the screening of the internal field by the photocreated carriers is observable over a wide excitation density range [55].

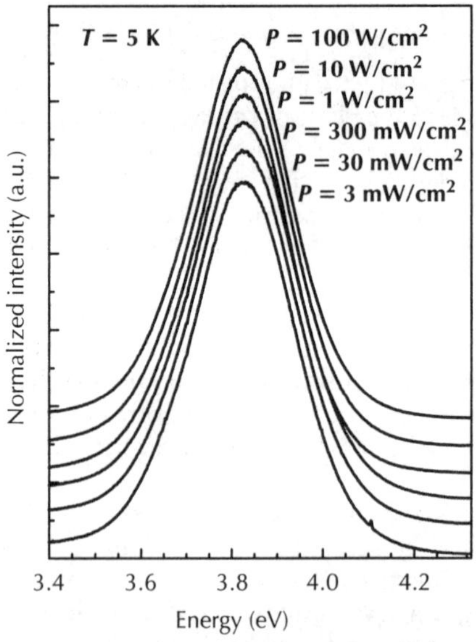

Figure 6.23   Photoluminescence spectrum at 5 K for various excitation power densities (at λ = 244 nm). The luminescence is above the GaN band gap, and no screening effect is observed in this excitation power range.

The most convincing experiment to probe the quantum confined Stark effect is to measure the radiative lifetime of the confined excitons, which is very sensitive to the spatial separation of electrons and holes. This is especially true for non-polar structures as the two charged planes are spatially well separated (by one QD diameter): in that case even a small internal electric field will separate the electron and hole wavefunctions and thus reduce drastically the transition oscillator strength, while the energy of the transition would be little affected. The time-resolved photoluminescence experiment was performed using a triple-pulsed Ti-sapphire laser emitting at 250 nm. The collected luminescence was analysed by a streak camera with 4 ps time resolution (Fig. 6.24).

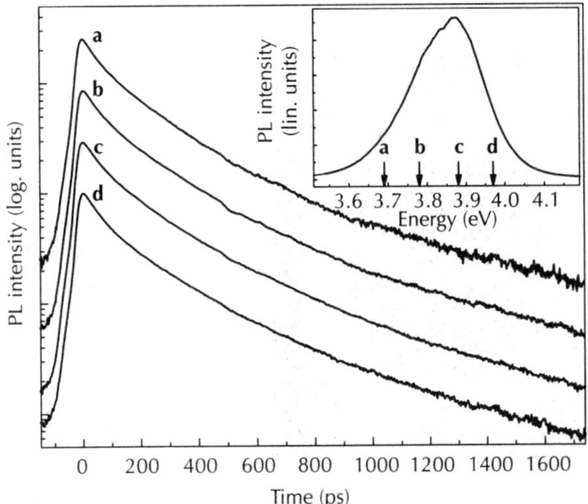

Figure 6.24   *Time-resolved photoluminescence at various emission wavelengths. The decay curves are independent of emission wavelength. The decay is not mono-exponential.*

It has to be noted that the time-resolved measurements were also performed as a function of sample temperature, showing no change up to 100 K. This indicates that non-radiative recombination can be neglected in this temperature range, and thus that the observed photoluminescence decay curves are characteristic of the radiative decay of the quantum dots. The first striking feature is the fact that the photoluminescence decay curves are independent of the emission energy. This means that the oscillator strength does not depend on the QD size. This again is a difference with the behaviour of polar QDs, for which the oscillator strength decreases exponentially with QD size [54, 55]. The second striking feature in the photoluminescence decay curves is that the decay is very fast as compared to the decay of polar GaN QDs [54, 55]. The decay in Fig. 6.24 is not mono-exponential, but an average decay time of around 280 ps can be extracted. This is to be compared with lifetimes ranging between a few ns and several hundred μs for polar QDs [54, 55].

From these first experimental observations in cw and time-resolved PL, it can be concluded that no quantum confined Stark effect is observed in these [11$\bar{2}$0] quantum dots [13, 56].

6.6.2   *Single dot PL – spectral diffusion and temperature broadening*

In order to gain more insights on the properties of confined carriers in non-polar GaN QDs, spatially resolved photoluminescence experiments were performed in order to analyse optical transitions stemming from single QDs. As compared to other QD systems for which single dot spectroscopy has been performed for several years, additional difficulties appear. A first difficulty is that the density of GaN QDs is difficult to master, and in this case to reduce (see earlier parts in this chapter). In the present case, the average density of QDs is 2×10^{11} cm^{-2}. However, the density is not homogeneous, so that locally the density can be reduced. Another difficulty is the emission wavelength in the UV, for which good optical components are scarce. In our case, we use an NA = 0.4 refractive microscope objective which can focus the 244 nm doubled argon laser on a ~1 μm diameter spot. This spot size is still too large to isolate a few quantum dots, so an additional processing of the sample is required. Two techniques were alternatively used to reduce the number of QDs probed under the excitation spot: (i) mesas defined by e-beam lithography were etched by inductively coupled plasma reactive ion etching, with sizes ranging between 200 nm and 5 μm and (ii) polystyrene nanobeads (diameter 300 nm) were deposited on the samples, covered by 100 nm aluminium; the beads were then removed, leaving apertures in the opaque aluminium mask. Both techniques allow the study of isolated transitions from single QDs and give similar results (Fig. 6.25)

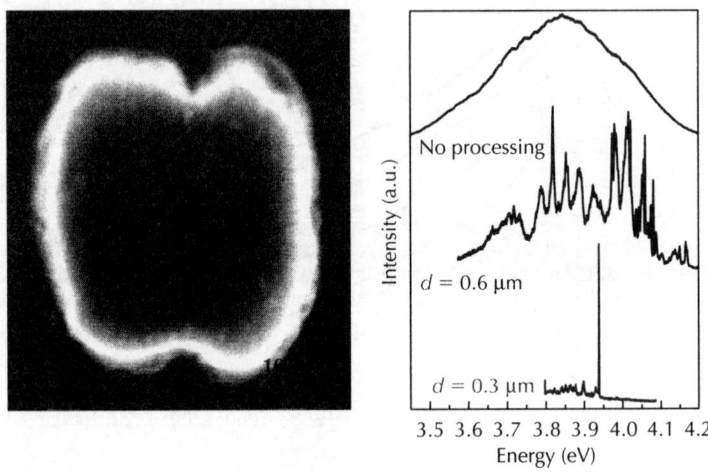

Figure 6.25   Left: SEM image of an etched mesa, the aligned surface QDs are visible on the image. Right: Microphotoluminescence spectra on an unprocessed region, on a 0.6 μm mesa and on a 0.3 μm mesa. Isolated sharp lines are observed on the high-energy side for some of the smallest mesas.

By studying several mesas and nanoapertures, we could isolate sharp spectrally isolated lines. One first remark is that these sharp lines were always isolated on the high-energy side of the quantum dot distribution (for energies above 3.8 eV). On the lower-energy side, structured features are observed, but are never as sharp and isolated as on the higher-energy side. The reasons behind this phenomenon are still unclear. As the sharp lines appear on the high-energy side, we first checked that they correspond to the recombination of single electron–hole pairs, and not to excited transitions of the QDs. This was done by performing power dependent measurements, in which the line intensities do have a linear dependency as a function of excitation power.

In such experiments, the first notable feature is the linewidth of the observed transitions. In our case, the linewidths range between 0.5 and 2 meV (the resolution of the set-up at these energies is around 0.4 meV). These linewidths are quite large compared to what is usually observed for QDs in other systems (usually in the 1 μeV to 100 μeV range). This is, however, less than what was reported for polar GaN/AlN QDs. At first, Kako and co-workers [57] reported linewidths ranging between 5 meV and 20 meV, i.e. one order of magnitude higher than our findings. Later, Bardoux and co-workers [58] reported linewidths as narrow as what we observe for small polar quantum dots; however, the observed lines always come as multiplets of lines spread over a few meVs. Both these results for polar QDs are explained by spectral diffusion: this phenomenon occurs when traps near a quantum dot are filled and emptied at a pace that is faster than the observation time. In that case, the time varying Coulomb force due to the presence or not of a charge in a trap leads to a time-dependent electrostatic potential inside the QD, which changes the energy of the discrete electronic transitions. In the work of Kako and co-workers [57], the volumic density of traps is such that the various electrostatic configurations found over the integration time lead to a large broadening of the transition. In the work of Bardoux and co-workers [58], the density of traps is less: a low number of traps are close enough to a given QD to induce large spectral shifts of the transition, giving rise to multiplets. Each line of the multiplet is also broadened through an interaction with traps that are farther away.

Returning to the case of [11$\bar{2}$0] quantum dots, the experimental situation is that narrow isolated lines or doublets are observed. Spectral diffusion is indeed observed when acquiring successive spectra with 1 s acquisition time (Fig. 6.26). It can be argued that the effect produced by the emptying and filling of a trap on a QD transition will be larger in the presence of an internal electric field, so that it is reasonable that non-polar QDs undergo less spectral diffusion than polar QDs. One, however, has to be cautious with such reasoning as the spectral diffusion depends much on the sample quality so that it is not simple to compare different samples.

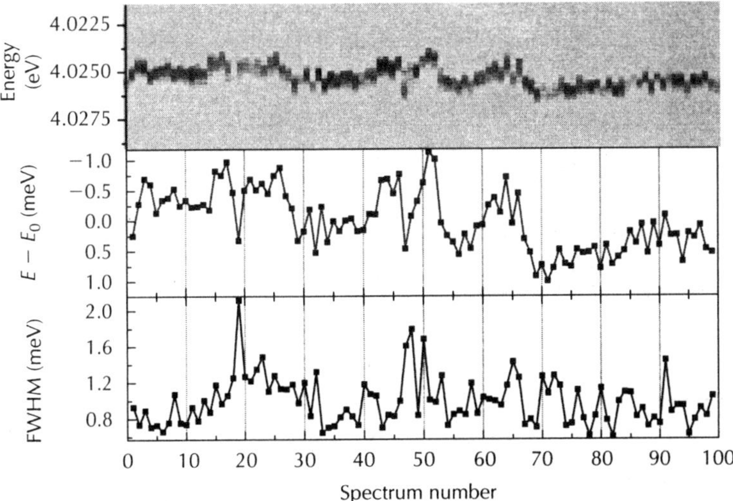

Figure 6.26   *Spectral diffusion for an isolated PL line. One spectrum is acquired every second. The fluctuations of the emission energy and of the FWHM are represented. The typical energy fluctuations are comparable with the typical FWHM.*

The next question that arises when studying single quantum dots is whether many particle states such as the biexciton and charged excitons can be observed. The biexciton was, for instance, observed by Kako and co-workers in polar GaN QDs; in that case the biexciton has a large negative binding energy ($\sim$30 meV). This unusual situation is due to the electron–hole spatial separation in polar heterostructures which enhances the ratio between the repulsive electron–electron and hole–hole interactions and the attractive electron–hole interaction, so that by adding a second exciton in the quantum dot is energy costly [57].

For polar quantum dots, the biexciton could not be observed. The reasons for this are experimental: when raising the excitation power, usually many lines grow in the vicinity of the main studied line. These new lines might correspond to excited states of other QDs lying lower in energy (recall that only the fundamental transitions of small dots with a high emission energy could be observed). The biexciton line of the studied QD might be among these new lines; however, it could so far not be clearly identified, so that the binding energy of the biexciton in non-polar GaN quantum dots is so far not known.

The narrow linewidth of the fundamental transition in non-polar GaN QDs allows the study of the coupling between the confined carriers and the acoustic phonons when raising the temperature of the sample. Indeed, when the temperature is increased, the acoustic phonon population builds up, leading to an increase of emission and absorption of acoustic phonons in the photoluminescence process. These absorption and emission events are observed as "phonon wings" above and below the main line (zero phonon line). We thus studied the single QD emission line as a function of sample temperature (Fig. 6.27). Below 60 K, the broadening of the line is dominated by the spectral diffusion so that no acoustic phonon broadening is observable. However, for higher temperature, high and low energy wings can be observed. The asymmetry between both wings is characteristic of acoustic phonon wings (the acoustic phonon emission processes are more likely than the absorption processes). These phonon wings were already observed in other

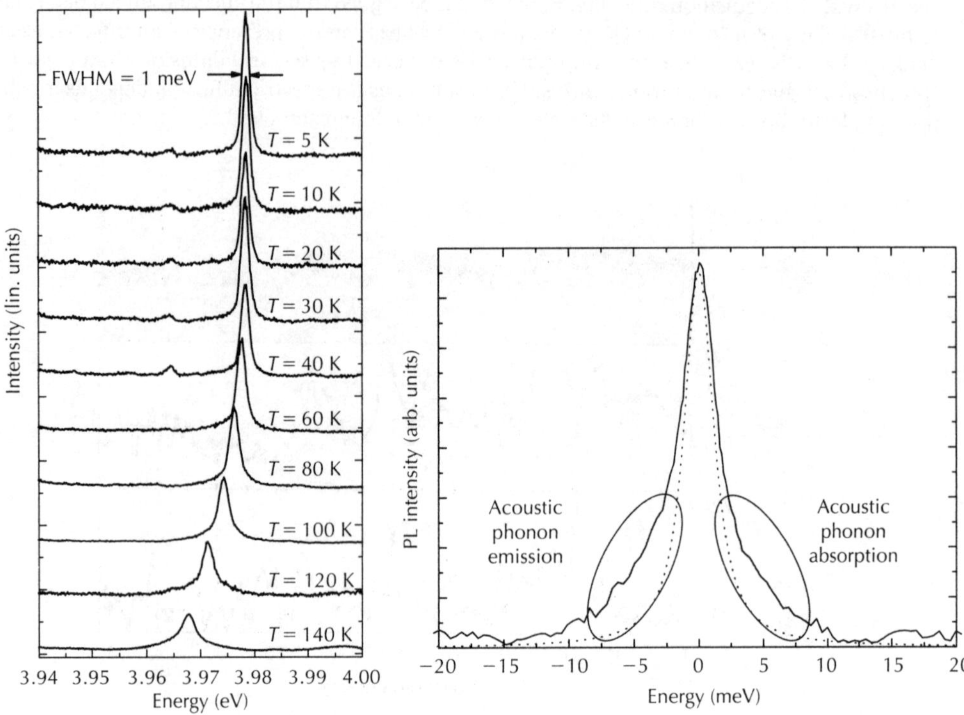

Figure 6.27   Left: *Broadening of a single QD emission line as a function of temperature.* Right: *Zoom on the curve at 140 K, the blue line is a Voigt fit to the zero phonon line.*

QD systems [59, 60, 61]. The important feature – which will be developed in the next section – is that this acoustic phonon coupling is a way of probing the confined carrier wavefunctions [59].

6.6.3   *Phonon coupling and oscillator strength: the localization hypothesis*

The basic idea behind the probing of confined wavefunctions through acoustic phonon coupling is that the maximum coupling occurs for acoustic phonons with a wavelength that matches the exciton spatial extension [59]. The properties of the acoustic phonons being known, the phonon wings can then be modelled with the exciton spatial extension as a fitting parameter, so that the fitting of experimental curves such as the one presented in Fig. 6.28 leads to the measurement of the exciton spatial extension.

In order to apply such a model, one has to propose a test wavefunction for the confined exciton. In the present study, we first of all propose to neglect any internal electric field. Indeed, we have shown previously that in macrophotoluminescence measurements, no quantum confined Stark effect is observed, so that the internal electric field in these non-polar quantum dots does not play an important role for the wavefunction geometry. In order to propose a test wavefunction, we then consider the typical length scales in our system. The typical QD size that we study is 2 nm height and 20 nm diameter. The Bohr radius being 2.8 nm in GaN, we see that the confinement due to the QD potential is strong in the vertical direction (growth axis), while it is weak in the in-plane dimensions. The wavefunction can then be factorized into a vertical component corresponding to the GaN/AlN quantum well of 2 nm thickness, and an in-plane component corresponding to a bound 2D exciton whose centre of mass is confined by the QD boundaries:

$$\psi_X(r_e, r_h) = \left(\frac{2}{L}\cos(\pi z_e/L_Z)\cos(\pi z_h/L_Z)\right)\left(\sqrt{\frac{2}{\pi a_{2D}^2}}e^{-r/a_{2D}}\right)\left(\frac{1}{\xi\sqrt{\pi}}e^{-R_p^2/2\xi^2}\right) \qquad (6.1)$$

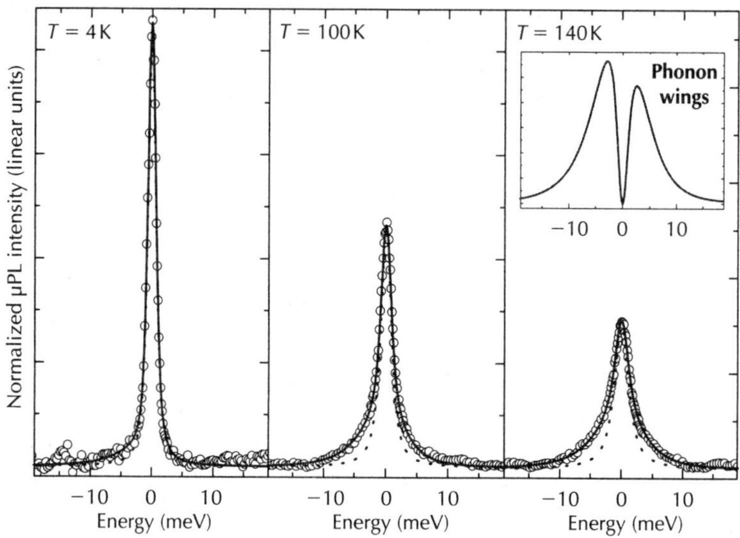

Figure 6.28   Experimental PL data (open dots), Voigt fit of the zero phonon line (dotted blue line) and fit with our phonon coupling model for $\xi = 2.1\,nm$ (red line). The inset for $140\,K$ shows the theoretical model without the ZPL, i.e. only the phonon wings, showing clearly the asymmetry between emission and absorption of acoustic phonons.

The first factor in the proposed wavefunction corresponds to the vertical confinement in a quantum well of thickness L_z with infinite barriers (which is here a very good approximation given the GaN/AlN huge band offsets). For each studied QD, we calculate the corresponding L_z from the QD transition energy from a simple QW model, assuming that the transition energy of the QD is mainly due to the vertical confinement. The second factor in Ψ corresponds to the 2D hydrogenoid exciton, r being the electron–hole relative coordinate $\mathbf{r} = \mathbf{r}_e - \mathbf{r}_h$ and a_{2D} the two-dimensional Bohr radius corresponding to the QW with infinite barriers of height L_z. The last factor is the confinement of the centre of mass of the exciton of coordinate $\mathbf{R}_p = (m_e\mathbf{r}_e + m_h\mathbf{r}_h)/(m_e + m_h)$, ξ being the parameter describing the lateral extension of the exciton.

Within this framework, the coupling to acoustic phonons can be modelled, with ξ as fitting parameter. The details of the model can be found in [56] and we shall here only describe its main features. The model is basically the Huang–Rhys model of interaction between a localized electron and phonons, applied here to a localized exciton in a quantum dot [62]. In general, one has to consider both the deformation potential and the piezoelectric potential couplings between the localized exciton and all acoustic phonon branches. For semiconductors with zinc-blende symmetry, it can be shown that only the deformation potential coupling to longitudinal acoustic phonons has to be taken into account [63]. In the present study, we are dealing with a wurtzite symmetry semiconductor so that the piezoelectric coupling potential might not be negligible. It can be shown that while the deformation potential coupling is mainly sensitive to the overall extent of the excitonic wavefunction, the piezoelectric coupling potential is large in the case when the electron and hole wavefunctions are spatially offset and thus result in a permanent dipole. For instance, for large enough polar GaN quantum dots, the piezoelectric coupling to acoustic phonons is large and dominates over the deformation potential coupling [64]. On the contrary, if the electron and hole wavefunctions are exactly superimposed so that the system is everywhere locally neutral, then the piezoelectric coupling is strictly zero [64]. Again in our case, the macrophotoluminescence experiments do not show any evidence of electron–hole spatial separation so that we choose to neglect the piezoelectric coupling of the confined exciton to acoustic phonons. This is justified *a posteriori* by the good fits obtained on the experimental data.

Another difficulty in our case is to be able to fit the zero phonon line. As already said, the zero phonon line is dominated by spectral diffusion, which we describe in a very phenomenological way. In particular, we do not have a modelling of the evolution of spectral diffusion with temperature.

In general, the line shape due to the spectral diffusion can be well fitted by a Voigt function. For each temperature, we thus use a Voigt fit of the zero phonon line, while the ξ parameter that fits the phonon wings is used as a temperature independent fitting parameter. As the zero phonon line is typically below 2 meV FWHM and the phonon wings extend up to 10 meV on each side of the zero phonon line, the fit of the ZPL and of the phonon wings can be considered as essentially independent.

The typical result for the fit is shown in Fig. 6.28, which displays the experimental data for a low, an intermediate and a high temperature. The fit of the zero phonon line is indicated in blue, while the model including the phonon coupling is displayed in red. It can be seen that the experimental curves can be very well adjusted over the entire temperature range. The fitting parameter ξ that describes the centre of mass localization is equal to 2.1 nm in this case. This corresponds to a full width at half maximum for the in-plane wavefunction of the centre of mass of 5 nm, to be compared with the diameter of 20 nm of the quantum dot. It can be shown that even with a GaN quantum dot with the shape of a cylinder with a diameter equal to the top facet of the dot and surrounded by infinite barriers (in other words, only the central cylinder of the actual QD is considered, while its sides are replaced by an infinite barrier), the lateral extension of the centre of mass would still be larger ($\xi = 3.4$ nm) than what we measure due to the phonon coupling experiment. This means that the exciton localization in the quantum dot is not solely governed by the barrier confinement, but that an additional mechanism leads to a supplementary lateral confinement.

This last conclusion can actually be probed in another way, namely by giving a closer look at the time-resolved data acquired on an ensemble of non-polar GaN QDs. Indeed, as described earlier in this article, the average decay time for an ensemble of QDs is around 280 ps. This seems to be rather short for quantum dots; however, one has to recall that a better way of describing the field–emitter interaction is the dimensionless oscillator strength for a given transition, the oscillator strength scales as the square of the emission wavelength. The oscillator strength deduced from the experimental data is in our case around 9. This is thus the same as for the typical InAs/GaAs quantum dot. However, the physics in both systems is quite different. The typical InAS /GaAs QD size is typically 3 nm height and 20 nm diameter – i.e. very similar to the GaN QDs considered here – while the Bohr radius in InAs is 28 nm, which is an order of magnitude larger than in GaN. In an InAs QD, the carriers are thus strongly confined in all directions: the confinement energy dominates over the carrier–carrier binding energy. As already described by wavefunction 7.1, for non-polar GaN QDs, the carriers are *a priori* expected to be strongly confined in the vertical direction and weakly confined in the in-plane directions. The corresponding physical situation is that of a bound two-dimensional exciton which is coherent over the in-plane surface of the QD. In that case, the regime of giant oscillator strength should be reached [65, 66]. This effect has, for instance, been clearly characterized for QDs formed by interface fluctuations of a GaAs/AlGaAs quantum well [67]. We can thus test various hypotheses for the confinement regime of the non-polar GaN QDs and see in each case which is the expected oscillator strength. We can examine a first case for which we consider the wavefunction 6.1, with a lateral confinement of the centre of mass limited only by the confining potential of the AlN barrier. As discussed earlier, in that case a lower limit for ξ can be taken as 3.4 nm. For such a wavefunction, the oscillator strength is given by:

$$f = \frac{8E_p}{E} \cdot \left(\frac{\xi}{a_{2D}}\right)^2 \tag{6.2}$$

where E is the transition energy and E_p is the interband matrix element (Kane energy). For E_p, the data found in the literature is scarce and scattered, we choose to use the value (for the conduction band to heavy-hole valence band transition) of 15.7 eV calculated by Chang and Chuang [68], which is close to the value of 14 eV recommended by Vurgaftman in a review of III-N parameters [69]. This then gives a value of the oscillator strength of 85, which corresponds to a radiative lifetime of 24 ps. This is clearly much lower than what we measure.

The opposite situation would be that of laterally strongly confined carriers, in which case the oscillator strength is given by [66]:

$$f = \frac{E_p}{E} \tag{6.3}$$

in the case of a perfect electron–hole overlap. This value is lowered if the electron and hole wavefunctions are different. The value for f given by the above formula is 4, corresponding to a radiative lifetime of 520 ps, which is larger than the mean decay time that we measure. The measured mean decay time is thus intermediate between an upper limit to the giant oscillator strength regime and a lower limit to the strongly confined case.

As already said, the Kane energy is not known precisely for GaN, and a value of 7.7 eV is reported in [70], but even with this value the precedent conclusion would still hold.

Let us now consider again the confined centre of mass wavefunction 6.1, but with this time the value for $\xi = 2.1$ nm that was deduced from the acoustic phonon coupling analysis. In that case, equation 6.2 (with $E_p = 15.7$ eV) yields $f = 33$, and correspondingly a decay time of 62 ps. This value is still much lower than what is achieved experimentally. As said earlier, the value of the Kane energy might be about half of the value with use. But, most importantly, it has to be pointed out that the enhancement of the oscillator strength in equation 6.2 is due to an excitonic effect. With $\xi = 2.1$ nm (to be compared with $a_{2D} = 1.8$ nm), the physical situation is not really that of a 2D bound exciton which is coherent over a large (much larger than a_{2D}) area. The enhancement of the oscillator strength thus should be reduced in equation 6.2.

The decay time analysis thus shows that the situation is intermediate between the strong confinement and the 2D giant oscillator strength regime, which is coherent with the acoustic phonon coupling analysis.

Going further along this analysis, one might reasonably consider that the localization length might vary from dot to dot, independently of QD size. In that case, for each emission energy (corresponding to a collection of QDs with a given size), there should be a distribution of decay times. This is indeed what is experimentally observed when looking in more detail at Fig. 6.24. Each decay curve is not mono-exponential, and is better fitted by a sum of exponentials with decay times ranging between 70 and 600 ps. In order to confirm this point, we performed time-resolved measurements on single QD lines. Figure 6.29 displays, for instance, the time resolved signal for three different quantum dots. The first striking feature is that in that case the decays are mono-exponential, which is not the case for an ensemble of QDs. The second important fact is that the decay times for the three different QDs are different, and vary non-monotonously as a function of emission energy. This supports our interpretation that the decay time is dominated by the localization length inside the QD, and not by the actual QD size [56].

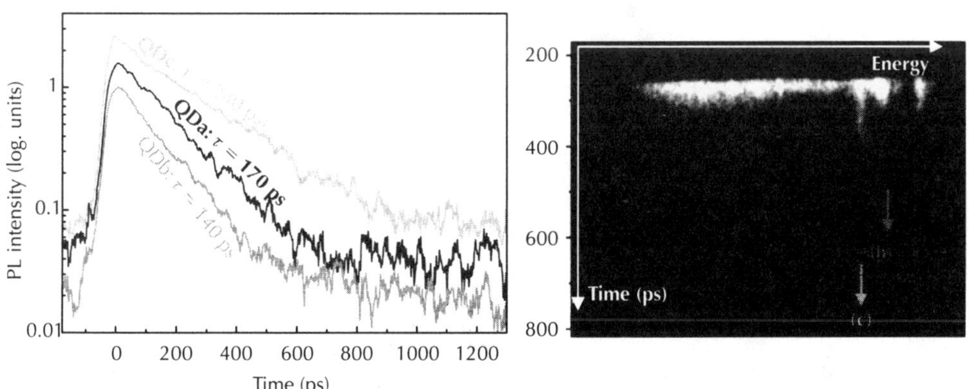

Figure 6.29   Left: *Time resolved data for single QD lines; the decay is mono-exponential and the decay time varies from dot to dot.* Right: *Same data as they appear on the streak camera CCD.*

6.7   Rare earth doping of GaN QDs

As a peculiar application of GaN QDs, their doping with rare earth ions has been achieved in view of realizing visible light-emitting diodes (LEDs). The idea of combining the RE luminescence with a semiconductor host came about by using Er and Yb (infrared) in GaAs, GaP, Si and SiC. Especially, Er-doped Si structures have been well investigated since the research was driven by the appeal of integration with Si microelectronics. For intense and thermally stable visible light emission from electroluminescence devices, wide band gap semiconductors are required since efficient energy transfers can only occur if the band gap of the host is wider than the energy difference between RE^{3+} ground states and excited state. Based on this idea, Steckl and his co-workers combined the RE^{3+} luminescent with the wide band gap of GaN and realized electroluminescence from the blue to the red spectral range [71, 72]. Furthermore the combination of different colours from RE^{3+} ions – red (Eu, Pr), green (Tb, Er, Ho) and blue (Tm, Dy, Ce) – would yield to white light emission. The choice of GaN as a semiconductor host can be further explained since it exhibits first a direct band gap, required for a good semiconductor host–RE^{3+} ion energy transfer probability and since it is a III–V semiconductor the RE ions are supposed to occupy metal sites with a 3+ charge, which was later experimentally established [73, 74].

However, the RE radiative quantum efficiency strongly depends on the carrier-mediated energy transfer process, which has to compete with fast non-radiative recombination channels. Non-radiative processes can occur, for example, at defects, which are found with a density of about $10^{10}/cm^2$ for GaN, which is a rather high number compared to other semiconductors. The result is a thermal quenching of the RE luminescence between one and two orders of magnitude from liquid helium to room temperature for REs in GaN films [75, 76].

One way to overcome this problem can be the doping of nitride quantum dots (QDs) with RE ions. Then the carrier-mediated energy transfer to RE ions should improve significantly since QDs are defect-free regions and act as carrier confinement centres so that the thermal stability of the observed RE luminescence should be enhanced drastically. In other words, the energy transfer to RE ions will be dominant over other possible non-radiative processes in the semiconductor host material.

Following the strategy described above, [0001] GaN QDs have been doped with Eu [74, 77], with Tm [78] and with Tb [79]. As expected, the quenching of RE photoluminescence is considerably decreased in the case of doped QDs with respect to the case of the bulk layer, as illustrated in Fig. 6.30.

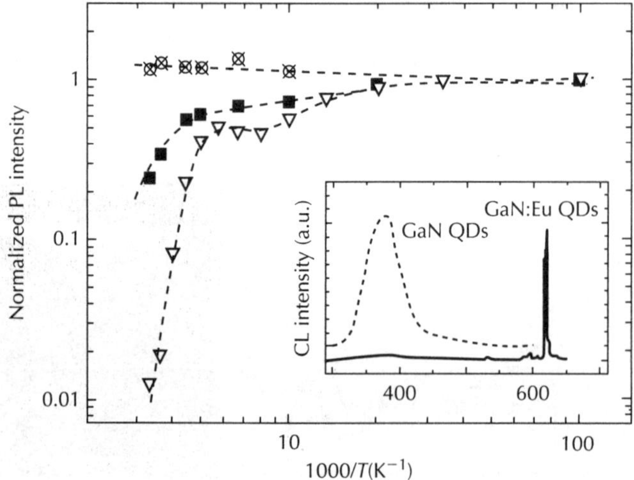

Figure 6.30 Photoluminescence intensity as a function of temperature for bulk GaN doped with Eu (open triangles), for GaN/AlN QDs (black squares) and for EU-doped GaN/AlN QDs (crossed squares). The luminescence of the EU-doped QDs is stable up to room temperature. The inset displays the cathodoluminescence of GaN QDs and Eu-doped GaN QDs. The quenching of the QD luminescence indicates an efficient energy transfer towards the Eu transitions.

In addition, confinement in QDs was found to make possible the excitation of levels in RE ions which are not accessible when using bulk material. As a matter of fact, transitions originating from $^1I_2$ and $^1I_6$ levels of Tm^{3+} ions have been observed in GaN QDs but not in bulk material [80].

6.8   Conclusion

Although being a latecomer in the family of self-assembled QD forming systems, GaN presents specificities which make it a fascinating one: its versatility to form polar and non-polar QDs, the presence of an electric field in polar QDs, the possibility, as concerns [0001] QDs, to fabricate them on a Ga-terminated or N-terminated face, the lack of interdiffusion between dots and barrier, including during the capping process, make this system rather unique. The lack of interdiffusion, in particular, makes it a model system to a certain extent and potentially allows one to probe the interplaying role of mismatch, surface and interface energy.

Optical properties of polar dots are dominated by the presence of an electric field contrary to the case of non-polar ones. However, in both cases, in-depth understanding of optical properties is far from being reached, due to a lack of control of residual doping level in both dots and barriers. Work is presently in progress to overcome these difficulties. Despite them, it is certain that optical properties of GaN QDs are as fascinating as structural ones: the large confinement resulting from the large band gap energy difference between GaN and AlN or AlGaN makes these nanostructures of potential interest for room temperature optoelectrical applications.

Acknowledgements

The authors acknowledge the contributions of the former and present students who participated in the work described in this chapter, namely C. Adelmann, B. Amstatt, T. Andreev, J. Coraux, S. Founta, N. Gogneau, E. Martinez-Guerrero, F. Rol, E. Sarigiannidou, and J. Simon. They also acknowledge the many contributions from their colleagues: E. Bellet-Amalric, J. Bleuse, C. Bougerol, V. Favre-Nicollin, J.-M. Gérard, Y. Hori, D. Jalabert, Le Si Dang, H. Mariette, E. Monroy, N. Pelekanos, D. Peyrade, M.G. Proietti, H. Renevier, J.L. Rouvière, and P. Venném gués. It is our pleasure to acknowledge the support of the NOVASiC company which provided high-quality polishing of SiC substrates.

References

1. See, for instance, the papers in the *Special Issue* of *Philosophical Magazine*, **87**, 1967 (2007).
2. F. Bernardini and V. Fiorentini, *Phys. Rev. B.* **57**, R9427 (1998).
3. T. Paskova, R. Kroeger, S. Figge, D. Hommel, V. Darakchieva, B. Monemar, E. Preble, A. Hanser, N.M. Williams, and M. Tutor, *Appl. Phys. Lett.* **89**, 051914 (2006).
4. M.D. Craven, S.H. Lim, F. Wu, J.S. Speck, and S.P. DenBaars, *Appl. Phys. Lett.* **81**, 469 (2002).
5. H.M. Ng, *Appl. Phys. Lett.* **80**, 4369 (2002).
6. C. Chen, J. Yang, H. Wang, J. Zhang, V. Adivarahan, M. Gaevski, E. Kuokstis, Z. Gong, M. Su, and M. Asif Khan, *Jpn. J. Appl. Phys.* **42**, L640 (2003).
7. P.P. Paskov, R. Schifano, B. Monemar, T. Paskova, S. Figge, and D. Hommel, *J. Appl. Phys.* **98**, 093519 (2005).
8. K. Kusakabe and K. Ohkawa, *Jpn. J. Appl. Phys.* **44**, 7931 (2005).
9. Y. Tsuchiya, Y. Okadome, A. Honshio, Y. Miyake, T. Kawashima, M. Iwaya, S. Kamiyama, H. Amano, and I. Akasaki, *Jpn. J. Appl. Phys.* **44**, L1516 (2005).
10. S. Chevtchenko, X. Ni, Q. Fan, A.A. Baski, and H. Morkoç, *Appl. Phys. Lett.* **88**, 122104 (2006).
11. N. Onojima, J. Suda, T. Kimoto, and H. Matsunami, *Appl. Phys. Lett.* **83**, 5208 (2003).
12. M.D. Craven, F. Wu, A. Chakraborty, B. Imer, U.K. Mishra, S.P. DenBaars, and J.S. Speck, *Appl. Phys. Lett.* **84**, 1281 (2004).

13. S. Founta, F. Rol, E. Bellet-Amalric, J. Bleuse, B. Daudin, B. Gayral, C. Moisson, and H. Mariette, *Appl. Phys. Lett.* **86**, 171901 (2005).
14. D.N. Zakharov, Z. Liliental-Weber, B. Wagner, Z.J. Reitmeier, E.A. Preble, and R.F. Davis, *Phys. Rev. B.* **71**, 235334 (2005).
15. M.D. Craven, S.H. Lim, F. Wu, J.S. Speck, and S.P. DenBaars, *Appl. Phys. Lett.* **81**, 1201 (2002).
16. E.J. Tarsa, B. Heying, X.H. Wu, P. Fini, S.P. DenBaars, and J.S. Speck, *J. Appl. Phys.* **82**, 5472 (1997).
17. G. Mula, C. Adelmann, S. Moehl, J. Oullier, and B. Daudin, *Phys. Rev. B.* **64**, 195406 (2001).
18. C. Adelmann, J. Brault, D. Jalabert, P. Gentile, H. Mariette, G. Mula, and B. Daudin, *J. Appl. Phys.* **91**, 9638 (2002).
19. J. Neugebauer, T.K. Zywietz, M. Scheffler, J.E. Northrup, H. Chen, and R.M. Feenstra, *Phys. Rev. Lett.* **90**, 056101 (2003).
20. B. Daudin, G. Feuillet, G. Mula, H. Mariette, J.L. Rouvière, N. Pelekanos, G. Fishman, C. Adelmann, and J. Simon, *Phys. stat. sol. (a)* **176**, 621 (1999).
21. J.E. Northrup, J. Neugebauer, R.M. Feenstra, and A.R. Smith, *Phys. Rev. B.* **61**, 9932 (2000).
22. M.D. Craven, P. Waltereit, J.S. Speck, and S.P. DenBaars, *Appl. Phys. Lett.* **84**, 496 (2004).
23. C. Adelmann, N. Gogneau, E. Sarigiannidou, J.L. Rouvière, and B. Daudin, *Appl. Phys. Lett.* **81**, 3064 (2002).
24. N. Gogneau, D. Jalabert, E. Monroy, T. Shibata, M. Tanaka, and B. Daudin, *J. Appl. Phys.* **94**, 2254 (2003).
25. S. Founta, C. Bougerol, H. Mariette, B. Daudin, and P. Vennéguès, *J. Appl. Phys.* **102**, 074304 (2007).
26. B. Daudin, F. Widmann, G. Feuillet, Y. Samson, M. Arlery, and J.L. Rouvière, *Phys. Rev. B.* **56**, R7069 (1997).
27. B. Damilano, N. Grandjean, F. Semond, J. Massies, and M. Leroux, *Appl. Phys. Lett.* **75**, 962 (1999).
28. S. Tanaka, S. Iwai, and Y. Aoyagi, *Appl. Phys. Lett.* **69**, 4096 (1996).
29. N. Grandjean and J. Massies, *Appl. Phys. Lett.* **71**, 1816 (1997).
30. M. Miyamura, K. Tachibana, and Y. Arakawa, *Appl. Phys. Lett.* **80**, 3937 (2002).
31. Y. Hori, O. Oda, E. Bellet-Amalric, and B. Daudin, *J. Appl. Phys.* **102**, 024311 (2007).
32. J.L. Rouvière, J. Simon, N. Pelekanos, B. Daudin, and G. Feuillet, *Appl. Phys. Lett.* **75**, 2632 (1999).
33. N. Gogneau, D. Jalabert, E. Monroy, E. Sarigiannidou, J.L. Rouvière, T. Shibata, M. Tanaka, J.-M. Gérard, and B. Daudin, *J. Appl. Phys.* **96**, 1104 (2004).
34. J. Coraux, B. Amstatt, J.A. Budagoski, E. Bellet-Amalric, Jean-Luc Rouvière, V. Favre-Nicolin, M.G. Proietti, H. Renevier, and B. Daudin, *Phys. Rev. B.* **74**, 195302 (2006).
35. B. Daudin, J.L. Rouvière, and M. Arlery, *Appl. Phys. Lett.* **69**, 2480 (1996).
36. J. Northrup and J. Neugebauer, *Phys. Rev. B.* **60**, R8473 (1999).
37. N. Gogneau, E. Sarigiannidou, E. Monroy, S. Monnoye, H. Mank, and B. Daudin, *Appl. Phys. Lett.* **85**, 1421 (2004).
38. D.W. Moon, H.I. Lee, B. Cho, Y.L. Foo, T. Spila, S. Hong, and J.E. Greene, *Appl. Phys. Lett.* **83**, 5298 (2003).
39. K. Sumitomo, H. Omi, Z. Zhang, and T. Ogino, *Phys. Rev. B.* **67**, 035319 (2003).
40. D. Jalabert, J. Coraux, H. Renevier, B. Daudin, C. Mann-Ho, C. Kwun-Bum, D.W. Moon, J.M. Llorens, N. Garro, A. Cros, and A. García-Cristóbal, *Phys. Rev. B.* **72**, 115301 (2005).
41. M.J. Hÿtch, E. Snoeck, and R. Kilaas, *Ultramicroscopy* **74**, 131 (1998).
42. J.L. Rouvière and E. Sarigiannidou, *Ultramicroscopy* **106**, 1 (2005).
43. S. Founta, J. Coraux, D. Jalabert, C. Bougerol, F. Rol, H. Mariette, H. Renevier, B. Daudin, R.A. Oliver, C.J. Humphreys, T.C.Q. Noakes, and P. Bailey, *J. Appl. Phys.* **101**, 063541 (2007).
44. J. Tersoff, C. Teichert, and M.G. Lagally, *Phys. Rev. Lett.* **76**, 1675 (1996).
45. F. Widmann, B. Daudin, G. Feuillet, Y. Samson, J.L. Rouvière, and N. Pelekanos, *J. Appl. Phys.* **83**, 7618 (1998).
46. A. Letoublon, V. Favre-Nicolin, H. Renevier, M.G. Proietti, C. Monat, M. Gendry, O. Marty, and C. Priester, *Phys. Rev. Lett.* **92**, 186101 (2004).
47. M.G. Proietti, H. Renevier, J.-L. Hodeau, J. Garcia, J.F. Bérar, and P. Wolfers, *Phys. Rev. B.* **59**, 5479 (1999).

48. J. Coraux, V. Favre-Nicolin, M.G. Proietti, B. Daudin, and H. Renevier, *Phys. Rev. B.* **75**, 235312 (2007).
49. J. Coraux, B. Amstatt, J.A. Budagoski, E. Bellet-Amalric, J.-L. Rouvière, V. Favre-Nicolin, M.G. Proietti, H. Renevier, and B. Daudin, *Phys. Rev. B.* **74**, 195302 (2006).
50. J. Coraux, H. Renevier, V. Favre-Nicolin, G. Renaud, and B. Daudin, *Appl. Phys. Lett.* **88**, 153125 (2006).
51. P. Waltereit, O. Brandt, A. Trampert, H.T. Grahn, J. Menniger, M. Ramsteiner, M. Reiche, and K.H. Ploog, *Nature* **406**, 865 (2000).
52. N. Akopian, G. Bahir, D. Gershoni, M.D. Craven, J.S. Speck, and S.P. DenBaars, *Appl. Phys. Lett.* **86**, 202104 (2005).
53. A. Cros, J.A. Budagosky, A. García-Cristóbal, N. Garro, A. Cantarero, S. Founta, H. Mariette, and B. Daudin, *Phys. Status Solidi B.* **243**, 1499 (2006).
54. J. Simon, N.T. Pelekanos, C. Adelmann, E. Martinez-Guerrero, R. André, B. Daudin, L.S. Dang, and H. Mariette, *Phys. Rev. B.* **68**, 035312 (2003).
55. T. Bretagnon, P. Lefebvre, P. Valvin, R. Bardoux, T. Guillet, T. Taliercio, B. Gil, N. Grandjean, F. Semond, B. Damilano, A. Dussaigne, and J. Massies, *Phys. Rev. B.* **73**, 113304 (2006).
56. F. Rol, S. Founta, H. Mariette, B. Daudin, L.S. Dang, J. Bleuse, D. Peyrade, J.-M. Gérard, and B. Gayral, *Phys. Rev. B.* **75**, 125306 (2007).
57. S. Kako, K. Hoshino, S. Iwamoto, S. Ishida, and Y. Arakawa, *Appl. Phys. Lett.* **85**, 64 (2004).
58. R. Bardoux, T. Guillet, P. Lefebvre, T. Taliercio, T. Bretagnon, S. Rousset, B. Gil, and F. Semond, *Phys. Rev. B.* **74**, 195319 (2006).
59. L. Besombes, K. Kheng, L. Marsal, and H. Mariette, *Phys. Rev. B.* **63**, 155307 (2001).
60. I. Favero, G. Cassabois, R. Ferreira, D. Darson, C. Voisin, J. Tignon, C. Delalande, G. Bastard, Ph. Roussignol, and J.-M. Gérard, *Phys. Rev. B.* **68**, 233301 (2003).
61. E. Peter, J. Hours, P. Senellart, A. Vasanelli, A. Cavanna, J. Bloch, and J.-M. Gérard, *Phys. Rev. B.* **69**, 041307(R) (2004).
62. K. Huang and A. Rhys, *Proc. R. Soc. London, Ser. A.* **204**, 404 (1950).
63. T. Takagahara, *Phys. Rev. B.* **60**, 2638 (1999).
64. B. Krummheuer, V.M. Axt, and T. Kuhn, *Phys. Rev. B.* **65**, 195313 (2002).
65. A.V. Kavokin, *Phys. Rev. B.* **50**, 8000 (1994).
66. L.C. Andreani, G. Panzarini, and J.-M. Gérard, *Phys. Rev. B.* **60**, 13276 (1999).
67. J. Hours, P. Senellart, E. Peter, A. Cavanna, and J. Bloch, *Phys. Rev. B.* **71**, 161306(R) (2005).
68. S.L. Chuang and C.S. Chang, *Phys. Rev. B.* **54**, 2491 (1996).
69. I. Vurgaftman, J.R. Meyer, and L.R. Ram-Mohan, *J. Appl. Phys.* **89**, 5815 (2001).
70. J.S. Im, A. Moritz, F. Steuber, V. Härle, F. Scholz, and A. Hangleiter, *Appl. Phys. Lett.* **70**, 631 (1996).
71. A.J. Steckl, M. Garter, D.S. Lee, J. Heikenfeld, and R. Birkhahn, *Appl. Phys. Lett.* **75**, 2184 (1999).
72. J. Heikenfeld, M. Garter, D.S. Lee, R. Birkhahn, and A.J. Steckl, *Appl. Phys. Lett.* **75**, 1189 (1999).
73. H. Bang, S. Morishima, Z. Li, K. Akimoto, M. Nomura, and E. Yagi, *J. Cryst. Growth* **237–239**, 1027 (2002).
74. Y. Hori, X. Biquard, E. Monroy, D. Jalabert, F. Enjalbert, L.S. Dang, M. Tanaka, O. Oda, and B. Daudin, *Appl. Phys. Lett.* **84**, 206 (2004).
75. E.E. Nyein, U. Hömmerich, J. Heikenfeld, D.S. Lee, A.J. Steckl, and J.M. Zavada, *Appl. Phys. Lett.* **82**, 1655 (2003).
76. T. Andreev, Y. Hori, X. Biquard, E. Monroy, D. Jalabert, A. Farchi, M. Tanaka, O. Oda, L.S. Dang, and B. Daudin, *Phys. Rev. B.* **71**, 115310 (2005).
77. Y. Hori, D. Jalabert, T. Andreev, E. Monroy, M. Tanaka, O. Oda, and B. Daudin, *Appl. Phys. Lett.* **84**, 2247 (2004).
78. T. Andreev, Y. Hori, X. Biquard, E. Monroy, D. Jalabert, A. Farchi, M. Tanaka, O. Oda, L.S. Dang, and B. Daudin, *Phys. Rev. B.* **71**, 115310 (2005).
79. Y. Hori, T. Andreev, D. Jalabert, E. Monroy, L.S. Dang, M. Tanaka, O. Oda, and B. Daudin, *Appl. Phys. Lett.* **88**, 053102 (2006).
80. T. Andreev, N.Q. Liem, Y. Hori, M. Tanaka, O. Oda, D.L.S. Dang, B. Daudin, and B. Gayral, *Phys. Rev. B.* **74**, 155310 (2006).

CHAPTER 7

Optical and Vibrational Properties of Self-assembled GaN Quantum Dots

Núria Garro, Ana Cros, Alberto García-Cristóbal, and Andrés Cantarero

Institute of Materials Science, University of Valencia (Spain)

7.1 Introduction

Quantum dots (QDs) based on group III nitrides (III-N) are expected to be the active medium of new optoelectronic devices operating at high powers and high temperatures. Besides the well-known advantages of their bulk and quantum well (QW) counterparts [1], III-N QDs provide strong confinement of carriers in nearly perfect zero-dimensional boxes. Consequently, their optical response is almost insensitive to the high dislocation density present in state-of-the-art epitaxial nitride semiconductors. Furthermore, quantum effects provide new degrees of freedom for the design of advanced devices. As an illustrative example of the potential of these nanostructures we highlight that triggered single-photon emission has been achieved from GaN QDs at 200 K, a temperature easily reachable with thermo-electric cooling and much higher than temperatures required for quantum information-processing schemes based on InAs QDs [2]. III-N QDs can be grown by a variety of techniques, ranging from molecular beam epitaxy and metal-organic vapour phase chemical deposition to other less extended methods, such as vapour–liquid–solid mechanism, reactive laser ablation or inorganic pyrolysis [3]. While the mentioned growth schemes result in the formation of QDs that nucleate randomly, other techniques aim for a more ordered disposition of the dots. Selective growth of InGaN QDs on hexagonal uniform GaN pyramids was demonstrated by Tachibana *et al.* [4]. Another approach takes advantage of the embedding of GaN quantum discs in AlGaN nanocolumns grown on Si(111) substrates by plasma-assisted molecular beam epitaxy [5]. More recently, QD arrays have been selectively fabricated by holographic lithography and subsequent ion etching of InGaN QW structures [6]. Other approaches envisage the one-dimensional alignment of GaN dots when grown on vicinal substrates [7]. In the following, we concentrate on those systems of dots that appear spontaneously during epitaxial growth without the need of artificial post-processing, and designate them as self-assembled or self-organized quantum dots regardless of the mechanism responsible for their appearance.

Nowadays, research efforts on III-N QDs are mainly focused on the understanding and control of their optical properties. These properties are strongly influenced by the existence of internal electric fields of the order of several MV/cm. As a result of such built-in fields, the photoluminescence spectrum of GaN QDs is often redshifted below the bulk GaN band gap energy. Taking advantage of this shift, the emission of III-N QDs can sweep most of the visible spectrum. Photo- and cathodoluminescence are widely spread techniques in the investigation of the optical properties of both QD ensembles and single QDs. The resolution of single GaN QD emission has only been possible very recently thanks to the improvement in the control of the QD density and the development of processing techniques [8, 9]. Raman spectroscopy has also been used in the investigation of III-N QDs. It allows the analysis of strain, composition, size and orientation of the constituting materials through the study of their atomic vibrations. The widespread use of Raman scattering for nanostructure characterization is, however, limited by the small scattering rates. As a consequence, the experiments are usually performed on samples where the structure of interest is repeated over several periods. One way to overcome this limitation takes advantage of the enhancement of the Raman scattering rate under resonant conditions. Experimental efforts have to be complemented with theoretical modelling which is essential for understanding the interplay between electric field and quantum confinement effects in these systems. The results point out that the strain state of the system contributes significantly to the internal electric field and must be considered in full detail.

Several authors have reviewed in the past the properties of III-N QDs. Huang *et al.* have compiled a general description of growth methods and optical characterization of InGaN and GaN QDs [3]. Other authors have given partial accounts summarizing their own work [10–13]. However, being a field of lively activity it is timely to present an updated review of the properties of self-assembled nitride quantum dots. The scope of this chapter is restricted to the studies on GaN QDs, which present distinctive features as compared to InGaN QDs such as the different emission spectral range and the absence of alloy and segregation effects. In particular, we focus on the optical and vibrational properties of the dots, and pay special attention to the interplay between the strain state of the dots, the magnitude of their internal electric field and its effects on their optical properties. We shall review results obtained mostly on wurtzite QDs, though the properties of zinc-blende GaN QDs will be mentioned briefly. The aim of this chapter is to give an overview of the different results obtained by the various research teams working on the topic rather than analyse in detail the peculiarities of one specific type of sample. In chapter 6, Daudin and co-workers perform such an exhaustive review of the growth and structural characteristics of the GaN QDs grown by plasma-assisted molecular beam epitaxy.

7.2 Growth and structural characterization

It is well known that one of the most important technical barriers that prevent achieving high-quality nitride materials is the lack of adequate substrates for homoepitaxy. Lattice mismatch and thermal expansion coefficient differences between nitride films and the available substrates induce large densities of structural defects (mainly threading dislocations, with densities of the order of 10^{10} cm^{-2}). Such difficulties have been overcome by the growth of high-quality QDs. In the following we give a brief overview of the most relevant growth techniques employed for the spontaneous or self-assembled growth of GaN QDs. Special attention will be paid to epitaxial growth categorized into two different modes: Volmer–Weber (VW) mode corresponding to three-dimensional (3D) island growth and Stranski–Krastanov (SK) mode, where the deposition of a strained two-dimensional (2D) wetting layer (WL) is followed by elastic relaxation through 3D islanding. The strain distribution in GaN QDs will be analysed in detail at the end of this section.

7.2.1 Growth and structural properties

Self-organized GaN QDs have been successfully grown by several epitaxial techniques like molecular beam epitaxy (MBE) [14, 15], metal-organic chemical vapour deposition [16, 17] (MOCVD) or a combination of both techniques [18, 19]. Depending on the substrate material and orientation, both wurtzite [14, 16, 19] and zinc-blende [20] dots have been obtained. Dots with the wurtzite structure are typically grown on a buffer layer of AlN or Al$_x$Ga$_{1-x}$N deposited on sapphire (Al$_2$O$_3$), 4H-SiC(0001), 6H-SiC(0001) or Si(111), while dots with the zinc-blende structure have been obtained starting from a 3C-SiC(001) template grown by MOCVD on Si(001). Characteristic values of QD sizes and densities obtained by the relevant research groups working in the field are displayed in Table 7.1. Varying the growth techniques and conditions, the dot density may cover a wide range (10^8–10^{11} cm^{-2}), although most of the samples obtained so far lie in the high-density end of the range (10^{10} 10^{11} cm^{-2}).

Gallium nitride QDs were grown for the first time by Tanaka and co-workers via MOCVD [16]. They demonstrated that it is possible to take advantage of the antisurfactant properties of Si on

Table 7.1   Typical parameters of GaN self-assembled QDs grown by several authors. When there is no direct report of the representative values of the dot height h and diameter D they have been estimated from other data specified by the authors.

Spacer	h/D (nm)	Dot density (cm^{-2})	Substrate	T_s (°C)	Growth technique/ mode	Ref.
AlGaN	6/40	3×10^9	6H-SiC	1100	MOCVD/VW	[16]
AlGaN	$h \sim 20\text{–}200$	10^8	Al$_2$O$_3$	1100	MOCVD/VW	[21]
AlN	1.6/35	$<1.6 \times 10^8$	6H-SiC	1080	MOCVD/SK	[17]
AlN	1.5–4/20	10^{10}	Al$_2$O$_3$	960	MOCVD/SK	[22]
AlN	2.5–4/20	10^{11}	Al$_2$O$_3$	700	PAMBE/SK	[23]
AlN	5/20	$1\text{–}5 \times 10^{11}$	Si(111)	800	MBE/SK	[15]
AlN	3/15	$10^8\text{–}10^{11}$	Al$_2$O$_3$	750	PAMBE/MSK	[24]
AlN	4/40	$10^{10}\text{–}10^{11}$	6H-SiC	750	PAMBE/MSK	[25]
AlN	2.2/20	10^{11}	a-plane 6H-SiC	700	PAMBE/SK	[26]
AlGaN	$D \sim 9.3$	$>3 \times 10^{11}$	6H-SiC	600	GSMBE	[18]
AlN	8/4–20	4×10^9	6H-SiC	700	VLS	[19]

Al$_x$Ga$_{1-x}$N surfaces ($x \sim 0.15$) to induce the formation of GaN islands. The introduction of the antisurfactant (tetraethylsilane, TESi, with H$_2$ as the carrier gas) modifies the nitride growth kinetics so that the surface-free energy can be artificially controlled to change the growth mode of GaN from 2D to 3D. The detailed study of the growth mechanism involved [27] indicates the formation of a one monolayer thick Si–N mask (nano-mask) that influences the morphology of the deposited GaN surface, giving rise to the formation of the quantum structures and contributing to the termination of threading dislocations in the GaN film. In Fig. 7.1 a high-resolution transmission electron microscopy (HRTEM) image illustrates the morphological characteristics of the resulting dots [28]. The absence of a WL indicates that they are apparently grown on a VW mode. They have a disk-like shape and approximate dimensions of 20 nm diameter and 5 nm height. The density and size of the dots may be controlled by varying the growth temperature, the feeding rate of TESi and the GaN coverage.

Figure 7.1   HRTEM image of GaN QDs grown epitaxially on an AlGaN layer by MOCVD [28]. With permission from AIP© 2000.

A modification of this method has been recently proposed by K. Pakula *et al.* [21, 29]. They combined the antisurfactanct and etching properties of SiH$_4$ on Al$_x$Ga$_{1-x}$N to obtain QDs of pyramidal shape with heights ranging from 20 to 200 nm and densities as low as 10^8 cm^{-2} which allow the study of the optical properties of single QDs without artificially masking the samples. The use of antisurfactants may be advantageous for the growth of QDs in systems where the lattice mismatch between the dot material and the substrate is very small and does not allow an SK growth mode. In this case the resulting dots are to a large extent relaxed.

Alternatively, in the system GaN/AlN, where the in-plane lattice mismatch between the dot and substrate materials is large enough (~2.4%), GaN QDs may be grown by MOCVD in the SK mode [17]. The formation of GaN QDs on a very flat surface of AlN results from a delicate balance between III/N ratio and growth temperature. Indeed, it has been found that the dot formation takes place under a very low III/N ratio, around 100 times smaller than that used in the conventional two-dimensional MOCVD GaN growth. The experiments indicate that substrate temperatures higher than 990°C inhibit the formation of dots due to the large migration and evaporation rates of Ga atoms. The typical diameter and height of the QDs were of the order of 20 and 2 nm, respectively. By this method it is possible to obtain dot densities as small as 10^8 cm^{-2}, and they can be increased up to 10^{10} cm^{-2} by changing the growth conditions. Further details can be found in the review paper of Arakawa and Sako [13]. More recently, MOCVD growth of SK GaN QDs has also been reported by Grandjean and his collaborators at EPFL [22].

The MBE growth of GaN QDs by the SK mode was initiated by Daudin and co-workers already in 1997 [14]. They demonstrated that, under particular growth conditions, the deposition of a few monolayers (MLs) of GaN on AlN(0001) by plasma-assisted MBE (PAMBE), where the active nitrogen is provided by an RF plasma source, is followed by the relaxation of strain in the 2D layer through 3D islanding, a process characteristic of the SK growth mode. Besides being governed by strain relaxation, the growth of QDs by MBE is extremely sensitive to the III/N ratio. Nitrogen-rich conditions lead to the formation of small islands on top of a 2 ML thick WL when the deposition thickness exceeds a critical value slightly over 2 ML (2.25 ML) [30]. The size and density of the islands increase with GaN coverage until the density saturates, with the subsequent appearance of a bimodal distribution of dot sizes. In this regime, dot densities around 10^{11} cm^{-2} are obtained. Metal-rich conditions, on the other hand, reveal the autosurfactant properties of Ga for GaN growth. In these conditions, a dynamically stable Ga bilayer accumulates on top of the GaN surface, inhibiting the 2D–3D transition [31], much as happens during the growth of InAs on GaAs where the island formation is suppressed under excess In conditions [32]. The formation of the GaN QDs takes place after the subsequent evaporation of the Ga layer by growth interruption under vacuum, a growth procedure known as modified Stranski–Krastanov (MSK) [25]. This method allows the control of the dot density in the 10^{10}–10^{11} cm^{-2} range. Much smaller densities have been achieved following a similar growth scheme by J. Brown *et al.* [24]. As shown in Fig. 7.2, a strong correlation between QD density and GaN coverage is observed for a growth rate of 0.23 ML/s when the Ga/N flux ratio is in the Ga droplet regime at 750°C. Medium to high GaN coverage (above 2.5 ML) results in a homogeneous density, whereas a density gradient across the sample, reaching values as low as 3×10^8 cm^{-2}, has been observed for low GaN coverage.

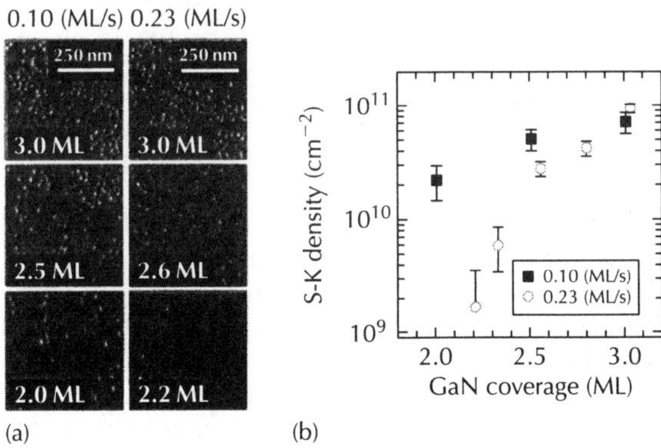

Figure 7.2   (a) Atomic force micrographs of GaN/AlN QDs grown by PAMBE for different GaN coverage and growth rates. (b) QD density as a function of GaN coverage [24]. With permission from AIP© 2004.

GaN QDs have also been grown using ammonia as the nitrogen active source for MBE on Si(111) or sapphire substrates covered by a thick AlN buffer layer [15]. Similar to the MSK growth, 3D islands are formed only if a growth interruption without ammonia flux is performed.

A critical thickness of 3 ML GaN was determined for this growth method. The QD size can then be easily controlled by varying the amount of GaN deposited with an upper limit of 12 ML to avoid plastic relaxation. Although typical QD densities are of a few times $10^{11}\,\mathrm{cm}^{-2}$, a recent report has shown that some samples may present a gradient of densities across the wafer going down to $10^{9}\,\mathrm{cm}^{-2}$ [9].

The shape of the QDs obtained by SK growth is similar for the different growth methods (MOCVD or MBE). They are truncated hexagonal pyramids with $\{1\bar{1}03\}$ facets inclined 32° with respect to the (0001) direction [13, 14, 33, 34]. In Fig. 7.3 the shape of MOCVD grown QDs obtained by atomic force microscopy (AFM) and cross-sectional HRTEM is presented. The aspect ratio (height/ diameter) of the dots depends on GaN coverage and growth temperature. While in other material systems the capping process strongly influences the shape and composition of the buried QDs, the GaN/AlN dots grown in the SK mode do not exhibit significant Ga–Al interdiffusion [36]. More recently, it has also been shown that during the capping process with AlN, the GaN/AlN QDs only experience a slight reduction of their volume, but do not change their shape [35].

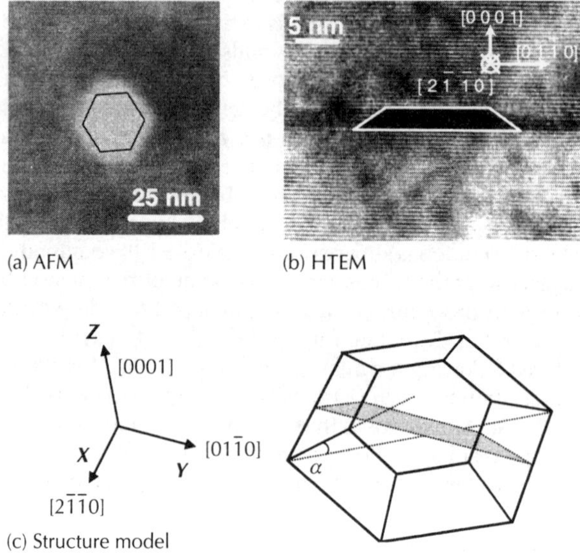

Figure 7.3   (a) AFM image of an uncovered GaN QD grown in the SK growth mode by MOCVD. (b) Cross-sectional HRTEM image of a covered QD. (c) Model for the QD structure. The angle α corresponds to 32° [13]. With permission from Wiley.

The smoothing effect of the AlN cap layer allows growing superlattices containing many layers of GaN QDs that preserve an excellent crystal quality. Furthermore, in QD multilayer structures with a thin enough spacer, QDs display a vertical alignment along the growth direction as shown in Fig. 7.4. The value of the critical AlN spacer thickness for which the vertical correlation between adjacent GaN dot layers is observed seems to depend not only on QD height, but also on its diameter [37, 38]. According to a model developed by Tersoff *et al.* [39], the vertical correlation between adjacent layers of dots is thought to be mediated by the strain field in the spacer that results from the presence of the buried QDs. The modifications in the strain in GaN/AlN QDs when going from an isolated dot to a multilayer sample will be analysed in section 7.2.2 in the light of the continuum elasticity model. Another effect of the piling-up process is the homogenization of the dot sizes in successive dot planes accompanied by an average increase of their diameter, see Fig. 7.4. The spacer layer acts like a bandpass filter that prevents the formation of QDs of smaller size. In Fig. 7.5 it is shown that, in correspondence to the diameter increase, a decrease of the dot density takes place as the number of QD layers increases [37].

Figure 7.4   HRTEM of vertically aligned GaN/AlN QDs grown by PAMBE in the SK mode at 720°C. The 2D GaN WL is clearly visible [23]. With permission from AIP© 1998.

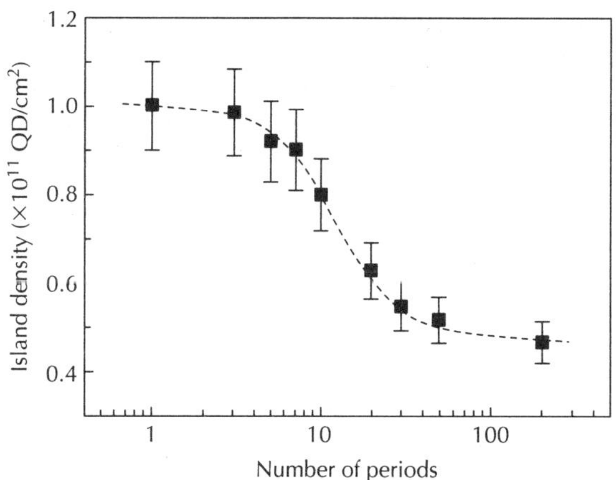

Figure 7.5   QD Areal density in the upper layer of GaN QD multilayer samples as a function of the number of layers [37]. With permission from AIP© 2004.

Until very recently, the (0001) wurtzite direction has been preferentially chosen for the growth of GaN/AlN QDs. Growth directions perpendicular to the polar c-axis may, however, pose interesting advantages for the optical performance of the QDs, as will be shown in section 7.4.5. Incidentally, these so-called "non-polar" GaN QDs have been grown on (112̄0)-oriented SiC substrates [26], and their growth dynamics are very different from those of the more common "polar" QDs grown along the c-axis. Extensive details on the growth of these structures can be found in chapter 6.

Other less extended growth methods have demonstrated the possibility of obtaining GaN quantum dots without the use of surfactants or the need of high lattice–mismatch between the GaN layer and the substrate. These techniques take advantage of the low melting point of Ga, and rely on the deposition of Ga droplets on the substrate surface, which are subsequently nitridated. Among other methods, we can mention the procedure developed by Kawasaki *et al.* [18] using a gas-source MBE (GSMBE) chamber, and the vapour–liquid–solid mechanism (VLS) proposed by Hu *et al.* [19], see Table 7.1.

7.2.2   Strain field distribution

The detailed understanding of the strain distribution in SK self-assembled QDs is of paramount importance for various reasons. In the first place, the strain relaxation acts as the main driving force in the self-organized growth itself. It is therefore important for understanding the equilibrium dot shape and the spatial correlation in the positioning of the islands. The local strain at the surface above the capped dots is also determinant for the vertical ordering of the subsequent dot layers [40]. Second, as in every pseudomorphic structure, the strain substantially modifies the band structure. And finally, in the particular case of III-N heterostructures, due to the large piezoelectric effect, the strain results in high internal electric fields, which in turn affect strongly the electronic and optical properties (see section 7.4.2). In addition, the study of the actual strain distribution in a given sample allows quantifying its deviation from the expected elastic relaxation, and therefore can shed some light on the presence of defects, such as dislocations, stacking faults, etc.

Information about the strain can be accessed experimentally through structural characterization techniques (electron diffraction and transmission, X-ray diffraction, ion scattering) and optical probe techniques (most notably Raman and infrared spectroscopy). In the case of GaN/AlN QDs, rather exhaustive studies of the strain by means of structural characterization techniques have been performed, as will be described in detail in chapter 6. The corresponding light scattering studies will be reviewed in section 7.3. On the theoretical side, two general approaches are used for the calculation of the elastic strain connected with self-assembled islands, namely, atomistic and continuum elasticity simulations [41]. In the former approach, two- and many-atom potentials are used for the calculation of the elastic energy stored in the sample; the strain is obtained by minimizing this energy. In the continuum elasticity calculations the actual structure is replaced by an elastic continuum and the strain distribution is obtained again by minimizing the elastic energy or by solving the elasticity equilibrium equation. The relevant parameters within this approach, besides the elastic constants, are the misfit strains:

$$\varepsilon_a = \frac{a_{AlN} - a_{GaN}}{a_{GaN}} = -0.024$$

$$\varepsilon_c = \frac{c_{AlN} - c_{GaN}}{c_{GaN}} = -0.039$$

(7.1)

The lattice constants of wurtzite GaN and AlN can be found in Table 7.2. Note that here the lattice mismatch is more moderate than in the case of InAs/GaAs dots, where it is around -0.07. Both modelling strategies, atomistic and continuum, have been applied to every kind of pseudomorphic structures, and they appear to give the same qualitative features and very similar quantitative results in many cases [42]. In the case of III-N wurtzite QDs, the atomistic valence-force-field method has been used to obtain the strain distribution in [43–46]. On the other hand, continuum elasticity models with various degrees of sophistication have also been applied to those nanostructures [47–51].

Table 7.2   Material parameters of bulk wurtzite GaN and AlN used to calculate the strain [52].

	GaN	AlN
a (Å) at $T = 300\,K$	3.189	3.112
c (Å) at $T = 300\,K$	5.185	4.982
C_{11} (GPa)	390	396
C_{33} (GPa)	398	373
C_{12} (GPa)	145	137
C_{13} (GPa)	106	108
C_{44} (GPa)	105	116

In order to illustrate the main features of the strain distribution of GaN/AlN(0001) QDs, we present here model calculations for a simplified situation, in which linear elasticity is assumed and the dot is considered as a misfitting inclusion in an otherwise infinite matrix. The model is based on the Eshelby's method of inclusions [53] generalized to account for the transversely isotropic elastic symmetry of the wurtzite crystal structure [54]. The calculations are similar to those presented in [47], except for the fact that a single inclusion in an infinite matrix is considered here, whereas a free-standing periodic array of dots is analysed in [47]. In order to avoid time-consuming three-dimensional calculations, the reported pyramidal shape of the dots (see section 7.2.1) was modelled with a truncated cone shape, with base radius R (diameter $D = 2R$), height h and an angle between the base and lateral surface equal to $30°$. Unless otherwise stated, this shape will be adopted for the remainder of the calculations shown in this work. The difference between the results for both shapes, pyramid and cone, is negligible except in the surface of the dot [47]. One important parameter for the specification of the dot geometry is the aspect ratio, $r = h/D$. The results for a QD of typical dimensions are shown in Fig. 7.6.

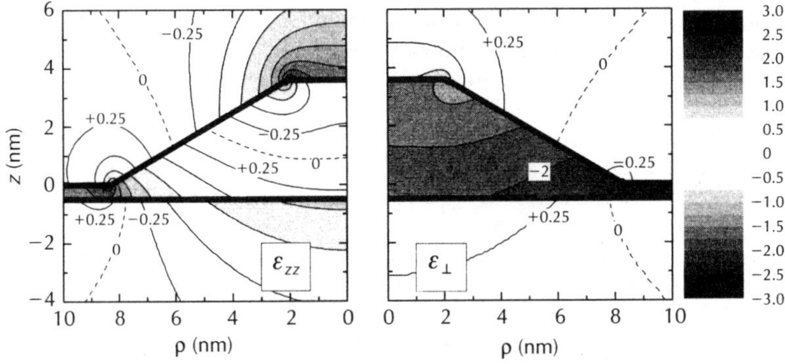

Figure 7.6   Contour plots of the strain components ε_{zz} and $\varepsilon_\perp = (\varepsilon_{xx} + \varepsilon_{yy})/2$ (in %) in a GaN QD with dimensions $R = 8.3\,nm$ and $h = 3.6\,nm$ (i.e. $r \approx 0.2$), laying on a WL of thickness $d = 0.5\,nm$. For the sake of simplicity, equal elastic constants (those of AlN, see Table 7.2) have been assumed throughout the whole structure. The dashed contour lines correspond to zero strain and the rest of the lines correspond to successive increments of $\pm 0.25\%$.

On account of its relative flatness, one approximation frequently used for the state of deformation of the QDs is to assume that it is equivalent to that of a GaN QW grown pseudomorphically on AlN (0001). This so-called biaxial strain is given by:

$$\varepsilon_{xx} = \varepsilon_{yy} = \varepsilon_a$$
$$\varepsilon_{zz} = -\frac{2C_{13,\text{GaN}}}{C_{33,\text{GaN}}}\varepsilon_{xx}.$$

(7.2)

The elastic constants of GaN and AlN can also be found in Table 7.2. The results in Fig. 7.6 reveal the limitations of the above simplified approximation. Due to the presence of the lateral surfaces, the strain in the GaN/AlN QD deviates considerably from the biaxial case. The general features are the following:

(i)   The strain is not homogeneous inside the dot, in contrast with the biaxial approximation. In general, the whole volume of the dot is compressed in the in-plane directions, i.e. $\varepsilon_\perp = (\varepsilon_{xx} + \varepsilon_{yy})/2 < 0$. On the contrary, as a consequence of the barrier constraint through the lateral surfaces, as one moves from the base to the top of the dot the strain ε_{zz} passes from tensile to compressive. In the case of Fig. 7.6, the average values at the QD base ($z = 0$) are $<\varepsilon_\perp> = -2.12\%$ (less negative than the misfit strain $\varepsilon_a = -2.4\%$) and $<\varepsilon_{zz}> = +0.75\%$ (much less than that predicted by the biaxial approximation $= +1.3\%$), and their ratio $\varepsilon_{zz}/\varepsilon_\perp = -0.35$ is clearly away from the biaxial value (-0.53). This different behaviour accounts for the partial relaxation of the stored elastic energy associated with the dot formation.

(ii) The strain is not zero in the barrier: in the regions above and below the dot, the AlN matrix is expanded in the plane and compressed along the z-direction. The deformation presents a maximum at the bottom and top surfaces and slowly decays with the distance.

(iii) The shear components of the strain (not shown in Fig. 7.6) are non-vanishing, though they are strongly localized to the perimeter of the bottom and top surfaces of the dot (in general, to the edges joining surfaces with different orientation), where they can reach rather large values.

It is to be noted, however, that the results shown in Fig. 7.6 change qualitatively when the geometrical parameters, most notably the QD aspect ratio r, are varied. Thus, for higher values of r (e.g. >0.25), ε_{zz} can become compressive through most of the dot volume, in radical contrast to the biaxial approximation, whereas for smaller values (e.g. $r < 0.15$), the biaxial approximation gives reasonable results. Fortunately, it is usually found that the aspect ratio of the dots within a given sample is approximately constant. For the sake of simplicity, some authors assume an isotropic elastic approximation in the study of the strain in wurtzite-type QDs [51, 55]. It must be warned that, while the general trends shown in Fig. 7.6 can be emulated with such approximation, it is impossible to reproduce accurately the quantitative values.

It is also instructive to comment about the modifications in the strain when going from an isolated dot to a series of stacked dots. These results are relevant for the interpretation of experiments performed on multiple dot layer samples, as will be seen in sections 7.3 and 7.4. In Fig. 7.7 we present the results of the on-axis ($\rho = 0$) strain for various stacks of varying number of QD layers N.

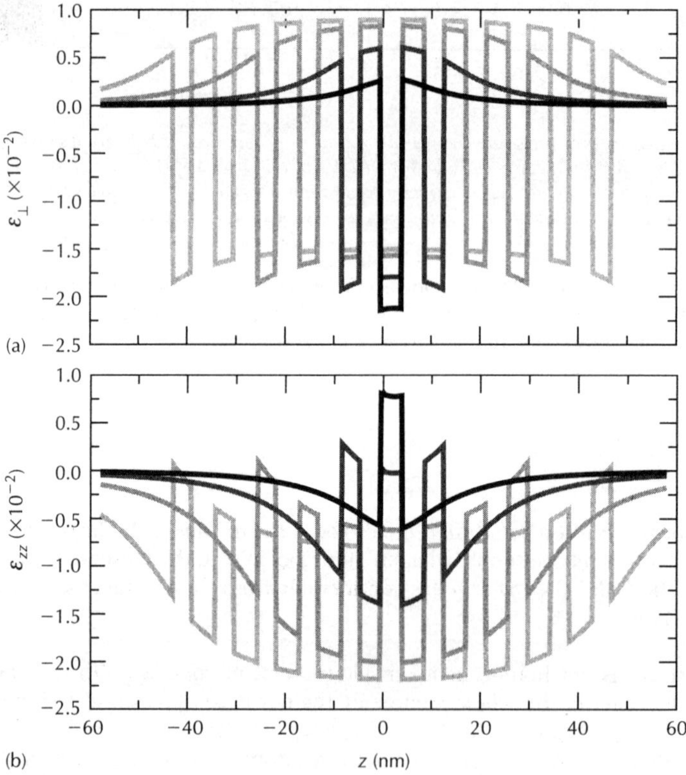

Figure 7.7   Strain components $\varepsilon_{\perp}$ (a) and ε_{zz} (b) along the axis of a stack of $N = 1, 3, 7, 11$ and 25 dots. The grey colour of the curves goes from darker to lighter as N increases. The dots lie on a WL of thickness $d = 0.5\,nm$. The sizes of the dots have been taken following the experimental results of [37], i.e. R = 17.5 nm and h = 3.8 nm (therefore, r = 0.11), with an interdot spacer thickness of 8.6 nm.

Within the volume of the central dot, the compressive in-plane deformation $\varepsilon_\perp$ is reduced when stacking an increasing number of dots, whereas ε_{zz} experiences a change of sign (from positive to negative). In the matrix region in between the dots, the in-plane expansion given by $\varepsilon_\perp$ increases, whereas the matrix becomes more and more compressed along the z-direction. It can be seen that the above trends saturate quickly as N increases: when going from $N = 11$ to $N = 25$ the strain varies very little, and can be taken as rather constant for $N > 25$, meaning that the interaction between the dots separated by more than 10 periods can be neglected. This is already apparent in Fig. 7.7 when looking at the extension into the barrier of the strain field created by a single dot.

Figure 7.8 shows a comparison of the volume averaged in-plane strain $<\varepsilon_\perp>$ in an isolated dot, on the left, and in the central dot of a stack of 21, on the right, as a function of the dot dimensions. These results have been employed in the interpretation of the effects of stacking on the Raman scattering [56] (see section 7.3). Whereas the strain in the isolated QD is independent of the dimensions for a given ratio, as determined by the straight lines in Fig. 7.8a, it presents noticeable differences in the QD of the stack. First, the strain is almost insensitive to the QD diameter provided it is larger than 20 nm. On the other hand, the values of strain change considerably with height. Even more pronounced changes were found for the strain component ε_{zz} (not shown).

Calculations of the elastic strain field of the above type have been successfully employed in the analysis of various experiments of X-ray diffraction [57, 35], TEM [58] and ion scattering [59]. To illustrate the relevance of the theoretical simulations we present in Fig. 7.9 a comparison

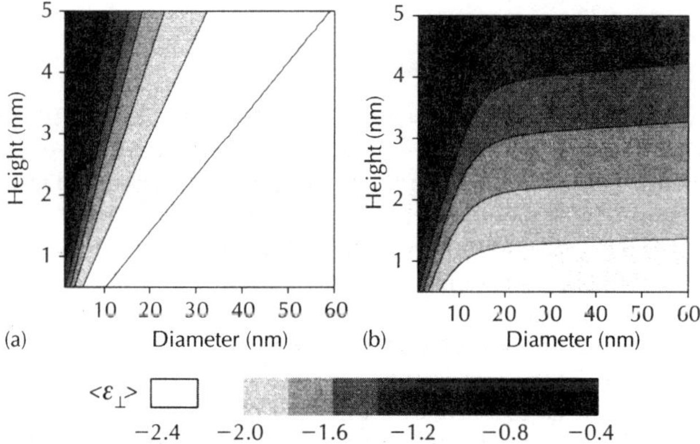

Figure 7.8   Averaged value of the in-plane strain $\varepsilon_\perp$ (in %) inside the QD as a function of its height and diameter for (a) an isolated QD and (b) the central dot of a stack of 21, separated by an 8 nm AlN spacer. With permission from APS© 2006.

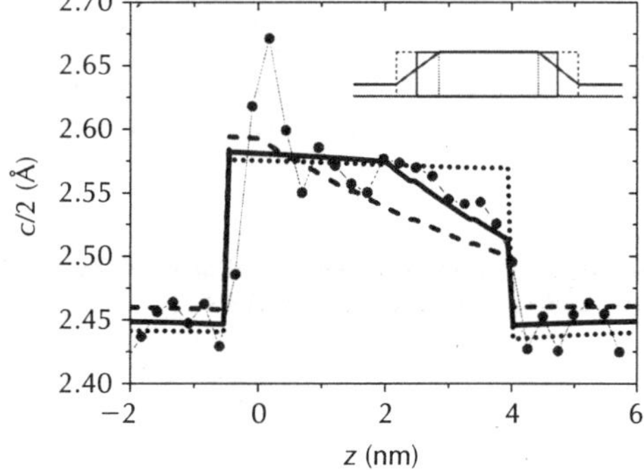

Figure 7.9   Comparison between the values of the strained lattice parameter c for a single GaN/AlN dot obtained after processing of the HRTEM images [60] (points), and the theoretically calculated (in-plane averaged) values. The different lines correspond to the different volumes employed for the strain average as indicated in the inset. The calculations correspond to a vertical QD superlattice of period 10 nm, the dot dimensions being R = 15 nm and h = 4 nm, with a WL of thickness d = 0.5 nm, as determined from the TEM images.

between the (0001) interlayer distances in a QD as extracted from HRTEM experiments [60] and those obtained from a calculation similar to that shown in Fig. 7.6. The very good agreement between the experimental and theoretical results highlights the adequacy of the continuum elastic model presented here for the description of the strain in self-assembled wurtzite QDs.

Although the above considerations have been made taking as example GaN/AlN(0001) QDs, a similar methodology has been applied for the study of dots grown along a non-polar axis [61]. Some related results will be presented in the following sections.

7.3   Raman scattering

In this section we will analyse the characteristics of the vibrational modes in GaN/AlN quantum dots and the structural information that can be obtained from their study. The section starts with a summary of the particularities of phonon modes in GaN and AlN wurtzite crystals, giving special attention to the dependence of their frequency with strain. Concerning the characteristics of the vibrational modes in the dots, we will perform an analysis of the influence of strain on non-polar and polar vibrations of (0001) GaN/AlN QDs considering results on resonant and non-resonant Raman scattering. The last part of this section is devoted to non-polar quantum dots.

7.3.1   Vibrational modes in bulk GaN and AlN

The symmetry of wurtzite GaN, AlN and their alloys belongs to the C_{6v}^4 (P63mc) space group, with two formula units per primitive cell and all atoms occupying C_{3v} sites. A group theory analysis [62, 63] of the lattice vibrations at the Γ point predicts six optical modes which decompose into the following representations of the C_{6v} (6mmm) point group: $A_1 + E_1 + 2E_2 + 2B_1$. The polar modes, A_1 and E_1, are both infrared and Raman active. The A_1 mode is polarized along the wurtzite c-axis (z-direction), while the E_1 mode is polarized in the xy plane. The non-polar E_2 modes are Raman active and the B_1 modes are silent. The uniaxial anisotropy of wurtzite crystals, together with the high ionic character of the crystal bonding, has a strong influence on the dynamic properties of these semiconductors and their nanostructures. The anisotropy of the short range atomic forces is responsible for the A_1–E_1 splitting while the long-range Coulomb field causes the longitudinal–transverse (LO–TO) splitting of the optical polar modes. In crystals with the wurtzite symmetry, the long-range forces prevail over the short-range ones. As a consequence, there is a dependence of the energy and polarization properties of the polar phonon modes on the direction of the phonon wavevector near the Γ point. These characteristics lead to phonon modes in wurtzite QDs that are strongly different from those of the more common zinc-blende semiconductors. Table 7.3 summarizes the symmetry, selection rules and frequencies of the Raman active phonon modes of bulk GaN and AlN.

The frequency of the vibrational modes is affected by strain, a fact that has been used to characterize strain in semiconductor nanostructures [65]. The relation between the phonon modes and

Table 7.3   Symmetry, selection rules and wavenumber of the zone-centre phonon modes of bulk GaN and AlN [64].

Mode	Selection rules	GaN (cm^{-1})	AlN (cm^{-1})
A_1(LO)	$z(x,x)z$	734.0	890.0
A_1(TO)	$x(z,z)x$, $x(y,y)x$	531.8	611.0
E_1(LO)	$x(y,z)y$	741.0	912.0
E_1(TO)	$x(y,z)y$, $x(y,z)x$	558.8	670.8
E_{2l}	$z(x,x)z$, $z(x,y)z$, $x(y,y)x$	144.0	248.6
E_{2h}	$z(x,x)z$, $z(x,y)z$, $x(y,y)x$	567.6	657.4

strain is established within the framework of the linear deformation potential theory. This approximation has been developed in detail for crystals with wurtzite symmetry by Briggs *et al.* [66]. The model describes the shift of the phonon modes, $\Delta\omega_\lambda = \omega_\lambda - \omega_0$, with respect to the unstrained bulk value, ω_0, by means of the phonon deformation potentials (PDP) of each mode λ. Under deformation, the energy of a mode belonging to the A_1 representation will shift by:

$$\Delta\omega_{A_1} = a_{A_1}(\varepsilon_{xx} + \varepsilon_{yy}) + b_{A_1}\varepsilon_{zz} \tag{7.3}$$

where a and b are the corresponding PDPs. The modes belonging to the E_1 and E_2 representations are doubly degenerate. The application of strain may break this degeneracy, giving rise to two separate modes. The resulting frequencies are given as a function of strain by:

$$\Delta\omega_\lambda = a_\lambda(\varepsilon_{xx} + \varepsilon_{yy}) + b_\lambda\varepsilon_{zz} \pm c_\lambda[(\varepsilon_{xx} - \varepsilon_{yy})^2 + 4\varepsilon_{xy}^2]^{1/2} \tag{7.4}$$

with $\lambda = E_1, E_2$. The values of the deformation potentials a_λ, b_λ and c_λ have been reported for various modes. They are displayed in Table 7.4.

Table 7.4   Deformation potentials for the different GaN and AlN phonon modes, expressed in cm^{-1}.

Mode λ	a_λ GaN	b_λ GaN	c_λ GaN [69]	a_λ AlN [70]	b_λ AlN [70]
E_{2l}	115 ± 25 [67]	-80 ± 35 [67]	–	–	–
A_1(LO)	-782 ± 174 [68]	-1181 ± 245 [68]	0	-643 ± 84	-1157 ± 136
A_1(TO)	-630 ± 40 [67]	-1290 ± 80 [67]	0	-930 ± 94	-904 ± 163
E_1(TO)	-820 ± 25 [67]	-680 ± 50 [67]	379 ± 43	-982 ± 83	-901 ± 145
E_1(LO)	–	–	678 ± 49	–	–
E_{2h}	850 ± 25 [67]	920 ± 60 [67]	379 ± 107	-1092 ± 91	-965 ± 161

7.3.2   Vibrational modes in GaN QDs

The interpretation of the Raman spectra of GaN QDs would require a theoretical model taking into account the peculiarities of the wurtzite nanostructures. These include not only the anisotropy characteristic of the wurtzite structure, but also the effect of strain on the vibrations, as well as the influence of the piezoelectric and pyroelectric properties of the constituting materials on the energy of the vibrational modes. A model incorporating all these effects has not been developed yet. However, calculations performed on GaN/AlN QWs [71] lead to the conclusion that the piezoelectric and pyroelectric effects may be considered as second-order corrections to the vibrational frequencies when compared with the influence of strain. Even so, few works have addressed the problem of the effect of strain on GaN/AlN quantum dot vibrations. Garro *et al.* [72] followed a two-step procedure for the quantification of the effect of strain on the frequencies of the QD vibrational modes. First, the atoms in the dot are considered to vibrate in the same manner as in a bulk sample, but under the influence of an average strain field. The changes in the vibration frequency due to strain can then be obtained by Eqs 7.3 and 7.4, both for the QD and the AlN barrier. The vibrational energies obtained by this calculation can be directly compared to experimental values for the non-polar E_2 mode, since confinement effects for this mode are very small. The polar vibrations, however, besides being affected by strain, change energy due to the presence of the QD interfaces. For the comparison of the energy of these modes with experimental values, a second step is necessary: the energy of the polar modes, modified by strain, must be recalculated within the anisotropic dielectric continuum model and the Loudon model of uniaxial crystals [73–75]. With this respect, Fonoberov *et al.* [74] developed a technique that allows the calculation of the polar phonon modes in wurtzite QDs of arbitrary shape, with the advantage of giving analytical solutions for QDs of ellipsoidal shape. The spectrum of polar optical phonons obtained from this model, however, does not depend on the absolute size of the QD, but only on its aspect ratio. This feature arises from the fact that the bulk optical phonon dispersion near the Γ point is neglected. Given the small phonon dispersion in GaN, this approximation is good for QDs with sizes over 3 nm. In smaller QDs, the energy of the vibrational modes will experience a small red shift due to the bending of the dispersion relations to lower frequencies for finite values of q.

Within the dielectric continuum model, the vibrational modes in a GaN quantum dot may be classified as quasi-confined TO, quasi-confined LO, and interface modes (TO, LO and mixed LO–TO modes). The frequency intervals corresponding to the modes of each character are summarized in Table 7.5. The prefix "quasi" used in reference to the confinement characteristics of the phonon modes arises from the fact that, unlike in zinc-blende nanostructures, in the wurtzite system the modes are not completely confined in the dot volume, but may penetrate to some extent into the surrounding material. Furthermore, all modes are dispersive and do not necessarily coincide with those of the bulk constituents. These two characteristics are peculiar to wurtzite semiconductors and arise as a consequence of the anisotropy of the dielectric function. On the other hand, the fulfilment of the boundary conditions imposed over the electric field associated to polar vibrations, combined with the anisotropy of the wurtzite dielectric function, originates a strong dependence of the energy of the QD vibrational modes on its aspect ratio, h/D, where h is the dot height and D its diameter.

Based on the model developed by Fonoberov *et al.* [74], it is possible to calculate analytically the quasi-confined and interface phonon modes of an oblate ellipsoidal dot of GaN embedded in an AlN matrix as a function of the QD aspect ratio [72]. The resulting vibrations can be classified

Table 7.5   Classification of the nature of the different vibrational modes in GaN/AlN QDs according to their frequency [72].

Interval	Character
$A_1^{GaN}(TO) < \omega < E_1^{GaN}(TO)$	Quasi-confined (TO)
$A_1^{GaN}(LO) < \omega < E_1^{GaN}(LO)$	Quasi-confined (LO)
$E_1^{GaN}(TO) < \omega < A_1^{AlN}(TO)$	Interface (TO)
$E_1^{GaN}(LO) < \omega < A_1^{AlN}(LO)$	Interface (LO)
$E_1^{AlN}(TO) < \omega < A_1^{GaN}(LO)$	Interface (LO–TO)

according to the angular and magnetic quantum numbers l and m and are shown in Fig. 7.10 for values of the QD aspect ratio ranging from 0 to 0.35. Unlike the results found in zinc-blende quantum dots, in the GaN/AlN system the interface modes cannot be ascribed to a particular material forming the heterostructure, and present in general mixed GaN–AlN character.

7.3.3 Non-resonant Raman scattering

The first micro-Raman studies performed on multilayers of GaN/AlN QDs were reported by Gleize *et al.* [76, 77]. In [77] the authors analysed the Raman spectra of a sample grown by MBE on an Si (111) substrate that consisted of 85 layers of GaN quantum dots. The features observed in the spectra were assigned to the GaN QDs and AlN spacer by a study of the dependence of their intensity with focus depth, their polarization selection rules and a detailed comparison of the spectra with that of an AlGaN alloy of similar composition. The authors identified the peak located at $606\,cm^{-1}$ as that of the E_{2h} of the GaN QDs. Two phonon modes were assigned to the AlN spacer: one with E_{2h} symmetry, at $645\,cm^{-1}$, and a second one with $E_1(TO)$ symmetry, at $657\,cm^{-1}$. A comparison of these values with those displayed in Table 7.3 for bulk GaN and AlN indicates a considerable blue shift of the QD related mode (almost $40\,cm^{-1}$), as a clue that the dots are under compressive stress. On the other hand, the modes corresponding to the spacer are red shifted by circa $13\,cm^{-1}$, as a consequence of the effect of tensile strain. Figure 7.11

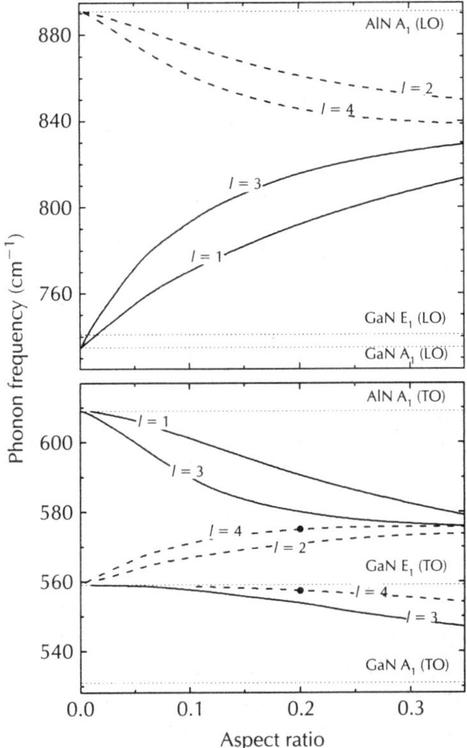

Figure 7.10   Frequencies of the polar optical vibrations for an oblate spheroid GaN QD embedded in an AlN matrix as a function of the QD aspect ratio. Each vibration is labelled by its corresponding value of l. Only modes with m − 0 have been plotted. The dotted lines indicate the energy of the bulk phonon modes of GaN and AlN [72]. With permission from APS© 2006.

shows the Raman spectra recorded in a sample consisting of 200 GaN/AlN QD layers grown on 6H-SiC, obtained for various polarization configurations. The peaks are identified by their symmetry and assigned to the QDs (GaN) or the spacer (AlN). Modes of A_1(TO), E_1(TO) and E_2 symmetries can be identified for both QDs and spacer. All the peaks related to the QDs are considerably blue shifted with respect to their bulk values, while those corresponding to AlN are red shifted. However, these shifts are found to be somewhat smaller than that reported for samples grown on (111) Si.

The identification of the phonon modes corresponding to the GaN/AlN dots paved the way for the analysis of structural changes in the QDs by means of Raman scattering. In this respect, it was found that the strain state of dots and spacer depended on the material constituting the substrate [56, 62, 76–78], its orientation [61, 79] and on the number of QD stacks [56, 78]. A systematic study of the evolution of strain with the number of dot layers was performed by Cros *et al.* [56]. The samples analysed were grown by MBE on 6H-SiC on a thin AlN buffer layer. Each period consisted of QDs formed after the deposition of 6 ML of GaN separated by 8 nm of AlN. The evolution of the E_{2h} phonon mode of QDs and spacer in stacks with 7, 10, 20, 50 and 200 QD layers is shown in Fig. 7.12. The linewidth of both spacer and QDs decreases, indicating increasing strain homogeneity in the samples. The compressive strain in the dots was found to remain almost constant in the different samples. The tensile strain of the spacer, on the other hand, increased with the number of layers. These changes seem to saturate for samples with more than 20 layers. This evolution of the phonon modes has been related to the homogenization of the QD sizes with stacking due to the vertical ordering of the dots. The detailed analysis of the strain can be performed by comparison of the Raman data concerning the non-polar E_{2h} mode with the result of a calculation of strain [56], in combination with Eq. 7.4.

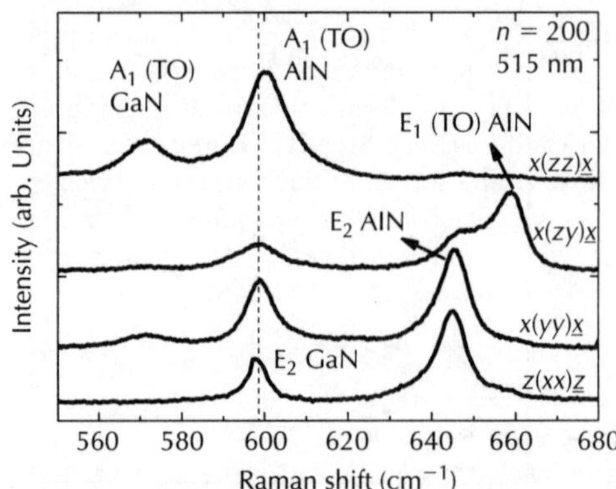

Figure 7.11   Micro-Raman spectra of a GaN/AlN QD sample (200 periods) recorded in various polarization configurations. The dots were obtained after the deposition of 6 ML GaN, and were separated by an 8 nm spacer. The spectra were excited at 514.5 nm. The dashed line indicates the frequency of the E$_2$ GaN mode.

The three curves included in Fig. 7.12 are the results of the calculations and differ in the way the values of the strain are obtained. For the solid line, strain was calculated considering stacks of QDs with variable diameter, with a distribution of sizes according to the AFM characterization of the samples [37]. In this calculation, the value of strain has been obtained by averaging to the QD material (shown in (a)) or to the AlN spacer (b) along the axis of the stacked structure. This is indicated schematically with the arrow in the picture beside the figure. The dashed lines, on the other hand, have been obtained considering stacks of QDs with equal diameter (45 nm), but averaging the value of strain to the volume of the QD (a) or spacer (b) located at the centre of the stack. In the calculations, the mean value of shear was found to be negligible. The dotted line in the figure corresponds to the shift calculated only with the in-plane strain (first term in Eq. 7.4). Concerning the QDs, the comparison of the calculations with the

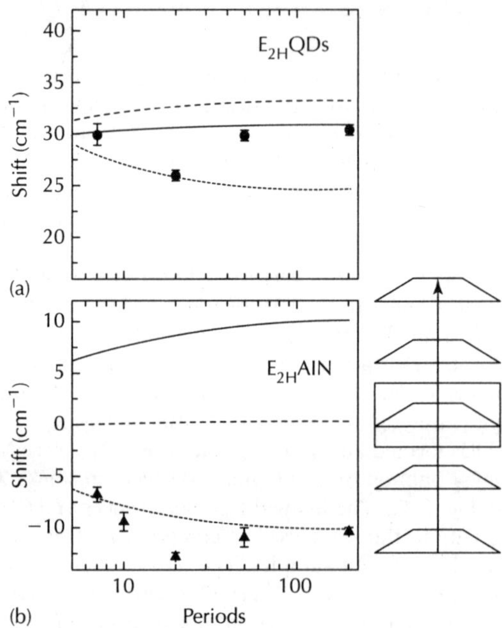

Figure 7.12   Strain induced shift of the E$_{2h}$ phonon mode of the GaN QDs (a) and AlN spacer (b) as a function of the number of periods. The curves are the result of a theoretical model based on the calculation of strain in the QD stack [56]. With permission from APS© 2006.

experimental Raman shift revealed a good agreement, reproducing the large positive shift of the E_{2h} phonon mode. The results confirmed the small dependence of strain with QD diameter, on the one hand, and its small variation with the number of stacks, in agreement with the results obtained by the model discussed in section 7.2.2 (see Fig. 7.8). It was found that the in-plane strain (dotted line) is responsible for 80% of the Raman shift observed in the GaN QDs. Regarding AlN, the fact that the best agreement with theory is obtained when only the in-plane strain component is taken into account (dotted line) suggests the existence of a relaxation mechanism along the growth direction that affects mainly the barrier. Other studies on Raman scattering in GaN/AlN multilayers grown on (111) Si [78] have found small shifts in the energy of the E_{2h} mode when comparing samples containing 40 and 85 dot periods. The dots in the thicker sample showed a slightly larger compressive stress, a tendency that was confirmed by power-dependent cathode luminescence (CL) experiments.

7.3.4   Resonant Raman scattering

As we have discussed before, it is possible to obtain information about the vibrational modes in QD multi-stacks by means of non-resonant Raman scattering. However, when the amount of material is very small, the Raman signal is too weak. The most efficient way to obtain information on GaN quantum dots is to work under resonant conditions. Resonant Raman scattering occurs when either the incident or the scattered photon energy approaches an electronic transition of the system under study [80]. Besides allowing the observation of the Raman signal from very small scattering volumes, the resonant enhancement enables the selective detection of different materials whenever their characteristic electronic energies can be independently tuned. Several experimental problems have to be confronted when measuring the Raman signal under resonance. First, the intense emission from the QDs may disguise the resonant Raman signal. This problem might be overcome by taking advantage of the quantum confined Stark effect, which shifts the fundamental transition of GaN QDs towards the visible. Under these conditions, the resonant excitation of the Raman signal with higher electronic states of the QD is possible far from the inconvenient influence of the PL signal. Second, the lack of tunable lasers in the UV may prevent the achievement of the resonant conditions in small quantum dots.

Gleize *et al.* [81] used various laser lines in the UV range to investigate resonance effects and multiphonon scattering in a GaN/AlN QD sample. The sample considered was grown along the [0001] direction on [111] Si and consisted of a stacking of 39 periods of GaN QDs. In Fig. 7.13 the spectra excited at 3.81 eV are displayed. The QD spectrum corresponds to that labelled as sample (b) in the figure, while sample (a) refers to the spectra obtained on a thick GaN layer under the same experimental conditions and has been included for comparison. The enhancement of the polar A_1(LO) mode due to Fröhlich electron–phonon interaction is visible in the spectra. Peaks corresponding to scattering by 1, 2, 3 and 4 LO-phonon modes are observed in the QD sample. The inset expands the region of the spectra corresponding to scattering by one phonon. The features observed at $602\,\text{cm}^{-1}$ and $744\,\text{cm}^{-1}$ are ascribed to the E_{2h} and A_1(LO),

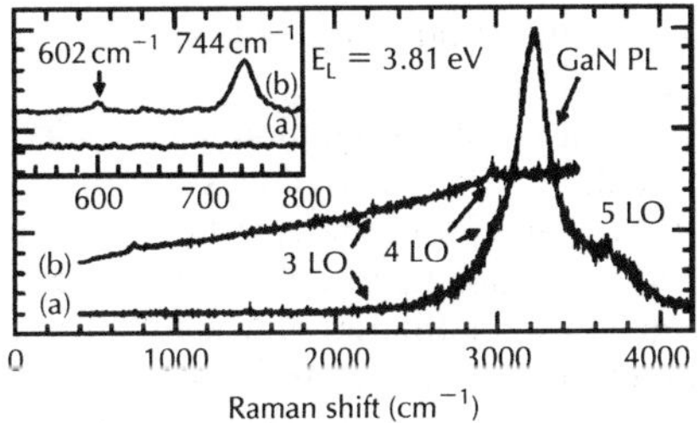

Figure 7.13   Raman spectra of a GaN layer (a) and a QD sample (b) excited at 3.81 eV. The multiphonon scattering by the A_1(LO) mode is indicated. The inset shows in detail the region corresponding to one phonon scattering [81]. With permission from AIP© 2001.

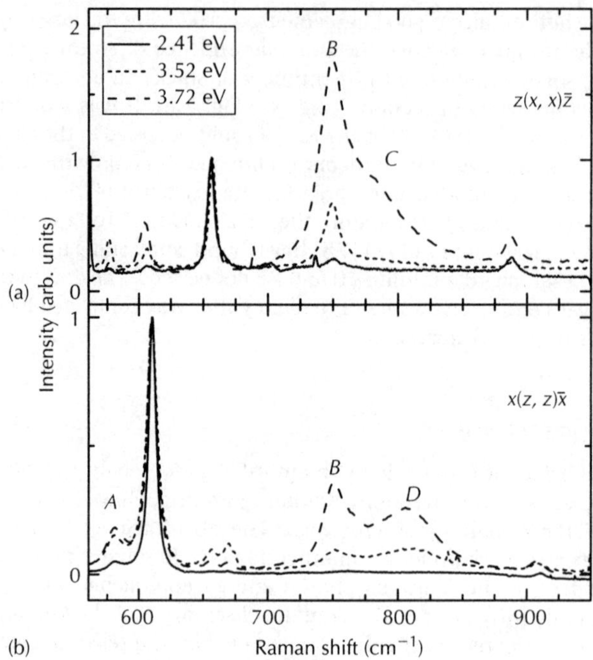

Figure 7.14   *Raman spectra of a GaN/AlN quantum dot stack measured for different excitation energies and polarization configurations [72]. With permission from APS© 2006.*

respectively. This assignment is supported by the detection of the E_{2h} mode under off-resonance excitation and by the presence of the $A_1(LO)$ multiphonon peaks at higher energies.

The characterization of QD stacks by means of resonant Raman scattering probed at various excitation energies in the UV and with different polarization configurations led to the identification in [72] of three longitudinal optical modes that were attributed to quasi-confined modes of the QDs (see Fig. 7.14). As explained before, the effect of strain in the QD vibrational modes is usually estimated by calculations of the average strain in the dot and the use of Eq. 7.4. In addition, confinement effects in polar phonon modes can be accounted for by means of a theoretical model based on the anisotropic dielectric continuum model [74]. This theoretical analysis, in combination with the experimentally observed selection rules, allowed the assignation of peak A at $582\,\mathrm{cm}^{-1}$ to a TO quasiconfined mode. On the other hand, peaks C ($780\,\mathrm{cm}^{-1}$) and D ($802\,\mathrm{cm}^{-1}$) were attributed to LO interface modes of the GaN type. Finally, the peak labelled as B ($751\,\mathrm{cm}^{-1}$) showed better agreement with the quasiconfined LO mode characteristic of a very flat (aspect ratio smaller than 0.02) quantum dot, and it is assigned to an interface mode localized at the flat bottom surface of the quantum dots. This last, more intense peak is the one observed in Fig. 7.13 [81] and identified on uncapped GaN QDs grown on AlGaN using Si as antisurfactant [82].

By means of Raman scattering at various excitation energies in the UV, it is also possible to probe selectively the phonon modes of QDs of various sizes in single layers of GaN QDs [82]. Smaller dots resonate at higher energies, and the vibrational frequencies of their polar modes may red shift as a consequence of confinement, reflecting novel shape of the phonon dispersion curves. Cros *et al.* [83] took advantage of the resonance enhancement of the Raman signal to study quantitatively the evolution of ε_{xx} and ε_{zz} strain components in only one layer of GaN/ AlN quantum dots as a function of AlN coverage. The QDs studied in this work were grown on 6H-SiC. The thickness of the AlN capping layer was varied from sample to sample, ranging from 0 to 14 ML. The resonance energy of the QDs was investigated by excitation with various UV laser lines. A resonant mode around $740\,\mathrm{cm}^{-1}$ was found in all the samples when exciting at

3.71 eV. The peak was identified as the quasi-confined A_1(LO)-phonon mode labelled as mode B in Fig. 7.6 [72]. Its energy increases with AlN coverage, indicating an increasing compressive strain with increasing AlN deposition. To obtain quantitative results of strain from the experimental data, the authors took advantage of the small aspect ratio of the investigated quantum dots (0.08) and combined Eq. 7.3 with the biaxial strain approximation. As discussed in section 7.2, this approximation strictly holds for 2D heterostructures (QWs and superlattices), but may be applied in flat isolated quantum dots. The samples were investigated as well by X-ray techniques (multiwavelength anomalous diffraction and diffraction anomalous fine structure) using grazing incidence geometry. These techniques are extensively described in chapter 6. The analysis of the diffracted intensities allowed the extraction of a and c lattice parameters of the GaN quantum dots as a function of AlN coverage.

Figure 7.15 [83] displays the comparison of the lattice parameters obtained by X-rays (triangles) and resonant Raman scattering (dots) as a function of coverage. It can be seen from the figure that the in-plane compression and axial expansion increase monotonically with capping thickness. It was found that the values of the in-plane lattice parameter obtained by both methods compare very well over the whole range of AlN capping investigated, in spite of the use of the biaxial approximation for the interpretation of the Raman data. Regarding the c parameter, the agreement was found to be good for small coverage (less than 4 ML). For thicker coverage the interpretation of the Raman results by means of the biaxial approximation underestimates the value of c by 0.4%. In spite of very good agreement found between the X-ray and Raman data, the quantitative determination of strain by a combination of resonant Raman measurements and the application of the biaxial approximation must be used with caution in other samples. It must be kept in mind that strain in dots with larger aspect ratios will be very far from the biaxial value, and that this approximation fails completely for correlated QD stacks, as was already pointed out in section 7.2.2.

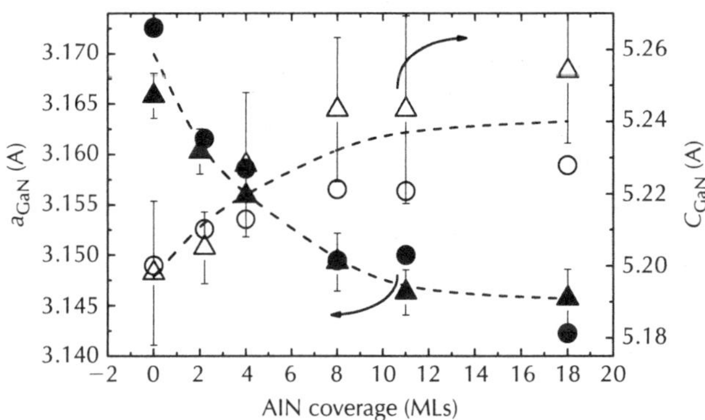

Figure 7.15   Comparison of a *and* c *lattice parameters of GaN QDs obtained by X-ray measurements (triangles) and micro-Raman scattering (dots). The dashed lines are guides to the eyes [83]. With permission from Wiley.*

7.3.5   QDs grown along non-polar directions

Raman scattering has been found to be very useful for the characterization of GaN/AlN quantum dots grown perpendicular to the wurtzite c-axis. As can be deduced from Table 7.3, one of the advantages found in the samples with this orientation is that the usual Raman experiments performed in backscattering geometry along the growth direction allow peaks related to the E_2, E_1(TO) and A_1(TO) modes to be obtained simultaneously.

Figure 7.16 shows the Raman spectra obtained from an a-plane QD sample at various polarization configurations [79], obtained with visible excitation. The sample was grown on a-plane 6H-SiC, and consisted of 18 layers of GaN quantum dots separated by 5.5 nm AlN. In the figure, the vibrational modes of GaN QDs and AlN spacer are classified according to their symmetry.

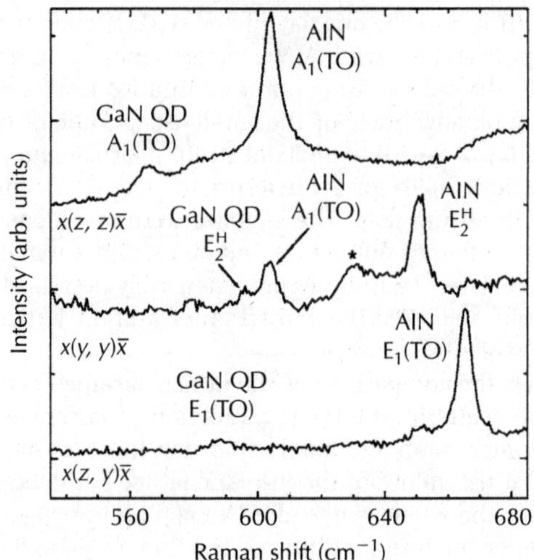

Figure 7.16   Raman spectra obtained for three polarization configurations under 514.5 nm excitation. The z-direction coincides with the c-axis of the wurtzite structure, while x corresponds to the growth direction. The vibrational modes of the QD and the AlN spacer have been labelled according to their symmetry [79]. With permission from AIP© 2005.

A comparison with the selection rules of the phonon modes of the SiC substrate confirmed that the *c*-axis of the QDs, spacer and substrate grow aligned [61].

Concerning strain in the sample, if *x* denotes the growth axis, and *z* is parallel to the wurtzite *c*-axis, it is expected that $\varepsilon_{xx} \neq \varepsilon_{yy} \neq \varepsilon_{zz}$. Inspection of Eq. 7.4 predicts under these conditions the splitting of the E_2 and E_1 vibrational modes. This splitting was not observed in the experiment, due to the small value of the c_λ deformation potential, reported already in Table 7.4. Values of strain may be extracted from the data combining the Raman shift measured for modes with different symmetry, and using Eqs 7.3 and 7.4. Values of $\varepsilon_{zz} = 0.015$ and $\varepsilon_{xx} + \varepsilon_{yy} = -0.009$ are obtained. Note that from the experiment it was not possible to obtain ε_{xx} and ε_{yy} independently. Strain in AlN was found to be in good agreement with that expected for the epitaxial growth of AlN on *a*-plane SiC, which for these samples can be related to the elastic constants and the SiC and AlN *c* lattice parameters through the expressions:

$$\varepsilon_{xx} = -\frac{C_{12}\varepsilon_{yy} + C_{13}\varepsilon_{zz}}{C_{11}} \quad \text{and} \quad \varepsilon_{zz} = \frac{C_{SiC}}{3c_{AlN}} - 1. \tag{7.5}$$

Here, the growth direction is identified with the *x*-axis, while *y* and *z* lie on the sample plane. For the GaN QDs the best agreement with the experimental data was obtained from the shift of the E_{2h} and A_1 modes, giving $\varepsilon_{zz} = -0.023$ and $\varepsilon_{xx} + \varepsilon_{yy} = -0.008$. However, this value was found to overestimate the strain induced frequency shift of the E_1 mode by 30%. The strain state of the QDs is also compatible with that expected for a biaxially strained layer grown on SiC (Eq. 7.5 applied to GaN), indicating that the deformation is largely imposed by the chosen substrate.

7.4   Luminescence

This section analyses the main characteristics of the interband electronic transitions in self-assembled GaN QDs as manifested in optical experiments (continuous wave and time-resolved photoluminescence (PL), cathode luminescence (CL), photoluminescence excitation (PLE)) and theoretical simulations.

7.4.1   Confinement effects

Increased thermal stability and improved emission efficiency at room temperature are well-established consequences of zero-dimensional confinement and common probes for identifying QD luminescence. In GaN/AlN heterostructures the 2.8 eV band gap mismatch between the dot and the barrier materials ensures a deep localization potential. Indeed, strong carrier confinement was already demonstrated in the first optical studies performed on GaN/AlN QDs grown in the SK mode by MBE [23] and by MOCVD [17, 84] and on GaN/Al$_x$Ga$_{1-x}$N ($x < 0.2$) QDs grown in the VW mode by using Si as a surfactant [16, 85]. Figure 7.17 shows the resonantly excited PL spectrum of a GaN/AlN QD superlattice grown by MBE on a commercial GaN substrate at two different temperatures. While the emission from the bulk GaN substrate is visible at low temperature only, a broader emission band centred at 3.75 eV and attributed to the GaN QDs remains almost unchanged, both in shape and intensity, with temperature. The blue shift of the QD emission with respect to the bulk band gap is a consequence of the confinement potential and the large inhomogeneous broadening reveals the wide distribution of dot sizes probed in this experiment. The integrated intensity of the QD emission stays constant up to room temperature, as can be seen in Fig. 7.18, because migration to non-radiative centres is hindered by strong confinement. On the contrary, a severe quenching in the PL intensity of GaN bulk and QW systems is observed as temperature increases due to the high density of threading dislocations and other defects that act as non-radiative traps.

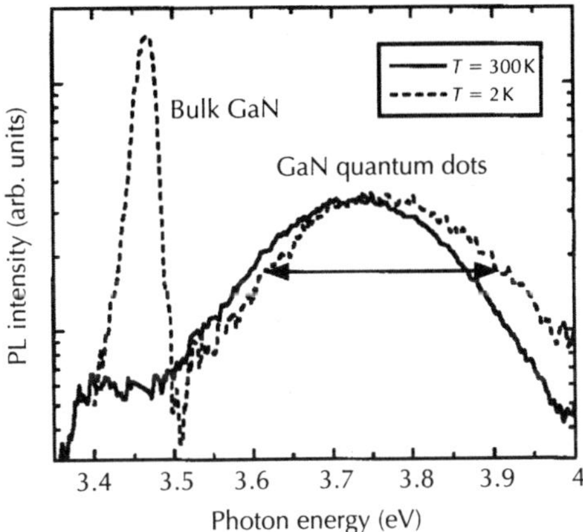

Figure 7.17   PL of a 20 period GaN/AlN QD superlattice grown by MBE on a GaN MOCVD-grown substrate. The excitation source was the 302 nm (4.1 eV) line of an argon laser which lies below the WL emission energy at 4.7 eV [23]. With permission from AIP© 1998.

In the same way as in other self-assembled systems, GaN QDs also present a large inhomogeneity in QD sizes resulting in large inhomogeneous broadening of the emission spectrum, see Fig. 7.17. A possible strategy for reducing the dispersion in island size is the growth of vertically correlated QD multilayers. As already described in section 7.2, when increasing the number of layers the vertical self-alignment is accompanied by the homogenization of the QD size distribution in successive dot planes and an average increase of their diameter [37, 38]. The spacer layer acts like a bandpass filter that prevents the formation of QDs of smaller size. Figure 7.19 shows the room temperature PL spectra of samples containing 1, 3, 5, and 10 layers of buried QDs grown by MOCVD [38]. Besides the emission from the QDs, a higher energy peak centred at 4.7 eV is attributed to the emission of the WL. In order to evidence the effects of the stacking in the multilayered samples, the PL spectrum of the reference single layer can be subtracted, see the inset of Fig. 7.19, and it is shown that the PL from the upper layers is red shifted in energy and shows a clear reduction of the spectral linewidth when compared to that of the single-layer sample. Another interesting observation is that the intensity of the emission increases with the number of layers

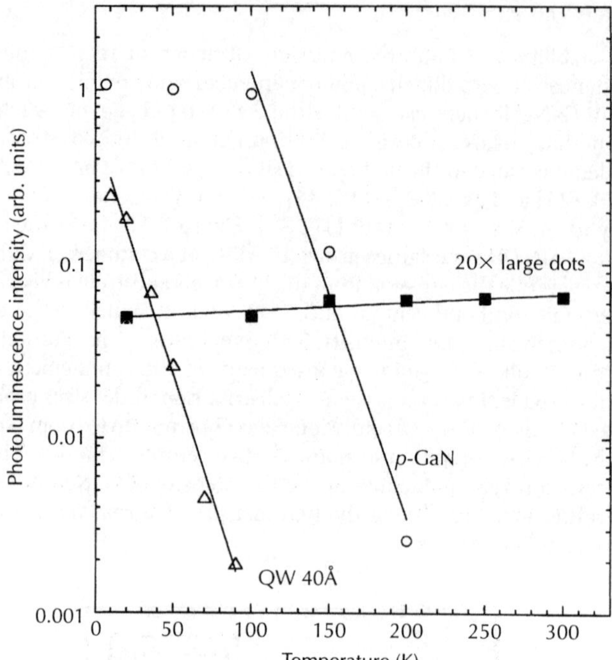

Figure 7.18   Comparison of the temperature dependence of the PL intensity of three different GaN systems: a 20 period GaN/AlN QD multilayer containing dots of 4.1 ± 0.4 nm height and 17 nm diameter; a single GaN QW with $Al_{0.15}Ga_{0.85}N$ barriers; and a p-type-doped GaN film [86].

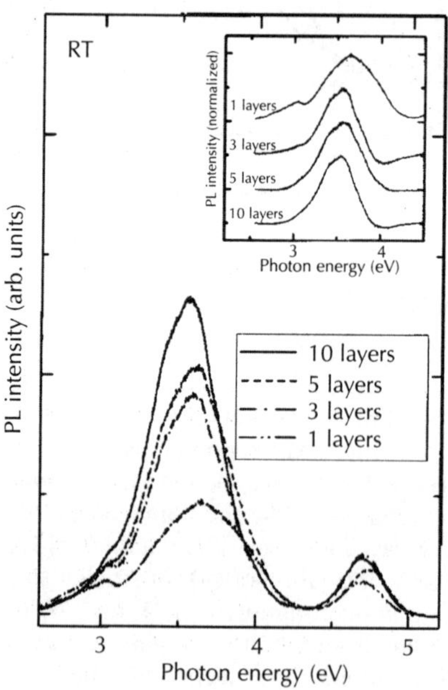

Figure 7.19   PL spectra of GaN QDs samples containing an increasing number of dot layers grown by MOCVD. The emission of the WL appears at 4.7 eV. The excitation wavelength is 193 nm (6.4 eV). The inset shows the normalized PL of the multilayer samples after subtracting the single-layer spectrum [38]. With permission from AIP© 2004.

indicating that new non-radiative recombination centres associated with structural defects are not introduced during the stacking process. This has important technological connotations since vertical stacking of the QDs is important to increase the total QD density and the subsequent increase of the gain for lasing operation.

The optical emission properties are determined by the electronic energy levels and wave functions of the system. Though there exist some atomistic approaches for the calculation of the electronic structure of GaN-based heterostructures [87, 88, 44, 45, 89–93, 46], the most common approach nowadays is the multiband envelope function approximation, which relies on the use of a multiband $\mathbf{k} \cdot \mathbf{p}$ Hamiltonian to describe the bulk band structure in combination with the standard envelope function theory to account for the confinement. The $\mathbf{k} \cdot \mathbf{p}$ Hamiltonian around the Γ point, including the effects of strain, allows the characterization of the wurtzite bandstructure in terms of a reduced number of parameters (energy gaps, effective masses, Luttinger-like $\mathbf{k} \cdot \mathbf{p}$ parameters, deformation potentials, etc.) [94]. Unfortunately, in the case of the III-N compounds the values of those parameters are not as well known as for other III–V semiconductors [47, 52, 95], and this situation hampers to some extent the comparison between theory and experiment in the III-N systems. For definiteness, in the calculations presented below we shall employ the same parameters used in [47].

In the case of lattice-mismatched heterostructures, such as the GaN/AlN systems considered here, the band-edge profiles are determined by the strained band offsets, as calculated from the electronic deformation potentials. This approach [96–98] has proved to be a convenient and reliable tool to describe quantum size effects in QWs [99, 100], as well as in QD structures [47, 101, 50, 102, 54]. To illustrate the impact of the deformation on the band structure, we show in Fig. 7.20a the band dispersion of GaN around the Γ point both without and with the effect of a biaxial strain (see section 7.2.2). In GaN, bands A and B are composed mostly of (p_x, p_y) atomic functions, whereas band C has dominantly p_z symmetry. The coupling between the conduction and valence bands in wide-band gap nitrides is relatively small over the energy range of interest, and can be

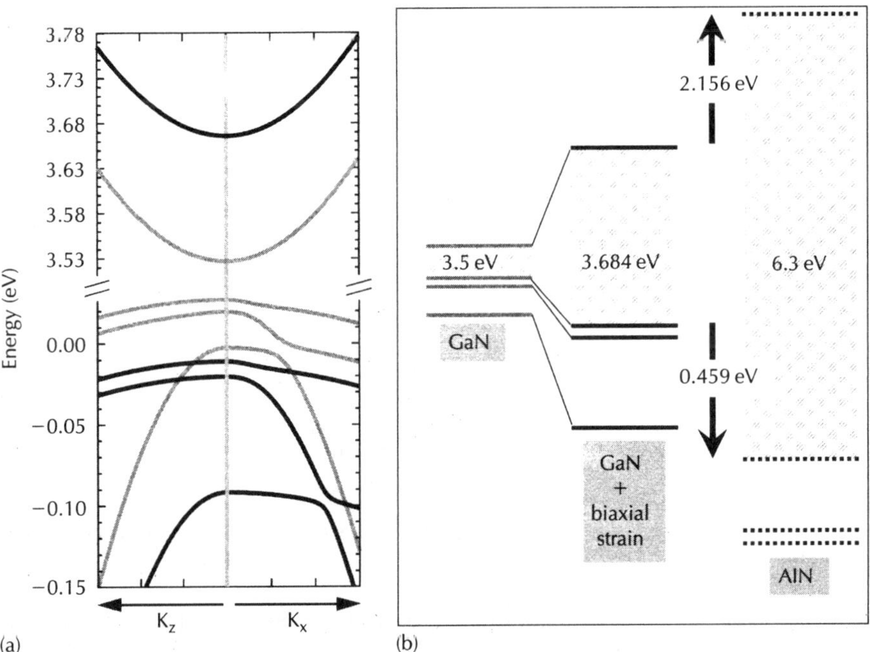

Figure 7.20   (a) Bulk bandstructure of GaN around the Γ point, along the in-plane direction (k_x) and [0001] direction (k_z), both without (grey line) and with (black line) the effect of a biaxial strain. (b) Band-edge profile of a GaN/AlN heterostructure both without and with the effect of a biaxial strain. The unstrained valence band offset between GaN and AlN has been taken to be 0.5 eV [47].

safely neglected. On the other hand, due to the small spin–orbit and crystal–field splittings, the mixing in the valence band is very important, especially in the k_x dispersion, leading to a strongly non-parabolic dispersion near the Γ point. Under deformation, the conduction band (CB) shifts to high energies and the valence bands (VBs) shift to low energies, thus increasing the value of the fundamental gap. In general, the curvature of the bands of GaN is not affected by strain as much as, e.g., in InAs. Figure 7.20b also presents the band-edge alignment in a heterojunction GaN/AlN both without and with the effect of a biaxial strain. As it can be seen, the combined effect of the confinement and the compressive strain in a GaN/AlN QW would produce a blue shift of the emission compared with the GaN bulk band gap larger than 200 meV. Nevertheless, we recall that the strain in the GaN/AlN QDs is far from homogeneous and deviates from the biaxial case, and this has to be fully taken into account for the quantitative modelling of the electronic properties of QDs.

7.4.2   Electric field effects: quantum-confined Stark effect

As a matter of fact, the most distinctive characteristic of III-nitride QD optical response comes from the existence of giant built-in electrostatic fields. On the one hand, the wurtzite crystalline structure is non-centrosymmetric, and therefore exhibits piezoelectric properties, i.e. develops electrical polarization when deformed [103]. The magnitude of the piezoelectric constants exceeds that of zinc-blende semiconductors by more than one order of magnitude [104, 52]. In addition, the [0001] direction or c-axis of the wurtzite structure is polar, meaning that the wurtzite crystals carry a spontaneous electrical polarization along [0001] even at equilibrium. These spontaneous polarizations are of the same order of magnitude as those found in ferroelectric materials [105]. Therefore, the strained heterostructures based on III-nitrides present polarization charges arising from spatial variations of the polarization that lead to huge internal fields and potentials, as shown in Fig. 7.21. The accurate values of the piezoelectric constants and spontaneous polarizations in III-N compounds are still the object of intense investigation. In Table 7.6 we compile the set of values used below to calculate the electric fields.

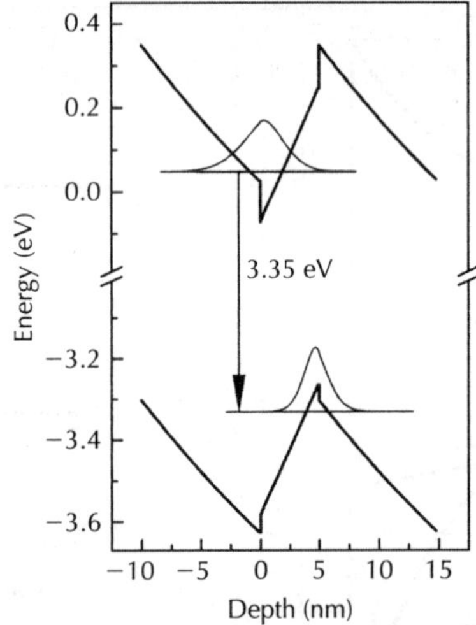

Figure 7.21   Impact of electrical polarization on the energy band diagrams and electron and hole ground states of a 5 mm width GaN/AlN(0001) QW [106].

The effects of the built-in electric fields have been widely reported in the optical properties of GaN and InGaN QWs [107, 108] and GaN/AlGaN heterostructures [109]. These effects are to be accounted for also in the study of the optical properties of GaN/AlN QDs, where the novelty is the non-trivial distribution of the electric field as a consequence of the multifaceted QD geometry and inhomogeneous strain distribution. The total potential is the sum of the piezoelectric plus spontaneous contributions, $\phi_{TOT} = \phi_{PZ} + \phi_{SP}$, which can be obtained by solving the Poisson

Table 7.6   Material parameters of bulk wurtzite GaN and AlN used to calculate the built-in electric potentials [105].

	GaN	AlN
e_{31} (C/m^2)	-0.49	-0.6
e_{33} (C/m^2)	0.73	1.46
e_{15} (C/m^2)	-0.49	-0.6
P_{SP} (C/m^2)	-0.029	-0.081
ε_r	9.6	9.6

equation with the appropriate polarization charges [47]. In Fig. 7.22a we show a map of the calculated total potential ϕ_{TOT} for a typical QD geometry. The potential profile along the [0001] dot axis is also displayed in Fig. 7.22b, together with the separate contributions of ϕ_{PZ} and ϕ_{SP}. Quantitatively ϕ_{PZ} is around one third of ϕ_{SP}, and both present the same polarity. Though the spontaneous contribution is dominant, the piezoelectric potential contribution cannot be neglected. Thus any realistic model of this system must include these two contributions, and therefore needs a previous evaluation of the strain. Qualitatively, it is clear that the variation of the potential inside the dot is considerably more pronounced along the [0001] direction than in the radial one. This fact has been used sometimes as a justification to approximate the electrostatic potential in the dot as equivalent to a parallel plate capacitor with a constant electric field. Figure 7.22b also displays the electric field calculated from ϕ_{TOT}, showing that it is approximately constant and can reach values as large as 5–7 MV/cm, which are nevertheless half of those found in a QW of the same height.

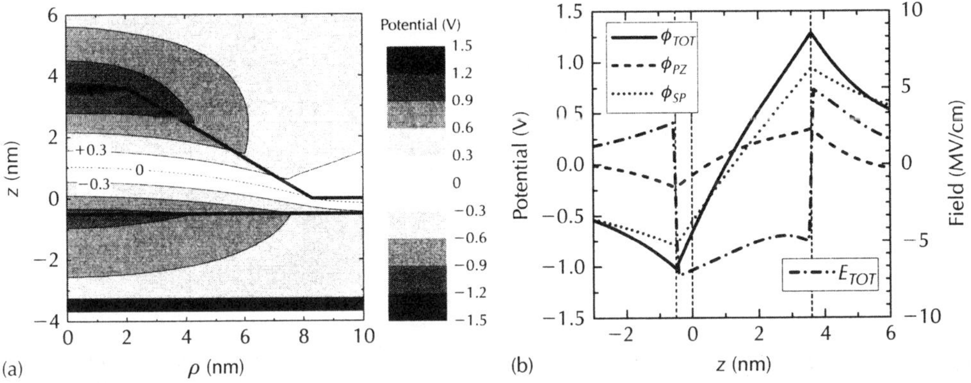

Figure 7.22   (a) Contour map of the internal electrostatic potential ϕ_{TOT} of a GaN/AlN QD, with the same dimensions as used in Fig. 7.6. The strain used to calculate ϕ_{PZ} is given in Fig. 7.6 and the same piezoelectric and dielectric constants (those of GaN) have been taken throughout the whole structure (see Table 7.6). The spontaneous polarization values used to calculate ϕ_{SP} are also given in Table 7.6. (b) Values of $\phi_{TOT}, \phi_{PZ}, \phi_{SP}$, and electric field E_{TOT} along the [0001] axis.

The impact of this internal field on the optical properties of a QD can be best substantiated by calculating the electronic states and wave functions. The electric potential ϕ_{TOT} has to be added to the confinement potential given by the (strain-modified) band edge profile. In Fig. 7.23a we show the dependence on the height of the dot of the first few CB and VB confined electronic states, as calculated with the multiband envelope function approximation [54]. It is seen that the energy of the CB states (VB states), decreases (increases) steeply with increasing dot height, unlike in other kinds of self-assembled QDs where the decrease is slower. This is due mostly to the presence of the huge potential difference between the top and bottom surfaces of the dot (see Fig. 7.22), which exhibits a quasi-linear dependence on the QD height. The electrons (holes) are mostly confined in the top (bottom) region of the dot, as is better visualized in Fig. 7.23b, where the squared envelope functions are represented.

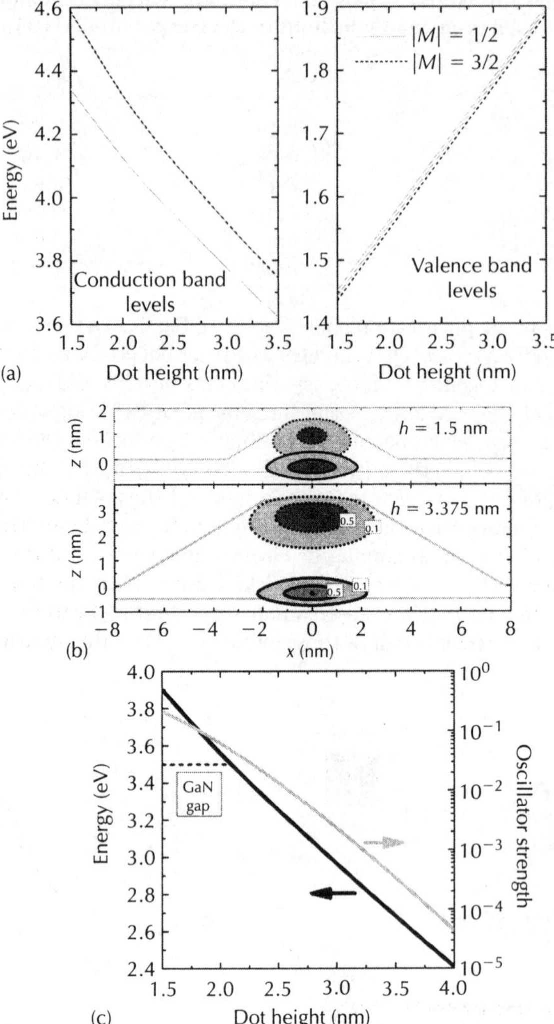

Figure 7.23 *(a) Three lowest CB and VB energy levels of GaN/AlN QDs as calculated within the multiband envelope function approximation [54]. The same aspect ratio as in Figs 7.6 and 7.22 has been employed. The coupling between the bulk CB and the VB is neglected, and the same band parameters (those of GaN [47]) have been taken throughout the whole structure. The zero of energy is taken on the VB top of unstrained AlN. (b) Probability density distribution for the electron (dashed line) and hole (solid line) ground states. The points indicate the maxima of the distributions, which have been scaled to unity. (c) Electron–hole transition energy and oscillator strength as a function of the QD height.*

Therefore, their energies follow the top (bottom) potential which decreases (increases) almost linearly with the QD height. It must be noted that the quantitative fitting of experimental measurements is at present difficult because small changes in the parameters that determine the magnitude of the built-in potential (i.e. the piezoelectric constants and spontaneous polarization values assumed for GaN and AlN) strongly modify the calculated energy levels [47]. Given the above results for the electron and hole energy levels, their difference, i.e. the electron–hole transition energy, also decreases rapidly with increasing the QD height. This decrease can eventually lead to an emission energy lower than the bulk GaN band gap, see Fig. 7.23c. Concomitantly, there is a decrease in the overlap of the electron and hole wavefunctions, and therefore a drastic reduction of the oscillator strength, as can also be seen in Fig. 7.23c. Thus, the most distinctive characteristic of III-nitride QD optical response derived from the intrinsic fields is a sizeable red shift of the ground-level transitions for increasing well width (quantum confined Stark effect, QCSE) accompanied by the reduction of the oscillator strength.

The theoretical picture presented in the above paragraphs is essentially consistent with the experimental findings on all types of polar GaN dots. For example, in Fig. 7.24, it is evidenced by the remarkable red shift experienced by the maximum of the QD PL spectrum (depicted by triangles) with increasing dot height [112]. As a matter of fact, the QD emission energy falls more than 1 eV below the strained GaN bulk band gap energy for heights higher than 4 nm. This is very different to the energy calculated assuming a vanishing internal field (see dotted line) and to the dependence of the emission of GaN/AlN QDs with zinc-blende structure (depicted by squares). The behaviour of cubic QDs will be analysed in more detail in section 7.4.5. The experimental results of Fig. 7.24 were in agreement with the theoretical calculations of Widmann *et al.* [110] for an electric field strength $F = 7$ MV/cm in good agreement with the theoretical predictions of Fig. 7.22.

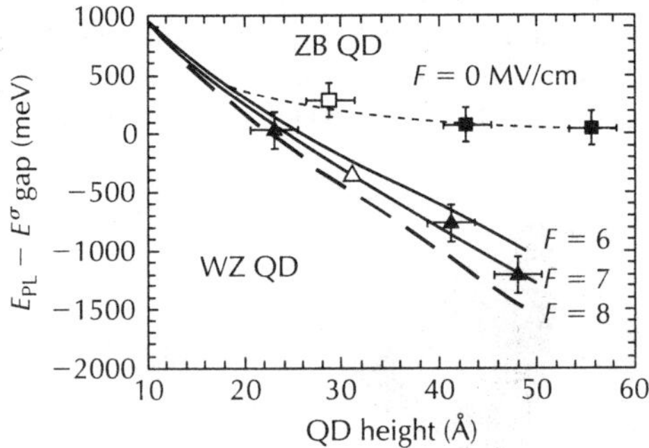

Figure 7.24   Energy shift of QD emission with respect to the GaN band gap energy as a function of the QD height. The GaN band gap energy has been corrected to take into account the biaxial compression of the dot material due to the AlN substrate [112]. With permission from APS© 2003.

Several authors have determined the strength of the built-in electric field in self-assembled GaN/AlN QDs employing a variety of experimental techniques and theoretical models. Table 7.7 summarizes some of the values published to date and shows a clear disagreement of the quoted figures. Uncertainties in the determination of the dot size, especially of the dot height often measured from AFM images and therefore subjected to considerable error, can explain partly the wide range

Table 7.7   Estimated value of the built-in electric field in self-assembled GaN/AlN QDs.

F (MV/cm)	As determined by:	Ref.
5.5	PL energy versus QD height	[110]
7	PL energy and radiative lifetime versus QD height	[112]
9	PL decay time versus PL energy	[113]
≪5	PL Stark shift induced by an external electric field[a]	[114]
6	Calculation of the strain distribution with a Green's function technique	[47]
3.8	Self-consistent tight-binding calculations	[89]

[a][114] predicts a 50% reduction in the AlN/GaN spontaneous polarization disagreement as compared to that calculated by Bernardini *et al.* [105] (see Table 7.2).

of electric field values. Also, as analysed in section 7.2.2, the piezoelectric contribution of QD multilayer stacks significantly differs from that of a single QD layer (see Fig. 7.7). Furthermore, there is an ongoing controversy about the magnitude of the difference in the spontaneous polarization between AlN and GaN that affects strongly the value of the internal fields [105, 111].

A side effect of the size dependence of the QD emission energy described above is the amplification of inhomogeneous broadening of the PL due to fluctuations in the QD size. Typical FWHM of GaN/AlN QDs fall in the 200–400 meV range much above the 60–80 meV of other non-polar self-assembled nanostructures.

Although the red shift of the QD emission is an undesired effect from the point of view of obtaining UV operating devices, this can also be used commercially for white light emission. It has been demonstrated that by controlling the QD size the emission can be continuously tuned from blue to orange. Figure 7.25 shows the room temperature PL of four GaN/AlN QD samples grown by MBE on Si(111). Three of them contain a single QD layer grown after the deposition of 7, 10, and 12 ML of GaN at a growth temperature of 800°C and the fourth one corresponds to a stacking of four dot planes of different deposition thickness [15]. As a consequence of the QCSE the addition of 1 ML of GaN induces a red shift of 150 meV in the emission energy. A proper combination of QD sizes can be used to obtain white light emission. Alternatively, rare earth-doped nitride QDs are being investigated as white light sources [115–117]. Vertically coupled GaN QDs have also been proposed as the material support for quantum computation by exploiting the existence of strong built-in electric fields to generate entangled states [118].

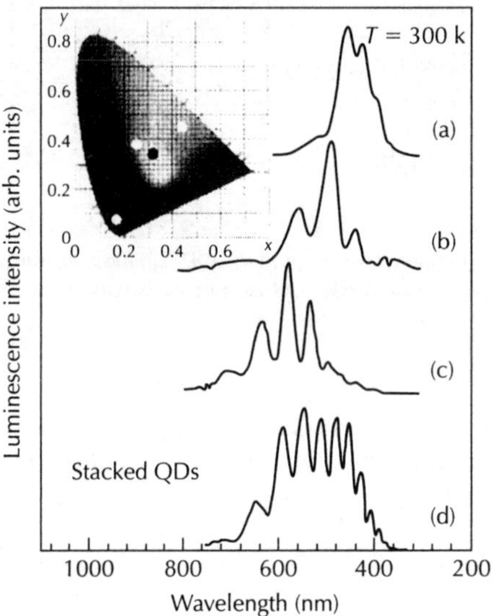

Figure 7.25   Emission of a single plane of GaN/AlN QDs grown on Si(111) by MBE for a GaN growth deposition of (a) 7; (b) 10; and (c) 12 MLs. Spectrum (d) corresponds to a four QD plane stack. All spectra present Fabry–Perot interference fringes due to the buffer layer [15]. With permission from AIP© 1999.

Another clear manifestation of the QCSE is observed when studying the dependence of the emission upon excitation power [110, 119, 120]. In this case a blue shift of the emission energy occurs for increasing excitation density, as shown in Fig. 7.26 for three different GaN QD samples grown by MBE on Si(111). Samples 1, 2, and 3 contained 86, 40, and 10 periods of GaN/AlN QDs and the average dot height was 3.7, 5, and 4.3 nm, respectively. More details on the sample growth are reported in [120]. The reported energy shifts are now understood as the result of the partial screening of the internal electric field by the field induced by the photogenerated electron–hole dipoles. These screening effects are larger for higher dots due to the amplification of the QCSE for increasing dot height, note that sample 2 presents the largest blue shift in the emission energy. The effect of screening by photogenerated carriers has been discussed

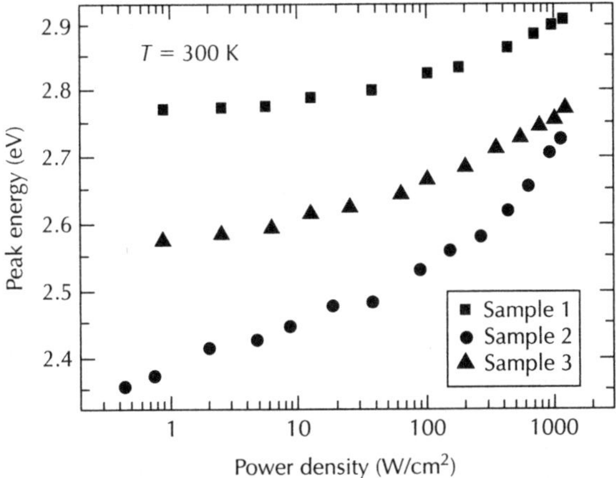

Figure 7.26   Evolution of the maximum energy of the PL spectra corresponding to three different GaN/AlN QD samples. The excitation source was the 266 nm fourth harmonic of an Nd:YAG laser which produced 5 ns pulses at a 10 Hz repetition rate [120]. With permission from AIP© 2004.

theoretically by Rajan *et al.* [89]. A linear blue shift of the emission energy with increasing electron–hole pair density is predicted. The calculated screening energy, defined as the energy shift per electron–hole pair, increases with the QD height as observed experimentally. The emission broadening is also affected by the excitation density but in a less straightforward manner. An increase in the FWHM for higher excitation powers can be attributed to the population of the excited states of the dots. However, a reduction of the emission width has been observed in some GaN QD multilayer stacks [120]. Kalliakos *et al.* pointed out that differences in the photo-generated carrier densities between different dot planes in QDs stacks could explain the observed reduction in the FWHM.

Cathodoluminescence has also been used in many investigations of the optical properties of III-N and their heterostructures. This technique is in many ways equivalent to PL but it allows a better control over the depth of the emission and, on the downside, it induces significantly higher excitation densities [78, 117]. Thus, in CL experiments the internal electric field is partly screened by the excitonic dipoles and the spectra appear blue shifted in energy. The CL and PL spectra of a GaN/AlN QD multi-stack are compared in Fig. 7.27. Both of them were recorded from the lateral section of the sample in order to prevent Fabry–Perot interference fringes as those observed in

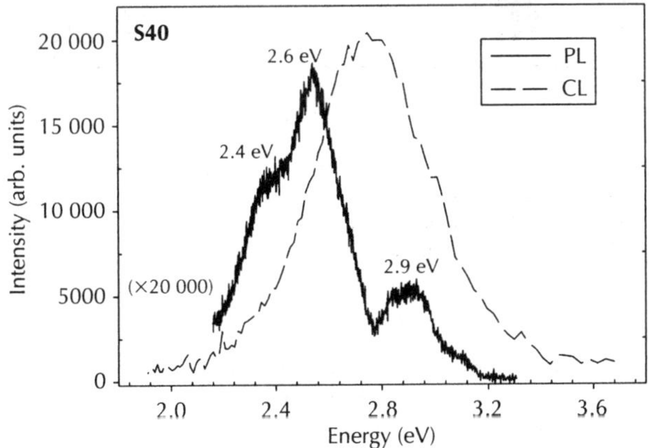

Figure 7.27   Comparison between the cross-sectional PL and CL of a sample containing 40 periods of GaN/AlN QDs grown on Si(111) by MBE. Measurements were carried out at room temperature and CL spectrum was recorded under an injection power of 5 nA and an electron beam energy of 20 keV [78]. With permission from IOP© 2002.

Fig. 7.25. The higher injection conditions not only give rise to the 200 meV blue shift but also to the higher intensities of the CL in comparison to the PL peaks. The screening of the internal field yields a strong increase of the oscillator strength and a more efficient absorption due to an increased overlap of the electron and hole wavefunctions.

The excited states of semiconductor nanostructures are often investigated by means of PLE. This technique has not been very much exploited in the study of group III nitrides due to the lack of tunable lasers in the UV range. One of the very few published PLE studies of a GaN/AlN QD sample is depicted in Fig. 7.28 together with the PL of the sample for two different excitation energies. The sample was grown by plasma-assisted MBE on 6H-SiC and consisted of five periods of GaN QDs separated by $Al_{0.25}Ga_{0.75}N$ spacers grown in between two 45 nm thick $Al_{0.25}Ga_{0.75}N$ optical waveguide layers. Several peaks appear in the PL spectrum of the sample excited at 4.7 eV (266 nm): the QD emission centred at 3.57 eV; the emission from the WL at 3.87 eV; and the emission from the AlGaN waveguide layers at 4.07 eV. The PLE maxima for the WL and the AlGaN layers is blue shifted by 100 meV, approximately, with respect to the emission maxima due to alloy potential fluctuations. Figure 7.28 also shows three PLE spectra of the GaN QDs measured for three detection energies within the emission peak. A maximum is observed in each PLE spectra, pointed by arrows, which appears more than 200 meV blue shifted with respect to the PL maximum, the so-called Stokes shift between the PL and PLE maxima. This increases for decreasing detection energy and therefore for increasing dot height. The severe reduction of the resonant absorption observed experimentally is attributed to a drop in the exciton oscillator strength as electrons and holes are separated spatially by the internal electric field.

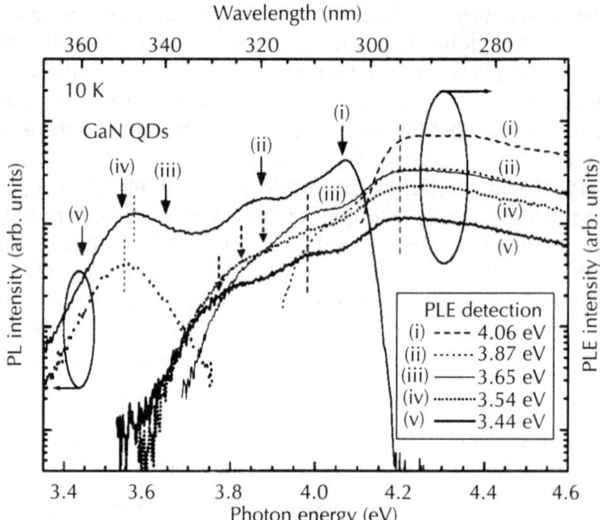

Figure 7.28   PLE spectra of GaN/AlGaN QDs measured at 10 K for different detection energies. A xenon lamp was used as excitation source in the PLE experiment. The detection energies from (i) to (v) are indicated on the PL spectrum excited at 266 nm (solid line). The dotted line corresponds to the PL spectrum excited at a lower energy of 3.81 eV (325 nm) [121]. With permission from AIP© 2002.

7.4.3   Electric field effects on the exciton dynamics

The spatial separation of electrons and holes inside nitride QDs reduces the overlap between the electron and the hole wavefunctions and the magnitude of the exciton oscillator strength, as discussed in the previous section and shown in Fig. 7.23b and c. In the time domain, the exciton radiative decay time is related to its oscillator strength f by:

$$\tau_R = \frac{3m_e c^3 2\pi\varepsilon_0}{ne^2\omega_0^2 f} \tag{7.6}$$

where ω_0 is the optical transition frequency, n is the refractive index, and m_e is the free electron mass. Hence, the presence of strong built-in electric fields should slow down the exciton radiative recombination. Time-resolved PL experiments have revealed that the decay time of the emission, τ_D, can be as high as tens of microseconds and presents a strong dependence on the QD height [112, 113, 122, 123]. Figure 7.29 shows the measured PL decay times of a series of samples containing a single period of GaN/AlN QDs with emission energies varying from 1.8 to 3.7 eV [113]. The experiments were conducted at 8 K and the repetition rate of the excitation source was either 10 Hz or adjusted in the range from 800 Hz to 82 MHz to avoid multiple excitations. At low temperatures, it is often assumed that τ_D is purely radiative and $\tau_D \approx \tau_R$ since the exciton non-radiative decay time, τ_{NR}, should be much longer. The validity of such an assumption for hexagonal GaN/AlN QDs will be re-examined later on in this section. The variation of τ_D is compatible with the QCSE: lower emission energies correspond to higher QDs where the separation of electrons and holes is more pronounced and τ_D is longer, and for shorter dots the emission energy is higher and so is the overlap between the electron and hole wavefunctions leading to shorter τ_D. Therefore, whereas piezoelectric effects are dominant for taller QDs, the dynamics of excitons in shorter dots are still governed by confinement effects. Solid lines in Fig. 7.29 depict the results of a simple theoretical model based on the effective mass and envelope function approximations that assumes a constant effective electric field inside the QD [113].

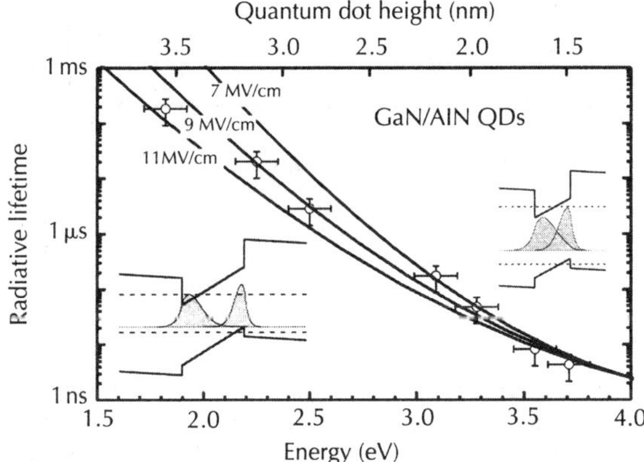

Figure 7.29   PL decay time as a function of the energy of the PL maximum and the calculated QD height for different GaN/AlN QDs with a constant effective electric field of 9 MV/cm [113]. With permission from APS© 2006.

Another unambiguous consequence of the existence of strong electric fields is the dynamic red shift of the time-resolved emission for increasing delay times after the arrival of the excitation pulse. This is shown in Fig. 7.30a where three normalized PL spectra corresponding to different delay times are displayed. At early times, the high density of photogenerated carriers (of the order of $10^{13}\,\mathrm{cm}^{-2}$ in this experiment) screens the internal electric field very effectively and reduces the QCSE (see the right-hand sketch of Fig. 7.30b). The peak shifts to the red as the photogenerated carriers recombine and the internal electric field is recovered. This red shift is faster at early times and becomes almost negligible after 10 μs when the "classical" radiative recombination regime is reached as reported in Fig. 7.30b. The radiative lifetime decreases exponentially with the density of electron–hole pairs [89] so that the dynamical descreening of the internal electric field also results in the observed non-exponential decay of the PL intensity (see Fig. 7.30b).

Although 3D confinement reduces the probability of non-radiative recombination, the presence of misfit dislocations, often acting as nucleation centres for the QDs, may still diminish the internal radiative efficiency of the QDs, η, defined as

$$\eta = \frac{\tau_{NR}}{\tau_{NR} + \tau_R} = \frac{\tau_D}{\tau_R}. \tag{7.7}$$

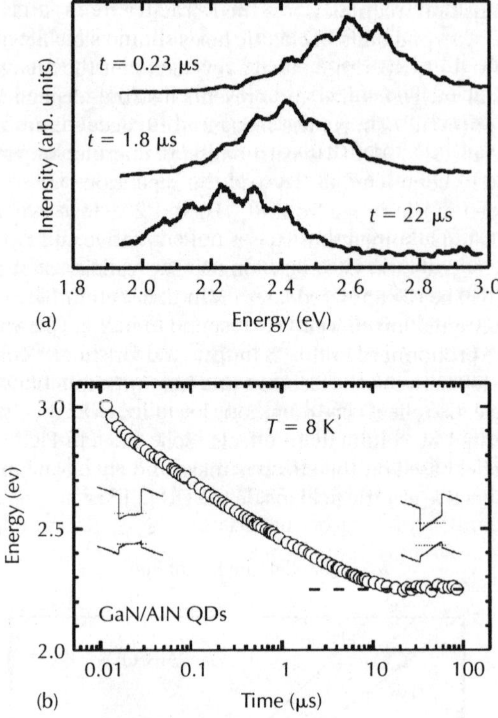

Figure 7.30   (a) Dynamic red shift of the time-resolved emission of a 3.3 nm height GaN/AlN QD. (b) Temporal evolution of the PL maximum. The excitation density was 100 μJ/cm² per 5 ns pulse. The inset sketches represent the dynamic descreening of the potential for electrons and holes along the QD axis [113]. With permission from APS© 2006.

In hexagonal GaN/AlN QDs η depends not only on the sample temperature but also on the dimensions of the dot. In particular, it is very much dependent on the QD height. It is reasonable to assume that τ_D is purely radiative and that η is close to 1 for the shortest dots at low temperatures. Fixing $\eta = 0.8$ for $h = 2.5$ nm, Simon *et al.* [112] have deduced the size dependence of η at low temperatures, finding that it falls down to 0.25 for $h = 3.6$ nm. Such a radical decrease follows from an almost size independent τ_{NR} (~10 ns) and a fast increase in τ_R for taller QDs due to the drop in the oscillator strength, see Eq. 7.6. As a result, non-radiative processes have a more significant contribution in the emission of taller QDs even at low temperatures. The temperature dependence of η shows that a decrease with increasing temperature ($\eta = 0.27$ for $h = 2.5$ nm at room temperature) occurs due to the reduction of τ_{NR} by an order of magnitude while τ_R remains almost constant [112]. The latter is expected for a 0D system. On the other hand, the decrease of τ_{NR} is more pronounced for shorter dots as the wavefunctions of the confined carriers penetrate more into the barrier making them more sensitive to nearby defects, such as threading dislocations.

7.4.4   Single quantum dot studies

Although optical spectroscopy on ensembles of QDs gives access to valuable information, the large inhomogeneous broadening of the spectral lines still hides many other fundamental aspects, such as the exciton fine structure, the homogeneous broadening of the emission lines, etc. Microscopic measurements where the emission of a single QD is resolved have proliferated for zinc-blende materials since the mid 1990s. Despite the interest that short-wavelength single-photon emitters convey [124] the study of self-assembled GaN QDs has evolved at a much slower pace. This is partly due to the poor control over the dot density described in section 7.2. The first μPL spectra showing single QD features were achieved in the group of Y. Arakawa in 2004 on samples grown by MOCVD [8]. Almost simultaneously, near-field scanning optical microscopy images were published showing the emission of relatively large (20 nm height and 150 nm diameter) single GaN QDs [125]. Previously, single QD CL measurements [126] were performed in uncapped GaN dots grown on AlGaN by MOCVD using TESi as antisurfactant. The density of dots in the samples grown by this method was small enough to ensure the observation of single dot emission without the need of isolation techniques. The position of a QD of a diameter of 100 nm and a height of 40 nm was determined by high-resolution field emission scanning electron microscopy. Its CL spectrum was characterized by a peak at 3.47 eV, ascribed to the dot, and a second peak centred at 3.71, characteristic of the AlGaN layer. The large width of the single QD emission (50 meV) was attributed to the presence of Si impurities introduced in the dot during growth. More recently, the combination of detailed CL mappings obtained at various emission wavelengths (Fig. 7.31) and AFM micrographs [127] allowed the establishment of a direct correlation between the emission characteristics of the individual dots and their size and density.

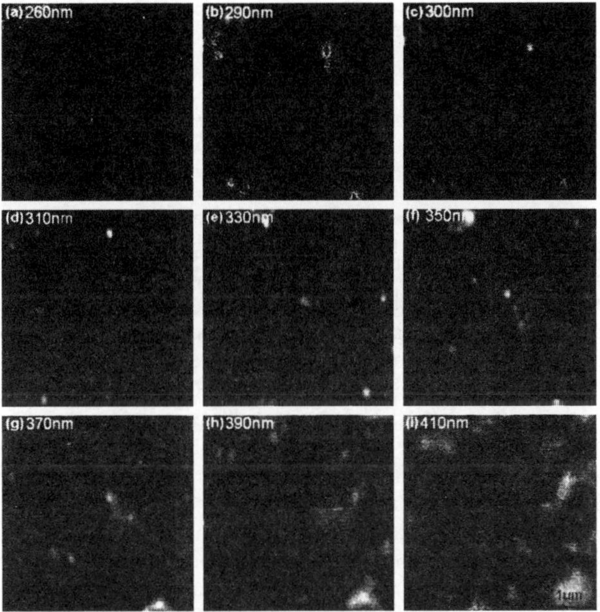

Figure 7.31   CL images of 4 × 4 µm² regions taken at (a) 260 nm, (b) 290 nm, (c) 300 nm (d) 330 nm, (f) 350 nm, (g) 370 nm, and (i) 410 nm at 300 K and V_{acc} = 3 kV of GaN QDs grown on an AlN buffer layer on (0001) 6H-SiC by MOCVD [127].

The isolation of a few QDs by µPL requires either etching submicron mesa patterns [8] or opening submicron holes in an opaque metallic mask [9] on a sample containing a low density of QDs ($<10^{10}$ cm^{-2}). Single QD PL lines measured on a 200 nm mesa are shown in Fig. 7.32. For low excitation powers, a single emission peak is observed and is attributed to the confined exciton (X). With increasing power, a second peak, labelled XX, appears on the high energy side. Two additional peaks of unknown nature appear in the P_{ex} = 9 mW spectrum. Notice that while the intensity of X scales linearly with P_{ex}, XX presents a superlinear (quadratic) power dependence. Hence, X and XX can be assigned to the exciton ground state and the biexciton transitions, respectively. According to that allocation, the biexciton binding energy is negative and of the order of 30 meV. Biexciton binding energies are most often positive. The negative binding energies observed in GaN QDs are a consequence of the spatial separation of electrons and holes (see Fig. 7.23b) which leads to, on the one hand, a reduction of the electron–hole binding energy and, on the other hand, an increase of the electron–electron and hole–hole binding energies [128].

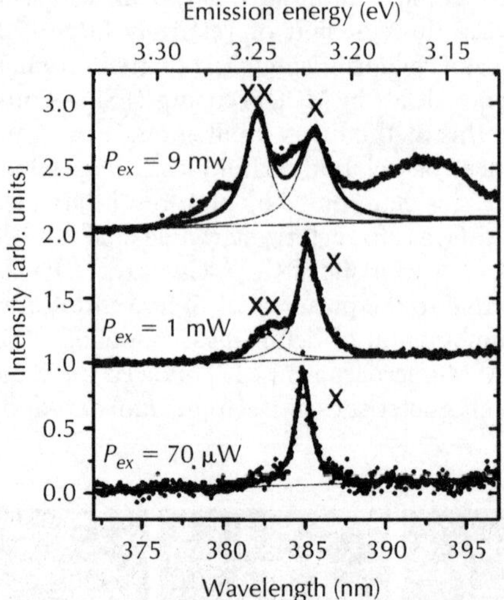

Figure 7.32   Single dot PL spectra of a GaN/AlN QD for increasing power of the 266 nm laser line focused on a 4 μm diameter spot. The dots were grown by MOCVD and had average height and diameter of 4 and 20 nm, respectively. Measurements were performed at 3.5 K [8]. With permission from AIP© 2004.

Another interesting characteristic of the single QD PL displayed in Fig. 7.32 is the broad linewidth of the X peaks. The FWHM of the emission lines is of the order of a few meV, almost three orders of magnitude larger than the typical values observed for InAs QDs and one order of magnitude larger than those for InGaN. Such large homogeneous broadening is far beyond the value expected for phonon dephasing and results from fluctuations in the exciton energy, the so-called spectral diffusion produced by the variations in the strength of the internal field caused by charge fluctuations in the vicinity of the QDs. The spectral diffusion has been observed experimentally following the time evolution of the continuous wave μPL spectrum of single GaN/AlN QDs over several minutes [9]. These studies have identified two types of spectral diffusion: a discrete one where the emission peak jumps abruptly in an energy range of the order of 10 meV; and a continuous shift of the transitions producing an inhomogeneous broadening of the spectral lines. Both scenarios can be explained by the changes in the QD internal electric field produced by charges trapped in defects in the vicinity of the dots. The discrete drift is produced by closer defects and more distant charges induce the continuous shift [9].

Single QD spectroscopy has allowed the demostration of QCSE by the application of an external electric field in different geometries [129, 114]. Figure 7.33a shows the evolution of the peak energy of a single exciton emission line as a function of the strength of an applied vertical electric field. The sample contained a single period of GaN/AlN QDs of 3 nm average height grown by MOCVD on *n*-doped 6H-SiC. A Cr/Au semitransparent Schottky contact was deposited on the top surface of the sample. The PL blue shifts linearly for increasing positive fields as electrons and holes are pushed towards the centre of the QD and the effect of the built-in field is partly compensated. Changing the geometry of the experiment and applying the electric field along the QD plane should induce a red shift of the emission line resulting from the polarization of the electron–hole pairs. The experimental observations reported in Fig. 7.33b correspond to an experiment performed on a similar sample with Al/Au interdigital electrodes with 200 nm spacing and prove that the emission is blue shifted for any direction of the field instead. There is a slight asymmetry in the energy shift due to the presence of a weak vertical component in the applied field which produces an asymmetric shift as seen in Fig. 7.33a. The increase of the emission energy in the case of in-plane external fields reflects the decrease in the exciton binding energy as electrons and holes are taken apart from each other and the cancellation of the energy shifts of the electron and hole confined state energies [114]. The latter occurs because there are different confinement potentials acting on electrons and holes since the internal electric field pushes their wavefunctions to the top and bottom part of the QD, respectively.

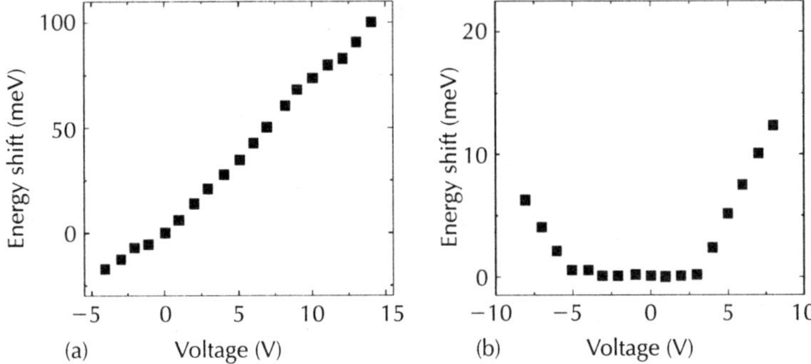

Figure 7.33   PL energy shift as a function of the voltage of an applied electric field parallel (a) and perpendicular (b) to the QD axis. In (a) positive voltage corresponds to electric fields pointing the sample top [129, 114]. With permission from APS© 2006.

7.4.5   QDs grown along non-polar directions

Avoiding the undesired effects of the built-in electric field present in polar III-nitride heterostructures is the main motivation for growing their non-polar counterparts. In the case of QDs two different strategies have proved a drastic suppression of the field effects: the growth of GaN QDs with zinc-blende structure and the growth of hexagonal dots along non-polar directions.

Zinc-blende GaN QDs embedded in an AlN matrix exhibit a strong emission in the UV without any thermal quenching up to room temperature [23]. The lack of an internal electric field also results in a faster and size independent recombination rate [112].

The situation of hexagonal non-polar QDs is more complex since the polar c-axis is now contained in the plane of the dot and still crosses GaN/AlN interfaces. Thus an electric field should still be present inside the QD. However, the optical properties of a-plane GaN/AlN QDs grown on $(11\bar{2}0)$ 6H-SiC indicate that the field is very much reduced. Figure 7.34 shows the room temperature PL of a sample containing 18 layers of non-polar dots for different excitation powers together with that of a polar sample. Among the differences between the two types of QDs, the lack of screening effects and the much narrower linewidth of the non-polar QD emission are

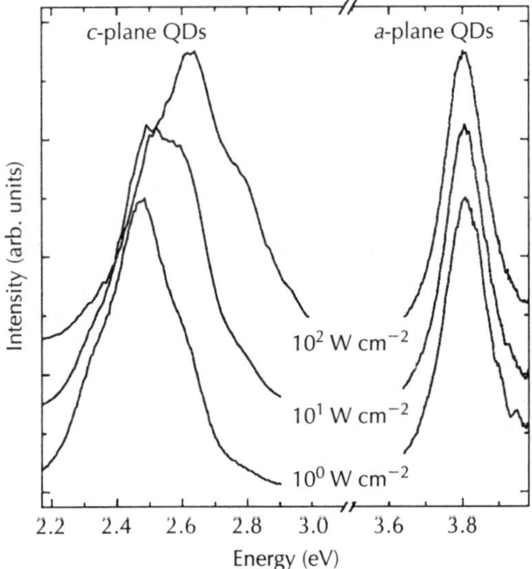

Figure 7.34   PL spectra of a-plane GaN/AlN QDs as a function of the average power density of a pulsed laser (2 ps pulse at 76 MHz repetition rate). The spectra of c-plane GaN QDs is also shown for comparison. The spectra were recorded at room temperature [79]. With permission from AIP© 2005.

highlighted. Indeed, single QD studies have reported linewidths which go down to 0.5 meV, i.e. one order of magnitude lower than those of polar QDs [130, 131].

The emission also evolves in a much faster dynamic than those described in section 7.4.3 for polar QDs. The *a*-plane QD PL decay curves for two different energies are depicted in Fig. 7.35. The curves can be fitted by single exponentials with characteristic decay times well below 1 ns. Since non-radiative recombination does not have a large impact in the low-temperature dynamics of these systems, the fast decay of the PL indicates a fast radiative lifetime of the exciton and, hence, an increased wavefunction overlap for electrons and holes in the dots.

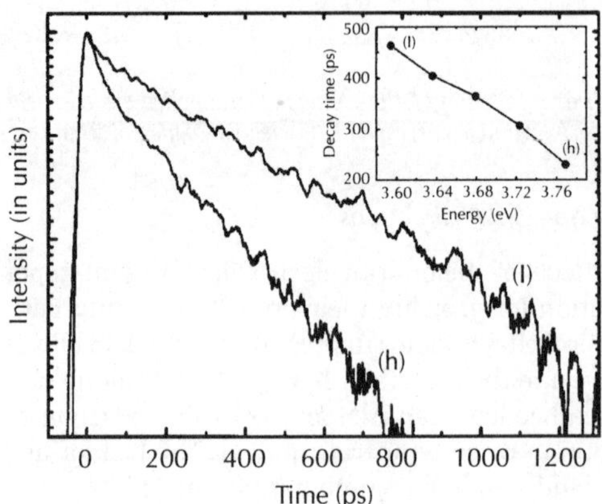

Figure 7.35   *Time evolution of the PL intensity for two energies at the high- and low-energy side of the* a-*plane GaN/AlN QD emission measured at low temperature [26]. With permission from AIP© 2005.*

The overall phenomenology just described points out a strong reduction of the internal electric field. A possible explanation for the lack of electric field effects becomes evident after the analysis of the strain-induced piezoelectric polarization. The determination of strain through Raman scattering presented in section 7.3.4 allows the estimation of the mean piezoelectric polarization, P_{pz}, for the GaN QD and the AlN barrier through the piezoelectric coefficients as

$$P_{pz} = e_{31}(\varepsilon_{xx} + \varepsilon_{yy}) + e_{33}\varepsilon_{zz}. \tag{7.9}$$

The values obtained are 0.027 C/m^2 and -0.013 C/m^2 for the barrier and the QD, respectively. These should be added to the spontaneous polarization calculated by Bernardini *et al.* [105] which are $P_{sp}^{AlN} = -0.081$ C/m^2 and $P_{sp}^{QD} = -0.029$ C/m^2. Finally, the total polarization difference between AlN and the *a*-plane QDs is $\Delta P = -0.012$ C/m^2, i.e. seven times smaller than that found in quantum dots grown along the *c*-axis [61]. Such a strong reduction would then result in a proportional reduction of the internal electric field. This is in very good agreement with the theoretical results of Fig. 7.36, where the electric field along the [0001] direction along the centre of a non-polar GaN/AlN QD is shown. The value of the electric field diminishes by a factor 10 with respect to that found in the polar QD of Fig. 7.22. A second striking difference between Figs 7.22 and 7.36 arises from the piezoelectric contribution to the potential induced by strain (dashed line) which now has an opposite sign. As a consequence, the total potential, obtained by adding both contributions, is considerably reduced.

7.5   Intraband absorption

The large conduction band offset characteristic of GaN/AlN heterostructures ($\sim$2 eV) offers the possibility of tuning intersubband transitions at very short wavelengths. Due to the high ionicity

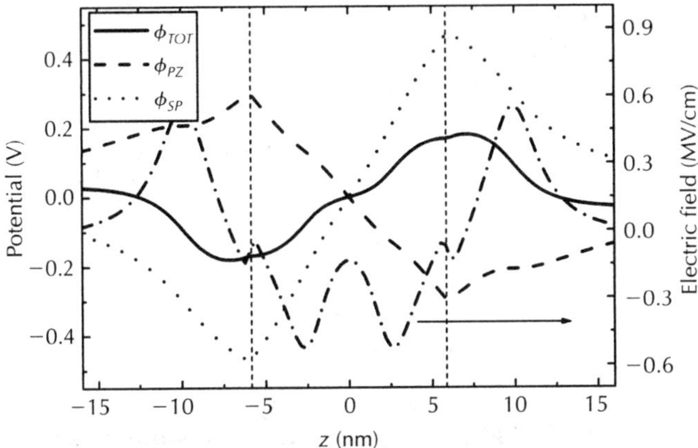

Figure 7.36   *Variation of the electrostatic potential components along the wurtzite c-axis of a non-polar QD. The strain-induced piezoelectric contribution is shown with a dashed line and spontaneous polarization term with a dotted line. Their sum corresponds to the total potential (solid line). The variation of the z-component of the electric field along the c-axis is represented by a dashed–dotted line.*

of GaN, the electron–LO-phonon interaction is largely enhanced, so that the absorption recovery time was found to be extremely short, of the order of 150 and 370 fs in QW heterostructures, for intersubband transitions at 4.5 and 1.7 μm, respectively [132, 133]. In contrast to QWs, the 3D confinement of electrons in quantum dots allows intraband transitions with a polarization component in the layer plane (TE) or along the growth axis (TM).

The first observation of intraband absorption in GaN/AlN QD superlattices was reported by Moumanis *et al.* [134]. Various absorption peaks were identified, with energies around 0.6 eV, which varied by several tens of meV with sample substrate.

The differential absorption under *s*-polarized radiation (electric field contained in the layer plane) was found to be a factor two or three smaller than that for *p*-polarized light. It amounts to 5.6×10^{-4} (1.7×10^{-4}) for the sample grown on Si (sapphire) determined for the main absorption peak. This absorption is attributed to the transition from the *s*-like to the *p*-like orbital states of the conduction band.

In order to achieve absorption at shorter wavelengths suitable for unipolar intersubband devices operating at fibre-optics telecommunication wavelengths (1.3 μm, 1.5 μm), Tchernycheva *et al.* [135] investigated samples with very small dot heights of approximately 1 nm. The samples studied in this work were grown by plasma-assisted MBE on AlN-on-sapphire. They consisted in 20 periods GaN QD layers, formed by the deposition of 4 ML GaN, with 3 nm thick AlN spacer. A detailed study of the different growth parameters and structural characteristics of the QDs that affect the intraband transitions and allow telecommunication wavelengths to be achieved was reported elsewhere [136]. The absorption observed for the different samples for *p*-polarization peaked between 0.81 eV (1.53 μm) and 0.878 eV (1.41 μm), and showed FWHM as small as 88 meV. The absorption characteristics was found to depend on the interruption growth time in Si-doped samples. On the other hand, nominally undoped samples showed residual intraband absorption as well as light-activated absorption. They were attributed to the detrapping of electrons in the AlN barriers and subsequent capture in the dots. No absorption was observed for *s*-polarization.

A communication wavelength photodetector based on intersubband transition in GaN/AlN self-assembled quantum dot heterostructures was reported in [137]. The photodetector is based on in-plane transport and has a room temperature spectral peak responsivity of 8 mA/W at a wavelength of 1.41 μm. The authors used photoluminescence, transmission, and photocurrent spectroscopy in order to explore the energy level scheme of the device. The sample investigated consisted of 20 periods of Si-doped GaN QD layers (1.2 nm QD height and 15 nm diameter) separated by 3 nm thick AlN barriers, and was grown on an AlN-on-sapphire template. Figure 7.37 presents the results of three techniques: photocurrent spectroscopy (PC), transmittance (T) and

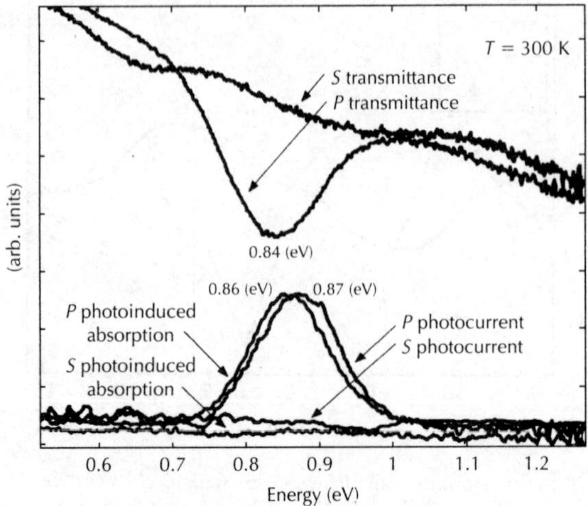

Figure 7.37   Photoconductive, transmission and photoinduced absorption spectra of the QD device in wedge illumination geometry for p- *and* s-*polarizations [137]. With permission from AIP© 2006.*

normalized photoinduced absorption ($\Delta T/T$). For the experiments, a multipass wavelength geometry was used, in which the sample was illuminated through the 45° wedge and a polarizer was used to select s- or p-polarization. The p-polarized PC peak takes place at 1.41 µm (0.88 eV), while that corresponding to the transmission and absorption spectra appear at slightly lower energies (20–40 meV). All the peaks are strongly p-polarized and show similar widths of about 150 meV.

In order to investigate the origin of the observed transitions and photocurrent spectral response, calculations of the electron energy levels in the QDs and in the WL were performed by means of an eight-band $\mathbf{k} \cdot \mathbf{p}$ model that takes into account lattice mismatch and strain, but neglects the in-plane confinement of the carriers. Good agreement between the calculated transition energies and the experiment was found. The calculations showed that the second QD electronic level (p_z) overlaps in energy with the first electronic level of the WL. There is therefore an efficient transfer of the excited carriers from the QDs into the WL. This overlap accounts as well for the observed blue shift of the PC peak relative to that measured by absorption (Fig. 7.37). The QDs of smaller size have a p_z energy larger than the first electronic state of the WL. They are therefore more likely to contribute to the lateral transport in the WL, resulting in a PC spectral peak of higher energy in comparison with that of absorption measurements.

Acknowledgements

The authors are thankful to the Ministry of Science and Education of Spain (Grant MAT 2006-01825, FEDER), European Network SANDIE (NMP4-CT-2004-500101) and the Generalitat Valenciana for their financial support.

References

1. H. Morkoç, *Nitride Semiconductors and Devices* (Springer-Verlag, Berlin, 1999).
2. S. Kako, C. Santori, K. Hoshino, S. Gotzinger, Y. Yamamoto, and Y. Arakawa, *Nature Mater.* **5**, 887 (2006).
3. D. Huang, Y. Fu, and H. Morkoç, Growth, structures, and optical properties of III-nitride quantum dots, in: T. Steiner, ed., *Semiconductor Nanostructures for Optoelectonics Applications* (Artech House, Boston, MA, 2004).
4. K. Tachibana, T. Someya, S. Ishida, and Y. Arakawa, *Appl. Phys. Lett.* **76**, 3212 (2000).
5. J. Ristic, E. Calleja, M.A. Sánchez-García, J.M. Ulloa, J. Sánchez-Páramo, J.M. Calleja, U. Jahn, A. Trampert, and K.H. Ploog, *Phys. Rev. B.* **68**, 125305 (2003).

6. S. Keller, C. Schaake, N.A. Fichtenbaum, C.J. Neufeld, Y. Wu, K. McGroddy, A. David, S.P. DenBaars, C. Weisbuch, J.S. Speck, and U.K. Mishra, *J. Appl. Phys.* **100**, 054314 (2006).

7. J. Brault, S. Tanaka, E. Sarigiannidou, J.L. Rouvière, B. Daudin, G. Feuillet, and H. Nakagawa, *J. Appl. Phys.* **93**, 3108 (2003).

8. S. Kako, K. Hoshino, S. Iwamoto, S. Ishida, and Y. Arakawa, *Appl. Phys. Lett.* **85**, 64 (2004).

9. R. Bardoux, T. Guillet, P. Lefebvre, T. Taliercio, T. Bretagnon, S. Rousset, B. Gil, and F. Semond, *Phys. Rev. B.* **74**, 195319 (2006).

10. N. Grandjean, B. Damilano, and J. Massies, *J. Phys.: Condens. Matter.* **13**, 6945 (2001).

11. N.T. Pelekanos, G.E. Dialynas, J. Simon, H. Mariette, and B. Daudin, *J. Phys.: Conf. Ser.* **10**, 61 (2005). L.S. Dang, G. Fishman, H. Mariette, C. Adelmann, E. Martinez, J. Simon, B. Daudin, E. Monroy, N. Pelekanos, J.L. Rouviere, and Y.H. Cho, *J. Korean Phys. Soc.* **42**, S657 (2003).

12. Y. Arakawa, T. Someya, and K. Tachibana, *Phys. Status Solidi (b).* **224**, 1 (2001).

13. Y. Arakawa and S. Kako, *Phys. Status Solidi (a).* **203**, 3512 (2006).

14. B. Daudin, F. Widmann, G. Feuillet, Y. Samson, M. Arlery, and J.L. Rouvière, *Phys. Rev. B.* **56**, R7069 (1997).

15. B. Damilano, N. Grandjean, F. Semond, J. Massies, and M. Leroux, *Appl. Phys. Lett.* **75**, 962 (1999).

16. S. Tanaka, S. Iwai, and Y. Aoyagi, *Appl. Phys. Lett.* **69**, 4096 (1996).

17. M. Miyamura, K. Tachibana, T. Someya, and Y. Arakawa, *J. Cryst. Growth* **237**, 1316 (2002).

18. K. Kawasaki, D. Yamazaki, A. Kinoshita, H. Hirayama, K. Tsutsui, and Y. Aoyagi, *Appl. Phys. Lett.* **79**, 2243 (2001).

19. C.W. Hu, A. Bell, F.A. Ponce, D.J. Smith, and I.S.T. Tsong, *Appl. Phys. Lett.* **81**, 3236 (2002).

20. E. Martínez-Guerrero, C. Adelmann, F. Chabuel, J. Simon, N.T. Pelekanos, G. Mula, B. Daudin, G. Feuillet, and H. Mariette, *Appl. Phys. Lett.* **77**, 809 (2000).

21. K. Pakuła, R. Bożek, K. Surowiecka, R. Stępniewski, A. Wysmolek, and J.M. Baranowski, *J. Cryst. Growth* **289**, 472 (2006).

22. D. Simeonov, E. Feltin, J.-F. Carlin, R. Butté, M. Ilegems, and N. Grandjean, *J. Appl. Phys.* **99**, 083509 (2006).

23. F. Widmann, B. Daudin, G. Feuillet, Y. Samson, J.L. Rouvière, and N. Pelekanos, *J. Appl. Phys.* **83**, 7618 (1998).

24. J. Brown, F. Wu, P.M. Petroff, and J.S. Speck, *Appl. Phys. Lett.* **84**, 690 (2004).

25. N. Gogneau, D. Jalabert, E. Monroy, T. Shibata, M. Tanaka, and B. Daudin, *J. Appl. Phys.* **94**, 2254 (2003).

26. S. Founta, F. Rol, E. Bellet-Amalric, J. Bleuse, B. Daudin, B. Gayral, H. Mariette, and C. Moisson, *Appl. Phys. Lett.* **86**, 171901 (2005).

27. S. Tanaka, M. Takeuchi, and Y. Aoyagi, *Jpn. J. Appl. Phys.* **39**, L831 (2000).

28. P. Ramvall, P. Ribet, S. Nomura, Y. Aoyagi, and S. Tanaka, *J. Appl. Phys.* **87**, 3883 (2000).

29. K. Pakuła, R. Bożek, K. Surowiecka, R. Stępniewski, A. Wysmolek, and J.M. Baranowski, *Phys. Status Solidi (b).* **243**, 1486 (2006).

30. C. Adelmann, B. Daudin, R.A. Oliver, G.A.D. Briggs, and R.E. Rudd, *Phys. Rev. B.* **70**, 125427 (2004).

31. G. Mula, C. Adelmann, S. Moehl, J. Oullier, and B. Daudin, *Phys. Rev. B.* **64**, 195406 (2001).

32. J.M. Gérard and J.Y. Marzin, *Appl. Phys. Lett.* **53**, 568 (1988).

33. B. Daudin, F. Widmann, J. Simon, G. Feuillet, J.L. Rouvière, N.T. Pelekanos, and G. Fishman, *MRS Internet J. Nitride Semicond. Res.* **4S1**, G9.2 (1999).

34. B. Damilano, N. Grandjean, F. Semond, J. Massies, and M. Leroux, *Phys. Status Solidi (b).* **216**, 451 (1999).

35. J. Coraux, B. Amstatt, J. Budagosky, E. Bellet-Amalric, J.L. Rouvière, V. Favre-Nicolin, M.G. Proietti, H. Renevier, and B. Daudin, *Phys. Rev. B.* **74**, 195302 (2006).

36. N. Gogneau, D. Jalabert, E. Monroy, E. Sarigiannidou, J.L. Rouvière, T. Shibata, and M. Tanaka, *J. Appl. Phys.* **96**, 1104 (2004).

37. N. Gogneau, F. Fossard, E. Monroy, S. Monnoye, H. Mank, and B. Daudin, *Appl. Phys. Lett.* **84**, 4224 (2004).

38. K. Hoshino, S. Kako, and Y. Arakawa, *Appl. Phys. Lett.* **85**, 1262 (2004).

39. J. Tersoff, C. Teichert, and M.G. Lagally, *Phys. Rev. Lett.* **76**, 1675 (1995).

40. V. Chamard, T.H. Metzger, M. Sztucki, V. Holý, M. Tolan, E. Bellet-Amalric, C. Adelmann, B. Daudin, and H. Mariette, *Europhys. Lett.* **63**, 268 (2003).
41. J. Stangl, V. Holý, and G. Bauer, *Rev. Mod. Phys.* **76**, 725 (2004).
42. C. Pryor, J. Kim, L.W. Wang, A.J. Williamson, and A. Zunger, *J. Appl. Phys.* **83**, 2548 (1998).
43. T. Saito and Y. Arakawa, *Inst. Phys. Conf. Ser.* **162**, 741 (1999).
44. T. Saito and Y. Arakawa, *Physica E.* **15**, 169 (2002).
45. T. Saito and Y. Arakawa, *Phys. Status Solidi (c).* **0**, 1169 (2003).
46. N. Skoulidis, V. Vargiamidis, and H.M. Polatoglou, *Superlatt. Microstruct.* **40**, 432 (2006).
47. A.D. Andreev and E.P. O'Reilly, *Phys. Rev. B.* **62**, 15851 (2000).
48. B. Jogai, *J. Appl. Phys.* **90**, 699 (2001).
49. E. Pan and B. Yang, *J. Appl. Phys.* **93**, 2435 (2003).
50. S.P. Lepkowski, J.A. Majewski, and G. Jurczak, *Phys. Rev. B.* **72**, 245201 (2005).
51. A.E. Romanov, P. Waltereit, and J.S. Speck, *J. Appl. Phys.* **97**, 043708 (2005).
52. I. Vurgaftman and J.R. Meyer, *J. Appl. Phys.* **94**, 3675 (2003).
53. J.D. Eshelby, *Proc. R. Soc. London Ser. A.* **241**, 376 (1957).
54. A. García-Cristóbal and J.M. Llorens, to be published.
55. D.P. Williams, A.D. Andreev, E.P. O'Reilly, and D.A. Faux, *Phys. Rev. B.* **72**, 235318 (2005).
56. A. Cros, N. Garro, J.M. Llorens, A. García-Cristóbal, A. Cantarero, N. Gogneau, E. Monroy, and B. Daudin, *Phys. Rev. B.* **73**, 115313 (2006).
57. A. Letoublon, V. Favre-Nicolin, H. Renevier, M.G. Proietti, C. Monat, M. Gendry, O. Marty, and C. Priester, *Phys. Rev. Lett.* **92**, 186101 (2004).
58. E. Sarigiannidou, E. Monroy, B. Daudin, J.-L. Rouvière, and A.D. Andreev, *Appl. Phys. Lett.* **87**, 203112 (2005).
59. D. Jalabert, J. Coraux, H. Renevier, B. Daudin, M.H. Cho, K.B. Chung, D.W. Moon, J.M. Llorens, N. Garro, A. Cros, and A. García-Cristóbal, *Phys. Rev. B.* **72**, 115301 (2005).
60. V. Chamard, T. Schülli, M. Sztucki, T.H. Metzger, E. Sarigiannidou, J.-L. Rouvière, M. Tolan, C. Adelmann, and B. Daudin, *Phys. Rev. B.* **69**, 125327 (2004).
61. A. Cros, J.A. Budagosky, A. García-Cristóbal, N. Garro, A. Cantarero, S. Founta, H. Mariette, and B. Daudin, *Phys. Status Solidi (b).* **243**, 1499 (2006).
62. C.A. Arguello, D.L. Rousseau, and S.P.S. Porto, *Phys. Rev. B.* **181**, 1251 (1969).
63. D.L. Rousseau, R.P. Bauman, and S.P.S. Porto, *J. Raman Spec.* **10**, 253 (1998).
64. V. Yu Davydov, Yu.E. Kitaev, I.N. Goncharuk, A.N. Smirnov, J. Graul, O. Semchinova, D. Uffmann, M.B. Smirnov, A.P. Mirgorodsky, and R.A. Evarestov, *Phys. Rev. B.* **58**, 12899 (1998).
65. E. Anastassakis and M. Cardona, in: T. Suski and W. Paul, eds., *High Pressure in Semiconductor Physics II* (Academic Press, San Diego, 1998).
66. R.J. Briggs and A.K. Ramdas, *Phys. Rev. B.* **13**, 5518 (1976).
67. V.Yu. Davydov, N.S. Averkiev, I.N. Goncharuk, D.K. Nelson, I.P. Nikitina, A.S. Polkovnikov, A.N. Smirnov, M.A. Jacobson, and O.K. Semchinova, *J. Appl. Phys.* **82**, 5097 (1997).
68. F. Demangeot, J. Frandon, P. Baules, F. Natali, F. Semond, and J. Massies, *Phys. Rev. B.* **69**, 155215 (2004).
69. V. Darakchieva, T. Paskova, M. Schubert, H. Arwin, P.P. Paskov, B. Monemar, D. Hommel, M. Heuken, J. Off, F. Scholz, B.A. Haskell, P.T. Fini, J.S. Speck, and S. Nakamura, unpublished.
70. J. Gleize, M.A. Renucci, J. Frandon, E. Bellet-Amalric, and B. Daudin, *J. Appl. Phys.* **93**, 2065 (2003).
71. J. Gleize, J. Frandon, M.A. Renucci, and F. Bechstedt, *Phys. Rev. B.* **63**, 073308 (2001).
72. N. Garro, A. Cros, J.M. Llorens, A. García-Cristóbal, A. Cantarero, N. Gogneau, E. Sarigiannidou, E. Monroy, and B. Daudin, *Phys. Rev. B.* **74**, 075305 (2006).
73. D.A. Romanov, V.V. Mitin, and M.A. Stroscio, *Phys. Rev. B.* **66**, 115321 (2002).
74. V.A. Fonoberov and A.A. Balandin, *Phys. Rev. B.* **70**, 233205 (2004).
75. L. Zhang, J. Shi, and H. Xie, *Solid Stat. Commun.* **140**, 549 (2006).
76. J. Gleize, F. Demangeot, J. Frandon, M.A. Renucci, M. Kuball, F. Widmann, and B. Daudin, *Phys. Status Solidi (b).* **216**, 457 (1999).
77. J. Gleize, J. Frandon, F. Demangeot, M.A. Renucci, C. Adelmann, B. Daudin, G. Feuillet, B. Damilano, N. Grandjean, and J. Massies, *Appl. Phys. Lett.* **77**, 2174 (2000).
78. G. Salviati, O. Martinez, M. Mazzoni, F. Rossi, N. Armani, P. Gucciardi, A. Vinattieri, D. Alderighi, M. Colocci, M.A. Gonzalez, L.F. Sanz-Santacruz, and J. Massies, *J. Phys.: Condens. Matter.* **14**, 13329 (2002).

79. N. Garro, A. Cros, J.A. Budagosky, A. Cantarero, A. Vinattieri, M. Gurioli, S. Founta, H. Mariette, and B. Daudin, *Appl. Phys. Lett.* **87**, 011101 (2005).

80. M. Cardona, in: M. Cardona and G. Günterodt, eds., *Light Scattering in Solids II*, (Springer, Berlin, 1982).

81. J. Gleize, F. Demangeot, J. Frandon, M.A. Renucci, M. Kuball, B. Damilano, N. Grandjean, and J. Massies, *Appl. Phys. Lett.* **79**, 686 (2001).

82. M. Kuball, J. Gleize, S. Tanaka, and Y. Aoyagi, *Appl. Phys. Lett.* **78**, 987 (2001).

83. A. Cros, J.A. Budagosky, N. Garro, A. Cantarero, J. Coraux, H. Renevier, M.G. Proietti, V. Favre-Nicolin, and B. Daudin, *Phys. Status Solidi (c)*, **4**, 2379 (2007).

84. M. Miyamura, K. Tachibana, and Y. Arakawa, *Appl. Phys. Lett.* **80**, 3937 (2002).

85. S. Tanaka, H. Hirayama, Y. Aoyagi, Y. Narukawa, Y. Kawakami, S. Fujita, and S. Fujita, *Appl. Phys. Lett.* **71**, 1299 (1997).

86. F. Widmann, J. Simon, N.T. Pelekanos, B. Daudin, G. Feuillet, J.L. Rouvière, and G. Fishman, *Microelect. J.* **30**, 353 (1999).

87. A. Niwa, T. Ohtoshi, and T. Kuroda, *Jpn. J. Appl. Phys.* **35**, L599 (1996).

88. R. Cingolani, A. Botchkarev, H. Tang, H. Morkoc, G. Traetta, G. Coli, M. Lomascolo, A. Di Carlo, F. Della Sala, and P. Lugli, *Phys. Rev. B.* **61**, 2711 (2000).

89. V. Ranjan, G. Allan, C. Priester, and C. Delerue, *Phys. Rev. B.* **68**, 115305 (2003).

90. V.R. Velasco, J. Tutor, and H. Rodriguez-Coppola, *Surf. Sci.* **565**, 259 (2004).

91. I. Lo, W.T. Wang, M.H. Gau, S.F. Tsay, and J.C. Chiang, *Phys. Rev. B.* **72**, 245329 (2005).

92. S. Schulz, S. Schumacher, and G. Czycholl, *Phys. Rev. B.* **73**, 245327 (2006).

93. A. Lakdja, B. Bouhafs, and P. Ruterana, *Phys. Status Solidi (a).* **203**, 2247 (2006).

94. S.L. Chuang and C.S. Chang, *Phys. Rev. B.* **54**, 2491 (1996).

95. D. Fritsch, H. Schmidt, and M. Grundmann, *Phys. Rev. B.* **67**, 235205 (2003).

96. Y.M. Sirenko, J.B. Jeon, K.W. Kim *et al.*, *Phys. Rev. B.* **53**, 1997 (1996).

97. S.L. Chuang and C.S. Chang, *Sem. Sci. Tech.* **12**, 252 (1997).

98. F. Mireles and S.E. Ulloa, *Phys. Rev. B.* **60**, 13659 (1999).

99. B. Gil, in: J.I. Pankove and T.D. Moustakas, eds., *Gallium Nitride II* (Academic Press, San Diego, 1999).

100. T. Taliercio, P. Lefebvre, M. Gallart, and A. Morel, *J. Phys.: Condens. Matter.* **13**, 7027 (2001).

101. V.A. Fonoberov and A.A. Balandin, *J. Appl. Phys.* **94**, 7178 (2003).

102. N. Vukmirovic, Z. Ikonic, D. Indjin, and P. Harrison, *J. Phys.: Condens. Matter.* **18**, 6249 (2006).

103. J.F. Nye, *Physical Properties of Crystals* (Oxford University Press, 1985).

104. M. Grundmann, O. Stier, and D. Bimberg, *Phys. Rev. B.* **52**, 11969 (1995).

105. F. Bernardini, V. Fiorentini, and D. Vanderbilt, *Phys. Rev. B.* **56**, R10024 (1997).

106. P. Waltereit, O. Brandt, A. Trampert, H.T. Grahn, J. Menniger, M. Ramsteiner, M. Reiche, and K.H. Ploog, *Nature* **406**, 865 (2000).

107. S.F. Chichibu, A.C. Abare, M.S. Minsky, S. Keller, S.B. Fleischer, J.E. Bowers, E. Hu, U.K. Mishra, L.A. Coldren, S.P. DenBaars, and T. Sota, *Appl. Phys. Lett.* **73**, 2006 (1998).

108. J.S. Im, H. Kollmer, J. Off, A. Sohmer, F. Scholz, and A. Hangleiter, *Phys. Rev. B.* **57**, R9435 (1998).

109. O. Ambacher, J. Smart, J.R. Shealy, N.G. Weimann, K. Chu, M. Murphy, W.J. Schaff, L.F. Eastman, R. Dimitrov, L. Wittmer, M. Stutzmann, W. Rieger, and J. Hilsenbeck, *J. Appl. Phys.* **85**, 3222 (1999).

110. F. Widmann, J. Simon, B. Daudin, G. Feuillet, J.L. Rouvière, and N.T. Pelekanos, *Phys. Rev. B.* **58**, R15989 (1998).

111. R.A. Hogg, C.E. Norman, A.J. Shields, M. Pepper, and M. Iizuka, *Appl. Phys. Lett.* **76**, 1428 (2000).

112. J. Simon, N.T. Pelekanos, C. Adelmann, E. Martínez-Guerrero, R. André, B. Daudin, L.S. Dang, and H. Mariette, *Phys. Rev. B.* **68**, 035312 (2003).

113. T. Bretagnon, P. Lefebvre, P. Valvin, R. Bardoux, T. Guillet, T. Taliercio, B. Gil, N. Grandjean, F. Semond, B. Damilano, A. Dussaigne, and J. Massies, *Phys. Rev. B.* **73**, 113304 (2006).

114. T. Nakaoka, S. Kako, and Y. Arakawa, *Phys. Rev. B.* **73**, 121305 (2006).

115. T. Andreev, Y. Hori, X. Biquard, E. Monroy, D. Jalabert, A. Farchi, M. Tanaka, O. Oda, L.S. Dang, and B. Daudin, *Phys. Rev. B.* **71**, 115310 (2005).

116. T. Andreev, N.Q. Liem, Y. Hori, M. Tanaka, O. Oda, L.S. Dang, and B. Daudin, *Phys. Rev. B.* **73**, 195203 (2006).

117. G. Salviati, F. Rossi, N. Armani, V. Grillo, O. Martinez, A. Vinattieri, B. Damilano, A. Matsuse, and N. Grandjean, *J. Phys.: Condens. Matter.* **16**, S115 (2004).

118. S. De Rinaldis, I. D'Amico, E. Biolatti, R. Rinaldi, R. Cingolani, and F. Rossi, *Phys. Rev. B.* **65**, 081309 (2002).

119. P. Riblet, S. Tanaka, P. Ramvall, S. Nomura, and Y. Aoyagi, *Solid State Comm.* **109**, 377 (1999).

120. S. Kalliakos, T. Bretagnon, P. Lefebvre, T. Taliercio, B. Gil, N. Grandjean, B. Damilano, A. Dussaigne, and J. Massies, *J. Appl. Phys.* **96**, 180 (2004).

121. Y.-H. Cho, B.J. Kwon, J. Barjon, J. Brault, B. Daudin, H. Mariette, and L.S. Dang, *Appl. Phys. Lett.* **81**, 4934 (2002).

122. S. Kako, M. Miyamura, K. Tachibana, K. Hoshino, and Y. Arakawa, *Appl. Phys. Lett.* **83**, 984 (2003).

123. T. Bretagnon, S. Kalliakos, P. Lefebvre, P. Valvin, B. Gil, N. Grandjean, A. Dussaigne, B. Damilano, and J. Massies, *Phys. Rev. B.* **68**, 205301 (2003).

124. C. Santori, S. Götzinger, Y. Yamamoto, S. Kako, K. Hoshino, and Y. Arakawa, *Appl. Phys. Lett.* **87**, 051916 (2005).

125. A. Neogi, H. Morkoç, T. Kuroda, A. Tackechi, T. Kawazoe, and M. Ohtsu, *Nano Lett.* **5**, 213 (2005).

126. A. Petersson, A. Gustafsson, L. Samuelson, S. Tanaka, and Y. Aoyagi, *Appl. Phys. Lett.* **74**, 3513 (1999).

127. Y. Yao, T. Sekiguchi, Y. Sakuma, M. Miyamura, and Y. Arakawa, *Scripta Materialia.* **55**, 679–682 (2006).

128. D.P. Williams, A.D. Andreev, D.A. Faux, and E.P. O'Reilly, *Physica E.* **21**, 358 (2004).

129. T. Nakaoka, S. Kako, and Y. Arakawa, *Physica E.* **32**, 148 (2006).

130. F. Rol, B. Gayral, S. Founta, B. Daudin, J. Eymery, J.-M. Gérard, H. Mariette, L.S. Dang, and D. Peyrade, *Phys. Status Solidi.* **243**, 1652 (2006).

131. F. Rol, B. Gayral, S. Founta, H. Mariette, B. Daudin, L.S. Dang, J. Bleuse, D. Peyrade, J.M. Gérard, and B. Gayral, *Phys. Rev. B.* **75**, 125306 (2007).

132. N. Iizuka, K. Kaneko, N. Suzuki, T. Asano, S. Noda, and O. Wada, *Appl. Phys. Lett.* **77**, 648 (2000).

133. C. Gmachl, S.V. Frolov, H.M. Ng, S.-N.G. Chu, and A.Y. Cho, *Electron. Lett.* **37**, 378 (2001).

134. Kh. Moumanis, A. Helman, F. Fossard, M. Tchernycheva, A. Lusson, F.H. Julien, B. Damilano, N. Grandjean, and J. Massies, *Appl. Phys. Lett.* **82**, 868 (2003).

135. M. Tchernycheva, L. Nevou, L. Doyennette, A. Helman, R. Colombelli, F.H. Julián, F. Guillot, E. Monroy, T. Shibata, and M. Tanaka, *Appl. Phys. Lett.* **87**, 101912 (2005).

136. F. Guillot, E. Bellet-Amalric, E. Monroy, M. Tchernycheva, L. Nevou, L. Doyennette, F.H. Julien, L.S. Dang, T. Remmele, M. Albrecht, T. Shibata, and M. Tanaka, *J. Appl. Phys.* **100**, 044326 (2006).

137. A. Vardi, N. Akopian, G. Bahir, L. Doyennette, M. Tchernycheva, L. Nevou, F.H. Julien, F. Guillot, and E. Monroy, *Appl. Phys. Lett.* **88**, 143101 (2006).

CHAPTER 8

GaSb/GaAs Quantum Nanostructures by Molecular Beam Epitaxy

Koichi Yamaguchi,[1] Shiro Tsukamoto,[2] and Kazunari Matsuda[3]

[1]*Department of Electronic Engineering, The University of Electro-Communications, 1-5-1 Chofugaoka, Chofu, Tokyo 182-8585, Japan*
[2]*Center for Collaborative Research, Anan National College of Technology, 265 Aoki, Minobayashi, Anan, Tokushima 774-0017, Japan*
[3]*International Research Center for Elements Science, Institute for Chemical Research, Kyoto University, Gokasho, Uji, Kyoto 611-0011, Japan*

8.1 Introduction

GaSb-based quantum nanostructures in a GaAs matrix have unique band alignment of a staggered type-II [1, 2] large bowing of the band gap [2] and strong hole localization [3]. Based on these properties, GaSb/GaAs nanostructures including quantum wells (QWs) and quantum dots (QDs) are of considerable interest for use in various applications, such as infrared emitting and detecting devices, and memory and spin-controlled devices. In the GaSb/GaAs heteroepitaxial growth, an exchange of group V elements and a large lattice mismatch of 7.8% between GaSb and GaAs should be noted to control the heterointerface. Chidley *et al.* first reported the growth of GaSb QWs on GaAs by metalorganic chemical vapour deposition (MOCVD) [4]. In their report, a transition from the pseudomorphic thin film growth to the island growth was observed at a thickness of above 1.5 nm. The growth of GaSb QDs on GaAs was first reported by Hatami *et al.* [1]. The GaSb/GaAs QDs were spontaneously formed by molecular beam epitaxy (MBE) using a Stranski–Krastanov (SK) growth technique, which has been developed in InAs/GaAs and Ge/Si heterostructures [5–7]. Currently, the SK growth technique of the self-assembled QDs is the most promising way for producing coherent QDs. Recently, the self-assembled GaSb QDs on GaAs have been attempted by many groups.

This chapter focuses on the GaSb/GaAs quantum nanostructures by MBE. Section 8.2 introduces the surface reconstruction of GaSb on GaAs and the heterointerface structure of GaSb/GaAs. Section 8.3 introduces the self-assembled GaSb/GaAs QDs and the optical properties of GaSb/GaAs QDs.

8.2 Surface and interface structures of GaSb/GaAs

In order to fabricate GaSb/GaAs quantum nanostructures, we have to control surface and interface structures. In this section, we will discuss the surface and interface structures for GaSb growth on GaAs and for GaAs growth on GaSb. First, we describe surface reconstructions of the GaSb grown on the GaAs(001) in Section 8.2.1. Then, we analyse a broadening of GaAs/GaSb heterointerface by using a kinetic model of surface exchange reaction between Sb and As atoms in Section 8.2.2.

8.2.1   Surface reconstruction of GaSb on GaAs and in situ STM observation

The fine deposition control of various semiconductor materials by MBE has enabled the fabrication of quantum-size effect devices. III–V semiconductor materials such as GaAs and InP have been used as diverse devices, in particular, optoelectronic device applications. Recently, the novel fabrication method that irradiates antimony (Sb) on a GaAs(001) surface and forms an Sb-terminated layer by As–Sb exchange reaction has been devised [8]. Sb irradiation brings those surface reconstruction changes from $c(4 \times 4)$ to $(n \times 3)$ $(n - 1, 2, 4)$. In the new application of Sb/GaAs(001) it has been proposed that InAs quantum dots grown on Sb-irradiated GaAs(001) by MBE obtained high density and high uniformity [9, 10]. However, a detailed structure of this surface has not been investigated. Although surface reconstruction of Sb-terminated GaAs(001) has been reported to exist $((2 \times 4)$ [11, 12] and $(2 \times 8))$ [12], $(n \times 3)$ surface reconstruction has not been reported yet. The $\times 3$ ordered reconstructions have also been studied for Sb-passivated GaAs (001) at room temperature [13–15]. Although scanning tunneling microscopy (STM) observation of Sb-passivated surface-(1×3) [15] and GaSb on GaAs(001)-$c(4 \times 4)$ [16] have been reported, any reconstruction model is not discussed for these $\times 3$ ordered reconstructions, including (2×3) and (4×3). Recently, using *in situ* STM and first-principles calculation, surface reconstructions were investigated on Sb-irradiated GaAs(001) formed by MBE and its structure models were proposed.

It was used a non-doped GaAs(001) just-oriented substrate. After oxide layer removal, a GaAs buffer layer of 200 nm was grown at 570°C under an As_4 pressure of 5×10^{-5} Pa and Ga pressure of 3.3×10^{-6} Pa. Next, under As_4 flux, the substrate was set up to an irradiation temperature of 435°C. The $c(4 \times 4)$ reconstruction of the As-rich surface was confirmed by reflection high energy electron diffraction (RHEED). When the Sb shutter was opened, simultaneously closing the As shutter, an Sb_4 flux of 1.0×10^{-6} Pa was irradiated on the As-terminated GaAs so that an As–Sb exchange reaction was carried out [8]. The irradiation amount depends on the RHEED pattern changing from $c(4 \times 4)$ to (2×3) or (4×3). The substrate was annealed for over 10 minutes at the same temperature after the irradiation. The (4×3) or (2×3) reconstructions were unchanged during substrate quenching to 200°C. The sample was observed at the same temperature on the same substrate heater by the *in situ* STM located inside the MBE growth chamber [17] STM experiments used a tungsten tip and were performed under the background pressure of 8×10^{-8} Pa. All STM images were acquired by a constant current of 0.1 nA, a scan speed of 250 nm/s and a tip bias of $+3$ V. Then, first-principles density-functional total-energy calculations were performed based on the density-functional theory using the Vienna *ab initio* simulation package [18]. The local-density approximation was used for the exchange correlation and the projector augmented wave [19, 20] pseudopotentials. The Ga 3d electrons were treated as part of the valence band. The cutoff energy for the plane-wave basis was 400 eV. We employed supercells containing six atomic layers of GaAs and two layers of GaSb where the bottom side of the slab was terminated with fractionally charged hydrogen in order to maintain bulk properties.

Figure 8.1 shows RHEED patterns in the [110] and [1$\bar{1}$0] azimuth and a filled-state STM image on an Sb-irradiated surface. The RHEED patterns indicate three-fold order distinctly in the [1$\bar{1}$0] azimuth (Fig. 8.1a) and weak $2\times$ and very weak $4\times$ in the [110] azimuth (Fig. 8.1b) so that the

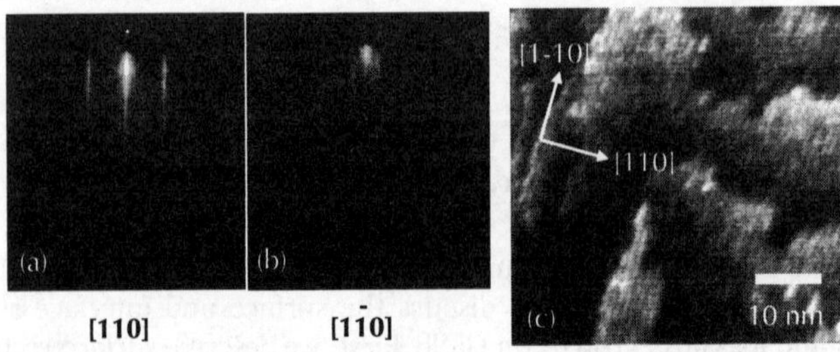

Figure 8.1   RHEED patterns and STM image of Sb-irradiated GaAs(001) [21]. (a) Slightly obscure two-fold in [110] azimuth and (b) three-fold in [1$\bar{1}$0] azimuth. (c) STM image at 200°C, tunnelling current of 0.1 nA, tip bias of +3 V, scan size of 50 × 50 nm, scan speed of 250 nm/s and background pressure of 8×10^{-8} Pa. The height range of the surface is approximately 1.0 nm. The periodicity perpendicular to the [1$\bar{1}$0] direction is 1.2 nm.

surface is dimerized (2×3) (or (4×3)). STM image (Fig. 8.1c) shows that anisotropic islands are formed with multilayer behaviour and kinks that result from steps. The step density of this surface, defined here as the total length of step edges per unit area, is $183\,\mu m^{-1}$, whereas that of GaAs(001)$-c(4 \times 4)$ is $40\,\mu m^{-1}$ [22]. This Sb-irradiated surface was thus found to be rough, which agreed with the ribbon-like structures reported by Bennett *et al.* [16]. They suggested that the additional growth of GaSb of 2.5 ML on GaSb of 1 ML (3.5 ML total)/GaAs(001) surface enhances the anisotropy of the surface attributed to the direction-dependent strain associated with the dimer-based surface reconstructions of the wetting layers. The result reinforces their suggestion because of the anisotropy along the $[1\bar{1}0]$ direction. The Sb irradiation causes an increase of surface step density. The same phenomenon is reported in the co-deposition of Sb and Si on Si(001) [23] and AlAs on GaAs(001) [24]. The high QD density can be achieved by using these high-step density surfaces since dangling bonds on the step edges plays a role of nucleation sites [10, 24]. However, the incorporation mechanism of InAs deposition into Sb/GaAs(001) and the segregation mechanism of Sb remain unclear. Therefore, it is important to know its initial surface which means the (2×3) (or (4×3)) reconstructed Sb-terminated GaAs(001).

Figure 8.2 shows a high-resolution empty-state image of Sb-irradiated GaAs(001). A $\times 3$ structure of approximately 1.2 nm periodicity was confirmed perpendicular to the $[1\bar{1}0]$ direction and many swinging dimer rows were observed along the $[1\bar{1}0]$ direction, similar to the disordered (1×3) observed by Moriarty *et al.* [15]. On the other hand, the RHEED pattern showed slightly obscure two-fold order in the [110] azimuth. Therefore, these surface reconstructions can be classed as disordered (2×3) structures with swinging dimer rows along the $[1\bar{1}0]$ direction. In order to confirm the interpretation of these structures, the models proposed by Righi *et al.* [25] are useful. They studied the respective stability for (1×3) [26] (1×3)/$c(2 \times 6)$ [27], as well as for (4×3) reconstructions for the Sb-rich surface on GaSb(001). Since the model of h0(4×3) proposed by Righi *et al.* [25] has the lowest surface energy, we consider three (2×3)-reconstructed structures for Sb/GaAs(001), as shown in Fig. 8.3, based on this h0(4×3) structure. It was considered that first- and second-layer atoms consisted of Sb only since a $\times 3$ RHEED pattern in the $[1\bar{1}0]$ azimuth can be achieved by about 2 ML Sb-irradiation on GaAs(001). Note that Sb–Sb bonds are stronger than Ga–Sb bonds [28, 29]. It is assumed that a little re-evaporation of Sb atoms from the top layer might occur during annealing at 435°C, but it does not affect reconstruction formations.

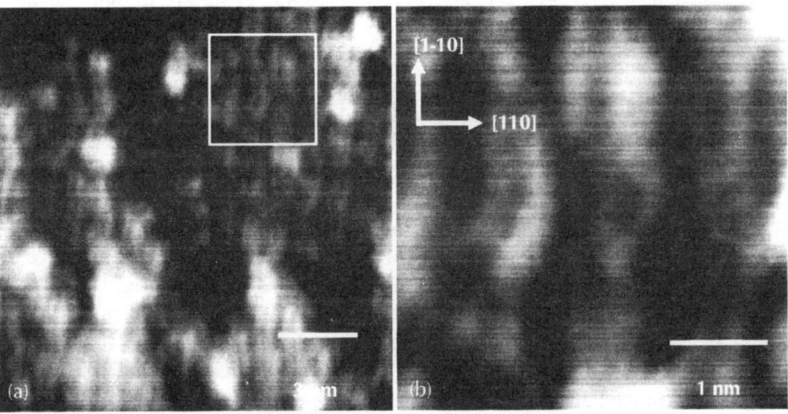

Figure 8.2   The high-resolution empty-state STM image of Sb/GaAs(001) [21], tip bias of $-3\,V$, tunnelling current of 0.1 nA, scan speed of 250 nm/s and background pressure of 8×10^{-8} Pa. The height range of the surface is approximately 1.0 nm. (b) is a magnified image as indicated by the white open square in (a).

In order to study the imbalance of charge, two bias-alternative STM images were observed in the identical region. Following a filled state (Fig. 8.3a), an empty-state image (Fig. 8.3b) was scanned. In empty state many contrasts are shown over the surface (C_1 shown in Fig. 8.3b, for example). It is confirmed that the identical regions in filled state of contrasts C_1 do not change

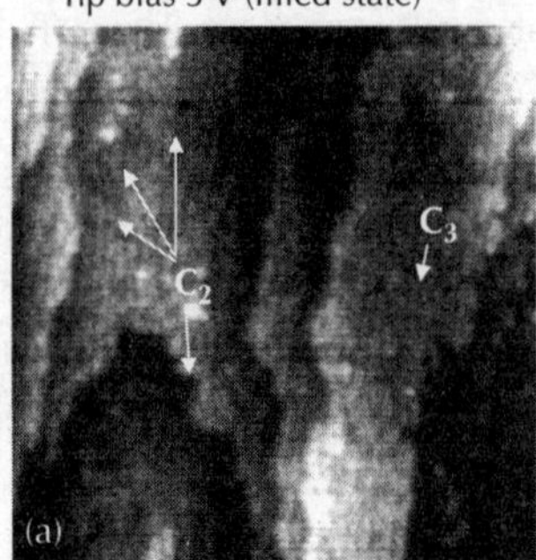

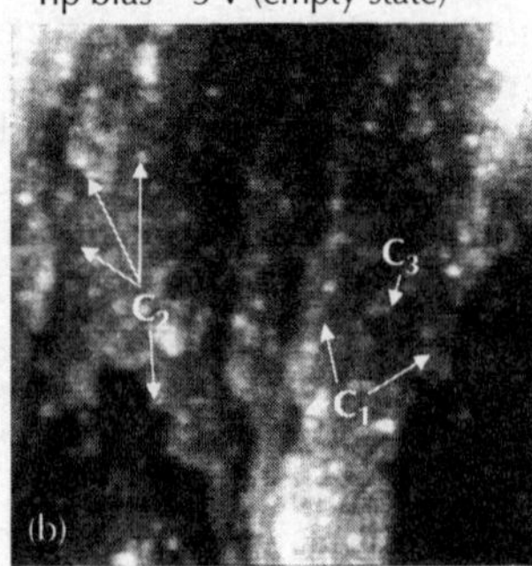

Figure 8.3   Bias-alternative STM images of Sb-irradiated GaAs(001) surface at the identical position [21]. The scan parameters are scan size of 50 nm × 50 nm and tunnelling current of 0.1 nA, respectively. The tip bias conditions are (a) a tip bias of 3 V (filled state) and (b) −3 V (empty state), respectively. The grey-scale height range is approximately 1.2 nm.

$3\times$ periodicity along $[1\,\bar{1}0]$ but have not detected a clear change perpendicular to the $[1\,\bar{1}0]$ direction. Since C_1 is sensitive in empty state, we might assume that C_1 originates not from contamination but from a segregated Ga or Sb cluster. If a III-rich or V-rich condition is matched locally, some reconstructed changes should arise from those concentrations [30]. On the other hand, C_2 and C_3 shown in Fig. 8.3a and 8.3b are probably contamination and lattice defect, respectively. The presence of such contrasts may relate strongly to the charge excess on this surface, but for detailed analysis it is assumed that first-principles calculation considering defect structures is needed. Moreover, there are still many possibilities of the structure model of this surface. We show here only some of them. For example, a recent report describes the possibility of Sb multilayers [31]. Finally, Sb irradiation on GaAs(001) causes high step density and has specific structures such as the swinging dimers and the contrasts C_1. InAs deposition on this surface causes drastic changes such as wetting layer and high density of dots. However, since surplus Sb supply gives rise to the deterioration of crystal quality [10], close attention must be paid to the Sb amount. In order to analyse the details of InAs deposition on this surface, *in situ* STM observations during growth [17] are suitable and powerful. Therefore, its details are described in next paragraphs.

Scanning tunnelling microscopy (STM) has been applied to GaAs and related materials since the 1980s. Somewhat more recently it has been applied to GaAs grown *in vacuo* by molecular beam epitaxy (MBE). The ability to image surface reconstructions and morphology as well as island structures and dopants has added substantially to the understanding of fundamental MBE processes. In a typical STM–MBE experiment, the ultra-high vacuum (UHV) chambers housing the MBE system and STM system are separate. After halting the MBE growth, the sample is moved through UHV to the STM for imaging. An early system of this type is described by Orr *et al.* [32] while a more recent design is detailed by Geng *et al.* [33]. These instruments enable one to interrupt growth and obtain "snapshots" of the growing surface. The length scales of interest range from 10^{-10} m (surface reconstruction, dopant incorporation, alloying) to 10^{-7} m (islands, domains, step-terrace structure). Of course, MBE is a dynamic process covering time scales from seconds or tens of seconds, which are typical monolayer (ML) completion times, to the vibrational periods of adatoms (10^{-13} s) which govern attempt frequencies for migration.

Two conditions are placed on the system if one wishes to faithfully image the state of the surface during growth. First, the surface should not change during imaging. Adsorption of contaminants and MBE materials is easily limited by the UHV conditions and the isolation of the MBE chamber from the STM system, respectively. However, the possibility of rearrangement of surface structures during imaging remains, so that the imaged surface may not represent the as-grown surface. MBE growth of III–V materials normally takes place at elevated surface temperatures ($>500°C$ for GaAs, $>400°C$ for InAs). In most GaAs STM–MBE systems, STM imaging occurs at room temperature, and this is sufficient to suppress adatom migration. Furthermore, the stability of surface structures is easily checked by repeated imaging.

The second condition relates to the process of quenching the sample from the MBE chamber. In the case of GaAs MBE, there is normally an excess flux of As_2 or As_4 molecules which stabilizes the surface at high temperatures. In some STM–MBE experiments, the substrate temperature is lowered at a fixed rate while the As flux is also lowered. The two rates must be balanced to avoid the surface becoming As deficient or As rich. Surface reconstructions have been imaged successfully at atomic resolution in this type of experiment: see, for example, Avery *et al.* [34] and LaBella *et al.* [35]. These surfaces are fairly uniform and can be stabilized without disrupting the surface reconstruction, which can be monitored throughout using reflection high energy electron diffraction (RHEED). Moreover, *discrete* structures such as islands can be affected by the continued annealing inherent in such a cooling protocol. A detailed cooling/growth scheme for the STM study of GaAs islands on GaAs(001) has been presented by Yang *et al.* [36].

An alternative to controlled cooling is to quench the surfaces to clean UHV as rapidly as possible. This technique has been applied to image discrete growth structures such as coexisting reconstruction domains [37], GaAs islands [38], and InAs quantum dots (QDs) [39]. In these studies, the small size of the samples required for STM combined with the relatively large thermal mass of the transfer mechanism produced an estimated cooling rate of around $50°C\ s^{-1}$. The samples could be quickly rotated away from the impinging As flux and the total time to remove samples from the MBE chamber was around 5 s by rapid-quench methods [37–39]. Comparing the sample with rapid-quenched immediately to the one allowed to anneal for 10 s at the growth temperature of $550°C$ under the As_2 flux prior to rapid quenching, in the latter case, the number density has dropped by nearly a factor of three with a corresponding increase in the mean island area. This is due to adatom migration during the short anneal and demonstrates that metrics such as the island size distribution can evolve markedly during the quench process. Note that, in both cases, approximately 0.1 ML GaAs was deposited on to $GaAs(001)-(2 \times 4)$ to produce an array of ML-height islands and both the islands and substrate are (2×4) reconstructed.

It is not possible to perform true dynamic imaging by using *in vacuo* remote STM–MBE. The development of a particular surface feature during growth cannot be followed since the tip cannot normally be returned to the same location after additional MBE growth. Of course, observations of "typical" features and metrics such as island size distributions are still extremely useful. In order to overcome these limitations and any uncertainties about the effects of sample quenching, it is desirable to place the STM inside the MBE chamber (STMBE) and perform true *in situ* imaging during MBE growth [17]. This technique presents significant experimental challenges and shows the possibility of STM observations on GaAs surfaces even under As_4 irradiation at $440°C$ [40]; it also shows key results for InAs–GaAs(001) heteroepitaxial growth [41] and the fundamental processes governing QD formation [42–44].

8.2.2 *Heterointerface structure of GaSb/GaAs*

In heteroepitaxy of GaSb/GaAs nanostructures, fabrication of abrupt heterointerface is requested for many device applications. However, a broadening of the GaAs/GaSb heterointerface is frequently induced by surface exchange reaction between Sb and As atoms during a GaAs growth on the GaSb layer. In particular, a large lattice mismatch of about 7% between GaSb and GaAs enhances Sb surface segregation due to the surface exchange reaction. The segregated Sb atoms are incorporated into the GaAs overgrown layer, and then GaAsSb alloy layer is formed at the heterointerface. As a result, the quantum confinement effect changes in the modified GaSb/GaAs quantum nanostructure. Here, in order to control the GaAs/GaSb heterointerface and to design quantum energy levels, we consider a kinetic model of the Sb–As surface exchange reaction [45], which is based on a kinetic model of indium surface segregation in heteroepitaxy of InGaAs/GaAs [46]. Figure 8.4 shows a schematic diagram of surface exchange reaction between Sb and As atoms.

The surface exchange reaction of Sb and As atoms between surface and underlying bulk layers is described as:

$$[As]^s + [Sb]^b \Leftrightarrow [As]^b + [Sb]^s \quad (\text{s: surface, b: bulk}). \tag{8.1}$$

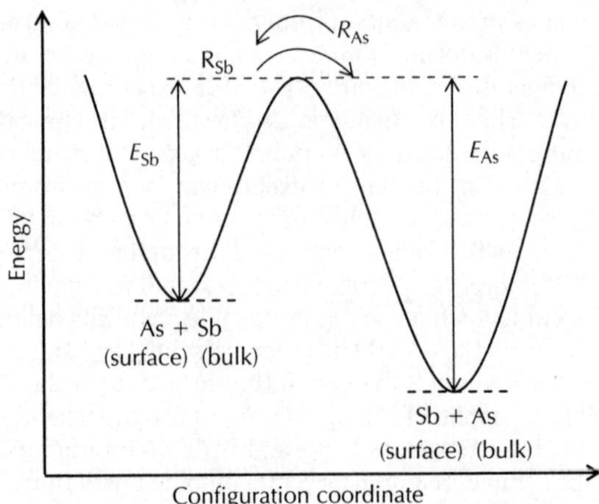

Figure 8.4   Schematic diagram of a surface exchange reaction model of Sb and As atoms.

[As] and [Sb] are concentrations of As and Sb atoms. Reaction rates of both forward and reverse exchange reactions can be respectively given by

$$R_{Sb} = v \exp(-E_{Sb}/kT_g) \tag{8.2}$$

and

$$R_{As} = v \exp(-E_{As}/kT_g). \tag{8.3}$$

Here, E_{Sb} and E_{As} are activation energies for the surface exchange reaction. T_g and v are the growth temperature of the GaAs capping layer and the vibration frequency (10^{-13} s), respectively. During GaAs capping growth on the GaSb layer, a variation of Sb surface concentration in a small time is given by the following continuous equation:

$$d[Sb]^s(t)/dt = R_{Sb}[Sb]^b(t)[As]^s(t) - R_{As}[Sb]^s(t)[As]^b(t). \tag{8.4}$$

Therefore, the Sb composition profile in the GaAs/GaSb heterostructure can be calculated using the above equations and mass conservation law of Sb and As. In this model, the growth parameters of growth rate and temperature are included. However, the impinging As flux was neglected because the sticking coefficient of As is very low on the V-group surface.

Figure 8.5 shows calculated Sb composition profiles as a function of the growth temperature (a) and the growth rate (b) of the GaAs capping layer on a 1 monolayer (ML) thick GaSb layer. In this calculation, $E_{Sb} = 1.68$ eV and $E_{As} = 1.82$ eV were used as an example. At a high growth temperature, the Sb surface exchange reaction becomes active, and the heterointerface broadens as shown in Fig. 8.5a. In addition, broadening of the heterointerface is similarly enhanced with a decreasing growth rate of the GaAs capping layer. Therefore, the low temperature and high growth rate conditions in the GaAs capping growth are requested to suppress broadening of the GaAs/GaSb heterointerface. However, such growth conditions are not suitable for keeping high crystal quality of the GaAs.

Next, we will consider prediction of transition energy in the broadened Ga(As)Sb/GaAs QWs, calculated by using this kinetic model. For GaAs$_{1-x}$Sb$_x$ alloys, the bowing effect of the band gap energy is used as:

$$E_g(x) = 1.52(1 - x) + 0.81x - 1.2x(1 - x) \quad \text{at 12 K.} \tag{8.5}$$

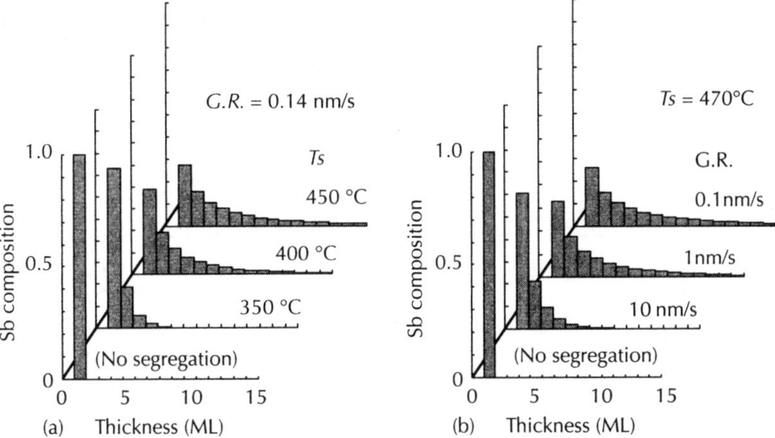

Figure 8.5 Sb composition profiles of GaSb(1 ML)/GaAs QWs, calculated using a kinetic model, as functions of the growth temperature (a) and the growth rate (b) of GaAs capping layer. Activation energies of the Sb–As surface exchange reaction, $E_{Sb} = 1.68\,eV$ and $E_{As} = 1.82\,eV$, were used.

Here, the band gap bowing effect in the valence band edge is represented as:

$$E_v(x) = 0.48(1-x) + 1.16x + b_v x(1-x) \tag{8.6}$$

where b_v is the added bowing parameter [47]. The conduction band edge is given by:

$$E_c(x) = E_v(x) + E_g(x). \tag{8.7}$$

Both $E_v(x)$ and $E_c(x)$ in $GaAs_{1-x}Sb_x$ are raised by compressive strain due to lattice mismatch [47]. For example, the energy band diagrams of the broadened Ga(As)Sb/GaAs QW with and without strain are shown in Fig. 8.6, where b_v of 0.67 was used tentatively. Parameters used for the calculation of the band line-up of $GaAs_{1-x}Sb_x$ are summarized in Table 8.1. The optical transition energy in the calculated QW structure was obtained from the difference in energy between the conduction band edge of the GaAs and the confined heavy-hole level in the strained $GaAs_{1-x}Sb_x$ QW. The bowing parameter b_v and activation energies E_{Sb} and E_{As} can be estimated from a fitting between experimental results (photoluminescence (PL)) and calculated ones, as follows. In this

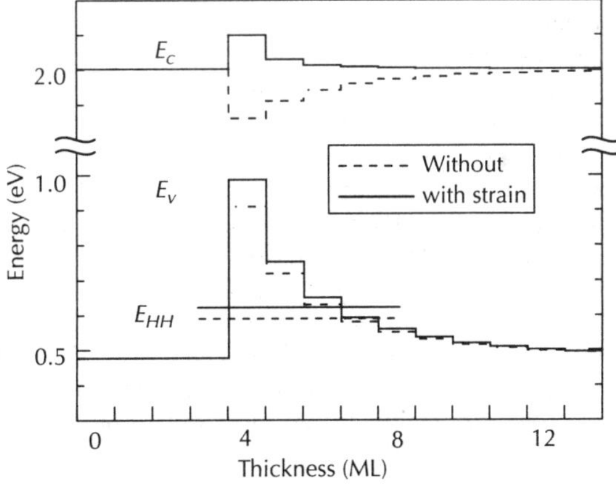

Figure 8.6 Energy band diagrams of conduction band edge (E_c) and valence band edge (E_v) for broadened Ga(As)Sb/GaAs QWs with strain and without strain. Bowing parameter $b_v = 0.67$ was used in this calculation.

Table 8.1 Parameters used for calculation of band line-up of GaAs$_{1-x}$Sb$_x$ by an empirical method [47] and the model solid theory [48].

Parameter	Symbol (unit)	GaAs	GaSb
Lattice constant	a (Å)	5.6533	6.0959
Energy gap	E_g (eV)	1.522	0.812
Elastic stiffness constant	C_{11}	1.223	0.908
	C_{12}	0.571	0.413
Hydrostatic deformation potential			
for conduction band	a_c (eV)	−7.17	−6.85
for valence band	a_v (eV)	1.16	0.79
Shear deformation potential			
for valence band	b (eV)	−1.70	−2.00
Model solid theory			
average valence band position	E_v, a_v (eV)	−6.92	−6.25
spin-orbit split-off energy	Δ (eV)	0.34	0.82
Empirical method			
conduction band position	E^E_{c0} (eV)	1.91	1.89
valence band position	E^E_{v0} (eV)	0.48	1.16
Electron effective mass	m_e/m_v	0.067	0.045
Heavy-hole effective mass	m_{HH}/m_o	0.50	0.40

analysis, it was assumed that activation energy of the surface exchange reaction is kept constant during GaAs capping growth, and a slight expansion of the QW thickness due to the strain effect was neglected.

Figure 8.7a shows PL experimental and calculated results of optical transition energy obtained from 1 ML thick GaSb/GaAs QWs as a function of the growth temperature of the GaAs capping layer. The samples were grown by MBE. For the ideal thickness of the GaSb QW of 1 ML, the calculated optical transition energy is 1.364 eV at 12 K. When the growth temperature decreases from 470°C to 350°C, the PL peak energy shifts toward the low-energy side in this experiment. Here, it should be noted that the PL peak energy decreases to below 1.364 eV (ideal 1 ML thick QW) at lower than 390°C. Furthermore, when the growth temperature is lower than 350°C, the peak energy oppositely shifts toward the high-energy side. That is, the minimum PL peak energy of 1.33 eV is obtained at 350°C. The broadening of the GaSb/GaAs heterointerface induces an increase in the QW width and an increase in the band gap of the GaAsSb, due to lowering the Sb composition. The former mainly induces a red shift of the transition energy and the latter mainly induces a blue shift. Thus both effects determine the transition energy of the broadened QW. Therefore, it is predicted that the dominant effect changes from the former to the latter at about 350°C. Consequently, in order to fit the PL experimental results to the calculated ones, we determine the bowing parameter (b_v) for the valence band edge in GaAsSb and the activation energies (E_{Sb} and E_{As}) of the surface exchange reaction in this kinetic model. Here, when the bowing effect for the valence band becomes strong (i.e. b_v increases), the minimum transition energy decreases. In contrast, for large E_{Sb}, the minimum transition energy appears at a higher growth temperature because of the suppression of the Sb–surface segregation effect. In addition, the optical transition energy shifts toward the low energy side in the higher temperature region. For large E_{As}, although the growth temperature indicating the minimum transition energy is almost maintained, the optical transition energy at the high-temperature side mainly increases. Based on these behaviours of the three parameters, we can fit the calculated results to the PL experimental results by changing the three parameters, as shown in Fig. 8.7a. In particular, the calculated minimum transition energy of 1.337 eV appears at 350°C, similar to the experimental result. From the above fitting, we could empirically obtain $b_v = 0.67 \pm 0.06$, $E_{Sb} = 1.68 \pm 0.01$ eV and $E_{As} = 1.80 \pm 0.01$ eV. In Fig. 8.7a, the calculated results, shown by the solid line, were obtained using $b_v = 0.67$, $E_{Sb} = 1.68$ eV and $E_{As} = 1.80$ eV.

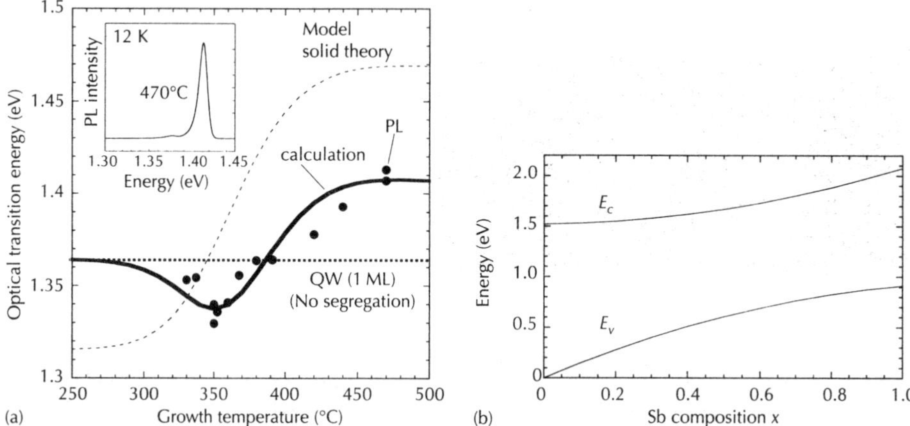

Figure 8.7   *(a) Experimental PL (12 K) results and calculated ones of optical transition energy obtained from the GaSb (1 ML)/GaAs QWs as a function of the growth temperature of the GaAs cap layer. The growth rate of the cap layer was 0.14 nm/s. The solid line indicates calculated transition energy of GaSb/GaAs QW structures, analysed using the kinetic model with fitting parameters of $b_v = 0.67$, $E_{Sb} = 1.68 eV$ and $E_{As} = 1.80 eV$. A broken line indicates the results calculated using a model solid theory [48]. (b) Calculated conduction and valence band edges (E_c and E_v) for strained $GaAs_{1-x}Sb_x$ (12 K) as a function of Sb composition. The obtained $b_v = 0.67$ was used in this calculation.*

For a comparison, the calculated result by using a model solid theory [48] is also indicated by the broken line in Fig. 8.7a. The optical transition energy calculated by the model solid theory increases with the temperature of cap growth and does not show the minimum value at about 350°C. The deviation between the calculated energy and experimental one was about 40 meV at 350°C.

Figure 8.7b shows the conduction band edge (E_c) and valence band edge (E_v), calculated from the obtained $b_v = 0.67$. Here, both band edges include a strain effect. The obtained $b_v = 0.67$ is smaller than that reported by Liu *et al.* [47] (1.08) and indicates a bowing effect in not only the valence band edge but also in the conduction band edge, as shown in Fig. 8.7b. This difference is mainly due to the different QW structures: the rectangular QW potential without the broadening of the Sb composition was used in the previous reports. In our analysis, the broadened QW structures due to the Sb–As segregation effect are taken into consideration, as explained above. As a result, we can empirically determine the added bowing parameter of b_v (0.67 ± 0.06), which is independent of the Sb composition and the QW structure.

Figure 8.8a shows a ($1\bar{1}0$) cross-sectional high-resolution transmission electron microscopy (HRTEM) image of the GaSb (1 ML)/GaAs QW, grown at 470°C. In this image, intermixing of Sb atoms in the GaAs capping layer is observed, and the width of the QW is roughly estimated to be about 3 ML. Figure 8.8b shows the Sb-composition profile of the GaSb (1 ML)/GaAs QW, calculated for 470°C using the presented kinetic model with the above fitting parameters ($b_v = 0.67$, $E_{Sb} = 1.68 eV$ and $E_{As} = 1.80 eV$). The Sb composition decreases to less than 0.1 for a thickness of more than 3 ML. The calculated QW width is comparable to the experimental one.

On the basis of these results, it is concluded that the presented kinetic model that includes the Sb–As surface exchange reaction is effective for analysing the MBE growth of the GaSb/GaAs heterointerface structures. For the activation energies of the Sb–As exchange reaction, Magri and Zunger obtained $E_{Sb} = 1.68 eV$ and $E_{As} = 1.75 eV$ from the analysis of the InAs/GaSb superlattice [49]. As presented above, for the GaAs/GaSb QWs, the almost same E_{Sb} (1.68 ± 0.01 eV) was obtained, and E_{As} was about 50 meV higher (1.80 ± 0.01 eV). It is possible that the difference in the value of E_{As} is caused by the different binding energies of InAs and GaSb.

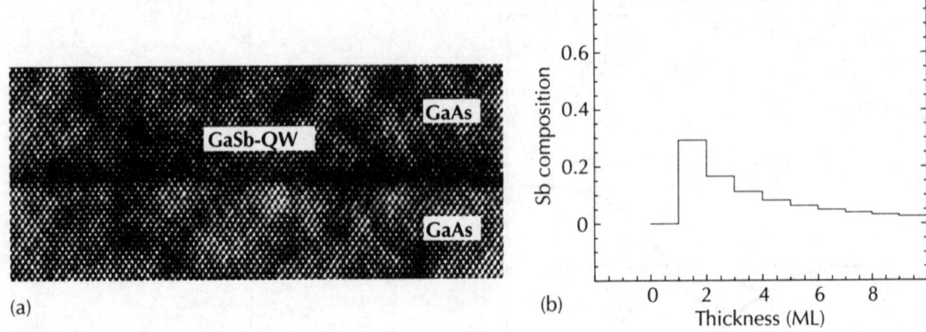

Figure 8.8   *(110) cross-sectional HRTEM image (a) and calculated Sb composition profile (b) of GaSb (1 ML)/ GaAs QW. The growth temperature and the growth rate of the GaAs capping layer were 470°C and 0.14 nm/s, respectively. Fitting parameters* $b_v = 0.67$, $E_{Sb} = 1.68\,eV$ *and* $E_{As} = 1.80\,eV$ *were used in this calculation.*

8.3   GaSb quantum dots on GaAs

GaSb quantum dots (QDs) on GaAs are spontaneously formed by Stranski–Krastanov (SK) growth mode because GaSb on GaAs has a lattice-mismatched interface ($\Delta a_0/a = 7.8\%$) similar to InAs on GaAs. In this section, we will discuss the self-assembled GaSb/GaAs QDs grown by molecular beam epitaxy (MBE) using the SK mode. The formation of three-dimensional (3D) GaSb islands on GaAs(001) will be described in Section 8.3.1. Then, optical properties of ensemble GaSb/GaAs QDs and single GaSb/GaAs QD will be discussed at in Section 8.3.2.

8.3.1   MBE growth of self-assembled GaSb/GaAs quantum dots

Thin film growth on a bulk substrate is typically classified by three growth modes; Frank–van der Merwe (FM) mode exhibiting layer-by-layer growth, Volmer–Weber (VM) mode forming 3D islands and Stranski–Krastanov (SK) mode. The SK mode is a combination of the FM and VW modes; the 3D island growth occurs followed by the 2D growth. The growth mode transition from 2D to 3D growth is induced by the difference in the surface energy between the grown material and the substrate and by strain energy due to lattice mismatch between both materials. In the case of GaSb growth on GaAs, the SK growth frequently occurs, and such 3D islands of GaSb/GaAs have been expected as type II QD structures. The self-assembled GaSb/GaAs QDs based on the SK mechanism have been studied recently [50–53].

The typical growth process of GaSb/GaAs QDs is described as follows. First, prior to the GaSb growth on GaAs, Sb_4 flux is irradiated on the GaAs buffer layer to fabricate initial GaSb surface layer. Figure 8.9 shows RHEED intensity of a specular beam as a function of irradiation time of Sb_4 flux. After the GaAs buffer layer growth at 580°C, the substrate temperature was dropped to 470°C. Then Sb_4 flux was irradiated on the GaAs surface instead of As_4 flux. During the Sb_4 irradiation, RHEED pattern changes from $c(4 \times 4)$ to (1×3) reconstruction, which reveals the Sb-terminated Ga(As)Sb surface. In Fig. 8.9, the RHEED specular beam intensity decreases and then saturates. This intensity saturation means the preparation of the stable Sb-terminated Ga(As)Sb surface layer. More recently, this surface layer structure has been analysed by *in situ* X-ray crystal truncation rod (CTR) scattering measurements [31].

Next, on this Ga(As)Sb surface layer, GaSb QDs were grown at 470°C. Figure 8.10 shows a relationship between RHEED diffraction beam intensity and GaSb growth time. As the GaSb growth proceeds, the RHEED pattern changes from streak to spot pattern. It means the growth mode transition from 2D to 3D growth; SK growth mode. The RHEED intensity is almost kept constant for the 2D growth and then rapidly increases for the 3D growth, as shown in Fig. 8.10. The critical thickness for 2D–3D transition can be estimated by using this time dependence of

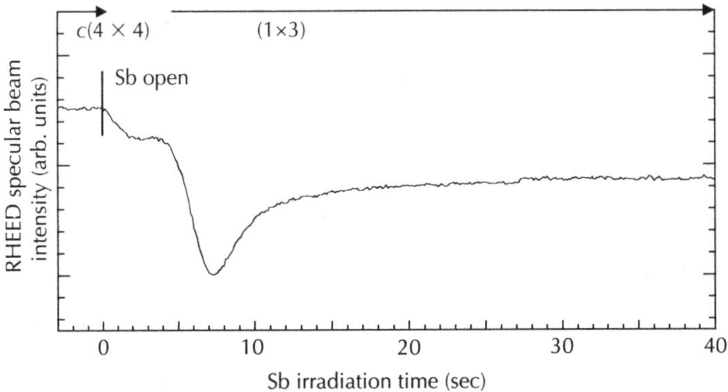

Figure 8.9   RHEED specular beam intensity as a function of irradiation time of Sb_4 flux on GaAs(001) surface.

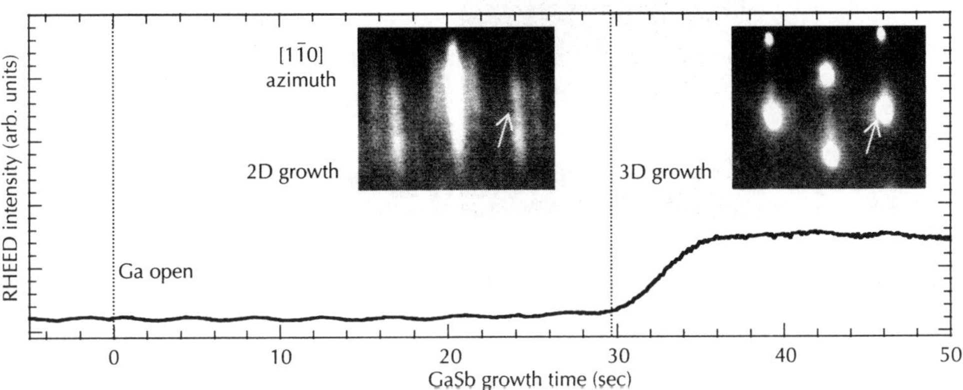

Figure 8.10   Two typical RHEED patterns for 2D and 3D growth stages and RHEED diffraction beam intensity as a function of the growth time during MBE growth of GaSb on GaAs(001).

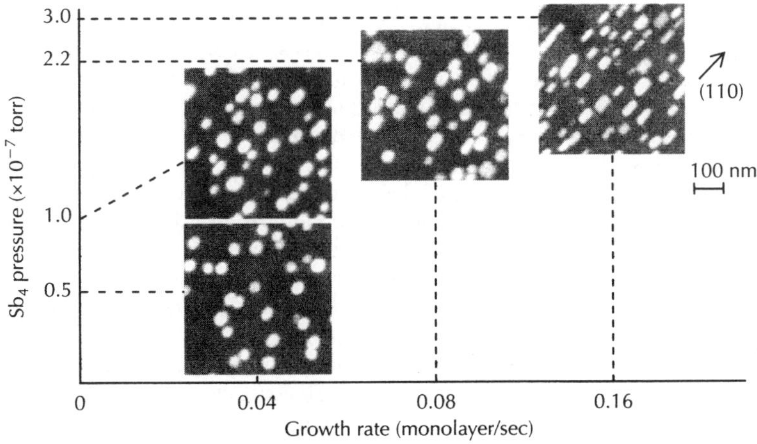

Figure 8.11   AFM images of GaSb QDs grown on Sb-initiated GaAs at various growth conditions.

RHEED intensity. The critical thickness usually depends on the GaSb growth condition. The difference in critical thickness is possible to relate to Sb–As intermixing during growth.

Figure 8.11 shows AFM images of GaSb QDs grown on Sb-initiated GaAs at various growth conditions. In the conventional SK growth, the change of the growth conditions modifies the

3D structures and density. For high growth rate and high Sb_4 pressure, the dot density increases, and size fluctuation becomes large. In addition, the dot shape is anisotropy: the lateral size elongates along the [110] direction. As the growth rate and the Sb_4 pressure decrease, the surface migration is enhanced. As a result, size fluctuation and dot density decrease. The QD structure reveals a dome-like shape. Figure 8.12 shows cross-sectional scanning transmission electron microscope (STEM) images of GaSb QDs grown at two typical growth conditions: growth rate of 0.16 ML/s and Sb_4 pressure of 2.2×10^{-7} torr (a), 0.04 ML/s and 0.5×10^{-7} torr (b). These QDs were embedded into GaAs matrix. STEM images of both samples show the presence of the wetting layer. In the anisotropic large dot (a), misfit dislocations are observed at the GaSb/GaAs heterointerface [54]. Hence, the strain of the QD is relaxed. Such interface misfit dislocations are frequently formed under high Sb pressure [55] and high growth rate conditions. On the contrary, in the dome-like dot with small size (b), no dislocations are observed. Low Sb pressure and low growth rate are preferable conditions for getting coherent GaSb QDs with isotropic shape and high size uniformity.

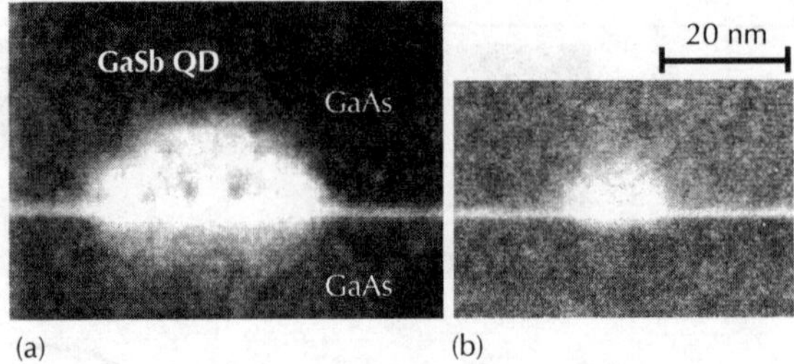

Figure 8.12   Cross-sectional STEM images of GaSb QDs grown at two typical growth conditions: growth rate of 0.16 ML/s and Sb_4 pressure of 2.2×10^{-7} torr (a), 0.04 ML/s and 0.5×10^{-7} torr (b). STEM images were acquired by a high-angle annular dark field (HAADF) mode.

Figure 8.13 shows GaSb coverage dependences of the QD size (height (a) and lateral size (b)) and density (c). GaSb growth conditions were low growth rate of 0.04 ML/s and low Sb_4 pressure of 0.5×10^{-7} torr. Here, the QD size was measured by using AFM, which shows large lateral size because of tip curvature. The lateral size and dot density tend to saturate at more than about 3 ML in coverage. In the case of InAs SK growth on GaAs, self size-limiting phenomena due to the strain and facet formation have been observed under low growth rate and low arsenic pressure conditions [56]. However, for the GaSb QDs on GaAs, the stable facet did not clearly appear on the side wall. In addition, as the growth proceeded, the dot height increased and did not saturate, as shown in Fig. 8.13a. It is possible that the saturation of the lateral size is caused by the strain at the island edge. At the step sites of the dot edge, the compressive strain increases with increasing dot size [57]. Consequently, the lateral size is limited by inhibition of adatom incorporation. On the contrary, the adatom incorporation occurs actively at the top of the dot because of the lattice relaxation at the top area. So, the dot height increases with the GaSb coverage. Although the self-limiting effect of height does not occur, size fluctuation in height can be suppressed by preparing high aspect ratio in the QD structure.

Figure 8.14 shows the average QD size (a), standard deviation (b) and density (c) as a function of the growth temperature of the GaSb QDs. The GaSb coverage was 4.5 ML. As the substrate temperature increases, the dot size increases and the dot density decreases. It is attributed to enhancement of the surface migration for high substrate temperature. The long surface migration is favourable for suppression of the size fluctuation. Indeed, the standard deviation reduces with increase in the substrate temperature, as shown in Fig. 8.14b. Figure 8.15 shows size distributions of the GaSb QDs grown under low growth rate of 0.04 ML/s, low Sb_4 pressure of 0.5×10^{-7} torr and substrate temperature of 470°C. The GaSb coverage was 5 ML, and the QD density was $8.9 \times 10^9 \, cm^{-2}$. The standard deviations were 9.2% for the lateral size and 14% for the height. After their GaSb QDs were embedded in the GaAs capping layer, PL properties were measured as follows.

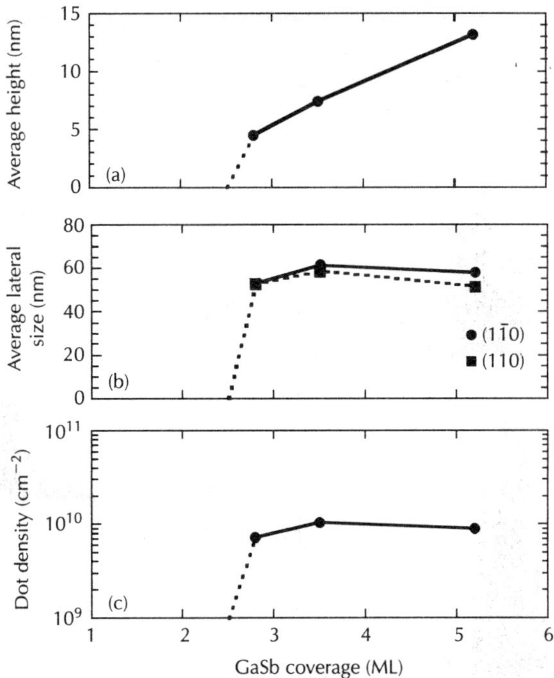

Figure 8.13 *GaSb coverage dependences of QD height (a), lateral size (b) and density (c). GaSb growth rate was 0.04 ML/s. Sb_4 pressure was 0.5×10^{-7} torr.*

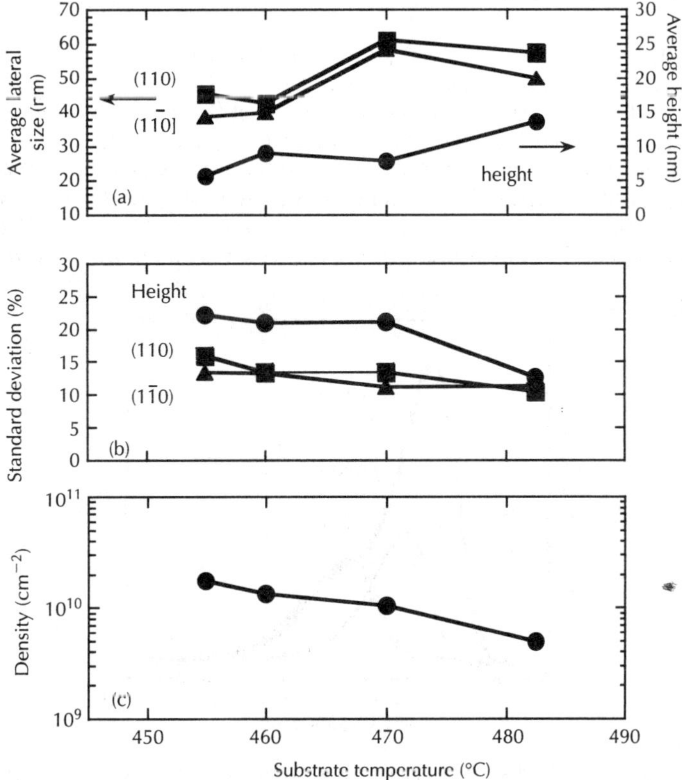

Figure 8.14 *Average QD size (a), standard deviation (b) and density (c) as a function of the growth temperature of GaSb QDs. GaSb coverage was 4.5 ML. GaSb growth rate was 0.04 ML/s. Sb_4 pressure was 0.5×10^{-7} torr.*

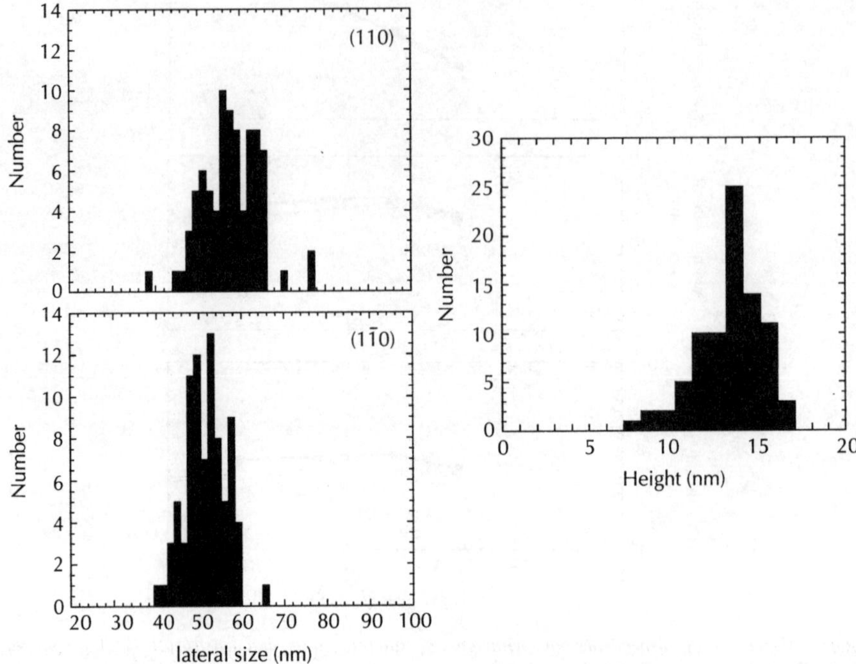

Figure 8.15   Size distributions of the GaSb QDs grown under growth rate of 0.04 ML/s, Sb$_4$ pressure of 0.5 × 10^{-7} torr and substrate temperature of 470°C. GaSb coverage was 5 ML.

Figure 8.16 shows PL spectra obtained from GaSb QDs as a function of the growth temperature of the GaAs capping layers (490°C, 470°C and 450°C). Growth conditions of the GaSb QDs were low growth rate of 0.04 ML/s, low Sb$_4$ pressure of 0.5 × 10^{-7} torr and 470°C. PL peak wavelength reveals a blue shift as the growth temperature increases. In the case of the 490°C growth, the RHEED pattern changed from spot pattern into streak one during the growth interruption between the SK growth and the capping growth. In short, the QD structure changes into QW-like structure during the growth interruption. For the growth below 470°C, the RHHED spot pattern was kept until the embedding in the GaAs capping layer. However, reduction of the

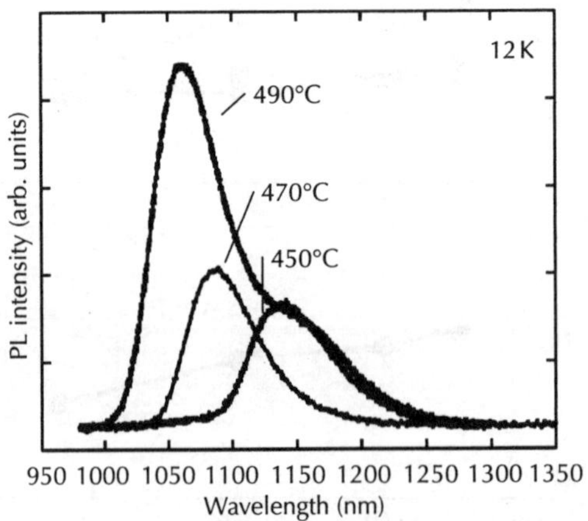

Figure 8.16   PL spectra of GaSb QDs as a function of growth temperature of GaAs capping layers (490°C, 470°C and 450°C). Growth conditions of GaSb QDs: growth rate of 0.04 ML/s, Sb$_4$ pressure of 0.5 × 10^{-7} torr and growth temperature of 470°C.

QD size and Sb–As intermixing at the heterointerface should be considered during the growth interruption and the capping growth. These effects induce the blue shift for higher temperature growth of the capping layer. As mentioned before, the GaSb growth conditions of low growth rate and low Sb_4 pressure suppress the size fluctuation of the QDs. In fact, PL spectrum for 470°C reveals narrow linewidth of 65 meV [53], as compared with that of previous reports.

8.3.2 Optical properties of GaSb/GaAs quantum dots

8.3.2.1 Energy structure of GaSb/GaAs quantum dots

Self-assembled quantum dots (QDs) are an ideal system for studying zero-dimensional quantum effects [58] while showing promise for future quantum devices [59–61]. So far, many experimental studies have been carried out mainly on self-assembled InAs/GaAs QDs. From the viewpoint of the energy level structure, the InAs/GaAs QD has a band alignment classified as type I, in which both electrons and holes are confined within the InAs QD as shown in Fig. 8.17. Their wavefunctions show a strong spatial overlap. In a staggered type II band structure, however, the lowest electron and hole energy states are concentrated on different layers [2, 51, 52, 62–67]. In a type II GaSb/GaAs QD, the spatial separation is between the electron wavefunction in the GaAs layer and the hole wavefunction in a GaSb QD. Consequently, the optical properties of a type II QD differ from those of a type I QD, as described later.

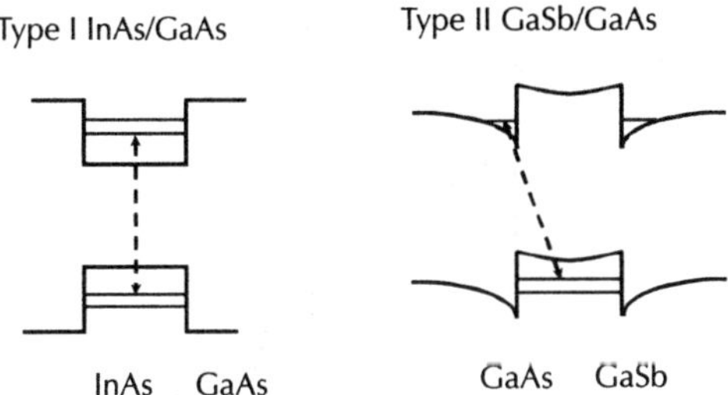

Figure 8.17   Energy structure of type I InAs/GaAs QDs and type II GaSb/GaAs QDs.

8.3.2.2 Fundamental optical properties of GaSb/GaAs QDs

The characteristic optical properties resulting from the type II band structure have been observed in various experiments. The inset of Fig. 8.18 shows the macroscopic (ensemble-averaged) PL spectra of GaSb/GaAs QDs at 10 K [62]. The PL spectra have a broad peak at 1.08 eV with a full width at half-maximum (FWHM) linewidth of about 80 meV. These broadened

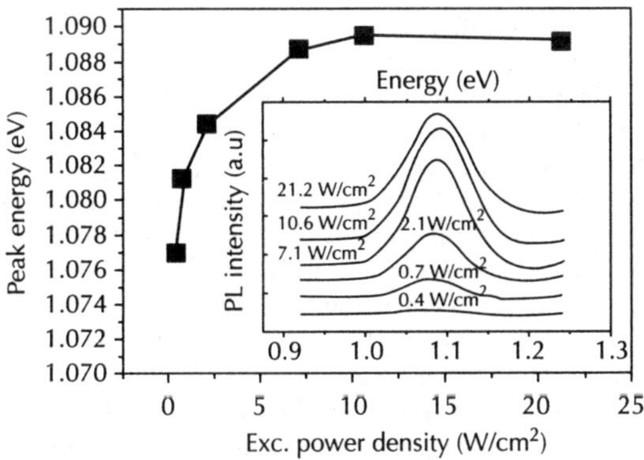

*Figure 8.18   Excitation power dependence of the PL peak energy [62]. The inset shows PL spectra at 10 K as a function of the excitation density. Reprinted with permission from Motlan et al., Appl. Phys. Lett. **79**, 2976 (2001). Copyright 2001, American Institute of Physics.*

PL spectra arise from the distribution of the optical transition (exciton) energies in the individual QDs due to inhomogeneities in properties such as size, strain, and composition. The resultant broadened spectral linewidth is therefore referred as "inhomogeneous linewidth". Figure 8.18 shows the excitation power dependence of the PL peak energy and the inset shows the PL spectra as a function of excitation density. A higher energy shift (blue shift) of the PL maximum ($\sim$12 meV) is observed with increasing excitation density from 0.4 to 10.6 W/cm^2. Such a blue shift is characteristic of type II structures [62, 64, 65]. In type I InAs/GaAs QDs, the lowest PL peak arising from exciton ground state shifts at lower energies with an increase in the excitation density [68]. Based on the macroscopic PL measurements, it was speculated that the blue shift of the peak arises from change of the band bending occurring at the interfaces where the electrons accumulate [62].

In addition to the characteristic blue shift of the PL peak, the recombination dynamics of the exciton in type II QDs is much different from that in the type I. Figure 8.19 presents the PL decay curves at 10 K, for different excitation intensities [2]. Under the lowest excitation condition (I/10), the PL dynamics can be viewed as a single-exponential decay with a time constant of 23 ns. This measured lifetime, arising from the reduced spatial overlap between electron and hole wavefunctions, is much longer than the typical radiative lifetime of type I InAs/GaAs QDs ($\sim$1 ns) [69]. This means the reduced recombination probability; that is, the exciton transition has a smaller oscillator strength.

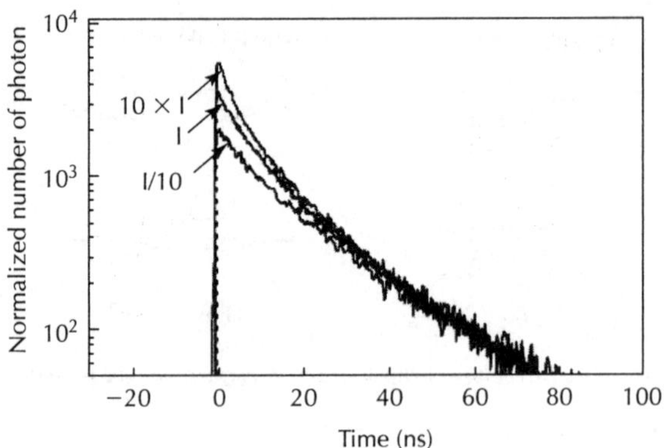

*Figure 8.19   PL decay curves of GaSb QDs at 10 K with different excitation intensities [2]. The data are normalized with the tail parts of the decay curve. Reprinted with permission from C.-K. Sun et al., Appl. Phys. Lett. **68**, 1543 (1996). Copyright 1996, American Institute of Physics.*

8.3.2.3   Exciton fine structures of GaSb/GaAs QDs

The ensemble-averaged optical spectrum has a broad feature that masks the intrinsic nature of the QD optical properties. Single QD optical spectroscopy, which is spectroscopic observation of a single QD, is one of the most useful tools in studying these intrinsic properties. The various approaches to realizing single QD PL spectroscopy have been applied using the conventional optical microscopes [70], near-field scanning optical microscopes [71], and so on [72]. Figure 8.20a shows typical PL spectra for a single GaSb QD at various excitation power densities, obtained using a near-field scanning optical microscope [73, 74]. A single and sharp emission peak in the PL spectrum, denoted by X, is observed at 1.2716 eV under lower excitation power conditions (less than several μW), while the ensemble-averaged PL spectrum has a broad peak with a 60 meV linewidth. The PL intensities of the X line as a function of excitation power density show an almost linear power dependence under lower excitation conditions (see Fig. 8.20b). The sharp X emission line, having a less than 1 meV linewidth (FWHM), is assigned to the radiative recombination of the exciton consisting of a hole confined in a GaSb QD and an electron in the surrounding GaAs barrier layer, which are weakly bound by an attractive Coulomb interaction.

Figure 8.21 shows the PL peaks for exciton emission from three different GaSb QDs on an expanded energy scale [74]. The linewidths of the three emission peaks are estimated to be less

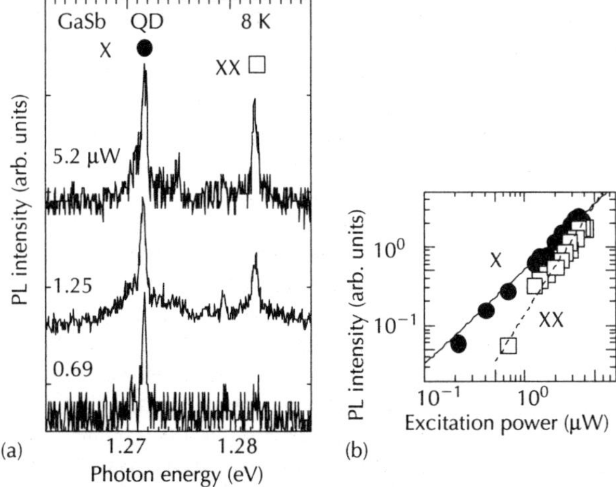

Figure 8.20 (a) PL spectra of a single GaSb QD at 8 K under various excitation power densities. The PL peaks at 1.2716 eV and 1.2824 eV are denoted as X and XX. (b) Excitation power dependence of PL intensities of the X and the XX line. The solid (dotted) line corresponds to the gradient associated with linear (quadratic) power dependence [74]. Reprinted with permission from K. Matsuda et al., Appl. Phys. Lett. **90**, 013101 (2007). Copyright 2007, American Institute of Physics.

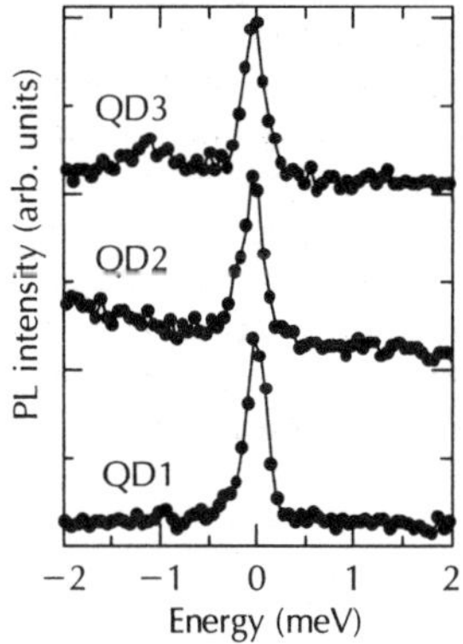

Figure 8.21 PL spectra of three different QDs in an expanded energy scale [74]. Reprinted with permission from K. Matsuda et al., Appl. Phys. Lett. **90**, 013101 (2007). Copyright 2007, American Institute of Physics.

than 250 μeV. As a consequence, the "homogeneous line width" (the intrinsic line width of a single QD) of an exciton state in a type II GaSb QD is evaluated as less than 250 μeV, narrower than the 280 μeV theoretically predicted for an ideal quantum well at 8 K [72].

Furthermore, as seen in Fig. 8.20a, an additional peak appears at 1.2824 eV in the PL spectra under higher excitation conditions (more than several μW). This peak, denoted by XX, is observed about 11 meV on the higher energy side of an exciton emission (X). Figure 8.20b shows the nearly quadratic power dependence of the XX line as a function of excitation power. This power dependence of the PL intensity suggests that the XX emission results from the radiative transition from a two-exciton state to the exciton ground state. The experimental result of the two-exciton emission (XX) on the higher energy side of the exciton emission (X) is in contrast the results commonly reported in type I structures such as in self-assembled In(Ga)As QDs [75, 76] and naturally occurring GaAs QDs [77, 78]. This is consistent with the behaviour of the macroscopic PL spectrum of an ensemble of GaSb QDs, which shows a blue shift of the PL peak with increasing excitation power (see Fig. 8.18) [62]. In type I QDs, the quadratic power dependent PL line usually observed at 3 to 5 meV on the lower energy side of the exciton emission is generally assigned to the bound biexciton emission.

The energy difference between the two exciton emission (XX) and the exciton emission (X) corresponds to the binding energy ($E_{bin} = 2E_X - E_{XX}$), where E_{XX} and E_X are the energies of the two exciton state and the exciton ground state, respectively. A negative binding energy implies that the exciton–exciton interaction is repulsive in these QDs. The binding energy of the two exciton state is determined by summing the repulsive Coulomb interactions between the electrons and between the holes, and the attractive Coulomb interaction between electrons and holes. Since both electron and hole wavefunctions are strictly confined in a type I QD, all three terms, the electron–electron, the hole–hole, and the electron–hole Coulomb interactions are of the same order of magnitude and the delicate balance between them leads to the binding of the two excitons (biexcitons). The biexcitons acquire a positive binding energy due to interparticle correlations in a way similar to the formation of a hydrogen molecule. The observed binding energy in type I QDs is typically 3–5 meV. The situation is quite different in type II GaSb QDs, in which only the holes are confined inside the QD, while the electron wavefunction is relatively delocalized in the GaAs barrier layer around the QDs. Consequently, because the strengths of the electron–hole and electron–electron interactions are smaller than that of the hole–hole interaction, it is reasonable for the Coulomb energy of the two exciton ground state to be dominated by the hole–hole repulsive Coulomb interaction, and thus the binding energy has a negative value.

8.3.2.4   Many exciton states of GaSb/GaAs QDs

To develop a quantitatively accurate understanding of the exciton and two exciton states in GaSb QDs, a theoretical calculation based on the empirical pseudopotential model (EPM) was applied [79]. The single particle states were obtained by solving the one electron Schrödinger equation in a potential $V(\vec{r})$ obtained from a superposition of atomic pseudopotentials $v_\alpha(\vec{r})$ centred at each atom location in a supercell containing the QD and its surrounding matrix, $\sum_{\alpha,n} v_\alpha(\vec{r} - \vec{d}_{\alpha,n})$. Here, α denotes the atom type and $\vec{d}_{\alpha,n}$ its equilibrium position, which was obtained by minimizing the strain energy [80]. Spin-orbit coupling is included as a similar sum of non-local potentials [79]. EPM parameters fitted to the bulk band structure parameters of GaSb and GaAs were used [81]. For the type II QDs, the energy structure was evaluated by applying a novel self-consistent mean field (SCF) calculation to the multiple electron–hole pair excitations within the EPM framework. The SCF Hamiltonian may then be written as:

$$-\frac{\hbar^2}{2m}\nabla^2 + V(\vec{r}) + \int d^3r'' d^3r' \frac{\rho_e(\vec{r}') - \rho_h(\vec{r}')}{\varepsilon(\vec{r} - \vec{r}'')\left|\vec{r}'' - \vec{r}''\right|} \tag{8.8}$$

where $\rho_e(\rho_h)$ is the density of the occupied conduction band (unoccupied valence band) states, excluding the state being calculated. The present approach treats the electron–electron and electron–hole interactions at the Hartree–Fock level for the one exciton and two exciton ground states.

Using a linear combination of bulk Bloch functions as the basis, the single particle energies and orbitals were calculated for a few of the lowest conduction and highest valence band states with zero, one, and two electron–hole pairs [82].

The exciton and two exciton energies were calculated as the sum of single particle energies, corrected for double counting of the Coulomb interaction. For example, the two exciton energy E_{XX} is given by:

$$E_{XX} = 2(E_e - E_h) - (V_e + V_h) \tag{8.9}$$

where E_e and E_h are the energies and V_e and V_h are the expectation values of the third term in Eq. 8.8 for the lowest conduction band and highest valence band states, respectively. The two exciton binding energy is given by $2E_X - E_{XX}$. A negative binding energy indicates that the two exciton-to-exciton emission in PL would appear on the higher energy side of the exciton emission.

The calculated binding energy as a function of QD size is shown in Fig. 8.22. This result corresponds to typical self-assembled GaSb QDs of heights in the range of 4.8 to 6.6 nm with the

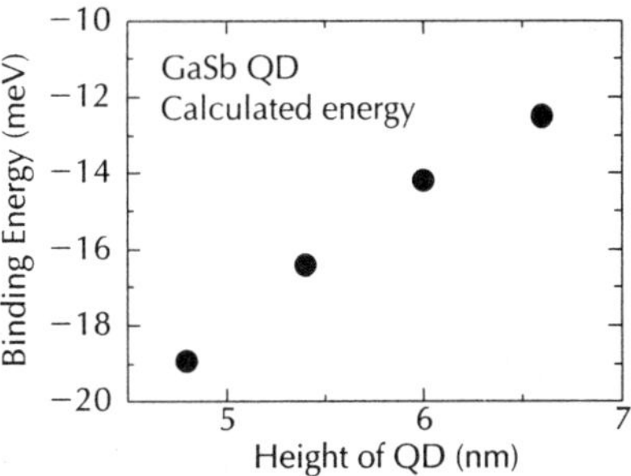

*Figure 8.22   Calculated two exciton binding energy as a function of lens-shaped GaSb QDs (height to diameter ratio 0.3) [74]. Reprinted with permission from K. Matsuda et al., Appl. Phys. Lett. **90**, 013101 (2007). Copyright 2007, American Institute of Physics.*

height to diameter ratio fixed at 0.3. The resultant binding energy ranged from a negative value of -12 to -19 meV. Although the two exciton energy shift relative to the exciton could qualitatively be attributed to the repulsion between the two confined holes, the contributions from the electron–hole attraction and electron–electron repulsion are not negligible. For example, for the 4.8 nm height QD, the two exciton binding energy of 19 meV includes 27 meV of the hole–hole repulsion, 5 meV of the electron–electron repulsion, and 12 meV of the electron–hole attraction. The strong correlation exists between a pair of excitons in GaSb type II QDs having an antibound nature due to these Coulomb interactions.

8.3.2.5   Exciton fine structures of GaSb/GaAs QDs under a magnetic field

Magnetic field is one of most useful and controllable parameters for investigating the intrinsic nature of the excitons. Figure 8.23a shows the PL spectra of a single GaSb QD under a magnetic field with Faraday configuration. The sharp exciton emission line splits gradually into two peaks with increasing magnetic field [83]. Figure 8.23a shows the centre energy of the two peaks, which corresponds to a diamagnetic shift [83]. The energy positions after subtracting the

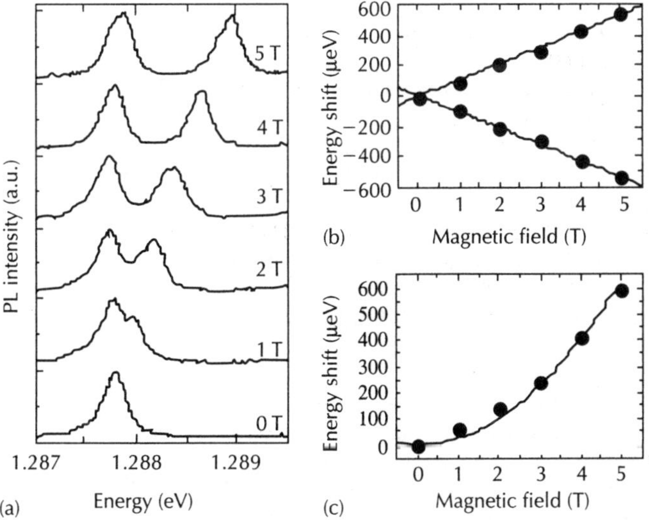

*Figure 8.23   (a) A single GaSb QD PL spectra under a magnetic field. (b) Diamagnetic shift and Zeeman splitting of an exciton state [83]. Reprinted with permission from T. Sato et al., Physica E **32**, 152 (2006). Copyright 2006, Elsevier.*

contribution of the diamagnetic shift, due to Zeeman splitting, are also given in Fig. 8.23c [83]. The energies of the split peaks are given by

$$E = E_0 \pm \frac{1}{2} g_{ex}\mu_B B + \beta B^2 \qquad (8.10)$$

where E_0 is the peak energy at 0T, μ_B is the Bohr magneton, g_{ex} is the g-factor of the exciton, and β is the diamagnetic coefficient. From the experimental data, $25\,\mu eV/T^2$ is obtained for the diamagnetic coefficient and 3.7 for the g-factor. The derived diamagnetic coefficient is larger than that typical of a type I In(Ga)As QD [84], and is attributable to the weak electron confinement outside the GaSb QD, because the diamagnetic coefficient is determined by the radius of the exciton wavefunction and the effective mass. The g-factor of GaSb QD is also slightly larger than that of In(Ga)As QDs. The exciton state in GaSb QD also has characteristic magneto-optical properties due to the type II band structure.

References

1. F. Hatami, N.N. Ledentsov, M. Grundmann, J. Bohrer, F. Heinrichsdorff, M. Beer, S.S. Ruvimov, P. Wermer, U. Gosele, J. Heydenreich, U. Richter, S.V. Ivanov, B.Ya. Meltser, P.S. Kop'ev, and Zh.I. Alferov, *Appl. Phys. Lett.* **67**, 656 (1995).
2. C.K. Sun, G. Wang, J.E. Bowers, B. Brar, H.R. Blank, H. Kroemer, and M.H. Pilkuhn, *Appl. Phys. Lett.* **68**, 1543 (1996).
3. M. Geller, C. Kapteyn, L. Muller-Krish, R. Heitz, and D. Bimberg, *Appl. Phys. Lett.* **82**, 2706 (2003).
4. E.T.R. Chidley, S.K. Haywood, R.E. Mallard, N.J. Mason, R.J. Nicholas, P.J. Walker, and R.J. Warburton, *Appl. Phys. Lett.* **54**, 1241 (1989).
5. L. Goldstein, F. Glas, J.Y. Marzin, M.N. Charasse, and G.L. Roux, *Appl. Phys. Lett.* **47**, 1099 (1985).
6. D.J. Eaglesham and M. Cerullo, *Phys. Rev. Lett.* **64**, 1943 (1990).
7. Y.W. Mo, D.E. Savage, B.S. Swartzentruber, and M.G. Legally, *Phys. Rev. Lett.* **65**, 1020 (1990).
8. T. Miura, T. Nakai, and K. Yamaguchi, *Appl. Surf. Sci.* **237**, 242 (2004).
9. K. Yamaguchi and T. Kanto, *J. Cryst. Growth* **275**, e2269 (2005).
10. M. Ohta, T. Kanto, and K. Yamaguchi, *Jpn. J. Appl. Phys.* **45**, 3427 (2006).
11. L.J. Whitman, B.R. Bennett, E.M. Kneedler, B.T. Jonker, and B.V. Shanabrook, *Surf. Sci.* **436**, L707 (1999).
12. P. Laukkanen, R.E. Perala, R.L. Vaara, I.J. Vayrynen, M. Kuzmin, and J. Sadowski, *Phys. Rev. B.* **69**, 205323 (2004).
13. F. Maeda, Y. Watanabe, and M. Oshima, *Phys. Rev. B.* **48**, 14733 (1993).
14. J.J. Zinck, E.J. Tarsa, B. Brar, and J.S. Speck, *J. Appl. Phys.* **82**, 6067 (1997).
15. P. Moriarty, P.H. Beton, Y.R. Ma, M. Henini, and D.A. Woolf, *Phys. Rev. B.* **53**, R16148 (1996).
16. B.R. Bennett, B.V. Shanabrook, P.M. Thibado, L.J. Whitman, and R. Magno, *J. Cryst. Growth* **175/176**, 888 (1997).
17. S. Tsukamoto and N. Koguchi, *J. Cryst. Growth* **201/202**, 118 (1999); **209**, 258 (2000).
18. G. Kresse and J. Heffner, *Phys. Rev. B.* **54**, 11169 (1996).
19. P.E. Bloechl, *Phys. Rev. B.* **50**, 17953 (1994).
20. G. Kresse and D. Joubert, *Phys. Rev. B.* **59**, 1758 (1999).
21. N. Kakuda, S. Tsukamoto, A. Ishii, K. Fujiwara, T. Ebisuzaki, K. Yamaguchi, and Y. Arakawa, *Microelectronics J.* **38**, 620 (2007).
22. T.J. Krzyzewski, P.B. Joyce, G.R. Bell, and T.S. Jones, *Surf. Sci.* **517**, 8 (2002).
23. G.G. Jernigan and P.E. Thompson, *Thin Solid Films* **380**, 114 (2000).
24. S.K. Park, J. Tatebayashi, and Y. Arakawa, *Appl. Phys. Lett.* **84**, 1877 (2004).
25. M.C. Righi, R. Magri, and C.M. Bertoni, *Phys. Rev. B.* **71**, 075323 (2005).
26. G.E. Franklin, D.H. Rich, A. Samsavar, E.S. Hirschorn, F.M. Leibsle, T. Miller, and T.-C. Chiang, *Phys. Rev. B.* **41**, 12619 (1990).
27. M.T. Sieger, T. Miller, and T.-C. Chiang, *Phys. Rev. B.* **52**, 8256 (1995).

28. The cohesion energies of Sb–Sb and Ga–Sb are 2.6 and 1.5 eV, respectively.
29. L.J. Whitman, P.M. Thibado, S.C. Erwin, B.R. Bennett, and B.V. Shanabrook, *Phys. Rev. Lett.* **79**, 693 (1997).
30. C. Kendrick, G. LeLay, and A. Kahn, *Phys. Rev. B.* **54**, 17877 (1996).
31. T. Kaizu, M. Takahasi, K. Yamaguchi, and J. Mizuki, *ISCS-2007* (2007).
32. B.G. Orr, C.W. Snyder, and M. Johnson, *Rev. Sci. Instrum.* **62**, 1400 (1991).
33. P. Geng, J. Marquez, L. Geelhar, J. Platen, C. Stezer, and K. Jacobi, *Rev. Sci. Instrum.* **71**, 504 (2000).
34. A.R. Avery, D.M. Holmes, J. Sudijono, T.S. Jones, and B.A. Joyce, *Surf. Sci.* **323**, 91 (1995).
35. V.P. LaBella, H. Yang, D.W. Bullock, P.M. Thibado, P. Kratzer, and M. Scheffler, *Phys. Rev. Lett.* **83**, 2989 (1999).
36. H. Yang, V.P. LaBella, D.W. Bullock, and P.M. Thibado, *J. Vac. Sci. Technol. B.* **17**, 1778 (1999).
37. G.R. Bell, J.G. Belk, C.F. McConville, and T.S. Jones, *Phys. Rev. B.* **59**, 2947 (1999).
38. M. Itoh, G.R. Bell, A.R. Avery, T.S. Jones, B.A. Joyce, and D.D. Vvedensky, *Phys. Rev. Lett.* **81**, 633 (1998).
39. P.B. Joyce, T.J. Krzyzewski, G.R. Bell, T.S. Jones, and B.A. Joyce, *Phys. Rev. B.* **58**, R15981 (1998).
40. S. Tsukamoto and N. Koguchi, *MRS Symposium Proceedings* **648**, P11.20 (2001).
41. G.R. Bell, M. Pristovsek, S. Tsukamoto, B.G. Orr, Y. Arakawa, and N. Koguchi, *Surf. Sci.* **544**, 234 (2003).
42. S. Tsukamoto, T. Honma, G.R. Bell, A. Ishii, and Y. Arakawa, *Small* **2**, 386 (2006).
43. T. Honma, S. Tsukamoto, and Y. Arakawa, *Jpn. J. Appl. Phys.* **45**, L777 (2006).
44. S. Tsukamoto, G.R. Bell, and Y. Arakawa, *Microelectronics J.* **37**, 1498 (2006).
45. T. Nakai and K. Yamaguchi, *Jpn. J. Appl. Phys.* **44**, 3803 (2005).
46. O. Dehaese, X. Wallart, and F. Mollot, *Appl. Phys. Lett.* **66**, 52 (1995).
47. G. Liu, S-L. Chuang, and S-H. Park, *J. Appl. Phys.* **88**, 5554 (2000).
48. C.G. Van der Walle, *Phys. Rev. B.* **39**, 1871 (1989).
49. R. Magri and A. Zunger, *Phys. Rev. B.* **64**, 81305 (2001).
50. B.R. Bennett, R. Magno, and B.V. Shanabrook, *Appl. Phys. Lett.* **68**, 505 (1996).
51. F. Hatami, M. Grundmann, N.N. Ledentsov, F. Heinrichsdorff, R. Heitz, J. Bohrer, D. Bimberg, S.S. Ruvimov, P. Werner, V.M. Ustinov, P.S. Kop'ev, and Zh.I. Alferov, *Phys. Rev. B.* **57**, 4635 (1998).
52. K. Suzuki, R.A. Hogg, and Y. Arakawa, *J. Appl. Phys.* **85**, 8349 (1999).
53. T. Nakai, S. Iwasaki, and K. Yamaguchi, *Jpn. J. Appl. Phys.* **43**, 2122 (2004).
54. S.H. Huang, G. Balakrishnan, A. Khoshkhlagh, A. Jallipalli, L.R. Dawson, and D.L. Huffaker, *Appl. Phys. Lett.* **88**, 131911 (2006).
55. G. Balakrishnan, J. Tatebayashi, A. Khoshkhlagh, S.H. Huang, A. Jallipalli, L.R. Dawson, and D.L. Huffaker, *Appl. Phys. Lett.* **89**, 161104 (2006).
56. T. Kaizu and K. Yamaguchi, *Jpn. J. Appl. Phys.* **40**, 1885 (2001).
57. A.-L. Barabasi, *Appl. Phys. Lett.* **59**, 3580 (1991).
58. D. Bimberg, M. Grundmann, and N.N. Ledentsov. *Quantum Dot Heterostructures* (John Wiley & Sons Ltd, New York, 1999).
59. A. Imamoglu, D.D. Awschalom, G. Burkard, D.P. DiVincenzo, D. Loss, M. Sherwin, and A. Small, *Phys. Rev. Lett.* **83**, 4204 (1999).
60. A. Zrenner, E. Beham, S. Stufler, F. Findels, M. Blchler, and G. Abstreiter, *Nature* **418**, 612 (2002).
61. Z. Yuan, B.E. Kardynal, R.M. Stevenson, A.J. Shields, C.J. Lobo, K. Cooper, N.S. Beattie, D.A. Ritchie, and M. Pepper, *Science* **295**, 102 (2002).
62. Motlan and E.M. Goldys, *Appl. Phys. Lett.* **79**, 2976 (2001).
63. R.A. Hogg, K. Suzuki, K. Tachibana, L. Finger, K. Hirakawa, and Y. Arakawa, *Appl. Phys. Lett.* **72**, 2856 (1998).
64. Ph. Lelong, K. Suzuki, G. Bastard, H. Sakaki, and Y. Arakawa, *Physics E.* **7**, 393 (2000).
65. L. Müller-Kirsch, R. Heitz, A. Schliwa, O. Stier, D. Bimberg, H. Kirmse, and W. Neumann, *Appl. Phys. Lett.* **78**, 1418 (2001).
66. M. Geller, C. Kapteyn, L. Müller-Kirsch, R. Heitz, and D. Bimberg, *Appl. Phys. Lett.* **82**, 2706 (2003).
67. E. Ribeiro, A.O. Govorov, W. Carvalho, Jr., and G. Medeiros-Ribeiro, *Phys. Rev. Lett.* **92**, 126402-1, (2004).

68. R. Heitz, F. Guffarth, I. Mukhametzhanov, M. Grundmann, A. Madhukar, and D. Bimberg, *Phys. Rev. B.* **62**, 016881 (2000).
69. E. Dekel, D.V. Regelman, D. Gershoni, E. Ehrenfreund, W.V. Schoenfeld, and P.M. Petroff, *Phys. Rev. B.* **62**, 11038 (2000).
70. E. Dekel, D. Gershoni, E. Ehrenfreund, D. Spektor, J.M. Garcia, and P.M. Petroff, *Phys. Rev. Lett.* **80**, 4991 (1998).
71. K. Matsuda, T. Saiki, S. Nomura, M. Mihara, Y. Aoyagi, S. Nair, and T. Takagahara, *Phys. Rev. Lett.* **91**, 177401-1 (2003).
72. D. Gammon, E.S. Snow, B.V. Shanabrook, D.S. Katzer, and D. Park, *Science* **76**, 3005 (1996).
73. T. Saiki and K. Matsuda, *Appl. Phys. Lett.* **74**, 2773 (1999).
74. K. Matsuda, S.V. Nair, H.E. Ruda, Y. Sugimoto, T. Saiki, and K. Yamaguchi, *Appl. Phys. Lett.* **90**, 013101 (2007).
75. D.V. Regelman, U. Mizrahi, D. Gershoni, E. Ehrenfreund, W.V. Schoenfeld, and P.M. Petroff, *Phys. Rev. Lett.* **87**, 257401 (2001).
76. L. Besombes, J.J. Baumberg, and J. Motohisa, *Phys. Rev. Lett.* **90**, 257402 (2003).
77. K. Brunner, G. Abstreiter, G. Böhm, G. Tränkle, and G. Weimann, *Phys. Rev. Lett.* **73**, 1138 (1994).
78. G. Chen, T.H. Stievater, E.T. Batteh, X. Li, D.G. Steel, D. Gammon, D.S. Katzer, D. Park, and L.J. Sham, *Phys. Rev. Lett.* **88**, 117901-1 (2002).
79. A.J. Williamson, L.W. Wang, and A. Zunger, *Phys. Rev. B.* **62**, 12963 (2000).
80. A. Zunger, *Phys. Stat. Sol (b)* **224**, 727 (2001).
81. R. Resta, *Phys. Rev. B.* **16**, 2717 (1977).
82. L.-W. Wang, J. Kim, and A. Zunger, *Phys. Rev. B.* **59**, 5678 (1999).
83. T. Sato, T. Nakaoka, M. Kubo, and Y. Arakawa, *Physica E.* **32**, 152 (2006).
84. Y. Toda, O. Moriwaki, M. Nishioka, and Y. Arakawa, *Phys. Rev. B.* **58**, 10147 (1998).

CHAPTER 9

Growth and Characterization of ZnO Nano- and Microstructures

Marius Grundmann,[1] Andreas Rahm,[1] Thomas Nobis,[1]
Michael Lorenz,[1] Christian Czekalla,[1] Evgeni M. Kaidashev,[1]
Jörg Lenzner,[1] Nikos Boukos,[2] and Anastasios Travlos[2]

[1]Universität Leipzig, Fakultät für Physik und Geowissenschaften, Institut für
Experimentelle Physik II, Linnéstr. 5, D-04103 Leipzig, Germany
[2]Institut of Materials Science, National Center for Scientific Research Demokritos,
GR 15310 Ag. Paraskevi Attikis, POB 60228 Athens, Greece

9.1 Introduction

Zinc oxide (ZnO) was subject to intensive research in the 1950s [1]. Since then it had been rediscovered several times. From the late 1990s of the last century there is again renewed interest in this wide-band gap II VI semiconductor. In particular, it is considered a promising candidate for use as visible and UV light emitters. The first report of lasing in ZnO by optical pumping at 2 K dates back to 1996 [2]. In 1997, lasing of the exciton was observed at room temperature by two groups independently [3, 4]. The material exhibits tunable defect and dopant-related luminescence properties in the visible region of the spectrum [1]. In contrast to gallium nitride (GaN), its main competitor for ultraviolet (UV) applications, ZnO exhibits favourable properties such as a high exciton binding energy of 60 meV (prospect of low laser threshold at room temperature, in contrast to 25 meV for GaN), unproblematic wet-chemical etching, radiation hardness (space applications) and the availability of substrates for homoepitaxy. They are, for instance, available from Tokyo Denpa [5]. Other advantages of ZnO include its biocompatibility [6], a high piezoelectric constant [7] and large optical gain [8]. This fuels high device-related expectations, not only for optoelectronics but also for biomedical and biosensor applications.

ZnO is already in everyday use as UV blockers in sun tan lotions, in surface acoustic wave devices (SAW), as gas sensors, in high-field varistors and as a catalyst, to name only a few of its well-established applications.

Furthermore, it has been shown that the band gap of zinc oxide can be controlled between 2.8 eV and 4 eV by alloying with CdO and MgO, respectively [9–11]. ZnO/$Mg_xZn_{1-x}O$ quantum wells [12, 13] and superlattices [14] as well as $Mg_xZn_{1-x}O/Cd_yZn_{1-y}O$ multiquantum wells have been studied [15].

However, many potential applications are still hampered by the lack of control over p-type conductivity. Typically, ZnO is n-type which is believed to be due to native point defects [16], and although there are several attempts to achieve p-type doping and conduction, reproducibility and temporal stability are currently serious issues [17]. Recent progress includes the observation of near UV to blue colour electroluminescence from a ZnO p-i-n homojunction light-emitting diode (LED) grown by laser molecular-beam epitaxy as reported by Tsukazaki et al. [18, 19]. Furthermore, ultraviolet electroluminescence from n-type ZnO/p-type polymer heterojunction

LEDs grown at low temperatures was demonstrated [20]. Also, it is expected that the spin degree of freedom will play a significant role in 21st century information technology [21]. Due to its predicted ferromagnetism above room temperature [22], heavily doped p-type ZnMnO is considered a promising material for spintronics [23] which could pave the way to exploit spin in addition to charge in semiconductor devices. It was shown that the incorporation of 3D transition metals can produce ferromagnetism in ZnO above room temperature [24].

Especially nanoscale one-dimensional zinc oxide (ZnO) structures have stimulated great interest during the past years because of their extraordinary structural diversity [25]. Nanowires, -tubes, -belts, -ribbons, -sheets and many more complex ZnO structures can be fabricated. There are reports on the synthesis of quasi-1D and 2D nanostructures [25] and combinations thereof [26], of tetrapod zinc oxide structures [27], of tapered tip morphologies [28] and of flowerlike ZnO [29]. At present, nanowires are the second hottest research topic in physics after carbon nanotubes [30, 31] according to ranking scientific fields based on the "Hirsch index" [32]. Due to its c-axis dominated growth behaviour, ZnO is of particular interest for self-assembled growth of nanowire structures. Such bottom-up architecture approaches are believed to lead to applications which are, at least to a certain amount, defect tolerant. A considerable amount of work has been done on catalyst-driven self-organization processes which are based on the vapour–liquid–solid (VLS) growth mechanism proposed by Wagner and Ellis in 1964 [33]. Recent progress in ZnO nanowires research is documented, for instance, in [34, 35].

ZnO whiskers can be utilized for the understanding of basic research phenomena [36], but primarily the unique physical properties were employed to demonstrate a variety of nanodimensional device prototypes. They include nanoscale lasers [37], nanocantilevers [38], piezoelectric nanogenerators based on ZnO nanowire arrays [39], ZnO nanorod logic circuits [40] and nanowire-field effect transistors [41]. Also, field-emission from ZnO nanowires was observed [42]. They can further be employed to detect various chemical species like oxygen [43, 44], ethanol [45], NO_2 or NH_3 [46]. ZnO/$Mg_xZn_{1-x}O$ nanorod multi-quantum-well structures (MQWs) were fabricated and quantum confinement effects were observed [47, 48]. Room-temperature ferromagnetism in well-aligned $Zn_{1-x}Co_xO$ nanorods [49] was also detected. However, detailed studies on the growth evolution and formation mechanisms, as well as on the lateral homogeneity of samples, are rare in the literature [50].

Pulsed laser deposition (PLD) has been proven to be a versatile technique for the growth of high-quality ZnO-based thin films [51, 52]. In this chapter, the growth of ZnO nanostructures by a unique high-pressure PLD process will be studied in detail. Those structures will be characterized structurally and optically with the main focus on so-called whispering gallery modes in which our nanostructures act as dielectric resonators. Furthermore, we present stimulated emission from ZnO microcrystals grown by carbothermal evaporation by spatially resolved photoluminescence (PL) and high excitation spectroscopy (HES). Electrical characterization of those crystals is also reported here.

9.2   Growth of nanowires PLD

9.2.1   Synthesis strategies for ZnO nanostructures

In the 1960s, the vapour–liquid–solid (VLS) mechanism was proposed by Wagner and Ellis as a model to explain the morphology and growth of silicon whiskers [33, 53]. It was observed that (a) an impurity was essential for whisker growth and (b) a small globule was present at the tip of each whisker [33]. In the mechanism, the impurity is thought to form a liquid alloy droplet which acts as a preferred site for deposition of particles from the vapour phase which then causes the droplet to become supersaturated. For instance, a small particle of Au can be intentionally used as such an impurity. At elevated temperature a small droplet of Si–Au formed in the case investigated by Wagner and Ellis. This alloy has a high sticking coefficient for arriving atoms and hence more Si enters the droplet, supersaturates it and finally freezes out at the substrate. This process results in whiskers with an alloy droplet on top. Typical more recent examples of semiconductor nanowire growth controlled by the VLS mechanism are Si [54, 55] and GaAs [56].

For Si nanowires the VLS synthesis can be directly observed at different growth stages [57]. The choice of the proper impurity/catalyst depends on a number of factors such as, the inertness to the reaction products and the formation of a liquid alloy at the growth temperature [33]. Gold as an impurity is very well suited to form an alloy with Zn at temperatures below 1000°C according to their binary phase diagram.

A detailed study of the fundamental aspects of VLS growth, like kinetics, temperature effects or the role of surface diffusion, was presented in 1975 by E.I. Givargizov [58]. Important results include the observation that thin whiskers grow slower than thick ones and that there is a minimum critical diameter at which growth stops completely due to a decrease of supersaturation [58]. Experimental proof of this theory has been observed for GaAs, GaP, InAs and InP [59]. The growth rate of whiskers depends on four mains steps [58]: (i) mass transport in the gas phase, (ii) chemical reaction on the vapour–liquid interface, (iii) diffusion in the liquid phase and (iv) the incorporation of the material in the crystal lattice.

Beyond those considerations for ideal case scenarios, there are other factors which can play a role in catalysed nanowire growth. Every model has its limits: contrary to the calculations of Givargizov [58], there are predictions that the attainable VLS nanowire minimum size is not limited *a priori* [60]. Also, there is a thermodynamic limit for the minimum radius of the metal–liquid clusters at high temperature [61]. A further deviation from the VLS growth model was observed by temperature dependent studies of the Au-assisted growth of InAs nanowires [62]. In the particular case of ZnO, several growth processes occur in parallel for appropriate conditions. It is observed that the whisker diameter can be far beyond the catalytic particle size [50] which is due to an additional vapour–solid mechanism. The growth models for ZnO nanowires and more complex two-dimensional ZnO nanostructures are large in number as is the structural diversity of ZnO. Next, we give a selective overview of the growth conditions for ZnO-based nanostructures.

ZnO whiskers, platelets, and dendritic crystals have already been reported before renewed interest arose. Earlier works include the vapour phase experiments by Sharma *et al.* (1971) [63]. He and his co-workers synthesized mm-sized crystals by a modified vapour–phase method at temperatures between 930 and 980°C. Furthermore, the growth of colourless and transparent ZnO needles with a length of up to 200 mm by oxidation of zinc vapour was observed (1974) [64]. In 1978, Iwanaga *et al.* [65] presented structures with different morphologies (hollow, needle, whisker, ribbon) depending on their starting material (ZnF_2, ZnS, ZnSe and others) for the reaction. They attributed the growth partly to the VLS mechanism. However, the typical size of all those crystals was beyond several μm. As already mentioned in the introduction, there is a tremendous interest in nanometre-sized ZnO crystals which has led to hundreds of publications during the past years.

Most commonly, physical vapour deposition or thermal evaporation are employed for the synthesis of ZnO-based nanostructures. But also metal-organic vapour phase epitaxy (MOVPE), electrochemical deposition or PLD was used. The growth temperatures span a very wide range: from only 70°C to above 1400°C. The use of a catalyst for VLS growth is a feature which a large part of the approaches share. Besides gold, which is most common, also liquid Zn droplets [66], NiO [67], tin [68] or copper [69] were used as nucleation sites. ZnO nanostructures grow on almost every substrate: sapphire, Si, glass, plastic, and so on. More special synthesis techniques are, for instance, the production of ZnO by the oxidation of Zn wires [70]. All those methods can lead to high-quality nm-sized crystals the shape of which depends strongly on the specific growth conditions. Some formation mechanisms are well understood while others need to be investigated further. For instance, for the formation of ZnO nanorings and nanohelices there are theoretical works on the necessary conditions upon which they form [71]. Also, the doping and alloying has been shown to be practicable. ZnO has been combined with MgZnO to form coaxial nanowire quantum wells (QW) [72] or multi-QWs [48, 73]. CdZnO [45, 74], ZnCoO [75] and ZnMnO [76] structures were presented. For a more comprehensive and complete overview, the reader is referred to corresponding review articles [25, 34, 77].

In our group, ZnO nanowires were grown by a special gold-catalysed high-pressure pulsed laser deposition process which is described in detail in the next subsection. Furthermore, we employ a vapour phase transport process proposed by Huang and Yao [61, 78] for the growth of

ZnO microcrystals. This method is conventionally used to synthesize one- and two-dimensional nanostructres [79, 80]. Here, our standard process was modified to yield whiskers with diameters in th µm range by thermal evaporation of a pressed ZnO–graphite (mass ratio 1:1) target at ambient pressure. The growth temperature was 1100°C. An argon downstream was applied as transport gas. The ZnO microwires were isolated on a carbon gluepad. SEM images of such a microwire are shown in Fig. 9.1.

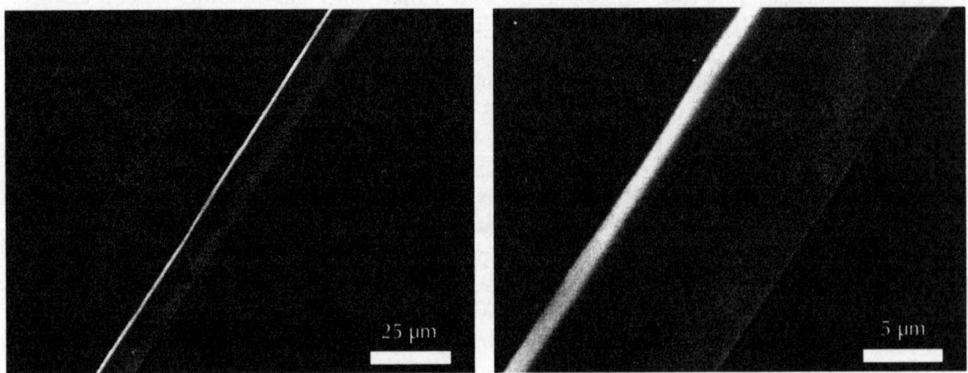

Figure 9.1   *SEM images of a ZnO microwire on a carbon gluepad.*

9.2.2   *High-pressure PLD process*

PLD belongs to the class of physical vapour deposition techniques. It was first demonstrated by Smith and Turner in 1965 [81]. However, it took more than two decades until PLD became a relevant growth technique. The breakthrough came with the development of high-power lasers and with new multi-element materials systems, such as high-T_c oxide superconductors, for which PLD is well suited. Nowadays, it is an established method for a variety of materials, especially oxides, but also organic polymers, metals and semiconductors [52]. Sankur *et al.* [82] first reported on the epitaxial growth of ZnO using PLD.

At the University of Leipzig we have established a high-pressure PLD process designed for nanostructure growth. The main parts of this system are a high-power pulsed KrF excimer laser from Lambda Physik (LPX 305), the laser optics and the deposition chamber [51]. The chamber consists of a T-shaped quartz tube with an outer diameter of 30 mm. The KrF excimer laser beam (wavelength 248 nm) is guided to the deposition chamber inside a protection shielding, enters along the centre bar of the T and is focused by a UV lens on the cylindrical surface of one of the rotating PLD targets. Up to three different targets fit onto the linear rotary feedthrough bar thus allowing *in situ* modulation of chemical composition. The rotation by the feedthrough ensures homogeneous ablation all around the poly-crystalline targets. They were prepared by mixing, pressing and sintering high-purity powders for 12 h at 1150°C. The growth from multi-component targets (ZnMgO, ZnMnO and ZnCoO) is discussed in [51] and [76].

The direction of the plasma plume is pointing above the substrate which is placed off-axis on a ceramic slat. Its surface normal is perpendicular to the symmetry axis of the plume and to the laser beam. The target to substrate distance was 5–35 mm. An encapsulated heater with an arrangement sof KANTHAL wire in ceramic tubes (patent pending DE 10255453.7) and FIBROTHAL isolation material is built around the quartz tube. A type K thermoelement (Ni–Cr–Ni) is used to measure the temperature at the outside wall of the tube. By this arrangement the temperature at the growth zone can be varied in the range between room temperature and 950°C. With this configuration the gas flow is directed from the target towards the substrate and, hence, potentially supports the transport of material from the plasma plume. The flow can be controlled from 5 sccm to 30 slm by Qualiflow AFC 50D digital mass flow controllers (MFCs). The

Table 9.1   Key parameters of the PLD system for nanowire growth at the University of Leipzig

PLD parameter	Value
Background gas	100% Ar or Ar + O_2 with up to 50% O_2
Background gas pressure	10–200 mbar
Gas flow	10–500 sccm
Growth temperature	400–950°C
Laser pulse energy	350–650 mJ
Laser wavelength	248 nm (KrF)
Laser focus size at target	$2 \times 5\,mm^2$
Pulse number	10–36 000
Pulse frequency	1–20 Hz
Target–substrate distance	5–35 mm

Ar background gas has a purity of 99.995%. In Table 9.1 the typical growth parameters for ZnO nanostructures are summarized.

Due to the increased collision probability between the laser-produced plume and the background gas, the plume dimension for the high-pressure PLD case presented here is considerably smaller than in conventional, i.e. high-vacuum, PLD. Furthermore, the ambient gas leads to a broadening of the angular distribution of the plume [83]. Dyer and co-workers have shown that the ratio between laser-pulse energy E and the background gas pressure p is the scaling parameter for the plume range L [84]:

$$L \propto \left(\frac{E}{p} \right)^{\gamma/3}$$

where γ is the ratio of specific heats of the elements in the plume [85]. For a typical laser pulse energy of 500 mJ and a pressure of 0.1 mbar the plume dimension of an ablated ZnO target is around 5 cm. Hence, the plume range for 100 mbar is well below 1 cm, a fact which can be directly verified by the observation of the optical emission from the excited plume, as it is seen through the viewport during ablation. The kinetic energies of species arriving at substrates located beyond the dimension of the plume will be very low, i.e. they are expected to be thermalized. As will be shown in this work, this versatile set-up allows the fabrication of ZnO-based nanostructures with tunable shape and chemical composition.

9.2.3   Growth morphology

In order to observe the nanowhisker growth at different stages, a series of five samples with different growth times or, correspondingly, a different number of laser pulses (1000–16 000) with a constant repetition frequency of 10 Hz has been produced (Fig. 9.2). Note that there is always a factor of two between the growth times of successive samples. All other PLD parameters remained unchanged. Typical as-grown ZnO nanowire structures are shown in the SEM images in Fig. 9.2. For the purpose of accurate determination of the nanostructure dimensions, the substrate has been cut along the axis of symmetry (*cf.* PLD chamber geometry [51]) by scratching the sapphire with a diamond needle tip. This facilitates an open view down to the nanowire base and the substrate cleaving edge. For better comparability, all SEM pictures shown in this section were taken exactly at the centre of the samples. Figure 9.2a–c represents tilted views for the samples grown from 10 nm Au dots with 1000, 4000 and 16 000 PLD pulses, respectively. Figure 9.2d is the corresponding top view of the sample in Fig. 9.2c. Nanowires with hexagonal cross-section are clearly resolved mimicking the wurtzite crystal structure of ZnO. The prism side faces are oriented parallel to each other indicating that there is a very good in-plane epitaxial

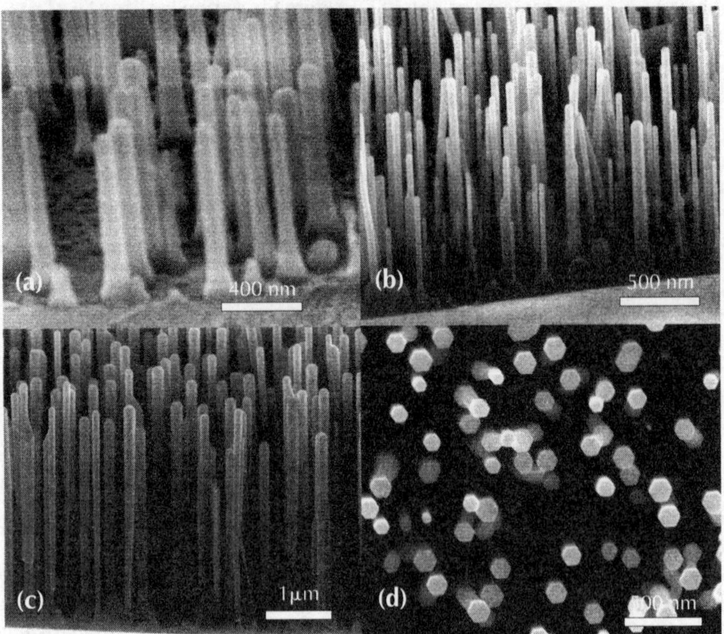

Figure 9.2   Typical SEM images of PLD-grown ZnO nanowires taken at different growth stages after application of: (a) 1000, (b) 4000 and (c) 16 000 laser pulses. (d) represents the top view of the sample shown in (c).

relationship. It is apparent from the pictures that the length of the whiskers increases upon increase of deposition time. The growth evolution is shown in more detail in Fig. 9.3a and b.

In the first growth steps the aspect ratio (length/diameter) increases almost linearly with deposition time (Fig. 9.3b). Note that the x-scale in the figure is logarithmic for better visualization. Most of the whiskers are between 20 and 70 nm in diameter (Fig. 9.3c and d). Figure 9.3c shows a typical corresponding histogram (sample with 8000 pulses). The sample with 16 000 pulses plays an exceptional role. Most of its nanowires are thicker than 70 nm while the diameter distribution functions (Fig. 9.3d) are similar for all other samples. In this last growth step the length and the diameter increase by similar relative amounts, resulting in a nearly constant aspect ratio. The lateral nanowire density of the first four samples scattered between 56 and 80 wires per μm^2 and for the last sample (16 000 pulses) it was as low as 23 wires/μm^2. It is evident from Fig. 9.3c and d that at later growth stages the whiskers tend to grow together thus forming structures with larger diameter but reduced lateral density.

For comparison, a sample based on colloids with 50 nm diameter was synthesized using 8000 laser pulses. The substrate in the PLD chamber is oriented in such a way that one side faces the PLD target, i.e. the surface is parallel to the symmetry axis of the plasma plume. In Fig. 9.4 the mean whisker diameter is plotted against the relative position on the sample. The inset of Fig. 9.4 shows schematically five different positions on a substrate where position 1 corresponds to the side with the smallest distance to the target and position 5 is 10 mm further apart from it. The diameter decreases systematically with increasing distance from the target. An analogous finding was presented earlier for different substrate positions [51]. The narrowest diameter variation (represented by the bars in Fig. 9.4) can be found in the middle of the sample, at position 3. Also the height of the nanowires varies over the substrate surface (Fig. 9.5). This difference of growth morphology on a single sample arises from the expansion dynamics of the plasma plume. The energy and density of the ablated species depend strongly on the distance from the target. The availability of ZnO whiskers with different and defined sizes on a single sample is of great use for basic research, as, for example, for optical cavity experiments (see section 9.3).

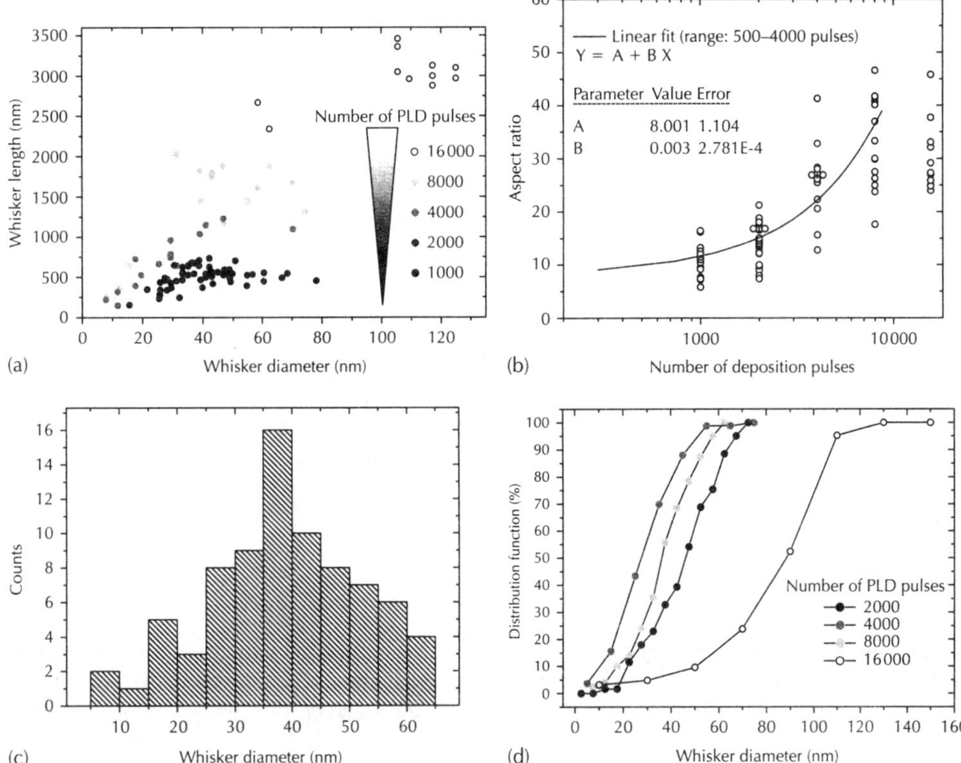

Figure 9.3   *Detailed growth evolution study of ZnO nanowires. (a), (b) Diameter, length and the aspect ratio in dependence of the number of PLD pulses. The linear fit in (b) includes the data points up to 8000 laser pulses. (c) Histogram of the whisker diameter of the sample with 8000 pulses. (d) Distribution functions of the diameter for different samples.*

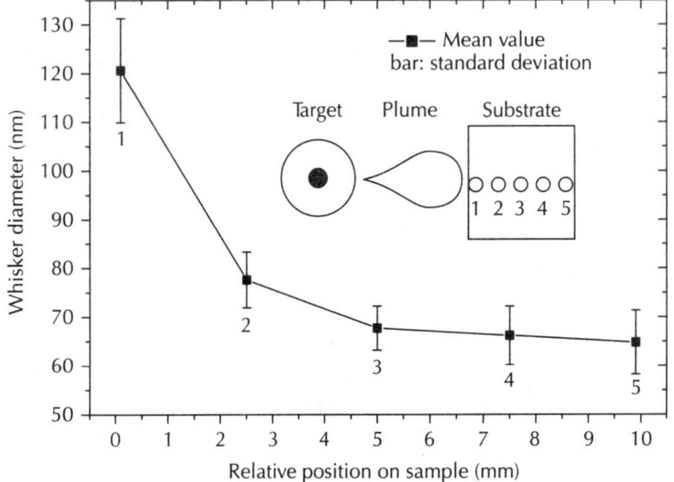

Figure 9.4   *Lateral homogeneity investigation over a sample. 1–5 represent different spots on the substrate, from front edge (near plasma, spot 1) to the back edge (spot 5, see inset). The bar indicates the standard deviation from the mean ZnO whisker diameter.*

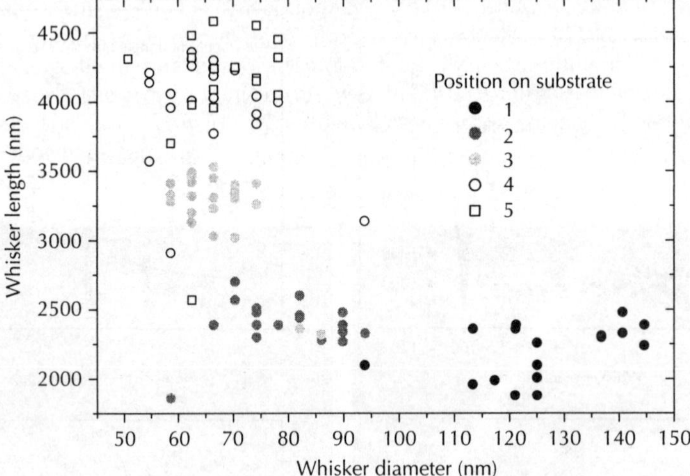

Figure 9.5   ZnO whisker length vs diameter for different positions on a single substrate. The positions 1–5 correspond to the inset of Fig. 9.4.

9.2.4   Structural characterization

XRD measurements show that the crystallographic c-axis of the ZnO nanowires is oriented perpendicular to the surface of the a-plane sapphire substrate. Figure 9.6 depicts an HRXRD diffractogram of the ZnO(0002) and the Al_2O_3 ($11\bar{2}0$) substrate reflection of the ZnO nanowire sample of Figs 9.4 and 9.5. The corresponding peak widths (FWHMs) are 22 arcsec and 15 arcsec, respectively. This underlines the superior crystalline quality of the whiskers as compared to ZnO thin films. The rocking curve of the ZnO (0002) reflection has an FWHM of 0.31° (inset of Fig. 9.6) which is comparable to the best values reported for ZnO nanowires and PLD thin films [10, 51, 86]. Hence, as it can be seen already in the SEM pictures, there is very little tilt between the whiskers. However, the crystal quality changes significantly with the growth temperature. In Fig. 9.7 the peak width of the ZnO(0002) reflection is shown for the rocking curve as well as for the 2θ–ω scan

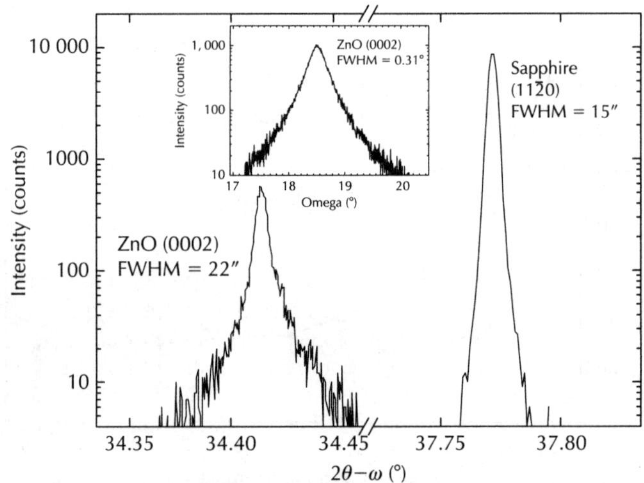

Figure 9.6   XRD diffractogram of a ZnO nanowire sample. The peak widths of the ZnO and the substrate reflection are in the range of the instrumental line broadening (12 arcsec, determined with sapphire and Si substrates). The rocking curve of the ZnO (0002) reflection with an FWHM of 0.31° is shown in the inset.

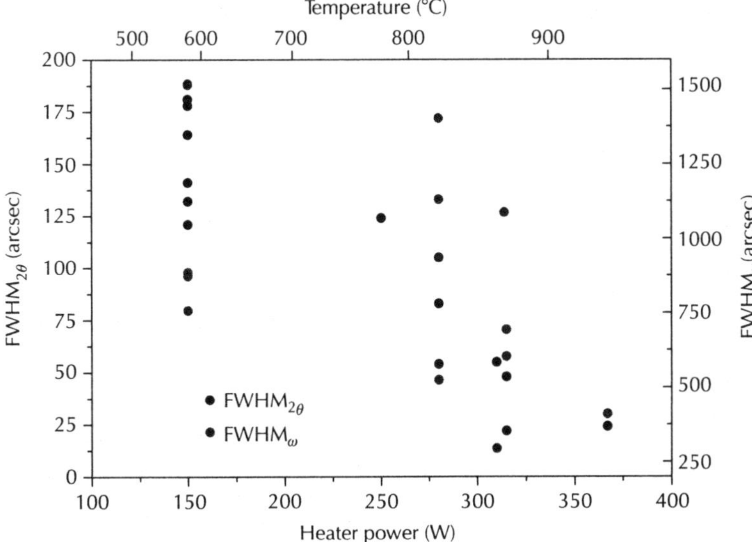

Figure 9.7   FWHMs of the ZnO(0002) reflection (XRD) vs growth temperature; black: 2θ–ω-scan width (left axis), red: rocking curve width (right axis).

in dependence of the heater power/temperature. Both widths are minimal for growth temperatures between 850 and 900°C though it should be noted that this trend is more pronounced for the 2θ–ω scans than for the ω scans. The data points at 150 W heater power come from samples which were grown on the bare substrate without employing any catalyst. Those samples are subject to ongoing investigations. Preliminary results indicated that catalyst-free ZnO growth is possible by the low-vacuum PLD set-up at sufficiently low temperature. The c-axis lattice constants for the pure ZnO samples was 5.207 Å which is again exactly the value for bulk ZnO.

The typical rod-like growth morphology is studied in more detail by TEM (Fig. 9.8). A single crystalline wurtzite ZnO whisker with a little gold droplet on top is revealed. However, the diameter of the whisker is much larger that the droplet size. The inset of Fig. 9.8a shows the SAD of ZnO. As already deduced from the XRD measurements, the growth direction is parallel to the c-axis of ZnO and in accordance with the very low FWHMs of the XRD ZnO(0002) reflection, dislocations in the nanowires are very rare. Figure 9.8b and c depicts steps on the side surfaces of the whiskers which are not periodic along the growth direction and not symmetric on both wire

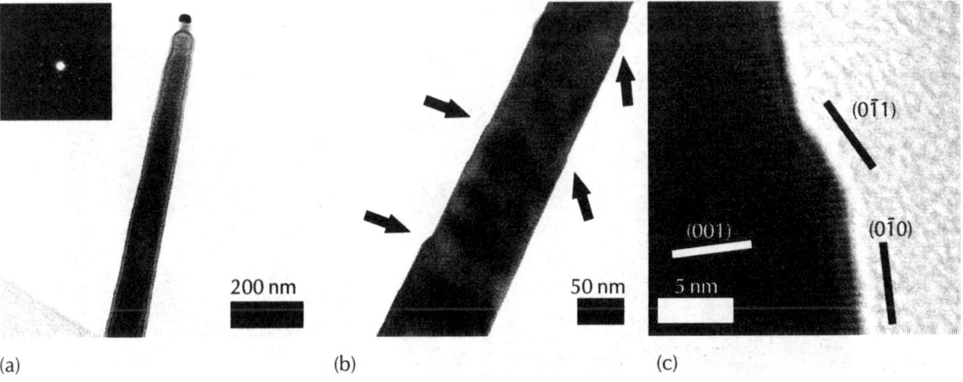

Figure 9.8   TEM images of a single ZnO nanowire. The inset in (a) shows the corresponding SAD pattern of the whisker. In (b) the arrows indicate the positions of steps on the side faces which are shown in more detail in (c). (b) and (c) were studied in the [100] zone axis.

sides. They exhibit various heights usually between 1 and 15 atomic steps. In Fig. 9.8c the important ZnO planes as well as the step habit plane are indicated. Angled $\{11\bar{2}3\}$ faces as reported in [87] were not observed here.

9.3 Optical properties I: whispering gallery modes

9.3.1 Nanowire luminescence

Depending on the respective growth conditions the PLD process provides a variety of differently shaped ZnO micro- and nanocrystals shown in Fig. 9.9. The room temperature luminescence of our PLD grown ZnO nanostructures typically consists of two bands, the first (UV) around 3.25 eV, the second (visible, VIS) around 2.35 eV. The UV band is associated with free-exciton emission. The latter is also called "green band" in the literature; its exact nature, involving deep levels, is still under debate. Figure 9.10 shows typical CL spectra of arrays of ZnO nanowires within the spectral region of donor-bound exciton (DBX) emission for $T = 10\,\mathrm{K}$. The size of the excited nanowire array has been increased from a few (1–3) wires up to 100–300 wires. All CL spectra are dominated by a strong emission line at 3.360 eV indicating the recombination of excitons bound to a neutral aluminium donor (I_6, [88]). The FWHM of this line varies from 1.4 to 2.0 meV and is thus comparable to values for high-quality PLD-grown thin films and single crystals [80, 89]. A transition at 3.364 eV possibly belongs to an ionized–donor bound exciton (I_3, [88]). The luminescence intensity of further DBXs (3.356 eV, indium, I_9 [88]) is weaker by about one order of magnitude. Free-exciton (FX) emission is barely visible at this temperature. The absence of significant differences between the CL spectra originating from different array sizes proves the lateral homogeneity of the nanowire sample on the μm scale.

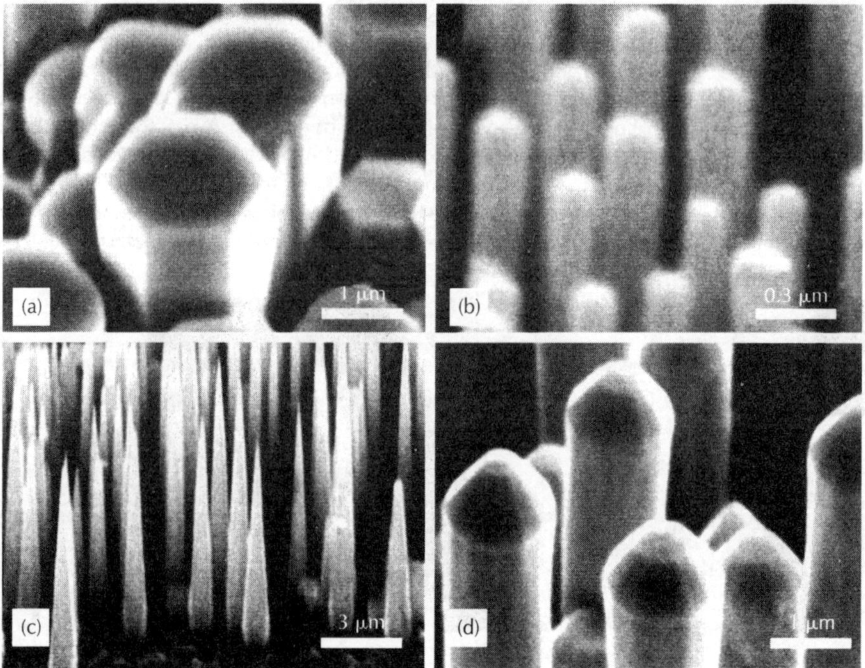

Figure 9.9 Different shapes of PLD grown ZnO micro- and nanocrystals. (a) Hexagonal microcrystals with diameters of about 2 μm. (b) ZnO nanowires with diameters around 100 nm. (c) Needle-like nanostructures. (d) Microcrystals exhibiting a dodecagonal cross-section.

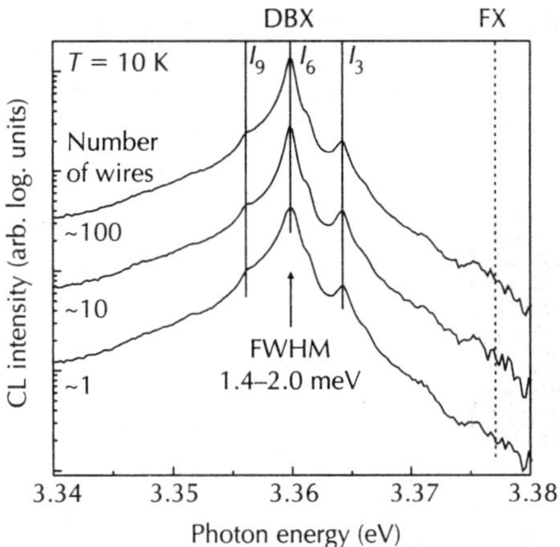

Figure 9.10   CL spectra (shifted vertically for clarity) of arrays of ZnO nanowires within the spectral region of donor-bound exciton (DBX) emission. The excited lateral area of the nanowire sample has been increased to include 1–3 nanowires (lower curve), 10–30 nanowires (middle curve) and 100–300 nanowires (upper curve). Vertical lines refer to dominating DBX transitions. The dashed vertical line marks the expected position of free exciton (FX) emission.

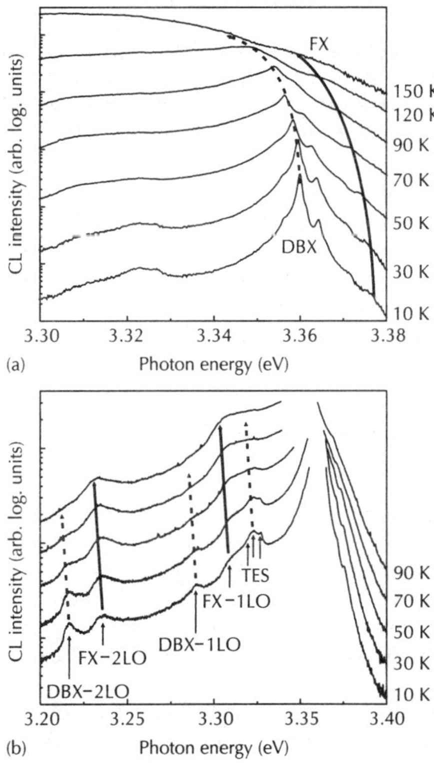

Figure 9.11   Temperature dependence of CL spectra of an array of about 10 nanowires. Results are given for the spectral region of DBX emission (a) and for the spectral region of phonon replica (b).

With increasing temperature (Fig. 9.11) all transitions broaden and shift to lower energies. DBX emission vanishes (dashed line in Fig. 9.11a), while FX emission gains in intensity (solid line). Figure 9.11b focuses on photon energies <3.34 eV. Phonon replicas and two electron satellites (TES) of the DBX lines can be observed [88]. Both types of transitions vanish with increasing temperature (dashed lines) due to their physical correlation to donor bound excitons. Additional peaks that can be detected even at elevated temperatures indicate phonon replicas of the FX transition (solid line). Due to the relative large diameter of ZnO wires presented here, a quantum confinement effect is neither expected nor visible in the spectra.

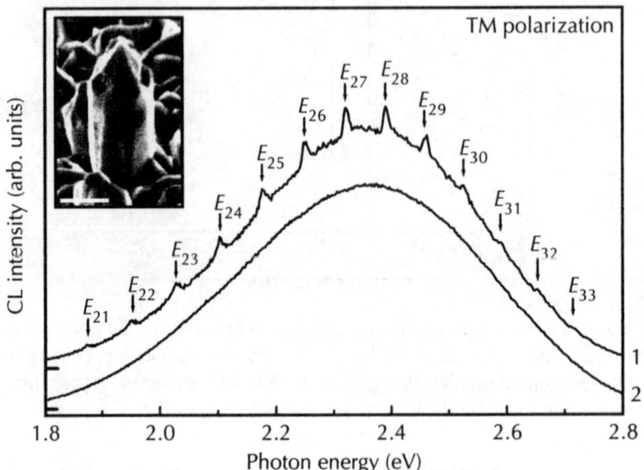

Figure 9.12   Room temperature VIS emission CL spectra of a ZnO microcrystal (curve 1) and of thin film material (curve 2) scaled to the same intensity and vertically shifted for clarity. Arrows mark theoretical energetic positions of WGMs due to Eq. 9.1 for TM polarization and D = 2.86 μm. The inset shows an SEM image of the investigated microcrystal. The scalebar has a length of 2 μm. The experimental cavity diameter is $D_{exp} = (2.90 \pm 0.06)$ μm.

Figure 9.12 shows the typical VIS emission CL spectrum from a ZnO microcrystal. An electron microscopy image of the crystal is given in the inset. In comparison to the broad and unstructured VIS emission of thin films or bulk material, VIS emission of the microcrystal is accompanied by a series of comparatively sharp peaks. This modulated broad band is visible and can be assigned to optical whispering gallery resonances. Within the VIS range ZnO is optically transparent and exhibits a refractive index of $n \approx 2$ [90], i.e. ZnO micro- and nanocrystals represent (hexagonal) dielectric resonators in a wide spectral range.

9.3.2   Theory of hexagonal whispering gallery modes

9.3.2.1   Dielectric resonators

From a general point of view a dielectric resonator is a transparent body of certain shape consisting of a substance with a high refractive index n, usually surrounded by air. Such a structure has the ability to confine light inside, e.g. utilized in any kind of optical fibres or waveguides. To explain this phenomenon in terms of classical electrodynamics, and any similar problem respectively, it is necessary to solve the three-dimensional wave equation for the given geometry. Steady-state solutions of this equation are then usually called eigenmodes of the considered resonator.

In the case of a hollow metallic resonator as an example for a so-called *closed* cavity, a discrete set of eigenmodes exists related to a discrete spectrum of eigenmode energies [91]. This is due to the boundary conditions for the electric and magnetic field that lead to a complete suppression of wave propagation outside the resonator. Hence, no energy is radiated from the hollow resonator; the whole radiation field is restricted inside the metallic cavity.

In contrary, a dielectric resonator is an *open* cavity. Because of the continuity conditions for the electric and magnetic field at the cavity's boundary, the radiation field is not exclusively restricted to the resonator itself, but actually extended to infinity. This results in a permanent radiation loss of energy and, hence, every imaginable field distribution decays exponentially with a lifetime τ, therefore not representing a steady-state solution [92]. The actual solutions of this problem form a continuum of so-called scattering states consisting of an incoming plane wave and an outgoing scattered wave. Therefore, the eigenmodes of a dielectric cavity appear as maxima within the continuous spectrum of the total scattering cross-section in dependence of the energy of the incoming wave. Those maxima, i.e. those incoming wave energies that exhibit the strongest interaction with the given resonator, are usually called resonance energies or *resonances* of the dielectric cavity [92]. Due to the above-mentioned finite lifetime, all resonances exhibit a finite linewidth. The larger the losses of a given resonance the shorter is the lifetime and the larger is the resonance's linewidth.

9.3.2.2   Two-dimensional hexagonal resonator

An important class of dielectric resonators are two-dimensional cavities, i.e. their in-plane dimensions are much larger compared to their height. In this case, wave propagation parallel to the longitudinal axis c is negligible. The light can be assumed to circulate around only within the resonator's cross-section. The respective mode patterns are so-called whispering gallery modes (WGMs), obtained by solving the wave equation in its two-dimensional form. Hence, two decoupled types of polarizations occur [92], namely TE polarized modes ($E \perp c$), and TM polarized modes ($E \parallel c$) as explained in Fig. 9.13b and c.

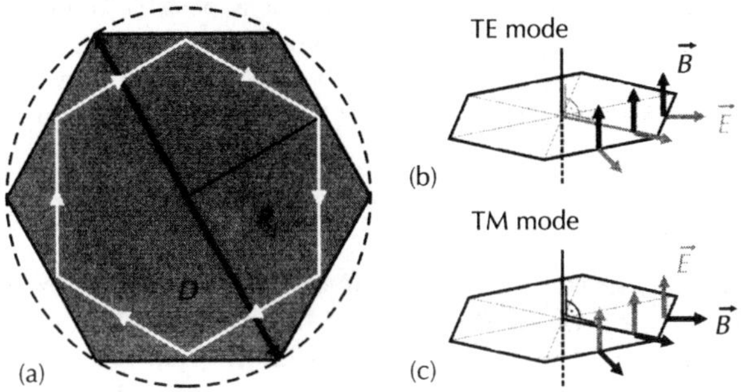

Figure 9.13   *(a) The light confined within a hexagonal cross-section can be assumed to circulate inside the cavity as indicated by the white arrows. The geometry of the cavity can be described by the radius of the incircle R_i, by the radius of the circumscribing circle R, or by its related diameter D = 2R using the geometric relation R = $2R_i/\sqrt{3}$. Note that the circumference of the inscribed white hexagon has a length of $6R_i$. (b), (c) Two possible types of polarization of two-dimensional whispering gallery modes, exemplarily given for a hexagonal resonator. For TE modes the electric field is transverse to the longitudinal axis of the resonator, for TM modes the magnetic field is.*

Eigenmodes of such two-dimensional whispering gallery resonators have been calculated in the past; the solutions for a circular cavity can be expressed in an analytical way [93]. Other kinds of geometries have been subject to present literature, as, e.g., deformed circles [94, 95], rectangles [96] or hexagons [97], which necessarily have to be calculated numerically [97, 98]. A scheme of the geometry of the two-dimensional hexagonal cavity is given in Fig. 9.13a.

The results of theoretical modelling [97] show that the eigenmodes of such a hexagonal cavity can be numbered by an integer mode number N. Additionally, the modes can be distinguished concerning their symmetry, e.g. leading to pairs of degenerated modes N^+ and N^- related to their respective chirality. In [16], hexagonal WGMs have been calculated for a fixed $n = 1.466$ and in the mode number range of about N = 20 to 70.

For large mode numbers N > 70 a simple plane wave model (PWM) has been deduced [97]. Its main idea is that the light wave circulates around exploiting the process of multiple total internal reflection (TIR, see Fig. 9.13a) and finally interferes with itself when having completed one full circulation within the resonator. To enforce constructive interference the total phase shift of the wave along its path has to be an integer multiple of 2π, i.e. only entire wave trains are allowed to perform multiple circulations generating a standing wave. Taking into account the polarization-dependent negative phase shift that occurs during the process of TIR [91], we obtain the following equation:

$$6R_i = \frac{hc}{nE}\left[N + \frac{6}{\pi}\arctan\left(\beta\sqrt{3n^2 - 4}\right)\right]. \tag{9.1}$$

The factor β is polarization; for TM polarization $\beta = \beta_{\mathrm{TM}} = n^{-1}$ has to be used; TE polarization leads to $\beta = \beta_{\mathrm{TE}} = n$. Due to the spectral dependence of the refractive index $n = n(E)$, Eq. 9.1 is an implicit equation to determine the discrete resonance energies $E = E_N(R_i)$ in terms of the geometric parameter R_i, Planck's constant h and vacuum speed of light c. The first factor of the right side of Eq. 9.1 corresponds to the wavelength in matter. The integer $N \geq 1$ characterizes the interference order of the resonance, which is in this case identical to the respective WGM number [97]. The following term containing β refers to the additional phase shift mentioned above. Furthermore, since ZnO is uniaxial, $n = n_{\parallel}(E)$ and $n = n_{\perp}(E)$ have to be applied for TM and TE polarization, respectively. We note that Eq. 9.1 for TM polarisation and $n = $ const. is identical to Eq. (17) of [16]. Neglecting the spectral dependency of n, Eq. 9.1 leads to:

$$E_N \propto \frac{1}{R_i} \tag{9.2}$$

and thus WGMs generally shift to higher energies with decreasing cavity diameter. Note that although TIR suggests that it is not possible for the light to leave the resonator, emission nevertheless occurs at the corners of the hexagon [97]. After all, radiation losses are always allowed due to the reasons given in Section 9.3.2.1. Therefore it is possible to optically detect WGMs. Since the geometrical model leading to Eq. 9.1 is a combination of basic ray and plane wave optics, it is expected to be valid only for $R \gg \lambda$, i.e. for $N \gg 1$ [97]. However, we show in this work that Eq. 9.1 describes the experimentally found WGMs even down to mode number $N = 1$.

9.3.3   Whispering gallery modes in hexagonal ZnO microcrystals

Using the PWM, those peaks in Fig. 9.12 can unambiguously be attributed to WGMs of a hexagonal cavity. Therefore it is necessary to determine the correct mode number N of every single detected peak. Provided that this mode number is known, Eq. 9.1 enables one to calculate the theoretical diameter of the crystal $D = 4R_i/\sqrt{3}$ out of every single detected peak energy. Fortunately, since the peaks have to be numbered in ascending order, one only has to find the correct mode number to start. The best fitting and therefore the final peak numbering is found if every single peak predicts the *same* radius as all the other resonance peaks, or at least if the variations in the predicted radius become smallest. This procedure is explained in more detailed in Fig. 9.14.

For the investigated crystal, the correct mode numbering is shown also in Fig. 9.3; the respective theoretical diameter yields to $D_{theory} = 2.86\,\mu\mathrm{m}$ with a small minimum–maximum spreading of less than $10\,\mathrm{nm}$. Although theory predicts both TM and TE polarization, all calculations have been performed for TM type, since polarization-dependent microphotoluminescence experiments showed that WGMs are preferentially TM polarized (see Fig. 9.7). The required data for $n_{\parallel}(E)$ were obtained from ellipsometry measurements on PLD-grown ZnO thin film samples [90].

The experimentally determined cavity diameter obtained from electron microscopy amounts to $D_{exp} = (2.90 \pm 0.06)\,\mu\mathrm{m}$. Hence, the deviations between theory and experiment are in the range of only 2%! This means that the simple PWM fits very well, even if N is in the range of only 20 to 30. To emphasize this fact, the theoretical values E_N calculated for fixed diameter $D = D_{theory}$ are given as black arrows in Fig. 9.3; they appear very close to the measured peaks. We note that if a constant value of the index of refraction n without spectral dispersion is used, the agreement between theory and experiment becomes worse.

9.3.4   Whispering gallery modes in ZnO nanostructures

Although the simple PWM obviously is valid for WGMs with mode numbers in the range of about $N = 20$–30, i.e. for optical cavities in the micrometre regime, it does not necessarily need to be valid for nanostructures. Furthermore, prism-shaped microcrystals only enable analysis of WGMs for fixed cavity diameter given by the cross-section of the particular microcolumn. Those restrictions can be vanquished by focusing on nanostructures with a *needle-like* shape.

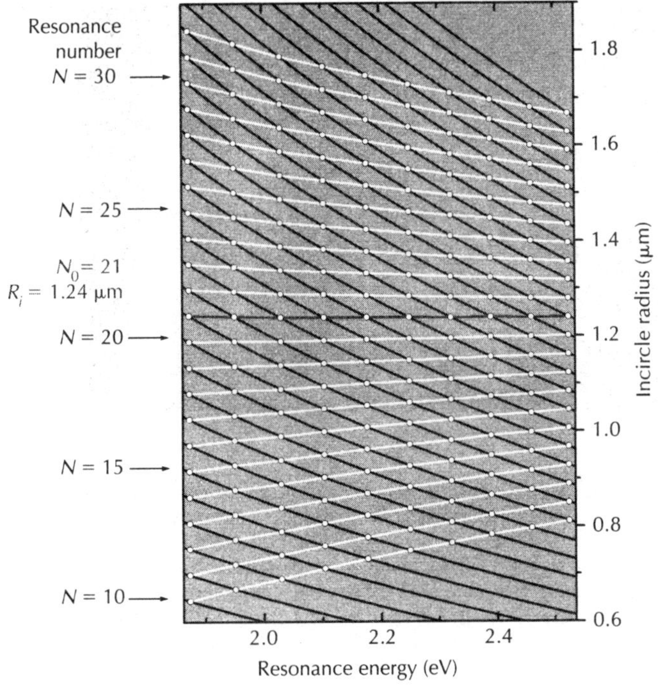

Figure 9.14   *Diagram $R_i(E)$ to determine the correct mode numbering for the spectrum of Fig. 9.12. Equation 9.1 enables one to calculate the radius of the cavity incircle $R_i = R_i(E,N)$ out of a given resonance number N and resonance energy E. The resulting set of curves is given here for N = 10...41 as a set of black lines approximately exhibiting a hyperbolic shape (as expected from Eq. 9.2). The index of the curves, i.e. the respective resonance number, is shown on the left side of the scheme marked by black arrows. The experimentally detected resonance energies are given as white circles. Since the resonances have to be numbered in ascending order, every white line represents a possible peak numbering. As can be seen from the diagram there exists only one peak numbering that results in an energy-independent incirlce radius, denoted by the horizontal line. Hence, the starting number for the peak numbering of the spectrum in Fig. 9.3 yields to $N_0 = 21$ with $R_i = 1.24\,\mu m$; this leads to a respective cavity diameter D = 2.86 \mu m.*

9.3.4.1   ZnO nanoneedles

An example of an array of PLD-grown ZnO nanoneedles is shown in Fig. 9.15a. As can be seen in Fig. 9.15b–d, the width of the hexagonal cross-section of a selected needle is continuously tapered when approaching the needle's top. Since the aspect ratio of this structure is very high, every plane perpendicular to the needle's longitudinal axis creates a hexagonal cavity and actually those structures allow for systematic investigations of WGMs in the nanometre regime from $D = 800\,nm$ down to $D = 0$!

9.3.4.2   Detection

For this purpose, we performed spatially resolved CL experiments [99, 100], i.e. pointwise CL spectra have been recorded across the whole needle using a scan grid of 16×150 points. Typical spectra and their interpretation are given in Fig. 9.16. As in the case of microcrystals, the broad VIS band of luminescence again is modulated due to a preferred constructive interference of light whose photon energy again fulfils the resonance condition of Eq. 9.1. For decreasing diameter, approaching the top of the needle, these maxima continuously shift to higher energies. This effect supplies the unambiguous proof that the measured spectral modulation actually originates from WGMs. When the dominating WGM with number N is blue shifted so far that it leaves the VIS range, the next lower resonance $N - 1$ occurs at lower energies indicated by a *discontinuous* red shift of the spectral maximum in Fig. 9.16c and d. This process continues until $N = 1$ is reached referring to the last resonance that can be observed.

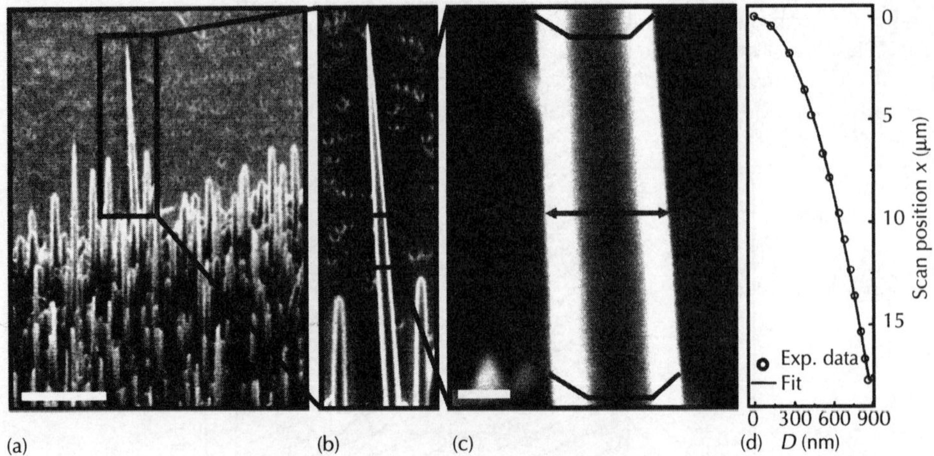

Figure 9.15   *SEM image of the investigated nanoneedle. All images have been obtained at 45° perspective. (a) SEM image of an array of ZnO nanoneedles containing the reported one marked by a black rectangle. The scalebar has a length of 10 µm. (b) Larger-scale SEM image of the investigated nanoneedle. The needle's diameter is continuously tapered approaching zero at the top. (c) High-resolution SEM image indicating the hexagonal cross-section of the needle. The scalebar has a length of 300 nm. (d) Experimentally determined shape of the needle. The obtained characteristic of the diameter D vs scan position x can be fitted using a potential law $D \propto x^{0.523 \pm 0.007}$, leading to a square root-like shape of the needle.*

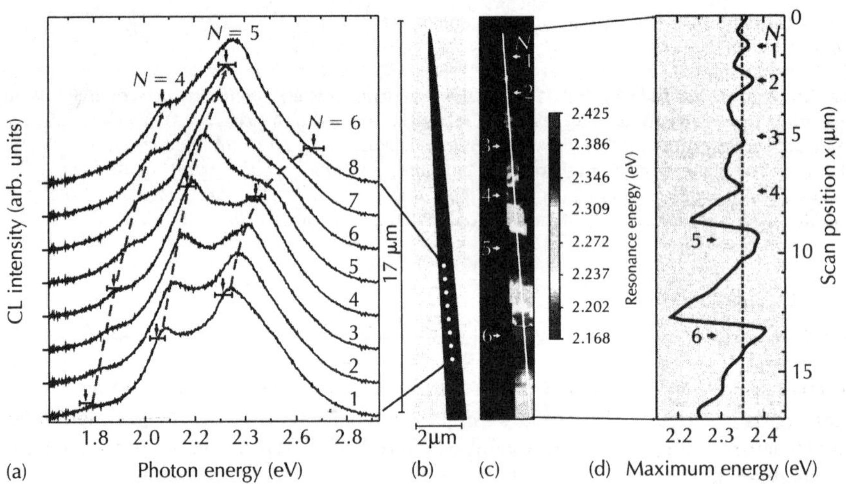

Figure 9.16   *Spatially resolved CL investigation of a single tapered ZnO nanoneedle. (a) CL spectra shifted vertically for clarity collected at eight equidistant locations marked on the needle's longitudinal axis by white dots in (b). The VIS band of ZnO is modulated such that maxima can clearly be distinguished from each other and attributed to WGMs labelled according to Eq. 9.1. Since the radius of the needle decreases along the longitudinal axis, the spectral maxima continuously shift to higher energies as indicated by the red dashed lines. Blue arrows and error bars mark selected TM resonance energies obtained from Eq. 9.1 and their error. (b) Experimentally determined shape of the needle. (c) Map of the energy of the spectral maximum within the visible spectral range. (d) Line scan along the white line shown in (c). The red dashed line gives the maximum position of the unstructured VIS band in bulk material.*

9.3.4.3   Polarization

To obtain essential information about the state of polarization of the detected WGMs we performed local polarization-dependent microphotoluminescence experiments. These show that the WGMs are preferentially TM polarized, since detecting TE polarization causes an almost complete

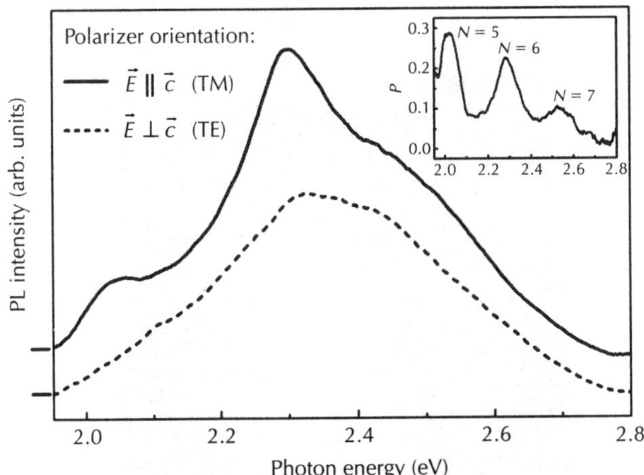

Figure 9.17   Micro-photoluminescence spectra of the nanoneedle of Fig. 9.15 for $D \approx 790\,nm$ at two different polarizer azimuth orientations shifted vertically for clarity. The modulation of the VIS band almost disappears when detecting the TE mode. Thus, the resonances are mainly TM polarized. The inset shows the ratio $P = (I_{TM} - I_{TE})/(I_{TM} + I_{TE})$, which visualizes the TM-WGMs as distinguishable peaks. The polarization effect has been found to be independent of the polarization of excitation.

suppression of WGM modulations (see Fig. 9.17). The remaining spectrally unstructured emission occurs due to the unpolarized character of ZnO photoluminescence originating from the needle's centre. However, this result is consistent with former investigations of lasing in hexagonal cavities within the micrometre regime that are reported to emit only TM modes [101]. Hence, in comparison with TE modes, TM modes obviously exhibit lower losses. This can be generally understood since, leaving the regime of TIR, the reflection coefficient for TM polarized waves due to Fresnel's equations is always larger than that for TE polarized ones.

9.3.4.4   Theory vs experiment

To compare the measured energies of the WGMs in nanostructures with theoretically predicted ones for both types of polarization, a spectral line scan along the needle's longitudinal axis is given in Fig. 9.18. In grey scales, the resonances appear as a set of bright lines that actually exhibit a curvature as expected from Eq. 9.2. The theoretical values are shown with dots and cross symbols in Fig. 9.18. They are directly calculated without free parameter using the experimentally determined shape of the needle, i.e. $R_i(x)$, see Fig. 9.15, and the above-mentioned thin film data for $n_{\parallel}(E)$ and $n_{\perp}(E)$ [90]. The dots (TM) match the measured resonance energies with a very good agreement. Additionally, the well fitting blue arrows within Fig. 9.16a confirm that the simple PWM gives a good description of the basic physics of WGMs even in nano-sized crystal geometries within the limits of the resonance linewidths.

9.3.4.5   Line broadening effects

As shown in Fig. 9.16, compared to microcavities, nano-sized crystals show broadened WGMs due to the increase of losses with decreasing cavity diameter. Nevertheless, CLI of the broad VIS band still enables the visualization of WGMs in the ZnO nanoneedle even at diameters $D < \lambda/n \cong 270\,nm$. For mode numbers $N \geq 4$ single TM-WGMs can be detected unambiguously as distinguishable maxima modulating the broad VIS band. For smaller mode numbers $N \leq 3$ WGMs can only be visualized as a continuous blue shift of the intrinsic VIS maximum, and the discontinuities in Fig. 9.16c and d become smooth. This is due to the extreme line broadening of both types of WGMs with decreasing cavity diameter. Theory [97] predicts line broadening by loss processes due to boundary waves that are scattered out of the resonator and due to light paths whose angle of incidence is slightly deviating from 60°. A relation $FWHM \propto E^{-1}R^{-2}$ is obtained, which explains the dramatic broadening of the measured resonances. For TM polarization and

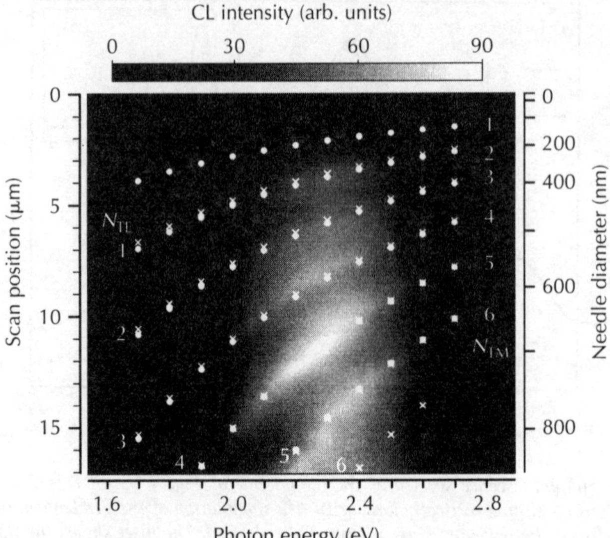

Figure 9.18 *Two-dimensional plot of spectra recorded along a line scan on the needle's longitudinal axis. The left vertical axis shows the line scan position x, the right one refers to the respective needle diameter D. The grey scales refer to the spectral CL intensity. The spectral maxima, i.e. the measured WGM energies, appear as bright belts going from the bottom left corner to the right upper one. With decreasing diameter all resonances shift systematically to higher energies. The white dots give theoretical TM resonance energy positions obtained from Eq. 9.1, white crosses give the same for TE polarization. Without adjusting free parameters there is a very good agreement between experiment and theory.*

$N = 6$ (spectrum 1 of Fig. 9.16a) a *FWHM* of about 350 meV is predicted. Considering that the underlying theory again is prepared for $N \gg 1$, the obtained spectrum verifies this value within the limits of resolution. For resonance $N = 3$ the theoretical *FWHM* is about 900 meV, which exceeds even the intrinsic linewidth of the VIS band. This explains the difficulty to obtain distinct WGM peaks for $N \leq 3$. Nevertheless, VIS emission is still affected when a WGM crosses the VIS range, as Fig. 9.18 clearly shows intensity modulations at the respective line scan positions. We note that losses due to surface roughness of the cavity faces could also lead to line broadening. However, as the high resolution SEM image of Fig. 9.15c shows well-shaped resonator faces, we do not adopt this loss process to be the dominating one.

9.3.4.6 *Discussion*

As mentioned above, there is an energy difference between predicted TM and TE modes. This difference is mostly due to β in Eq. 9.1, since the ZnO birefringence $(n_\parallel - n_\perp)/(n_\parallel + n_\perp) \approx 1.2\%$ is small [90]. However, TE mode with mode number N is predicted (coincidentally due to the particular value of n) to appear very close to the spectral position of TM mode $N + 1$ (Fig. 9.18), and, hence, this could be one reason for the missing TE mode series. Possible TE-WGM maxima should always lie beyond higher-ordered TM maxima, and as the former ones suffer larger losses, they lead to much broader resonance peaks hardly detectable in the VIS band.

Certainly there are, however, small deviations between the detected peak energies and the predicted TM-WGMs according to Eq. 9.1. (see Figs 9.16a and 9.18). These deviations probably reveal the limits of the simple PWM. To discuss this fact more in detail we note that the application of numerical methods to predict resonance energies at the end always depends on a precise determination of the considered geometry, e.g. R_i, and a good knowledge of the refractive index n. SE microscopy measurements of the cavity diameter D are performed with an error of at least 1%. The ellipsometrically determined values for $n_\parallel$ for bulk material yield an error in n of about 2%. The actual refractive index of the nanocrystals may possibly differ additionally. These effects lead to an error in E of about 3%, i.e. 50–80 meV in the considered spectral range, which is indicated as error bars in Fig. 9.16a. Compared to the linewidths of the WGMs, Eq. 9.1 for TM polarization gives a very good description of the spectral position of the resonant modes.

9.4 Optical properties II: stimulated emission from ZnO microcrystals

In the last section we showed that micrometre-sized ZnO structures can be employed as optical resonators are hence promising candidates for laser devices and should allow fundamental investigations of light–matter coupling. The microcrystal structures show a high photoluminescence (PL) signal under low excitation conditions.

If ZnO is optically excited under high intensities, typically an additional peak arises in the ZnO PL spectrum. This peak is most probably caused by an inelastic exciton–exciton scattering process and exhibits a fine structure. The emitted photons have the energies [102]:

$$E_n = E_{ex} - E_x^{\,b}\left(1 - \frac{1}{n^2}\right) - \frac{3}{2}k_B T \tag{9.3}$$

where E_{ex} is the energy of the free exciton luminescence, $E_x^{\,b}$ is the binding energy of the excitons and $k_B T$ the thermal energy. The parameter n is the quantum number of the problem. The respective peak is called P-band and in many cases stimulated emission related to this band can be observed. Under even higher excitation, the carrier density in the sample increases and the electron–hole plasma (EHP) peak at even lower energies than the P-band becomes observable.

A laser is a device emitting stimulated optical radiation. For practical purposes, one uses a cavity for optical feedback. A ZnO microcrystal *per se* cannot be expected to act as such a cavity due to the high losses caused by the short distance of the mirrors. The mirror loss for reflectivities R_1 and R_2 in a cavity of length L is given by [103]:

$$\alpha = \frac{2}{L}\ln\frac{1}{R_1 R_2}. \tag{9.4}$$

The reflectivity of the ZnO/air interface is approximately 0.2. For a $10\,\mu m$ cavity, Eq. 9.4 gives a loss of $1600\,cm^{-1}$. In a real structure, other losses due to scattering and absorption will additionally play a role. Typical gain reported for ZnO ($150\,cm^{-1}$ in [104], $600\,cm^{-1}$ in [105]) cannot overcome these losses. The observation of Fabry–Perot modes in lasing thus seems unlikely in our structures. However, we will demonstrate stimulated emission as result of a single pass gain.

9.4.1 Experiments

Figure 9.19 shows a scheme of the set-up used for the PL and high excitation spectroscopy (HES) measurements. It allows spatially resolved PL measurements and the observation of emitted light from single microstructures. For PL, the excitation source was a continuous-wave He–Cd Kimmon laser with a wavelength of 325 nm and a power of 20 mW. For HES measurements, a pulsed Thales DIVA II Nd:YAG laser beam (266 nm, 10 ns, 20 Hz) is focused on the sample. The energy per pulse is 2 mJ. Using these two lasers, the excitation intensity has been varied in a range from 1 to $2500\,kW/cm^2$. The angle of incidence of the pump beam is approximately $60°$ with respect to the surface normal as depicted in the inset of Fig. 9.19. Excitation spots smaller than the diameter of the pump laser beam are possible using a focusing lens. Furthermore, the excitation intensity can be controlled by a reflective Newport attenuator for high power lasers.

The light emitted from the sample can be deflected into a digital camera by a movable mirror making the set-up work as an optical microscope. If the mirror is removed, the emitted light is focused onto an optical fused silica fibre and directed to the monochromator. It is then spectrally dispersed with a 2400 lines/mm grating (blaze 330 nm) and detected by a liquid nitrogen cooled back-illuminated CCD camera. The set-up is diffraction limited and confocal and allows a spatial resolution of $1\,\mu m$.

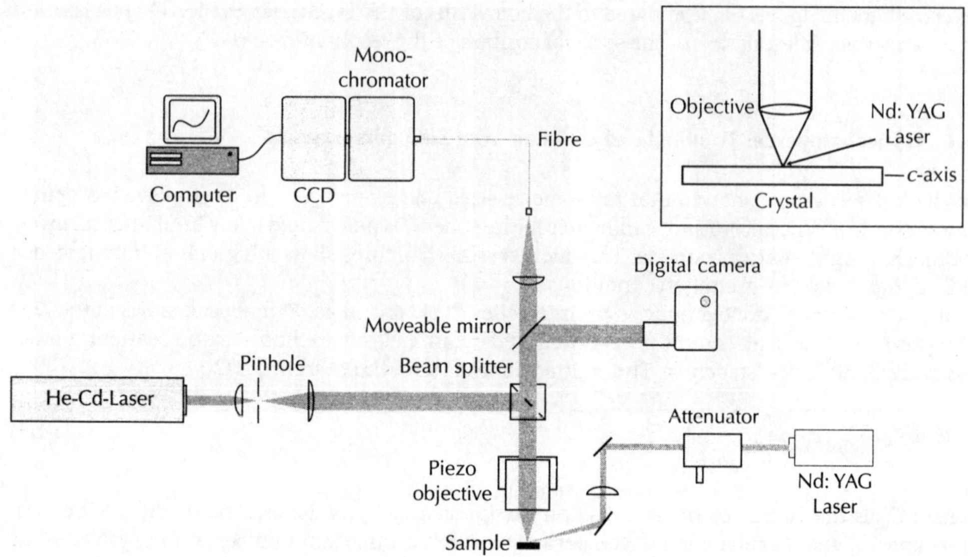

Figure 9.19 Set-up for spatially resolved PL measurements. The inset shows the orientation of the wire with respect to the laser beam and the objective.

9.4.2 Results and discussion

Figure 9.20 shows the low and high excitation spectra of a ZnO microcrystal in a semilogarithmic plot at room temperature. The dashed cw spectrum was shifted for clarity and does not fit to the same scale. The excitation intensities were chosen between 45 and 250 kW/cm^2.

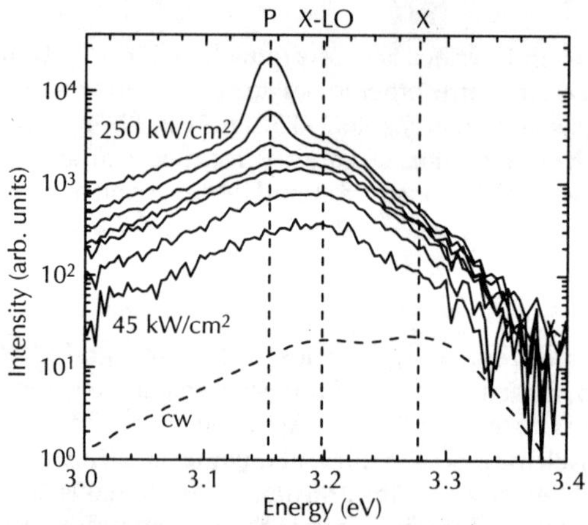

Figure 9.20 Room temperature PL spectra of a ZnO microcrystal arranged on a carbon gluepad under low (dashed curve) and high (solid curves) excitation intensities varied between 45 and 250 kW/cm^2. At higher excitation intensities, the P-band arises from the free exciton emission band. The dashed curve of the cw spectrum was shifted for clarity.

Two peaks at 3.20 and 3.28 eV are observed in the cw spectrum. The lower energy peak is most probably related to the first LO-phonon replica of the 3.28 eV maximum. Under higher excitation only the 3.20 eV peak is visible. It can be seen from Fig. 9.20 that a sharp peak at 3.154 eV is predominant in the spectrum for higher excitation intensities. The FWHM is 88 meV for 85 kW/cm^2 and 29 meV for 250 kW/cm^2, indicating spectral narrowing as depicted in Fig. 9.21.

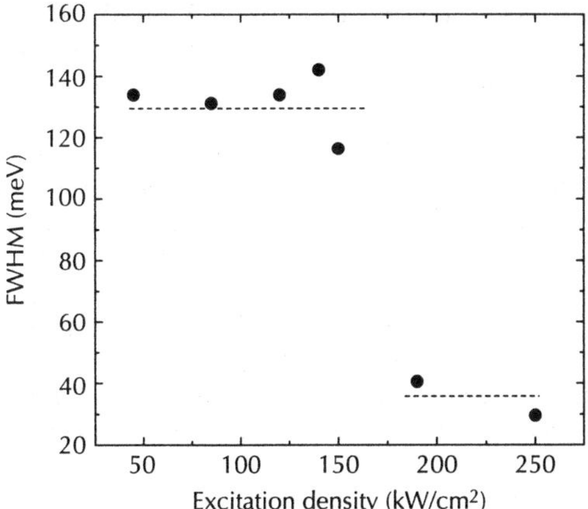

Figure 9.21 FWHM of the PL peak as a function of the excitation intensity. Dashed lines are guides to the eye.

The peak intensity increases superlinearly with increasing excitation intensity as depicted in Fig. 9.22. The spectral narrowing (Fig. 9.21) and the superlinear characteristic curve in Fig. 9.22 implies the presence of stimulated emission from the microcrystal. The threshold pump intensity is approximately $170\,\mathrm{kW/cm^2}$.

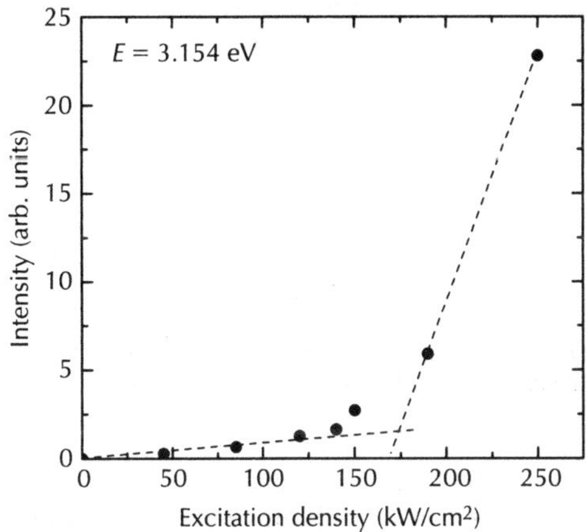

Figure 9.22 Dependence of the peak intensity on the excitation intensity. The superlinear dependency is clearly visible. The treshold intensity for the stimulated emission is approximately $170\,kW/cm^2$.

For $T = 300\,\mathrm{K}$, Eq. 9.3 gives values between 3.234 ($n = 2$) and 3.219 eV ($n = \infty$). The observed values are smaller than those expected from Eq. 9.3. We tentatively attribute the laser peak to the P-band and the energy shift of about $70\,\mathrm{meV}$ from the expected room temperature position to a local increase of the sample temperature by approximately $100\,\mathrm{K}$, taking into account the temperature dependence of the band gap energy [106]. Further excitation density and temperature dependence measurements will clarify this point.

9.5 Electrical characterization of ZnO microcrystals

Nanowires and nanotubes carry charge efficiently and are potentially ideal building blocks for future (opto)electronics. Knowledge about contacts and electrical conduction properties of

ZnO is essential for practical device applications. In 1965, Mead [107] investigated surface barrier heights of several metals on vacuum cleaved ZnO surfaces. For thin films and bulk material there are several recent reports on electrical characterization of ZnO, e.g. by H. von Wenckstern *et al.* [108] or D.C. Look *et al.* [16]. However, for single nano- and microsized ZnO crystals there are only a few experimental results on electrical transport and contacts, e.g. [109, 110]. For instance, Park *et al.* investigated the electrical transport mechanism of single ZnO nanorods [111]. Also some prototype devices have been demonstrated including field effect transistors (FET), p–n junctions and electroluminescent nanodevices (grown on p-GaN) on the basis of ZnO nanostructures [41, 112]. The large surface-to-volume ratio of such structures is expected to cause a strong dependence of the transport properties on environmental conditions, a fact that makes ZnO nanowires ideal candidates for sensing applications [43, 44].

In this section the results of rectifying and ohmic contacts on ZnO microwires will be presented and the temperature dependence of the conductivity will be discussed.

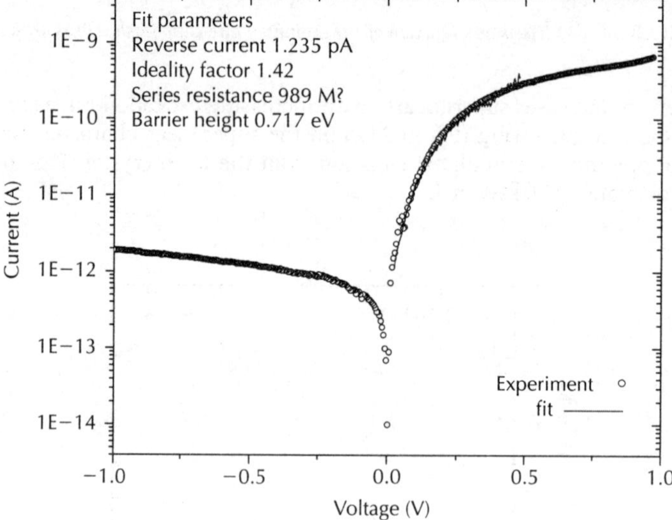

Figure 9.23 *Current vs voltage characteristic of a ZnO microwire with an Epo-Tek H20E rectifying contact. The parameters of the fit according to Eq. 9.5 (red curve) are shown in the inset.*

The dependence of the current *I* on the applied bias of a carbothermally grown microwire with one rectifying contact (Epo-Tek H20E) and one ohmic contact (GaIn) was measured at room temperature with the semiconductor parameter analyser and a wafer prober. A typical rectifying diode-like behaviour with a forward/reverse current ratio around 350 is observed. According to the Schottky–Mott model, the barrier formation can be described by the barrier height which is given by the difference of the Fermi level of the contact metal and the electron affinity of the semiconductor [113]. Reports on investigations in the temperature range between 300 and 500 K from several groups reveal a transport mechanism according to the thermionic emission model [43, 110]. In order to determine the diode key figures, the forward current was fitted under the standard assumption that thermionic emission is the dominating current transport process at room temperature [114]:

$$I = AA^*T^2 \exp\left(\frac{-\Phi_{B.eff}}{kT}\right)\left[\exp\left(\frac{e(V_a - IR_s)}{nk_BT}\right) - 1\right] \tag{9.5}$$

where A is the contact area, A^* is the effective Richardson constant having a theoretical value of 32 Acm^{-2}K^{-2} (for $m_e^* = 0.27m_0$ [103, 114], T is the absolute temperature, $\Phi_{B.eff}$ is the effective barrier height at zero bias, e is the elementary charge, V_a is the applied voltage, R_S is the series resistance, n is the ideality factor of the diode and k_B is Boltzmann's constant. The effective Schottky barrier height was determined to be $\Phi_{B.eff} = 0.72\,eV$ and the ideality factor was 1.42. For ZnO thin films and single crystals similar values have been reported for Ag as a Schottky contact metal: a barrier height of 0.56–0.84 eV [115, 116] and ideality factors between 1.33 and 1.5 were obtained from I–V and C–V (capacitance–voltage) measurements [115, 117].

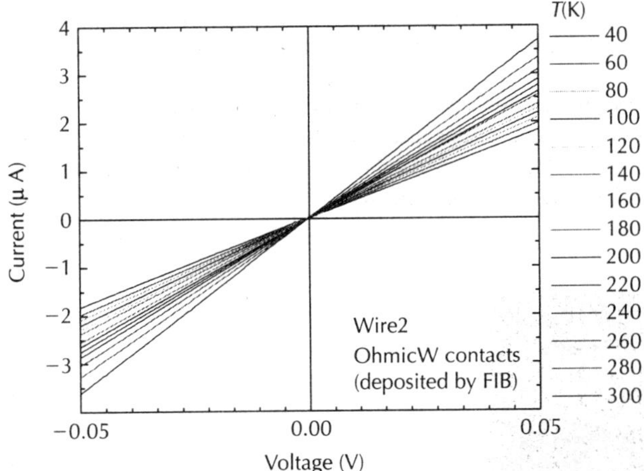

Figure 9.24   Temperature-dependent current–voltage characteristics of wire 2. It shows an ohmic behaviour due to the two tungsten contacts which were attached to the microwire with the FIB.

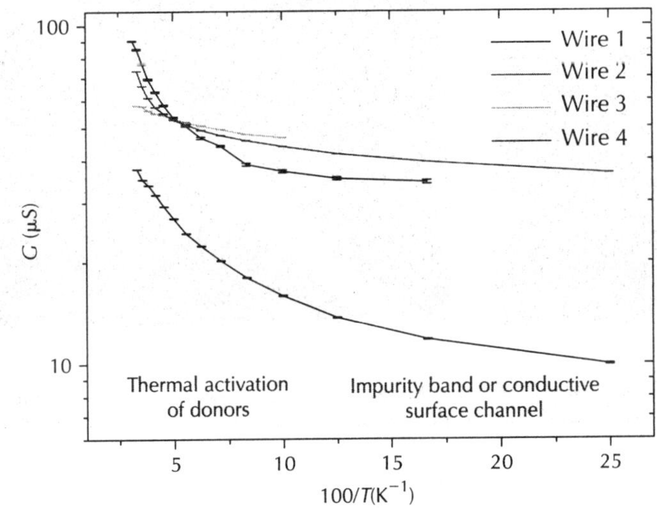

Figure 9.25   Temperature dependence of the conductivity of four different ZnO microwires. At low temperature the conduction is very weakly temperature dependent and apparently governed by a surface channel or an impurity band (see text).

The temperature dependence of ohmic I–V characteristics has been measured for four different microwires in the dark in a closed cycle helium cryostat with a semiconductor parameter analyser (Agilent 4156C). Wires 1 and 2 were prepared with two FIB-deposited tungsten contacts and wires 3 and 4 with two InGa contacts each. Figure 9.24 shows the result of the temperature range from 40 up to 300 K for wire number 2. As expected, it shows an ohmic behaviour and the resistance (inverse slope) decreases with increasing temperature. The other three crystals showed qualitatively exactly the same response. This dependence was analysed in more detail by linear fitting of the *I–V* curves. Figure 9.25 shows the Arrhenius plot of the corresponding conductivities $G = 1/R$. The very small vertical bars in the plot indicate the error obtained from the linear fitting of the slope of the *I–V* curves. In the temperature regime from 200 to 300 K all samples show a strongly temperature dependent conduction. At low temperatures, i.e. in the freeze-out range, the conductivity is almost temperature independent, which is especially pronounced for wires 2 and 4. This can be attributed either to a conductive surface channel or to an impurity band conduction process. In order to provide the fundamentals for nanowire device applications, it is essential to study the crucial electrical parameters, like, for example, the carrier concentration (defined by the doping level) and carrier mobility. However, to do so independently, experiments such as temperature-dependent Hall measurements need to be performed which are connected with obvious difficulties for micro- and nanosized structures. But also from the present data the carrier concentration can be estimated.

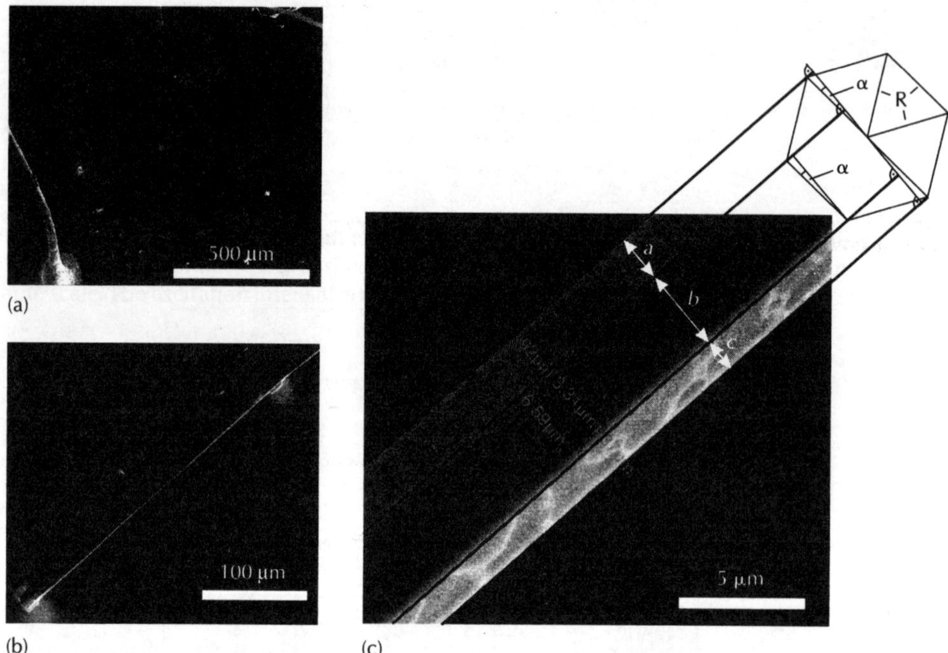

Figure 9.26   *(a), (b) SEM pictures of a carbothermally grown ZnO microwire with two tungsten contacts at different magnifications. (c) illustrates the geometrical considerations for the determination of* R *which account for the tilt of the crystal.*

The specific conductivity for electron and hole conduction can be written as [103]:

$$\sigma = -en\mu_n + ep\mu_p \tag{9.6}$$

where e is the electron charge, n and p are the electron and hole densities, respectively, and μ_i is the corresponding charge carrier mobility. Since ZnO is intrinsically *n*-conducting [118], the *p*-conducting part is negligible here and the electron density can be estimated by:

$$n \approx -\frac{\sigma}{e\mu_n} = \frac{IG}{Ae\mu_n} \tag{9.7}$$

where *l* is the length and A is the area of the hexagonal cross-section. It is given by:

$$A = \frac{3\sqrt{3}}{2} R^2 \tag{9.8}$$

where R is the radius of the circumscribing circle as indicated in Fig. 9.26. The length, as well as the projected distances between the edges of the hexagons:

$$\begin{aligned} a &= R\cos(60° - \alpha) \\ b &= R\cos\alpha \\ c &= R\sin(30° - \alpha) \end{aligned} \tag{9.9}$$

of the ZnO microwires, can be determined from top-view SEM pictures (see Fig. 9.26). Each pair of those equations can be solved for the tilt angle α by eliminating R:

$$\begin{aligned} \alpha &= \arctan \frac{\sqrt{3}(2a - b)}{3b} \\[6pt] \alpha &= \arctan \frac{\sqrt{3}(a - c)}{3(a + b)} \\[6pt] \alpha &= \arctan \frac{\sqrt{3}(b - 2c)}{3b} \end{aligned} \tag{9.10}$$

Once α is known, R can be calculated:

$$a + b + c = 2R\cos\alpha \Leftrightarrow R = \frac{a + b + c}{2\cos\alpha} \tag{9.11}$$

The mobility of ZnO above 200 K is limited mainly by polar optical phonon scattering [103]. Since the crystal structure of the ZnO micro- and nanowhiskers investigated here is very good, the mobility is not limited by grain boundary scattering, as is usually the case for PLD-grown ZnO thin films [108]. Hence, the temperature dependence of the Hall mobility of commercially available single ZnO bulk crystals grown by seeded chemical vapour deposition (Eagle Picher Inc.) can be taken as a good appoximation for the microwires in the temperature regime above 200 K. The values from sample A in [108] have been used to determine the carrier density according to

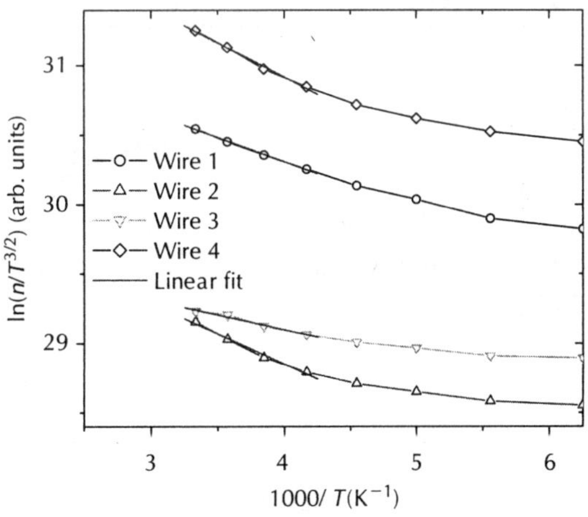

Figure 9.27 *Temperature dependence of the carrier concentration of ZnO microwires. The carrier activation energy for 240–300 K can be extracted from the slope (cf. Eq. 9.12).*

Eq. 9.7. It should be noted that a considerably lower room temperature mobility ($17\,\text{cm}^2/\text{Vs}$) was estimated from the transconductance of nanowire FETs [43]. The room temperature results are summarized in Table 9.2. The carrier concentration varies between $1.22 \times 10^{16}\,\text{cm}^{-3}$ (wire 2) and $1.94 \times 10^{17}\,\text{cm}^{-3}$ (wire 4).

Table 9.2   Tilt angle, length and radius, as well as room temperature conductivity, the estimated carrier concentration and the thermal carrier activation energy of the four microcrystal samples

Wire	α (°)	l (μm)	R (μm)	G (300 K) (S)	n (300 K) (cm^{-3})	ΔE (meV)
1	5.49	234.06	3.31	3.77×10^{-5}	9.58×10^{16}	30.09
2	−0.22	138.60	7.13	7.34×10^{-5}	1.22×10^{16}	37.60
3	−12.79	330.83	9.42	5.80×10^{-5}	2.57×10^{16}	18.41
4	1.47	2240.00	10.80	8.51×10^{-5}	1.94×10^{17}	39.59

The temperature dependence of the electron concentration in the Boltzmann approximation can be written as [103]:

$$n = N_C \exp \frac{E_F - E_C}{k_B T} \tag{9.12}$$

where E_F is the Fermi energy, E_C is the conduction band energy and N_C is the conduction band edge density of states which is proportional to $T^{3/2}$. Hence, the thermal activation energy $E_F - E_C$ can be determined from the slope in a $\ln(n/T^{3/2})$ vs $1/T$ plot. The results are shown in Fig. 9.27 with n determined by Eq. 9.7 using again the temperature dependent mobility from [108]. The activation energies were determined from the linear fit of the range from 240 to 300 K and are summarized in Table 9.2. The values are between 18 and 40 meV. It was reported for undoped ZnO that native defects of zinc interstitials contribute to the n-type semiconducting behaviour, and these defects serve as shallow donors with a binding energy of 30–60 meV [16]. Other studies based on PL spectroscopy, temperature-dependent Hall experiments or electrical conductivity measurements report on typical activation energies between 11 and 53 meV for nominally undoped ZnO [119–121]. Hence, the microwires seem to have similar contact properties and the conduction mechanism is, apart from surface effects, comparable to bulk ZnO.

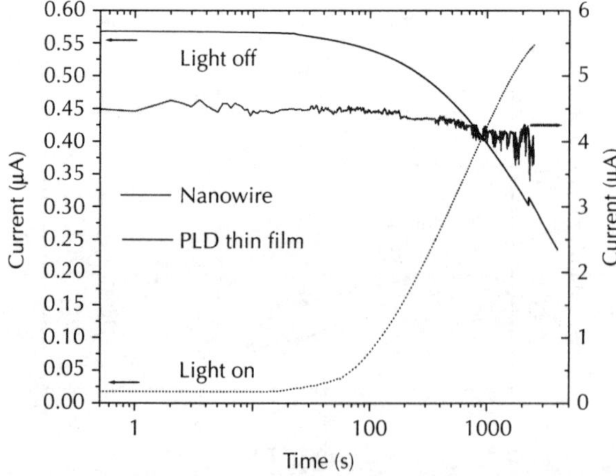

Figure 9.28   Time dependence of the current through a nanowire. As a comparison the persistent photocurrent of a PLD thin film is shown (black solid line) as well as the photoresponse (black dotted line).

Figure 9.28 shows the time dependence of the current of a ZnO nanowire with two ohmic contacts under constant bias (red curve) measured at the wafer–prober in comparison with that of a PLD thin film (black solid curve). The samples were exposed several days to ambient light prior to the experiments and at $t = 10\,\text{s}$ the wafer–prober was completely darkened. Unlike the thin film, the nanowire exhibits almost time-independent conductivity. For the thin film the inverse effect (switching on the light) can also be observed (dotted black line). ZnO is known to exhibit persistent n-type photoconductivity [118] and a slow photoresponse [122]. Typically, the photosensitivity factor [120]

$$S = \frac{\sigma_{\text{photo}} - \sigma_{\text{dark}}}{\sigma_{\text{dark}}} \tag{9.13}$$

is in the order of 10–100 and in some cases even values above 10^4 have been reported [120] with exponential decay constants in the 10 s range. This effect is commonly attributed to oxygen vacancies [118] but the exact mechanism is still under debate since computational studies [17, 123] have cast doubt on the categorization of oxygen vacancies as shallow donors. Already in the 1950s, oxygen adsorption at the surface was reported to play an important role for time-dependent photoconductivity of ZnO [124]. In this model, adsorbed oxygen molecules can bind one electron each, the surface is therefore charged negatively, a depletion layer forms and the conductivity is reduced. Upon illumination the depletion layer vanishes. Also the free carrier concentration as well as the degree of compensation have a strong influence [125]. Due to the larger surface-to-volume ratio and good crystallinity, surface effects dominate the photoconduction transient in nanowires [122]. For instance, Li *et al.* have shown that the transient photocurrent decay time depends strongly on the ambient atmosphere. This is also the reason why ZnO nanowires exhibit good gas sensing properties.

9.6 Conclusion

In summary, ZnO nanowires have been grown by high-pressure PLD on sapphire substrates covered with gold colloidal particles as nucleation sites. A detailed study of the growth evolution and of the nanowire size and length distribution was presented. It was found that the aspect ratio varies linearly with the deposition time.

We investigated the resonant optical behaviour of ZnO micro- and nanocrystals, giving detailed insight into optical whispering gallery modes within the micro *and* nanometer regime. In particular, we analysed hexagonal WGMs in nanocavities for mode numbers in the range from $N = 1$ to $N = 6$. The energy shift and broadening of the modes for cavity diameter decreasing to zero is well described without free parameter by Eq. 9.1 from a plane wave model. Furthermore, we have found RT stimulated emission in a ZnO microcrystal grown by carbothermal evaporation.

Moreover, rectifying as well as ohmic contacts attached to single microwires were characterized. We find that the conductivity is strongly temperature dependent in the range from 200 to 300 K. From this, carrier activation energies were determined by taking the sample geometry into account. The values are between 18 and 40 meV while the estimated carrier concentration is around $10^{16}\,\text{cm}^{-3}$ at room temperature.

Acknowledgements

This work was supported by Deutsche Forschungsgemeinschaft within FOR 522 (Project Gr 1011/12-1), by the European Network of Excellence SANDIE (NoE 500101) and by the EU STREP NANDOS (FP6-016924).

References

1. C. Klingshirn, M. Grundmann, A. Hoffmann, B. Meyer, and A. Waag, *Physik-Journal* **1**, 33 (2006).
2. D.C. Reynolds, D.C. Look, and B. Jogai, *Solid State Comm.* **99**, 873 (1996).

3. D.M. Bagnall, Y.F. Chen, Z. Zhu, T. Yao, S. Koyama, M.Y. Shen, and T. Goto, *Appl. Phys. Lett.* **70**, 2230 (1997).
4. Y. Segawa, A. Ohtomo, M. Kawasaki, H. Koinuma, Z. Tang, P. Yu, and G. Wong, *Physica Status Solidi (b)* **202**, 669 (1997).
5. K. Maeda, M. Sato, I. Niikura, and T. Fukuda, *Semi. Sci. Tech.* **20**, S49 (2005).
6. S.C. Minne, S.R. Manalis, and C.F. Quate, *Appl. Phys. Lett.* **67**, 3918 (1995).
7. P.M. Verghese and D.R. Clarke, *J. Appl. Phys.* **87**, 4430 (2000).
8. A. Yamamoto, T. Kido, T. Goto, Y. Chen, T. Yao, and A. Kasuya, *Appl. Phys. Lett.* **75**, 469 (1999).
9. A.K. Sharma, J. Narayan, J.F. Muth, C.W. Teng, C. Jin, A. Kvit, R.M. Kolbas, and O.W. Holland, *Appl. Phys. Lett.* **75**, 3327 (1999).
10. A. Ohtomo, M. Kawasaki, T. Koida, K. Masubuchi, H. Koinuma, Y. Sakurai, Y. Yoshida, T. Yasuda, and Y. Segawa, *Appl. Phys. Lett.* **72**, 2466 (1998).
11. T. Makino, Y. Segawa, M. Kawasaki, A. Ohtomo, R. Shiroki, K. Tamura, T. Yasuda, and H. Koinuma, *Appl. Phys. Lett.* **78**, 1237 (2001).
12. T. Maemoto, N. Ichiba, S. Sasa, and M. Inoue, *Thin Solid Films* **486**, 174 (2005).
13. T. Gruber, C. Kirchner, R. Kling, F. Reuss, and A. Waag, *Appl. Phys. Lett.* **84**, 5359 (2004).
14. A. Ohtomo, K. Tamura, M. Kawasaki, T. Makino, Y. Segawa, Z.K. Tang, G.K.L. Wong, Y. Matsumoto, and H. Koinuma, *Appl. Phys. Lett.* **77**, 2204 (2000).
15. T. Makino, C.H. Chia, N.T. Tuan, Y. Segawa, M. Kawasaki, A. Ohtomo, K. Tamura, and H. Koinuma, *Appl. Phys. Lett.* **77**, 1632 (2000).
16. D.C. Look, J.W. Hemsky, and J.R. Sizelove, *Phys. Rev. Lett.* **82**, 2552 (1999).
17. A. Janotti and C.G.V. de Walle, *Appl. Phys. Lett.* **87**, 122102 (2005).
18. A. Tsukazaki, A. Ohtomo, T. Onuma, M. Ohtani, T. Makino, M. Sumiya, K. Ohtani, S.F. Chichibu, S. Fuke, Y. Segawa, H. Ohno, H. Koinuma, and M. Kawasaki, *Nature Materials* **4**, 42 (2005).
19. A. Tsukazaki, M. Kubota, A. Ohtomo, T. Onuma, K. Ohtani, H. Ohno, S.F. Chichibu, and M. Kawasaki, Blue light-emitting diode based on ZnO, *Jpn. J. Appl. Phys.* **44**, L643 (2005).
20. R. Konenkamp, R. Word, and M. Godinez, *Nano Lett.* **5**, 2005 (2005).
21. D.D. Awschalom, D. Loss, N. Samarth, eds, *semiconductor Spintronics and Quantum Computation* (Springer Verlag Berlin, Heidelberg – New York, 2002).
22. T. Dietl, H. Ohno, F. Matsukura, J. Cibert, and D. Ferrand, *Science* **287**, 1019 (2000).
23. S.A. Wolf, D.D. Awschalom, R.A. Buhrman, J.M. Daughton, S. von Molnar, M.L. Roukes, A.Y. Chtchelkanova, and D.M. Treger, *Science* **294**, 1488 (2001).
24. M. Diaconu, H. Schmidt, A. Poppl, R. Bottcher, J. Hoentsch, A. Klunker, D. Spemann, H. Hochmuth, M. Lorenz, and M. Grundmann, *Phys. Rev. B (Cond. Mat. Mat. Phys.)* **72**, 085214 (2005).
25. Z.L. Wang, *J. Phys.: Cond. Matt.* **16**, R829 (2004).
26. P. Hu, Y. Liu, L. Fu, X. Wang, and D. Zhu, *Appl. Phys. A: Mat. Sci. Proc.* **V80**, 35 (2005).
27. H. Yan, R. He, J. Pham, and P. Yang, *Adv. Mat.* **15**, 402 (2003).
28. W. Park, G.-C. Yi, M. Kim, and S. Pennycook, *Adv. Mat.* **14**, 1841 (2002).
29. X. Gao, X. Li, and W. Yu, *J. Phys. Chem. B.* **109**, 1155 (2005).
30. M.G. Banks, *Scientometrics* **V69**, 161 (2006).
31. B. Dume, http://physicsweb.org/articles/news/10/5/4/1, 2006.
32. J.E. Hirsch, *Proc. Nat. Acad. Sci.* **102**, 16569 (2005).
33. R.S. Wagner and W.C. Ellis, *Appl. Phys. Lett.* **4**, 89 (1964).
34. G.-C. Yi, C. Wang, and W.I. Park, *Semicond. Sci. Techn.* **20**, S22 (2005).
35. Z.L. Wang, *Mat. Today* **7**, 26 (2004).
36. T. Nobis, E.M. Kaidashev, A. Rahm, M. Lorenz, and M. Grundmann, *Phys. Rev. Lett.* **93**, 103903 (2004).
37. J. Johnson, H. Yan, R. Schaller, L. Haber, R. Saykally, and P. Yang, *J. Phys. Chem. B.* **105**, 11387 (2001).
38. W.L. Hughes and Z.L. Wang, *Appl. Phys. Lett.* **82**, 2886 (2003).
39. Z.L. Wang and J. Song, *Science* **312**, 242 (2006).
40. W. Park, J. Kim, G.-C. Yi, and H.-J. Lee, *Adv. Mat.* **17**, 1393 (2005).
41. M. Arnold, P. Avouris, Z. Pan, and Z. Wang, *J. Phys. Chem. B.* **107**, 659 (2003).

42. S.H. Jo, J.Y. Lao, Z.F. Ren, R.A. Farrer, T. Baldacchini, and J.T. Fourkas, *Appl. Phys. Lett.* **83**, 4821 (2003).

43. Z. Fan, D. Wang, P.-C. Chang, W.-Y. Tseng, and J.G. Lu, *Appl. Phys. Lett.* **85**, 5923 (2004).

44. Q.H. Li, Y.X. Liang, Q. Wan, and T.H. Wang, *Appl. Phys. Lett.* **85**, 6389 (2004).

45. Q. Wan, Q.H. Li, Y.J. Chen, T.H. Wang, X.L. He, X.G. Gao, and J.P. Li, *Appl. Phys. Lett.* **84**, 3085 (2004).

46. Z. Fan and J.G. Lu, *Appl. Phys. Lett.* **86**, 123510 (2005).

47. T. Yatsui, J. Lim, M. Ohtsu, S.J. An, and G.-C. Yi, *Appl. Phys. Lett.* **85**, 727 (2004).

48. W. Park, G.-C. Yi, M. Kim, and S. Pennycook, *Adv. Mat.* **15**, 526 (2003).

49. J.-J. Wu, S.-C. Liu, and M.-H. Yang, *Appl. Phys. Lett.* **85**, 1027 (2004).

50. H. Fan, W. Lee, R. Hauschild, M. Alexe, G. Le Rhun, R. Scholz, A. Dadgar, K. Nielsch, H. Kalt, A. Krost, M. Zacharias, and U. Gösele, *Small* **2**, 561 (2006).

51. M. Lorenz, E.M. Kaidashev, A. Rahm, T. Nobis, J. Lenzner, G. Wagner, D. Spemann, H. Hochmuth, and M. Grundmann, *Appl. Phys. Lett.* **86**, 143113 (2005).

52. M. Lorenz, Pulsed laser deposition of ZnO-based thin films, in: *Zinc Oxide as Transparent Electronic Material and its Application in Thin Film Solar Cells*, A.K.K. Ellmerm, ed. (Springer, 2006).

53. R. Wagner, VLS mechanism of crystal growth. In: *Whisker Technology* (Wiley-Interscience, New York, 1970), Chap. 3, pp. 47–120.

54. D.P. Yu, C.S. Lee, I. Bello, X.S. Sun, Y.H. Tang, G.W. Zhou, Z.G. Bai, Z. Zhang, and S.Q. Feng, *Solid State Comm.* **105**, 403 (1998).

55. A.M. Morales and C.M. Lieber, *Science* **279**, 208 (1998).

56. K. Hiruma, M. Yazawa, T. Katsuyama, K. Ogawa, K. Haraguchi, M. Koguchi, and H. Kakibayashi, *J. Appl. Phys.* **77**, 447 (1995).

57. Y. Wu, H. Yan, and P. Yang, *Topics in Catalysis* **V19**, 197 (2002).

58. E.I. Givargizov, *J. Crys. Growth* **31**, 20 (1975).

59. W. Seifert, M. Borgstrom, K. Deppert, K.A. Dick, J. Johansson, M.W. Larsson, T. Martensson, N. Skold, C. Patrik T. Svensson, and B.A. Wacaser, *J. Crys. Growth* **272**, 211 (2004).

60. T.Y. Tan, N. Li, and U. Gosele, *Appl. Phys. Lett.* **83**, 1199 (2003).

61. M.H. Huang, Y. Wu, H. Feick, N. Tran, E. Weber, and P. Yang, *Adv. Mat.* **13**, 113 (2001).

62. K. Dick, K. Deppert, T. Martensson, B. Mandl, L. Samuelson, and W. Seifert, *Nano Lett.* **5**, 761 (2005).

63. S.D. Sharma and S.C. Kashyap, *J. Appl. Phys.* **42**, 5302 (1971).

64. T. Matsushita, K. Kodaira, J. Saito, and R. Yoshida, *J. Crys. Growth* **26**, 147 (1974).

65. H. Iwanaga, N. Shibata, O. Nittono, and M. Kasuga, *J. Crys. Growth* **45**, 228 (1978).

66. M. Wei, D. Zhi, and J.L. MacManus-Driscoll, *Nanotechnology* **16**, 1364 (2005).

67. S.C. Lyu, Y. Zhang, H. Ruh, H.-J. Lee, H.-W. Shim, E.-K. Suh, and C.J. Lee, *Chem. Phys. Lett.* **363**, 134 (2002).

68. P. Gao, Y. Ding, and Z. Wang, *Nano Lett.* **3**, 1315 (2003).

69. S.Y. Li, C.Y. Lee, and T.Y. Tseng, *J. Crys. Growth* **247**, 357 (2003).

70. J. Liu, Z. Zhang, X. Su, and Y. Zhao, *J. Phys. D: Appl. Phys.* **38**, 1068 (2005).

71. Z.C. Tu, Q.X. Li, and X. Hu, *Phys. Rev. B (Cond. Mat. Mat. Phys.)* **73**, 115402 (2006).

72. J.Y. Bae, J. Yoo, and G.-C. Yi, *Appl. Phys. Lett.* **89**, 173114 (2006).

73. C. Kim, W.I. Park, G.-C. Yi, and M. Kim, *Appl. Phys. Lett.* **89**, 113106 (2006).

74. F.Z. Wang, Z.Z. Ye, D.W. Ma, L.P. Zhu, F. Zhuge, and H.P. He, *Appl. Phys. Lett.* **87**, 143101 (2005).

75. L.W. Yang, X.L. Wu, T. Qiu, G.G. Siu, and P.K. Chu, *J. Appl. Phys.* **99**, 074303 (2006).

76. A. Rahm, E.M. Kaidashev, H. Schmidt, M. Diaconu, A. Pöppl, R. Böttcher, C. Meinecke, T. Butz, M. Lorenz, and M. Grundmann, *Microchimica Acta.* **156**, 21 (2007).

77. C.N.R. Rao, F.L. Deepak, G. Gundiah, and A. Govindaraj, *Prog. Solid State Chem.* **31**, 5 (2003).

78. B.D. Yao, Y.F. Chan, and N. Wang, *Appl. Phys. Lett.* **81**, 757 (2002).

79. A. Rahm, G. Yang, M. Lorenz, T. Nobis, J. Lenzner, G. Wagner, and M. Grundmann, *Thin Solid Films* **486**, 191 (2005).

80. M. Lorenz, J. Lenzner, E. Kaidashev, H. Hochmuth, and M. Grundmann, *Ann. Phys.* **13**, 39 (2004).

81. H.M. Smith and A.F. Turner, *Appl. Pot.* **4**, 147 (1965).

82. H. Sankur and J.T. Cheung, *J. Vac. Sci. Tech. A: Vac. Surf. Films* **1**, 1806 (1983).

83. K.L. Saenger, Angular distribution of ablated material. In: *Pulsed Laser Deposition of Thin Films* (John Wiley & Sons, New York, 1994), Chap. 7, pp. 199–228.

84. P.E. Dyer, A. Issa, and P.H. Key, *Appl. Phys. Lett.* **57**, 186 (1990).

85. L. Chen, Particulates generated by pulsed laser ablation. In: *Pulsed Laser Deposition of Thin Films* (John Wiley & Sons, New York, 1994), Chap. 6, pp. 167–198.

86. S. Choopun, R.D. Vispute, W. Yang, R.P. Sharma, T. Venkatesan, and H. Shen, *Appl. Phys. Lett.* **80**, 1529 (2002).

87. J.B. Baxter, F. Wu, and E.S. Aydil, *Appl. Phys. Lett.* **83**, 3797 (2003).

88. B.K. Meyer, H. Alves, D.M. Hofmann, W. Kriegseis, D. Forster, F. Bertram, J. Christen, A. Hoffmann, M. Straßburg, M. Dworzak, U. Haboeck, and A.V. Rodina, *Physica Status Solidi (b)* **241**, 231 (2004).

89. E.M. Kaidashev, M. Lorenz, H. von Wenckstern, A. Rahm, H.-C. Semmelhack, K.-H. Han, G. Benndorf, C. Bundesmann, H. Hochmuth, and M. Grundmann, *Appl. Phys. Lett.* **82**, 3901 (2003).

90. R. Schmidt, B. Rheinlander, M. Schubert, D. Spemann, T. Butz, J. Lenzner, E.M. Kaidashev, M. Lorenz, A. Rahm, H.C. Semmelhack, and M. Grundmann, *Appl. Phys. Lett.* **82**, 2260 (2003).

91. J.D. Jackson, *Classical Electrodynamics*, (Wiley, New York, 1999).

92. J.U. Nöckel, 2-d microcavities: theory and experiments, in: *Cavity-Enhanced Spectroscopies*, edited by R.D. van Zee and J.P. Looney (Academic Press, San Diego, 2002) pp. 185–226.

93. N.S. Kapany and J.J. Burke. *Optical Waveguides*, (Academic Press, New York and London, 1972).

94. S.-B. Lee, J.H. Lee, J.S. Chang, H.J. Moon, S.W. Kim, and K. An, *Phys. Rev. Lett.* **88**, 033901–033903 (2002).

95. C. Gmachl, F. Capasso, E.E. Narimanov, J.U. Nöckel, A.D. Stone, J. Faist, D.L. Sivco, and A.Y. Cho, *Science* **280**, 1556 (1998).

96. M. Lohmeyer, *Opt. Quant. Electron.* **34**, 541 (2002).

97. J. Wiersig, *Phys. Rev. A.* **67**, 023807 (2003).

98. J. Wiersig, *J. Opt. A.* **5**, 53 (2003).

99. J. Christen, M. Grundmann, and D. Bimberg, *J. Vac. Sci. Technol. B.* **4**, 2358–2368, (1991).

100. Th. Nobis, E.M. Kaidashev, A. Rahm, M. Lorenz, J. Lenzner, and M. Grundmann, *Nano Lett.* **4**, 797 (2004).

101. I. Braun, G. Ihlein, F. Laeri, J.U. Nöckel, G. Schulz-Ekloff, F. Schüth, U. Vietze, Ö. Weiss, and D. Wöhrle, *Appl. Phys. B.* **70**, 335 (2000).

102. C. Klingshirn, *Phys. Status Solidi B.* **71**, 547 (1975).

103. M. Grundmann. *The Physics of Semiconductors*, (Springer Verlag, 2006).

104. J. Cui, S. Sadofev, S. Blumstengel, J. Puls, and F. Henneberger, *Appl. Phys. Lett.* **89**, 051108 (2006).

105. Y. Chen, Y. Segawa, N.T. Tuan, S.-K. Hong, and T. Yao, *Appl. Phys. Lett.* **78**, 11 (2001).

106. R. Schmidt-Grund, N. Ashkenov, M. Schubert, W. Czakai, D. Faltermeier, G. Benndorf, H. Hochmuth, M. Lorenz, J. Bläsig, M. Grundmann, ICPS Poster 2006.

107. C.A. Mead, *Phys. Lett.* **18**, 218 (1965).

108. H. von Wenckstern, S. Weinhold, G. Biehne, R. Pickenhain, H. Schmidt, H. Hochmuth, M. Grundmann, donor levels in ZnO in: *Advances in Solid State Physics*, vol. 45 (Springer Berlin, 2005).

109. C.H. Liu, W.C. Yiu, F.C.K. Au, J.X. Ding, C.S. Lee, and S.T. Lee, *Appl. Phys. Lett.* **83**, 3168 (2003).

110. Q.H. Li, Q. Wan, Y.X. Liang, and T.H. Wang, *Appl. Phys. Lett.* **84**, 4556 (2004).

111. J.Y. Park, H. Oh, J.-J. Kim, and S.S. Kim, *Nanotechnology* **17**, 1255 (2006).

112. W. Park and G.-C. Yi, *Adv. Mat.* **16**, 87 (2004).

113. S. Kurtin, T.C. McGill, and C.A. Mead, *Phys. Rev. Lett.* **22**, 1433 (1969).

114. H. von Wenckstern, G. Biehne, R.A. Rahman, H. Hochmuth, M. Lorenz, and M. Grundmann, *Appl. Phys. Lett.* **88**, 092102 (2006).

115. H. von Wenckstern, E.M. Kaidashev, M. Lorenz, H. Hochmuth, G. Biehne, J. Lenzner, V. Gottschalch, R. Pickenhain, and M. Grundmann, *Appl. Phys. Lett.* **84**, 79 (2004).

116. S. Liang, H. Sheng, Y. Liu, Z. Huo, Y. Lu, and H. Shen, *J. Crys. Growth* **225**, 110 (2001).

117. H. Sheng, S. Muthukumar, N.W. Emanetoglu, and Y. Lu, *Appl. Phys. Lett.* **80**, 2132 (2002).

118. B. Claflin, D. Look, S. Park, and G. Cantwell, *J. Crys. Growth* **287**, 16 (2006).

119. E. Ziegler, A. Heinrich, H. Oppermann, and G. Stöver, *Phys. Status Solidi (a)* **66**, 635 (1981).

120. A.E. Jimenez-Gonzalez, J.A. Soto Urueta, and R. Suarez-Parra, *J. Crys. Growth* **192**, 430 (1998).

121. K. Thonke, T. Gruber, N. Teofilov, R. Schonfelder, A. Waag, and R. Sauer, *Phys. B: Cond. Mat.* **308–310**, 945 (2001).

122. Q.H. Li, T. Gao, Y.G. Wang, and T.H. Wang, *Appl. Phys. Lett.* **86**, 123117 (2005).

123. S.B. Zhang, S.-H. Wei, and A. Zunger, *Phys. Rev. B.* **63**, 075205 (2001).

124. D.A. Melnick, *J. Chem. Phys.* **26**, 1136 (1957).

125. Y. Li, H.-S. Kim, J.-M. Erie, F. Ren, S.J. Pearton, and D.P. Norton, *SPIE*, vol. **6337**, 633708 (2006).

CHAPTER 10

Miniband-related 1.4–1.8 μm Luminescence of Ge/Si Quantum Dot Superlattices

V.G. Talalaev,[1,2,3] G.E. Cirlin,[1,4] A.A. Tonkikh,[1,4] N.D. Zakharov,[1] P. Werner,[1] U. Gösele,[1] J.W. Tomm,[2] and T. Elsaesser[2]

[1]Max-Planck-Institut für Mikrostrukturphysik, Weinberg 2, 06120 Halle/Saale, Germany;
[2]Max-Born-Institut für Nichtlineare Optik und Kurzzeitspektroskopie,
Max-Born-Strasse 2A, 12489 Berlin, Germany;
[3]V.A. Fock Institute of Physics, St Petersburg State University, Ulyanovskaya 1,
198504 Petrodvorets, St Petersburg, Russia;
[4]Ioffe Physico-Technical Institute RAS, 194021 Polytekhnicheskaya 26, St Petersburg, Russia

10.1 Introduction

The development of efficient silicon-based light-emitting devices is a challenging task in modern semiconductor physics and optoelectronics. Due to the indirect nature of the band gap, bulk silicon has a poor luminescence. One of the promising approaches to increase the luminescence efficiency from Si-based materials is to apply a concept of nanostructures based on Ge inclusions into an Si matrix. The development of epitaxial methods enabled the growth of ultra-thin Ge/Si layer superlattices. Structures that are based on the Brillouin folding concept can result in quasi-direct optical transitions near $1.55\,\mu m$ wavelength [1–4]. However, due to the small localization potential the photoluminescence (PL) from Ge/Si layer superlattices is observed only at low temperatures [3–6]. The concept of a quantum-cascade Ge/Si-based laser has also been demonstrated. However, it emits only in the mid-infrared region and shows electroluminescence at low temperatures [7].

Several attempts have been made to obtain stronger localization and to achieve a room temperature luminescence at $1.55\,\mu m$ (0.8 eV) in Ge quantum dots (QDs) embedded into an Si matrix. Although arrays of self-assembled Ge islands grown by the Stranski–Krastanov mode on silicon have been studied quite intensively, just a few research teams have reported about PL [8–9] and electroluminescence (EL) data [11–14]. Probably, due to the low emission intensity, the *internal* quantum efficiency (QE) measured at room temperature for a Ge/Si light-emitting diode (LED) at $1.42\,\mu m$ has been reported only in two studies with the following results: $5 \times 10^{-4}\%$ [12] and 0.015% [14]. In one case [14] an *external* QE of $3.4 \times 10^{-4}\%$ was measured for EL close to $1.5\,\mu m$. It should be mentioned that all publications mostly concentrate on QDs, i.e. on the hole subsystem. It is well known [15] that Ge/Si heterostructures have a type-II band alignment, in which both carriers have opposite locations in relation to the heterointerface: electrons in Si and holes in Ge. Due to the small overlap of electron and hole wavefunctions (not more than 15%) the oscillator strength of indirect excitons in a Ge/Si heterostructure is quite low [16]. This is one of the key factors for the weak near-infrared luminescence of Ge/Si heterostructures.

In this work a novel approach is presented, which overcomes the disadvantages mentioned above and allows efficient luminescence at room temperature from Ge/Si heterostructures to be achieved. Special regimes of molecular beam epitaxy (MBE) with Sb-doping enabled the growth of highly strained dislocation-free Ge/Si quantum dot multilayer structures. Our samples consist of stacked QD arrays, in which the columns of Ge QDs are well correlated vertically. We characterize such structures as Ge/Si QD superlattices (QDSLs), because there exists a confinement of holes in the Ge QDs, as well as a confinement for electrons in the Si spacer layers. The electron states in the Si layers, which behave as real quantum wells (QWs), are characterized by activation energies in the range of 45 to 85 meV, i.e. are stable up to room temperature. The optimization of QD columns have provided conditions, which are favourable for vertical electron tunnelling and for the formation of a conduction miniband. The different investigations presented in the following give strong indications that a conversion to quasi-direct excitons occurs. This conversion enables an *external* QE of 0.04% for 1.55 µm EL maximum at room temperature.

10.2   Experimental techniques

The Ge/Si structures are grown on Si(001) 5-inch substrates using the MBE set-up Riber SIVA 45. Samples consist of an undoped Si buffer layer with 100 nm thickness, a Ge/Si multilayer structure and a 50 nm Si cap layer. The growth temperature is 600°C. The growth rates of Si and Ge are 1.0 and 0.2 Ås^{-1}, respectively. As the nominal thickness of the single Ge layers values between 0.7 nm and 0.9 nm are chosen. The MBE growth of the heterostructures is monitored *in situ* by a reflection high-energy electron diffraction system (RHEED).

Due to the difference in the lattice parameters between the bulk of Ge and of Si ($\sim$4%) the initial stage of the Ge deposition is accompanied by the formation of nano-size islands known in the literature as Stranski–Krastanov QDs. The formation of Ge nano-islands in all layers of each sample is evident from RHEED by observing the spot patterns. These measurements document the transition from the 2D growth mode of layers to the 3D growth of islands (QDs). The effective thickness of an Si spacer is varied in the range 5.5–9.5 nm. A doping of the Si spacer layer by antimony is performed in the following way: a growth interruption after Ge QD formation is used. During the growth interruption the Ge QD array is exposed to the Sb flux. A Sb deposition rate was 2×10^{-4} nms^{-1}. The time of the exposition is tuned between 0 and 40 s. After exposition the first part of the Si spacer layer (2 nm) is also doped by Sb, whereas the remaining part of the spacer layer is kept undoped. Sb concentration profiling in QDSLs is carried out by secondary ion mass spectroscopy (SIMS) using a Cameca set-up. In order to optimize the structure for getting the most effective luminescence at 1.55 µm the Ge and Si thickness as well as the time of Sb exposition have to be optimized [17].

Two types of boron-doped Si substrates are used: p-type ($\rho \sim 5\,\Omega$cm) for PL investigations and p^+-type ($\rho \sim 0.015\,\Omega$cm) for EL measurements and LED fabrication. The capping layer in PL structures is undoped, in LED structures it is doped with Sb. SIMS data for optimized QDSLs show that due to diffusion of Sb atoms to the growth surface, both structures have an increased Sb concentration in the cap layer: 2×10^{17} cm^{-3} for PL and 2×10^{18} cm^{-3} for LED structures (Fig. 10.1c). Thus we deal with QDSLs embedded into a p–n-junction (PL structures) and into a p^+–n^+-junction (LED structures). Ohmic contact to the cap layer is formed using Al/Au deposited by magnetron sputtering. Indium is used to form the backside metal contact. The LED is glued onto a copper heat-sink and the top contacts are bonded with gold wires using standard thermo-compression bonding.

The structure of the grown samples is investigated by transmission electron microscopy (TEM) and selected area electron diffractometry using JEM 4010 and CM 20 microscopes operating at acceleration voltages 400 kV and 200 kV, respectively. The Ge concentration profile is determined by applying an image analysing technique of bright-field cross-section TEM micrographs recorded on a slow scan charge-coupled-device (CCD) camera to keep the linearity between electron intensity and contrast.

Raman spectra are recorded in backscattering geometry at ambient temperature. The Raman scattered light is collected by a microscope and subsequently analysed by a Dilor X-Y triple spectrometer equipped with a liquid nitrogen-cooled Si CCD camera. The 488 nm line of an Ar^+ laser is used for excitation with a typical power below 1 mW. The spectra are taken for incident and scattering light polarized both parallel or perpendicular to each other. The excitation laser spot has a diameter of 1 μm. The spectral resolution is $0.5\,cm^{-1}$.

Steady-state (cw) PL measurements are carried out in a standard lock-in configuration. The PL spectra are excited by the 488 nm line of an Ar^+ laser. The laser spot on the sample has a diameter of 1.5 mm. For studying the PL intensity versus excitation power density the laser beam is focused down to a 100 μm diameter and adjusted by neutral filters. For low-temperature PL measurements the samples are cooled in a continuous-flow He cryostat. The PL signal is collected by mirror optics and dispersed by a single 0.5 m grating monochromator coupled with a liquid nitrogen-cooled Ge detector (Edinburgh Instruments) having the photoelectric threshold at 1.7 μm (0.73 eV). In order to register luminescence at longer wavelengths the monochromator exit is equipped with a liquid nitrogen-cooled InGaAs photodiode G7754-01 (Hamamatsu) having a cut-off wavelength of 2.4 μm (0.5 eV). PL spectra are always normalized to the spectral photo-detector sensitivity. EL spectra are obtained for in-plane and edge-emitting geometries in the constant current mode using the same set-up. The diode chip is mounted on a water-cooled pedestal in order to prevent EL degradation by heating of the active zone during electrical excitation.

EL spectra, current–voltage characteristics and photo-voltage saturation measurements are performed using a Keithley 2400 source measure unit. All temperature dependent PL measurements are carried out in the short-circuit regime of the p–n-junction.

Time-resolved PL (TRPL) measurements are performed in an He-closed-cycle cryostat at 15 K. The excitation wavelength is 395 nm, the pulse width is 1 ps, and the repetition rate is 1 kHz. These excitation pulses are generated by a system consisting of a Spectra Physics Tsunami seed laser (mode locked Ti–sapphire, 82 MHz, 100 fs) followed by a Spectra Physics Spitfire regenerative amplifier (Ti–sapphire). An optical BBO crystal is used for the second harmonic generation. The laser beam is focused onto the sample down to a 100 μm diameter. The resulting PL is dispersed by a 0.25 m grating monochromator and detected by a Judson Technologies J16D-M204 Ge detector having a cut-off wavelength of 1.6 μm. The signal is analysed by a digital oscilloscope Agilent Infinium 54833A. The temporal resolution of the total system is limited to ∼10 ns by the response time of the detector–preamplifier (Femto HCA-100) combination. Neutral density filters are used for adjusting the excitation power density.

In order to determine the external efficiency of EL from Ge/Si QDSLs at 1.55 μm, we developed an absolute calibration method based on the steady-state PL set-up. An integrating 115 mm diameter sphere is used. The diode is mounted on the entrance port. The exit port has a detachable opal–glass diffuser and is always positioned in the monochromator focal plane. Before these measurements a calibration of photodetector spectral sensitivity is performed using the black-body simulator. Subsequently, a power normalization of the photodetector is carried out using the 1.55 μm line from a calibrated laser diode (0.9 mW).

10.3 Experimental data and interpretation

10.3.1 Structural properties of Ge/Si QDSL

Cross-section TEM images of undoped and Sb-doped multilayer structures are compared in Fig. 10.1a and b. The undoped structure (Fig. 10.1a) shows smeared heterointerfaces and a relatively flat upper surface. The lateral size of the Ge islands increases with the layer number from bottom to the top. The Sb-doped structure (Fig. 10.1b) is characterized by sharp heterointerfaces and a highly corrugated upper surface. The lateral size of the Ge clusters is nearly the same for the top and bottom layers. No misfit dislocations are observed in the samples.

A plan-view TEM image of the Sb-doped 20-layer Ge/Si structure is shown in Fig. 10.2. The Ge clusters look like squares with the edges oriented along <100> crystallographic directions and are assembled into a kind of square lattice. The average base size of the squares is $60 \times 60\,nm^2$. The array surface density is $1.1 \times 10^{10}\,cm^{-2}$. The array is characterized by high uniformity of the Ge cluster sizes. It should be mentioned that every surface cluster is related to a single Ge/Si column.

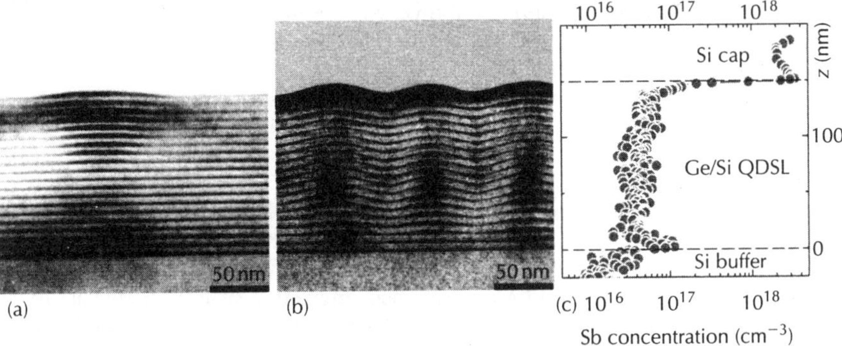

Figure 10.1   TEM cross-section images of Ge/Si heterostructures with 20 Ge layers: (a) undoped; (b) Sb doped. The dark regions correspond to Ge layers. (c) SIMS data of the Sb concentration profile in QDSL similar to (b) embedded into a p^+–n^+-junction. z-growth direction.

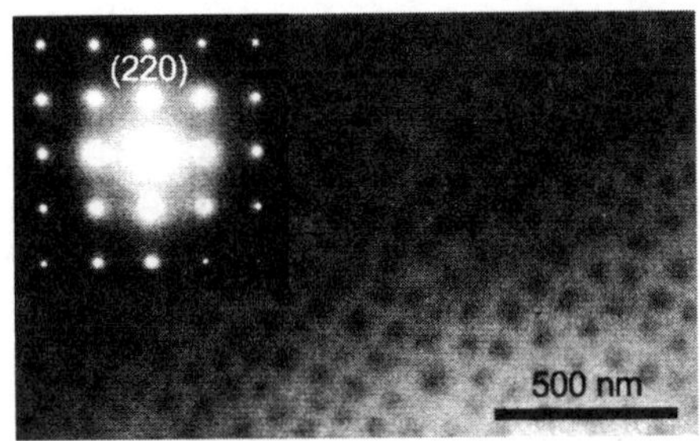

Figure 10.2   [001] TEM plan-view image of the Ge/Si QDSL. The insert shows the electron diffraction pattern taken along [001].

A cross-section TEM image of a Ge/Si multilayer column is shown in detail in Fig. 10.3 (left). Lens-shaped vertically correlated Ge clusters separated by Si spacers are seen. The column is characterized by a high uniformity of the lateral sizes (55 nm ± 5%). It was established that the height of the clusters in the column has a ±10% deviation from the average value ($B = 4.5$ nm). We will follow the traditional name of these clusters as *quantum dots* taking into account that the quantum confinement of the carriers is valid in the growth direction while in the in-plane direction only the carrier localization takes place. The Si spacer thickness in the column (W) has a still smaller deviation (±5%).

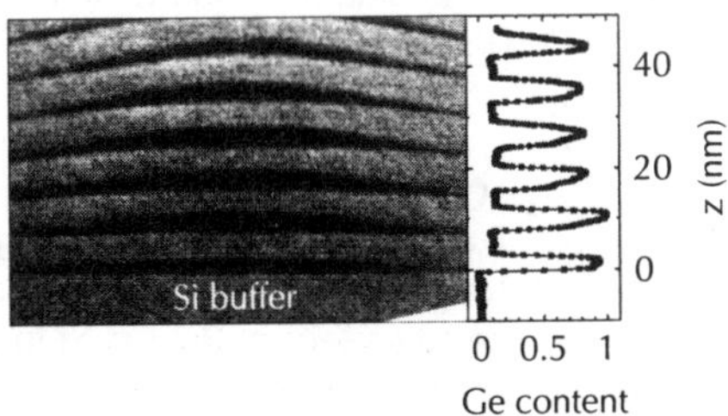

Figure 10.3   Cross-section TEM image of the Ge/Si QDSL. The average height of QD in column B = 4.5 nm (left). Right: Ge content profile along the growth direction z for the same sample.

The Ge content distribution across the QDSL in the growth direction is demonstrated in Fig. 10.3 (right). Ge/Si interfaces are rather sharp. Very little intermixing has taken place and the Ge content in the QDs is $x = 0.8$ (±10%), and in the Si spacer layers it is $x = 0.1$ (±10%). It should be pointed out that all TEM images show a thin Ge wetting layer (WL), which is typical for the QD arrays grown by the Stranski–Krastanov mode. TEM data show that the Sb-doped structures under investigation are mostly defect free.

Figure 10.4 shows Raman scattering spectra for the Sb-doped Ge/Si QDSL with $B = 3.6\,nm$ measured in two different configurations. The quasi-periodical structure in the $20-200\,cm^{-1}$ spectral range, see Fig. 10.4 (bottom), is attributed to folded longitudinal acoustic (FLA) phonons, which have already been reported for Ge/Si QDSLs in [18]. According to the selection rules [19], our geometry, namely $z(x'y')\check{z}$, is not Raman active for FLA phonons in flat Ge/Si layers. Their presence in layers with QDs, however, is explained by symmetry lowering [20]. The high structural quality of the studied QDSLs, in particular the presence of sharp Ge/Si interfaces, is confirmed by the observation of 20 FLA modes. Their intensity distribution appears to be non-monotonous. The observed beats at $7-9\,meV$ and $14\,meV$ are tentatively explained by electron–phonon interaction in the QDSL.

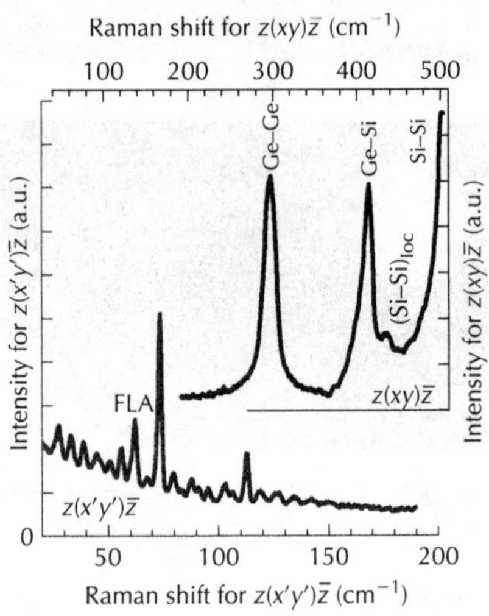

Figure 10.4   Raman spectra of a Ge/Si QDSL measured in two backscattering configurations. The terms z, x, y, x' and y' refer to the [001], [100], [010], [1̄10] and [110] directions, respectively.

Raman spectra in the $200-500\,cm^{-1}$ spectral range, see Fig. 10.4 (top), are recorded in $z(xy)\check{z}$ geometry. According to the selection rules for this geometry all features are assigned to longitudinal optical (LO) phonon modes: Ge–Ge, Ge–Si, (Si–Si)$_{loc}$, and Si–Si. The frequency of the Ge–Ge mode ($297.5\,cm^{-1}$) is below the typical Ge bulk value ($301\,cm^{-1}$) [21]. This shift of the Ge–Ge LO mode in the QDSL is likely to be caused by phonon confinement [22]. It should be noted that in those structures with the thicker Ge QDs, e.g. $B = 5.8\,nm$, the Ge–Ge mode frequency agrees well with the bulk value.

The Ge-Si interface mode at $415\,cm^{-1}$ has an amplitude comparable to that of the Ge–Ge LO-mode. The additional features (Si–Si)$_{loc}$ at 435, 450 and $465\,cm^{-1}$ are attributed to local vibrations (Si–Si)$_{loc}$ under the influence of Ge atoms in their vicinity [23]. These features reflect the presence of a near-range order at the interfaces between the Ge QDs and the adjoining Si layers. Thus, the Raman spectra indicate an ordered surface of the Ge–Si interfaces in Sb-doped QDSLs. This does not contradict the observed sharpness, since this order is still on atomic-scale dimension. Raman spectra of undoped Ge–Si QDSL look quite different: the FLA and Si–Si local modes are absent, and the Ge–Si interface mode is always weaker than the Ge–Ge mode.

10.3.2 *Luminescence properties and initial electronic structure*

The influence of Sb doping on the PL spectrum of Ge/Si QDSLs at room temperature is demonstrated in Fig. 10.5. Undoped structures had always a poor PL – in the low-energy part of the

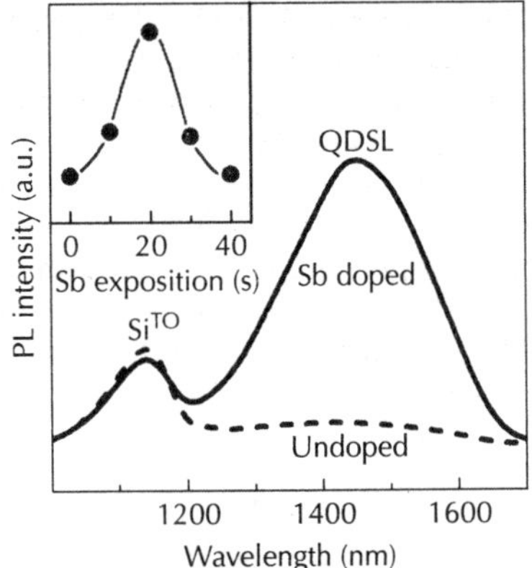

Figure 10.5   Room temperature PL spectra of two QDSLs: undoped and Sb doped with 20 s exposition. The inset shows the influence of Sb exposition time on the PL intensity of the QDSL band. Excitation power density P = 6 Wcm^{-2}. The QDSLs are grown on a p-type substrate.

PL spectrum the QDSL band is very weak. Sb doping leads to the noticeable improvement of PL properties in the spectral range (1.4–1.8)μm – QDSL band becomes dominant. The maximum effect is observed for 20 s exposition (the inset in Fig. 10.5). At higher Sb doses the QDSL PL is quenched again. All results presented below are produced on the Ge/Si QDSLs doped with a 20 s Sb exposition.

The PL spectrum of Ge/Si QDSL for different measurement temperatures between 5 and 80 K is shown in Fig. 10.6. The high-energy part of the low-temperature PL spectrum contains a group of narrow lines related to the carrier radiative recombination in the Si matrix. Basically, it is the band–band recombination assisted by the TA, TO and (TO + OΓ) phonons, and the lines of bound excitons. The fundamental Si^{TO} line is detected in the PL spectrum up to room temperature (Fig. 10.5). At temperatures below 20 K two bands marked as WL^{NP} and WL^{TO} are observed in the middle part of the PL spectrum. Most authors explain these bands by the carrier recombination in the Ge WL: non-phonon and TO–phonon assisted, respectively. It should be noted that dislocation PL lines D3 and D4 could be found in the same spectral region [24, 25]. In our case, however, the WL^{NP} and WL^{TO} bands cannot be associated with a D3 and D4 dislocation PL. We have not observed D3 and D4 lines even for specially dislocated structures [26]. Besides, the spectral positions of WL^{NP} and WL^{TO} bands change depending on the WL thickness, namely in a structure with a thicker Ge layer both WL bands are shifted in the low-energy direction.

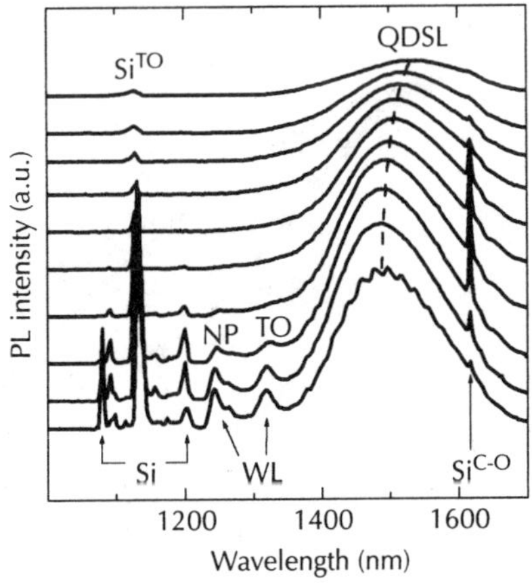

Figure 10.6   PL spectra of QDSL at different measurement temperatures: 5, 10, 15, 20, 25, 30, 40, 50, 60, 80 K, from the bottom to the top. P = 6 Wcm^{-2}. The QDSL is grown on p-type substrate.

The broad QDSL band in the low-energy part of the PL spectrum is attributed to the optical transitions in the Ge/Si columns between holes, localized in the QDs, and electrons, tied to the interface by Coulomb interaction [8–10, 27–32]. At temperatures $T \leq 10\,\mathrm{K}$ the fine periodic structure of the QDSL band is distinctly observed. The deconvolution into 10 Lorentzians is the best fit of the observed multi-modal structure [−5; +5] (Fig. 10.7a). The average distance between neighbouring maxima is $\delta_m = 10\,\mathrm{meV}$. The full width at half maximum of the component is FWHM = 15 meV. The QDSL embedded into a p^+–n^+-junction has the structure of the QDSL PL band too (Fig. 10.7b), but the number of components is only 5 [−3; +2], the distance δ_m is larger (about 20 meV) and the FWHM is 30 meV. On the other hand, this multi-modal structure was kept in the PL spectrum of an LED up to 150 K.

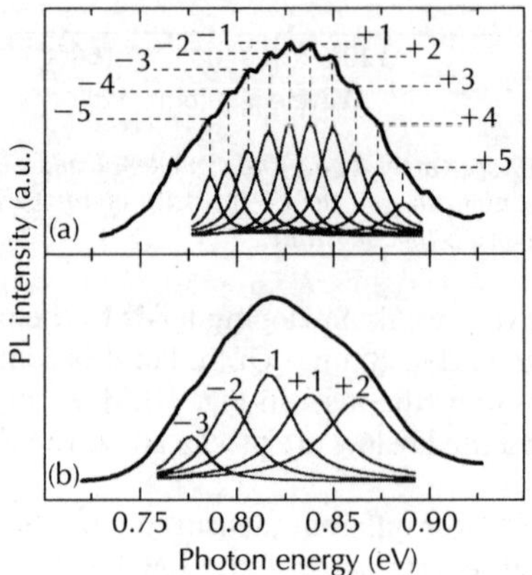

Figure 10.7 QDSL PL band measured at 5 K and an excitation density of 6 Wcm^{-2} for two Ge/Si QD structures: (a) grown on p-type substrate (see Fig. 10.6) and (b) on p$^+$-type substrate. Deconvolution into Lorentzians is also given.

The temperature dependence of the QDSL PL peak energy (E_m) for a p^+–(i)–n^+-structure is more informative and is shown in detail in Fig. 10.8. A pronounced red shift of E_m occurs at lower temperatures than the corresponding band gap narrowing of bulk GeSi. The deviation starts as early as 20 K, reaches a maximum in the range 150–200 K and disappears at 300 K. The pronounced red shift of the PL peak is typical for In(Ga)As/GaAs QDs, but not earlier than at 100 K. It has been attributed to carrier redistribution between small and large QDs [33–35].

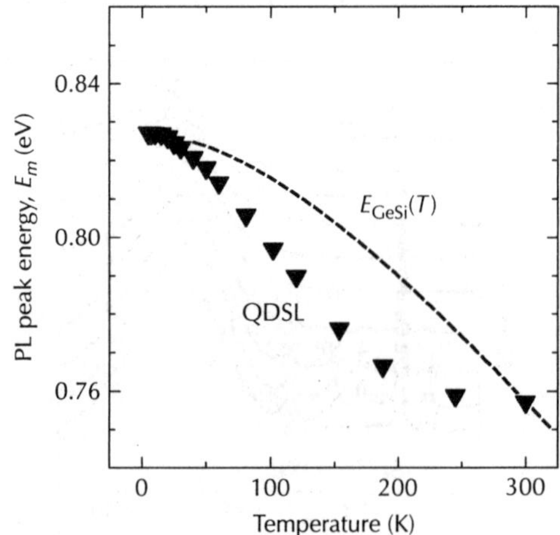

Figure 10.8 Temperature dependence of the QDSL PL peak energy E$_m$. Excitation power density amounts to 6 Wcm^{-2}. The dashed line shows the temperature dependence of the bulk Ge$_{0.8}$Si$_{0.2}$ band gap. Composition x = 0.8 is equivalent to the Ge content in QDs. The QDSL is grown on p$^+$-type substrate.

In our system, the early red shift is caused by other mechanisms which will be discussed on the basis of an energy band model below.

The temperature dependence of the QDSL integrated intensity (J) is presented in Fig. 10.9 for both types of structures. For an analysis, an energy E_A for thermal-activated electrons leaving the states contributing to PL is introduced. The activation energy E_A was calculated from an Arrhenius plot. We used extended Arrhenius analysis. Beside the main quenching mechanism of PL with E_A we took into account an additional competing transition with E_{A2}. Experimental points were fitted using the following expression:

$$\frac{J(T)}{J(0)} = \left[1 + A_1 \exp\left(-\frac{E_A}{kT}\right) + A_2 \exp\left(-\frac{E_{A2}}{kT}\right) \right]^{-1}$$

(10.1)

where $J(0)$ is maximal PL intensity, A_1, A_2 are fitting parameters and k is the Boltzmann's constant. The values of the main activation energy E_A for the two samples are shown in Fig. 10.9.

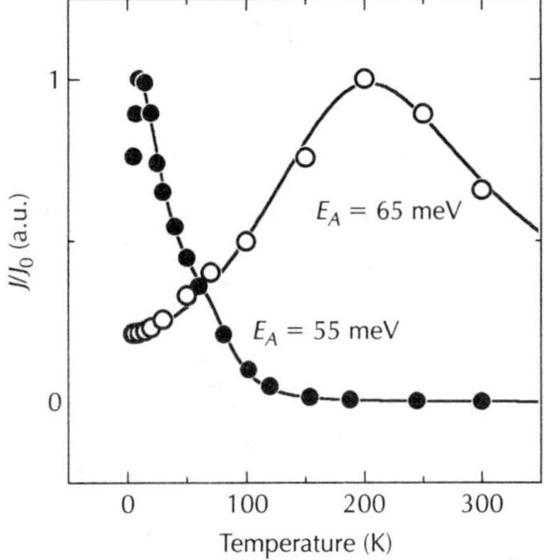

Figure 10.9   Temperature dependence of the QDSL PL integrated intensity for two structures. Filled circles – QDSL on the p-type substrate. Open circles – QDSL on the p⁺-type substrate. PL excitation density – 6 Wcm⁻². Solid lines – fit using Eq. 10.1. Activation energies E_A deduced from the fit are also shown.

The Arrhenius analysis was applied to obtain the activation energy for a set of structures having different values of the QD height B and the spacer thickness W. Fig. 10.10 shows that the activation energy in QDSLs does not depend on the QD size B (Fig. 10.10c), but on the spacer thickness W (Fig. 10.10d). This means that the Si spacer in the column acts as a real QW with a discrete level for electrons (Fig. 10.11). In fact, the main activation energy E_A is the barrier height for electrons on this level, and it is determined as the difference between the QW depth (conduction band offset U_e) and confinement energy of electron ground state E_e. As W increases (3.0–6.6 nm), the electron 1e level goes down, and main activation energy increases (45–85 meV). The competing activation energy E_{A2} has been kept between 6 and 10 meV. If parameter A_2 is positive, the second term is responsible for the early but slow temperature quenching of the QDSL PL, for example at relatively low excitation of QDSLs embedded into a p–n-junction. In case of a negative parameter A_2 the second term in (10.1) is responsible for the temperature-induced enhancement of the QDSL PL, for example at relatively low excitation of QDSLs embedded into a p^+–n^+-junction. The PL intensity increase in a certain temperature interval is typical for the majority of QDSLs. Figure 10.9 shows how the QDSL band intensity grows in the range 5–20 K for the p–(QDSL)–n-structure and in the range 5–200 K for the p^+–(QDSL)–n^+-structure.

The upper part of Fig. 10.10 presents the QDSL peak position E_m as a function of QD height B and spacer thickness W for a set of structures investigated. It should be noted that E_m in the PL spectrum obviously does not depend on the Si spacer thickness W in the column (Fig. 10.10b),

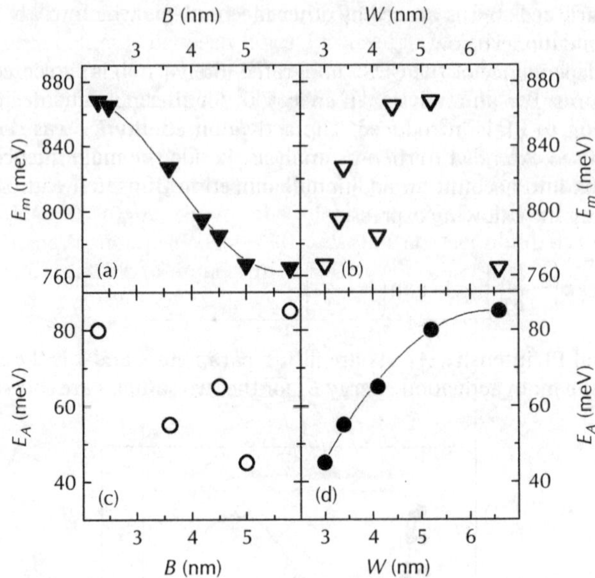

Figure 10.10   *QDSL peak position in the low-temperature PL spectrum (E_m) and the main activation energy (E_A) of the QDSLs as a function of QD height B and spacer thickness W. The values of B and W are measured on the column axis using TEM data. The PL excitation power density is 6 Wcm^{-2}. The solid lines are given for clarity.*

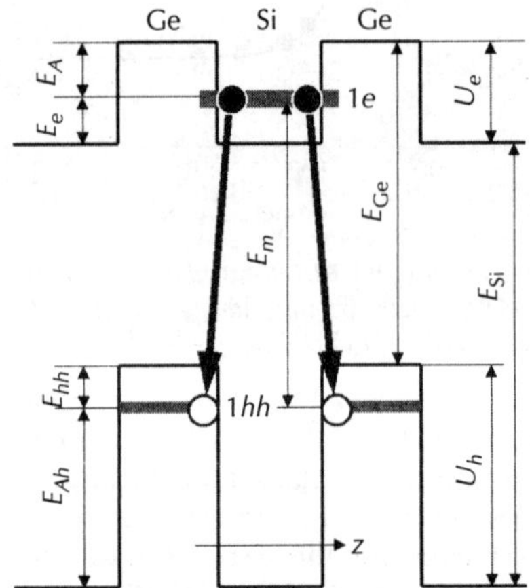

Figure 10.11   *Scheme of type-II heterostructure with QWs for holes in the Ge layer and for electrons in the Si layer. U – band offset, E_e – confinement energy of electron ground state, $E_A = (U_e - E_e)$ – activation energy of this state, E_m – optical transition energy (QDSL peak position). E_{Ge} and E_{Si} – band gaps of bulk Ge (QD) and Si (spacer) for Δ valley with intermixing.*

but on the Ge QD height B (Fig. 10.10a). Specifically, as B increases, the QDSL band has a lower energy position. This phenomenon is in full agreement with the assumption that in the Ge/Si heterostructures the radiative recombination energy of the electron–hole pair mostly depends on the heavy-hole level in the deep QW of the Ge layer [36]. As the Ge QD size increases, the hole confinement energy E_{hh} decreases, which results in a lower transition energy E_m.

The energy band line-up shown in Fig. 10.11 can be considered as the initial approximation for the studied Ge/Si QDSLs. Potential wells are formed for the holes in the Ge QDs as well as for the electrons in the Si spacer layers. The Ge QW for holes is characterized by a depth of up to several hundreds of meV and, thus, the thermal redistribution of holes is negligible for the Ge/Si structures. It is evident that the rapid red shift of the QDSL peak with temperature (Fig. 10.8) is explained by the processes in the electron subsystem. The localization of electrons in the Ge/Si interface is mainly determined by the Coulomb interaction, i.e. by the indirect exciton binding energy, which is constant and equals 25 meV [36]. The latter value is close to kT at room temperature, making an observation of the QDSL band at room temperature quite difficult [8–14]. In our highly strained Sb-doped QDSLs the potential well for electrons is deeper due to the tensile strain–induced lowering of the conduction band. This leads to higher activation energies for electrons (45–85 meV). This fact, however, cannot be the reason for the observed red shift of the QDSL PL peak either.

Available experimental data on the activation energy for the QD PL band in Ge/Si structures are very controversial and scattered between 15 and 183 meV [9, 10, 13, 37]. It is noteworthy that only values between 21 and 46 meV are attributed to the electron subsystem in [13]. All other reported values are interpreted as the hole escaping from the Ge QDs into the WL or barrier lowered by intermixing. Close to 1.55 μm the well-known D1 PL line can also be found, and this line is attributed to dislocations in Si and has an activation energy of 170 meV [38, 39]. It should be noted that our QDSL-related PL band has nothing in common with the D1 line (except the spectral position) [40]. The QDSL PL peak position could be controlled by choosing the growth parameters, which can change the Ge QD sizes. Figure 10.12 presents the PL and the EL spectra of the Ge/Si QDSLs, which were produced by combining different parameters. The QDSL PL peak dominates at room temperature. Its integrated intensity exceeds the Si^{TO} fundamental emission by a factor of 10 to 1000. The experimentally obtained positions of the QDSL maxima are denoted by circles. They correspond to the spectral region of 1.4 μm (0.89 eV) to 1.8 μm (0.69 eV).

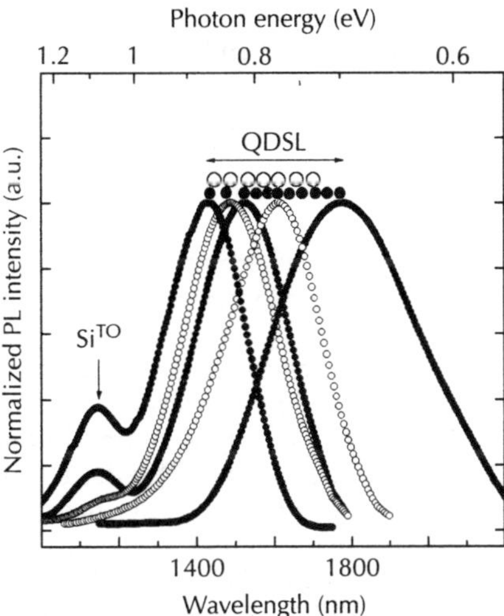

Figure 10.12   PL and EL spectra of Ge/Si QDSLs at room temperature. Full circles – PL, empty circles – EL. Circles on the top denote the QDSL peak positions reached in the experiments.

The LED structures have a p^+-type substrate and an Sb concentration profile shown in Fig. 10.1c. In fact, during the carrier injection the Ge/Si QDSL having 20 periods acts as the active zone. The EL spectra for different current densities at room temperature (300 K) are presented in Fig. 10.13. The current–voltage characteristics of LEDs (the left inset of Fig. 10.13) demonstrates the high-quality p^+–n^+-junction with a low dark current. At an increase of the current density (j) the integrated EL intensity (J) grows superlinearly. The right inset of Fig. 10.13 presents this dependence ($J = j^m$) in a double logarithmic plot. In the range of current densities $j = (0.9–1.8)$ Acm^{-2} a factor $m = 4.8$ is derived, at which the EL spectra in Fig. 10.13 are measured. For optical pumping the value of m-factor does not exceed 1.65 at room temperature [41, 42]. We have

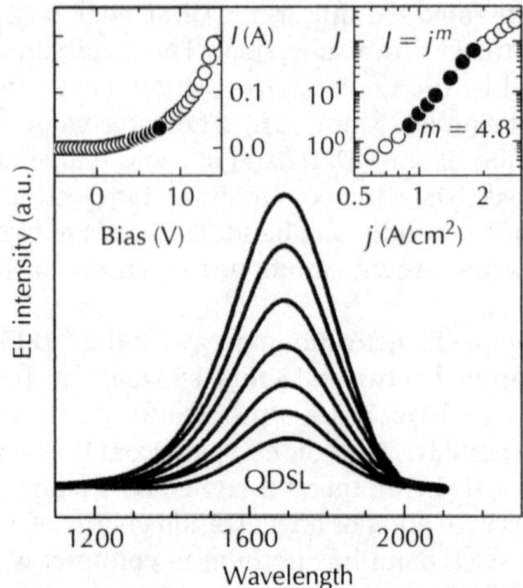

Figure 10.13   Ge/Si EL spectra measured at room temperature for QDSL, having B = 3.8 nm, W = 2.5 nm. Current densities j (Acm⁻²): 0.9; 1.0; 1.1; 1.2; 1.4; 1.6 and 1.8, from the bottom to the top. Left inset: Dark current–voltage characteristics. Right inset: Double logarithmic plot for EL integrated intensity J versus current density j. Factor m is deduced from fit J = jᵐ. Full circles correspond to the EL measurement points shown in the main graph.

reported [43] such an unusually large m-factor for Ge/Si QDSL EL. It has the same nature as the anomalous temperature dependence in Figs 10.8 and 10.9 and will be discussed below after consideration of the energy band model of QDSL. Concerning all available publications, only in [14] the $J(j)$ dependence in Ge/Si QD multilayer structures at 300 K was measured. At $j < 20$ Acm⁻² the dependence was also found to be superlinear with a factor of 1.3. The authors of [11] measured the EL signal from a Ge/Si QD array up to 290 K. It is noteworthy that the QDSL EL intensity was maximal at 225 K.

Our result ($m = 4.8$) demonstrates a high efficiency for an electrical pumping of Ge/Si QDSLs [43]. The external QE of the EL was measured for the QDSL band with a maximum at 1.55 μm. At a current density of 2 Acm⁻², the external efficiency was $\eta = 4 \times 10^{-4}$. To the knowledge of the authors this achieved value is the highest for Ge/Si structures in this spectral region at ambient temperature. This value is higher than the external efficiency reported for the QD-based Ge/Si LEDs ($\eta = 10^{-6}$ for $\lambda = 1.4$ μm, [44]). In [12], the same authors report the following values of internal QE in the Ge QD-based structures: 10^{-5} for the 10 layer structure and 5×10^{-6} ($\lambda = 1.42$ μm) for the one layer QD array. Normally, the efficiency of LEDs based on band-to-band luminescence in bulk silicon ($\lambda = 1.12$ μm) is 10^{-4}–10^{-5} [45]. The LEDs based on dislocation-rich silicon are characterized by an external QE of 10^{-6} ($\lambda = 1.6$ μm) [46]. Only a special surface treatment of the highly purified silicon wafer is allowed to reach $\eta = (1 - 2) \times 10^{-3}$ for the dislocation luminescence [47].

10.3.3   Effect of Sb doping

The unusually high localization potential for electrons in QDSLs (up to 85 meV) is related to the extremely strained Ge/Si columns. It is known [48] that Sb is a perfect surfactant for the growth of the Ge/Si heterostructures. The Sb predeposition leads to a decrease of the Si adatom migration. In this way antimony blocks possible channels of elastic strain relaxation in the Si spacer layers, namely it suppresses intermixing and prevents the nucleation of dislocations. The accumulation of tensile strain in the Si layers and a compressive strain in the Ge QDs leads to an increase of U_e, i.e. to the deepening of the electron QW. Values of the activation energy become 2–3 times larger than the thermal energy at room temperature ($kT \sim 25$ meV) and account for its dependence on thickness W of the Si spacer (Fig. 10.10). In [49, 50] it was possible to get room temperature luminescence from Ge/Si strained layer superlattices probably due to the Sb predeposition during the MBE growth. In [51] the post-growth Sb modulation doping of the Ge/Si superlattice resulted in an electron mobility enhancement at room temperature. Electron localization with a band offset of $U_e \geq 100\,\mathrm{meV}$ was also reported in [9] for undoped Ge/Si nanostructures.

Our undoped Ge/Si multilayer structures have less sharp interfaces (Fig. 10.1a in comparison to Fig. 10.1b) and are characterized by a poor near-infrared PL (Fig. 10.5). The same result was produced by a number of special methods directed towards the improvement of intermixing (smearing of interfaces): an increase of the growth temperature, a decrease of the growth rate [26] and a post-growth annealing [52].

Below we will provide a qualitative analysis of the Si QW profile in Sb-doped QDSLs. It is evident that due to the well-defined interfaces the QW energy walls are practically vertical. The QW energy bottom is likely to be non-symmetrical, because the tensile strain in the Si spacer is distributed inhomogeneously. Following the scheme for a single Ge QD in an Si matrix [36], a higher tensile strain exists in the vicinity of the QD apex than near the base. It is probable that the Si spacer thickness strongly influences the Si QW bottom profile in our QDSLs. However, the main activation energy E_A is primarily determined by the QW depth. The competing activation energy ($E_{A2} = 6 - 10\,\mathrm{meV}$, $A_2 > 0$) depends on the QW bottom profile. The authors of [13] found $E_{A2} = (5-6)\,\mathrm{meV}$ in the undoped structures and attributed this energy to the electron transitions between Δ_2 valleys in the inhomogeneously strained Si spacer. It is known [53] that the tensile strain results in a splitting of the six-fold degenerated Δ valleys into the four-fold degenerated Δ_4 and two-fold degenerated Δ_2 valleys. The latter forms the absolute minimum of conduction band in the momentum space. Due to the asymmetric strain profile in the Si spacer the Δ_2 valley near the QD apex is shifted lower than Δ_2 near the QD base. Further, we assume that in our thin spacer QDSLs the slope of the QW bottom can still be steep enough to cause the splitting of the two-fold degenerated electron level in the Si QW. Due to the entanglement of states only the lower split $1e$ state is active in the PL. Thermalization of electrons from the $1e$ state into the "dark" $2e$ state can explain the appearance of a competing activation energy E_{A2} ($A_2 > 0$). We have found that the competing process disappears at a rise of the excitation level ($\geq 12\,\mathrm{Wcm}^{-2}$), i.e. after filling of the $2e$ state.

The Sb-doping parameters are optimized by applying SIMS, TEM and PL. The highest intensity of the QDSL PL band is reached at a medium level of doping in the active zone for $n \sim 5 \times 10^{16}\,\mathrm{cm}^{-3}$ (Fig. 10.1) [54]. This concentration corresponds to a Sb exposition of $20\,\mathrm{s}$ (Fig. 10.5). For this value of n the sharp interfaces and high strains in QD columns are observed. A further increase of the doping level results in PL degradation. In [42] we showed that at a high Sb concentration the segregation takes place and amorphous clusters appear in the Si spacer layers. We do not assume that the clusters themselves and/or their surfaces are the effective channels of the non-radiative recombination. But they are the agents of stress relaxation in the columns. And this is detrimental for the depth U_e of the Si QWs. For a small Si QW area ($\int E(z)dz = U_e \times W$) the $1e$ state is pushed into the continuum.

Besides the nanoscale impact on the Si QWs, Sb doping also results in a microscale transformation of energy line-up in the whole Ge/Si structure; it actually brings the cap layer to n-type, and QDSL (with buffer) becomes an i-region inside the p-n- or p^+-n^+-junction (Fig. 10.1). We have measured a built-in band bending (Φ) by photo-voltage saturation at $5\,\mathrm{K}$ and room temperature. Φ values, as well as built-in electric field strength (F) and the voltage drop per period of QDSL (U_C), which are calculated from these measured values, are presented in Table 10.1 for two samples, the PL temperature dependencies of which are shown in Figs 10.8 and 10.9. The decrease in built-in voltage with temperature growth is probably related to an increase in the free carrier concentration due to the thermo-ionization of shallow impurities in the Si cap (donor $\mathrm{Sb} - 43\,\mathrm{meV}$) and in the Si substrate (acceptor $\mathrm{B} - 45\,\mathrm{meV}$). Thus, Sb doping stimulates a temperature dependence of the built-in field.

The observation of the QDSL fine structure in low-temperature PL spectra (Fig. 10.6) became possible also due to the impact of Sb. It was shown [48, 55] that the Sb surfactant homogenized the QD size and shape. An Sb doped InAs/GaAs structure with QDs [56], which were monolayer-stepwise different in the height, had a similar shape of the PL band. In the case of Ge/Si QDSLs we also found a very narrow QD height distribution in each Ge layer and each Ge/Si column

Table 10.1   Parameters of the band line-up and the built-in voltage for two samples with 20 periods at 5 K and 300 K. $C = (B + W) -$ QDSL period, Φ – dark band bending in junction, $F = \Phi/(20 \times C + L)$ – built-in electric field strength (L – Si buffer thickness), U_C – voltage drop per period

Type of structure	p–(i)–n	p^+–(i)–n^+
C (nm)	7.0	8.6
Temperature (K)	5	5
Φ (eV)	0.35	0.6
F (kV·cm^{-1})	14.6	22.1
U_C (meV)	10.2	19.0
Temperature (K)	300	300
Φ (eV)	0.2	0.25
F (kV·cm^{-1})	8.3	9.2
U_C (meV)	5.8	7.9

(FWHM = 15 meV). But this does not explain the temperature sensitivity of the fine structure. Its temperature-induced disappearance is explained below after consideration of the Ge/Si QDSL energy band model.

10.4   Miniband model for the Ge/Si QDSL

10.4.1   PL excitation power dependence

Following the scheme for a single Ge QD in an Si matrix [36], the first spatially indirect exciton should be localized in the vicinity of the QD apex, i.e. in the region of maximum inhomogeneous strain. If the number of free carriers is sufficiently large, a second exciton can be formed on the opposite heterointerface, near the QD base. Due to the asymmetric strain profile this second local minimum for electrons is shallower than the first one. This difference results in the 20 meV blue shift of the exciton emission maximum [36]. In our case the QDSL blue shift (ΔE_m) is caused by increasing the optical pumping up to 6 Wcm^{-2} which only amounts to 4 meV and at 12 Wcm$^{-2}\Delta E_m = 7$ meV (Fig. 10.14). We have established that at 6 Wcm^{-2} excitation the $1e$ state is already occupied, and up to 12 Wcm^{-2} the $2e$ state is filled up. In this way, such very

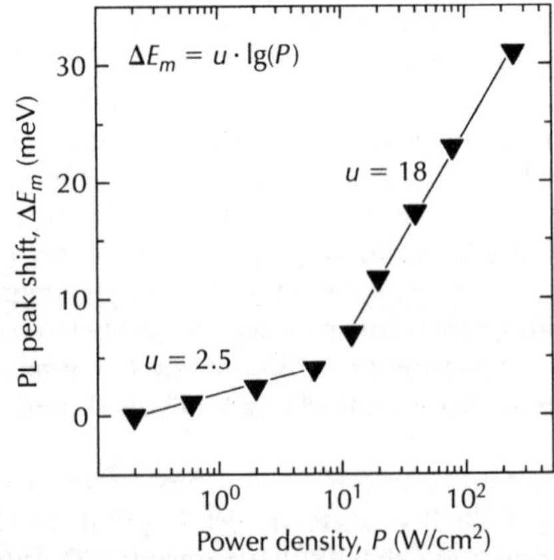

Figure 10.14   Influence of the excitation power density on the QDSL PL peak shift for p–(QDSL)–n-structure at 15 K. Solid lines – linear fit for the deduced factor u.

small E_m shifts confirm (i) the identity of QWs and (ii) the similarity (resonance) of E_e energies at the opposite sides of a Ge QD, as shown in Fig. 10.11.

It is well known [57–59] that resonant tunnel coupling the identical QWs separated by potential barriers can form an energetic miniband from the separate levels. The QW wavefunctions in a miniband are delocalized and shared by the whole structure (superlattice). The miniband transport mode in the superlattices was first investigated by Esaki and Tsu [60]. Theoretical studies [61–63] and experimental observations of resonant tunnelling [64, 65] gave an impetus for a whole series of investigations of III–V layer-based superlattices. Experimentally, the miniband formation was found in type-I structures with III–V-based QDs by PL spectroscopy [66–70] and by photoconductivity [71]. The authors of [72] performed the calculations for type-II Ge QD-based Ge/Si structures and showed that a hole miniband could be established. Measurements of the Hall mobility [73] confirm that the carrier transport in Ge/Si QDSLs is likely to have a miniband conduction type. An important feature of the miniband formation in type-II heterostructures is the conversion to spatially quasi-direct excitons. We attribute the high efficiency of luminescence in our Ge/Si QDSLs to this phenomenon, as will be discussed in the following.

10.4.2 Miniband calculation

In order to calculate the energetic structure of QDSL we used a modified Kronig–Penney approach (KPA). Primitive KPA is limited to the infinite superlattice and to the same effective mass m^* for the QW as well as for the barrier. Nevertheless, the KPA is the most widely used instrument for energy calculations in periodic structures. It was demonstrated that the KPA gives reliable results for III–V-based superlattices [74–76]. Calculations performed by the authors of [72] for the Ge/Si system took into account the different longitudinal bulk effective masses in the QW and in the barrier. It is known [57] that a correct choice of the effective mass in KPA requires the accounting of non-parabolicity of energy versus wave vector $E(q)$. For the Ge/Si system the choice of a proper m^* for electrons is complicated by the following: the absolute minima of the conduction bands in the QW (Si) and in the barrier (Ge) are not only located at different q, but also in different crystallographic directions. An extremely small effective mass of the exciton for QDSLs has been predicted in [77]. We have used the results of calculations [72] with effective masses of the state density in the conduction band of Si and Ge:

$$
\begin{aligned}
\cos(qC) = {} & \cos\!\left(\frac{W}{\hbar}\sqrt{2m_w^* E}\right)\cosh\!\left(\frac{B}{\hbar}\sqrt{2m_B^*(U_e - E)}\right) \\
& -\frac{1}{2}\left(\frac{m_B^*}{m_W^*}\sqrt{\frac{m_W^* E}{m_B^*(U_e - E)}} - \frac{m_W^*}{m_B^*}\sqrt{\frac{m_B^*(U_e - E)}{m_W^* E}}\right) \\
& \times \sin\!\left(\frac{W}{\hbar}\sqrt{2m_W^* E}\right)\sinh\!\left(\frac{B}{\hbar}\sqrt{2m_B^*(U_e - E)}\right)
\end{aligned}
\tag{10.2}
$$

where m_W^* refer to an electron effective mass in Si QW ($0.32m_0$) and m_B^* to the electron effective mass in the Ge barrier ($0.22m_0$).

Iterations applying Eq. 10.2 were performed to fit calculated values of $(U_e - E)$ to the experimental data of the electron activation energy E_A (Fig. 10.10). Simultaneously, the electron confinement energy E_e and Si QW depth U_e were also calculated. For different structures E_e values ranged between 30 and 60 meV. The Si QW depth U_e for all QDSLs was assumed 110 ($\pm 5\%$) meV. As the result of this simulation, an energy spectrum of electrons in QDSL represented a *miniband*, formed by the ground electron state ($1e$). The calculated width of the electron miniband Δ_{MB} as the function of the period C for the studied Ge/Si QDSLs is presented in Fig. 10.15. The miniband width varies from 3 to 34 meV. This qualitatively reflects a relation with the effective mass in the miniband: $m_{MB}^* \sim \Delta_{MB}^{-1} \times C^{-2}$ [59]. In the calculations of Δ_{MB} we assumed that the maximum of electron state density in the miniband was located in its bottom quarter (low temperatures).

Calculations also show that in the studied QDSLs having a QD array density of $1.1 \times 10^{10}\,\text{cm}^{-2}$ the electron miniband is not formed for the layer plane directions ($\Delta_{MB} < 0.1\,\text{meV}$). The same calculation for the hole subsystem has proved that the QW depth for holes in Ge amounts to

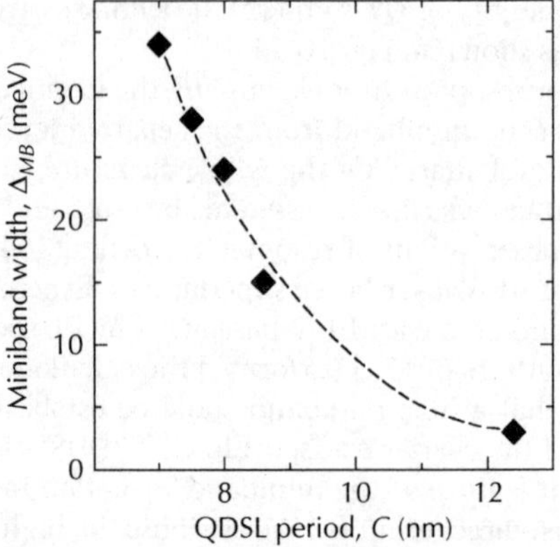

Figure 10.15   *Electron miniband width* Δ_{MB} *versus period C in the QDSL column. Diamonds – value of* Δ_{MB} *calculated on the basis of the experimental data* B, W *and* E_A *using Eq. 10.2 and effective mass of the state density in Ge/Si QDSL. Dashed line is a guide for the eye.*

440 meV (±5%). Nevertheless, the miniband width for the hole ground state does not exceed 1 meV.

We also performed the calculation of the QDSL energy spectrum using the solution of Schrödinger's equation by the transfer matrix method [78]. Taking into account the thickness fluctuations of the layers in the QD column and the exciton–phonon interaction it was determined that electron states form the miniband. Its width lies between 15 and 65 meV at 300 K. The wavefunctions of the holes remain localized in the QDs. Calculations [78] provide a new indicator of miniband existence in Ge/Si QDSLs. Specifically, the squared integral of the electron and hole wavefunction overlap is described by a quadratic dependence on the number of periods.

10.4.3   PL dependence on the number of periods

The measured integrated intensity J of the QDSL band in dependence on the number N of the Ge layers (Fig. 10.16) follows a quadratic dependence $J \sim N^2$ rather than a linear additive rule $J \sim N$. This quadratic dependence was predicted by our calculations [78]. The superlinear behaviour indicates the presence of a new recombination mechanism in the studied superlattices. We believe that it is a more efficient type-I recombination, which is provided by the electron miniband in the Ge/Si QDSLs. In other words, the recombination of electron and hole takes place within

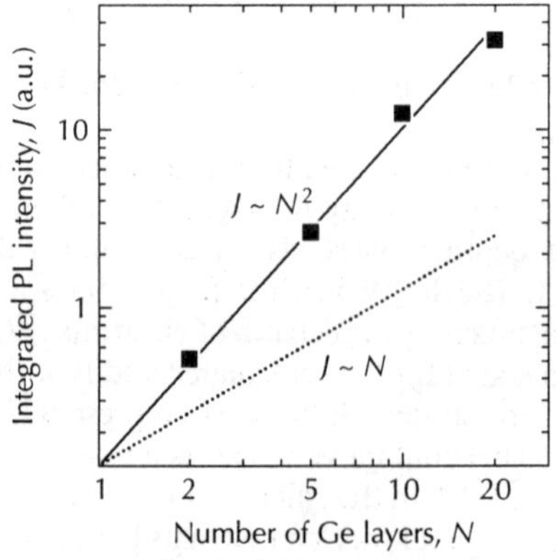

Figure 10.16   *Room temperature PL integrated intensity of QDSL band (J) versus the number* N *of Ge layers. Excitation power density – 6 Wcm^{-2}. Squares – experimental data; solid line is a fit to the quadratic dependence on* N; *dotted line – linear dependence on* N. *Both curves converge for* N = 1.*

the same nanocluster. For a sufficient concentration of non-equilibrium carriers this results in the observed dependence $J \sim N^2$.

What benefit for the PL intensity can be expected from the conversion to the type-I recombination? In the type-I InAs/GaAs structure the electron and the hole are in the same QD, and the overlap of their wavefunctions reaches 80% (direct exciton) [79]. In the Ge/Si structure this value for the electron in Si and for the hole in Ge does not reach 15% (indirect exciton) [16]. Therefore, the exciton oscillator strength due to the miniband should by more than one order larger than without miniband. We have obtained a ratio $J/J_1 = 10$ for the periods number of $N = 10$ and $J/J_1 = 12$ for $N = 20$, where J corresponds to the measured integrated intensity of the QDSL band, J_1 – QDSL intensity, extrapolated for the linear dependence on N (Fig. 10.16). Thus the recombination transition becomes quasi-direct in real space, but probably remains indirect in the momentum (q) space. Though for the strongly localized holes the relaxation of q-conservation and, consequently, non-phonon recombination can be expected. The attempts of an indirect–direct conversion in q-space in a Ge/Si structure have been taken for small Ge QDs [48], for thin Ge WLs [80] and for thin Si spacers [9]. Unfortunately, a hybrid situation, in which the heavy holes are localized, but the electrons are not, can cause a reduced miniband transport.

10.4.4   Temporal profile of QDSL PL

If one presumes an indirect–direct conversion of the exciton in our Ge/Si QDSLs, an increase in oscillator strength and, consequently, a decrease in the exciton lifetime is expected. The typical PL decay time for conventional Ge/Si QD structures is in the range 2–5 µs [26, 81]. In this case, the slow PL decay is due to the indirect character of transitions in real and momentum space. The elimination of at least the first factor could accelerate the transients by more than one order of magnitude. This expectation is based on the following data. The PL of InAs QDs in an Si matrix (type-II line-up, but q-direct exciton) is characterized by a decay time of about 440 ns [82]. The TRPL from InAs QDs in a GaAs matrix (type-I line-up and q-direct exciton), which we measured at the same wavelength (1.3 µm), exhibited a time constant of 2 ns [83].

Figure 10.17 presents the PL kinetics for a Ge/Si QDSL at 15 K. This temporal profile was measured for an excitation with a flux density of 5×10^{10} photons $\times$ pulse^{-1}cm^{-2}. For the QD array density of 1.1×10^{10} cm^{-2} in the 20 layer structure it corresponds to an incomplete occupation of $1e$ states in Si QWs (see region of $u = 2.5$ in Fig. 10.14). The PL decay profile is well described by one exponent with a time constant $t_L = 95$ ns (Fig. 10.17): $I(t) = I_0 \exp[-(t/t_L)^s]$. The factor of QD size dispersion $s \approx 1$, i.e. all nanoclusters giving a contribution into the QDSL PL band have similar sizes. Analysis of the full transient PL profile is based on a three-level system, which

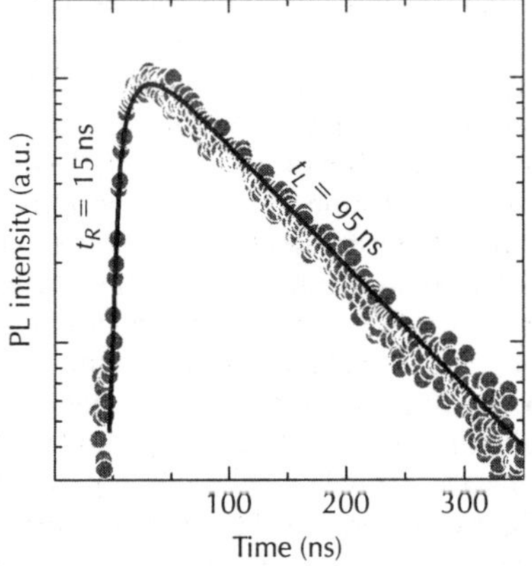

Figure 10.17   PL decay profile of the QDSL band at T = 15 K for the p–(QDSL)–n-structure with 20 Ge/Si periods. Time zero corresponds to a maximum of the laser pulse. Excitation – 5×10^{10} photons pulse^{-1}cm^{-2}. The solid line shows a fit to the temporal evolution of PL, based on the three-level system (10.3).

includes both radiative recombination with decay time t_L and carrier relaxation into the $1e$ states with a PL rise time t_R (fitted line in Fig. 10.17):

$$I(t) \propto \frac{t_L}{t_L - t_R} \times \left[\exp(-t/t_L) - \exp(-t/t_R)\right] \tag{10.3}$$

Therefore, the observed decay time is assigned to the radiative lifetime of the excitons in the QDSL. This value is much smaller than the time relevant for type-II recombination, where the carriers are spatially separated. The 95 ns lifetime indicates the conversion to type-I recombination, which is likely to be caused by the creation of the electron miniband in the Ge/Si QDSL.

The short PL rise time of 15 ns (Fig. 10.17) is explained by the fast capture of carriers from the Si matrix into the miniband. The time constant of $t_R = 15$ ns corresponds to the resonant tunnelling time within the QDSL. The time constant of non-resonant tunnelling between Ge/Si QWs is considerably longer (about 325 ns [84]). The large difference between these time constants (more than one order of magnitude) has been observed in GaAs QWs [85] and CdTe QWs [86].

At a higher excitation level, corresponding to the region with a factor $u = 18$ meV $\times$ decade^{-1} (Fig. 10.14), the shape of the transient PL splits into two components – a fast one, not resolved in our experiment, and a following slower one with a decay time of a few μs. We suppose that the slow component is typical for the lifetime of a spatially indirect exciton in Ge/Si heterostructures. It means that a high excitation power leads to a collapse of the miniband and a recovery of the type-II recombination. The fast component appears to be due to high densities of non-equilibrium carriers and most likely originates from such relaxation processes as Auger recombination and exciton–exciton scattering. The latter factor restricts the coherence length of the excitons in the QDSL and makes the miniband unstable. The destruction of a miniband under high PL excitation densities by exciton–exciton scattering was observed in InGaAs/GaAs QDSLs by [70].

10.4.5   Temperature and power dependence of miniband

A description based on discrete levels in Si QWs (Fig. 10.11) does not account for the following experimental findings: (i) multi-modal structure (Figs 10.6 and 10.7); (ii) intensity rise (Fig. 10.9) and (iii) rapid red shift (Fig. 10.8) of QDSL PL as a function of temperature; (iv) superlinearity for the integrated intensity (Fig. 10.13); and (v) small blue shift (Fig. 10.14) of the QDSL band in the power dependence. However, these facts may be explained within the miniband model.

The electron subsystem of the studied Ge/Si QDSLs is made up from QWs with one confined state. The possible transport regimes of such a superlattice under applied field are a miniband (MB), a Wannier–Stark (WS) ladder, and multi-QWs with spatially separated subbands and their crossovers. Four main energy parameters play a role for a realistic description: the miniband width (Δ_{MB}), the WS energy (eFC), the thermal energy (kT), and the structural and dynamic disorder broadening (δ). For a given QDSL, the Δ_{MB} is determined by the overlap of the wavefunctions of electrons in the adjacent identical QWs and the coherent length (L_C), in other words, by the probability of resonant tunnelling. If the neighbouring QWs are not identical and/or an electric field is applied the MB regime can be supported by temperature (kT) and dynamic broadening (scattering). Simultaneously, the scattering processes (phonon, interface roughness, exciton–exciton) shorten the coherence length in QDSL. Under an electric field with $eFC \geq \delta$ the miniband breaks up into a series of WS states (WS ladder). The localization degree of an electron in the WS state is determined also by the probability of tunnelling through the barrier. Depending on this the electrons may be transported between WS states via hopping (moderate fields, $\delta \leq eFC < \Delta_{MB}$) or sequential non-resonant tunnelling (high fields, $eFC \geq \Delta_{MB}$). In the last extreme case, the wavefunctions are strongly localized in separate QWs. Such structures behave like the uncoupled multi-QWs. At moderate fields the wavefunction of WS states can be well delocalized throughout several periods and can give the additional optical transitions [87].

We believe that the multi-modal structure of the QDSL PL band reflects the presence of a WS ladder in the QDSL at low temperatures (Figs 10.6 and 10.7). The interval between PL neighbouring components ($\delta_m = 10$ and 20 meV) correlates well with the voltage drop per period (Table 10.1) for $p(i)n$- and $p^+(i)n^+$-structures: $U_C = 10.2$ and $19\,\text{meV}$, respectively. In Fig. 10.18c we present the sketched band line-up and optical transitions for a QDSL in the WS regime. In fact, the multimodal QDSL PL band reflects the distribution of squared wavefunctions in the WS states of the ladder. Thus the electron wavefunction is extended up to 10 periods in $p(i)n$-structures, i.e. a coherence length $L_C \approx 70\,\text{nm}$. Due to the stronger built-in field (Table 10.1) only five components are distinguished in $p^+(i)n^+$-structure ($L_C \approx 43\,\text{nm}$). Two WS ladders with $L_C = 32.5$ and $60\,\text{nm}$ were observed by the authors of [87, 88] in GaAs/AlGaAs superlattices. The periodicity of optical oscillations with F^{-1}-dependence was predicted in [89]. Hopping conduction was identified in the plane of QD arrays at low temperature in [71] for InGaAs and in [90, 91] for Ge QDs.

From this point of view the disappearance of the multi-modal structure in the QDSL PL spectrum can only be the result of miniband formation in the column due to temperature-induced weakening of the built-in field (Table 10.1). The miniband develops gradually from separate WS states, the electron wavefunctions extend throughout the QDSL (Fig. 10.18a) having an average thickness of 140–$250\,\text{nm}$. The authors of [92] obtained a coherence length equal to $150\,\text{nm}$ for a miniband in a 20 period GaAs/AlGaAs superlattice. The miniband formation in the $p(i)n$-structure occurs already at $10\,\text{K}$ and in the $p^+(i)n^+$-structure above $150\,\text{K}$ due to differences in the built-in field strength in these structures (Table 10.1). The $p(i)n$-field at low temperatures ($F \approx 14.6\,\text{kV} \times \text{cm}^{-1}$) can be considered a miniband breaking field $F = \delta/eC$. In that case the broadening energy of the electron level in the Ge/Si QDSL at low temperature should be around $10\,\text{meV}$. It is evident that such noticeable broadening will effectively support the MB regime in QDSLs with narrow minibands and/or under strong electric fields, e.g. in $p^+(i)n^+$-structures. Acoustic-phonon scattering and scattering via interface roughness can give the disorder broadening $\delta \approx 10\,\text{meV}$ at low temperatures. A short scattering time $t = \hbar/\delta \approx 70\,\text{fs}$, as well as some modes in the Raman spectra (Fig. 10.4) are an indication of both processes in Sb-doped QDSLs. MB transport controlled acoustic-phonon scattering was found in [71]. The existence of true miniband conduction via interface roughness scattering was highlighted by the authors of [93].

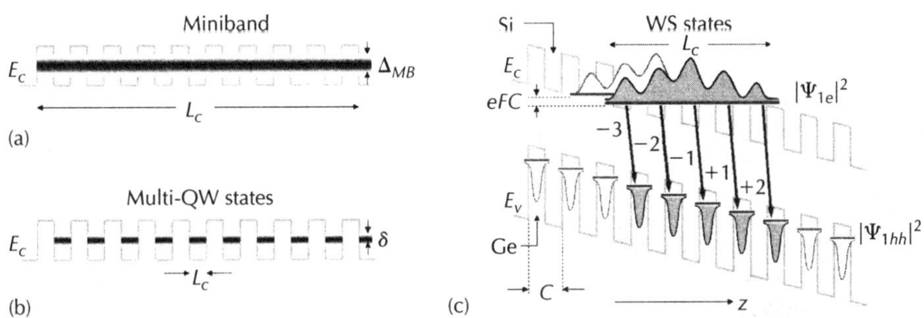

Figure 10.18 *Scheme of band edge profile of Ge/Si QDSL in different regimes: (a) miniband (low electric field, $L_C \approx C \times N$ – for not too low temperatures, low excitation densities); (b) multi-QW structure (low field, $L_C \approx C$ – high excitation level); (c) WS ladder (moderate field, $L_C \approx 5 \times C$ – low temperature, low excitation level), squared wavefunctions and optical transitions $[-3; +2]$ are presented as well.*

The miniband development in Ge/Si QDSLs results in spatially direct recombination. Due to indirect–direct conversion the QDSL PL intensity increases as a function of temperature (Fig. 10.9). The maximum is reached at different temperature values depending on the original built-in field: $20\,\text{K}$ for the p–(QDSL)–n-structure and $200\,\text{K}$ for the p^+–(QDSL)–n^+. The magnitude of this effect excludes its explanation by the carrier transfer from WL to the Ge QDs [80]. A further intensity decrease is caused by the thermal activation of electrons from the miniband. A slope with activation energy E_{A1} appears in the Arrhenius plot. Such behaviour was observed in all QDSLs embedded into p^+–n^+-junctions. Typically, intensity increases by a factor of 3–4 in the range of 5–200 K. As a rule, the QDSL intensity at room temperature is higher than at low temperatures. For $p(i)n$-structures this ratio is always less than 1, namely from 0.1 to 0.01 depending on the value of E_{A1}. The authors of [93] observed miniband transport between ~77 and 400 K in GaAs/AlAs superlattices.

The miniband formation is accompanied also by the transformation of a random electron distribution on the WS levels in separate Si QWs to the Fermi–Dirac distribution within a miniband. The maximum of electron state density shifts towards the miniband bottom and deviates from $E_{GeSi}(T)$ (Fig. 10.8). Further evolution of $E_m(T)$ is related to the "traffic" of Fermi–Dirac electron gas within the miniband, namely to its middle. $E_m(T)$ returns to $E_{GeSi}(T)$. A similar, unusually rapid temperature shift of the PL peak was observed for a planar InAlAs QD array [69] below 105 K, and for stacked InAs QDs [68] below 250 K. The phenomenon has been attributed to the effective thermalization of carriers at strong coupling of InAs QDs and miniband formation.

The superlinear behaviour of PL and EL intensity ([41, 42] and Fig. 10.13) has the same nature as the temperature dependence – the conversion to the MB regime due to straightening of a band bending. The lowering of the built-in electric field in EL measurements is achieved by a sequential growing of the forward bias. In PL intensity versus excitation power this is achieved by an increase of the non-equilibrium carrier concentration (open-circuit regime).

A typical sign of type-I band alignment and of spatial direct exciton is a small u-factor of the $\Delta E_m(\lg P)$-dependence, which is caused by the well-known state filling effect. In contrast to this, a large blue shift is typical for a type-II structure, in which electrons and holes are spatially separated by the interface. A dipole layer is formed there due to the photo-excitation. A high Hartree potential of this layer raises the confinement energy of electrons and holes and causes a considerable blue shift in the PL spectrum [81, 94]. The value of that shift by one order exceeds the shift, caused by the state filling effect. Therefore, the $\Delta E_m(\lg P)$-dependence can serve as an indicator of the band alignment type. The available data on Ge/Si QD structures deal with low temperatures (4.5–10 K) and quite different power ranges within 0.6–400 Wcm^{-2}. For the sake of comparability we introduced the u-factor and reduced all data to it. For the type-II Ge/Si heterostructures the values of u ranged from 13 to 22 meV decade^{-1} [37, 80, 81, 94]. The II–I-type conversion has been reported for the GeSi structures having strong intermixing in the annealed QDs [37] and in WL [80] with $u \approx 1$ meV decade^{-1}. Our QDSLs had u-factors less than 3 meV decade^{-1} for low excitation densities ($u = 2.5$ meV decade^{-1} for $P \leq 6$ Wcm^{-2} in Fig. 10.14). This value agrees very well with type-I recombination. At higher powers the u-factor reached 20 meV decade^{-1}. In Fig. 10.14 the value $u = 18$ meV decade^{-1} was observed for $P \geq 12$ Wcm^{-2}, where the filling of the $2e$ states occurred. By TRPL measurements we showed that a high excitation power leads to the collapse of the miniband and the recovery of type-II recombination. At high density of non-equilibrium carriers the exciton–exciton scattering restricts the coherence length L_C in QDSLs. If L_C becomes extremely short ($L_C \approx W$), the miniband and WS ladder break up into localized states of separate QWs. Sequential phonon-assisted tunnelling in such decoupled multi-QWs becomes the dominant process (Fig. 10.18b).

Arguments given above prove the presence of regimes typical for coupled superlattices (miniband and WS ladder) and uncoupled multi-QWs structures (sequential tunnelling) in Ge/Si QDSLs. The crossovers between transport modes were investigated using two experimental dependencies of PL, namely the temperature influence between 5 and 300 K at $P = 6$ Wcm^{-2} in the short-circuit regime and the power density dependence between 0.2 and 250 Wcm^{-2} at $T = 15$ K in the open-circuit regime, i.e. we actually have studied two cross-sections in the parametric space $\{T;P\}$. Both parameters (each in its own regime) induced the change of built-in electric field without external bias. It demonstrates that the Ge/Si QDSL is a promising system not only for Si-based optoelectronics, but also for other integrated optics.

10.5 Conclusions

Structural and luminescence properties of Sb-doped Ge/Si QDSLs grown by MBE were studied. Highly strained columns and sharp Ge/Si interfaces were found to form a large conduction band offset of 110 meV. The Si spacer QWs were found to have one single confined electron state only. Kronig–Penney calculations showed that an electron miniband is formed in the QDSLs rather than discrete electronic levels. The miniband width was found to be about 30 meV. The miniband formation leads to a conversion of the band structure from indirect to quasi-direct, i.e. towards type-I recombination. The proposed miniband model gives an adequate explanation of the experimental results, i.e. the high efficiency of photo- and electroluminescence (0.04%) in the 1.55 µm region at room temperature; the square dependence of PL intensity on the number of periods; the substantially decreased exciton lifetime of 95 ns; the temperature-induced growth of the photoluminescence intensity and the superlinear dependence of luminescence on excitation power.

Acknowledgements

We gratefully acknowledge helpful discussions with V. Kveder, D. Kovalev, G. Abstreiter and D. Grützmacher. We would like to thank A. Frommfeld for supporting the MBE growth and S. Schwirzke-Schaaf for contributions to Raman measurements.

References

1. S. Satpathy, R.M. Martin, and C.G. Van de Walle, *Phys. Rev. B* **38**, 13237 (1988).
2. R.J. Turton and M. Jaros, *Mat. Sci. Eng. B* **7**, 37 (1990).
3. R. Zachai, K. Eberl, G. Abstreiter, E. Kasper, and H. Kibbel, *Phys. Rev. Lett.* **64**, 1055 (1990).
4. H. Presting, H. Kibbel, M. Jaros, R.M. Turton, U. Menczigar, G. Abstreiter, and H.G. Grimmeiss, *Semicond. Sci. Technol.* **7**, 1127 (1992).
5. S. Ghosh, J. Weber, and H. Presting, *Phys. Rev. B* **61**, 15625 (2000).
6. S. Fukatsu, H. Akiyama, Y. Shiraki, and H. Sakaki, *J. Cryst. Growth* **157**, 1 (1995).
7. G. Dehlinger, L. Diehl, U. Gennser, H. Sigg, J. Faist, K. Ensslin, D. Grützmacher, and E. Müller, *Science* **290**, 2277 (2000).
8. H. Sunamura, S. Fukatsu, N. Usami, and Y. Shiraki, *J. Cryst. Growth* **157**, 265 (1995).
9. K. Eberl, O.G. Schmidt, R. Duschl, O. Kienzle, E. Ernst, and Y. Rau, *Thin Solid Films* **369**, 33 (2000).
10. W.-H. Chang, W.-Y. Chen, A.-T. Chou, T.-M. Hsu, P.-S. Chen, Z. Pei, and L.-S. Lai, *J. Appl. Phys.* **93**, 4999 (2003).
11. T. Brunhes, P. Boucaud, S. Sauvage, J.-M. Lourtioz, C. Hernandez, Y. Campidelli, O. Kermarrec, D. Bensahel, G. Faini, and I. Sagnes, *Appl. Phys. Lett.* **77**, 1822 (2000).
12. L. Vescan, O. Chretien, T. Stoica, E. Mateeva, and A. Mück, *Mat. Sci. Semicond. Proc.* **3**, 383 (2000).
13. M. Stoffel, U. Denker, and O.G. Schmidt, *Appl. Phys. Lett.* **82**, 3236 (2003).
14. W.-H. Chang, A.T. Chou, W.Y. Chen, H.S. Chang, T.M. Hsu, Z. Pei, P.S. Chen, S.W. Lee, L.S. Lai, S.C. Lu, and M.-J. Tsai, *Appl. Phys. Lett.* **83**, 2958 (2003).
15. M.M. Rieger and P. Vogl, *Phys. Rev. B* **48**, 14276 (1993).
16. A.I. Yakimov, N.P. Stepina, A.V. Dvurechenskii, A.I. Nikiforov, and A.V. Nenashev, *Semicond. Sci. Technol.* **15**, 1125 (2000).
17. P. Werner, V. Egorov, V. Talalaev, G. Cirlin, N. Zakharov. Patent granted USA, US 7,119,358 B2 (10 October 2006).
18. P.H. Tan, D. Bougeard, G. Abstreiter, and K. Brunner, *Appl. Phys. Lett.* **84**, 2632 (2004).
19. R. Schorer, G. Abstreiter, S. de Gironcoli, E. Molinari, H. Kibbel, and H. Presting, *Phys. Rev. B* **49**, 5406 (1994).
20. A.G. Milekhin, A.I. Nikiforov, O.P. Pchelyakov, S. Schulze, and D.R.T. Zahn, *JETP Letters* **73**, 461 (2001).
21. S.H. Kwok, P.Y. Yu, C.H. Tung, Y.H. Zhang, M.F. Li, C.S. Peng, and J.M. Zhou, *Phys. Rev. B* **59**, 4980 (1999).
22. J.L. Liu, G. Jin, Y.S. Tang, Y.H. Luo, and K.L. Wang, *Appl. Phys. Lett.* **76**, 586 (2000).
23. V.I. Alonso and K. Winer, *Phys. Rev. B* **39**, 10056 (1989).
24. W. Schröter and H. Cerva, *Sol. St. Phenom.* **85–86**, 67 (2002).
25. V.V. Kveder, E.A. Steinman, S.A. Shevchenko, and H.G. Grimmeiss, *Phys. Rev. B* **51**, 10520 (1995).

26. N.D. Zakharov, V.G. Talalaev, P. Werner, A.A. Tonkikh, and Cirlin, *Appl. Phys. Lett.* **83**, 3084 (2003).
27. S. Fukatsu, H. Sunamura, Y. Shiraki, and S. Komiyama, *Appl. Phys. Lett.* **71**, 258 (1997).
28. P. Schittenhelm, C. Engel, F. Findeis, G. Abstreiter, A.A. Darhuber, G. Bauer, A.O. Kosogov, and P. Werner, *J. Vac. Sci. Technol. B* **16**, 1575 (1998).
29. O.G. Schmidt, O. Kienzle, Y. Hao, K. Eberl, and F. Ernst, *Appl. Phys. Lett.* **74**, 1272 (1999).
30. L. Vescan, *Mat. Sci. Eng. A.* **302**, 6 (2001).
31. R. Loo, P. Meunier-Beillard, D. Vanhaeren, H. Bender, M. Caymax, W. Vandervorst, D. Dentel, M. Goryll, and L. Vescan, *J. Appl. Phys.* **90**, 2565 (2001).
32. A.V. Novikov, D.N. Lobanov, A.N. Yablonsky, Y.N. Drozdov, N.V. Vostokov, and Z.F. Krasilnik, *Physica E.* **16**, 467 (2003).
33. D.I. Lubyshev, P.P. González-Borrero, E. Marega Jr, E. Petitprez, N. La Scala Jr, and P. Basmaji, *Appl. Phys. Lett.* **68**, 205 (1996).
34. C. Lobo, R. Leon, S. Marcinkevicius, W. Yang, P.C. Sercel, X.Z. Liao, J. Zou, and D.J.H. Cockayne, *Phys. Rev. B* **60**, 16647 (1999).
35. V.G. Talalaev, B.V. Novikov, M.A. Smirnov, V.V. Kachkanov, G. Gobsch, R. Goldhahn, A. Winzer, G.E. Cirlin, V.A. Egorov, and V.M. Ustinov, *Nanotechnology* **13**, 143 (2002).
36. A.V. Dvurechenskii and A.I. Yakimov, *Semiconductors* **35**, 1143 (2001).
37. J. Wan, Y.H. Luo, Z.M. Jiang, G. Jin, J.L. Liu, K.L. Wang, X.Z. Liao, and J. Zou, *Appl. Phys. Lett.* **79**, 1980 (2001).
38. V.V. Kveder, E.A. Steinman, S.A. Shevchenko, and H.G. Grimmeiss, *Phys. Rev. B* **51**, 10520 (1995).
39. W. Schröter and H. Cerva, *Sol. St. Phenom.* **85–86**, 67 (2002).
40. Discussion with V.V. Kveder and M. Seibt. Institute of Physics, Georg-August-University Göttingen, 25 November 2004.
41. G.E. Cirlin, V.G. Talalaev, N.D. Zakharov, V.A. Egorov, and P. Werner, *Phys. Stat. Sol. (b)* **232**, R1 (2002).
42. G. Cirlin, A. Tonkikh, V. Talalaev, N. Zakharov, P. Werner, Proceedings of SPIE, 5946 "Optical Materials and Applications", 266 (2005).
43. V.G. Talalaev, G.E. Cirlin, A.A. Tonkikh, N.D. Zakharov, and P. Werner, *Phys. Stat. Sol. (a)* **198**, R4 (2003).
44. L. Vescan, T. Stoica, O. Chretien, M. Goryll, E. Mateeva, and A. Mück, *J. Appl. Phys.* **87**, 7275 (2000).
45. T.C. Ong, K.W. Terrill, S. Tam, and C. Hu, *IEEE Elec. Dev. Lett.* **4**, 460 (1983).
46. E.Ö. Sveinbjörnsson and J. Weber, *Appl. Phys. Lett.* **69**, 2686 (1996).
47. V. Kveder, M. Badylevich, E. Steinman, A. Izotov, M. Seibt, and W. Schröter, *Appl. Phys. Lett.* **84**, 2106 (2004).
48. C.S. Peng, Q. Huang, W.Q. Cheng, J.M. Zhou, Y.H. Zhang, T.T. Sheng, and C.H. Tung. *Appl. Phys. Lett.* **72**, 2541 (1998).
49. J. Engvall, J. Olajos, H.G. Grimmeiss, H. Presting, H. Kibbel, and E. Kasper, *Appl. Phys. Lett.* **63**, 491 (1993).
50. H. Presting, T. Zinke, O. Brux, M. Gail, G. Abstreiter, H. Kibbel, and M. Jaros, *J. Cryst. Growth* **157**, 15 (1995).
51. E. Kasper, *Surf. Sci.* **174**, 630 (1986).
52. A.A. Tonkikh, V.G. Talalaev, N.D. Zakharov, G.E. Cirlin, V.M. Ustinov, and P. Werner, *Tech. Phys. Lett.* **29**, 739 (2003).
53. O.G. Scmidt, K. Eberl, and Y. Rau, *Phys. Rev. B* **62**, 16715 (2000).
54. V.A. Egorov, G.E. Cirlin, A.A. Tonkikh, V.G. Talalaev, A.G. Makarov, N.N. Ledentsov, V.M. Ustinov, N.D. Zakharov, and P. Werner, *Phys. Sol. State* **46**, 49 (2004).
55. G.E. Cirlin, A.A. Tonkikh, V.E. Ptitsyn, V.G. Dubrovskii, S.A. Masalov, V.P. Evtikhiev, D.V. Denisov, V.M. Ustinov, and P. Werner, *Phys. Sol. State* **47**, 58 (2005).
56. F. Guffarth, R. Heitz, A. Schliwa, K. Pötschke, and D. Bimberg, *Physica E.* **21**, 326 (2004).
57. H.T. Grahn, ed., *Semiconductor Superlattices, Growth and Electronic Properties*, (World Scientific, Singapore (1995).
58. A. Sibille, C. Minot, and F. Laruelle, *Int. J. Mod. Phys. B* **14**, 909 (2000).

59. A. Wacker, *Phys. Rep.* **357**, 1 (2002).

60. L. Esaki and R. Tsu, *IBM J. Res. Develop.* **14**, 61 (1970).

61. R.F. Kazarinov and R.A. Suris, *Semiconductors* **6**, 120 (1972).

62. R. Tsu and L. Esaki, *Appl. Phys. Lett.* **22**, 562 (1973).

63. R. Tsu and G. Döhler, *Phys. Rev. B* **12**, 680 (1975).

64. L.L. Chang, L. Esaki, and R. Tsu, *Appl. Phys. Lett.* **24**, 593 (1974).

65. L. Esaki and L.L. Chang, *Phys. Rev. Lett.* **33**, 495 (1974).

66. M. Colocci, A. Vinattieri, L. Lippi, F. Bogani, M. Rosa-Clot, S. Taddei, A. Bosacchi, S. Franchi, and P. Frigeri, *Appl. Phys. Lett.* **74**, 564 (1999).

67. G.S. Solomon, J.A. Trezza, A.F. Marshall, and J.S. Harris, Jr, *Phys. Rev. Lett.* **76**, 952 (1996).

68. B. Ilahi, L. Sfaxi, F. Hassen, L. Bouzaiene, H. Maaref, B. Salem, G. Bremond, and O. Marty, *Phys. Stat. Sol. (a)* **199**, 457 (2003).

69. B.L. Liu, Z.Y. Xu, H.Y. Liu, and Z.G. Wang, *J. Cryst. Growth* **220**, 51 (2000).

70. S. Lan, K. Akahane, K.-Y. Jang, T. Kawamura, Y. Okada, M. Kawabe, T. Nishimura, and O. Wada, *Jpn. J. Appl. Phys.* **38**, 2934 (1999).

71. H.Z. Song, K. Akahane, S. Lan, H.Z. Xu, Y. Okada, and M. Kawabe, *Phys. Rev. B* **64**, 085303 (2001).

72. O.L. Lazarenkova and A.A. Balandin, *J. Appl. Phys.* **89**, 5509 (2001).

73. Y. Bao, A.A. Balandin, J.L. Liu, J. Liu, and Y.H. Xie, *Appl. Phys. Lett.* **84**, 3355 (2004).

74. Á. Perales, L.L. Bonilla, and R. Escobedo, *Nanotechnology* **15**, S229 (2004).

75. K. Kawashima, M. Takata, K. Fujiwara, T. Kagawa, O. Tadanaga, and H. Iwamura, *Microelectronic Engineering* **51–52**, 143 (2000).

76. Y. Shimada, N. Sekine, and K. Hirakawa, *Physica E.* **21**, 661 (2004).

77. T. Takagahara, *Surf. Sci.* **267**, 310 (1992).

78. N.V. Sibirev, V.G. Talalaev, A.A. Tonkikh, G.E. Cirlin, V.G. Dubrovskii, N.D. Zakharov, and P. Werner, *Semiconductors* **40**, 224 (2006).

79. M. Grundmann, O. Stier, and D. Bimberg, *Phys. Rev. B* **52**, 11969 (1995).

80. J. Wan, G.L. Jin, Z.M. Jiang, Y.H. Luo, J.L. Liu, and K.L. Wang, *Appl. Phys. Lett.* **78**, 1763 (2001).

81. G. Bremond, M. Serpentini, A. Souifi, G. Guillot, B. Jacquier, M. Abdallah, I. Berbezier, and B. Joyce, *Microelectron. J.* **30**, 357 (1999).

82. R. Heitz, N.N. Ledentsov, D. Bimberg, A.Y. Egorov, M.V. Maximov, V.M. Ustinov, A.E. Zhukov, Zh.I. Alferov, G.E. Cirlin, I.P. Soshnikov, N.D. Zakharov, P. Werner, and U. Gösele, *Appl. Phys. Lett.* **74**, 1701 (1999).

83. V.G. Talalaev, J.W. Tomm, A.S. Sokolov, I.V. Shtrom, B.V. Novikov, A.T. Winzer, R. Goldhahn, G. Gobsch, N.D. Zakharov, P. Werner, U. Goesele, G.E. Cirlin, A.A. Tonkikh, V.M. Ustinov, and G.G. Tarasov, *J. Appl. Phys.* **100**, 083704 (2006).

84. L.C. Lenchyshyn, M.L.W. Thewalt, J.C. Sturm, and X. Xiao, *Phys. Rev. B* **47**, 16659 (1993).

85. K. Leo, J. Shah, E.O. Göbel, J.P. Gordon, and S. Schmitt-Rink, *Semicond. Sci. Technol.* **7**, B394 (1992).

86. S. Haacke, N.T. Pelekanos, H. Mariette, M. Zigone, A.P. Heberle, and W.W. Rühle, *Phys. Rev. B* **47**, 16643 (1993).

87. E.E. Mendez, F. Agulló-Rueda, and J.M. Hong, *Phys. Rev. Lett.* **60**, 2426 (1988).

88. F. Agulló-Rueda, E.E. Mendez, and J.M. Hong, *Phys. Rev. B* **40**, 1357 (1989).

89. J. Bleuse, G. Bastard, and P. Voisin, *Phys. Rev. Lett.* **60**, 220 (1988).

90. A.I. Yakimov, C.J. Adkins, R. Boucher, A.V. Dvurechenskii, A.I. Nikiforov, O.P. Pchelyakov, and G. Biskupski, *Phys. Rev. B* **59**, 12598 (1999).

91. A.I. Yakimov, A.V. Dvurechenskii, A.I. Nikiforov, and C.J. Adkins, *Phys. Stat. Sol. (b)* **218**, 99 (2000).

92. C. Rauch, G. Strasser, K. Unterrainer, W. Boxleitner, E. Gornik, and A. Wacker, *Phys. Rev. Lett.* **81**, 3495 (1998).

93. A. Sibille, J.F. Palmier, M. Hadjazi, H. Wang, G. Etemadi, E. Dutisseuil, and F. Mollot, *Superlattices and Microstructures* **13**, 247 (1993).

94. M. Larsson, A. Elfving, P.-O. Holtz, G.V. Hansson, and W.-X. Ni, *Physica E.* **16**, 476 (2003).

CHAPTER 11

Effects of the Electron–Phonon Interaction in Semiconductor Quantum Dots

J.T. Devreese,[1,2] V.M. Fomin,[1,2,3] and V.N. Gladilin[1,3]

[1]*TFVS, Universiteit Antwerpen, Belgium;*
[2]*PSN, COBRA, TU Eindhoven, The Netherlands;*
[3]*PMS, State University of Moldova, Chişinău, Moldova*

11.1 Introduction

The state-of-the-art nanodevices based on quantum dots have provided new insights in the fundamental physical properties of confined semiconductors as well as advancement of novel applications in microelectronics and optoelectronics and for manipulation of materials at the nanoscale [1–3].

The study of *excitons* in nanoscale systems has been triggered by the advancement of spectroscopic methods sensitive to the electronic structure of nanomaterials and dynamic processes of charge carriers, including absorption, redistribution and emission of excitation energy. Features of excitons in semiconductor quantum dots constitute a subject of an active cross-disciplinary research [4] combining structural characterization, chemistry and physics. The quantum confinement effects on the exciton energy spectra in semiconductor quantum dots have been devoted an extensive literature, see, e.g., monographs and reviews [5–9] and references therein.

Non-adiabaticity [10, 11] is an inherent property of exciton–phonon systems in various quantum dot structures. The present authors and their co-authors found that non-adiabaticity can *drastically enhance the efficiency of the exciton–phonon interaction*. The effects of non-adiabaticity are important to interpret the surprisingly high intensities of the phonon "sidebands" observed in the optical absorption, the photoluminescence and the Raman spectra of quantum dots, in particular, an enhancement of these intensities with decreasing the quantum dot size (see, e.g., [12, 13]). Deviations of intensities of the phonon-peak sidebands, observed in some experimental optical spectra, from the Franck-Condon progression, which is prescribed by the commonly used adiabatic approximation, find a natural explanation within our non-adiabatic approach [9–11, 14–18].

Photoluminescence excitation (PLE) experiments for self-organized InAs/GaAs quantum dots [19] revealed size-dependent excitation resonances. The observed PLE resonances are identified based on their energy, relative intensity, and sensitivity to size variations in comparison to results of eight-band $\mathbf{k} \cdot \mathbf{p}$ calculations of exciton transitions for pyramidal InAs/GaAs QDs. Band mixing, strain, and the particular geometry of the three-dimensional confinement lead to a rich fine structure with a variety of "forbidden" excitonic transitions. A good agreement between experiment and theory is found for large QDs (with a ground-state transition energy $E \lesssim 1.1\,\mathrm{eV}$), whereas the agreement becomes worse for smaller QDs because of the uncertainties in the size- and growth-dependent variations of the QD shape and composition as well as Coulomb-induced localized wetting layer states.

We overview in this chapter the present understanding of the effects of non-adiabaticity of the exciton–phonon systems (section 11.2), which allow for the interpretation of the surprisingly high intensities of the phonon sidebands observed in the optical absorption, the photolumines-cence, the PLE and the Raman spectra of some types of quantum dots. We start with the theory of non-adiabatic processes in spherical quantum dots (section 11.3). Next we analyse stacked quantum dots, modelling their shape by parallelepipeds (section 11.4) and discuss the notion of excitonic polarons in quantum dots and their effect on the optical properties of quantum dots (section 11.5). Finally, we consider some recent studies of the effects of the electron–phonon interaction in semiconductor quantum dots (section 11.6).

11.2   Non-adiabaticity of the exciton–phonon systems in spherical quantum dots

In the last two decades, there has been an increasing experimental interest in multiphonon pho-toluminescence [12, 21–24] and Raman scattering [25–30] of nanosize quantum dots.

Photoluminescence peaks due to phonon-assisted processes were first observed in colloidal quantum dots (nanocrystals) under size-selective excitation [21, 22]. Observation of multipho-non photoluminescence of quantum dots embedded in glass is complicated because of a quick trapping of holes onto the local surface levels [31]. Overlapping of photoluminescence bands related to recombination of the electron–hole pairs, which are present in different states local-ized at the surface, smears the features of the spectrum due to the phonon-assisted processes [12, 32]. Distinct phonon-line progressions, which are caused by recombination of the elec-tron–hole pairs in "interior" states of a quantum dot (i.e. the states spatially quantized due to the confinement potential), were observed in the fast components of photoluminescence of CdSe quantum dots embedded in glass [12, 23] using time-resolved spectroscopy.

Existing attempts to interpret [12, 24, 26, 33, 34] the aforementioned experiments on the basis of the *adiabatic* theory of multiphonon transitions in deep centres by Pekar [35] and Huang–Rhys [36] meet considerable difficulties. In the framework of the adiabatic theory, the values of the Huang–Rhys parameter were obtained in [33, 34] using a spherical model Hamiltonian for the electron–hole pair in a quantum dot [37–39] and taking into account the valence-band mix-ing. The values of the Huang-Rhys parameter obtained in this way are significantly (by one–two orders of magnitude) smaller as compared with those derived from experiment. This discrepancy is due to the fact that under a strong confinement the charge density of the electron–hole pair in the ground state is small everywhere in the quantum dot. In order to achieve agreement with the experimental data, additional mechanisms have been intuitively introduced, which ensure sepa-ration of the electron and hole charges in space: (i) additional built-in charge [26, 33, 40]; (ii) traps which would localize holes [22]; and (iii) different boundary conditions for electrons and holes [25].

The main idea of the adiabatic approach consists in the assumption that a stationary state of charge carriers is formed for each *instantaneous* position of ions (i.e. charge carriers follow the ion motion adiabatically). In the framework of this approach, the exciton–phonon interaction leads only to a modification of the exciton wave function, but does not give rise to transitions between different exciton states.

However, two circumstances are to be mentioned, which imply that the exciton–phonon sys-tems in quantum dots are essentially non-adiabatic.

1.   The states of an exciton in a quantum dot including the ground state are, generally speak-ing, *degenerate*. In this connection, it is worthwhile recalling [41–46] that the electron–vibrational interaction in the impurity centres with a degenerate electron level may cause dynamic Jahn–Teller effect [47], or, equivalently, so-called *internal non-adiabaticity*. Namely, if the electron interaction with some vibrational modes is described by non-commutative matrices calculated on the basis of a degenerate electron level (those vibrational modes are usually called the non-adiabatic or Jahn–Teller modes), there is a non-adiabatic mixing

of electron states belonging to this level. In a direct analogy to this scenario of the proper Jahn–Teller effect for the impurity centres, transitions do occur between different states of a degenerate exciton level in a quantum dot, provided that the exciton–phonon interaction in the basis of this degenerate level is characterized by non-commutative matrices.

2. Differences between energy levels of an exciton in a quantum dot can become comparable with the optical phonon energy in the experiments [12, 21–29]. Therefore, the resulting effects of *external non-adiabaticity*, or so-called pseudo Jahn–Teller effect [45], in the phonon-assisted optical transitions are of crucial importance. The term "external non-adiabaticity" is used in order to make a distinction between this class of non-adiabatic phenomena and the above-described proper Jahn–Teller effect relevant to a degenerate exciton level.

11.3 Photoluminescence of spherical quantum dots

11.3.1 The light absorption by quantum dots

Quantum dots are considered to be embedded in a medium with a weak dispersion of refractive index. In this section, the quantum dots are supposed to be identical. On the basis of the Kubo formula [48], the linear coefficient of absorption by an ensemble of quantum dots in the electric dipole approximation [49] can be represented as:

$$\alpha(\Omega) = \frac{8\pi\Omega\mathcal{N}}{3\hbar cn(\Omega)}(1 - e^{-\frac{\hbar\Omega}{k_B T}})\,\mathrm{Re}\int_0^\infty dt\, e^{i(\Omega+i\epsilon)t}\langle e^{\frac{i}{\hbar}Ht}d^\dagger e^{-\frac{i}{\hbar}Ht}d\rangle, \quad \epsilon \to +0, \tag{11.1}$$

where Ω is the light frequency, $\mathcal{N}$ is the concentration of quantum dots, $n(\Omega)$ is the refractive index of the medium, d is the dipole moment operator, and $\langle ... \rangle$ denotes the averaging over the statistical ensemble of the exciton–phonon systems. The total Hamiltonian of the exciton–phonon system:

$$H = H_{ex} + H_{ph} + H_{int} \tag{11.2}$$

contains the exciton Hamiltonian H_{ex}, the phonon Hamiltonian H_{ph}, and the exciton–phonon interaction Hamiltonian:

$$H_{int} = \sum_\lambda (\gamma_\lambda a_\lambda + \gamma_\lambda^\dagger a_\lambda^\dagger) \tag{11.3}$$

where $a_\lambda^\dagger$ and a_λ are the creation and annihilation operators for the phonons of the λth vibrational mode, the interaction operators γ_λ will be specified below.

Note that in the experiments [12, 21–24], the exciton energy is much larger than both the phonon energy and the value $k_B T$. This means, in particular, that the probability of thermal generation of an exciton is vanishing. For optical transitions leading to generation of an exciton and starting from the exciton vacuum $|0\rangle$ whose energy is chosen as zero, the initial states of the exciton–phonon system are described by the wave functions $|0\rangle|n\rangle$, where $|n\rangle$ are eigenstates of the phonon Hamiltonian H_{ph}. Further, the final states of the exciton–phonon system are not occupied, so that in Eq. (11.1), one can omit the term which is proportional to $\exp(-\hbar\Omega/k_B T)$ and describes a correction due to stimulated emission of light with frequency Ω. Finally, the frequency of the exciting radiation is much larger than the phonon frequency; hence, the interaction of this radiation with free phonons can be neglected, i.e. $d|n\rangle = |n\rangle d$. Hence, the transformation of the average follows:

$$\left\langle e^{\frac{i}{\hbar}Ht}d^\dagger e^{-\frac{i}{\hbar}Ht}d\right\rangle = \sum_n \rho_n\langle n|e^{\frac{i}{\hbar}H_{ph}t}\langle 0|d^\dagger e^{-\frac{i}{\hbar}Ht}d|0\rangle|n\rangle$$

$$= \langle 0|d^\dagger|\beta\rangle\langle\beta|\left(\sum_n \rho_n\langle n|e^{\frac{i}{\hbar}H_{ph}t}e^{-\frac{i}{\hbar}Ht}|n\rangle\right)|\beta'\rangle\langle\beta'|d|0\rangle \tag{11.4}$$

where $|\beta\rangle$ labels eigenstates of the exciton Hamiltonian and ρ_n is the probability of the phonon state $|n\rangle$ in the equilibrium statistical ensemble of the phonon systems:

$$\rho_n = \frac{1}{\mathcal{Z}_{ph}} \exp\left(-\frac{\mathcal{E}_n}{k_B T}\right)$$

$\mathcal{Z}_{ph}$ is the partition function of the free phonon subsystem and $\mathcal{E}_n$ is the energy of the phonon state $|n\rangle$. Using the well-known Feynman relations [50] for operator exponentials, one obtains

$$e^{\frac{i}{\hbar}H_{ph}t}e^{-\frac{i}{\hbar}Ht} = e^{-\frac{i}{\hbar}H_{ex}t}U(t) \tag{11.5}$$

where the evolution operator:

$$U(t) = T\exp\left(-\frac{i}{\hbar}\int_0^t dt_1 H_{int}(t_1)\right) \tag{11.6}$$

is represented in terms of the chronological ordering operator T, while

$$A(t) = e^{\frac{i}{\hbar}(H_{ex}+H_{ph})t} A\, e^{-\frac{i}{\hbar}(H_{ex}+H_{ph})t}$$

denotes the operator A in the interaction representation. Hence, Eq. (11.1) takes the form:

$$\alpha(\Omega) = \frac{8\pi\Omega\mathcal{N}}{3\hbar c n(\Omega)} \text{Re} \sum_{\beta,\beta'} d_\beta^* d_{\beta'} \int_0^\infty dt\, e^{i(\Omega-\Omega_3+i\epsilon)t}\langle\beta|\langle U(t)\rangle_{ph}|\beta'\rangle \tag{11.7}$$

where Ω_β is the Franck–Condon frequency of generation of an exciton in the state $|\beta\rangle$ and $d_\beta \equiv \langle\beta|d|0\rangle$ is the dipole matrix element of a transition between the exciton vacuum state and the state $|\beta\rangle$,

$$\langle\ldots\rangle_{ph} \equiv \sum_n \rho_n\langle n|\ldots|n\rangle$$

denotes the averaging over the phonon ensemble.

Using Eq. (11.3) and neglecting the dispersion of the frequency of the optical phonons ω_0, one obtains the following result of the averaging over the equilibrium phonon ensemble in Eq. (11.7):

$$\langle U(t)\rangle_{ph} = T\exp\left\{-\frac{1}{\hbar^2}\sum_\lambda \int_0^t dt_1 \int_0^{t_1} dt_2 \left[(\bar{n}+1)e^{-i\omega_0(t_1-t_2)}\gamma_\lambda(t_1)\gamma_\lambda^\dagger(t_2)\right.\right.$$
$$\left.\left. +\bar{n}\,e^{i\omega_0(t_1-t_2)}\gamma_\lambda^\dagger(t_1)\gamma_\lambda(t_2)\right]\right\} \tag{11.8}$$

where:

$$\bar{n} = \left(e^{\frac{\hbar\omega_0}{k_B T}} - 1\right)^{-1}. \tag{11.9}$$

The absorption coefficient $\alpha(\Omega)$ of Eq. (11.7) with Eq. (11.8) can be analytically calculated at arbitrary exciton–phonon coupling (i) when the exciton levels are non-degenerate (this case was analysed by Pekar [35] and Huang-Rhys [36]) and (ii) when the vibrational modes interacting with an exciton are adiabatic, i.e. the matrices $||\langle\beta|\gamma_\lambda|\beta'\rangle||$ in the basis of the degenerate level can be simultaneously diagonalized (this case is referred to as a static Jahn–Teller effect [41–44]).

However, these analytical approaches cannot be applied for the exciton–phonon systems in quantum dots, because these systems are essentially non-adiabatic as already stated in the Introduction. The fact that the exciton–phonon interaction in quantum dots is weak enables to adequately describe the effect of the non-adiabaticity on phonon-assisted optical transitions. Here we will concentrate on the intensity of phonon satellites in the optical spectra; the influence of the Jahn–Teller effect on the energy spectrum of the system will be considered in section 11.4.

Under the condition of weak exciton–phonon interaction:

$$\eta \equiv \frac{\max |\langle \beta| \gamma_\lambda| \beta'\rangle|}{\hbar\omega_0} \ll 1 \tag{11.10}$$

contributions to the absorption coefficient from transitions, accompanied by a change of the number of phonons by K, can be calculated to leading (Kth) order in the small parameter η^2. For the details, we refer to [11].

The average absorption coefficient on the narrow frequency intervals of width $\Delta\Omega$ is represented as (*cf.* [11]):

$$\alpha(\Omega) = \frac{8\pi^2 \Omega \mathcal{N}}{3\hbar c n(\Omega)} \sum_\beta \left[\sum_{K=0}^\infty (\bar{n}+1)^K f_{\beta K}^{(+)} \delta(\Omega_\beta + K\omega_0 - \Omega) \right.$$

$$\left. + \sum_{K=1}^\infty \bar{n}^K f_{\beta K}^{(-)} \delta(\Omega_\beta - K\omega_0 - \Omega) \right]. \tag{11.11}$$

Here, the coefficients $f_{\beta K}^{(+)}$ ($f_{\beta K}^{(-)}$) defined by:

$$f_{\beta 0}^{(\pm)} = |d_\beta|^2 \tag{11.12}$$

$$f_{\beta K}^{(\pm)} = \sum_{\beta_{-K},\ldots,\beta_K} \delta_{\beta_0 \beta} \xi_K(\beta_{-K},\ldots,\beta_K)$$

$$\times \prod_{k=1}^K \frac{\theta(|\Omega_{\beta\beta_k} \pm k\omega_0| - \Delta\Omega)\theta(|\Omega_{\beta\beta_{-k}} \pm k\omega_0| - \Delta\Omega)}{(\Omega_{\beta\beta_k} \pm k\omega_0)(\Omega_{\beta\beta_{-k}} \pm k\omega_0)}, \quad K \geq 1 \tag{11.13}$$

are proportional to the probabilities for the generation of an exciton in the state $|\beta\rangle$ with simultaneous emission (absorption) of K phonons;

$$\theta(x) = \begin{cases} 1, & x \geq 0 \\ 0, & x < 0 \end{cases} \tag{11.14}$$

$$\Omega_{\beta\beta'} = \Omega_\beta - \Omega_{\beta'}. \tag{11.15}$$

In Eq. (11.13), there are three kinds of terms which contain non-diagonal matrix elements $\langle \beta_i |\gamma_\lambda| \beta_\kappa\rangle$ in the coefficients ξ_K:

1. terms with all the states $|\beta_{-K}\rangle,\ldots,|\beta_K\rangle$ belonging to the *same* energy level. Those terms describe the influence of the internal non-adiabaticity (or the proper Jahn–Teller effect) on the optical transition probabilities;
2. terms with $|\beta_{-K}\rangle,\ldots,|\beta_K\rangle$ pertaining to *different* energy levels; they take into account the effects of the external non-adiabaticity (or the so-called pseudo Jahn–Teller effect);
3. terms proportional to $d_{\beta_{-K}}^* d_{\beta_K}$ with $\beta_{-K} \neq \beta_K$; such terms describe the *intermultiplet mixing* between the exciton states with the same symmetry. This mixing occurs due to the exciton–phonon interaction.

When an exciton energy level is degenerate, Eq. (11.13) differs from the result of the adiabatic theory [35, 36], which gives for weak exciton–phonon interaction

$$\tilde{f}_{\beta K} = \frac{|d_\beta|^2}{K!} \left(\sum_\lambda \left| \frac{\langle \beta| \gamma_\lambda| \beta\rangle}{\hbar\omega_0} \right|^2 \right)^K, \tag{11.16}$$

the difference taking place even if one neglects the effects mentioned in points (2) and (3) above. For example, the shape of the absorption band edge of a spherical quantum dot can be obtained using the spherical model [37–39] for the exciton Hamiltonian (see appendix to [11]). The coefficients $f_{\beta K}^{(\pm)}$, which are proportional to the probabilities of generation of an exciton in the ground state with emission or absorption of K phonons of non-adiabatic (or, equivalently, Jahn–Teller) modes, are then given by:

$$\frac{f_{\beta K}^{(\pm)}}{\tilde{f}_{\beta K}} = \frac{4}{3}\left(\left[\frac{K+1}{2}\right] + \frac{1}{2}\right)\left(\left[\frac{K+1}{2}\right] + \frac{3}{2}\right) \tag{11.17}$$

where $[x]$ denotes the integer part of x. The adequate account of the exciton interaction with the Jahn–Teller vibrations leads to a significant increase in the multiphonon transition probabilities and to a more complicated dependence of these probabilities on the number of emitted or absorbed phonons in comparison with the results of the theory by Pekar [35] and Huang-Rhys [36]. Therefore, the multiphonon optical spectrum determined by Eq. (11.11) with Eq. (11.17) is considerably different from the Franck–Condon progression described by Eq. (11.11), where the $f_{\beta K}^{(\pm)}$ are replaced by the $\tilde{f}_{\beta K}$ given through Eq. (11.16).

11.3.2   The photoluminescence spectrum

The relaxation processes during the time interval t between the generation and the recombination of an exciton substantially influence the photoluminescence spectrum. Two limiting cases are examined here: (i) the thermodynamic equilibrium photoluminescence which takes place when τ_0 the time of the relaxation between the exciton energy levels is much smaller than the radiative lifetime of an exciton τ and (ii) the opposite case relevant to slow relaxation, $\tau_0 \gg \tau$. The energy level broadening due to the finite values of τ_0 and τ is disregarded in [11], i.e. the inequalities $\tau_0^{-1}, \tau^{-1} \ll \omega_0$ are supposed to be satisfied. This supposition is in agreement with the theoretical estimations of the exciton lifetime (for example, the value $\tau \sim 1\,\mathrm{ns}$ was obtained in [12] for CdSe quantum dots) and with the experimental observation of ultra-narrow ($<0.15\,\mathrm{meV}$) luminescence lines from *single* quantum dots [51]. The spectral broadening of the multiphonon photoluminescence lines will be considered below in section 11.3.4.

The spectra of equilibrium luminescence and of the slow relaxation photoluminescence were obtained in [11] on the basis of the optical absorption spectrum analysed above.

11.3.3   Models for quantum dots

Both the quantum dots in glass and the colloidal quantum dots of small size (1 to 2 nm) are known [12, 22–24] to have a geometrical shape close to spherical. For the samples investigated in [12, 23], small-angle X-ray measurements have indicated that the function $\mathcal{N}(R)$ which describes the distribution of quantum dots over radii is close to the logarithmic standard distribution

$$\mathcal{N}(R) = \frac{\mathcal{N}_{\mathrm{tot}}}{\sqrt{2\pi\varrho}R}\exp\left\{-\left[\frac{\ln(R/R_0)}{\sqrt{2}\varrho}\right]^2\right\}, \tag{11.18}$$

the average radius of quantum dots being determined as

$$\langle R \rangle = R_0 \exp\left(\frac{\varrho^2}{2}\right). \tag{11.19}$$

Here R_0 and ϱ are distribution parameters, $\mathcal{N}_{\mathrm{tot}}$ is the total concentration of quantum dots:

$$\mathcal{N}_{\mathrm{tot}} = \int_0^\infty \mathcal{N}(R)\,dR.$$

In section 11.3.4, the function $\mathcal{N}(R)$ (Eq. 11.18) is used to model the distribution of quantum dots over radii.

In the exciton–phonon interaction Hamiltonian H_{int} (Eq. 11.3), the operators γ_λ are

$$\gamma_\lambda = \gamma_\lambda(\mathbf{r_e}) - \gamma_\lambda(\mathbf{r}_h) \tag{11.20}$$

where $\mathbf{r}_e$ and $\mathbf{r}_h$ are the coordinates of an electron and a hole, respectively. In a spherical quantum dot the interaction operators $\gamma_\lambda^B(\mathbf{r})$ and $\gamma_\lambda^I(\mathbf{r})$ describing the electron interaction, respectively, with the bulk and interface phonons are expressed according to [52] by

$$\gamma_\lambda^{(B)}(\mathbf{r}) \equiv \gamma_{nlm}^{(B)}(\mathbf{r}) = v_{nl}^{(B)}(r)i^{-l-m-|m|}Y_{lm}(\vartheta,\varphi) \tag{11.21}$$

$$\gamma_\lambda^{(I)}(\mathbf{r}) \equiv \gamma_{lm}^{(I)}(\mathbf{r}) = v_l^{(I)}(r)i^{-l-m-|m|}Y_{lm}(\vartheta,\varphi) \tag{11.22}$$

where $Y_{lm}(\vartheta, \varphi)$ are the spherical harmonics. The radial parts of the interaction operators are described by

$$v_{nl}^{(B)}(r) = e\sqrt{\frac{\hbar\omega_0}{\epsilon_0 R}\left(\frac{1}{\epsilon(\infty)} - \frac{1}{\epsilon(0)}\right)}\frac{j_l(z_{ln}r/R)}{z_{ln}j_l'(z_{ln})} \tag{11.23}$$

$$v_l^{(I)}(r) = e\sqrt{\frac{\hbar\omega_0}{\epsilon_0 R}\left(\frac{1}{\epsilon(\infty)} - \frac{1}{\epsilon(0)}\right)}\frac{\sqrt{l}\epsilon(\infty)}{l\epsilon(\infty) + (l+1)\tilde{\epsilon}(\infty)}\left(\frac{r}{R}\right)^l \tag{11.24}$$

where $j_l(z)$ are the spherical Bessel functions, z_{ln} is the nth zero of the function $j_l(z)$, ε_0 is the permittivity of vacuum, $\epsilon(\infty)$ and $\tilde{\epsilon}(\infty)$ are the optical dielectric constants, respectively, of a quantum dot and of its surrounding medium, and $\epsilon(0)$ is the static dielectric constant of a quantum dot.

The exciton Hamiltonian for the spherical model [37–39] supplemented to account for the electron–hole exchange interaction [53, 54] is as follows:

$$H_{\text{ex}} = \frac{1}{2m_e}\mathbf{p}_e^2 + \frac{\gamma_1}{2m_0}\mathbf{p}_h^2 - \frac{\gamma_2}{9m_0}(\mathbf{P}^{(2)} \cdot \mathbf{J}^{(2)})$$
$$+ V_C(\mathbf{r}_e,\mathbf{r}_h) - \frac{2}{3}\varepsilon_{\text{exch}}a_0^3\delta(\mathbf{r}_e - \mathbf{r}_h)(\sigma \cdot \mathbf{J}). \tag{11.25}$$

Here $\mathbf{p}_e$ and $\mathbf{p}_h$ are the momenta of an electron and a hole, respectively, γ_1 and γ_2 are the Luttinger parameters, $\mathbf{P}^{(2)}$ and $\mathbf{J}^{(2)}$ are the irreducible second-rank tensor operators of the momentum and the spin-$\frac{3}{2}$ angular momentum of a hole, σ and $\mathbf{J}$ are the spin operators of an electron and a hole, a_0 is the lattice constant, and ϵ_{exch} is the exchange strength constant [55], which is equal to $320\,\text{meV}$ in CdSe [54]. In a spherical quantum dot, the Coulomb electron–hole interaction may be approximately described by the Hamiltonian [52]

$$V_C(\mathbf{r}_e, \mathbf{r}_h) = -\frac{e^2}{4\pi\epsilon_0\epsilon(\infty)}\sum_{l=0}^{\infty}P_l(\cos\vartheta_{eh})\left\{\frac{r_e^l}{r_h^{l+1}}\theta(r_h - r_e) + \frac{r_h^l}{r_e^{l+1}}\theta(r_e - r_h)\right.$$
$$\left. + \frac{(r_e r_h)^l}{R^{2l+1}}\frac{[\epsilon(\infty) - \tilde{\epsilon}(\infty)](l+1)}{\epsilon(\infty) + \tilde{\epsilon}(\infty)(l+1)}\right\} \tag{11.26}$$

where $P_l(x)$ is a Legendre polynomial of degree l and ϑ_{eh} is the angle between $\mathbf{r}_e$ and $\mathbf{r}_h$. The Franck–Condon frequency for generation of an exciton in the eigenstate $|\beta\rangle$ is determined by the corresponding eigenvalue $\mathcal{E}_\beta$ of the Hamiltonian (Eq. 11.25) as follows:

$$\Omega_\beta = \frac{E_g + \mathcal{E}_\beta}{\hbar} \tag{11.27}$$

where E_g is the energy band gap for the substance of quantum dot.

For typical [12, 22, 23] quantum dot radii, the exciton states are determined mainly by confinement while the electron–hole interaction can be treated as a perturbation. To zeroth order in this perturbation, the hole states [39], designated as $nS_{3/2}$, $nP_{1/2}$, $nP_{3/2}$, $nP_{5/2}$, $nD_{1/2}$, $nD_{5/2}$, ..., are characterized by definite values of the hole angular momentum $\mathbf{F} = \mathbf{L} + \mathbf{J}$ (where $\mathbf{L}$ is the orbital angular momentum of a hole) and the parity; n labels the solutions of the equations for the radial components of the hole wave function [39].

Under size-selective excitation, the excitation frequency Ω_{exc} is chosen to lie in the absorption band tail. Thus, only the largest quantum dots are involved in the absorption processes and excitons are generated in the lowest states. The exciton ground state $(1S,1S_{3/2})$ and the state $(1S,1P_{3/2})$ separated from the ground state by an energy comparable with the phonon energy $\hbar\omega_0$ play the main role in the absorption and photoluminescence. (As for an electron, the energy of its size quantization is large and only the state 1S must be taken into account.)

A detailed analysis of the lowest energy levels of an exciton as a function of the radius in a spherical CdSe quantum dot with zinc-blende (cubic) structure and with wurtzite (hexagonal) structure was performed in [11]. In quantum dots with both the zinc-blende structure and the wurtzite structure, the zero-phonon generation of an exciton in the ground state and the radiative recombination from this state are forbidden in the dipole approximation. Therefore, the intensity of zero-phonon line is due to transitions from high energy levels. As a consequence, a decrease of temperature must result in a decrease of the zero-phonon-line intensity in the equilibrium-luminescence spectra. Such behaviour is in agreement with the experimental data [22].

11.3.4   Numerical results and comparison with the experiment

In Fig. 11.1, the calculated fluorescence spectra $I(\Omega)$ are presented together with the experimental data of [22] on colloidal CdSe quantum dots of wurtzite structure, respectively, at different temperatures T. The following values of the parameters were used: $\epsilon(\infty) = 6.23$, $\epsilon(0) = 9.56$, $m_e = 0.13m_0$, $E_g = 1.75\,\text{eV}$ [33]; $\gamma_1 = 2.04$, $\gamma_2 = 0.58$ [56]. From a comparison of the calculated absorption spectrum (Eq. 11.11) with the experimental absorption spectrum of [22], the value 0.06 was obtained for the parameter ϱ of the distribution function $\mathcal{N}(R)$ (Eq. 11.18). Because of the chaotic nature of the energy spectrum of excitons in real quantum dots (e.g. owing to the strain or non-perfect geometrical form) and also due to a limited spectral resolution of experimental equipment, photoluminescence lines described by δ-functions turn into broadened peaks. The shape of these peaks was modelled by Lorentzians of finite width $\Gamma = 15\,\text{meV}$.

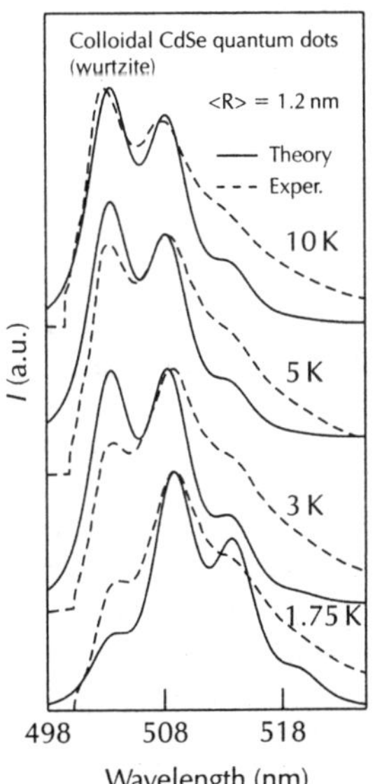

Figure 11.1   Fluorescence spectra of CdSe quantum dots with wurtzite structure at various temperatures. Solid lines were calculated by using the non-adiabatic approach, dashed lines represent the experimental data of [22]. Different lines are shifted along the vertical axis for clarity. (Reprinted with permission after [11]. © 1998 by the American Physical Society.)

It is worthwhile noting that in CdSe quantum dots of wurtzite structure, the magnitude of the $(1S, 1S_{3/2})$-level splitting by the crystal field for a wide range of radii R is close to the optical phonon energy $\hbar\omega_0 = 25\,\text{meV}$. Therefore, the observed peak of the equilibrium photoluminescence in the spectral region $\Omega \approx \Omega_{\text{exc}} - K\omega_0$ is caused not only by K-phonon processes (with probability $\sim\eta^{2K}$), but also by $(K-1)$-phonon processes (with probability $\sim \eta^{2(K-1)}$). The latter processes are related to generation of an exciton in the *upper* group of states (resulting from the splitting of the $(1S, 1S_{3/2})$ level) and the subsequent radiationless relaxation to the *lower* group of states. This feature of the energy spectrum of the exciton–phonon system in quantum dots, in combination with the pseudo Jahn–Teller effect, leads to a substantial difference of the observed spectrum of multiphonon photoluminescence from the Franck–Condon progression, as shown in Fig. 11.2. It is important to note that a straightforward calculation in the framework of the adiabatic theory leads to intensities of the phonon satellites, which do not fit the observed spectrum well for any value of the Huang-Rhys parameter.

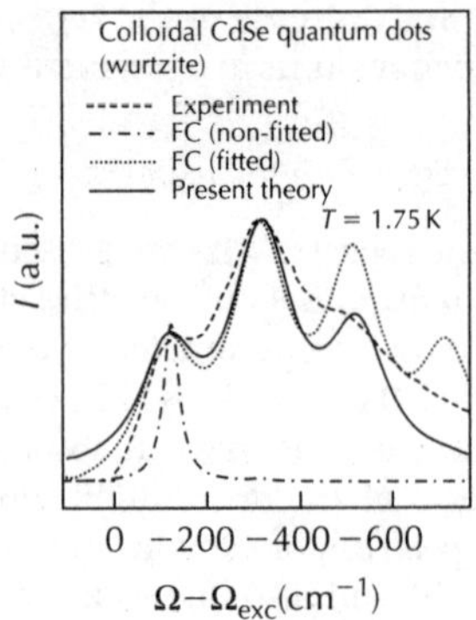

$$\Omega - \Omega_{\text{exc}}(\text{cm}^{-1})$$

Figure 11.2   Fluorescence spectra of CdSe quantum dots with wurtzite structure at the average radius $\langle R \rangle = 1.25\,nm$. Dashed line represents the experimental data of [22], dot–dashed line displays a Franck–Condon progression with the Huang-Rhys parameter $S = 0.06$ calculated in [33], dotted line shows another Franck–Condon progression with the Huang-Rhys parameter $S = 1.7$, which is obtained by fitting the ratio of one-phonon and zero-phonon peak heights to the experimental value, solid line results from the non-adiabatic approach. (Reprinted with permission after [11]. © 1998 by the American Physical Society.)

In Fig. 11.3, the photoluminescence spectra $I(\Omega)$ calculated for a zinc-blende-type crystal lattice are given and compared with the experimental data of [12] on photoluminescence of CdSe quantum dots in borosilicate glass. The parameter values $m_e = 0.11m_0$, $E_g = 1.9\,\text{eV}$ [57] were used for the zinc-blende lattice. The theoretical curves are calculated for the case of the equilibrium photoluminescence [11]. The shape of photoluminescence peaks was modelled by Lorentzians of width $\Gamma = 2\,\text{meV}$ relevant to the limited spectral resolution of measurements (better than $5\,\text{meV}$ in [12]). From a comparison of the calculated absorption spectra (Eq. 11.11) with the experimental absorption spectra of [12], the values 0.15, 0.18, and 0.20 of the parameter ϱ were obtained for average radii $\langle R \rangle$ equal to 1.4 nm, 1.8 nm, and 2.7 nm, respectively. The experimental photoluminescence spectra at each value of the average radius $\langle R \rangle$ refer to different time intervals between the pumping pulse (of duration 2.5 ps) and the measurement, the upper curve corresponding to the time interval equal to 0. According to [12] the decay of fast photoluminescence components is caused by trapping of holes onto deep (binding energy $\sim200\,\text{meV}$) local surface levels. Depending on the average radius of the quantum dots, the decay time varies from some tens to some hundreds of ps. The separation between neighbouring energy levels of a non-trapped exciton is smaller than the depth of local levels. Therefore, the relaxation in the system of "interior" exciton states, which were discussed in the Introduction, can be expected to proceed more quickly than the hole trapping. This supposition is in agreement with the behaviour of fast photoluminescence components under the excitation by light whose frequency lies near the absorption band maximum (no size-selective excitation). In this case a significant shift of the photoluminescence intensity maximum from the excitation frequency appears 10 ps after the excitation pulse [12]. Therefore, it is the limiting case of the equilibrium photoluminescence rather than the opposite case of slow relaxation which seems to be relevant to the experimental data of [12].

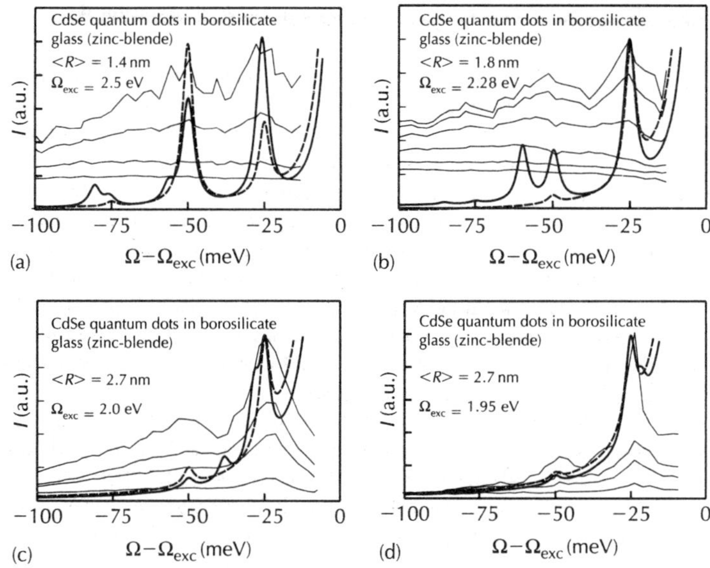

Figure 11.3   Photoluminescence spectra of CdSe quantum dots embedded in borosilicate glass at different average radii and excitation energies: $\langle R \rangle = 1.4\,nm$, $\hbar\Omega_{exc} = 2.50\,eV$ (a), $\langle R \rangle = 1.8\,nm$, $\hbar\Omega_{exc} = 2.28\,eV$ (b), $\langle R \rangle = 2.7\,nm$, $\hbar\Omega_{exc} = 2.00\,eV$ (c), $\langle R \rangle = 2.7\,nm$, $\hbar\Omega_{exc} = 1.95\,eV$ (d). Thin solid lines represent the families of experimental time-resolved photoluminescence spectra of [12] measured at different time intervals between the pumping pulse and the measurement, the upper curve corresponding to the time interval equal to 0. Theoretical results are displayed for the equilibrium–photoluminescence spectra (heavy solid lines) and for the photoluminescence spectra in the case of slow relaxation of the exciton energy (dashed lines). (Reprinted with permission after [11]. © 1998 by the American Physical Society.)

The quantum dots in glass are characterized by broad distribution functions over the radii R [31, 32]. Therefore, even under size-selective excitation some excitons are generated in states with relatively high energy (first of all, in the states $(1S,2S_{3/2})$, $(1S,1P_{5/2})$, $(1S,1D_{5/2})$ whose energies differ slightly from each other). In the case of the equilibrium photoluminescence, the radiationless relaxation of such excitons into the lowest states $(1S,1S_{3/2})$ leads to the appearance of zero-phonon luminescence peaks shifted from Ω_{exc} to lower frequencies. These peaks hide the phonon satellites (see Fig. 11.3c). This effect becomes less pronounced when the excitation is deeper in the absorption band tail (see Fig. 11.3d).

The observed photoluminescence spectra in InAs/GaAs self-assembled quantum dots were attributed to the large efficiency of the Fröhlich interaction between a strain-induced polarized exciton and the longitudinal optical phonons [58, 59]. The theoretical analysis in [60] has shown that the phonon-assisted photoluminescence due to the intraband transitions of an electron between the size-quantized states in self-assembled quantum dots modelled as rectangular parallelepiped InAs quantum dots ("quantum bricks") embedded into GaAs is strongly enhanced by two processes. First, the efficiency of the electron–phonon interaction in an individual quantum dot is enhanced in sufficiently small dots. Second, the ratio between the intensities of the zero-phonon line and the one-phonon line in the photoluminescence spectrum is efficiently controlled by both the shape and the size distribution of those quantum dots. The relative enhancement of the phonon-assisted photoluminescence for narrow size distributions is demonstrated to provide an effective method to characterize vibrational states in single InAs/GaAs quantum dots [60]. Recently, the size, shape, and composition of self-assembled InAs/GaAs and InAs/AlAs quantum dots have been accurately determined by cross-sectional scanning tunnelling microscopy [61]. Those structural results provide a basis for the further, more detailed, analysis of the phonon-assisted transitions in the photoluminescence spectra of self-assembled quantum dots.

11.4 Non-adiabaticity of the exciton–phonon systems in stacked quantum dots

Recently, *stacked* quantum dots have received increasing attention (see, e.g., [62–68]) due to the possibility of finely controlling their energy spectra. This makes stacked quantum dots very promising for future nanodevices [67, 68]. In the present work, the non-adiabatic approach is applied to stacked InAs/GaAs quantum dots, which reveal a richer structure of phonon and exciton spectra in comparison to those for a single quantum dot. Namely, in the stacked quantum dots, as distinct from a single quantum dot, the exciton (phonon) spectra contain groups of states (modes) of different symmetry with close energies (frequencies). Owing to small energy differences between the exciton energy levels within these groups, additional channels for non-adiabatic transitions as compared to a single quantum dot open in stacked quantum dots. Therefore, optical absorption spectra of stacked quantum dots contain more phonon satellites than those of a single quantum dot. These features of the optical absorption can be experimentally revealed, e.g., in the photoluminescence excitation measurements.

In order to model coupled self-assembled InAs/GaAs quantum dots, we consider a stack of N parallelepiped-shaped quantum dots with heights l_{nz} ($n = 2, 4, ..., 2N$) and with the interdot distances l_{nz} ($n = 3, 5, ... , 2N - 1$) along the z-axis. Within the present approach, the lateral sizes of each quantum dot in a stack are supposed to be much larger than its size l_{nz} along the growth axis. The stack is a system of $(2N + 1)$ layers ($n = 1, ... , 2N + 1$) with parameters:

$$\begin{cases} l_{nz}, \varepsilon_n = \varepsilon_{\text{InAs}} & \text{for } n = 2, 4, ..., 2N \\ l_{nz}, \varepsilon_n = \varepsilon_{\text{GaAs}} & \text{for } n = 3, 5, ..., 2N - 1 \\ l_{nz} \to \infty, \varepsilon_n = \varepsilon_{\text{GaAs}} & \text{for } n = 1, 2N + 1. \end{cases}$$

The bulk-like optical-phonon frequencies in InAs and GaAs layers of the stacked InAs/GaAs quantum dots coincide with the LO-phonon frequencies in InAs and GaAs, respectively. The interface frequencies belong to the stacked quantum dots as a whole and satisfy the dispersion equation:

$$\det \| a_{kn}(\omega) \| = 0 \quad (k, n = 1, ... 2N) \tag{11.28}$$

where $a_{kn}(\omega)$ is the dynamic matrix of the interface vibrations with the matrix elements:

$$\begin{cases} a_{nn}(\omega) = \varepsilon_n(\omega) \coth q_{\|} l_n + \varepsilon_{n+1}(\omega) \coth q_{\|} l_{n+1} \\ a_{n,n-1}(\omega) = a_{n-1,n}(\omega) = -\varepsilon_n(\omega) / \sinh q_{\|} l_n \end{cases} \tag{11.29}$$

and all other matrix elements are equal to zero.

In Fig. 11.4, typical interface-phonon spectra are represented for stacked InAs/GaAs quantum dots formed by two InAs parallelepipeds. The frequencies are plotted as a function of the in-plane wave number $q_{\|}$, which takes discrete values due to the quantization of the phonons in the xy-plane. In a stack of N quantum dots, each interface-phonon frequency of a single quantum dot splits into N branches. The splitting of the interface-phonon frequencies is due to the electrostatic interaction between the optical polar vibrations of the different quantum dots.

These features of the optical-phonon spectrum of stacked quantum dots are manifested in their optical properties. We calculate the optical absorption spectrum of polaronic excitons in stacked quantum dots starting from the Kubo formula. Within the non-adiabatic approach [11] the following expression results for the linear coefficient of the optical absorption (Eq. 11.7) by the exciton–phonon system in a quantum-dot structure:

$$\alpha(\Omega) \propto \text{Re} \sum_{\beta, \beta'} d_{\beta}^* d_{\beta'} \int_0^{\infty} dt \, e^{i(\Omega - \Omega_\beta + i0^+)t} \langle \beta | \bar{U}(t) | \beta' \rangle \tag{11.30}$$

where Ω is the frequency of the incident light and d_β and Ω_β are, respectively, the electric dipole matrix element and the Franck–Condon frequency of a transition between the exciton vacuum

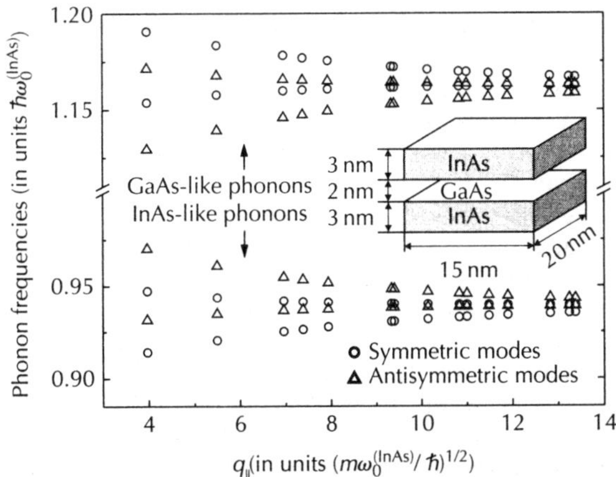

Figure 11.4   *Interface-phonon frequencies for two stacked parallelepiped-shaped InAs/GaAs quantum dots.* $\omega_0^{(InAs)}$ *is the frequency of LO phonons in InAs at the centre of the Brillouin zone. (Reprinted with permission from [20]. © 2004 by the American Physical Society.)*

state and the one-exciton state $|\beta\rangle$. Exciton states in stacked InAs/GaAs quantum dots are determined using an exact diagonalization of the exciton Hamiltonian with simple parabolic valence and conduction bands within a finite-dimensional basis of the electron–hole states. The evolution operator averaged over the phonon ensemble, $\bar{U}(t)$, Eq. (11.8), can be represented as:

$$\bar{U}(t) = \mathrm{T} \exp\left\{-\frac{1}{\hbar^2}\sum_\lambda \int_0^t dt_1 \int_0^{t_1} dt_2 \left[\frac{e^{-i\omega_\lambda(t_1-t_2)}}{1-y_\lambda}\right.\right.$$
$$\left.\left. \times \gamma_\lambda(t_1)\gamma_\lambda^\dagger(t_2) + \frac{y_\lambda e^{i\omega_\lambda(t_1-t_2)}}{1-y_\lambda}\gamma_\lambda^\dagger(t_1)\gamma_\lambda(t_2)\right]\right\}. \tag{11.31}$$

In Eq. (11.31), T is the time ordering operator, the index λ labels the phonon modes specific for the quantum-dot structure under consideration, ω_λ are phonon frequencies, $\gamma_\lambda(t)$ are the exciton–phonon interaction amplitudes in the interaction representation, and $y_\lambda = \exp(-\hbar\omega_\lambda/k_BT)$.

Within the adiabatic approximation, which has been widely used to calculate the optical spectra of quantum dots, non-diagonal matrix elements of the exciton–phonon interaction are neglected when calculating $\alpha(\Omega)$ as given by Eq. (11.30) with Eq. (11.31). In the adiabatic approach [69, 70] one supposes that (i) both the initial and the final states of a quantum transition are non-degenerate and (ii) the energy differences between the exciton states are much larger than the phonon energies. It has been shown in [11, 15–17] that these conditions are often violated for optical transitions in small quantum dots, which have sizes less than the bulk exciton radius.

In general, the efficiency of the exciton–phonon interaction, as revealed through the optical spectra of quantum dots, depends on their geometric and material parameters in a complicated way. On the one hand, an increase of the confinement strength, which results from a decrease of the quantum-dot size or from a rise of potential barriers at the quantum-dot boundary, enhances the efficiency of the interaction with phonons for a *solitary electron (a solitary hole)*. On the other hand, strengthening confinement leads to an increasing overlap between the electron and hole wave functions so that the net charge of the electron–hole pair and, correspondingly, the efficiency of the *exciton–phonon* interaction tend to decrease. As a result, the efficiency of the exciton–phonon interaction appears a non-monotonous function of the confinement strength even within the adiabatic approximation (see, e.g., [71]). Effects of non-adiabaticity, related to phonon-assisted transitions between *different* exciton states, drastically enhance intensities of phonon satellites in the optical spectra of quantum dots and, at the same time, make these intensities dependent on a vast set of parameters, which characterize the phonon-induced coupling between *all* the involved exciton states. Thus, side by side with the Fröhlich coupling constant α, energy spacing between exciton states strongly influences the aforementioned intensities. When the above spacing is comparable with the LO-phonon energy, non-adiabatic effects can strongly affect the optical spectra even in quantum dots of III–V semiconductor compounds, where the Fröhlich coupling constant is significantly smaller ($\alpha = 0.0504$ for InAs) than the values $\alpha \sim 0.5$ typical for II–VI semiconductors considered in [11, 16].

In section 11.3, we described a method to calculate the absorption spectrum given by Eqs (11.30) and (11.31) taking into account the effect of non-adiabaticity on the probabilities of phonon-assisted optical transitions. The key step is the calculation of the matrix elements of the evolution operator $\langle \beta | \bar{U}(t) | \beta' \rangle$. In order to describe the effect of non-adiabaticity both on the intensities and on the positions of the absorption peaks, a diagrammatic approach can be used. When calculating these matrix elements we take into account that in a quantum dot, due to the absence of momentum conservation, the product $\langle \beta_1 | \gamma_\lambda | \beta_2 \rangle \langle \beta_2 | \gamma_\lambda^* | \beta_3 \rangle$ can be non-zero for $\beta_1 \neq \beta_3$, as distinct from the bulk case. Consequently, the evolution operator is in general non-diagonal in the basis of one-exciton wave functions $| \beta \rangle$. For the absorption coefficient we obtain:

$$\alpha(\Omega) \propto -\mathrm{Im}\sum_\beta | d_\beta |^2 G_\beta(\Omega + i0^+) - \mathrm{Im}\sum_{\beta,\beta'} d_\beta d_{\beta'}^*$$
$$\times \left[Q_{\beta\beta'}^{(1)}(\Omega + i0^+) + Q_{\beta\beta'}^{(2)}(\Omega + i0^+) \right] \tag{11.32}$$

where:

$$G_\beta(\Omega) = \sum_{\{j_\lambda = -\infty\}}^{\{\infty\}} C_{\{j_\lambda\}\beta}^{(+)} \left[\Omega - \Omega_\beta + \sum_\lambda S_{\lambda,\beta}\omega_\lambda - \sum_\lambda j_\lambda \omega_\lambda \right.$$
$$\left. - \Sigma_\beta^{(1)}\left(\Omega - \sum_\lambda j_\lambda \omega_\lambda \right) - \Sigma_\beta^{(2)}\left(\Omega - \sum_\lambda j_\lambda \omega_\lambda \right) \right]^{-1} \tag{11.33}$$

$$C_{\{j_\lambda\}\beta}^{(\pm)} = \prod_\lambda (\pm 1)^{j_\lambda} \exp\left[\mp(2\bar{n}_\lambda + 1)S_{\lambda\beta} + \frac{j_\lambda \hbar\omega_\lambda}{2k_B T} \right]$$
$$\times I_{|j_\lambda|}\left(S_{\lambda\beta}\left[2\sinh\left(\frac{\hbar\omega_\lambda}{2k_B T} \right) \right]^{-1} \right) \tag{11.34}$$

$I_n(x)$ is a modified Bessel function of the first kind and $S_{\lambda\beta}$ is the Huang-Rhys parameter, which is related to the interaction of the exciton in the state β with phonons of the λth mode:

$$S_{\lambda\beta} = \left| \frac{\langle \beta | \gamma_\lambda | \beta \rangle}{\hbar\omega_\lambda} \right|^2 . \tag{11.35}$$

The self-energy terms $\Sigma_\beta^{(1)}(\Omega)$ and $\Sigma_\beta^{(2)}(\Omega)$ in Eq. (11.33) are obtained by summing diagrams, which describe one- and two-phonon non-adiabatic contributions:

$$\Sigma_\beta^{(1)}(\Omega) = \sum_{j=\pm 1} \sum_{\lambda,\beta_1} F_{\beta\beta_1}(\Omega - j\omega_\lambda) M_{\lambda\beta\beta_1\beta_1\beta}^{(j)} \tag{11.36}$$

and

$$\Sigma_\beta^{(2)}(\Omega) = \sum_{j_1,j_2=\pm 1} \sum_{\lambda_1,\lambda_2} \sum_{\beta_1,\beta_2,\beta_3} F_{\beta\beta_1}(\Omega - j_1\omega_{\lambda_1})$$
$$\times F_{\beta\beta_3}(\Omega - j_2\omega_{\lambda_2})\left[F_{\beta\beta_2}(\Omega - j_1\omega_{\lambda_1} - j_2\omega_{\lambda_2}) \right.$$
$$\times M_{\lambda_1\beta\beta_1\beta_2\beta_3}^{(j_1)} M_{\lambda_2\beta_1\beta_2\beta_3\beta}^{(j_2)} + F_{\beta\beta_2}(\Omega)$$
$$\left. \times M_{\lambda_1\beta\beta_1\beta_1\beta_2}^{(j_1)} M_{\lambda_2\beta_2\beta_3\beta_3\beta}^{(j_2)} (1 - \delta_{\beta_2\beta}) \right] \tag{11.37}$$

where

$$F_{\beta\beta_1}(\Omega) = \sum_{\{j_\lambda=-\infty\}}^{\{\infty\}} C^{(-)}_{\{j_\lambda\}\beta} G_{\beta_1}\left(\Omega - \sum_\lambda j_\lambda \omega_\lambda\right) \tag{11.38}$$

$$M^{(j)}_{\lambda\beta_1\beta_2\beta_3\beta_4} = m^{(j)}_{\lambda\beta_1\beta_2\beta_3\beta_4} - m^{(j)}_{\lambda\beta_1\beta_1\beta_1\beta_1}\delta_{\beta_1\beta_2}\delta_{\beta_3\beta_4} \tag{11.39}$$

$$m^{(j)}_{\lambda\beta_1\beta_2\beta_3\beta_4} = \frac{j\langle\beta_{2-j}|\gamma_\lambda|\beta_{3-j}\rangle\langle\beta_{2+j}|\gamma^\dagger_\lambda|\beta_{3+j}\rangle}{\hbar^2(1-y^j_\lambda)}. \tag{11.40}$$

The above restriction to one- and two-phonon contributions is justified when the non-adiabatic exciton–phonon interaction is weak: $|\langle\beta|\gamma_\lambda|\beta'\rangle|^2 << \hbar\omega_\lambda$ for $\beta' \neq \beta$. The functions $Q^{(1)}_{\beta\beta'}(\Omega)$ and $Q^{(2)}_{\beta\beta'}(\Omega)$ in Eq. (11.32), which describe contributions of one- and two-phonon processes to non-diagonal matrix elements of the evolution operator, take the form

$$Q^{(1)}_{\beta\beta'}(\Omega) = G_\beta(\Omega)G_{\beta'}(\Omega)(1-\delta_{\beta\beta'})$$
$$\times \sum_{j=\pm 1}\sum_{\lambda,\beta_1} G_{\beta_1}(\Omega - j\omega_\lambda)m^{(j)}_{\lambda\beta\beta_1\beta_1\beta'} \tag{11.41}$$

$$Q^{(2)}_{\beta\beta'}(\Omega) = G_\beta(\Omega)G_{\beta'}(\Omega)(1-\delta_{\beta\beta'})$$
$$\times \sum_{j_1,j_2=\pm 1}\sum_{\lambda_1,\lambda_2}\sum_{\beta_1,\beta_2,\beta_3} G_{\beta_1}(\Omega - j_1\omega_{\lambda_1})$$
$$\times G_{\beta_3}(\Omega - j_2\omega_{\lambda_2})\Big\{G_{\beta_2}(\Omega - j_1\omega_{\lambda_1} - j_2\omega_{\lambda_2})$$
$$\times\Big[(1-\delta_{\beta_3\beta_1})m^{(j_1)}_{\lambda_1\beta\beta_1\beta_2\beta_3\beta'}m^{(j_2)}_{\lambda_2\beta_1\beta_2\beta_3\beta_3}$$
$$+ m^{(j_1)}_{\lambda_1\beta\beta_1\beta_2\beta_3}m^{(j_2)}_{\lambda_2\beta_1\beta_2\beta_3\beta'}\Big] + G_{\beta_2}(\Omega)(1-\delta_{\beta_2\beta})$$
$$\times(1-\delta_{\beta_2\beta'})m^{(j_1)}_{\lambda_1\beta\beta_1\beta_1\beta_2}m^{(j_2)}_{\lambda_2\beta_2\beta_3\beta_3\beta'}\Big\}. \tag{11.42}$$

The absorption spectrum is thus expressed through the functions $G_\beta(\Omega)$, which in turn are determined by a closed set of equations (11.33) and (11.36) to (11.38).

In Figs 11.5, 11.6, and 11.7 the calculated optical absorption spectra are shown for a single quantum dot, two stacked identical quantum dots, and two stacked dots with somewhat different heights, respectively. The calculations were performed for low temperatures, $\{y_\lambda << 1\}$, when the absorption-line broadening due to the exciton–LO-phonon interaction is negligible. The broadening shown in Figs 11.5 to 11.7 is introduced only to enhance visualization. From the comparison of the spectra obtained in the adiabatic approximation with those resulting from the non-adiabatic approach, the following effects of non-adiabaticity are revealed. First, the *polaron shift* of the zero-phonon lines with respect to the bare-exciton levels is larger in the non-adiabatic approach than in the adiabatic approximation. Second, there is a strong *increase of the intensities of the phonon satellites* compared to those given by the adiabatic approximation. This increase can be by more than two orders of magnitude. Third, in the optical absorption spectra found within the non-adiabatic approach, there appear phonon satellites related to *non-active bare-exciton states*.

Fourth, the optical-absorption spectra demonstrate the crucial role of *non-adiabatic mixing* of different exciton and phonon states in quantum dots. This results in a rich structure of the absorption spectrum of the exciton–phonon system [14, 15, 17]. For the stacked quantum dots, this effect is significantly enhanced when the exciton-level splitting, caused by the coupling between quantum dots, is comparable with an LO-phonon energy [see Figs 11.6b and 11.7b]. Similar conclusions about the influence of the exciton–phonon interaction on the optical spectra of quantum dots have been formulated in [72] for the "strong coupling regime" for excitons and LO-phonons. Such a "strong coupling regime" is a particular case of the non-adiabatic

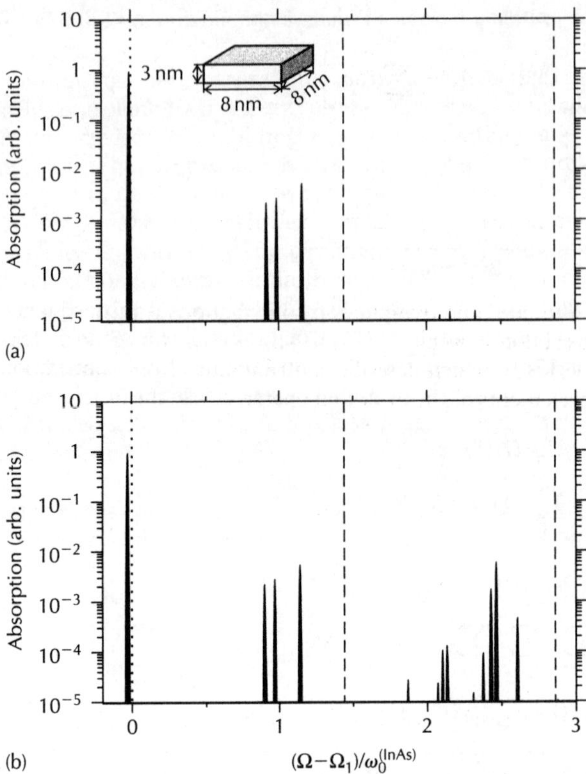

Figure 11.5   Absorption spectra, calculated with the adiabatic approximation (panel (a)) and with the non-adiabatic approach (panel (b)) for a single quantum dot. Optically active and non-active energy levels of a bare exciton are shown as dotted and dashed lines, respectively. Ω_1 is the transition frequency for the lowest state of a bare exciton. (Reprinted with permission from [20]. © 2004 by the American Physical Society.)

mixing related to a resonance, which arises when the spacing between exciton levels is close to the LO-phonon energy. As seen from Figs 11.6 and 11.7, effects of non-adiabaticity lead to a significant increase of absorption peaks in the spectral ranges, characteristic for one- and two-phonon satellites, even in the case when the exciton-level spacing is relatively far from satisfying the aforementioned resonant conditions.

In some cases (see, e.g., [11]) the luminescence spectrum of a quantum dot can be easily derived from its absorption spectrum. For example, under the assumption that the distribution function of the states of an exciton coupled to the phonon field, $f(\Omega)$, depends only on the energy of a state, the luminescence intensity at low temperatures ($\{y_\lambda \ll 1\}$) can be represented as:

$$I(\Omega) \propto \sum_{K=0}^{\infty} \frac{1}{K!} \sum_{\lambda_1,\dots,\lambda_K} f\left(\Omega + \sum_{k=0}^{K} \omega_{\lambda_k}\right) \left(\prod_{k=1}^{K} \frac{\partial}{\partial y_{\lambda_k}}\right) \alpha(\Omega)\Bigg|_{\{y_\lambda \to 0\}} . \tag{11.43}$$

Equation (11.43) is applicable, for instance, for the thermodynamic equilibrium photoluminescence. In this case the radiative lifetime of an exciton is much larger than the time characteristic of radiationless relaxation between one-exciton states.

We have calculated the spectra of the thermodynamic equilibrium luminescence at $T \to 0$ (not shown here) for quantum dots with parameters indicated above. Both for single and for coupled quantum dots, the intensities of phonon satellites in these spectra are significantly smaller than in the absorption spectrum. This is because in quantum dots under consideration the lowest one-exciton energy level (with $\beta = 1$) is less affected by the phonon-induced non-adiabatic mixing of states than higher levels. The luminescence spectrum at zero temperature is determined by

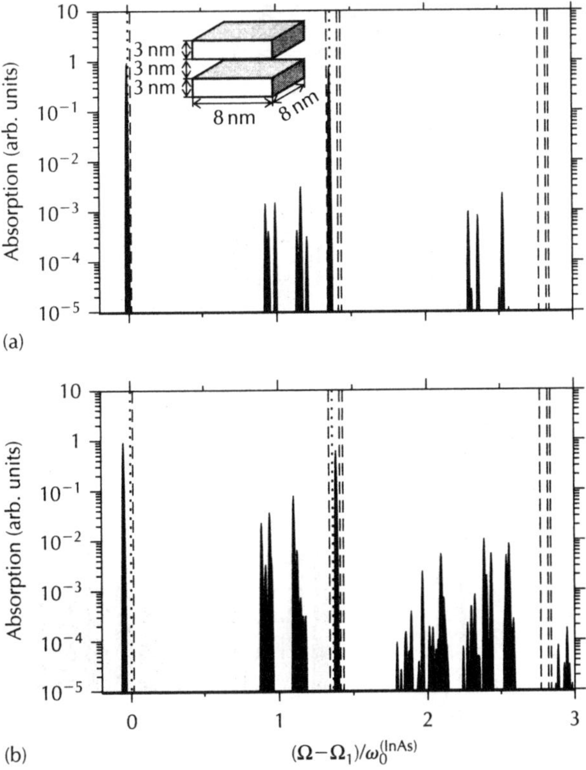

Figure 11.6 *Absorption spectra, calculated with the adiabatic approximation (panel (a)) and with the non-adiabatic approach (panel (b)) for two stacked identical quantum dots. Notations are the same as in Fig. 11.5. (Reprinted with permission from [20]. © 2004 by the American Physical Society.)*

transitions from the state with $\beta = 1$, while the absorption spectrum contains appreciable contributions due to transitions to higher exciton-phonon states.

11.5 Excitonic polarons in quantum dots: modification of the optical spectra

The conclusion about the large enhancement of the two-phonon sidebands in the luminescence spectra as compared to those predicted by the Huang-Rhys formula, which was explained in [11, 14] by non-adiabaticity of the exciton–phonon system in certain quantum dots, has been reformulated in [72] in terms of the Fröhlich coupling between product states with different electron and/or hole states.

A lens-shaped InAs/GaAs quantum dot is modelled in [72] as a truncated cone with basis radius R and height h. The exciton states in this cylindrically symmetric system are classified in terms of the z-projection of the total angular momentum of the electron–hole pair L_z. The Coulomb term

$$H_C = -\frac{e^2}{4\pi\epsilon_0\kappa|\mathbf{r}_e - \mathbf{r}_h|} \tag{11.44}$$

with the dielectric constant κ mixes electron and hole pair states with the same L_z (for example, the states with $L_z = 0$). In [72], the case $T = 0$ is analysed.

The exciton–phonon interaction Hamiltonian Eq. (11.3) with Eq. (11.20), leads to the direct Fröhlich coupling between the factorized (adiabatic) exciton–phonon states, which differ by one

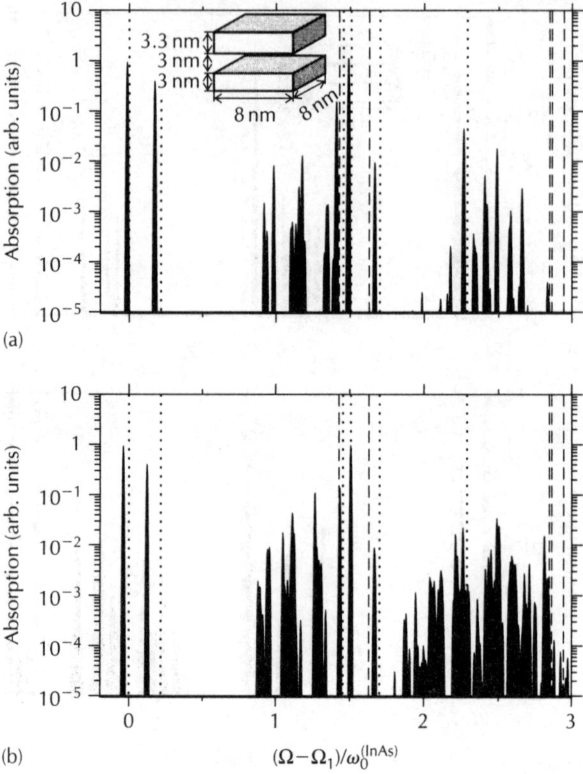

(a)

(b)

$(\Omega - \Omega_1)/\omega_0^{(\mathrm{InAs})}$

Figure 11.7   Absorption spectra, calculated with the adiabatic approximation (panel (a)) and with the non-adiabatic approach (panel (b)) for two stacked quantum dots of different height. Notations are the same as in Fig. 11.5. (Reprinted with permission from [20]. © 2004 by the American Physical Society.)

phonon and for which either the electron on the hole quantum number changes:

$$\langle n'_e, n'_h, n'_{\mathbf{q}} | H_{int} | n_e, n_h, n_{\mathbf{q}} \rangle \propto$$
$$\delta_{n'_{\mathbf{q}}, n_{\mathbf{q}} \pm 1} \left[\delta_{n'_h, n_h} \langle n'_e | \exp(\mp i\mathbf{q} \cdot \mathbf{r}_e | n_e) \rangle - \delta_{n'_e, n_e} \langle n'_h | \exp(\mp i\mathbf{q} \cdot \mathbf{r}_h | n_h) \rangle \right]. \tag{11.45}$$

The Hamiltonian of the exciton–phonon systems was diagonalized numerically in the basis, which consisted of the zero-phonon states $|S_e, S_h, 0\rangle$, $|S_e, P_{h\pm}, 0\rangle$, $|P_{e\pm}, S_h, 0\rangle$, $|P_{e\pm}, P_{h\pm}, 0\rangle$; the one-phonon continuum $|S_e, S_h, 1_{\mathbf{q}}\rangle$, $|S_e, P_{h\pm}, 1_{\mathbf{q}}\rangle$, $|P_{e\pm}, S_h, 1_{\mathbf{q}}\rangle$, $|P_{e\pm}, P_{h\pm}, 1_{\mathbf{q}}\rangle$; and the two-phononcontinuum $|S_e, S_h, 1_{\mathbf{q}}, 1_{\mathbf{q}'}\rangle$, $|S_e, P_{h\pm}, 1_{\mathbf{q}}, 1_{\mathbf{q}'}\rangle$, $|P_{e\pm}, S_h, 1_{\mathbf{q}}, 1_{\mathbf{q}'}\rangle$, $|P_{e\pm}, P_{h\pm}, 1_{\mathbf{q}}, 1_{\mathbf{q}'}\rangle$. This diagonalization led to entangled exciton–phonon states, which are called "excitonic polaron states" in [72] and which realize – in the aforementioned basis – a representation of the non-adiabatic exciton–phonon states introduced in [11].

Their energies are shown in Fig. 11.8 as a function of the quantum dot size. The part of the spectrum with $L_z = \pm 1$ reveals a double anticrossing between $|S_e, P_{h\pm}, 0\rangle$, $|P_{e\pm}, S_{h\pm}, 0\rangle$ and $|S_e, P_{h\pm}, 1_{\mathbf{q}}\rangle$. The magnitude of the anticrossings constitutes $\approx 7(10)$ meV for $R \approx 7(12.5)$ nm. The symmetrical [73] $L_z = 0$ states reveal anticrossings as large as ≈ 7 meV when a factorized symmetrical zero-phonon state is symmetrical with a factorized one-phonon continuum. When a factorized symmetrical zero-phonon state crosses a two-phonon polaron continuum, weaker anticrossings occur through the one-phonon continuum states $|S_e, P_{h\pm}, 1_{\mathbf{q}}\rangle|$ and $|P_{e\pm}, S_h, 1_{\mathbf{q}}\rangle$. In the excitonic ground state, the electron and hole envelope functions are very similar. Correspondingly, the excitonic polaron behaves as a neutral particle and manifests a weak polaronic red shift by about 0.9 meV with respect to the unperturbed factorized state $|S_e, S_h, 0\rangle$.

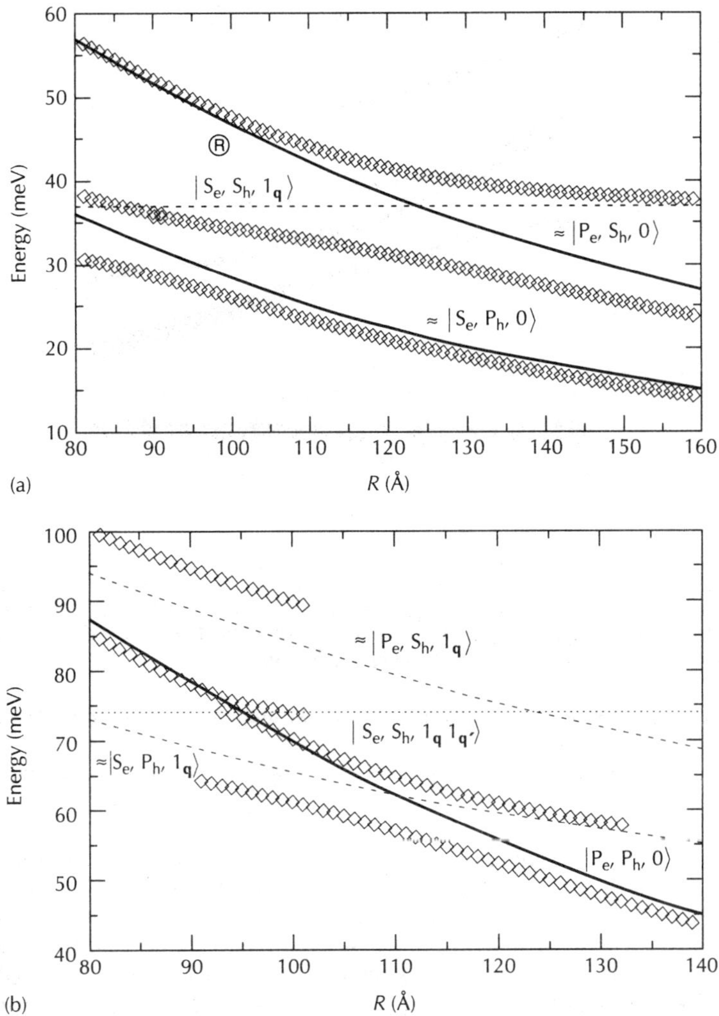

Figure 11.8   Excitonic polaron energies versus R (h/R = 0.1) in truncated cone-shaped InAs/GaAs quantum dots. (a) $L_z = \pm 1$ states. (b) Symmetrical $L_z = 0$ states. Only those states with more than 10% of zero phonon parts are shown. The energy zero is taken at the energy of the factorized $|S_e, S_h, 0\rangle$ state. Solid lines: zero-phonon factorized state. Dashed lines: one-phonon factorized states. Dotted lines: two-phonon factorized states. Symbols: excitonic polaron energies. (Reprinted with permission after [72]. © 2002 by the American Physical Society.)

For absorption at $T = 0$, the initial state is the vacuum with zero phonons. Therefore, the final state should contain an optically active zero-phonon component. As shown in Fig. 11.9, the calculated energies and intensities of interband absorption for excitonic polaron are significantly different from the purely excitonic predictions. As in [11], it is found that there are more absorption lines within the excitonic polaron description than in the adiabatic exciton approximation. The latter, in the range of quantum dot radii under consideration, gives two lines of comparable intensity, which correspond to the $L_z = 0$ states (corresponding to the $|S_e, S_h\rangle$ and the symmetrical combination of the $|P_e, P_h\rangle$ as shown in Fig. 11.8b). The excitonic polaron absorption displays up to five lines of different strength. Except the ground-state transition, which mostly relates to the $|S_e, S_h, 0\rangle$ state, their energy positions and intensities vary substantially with radius. The microscopic interpretation of the excitonic polaron absorption lines follows directly from Fig. 11.8b. In terms of the earlier works [11, 14], those absorption lines are due to the non-adiabatic transitions in the exciton–phonon system. The excitonic polaron reveals also an absorption

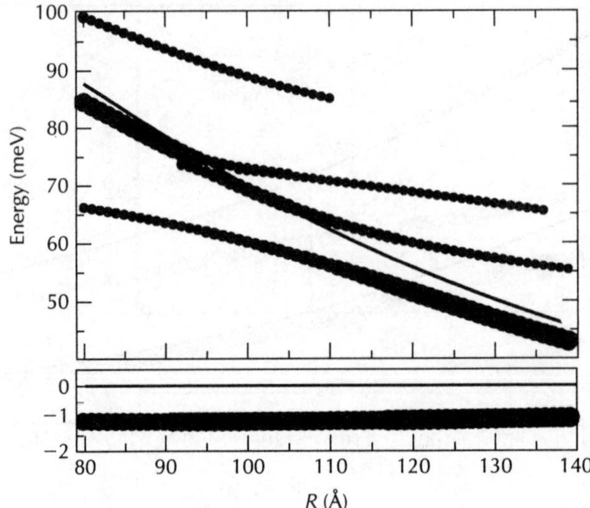

Figure 11.9   Calculated energies and intensities of interband optical transitions in truncated cone-shaped InAs/ GaAs quantum dots versus R (h/R = 0.1). Solid lines: excitonic picture. Symbols: excitonic polaron description. The radius of each symbol is proportional to the corresponding oscillator strength. Only symbols with radii larger than 5% of the ground transition one are displayed. The energy zero is taken at the ground excitonic transition. (Reprinted with permission after [72]. © 2002 by the American Physical Society.)

line at $\hbar\omega_{LO}$ above the ground-state transition, but for parameters used in [72] it is too weak to be shown in Fig. 11.9.

The excitonic polaron model of [72] has a restricted basis and therefore could not explain the totality of the absorption lines in the energy range of the expected excitonic transitions observed in [74].

The key difference between luminescence and absorption in the excitonic polaron picture is that even at $T = 0$ there exists a possibility of recombining to the vacuum state with 0,1,2... phonons. This gives rise to 1-, 2-... phonon satellite lines, which are red shifted from the zero-phonon line by $\hbar\omega_{LO}$, $2\hbar\omega_{LO}$.... The calculation in [72] showed that, e.g., for a quantum dot with $R = 8\,\text{nm}$ a two-photon satellite has intensity, which is about 100 times larger than that predicted within the adiabatic Pekar–Huang-Rhys approach. Thus the prediction of large enhancement of the phonon satellites in the luminescence spectra of exciton–phonon systems obtained within the non-adiabatic approach [11] has been rederived using the Fröhlich coupling between the adiabatic product states [72].

11.6   Recent studies

Due to non-adiabaticity, multiple absorption peaks appear in spectral ranges characteristic for phonon satellites. From the states which correspond to these peaks, the system can rapidly relax to the lowest emitting state. Therefore, in the photoluminescence excitation (PLE) spectra of quantum dots, pronounced peaks can be expected in spectral ranges characteristic for phonon satellites. Experimental evidence of the enhanced phonon-assisted absorption due to effects of non-adiabaticity has been provided by PLE measurements on single self-assembled InAs/GaAs [75] and InGaAs/GaAs [76] quantum dots. The polaron concept was also invoked for the explanation of the PLE measurements on self-organized $In_xGa_{1-x}As$/GaAs [77] and CdSe/ZnSe [78] quantum dots. Recently [79], the influence of quantum dot occupancy and incident light polarization on the carrier–lattice interactions in electron-doped InAs self-assembled quantum dots grown on GaAs has been investigated up to 28 T by far-infrared magnetotransmission spectroscopy. An enhancement of the efficiency of the electron–phonon coupling with increased electron population from one to two electrons per dot is found, in good agreement with the predicted $\sqrt{2}$

increase due to the antisymmetrization of the electron wave function in the two-electron quantum dot. The authors of [80] have reported the direct observation of polarons in InAs/GaAs self-assembled quantum dots populated by a few electrons. The degree of coupling is varied in a systematic way with a set of samples having electron intersublevel spacing changing from larger to smaller than the longitudinal-optical-phonon energy. The signature of polarons is manifested clearly by the observation of a large (12–20 meV) anticrossing for both InAs and GaAs-like QD phonons using resonant Raman spectroscopy.

Preparation of the CdS quantum dots with HgS quantum wells (QDQWs) was described in [81]. Such structures possess greater photochemical stability as compared to ordinary quantum dots. Excitons, localized in the HgS quantum well, are separated from the localized surface levels by the CdS shell and, as a result, have longer lifetime and improved quantum yield/photoresponse. It has been revealed that QDQWs are preferentially tetrahedral particles [82] with zinc-blende crystal lattice. An insight into the nature of the optical response of CdS/HgS/CdS QDQWs has been provided by a simultaneous consideration of the tetrahedral shape of a QDQW, interface optical phonons, and non-adiabatic phonon-assisted transitions [18, 83]. The exciton states and the photoluminescence spectra appear to be very sensitive to the shape of the QDQWs.

Exciton energies and wave functions have been found numerically, within a finite-difference scheme [18], as eigenstates of the Hamiltonian:

$$\hat{H}_{ex} = [\hat{H}_e + V_{s-a}(\mathbf{r}_e)] - [\hat{H}_h - V_{s-a}(\mathbf{r}_h)] + V_{int}(\mathbf{r}_e, \mathbf{r}_h) \tag{11.46}$$

where $\hat{H}_e$ is the two-band electron Hamiltonian and $\hat{H}_h$ is the six-band hole Hamiltonian derived from a general eight-band Hamiltonian for heterostructures [85]. The dielectric mismatch at the boundaries of QDQW's shells leads to the appearance of the electron and hole self-interaction potentials V_{s-a} and significantly changes the potential of the electron–hole interaction V_{int}.

While only a few lowest electron states lie below the CdS bulk conduction band edge and are localized near the HgS shell, many hole states are trapped in the HgS shell and lie above the CdS bulk valence band edge. As shown in [18], the electron in the ground state is practically uniformly distributed in the HgS shell, while the hole in the ground state is localized near the edges of the tetrahedral HgS shell. Unlike ground state, the hole in the first excited state substantially penetrates the region of facets of the HgS shell, where the electron density in the ground state is high. The electron states are widely separated by energy and the Coulomb mixing between the electron–hole pair states with different electron states can be ignored. As distinct from that, the hole states are closely spaced and the Coulomb mixing between the pair states with different hole states is taken into account [18].

In the QDQW's absorption spectrum (Eq. 11.11), the intensity of a zero-phonon transition to the exciton level β is $|d_\beta|^2$ (cp. Eq. 11.12). Figure 11.10 shows that the intensity of the second

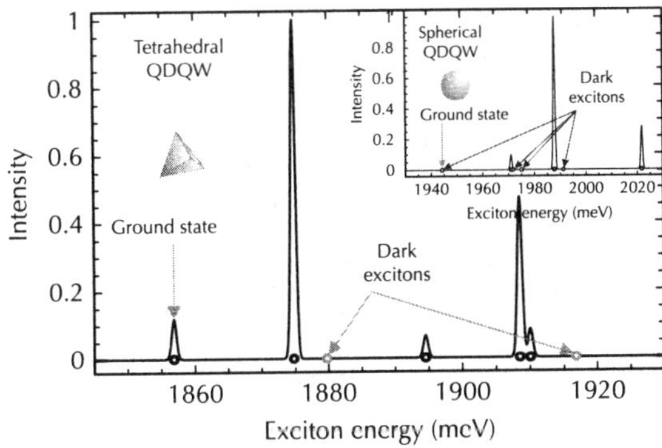

Figure 11.10 Normalized intensities of the lowest exciton levels in the absorption spectrum of the tetrahedral QDQW (the inset gives the intensities for the spherical QDQW considered in [86]). Spectral lines are broadened by a Gaussian to enhance visualization. (Reprinted with permission after [18]. © 2004 by the American Physical Society.)

exciton level is 8.5 times larger than that of the ground state level. Since the exciton ground state is luminescent and the second exciton state is absorptive, a Stokes shift should be observed between the excitation and the zero-phonon line in the equilibrium photoluminescence spectrum. In addition to the fact that the calculated exciton ground state energy of 1857 meV in [18] is in excellent agreement with experimental data of [82, 84], the obtained energy difference of 18 meV between the second and ground states of the exciton is very close to the value of 19 meV of the Stokes shift found in the fluorescence line narrowing spectrum [82]. The inset to Fig. 11.10 shows the result of an analogous finite-difference calculation of the exciton states for a spherical QDQW [86] with the same thickness of the HgS shell as that for the tetrahedral QDQW. The exciton ground state is eight-fold degenerate for both spherical and tetrahedral QDQWs. The remarkable effect of the QDQW's shape is that the exciton in the ground state is *bright* for the tetrahedral QDQW shape, while it is *dark* for the spherical QDQW.

The bulk-like and interface phonon modes and the amplitudes of the electron–phonon interaction for the tetrahedral QDQW are theoretically derived in [18, 83] using the same finite-difference scheme as for the excitons.

Figure 11.11 represents a comparison of the photoluminescence spectrum calculated using the non-adiabatic theory of [11] for a single tetrahedral QDQW with two experimental spectra measured at different excitation wavelengths and intensities [84, 87]. Spectral diffusion, which has been explained recently by a reorganization of local electric fields in or around the particles [88], is most likely responsible for rather broad lines in the photoluminescence spectra of a single QDQW. Experimentally defined average phonon frequency is 35.3 ± 0.6 meV and the Huang-Rhys parameter is 0.25 ± 0.05 [84]. Because of the localization of an exciton in the HgS well, the amplitudes of bulk-like phonon modes are much smaller than those of interface-phonon modes. The dominant contribution to the intensity of the one-phonon lines in the calculated photoluminescence spectrum is due to p-like interface-phonon modes. Two highest one-phonon peaks correspond to the phonon modes with frequencies 34.3 and 36.7 meV. The average frequency, weighted with the heights of the two peaks, is 35.5 meV, which is in a good agreement with the experimentally defined value. The intensity of the calculated two-phonon lines can hardly be compared with experimental spectra because of the high background level in this spectral region. The inset to Fig. 11.11 shows one- and two-phonon bands in the photoluminescence spectrum calculated within the adiabatic approximation. Since the exciton ground state β_0 is eight-fold degenerate and its intensity $|d_{\beta_0}|^2$ is small compared to the intensities of the higher lying exciton states (see Fig. 11.10), the heights of phonon peaks in the adiabatic approximation are much lower than those calculated using the non-adiabatic theory. Only the simultaneous consideration of the tetrahedral

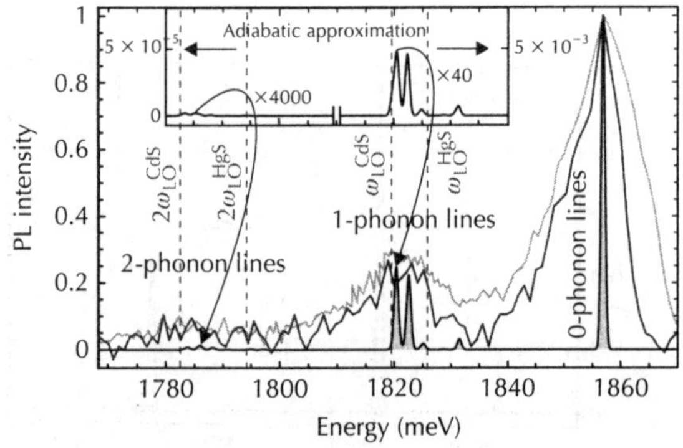

Figure 11.11   Photoluminescence spectrum of a tetrahedral QDQW normalized to the zero-phonon line's height. Calculated spectral lines (black curve) are broadened by a Gaussian to enhance visualization. Experimental spectra are measured at T = 10 K with excitation wavelength 633 nm and intensity 15 kW/cm² (light grey curve) [84] and with excitation wavelength 442 nm and intensity 5 kW/cm² (dark grey curve) [87]. The insert gives one- and two-phonon bands (magnified by factors 40 and 4000, respectively) as calculated within the adiabatic approximation. (Reprinted with permission after [18]. © 2004 by the American Physical Society.)

shape of a QDQW, interface optical phonons, and non-adiabatic phonon-assisted transitions allows for a quantitative interpretation of the observed photoluminescence spectra of a CdS/HgS/CdS QDQW [18, 83].

11.7 Conclusions

Due to the non-adiabaticity of the exciton–phonon system in several types of quantum dots, different *new channels* of the phonon–assisted optical transitions open. As distinct from the adiabatic approximation, which describes only phonon transitions through intermediate exciton states coinciding either with its initial or its final state, the non-adiabatic approach also takes into consideration *additional* phonon transitions: (i) between different exciton states belonging to the same degenerate level (proper Jahn–Teller effect), as well as (ii) between exciton states of different energy (pseudo Jahn–Teller effect). In contrast to the works based on the theory of Pekar and Huang-Rhys, which adequately account for the adiabatic phonons only, the non-adiabatic approach enables describing correctly the transitions involving *all* types of phonons, including Jahn–Teller phonons.

The effect of the channels leads to a considerable enhancement of the phonon-assisted transition probabilities in the photoluminescence of quantum dots even with relatively weak electron–phonon coupling strength. The resulting multiphonon optical spectra are considerably different from the Franck–Condon progression. In order to obtain agreement between theory and experiment, in particular regarding the optical properties of self-assembled quantum dots, it is necessary to take into account these new channels of the phonon-assisted optical transitions.

Acknowledgements

We acknowledge discussions with Evghenii Pokatilov, Serghei Klimin, Vladimir Fonoberov, Serghei Balaban, Fritz Henneberger, Jean-Pierre Leburton, Dieter Bimberg, the late Robert Heitz, Andrei Schliwa, Marius Grundmann, Mario Capizzi, Paul Koenraad, Andrei Silov, Jos Haverkort, Alberto García-Cristóbal, Joachim Wolter, Maurice Skolnick, Laurence Eaves, Mohamed Henini, Jörg Kotthaus, Victor Moshchalkov, Manus Hayne, Frank Wise, Thomas Basché, and Isaac Bersuker. This work was supported by the FWO-V project G.0435.03, the WOG WO.035.04N (Belgium) and the European Commission SANDiE Network of Excellence, contract No. NMP4-CT-2004-500101.

References

1. D. Bimberg, M. Grundmann, and N.N. Ledentsov. *Quantum Dot Heterostructures* (Wiley, Chichester, 1999).
2. P. Michler. *Single Quantum Dots: Fundamentals, Applications, and New Concepts* (Springer, Berlin, 2003).
3. M.S. Skolnick and D.J. Mowbray, *Ann. Rev. Mat. Res.* **34**, 181 (2004).
4. G.D. Scholes and G. Rumbles, *Nature Materials* **5**, 683 (2006).
5. L. Jacak, P. Hawrylak, and A. Wójs. *Quantum Dots*, (Springer, Berlin, 1998).
6. C. Delerue and M. Lannoo. *Nanostructures. Theory and Modeling*, (Springer, Berlin, 1998).
7. J.T. Devreese, in: *Encyclopedia of Physics*, edited by R.G. Lerner and G.L. Trigg (Wiley-VCH, 2005), Vol. 2, pp. 2004–2027.
8. J.T. Devreese, in: *Handbook of Semiconductor Nanostructures and Nanodevices*, edited by A.A. Balandin and K.L. Wang (American Scientific Publishers, Los Angeles, 2006), Vol. 4, pp. 311–337.
9. J.T. Devreese, V.M. Fomin, and E.P. Pokatilov, in: *Handbook of Semiconductor Nanostructures and Nanodevices*, edited by A.A. Balandin and K.L. Wang (American Scientific Publishers, Los Angeles, 2006), Vol. 4, pp. 339–407.

10. V.M. Fomin, E.P. Pokatilov, J.T. Devreese, S.N. Klimin, S.N. Balaban, and V.N. Gladilin, in: *Proceedings of the 23rd International Conference on the Physics of Semiconductors*, edited by M. Scheffler and R. Zimmermann (World Scientific, Singapore, 1996), pp. 1461–1464.

11. V.M. Fomin, V.N. Gladilin, J.T. Devreese, E.P. Pokatilov, S.N. Balaban, and S.N. Klimin, *Phys. Rev. B* **57**, 2415 (1998).

12. V. Jungnickel and F. Henneberger, *J. Lumin.* **70**, 238 (1996).

13. M. Bissiri, G. Baldassarri Höger von Högersthal, A.S. Bhatti, M. Capizzi, A. Frova, P. Frigeri, and S. Franchi, *Phys. Rev. B* **62**, 4642 (2000).

14. V.N. Gladilin, S.N. Balaban, V.M. Fomin, and J.T. Devreese, in: *Proceedings of the 25th International Conference on the Physics of Semiconductors, Osaka, Japan, 2000* (Springer, Berlin, 2001), Part II, pp. 1243–1244.

15. J.T. Devreese, V.M. Fomin, V.N. Gladilin, E.P. Pokatilov, and S.N. Klimin, *Nanotechnology* **13**, 163 (2002).

16. E.P. Pokatilov, S.N. Klimin, V.M. Fomin, J.T. Devreese, and F.W. Wise, *Phys. Rev. B* **65**, 075316 (2002).

17. J.T. Devreese, V.M. Fomin, E.P. Pokatilov, V.N. Gladilin, and S.N. Klimin, *Phys. Stat. Sol. (c)* **0**, 1189 (2003).

18. V.A. Fonoberov, E.P. Pokatilov, V.M. Fomin, and J.T. Devreese, *Phys. Rev. Lett.* **92**, 127402 (2004).

19. R. Heitz, O. Stier, I. Mukhametzhanov, A. Madhukar, and D. Bimberg, *Phys. Rev. B* **62**, 11017 (2000).

20. V.N. Gladilin, S.N. Klimin, V.M. Fomin, and J.T. Devreese, *Phys. Rev. B* **69**, 155325 (2004).

21. M.G. Bawendi, L. Wilson, L. Rothberg, P.J. Caroll, T.M. Jedju, M.L. Steigerwald, and L.E. Brus, *Phys. Rev. Lett.* **65**, 1623 (1990).

22. M. Nirmal, C.B. Murray, D.J. Norris, and M.G. Bawendi, *Z. Phys. D* **26**, 361 (1993).

23. V. Jungnickel, F. Henneberger, and J. Puls, in: *22nd International Conference on the Physics of Semiconductors* (World Scientific, Singapore, 1994), Vol. 3, p. 2011.

24. D.J. Norris, Al.L. Efros, M. Rosen, and M.G. Bawendi, *Phys. Rev. B* **53**, 16347 (1996).

25. A.P. Alivisatos, T.D. Harris, P.J. Carroll, M.L. Steigerwald, and L.E. Brus, *J. Chem. Phys.* **90**, 3463 (1989).

26. M.C. Klein, F. Hache, D. Ricard, and C. Flytzanis, *Phys. Rev. B* **42**, 11123 (1990).

27. Al.L. Efros, A.I. Ekimov, F. Kozlowski, V. Petrova-Koch, H. Schmidbaur, and S. Shulimov, *Solid State Commun.* **78**, 853 (1991).

28. J.J. Shiang, S.H. Risbud, and A.P. Alivisatos, *J. Chem. Phys.* **98**, 8432 (1993).

29. A. Mlayah, A.M. Brugman, R. Carles, J.B. Renucci, M.Ya. Valakh, and A.V. Pogorelov, *Solid State Commun.* **90**, 567 (1994).

30. G. Scamarcio, V. Spagnolo, G. Ventruti, M. Lugará, and G.C. Righini, *Phys. Rev. B* **53**, R10489 (1996).

31. S.V. Gaponenko, *Fiz. Tverd. Tela.* **30**, 577 (1996).

32. F. Henneberger and J. Puls, in: *Optics of Semiconductor Nanocrystals*, edited by F. Henneberger, S. Schmitt Rink, and E.O. Göbel (Akademie-Verlag, Berlin, 1993), p. 447.

33. S. Nomura and T. Kobayashi, *Phys. Rev. B* **45**, 1305 (1992).

34. Y. Chen, S. Huang, J. Yu, and Y. Chen, *J. Lumin.* **60 & 61**, 786 (1994).

35. S.I. Pekar, *Zh. Eksp. Teor. Fiz.* **20**, 267 (1950).

36. K. Huang and A. Rhys, *Proc. R. Soc. London Ser. A* **204**, 406 (1950).

37. N.O. Lipari and A. Baldereschi, *Phys. Rev. Lett.* **42**, 1660 (1970).

38. A. Baldereschi and O. Lipari, *Phys. Rev. B* **8**, 2697 (1973).

39. J.B. Xia, *Phys. Rev. B* **40**, 8500 (1989).

40. D.M. Mittleman, R.W. Schoenlein, J.J. Shiang, V.L. Colvin, A.P. Alivisatos, and C.V. Shank, *Phys. Rev. B* **49**, 14435 (1994).

41. R. Englman. *The Jahn-Teller Effect in Molecules and Crystals*, (Wiley, New York, 1972).

42. A.M. Stoneham. *Theory of Defects in Solids: Electronic Structure of Defects in Insulators and Semiconductors*, (Clarendon, Oxford, 1975), Chap. 17.

43. Yu.E. Perlin and B.S. Tsukerblat. *The Effects of Electron–Vibrational Interaction on the Optical Spectra of Paramagnetic Impurity Ions* (Shtiintsa, Kishinev, 1974), (in Russian).

44. Yu.E. Perlin and B.S. Tsukerblat, in: *The Dynamical Jahn–Teller Effect in Localized Systems*, edited by Yu.E. Perlin and M. Wagner (Elsevier, Amsterdam, 1984), p. 251.

45. I.B. Bersuker and V.Z. Polinger. *Vibronic Interactions in Molecules and Crystals* (Springer, Berlin, 1989).

46. I.B. Bersuker. *The Jahn–Teller Effect*, (Cambridge University Press, Cambridge, 2006), p. 495.

47. H.A. Jahn and E. Teller, *Proc. R. Soc. London A* **161**, 220 (1937).

48. R. Kubo, *J. Phys. Soc. Japan* **12**, 570 (1957).

49. V.B. Berestetskii, E.M. Lifshitz, and L.P. Pitaevskii. *Relativistic Quantum Theory* (Oxford, 1971).

50. R.P. Feynman, *Phys. Rev.* **84**, 108 (1951).

51. M. Grundmann, J. Christen, N.N. Ledentsov, J. Böhrer, D. Bimberg, S.S. Ruvimov, P. Werner, U. Richter, U. Gösele, J. Heydenreih, V.M. Ustinov, A.Yu. Egorov, A.E. Zhukov, P.S. Kop'ev, and Zh.I. Alferov, *Phys. Rev. Lett.* **74**, 4043 (1995).

52. S.N. Klimin, E.P. Pokatilov, and V.M. Fomin, *Phys. Status Solidi (b)* **184**, 373 (1994).

53. T. Takagahara, *Phys. Rev. B* **47**, 4569 (1993).

54. M. Nirmal, D.J. Norris, M. Kuno, M.G. Bawendi, A.l.L. Efros, and M. Rosen, *Phys. Rev. Lett.* **75**, 3728 (1995).

55. G.L. Bir and G.E. Pikus. *Symmetry and Strain-Induced Effects in Semiconductors*, (Wiley, New York, 1975).

56. D.J. Norris and M.G. Bawendi, *Phys. Rev. B* **53**, 16338 (1996).

57. *Semiconductors. Physics of II–VI and I–VII Compounds, Semimagnetic Semiconductors*, edited by K.H. Hellwege, Landolt-Börnstein, New Series, Group III, Vol. 17, Pt b (Springer, Berlin, 1982).

58. A.W.E. Minnaert, A.Yu. Silov, W. van der Vleuten, J.E.M. Haverkort, J.H. Wolter, A. García-Cristóbal, V.N. Gladilin, V.M. Fomin, and J.T. Devreese, in: *Proceedings of the 24th International Conference on the Physics of the Semiconductors*, edited by, D. Gershoni (World Scientific, Singapore, 1999) pp. Mo–P139.

59. A. García-Cristóbal, A.W.E. Minnaert, V.M. Fomin, J.T. Devreese, A.Yu. Silov, J.E.M. Haverkort, and J.H. Wolter, *Phys. Stat. Sol. (b)* **215**, 331 (1999).

60. V.M. Fomin, V.N. Gladilin, S.N. Klimin, J.T. Devreese, P.M. Koenraad, and J.H. Wolter, *Phys. Rev. B* **61**, R2436 (2000).

61. P. Offermans, P.M. Koenraad, J.H. Wolter, K. Pierz, M. Roy, and P.A. Maksym, *Phys. Rev. B* **72**, 165332 (2005).

62. T. Schmidt, R.J. Haug, K. von Klitzing, A. Förster, and H. Lüth, *Phys. Rev. Lett.* **78**, 1544 (1997).

63. R.J. Luyken, A. Lorke, M. Fricke, J.P. Kotthaus, G. Medeiros-Ribeiro, and P.M. Petroff, *Nanotechnology* **10**, 14 (1999).

64. B. Partoens and F.M. Peeters, *Phys. Rev. Lett.* **84**, 4433 (2000).

65. M. Pi, A. Emperador, M. Barranco, F. Garcias, K. Muraki, S. Tarucha, and D.G. Austing, *Phys. Rev. Lett.* **87**, 066801 (2001).

66. M. Bayer, P. Hawrylak, K. Hinzer, S. Fafard, M. Korkusinski, Z.P. Wasilewski, O. Stern, and A. Forchel, *Science* **297**, 1313 (2002).

67. L. Rebohle, F.F. Schrey, S. Hofer, G. Strasser, and K. Unterrainer, *Appl. Phys. Lett.* **81**, 2079 (2002).

68. S. Bednarek, T. Chwiej, J. Adamowski, and B. Szafran, *Phys. Rev. B* **67**, 205316 (2003).

69. S.I. Pekar, *Zh. Eksp. Teor. Fiz.* **20**, 267 (1950).

70. K. Huang and A. Rhys, *Proc. R. Soc. London, Ser. A* **204**, 406 (1950).

71. S. Nomura and T. Kobayashi, *Phys. Rev. B* **45**, 1305 (1992).

72. O. Verzelen, R. Ferreira, and G. Bastard, *Phys. Rev. Lett.* **88**, 146803 (2002).

73. The states with $L_z = 0$ can be split into symmetrical and antisymmetrical states under the transformation of the polar angles $\theta_{e,h} \rightarrow -\theta_{e,h}$ and $\phi_{\mathbf{q}} \rightarrow -\phi_{\mathbf{q}}$ of the vectors $\vec{r}_{e,h}$ and $\vec{q}$, respectively.

74. Y. Toda, O. Moriwaki, M. Nishioka, and Y. Arakawa, *Phys. Rev. Lett.* **82**, 4114 (1999).

75. A. Lemaître, A.D. Ashmore, J.J. Finley, D.J. Mowbray, M.S. Skolnick, M. Hopkinson, and T.F. Krauss, *Phys. Rev. B* **63**, 161309(R) (2001).

76. A. Zrenner, F. Findeis, M. Baier, M. Bichler, and G. Abstreiter, *Physica B* **298**, 239 (2001).

77. R. Heitz, H. Born, F. Guffarth, O. Stier, A. Schliwa, A. Hoffmann, and D. Bimberg, *Phys. Rev. B* **64**, 241305 (2001).

78. U. Woggon, D. Miller, F. Kalina *et al.*, *Phys. Rev. B* **67**, 045204 (2003).

79. B.A. Carpenter, E.A. Zibik, M.L. Sadowski, L.R. Wilson, D.M. Whittaker, J.W. Cockburn, M.S. Skolnick, M. Potemski, M.J. Steer, and M. Hopkinson, *Phys. Rev. B* **74**, 161302(R) (2006).
80. B. Aslan, H.C. Liu, M. Korkusinski, P. Hawrylak, and D.J. Lockwood, *Phys. Rev. B* **73**, 233311 (2006).
81. A. Eychmüller, A. Mews, and H. Weller, *Chem. Phys. Lett.* **208**, 59 (1993).
82. A. Mews, A.V. Kadavanich, U. Banin, and A.P. Alivisatos, *Phys. Rev. B* **53**, R13242 (1996).
83. V.A. Fonoberov, E.P. Pokatilov, V.M. Fomin, and J.T. Devreese, *Physica E* **26**, 63 (2005).
84. F. Koberling, A. Mews, and T. Basché, *Phys. Rev. B* **60**, 1921 (1999).
85. E.P. Pokatilov, V.A. Fonoberov, V.M. Fomin, and J.T. Devreese, *Phys. Rev. B* **64**, 245328 (2001).
86. E.P. Pokatilov, V.A. Fonoberov, V.M. Fomin, and J.T. Devreese, *Phys. Rev. B* **64**, 245329 (2001).
87. T. Basché (private communications).
88. R.G. Neuhauser, K.T. Shimizu, W.K. Woo, S.A. Empedocles, and M.G. Bawendi, *Phys. Rev. Lett.* **85**, 3301 (2000).

CHAPTER 12

Slow Oscillation and Random Fluctuation in Quantum Dots: Can we Overcome?

Raphael Tsu

University of North Carolina at Charlotte, Charlotte NC 28223, USA

12.1 Introduction

I would like to place the development of man-made solids, such as the introduction of superlattice [1] followed by quantum wells [2, 3] more than 30 years ago, evolving into the present with quantum dots, into the category of *quantum composites* which is in contrast to *classical composites* where mean-field approximation is adequate. First of all, let us define a size a when quantum interference is important. Surely, it is dictated by the condition that $a < \Lambda$, the mean-free path in bulk or more precisely the coherence length of an electron. On the other hand, one often encounters the notion that quantum confinement is important whenever $a < R_B$, the Bohr radius. The latter condition is doubtful because only in bulk can the dielectric constant representing screening be uniquely defined. At low temperatures, Λ may be 100 nm.So that for $a \sim 100$ nm, R_B is practically the bulk value, therefore quantum interference is very much operative in a QD of 100 nm. In any case, we shall use the former criterion to define the existence of a QD. Nicollian challenged me to place many quantum dots under one large contact. Our first attempt was to crystallize a thin layer of a Si followed by oxidation resulting in many islands of Si QD embedded in an oxide layer. Our plan was to construct a functional device by utilizing several wells with separated discrete quantum states produced by the high oxide barrier, being nearly 3.2 eV above the conduction band of silicon. We were pleased by our initial success in 1991 [4] of observing discrete structures having a metal contact, indicating that there were several distinct quantum states of the Si QD. However, strange conductance oscillations and switching were then observed [5].

Typically in a contact of $40 \times 40 \, \mu m^2$, there were ~ 1600 Si QDs of lateral dimension ~ 5–10 nm, so that these QDs are well separated. The conductance jump is in multiples of $39 \, \mu S$, or $\Delta G \sim ge^2/h$, with g, a degeneracy factor, usually 2, 4, etc. At this stage, we were sure that we were dealing with complicated trappings, particularly the oscillation period, and switching time is typically varying from a fraction of a second to as long as 20 s. This complexity discouraged us from pursuing the study in a more aggressive manner, partly because we lost the NSF support because our programme manager did not appreciate the complexity. I was also pursuing other topics, e.g. porous silicon, very fashionable at the time, although I continued to analyze the vast amount of data and applied a deeper interpretation and understanding. For example, it is only recently that I discovered an understanding on the conductance spectrum. Meanwhile, I learned about optical blinking in QDs which is quite similar to conductance switching and have decided to present both cases in this chapter to show how similar the data are on electrical switching and optical blinking.

12.2   What is a quantum dot?

Let us first describe what a quantum dot is:

1. A quantum dot should have the same coordination number. For example, silicon as a solid has a coordination number of four, but six in the molten state. Clearly, a molecule of two silicon atoms would have a coordination number of two. Taking the coordination number of four as in a crystalline solid of silicon implies an Si QD defined with four coordination numbers, this limits our consideration of a cluster of silicon atoms resembling solid silicon.

2. To characterize solid silicon, regardless of how good or, more precisely, how high the mobility is, the energy band gap and the effective masses for the valence and conduction bands are basically identical. Whereas for a quantum dot of silicon, apart from the separation of discrete states dictated by size, there is really no standard defining two Si QDs. The question is how can we choose a standard? For example, because of the tensor nature of the effective masses, the states are far from the usual 1S, 2S, 2P, etc. quantum numbers we use for the central field atomic states. Simple computation with two masses, one along the longitudinal ellipsoid and the other transverse to the axis of the ellipsoid, leads to a different set of energy degeneracy factors for an Si QD. As size is reduced to $\sim 1\,$nm, clearly the indirect nature of the band gap would have changed, leading to another complication for the classification of Si QDs. Does it make sense to rely on *ab initio* computation? The answer is no, because the computed energy structure, even if correct, gives us little physical meaning to serve as a guide for understanding the physics of an Si QD. My view of the importance of *ab initio* computation lies in the confirmation of conventional approaches with parameters. Once these parameters are in place, we can rely on the usual physical models: energy bands, symmetry, effective masses, etc.

3. Let us see what we can do with the spectral linewidth of photoluminescence of a QD. Before we delve into it, let us digress by considering what the linewidth is of optical transitions involved in a QD. In solid, thermal broadening, $k_B T$, represents the most important factor in addition to other energy non-conserving processes. It is not quite recognized by many workers in QDs that the primary broadening factor is still $k_B T$, because while energy barriers may serve to isolate the energy states of a QD, because of the similarity of the dynamic force matrix, or simply the small difference in elastic constants of the barrier confining the QD from that of the QD, phonons are much less confined, leading to efficient energy losses in terms of phonons. At this stage of the discussion, I want to point out that in an atom, unlike the vibronic states in a molecule, there is no broadening factor other than the perturbation presented by the application of a vector potential to the atomic Hamiltonian. Even at very low frequency such as near a dc electric field, the Stark effect is sufficient to induce coupling between states allowing energy to be dispersed to states having a probability of escaping from the atom, even though the probability is very small. In short, the standard or the Figure of merit of a QD may be tied to the linewidth of the optical transition. Measuring the spectrum of a single InP QD, Bertram *et al.* [6] observed a linewidth of $0.6\,$meV at $1.645\,$eV being equal to $k_B T$ at $T = 6.9\,$K, indicating the absence of other energy non-conserving processes. Figure 12.1 shows their measurement.

4. Other parameters generally used to characterize a solid, such as size with AFM, XRD, high-resolution TEM, and Raman scattering, are useful, but generally do not tell you the quality of the QD. Most researchers in QDs still cling onto the terminology – quantum transport. In fact, there is no such thing with QDs because only tunnelling and hopping between dots can transport an electron through the dots.

5. Let us discuss the possibility of forming a 3D superlattice structure with QDs. Unlike photonic crystals [7] where the periodic structure is characterized by the wavelength of photons with the first Brillouin zone 50 times larger than the de Broglie wavelength of electrons in typical QDs, allowing successful fabrication using available lithographic tools as well as various schemes such as stacking of spheres [8]. Most QDs utilize the strain-induced Stranski–Krastanov growth techniques, which is not possible at present to produce a periodic 3D structure of QDs. A single QD has the problem of contact for electronic

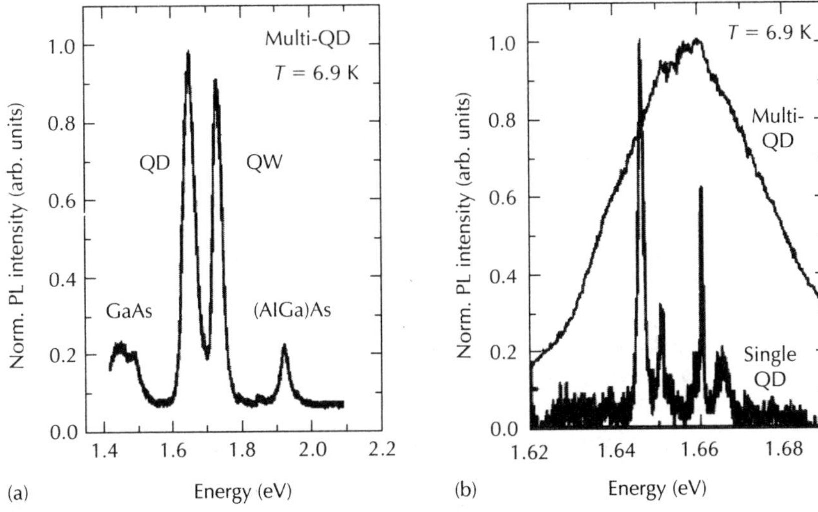

Figure 12.1 *Photoluminescence at 5 K of InP SK QDs on 3 nm GaAs QW with* $Al_{0.3}Ga_{0.7}As$ *barrier forming QDs of 140 nm lateral dimension with an average density of 5 QDs/μm^2. (a) Many QDs with line width ~435 meV, and (b) single QD with linewidth ~0.6 meV at 1.66 eV, which is K_BT at 6.9 K. From [6].*

applications [9]. The fundamental issue is the lack of an equal potential surface for defining a voltage for input/output. Is there any advantage in having a distribution of QDs either in electronic applications or in optical devices? Thus far it is doubtful that such arrangement can outperform the present-day superlattice and quantum well structures. After the presentation of random fluctuation, I shall return to make some comments and judgements.

6. When the size of the QD is reduced to ~1 nm, several things happen. First, the capacitance is reduced [10] and the dielectric constant is lowered [11] to the point that Bohr radius is much reduced with much increased ionization energy for dopants [12]. Even if we can overcome the statistical problem of accurately introducing dopants into a QD, the QD remains intrinsic at room temperature and below. On the other hand, since quantum transport has no meaning, there is no need for doping. QD devices depend on carrier injection via tunnelling.

12.3 Slow oscillation and random switching instability in a distribution of QDs

There is a significant increase in interaction when the individual QD states are occupied consistent with Pauli's exclusion principle. Coupling of these QDs form a 2D-like system giving rise to steps in conductance. This important many-body effect which explains why conductance steps are usually led by conductance peaks was recognized only recently after LaFave and I [13a,b] solved the effect of the discrete nature of electrons on the capacitance of a dielectric sphere, a problem similar to the QD. In 1990 Nicollian and I decided to crystallize *a*-Si into Si QDs, separated by thin oxide barriers and metallic contacts for I–V and G–V measurements [4].

Figure 12.2 shows a schematic drawing of the modified MOSFET structure with nanocrystallites shown as shaded circles embedded in an a-SiO_2 matrix. The substrate used has $n \sim 3.5 \times 10^{16} cm^{-3}$ silicon wafers. Starting with a thermally grown field of oxide 100 nm thick at 1050°C in dry oxygen, the active device, varying in size from $40 \times 40\,\mu m^2$ down to $10 \times 10\,\mu m^2$, is formed by etching a window photolithographically. A thin undoped a-Si layer, ~15−18 nm, is deposited by e-beam, followed by crystallization and oxidation at 800−900°C in a 3:1 dry $N_2 + O_2$ at atmospheric pressure. The reason we did not anneal at 1000°C is to avoid oxidation leading to deleterious effects. Typically, our samples consist of ~5−10 nm crystallites surrounded by 2.5 − 3 nm of oxide serving as barriers.

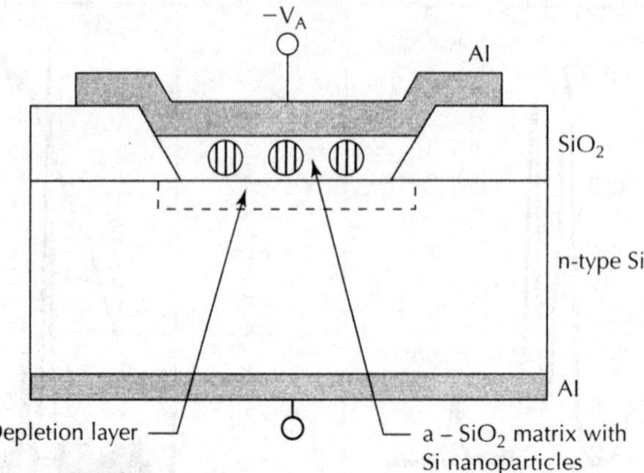

Figure 12.2   *Modified MOSFET with nanocrystallites shown shaded circles in an a-SiO$_2$ matrix.*

The device is electrically characterized by measuring dc I−V, capacitance and conductance at 1 MHz as functions of gate bias V_G using a voltage ramp generator, Model 410 C-V plotter, at a rate of 1−10 mV/s, with a lock-in amplifier. An impedance transforming circuit with $Z_{in} = 10$ KΩ, and $Z_{out} = 50\,\Omega$, having an op-amp AD540K is designed to insure the output of the voltage ramp capable of keeping pace with the demand at resonance. In usual resonant tunnelling via quantum wells with n-doped contact, negative differential conductance, NDC, appears whenever the applied voltage is such that the quantum state of the well moves below the source of electrons from the contact into the forbidden gap. Current drops because source is cut off. However, with a metal contact, the Fermi sphere is very large compared to all the quantum states involved.

As shown in Fig. 12.3, as the bias voltage sweeps the state E_1 below the conduction band minimum, the supply of the electrons is always present with a metal contact and almost constant. Therefore I−V is a step rather than a peak, tracing out the DOS.

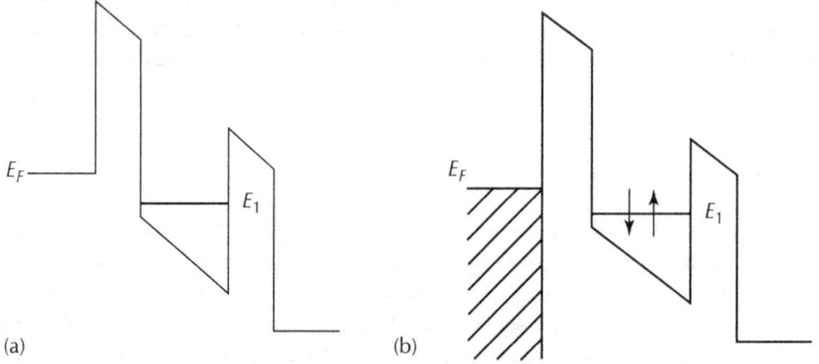

Figure 12.3   *Current peak with n-contact (a), conductance peak or current step with metal (b).*

Empty lattice energy states of an Si QD were computed for a cube oriented in various symmetry directions as well as for a sphere using variational technique as shown in Fig. 12.4 [14]. Resonant tunnelling via nano-crystalline silicon, nc-Si, embedded in an amorphous silicon dioxide matrix has been exploited [4, 14, 15]. The smallest grain size, ~ 300 nm, is obtained by annealing at $T \sim 600°$C of a thin layer of deposited a-Si at 30°C. Further size reduction is possible at lower deposition temperature. High resolution TEM showed almost perfect spherical shape for the nano-crystallites [16]. However, with interfacial inhibition, a-SiO$_2$ crystallizes at the annealing temperature above 900°C.

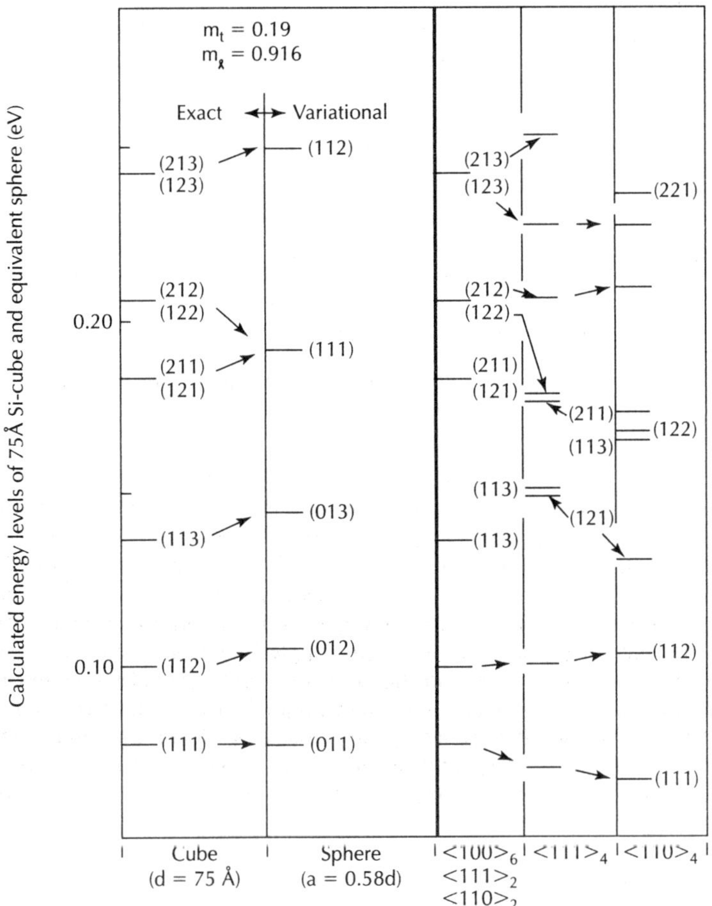

Figure 12.4   Calculated energy state for a d = 7.5 nm Si cube, compared with a sphere of radius a = 0.58d for aligning the ground staes. Right side gives the cube for other directions. The notation <100>₆ indicates the orientation in <100> with degeneracy factor 6.

For a constant transition matrix, conductance is proportional to the density of states, DOS, resulting in peaks for 0D and steps for 2D as shown in Fig. 12.5. Let us first discuss why conductance plays a leading role in the characterization of quantum dots, QDs, and quantum wells, QWs. If the matrix element is constant, then the conductance is proportional to the density of states, $N(E)$ shown in Fig. 12.5, for QD with 0D, Qwire with 1D and QW with 2D. For the parabolic energy band in 2D, DOS is a constant, m^*/π, with m^* being the appropriate transverse effective masses, when the energy of the level is reached for multiple levels in a QW, a series of steps appears.

Let E_1, E_2, ..., be the first, second ... states in a QW. Since the confinement is in one direction, the other two directions form a 2D system. Therefore the DOS, or $N(E)$ consists of a staircase, or steps, at the voltages corresponding to the energies E_1, E_2, ... With QD, the steps are replaced by a series of δ function-like peaks. We need to be extremely cautious in dealing with input/output for tunnelling via a QD. For layered superlattices, the equal potential surfaces representing the input and output contacts are planes, therefore it is simple to define the Fermi levels for the two contacts, and the incident and transmitted electrons are represented by plan waves. In microwave transmission, the cavity resonator is basically a QD coupled to a waveguide, and via a horn coupled to free space. What happens to radar systems? Take, for example, the scattering of plane waves from a sphere; depending on the frequency, it may be that only the lowest dipole term is significant. There is no such thing as a "universal conductance" or universal impedance because

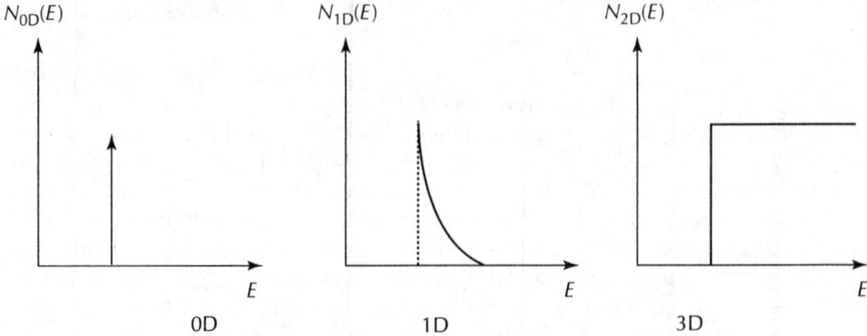

Figure 12.5   N(E) for QD with 0D, 1D and QW with 2D.

the scattering matrix determines which incoming mode is scattered into which outgoing mode. It is entirely similar for electron scattering, only that the problem is simpler because photons have to contend with polarization. However, the bulk of the situation is identical; the most important point is to recognize that no single number may serve as an impedance function. Even for dipole terms at low frequency, each scattering term has at least four components to specify, the incident and outgoing polarizations, as the incident and outgoing angles consist of two sets of (θ, ϕ). Therefore one must set up the problem as in waveguides uniquely specifying the input and output, with the QD as a plug in the waveguide. However, there is a further complication. Even in a waveguide it is generally not possible to define the input voltage and current. However, one can use the Poynting vector at some plane considered as the input end, and another plane at the output end. As shown on p. 301 of [15], the wave impedance of the electron waveguide is:

$$Z = \frac{\hbar}{2e^2} \frac{k_0^2 L}{k_z}$$
(12.1)

with

$$k_z = \sqrt{k_0^2 - k_c^2}$$

and

$$E = \frac{\hbar^2 k_0^2}{2m}$$

where the cutoff k_c is the momentum in the transverse degree of freedom given by $k_c^2 \equiv (m\pi/a)^2 + (n\pi/b)^2$ for dimension a and b. Equation 12.1 looks similar to but not quite the same as the waveguide case for photons. Furthermore, along the direction of propagation, unlike the sine dependences in the x and y directions, the wavefunction dependence on z is $\exp(ik_z z)$, a propagating function, therefore periodic boundary conditions must be applied, i.e. $k_z L = 2\ell\pi$, with ℓ being any integer. Then Eq. 12.1 becomes:

$$Z = Z_0 \, \ell \, [1 - k_c^2/k_0^2]^{-1}$$
(12.2)

with

$$Z_0 = h/2e^2.$$
(12.3)

At this point one might be tempted to call Z_0 the universal impedance. In Eq. 12.2 for $a = b = \infty$, and for $\ell = 1$, obviously $Z = Z_0$. Similarly, there is no such basis on which to call G_0 the universal conductance. It is simply the prefactor. Apart from what I have discussed thus far, there is another very important consideration, the unique Fermi energy serving to characterize conduction and propagation through a quantum dot. As discussed, without a definite equal potential, how do we assign the energy in the Fermi distribution function? We know that the potential varies from point to point. This point is underlined when LaFave and I decided to use an average potential [13b]. To help the reader understand, we need to realize the fact that the geometrical boundary of a QD is not an equal potential surface except for a sphere with only one electron! With this point made, let us move on to the data.

To arrive at very interesting data, I want to emphasize the need for electrical forming. The idea came from the fact that dull I–V, after repeated measuring of the I–V, produced interesting structures. Because of the Schottky barrier in the device structure, all our data were taken in the reverse bias, because with forward bias, all that is present is a large current as in any Schottky diode. However, the forward current is very repeatable, and Nicollian and I decided to past a current in the forward direction until structures showed up in the reverse direction. This is precisely what is referred to as electrical forming. Incidentally, almost all electroluminescent devices are electrically formed! Forming may have resulted in the establishment of a filamentary conduction path. Figure 12.6 shows how Nicollian chose to electrically form in the forward direction where I–V is represented by the typical forward direction of a Schottky diode.

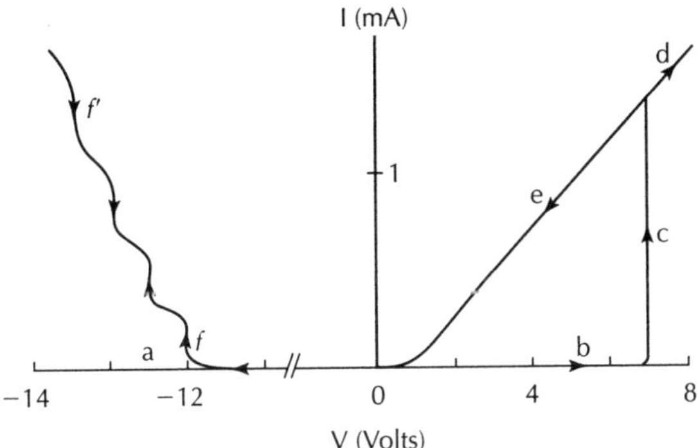

Figure 12.6   Electrical forming used to eliminate the unwanted tissue region of an annealed crystallized Si QD. The path (a) and (b) indicates only minute conduction until path (c) is reached. Subsequent paths, (d)–(f), indicate structures after forming with (f´) often showing hysteretics. After [5].

In Figure 12.7, two sharp conductance peaks at ~−11.1 V and −11.35 V are separated by $\Delta V = 0.25$ V. *Without electrical forming, usually no sharp peaks and steps were observed.* The difference between the voltage V_{n+1} and V_n per electron is

$$e(V_{n+1} - V_n) \equiv e\Delta V = E_{n+1} - E_n + e^2/2C \tag{12.4}$$

where C is the capacitance [17]. For QDs of ~6 nm, $E_{n+1} - E_n \sim 0.06$ eV, and $e^2/C \sim 0.16$ eV, or ~0.22 eV per electron, which is close to $\Delta V = 0.25$ V. This indicates that bulk of the voltage ~1 V represents the voltage across the deep depletion. (Unlike the oxide gate, holes are trapped at the interface resulting in inversion.) Thus we know that only 20% of the applied voltage appears across the quantum dot device, and the rest appears across the substrate due to deep depletion as first pointed out in [4]. Note also in Fig. 12.7 that the flat part of the G–V gives $\Delta G \sim 160\,\mu$S, close to $4G_0$, showing that the degeneracy factor is 2, one for each spin for the first excited state. Detailed discussions on peaks and steps will be given later. Figure 12.8 shows that the conductance peak has thermal linewidth, indicating that the majority of the QDs are nearly what they are supposed to be.

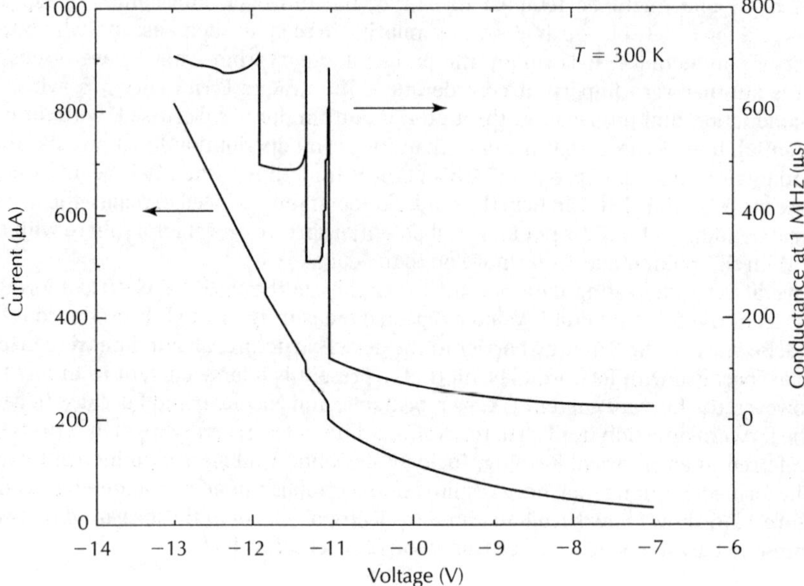

Figure 12.7   Typical I−V and G−V for devices after electrical forming. Forming, a sort of controlled spot heating, is performed with a current limiter usually for 15 minutes. Note the typical conductance steps with ΔG ∼ 160 µS, led by conductance peaks.

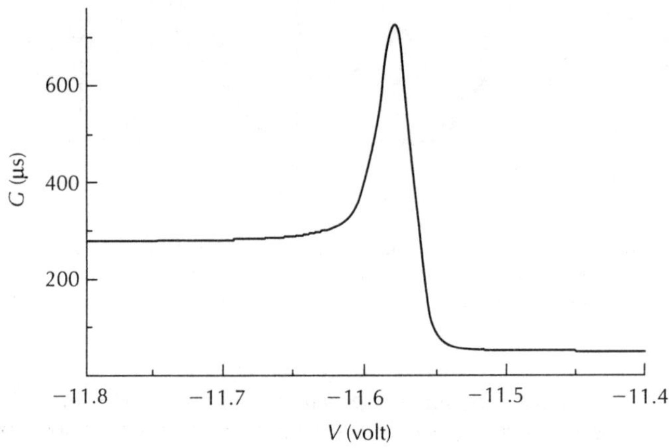

Figure 12.8   Linewidth of a typical G–V is ∼k$_B$T, at T = 300 K.

Figure 12.9 shows hyteresis, indicating the presence of trapping. We have identical G–V without hysteresis indicating that hysteresis is from traps. Figure 12.10 shows that conductance steps and peaks are not affected by size variation of the contact, indicating filamentary rather than area conduction. We know how to use electrical forming to select peaks and steps. Slower forming results in steps, characterized by pseudo-2D coupled QDs, while rapid forming results in peaks, typical for 0D QDs. Usually at lower bias, mostly below −15 V, conductance steps are preceded by conductance peaks, while at high reverse bias, only conductance steps without the leading conductance peak. As pointed out in [4] our interpretation is that at high bias involving higher energy states, the wave-functions of adjacent QDs are not well localized which result in coupling resembling a 2D system. In essence, the filamentary conduction path gives us an ideal

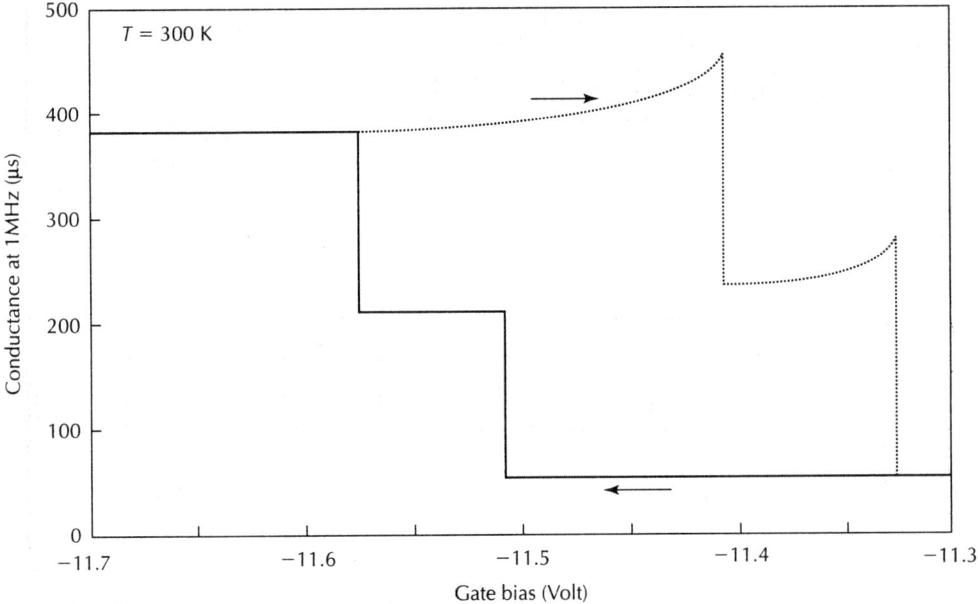

Figure 12.9   *Typical step-like G–V with hysteresis at 300 K. The conductances 40, 205, 380 μS correspond to* $\Delta G = 40, 165, 185\ \mu S \sim G_0, 4G_0, 5G_0$ *[5], [15], and Li's thesis [16].*

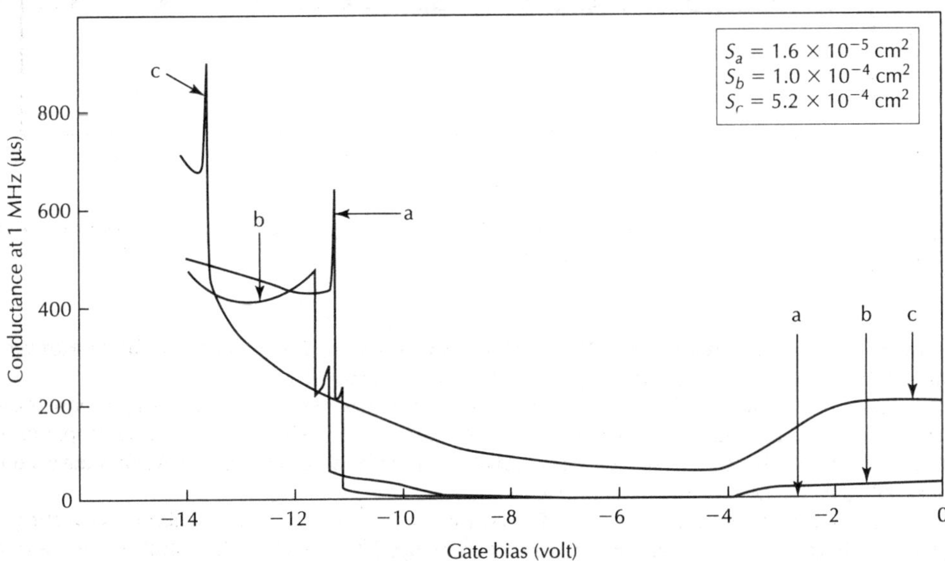

Figure 12.10   *G–V with three contact areas: lower, 1.6, middle, 10, and upper* $52 \times 10^{-5}\ cm^2$. *Conductance for all peaks and steps are same for all contacts. Conductance near V = 0 is caused by trappings.*

opportunity to study G–V in detail without the need to zero in on a QD as it is done in photoluminescence studies; this will be detailed in the next section.

Figure 12.11 shows slow conductance oscillation [5]. The top Figure shows steps with oscillatory sharp peaks vs applied bias voltage. In the middle Figure, at bias $V = -11.95$ V, closer to the top step, the on-time is longer. G oscillates between 260 and 420 with $\Delta G = 160\,\mu S$. In the bottom figure, at bias $V = -11.85$, closer to the bottom step, on-time is very short, G oscillates

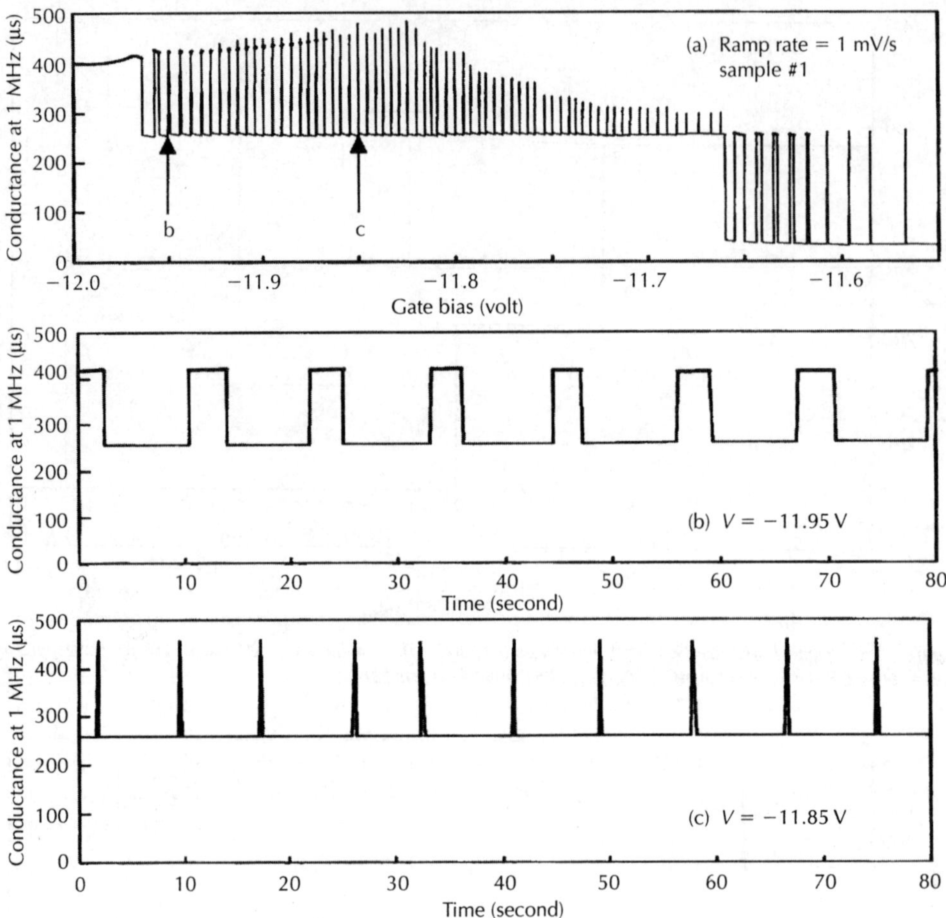

Figure 12.11   Slow conductance oscillation. Top: G vs V, Middle: G vs time at V = −11.95 V, closer to the next step Bottom: V = −11.85 V, closer to the previous step.

between 260 and 460, with $\Delta G = 200\,\mu S$, *this indicates that the origin of the oscillation with periods ~0.2 s to more than 20 s is caused by trap-induced Coulomb blockade.*

Figure 12.12 shows another sample with oscillation period ~0.3 s [18, 19]. In the past these data were put to one side because the oscillations indicated a serious problem with trapping in the QD structure. However, it is now known that even the best InP QDs and InAs QDs have similar problems of instability due to trapping.

Before I deal with optical instability of QDs, I want to show the effect of light, converting a small conductance peak to a huge peak as shown in Fig. 12.13 with a slight shift in the position of the steps [20, 21]. The shift may be due to heating; however, the appearance of a very large conductance peak may be caused by light-induced passivation of defects. Using various filters and a tungsten light, it was discovered that IR below the band gap of Si has no effect.

Figure 12.14 shows a random telegraph-like conductance spectrum obtained first by Ding [20]. As pointed out before, we eventually learn how to control various types of QDs by controlling the power and time of electrical forming in the forward part of the I–V.

The type of samples we fabricated using various annealing and electrical forming results in selecting a simple configuration having a current filament consisting of only one silicon dot. As pointed out in [5] whenever eV_1 in Eq. 12.4 is aligned with E_F, the energy $E_1 + e^2/2C$, $E_1 + 2e^2/2C$, etc. includes charging the capacity C, referred to as Coulomb blockade by Likharev [23],

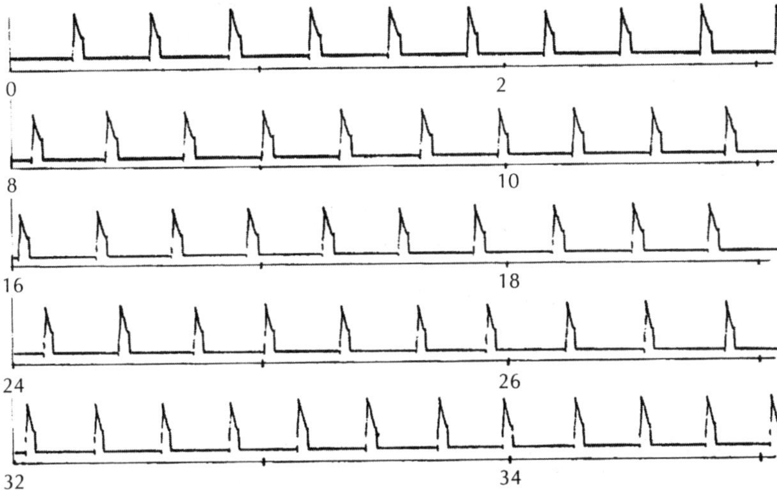

Figure 12.12 *Typical conductance oscillation shows up in less than 5% of the devices with Si QD formed by crystallization from the a-Si phase.*

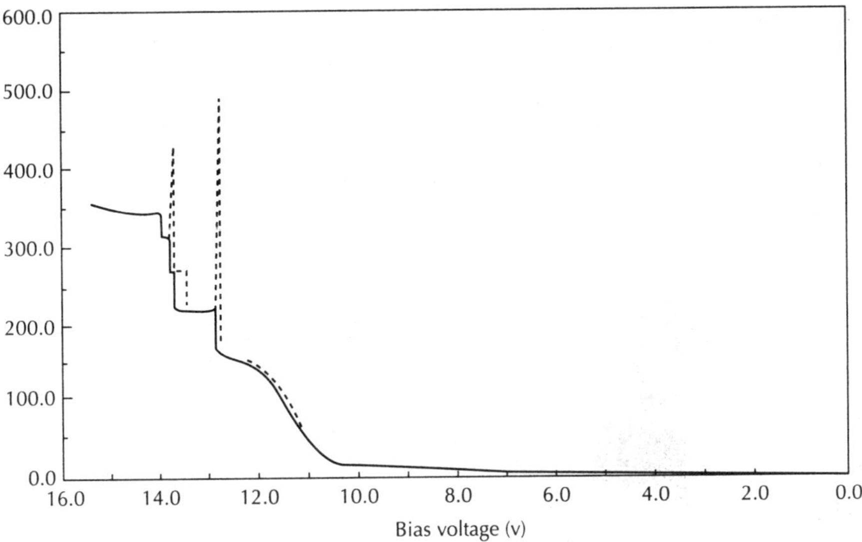

Figure 12.13 *Conductance versus bias voltage without light (solid) and with light (dotted). IR irradiation below the band gap of silicon has no effect.*

which is simply electrostatics. When trapping by a defect is present, the first current jump occurs at an applied voltage $V_a = V_1 + Q/C$. Suppose an electron is captured resulting in $Q = e(n + 1)$, an additional voltage of e/C is necessary to maintain resonant tunnelling. Because the applied bias is fixed, the conductance will jump down to a lower value. Conversely, whenever an electron is emitted from a trap, the charge Q returns to $Q = en$, the potential at the QD drops back so that the energy state involved is again aligned with the Fermi level of the contact at V_1, causing the conductance to jump back to a higher value. Therefore the period is the sum of the electron capture and emission time constants. In this picture, oscillation is the result of a flip-flop between two charge states involving exchange with a defect. Alternately, by assuming that a non-conducting state is weakly coupled to a conducting state via an oxide barrier 3.2 eV, for a period of

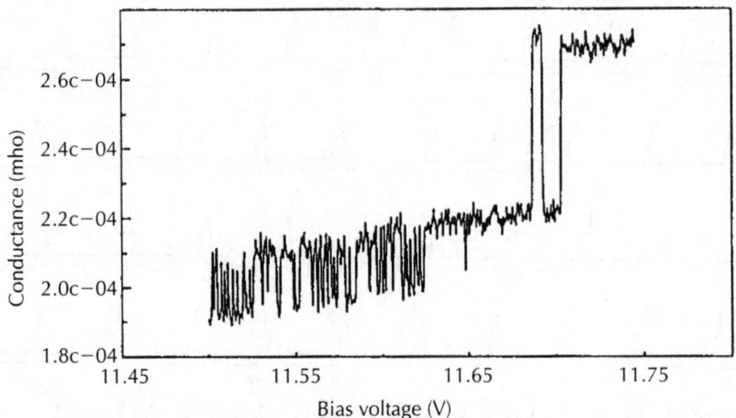

Figure 12.14   Conductance versus bias voltage shows telegraph-like noise. At a fixed voltage, the variation in time is an like all those we have shown as oscillations or oscillatory switching, rather it is a typical noise-like spectrum. Unpublished thesis by Chen Ding (1994).

10 s, it is necessary that the barrier width $\sim$15 nm. Since the total layer thickness is $\sim$15 nm, the origin of the switching is more likely due to a defect state located near a conducting QD rather than a similar but non-conducting QD state. Conduction peaks have been reported by resonant tunnelling via bound states of a single donor in a quantum well [24], which is not too different from our cases.

In Figure 12.15a the sweeping rate of 1 mV/s for the applied bias is used for conductance versus the bias voltage near and before a peak. The voltage separation between points b and d in (a) is $\sim$50 mV corresponding to $\sim$50 s. At a period varying from 0.2 s to 5 s, there should be 250 to 10 oscillations. In other words, most of the oscillatory peaks shown in Figure 12.15a are oscillations

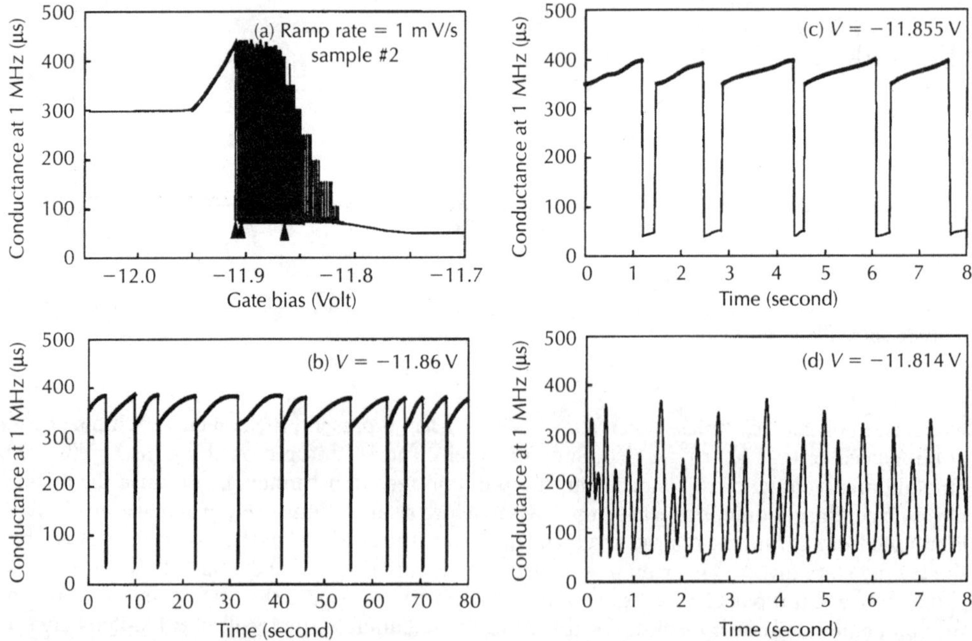

Figure 12.15   Conductance spectrum near and before a peak is shown in (a). At fixed voltages, (b)–(d), oscillation with time is detailed. Taken by X. Li, unpublished [25].

in time rather than with bias voltage! The period of oscillation is reduced when approaching the conductance peak as shown in (b) and (c). Oscillation is more complex in (d) where the number of electrons involved in jumping back and forth varies between 4 and 8 while in (b) nd (c) the number appears to be fixed at 8 [25]. At this stage, emphasize should be that the energy states are rather different from the central field problem typified in atomic hydrogen, having 1 s, 2 s, 2 p, etc. As shown in Figure 12.4, the lowest state of a quantum cube has a set of three quantum numbers, (111), but a quantum sphere of Si has (011), in addition to the other complications arising from the directional effective masses. These considerations, together with the degeneracy factor, are very complex.

Figure 12.16 shows telegraph-like noise spectra of complex current fluctuation prior to oxide breakdown due to discrete multilevel switching [22]. Note that the spectra are also voltage dependent. Actually, I do not like to use the word noise, because the broad feature of the signal is very different from random thermal noise. Instability is probably a better description. As long as the majority of samples are bad samples and are rejected, our selection process does have merits, which is not so different from annealing in steam to reduce the interface density of the MOS capacitors, or even something more current, such as picking a single thread of carbon nanotube.

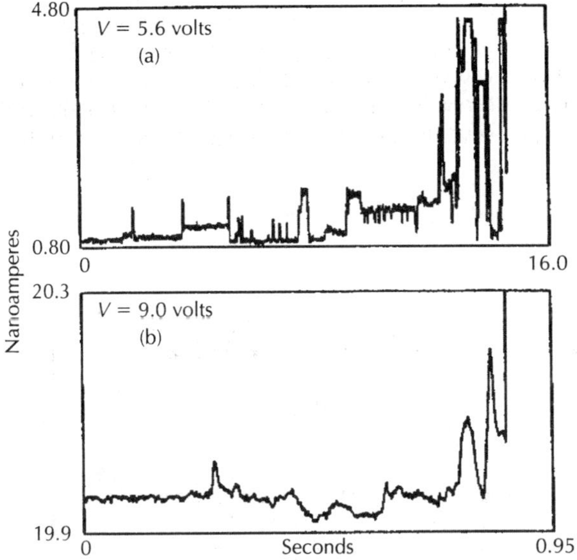

Figure 12.16   Telegraph-like current fluctuation prior to oxide breakdown. After Farmer et al. [22].

12.4 Many-body effects in coupled quantum dots

Because in 2D systems, DOS is proportional to the conductance G, and taking $G = gG_0$ where $G_0 = 39\,\mu S$, the degeneracy factor $g = 1, 2, 3, \ldots$ For a state without any other symmetry-induced degeneracy, $g = 1$ per spin, so that for $+1/2$ and $-1/2$ spins, $g = 2$. I have put together several typical cases representing $g = 2, 4,$ and 6 [26].

After the initial peak, the majority of $\Delta G \sim 160\,\mu S$, with $\Delta g = 4$ for the step, or involving two pairs of $+$ and $-$ spins, with the chemical bonds with two electrons per bond with coordination number 4 – although there are also $g = 2, 6$ or even 8, with higher numbers representing degeneracy when symmetry is introduced Occasionally, in some 10% of our sample, we have seen odd numbers, for example $g = 5$ with $\Delta G = 200\,\mu S$, and usually associated with telegraph-like noise. But some samples show 1D conductance, which may be due to the two coupled QDs arranged in line with the current path mimicking a Qwire. *However, it is little understood why peaks always precede steps, although conductance involving higher energy states of the Si QD shows steps. Since I have* been involved with LaFave in the study of the discrete nature of electrons in capacitance using a dielectric sphere model [13b]. I have acquired a better understanding, or better respect, for electron–electron Coulomb interactions. Basically, the coupling of the QDs into a 2D-like system is enhanced by the occupation of given states by electrons, with interaction terms including the direct Coulomb term, $e^2/\mathbf{r}_{ij}$, as well as all the induced polarization terms on the individual QDs, inside and outside as well as on the interface. In short, it is the many-body effect that creates enhanced coupling in forming the 2D-like system from 0D QD states and results in creating a peak leading to step, a model see the Fig. 12.16. As the applied voltage is increased, electrons tunnel into the QD and occupy the empty states, resulting in all the induced terms as the basis of our calculation for capacitance [13a,b]. The net result is to enhance the interaction between neighbouring QDs, essentially by lowering the tunnelling barriers with respect to the self-consistent potential of the electrons occupying these states. Therefore the coupling would be much weaker without the electron occupation. *To summarize the many-body effect may be simply described by the self-consistent potential from occupation, raising the potential of the individual QD state with respect to the barriers separating these QDs, creating an enhanced coupling between neighbouring dots.*

This schematic representation of the enhanced coupling shown in Figure 12.17 is based on the coupling of two Si QDs, with an electron in one and a neighbouring QD. The top shows singly occupied individual QDs, the middle shows doubly occupied QDs, and the bottom shows exchanging occupations leading to oscillations. However, the exchange should be very fast. Only when a trap replaces a regular QD, serving as an imposter with very slow emission and capture rates, does telegraph-like slow oscillation occur. If triply occupied, $g = 4, 8, 12$. It is more complicated for the coupling of three. Nevertheless, using the chemical approach to the formation of molecules, it is clear that coupling dictates occupation of states by electrons leading to initial coupling.

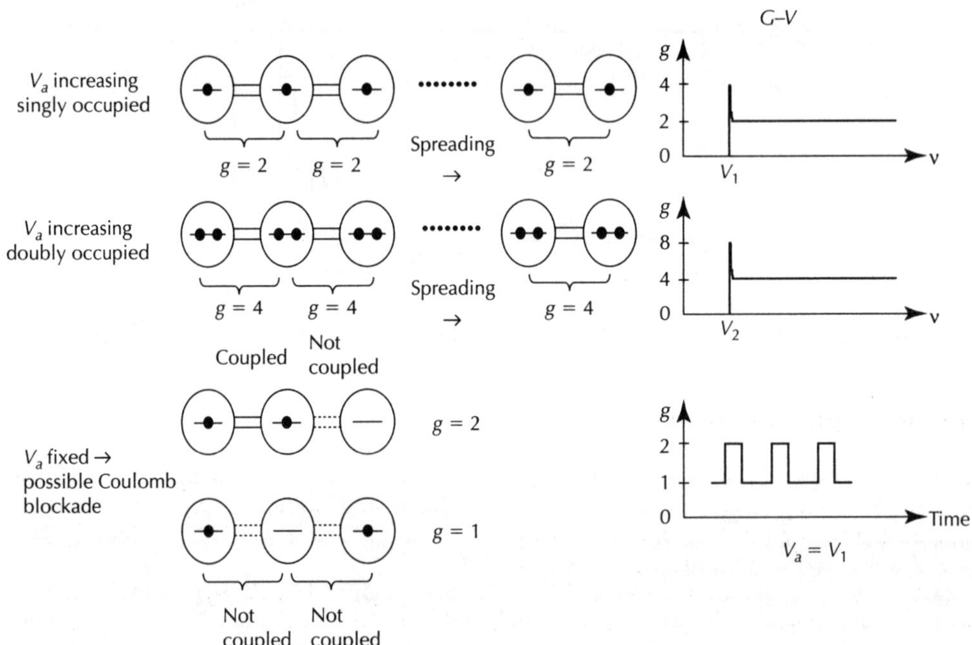

Figure 12.17   A model for the enhanced coupling between QDs with an electron in one and in a neighbouring QD. The top shows singly occupied individual QDs, the middle shows doubly occupied QDs, and the bottom shows exchanging occupations leading to oscillations. However, the exchange should be very fast. When a trap replaces a regular QD, serving as an imposter with very slow emission and capture rates, telegraph-like slow oscillation occurs. If triply occupied, g = 4, 8, 12.

12.5   Conventional optical study of QDs

Before we deal with the study of instabilities in the photo-luminescence of QDs, we shall present the management of absorption and luminescence collected from a distribution of QDs with conventional spectroscopic techniques, i.e. measurement involves many QDs having a distribution of sizes. Recently, Liu *et al.* [27], undertook placing CdSe QDs, ~2.4 nm, in a photonic crystal consisting of silica spheres with a diameter of ~300 nm. The purpose is to match the photo-emission to the forbidden gap of the photonic crystal for enhanced interaction. In one dimension, interaction is enhanced by placing optically active transitions within the resonance of a Fabry–Perrot interferometer.

The CdSe/ZnS core-shell QDs with ~2.4 nm size were commercially available from Evident Technologies Inc. To study the thiolation effect on the PL properties of QDs, several substrate surfaces, e.g. Si substrate, Au film on Si, and fused silica, with and without sulphur were introduced, and the root-mean-square (RMS) surface roughness of these samples was measured by a Veeco Dimension 3100 atomic force microscope (AFM). Substrates were first immersed in a solution of 5 mM (3-mercaptopropyl)-trimethoxysilane (MPTMS) in ethanol for 24 h, followed by a thiolation process for introducing sulphur onto a substrate surface. To study the interaction among the QDs themselves, low to high concentrations of QDs in toluene solution were used to deposit the thin film with QDs onto the above substrates via evaporation of the toluene solvent. 3D photonic crystals of silica spheres were self-assembled onto the substrate via evaporation of the ethanol, and the film thickness of more than 15 layers was obtained. Thermal treatment was used to stabilize the stop-band of the photonic crystals, involving annealing for 3 h in a quartz tube in the range of 300 ~ 1000°C. A Cary 300 Bio UV-Visible spectrophotometer was used for the measurements of the transmission spectrum at normal incidence. At this point one must recognize that the experimental technique used for this study is no different from 50 years ago, except AFM was used to characterize the structure and surface roughness.

To study the interaction between QDs and between QDs and the matrix, we start from a simple but fundamental scheme. There are two types of shifts of the emission peak, a one-way shift, for example caused by pressure, temperature, etc., and a two-way shift, one up-shift of the higher peak and another down-shift of the lower peak caused by coupling. Let the electron wavefunctions $|1\rangle$ and $|2\rangle$ represent two uncoupled states, with $H_o|1\rangle = E_1|1\rangle$ and $H_o|2\rangle = E_2|2\rangle$. In $H\Psi = E\Psi$, $H = H_o + H_1$, where H_1 represents interaction. Defining a coupling term, $C = \langle 1|H_1|2\rangle$ and $C^+ = \langle 2|H_1|1\rangle$. The wavefunction Ψ can be expressed by a linear combination of the two uncoupled states, $\Psi = a|1\rangle + b|2\rangle$, where a and b are two constants. The secular determinant $(E - E_1)(E - E_2) - C^2 = 0$. The roots give the two shifted energy states:

$$E_\pm = \frac{E_1 + E_2}{2} \pm \sqrt{\left(\frac{E_1 - E_2}{2}\right)^2 + C^2}. \tag{12.5}$$

For two identical QDs, $E_1 = E_2$, so that $E_+ = E_1 + C$, and $E_- = E_1 - C$. Since electrons at the upper state can relax to the lower state via phonons, the observed transition is from the lower state to the ground states, assumed not coupled. In this process, the observed PL is red shifted as observed. For QDs coupled to a matrix such as some sort of molecular surface complex, the observed shift may go up or down depending on whether the interaction molecules are located at lower energy or high energy, respectively. For convenience to the reader, we point out that interaction with the matrix indicated that the molecule in question is located at a lower energy because the PL is blue shifted. Before we show the measured shifts, let us point out the problem when a distribution of particle size is present. For two adjacent QDs with different sizes, the full expression represented by Eq. 12.5 should be involved. Not only are the uncoupled spectra broadened by distribution, the interaction, as shown in Eq. 12.5, has additional spreading. The particle size distribution resulted in a PL linewidth being much more than the intrinsic thermal broadening. Therefore the measured data we obtained do not land themselves to simple modelling; nevertheless, from an application point of view, what really matters is what we measured. Furthermore, we did not know the details of compacting the QDs with dilution followed by subsequent removing of the toluene solvent. We estimated an increase in packing density of perhaps several per cent, from low to high concentration. With these uncertainties in mind, we present the measured data as shown in Fig. 12.18.

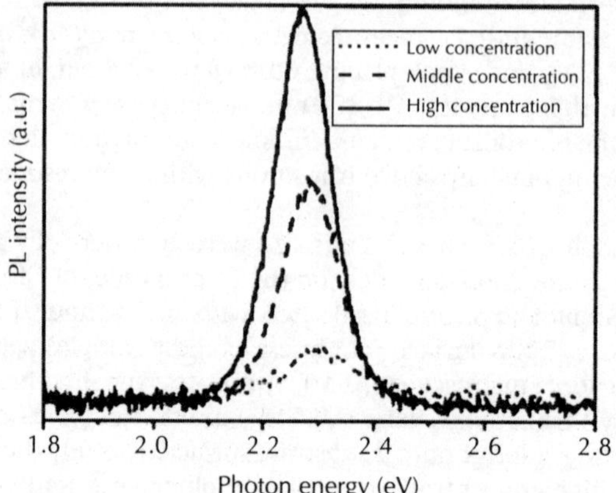

Figure 12.18   PL spectra of CdSe/ZnS QDs deposited on an Si surface with native oxide for various dot concentrations. The degree of stacking probably has increased by several percentage points. Note that there is a 1.3% down-shift of the centre peak.

By correlating with AFM, we discovered that PL is much enhanced from surface roughness resulting in enhanced surface interaction due to the increase of surface areas and effective surface electric field, which is very similar to the surface enhanced Raman. The roughness of the fused silica is ~20 times greater than that of the Si substrate with native oxide coverage, which explains why we have an extra peak on the fused silica, but not on the Si wafer with thin and smooth native oxide.

Figures 12.18 and 12.19 illustrate the acquisition of information regarding the coupling between the quantum dots and between the quantum dots with surface complexes. Much of

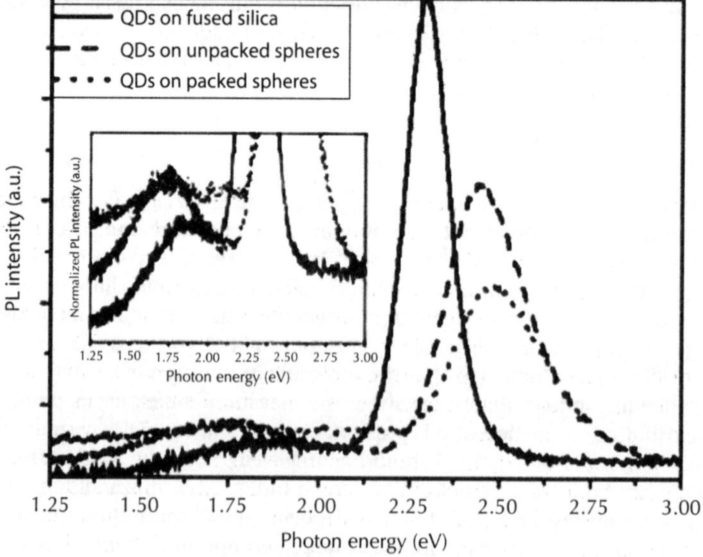

Figure 12.19   PL spectra of CdSe/ZnS QDs deposited on: (1) fused silica, (2) unpacked silica spheres, and (3) packed crystals of silica spheres forming closed packed structures. Note that the main peak is now pushed up by the molecular surface complex located below the main peak. The inset shows the portion of the surface peaks expanded.

the understanding is derived from conventional spectroscopy. From the point of view of device optimization, these conventional techniques may be adequate. However, the detailed interaction must come from spectroscopy on a single quantum dot, which is presented in the next section.

12.6 Single quantum dot spectroscopy

As mentioned, only single QD, SQD, can display the correct linewidth. In addition to Fig. 12.1 showing the linewidth of InP on GaAs [6] being $K_B T$, another excellent example is shown in Fig. 12.20 of InAs QD from 1.7 ML on 100 nm of GaAs on semi-insulating GaAs(100), capped by 100 nm of GaAs for protection. A cw Ar laser-pumped Ti–Sp laser between 700 and 900 nm was used and focused to 2 μm spot spatial resolution, and 0.15 meV of spectral resolution corresponding to 1.7 K. PL from 1.515 eV of GaAs shows LO-phonon replicas, allowing them to account for varying diffusion: hitting an LO-phonon line allowing the creation of phonons leading to phonon-assisted carrier relaxation (see [25, 26]). Without it, diffusivity would be much lower. Varying diffusivity by selecting the excitation energy explains why PL at 1.447 eV in the thin (<1 nm) wetting layer is two orders of magnitude above the emission in the thick (200 nm) GaAs. The two vertical dotted lines draw attention to oscillatory behaviour in P_{IR}. In brief, SQD spectroscopy allows far more detailed understanding, hitherto not possible with conventional spectroscopy. Even if the QDs are oriented the same, symmetry and selection rules in optical transition cannot be accurately obtained without having SQD. For example, although most researchers assume the transition in QD with SS for ground state and PP for the first excited state, as shown in Fig. 12.4, these atomic quantum numbers do not apply to any quantum dots, not even for a sphere.

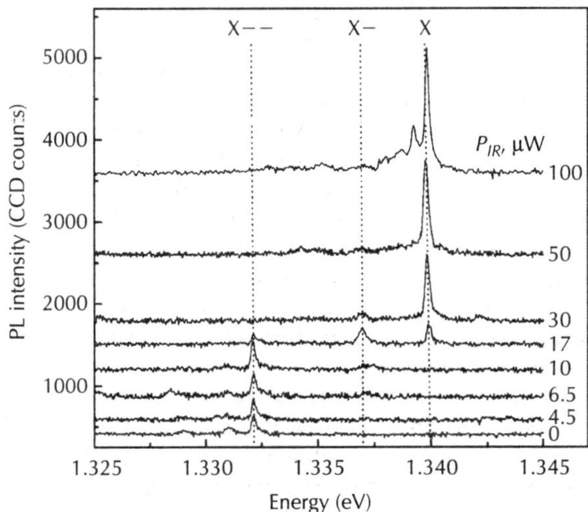

Figure 12.20   Micro-PL of the single InAs QDs on GaAs, with dual-laser excitation at T = 5 K with P_{ex} = 20 nW, varying P_{IR}, $\hbar\omega_{ex}$ = 1.503 eV, $\hbar\omega_{IR}$ = 1.240 eV. For details, see [28, 29].

12.7 Instability in the PL of quantum dots

Optical blinking, similar to "telegraph instability" discussed with respect to tunnelling, has been observed in many small systems. The first time I noticed blinking was in porous silicon, PS. Since the period was in seconds, I dismissed it as totally uninteresting, useless and annoying consequences of trapping. In fact I have even seen it with a broken silicon wafer by bending. Slow oscillation in PL output has been reported in CdSe QDs [30], in InP QD/GaInP SK QDs [31]. Figure 12.21 shows bi-level oscillation with a period ~5 s [6].

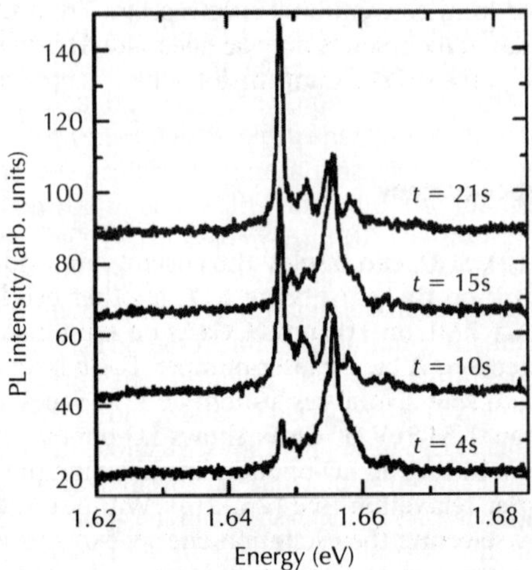

Figure 12.21   Slow oscillation in time of a single QD at low excitation. Note that there are two opposing out-of-phase oscillations, one at 1.649 eV (1.1 meV) and the other at 1.656 eV (FWHM = 1.27 meV), at the same period of oscillation ~5 s. After Bertram [6].

Figure 12.22 shows PL in InP QD on $Ga_xIn_{1-x}P$ involving switching with single, double and even triple states. In the case of conductance switching how do processes with single, double, etc states work? Let us go back to the top of Fig. 12.11. Note that the difference of G involves $g = 1$, going from $G = 260 - 300\,\mu S$, $-340\,\mu S$, $-380\,\mu S$, and finally $-420\,\mu S$, corresponding to $g = 1, 2, 3$, and 4. Note that even $g = 6$ or 8 have been observed [26]. These higher values of g correspond to coupling to even more dots with a given defect, or even more defects surrounding a given dot. In Fig. 5 of [28], it was pointed out that switching seems to be faster at higher temperature. At least in the case of conductance switching, temperature seems to play a minor role. In other words, phonons may not play a direct role in switching! Could that be due to some kind of relaxation oscillation? In my book, I presented a summary of the thesis by my student Sen [15], solving the time-dependent Schrödinger equation. He found that the relaxation oscillation dominates whenever the incident electron energy is not an eigen-state of the quantum system. As I have mentioned, switching is present even in quantum wells. However, everything seems to be magnified in QDs. By the same token, switching is present in bulk; however, the overwhelming DOS of the band states totally suppresses the effects of a few defects except in devices involving oxide gates with charging and discharging [32].

Figure 12.22 shows telegraph-like time dependence of PL from an $InP/Ga_xIn_{1-x}P$ single QD at 10 K. These nominally undoped QDs are separated, ~10 μm, having 5 nm high by 25 nm laterally determined by AFM. The area of illumination is ~$100 \times 100\,\mu m^2$, with collection resolution of 1.5 μm and spectral resolution of 0.1 meV. I purposely detail the optical set-up to illustrate that such a system is exactly what is needed to obtain detailed property on these QDs. For example, this set-up can avoid the problem Ke Liu and I encountered during our study of the interaction between QDs with a distribution of particle sizes. Note that the observed single-level and bi-level switching are same as the cases in the conductance switching involving $g = 2, 4, 8$, etc. The ideal set-up is a combined capability of both single QD optical and electrical characterization, capable of measurements *in situ* to avoid contamination and the oxidation problem. In some sense, it should be easier for QDs because we know where to look. Imagine looking for detailed features of finding the trap of perhaps less than 1 nm in a conventional MOSFET with a dimension tens or hundreds of nm.

In tunnelling via QW, the conservation of the longitudinal energy and transverse momentum means that the transverse energy grows with the longitudinal energy fixed, resulting in constant conductance. *Quantum wells with longitudinal separation by tens or hundreds of lattice constants, but coupling in the transverse direction, are same as in usual solids. Quantum dots, on the other hand,*

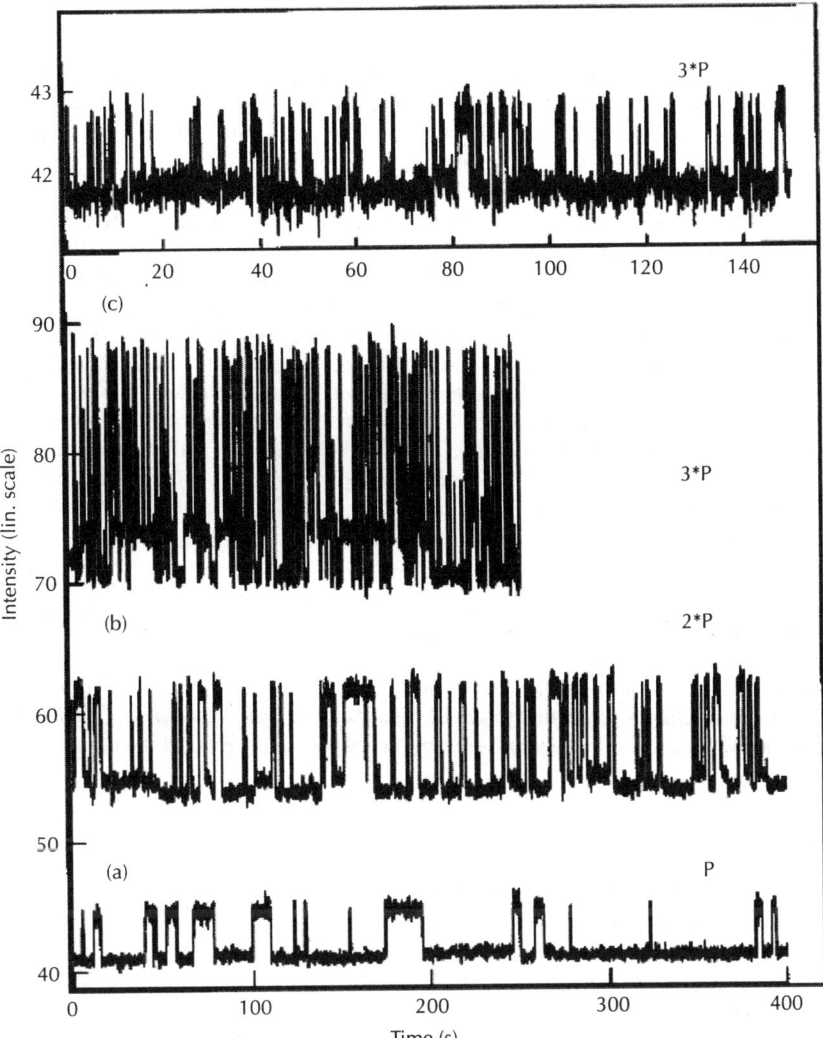

Figure 12.22   Telegraph switching of InP in $Ga_xIn_{1-x}P$: (a) low excitation – switching between two levels; (b) high excitation – switching between three levels; (c) after partial quenching [33].

having separation in the transverse direction comparable to the QD size, and coupling in the transverse direction, may be enhanced by electron occupation forming "molecular bonds", resulting a 2D sheet. Spreading the transverse momentum means having the incident electron at an angle, which is more affected by interface roughness and trapping sites. In classical systems, individual contributions without phase differences are simply summed. In waves, phase differences of individual entities produce interference and possible instabilities. *Therefore inhomogeneity in QDs magnifies problem in QWs.* Using common sense, a parallel system is more subjected to instability caused by localized deviations. I conclude after many years of thinking about the problem of quantum electronic devices, that our saviour is the tolerance brought by the introduction of phase randomization in a large N-body coupled system that overwhelms the "devil", Pauli's exclusion principle. Too bad that electrons, mainly because of their charge, cannot be allowed in the same state as Bosons.

12.8 Summary

Let us look more closely at conductance consisting of delta function-like peaks and steps, as a function of the applied voltage. Whenever the first jump in conductance involves two electrons, our data are more predictable, whereas the first jump involves only one electron, it is more common to observe telegraph-like G–V. This fact indicates that a defect traps an electron and holds onto it by drastically changing the potential and blocking other electrons. Except in SK QDs, I am inclined to think that QDs in general have more defects than QWs simply because the surface volume ratio in QDs is much higher than in QWs. Second, the state in a nanoscale QD is similar to a defect. After we discovered the effects from light, we realized that the process seems to be caused by a variety of trappings, although the data did not rule out trapping from non-conducting dots, instead of some unknown defects. The main many-body effects can be lumped into one, the Coulomb term normally taken as a small perturbation in atomic and even solid-state physics. *There is no doubt in my mind that some sort of general usage of QDs will be found and developed. However, I also predict that nanoelectronics will never be developed in computers. I have mentioned that living organism utilize ions instead of electrons. The ratio of weight between a sodium atom and electrons is the same as that between a 200 kg lead ball and a ping pong ball. This is why nature, with a few hundred millions of years of evolution, perfected the ionic instead of electronic system, in what may be called the single ion transistor.*

Acknowledgements

I take this opportunity to acknowledge our original funding from NSF, followed by ONR, ARO and DARPA. Most of all thanks go to my students J. Clay Lofgren, Daniel Boeringer, Xiaolei Li, Amanda Bowhill, Jonder Morais, Chen Ding, Quiyi Ye and my colleague the late E.H. Nicollian.

References

1.   L. Esaki and R. Tsu, *IBM Res. Develop.* **14**, 61 (1970).
2.   R. Tsu and L. Esaki, *Appl. Phys. Lett.* **22**, 562 (1973).
3.   L.L. Chang, L. Esaki, and R. Tsu, *Appl. Phys. Lett.* **24**, 593 (1974).
4.   R. Tsu, Q.-Y. Ye, and E.H. Nicollian, *SPIE* **1361**, 232 (1990). Q.-Y. Ye, R. Tsu, and E.H., Nicollian, *Phys. Rev. B.* **44**, 1806 (1991).
5.   R. Tsu, X.-L. Li, and E.H. Nicollian, *Appl. Phys. Lett.* **65**, 842 (1994).
6.   D. Bertram, M.C. Hanna, and A.J. Nosik, *Appl. Phys. Lett.* **74**, 2666 (1999).
7.   J.D. Joannopoulos, R.D. Meade, and J.N. Winn. *Photonic Crystals* (Princeton U Press, 1995).
8.   K. Liu, T.A. Schmedake, K. Daneshvar, and R. Tsu, *Microelectronic J.* (2007).
9.   R. Tsu, *Nanotechnology* **12**, 1 (2001).
10. D. Babic, R. Tsu, and R.F. Greene, *Phys. Rev. B.* **45**, 14150 (1992).
11. R. Tsu, D. Babic, and L. Ioriatti, *J. Appl. Phys.* **82**, 1327 (1997).
12. R. Tsu and D. Babic, *Appl. Phys. Lett.* **64**, 1806 (1994).
13a. J. Zhu, T.J. LaFave, and R. Tsu, *Microelectronic J.* **37**, 1296 (2006).
13b. T.J. LaFave and R. Tsu to be published.
14. R. Tsu, *SPIE* **1361**, 313 (1990).
15. R. Tsu. *Superlattice to Nanoelectronics*, (Elsevier, Amsterdam, 2005).
16. R. Tsu, J.G. Hernandez, S.S. Chow, and D. Martin, *Appl. Phys. Lett.* **48**, 647 (1986).
17. R. Tsu, *Physica B.* **189**, 235 (1993) where additional trap-charge is included.
18. R. Tsu, *Microelectronics J.* **34**, 329 (2003).
19. X.-L. Li (1993) MS thesis, Department of Electrical Engineering, UNC-Charlotte.
20. C. Ding (1994) MS thesis, ECE, UNCC.
21. C. Ding and R. Tsu, *Mat. Res. Soc. Symp. Proc.* **378**, 757–760 (1995).
22. K.R. Farmer, R. Saletti, and R.A. Buhrman, *Appl. Phys. Lett.* **52**, 1749 (1988).

23. K.K. Likharev, *IBM J. Res. Dev.* **32**, 114 (1988).
24. M.W. Dellow, P.H. Beton, C.J.G.M. Langerak, T.J. Foster, P.C. Main, L. Eaves, M. Henini, S.P. Beaumont, and C.D.W. Wilkinson, *Phys. Rev. Lett.* **68**, 1754 (1992).
25. X.L. Li (1994), MS thesis, unpublished, UNC-Charlotte.
26. R. Tsu, *Microelectronic J.* **39**, 335 (2008).
27. K. Liu, T.A. Schmedake, K. Daneshvar, and R. Tsu, *Microelectronic J.* **38**, 700 (2007).
28. E. Moskalenko, K.F. Karlsson, T. Donchev, P.O. Holtz, B. Monemar, W. Shoenfeld, and P.M. Petroff, *Nano Lett.* **5**, 2118 (2005).
29. K.F. Karlsson, E. Moskalenko, P.O. Holtz, B. Monemar, W.V. Shoenfeld, J.M. Garcia, and P.M. Petroff, *Appl. Phys. Lett.* **78**, 2952 (2001).
30. M. Nirmal, B.O. Dabbousi, M.G. Bawendi, J.J. Macklin, J.K. Trautman, T.D. Harris, and L.E. Brus, *Nature* **383**, 803 (1996).
31. P. Castrillo, D. Hessman, M.E. Pisto, J.A. Prieto, C. Pryor, and L. Samuelson, *Jpn. J. Appl. Phys.* Part 1. **36**, 4188 (1997).
32. M.J. Kirton and M.J. Uren, *Adv. In Physics* **38**, 367 (1989).
33. M.-E. Pisto, *Phys. Rev. B.* **63**, 113301–1 (2001).

CHAPTER 13

Radiation Effects in Quantum Dot Structures

Nikolai A. Sobolev

I3N-Institute for Nanostructures, Nanomodelling and Nanomanufacturing, and
Departamento de Física, Universidade de Aveiro, 3810-193 Aveiro, Portugal

13.1 Introduction

The tolerance of materials and devices to radiation-induced defects (radiation defects, RDs) is of crucial importance in atomic energy and space applications. In a nuclear reactor, the samples are exposed to neutrons and gamma-quanta. The space-radiation environment accompanying most useful orbits consists of energetic electrons (energies up to ~ 7 MeV), protons (energies extending to hundreds of MeV) and small amounts of low energy heavy ions [1]. The predicted proton and electron fluxes for low earth orbit (LEO) missions are presented in Figs 13.1 and 13.2. The LEO proton spectrum is especially hard: between 50 and 500 MeV the proton flux decreases only by a factor of 4.

Furthermore, the creation of RDs is an inevitable collateral effect in ion implantation that is a well-established technique of materials modification. Finally, there is a possibility of using RDs

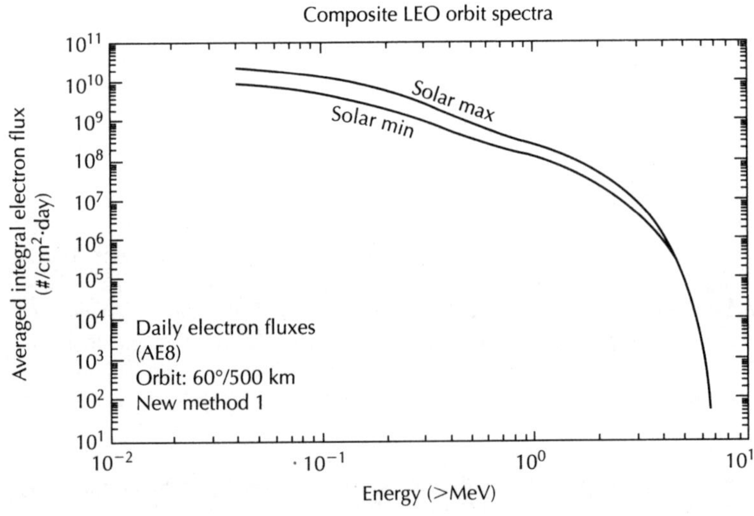

Figure 13.1   Low earth orbit (LEO) proton fluxes [1].

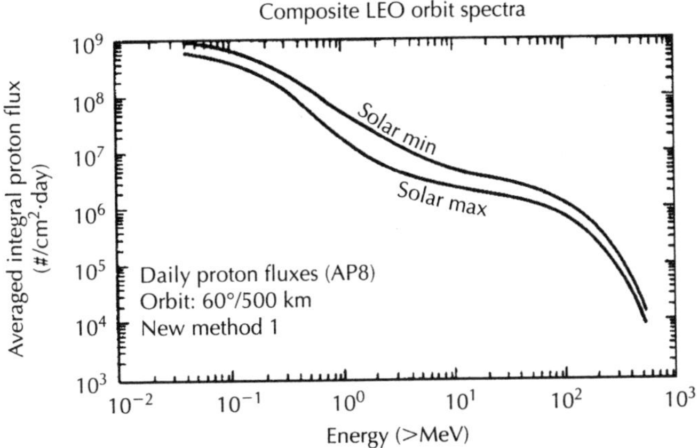

Figure 13.2   Low earth orbit (LEO) electron fluxes [1].

themselves in the device technology. With the onrushing advent of quantum-size semiconductor structures (QSSS), the studies of RDs in them rapidly grow in importance.

Before discussing the physics of the interaction of radiation with matter, it is worth commenting on the terminology. The energy absorbed by a specimen per mass unit is dubbed *dose* and is measured in Gray ($Gy = J \cdot kg^{-1}$). One Gy is equal to 100 rad. The *flux* is the number of particles passing during 1 s through a 1 m² area ($m^{-2} \cdot s^{-1}$); integrating over time gives the *fluence* measured in m^{-2}. It has to be noted that physicists often confuse the terminology in their publications, calling *dose* what in reality is *fluence* and measuring it, for convenience, in cm^{-2}.

The result of irradiating a semiconductor material will depend on the type of radiation, its mode (pulsed, continuous) and type of interaction with the material, as well as the type of material. The two main types of interaction of radiation with materials are atomic displacements and ionization. All particles (electrons, protons, heavy ions and photons ranging from UV to gamma energies) except neutrons produce ionization effects in materials. Besides, the radiation effects in solid-state devices include single-event upsets.

When an energetic ion penetrates a material, it loses energy mainly by two nearly independent processes: (i) elastic collisions with the nuclei known as nuclear energy loss $(dE/dx)_n$, which dominates at an energy of about 1 keV/amu; and (ii) inelastic collisions of the highly charged projectile ion with the atomic electrons of the matter known as electronic energy loss $(dE/dx)_e$, which dominates at an energy of about 1 MeV/amu or more. In the inelastic collision (cross-section $\sim 10^{-16} cm^2$) the energy is transferred from the projectile to the atoms through excitation and ionization of the surrounding electrons. The amount of electronic loss in each collision varies from tens of eV to a few keV per Å [2].

The atomic displacements occur due to the transfer of momentum of the incident particle to the atoms of the target material (nuclear energy loss). Provided an atom subjected to such a collision receives sufficient kinetic energy, it will be removed from its position and leave behind a vacancy. The removed atom may meet another such vacancy and recombine or lodge in an interstitial position in the lattice (a self-interstitial) or be trapped by an impurity atom. The vacancies may be mobile, too, and either combine with impurity atoms or/and cluster with other vacancies. Defects that are stable at the irradiation temperature may become mobile upon subsequent heating. For the evaluation of the radiation damage in solids the mobility of the defects is of paramount importance. The self-interstitials in silicon are mobile even at 0.5 K [3]. In semiconductors such as Si, Ge, GaAs, GaSb, InP, InAs, AlAs, and InSb a large part of the primary defects undergoes annihilation even below room temperature (RT). The RDs in these materials, which are found at RT, consist mainly of secondary and tertiary complexes formed by migration and agglomeration of vacancies and interstitials with each other and with impurities. The resulting complexes are usually electronically active.

When the energy of the primary recoil atom is high, which is especially the case upon ion implantation or neutron irradiation, a collision cascade develops, where the defect density is much higher than upon electron or proton irradiation. This high defect density may lead to an amorphization of the implanted layer. The accompanying defect rearrangement processes are usually quite complex and not yet understood in all details even in elemental semiconductors such as Si.

The fundamentals of the radiation defect creation in solids can be briefly described as follows [4]. The total cross-section for the displacement of an atom from its regular lattice site as a result of an elastic collision is given by:

$$\sigma(E) = \int_{T_d}^{T_m} d\sigma(E,T) \tag{13.1}$$

where E is the kinetic energy of the incident particle, T is the kinetic energy transmitted to the lattice atom, and $d\sigma(E, T)$ is the differential cross-section of the corresponding interaction. The integration is performed over the energy T from the minimum energy T_d necessary for the displacement of a lattice atom into an interstitial position to the maximum energy T_m that the incident particle can transmit to the target atom. The energy T_d is called threshold energy. The differential cross-section depends on the interaction potential. In the case of charged particles (electrons, protons, ions) the interaction can be described by the Coulomb potential; for the neutral particles (neutrons) it is rather similar to the collision of hard spheres. Usually the threshold energy is taken as isotropic, because it is difficult to observe an appreciable orientation dependence of T_d in the experiment. The experimental values of T_d for various semiconductor materials are given in Table 13.1 [4, 5].

Table 13.1   Experimental threshold energy values for the atomic displacement damage for various semiconductors (data taken from [4, 5]).

Crystal	T_d, eV	Crystal	T_d, eV
Diamond	35±5	InSb	6.4–9.9
Si	21	InAs	6.7–8.3
Ge	27.5	GaSb	6.2–7.5
GaAs	7–11	InP	3–4 (In) 8 (P)

To calculate the number of displacements produced by an incident particle, one has to solve the integral in Eq. (13.1), taking into account the type of interaction.

For heavy charged particles, the value of $d\sigma(E, T)$ is given by the Rutherford formula, and the result of the integration is:

$$\sigma(E) = a \left(\frac{1}{T_d} - \frac{1}{T_m} \right) \tag{13.2}$$

with

$$a = \frac{\pi Z_1^2 Z_2^2 e^4}{E} \cdot \frac{M_1}{M_2}. \tag{13.3}$$

The index 1 designates the incident particle and 2 belongs to the target atom.

For neutron irradiation, the hard spheres collision model may be adopted, so that one obtains:

$$d\sigma = \pi (R_1 + R_2)^2 \frac{dT}{T_m} \tag{13.4}$$

where R_1 and R_2 are the radii of the incident particle and the target atom, respectively. Taking into account that $M_1 \ll M_2$, the result is:

$$\sigma(E) = \pi(R_1 + R_2)^2 \left(1 - \frac{T_d}{T_m}\right). \tag{13.5}$$

For the electrons with MeV energies, the Rutherford formula has to include relativistic corrections. The exact calculation has been performed by N. Mott [6]; an approximate solution has been obtained by McKinley and Feshbach [7]. However, even the latter formula is relatively sophisticated.

Figure 13.3 gives an idea of the energy transferred by an incident particle to the target atoms as a function of the energy for electrons, protons and neutrons [4]. Note that collisions with neutrons are much harder in terms of the average energy transmitted to a target atom than those with protons of equal energy. In Fig. 13.4 a comparison of experimental damage coefficients in

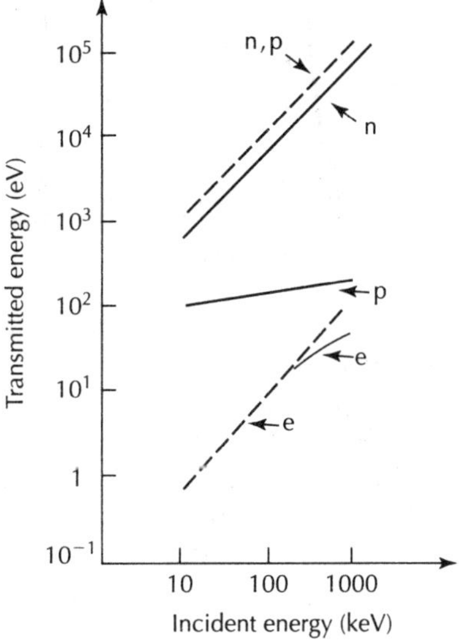

Figure 13.3   Maximum (dashed line) and average (full line) energy transmitted to a silicon atom as a function of the incident energy for electrons (e), protons (p), and neutrons (n). [4]

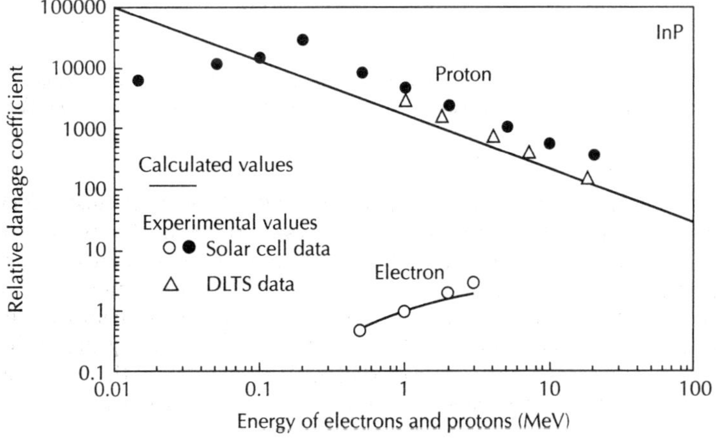

Figure 13.4   Relative damage coefficients for InP between electrons with energy E_e and protons with energy E_p normalized to 1 MeV electron irradiation results, determined from the solar cell property degradation and the DLTS method, in comparison with normalized values of the non-ionizing energy loss (NIEL) [10].

InP between electrons and protons in a wide range of energies is shown. A monotonic decrease in damage constant with proton energy increasing up to 500 MeV has been found experimentally in GaAs light-emitting diodes (LEDs) [8] and InGaAs/GaAs quantum well (QW) LEDs [9]. It is important to note that the proton-induced damage profile is highly non-uniform with a sharp maximum near the projected range R_p (penetration depth) so that the damage *density* at depths well below R_p can *decrease* with increasing proton energy, despite the increase of the total energy deposited in elastic collisions (see Fig. 13.5). This fact must be taken into account when irradiating nanometre thick layers containing, e.g., QDs and situated near the sample surface.

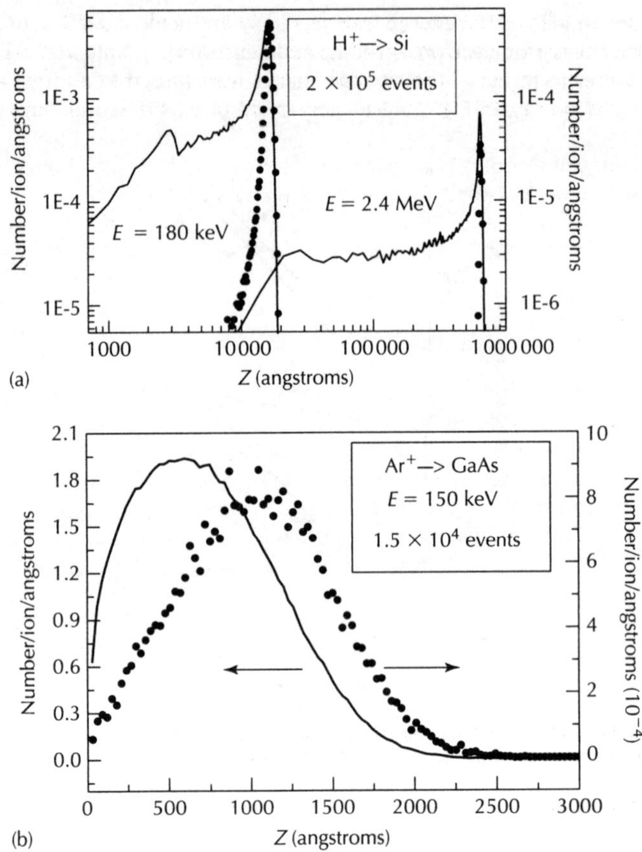

Figure 13.5   *TRIM simulations of the depth distribution of the displaced target atoms (solid lines, left scale) and implanted atoms (dots, right scale) for the implantation of 180 keV and 2.4 MeV H$^+$ in Si (a) and 150 keV Ar$^+$ in GaAs (b).*

It can be immediately seen from Fig. 13.3 that the energy of electrons must achieve hundreds of keV in order to implement transfer of an energy amount exceeding T_d. On the contrary, in the case of the ions, the masses of the incident particle and the target atom are comparable, so that the energy transfer is very efficient, and even in the case of the ion etching of the surface with energies of hundreds of eV the creation of radiation defects must be taken into account.

Fast neutron irradiation produces energetic recoil atoms and in terms of the produced damage can be understood as "internal" self-ion implantation. However, due to the small collision cross-section, the displacement cascades are well separated in the crystal volume even at medium irradiation doses.

On the contrary, implantation of medium and heavy mass ions produces a very dense damage within a thin subsurface layer of a solid target, so that amorphization of this layer can be readily achieved. The critical fluence needed for the amorphization of a given crystal depends on the ion mass and the target temperature. For each ion–target combination, there is a critical temperature above which the amorphization becomes impossible due to dynamical defect annealing [11, 12]. (At cryogenic temperatures, the amorphization of silicon was induced even under MeV electron irradiation, but the required fluences were exceedingly high [13, 14].) The energy dependence of the critical ion fluence exists but is less pronounced. The theoretical description of the crystalline-to-amorphous transition upon ion irradiation is still a matter of debate [15]. Another important peculiarity of the ion irradiation is the sputtering of the target [16]. This phenomenon has important applications in the radiation technology of nanostuctures, see Section 13.4.2.

In practical terms, the SRIM/TRIM Monte Carlo simulation code [17] can be applied to calculate the ranges and primary displacement defects created by energetic ions in matter. To apply SRIM/TRIM to the calculation of the effects of other types of radiation, one additionally needs the "Integrated TIGER Series" (ITS code) for electrons and photons, or the "Monte Carlo Neutron Program" (MCNP code) for neutrons [18]. Electron trajectories and energy loss profiles can also be simulated using the CASINO code [19]. A few examples of the primary damage and implanted ion concentration profiles calculated by TRIM are given in Fig. 13.5.

The effect of ionization on the defect production in the common semiconductors exists but mostly is a minor one. Anyway, the ionization (formation of electron–hole pairs) alone does not produce RDs in these semiconductors as it is the case in wide-gap insulators. However, the degradation of a MOS device, especially at low irradiation doses, is almost entirely due to the long-lived effects of ionization in the dielectric subelement, i.e. in the gate insulator [20]. As to the ionization effects in devices, one has to distinguish between the effect of the total ionizing dose (TID) and single event effects (SEE) such as single event upset (SEU) and others. SEUs are defined by NASA as "radiation-induced errors in microelectronic circuits caused when charged particles (usually from the radiation belts or from cosmic rays) lose energy by ionizing the medium through which they pass, leaving behind a wake of electron–hole pairs"[21]. The SEUs are transient soft errors, and are non-destructive. A reset or rewriting of the device results in normal device behaviour thereafter. The effects like SEUs are out of the scope of this chapter that is dedicated to the effects of persisting damage. In bipolar devices the primary effect of ionizing radiation is gain reduction. This is usually due to an increase in surface recombination near the emitter–base region. Ionization damage also causes leakage current to increase [23]. As we shall see in Section 13.2.7, there are pronounced effects of irradiation with electrons of subthreshold energies and X-rays on some quantum dot structures.

The sensitivity of the device parameters to irradiation is further determined by material properties, such as the threshold energy for atomic displacement (see Table 13.1), probability of the annihilation of the self-interstitials and vacancies, type and level of doping, and position of the defect-induced energy levels in the gap. Let us cite a few instances. GaN is about two orders of magnitude more resistant to 2 MeV proton irradiation than GaAs [24, 25]. The damage constant of p-GaAs is smaller than that of n-GaAs [26]. ZnO is generally considered to be radiation hard [27–29]; it is about two orders of magnitude harder than GaN [30–32]. A very informative comparison can be made using solar cells made out of various materials (the latter may be present in one multi-junction cell [33]). So, e.g., the InP and $CuInSe_2$ solar cells have been found to be more radiation resistant than those made of Si and GaAs [22]. A confrontation of results for some III–V compounds is given in Figure 13.6. Ample damage correlations in Si, GaAs, and InP exposed to gamma, electron and proton radiations have been undertaken, e.g., in [34]. It has recently been shown that improved resistance against amorphization is to be found in compounds that have a natural tendency to accommodate lattice disorder [35]. There is a comprehensive literature on the subject of the radiation hardness of semiconductors and semiconductor devices [36].

A very special consequence of high-energy irradiation is the activation of materials due to nuclear reactions. It is certainly unimportant on board a telecom satellite, but the fluence has to be restricted in laboratory tests to avoid handling parts where the induced radioactivity creates unacceptable personnel hazards. Luckily, the threshold energy of nuclear reactions is usually quite high, so that electrons and protons with energies of several MeV do not induce any persistent radioactivity. Importantly, thermal neutrons induce transmutation reactions often producing long-lived isotopes.

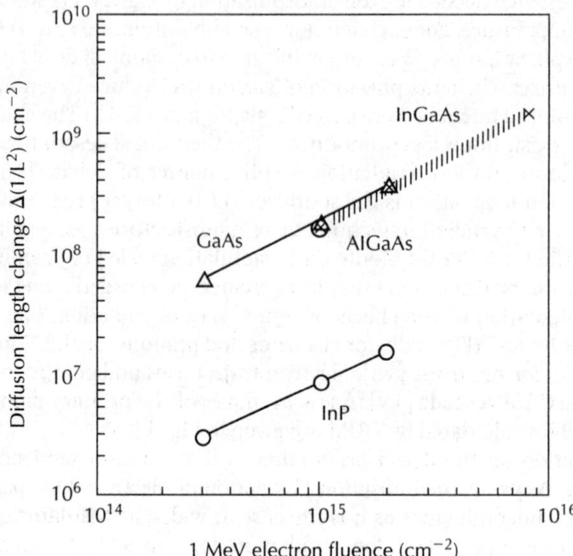

Figure 13.6   Changes in minority-carrier diffusion length L, determined by the EBIC method, for $Al_{0.35}Ga_{0.65}As$, $In_{0.15}Ga_{0.85}As$ and GaAs due to 1 MeV electron irradiations as a function of electron fluence in comparison with those of InP [22].

Summarizing, in order to predict the radiation damage in quantum-size semiconductor structures (QSSS), first of all one needs knowledge of the creation, transformation, and annihilation processes of RDs in corresponding bulk materials including alloys. Whereas these processes in Si are well understood, the information concerning Ge, GaAs and InP is much less detailed. The worst situation is to be stated for the other III–V compounds and alloys, let alone the II–VI semiconductors [37, 38], though, e.g., CdTe is one of the most used semiconductors for nuclear detectors [39]. To solve the problem of the radiation hardness of a device, one has to establish which layer (or layers) in a concrete, probably very complicated, structure predominantly determines the device parameters degradation. In devices like light-emitting diodes (LEDs) or lasers containing low-dimensional active layers, it is important to know which is the volume sampled by the wavefunction of the electrons and holes participating in the radiative recombination. Finally, the role of the Fermi level, hetero-interfaces and strain in the defect evolution and defect reactions, the mutual influence of the adjacent layers, and the impact of the quantum confinement on the structure and properties of local defects, which are supposed to be already known from the studies of the corresponding bulk semiconductors, have to be elucidated.

On the other hand, what useful information can we learn from the irradiation studies?

- Influence of the defects on the electronic properties of QSSS and on the corresponding device parameters;
- Elucidation of the electronic structure of QSSS as well as their carrier transport, relaxation and recombination processes using defects as microprobes;
- Diffusion processes in QSSS;
- Novel technological processes of micro-, nano- and optoelectronics.

The chapter presents a survey of effects occurring in Si–Ge, III–V and II–VI quantum dots (QDs) and, for comparison, in quantum wells (QWs) and superlattices (SLs) upon electron and proton irradiation as well as upon ion implantation. **Section 13.2** is dedicated to the important issue of the radiation hardness. It is shown that QD-based devices can withstand much higher radiation fluences than corresponding 2D and bulk structures. The physical mechanisms of this phenomenon are discussed. In **Section 13.3**, the influence of irradiation on the QD lasers is considered. **Section 13.4** demonstrates examples of the application of particle irradiation to the device technology, especially QD intermixing and ion-beam synthesis. Emphasis is given to the synthesis of magnetic nanocrystals in different solid matrices. **Section 13.5** contains concise conclusions.

13.2   Radiation hardness of quantum dot heterostructures

13.2.1   General remarks

The term "radiation hardness" (the same as "radiation resistance") describes the ability of a structure's property to withstand the deteriorating action of radiation. The recombination parameters of semiconductors are much more sensitive to RDs than, e.g., equilibrium carrier concentration or mobility. RDs with deep levels in the band gap act as non-radiative recombination centres (lifetime killers) limiting the photoluminescence (PL) and electroluminescence (EL) intensity as well as photosensitivity.

Before starting to treat the radiation hardness of QD heterostructures, it is worth noting that there apparently is a general trend for low-dimensional structures to be more defect free than corresponding bulk materials. As pointed out in [40], Turnbull was probably the first to propose, as early as 1950, that small crystals will contain fewer defects [41]. This "self-purification" was recently shown [40] to be an intrinsic property of defects in semiconductor nanocrystals, for the formation energies of defects increase as the size of the nanocrystal decreases, see Fig. 13.7.

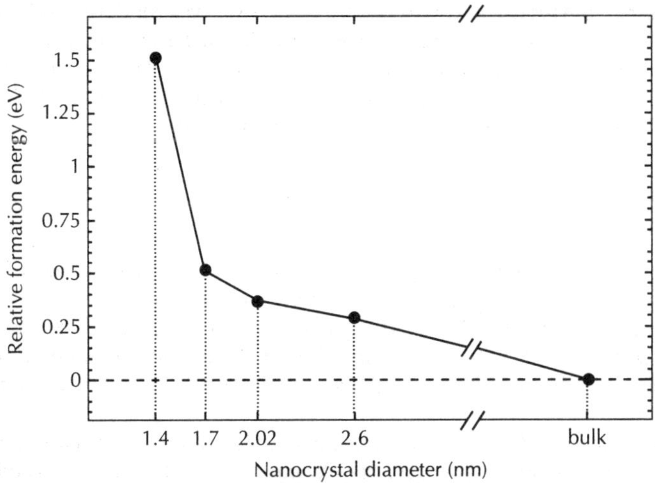

Figure 13.7   Variation of the formation energy of a substitutional Mn impurity in a CdSe nanocrystal as a function of the nanocrystal diameter. The increase in the formation energy is important to explain self-purification in nanocrystals [40].

There are several experimental corroborations of this trend. The correlation of defect density to the grain size was examined in nanocrystalline Pd and ZrO_2 upon irradiation with 4 MeV Kr ions [42]. A drastic reduction of defect clusters in the small grains below 50 nm was found. In the smallest ZrO_2 (<15 nm) and Pd grains (<30 nm) no defects could be detected. PL experiments performed on InAs QDs embedded in low-temperature (LT) grown GaAs led to the conclusion that the PL quenching centres are located only in the GaAs barrier and not inside the QDs [43]. In another work it was concluded from the structural characterization of self-organized InGaAs/GaAs QDs grown on Si substrates that the dots themselves may be defect free [44]. However, only relatively small dots can be defect free. So, e.g., when the InAs or InGaAs QDs exceeded certain dimensions, dislocations were observed within the dots [45–47].

There also are many examples of experimentally found enhanced defect tolerance of the low-dimensional structures. (It has to be noted that the role of native, technological and radiation-induced defects on the performance of devices is the same.) The amphoterically doped GaAs or single heterojunction GaAs/AlGaAs light-emitting diodes (LEDs) are much more sensitive to proton-induced damage than the double heterojunction devices [48, 49]. Nanostructuring leads to one order of magnitude enhancement of radiation hardness against high-energy heavy ion bombardment in GaN layers [50]. The effect was tentatively explained as due to the increase in the specific surface of the nanostructured sample which allows the migration of the defects formed during the ion bombardment process towards the surface (see also [51]), and due to the enhanced dynamic annealing of these defects.

Furthermore, it was predicted theoretically that the high recombination rate of non-radiative centres (lifetime killers) could be reduced by the low mobility of carriers and that (which is important in the context of this chapter) the effect should be more pronounced in low-dimensional structures [52]. Indeed, a strong reduction in the carrier diffusion length is observed going from InGaAs QWs ($L_d \approx 2.7\,\mu\mathrm{m}$) to InAs QDs ($L_d < 100\,\mathrm{nm}$) [53].

The increased tolerance of defects is one of the most important promises of the self-assembled QD nanotechnology [54]. The basic argument is that more strongly localized carriers exhibit reduced migration to non-radiative centres. The first work on the defect tolerance of InAs QDs has been reported in [55, 56]. The authors compared the PL of an array of self-assembled InAs/GaAs QDs and of a single high-quality InGaAs QW. On GaAs substrates, the radiative quantum efficiency η was essentially the same for both structures. The growth on a commercial GaAs-on-Si substrate with a high dislocation density entailed drastic quenching of the integrated PL intensity and shortening of the carrier lifetime τ for the InGaAs QW, whereas both τ and η were not modified for the QD array. The authors came to the conclusion that the efficient carrier capture by InAs QDs, combined with the localized nature of QD excitons, hindered in this case the carrier diffusion toward dislocations. Coexistence of growth-induced defect-related deep levels with InAs QDs emitting bright luminescence was reported in [57]. InGaAs QD lasers grown on Si were successfully fabricated [44, 431, 432]. Interestingly, a QD layer was utilized in [431] to suppress the propagation of dislocations present in the GaAs buffer and to fabricate a low defect density active QD region.

13.2.2   In(Ga)As/GaAs quantum dots

Irradiation with low- or medium-energy ions of In(Ga)As/GaAs QDs has been found to quench the PL intensity at least one order of magnitude more slowly than in comparable QW structures [58, 59]. The radiation hardness of the QD PL was observed also in experiments employing electron [60, 61] and proton [62–66, 81, 435] irradiation. An enhanced radiation resistance of the electrical properties of the InGaAs/GaAs QD structures upon ion implantation [67] has been shown, too.

Let us consider in more detail the behaviour upon 2 MeV electron irradiation of three samples containing one QD layer (sample 1 × QD), five QD layers (5 × QD) and two coupled QWs (CQW) [60]. The PL spectra of the as-grown samples taken at the measurement temperature $T_m = 10\,\mathrm{K}$ upon excitation with an Ar^+ laser and their evolution versus electron irradiation fluence are shown in Fig. 13.8a–c. With increasing irradiation fluence all samples exhibit a decrease in the PL intensity. However, whereas in samples 1 × QD and 5 × QD the QD-related PL peaks could be observed up to the fluences $\Phi = 1 \times 10^{17}$ and $2 \times 10^{17}\,\mathrm{cm}^{-2}$, respectively, the QW-related PL in sample CQW was quenched already between $\Phi = 2 \times 10^{16}$ and $5 \times 10^{16}\,\mathrm{cm}^{-2}$. This behaviour is essentially the same at 10, 77 and 300 K. At electron fluences $\Phi \geq 1 \times 10^{17}\,\mathrm{cm}^{-2}$ the PL of bulk n-GaAs with a doping level of $1 \times 10^{17}\,\mathrm{cm}^{-3}$ is completely suppressed [68]. Thus, Fig. 13.8 demonstrates a much higher tolerance of QDs with respect to the electron irradiation as compared to bulk GaAs or QWs.

In order to separate the defect-related recombination processes in the GaAs barrier from those in the QDs and QWs themselves, the samples were investigated using resonant excitation by means of a Ti^+–sapphire laser. Selected spectra taken at 12 K are shown in Fig. 13.9. The excitation energy 1.318 eV lies below the PL peak of the WL centred at 1.35–1.36 eV. As in the case of

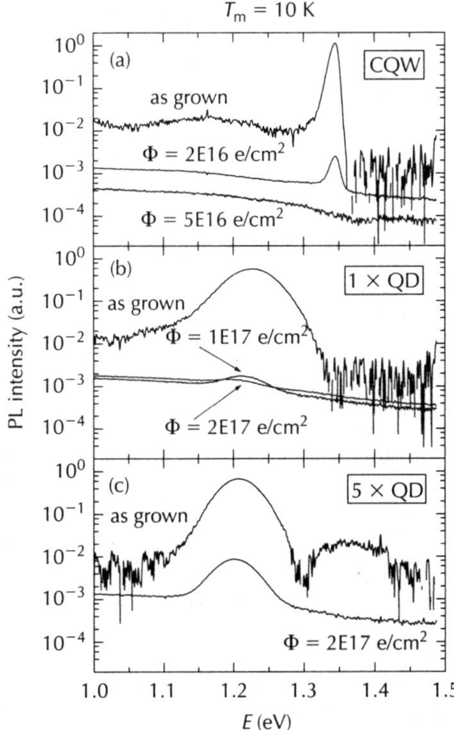

Figure 13.8   FTIR PL spectra of samples $1 \times$ QD, CQW and $5 \times$ QD measured at $10\,K$ upon excitation with the $457.9\,nm$ line of an Ar^+ laser. The electron irradiation fluence Φ is indicated at the spectra. The PL intensity is normalized to the excitation power. The latter did not exceed $\sim 10^{-2}\,W/cm^2$ for as-grown samples to avoid PL and detector saturation. For irradiated samples with a low PL intensity, the excitation power was increased to maximize the signal [60].

the above-band gap excitation, the PL intensity degradation of the CQW sample is much stronger than that of the QD samples. The behaviour observed upon 1 MeV proton irradiation [62] and low-energy Ar ion bombardment [58] is completely analogous, as illustrated in Fig. 13.10.

Possible processes leading to the quenching of the PL intensity upon electron irradiation are:

1. Capture and non-radiative recombination of photoexcited carriers at defects in the GaAs barrier. This process should influence the QW and QD luminescence to one and the same extent, provided a linear recombination regime (no saturation) is ensured. In fact, under these conditions the PL intensity of the kth centre I_k is given by the expression $I_k \propto g\sigma_k N_k/(\sum_i \sigma_i N_i)$, where g is the carrier generation rate, σ_i is the capture cross-section and N_i is the concentration of the ith centre, and the summation is made over all recombination channels. However, upon above-band gap excitation with the Ar^+ laser very different PL quenching rates for QD and CQW samples which had similar PL intensities prior to irradiation were observed. This means that (i) there is an additional irradiation-induced "QW- and QD-internal" non-radiative recombination channel that has a much larger cross-section in the QW than in the QDs and (ii) QDs can efficiently capture carriers directly from the barrier.

2. If the QDs are fed by carriers also through the WL, the defect-induced non-radiative recombination in the WL should influence the quenching rate of the QD PL. Therefore, an experiment with QD PL excitation below the WL band edge is essential. The QD PL quenching observed upon subband gap excitation (Figs. 13.9 and 13.10 (bottom)) confirms very well the existence of an internal defect-related recombination mechanism in the QDs and the QW and its greater influence in the latter.

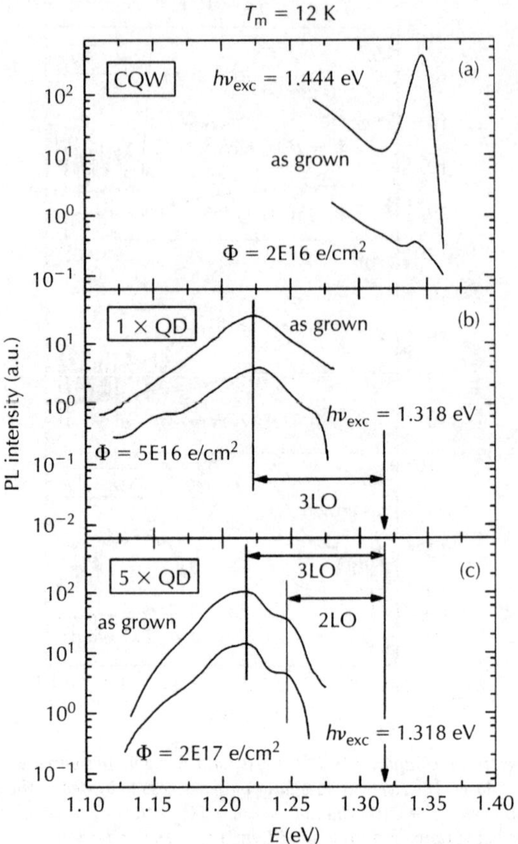

Figure 13.9   PL spectra of samples $1 \times$ QD, CQW and $5 \times$ QD measured at 12 K upon excitation with a Ti$^+$–sapphire laser. The PL intensity is normalized to the excitation power. With the excitation energy $h\nu_{exc}$ lying near enough to the QD PL band, resonances within the latter at energies being a multiple of the LO-phonon energy can be observed so that a subset of the QD ensemble defined by the ground-state transition energy is probed [74]. The energies of the phonons involve range from 29.6 eV (InAs LO-phonon in the WL) to 37.6 meV (GaAs LO-phonon in the strained barrier layers adjacent to the QDs) [60].

3. The internal irradiation-induced non-radiative recombination mentioned above may occur at defects created inside the QW (WL) and QDs as well as at the interfaces QW (WL)/ GaAs barrier, QD/WL and QD/GaAs barrier. The delocalization of carriers in the QW (WL) plane perpendicular to the growth direction makes them interact with a much greater number of defects than in QDs [58]. Thus, *a priori* a higher radiation hardness of the QD PL is expected. The role of inhomogeneous strain in defect trapping at interfaces with possible subsequent separation or annihilation is difficult to evaluate. It cannot be excluded that the differing strain gradients influence the defect reactions in the QW and QD to a different extent.

4. Another mechanism of the "internal" defect-induced PL quenching proposed in [69]and corroborated by detailed calculations in [70] is the carrier tunnelling out of the QW or QDs to defects created in the GaAs barrier. In fact, QDs and QWs have a different sensitivity to defects in the adjacent barrier regions. This is due to a differing degree of the wavefunction penetration into the barrier. So, e.g., In$_{0.13}$Ga$_{0.87}$As QWs start to "feel" a free surface located at distances less than about 20 nm [71]. The PL intensity of InAs/GaAs QDs, however, only degrades for a cap thickness of 9 nm or smaller [72].

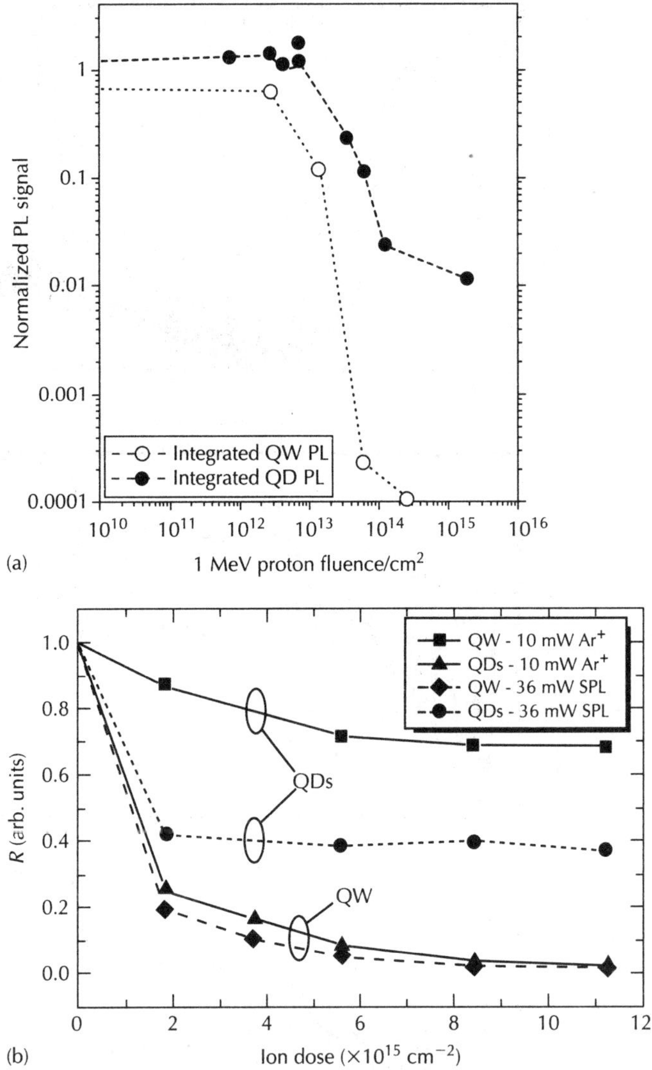

Figure 13.10 Top: *PL intensity of a QD and a reference QW sample vs the 1 MeV proton fluence upon above-band gap excitation [62]*. Bottom: *Dependence of the PL intensity of a QD and a reference QW sample on the 300 eV Ar ion fluence. The measurements were done upon above-band gap (labelled 10 mW Ar$^+$) or resonant excitation (labelled 36 mW SPL) [58]*.

Further insights into the mechanism of the radiation damage of the QD heterostructures were obtained by using the PL excitation (PLE) and time-resolved PL (TRPL) spectroscopy [61, 63–66, 73]. In samples A and B used in [65, 66], the active layer together with GaAs cladding layers were placed between two AlGaAs barriers, with a GaAs capping layer on top of the whole structure. Sample A had an active layer composed by the QD layer, with a dot density of $\approx(3–5) \times 10^{10}$ cm^{-2}, overgrown by a 2 nm In$_{0.25}$Ga$_{0.75}$As QW. Sample B had a dot density of $\approx 10^9$ cm^{-2} in an active layer composed by a single QD layer. The irradiation by 2.4 MeV protons with fluences in the range from 1×10^{12} to 1×10^{14} cm^{-2} was performed at room temperature.

Usually the defects reduce the lifetime of non-equilibrium carriers and, consequently, their diffusion length, thus limiting the carrier supply to the radiative recombination centres and killing the PL intensity. However, because of the presence of the closely spaced AlGaAs barriers, the carrier capture by the QDs in the samples used in [65, 66] was not diffusion limited. That is why no

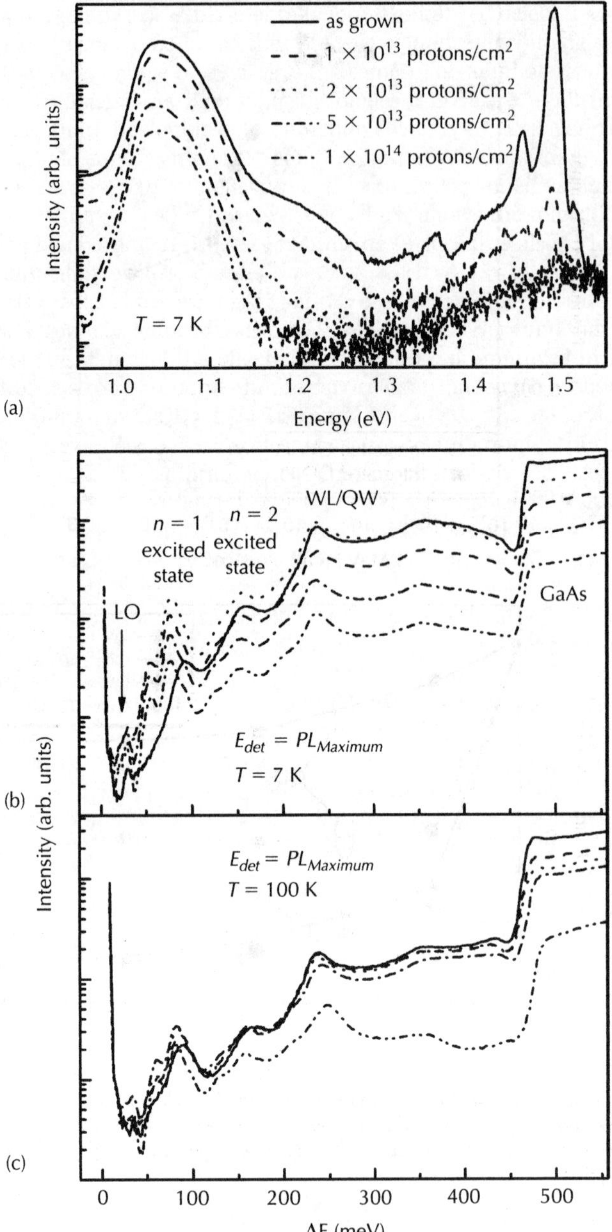

Figure 13.11   PL (a) and PLE (b, c) spectra of sample A with a dot density of ≈(3–5) × 10¹⁰ cm⁻² for various 2.4 MeV proton irradiation doses, measured at 7 K (a, b) and 100 K (c) [65].

difference in the quenching factor of the PL intensity at a given irradiation dose was observed at 7 K for the above-band gap and below-band gap excitation for all energies above the $n = 2$ QD excited state (Fig. 13.11b). Thus, the loss of carriers occurs mainly in the dots themselves. A probable reason of this effect is a tunnelling out of the dots to adjacent radiation-induced non-radiative recombination centres. The situation changes dramatically, when the temperature is increased to ~100 K (Fig. 13.11c). The PL intensity in the irradiated samples is recovered (within ~10%) to the undamaged case. The channel leading to non-radiative recombination is obviously not relevant at elevated temperatures. This observation was attributed to the

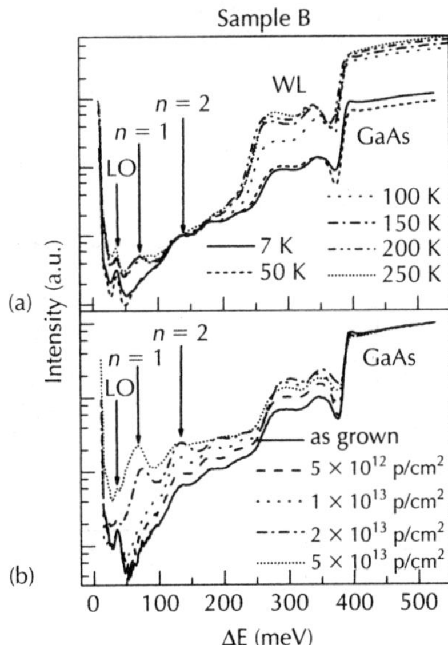

Figure 13.12   (a) PLE spectra vs the excess excitation energy $\Delta E = E_{exc} - E_{det}$ of the as-grown sample B with a dot density of $\approx 10^9 \, cm^{-2}$ taken at different temperatures. The spectra are normalized to the detection energy and to the second excited state intensity. (b) PLE spectra of the same sample after irradiation with various 2.4 MeV proton doses, recorded at the QD PL maximum at 7 K. The spectra are normalized to the intensity above the GaAs band gap [66].

temperature-dependent competition of intradot relaxation and non-radiative recombination in the first excited state [74]. With increasing temperature intradot relaxation becomes more efficient due to the growth of the phonon density [75]. Consequently, the captured carriers reach the unperturbed ground state more quickly, thus overcoming the phonon bottleneck and bypassing tunnelling. Upon further increasing the temperature above 100 K, the intensity decreases dramatically with increasing dose [65]. This probably occurs because thermal activation into the barrier allows the carriers to reach defect states in the latter without tunnelling. An important conclusion must be drawn from this behaviour: the radiation hardness is temperature dependent.

Another striking feature of the PLE spectra is the increase upon irradiation of the PLE intensity at energies corresponding to the low-lying QD excited states (Fig. 13.11b and Fig. 13.12b). The effect is particularly well seen in Fig. 13.12b, where the PLE intensities for all doses have been normalized to that of the as-grown sample at energies exceeding the GaAs band gap, and in the contour plots (Fig. 13.13). Qualitatively similar effects are observed when the temperature is increased (Fig. 13.12a) or a reverse bias is applied to a diode structure with the dots located in the space charge region (Fig. 13.14) [65]. The radiation defects shift the Fermi level towards the middle of the band gap, thus emptying the low-lying QD states from spectator carriers and permitting the resonant absorption.

The results of the time-resolved PL (TRPL) measurements performed on sample A as a function of the irradiation dose, for resonant and non-resonant excitation, corroborate the PLE data. No influence of the irradiation on the PL decay kinetics from the ground state is observed (Fig. 13.15). However, the rise time shortens by about a factor of four for the dose $1 \times 10^{14} \, p/cm^2$, both for the excitation into the barrier (Fig. 13.15a) and into the sublevels of the first excited state at 1.140 and 1.123 eV (Fig. 13.15b and c). This means that the rise time shortening upon above-band gap excitation is caused by a carrier (exciton) loss in the QDs and not by any reduction of the diffusion length in the barrier or the WL. As discussed above, the effect can be explained by the tunnel escape of the carriers to adjacent defects. The ground state, having a more localized wavefunction than the excited ones, remains essentially "undamaged". Contrary to measurements on electron irradiated QDs of another type [61], no development of a second,

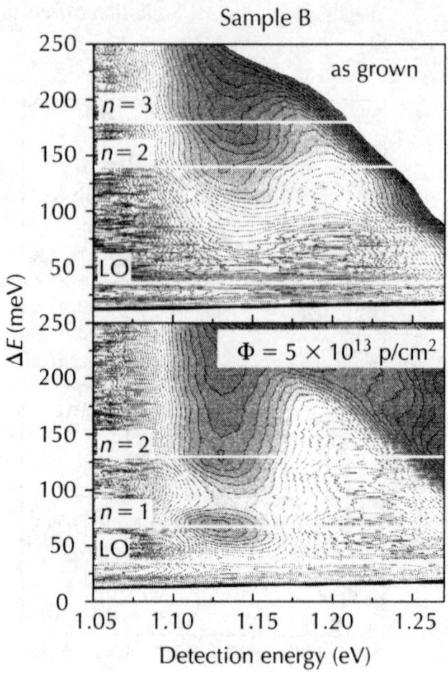

Figure 13.13 Contour plots of the QD PL intensity as a function of the detection energy and the excess excitation energy $\Delta E = E_{exc} - E_{det}$ for sample B at 7 K. The white lines denote PLE resonances from the excited states and LO-phonons [66].

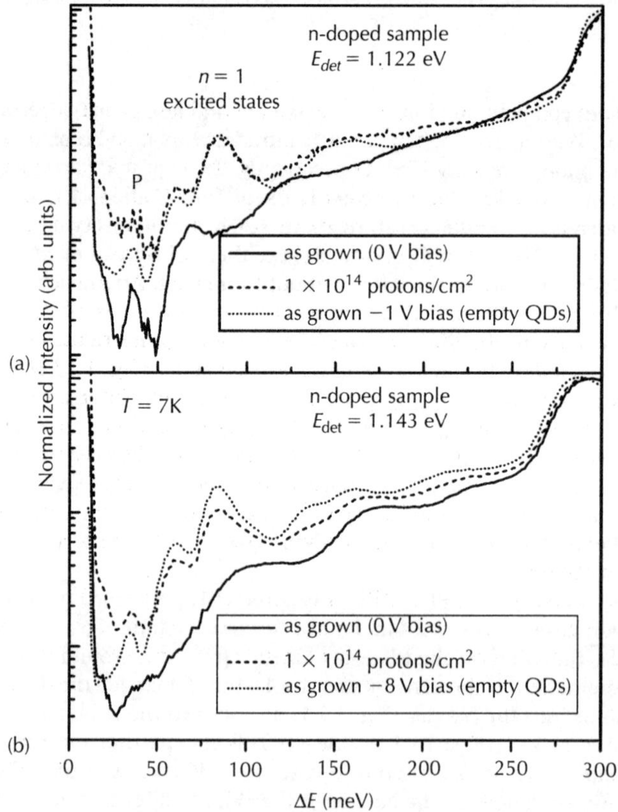

Figure 13.14 PLE spectra normalized to the wetting layer peak for as-grown and proton irradiated diode structures with the dots located in the space charge region. (a) An n-doped sample and (b) a p-doped sample. The solid and dotted lines show PLE spectra of charged and neutral QDs in the as-grown samples without and with bias, respectively. The dashed lines show the PLE spectra of the QDs in the irradiated samples [65].

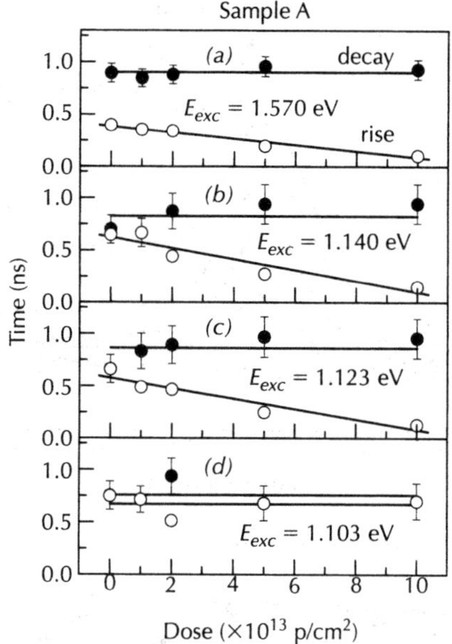

Figure 13.15   *Rise (open circles) and decay (solid circles) times from TRPL measurements of sample A. The detection energy was at the ground state transition. (a) E_{exc} = 1.570 eV (above the GaAs band gap); (b) E_{exc} = 1.140 eV (a sublevel of the 1st excited state); (c) E_{exc} = 1.123 eV (another sublevel of the 1st excited state) [66].*

shorter decay time was observed. This difference was tentatively attributed to the heavier damage caused by the electron doses up to 10^{17} e/cm^2 used in [61]. Similarly, though a slight decrease of the carrier lifetime was observed in irradiated QDs in [63, 64, 73], it was much smaller than in reference QW samples.

The above-mentioned facts bring up an important question. Do stable point defects created by atomic displacements at room temperature (RT) exist inside the In(Ga)As QDs? In fact, their existence has never been proven. Since, as discussed in Section 13.1, the primary defects (vacancies and interstitial atoms) are mobile at RT in GaAs and InAs, it is very likely that they are captured at the interfaces (cf. [76]). Moreover, the defects raise the free energy of the crystal, so that it is only natural that the QDs expulse mobile defect components into the matrix. A related effect is the "self-purification" of nanocrystals [40]. The high radiation hardness of the ground state of the QDs described above leads to the conclusion that the (small enough) QDs are defect free, at least up to moderate irradiation fluences. This, along with the localization of the wave function, is a reason of the high radiation hardness of the QD structures.

The behaviour upon 3 MeV proton irradiation and subsequent annealing of the optical absorbance spectra of the intersubband transitions in In$_{0.3}$Ga$_{0.7}$As/GaAs QD samples and in GaAs/AlGaAs and InGaAs/InAlAs multiple QW samples was investigated in [77], see Fig. 13.16. The intensity of the intersubband transition was observed to decrease with increasing fluence, and the transition in QDs was completely depleted at a fluence of 5×10^{14} cm^{-2}. The comparison of the radiation hardness of the QDs and the QWs could not be conclusive as the samples were based on different materials systems. However, interesting differences in the annealing behaviour were observed: whereas thermal annealing recovery of the depleted intersubband transitions was observed in irradiated multiple QW samples, in the QD samples the transition was not recovered.

13.2.3   Other quantum dots

An enhanced tolerance of the PL to proton irradiation was also observed in the Ge/Si QDs [78]. The underlying mechanisms should be similar to those in the InAs/GaAs QDs. However, the Ge islands were very large (basis diameter ~200 nm), so that any effects of self-purification or expulsion of defects out of the dots could not be expected.

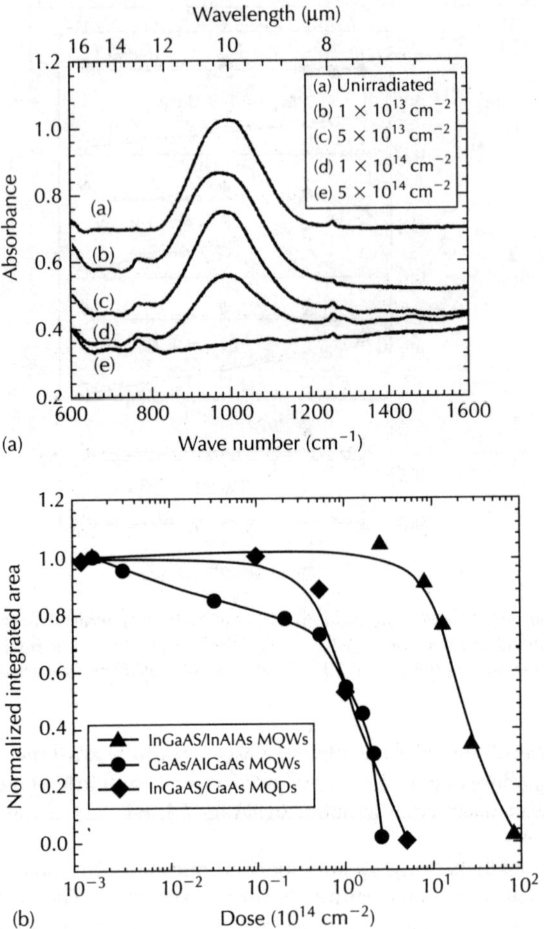

Figure 13.16 Top: *Absorbance spectra of the intersubband transition in 3 MeV proton irradiated In$_{0.3}$Ga$_{0.7}$As/ GaAs QD samples. The spectra were obtained at RT after the samples were irradiated with different fluences. Bottom: Normalized integrated area of the intersubband transition as a function of irradiation fluence obtained for In$_{0.3}$Ga$_{0.7}$As/GaAs QD samples (diamonds), In$_{0.52}$Ga$_{0.48}$As/In$_{0.52}$Al$_{0.48}$As multiple QW samples (triangles) and GaAs/Al$_{0.3}$Ga$_{0.7}$As multiple QW samples (circles)* [77].

There is, to the best of our knowledge, only one thorough report on the particle irradiation of CdS$_{1-x}$Se$_x$ QDs [79] (10 MeV electron irradiation was briefly mentioned also in [80]). The effect of 5 MeV and 10 MeV electron irradiation and subsequent annealing on the optical absorption of the dots embedded in a borosilicate glass matrix was studied. Gradual smearing and vanishing of the confinement-related maxima and a considerable (up to 0.13 eV) blue shift of the absorption edge were observed (Fig. 13.17). Contrary to CdS$_{0.4}$Se$_{0.6}$, where the confinement-related absorption maxima vanished already after irradiation by 10¹³ cm⁻² (Fig. 13.17a), in CdS$_{0.22}$Se$_{0.78}$ QDs the maxima, though smaller in intensity, were still visible even at 10¹⁵ cm⁻² (Fig. 13.17b). Annealing at 425–550 K resulted in the recovery of the initial absorption edge and the confinement-related maxima. The gradual smearing of the quantum size singularities was attributed to the irradiation-induced ionization of the QDs due to electron (hole) transfer between the NCs and irradiation-activated electron (hole) traps in the host matrix: with the increase of the irradiation fluence the transferred charge carriers occupy the confinement-related levels in the dots, gradually disabling the lower-energy transitions. The possibility of the creation of radiation defects inside the QDs themselves was discussed [79] and found improbable: as the optical absorption

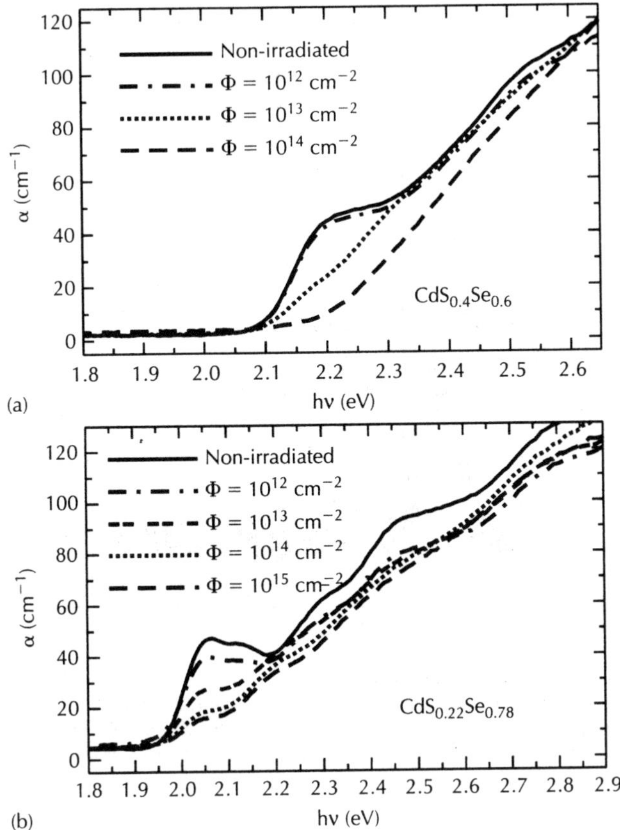

Figure 13.17 *Optical absorption spectra of $CdS_{0.4}Se_{0.6}$ (a) and $CdS_{0.22}Se_{0.78}$ (b) QDs in borosilicate glass matrix, irradiated at RT with 10 MeV electrons [79].*

spectra of the bulk $CdS_{1-x}Se_x$ crystals remain unchanged even after irradiation with an electron fluence as high as $10^{17}\,cm^{-2}$, it is hardly possible that the considerable changes of the spectra observed in the QDs samples could result from the radiation defects created at fluences of 10^{13}–$10^{14}\,cm^{-2}$.

13.2.4 Quantum dots embedded in a superlattice

Recently, the radiation hardness of InAs QDs embedded in a short-period AlAs/GaAs SL and of SiGe/Si QDs embedded in a Si/Ge SL was shown to be even higher than that of QDs in a bulk GaAs or Si barrier, respectively [81, 82].

To understand these results, it is important to consider that thin-layer Si/Ge SLs exhibit an enhanced radiation hardness of the PL as compared to Si/Ge QWs and to bulk Si. In the experiments described in [76], SLs with periods of 10 or 15 monolayers (Si_6Ge_4 and Si_9Ge_6) and double QWs $Ge_nSi_{20}Ge_n$ ($n = 2$ or 4) were used. However, the exciting laser light was absorbed mainly in the SiGe waveguide layer situated on top of the samples, so that its degradation upon irradiation might have affected the observed result. Therefore, analogous experiments were repeated using electrical injection. In Fig. 13.18, the relative degradation of the EL from fully processed QW and SL diodes is shown. The diodes were exposed to one and the same electron fluence and the current density employed in the measurement was kept the same before and after irradiation. The Figure corroborates the PL data of [76]: whereas the EL intensity of the QW diode drops below the noise level after electron irradiation, the SL emission is still quite detectable. This means

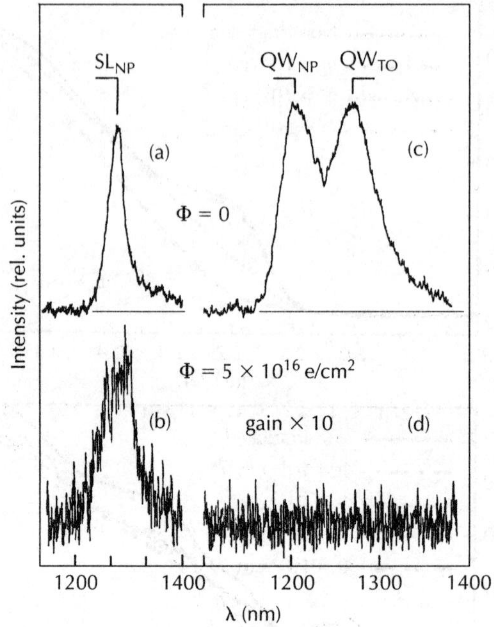

Figure 13.18   EL spectra of an Si$_6$Ge$_4$ SL and an Ge$_2$Si$_{20}$Ge$_2$ QW prior to (a, c) and after (b, d) irradiation with 5×10^{16} cm^{-2} of 3–4 MeV electrons. The spectra after irradiation were recorded with a gain $\times$ 10. T$_{meas}$ = 4.2 K. (a, b) SL, current density j = 6.4 A/mm^2. (c, d) QW, j = 2.8 A/mm^2. SL and QW are characteristic luminescence bands of the SL and QW, respectively. The NP and TO indices refer to no-phonon transitions and their TO-replica, respectively [83].

that we observe an enhanced radiation hardness of the very SL. The following model has been proposed to explain the effect [76]. As mentioned above, when Si is irradiated with 3–4 MeV electrons at room temperature, most of the stable radiation defects are formed after long-range migration of the primary radiation defects, i.e. vacancies and self-interstitials, by subsequent interaction and formation of complexes with impurities. At doping levels above $\sim 10^{17}$ cm^{-3} the creation rate of the stable defects becomes independent of the doping level and approaches the primary displacement rate since each primary defect is then captured by an impurity atom [84]. An analogous picture may basically be assumed for bulk Ge. Hence, even for an impurity concentration as high as 10^{19} cm^{-3} the average distance between neighbouring impurities is still of the order of 40 Å, and this determines the mean migration length of a primary defect to form a non-radiative centre. At the same time, the mean diffusion length to reach an interface in a short-period SL is a few Ångströms only. The interfaces act as sinks and annihilation centres for the mobile primary radiation defects, thus leading to a lower concentration of non-radiative centres than in the bulk material.

It should be noted that the role of an SL as a diffusion stop layer in MBE growth is well known. An SL buffer can effectively getter defects from the substrate and prevent their propagation into the overlying epitaxial material [433]. Quite recently, a GaAs/AlGaAs SL was again found to minimize the defect diffusion into the QD active region during overgrowth at moderate temperatures [436]. A damage-reducing effect of an intervening SL placed in the surface barrier region was observed in QW samples subjected to low-energy ion bombardment [434].

13.2.5   Amorphizing damage

The influence on the radiation hardness of reducing the size to nanodimensions is not so unambiguous in the case of heavy damage leading to amorphization (usually this is the case of ion irradiation). So, bulk zirconia ZrO$_2$ exhibits no evidence of irradiation-induced amorphization at doses as high as 110 displacements per atom (dpa). However, ~ 3 nm diameter ZrO_2 nanocrystals (NCs) can be amorphized at a dose corresponding to only 0.9 dpa [85]. Interestingly, in a control experiment the authors failed to amorphize 3 nm precipitates of Au embedded in amorphous SiO_2. The results have been explained by the delicate balance between the bulk free energy and surface free energy of various zirconia polymorphs and the defect free energy introduced by energetic ions. The excess surface free energy in nanocrystalline zirconia tips the balance in favour of radiation-induced amorphization. The thermodynamic properties of Au are very different from those of ZrO_2: high-temperature metastable phases do not exist, the surface free energy is relatively low, and besides defects can readily recombine due to lattice site equivalence.

The behaviour of Au NCs is not surprising as elemental metals usually cannot be amorphized via radiation damage, even at low temperatures. However, very recently the authors of [441] succeeded in amorphizing, at 77 K, ~ 2.5 nm small Cu NCs embedded in SiO_2, after they failed to do so with ~ 8 nm NCs [442]. The different behaviour of the NCs has been partly attributed to their different size: in the first case half the Cu atoms and in the second case less than 15% of them reside on the NC–matrix interface.

Similarly, Ge NCs embedded in SiO_2 turned up to be extremely sensitive to ion irradiation and are rendered amorphous at an ion fluence ~ 40 times less than that required to amorphize bulk, crystalline standards. This rapid amorphization was attributed to the higher-energy nanocrystalline structural state prior to irradiation, inhibited Frenkel pair (i.e. vacancy plus interstitial) recombination when Ge interstitials are recoiled into the SiO_2 matrix and preferential nucleation of the amorphous phase at the nanocrystal/matrix interface [86].

Si NCs in SiO_2 could be amorphized by irradiation at RT either with 30 or 130 keV He ions or with 400 keV electrons, contrary to bulk Si, in which electron and very light ion irradiation never leads to amorphization at RT [87]. The effect was explained by an accumulation of point defects produced by irradiation at the NC surface, leading to amorphization at low displacement rates (0.1–0.2 dpa). Subsequent publications confirmed the particular sensitivity of the Si NCs in SiO_2 to irradiation [88, 89].

Upon 60 MeV carbon ion irradiation, the defect accumulation rate in nano-Au at 15 K is larger than that in poly-Au, but for irradiation at 300 K it becomes much smaller than that in poly-Au [51]. In a subsequent work the authors came to the conclusion that both the cross-sections for defect annihilation and production in nano-Au increase monotonically as the grain size decreases [90].

13.2.6 Low-dose effect

There are several reports on an enhancement of the PL intensity in QD structures or of the QD laser performance due to irradiation with low fluences of electrons, protons or ions with energies being high enough to cause the creation of displacement-type defects [62–64, 91–94]. As an explanation, an increase in the efficiency of photocarrier transfer into the QDs via defect-induced energy levels [63], [64], reduction of grown-in defects around the QD regions [92] or even removal of the cap layer due to sputtering [94] have been proposed. A similar "low-dose effect" has been formerly observed in Si [95], $CuInS_2$[96] and GaAs [97] and interpreted as a result of the dissociation of grown-in defects assisted by the creation of mobile point defects by the irradiation. From the viewpoint of the thermodynamics, the low-dose effect manifests a shift of an imperfect crystal that is in a metastable state after growth to the thermodynamic equilibrium. On the contrary, high doses of radiation push the crystal off the equilibrium, whereas post-radiation annealing restores the equilibrium. As the low-dose effect is usually observed in a rather narrow fluence range, the following method of the crystal quality improvement of $CuInS_2$ crystals has been proposed in [96]: "high dose" irradiation with a posterior annealing at the optimum temperature. As we shall see in Section 13.4.1, such a method is successfully used to improve the characteristics of QD and QW structures.

13.2.7 Irradiation with electrons of subthreshold energies and x-rays

Irradiation that is unable to create damage in elastic collisions causes, nevertheless, persistent radiation effects in QDs based on wide-gap semiconductors.

The luminescence intensity of CdS NCs embedded in a glass matrix degrades upon X-ray and even light irradiation. The photoinduced defects were found to be different from those induced by X-rays and were supposed to be located in glass near the NCs [98]. The PL intensity of CdSe QDs gradually decreases upon laser illumination up to a certain exposure and then remains constant, contrary to the PL intensity of ZnCdSe QWs which keeps decreasing [99]. The absorption edge blue shift was observed in glass-embedded CdS [100] and $CdS_{1-x}Se_x$[101, 102] QDs under X-ray irradiation as well as in CdSe QDs under intense light illumination [100], though the transformation of the confinement-related optical absorption maxima was reported only in [101] and [102] and resembled the behaviour under MeV electron irradiation described above (see Section 13.2.3). A strong PL and CL degradation of semiconductor QD composites, formed by highly luminescent (CdSe)ZnS core–shell NCs embedded in a ZnS matrix, was observed with time [103]. PL experiments carried out at high laser fluences ($0.5–10$ mJ/cm^2 per pulse) showed that the PL intensity decay with illumination time depended on the size of the NCs and the nature of the surrounding matrix. For instance, close-packed films showed a much slower decay than composite films. The CL intensity degradation is enhanced at lower temperatures. Partial recovery of the CL signal could be achieved after thermal annealing at temperatures around 120°C, which indicates that activation of trapped carriers can be induced by thermal stimulation. The CL and PL decay in the composite films was attributed to photo- and electroionization of the NCs and subsequent trapping of the ejected electrons in the surrounding semiconductor matrix [103].

Electron irradiation with an energy as low as 10 keV quenches the CL in GaN/AlGaN QD structures [104–107]. It should be noted that it is not surprising because the CL degradation had formerly been observed in regions with zero dislocations of epitaxial laterally overgrown GaN (ELOG) layers also under 10 keV electron excitation [108]. The QD CL intensity degradation was attributed to an increase of the non-radiative recombination due to a recombination-induced diffusion of pre-existing defects [105]. A comparison of the radiation hardness of heterostructures containing GaN/AlGaN QDs, QWs and AlGaN barriers together with that of a thick ELOG layer revealed a much higher resistance of the QDs, especially at high electron beam current densities [106].

13.2.8 Hydrogen passivation

An effect that is in its essence opposite to the defect introduction and is often exploited in defect studies is the hydrogen passivation of the grown-in non-radiative recombination centres like dislocations and point defects [109]. A PL intensity enhancement due to the hydrogen passivation has been reported for the In(Ga)As/GaAs [110–116], InAs/AlAs [117], Ge/Si QDs [78], as well as for Si nanocrystals embedded in fused silica [118–121] or silicon nitride [122]. This effect has to be taken into account when, upon proton irradiation, the implanted hydrogen profile is placed over the active device layers.

13.2.9 Influence of defects on the thermal stability of the luminescence

It is well known that the thermal stability of the QD luminescence suffers from the presence of defects that act as non-radiative recombination centres (see, e.g., [110]). The temperature dependence of the PL intensity in Ge/Si QDs without any treatment (as grown) as well as subjected to 2.4 MeV proton irradiation or atomic hydrogen passivation is illustrated in Fig. 13.19. It is clearly seen that the irradiation deteriorates, and the passivation improves the thermal stability of the PL. The latter is only possible if the as-grown dots contain defects. Indeed, the Ge/Si QDs studied in [78] had a basis diameter of $\sim$200 nm, so that the effect of "self-purification"[40] mentioned in Section 2.1 could not occur. In InAs/GaAs QDs, a growth treatment using tetrachloromethane eliminated the quenching of the PL intensity up to RT [123]. A photocarrier statistical model taking into account the variations of the quasi-Fermi level position of the minority carriers, which are related to the concentration of trapping centres in the GaAs matrix, was developed [125]. When defects were introduced through proton irradiation, the PL quenching at RT appeared again [124, 125]. The calculated results for the PL intensity reproduced well the experimentally observed trends [125].

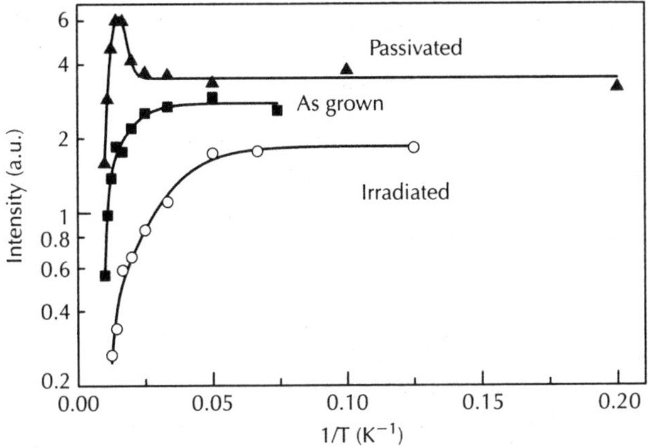

Figure 13.19   Temperature dependences of the PL intensity in the as-grown, passivated and 2.4 MeV proton irradiated ($2 \times 10^{12}\,cm^{-2}$) Ge/Si QD samples. The curves have been vertically shifted to avoid superposition of the points [78].

13.3   Radiation hardness of QD lasers

There is an extensive literature on radiation effects in heterostructure and QW laser diodes (see, e.g., [48, 126–136]), so that the logic of the radiation damage in these devices is well understood. From an application standpoint the most important parameters for a semiconductor laser diode are threshold current and slope efficiency. Figure 13.20 shows the effect of 5.5 MeV proton irradiation on the optical power of the strained InGaAs/GaAs QW laser diode operating near 980 nm as the forward current increases from very low currents to the region where laser operation begins. The optical power increases abruptly after the forward current reaches the threshold current. The latter grows with increasing proton fluence. On the contrary, the slope efficiency above threshold is independent of fluence [127].

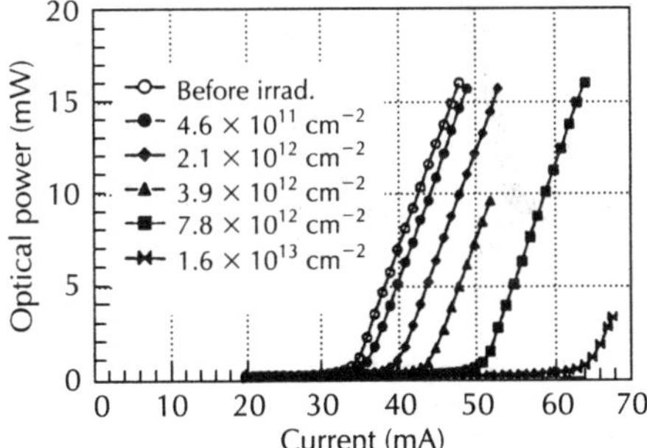

Figure 13.20   Degradation of L–I characteristics of a strained InGaAs/GaAs QW laser diode during 5.5 MeV proton irradiation [127].

Degradation in lasers is usually much greater at low injection conditions, when the laser operates in the LED mode. The light output (due to radiative recombination below threshold) is reduced far more at currents below the threshold current compared to the changes that occur in threshold current when the device enters the laser mode, and light output is dominated by stimulated emission [136].

The radiation hardness of In(Ga)As/GaAs QD lasers was investigated for the irradiation with phosphorous ions [91] and protons [137] in comparison with QW lasers. It was found that laser diodes can sustain significantly greater 8.56 MeV P ion fluences ($\times 100$) than comparable QW-based structures prior to the onset of failure.

Let us consider the influence of proton irradiation on laser-device characteristics, i.e. threshold, slope efficiency, internal differential efficiency and optical loss following [137].

The active region of the structures under investigation consisted either of a single $In_xGa_{1-x}As$ QW ($x \approx 0.2$) or of a triple stack of InGaAs/GaAs QDs with a dot density of $(3–5) \times 10^{10}\,cm^{-2}$. The lasing wavelength of the processed broad-mesa ridge waveguide lasers was 1096 nm for the QW laser and 1155 nm for the QD laser. The mesas were etched down to 400 nm above the active layer. The stripe width was 200 μm. For a waveguide thickness of 300 nm this resulted in weak index guiding of the optical modes and also in the suppression of current spreading. Cavity lengths of the investigated devices were 1.0 and 1.6 mm. The samples were irradiated with 2.4 MeV protons to a fluence of $2 \times 10^{13}\,cm^{-2}$.

Before and after irradiation, the QD laser operated on the ground state. The lasing wavelength of the QW devices also remained unchanged after irradiation. Figure 13.21 shows the L–I curves for the QW and QD lasers for 1.6 mm cavity length before and after irradiation. The extracted device parameters are summarized in Table 13.2. As shown in Fig. 13.21, for these two lasers the threshold current density and the external differential quantum efficiency (QE) before irradiation were very similar: 83 A/cm², 68% and 89 A/cm², 61%, respectively. After irradiation the differential external QE was reduced in both cases by ~43% for the QW and by 50% for the QD laser. The change in threshold current density after irradiation was, however, quite different: the value for the QW laser worsened by about twice as much as for the QD laser, 950 A/cm² and 550 A/cm², respectively. It was naturally concluded from these results that in the regime of spontaneous emission incorporated defects introduced by the proton irradiation result in non-radiative relaxation channels. These defects in or in the vicinity of the active region are much more critical in the QW case owing to the higher in-plane diffusion of the carriers, contrasted to QDs. As already discussed in Section 13.2.2, the localized carriers in the QDs have a reduced interaction

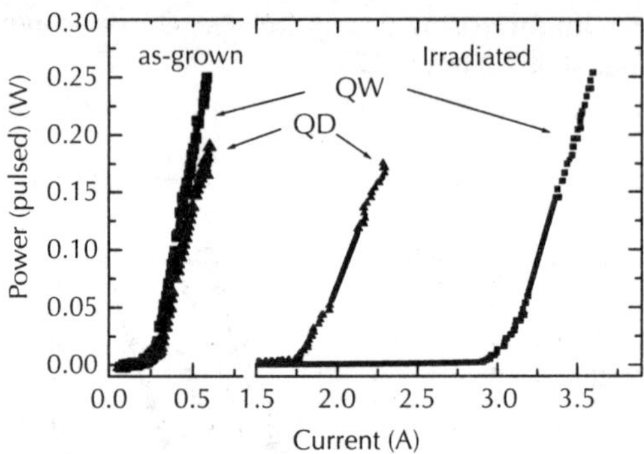

Figure 13.21   Optical output power against current in pulsed operation for the same QW and QD lasers before and after irradiation (1.6 mm cavity length, 200 μm stripe width) [137].

Table 13.2   Threshold current densities and external differential quantum efficiencies for three laser devices (1 QD and 2 QWs) before (as-grown) and after irradiation with high energy protons [137].

	QW laser (L = 1.0 mm)	QW laser (L = 1.6 mm)	QD laser (L = 1.6 mm)
j_{thr} (as-gr.)	92 A/cm²	83 A/cm²	89 A/cm²
η_{ext} (as-gr.)	76%	68%	61%
j_{thr} (irrad.)	1.2 kA/cm²	950 A/cm²	550 A/cm²
η_{ext} (irrad.)	44%	39%	30%

with the defects. Once the devices lase, the differential quantum efficiency is reduced owing to leakage currents in the barrier that are similar for QWs and QDs. The fast stimulated emission in the active media bypasses non-radiative recombination of localized carriers, resulting in similar differential quantum efficiencies for QWs and QDs.

From the external differential QE for both cavity lengths of the QW device, the internal quantum efficiency and the internal optical losses before and after irradiation can be estimated. Before irradiation, an internal QE η_{int} of 95% and an internal optical loss of 2.9 cm^{-1} were found to be in agreement with results obtained from a larger set of samples (η_{int} = 84% with a 10% error, α_{int} = 5 cm^{-1}). After irradiation, 56% and 3.3 cm^{-1} were found, respectively. Since the internal optical loss had a similar value before and after irradiation, the reduction in external differential QE after irradiation was attributed to the reduced internal QE [137].

13.4 Radiation technology

Radiation treatment can be used to improve the performance of QSSS-based devices or to modify their characteristics in a desired manner. New structures can be fabricated due to self-organization upon irradiation.

13.4.1 Intermixing

Spatially selective intermixing is particularly important in the QW laser fabrication (e.g. in GRINSCH: graded-index separate confinement heterostructure) [138]. Intermixing the wells and barriers of QW heterostructures generally results in an increase in the band gap energy (blue shift of the luminescence band) and is accompanied by changes in the refractive index.

A very efficient QW intermixing is achieved using particle irradiation with subsequent annealing [139, 140]. Independently of the amount of intermixing in the collision cascades (which is virtually absent when only isolated point defects are created, e.g., by proton or electron irradiation), the mixing can be strongly enhanced by post-irradiation heat treatment and occurs due to defect-enhanced diffusion [141]. The ion irradiation-induced intermixing has several advantages over the other intermixing techniques: the amount of defects can easily be controlled by the ion dose and mass, the defects can be placed at a certain depth by the choice of the ion energy, and, last but not least, the intermixing occurs at lower annealing temperatures than in the as-grown structures. The latter fact allows a high lateral band gap contrast using focused ion-beam implantation or selective ion implantation using masks to be achieved.

It is important to note that in order to realize advanced device parameters, the radiation-induced damage has to be fully eliminated by annealing, which requires a sufficiently high annealing temperature. On the other hand, as mentioned above, the temperature should be low enough to avoid thermally activated band gap shifts in unimplanted areas. As a compromise, an annealing temperature close to the onset of purely thermally activated band gap shifts is used [140]. However, this compromise can only be achieved when the implantation damage is not too heavy, as heavier damage requires, as a rule, higher annealing temperatures. This fact puts an upper limit to the useful implantation fluences.

It is well known that compound semiconductors tend to undergo stoichiometric changes and defect generation upon prolonged heating. To suppress these detrimental processes, the rapid thermal annealing (RTA) in an inert or forming atmosphere, with the sample surface protected against effusion by an oxide or nitride layer or proximity caps, is usually applied. It is noteworthy that the protection by oxide or nitride layers can enhance or retard the intermixing by itself [142]. Typical annealing conditions for III–V material systems are rapid heating to 600–900°C for times ranging from several seconds to a few minutes [139]. In some papers, even an improvement of the luminescent parameters of III–V QW samples subjected to As ion [143], alpha particle [144], and electron [145] irradiation or Ar plasma exposure [93] with subsequent annealing has been reported.

The ion mass determines the energy transfer in the collisions (see Section 13.1) and, therefore, the defect density. Whereas light ions tend to create isolated point defects or point defect clusters, heavy ones may produce amorphous regions in a single ion track [146]. As a result, the intermixing depends on the ion mass [147].

Besides the implantation dose and ion mass, the ion energy is an important technological parameter, because it governs the profiles of the damage and of the implanted impurity concentration. Depending on the target material, especially on the defect diffusivities in it, either the total number of created vacancies or the vacancy concentration in the very QW region may be important [139].

The target temperature during the implantation may have a profound effect on the intermixing because of the dynamic defect annealing that is mediated by defect diffusion and annihilation [148]. So, e.g., ion-beam mixing in Al_xGa_{1-x}As and InP matrices was measured as a function of target temperature upon 1 MeV Kr ion bombardment [149]. The mixing increased with temperature up to a critical temperature T_c at which point it precipitously dropped. T_c was identified as the highest temperature at which the matrix could be amorphized by 1 MeV Kr irradiation. Earlier the ion implantation into heated InGaAs/InP and AlGaAs/GaAs multilayers was found to induce compositional disordering at significantly lower temperatures than implantation at RT with subsequent annealing [150].

The dose rate (ion flux) and the channelling effect may also affect the intermixing; the first one as a result of both temporal and spatial overlap of defects within collision cascades [151] and the second one due to a modification of the density and depth distribution of the radiation damage [152]. Besides, channelling allows reducing the lateral straggling of implanted ions thus improving the steepness of the resulting lateral potentials [153, 154].

And last but not least, it is evident that the intermixing is strongly material dependent [140, 141]. In addition, in a quaternary system like InGaAs(P)/InP the intermixing is more complicated than that in a ternary one, e.g. AlGaAs/GaAs, as interdiffusion can occur on both group III and V sublattices, which are characterized by their diffusion lengths. Not only a blue shift but also a red shift in the band gap was observed in InGaAs(P)/InP QWs, depending on the diffusion length ratio governed by the RTA temperature after plasma exposure [155].

Whereas the radiation defect-induced QW intermixing is a mature technology, its application to QD structures is actively studied at the time. The QD intermixing should differ from that in QWs due to the large surface area-to-volume ratio, composition gradients inside the dots, and a peculiar configuration of strain fields around the dots. There are two different aspects of the chemical disorder via interdiffusion across the QD interfaces, namely: (i) the effect of the strain relief inside the QDs and (ii) the purely chemical effect due to the group III and group V atomic species interdiffusion. According to [156], these effects may be quantitatively comparable, significantly affecting the electronic and optical properties of the dots.

In the first work on ion implantation-induced intermixing of InAs/GaAs QDs [59], a blue shift of the ground state transition of up to 150 meV was achieved after implantation of $1 \times 10^{13}\,cm^{-2}$ of 50 keV Mn ions and annealing (Fig. 13.22). It is worth noting that a blue shift as large as 100 meV was obtained after implantation of $1 \times 10^{15}\,cm^{-2}$ prior to any heat treatment, however, the pristine PL intensity could not be restored after annealing at 700°C, and at 800°C the intermixing began also in the unimplanted sample, thus preventing the implementation of a high lateral band gap contrast required in device applications. Furthermore, the additional blue shift due to annealing had a larger effect on samples with low implantation dose.

Beside the blue shift, a partial suppression of the dot emission was observed in InAs/(AlGa)As QD structures implanted with 100 keV Cr ions and annealed at 700°C for 1 min [157]. This suppression was accompanied by an enhancement of the wetting layer (WL) emission and its red shift (Fig. 13.23). It was concluded that the used treatment drives the system towards a predominantly two-dimensional character. The unusual behaviour of the WL luminescence was interpreted as being due to a compensation of the Ga and/or Al diffusion from the GaAs or AlGaAs barriers to the WL by the In diffusion from the QDs to the WL (see the inset of Fig. 13.23). Incorporation of more In into the WL would lead to an In-rich and/or a thicker two-dimensional layer, thus causing a red shift of the WL emission. The same process can lead to the partial suppression and a blue shift of the dot PL in the annealed and/or implanted samples [157].

A comparison of the InGaAs/GaAs QD intermixing accomplished by means of the 360 keV P and As ion implantation at 200°C and of the impurity-free vacancy disordering (IFVD) using

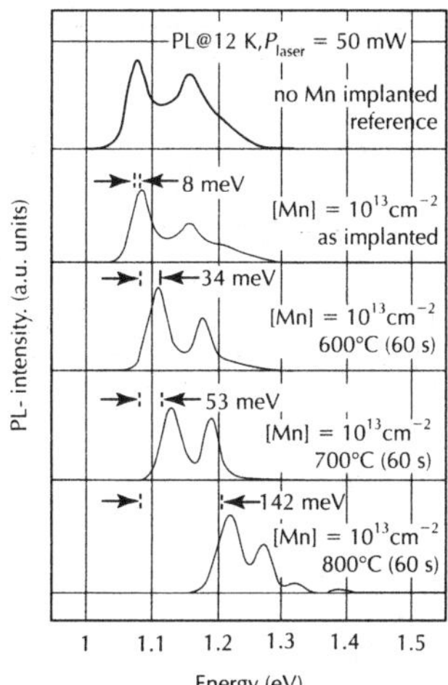

Figure 13.22   Evolution of the PL spectra as a function of annealing temperature for an Mn implantation dose of 1×10^{13} cm^{-2} *at* T = 12 K *[59].*

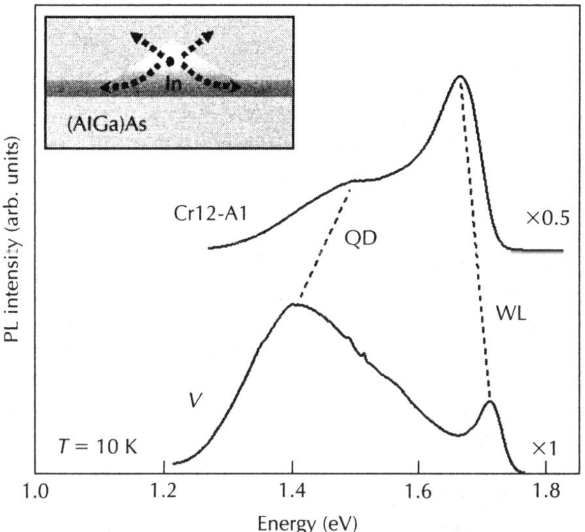

Figure 13.23   Photoluminescence spectra at 10 K for InAs/Al0.3Ga0.7As quantum dots. The spectra refer to the as-grown (V) sample and that subjected to the implantation of 4×10^{12} cm^{-2} *Cr ions and 60 s annealing at 700°C. Inset: Sketch showing the In interdiffusion in the dot [157].*

capping with dielectric SiO_2 or Si_xN_y films, always followed by RTA, has shown that in the first case high spatially selective intermixing can be achieved at a significantly lower annealing temperature than for typical dielectric cap techniques [158, 159].

In Fig. 13.24, the implantation-induced energy shift and the full width at half maximum (FWHM) of the PL spectrum of InAs/InP QDs are shown as a function of the 450 keV P ion fluence for two RTA temperatures. For low doses the FWHM for the implanted and annealed sample decreases noticeably compared to that of the annealed but unimplanted QD sample. It was inferred that ion implantation-induced intermixing reduces the overall variation in QD size, shape, strain, and composition profile [160].

As pointed out above, whatever the technique (IFVD using dielectric caps or implantation-induced), the intermixing mechanism is the defect-enhanced diffusion requiring the presence of

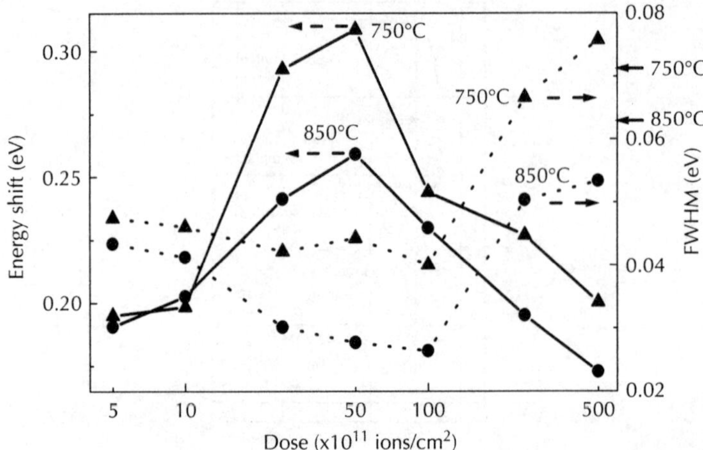

Figure 13.24   The implantation-induced energy shift (defined as the difference between the PL energy peaks of the implanted and unimplanted samples which are annealed at the same temperature) and the full width at half maximum (FWHM) of the PL spectrum of the InAs/InP QDs vs P ion fluence. The FWHMs of the PL spectra of the unimplanted and annealed QDs are also shown (by solid arrows) for comparison. All samples are annealed at either 750 or 850°C for 30 s [160].

point defects. In this sense, the existence of an optimum ion fluence for the achievement of the maximum intermixing (see, e.g., Fig. 13.24) is usually interpreted as a result of the formation of more complex defect clusters at higher fluences, which reduces the availability of point defects that cause intermixing. On the other hand, the virgin PL intensity usually cannot be restored after heavy ion implantation and subsequent annealing at temperatures below the onset of inter-mixing in unimplanted samples, obviously due to incomplete damage annealing. It follows from these considerations that any other technique producing a high concentration of point defects without heavily disordering the crystal lattice should lead to an efficient intermixing without degradation of the PL intensity. Two such radiation-based techniques are the high-energy proton irradiation and the low-energy ion implantation or plasma treatment of the surface.

Due to the large penetration depth as well as low lateral straggling and uniform damage along the major part of the trajectory (see Fig. 13.5), the protons of sub-MeV and MeV ener-gies are well suited for the treatment of multilayer QD structures. That is why proton beam writ-ing is a promising technique also in nanolithography: e.g. the energy distribution profile from a 2 MeV proton beam trajectory over the first 2 μm of penetration into a photoresist is essentially contained within a 10 nm diameter [161]. Using the proton irradiation, it is important to take into consideration the above-mentioned possibility of the hydrogen passivation of non-radiative recombination centres like dislocations and point defects [109, 162], see Section 13.2.8.

In the first investigation on intermixing a ten-layer InAs/GaAs QD structure by means of proton implantation with subsequent RTA [163], a blue shift of the PL peak energy up to 94.3 meV was achieved; however, the virgin PL intensity could not be restored, presumably due to the choice of too high 100 keV proton fluences. In a subsequent publication of the same group [115], the use of smaller fluences and energies (50 and 80 keV) led, beside a blue shift of up to 100 meV, to a PL intensity enhancement by a factor of six. However, as was elucidated by control experiments, the improvement occurred due to hydrogen passivation of non-radiative centres. In a parallel work on InGaAs/GaAs QDs [164], a linewidth reduction was found to be proportional to the amount of blue shift, and the latter decreased with the implantation temperature increasing from 20 to 200°C due to the dynamic annealing effect during implantation. A non-monotonous depend-ence of the FWHM on the hydrogen fluence was observed in [165]. A pronounced enhance-ment of room temperature PL up to 80-fold induced by 50 and 70 keV proton implantation and RTA in a multilayer InAs/GaAs QD structure was reported in [114]. As in [115], secondary ion mass spectroscopy (SIMS) showed a homogeneous hydrogen distribution over the QD multilayer. The huge PL intensity increase was only possible because RTA-treated unimplanted samples suffered relaxation by dislocation formation and thus exhibited a very weak PL. The dislocations were then passivated by the implanted hydrogen. An additional intensity enhancement mechanism was proposed due to an increase of the capture rate caused by an intermixing-induced change in QD shape. Interestingly, the authors of [166] compared the time-integrated PL and the time-resolved PL of several lattice-matched InGaAs/InP QWs intermixed either by ion implantation or by an impurity-free method. They have found that the carrier capture rates into QWs and carrier relaxation from the wells depended on the type of intermixing used. Their results indicated that the carrier lifetimes were significantly longer in samples intermixed by the impurity-free method, while the carrier collection efficiency of the QWs was more efficient in samples intermixed by ion implantation.

The influence of the embedding material (InP, GaInAsP, InP and InGaAs) on the proton implantation-induced intermixing of InAs QDs was studied by PL in [167]. The particle energy of 40 keV placed the damage peak over the QD region. The QDs capped with InP layers exhibited the highest implantation-induced energy shift due to strong group V interdiffusion whereas the QDs grown on and capped with GaInAsP layers showed the least implantation-induced energy shift due to weak group V and group III interdiffusion. The QDs capped with InP and InGaAs layers showed intermediate implantation-induced energy shift and were less thermally stable as compared to the QDs grown on and capped with GaInAsP layers. The QDs capped with InP layers demonstrated enhanced PL intensity when implanted with proton fluences less than $5 \times 10^{14}\,\mathrm{cm}^{-2}$. On the other hand, proton fluences above $1 \times 10^{14}\,\mathrm{cm}^{-2}$ reduced the PL linewidth in all samples.

Whereas the PL measurements give only indirect information on the damage and intermixing in irradiated QDs, such techniques as the Rutherford backscattering/channelling of light ions (RBS/C), X-ray diffraction and X-ray diffuse scattering, high-resolution transmission electron microscopy (HRTEM), high-angle annular dark-field (HAADF) imaging in a scanning TEM (STEM), and cross-sectional scanning tunnelling microscopy (XSTM) yield more direct evidence on structural and compositional changes in the samples. Unfortunately, applications of these techniques to the irradiation-induced intermixing of QDs are almost absent so far. The only work we are aware of is [168], where RBS/C was used to investigate the 1.0 MeV proton-induced intermixing in InAs/GaAs QD structures. An activation energy as low as 0.2 eV for the In–Ga intermixing was concluded from the experiments and ascribed to the presence of a non-equilibrium concentration of point defects produced by proton irradiation. Besides, evidence for mass transport of group III elements along or perpendicular to the growth <100> direction was offered.

In the implantation-induced intermixing experiments described above, the QD region was directly damaged by the ion beam. However, in the case of the high-energy proton-induced intermixing the concentration of point defects created at the site of the dots is low (cf., e.g., the defect profiles shown in Fig. 13.5a with typical layer thicknesses in QD structures), and the intermixing occurs mainly due to defects that experience a long-range migration to the dots from the adjacent barrier layers. This phenomenon is also exploited in intermixing experiments using low-energy ion implantation or plasma etching, in which the defects are created in a subsurface region. So, e.g., an exposure of InAs/Gas QDs to the 300 eV Ar plasma led to a degradation of the PL intensity though the dots were situated at a depth free of primary displacement damage [58]. In this sense, such intermixing experiments are similar to those were the defect flux from the surface is caused by a capping layer (see, e.g., [142, 169]), with the important difference that the irradiation technique allows the defect concentration and the heat treatment temperature to be chosen separately. Thus, in [94] a blue shift of 160 nm and an increase in the PL intensity of InAs/InGaAs/InP QD samples were observed immediately after Ar plasma exposure for 90 s, and a further increase in the blue shift of 330 nm, accompanied by a 2.5 times increase in the PL intensity and a narrowing in the PL linewidth, was achieved after subsequent RTA at 720°C. In another work [170], the use of an annealing temperature as low as 600°C after Ar plasma treatment allowed a blue shift of the InAs/InP QD PL greater than 100 nm to be achieved, limiting the thermal shift to only 10 nm. A halftoning of the band gap was performed using wet chemical etching to remove defects partially from the InP cap so that defect creation, defect amount control, and intermixing promotion by defects became three thoroughly independent processes.

A very large blue shift up to 350 nm (280 meV) and a narrow emission linewidth down to 30 nm in intermixed InAs/InP quantum sticks (QS) were reported after low-energy (18 keV) P ion implantation at 200°C and RTA at only 650°C [171], see Fig. 13.25. As in the case of Ar plasma treatment, the QD layer was not directly damaged by the ions. The choice of a low RTA temperature allowed a very small thermal shift (less than 3 nm) to be kept. Importantly, the PL intensity was not sacrificed up to an ion fluence of $5 \times 10^{13}\,\mathrm{cm}^{-2}$ at which the blue shift approached its maximum and the linewidth its minimum.

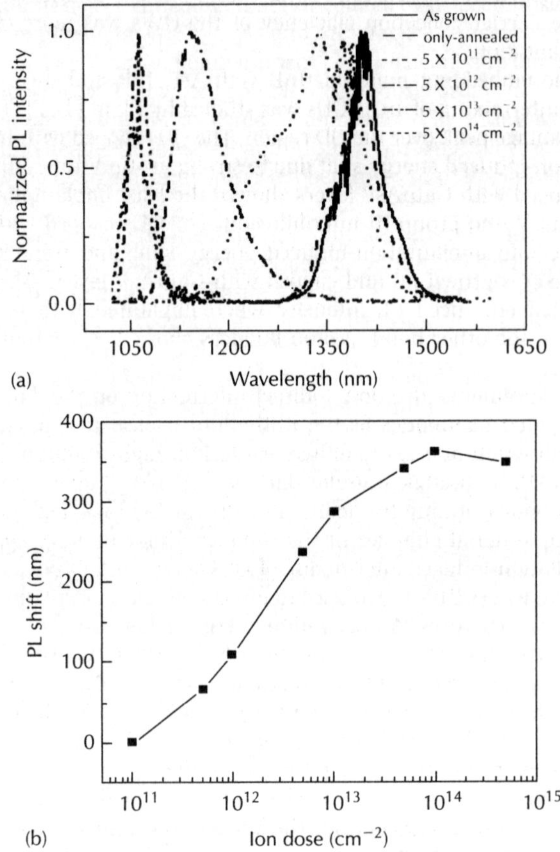

Figure 13.25   (a) Normalized PL spectra at 20 K of intermixed InAs/InP quantum stick samples as a function of the phosphorus ion implantation dose. The samples were annealed at 650°C for 120 s. (b) Photoluminescence peak wavelength shift vs the ion dose [171].

A direct comparison of proton and phosphorous ion implantation-induced intermixing of InAs/InP QDs was reported in [172]. Proton irradiation induced less energy shift than P ion implantation for a given concentration of atomic displacements, which was ascribed to the more efficient dynamic annealing of the defects created by protons. The implantation-induced energy shift reached a maximum value of about 260 meV for a fluence of $5 \times 10^{12}\,\mathrm{cm}^{-2}$ in the P ion implanted QDs, which also showed narrower PL linewidths compared to the proton implanted QDs. Defect production and annihilation processes evolved differently in samples with InGaAs and InP top cap layers and varied with the implantation temperature. When the implantation was performed at higher temperatures, the energy shift of the P ion implanted QDs capped with an InP layer increased, probably due to the reduction in larger defect cluster formation at higher temperatures, while the energy shift of the proton implanted QDs decreased due to increased dynamic annealing irrespectively of their cap layers.

The intermixing also opens the way to the *lowering of the dimensionality* of semiconductor heterostructures. So, e.g., *ion implantation through lithographically defined masks* and *focused ion beams* (FIB) [250] were used very early to implant quantum well structures and interdiffuse III–V heterojunctions thus creating quantum wires, dots and antidotes [173–184]. This is now a quickly developing area of technology [185–188]. At the present time, sub-5 nm FIB direct patterning of nanodevices is possible [189].

13.4.2 Self-organization upon irradiation

A very interesting and useful phenomenon is the self-organized creation of nanopatterns on the surfaces of targets irradiated by ion beams at low and intermediate energies. So, submicron ripples on various surfaces [190, 191] and nanometric dots on GaSb [192] can be produced by *ion-beam sputtering*. The dots even form a highly ordered array with hexagonal symmetry [192] (Fig. 13.26). Later on, nanodot production was reported in other materials such as InP [193], Si [194], Ge [195], InAs [196], InSb [197], MgO [198], SiC [199]. The theoretical description of the sputtering phenomenon is made by means of the molecular dynamics, Monte Carlo (e.g. TRIM), and continuum methods. The effect of the sputtering treatment depends on the ion–target combination, target temperature, ion energy and incidence angle, treatment time and ion flux. A crucial aspect is the surface morphology (roughness). Topographical features may cause differences in the energy deposition (e.g. shadow effect) and, thus, significant variations of the local sputtering yield. Accordingly, small irregularities on a relatively smooth surface may be enhanced by ion bombardment. This implies that microscopically flat surfaces are unstable under

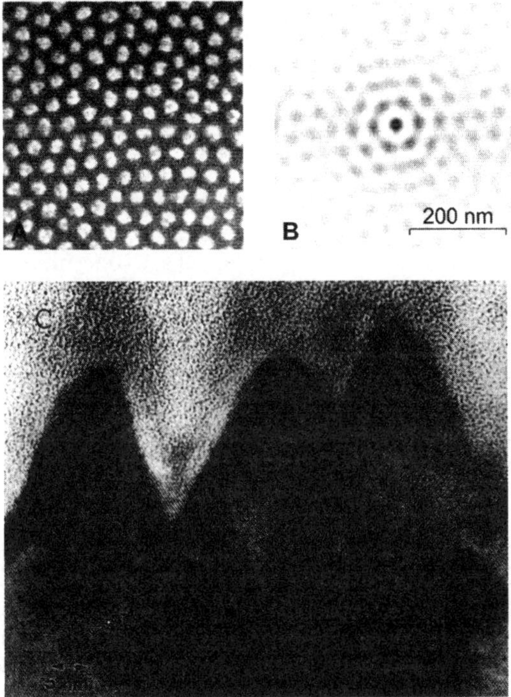

Figure 13.26   (A) The extract of an SEM image of QDs forming on a GaSb surface during bombardment by 420 eV Ar^+ ions at normal incidence and (B) the corresponding two-dimensional autocorrelation reveals the regularity and hexagonal ordering of the dots, which extend over more than six periods. (C) Cross-sectional TEM image of the nanostructures. The dots are crystalline without a disruption to the crystal lattice of the GaSb bulk. An amorphous layer of ~2 nm, which is the penetration depth of the low-energy ions into the material, covers the dots. The base and the height of the dots measure 30 nm [192].

high-dose ion bombardment unless atom migration acts as a dominating smoothing effect [200]. There are very recent reviews [191, 200] of the self-organized surface nanopatterning by ion beam sputtering.

When a SiGe buffer layer is grown on (001)Si upon *in situ* irradiation by 1 keV As ions, the implantation-induced defects represent effective channels for stress relaxation and create a nanopatterned surface. When several layers of Ge QDs are then grown by MBE on top of such a buffer, the nucleation of the Ge islands is enhanced in the strain field from the buried ion-induced pattern, and the dots grow in a vertically correlated manner [201]. Creation of high-density arrays of small-size Ge/Si QDs with a narrower size distribution of islands in comparison with conventional MBE based on the use of pulsed irradiation with low-energy Ge^+ ions during the Ge/Si heteroepitaxy was proposed in [202] and further investigated in [203–206]. It was concluded that *in situ* high-temperature ion bombardment during the QD growth pushes the epitaxial layer system towards equilibrium, thus enhancing the QD PL intensity.

13.4.3   Ion-beam synthesis

Another ample field is the ion-beam synthesis (IBS) of metal or semiconductor nanocrystals (NCs) in amorphous and crystalline matrices (see, e.g., [207–214]). These NCs possess unique physical properties and can be used in novel devices, which exploit size effects on a nanometre scale. They can be the basic unit in light-emitting diodes, photovoltaic devices and single nanocrystal–single electron transistors. The non-linear optical properties of metallic nanoclusters in insulators are based on the size dependence of the plasmon frequency and third-order optical non-linearity [215]. IBS of Si and Ge NCs in the gate oxide of MOS structures is now a standard technology [216, 217]. It was argued that Si NC non-volatile memory technology possesses intrinsic radiation tolerance to both total ionizing doses of ^{60}Co gamma-irradiation and single event effects, as well as built-in redundancy that is non-existent in flash memory. This originates from the ability of the cell to maintain functionality with only a fraction of intact charge storage nodes, in contrast to a monolithic floating gate [218]. Especially interesting is the achievement of the optical gain in structures containing Si NCs in SiO_2 that emit strong luminescence [219, 220]. The IBS of NCs is a rapidly developing technology, so that a review can hardly be exhaustive and up to date. Therefore only a bird's-eye view of the research area will be provided in this section. Magnetic NCs in semiconductors and isolating crystals will be discussed in more detail in Section 13.4.4.

In the IBS technology, an impurity is usually implanted at a concentration far exceeding its solubility limit, and a phase separation occurs, leading to the formation of NCs. The latter can happen during the implantation process or during the subsequent heat treatment. Heat treatment is usually necessary also for the damage annealing, when the irradiation temperature was not too high. More than one impurity can be implanted in order to synthesize compound NCs [207, 209, 221–230, 438] or even doped compound NCs [209, 231, 232]. However, via implantation of only one ion species in a compound or heavily doped target, NCs of compounds different from the matrix can be synthesized [233–240]. Three different cluster morphologies can be observed: separated systems, alloy clusters and core–shell clusters [241, 242]. Even rare gas nanocrystals can be formed in solid matrices by ion implantation [243, 244].

The reports on the formation of metal NCs in insulator crystals and glasses are especially numerous. A few examples illustrating important physical processes will be cited here. There are recent comprehensive reviews on the potentialities of ion implantation for the synthesis and modification of metal nanoclusters in silica and silica glasses [211, 245] and in sapphire [246]. Some recent results not mentioned above can be found in [247–251, 439, 440].

The IBS process forms a strongly non-equilibrium state that relaxes towards equilibrium during annealing. The driving force is the supersaturation chemical potential. It is noteworthy that, depending on the implanted species–host combination, supersaturation may also occur at near-zero concentrations, so that as soon as the temperature is high enough for atomic mobility to set in, the solution becomes unstable against very small concentration fluctuations and there is no nucleation barrier [212]. Although the self-organization of NCs occurs during relaxation towards equilibrium, it is difficult to achieve a process control which is sufficient for technological applications [252]. Post-implantation annealing inevitably leads to a broad NC size distribution, which is typical for conventional Ostwald ripening: the larger clusters grow at the expense of the smaller ones. The process has been analytically described in a mean field approximation, in the long-term limit of ripening, and under several other assumptions (LSW model) [253, 254]. A variation of temperature and ion flux during ion implantation can control the size distribution of NCs, at least to a certain extent [255].

On the other hand, the size and size distribution of NCs can be efficiently tailored by high-energy ion or electron irradiation through a layer of NCs [256–258], even when the irradiation is made at a temperature as low as $93\,K$ [259] or $77\,K$ [260]. Under the steady-state condition of an appropriate ion irradiation, an effective negative interface tension or capillary length appears in the Gibbs–Thomson relation for NCs embedded in a matrix. Therefore, the system of NCs evolves towards a maximum interface area, which is reached for a monodisperse size distribution (if nucleation of additional NCs is not allowed), i.e. inverse Ostwald ripening occurs [252]. If collisional mixing at interfaces exceeds a certain intensity, then new NCs can nucleate during ion irradiation [261]. The collisional mixing can be used to create well-controlled NC distributions also in flat layered structures. So, over-stoichiometric Si can be introduced to an SiO_2 layer or layers from surrounding Si layers by intermixing with high-energy Si irradiation when the implanted Si ions come to rest in the substrate well below the active device region [252, 262]. Thus, sandwiched between the stable phases of SiO_2 and Si, unstable non-stoichiometric SiO_x ($x < 2$) phases are formed. Annealing restores the upper and lower SiO_2/Si interfaces by phase separation. However, the tails of the Si atom mixing profiles do not reach the recovered interfaces by diffusion, so that the phase separation proceeds via nucleation and growth of Si NCs in SiO_2. The competition between interface restoration and nucleation self-aligns δ-layers of Si NCs in SiO_2 along the two interfaces [262]. A selective dealloying in bimetallic Au_xCu_{1-x} nanoclusters prepared by ion implantation in fused silica upon subsequent irradiation with light ions was reported in [263]. In this process the irradiation with Ne ions promoted a preferential extraction of Au from the alloy, resulting in the formation of Au-enriched "satellite" nanoparticles around the original cluster.

Let us briefly consider also a few selected examples of irradiation-induced *cluster precipitation*. Irradiation can promote nucleation and growth of elements already present in the substrate (both introduced by a preceding implantation or during the growth). A phase separation leading to the formation of Si NCs was induced in silica by electron irradiation with energies above $150\,keV$ without any need for introducing additional Si atoms by implantation or mixing [264]. According to the authors the transformation of amorphous SiO_2 into crystalline Si takes place in two steps. The first step involves transformation of amorphous SiO_2 into amorphous Si, while the second step is the crystallization of amorphous Si. Valence electron ionization was determined as the key factor for the transformation from SiO_2 to amorphous Si; beam heating and knock-on displacement were found to be responsible for the transformation from amorphous Si to crystalline Si.

Room temperature MeV ion irradiation of a glass containing copper oxide can initiate nucleation of pure Cu clusters via the inelastic "electronic" component of the ion energy loss, when the latter is above a threshold value [265]. Then the clusters grow under subsequent thermal annealing, following the LSW kinetics. The authors claimed that the decoupling of nucleation and growth allowed total control over the cluster density, average size, and size distribution.

A double-implantation technique was applied in [16]. First, Au was introduced into suprasil by ion implantation at fluences below the threshold fluence for spontaneous cluster formation; then the samples were irradiated with an MeV Si beam, thus assisting the formation of Au nanoclusters. A relationship between the energy deposited in electronic excitations by the MeV Si ions in the Au-implanted layer with the size of these nanoclusters as well as with the optical absorption band due to formation of Au nanoclusters was demonstrated.

Nanocrystals of tin oxide were formed in e-beam evaporated amorphous films by high-energy heavy ion irradiation without subsequent annealing [266]. The nucleation of nanocrystals was ascribed to electronic excitation by the ions.

In the case of the ion bombardment of zircon ($ZrSiO_4$), a phase decomposition occurs under irradiation at temperatures above $600°C$. Between 600 and $750°C$, zircon first becomes amorphous, but with increasing dose, it gradually decomposes into an assemblage of randomly oriented cubic or tetragonal ZrO_2 NCs that are embedded in a matrix of amorphous SiO_2. The ZrO_2 precipitates are approximately 3 nm in diameter. Irradiation above 750°C leads to direct decomposition of zircon into the component oxides without the formation of an intermediate amorphous phase [267].

The number of publications in this highly interesting area is rapidly growing, so that a separate review would be necessary to give an account of the state of the art within it.

Gettering of metal impurities in semiconductors by implantation-induced defects has been known for decades [268] and is widely used in the microelectronics technology. Furthermore, the formation of voids or bubbles in silicon implanted with H or He is a well-known phenomenon [269]. After annealing, He-implanted Si specimens exhibit voids that can be highly efficient gettering sites for metal impurities [270]. Self-assembled formation of spherically shaped nanometer-sized voids in an Si/SiGe layered structure after Ge ion implantation followed by RTA was reported [271, 272]. As we shall see, these processes are immediately related to the NC formation upon ion implantation. It has been directly shown by high-angle annular dark-field (HAADF) imaging in a scanning TEM (STEM) that the Ge and Er nanocrystal growth in ion-implanted SiC is defect mediated: upon high-dose implantation at 700°C the dislocation loops are created, and the precipitates segregate to the dislocation cores during subsequent annealing at 1600°C for 180 s [273]. A defect-mediated NC nucleation upon Pt implantation was implemented by pre-irradiating thin SiO_2 films with high-energy Ge ions [274]. A striking feature was observed by TEM in [275–279]: the formation of CdS and Cu nanoparticles in amorphous SiO_2 and of S, As, Se, Zn, Cd and Ag nanocrystals in crystalline Si by ion implantation and subsequent annealing was tightly related to the formation of intra- and interparticle voids in the implanted layers. The voids were formed either at the implant temperature or when heated during annealing to temperatures above the melting or boiling points of the implanted species. The authors indicated four processes that may occur either individually or in concert and can lead to void formation: (i) volume contraction of the precipitated material upon solidification from a liquid; (ii) evaporation of previously existing particles to leave behind a large void; (iii) diffusion and aggregation of gaseous species to create voids in a process similar to that proposed for He-implanted Si [270]; and (iv) gettering of the implanted material to partially fill voids that were previously formed in the host crystal. A combination of these processes may result in different and complex morphologies [277]. *In situ* TEM observations of Mn- and As-implanted silicon have proven even more clearly that the voids are first formed upon annealing and then Mn and MnAs nanocrystals are formed inside them [280].

13.4.4 Creation of magnetic nanocrystals

Creation of magnetic NCs embedded in semiconductor matrices opens the way to diverse spintronic applications. Ferromagnetic nanodots are basic elements for fabrication of various devices, for detection of magnetic field and for information recording [281]. As in the other cases described above, the ion implantation offers a versatile tool for nanofabrication. Despite the importance of the technological area, it has never been reviewed. Therefore, it will be discussed here in more detail.

13.4.4.1 III–V compounds

GaAs: Submicrometre magnetic structures were developed by FIB implantation of Mn in GaAs epilayers and QWs in 1994 [282]. At the same time, formation of metal/semiconductor composites by ion implantation of Fe and Ni into GaAs and a subsequent anneal to nucleate clusters was tried [283]. By means of the electron diffraction and HRTEM the precipitates were identified as hexagonal and metallic Fe_3GaAs or Ni_3GaAs. The orientation relationship to GaAs was determined. However, magnetic measurements were not performed. After a while, magnetic nanoclusters were successfully synthesized by Mn implantation in GaAs [284, 285]. This technology has been thoroughly explored in a series of forthcoming publications by various groups [286–299]. Implantation of Mn ions into MBE-grown or semi-insulating GaAs with subsequent RTA leads to the formation of GaMn precipitates from 200 to 400 nm in diameter, composed of Ga and Mn at a ratio of about 60:40 [284, 286, 287]. The precipitates are ferro-magnets with a Curie temperature well above 400 K. Considering that bulk GaMn phases at or near this composition are ferromagnetic with lower Curie temperatures or simply non-magnetic, an important role of the crystalline structure of the precipitates in their magnetism was concluded [287]. The easy magnetization axes of 80% of the GaMn particles align preferentially along three equivalent <001> directions of the GaAs host. The GaMn particles were identified by electron diffraction and HRTEM as icosahedral quasicrystals having a definite orientation relationship to the GaAs host matrix. Nearly 30% of the measured particles were identified as 3D quasicrystals, and more than 60% of the remaining crystals showed quasicrystal-related features such as quasicrystal approximants and incommensurate structures, from which one- or two-dimensional Fibonacci chains could be seen in electron diffraction patterns [288].

Low-temperature (LT) GaAs contains an excess of interstitial As, therefore implantation of Mn in such layers with subsequent RTA leads to the formation of hexagonal MnAs precipitates with particle sizes ranging from 10 to 25 nm (Fig. 13.27a) [289]. The diffraction pattern of these particles indicates an epitaxial relationship with the GaAs matrix (Fig. 13.27b): $(0001)MnAs \parallel (1\bar{1}1)$ GaAs and $(01\bar{1}1)MnAs \parallel (002)GaAs$. The α-MnAs phase is ferromagnetic with NiAs-type hexagonal structure ($a = 0.3724$ nm, $c = 0.5706$ nm). In the bulk, a reversible α-MnAs–β-MnAs (paramagnetic, orthorombic) phase transition takes place at 318 K [300]. The maximum Curie temperature obtained from SQUID measurements in [289] was $T_c = 360$ K, i.e. slightly higher than that of bulk MnAs. This behaviour was attributed to an admixture of Ga to the MnAs

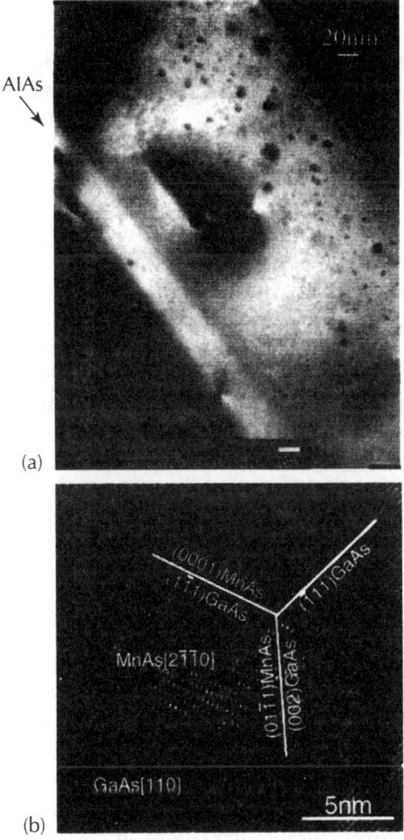

Figure 13.27   (a) Cross-sectional TEM image of LT-GaAs:Mn annealed at 750°C (5 s) (Mn = 1 × 10^{16} cm^{-2}). (b) High-resolution TEM image of a single MnAs crystallite in LT-GaAs:Mn annealed at 750°C (5 s) ((Mn)= 1 × 10^{16} cm^{-2}) [289].

precipitates, so that $MnAs_nGa_m$ alloy precipitates may have formed at high annealing temperatures. Similar results were obtained in [293, 294], where hexagonal MnAs nanoparticles with diameters ranging from 9 nm to 13 nm were synthesized in (001)GaAs by implantation of Mn alone and by co-implantation of Mn and As, in both cases followed by RTA. The crucial role of As in the retention of Mn was shown. The same epitaxial relation as in [289] was determined: $(0002)MnAs \parallel (\bar{1}\bar{1}1)$ GaAs and $[1\bar{2}10]MnAs \parallel [011]GaAs$, the highest attainable through the association of these two materials. There are four different but equivalent orientations of the hexagonal prisms, as their basal plane can lie on any of the four $\{111\}$ planes of GaAs. The formation of MnAs nanocrystals in GaAs co-implanted with Mn and As is facilitated by a pre-annealing at 600°C, just below the congruent temperature of GaAs $[294]T_G = 630°C$. Ferromagnetic resonance (FMR) spectra measured on Mn and As co-implanted GaAs after RTA treatment [301] were identical to those formerly reported in MO CVD-grown, cluster-rich GaMnAs layers and interpreted as resulting from hexagonal MnAs nanocrystals with the [0001] axis parallel to any of the $<111>$ axes of the GaAs host [302]. Synthesis of MnAs nanocrystals with a Curie temperature of approx. 320 K by Mn implantation in semi-isolating (001)GaAs wafers with subsequent RTA in a nitrogen gas with a silicon proximity cap was reported in [291]. Contrary to [289], the authors of [291] concluded from their magnetic force microscopy (MFM) measurements that the precipitates' [0001] axis (which is the hard magnetization axis) is normal to the (001) wafer surface, and the $[11\bar{2}0]_{MnAs}$ (easy magnetization axis) is parallel to $[110]_{GaAs}$. When the RTA was performed in a forming gas, GaMn precipitates were formed. MnAs and $MnAs_nGa_m$ precipitates were simultaneously observed after Mn implantation in semi-insulating GaAs and RTA in a nitrogen atmosphere [33].

Though the MnAs nanoclusters are embedded in the GaAs matrix, they possess a residual degree of freedom caused by lattice defects surrounding them, so that the first-order magneto-structural phase transition at approximately 40°C, from the hexagonal–ferromagnetic α-phase to the orthorhombic–paramagnetic β-phase, could be observed [298]. The nanoclusters' lattice parameter presents a thermal hysteresis, i.e. the phase transition occurs at higher temperatures during heating as compared to a cooling cycle, and is not very abrupt, which was interpreted as due to the coexistence of the two phases. By analogy with literature results on thin MnAs films grown on GaAs, both the coexistence and the hysteresis were attributed to strain effects.

Granular films, consisting of Co granules embedded in GaAs, were prepared by recoil implantation of Co in n-type (100)GaAs single crystals [303]. The GaAs surface, on which a 30 nm thick Co layer had been deposited prior to the implantation by RF magnetron sputtering, was bombarded with Xe ions. The samples showed large extraordinary Hall effect.

Giant magnetoresistance [290] and enhanced positive magnetoresistance [292, 304] were observed in GaAs layers containing MnAs NCs synthesized by Mn implantation. Several other interesting effects were reported in GaAs/MnAs and other hybrid semiconductor–nanomagnet systems fabricated by epitaxial techniques [305–312]. Obviously, similar results can be obtained using ion implantation, especially FIB.

Though magnetic nanocluster synthesis by ion implantation is best studied in GaAs, there are numerous reports on the implantation-induced formation of magnetic precipitates in other semiconductors and isolators.

GaP: Very small FM NCs containing only ten spins were detected in Mn-implanted *p*-type GaP [313]. The magnetism was suppressed when *n*-type GaP substrates were used. The presence of ferromagnetic clusters and hysteresis to temperatures of at least 330 K was attributed to disorder and proximity to a metal-insulating transition. The experimental data suggest a percolation type of picture in which isolated ferromagnetic clusters, immersed in a background of paramagnetic moments, grow in size as the temperature is lowered until, at $T = T_C$, long-range order extends through the whole system.

InP: MnP and $InMn_3$ crystallites with a size of ~ 20 nm and Curie temperatures of 291 K and well above RT, respectively, were observed in Mn-implanted InP:Zn [314, 315].

GaN: Mn and Fe were also implanted in GaN [316, 317]. After appropriate annealing, the samples showed signatures of ferromagnetism with Curie temperatures <250 K for Mn and <150 K for Fe implantation. The structural analysis of the Mn-implanted GaN showed regions consistent with the formation of $Ga_xMn_{1-x}N$ platelets occupying $\sim 5\%$ of the implanted volume. In [443] RT ferromagnetism was observed in GaN after Fe implantation at 623 K with different fluences and was mainly ascribed to α-Fe precipitations. The formation of secondary phases such as Mn_xGa_y or Mn_xN_y was excluded by careful diffraction analysis. Creation of Co or CoGa magnetic clusters as a result of the Co implantation in GaN was reported in [318].

13.4.4.2   Group IV semiconductors

SiC: $SmSi_2$, Sm_5C_2, Co_2Si and SmCo nanoinclusions were created by Sm and Co implantation in 4H–SiC [319] (see Fig. 13.28). Various types of epitaxial relationship between the nanocrystals and the SiC matrix were found. The magnetism of the Co-related nanocrystals was checked by the Lorentz microscopy, see Fig. 13.29 (as to the measurement technique, see the recent reviews [320, 321]). Magnetic fields ranging from 2 to 4 T were measured at RT.

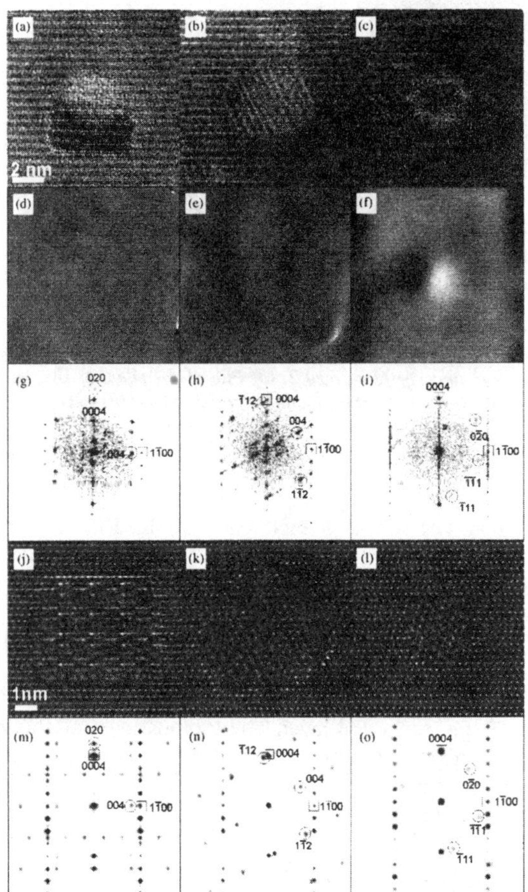

Figure 13.28   *Three typical types of nanocrystals created in 4H–SiC after Sm–ion implantation and annealing viewed along the [11$\bar{2}$0]-4H–SiC direction; experimental HRTEM images (a–c), corresponding lattice banding maps (d–f), diffractograms (g–i) and corresponding calculations: HRTEM images (j–l) and diffraction patterns (m–o) for [100]-SmSi$_2$-nanocrystal (a, d, g, j, m), [110]-SmSi$_2$-nanocrystal (b, e, h, k, n) and [110]-Sm$_5$Co$_2$-nanocrystal (c, f, i, l, o). Reflections of the nanocrystals are encircled; reflections of the 4H–SiC matrix are boxed [319].*

Ge: In Ge, ferromagnetic nanoparticles were synthesized by implantation of Mn and Fe. After an Mn implantation in (100)Ge substrates at elevated temperatures (240–300°C), with and without subsequent annealing at 400°C, X-ray diffraction (XRD), TEM, X-ray photoemission (XPS), electron energy loss spectroscopy (EELS), scanning tunnelling spectroscopy (STM) and magneto-optical Kerr effect (MOKE) measurements revealed the coexistence of Mn dilution into

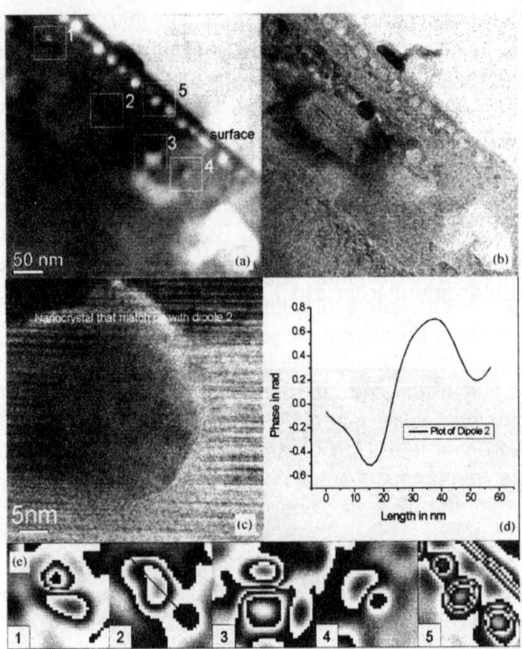

Figure 13.29   *(a) Reconstructed phase image of Co–ion implanted 4H–SiC. Nanocrystals marked 1–4 show magnetic activity, those marked 5 show electric activity. (b) Bright field images showing the same sample area as in (a). (c) HRTEM image of the nanocrystal marked 2. (d) Plot of the phase of the electron wave along the region marked 2. The distance of 13 nm between minimum and maximum phase fits exactly to the diameter of the round-shaped nanocrystal 2. (e) Magnified and amplified phase images of the magnetic dipoles marked 1–4 and the monopoles marked 5 in (a), the dotted line in 2 marks the trace of the plot of the dipole shown in (d) [319].*

the crystalline Ge matrix and of MnGe precipitates embedded in it [214, 322–325]. The precipitates with smaller size (3–10 nm) were amorphous, whereas the bigger ones (8–15 nm) were mainly composed of the metallic Mn_5Ge_3 hexagonal crystalline phase ($T_C \sim 296$ K, [326]). The small amorphous precipitates have semiconducting properties with a band gap of 0.45 ± 0.05 eV [327], and their contribution is believed to be predominant in determining the magnetic properties of the system [328, 329].

Mn implantation in Ge at RT at fluences used for magnetic functionalization ($>10^{15}$ cm^{-2}) leads to amorphization of the implanted layer. The topmost part of the implanted layer is also porous and oxidized [330]. Subsequent annealing at 400°C leaves this layer amorphous, whereas the deeper regions reconstruct into Ge nanocrystals (~ 10 nm in diameter). Though magnetic hysteresis was observed at the annealed samples up to above 250 K, such a structure appears less promising for electronic applications.

An important step in implementing a spin field-effect transistor [331] was done in [332]. The RT Mn implantation was undertaken through a periodic array of 20 nm diameter holes in an SiO_2 layer deposited on top of n- and p-type (111)Ge wafers. Following implantation, RTA was performed on samples at 450°C in an N_2 atmosphere for 20 min. The TEM pictures showed that the Mn_xGe_{1-x} DMS nanodots were coherently incorporated in the Ge matrix. XRD revealed the presence of the Mn_5Ge_3 and $Mn_{11}G_8$ ($T_C \sim 270$ K, [326]) ferromagnetic phases, the latter in a negligible concentration. More importantly, the structures formed on n- and p-type substrates exhibited different magnetic properties. Hysteresis was observed only in p-type, suggesting the existence of hole-mediated exchange coupling. Using gate-biased MOS structures, the authors succeeded in controlling the hysteresis by varying the hole concentration in the Mn_xGe_{1-x} channel.

Fe nanoclusters were also successfully incorporated in p-type (110)Ge by ion implantation [333, 334]. At the lower fluence used (2×10^{16} cm^{-2}) dilute Fe_3Ge precipitates of irregular shape with sizes varying from 13 to 25 nm were detected. The cubic Fe_3Ge is a ferromagnetic compound with Curie temperature equal to 755 K. As the size of the precipitates was large, no superparamagnetic relaxation effects were observed. With the fluence increasing up to $2 \times 10^{17}\,cm^{-2}$, the situation changed. As determined by XPS data, the majority of the implanted Fe ions were in the metallic phase, and a small amount was in the Fe–Ge, Ge–O, and Fe–O phases. From an analysis of the magnetometry results, the authors obtained a mean diameter of the superparamagnetic clusters of $\sim$4 nm. Negative magnetoresistance was observed and attributed to spin-dependent scattering of charge carriers by these clusters. Ion-beam synthesis of Fe nano-clusters in Ge was also reported in [335].

Ge/Si: Epitaxially grown Ge/Si QD heterostructures were subjected to Mn implantation and RTA [336]. Magnetization with $T_C = 160$ K was measured in a SQUID and ascribed to diluted magnetic QDs. Some intermixing was acknowledged as a result of Raman scattering measurements. However, no structural analysis in order to access the degree of intermixing and to prove the existence of QDs after implantation and annealing was reported.

Si: Though the most important semiconductor is still silicon, the research efforts to produce magnetic phases in it have been surprisingly weak. In an analogy to Ge, an obvious way to achieve magnetic functionalization of Si appears to be doping with transition metals (TMs). When TMs react with Si, silicides are readily formed upon heating [337–339]. $PtSi$, $TiSi_2$, $CoSi_2$, $NiSi$ have metallic conductivity; $CrSi_2$, β-$FeSi_2$, $ReSi_{1.75}$, Ru_2Si_3, $OsSi_2$, Ir_3Si_5 and $MnSi_{1.7}$ are semiconductors. Manganese silicides are especially numerous: besides $MnSi$ [340] and Mn_5Si_3[341], there is an entire family of semiconducting higher manganese silicides (HMS) $MnSi_x$ with a tetragonal lattice and composition x varying from 1.67 to 1.75, which are usually labelled $MnSi_{1.7}$ (Mn_4Si_7, $Mn_{11}Si_{19}$, $Mn_{15}Si_{26}$, $Mn_{26}Si_{45}$, and $Mn_{27}Si_{47}$) [342].

Unfortunately, most silicides either have a low Curie temperature T_C or are not ferromagnetic at all (iron silicide Fe_3Si is an exception [339]). In the absence of a magnetic field, MnSi orders magnetically with a helical structure below 30 K. However, in an applied field of 0.6 T the moments align in an induced ferromagnetic state. Thus, in a field MnSi is often regarded as a good example of a very weak itinerant ferromagnet [343–346]. In very thin layers ($<$60 Å of Mn deposited) MnSi exhibits ferromagnetism at 10 K [347]. Mn_4Si_7 has a Curie temperature of 40–43 K [348, 349]. Mn_5Si_3[350–352] and Mn_3Si [353, 354] are antiferromagnetic at low temperatures.

So it was quite surprising when room temperature ferromagnetism was reported in Si doped with Mn by ion implantation [355, 356], though there had been reports on magnetic thin SiMn films grown by other techniques [357–361]. However, it has not been shown which phase was responsible for the effect. Other works followed where the existence of nanoclusters in the implanted and annealed layers was proven with the most probable phase of the crystallites being $MnSi_{1.7}$[362, 363]. It was concluded from the XANES and EXAFS spectra that Mn ions implanted into bulk silicon do not form either metallic or oxide inclusions [364]. Both XANES and EXAFS spectra are in reasonable agreement with the model, which assumes the formation of clusters with short-range order close to strained and defected Mn–Si compounds with five to eight near-neighbour atoms. Clusters formed in the sample with Mn implantation performed on a heated substrate showed the local order closer to that in the Mn–Si phases.

Thus, although the high-temperature ferromagnetism observed in Mn implanted Si is most probably caused by clusters, it is not yet clear which is the crystallographic phase responsible for the magnetism.

It has to be noted that density functional theory (DFT) calculations have predicted for the monosilicides of Mn, Fe, Co, and Ni a CsCl-like structure that has not yet been observed as bulk compound but may be stabilized epitaxially on Si(001). For very thin films of CoSi and MnSi grown in this structure, a ferromagnetic ground state was obtained. It was found that the atomic structure of MnSi films on Si(111) is close to the natural crystal structure of bulk MnSi (B20), and also shows large magnetic moments of the Mn atoms at the surface and interface [365]. DFT calculations also predicted that Si-based heterostructures with one quarter to one layer δ-doping of interstitial Mn display high spin polarization [366].

Another approach was chosen in [280, 301, 367]. Mn and As ions were co-implanted at 350°C in (001)Si substrates that were subsequently subjected to RTA at various temperatures. In ferromagnetic resonance (FMR) measurements, highly anisotropic spectra were detected

(Fig. 13.30) that could be interpreted as arising from hexagonal magnetic clusters with the hard magnetization axis aligned along the four <111> Si axes. The spectra were strikingly similar to those measured in MO CVD-grown, MnAs cluster-rich GaMnAs layers; these spectra had been interpreted as arising from hexagonal MnAs clusters with MnAs[0001] ‖ GaAs[111] [302]. FMR spectra and magnetization measured in a SQUID persisted above RT. *Ex situ* and *in situ* energy-filtered cross-sectional TEM measurements revealed the existence of two main types of NCs. (i) The smaller ones with diameters up to 20 nm consist of Mn and As and are crystalline; no phase separation could be imaged. (ii) The larger ones are phase separated. They predominantly consist of manganese silicides and are surrounded by a shell of amorphous As. During the annealing voids are formed that are then filled by diffusing Mn and As atoms [280]. The magnetism and the FMR spectra are supposed to stem from the MnAs nanocrystals.

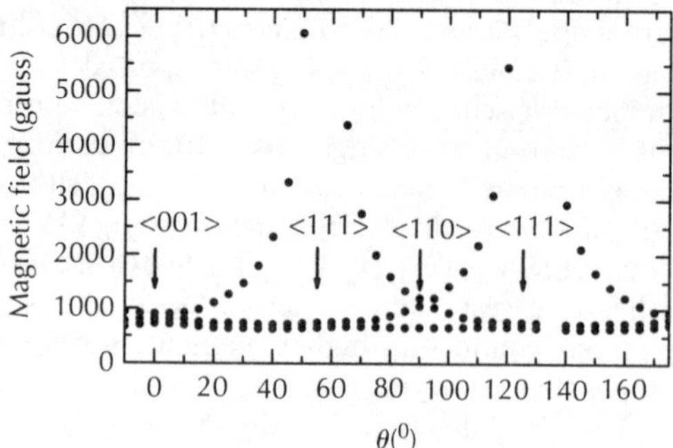

Figure 13.30   *Angular dependence of the FMR line positions in an Mn- and As-implanted (001)Si sample after RTA at 1100°C obtained for the magnetic field rotation in the (110) plane [301].*

13.4.4.3   Oxides

ZnO: Formation of magnetic nanoclusters was reported in ZnO implanted with Co [368–371, 445], Fe [370, 372–374, 444] and Ni ions [375, 445]. In the case of Co implantation, X-ray diffraction results indicated the presence of (110)-oriented hexagonal phase Co in the ZnO matrix. The nanocrystal size was estimated to be ∼3.5 nm. In-plane X-ray diffraction results showed that the nanocrystals were epitaxial with respect to the ZnO host matrix [369]. The orientation relationship is as follows [445]: hcp-Co(0001)[10–10] ‖ ZnO(0001)[10–10] and Ni (111) [112] ‖ ZnO(0001)[10–10]. The existence of small superparamagnetic clusters in ZnO implanted with Co at ∼350°C was confirmed in [371] by magnetic measurements. The structure and magnetic properties of Co or Ni NCs embedded inside ZnO could be tuned by postannealing. For the Co case, 823 K annealing resulted in the co-existence of fcc-Co and hcp-Co. The magnetic anisotropy was changing from out-of plane to in-plane. Especially for the Ni NCs, the magnetic anisotropy was different from the bulk crystals [445].

Tiny bcc-Fe particles, formed inside the host matrix, were found to be responsible for the ferromagnetic properties of ZnO implanted with Fe at 350°C [372]. The particles were identified using synchrotron X-ray diffraction and conversion electron Mössbauer spectroscopy (CEMS). Evidence for the existence of other, unidentified, superparamagnetic NCs was brought by the zero field-cooled (ZFC) and field-cooled (FC) magnetization measurements as a function of temperature in Fe-implanted ZnO subjected to flash lamp annealing [373]. In a subsequent work of the same group [374], crystallographically oriented Zn ferrite (ZnFe$_2$O$_4$) nanoparticles with the orientation relationship of ZnFe$_2$O$_4$(111)[110] ‖ ZnO(0001)[11$\bar{2}$0] were detected inside high fluence Fe-implanted ZnO after a long-term annealing at 800°C. These ZnFe$_2$O$_4$ nanoparticles showed a hysteretic behaviour upon magnetization reversal at 5 K. The influence of the implantation parameters on the state of the implanted Fe and the appearance of the ferromagnetism was thoroughly investigated in [444].

Also in ZnO implanted with Ni at 350°C crystalline fcc-Ni NCs were detected [375]. It was concluded from magnetic measurements that around 27% of implanted Ni atoms were in the metallic state.

MgO: In early experiments on the room temperature Fe implantation in MgO at fluences ranging from 10^{15} to 10^{17} cm^{-2}[376–379], mainly CEMS supported by RBS/C, optical absorption, electron microscopy and electrical conductivity measurements was implemented to investigate the state of the implanted ions. It was found that implantation introduces iron in three charge states: Fe^{2+}, Fe^{3+}, and metallic precipitates (Fe^0) with the dominant role of Fe^{3+} at low fluences, Fe^{2+} at medium fluences, and metallic iron clusters at the highest fluences. The isochronal thermal annealings in air at temperatures between 300 and 700°C gradually caused the oxidation and the nucleation of highly dispersed spinel-like clusters and then, at about 800–900°C, the growth of magnesioferrite particles. In contrast, the heat treatment in vacuum converted all iron into well-diluted Fe^{2+} in MgO phase. In [380, 381], a superparamagnetic behaviour of the magnetization was found after Fe implantation at RT until a fluence of 2.0×10^{17} cm^{-2}, suggesting the precipitation of implanted Fe ions as nanogranules in the MgO matrix. The GMR effect with significant MR ratio of 3.5% for fields of 0–12 kOe was observed only for the sample implanted at this dose and post-annealed at 300°C in argon. The superparamagnetic NCs changed into ferromagnetic ones by annealing at 300°C and/or with increasing fluence due to the growth in their size and the consequent blocking of superparamagnetic relaxation. CEMS, MOKE, XRD, TEM including electron holography [320, 321], and RBS/C were applied in [382] to investigate the Fe NC formation in (001)MgO implanted with Fe in a wide temperature range (and also in yttrium-stabilized zirconia ZrO$_2$, see below). Non-magnetic metallic γ-phase Fe NCs were synthesized inside MgO. The formation of the γ-Fe (fcc) phase was surprising, since it is usually not stable at RT. The fraction of the γ-phase Fe^0 increased from 28% (nominally RT implantation) to 60% ($T_i = 800$°C) as a function of implantation temperature T_i at the expense of Fe^{2+} states. At $T_i = 600$°C and 800°C the Fe^{2+} states were not present but were transformed into Fe^0 or Fe^{3+} states. It was found that at $T_i = 800$°C the NCs show an exclusive orientation relationship of γ-Fe(111) $\|$ MgO(111), respectively, γ-Fe[220] $\|$ MgO[220]. The mean diameter, as calculated by the Scherrer formula, amounted to 5 nm. It was shown recently that ferromagnetic Fe NCs can be synthesized by ion implantation in differently oriented MgO single crystals [(100), (110), and (111)] [383]. They all exhibited ferromagnetic behaviour at 5 and 300 K. The coercive field H_C strongly depended on the crystal orientations and obeyed the relation of $H_C(110) < H_C(100)$ $<H_C(111)$ at different measurement temperatures.

Implantation of Co and Ni in (001)MgO crystals with subsequent heat treatment up to 1280 K in vacuum was used to create colloidal dispersions of superparamagnetic Co and Ni particles [384–387]. The number and size of the aggregates depended on the atomic concentration of the implanted ions and the subsequent annealing treatments. From magnetic measurements, average diameters of fcc-Co NCs from 2.3 to 6.8 nm and that of fcc-Ni NCs from 4.1 to 6.7 nm as well as the anisotropy constants were calculated; the particles behaved as single domains. The presence of an antiferromagnetic layer surrounding the Ni aggregates, possibly a nickel oxide layer, was concluded from the asymmetric shape of the hysteresis curves and the non-saturating behaviour of the high-field magnetic susceptibility [385]. As evidenced by the Mie scattering from the colloids, Co implanted at an energy of 250 keV diffuses out of the precipitates, when the annealing temperature is increased from 1250 to 1600 K, and probably occupies substitutional sites in the MgO lattice [387]. For the implantation energy of 100 keV, the outdiffusion starts at 1050 K, and at 1250 K the precipitates are already dissolved. Very recently, the metal and alloy NC formation in MgO, LaAlO$_3$ and YSZ (see below) targets covered by Pt and FePt thin films by means of ion-induced mixing upon Pt ion irradiation was achieved [446].

TiO$_2$: Titanium dioxide occurs in several forms, such as rutile, anatase and brookite. Rutile is tetragonal ($a = 4.59$ Å and $c = 2.96$ Å) with a band gap of 3 eV, while the anatase phase also crystallizes in a tetragonal structure ($a = 3.78$ Å and $c = 9.52$ Å) with a band gap of 3.2 eV [388, 389]. Thin films of cobalt-doped anatase were reported as the first semiconductor exhibiting ferromagnetism at RT [390]. The nucleation of cobalt or cobalt-rich magnetic NCs was claimed to be responsible for this ferromagnetic behaviour [391, 392]. Thus, most of the publications on the transition metal ion implantation in TiO$_2$ were dedicated to cobalt [393–400] and only a few to nickel [395, 401].

Anatase films implanted with Co showed superparamagnetic properties due to the presence of magnetic NCs consisting of Co or a ferromagnetic Co compound, whose existence was proved by scanning TEM. Upon annealing, the particle size and the blocking temperature were increased by raising the annealing temperature and/or decreasing the oxygen pressure [393]. In Co-implanted rutile, the ordered magnetic behaviour due to the presence of Co NCs was observed even above RT [395].

Strongly anisotropic FMR spectra were observed in the Co-implanted (100) and (001) rutile samples and attributed to the metallic cobalt NCs synthesized during the ion implantation procedure. Dominant four-fold in-plane anisotropy of the FMR signal was interpreted as indicating the cubic (probably, fcc) phase of metal Co, while the additional axial anisotropy was ascribed to the influence of the host rutile structure. The annealing of implanted samples resulted in disappearance of the ferromagnetism and formation of paramagnetic Co^{2+} centres [394]. Magnetic anisotropy of similarly prepared samples was studied by means of MOKE and SQUID magnetometry. Room temperature magnetism was observed. A strong angular dependence of the remanent magnetization and coercive field in the plane of the implanted surface was observed: two-fold anisotropy for the (100) and (110) substrate and four-fold anisotropy for the (001) substrate samples [397, 398]. Upon varying the Co fluence in the range from $0.15–1.50 \times 10^{17}$ cm^{-2}, sequentially paramagnetic, superparamagnetic, weak ferromagnetic and, eventually, strongly anisotropic ferromagnetic response at RT was found [399, 400]. The thermomagnetic analysis revealed in the ferromagnetic samples two magnetic transitions with ordering temperatures $T_{C1} \sim 700$ K and $T_{C2} \sim 850$ K. Annealing of the samples in air strongly suppressed the ferromagnetic phases if the heating temperature exceeded the corresponding transition temperature. Subsequent high-vacuum annealing restored only the low-temperature ferromagnetic phase. The following model suggesting coexistence of two different magnetic phases in Co-implanted rutile was proposed to explain the temperature dependence of the magnetization and the anisotropy of the hysteresis loops [400]. Cobalt in a form of NCs in the very near-surface layer (the first magnetic phase) determines the superparamagnetic behaviour of the samples implanted with low and medium fluences. Cobalt dopant in the form of Co^{2+} ions substituting the host Ti^{4+} ions is the base of the second magnetic phase. The oxygen vacancies formed by implantation provide the charge compensation and serve as mediators for the exchange interaction between the Co^{2+} ions. At high fluences, the bulk-like ferromagnetic order is developed due to ion accumulation and indirect exchange interaction between the Co^{2+} ions.

Ni NCs with dimensions ranging from 3 to 20 nm were observed by high-angle annular dark-field (HAADF) Z-contrast scanning TEM and HRTEM in Ni-implanted rutile [401]. Accordingly, superparamagnetic behaviour was observed in the magnetometry measurements, exhibiting a blocking temperature of 50 K [395] or 85 K [401].

Al$_2$O$_3$: To realize perpendicular magnetic recording media with a high density, much attention has been paid to $L1_0$-type FePt (fct phase). One of the reasons is its high magnetocrystalline anisotropy (above 10^7 erg/cm^3), enabling a small fct-FePt particle to overcome thermal instability that causes superparamagnetism [402]. Formation of oriented ferromagnetic FePt NCs was achieved in Al$_2$O$_3$ also by ion implantation [403–405]. The NCs had coercivity values in excess of 20 kOe.

The production of buried granular magnetic layers was implemented through implantation of Fe [406–408] and Fe + Co ions [251, 409] into α-Al$_2$O$_3$ single crystals. Upon annealing in vacuum Fe ions precipitate as oriented α-Fe NCs, and the orientational relations are of (111) ‖ (0001)Al$_2$O$_3$ and [1$\bar{1}$0]Fe ‖ [11$\bar{2}$0]Al$_2$O$_3$, or (110)Fe ‖ (0001)Al$_2$O$_3$[406]. Magnetization curves exhibited superparamagnetic or ferromagnetic behaviours depending on the diameters of the iron NCs. The granular layers exhibited an eminent giant magnetoresistance (GMR) [251, 407, 410].

A full bibliography on metal-implanted sapphire till early 2005 can be found in [246].

SiO$_2$: There are numerous reports also on the synthesis of magnetic NCs in SiO$_2$ by implantation of Co [411–415], Fe [416–418], Ni [419], Co + Ni [420, 421], and Fe + Pd [422] ions.

In contrast to the IBS of magnetic NCs out of the implanted components, ion irradiation of silicon-based gels with the stoichiometry $SiO_{1.5}H_1$, prepared from triethoxysilane, and containing Fe^{3+} or Ni^{2+} ions in solid solution, was used to obtain the precipitation of metallic NCs in [423–426]. The films were irradiated with 1.5 MeV He or 3 MeV Au ions. It was concluded that electronic excitations were responsible for the reduction and precipitation processes, while atomic displacements induced a partial re-dissolution by mixing. The nucleation occurred for a critical value of the energy density deposited by ions in electronic excitation. The NCs exhibited a narrower range of sizes than in films submitted to heat treatments in vacuum. The hammering effect of ions on the silica matrix caused a tilt of easy magnetization axis when the volume fraction of particles was of a few per cents, but this axis remained in the film plane for over 10%.

The synthesis and modification of metal nanoclusters in silica and silica glasses through ion implantation has recently been reviewed in [211, 245].

Yttrium-stabilized ZrO$_2$: Precipitated layers of ferromagnetic α-Fe or ferrimagnetic Fe$_3$O$_4$ embedded in the near-surface region of (100)-oriented yttrium-stabilized zirconia ZrO$_2$ (YSZ) and exhibiting a magneto-optical activity were created by using Fe ion implantation and thermal processing [382, 427, 428]. TEM and XRD studies established that each α-Fe or Fe$_3$O$_4$ NC was a single crystal that was crystallographically aligned with respect to the YSZ host. The formation of either α-Fe (bcc) or Fe$_3$O$_4$ (or even γ-Fe) NCs depended on the annealing conditions. The α-Fe precipitates exhibited two different, but symmetry related, orientations relative to the YSZ host. The predominant orientation was: α-Fe(110) $\parallel$ YSZ(001) out-of-plane and α-Fe[001] $\parallel$ YSZ[100] in-plane. The less common relative orientation was found to be: α-Fe(001) $\parallel$ YSZ(001) out-of-plane and α-Fe[100] $\parallel$ YSZ[110] in-plane. The particles of the small γ-Fe component had only one relative orientation that was determined to be: γ-Fe(001) $\parallel$ YSZ(001) out-of-plane and γ-Fe[100] $\parallel$ YSZ[100] in-plane [382, 427]. Electron holography [320, 321] clearly indicated a magnetic dipole associated with a single Fe nanoparticle [382].

Mg$_2$SiO$_4$: The implantation of Fe ions at RT into forsterite (Mg$_2$SiO$_4$) also leads to the formation of metallic Fe NCs, especially at high implantation fluences [437]. The subsequent annealing in air at $T = 350$–$700°C$ gradually causes the oxidation of iron and then, at about 800°C, the growth of γ-Fe$_2$O$_3$ particles with a size of about 15 nm. At higher temperatures (900–1200°C) the transformation of γ-Fe$_2$O$_3$ into α-Fe$_2$O$_3$ was observed simultaneously with the large outdiffuison of iron.

In this chapter, we have only reviewed the cases where magnetic NCs had been deliberately synthesized. However, it has to be considered that in many reports claiming the creation of true diluted magnetic semiconductors (DMS) the real source of magnetism were magnetic nanocrystals, so the spectrum of existing hybrid nanomagnet/semiconductor systems is certainly much broader than described in this chapter.

13.4.5 Other nanotechnological applications of radiation

Swift heavy ions (SHI, i.e. ions moving at a velocity comparable to the Bohr velocity of the electron) lose energy in materials mainly through inelastic collisions with the atomic electrons. Along the trajectory, a trail of defects (point defects, defect clusters, structural phase transformation) known as latent track may be formed depending on the type of ion and its energy as well as the physical properties of the target material. This damage is always created in the close vicinity of the trajectory of the projectile [2]. The latent tracks can be transformed via etching into open nanochannels with a high aspect ratio ($\sim$1000), which allows the controlled production of nanoporous membranes that can be filled with the required material, thus producing nanowires, nanotubules and chains of nanoparticles in insulators, semiconductors and polymers [429]. To date, the tracks have been used as molecular filters, bio-sensors (e.g. for DNA detection) or to produce metallic nanowires, but single ion irradiation opens the door even more for the creation of well-ordered semiconductor nanostructures.

An effect similar to that produced by SHI in insulators, though predominantly potential energy driven, is provoked by the impact of slow highly charged ions [430]. There is a well-defined threshold of potential energy carried into the collision of about 14 keV for nano-sized hillock formation on the (111) surface of CaF$_2$ single crystals. Estimates of the deposited energy density suggest that the threshold is linked to a solid–liquid phase transition ("melting") on the nanoscale. This method makes it possible for regular nanostructures to form on surfaces without damaging deeper-lying crystal layers.

13.5 Conclusions

In this chapter, several radiation effects and their applications to quantum-size semiconductor structures, ante omnia quantum dots, have been considered. Numerous instances have been cited illustrating the main topics of interest of radiation effect studies as formulated at the end of the Introduction (Section 13.1). The QD heterostructures and QD lasers are generically more resistant to radiation damage ("radiation hard") than their bulk and 2D counterparts, which is caused not only by the localization of the wavefunction of the confined carriers but also by the expulsion of the mobile defect components to the surface/interface of the nanocrystals. There are many exciting applications of particle irradiation to QD technology, such as intermixing, self-organized formation of surface nanostructures, ion-beam synthesis of nanocrystals in solid matrices, as well as cluster precipitation in irradiated media. An emphasis has been given to the ion-beam synthesis of magnetic nanocrystals in solid hosts.

Acknowledgements

I had the luck to work with many outstanding colleagues and gifted students from Belarus, Germany, Portugal, Russia and other countries. They are too numerous to be particularized here, but I am deeply grateful to all of them for fruitful collaboration and for inspiration and impulses they have given me. This work was supported in part by the SANDiE Network of Excellence of the EU and by the project POCI/FIS/61462/2004 of the FCT of Portugal. The TRIM simulations made by N.M. Santos are gratefully acknowledged.

References

1. E.G. Stassinopoulos and J.P. Raymond, *Proc. IEEE* **76**, 1423 (1988).
2. D. Kanjilal, *Curr. Sci.* **80**, 1560 (2001).
3. P.S. Gwozdz and J.S. Koehler, *Phys. Rev. B* **6**, 4571 (1972).
4. J. Bourgoin, M. Lannoo, *Point Defects in Semiconductors. II. Experimental Aspects. Springer Series in Solid State Science*, Vol. **35**. (Springer, Berlin *et al.*, 1983).
5. B. Massarani and J.C. Bourgoin, *Phys. Rev. B* **34**, 2470 (1986).
6. N. Mott, *Proc. Roy. Soc.* **A124**, 426 (1929); **A135**, 429 (1932).
7. W.A. McKinley, Jr. and H. Feshbach, *Phys. Rev.* **74**, 1759 (1948).
8. A.L. Barry, A.J. Houdayer, P.F. Hinrichsen, W.G. Letourneau, and J. Vincent, *IEEE Trans. Nucl. Sci.* **42**, 2104 (1995).
9. R.J. Walters, S.R. Messenger, G.P. Summers, E.A. Burke, S.M. Khanna, D. Estan, L.S. Erhardt, H.C. Liu, M. Gao, M. Buchanan, A.J. Spring Thorpe, A. Houdayer, and C. Carlone, *IEEE Trans. Nucl. Sci.* **48**, 1773 (2001).
10. M. Yamaguchi, *J. Appl. Phys.* **81**, 6013 (1997).
11. F.F. Morehead and B.L. Crowder, *Rad. Eff. Def. Sol.* **6**, 27 (1970).
12. J.F. Gibbons, *Proc. IEEE*, **60**, 1062 (1972).
13. S. Takeda and J. Yamasaki, *Phys. Rev. Lett.* **83**, 320 (1999).
14. J. Yamasaki, S. Takeda, and K. Tsuda, *Phys. Rev. B* **65**, 115213 (2002).
15. H. Bernas, in: *Radiation Effects in Solids*, ed. by K.E. Sickafus *et al.* (Springer, Dordrecht, 2007), Chap. 12, p. 353, and references therein.
16. P. Sigmund, in: *Sputtering by Particle Bombardment I*, ed. by R. Behrisch (Springer, Berlin, 1981).
17. J.F. Ziegler, J.P. Biersack, and U. Littmark, *The Stopping and Range of Ions in Solids* (Pergamon, New York, 1985); http://www.srim.org/.
18. Available from the Radiation Shielding Information Center, Oak Ridge National Laboratory, P.O. Box 2008, Oak Ridge, TN, 37831-6362, USA, pdc@epic.epm.ornl.gov.
19. P. Hovington, *Scanning* **19**, 29 (1997).
20. *Ionizing Radiation Effects in MOS Devices and Circuits*, ed. by T.P. Ma and P.V. Dressendorfer (John Wiley, New York, 1989).

21. *NASA Thesaurus*: NASA/SP-2004-7501/Vol. **1**, p. 902.
22. M. Yamaguchi, *J. Appl. Phys.* **78**, 1476 (1995). and references therein.
23. A.H. Johnston, Talk presented at the 4th International Workshop on Radiation Effects on Semiconductor Devices for Space Application, Tsukuba, Japan, October 11–13, 2000.
24. S.M. Khanna, J. Webb, H. Tang, A.J. Houdayer, and C. Carlone, *IEEE Trans. Nucl. Sci.* **47**, (2322) (2000).
25. F. Gaudreau, C. Carlone, A. Houdayer, and S.M. Khanna, *IEEE Trans. Nucl. Sci.* **48**, 1778 (2001).
26. M. Yamaguchi and C. Amano, *J. Appl. Phys.* **54**, 50213 (1983).
27. C. Coskun, D.C. Look, G.C. Farlow, and J.R. Sizelove, *Semicond. Sci. Technol.* **19**, 752 (2004).
28. J.F. Cordaro, C.E. Shipway, and J.T. Schott, *J. Appl. Phys.* **61**, 429 (1987).
29. R. Khanna, K. Ip, K.K. Allums, K. Baik, C.R. Abernathy, S.J. Pearton, Y.W. Heo, D.P. Norton, F. Ren, S. Shojah-Ardalan, and R. Wilkins, *J. Electron. Mater.* **34**, 395 (2005).
30. F.D. Auret, M. Hayes, W.E. Meyer, J.M. Nel, L. Wu, and M.J. Legodi, 13th International Conference on Semiconductor and Insulating Materials, 2004 (SIMC-XIII-2004), p. 217.
31. D.C. Look, D.C. Reynolds, J.W. Hemsky, R.L. Jones, and J.R. Sizelove, *Appl. Phys. Lett.* **75**, 811 (1999).
32. A.Y. Polyakov, N.B. Smirnov, A.V. Govorkov, E.A. Kozhukhova, V.I. Vdovin, K. Ip, M.E. Overberg, Y.W. Heo, D.P. Norton, S.J. Pearton, J.M. Zavada, and V.A. Dravin, *J. Appl. Phys.* **94**, 2895 (2003).
33. T. Sumita, M. Imaizumi, S. Matsuda, T. Ohshima, A. Ohi, and H. Itoh, *Nucl. Instr. Methods B* **206**, 448 (2003).
34. G.P. Summers, E.A. Burke, P. Shapiro, S.R. Messenger, and R.J. Walters, *IEEE Trans. Nucl. Sci.* **40**, 1372 (1993).
35. K.E. Sickafus, R.W. Grimes, J.A. Valdez, A. Cleave, M. Tang, M. Ishimaru, S.M. Corish, C.R. Stanek, and B.P. Uberuaga, *Nature Materials*, **6**, 217 (2007).
36. See, e.g., *IEEE Trans. Nucl. Sci.*
37. For a review, see, e.g., *Landolt-Börnstein. Vol. 22 "Semiconductors", subvol. b "Impurities and Defects in Group IV Elements and III–V Compounds"*, ed. by M. Schulz (Springer, Berlin *et al.*).
38. For the most recent update, see *Proceeding of the 24th International Conference on Defects in Semiconductors* (22-27 July 2007, Albuquerque, New Mexico, USA), in: *Physica B* 401–402.
39. *Semiconductors for Room Temperature Nuclear Detector Applications (Semiconductors and Semimetals)*, Vol. **43**, ed. by A. Beer *et al.* (Elsevier, 1995).
40. G.M. Dalpian and J.R. Chelikowsky, *Phys. Rev. Lett.* **96**, 226802 (2006).
41. D. Turnbull, *J. Appl. Phys.* **21**, 1022 (1950).
42. M. Rose, A.G. Balogh, and H. Hahn, *Nucl. Instr. Methods B* **127–128**, 119 (1997).
43. D. Sreenivasan, J.E.M. Haverkort, T.J. Eijkemans, and R. Nötzel, *Appl. Phys. Lett.* **90**, 112109 (2007).
44. K.K. Lindner, J. Phillips, O. Qasaimeh, X.F. Liu, S. Krishna, and P. Bhattacharya, *J. Vac. Sci. Technol. B* **17**, 1116 (1999).
45. J.S. Wang, J.F. Chen, J.L. Huang, P.Y. Wang, and X.J. Guo, *Appl. Phys. Lett.* **77**, 3027 (2000).
46. Y.C. Chua, E.A. Decuir, Jr, B.S. Passmore, K.H. Sharf, M.O. Manasreh, Z.M. Wang, and G.J. Salamo, *Appl. Phys. Lett.* **85**, 1003 (2004).
47. M.V. Maximov, D.S. Sizov, A.G. Makarov, I.N. Kaiander, L.V. Asryan, A.E. Zhukov, V.M. Ustinov, N.A. Cherkashin, N.A. Bert, N.N. Ledentsov, and D. Bimberg, *Fiz. Tekhn. Poluprov.* **38**, 1245 (2004).
48. A.H. Johnston, *IEEE Trans. Nucl. Sci.* **48**, 1713 (2001).
49. R.A. Reed, P.W. Marshall, C.J. Marshall, R.L. Ladbury, H.S. Kim, L.X. Nguyen, J.L. Barth, and K.A. LaBel, *IEEE Trans. Nucl. Sci.* **47**, 2492 (2000). and references therein.
50. V.V. Ursaki, I.M. Tiginyanu, O. Volciuc, V. Popa, V.A. Skuratov, and H. Morkoç, *Appl. Phys. Lett.* **90**, 161908 (2007).
51. Y. Chimi, A. Iwase, N. Ishikawa, M. Kobiyama, T. Inami, and S. Okuda, *J. Nucl. Mater.* **297**, 355 (2001).
52. Al.L. Efros and V.N. Prigodin, *Appl. Phys. Lett.* **62**, 3013 (1993).
53. A. Fiore, M. Rosetti, B. Alloing, C. Paranthoen, J.X. Chen, L. Geelhaar, and H. Riechert, *Phys. Rev. B* **70**, 205311 (2004).

54. C. Weisbuch and J. Nagle, in: *Science and Engineering of 1D and OD Semiconductor Systems*, ed. by C.M. Sotomayor-Torres and S.P. Beaumont, NATO ASI Series B124 (Plenum, New York, 1990), p. 319.
55. J.-M. Gérard, O. Cabrol, and B. Sermage, *Appl. Phys. Lett.* **68**, 3123 (1996).
56. D. Lacombe, A. Pochet, J.-M. Gérard, and O. Cabrol, *Appl. Phys. Lett.* **70**, 2398 (1997).
57. S.W. Lin, C. Balocco, M. Missous, A.R. Peaker, and A.M. Song, *Phys. Rev. B* **72**, 165302 (2005).
58. W.V. Schoenfeldt, C.-H. Chen, P.M. Petroff, and E.L. Hu, *Appl. Phys. Lett.* **73**, 2935 (1998).
59. P.J. Wellmann, W. Schoenfeld, J.M. Garcia, and P.M. Petroff, *J. Electron. Mater.* **27**, 1030 (1998).
60. N.A. Sobolev, A. Cavaco, M.C. Carmo, M. Grundmann, F. Heinrichsdorff, and D. Bimberg, *Phys. Stat. Solidi (b)*, **224**, 93 (2001).
61. N.A. Sobolev, A. Cavaco, M.C. Carmo, H. Born, M. Grundmann, F. Heinrichsdorff, R. Heitz, A. Hoffmann, and D. Bimberg in: *Physics, Chemistry and Application of Nanostructures*, ed. by V.E. Borisenko *et al.* (World Scientific, Singapore, *et al.* 2001), p. 146.
62. R. Leon, G.M. Swift, B. Magness, W.A. Taylor, Y.S. Tang, K.L. Wang, P. Dowd, and Y.H. Zhang, *Appl. Phys. Lett.* **76**, 2074 (2000).
63. R. Leon, S. Marcinkevicius, J. Siegert, B. Cechavicius, B. Magness, W. Taylor, and C. Lobo, *IEEE Trans. Nucl. Sci.* **49**, 2844 (2002).
64. S. Marcinkevicius, J. Siegert, R. Leon, B. Cechavicius, B. Magness, W. Taylor, and C. Lobo, *Phys. Rev. B* **66**, 235314 (2002).
65. F. Guffarth, R. Heitz, M. Geller, C.M.A. Kapteyn, H. Born, R. Sellin, A. Hoffmann, D. Bimberg, N.A. Sobolev, and M.C. Carmo, *Appl. Phys. Lett.* **82**, 1941 (2003).
66. A. Cavaco, N.A. Sobolev, M.C. Carmo, F. Guffarth, H. Born, R. Heitz, A. Hoffmann, and D. Bimberg, *Phys. Stat. Solidi (c)*, **0**, 1177 (2003).
67. N.V. Baidus, V.K. Vasilev, Yu.A. Danilov, B.N. Zvonkov, P.B. Mokeeva, and N.A. Sobolev, in: *Design of Radiation-hard Semiconductor Devices for Communication Systems and Precision Measurements using Noise Analysis* (TALAM, Nizhni Novgorod, 2003), p. 59 (in Russian).
68. F.P. Korshunov, T.A. Prokhorenko, N.A. Sobolev, and E.A. Kudriavtseva, *Mater. Res. Soc. Symp. Proc.* **262**, 603 (1992).
69. P.C. Sercel, *Phys. Rev. B* **51**, 14532 (1995).
70. X.-Q. Li and Y. Arakawa, *Phys. Rev. B* **56**, 10423 (1997).
71. Y.-L. Chang, I.-H. Tan, Y.-H. Zhang, D. Bimberg, J. Merz, and E. Hu, *J. Appl. Phys.* **74**, 5144 (1993).
72. S. Fafard, *Appl. Phys. Lett.* **76**, 7270 (2000).
73. N. Žurauskiene, S. Ašmontas, A. Dargys, J. Kundrotas, G. Janssen, E. Goovaerts, S. Marcinkevičius, P.M. Koenraad, J.H. Wolter, and R. Leon, *Sol. State Phenomena* **99–100**, 99 (2004).
74. R. Heitz, M. Veith, N.N. Ledentsov, A. Hoffmann, D. Bimberg, M. Ustinov, P.S. Kopev, and Zh.I. Alferov, *Phys. Rev. B* **56**, 10435 (1997).
75. R. Heitz, H. Born, F. Guffarth, O. Stier, A. Schliwa, A. Hoffmann, and D. Bimberg, *Phys. Rev. B* **64**, 241305(R) (2001).
76. N.A. Sobolev, F.P. Korshunov, R. Sauer, K. Thonke, U. König, and H. Presting, *J. Cryst. Growth*, **167**, 502 (1996).
77. Y.C. Chua, E.A. Decuir Jr, M.O. Manasreh, and B.D. Weaver, *Appl. Phys. Lett.* **87**, 091905 (2005).
78. N.A. Sobolev, A. Fonseca, J.P. Leitão, M.C. Carmo, H. Presting, and H. Kibbel, *Phys. Stat. Sol. (c)*, **0**, 1267 (2003).
79. A.V. Gomonnai, Yu.M. Azhniuk, V.V. Lopushansky, I.G. Megela, I.I. Turok, M. Kranjčec, and V.O. Yukhymchuk, *Phys. Rev. B* **65**, 245327 (2002).
80. M. Silvestri and J. Schroeder, *J. Phys.: Condens. Matter* **7**, 8519 (1995).
81. M.B. Huang, J. Zhu, and S. Oktyabrsky, *Nucl. Instr. Methods B* **211**, 505 (2003).
82. A. Fonseca, N.A. Sobolev, J.P. Leitão, E. Alves, M.C. Carmo, N.D. Zakharov, P. Werner, A.A. Tonkikh, and G.E. Cirlin, *J. Luminescence* **121**, 417 (2006).
83. N.A. Sobolev and H. Presting, unpublished.
84. V.I. Gubskaya, P.V. Kuchinskii, and V.M. Lomako, *Radiation Effects* **55**, 35 (1981).
85. A. Meldrum, L.A. Boatner, and R.C. Ewing, *Phys. Rev. Lett.* **88**, 025503 (2002).
86. M.C. Ridgway, G. de M. Azevedo, R.G. Elliman, W. Wesch, C.J. Glover, R. Miller, D.J. Llewellyn, G.J. Foran, J.L. Hansen, and A. Nylandsted Larsen, *Nucl. Instr. Methods B* **242**, 121 (2006).

87. G.A. Kachurin, M.-O. Ruault, A.K. Gutakovsky, O. Kaitasov, S.G. Yanovskaya, K.S. Zhuravlev, and H. Bernas, *Nucl. Instr. Methods B* **147**, 356 (1999).
88. S. Cheylan, N. Langford, and R.G. Elliman, *Nucl. Instr. Methods B* **166–167**, 851 (2000).
89. D. Pacifici, E.C. Moreira, G. Franzo, V. Martorino, F. Priolo, and F. Iacona, *Phys. Rev. B* **65**, 144109 (2002).
90. Y. Chimi, A. Iwase, N. Ishikawa, M. Kobiyama, T. Inami, T. Kambara, and S. Okuda, *Nucl. Instr. Methods B* **245**, 171 (2006).
91. P.G. Piva, R.D. Goldberg, I.V. Mitchell, D. Labrie, R. Leon, S. Charbonneau, Z.R. Wasilewski, and S. Fafard, *Appl. Phys Lett.* **77**, 624 (2000).
92. D. Nie, T. Mei, H.S. Djie, B.S. Ooi, and X.H. Zhang, *Appl. Phys. Lett.* **88**, 251102 (2006).
93. H.S. Djie, T. Mei, and J. Arokiaraj, *Appl. Phys. Lett.* **83**, 60 (2003).
94. Z. Yin, X. Tang, C.-W. Lee, J. Zhao, S. Deny, and M.-K. Chin, *Nanotechnology* **17**, 4664 (2006).
95. N.N. Sirota, F.P. Korshunov, and L.Yu. Raines, in: *Radiation Defects in Semiconductors*, (University Press, Minsk, 1972), p. 259 (in Russian).
96. I.A. Aksenov, N.A. Sobolev, and V.A. Sheraukhov, *Phys. Stat. Sol. (a)*, **123**, K171 (1991).
97. A. Jorio, C. Rejeb, M. Parenteau, C. Carlone, and S.M. Khanna, *J. Appl. Phys.* **74**, 2310 (1993).
98. T. Miyoshi, H. Furukawa, K. Nitta, H. Okuni, and N. Matsuo, *Solid State Commun.* **104**, 451 (1997).
99. T. Ota, K. Maehashi, H. Nakashima, K. Oto, and K. Murase, *Phys. Stat. Sol (b)*, **224**, 169 (2001).
100. V.Ya. Grabovskis, Ya.Ya. Dzenis, A.I. Ekimov, A.I. Kudryavtsev, M.N. Tolstoi, and U.T. Rogulis, *Fiz. Tverd. Tela (Leningrad)* **31**, 272 (1989).
101. A.V. Gomonnai, Yu.M. Azhniuk, M. Kranjčec, A.M. Solomon, V.V. Lopushansky, and I.G. Megela, *Solid State Commun,* **119**, 447 (2001).
102. A.V. Gomonnai, A.M. Solomon, Yu.M. Azhniuk, M. Kranjčec, I.G. Megela, and V.V. Lopushansky, *J. Optoelectron. Adv. Mater.* **3**, 509 (2001).
103. J. Rodríguez-Viejo, H. Mattoussi, J.R. Heine, M.K. Kuno, J. Michel, M.G. Bawendi, and K.F. Jensen, *J. Appl. Phys.* **87**, 8526 (2000).
104. J. Verbert, J. Barjon, E. Monroy, B. Daudin, and B. Sieber, *J. Phys.: Condens. Matter.* **16**, S243 (2004).
105. B. Sieber, *J. Appl. Phys.* **98**, 083520 (2005).
106. B. Sieber, *J. Phys.: Condens. Matter,* **18**, 1033 (2006).
107. B. Sieber, *Phys. Stat. Sol. (c)*, **4**, 1517 (2007).
108. S. Dassonneville, A. Amokrane, B. Sieber, J.-L. Farvacque, B. Beaumont, P. Gibart, J.-D. Ganière, and K. Leifer, *J. Appl. Phys.* **89**, 7966 (2001).
109. *Hydrogen in Semiconductors II (Semiconductors and Semimetals)*, Vol. **61**, ed. by N.H. Nickel *et al.* (Academic Press, San Diego and London, 1999), and references therein.
110. E.C. Le Ru, P.D. Siverns, and R. Murray, *Appl. Phys. Lett.* **77**, 2446 (2000).
111. A.P. Jacob, Q.X. Zhao, M. Willander, F. Ferdos, M. Sadeghi, and S.M. Wang, *J. Appl. Phys.* **92**, 6794 (2002).
112. S. Mazzucato, D. Nardin, M. Capizzi, A. Polimeni, A. Frova, L. Seravalli, and S. Franchi, *Mater. Sci. Eng. C*, **25**, 830 (2005).
113. G. Saint-Girons, A. Lemaître, V. Navarro-Paredes, G. Patriarche, E.V.K. Rao, I. Sagnes, and B. Theys, *J. Cryst. Growth*, **264**, 334 (2004).
114. W. Lu, Y.L. Ji, G.B. Chen, N.Y. Tang, X.S. Chen, S.C. Shen, Q.X. Zhao, and M. Willander, *Appl. Phys. Lett.* **82**, 4300 (2003).
115. Y. Ji, G. Chen, N. Tang, Q. Wang, X.G. Wang, J. Shao, X.S. Chen, and W. Lu, *Appl. Phys. Lett.* **82**, 2802 (2003).
116. M. Gurioli, M. Zamfirescu, A. Vinattieri, S. Sanguinetti, E. Grilli, M. Guzzi, S. Mazzucato, A. Polimeni, M. Capizzi, L. Seravalli, P. Frigeri, and S. Franchi, *J. Appl. Phys.* **100**, 084313 (2006).
117. S.P. Park, J. Tatebayashi, T. Nakaoka, T. Sato, Y.J. Park, and Y. Arakawa, *Jpn. J. Appl. Phys. Part 1*, **43**, 2118 (2004).
118. K.S. Min, K.V. Shcheglov, C.M. Yang, H.A. Atwater, M.L. Brongersma, and A. Polman, *Appl. Phys. Lett.* **69**, 2033 (1996).
119. S.P. Withrow, C.W. White, A. Meldrum, J.D. Budai, D.M. Hembree Jr., and J.C. Barbour, *J. Appl. Phys.* **86**, 396 (1999).

120. S. Cheylan and R.G. Elliman, *Nucl. Instr. Methods, B* **175–177**, 422 (2001).
121. S. Cheylan and R.G. Elliman, *Appl. Phys. Lett.* **78**, 1912 (2001).
122. C.H. Cho, B.-H. Kim, T.-W. Kim, S.-J. Park, N.-M. Park, and G.-Y. Sung, *Appl. Phys. Lett.* **86**, 143107 (2005).
123. N.V. Baidus, A. Chahboun, M.J.M. Gomes, M.I. Vasilevskiy, P.B. Demina, E.A. Uskova, and B.N. Zvonkov, *Appl. Phys. Lett.* **87**, 053109 (2005).
124. A. Chahboun, N.V. Baidus, P.B. Demina, B.N. Zvonkov, M.J.M. Gomes, A. Cavaco, N.A. Sobolev, M.C. Carmo, and M.I. Vasilevskiy, *Phys. Stat. Sol. (a)* **203**, 1348 (2006).
125. A. Chahboun, M.I. Vasilevskiy, N.V. Baidus, A. Cavaco, N.A. Sobolev, M.C. Carmo, E. Alves, and B.N. Zvonkov, *J. Appl. Phys.* **103**, 083548 (2008).
126. H. Lischka, H. Henschel, W. Lennartz, and K.U. Schmidt, *IEEE Trans. Nucl. Sci.* **39**, 423 (1992).
127. B.D. Evans, H.E. Hager, and B.W. Hughlock, *IEEE Trans. Nucl. Sci.* **40**, 1645 (1993).
128. Y.M. Cheng, R.W. Herrick, P.M. Petroff, M.K. Hibbs-Brenner, and R.A. Morgan, *Appl. Phys. Lett.* **67**, 1648 (1995).
129. H. Schone, R.F. Carson, A.H. Paxton, and E.W. Taylor, *IEEE Photonics Technol. Lett.* **9**, 1552 (1997).
130. A.H. Paxton, R.F. Carson, H. Schone, E.W. Taylor, K.D. Choquette, H.Q. Hou, K.L. Lear, and M.E. Warren, *IEEE Trans. Nucl. Sci.* **44**, 1893 (1997).
131. Y.F. Zhao, A.R. Patwary, R.D. Schrimpf, M.A. Neifeld, and K.F. Galloway, *IEEE Trans. Nucl. Sci.* **44**, 1898 (1997).
132. H. Ohyama, E. Simoen, C. Claeys, T. Hakata, T. Kudou, M. Yoneoka, K. Kobayashi, M. Nakabayashi, Y. Takami, and H. Sunaga, *Physica, B* **273–274**, 1031 (1999).
133. A.H. Johnston, T.F. Miyahira, and B.G. Rax, *IEEE Trans. Nucl. Sci.* **48**, 1764 (2001).
134. A.H. Johnston, *IEEE Trans. Nucl. Sci.* **50**, 689 (2003).
135. O. Gilard, *J. Appl. Phys.* **93**, 1884 (2003).
136. A.H. Johnston and T.F. Miyahira, *IEEE Trans. Nucl. Sci.* **51**, 3564 (2004).
137. Ch. Ribbat, R. Sellin, M. Grundmann, D. Bimberg, N.A. Sobolev, and M.C. Carmo, *Electron. Lett.* **37**, 174 (2001).
138. For a review of QW intermixing, see, e.g., *IEEE J. Selected Topics Quant. Electron.* **4**, No. 4 (1998), and references therein.
139. S. Charbonneau, E.S. Koteles, P.J. Poole, J.J. He, G.C. Aers, J. Haysom, M. Buchanan, Y. Feng, A. Delage, F. Yang, M. Davies, R.D. Goldberg, P.G. Piva, and I.V. Mitchell, *IEEE J. Selected Topics Quant. Electron.* **4**, 772 (1998). and references therein.
140. J.P. Reithmaier and A. Forchel, *IEEE J. Selected Topics Quant. Electron.* **4**, 595 (1998). and references therein.
141. P.J. Poole, S. Charbonneau, G.C. Aers, T.E. Jackman, M. Buchanan, M. Dion, R.D. Goldberg, and I.V. Mitchell, *J. Appl. Phys.* **78**, 2367 (1995).
142. L. Fu, P. Lever, H.H. Tan, C. Jagadish, P. Reece, and M. Gal, *Appl. Phys. Lett.* **82**, 2613 (2003).
143. P.G. Piva, S. Charbonneau, I.V. Mitchell, and R.D. Goldberg, *Appl. Phys. Lett.* **68**, 2252 (1996).
144. J. Kundrotas, A. Dargys, G. Valusis, S. Asmontas, K. Köhler, and C. Leroy, *J. Appl. Phys.* **89**, 6007 (2001).
145. E.-M. Pavelescu, A. Gheorghiu, M. Dumitrescu, A. Tukiainen, T. Jouhti, T. Hakkarainen, R. Kudrawiec, J. Andrzejewski, J. Misiewicz, N. Tkachenko, V.D.S. Dhaka, H. Lemmetyinen, and M. Pessa, *Appl. Phys Lett.* **85**, 6158 (2004).
146. I. Jencic, M.W. Bench, I.M. Robertson, and M.A. Kirk, *J. Appl. Phys.* **78**, 974 (1995).
147. H. Leier, A. Forchel, G. Hörcher, J. Hommel, S. Bayer, H. Rothfritz, G. Weimann, and W. Schlapp, *J. Appl. Phys.* **67**, 1805 (1990).
148. K.K. Anderson, J.P. Donnelly, C.A. Wang, J.D. Woodhouse, and H.A. Haus, *Appl. Phys. Lett.* **53**, 1632 (1988).
149. D.V. Forbes, J.J. Coleman, J.L. Klatt, and R.S. Averback, *J. Appl. Phys.* **77**, 3543 (1995).
150. W. Xia, S.A. Pappert, B. Zhu, A.R. Clawson, P.K.L. Yu, S.S. Lau, D.B. Poker, C.W. White, and S.A. Schwarz, *J. Appl. Phys.* **71**, 2602 (1992).
151. S. Charbonneau, P.J. Poole, P.G. Piva, G.C. Aers, E.S. Koteles, M. Fallahi, J.-J. He, J.P. McCaffrey, M. Buchanan, M. Dion, R.D. Goldberg, and I.V. Mitchell, *J. Appl. Phys.* **78**, 3697 (1995).
152. D. Barba, B. Salem, D. Morris, V. Aimez, J. Beauvais, M. Chicoine, and F. Schiettekatte, *J. Appl. Phys.* **98**, 054904 (2005).

153. F. Laruelle, A. Bagchi, M. Tsuchiya, J. Merz, and P.M. Petroff, *Appl. Phys. Lett.* **56**, 1561 (1990).
154. F.E. Prins, G. Lehr, M. Burkard, H. Schweizer, and G.W. Smith, *Nucl. Instr. Methods*, B **80–81**, 827 (1993).
155. D. Nie, T. Mei, X.H. Tang, M.K. Chin, H.S. Djie, and Y.X. Wang, *J. Appl. Phys.* **100**, 046103 (2006).
156. R. Santoprete, P. Kratzer, M. Scheffler, R.B. Capaz, and B. Koiller, *J. Appl. Phys.* **102**, 023711 (2007).
157. T. Surkova, A. Patanè, L. Eaves, P.C. Main, M. Henini, A. Polimeni, A.P. Knights, and C. Jeynes, *J. Appl. Phys.* **89**, 6044 (2001).
158. H.S. Djie, B.S. Ooi, and V. Aimez, *Appl. Phys. Lett.* **87**, 261102 (2005).
159. H.S. Djie, B.S. Ooi, and V. Aimez, *J. Cryst. Growth*, **288**, 40 (2006).
160. S. Barik, H.H. Tan, and C. Jagadish, *Appl. Phys. Lett.* **90**, 093106 (2007).
161. F. Watt, M.B.H. Breese, A.A. Bettiol, and J.A. van Kan, *Materials Today* **10**, 20 (2007).
162. C.G. van de Walle and J. Neugebauer, *Annu. Rev. Mater. Res.* **36**, 179 (2006).
163. Y. Ji, W. Lu, G. Chen, X. Chen, and Q. Wang, *J. Appl. Phys.* **93**, 1208 (2003).
164. P. Lever, H.H. Tan, C. Jagadish, P. Reece, and M. Gal, *Appl. Phys. Lett.* **82**, 2053 (2003).
165. B. Ilahi, B. Salem, V. Aimez, L. Sfaxi, H. Maaref, and D. Morris, *Nanotechnology* **17**, 3707 (2006).
166. L.V. Dao, M. Gal, C. Carmody, H.H. Tan, and C. Jagadish, *J. Appl. Phys.* **88**, 5252 (2000).
167. S. Barik, H.H. Tan, and C. Jagadish, *Appl. Phys. Lett.* **88**, 223101 (2006).
168. J. Zhu, S. Oktyabrsky, and M.B. Huang, *J. Appl. Phys.* **100**, 104312 (2006).
169. H.S. Djie, O. Gunawan, D.-N. Wang, B.S. Ooi, and J.C.M. Hwang, *Phys. Rev. B* **73**, 155324 (2006).
170. D. Nie, T. Mei, C.D. Xu, and J.R. Dong, *Appl. Phys. Lett.* **89**, 131103 (2006).
171. B. Salem, V. Aimez, D. Morris, A. Turala, P. Regreny, and M. Gendry, *Appl. Phys. Lett.* **87**, 241115 (2005).
172. S. Barik, H.H. Tan, and C. Jagadish, *Nanotechnology* **18**, 175305 (2007).
173. J. Cibert, P.M. Petroff, D.J. Werder, S.J. Pearton, A.C. Gossard, and J.H. English, *Appl. Phys. Lett.* **49**, 223 (1986).
174. J. Cibert, P.M. Petroff, G.J. Dolan, S.J. Pearton, A.C. Gossard, and J.H. English, *Appl. Phys. Lett.* **49**, 1275 (1986).
175. Y. Hirayama, S. Tarucha, Y. Suzuki, and H. Okamoto, *Phys. Rev. B* **37**, 2774 (1988).
176. T. Hiramoto, K. Hirakawa, Y. Iye, and T. Ikoma, *Appl. Phys. Lett.* **54**, 2103 (1989).
177. F. Laruelle, P. Hu, R. Simes, R. Kubena, W. Robinson, J. Merz, and P.M. Petroff, *J. Vac. Sci. Technol. B* **7**, 2034 (1989).
178. F. Laruelle, Y.P. Hu, R. Simes, W. Robinson, J. Merz, and P.M. Petroff, *Surf. Sci.* **228**, 306 (1990).
179. H. Leier, A. Forchel, B.E. Maile, G. Mayer, J. Hommel, G. Weimann, and W. Schlapp, *Appl. Phys. Lett.* **56**, 48 (1990).
180. K. Ensslin and P.M. Petroff, *Phys. Rev. B* **41**, 12307 (1990).
181. P.M. Petroff, Y.J. Li, Z. Xu, W. Beinstingl, S. Sasa, and K. Ensslin, *J. Vac. Sci. Technol. B* **9**, 3074 (1991).
182. F.E. Prins, G. Lehr, H. Schweizer, and G.W. Smith, *Appl. Phys. Lett.* **63**, 1402 (1993).
183. F.E. Prins, F. Adler, G. Lehr, S.Y. Nikitin, H. Schweizer, and G.W. Smith, *J. Phys. IV*, **3**, 115 (1993).
184. Z. Xu, M. Wassermeier, Y.J. Li, and P.M. Petroff, *Appl. Phys. Lett.* **60**, 586 (1992).
185. N. Marthur and P. Littlewood, *Nature Materials* **3**, 207 (2004).
186. R.M. Langford, *Microsc. Res. and Techn.* **69**, 538 (2006).
187. A.A. Tseng, *Small*, **1**, 924 (2005).
188. For a recent update, see, e.g., *Proc. 32nd International Conference on Micro- and Nano-Engineering, Microelectronis Engineering* **84**, issues 5–8 (2007); *Ion-Beam-Based Nanofabrication*, ed. by D. Ila *et al.* (*Mater. Res. Soc. Symp. Proc.*, Vol. **1020**, Warrendale, PA, 2007).
189. J. Gierak, A. Madouri, A.L. Biance, E. Bourhis, G. Patriarche, C. Ulysse, D. Lucot, X. Lafosse, L. Auvray, L. Bruchhaus, and R. Jede, *Microelectron. Engin.* **84**, 779 (2007).
190. M. Navez, D. Chaperot, and C. Sella, *Compt. Rend. Acad. Sci.* **254**, 240 (1962).
191. W.L. Chan and E. Chason, *J. Appl. Phys.* **101**, 121301 (2007).
192. S. Facsko, T. Dekorsy, C. Koerdt, C. Trappe, H. Kurz, A. Vogt, and H.L. Hartnagel, *Science* **285**, 1551 (1999).

193. F. Frost, A. Schindler, and F. Bigl, *Phys. Rev. Lett.* **85**, 4116 (2000).

194. R. Gago, L. Vazquez, R. Cuerno, M. Varela, C. Ballesteros, and J.M. Albella, *Appl. Phys. Lett.* **78**, 3316 (2001).

195. B. Ziberi, F. Frost, and B. Rauschenbach, *Appl. Phys. Lett.* **88**, 173115 (2006).

196. F. Frost and B. Rauschenbach, *Appl. Phys. A* **77**, 1 (2003).

197. S. Facsko, H. Kurz, and T. Dekorsy, *Phys. Rev. B* **63**, 165329 (2001).

198. M. Lu, X.J. Yang, S.S. Perry, and J.W. Rabalais, *Appl. Phys. Lett.* **80**, 2096 (2002).

199. Y.S. Katharria, S. Kumar, P.S. Lakshmy, D. Kanjilal, and A.T. Sharma, *J. Appl. Phys.* **102**, 044301 (2007).

200. J. Muñoz-García, L. Vázquez, R. Cuerno, J.A. Sánchez-García, M. Castro, and R. Gago, in: *Lecture Notes on Nanoscale Science and Technology*, ed. by Z.M. Wang (Springer, Heidelberg, 2007). in press (arXiv:0706.2625v1 [cond-mat.mtrl-sci]).

201. P.I. Gaiduk, A. Nylandsted Larsen, J. Lundsgaard Hansen, A.V. Mudryj, M.P. Samtsov, and A.N. Demenschenok, *Appl. Phys. Lett.* **79**, 4025 (2001).

202. A.V. Dvurechenskii, V.A. Zinovyev, V.A. Kudryavtsev, and J.V. Smagina, *JETP Lett.* **72**, 131 (2000).

203. A.V. Dvurechenskii, V.A. Zinoviev, and Zh.V. Smagina, *JETP Lett*, **74**, 267 (2001).

204. A.V. Dvurechenskii, V.A. Zinovyev, V.A. Kudryavtsev, Zh.V. Smagina, P.L. Novikov, and S.A. Teys, *Phys. Low-Dim. Struct.* **1–2**, 303 (2002).

205. A.V. Dvurechenskii, J.V. Smagina, V.A. Zinovyev, S.A. Teys, and A.K. Gutakovskii, *Intern. J. Nanoscience*, **3**, 19 (2004).

206. N.P. Stepina, A.V. Dvurechenskii, V.A. Ambrister, J.V. Smagina, V.A. Volodin, A.V. Nenashev, J.P. Leitão, M.C. do Carmo, and N.A. Sobolev, to be published.

207. C.W. White, J.D. Budai, S.P. Withrow, J.G. Zhu, E. Sonder, R.A. Zuhr, A. Meldrum, D.M. Hembree Jr, D.O. Henderson, and S. Prawer, *Nucl. Instr. Methods B* **141**, 228 (1998).

208. For a recent update see, e.g., *Proceeding of the 15th International Conference on Ion Beam Modification of Materials, Nucl. Instr. Methods B* **257**, issues 1–2 (2007).

209. A. Pérez-Rodríguez, B. Garrido, C. Bonafos, M. López, O. González-Varona, J.R. Monrante, J. Montserrat, and R. Rodríguez, *J. Mater. Sci.: Mater. in Electronics.* **10**, 385 (1999).

210. A. Meldrum, R.F. Haglund Jr, L.A. Boatner, and C.W. White, *Adv. Mater.* **13**, 1431 (2001). and references therein.

211. P. Mazzoldi and G. Mattei, *Riv. Nuovo Cimento.* **28**, 1 (2005). and references therein.

212. H. Bernas and R. Espiau de Lamaestre, in: *Radiation Effects in Solids*, Ed. by K.E. Sickafus *et al.* (Springer, Dordrecht, 2007). Chap. 16, p. 449, and references therein.

213. B. Pécz, H. Weishart, V. Heer, and L. Tóth, *Appl. Phys. Lett.* **82**, 46 (2003).

214. Yu.N. Parkhomenko, A.I. Belogorokhov, N.N. Gerasimenko, A.V. Irzhak, and M.G. Lisachenko, *Semiconductors* **38**, 572 (2004).

215. C. Flytzanis, F. Hache, M.C. Klein, D. Richard, and P. Roussignol, in: *Semiconductor and Metal Crystallites in Dielectrics*, Ed. by E. Wolf (Elsevier Science, Amsterdam, 1991), p. 321.

216. V. Beyer and J. von Borany, in: *Materials for Information Technology. Devices, Interconnects and Packaging*, Ed. by E. Zschech *et al.* (Springer, London, 2005), p. 139.

217. P. Normand, E. Kapetanakis, P. Dimitrakis, D. Skarlatos, K. Beltsios, D. Tsoukalas, C. Bonafos, G. Ben Assayag, N. Cherkashin, A. Claverie *et al.*, *Nucl. Instr. Methods, B* **216**, 228 (2004).

218. M.P. Petkov, L.D. Bell, and H.A. Atwater, *IEEE Trans. Nucl. Sci.* **51**, 3822 (2004).

219. L. Pavesi, L. Dal Negro, C. Mozzoleni, G. Granzò, and F. Priolo, *Nature* **408**, 440 (2000).

220. P.M. Fauchet, J. Ruan, H. Chen, L. Pavesi, L. Dal Negro, M. Cazzaneli, R.G. Elliman, N. Smith, M. Samoc, and B. Luther-Davies, *Opt. Mater.* **27**, 745 (2005).

221. Y. Kanemitsu, H. Tanaka, Y. Fukunishi, T. Kushida, K.S. Min, and H.A. Atwater, *Phys. Rev. B* **62**, 5100 (2000).

222. C.W. White, J.D. Budai, J.G. Zhu, S.P. Withrow, and M.J. Aziz, *Appl. Phys. Lett.* **68**, 2389 (1996).

223. J.A. Wolk, K.M. Yu, E.D. Bourret-Courchesne, and E. Johnson, *Appl. Phys. Lett.* **70**, 2268 (1997).

224. E. Borsella, S. Dal Toé, G. Mattei, C. Maurizio, P. Mazzoldi, A. Saber, G. Battaglin, E. Cattaruzza, F. Gonella, A. Quaranta, and F. D'Acapito, *Mater. Sci. Eng. B* **82**, 148 (2001).

225. E. Borsella, M. Garcia, G. Mattei, C. Maurizio, P. Mazzoldi, E. Cattaruzza, F. Gonella, G. Battaglin, A. Quaranta, and F. D'Acapito, *J. Appl. Phys.* **90**, 4467 (2001).

226. E. Borsella, C. de Julián Fernández, M. Garcia, G. Mattei, C. Maurizio, P. Mazzoldi, S. Padovani, C. Sada, G. Battaglin, E. Cattaruzza *et al.*, , *Nucl. Instr. Methods, B* **191**, 447 (2002).

227. A. Tchebotareva, J.L. Brebner, S. Roorda, P. Desjardins, and C.W. White, *J. Appl. Phys.* **92**, 4664 (2002).

228. I.D. Desnica-Frankovic, U.V. Desnica, P. Dubcek, M. Buljan, S. Bernstorff, H. Karl, I. Grosshans, and B. Stritzker, *J. Appl. Crystallography*, **36**, 439 (2003).

229. R. Espiau de Lamaestre, H. Bernas, Ch. Ricolleau, and J. Majimel, *Nucl. Instr. Methods B* **242**, 214 (2006).

230. R. Espiau de Lamaestre, J. Majimel, F. Jomard, and H. Bernas, *J. Phys. Chem. B* **109**, 19148 (2005).

231. Y. Kanemitsu, H. Matsubara, and C.W. White, *Appl. Phys. Lett.* **81**, 535 (2002).

232. A. Ishizumi, K. Matsuda, T. Saiki, C.W. White, and Y. Kanemitsu, *Appl. Phys. Lett.* **87**, 133104 (2005).

233. P.I. Gaiduk, A. Nylandsted Larsen, and J. Lundsgaard Hansen, *Appl. Phys. Lett.* **79**, 3494 (2001).

234. W.M. Tsang, S.P. Wong, and J.K.N. Lindner, *Proc. SPIE* **6037**, 9 (2006).

235. P.K. Kuiri, H.P. Lenka, J. Ghatak, G. Sahu, B. Joseph, and D.P. Mahapatra, *J. Appl. Phys.* **102**, 024315 (2007).

236. X. Xiang, X.T. Zu, S. Zhu, Q.M. Wei, C.F. Zhang, K. Sun, and L.M. Wang, *Nanotechnology* **17**, 2636 (2006).

237. H. Amekura, N. Umeda, Y. Sakuma, N. Kishimoto, and Ch. Buchal, *Appl. Phys. Lett.* **87**, 013109 (2005).

238. C. Marques, N. Franco, L.C. Alves, R.C. da Silva, E. Alves, G. Safran, and C.J. McHargue, *Nucl. Instr. Methods, B* **257**, 515 (2007).

239. R. Espiau de Lamaestre and H. Bernas, *J. Appl. Phys.* **98**, 104310 (2005).

240. R. Espiau de Lamaestre, H. Bernas, D. Pacifici, G. Franzó, and F. Priolo, *Appl. Phys. Lett.* **88**, 181115 (2006).

241. G. Mattei, *Nucl. Instr. Methods, B* **191**, 323 (2002).

242. P. Kluth, B. Hoy, B. Johannessen, S.G. Dunn, G.J. Foran, and M.C. Ridgway, *Appl. Phys. Lett.* **89**, 153118 (2006).

243. G. Faraci, A.R. Pennisi, F. Zontone, B. Li, and I. Petrov, *Phys. Rev. B* **74**, 235436 (2006). and references therein.

244. G. Faraci, A.R. Pennisi, and F. Zontone, *Phys. Rev. B* **76**, 035423 (2007) and references therein.

245. P. Mazzoldi and G. Mattei, *Phys. Stat. Sol. (a)* **204**, 621 (2007).

246. A.L. Stepanov and I.B. Khaibullin, *Rev. Adv. Mater. Sci.* **9**, 109 (2005). and references therein.

247. R.A. Ganeev, A.I. Ryasnyansky, A.L. Stepanov, C. Marques, R.C. da Silva, and E. Alves, *Opt. Commun.* **253**, 205 (2005).

248. C. Marques, R.C. da Silva, A. Wemans, M.J.P. Maneira, A. Kozanecki, and E. Alves, *Nucl. Instr. Methods, B* **242**, 104 (2006).

249. C. Marques, N. Franco, L.C. Alves, R.C. da Silva, and E. Alves, *Surf. Coat. Technol.* **201**, 8190 (2007).

250. See, e.g., K. Gamo, *Semi Cond. Sci. Technol.* **8**, 1118 (1993).

251. N. Hayashi, T. Toriyama, M. Yamashiro, T. Moriwaki, Y. Oguri, K. Fukuda, I. Sakamoto, and M. Hasegawa, *Nucl. Instr. Methods B* **257**, 25 (2007).

252. K.H. Heinig, T. Müller, B. Schmidt, M. Strobel, and W. Möller, *Appl. Phys. A* **77**, 17 (2003). and references therein.

253. I.M. Lifshitz and V.V. Slyozov, *J. Phys. Chem. Solids* **19**, 35 (1961).

254. C. Wagner, *Z. Elektrochem.* **65**, 581 (1961).

255. M. Strobel, K.H. Heinig, W. Moeller, A. Meldrum, D.S. Zhou, C.W. White, and R.A. Zuhr, *Nucl. Instr. Meth. Phys. Res. B* **147**, 343 (1999).

256. A. Meldrum, S.J. Zinkle, L.A. Boatner, and R.C. Ewing, *Nature*, **395**, 56 (1998).

257. M. Strobel, K.H. Heinig, and W. Möller, *Phys. Rev. B* **64**, 245422 (2001).

258. G. Rizza, H. Cheverry, T. Gacoin, A. Lamasson, and S. Henry, *J. Appl. Phys.* **101**, 014321 (2007).

259. P. Kluth and M.C. Ridgway, *Nucl. Instr. Methods, B* **242**, 458 (2006).

260. P. Kluth, B. Johannessen, G.J. Foran, D.J. Cookson, S.M. Kluth, and M.C. Ridgway, *Phys. Rev. B* **74**, 014202 (2006).

261. G.C. Rizza, M. Strobel, K.H. Heinig, and H. Bernas, *Nucl. Instr. Methods, B* **178**, 78 (2001).

262. B. Schmidt, K.-H. Heinig, L. Röntzsch, T. Müller, K.-H. Stegemann, and E. Votintseva, *Nucl Instr. Methods B* **242**, 146 (2006).

263. G. Mattei, G. De Marchi, C. Maurizio, P. Mazzoldi, C. Sada, V. Bello, and G. Battaglin, *Phys. Rev. Lett.* **90**, 085502 (2003).

264. X.-W. Du, M. Takeguchi, M. Tanaka, and K. Furuya, *Appl. Phys. Lett.* **82**, 1108 (2003).

265. E. Valentin, H. Bernas, C. Ricolleau, and F. Creuzet, *Phys. Rev. Lett.* **86**, 99 (2001).

266. T. Mohanty, P.V. Satyam, and D. Kanjilal, *J. Nanosci. Nanotechnol.* **6**, 2554 (2006).

267. A. Meldrum, L.A. Boatner, C.W. White, and R.C. Ewing, *Mater. Res. Innov.* **3**, 190 (2000). and references therein.

268. See, eg., T.M. Buck, K.A. Pickar, J.M. Poate, and C.-M. Hsieh, *Appl. Phys. Lett.* **21**, 485 (1972).

269. R.E. Hurley, H.S. Gamble, and S. Suder, in: *Nanostructured and Advanced Materials for Applications in Sensor, Optoelectronic and Photovoltaic Technology*, NATO Science Series, Vol. **204** (Springer, Netherlands, 2005), Pt 2, p. 299, and references therein.

270. D.M. Follstaedt, S.M. Myers, G.A. Petersen, and J.W. Medernach, *J. Electron. Mater.* **25**, 151 (1996).

271. P.I. Gaiduk, J. Lundsgaard Hansen, A. Nylandsted Larsen, and E.A. Steinman, *Phys. Rev. B* **67**, 235310 (2003).

272. P.I. Gaiduk, A. Nylandsted Larsen, J. Lundsgaard Hansen, E. Wendler, and W. Wesch, *Phys. Rev. B* **67**, 235311 (2003).

273. U. Kaiser, D.A. Muller, J.L. Grazul, A. Chuvilin, and M. Kawasaki, *Nature Materials* **1**, 102 (2002).

274. R. Giulian, P. Kluth, L.L. Araujo, D.J. Llewellyn, and M.C. Ridgway, *Appl. Phys. Lett.* **91**, 093115 (2007).

275. A. Meldrum, C.W. White, L.A. Boatner, I.M. Anderson, R.A. Zuhr, E. Sonder, J.D. Budai, and D.O. Henderson, *Nucl. Instr. Methods B* **148**, 957 (1999).

276. A. Meldrum, E. Sonder, R.A. Zuhr, I.M. Anderson, J.D. Budai, C.W. White, L.A. Boatner, and D.O. Henderson, *J. Mater. Res.* **14**, 4502 (1999).

277. A. Meldrum, S. Honda, C.W. White, R.A. Zuhr, and L.A. Boatner, *J. Mater. Res.* **16**, 2670 (2001).

278. F. Ren, C.Z. Jiang, Y.H. Wang, Q.Q. Wang, and J.B. Wang, *Nucl. Instr. Methods B* **245**, 427 (2006).

279. F. Ren, C.Z. Jiang, G.X. Cai, Q. Fu, and Y. Shi, *Nucl. Instr. Methods B* **262**, 201 (2007).

280. J. Biskupek, A. Chuvilin, U. Kaiser, O. Picht, and W. Wesch, In-situ growth of MnAs nanocrystals in Si studied by transmission electron microscopy, in: *Proc. Internat. Microscopy Conf.* (Sapporo, Japan, 2006).

281. J.I. Martín, J. Nogués, K. Liuc, J.L. Vicente, and I.K. Schuller, *J. Magn. Magn. Mater.* **256**, 449 (2003).

282. J.M. Kikkawa, J.J. Baumberg, D.D. Awschalom, D. Leonard, and P.M. Petroff, *Phys. Rev. B* **50**, 2003 (1994).

283. J.C.P. Chang, N. Otsuka, E.S. Harmon, M.R. Melloch, and J.M. Woodall, *Appl. Phys. Lett.* **65**, 2801 (1994).

284. J. Shi, J.M. Kikkawa, R. Proksch, T. Schäffer, D.D. Awschalom, G. Medeiros-Ribeiro, and P.M. Petroff, *Nature* **377**, 707 (1995).

285. J. Shi, S. Gider, K. Babcock, and D.D. Awschalom, *Science* **271**, 937 (1996).

286. J. Shi, J.M. Kikkawa, D.D. Awschalom, G. Medeiros-Ribeiro, P.M. Petroff, and K. Babcock, *J. Appl. Phys.* **79**, 5296 (1996).

287. J. Shi, D.D. Awschalom, P.M. Petroff, and K. Babcock, *J. Appl. Phys.* **81**, 4331 (1997).

288. J.P. Zhang, A.K. Cheetham, K. Sun, J.S. Wu, K.H. Kuo, J. Shi, and D.D. Awschalom, *Appl. Phys. Lett.* **71**, 143 (1997).

289. P.J. Wellmann, J.M. Garcia, J.-L. Feng, and P.M. Petroff, *Appl. Phys. Lett.* **71**, 2532 (1997).

290. P.J. Wellmann, J.M. Garcia, J.-L. Feng, and P.M. Petroff, *Appl. Phys. Lett.* **73**, 3291 (1998).

291. K. Ando, A. Chiba, and H. Tanoue, *Appl. Phys. Lett.* **73**, 387 (1998).

292. Sh.U. Yuldashev, Y. Shon, Y.H. Kwon, D.J. Fu, D.Y. Kim, H.J. Kim, T.W. Kang, and X. Fan, *J. Appl. Phys.* **90**, 3004 (2001).

293. A. Serres, G. Benassayag, M. Respaud, C. Armand, J.C. Pesant, A. Mari, Z. Liliental-Weber, and A. Claverie, *Mater. Sci. Eng. B* **101**, 119 (2003).

294. A. Serres, M. Respaud, G. BenAssayag, J.-C. Pesant, C. Armand, and A. Claverie, *J. Magn. Magn. Mater.* **295**, 183 (2005).

295. W. Chunhua, C. Yonghai, G. Yu, and W. Zhanguo, *J. Cryst. Growth* **268**, 12 (2004).

296. Y. Shon, Y.S. Park, K.J. Chung, D.J. Fu, D.Y. Kim, H.S. Kim, H.J. Kim, T.W. Kang, Y. Kim, X.J. Fan, and Y.J. Park, *J. Appl. Phys.* **96**, 7022 (2004).

297. Yu.A. Danilov, A.V. Kruglov, M. Behar, M.C. dos Santos, L.G. Pereira, and J.E. Schmidt, *Phys. Sol. State* **47**, 1626 (2005).

298. O.D.D. Couto Jr, M.J.S.P. Brasil, F. Iikawa, C. Giles, C. Adriano, J.R.R. Bortoleto, M.A.A. Pudenzi, H.R. Gutierrez, and I. Danilov, *Appl. Phys. Lett.* **86**, 071906 (2005).

299. M. Kasai, J. Yanagisawa, H. Tanaka, S. Hasegawa, H. Asahi, K. Gamo, and Y. Akasaka, *Nucl. Instr. Methods B* **242**, 240 (2006).

300. H. Okamoto, *Bull. Alloy Phase Diagrams* **10**, 549 (1989).

301. N.A. Sobolev, M.A. Oliveira, V.S. Amaral, A. Neves, M.C. Carmo, W. Wesch, O. Picht, E. Wendler, U. Kaiser, and J. Heinrich, *Mater. Sci. Eng. B* **126**, 148 (2006).

302. Th. Hartmann, M. Lampalzer, W. Stolz, W. Heimbrodt, H.-A. Krug von Nidda, A. Loidl, and L. Svistov, *Physica E,* **13**, 572 (2002).

303. S. Honda, I. Sakamoto, B. Shibayama, T. Sugiki, K. Kawai, and M. Nawate, *J. Appl. Phys.* **95**, 7396 (2004).

304. V.A. Kulbachinskii, R.A. Lunin, P.V. Gurin, N.S. Perov, P.M. Sheverdyaeva, and Yu.A. Danilov, *J. Magn. Magn. Mater.* **300**, e20 (2006).

305. H. Akinaga, S. Miyanishi, K. Tanaka, W. Van Roy, and K. Onodera, *Appl. Phys. Lett.* **76**, 97 (2000).

306. M.V. Sapozhnikov, A.A. Fraerman, S.N. Vdovichev, B.A. Gribkov, S.A. Gusev, A.Yu. Klimov, V.V. Rogov, J. Chang, H. Kim, H.C. Koo, S.-H. Han, and S.H. Chun, *Appl. Phys. Lett.* **91**, 062513 (2007).

307. M. Sawicki, K.Y. Wang, K.W. Edmonds, R.P. Campion, A.W. Rushforth, C.T. Foxon, B.L. Gallagher, and T. Dietl, *J. Magn. Magn. Res.* **310**, 2126 (2007).

308. K.Y. Wang, M. Sawicki, K.W. Edmonds, R.P. Campion, A.W. Rushforth, A.A. Freeman, C.T. Foxon, B.L. Gallagher, and T. Dietl, *Appl. Phys. Lett.* **88**, 022510 (2006).

309. K. Sato, H. Katayama-Yoshida, and P.H. Dederichs, *Jpn. J. Appl. Phys.* **44**, L948 (2005).

310. S. Halm, G. Bacher, E. Schuster, W. Keune, M. Sperl, J. Puls, and F. Henneberger, *Appl. Phys. Lett.* **90**, 051916 (2007).

311. H. Shimizu, M. Miyamura, and M. Tanaka, *Appl. Phys. Lett.* **78**, 1523 (2001).

312. M. Tanaka, H. Shimizu, and M. Miyamura, *J. Cryst. Growth,* **227–228**, 839 (2001).

313. N. Theodoropoulou, A.F. Hebard, M.E. Overberg, C.R. Abernathy, S.J. Pearton, S.N.G. Chu, and R.G. Wilson, *Phys. Rev. Lett.* **89**, 107203 (2002).

314. Y. Shon, S. Lee, H.C. Jeon, S.-W. Lee, D.Y. Kim, T.W. Kang, E.K. Kim, D.J. Fu, X.J. Fan, C.S. Yoon, and C.K. Kim, *Appl. Phys. Lett.* **88**, 232511 (2006).

315. Y. Shon, S.-W. Lee, C.S. Park, S. Lee, H.C. Jeon, D.Y. Kim, T.W. Kang, C.S. Yoon, E.K. Kim, and J.J. Lee, *Appl. Surf. Sci.* (2007). doi:10.1016/ j.apsusc.2007.06.020.

316. N. Theodoropoulou, M.E. Overber, S.N.G. Chu, A.F. Hebard, C.R. Abernathy, R.G. Wilson, J.M. Zavada, K.P. Lee, and S.J. Pearton, *Phys. Stat. Sol. (b),* **228**, 337 (2001).

317. N. Theodoropoulou, A.F. Hebard, M.E. Overberg, C.R. Abernathy, S.J. Pearton, S.N.G. Chu, and R.G. Wilson, *Appl. Phys. Lett.* **78**, 3475 (2001).

318. W. Kim, H.J. Kang, S.K. Noh, J. Song, and C.S. Kim, *J. Magn. Magn. Mater.* **310**, e729 (2007).

319. J. Biskupek, U. Kaiserb, H. Lichte, A. Lenk, T. Gemming, G. Pasold, and W. Witthuhn, *J. Magn. Magn. Mater.* **293**, 924 (2005).

320. M.R. McCartney and D.J. Smith, *Annu. Rev. Mater. Res.* **37**, 729 (2007).

321. H. Lichte, P. Formanek, A. Lenk, M. Linck, C. Matzeck, M. Lehmann, and P. Simon, *Annu. Rev. Mater. Res.* **37**, 539 (2007).

322. L. Ottaviano, M. Passacantando, S. Picozzi, A. Continenza, R. Gunnella, A. Verna, G. Bihlmayer, G. Impellizzeri, and F. Priolo, *Appl. Phys. Lett.* **88**, 061907 (2006).

323. A. Verna, L. Ottaviano, M. Passacantando, S. Santucci, P. Picozzi, F. D'Orazio, F. Lucari, M. De Biase, R. Gunnella, M. Berti, A. Gasparotto, G. Impellizzeri, and F. Priolo, *Phys. Rev. B* **73**, 195207 (2006).

324. F. D'Orazio, F. Lucari, M. Passacantando, L. Ottaviano, A. Verna, G. Impellizzeri, and F. Priolo, *J. Magn. Magn. Mater.* **310**, 2150 (2007).

325. L. Ottaviano, M. Passacantando, A. Verna, P. Parisse, S. Picozzi, G. Impellizzeri, and F. Priolo, *Phys. Stat. Sol. (a)*, **204**, 136 (2007).

326. *Binary Alloy Phase Diagrams*, 2nd edn, Vol. **2**, ed. by T.B. Massalski (American Society for Metals, Metals Park, Ohio, 1990).

327. L. Ottaviano, P. Parisse, M. Passacantando, S. Picozzi, A. Verna, G. Impellizzeri, and F. Priolo, *Surf. Sci.* **600**, 4723 (2006).

328. S. Sugahara, K.L. Lee, S. Yada, and M. Tanaka, *Jpn. J. Appl. Phys.* **44**, L1426 (2005).

329. A. Verna, F. D'Orazio, L. Ottaviano, M. Passacantando, F. Lucari, G. Impellizzeri, and F. Priolo, *Phys. Stat. Sol. (a)* **204**, 145 (2007).

330. A. Verna, L. Ottaviano, M. Passacantando, S. Santucci, P. Picozzi, F. D'Orazio, F. Lucari, M. De Biase, R. Gunnella, M. Berti, A. Gasparotto, G. Impellizzeri, and F. Priolo, *Phys. Rev. B* **74**, 085204 (2006).

331. D.E. Nikonov and G.I. Bourianoff, *IEEE Trans. Nanotechnol.* **4**, 206 (2005).

332. J. Chen, K.L. Wang, and K. Galatsis, *Appl. Phys. Lett.* **90**, 012501 (2007).

333. R. Venugopal, B. Sundaravel, W.Y. Cheung, I.H. Wilson, F.W. Wang, and X.X. Zhang, *Phys. Rev. B* **65**, 014418 (2001).

334. R. Venugopal, B. Sundaravel, I.H. Wilson, F.W. Wang, and X.X. Zhang, *J. Appl. Phys.* **91**, 1410 (2002).

335. S. Amirthapandian, B.K. Panigrahi, M. Rajalakshmi, S. Kalavathi, B. Sundaravel, D. Datta, and A. Narayanasamy, *Nucl. Instr. Methods B* **244**, 52 (2006).

336. I.T. Yoon, C.J. Park, S.W. Lee, T.W. Kang, D.J. Fu, and X.J. Fan, *Solid State Commun.* **140**, 185 (2006).

337. L. Miglio and F. d'Heurle, *Silicides: Fundamentals and Applications* (ISBN 981-02-4452-5, 2000).

338. *Properties of Metal Silicides*, ed. by K. Maex and M. van Rossum (INSPEC, London, 1995).

339. V.E. Borisenko. *Semiconducting Silicides*, (Springer-Verlag, ISBN 3-540-66111-5, 2000).

340. Q. Zhang, M. Tanaka, M. Takeguchi, and K. Furuya, *Surf. Sci.* **507–510**, 453 (2002).

341. G. Ctistis, U. Deffke, J.J. Paggel, and P. Fumagalli, *J. Magn. Magn. Mater.* **240**, 420 (2002).

342. A. Mogilatenko, Ph.D. thesis (Technische Universität Chemnitz, 2003).

343. D. Bloch, V. Jaccarino, J. Voiron, and J.H. Wernick, *Phys. Lett.* **51A**, 259 (1975).

344. Y. Ishikawa, K. Tajima, D. Bloch, and M. Roth, *Solid State Commun.* **19**, 525 (1976).

345. Y. Ishikawa, Y. Noda, C. Fincher, and G. Shirane, *Phys. Rev. B* **25**, 254 (1982); *Phys. Rev. B* **16**, 4956 (1977).

346. Y. Ishikawa, G. Shirane, J.A. Tarvin, and M. Kohgi, *Phys. Rev. B* **16**, 4956 (1977).

347. K. Schwinge, C. Müller, A. Mogilatenko, J.J. Paggel, and P. Fumagalli, *J. Appl. Phys.* **97**, 103913 (2005).

348. U. Gottlieb, A. Sulpice, B. Lambert-Andron, and O. Laborde, *J. Alloys and Compounds* **361**, 13 (2003).

349. A. Sulpice, U. Gottlieb, M. Affronte, and O. Laborde, *J. Magn. Magn. Mater.* **272–276**, 519 (2004).

350. G.H. Lander, P.J. Brown, and J.B. Forsyth, *Proc. Phys. Soc.* **91**, 332 (1967).

351. P.J. Brown, J.B. Forsyth, V. Nunez, and F. Tasset, *J. Phys.: Cond. Mat.* **4**, 10025 (1992).

352. P.J. Brown and J.B. Forsyth, *J. Phys.: Cond. Mat.* **7**, 7619 (1995).

353. E.N. Babanova, F.A. Sidorenko, and P.V. Geld, *Sov. Phys. Solid State* **17**, 616 (1975).

354. S. Tomiyoshi and H. Watanabe, *J. Phys. Soc. Japan*, **39**, 295 (1975).

355. L. Liu, N. Chen, S. Song, Z. Yin, F. Yang, C. Chaia, S. Yanga, and Z. Liu, *J. Cryst. Growth*, **273**, 458 (2005).

356. M. Bolduc, C. Awo-Affouda, A. Stollenwerk, M.B. Huang, F.G. Ramos, G. Agnello, and V.P. LaBella, *Phys. Rev. B* **71**, 033302 (2005).

357. H. Nakayama, H. Ohta, and E. Kulatov, *Physica, B* **302–303**, 419 (2001).

358. H.-M. Kim, N.M. Kim, C.S. Park, S.U. Yuldashev, T.W. Kang, and K.S. Chung, *Chem. Mater.* **15**, 3964 (2003).

359. F.M. Zhang, Y. Zeng, J. Gao, X.C. Liu, X.S. Wu, and Y.W. Du, *J. Magn. Magn. Mater.* **282**, 216 (2004).

360. F.M. Zhang, X.C. Liu, J. Gao, X.S. Wu, Y.W. Du, H. Zhu, J.Q. Xiao, and P. Chen, *Appl. Phys. Lett.* **85**, 786 (2004).

361. H.M. Kim, T.W. Kang, and K.S. Chung, *J. Ceram. Proc. Res.* **5**, 238 (2004).

362. C. Awo-Affouda, M. Bolduc, M.B. Huang, F. Ramos, K.A. Dunn, B. Thiel, G. Agnello, and V.P. LaBella, *J. Vac. Sci. Technol. A* **24**, 1644 (2006).

363. S. Zhou, K. Potzger, G. Zhang, A. Mücklich, F. Eichhorn, N. Schell, R. Grötzschel, B. Schmidt, W. Skorupa, M. Helm, J. Fassbender, and D. Geiger, *Phys. Rev. B* **75**, 085203 (2007).

364. A. Wolska, K. Lawniczak-Jablonska, M. Klepka, M.S. Walczak, and A. Misiuk, *Phys. Rev. B* **75**, 113201 (2007).

365. P. Kratzer, S.J. Hashemifar, H. Wu, M. Hortamani, and M. Scheffler, *J. Appl. Phys.* **101**, 081725 (2007).

366. H. Wu, P. Kratzer, and M. Scheffler, *Phys. Rev. Lett.* **98**, 117202 (2007).

367. O. Picht, W. Wesch, J. Biskupek, U. Kaiser, M.A. Oliveira, A. Neves, and N.A. Sobolev, *Nucl. Instr. Methods, B* **257**, 90 (2007).

368. N.A. Theodoropoulou, A.F. Hebard, D.P. Norton, J.D. Budai, L.A. Boatner, J.S. Lee, Z.G. Khim, Y.D. Park, M.E. Overberg, S.J. Pearton, and R.G. Wilson, *Solid-State. Electron.* **47**, 231 (2003).

369. D.P. Norton, M.E. Overberg, S.J. Pearton, K. Pruessner, J.D. Budai, L.A. Boatner, M.F. Chisholm, J.S. Lee, Z.G. Khim, Y.D. Park, and R.G. Wilson, *Appl. Phys. Lett.* **82**, 5488 (2003).

370. W.K. Choi, B. Angadi, H.C. Park, J.H. Lee, J.H. Song, and R. Kumar, *Adv. Sci. Technol.* **52**, 42 (2006).

371. R.P. Borges, J.V. Pinto, R.C. da Silva, A.P. Gonçalves, M.M. Cruz, and M. Godinho, *J. Magn. Magn. Mater.* **316**, e191 (2007).

372. K. Potzger, Sh. Zhou, H. Reuther, A. Mücklich, F. Eichhorn, N. Schell, W. Skorupa, M. Helm, J. Fassbender, T. Herrmannsdörfer, and T.P. Papageorgiou, *Appl. Phys. Lett.* **88**, 052508 (2006).

373. K. Potzger, W. Anwand, H. Reuther, Sh. Zhou, G. Talut, G. Brauer, W. Skorupa, and J. Fassbender, *J. Appl. Phys.* **101**, 033906 (2007).

374. Sh. Zhou, K. Potzger, H. Reuther, G. Talut, F. Eichhorn, J. von Borany, W. Skorupa, M. Helm, and J. Fassbender, *J. Phys. D: Appl. Phys.* **40**, 964 (2007).

375. Sh. Zhou, K. Potzger, G. Zhang, F. Eichhorn, W. Skorupa, M. Helm, and J. Fassbender, *J. Appl. Phys.* **100**, 114304 (2006).

376. A. Perez, J.P. Dupin, O. Massenet, G. Marest, and P. Bussière, *Rad. Eff.* **52**, 127 (1980).

377. A. Perez, M. Treilleux, L. Fritsch, and G. Marest, *Nucl. Instr. Methods in Phys. Res.* **182–183**, 747 (1981).

378. A. Perez, J. Bert, G. Marest, B. Sawicka, and J. Sawicki, *Nucl. Instr. Methods in Phys. Res.* **209–210** Pt. 1, 281 (1983).

379. A. Perez, G. Marest, B.D. Sawicka, J.A. Sawicki, and T. Tyliszczak, *Phys. Rev. B* **28**, 1227 (1983).

380. N. Hayashi, I. Sakamoto, T. Okada, H. Tanoue, H. Wakabayashi, and T. Toriyama, *Phys. Stat. Sol. (a)*, **189**, 775 (2002).

381. N. Hayashi, I. Sakamoto, T. Toriyama, H. Wakabayashi, T. Okada, and K. Kuriyama, *Surf. Coat. Technol.* **169–170**, 540 (2003).

382. K. Potzger, H. Reuther, Sh. Zhou, A. Mücklich, R. Grötzschel, F. Eichhorn, M.O. Liedke, J. Fassbender, H. Lichte, and A. Lenk, *J. Appl. Phys.* **99**, 08N701 (2006).

383. Z. Mao, Z. He, D. Chen, W.Y. Cheung, and S.P. Wong, *Solid State Commun.* **142**, 329 (2007).

384. M.M. Cruz, R.C. da Silva, J.V. Pinto, R. González, E. Alves, and M. Godinho, *J. Magn. Magn. Mater.* **272–276**, 840 (2004).

385. J.V. Pinto, M.M. Cruz, R.C. da Silva, E. Alves, R. González, and M. Godinho, *Eur. Phys. J. B* **45**, 331 (2005).

386. M.M. Cruz, J.V. Pinto, R.C. da Silva, E. Alves, R. González, and M. Godinho, *J. Magn. Magn. Mater.* **316**, e776 (2007).

387. B. Savoini, R. González, J.V. Pinto, R.C. da Silva, and E. Alves, *Nucl. Instr. Methods B* **257**, 563 (2007).

388. H. Tang, H. Berger, P.E. Schmid, and F. Levy, *Solid State Commun.* **92**, 267 (1994).

389. W. Prellier, A. Fouchet, and B. Mercey, *J. Phys.: Condens. Matter*, **15**, R1583 (2003).

390. Y. Matsumoto, M. Murakami, T. Shono, T. Hasegawa, T. Fukumura, M. Kawasaki, P. Ahmet, T. Chikyow, S. Koshihara, and H. Koinuma, *Science* **291**, 854 (2001).

391. D.H. Kim, J.S. Yang, K.W. Lee, S.D. Bu, T.W. Noh, S.-J. Oh, Y.-W. Kim, J.-S. Chung, H. Tanaka, H.Y. Lee, and T. Kawai, *Appl. Phys. Lett.* **81**, 2421 (2002).

392. S.A. Chambers, T. Doubray, C.M. Wang, A.S. Lea, R.F.C. Farrow, L. Folks, V. Deline, and S. Anders, *Appl. Phys. Lett.* **82**, 1257 (2003).

393. D.H. Kim, J.S. Yang, Y.S. Kim, D.-W. Kim, T.W. Noh, S.D. Bu, Y.-W. Kim, Y.D. Park, S.J. Pearton, Y. Jo, and J.-G. Park, *Appl. Phys. Lett.* **83**, 4574 (2003).

394. B. Aktaş, F. Yildiz, B. Rameev, R. Khaibullin, L. Tagirov, and M. Özdemir, *Phys. Stat. Sol. (c)* **1**, 3319 (2004).

395. J.V. Pinto, M.M. Cruza, R.C. da Silva, E. Alves, and M. Godinho, *J. Magn. Magn. Mater.* **294**, e73 (2005).

396. R.I. Khaibullin, L.R. Tagirov, B.Z. Rameev, Sh.Z. Ibragimov, F. Yildiz, and B. Aktaş, *J. Phys.: Condens. Matter*, **16**, L443 (2004).

397. N. Akdogan, B.Z. Rameev, L. Dorosinsky, H. Sozeri, R.I. Khaibullin, B. Aktaş, L.R. Tagirov, A. Westphalen, and H. Zabel, *J. Phys.: Condens. Matter* **17**, L359 (2005).

398. N. Akdogan, B.Z. Rameev, R.I. Khaibullin, A. Westphalen, L.R. Tagirov, B. Aktaş, and H. Zabel, *J. Magn. Magn. Mater.* **300**, e4 (2006).

399. N. Akdogan, A. Nefedov, A. Westphalen, H. Zabel, R.I. Khaibullin, and L.R. Tagirov, *Superl. Microstr.* **41**, 132 (2007).

400. R.I. Khaibullin, Sh.Z. Ibragimov, L.R. Tagirov, V.N. Popok, and I.B. Khaibullin, *Nucl. Instr. Methods*, B **257**, 369 (2007).

401. S. Zhu, L.M. Wang, X.T. Zu, and X. Xiang, *Appl. Phys. Lett.* **88**, 043107 (2006).

402. O.A. Ivanov, L.V. Solina, V.A. Denishina, and L.M. Magat, *Phys. Met. Metallogr.* **35**, 81 (1973).

403. C.W. White, S.P. Withrow, J.D. Budai, L.A. Boatner, K.D. Sorge, J.R. Thompson, K.S. Beaty, and A. Meldrum, *Nucl. Instr. Methods B* **191**, 437 (2002).

404. C.W. White, S.P. Withrow, K.D. Sorge, A. Meldrum, J.D. Budai, J.R. Thompson, and L.A. Boatner, *J. Appl. Phys.* **93**, 5656 (2003).

405. C.W. White, S.P. Withrow, J.M. Williams, J.D. Budai, A. Meldrum, K.D. Sorge, J.R. Thompson, and L.A. Boatner, *J. Appl. Phys.* **95**, 8160 (2004).

406. M. Ohkubo, T. Hioki, N. Suzuki, T. Ishiguro, and J. Kawamoto, *Nucl. Instr. Methods B* **39**, 675 (1989).

407. I. Sakamoto, S. Honda, H. Tanoue, N. Hayashi, and Y. Yamane, *Nucl. Instr. Methods B* **148**, 1039 (1999).

408. E. Alves, C. MacHargue, R.C. Silva, C. Jesus, O. Conde, M.F. da Silva, and J.C. Soares, *Surf. Coat. Technol.* **128–129**, 434 (2000).

409. N. Hayashi, I. Sakamoto, H. Wakabayashi, T. Toriyama, and S. Honda, *J. Appl. Phys.* **94**, 2597 (2003).

410. N. Hayashi, T. Toriyama, H. Tanoue, H. Wakabayashi, and I. Sakamoto, *Hyperfine Interact.* **141–142**, 163 (2002).

411. E. Cattaruzza, F. Gonella, G. Mattei, P. Mazzoldi, D. Gatteschi, C. Sangregorio, M. Falconieri, G. Salvetti, and G. Battaglin, *Appl. Phys. Lett.* **73**, 1176 (1998).

412. M. Klimenkov, J. von Borany, W. Matz, D. Eckert, M. Wolf, and K.-H. Müller, *Appl. Phys. A* **74**, 571 (2002).

413. P. Gangopadhyay, T.R. Ravindran, K.G.M. Nair, S. Kalavathi, B. Sundaravel, and B.K. Panigrahi, *Appl. Phys. Lett.* **90**, 063108 (2007).

414. T. Haeiwa, S. Kazuhiro, and K. Konishi, *J. Magn. Magn. Mater.* **310**, e809 (2007).

415. W.M. Tsang, V. Stolojan, B.J. Sealy, S.P. Wong, and S.R.P. Silva, *Ultramicroscopy* **107**, 819 (2007).

416. Xi.-Z. Ding, B.K. Tay, X. Shi, M.F. Chiah, W.Y. Cheung, S.P. Wong, J.B. Xu, and I.H. Wilson, *J. Appl. Phys.* **88**, 2745 (2000).

417. K.S. Buchanan, A. Krichevsky, M.R. Freeman, and A. Meldrum, *Phys. Rev. B* **70**, 174436 (2004).

418. V.B. Guseva, A.F. Zatsepin, V.A. Vazhenin, B. Schmidt, N.V. Gavrilov, and S.O. Cholakh, *Phys. Solid State*, **47**, 674 (2005).

419. O. Cíntora-González, C. Estournès, D. Muller, J. Guille, and J.J. Grob, *Nucl. Instr. Methods B* **147**, 422 (1999).
420. C. de Julián Fernández, C. Sangregorio, G. Mattei, G. De, A. Saber, S.L. Russo, G. Battaglin, M. Catalano, E. Cattaruzza, F. Gonella, D. Gatteschi, and P. Mazzoldi, *Mater. Sci. Eng. C* **15**, 59 (2001).
421. C. de Julián, C. Sangregorio, G. Mattei, G. Battaglin, E. Cattaruzza, F. Gonella, S.L. Russo, F. D'Orazio, F. Lucari, G. De, D. Gatteschi, and P. Mazzoldi, *J. Magn. Magn. Mater.* **226–230**, 1912 (2001).
422. C. de Julián Fernández, G. Mattei, C. Sangregorio, M.A. Tagliente, V. Bello, G. Battaglin, C. Sada, L. Tapfer, D. Gatteschi, and P. Mazzoldi, *J. Non-Cryst. Solids* **345–346**, 681 (2004).
423. J.C. Pivin, E. Vincent, S. Esnouf, and M. Dubus, *Eur. Phys. J. B* **37**, 329 (2004).
424. J.C. Pivin, *Nucl. Instr. Methods, B* **216**, 239 (2004).
425. J.C. Pivin, S. Esnouf, F. Singh, and D.K. Avasthi, *J. Appl. Phys.* **98**, 023908 (2005).
426. F. Singh, D.K. Avasthia, O. Angelovb, P. Berthetc, and J.C. Pivin, *Nucl. Instr. Methods B* **245**, 214 (2006).
427. S. Honda, F.A. Modine, A. Meldrum, J.D. Budai, T.E. Haynes, and L.A. Boatner, *Appl. Phys. Lett.* **77**, 711 (2000).
428. Xi. Zhu, C. Blois, K.S. Buchanan, Z. Liu, A. Meldrum, and M.R. Freeman, *J. Appl. Phys.* **97**, 10A720 (2005).
429. D. Fink, A.V. Petrov, W.R. Fahrner, K. Hoppe, R.M. Papaleo, A.S. Berdinsky, A. Chandra, A. Zrineh, and L.T. Chadderton, *Int. J. Nanoscience* **4**, 965 (2005). and references therein.
430. A.S. El-Said, W. Meissl, M.C. Simon, J.R. Crespo López-Urrutia, C. Lemell, J. Burgdörfer, I.C. Gebeshuber, H.P. Winter, J. Ullrich, C. Trautmann *et al.*, *Nucl. Instr. Methods B* **258**, 167 (2007), and references therein.
431. Z. Mi, J. Yang, P. Bhattacharya, P.K.L. Chan, and K.P. Pipe, *J. Vac. Sci. Technol. B* **24**, 1519 (2006).
432. Z.I. Kazi, T. Egawa, Ma. Umeno, and T. Jimbo, *J. Appl. Phys.* **90**, 5463 (2001).
433. P.M. Petroff, R.C. Miller, A.C. Gossard, and W. Wiegmann, *Appl. Phys. Lett.* **44**, 217 (1984).
434. D.L. Green, E.L. Hu, P.M. Petroff, V. Liberman, M. Nooney, and R. Martin, *J. Vac. Sci. Technol.* **11**, 2249 (1993).
435. S. Oktyabrsky, M. Lamberti, V. Tokranov, G. Agnello, and M. Yakimov, *J. Appl. Phys.* **98**, 053512 (2005).
436. H.S. Djie, D.-N. Wang, B.S. Ooi, J.C.M. Hwang, X.-M. Fang, Y. Wu, J.M. Fastenau, and W.K. Liu, *Thin Solid Films*, **515**, 4344 (2007).
437. S. Massouh, A. Perez, and J. Serughetti, *Phys. Rev. B* **33**, 3083 (1986).
438. Y.-K. Huang, C.-P. Liu, Y.-L. Lai, C.-Y. Wang, Y.-F. Lai, and H.-C. Chung, *Appl. Phys. Lett.* **91**, 091921 (2007).
439. J.P. Zhao, Y. Meng, D.X. Huang, R.K. Rayabarapu, and J.W. Rabalais, *J. Appl. Phys.* **100**, 084308 (2006).
440. N. Arai, H. Tsuji, N. Gotoh, T. Minotani, T. Ishibashi, K. Adachi, H. Kotaki, Y. Gotoh, and J. Ishikawa, *J. Phys.: Conf. Ser.* **61**, 41 (2007).
441. B. Johannessen, P. Kluth, D.J. Llewellyn, G.J. Foran, D.J. Cookson, and M.C. Ridgway, *Appl. Phys. Lett.* **90**, 073119 (2007).
442. B. Johannessen, P. Kluth, C.J. Glover, S.M. Kluth, G.J. Foran, D.J. Cookson, D.J. Llewellyn, and M.C. Ridgway, *Nucl. Instr. Methods B* **250**, 210 (2006).
443. G. Talut, H. Reuther, Sh. Zhou, K. Potzger, F. Eichhorn, and F. Stromberg, *J. Appl. Phys.* **102**, 083909 (2007).
444. Sh. Zhou, K. Potzger, G. Talut, H. Reuter, J. von Borany, R. Grötzschel, W. Skorupa, M. Helm, J. Fassbender, N. Volbers, M. Lorenz, and T. Herrmannsdörfer, *J. Appl. Phys.* **103**, 023902 (2008).
445. Sh. Zhou, K. Potzger, J. von Borany, R Grötzschel, W. Skorupa, M. Helm, and J. Fassbender, *Phys. Rev. B* **77**, 035209 (2008).
446. G. Talut, K. Potzger, A. Mücklich, and Sh. Zhou, *J. Appl. Phys.* **103**, 07D505 (2008).

CHAPTER 14

Probing and Controlling the Spin State of Single Magnetic Atoms in an Individual Quantum Dot

L. Besombes, Y. Leger, L. Maingault, D. Ferrand, J. Cibert, and H. Mariette

CEA-CNRS group "Nanophysique et Semiconducteurs", Néel Institute, CNRS and University J. Fourier, 25 avenue des Martyrs, 38042 Grenoble, France

14.1 Introduction

As the size of magneto-electronic devices scales down, it becomes increasingly important to understand the properties of a single magnetic atom in a solid-state environement [1–5]. Atomic scale surface probes have been succesfully used in this regard [1–3]. More recently, optical probing of both magnetic [4] and non-magnetic [6] atoms in semiconductors has been demonstrated. Magnetically doped semiconductors have been used in the fabrication of electrically active devices that control the magnetic properties like transition temperature and coercitive field [7–9]. In these devices a macroscopic number of magnetic atoms was manipulated. We will present here electrically active devices that control the charge state of an individual II–VI quantum dot (QD) doped with a single Mn atom. Single dot micro-photoluminescence measurements reveal that the magnetic anisotropy and spin configuration of the single Mn atom are very different depending on the charge state of the dot, which can be 0, or $\pm 1e$. Thereby, these devices are able to tune the magnetic properties of a single Mn atom embedded in a QD and represents a first step in the implementation of several proposals of electrical control of the magnetism in Mn-doped quantum dots [10–13].

Quantum dots doped with magnetic atoms and filled with a tunable number of carriers can behave like tunable nanomagnets. Indeed, as it has been theoretically shown, the parity of the number of confined carriers can drastically change the magnetic properties of the localized magnetic moments [11–15].

In a magnetic QD, the sp–d interaction takes place with a single carrier or a single electron–hole pair. However, besides effects related to the carrier–Mn-ion exchange interaction such as giant Faraday rotation and strong Zeeman shift, it was found that even a small content of Mn introduced in a II–VI semiconductor material can strongly suppress photoluminescence if the energy gap E_g exceeds the energy of the Mn internal transition. This strongly limits the study of individual diluted magnetic semiconductor quantum dots [16]. The first studies of individual quantum dots doped with Mn atoms were reported by Maksimov *et al.* [17]. They studied CdMnTe QDs inserted in CdMgTe barriers in which the optical transition energies are lower than the energy of the internal transition of the Mn atom. This suppresses the non-radiative losses due to the transfer of confined carriers to the Mn energy levels. This system allowed the formation of a quasi zero-dimensional magnetic polaron to be observed. Moreover, the broadening of the emission lines caused by the magnetic fluctuations in the environment of the recombining electron–hole pair was controlled by an external magnetic field.

Another way to reduce the non-radiative losses was to introduce the magnetic atoms in the quantum dots barriers. This has been realized for CdSe dots embedded in ZnMnSe barriers by Seufert *et al.* [18]. In this system, the interaction between the confined exciton and the magnetic ions is due to the spread of the wave function in the barriers and to a small diffusion of the magnetic atoms in the quantum dots. In these diluted magnetic semiconductor (DMS) structures, the response time of the paramagnetic Mn spin was extracted from the transient spectral shift of the photoluminescence caused by the dynamical spin alignment of magnetic ions incorporated into the crystal matrix. The formation of a ferromagnetically aligned spin complex was demonstrated to be surprisingly stable as compared to bulk magnetic polaron [19, 20], even at elevated temperature and high magnetic fields. The photoluminescence of a single electron–hole pair confined in one magnetic QD, which sensitively depends on the alignment of the magnetic ions spins, allowed measurement of the statistical fluctuation of the magnetization on the nanometre scale. Quantitative access to statistical magnetic fluctuations was obtained by analysing the linewidth broadening of the single dot emission. This all-optical technique allowed a magnetic moment of about $100\,\mu_B$ to be addressed and changes in the order of a few μ_B [21–23] to be resolved.

A huge effort has also been done to incorporate magnetic ions in chemically synthesized II–VI nanocrystals [24]. The incorporation of the magnetic impurities is strongly dependent on the growth conditions and controlled by the adsorption of impurities on the nanocrystal surface during growth [25]. The doping of nanocrystals with magnetic impurities also leads to interesting magneto-optical properties [26] but once again, in these highly confined systems, the transfer of confined carriers to the Mn electronic levels strongly reduces their quantum efficiency and prevents the optical study of individual Mn-doped nanocrystals. However, by looking to MCD absorption spectra it is possible to observe a giant excitonic Zeeman splitting and to deduce directly the sp–d exchange interaction [27].

CdTe/ZnTe self-assembled QDs usually present an emission energy below the internal transition of the Mn atom. The incorporation of magnetic atoms is then possible without losing the good optical properties of these QDs. Up to now, however, all the experimental studies on these diluted magnetic QDs were focused on the interaction of a single carrier spin with its paramagnetic environment (large number of magnetic atoms) [28]. We will see that CdTe/ZnTe quantum dot structures doped with a low density of Mn atoms allow the optical control of spin states of a single magnetic ion interacting with a single electron–hole pair or a single carrier.

14.2 II–VI diluted magnetic semiconductors quantum dots

14.2.1 *Carrier–Mn coupling*

Carrier–Mn coupling was mostly studied in bulk DMS made of II–VI semiconductors in which Mn impurities were introduced (see the review papers [29, 30]). These semiconductors are formed with a cation from column II (Zn, Cd or Hg) and an anion from column VI (Te, Se, S and more recently even O). These compounds assume the zinc-blende structure or, for the most ionic ones, the closely related wurtzite structure. Manganese impurities have the d^5 electronic configuration and substitute the cations up to 100%. An important point is that Mn substitutes the column II cation as an isoelectronic impurity – by contrast to the acceptor character observed in GaAs and similar III–Vs [31, 32]. The Mn ground state is 6S (or 6A1 in cubic or hexagonal symmetry), introducing localized, isotropic spins with $S = 5/2$. If not interacting, these localized spins follow Maxwell–Boltzmann statistics, resulting in a magnetization M induced by an applied field H at temperature T given by a Brillouin function of H/T:

$$M = -xN_0 g_{(Mn)}\mu_B \frac{5}{2} B_{5/2}\left(\frac{5}{2}\frac{g_{(Mn)}\mu_B\mu_0 H}{k_B T}\right) \tag{14.1}$$

where x is the proportion of cations substituted by Mn, N_0 the density of cations in the zinc-blende or wurtzite structure, $g_{(Mn)} = 2$ (to a good approximation) is the Mn Lande factor, μ_B is the

Bohr magneton, and k_B the Boltzmann constant. This would apply for non-interacting spins. In the actual material, antiferromagnetic "superexchange" interactions appear as soon as the Mn density is not vanishingly small. These interactions result in a reduction of the magnetization: this is quantitatively described by using a "modified Brillouin function" where x is replaced by a number of free spins x_{eff}, and the argument is $H/(T + T_{AF})$. The values of the two phenomenological parameters x_{eff} and T_{AF} are experimentally well documented in the most usual DMSs, such as $Cd_{1x}Mn_xTe$ [33]. The number of free spins has a direct physical meaning: nearest-neighbour Mn pairs are blocked antiparallel by the strong superexchange interaction – it can be calculated from statistics over Mn pairs and clusters either in the homogeneous material [34, 35], or even at an interface with the non-magnetic semiconductor [36]. The description by a modified Brillouin function has proven to be very efficient at moderately low temperatures (1 to 30 K), moderate field (up to 5 T), and composition up to $x = 0.2$ [33]. A spin glass behaviour appears at higher Mn contents in this temperature range, and at lower temperature even for a low Mn content. At higher temperatures or higher field nearest-neighbour pairs are no longer blocked antiparallel. It was recognized early that the key property of a DMS is the strong coupling between the bands of the semiconductor (conduction band and, even more strongly, valence band) and the localized spins. Optical spectroscopy around the band gap reveals the so-called "giant Zeeman effect", with a spin splitting proportional to the Mn magnetization [37]. Several studies have demonstrated this proportionality and measured the strength of the coupling [38] (see Fig. 14.1).

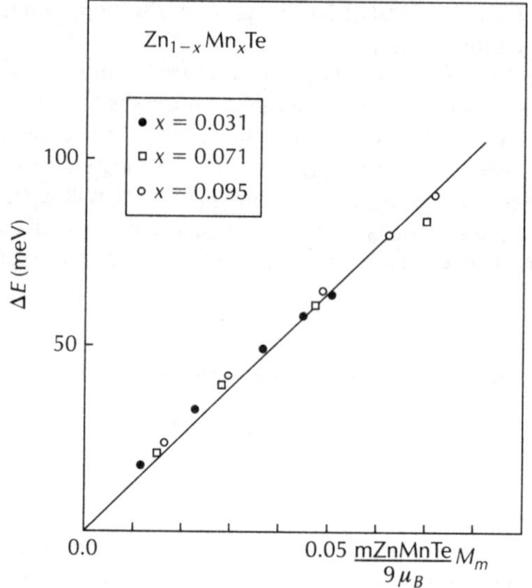

Figure 14.1 *Splitting of the free exciton line. The straight line is the theoretical dependence calculated for* $N_0(\alpha - \beta) = 1.29$ eV *(from A. Twardowski et al., Solid State Com.,* **50**, *509 (1984)).*

As a result, magneto-optical spectroscopy is now a very sensitive method for measuring locally the magnetization of the Mn system. For example, it was applied [33] to measure the diffusion of Mn across a (Cd,Mn)Te–CdTe interface during the growth by molecular beam epitaxy (MBE). Actually, the method is quite direct in semiconductors with intermediate values of the band gap width (tellurides, selenides). In wide band gap semiconductors, ex-citonic effects are important so that the splitting directly measured on the spectra is not proportional to the spin splitting of the bands [39]. Altogether, this excellent knowledge of the magnetic properties in (electrically) undoped II–VI DMSs, and of the coupling between the localized spins and the conduction/ valence band, constitutes a very firm basis for the further studies described below.

14.2.2 CdTe:Mn/ZnTe quantum dots

The CdTe/ZnTe QDs samples used in this study are grown by atomic layer epitaxy (ALE) with a deposition of six monolayers of CdTe on a ZnTe buffer. One should note that the vapour pressure above the column II or column VI elements in an ultra-vacuum chamber is much higher than above their compounds. As a result, stable surfaces are obtained under a flux of either the cation or the anion species – another difference with the III–Vs. That implies that (ALE) is feasible [40], as well as MBE under both cation-rich and anion-rich conditions. Cation-rich MBE growth on a (001) surface proceeds in a 2D mode (layer by layer) up to a sharply defined critical thickness [41] which depends on the lattice mismatch and which corresponds to a plastic relaxation. As far as the II–VI (QDs) are concerned, the first works published for CdSe QDs in ZnSe [42] and for CdTe in ZnTe [43] evidence zero-dimensional excitonic properties but did not reported *in situ* direct evidence of a spontaneous 2D–3D growth transition (the so-called Stranski–Krastanov mode). A modified procedure, which involves the re-evaporation of an amorphous layer of the anion, was developed later [44], in order to observe *in situ* a clear 2D–3D transition and to get well-formed CdTe QDs [45], and CdSe QDs [46]. With this process, the 2D–3D transition is mainly induced by a surface energy variation and is revealed by a clear, spotty, reflection, high-energy electron diffraction pattern [45, 47].

Let us explain now how we optimize the growth process in order to get one single Mn into the QD. Our criterion for the occurrence of such configuration is a special feature observed in the photoluminescence (PL) spectra: six narrow lines appear (see below). This corresponds to the interaction between an exciton with a hole quantized along the growth axis z and a single Mn ion, whose spin has six projections along this axis [4]. This is the proof that only one Mn interacts with the exciton. When more Mn ions interact with the exciton confined in the QD, only broad lines appear due to statistical magnetic fluctuations of all the spins of the Mn ions present in the QD [23, 48].

A simple geometric modellization shows that if we optimize the Mn density as compared to the QD one, we could get almost half of all the QDs that contain only one Mn. On a surface with only one Mn ion and one QD, the probability to have this Mn ion into the QD is $a = S_{QD} \times d_{QD}$, where d_{QD} is the QD density and S_{QD} is the area of a QD. With different numbers of QDs and Mn ions, N and p, respectively, this probability becomes $p(a/N)(1 - a/N)^{p-1}$. This probability is based on a random distribution of both QDs and Mn ions and does not assume any correlation between the QD nucleation and the presence of Mn ions. Then we found that the maximum probability of QDs having only one Mn is always about 40%, whatever the QD size, for an appropriate Mn density. This model was applied with the parameters of our CdTe QDs. Atomic force microscopy gives a density of about $d_{QD} = 3 \times 10^{10}$ QDs/cm^2 (Fig. 14.2a). The size and shape of our QDs are deduced from high-resolution transmission electron microscopy images. QDs are lens shaped with a typical radius of 5 nm and height of 3 nm (Fig. 14.2a). Thus the parameter a is found to be around 0.02. To go one step further, a single Mn is detected with six well-separated lines if the Mn ion is near the centre of the QD and also if all the other Mn ions are not in the vicinity of this QD. The influence of these other Mn ions was calculated in [49]. Taking into account such additional conditions induces a strong reduction of the maximum probability to get only one Mn into the QD and also reduces the number of Mn ions needed to get this maximum to a very low value [55].

From this simulation, a concentration of Mn as low as 0.02% is needed for the optimal probability ($d_{Mn}/d_{QD} = 4$). Such an Mn concentration corresponds to an Mn flux too low to be controlled precisely. Thus a specific approach to get a very precise composition of Mn over the ZnTe surface is needed. After a ZnTe buffer growth in order to get a flat 2D surface, a thin $Zn_{0.94}Mn_{0.06}Te$ buffer is grown. The Mn composition in this buffer is high enough to be determined very precisely with reflection high-energy electron diffraction oscillation calibration [50]. Then a ZnTe spacer of few monolayers is grown. Segregation of Mn ions into this spacer allows the control of the Mn density with the number of spacer monolayers. Magneto-optical spectroscopy experiments [36] showed that every additional monolayer has half the Mn composition of the former one. This model of interface is currently used to describe segregation at the growing interface [51]. Then the 2D growth control of the ZnTe spacer layer with a 1 ML accuracy allows us to adjust exactly the Mn concentration in the QD plane. We have performed a simulation assuming that Mn ions and QDs

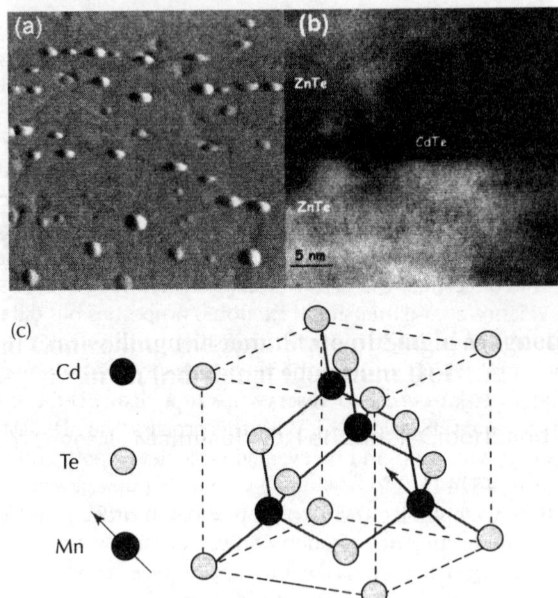

Figure 14.2   (a) AFM image of a CdTe surface deposited on a ZnTe substrate before deposition of a ZnTe capping layer. (b) High resolution TEM image showing the structure of a CdTe/ZnTe quantum dot. (c) Crystallin structure of CdTe with the substitution of a Cd atom by a magnetic Mn atom.

are randomly distributed over the surface (in plane) and that Mn ions segregate into the ZnTe spacer (growth direction). For every carrier in the QDs, the value of the interaction between Mn ions and the electron (hole) confined in the QD is calculated using the wave function as described later. If these interactions are such that the energy spacing between two successive emission lines is larger than the spectral resolution of our optical spectroscopy set-up ($50\,\mu eV$), a single Mn ion is present in the dot and can be detected with a comb of six lines. The maximum probability is reached when there are four spacer monolayers as shown in Fig. 14.3 (triangle). However, PL

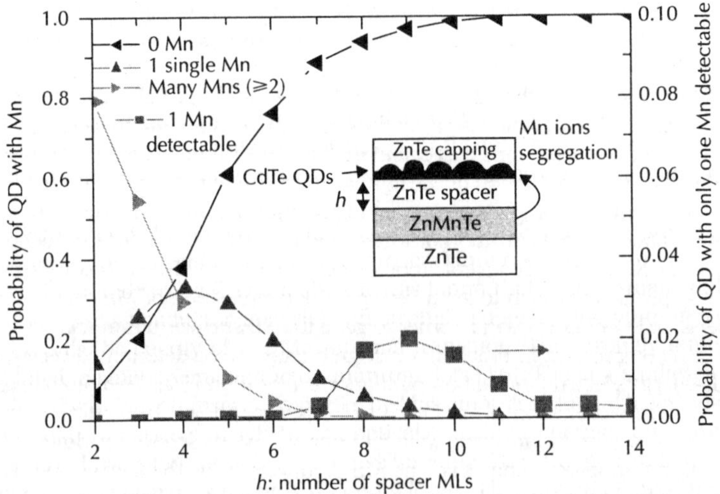

Figure 14.3   Random simulation of the incorporation of Mn ions into the dot as a function of the number of spacer monolayers. The maximum of single Mn QDs is reached (square) around 9 ML. The inset is a sample cross-section profile.

lines are detected only if no other Mn is close to the QD, i.e. if the interaction between all other Mn and the exciton does not induce an energy spacing above 20 µeV. With this additional condition the optimal probability to detect one single Mn ion is reached for a spacer thickness of 9 ML (square on Fig. 14.3). Moreover, this probability is now very low (see the right scale on Fig. 14.3) which shows the limit of random incorporation of Mn ions. Such results guide us efficiently in our approach to grow QDs doped with a single Mn ion.

14.3   Optical probing of the spin state of a single magnetic atom in a QD

14.3.1   Confined carriers–Mn exchange interaction

Microspectroscopy was used to study the magneto-optical properties of individual CdTe/ZnTe QDs. The low temperature (5 K) photoluminescence (PL) of single QDs is excited with the 514.5 nm line of an argon laser or a tunable dye laser and collected through a large numerical-aperture objective and aluminium shadow masks with 0.5–1.0 m apertures. The experiments are carried out in the backward geometry with the propagation direction of the incident and emitted light parallel to the [001] growth axis. Superconductive coils are used to apply a magnetic field up to 11 T in Voigt or Faraday configuration. The PL is then dispersed by a 2 m double monochromator and detected by a nitrogen-cooled Si charged-coupled device camera or an Si avalanche photodiode. Typical photoluminescence and excitation spectra of an exciton in a single CdTe/ZnTe QD are shown on Fig. 14.4.

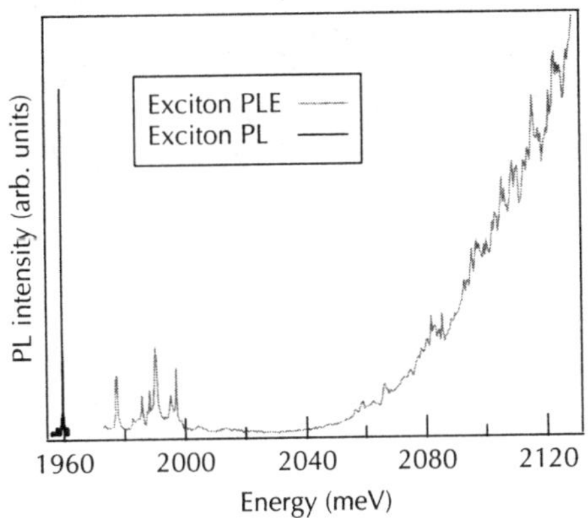

Figure. 14.4   Photoluminescence and photoluminescence excitation spectra of an exciton in a single CdTe/ZnTe quantum dot.

In Fig. 14.5, PL spectra of an individual Mn-doped QD are compared to those of a non-magnetic CdTe/ZnTe reference sample. In non-magnetic samples, narrow PL peaks (limited by the spectrometer resolution of about 50 µeV) can be resolved, each attributed to the recombination of a single electron–hole pair in a single QD. The emission of neutral QDs is split by the e–h exchange interaction and usually a linearly polarized doublet is observed [56]. On the other hand, most of the individual emission peaks of magnetic single QDs are characterized by a rather large linewidth of about 0.5 meV. For some of these QDs, a fine structure can be resolved and six emission lines are clearly observed at zero magnetic field. The measured splitting changes from dot to dot. This fine structure splitting as well as the broadening is obviously related to the influence of the magnetic ions located within the spatial extent of the exciton wave function. The broadening observed in magnetic QDs has been attributed by Bacher *et al.* to the magnetic fluctuations of the spin projection of a *large number* of Mn spins interacting with the confined exciton [23]. In the low concentration Mn-doped samples, the observation of a fine structure shows that the QD exciton interacts with a single Mn spin. In time-averaged experiments, the statistical fluctuations of a

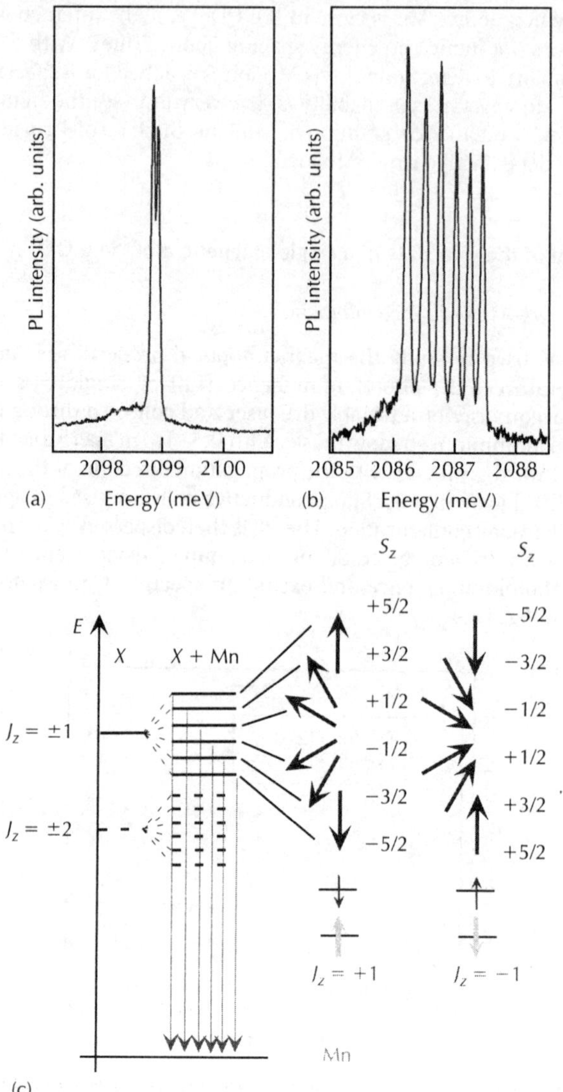

Figure 14.5   Low temperature (T = 5 K) PL spectra obtained at B = 0 T for an individual CdTe/ZnTe QD (a) and a Mn-doped QD (b). (c) Scheme of the energy levels of the Mn–exciton coupled system at zero magnetic field. The exciton–Mn exchange interaction shifts the energy of the exciton depending on the S_z component of the Mn spin projection.

single Mn spin ($S = 5/2$) can be described in terms of populations of its six spin states quantized along the direction normal to the QD plane. The exchange interaction of the confined exciton with the Mn atom shifts its energy depending on the Mn spin projection, resulting in the observation of six emission lines.

QDs doped with single Mn atoms were considered theoretically in the case of spherical nano-crystals with a strong confinement [26]. The eigenstates resulting from the exchange coupling between the exciton and the magnetic ion were obtained by a combination of the electron, hole and Mn magnetic moments. In flat self-assembled QDs with a relatively weak confinement, the bi-axial strains in the plane of the QD lift the degeneracy of the hole spin projections (heavy-hole/light-hole splitting). In a first approximation, this system can be described by a heavy-hole exciton confined in a symmetric QD, in interaction with the six spin projections of the manganese ion [52]. The spin interaction part of the Hamiltonian is given by:

$$H_{int} = I_e \sigma \cdot S + I_h j \cdot S + I_{eh} \sigma \cdot j \tag{14.2}$$

where I_e (I_h) is the Mn–electron (–hole) exchange integral, I_{eh} the electron–hole exchange interaction and σ (j, S) the magnetic moment of the electron (hole, Mn). The initial states of the optical transitions are obtained from the diagonalization of the spin Hamiltonian and Zeeman Hamiltonian in the subspace of the heavy-hole exciton and Mn spin components $|\pm 1/2\rangle_e |\pm 3/2\rangle_h |S_z\rangle_{Mn}$, with $S_z = \pm 5/2, \pm 3/2, \pm 1/2$. Since the dipolar interaction operator does not affect the Mn d electrons, the final states involve only the Mn states $|S_z\rangle_{Mn}$ with the same spin component [4].

In this framework, at zero magnetic field, the QD emission presents a fine structure composed of six doubly degenerate transitions roughly equally spaced in energy. The lower energy bright states, $|+1/2\rangle_e |-3/2\rangle_h |+5/2\rangle_{Mn}$ and $|-1/2\rangle_e |+3/2\rangle_h |-5/2\rangle_{Mn}$ are characterized by an antiferromagnetic coupling between the hole and the Mn atom. The following states are associated with the Mn spin projections: $S_z = \pm 3/2, \pm 1/2$ until the higher energy states $|-1/2\rangle_e |+3/2\rangle_h |+5/2\rangle_{Mn}$ and $|+1/2\rangle_e |-3/2\rangle_h |-5/2\rangle_{Mn}$ corresponding to ferromagnetically coupled hole and manganese. In this simple model the zero field splitting $\delta_{Mn} = \frac{1}{2}(I_e - 3I_h)$ depends only on the exchange integrals I_e and I_h and is thus related to the position of the Mn atom within the exciton wave function.

When an external magnetic field is applied in the Faraday geometry (Fig. 14.6), each PL peak is further split and 12 lines are observed, six in each circular polarization. As presented in Fig. 14.7, if the Zeeman effect of the Mn states is identical in the initial and final states of the optical transitions then the six lines in a given polarization follow the Zeeman and diamagnetic shift of the exciton, as in a non-magnetic QD. The parallel evolution of six lines is perturbed around 7 T in $\sigma-$ polarization by anticrossings observed for five of the lines. In addition, as the magnetic field increases, one line in each circular polarization increases in intensity and progressively dominates the others.

The electron–Mn part of the interaction Hamiltonian $I_e(\sigma \cdot S)$ couples the dark ($J_z = \pm 2$) and bright ($J_z = \pm 1$) heavy-hole exciton states. This coupling corresponds to a simultaneous electron and Mn spin flip changing a bright exciton into a dark exciton. Because of the strain-induced splitting of light-hole and heavy-hole levels, a similar Mn–hole spin flip scattering is not allowed. The electron–Mn spin flip is enhanced as the corresponding levels of bright and dark excitons are brought into coincidence by the Zeeman effect. An anticrossing is observed around 7 T for five of the bright states in $\sigma-$ polarization (experiment: Fig. 14.6 and theory: Fig. 14.7). It induces a transfer of oscillator strength to the dark states. In agreement with the experimental results, in the calculations the lower energy state in $\sigma-$ polarization $(|+1/2\rangle_e |-3/2\rangle_h |+5/2\rangle_{Mn})$ does not present any anticrossing. In this spin configuration, both the electron and the Mn atom have maximum spin projection and a spin flip is not possible.

The minimum energy splitting at the anticrossing is directly related to the electron–Mn exchange integral I_e. For instance, the splitting measured for the higher energy line in $\sigma-$ polarization (Fig. 14.6), $\Delta E = 150\,\mu eV$ gives $I_e \approx 70\,\mu eV$. From the overall splitting measured at zero field (1.3 meV) and with this value of I_e, we obtain $I_h \approx -150\,\mu eV$. These values are in good agreement with values estimated from a modelling of the QD confinement by a square quantum well in the growth direction and a truncated parabolic potential in the QD plane. With a quantum well thickness $L_z = 3\,nm$ and a Gaussian wave function characterized by an in-plane localization parameter $\xi = 5\,nm$ we obtain $I_e \approx 65\,\mu eV$ for an Mn atom placed at the centre of the QD.

However, the ratio of the exchange integral, $(3I_h)/I_e \approx -6$, for the QD presented in Fig. 14.6, does not directly reflect the ratio of the sp–d exchange constants $\beta/\alpha \approx -4$ measured in bulk CdMnTe alloys [29]. Such deviation likely comes from the difference in the electron–Mn and hole–Mn overlap expected from the difference in the electron and hole confinement length but it could also be due to a change of the exchange parameters induced by the confinement [57]. A dispersion of the zero field energy splitting observed from dot to dot is then due to a variation of the Mn–exciton overlap for different QDs. However, the spin Hamiltonian (2) does not reproduce the observed non-uniform zero field splitting between consecutive lines (Fig. 14.5b). As we

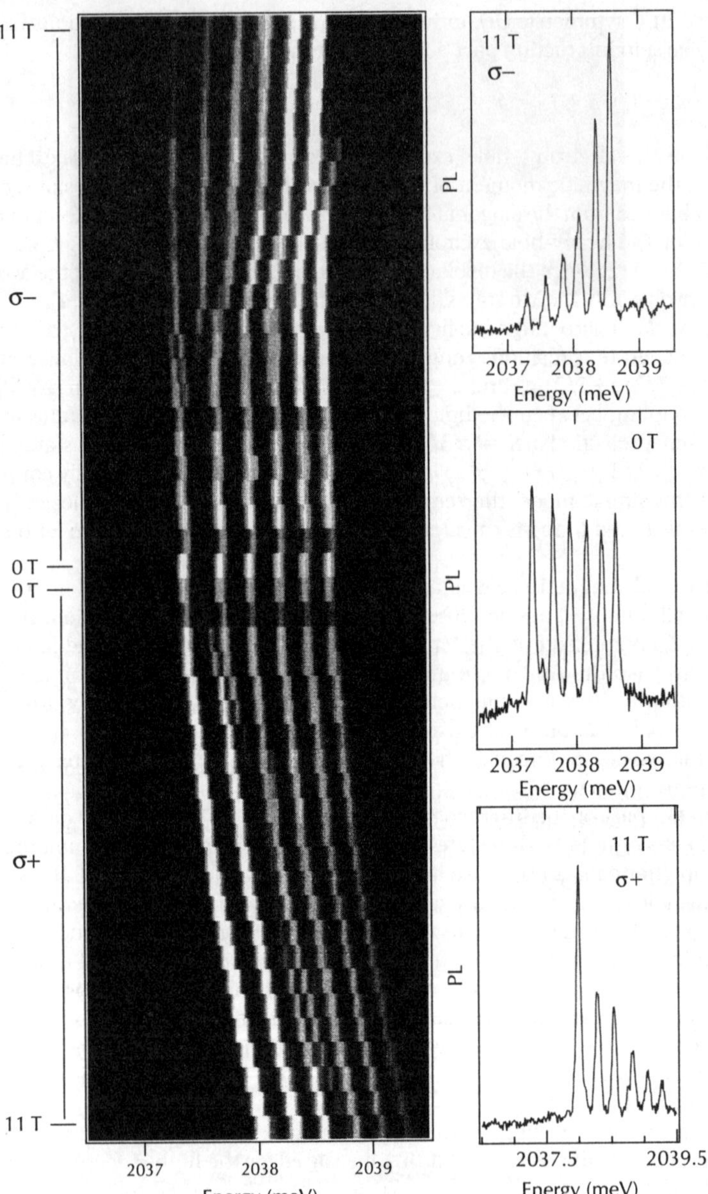

Figure 14.6   *Magnetic field dependence of the emission of an Mn-doped QD recorded in σ+ and σ− polarization. Anticrossing of the bright and dark states appears around 7 T in σ− polarization.*

will see in the following, a more accurate model has to take into account the full valence band structure and the heavy-hole/light-hole mixing.

14.3.2   The exciton as a probe of the Mn spin state

As illustrated in Fig. 14.6, the relative intensities of the six emission lines observed in each circular polarization depends strongly on the applied magnetic field. The emission intensity, which is almost equally distributed over the six emission lines at zero field, is concentrated on the high

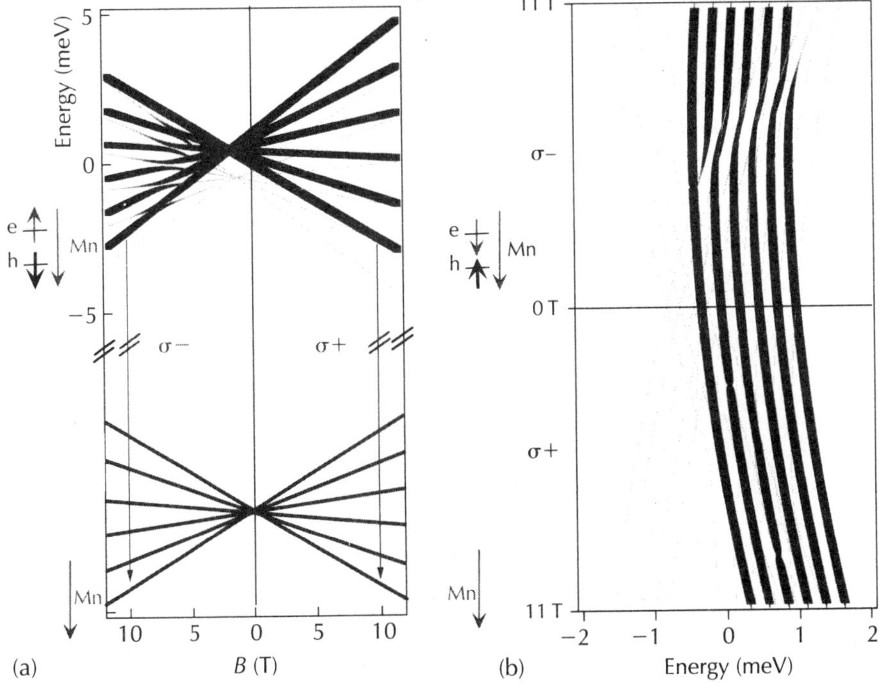

Figure 14.7 *(a) Scheme of the energy levels of the initial and final state involved in the optical transitions of a quantum dot containing an Mn atom. (b) Modellization of the optical transition obtained from the diagonalization of an effective spin Hamiltonian including the e–h exchange interaction, the exciton–Mn exchange interaction, the Zeeman and the diamagnetic energies. The contribution of the dark states appears in green.*

energy line for the $\sigma-$ emission and on the low energy line for the $\sigma+$ emission at high magnetic field. As the magnetic field increases, the Mn atom is progressively polarized. In time-averaged experiments, the probability of observing the recombination of the bright excitons coupled with the $S_z = -5/2$ spin projection is then enhanced. Two states dominate the spectra: $|-1/2\rangle_e|+3/2\rangle_h|-5/2\rangle_{Mn}$ in the low-energy side of the $\sigma+$ emission and $|+1/2\rangle_e|-3/2\rangle_h|-5/2\rangle_{Mn}$ in the high-energy side of the $\sigma-$ polarization. Changing the temperature of the Mn ion will affect the distribution of the exciton emission intensities. The PL of the exciton is then a direct probe of the magnetic state of the Mn ion.

14.3.3 Exciton–Mn thermalization process

The effective temperature of the manganese ion in the presence of the exciton, T_{Mn}, is found to depend of course on the lattice temperature but also on the laser excitation density (Fig. 14.8). For a fixed temperature and a fixed magnetic field, the asymmetry observed in the emission intensity distribution progressively disappears as the excitation intensity is increased (Fig. 14.8a). The variation of T_{Mn} deduced from the emission rates is presented in the inset of Fig. 14.8c (7 T) and Fig. 14.8d (0 T) as a function of the excitation density. A similar excitation intensity dependence of T_{Mn} was previously observed in DMS quantum wells and was attributed to the heating of the Mn^{2+} ions through their spin–spin coupling with the photo-created carriers [58]. The photo-carriers have excess energy. Via spin–flip exchange scattering they pass their energy to the Mn^{2+} ions and elevate their spin temperature. The energy flux from the Mn to the lattice, determined by the spin lattice relaxation, will tend to dissipate this excess energy. Under steady-state photo-excitation, the resulting temperature of the magnetic ions T_{Mn} exceeds the lattice temperature. The effect of this spin–spin coupling is strongly enhanced in our system since the isolated Mn^{2+} ion is only weakly coupled to the lattice and hardly thermalized with the phonon bath [59].

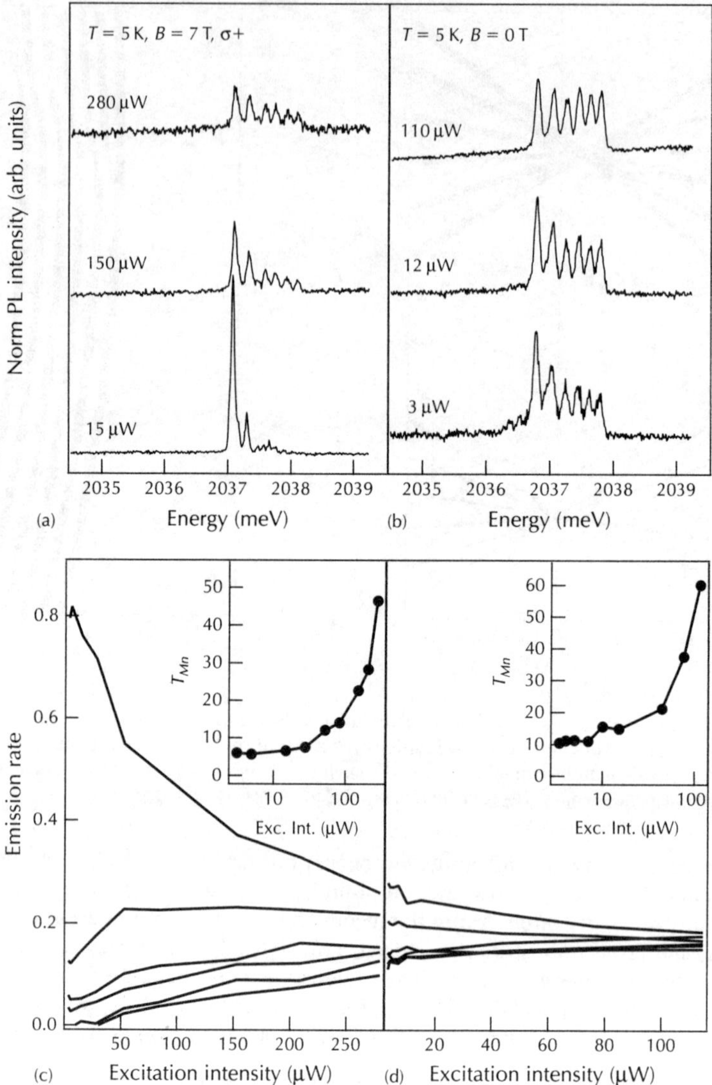

Figure 14.8   (a) Normalized PL (PL spectra divided by the total integrated intensity) in σ+ polarization versus excitation intensity for a fixed temperature (5 K) and magnetic field (7 T). (b) Excitation intensity dependence of the zero magnetic field emission of a single Mn-doped QD for a fixed lattice temperature T = 5 K. (c) and (d) Extracted emission rates of each PL line as a function of the excitation intensity at 7 T (c) and 0 T (d). The inset plots the extracted Mn effective temperature.

Under non-resonant excitation, the injection of an exciton changes the spin distribution of the magnetic ion. As illustrated in Fig. 14.8b, at 0 T and at low excitation intensity, an asymmetry is observed in the emission intensity distribution. This polarization shows that a spin flip of the exciton–Mn system can occur during the lifetime of the exciton. The exchange interaction with the exciton acts as an effective magnetic field which splits the Mn^{2+} levels in zero applied field, allowing a progressive polarization of its spin distribution.

Resonant excitation of electron–hole pairs directly in the QD limits the exciton–Mn spin relaxation [75]. This is illustrated in Fig. 14.9b and c where the emission intensity of the ground state of an Mn-doped QD is presented as a function of the detection energy when the laser excitation energy is scanned through the resonant absorption of an excited state identified in a PLE spectra (Fig 14.9a). This excitation energy scan reveals that the intensity distribution of the emission

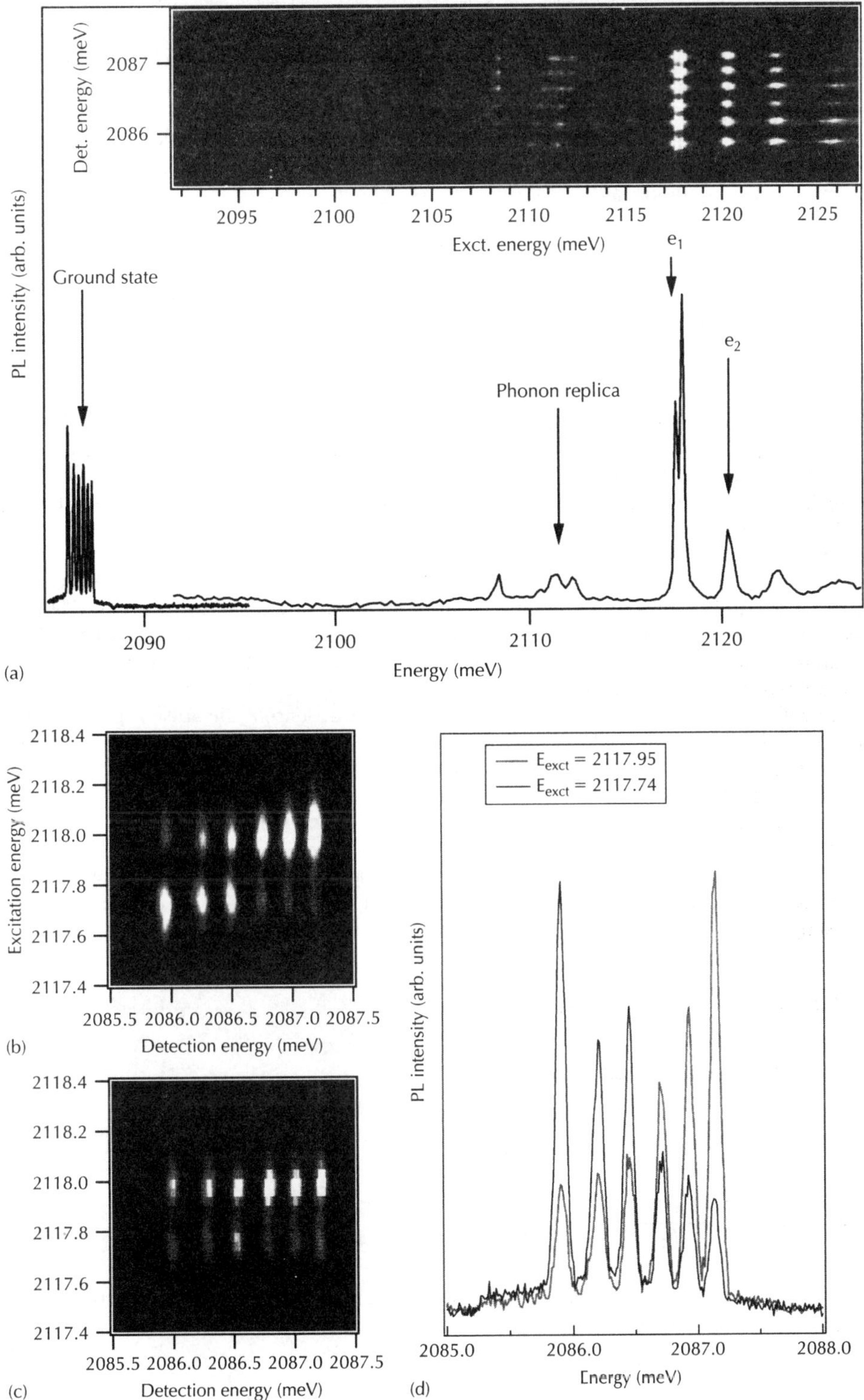

Figure 14.9 (a) Experimental PL and PLE spectra of an Mn-doped quantum dot exciton. The inset shows the contour plot of the multichannel PLE. (b) and (c) PLE contour plots for excited state e_1 obtained for co-polarized (b) and cross-polarized (c) circular excitation and detection. (d) Resonant PL spectra obtained in co-polarized circular excitation and detection for two different excitation wavelengths on e_1.

strongly depends on both the wavelength and on the polarization of the excitation laser. This dependence shows that under resonant excitation there is no complete spin relaxation of the exciton–Mn complex during the lifetime of the exciton. As shown in Fig. 14.9d, this long spin relaxation time combined with the fine structure of the excited states permits the selective creation of a given spin configuration of the exciton–Mn complex by tuning the polarization and wavelength of the excitation laser.

14.4   Geometrical effects on the optical properties of quantum dots doped with a single magnetic atom

14.4.1   Influence of an anisotropic strain distribution

Several phenomena can lead to the mixing between light and heavy holes in QDs. First, the symmetry reduction due to the confinement geometry of the dot has to be considered. In this case, a hole band mixing appears through the non-diagonal terms of the Luttinger–Kohn Hamiltonian. This mixing is responsible for the linear polarization rate observed in strongly confined quantum wires [77]. A large wave function anisotropy is needed to reproduce the observed linear polarization [78]. Such anisotropy can only be obtained in a very elongated confining potential for the holes with large barriers. This is inconsistent with the observation of huge linear polarization rates in shallow CdTe/ZnTe QDs presenting a quite weak in-plane asymmetry as revealed by AFM measurements [55]. Another origin of hh–lh mixing is the coupling of the X and Y valence band states produced by the microscopic arrangement of chemical bonds at heterointerfaces [79]. This contribution is expected to be weak in flat self-assembled QDs with almost symmetric interfaces.

As we will see in the following, an efficient mixing can arise from the anisotropic relaxation of strains in the QDs plane. Influence of strains on hole states has been intensively studied in quantum wells. Two main effects have been demonstrated, namely the variation of the gap due to hydrostatic strains and the hh–lh degeneracy lift due to biaxial strains. We will show that in QDs the strain distribution can also be responsible for a strong lh–hh mixing. This mixing has striking effects on the hole spin anisotropy.

In bulk semiconductors, the spin orbit interaction is responsible for splitting the hole states. We only consider here the lowest energy holes with angular momentum $j = 3/2$. These $|j, j_z\rangle$ states can be simply defined using orbital (X, Y, Z) and spin $(\uparrow, \downarrow)$ eigenvectors:

$$|3/2, +3/2\rangle = -\uparrow \frac{X + iY}{\sqrt{2}}$$

$$|3/2, +1/2\rangle = \sqrt{\frac{2}{3}} \uparrow Z - \downarrow \frac{X + iY}{\sqrt{6}}$$

$$|3/2, -1/2\rangle = \sqrt{\frac{2}{3}} \downarrow Z + \uparrow \frac{X - iY}{\sqrt{6}} \tag{14.3}$$

$$|3/2, -3/2\rangle = \downarrow \frac{X - iY}{\sqrt{2}}$$

Using these notations, the Bir–Pikus Hamiltonian describing the influence of the strain on the valence band structure is written:

$$H_{BP} = -a_v \sum_i \varepsilon_{ii} - b \sum_i \varepsilon_{ii}(j_i^2 - 1/3j^2) - \frac{2d}{\sqrt{3}}\left(\varepsilon_{xy} \frac{j_x j_y + j_y j_x}{2} + c.p. \right) \tag{14.4}$$

where the abbreviation $c.p.$ is used for circular permutation over x, y and z corresponding to the crystal axes. The matrices of the angular momentum operator j can be found in [69]. ε_{ij} denotes the ij component of the strain tensor and a_v, b and d are the deformation potentials for the valence band. Using the hole formalism, J. Allègre *et al.* found $a_v = 0.91\,\text{eV}$, $b = 0.99\,\text{eV}$ and $d = 2.76\,\text{eV}$ for CdTe [73].

While growing CdTe on ZnTe, CdTe is compressed in the growth plane and distended in the growth direction. CdTe QDs form by deformation of a two-dimensional layer induced by strain relaxation [45]. In this formation mechanism, both elastic and inelastic strain relaxations are involved and one expects each QD to probe a different local in-plane strain distribution. When studying a particular QD, one has to consider a particular set of values for ε_{ij}. In the following, we will mainly consider the effects of strain anisotropy in the growth plane and describe the strain on each QD by an average value of ε_{xy} and $\varepsilon_{xx} - \varepsilon_{yy}$. A non-zero value of the volume average ε_{ij} can result from a local deformation of the lattice induced by neighbouring QDs or dislocations. In this approximation, the Bir and Pikus Himiltonian is reduced to a block-diagonal matrix in the $(+3/2, -1/2, +1/2, -3/2)$ basis:

$$H_{BP} = \begin{pmatrix} P+Q & R & & \\ R^\dagger & P-Q & & \\ & & P-Q & R \\ & & R^\dagger & P+Q \end{pmatrix} \tag{14.5}$$

with

$$P = a_v \sum_i \varepsilon_{ii}$$

$$Q = b\left(\frac{\varepsilon_{xx} + \varepsilon_{yy}}{2} - \varepsilon_{zz} \right) \tag{14.6}$$

$$R = id\varepsilon_{xy} - b\frac{\sqrt{3}}{2}\left(\varepsilon_{xx} - \varepsilon_{yy} \right).$$

Using the *hh* band as the origin of the energies in the valence band, the strain Hamiltonian can be rewritten as:

$$H_{BP} = \begin{pmatrix} 0 & \rho_s e^{-2i\theta'_s} & & \\ \rho e^{2i\theta'_s} & \Delta_{lh} & & \\ & & \Delta_{lh} & \rho e^{-2i\theta'_s} \\ & & \rho_s e^{2i\theta'_s} & 0 \end{pmatrix}. \tag{14.7}$$

This notation allows us to introduce useful parameters to describe the strain effects, namely the light–heavy hole splitting Δ_{lh}, the strain coupling amplitude ρ_s and the strain-induced anisotropy axis in the QD plane defined by the angle θ'_s with respect to the x (100) axis. In the following, we will define angles relative to the (110) axis corresponding to the cleaved edge of the sample. We thus introduce the new angle $\theta_s = \theta'_s - 45°$. Because of the strain-induced VBM described by the Hamiltonian (Eq. 14.7), the general form for the lowest energy hole states is:

$$|\varphi_h^\pm\rangle \propto \chi_{hh}(\vec{r})\,|\pm 3/2\rangle - \frac{w_{lh}}{w_{hh}}\chi_{lh}(\vec{r})\,|\mp 1/2\rangle \tag{14.8}$$

where $w_{lh(hh)}$ is the probability of the hole being light (heavy) and $X_{lh(hh)}(\vec{r})$ its envelope function.

Considering now single Mn-doped QDs, the sp–d exchange interactions between the confined carriers and the single Mn atom are responsible for a peculiar fine structure of the emission [4]. It has been demonstrated that the hole–Mn spin interaction is of prime importance in this fine structure, because of its energy range and above all because of the spin anisotropy of the hole [67, 68]. It appears clearly that a valence band mixing will have dramatic consequences on the emission fine structures of Mn-doped QDs. In the heavy-hole approximation, a hole acts on the

Mn spin as an effective magnetic field applied along the growth axis, lifting its spin degeneracy. If a mixing of heavy and light holes is introduced, one has to consider a system of coupled spins in which hole–Mn spin flips become possible.

One can find a signature of the valence band mixing in the emission spectra of neutral single Mn-doped QDs [69]. In an Mn-doped QD, the ground state of the exciton is composed of six bright energy levels and six dark energy levels resulting from the coupling of bright excitons and dark excitons with the six projections of the Mn spin. For a given exciton spin (± 1 or ± 2), the six Mn spin projections have different energies. In this system, the hole spin interacts with both the Mn spin and the electron spin. The effect of an increasing valence band mixing on the excitonic emission spectrum of a QD is presented in Fig. 14.10a. Parameters are chosen to reproduce the features of QD5 emission spectrum plotted on Fig. 14.10b: the e-Mn exchange integral I_{e-Mn} is set to $70\,\mu eV$; the splitting between bright and dark exciton δ_0 is chosen equal to $550\,\mu eV$, equally divided among the long range and short range e–h exchange interaction. Other parameters are the same as in the previous charged exciton calculation.

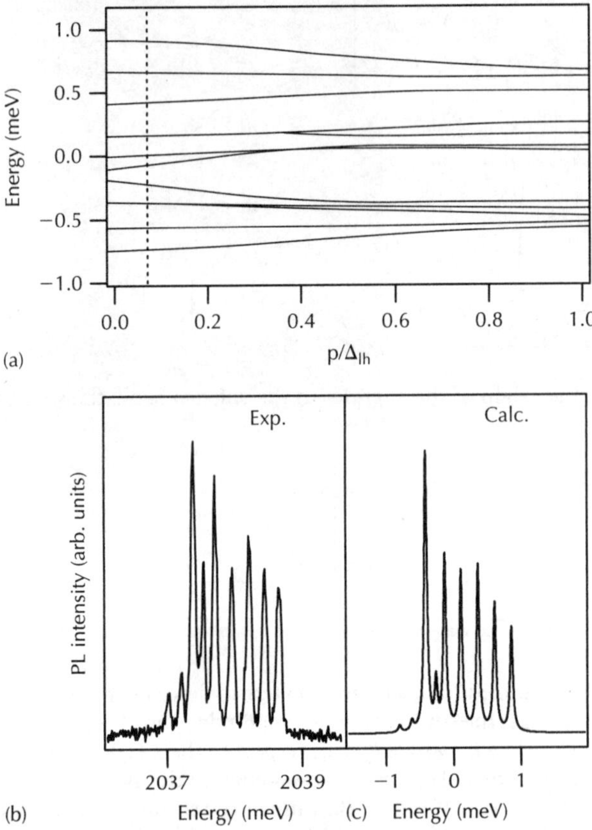

Figure 14.10   (a) Dependence of the exciton–Mn energy spectrum on the strength of the valence band mixing ρ/Δ_{lh}. The dotted line corresponds to the situation of QD6 ($\rho/\Delta_{lh} = 0.07$). (b) and (c) are the experimental and calculated emission spectra of QD6.

14.4.2   Influence of the shape anisotropy

Let us consider now the influence of the electron–hole exchange interaction. We have seen in the case of non-magnetic QDs that, in first approximation, the only effect of the valence band mixing on the short range e–h exchange is to couple bright exciton states together. For magnetic QDs, bright states associated with the same Mn spin projection are thus coupled. This coupling mainly concerns the energy levels associated with Mn spin projections $\pm 1/2$ because they are only slightly split by the exciton–Mn exchange interaction. An increase of the valence band mixing will thus open a slight gap in the middle of the bright exciton fine structure (around 0.25 meV in Fig. 14.10a).

Another gap appears on the low energy side of the structure because of the hole–Mn interaction. As a matter of fact, simultaneous hole–Mn spin flips couple bright and dark states associated with consecutive Mn spin projections. The exciton–Mn exchange interaction induces an overlap of the bright and dark exciton fine structure. States coupled by h–Mn spin flips are therefore quite close to each other and an increase in the VBM, that is to say an increase in the h–Mn spin flip efficiency, will open a gap between them. This can be observed in Fig. 14.10a around -0.1 meV. Finally, one may note once again that the overall splitting of the structure decreases as the VBM increases because of the decrease of the weight of the heavy-hole in the exciton ground state.

Such coupling between bright and dark excitons is observed in the spectrum of QD6 presented in Fig. 14.10b. In this QD, the valence band mixing is quite weak and the gaps induced by the e–h exchange and the h–Mn exchange can be hardly resolved in the exciton spectrum. Nevertheless, the h–Mn spin flips-induced bright–dark coupling is large enough to give non-negligible oscillator strengths to the dark states. Three "dark" states are then observed in the emission spectrum. The value ρ_s/Δ_{lh} can be estimated at 0.07. Using this parameter we can calculate the QD6 exciton emission spectrum. The result is plotted in Fig. 14.10c. It has been shown in previous work [4] that the exciton–Mn system is partially thermalized during the exciton lifetime. This results in an increase in the intensity of the lines as their energy decreases. In the calculation we thus consider that the system has relaxed and that it is characterized by an effective spin temperature T_{eff} depending both on the phonon bath and on the hot carrier gas produced by the laser excitation. In Fig. 14.10c $T_{eff} = 15$ K. With this effective temperature, "dark" states appear in the calculated spectrum but their intensity is weaker than in the experiment. Moreover, comparing the intensities of the bright lines in the calculation and in the experiment, we note that we cannot fit the data with an effective temperature smaller than 15 K in order to increase the "dark" state's intensity. These features show that we should not consider a total thermalization in the exciton–Mn system: if the spin relaxation is partially blocked, it is difficult to thermalize the system. As bright and dark excitons are created with the same generation rate, the intensities of the associated transitions can be comparable even though they have different oscillator strengths. This is particularly the case under weak excitation conditions where the influence of the hot carrier gas spin reservoir is reduced.

We finally consider the neutral exciton fine structure of a single Mn-doped QD that presents an anisotropic in-plane shape (QD7 presented in Fig. 14.11). Effects of the long range electron–hole exchange interaction on the emission spectrum of a single Mn-doped QD were discussed in a former letter [70]. Nevertheless, details of the polarization properties were not analysed. Here, we detail these polarization features and show that the VBM is once again necessary to explain them. Figure 14.11 presents the linear polarization dependence of the emission spectrum of QD7. The experimental spectrum shows the main characteristics of a quite strong VBM as detailed in Fig. 14.11a. The spectrum presents an overall linear polarization rate of about 25% orientated at $\theta_s = 110°$ from the cleaved edge of the sample and dark states appear on the low energy side of the structure. Two gaps can also be observed: one between the two first main lines and the other between the third and the fourth lines. However, if these peculiarities were only due to the combined effect of the VBM and short range e–h exchange, one could expect the third and the fourth lines to be polarized parallel and perpendicular to the strain direction (110°). It is not the case and, surprisingly, the emission spectrum also presents non-orthogonal linear polarization directions.

These polarization directions are the signature of a competition between a valence band mixing and long range electron–hole exchange interaction. The calculation of the emission spectrum of QD7 (right panel of Fig 14.11) gives a good agreement with the experiment using the following parameters: $I_{e-Mn} = 60\,\mu eV$, $I_{h-Mn} = 135\,\mu eV$, $\delta_0^{lr} = \delta_0^{sr} = 300\,\mu eV$, $\delta_2 = 450\,\mu eV$, $\Delta_{lh} = 30$ meV, $\rho_s/\Delta_{lh} = 0.25$ and $T_{eff} = 20$ K. The strain direction and the dot shape direction must be roughly perpendicular (80°) to fit the experiment. The main features of QD7 emission are well reproduced as presented in Fig. 14.11.

In anisotropic QDs, the interplay between the electron–hole and exciton–Mn exchange interactions is confirmed by magneto-optical measurements (Fig. 14.12). The typical Zeeman splitting of the six lines is clearly observed in the data at all fields, with a strong intensity gradient at the highest fields resulting from a rather strong Mn spin polarization. For this clearly anisotropic

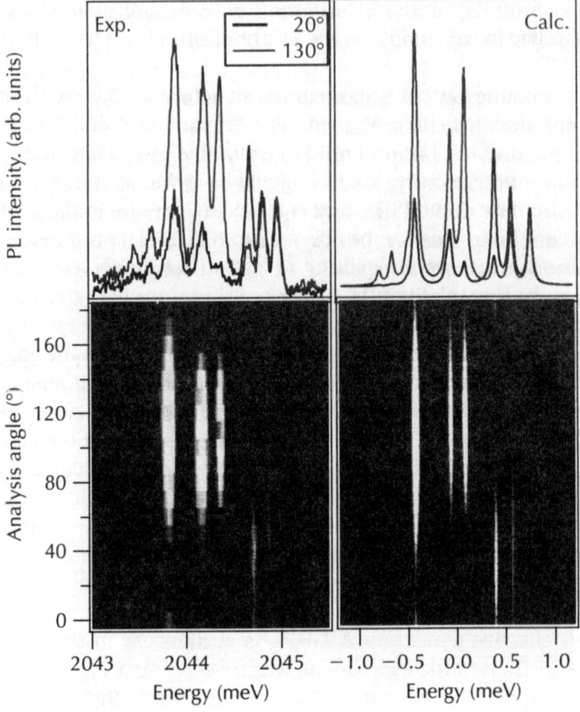

Figure 14.11 *Experimental and calculated polarization resolved photoluminescence of QD6. The intensity maps present the dependence of the emission on the analyser angle. The emission spectrum are presented for orthogonal linear analyser directions (red and black curves).*

dot, the central gap in the emission structure is maintained in both circular polarizations, with a small quadratic diamagnetic energy shift. This behaviour is explained as follows: the dot anisotropy leads to successive anticrossings of the ± 1 bright exciton states associated with given Mn spin projections ($-1/2$, $-3/2$ and $-5/2$) as a function of magnetic field – As B increases, transitions associated with the $J_z = +1$ exciton shift up in energy whereas the $J_z = -1$ transitions shift down. The anisotropic part of the electron–hole exchange interaction mixes successively the $J_z = \pm 1$ exciton states associated with $S_z = -1/2$, then with $S_z = -3/2$ and finally with $S_z = -5/2$ at successively higher B. For QD3, these anticrossings are observed successively at 2.5, 7 and 11 T.

To understand fully the rich magnetic behaviour of these anisotropic QDs, we calculated the optical transitions under magnetic field by diagonalizing the complete Hamiltonian of the electron-heavy hole–Mn system (including the exchange, Zeeman and diamagnetism Hamiltonians). Calculated transitions are presented in Fig. 14.12b. The fitted Landé factors of the electron ($g_e = -1.1$), the hole ($g_h = 0.3$) and the Mn atom ($g_{Mn} = 2.0$), the splitting between $J_z = \pm 1$ and $J_z = \pm 2$ excitons ($=1\,\text{meV}$) and the diamagnetic factor ($\gamma = 2.45\,\mu\text{eV} \cdot \text{T}^{-2}$) agree well with previous work [4, 29, 56]. Parameters δ_2 and δ_{Mn} were adjusted to fit the zero field data, as explained earlier.

Comparison between calculation and data explains most of the details of the magneto-optic properties. In particular, around 7 T, the central gap is perturbed in both circular polarizations. In $\sigma-$, this is due to anticrossings induced by the mixing of $|s_e = 1/2, j_{hz} = -3/2, S_z>$ states and $|-1/2, -3/2, S_z + 1>$ states by the *electron*–Mn exchange [4], i.e. corresponding to simultaneous spin-flips of electron and manganese spins. In $\sigma+$ polarization, Fig. 14.12b shows that the line of second lowest energy crosses the central gap as an essentially non-radiative transition. This implies a mixing of $|-1/2, 3/2, -3/2 >$ and $|-1/2, -3/2, -1/2 >$. This is a second-order mixing involving both mixing of $|-1/2, -3/2, -1/2>$ and $|1/2, -3/2, -3/2>$ by the e–Mn

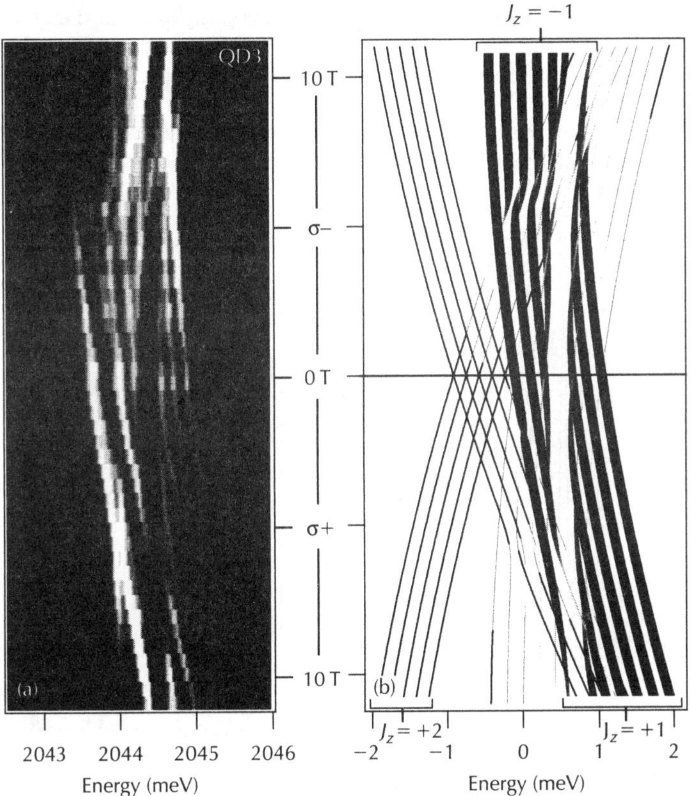

Figure 14.12   (a) Intensity map of magnetic field dependence of the emission spectrum of an asymmetric Mn-doped dot QD3, for circular polarization $\sigma+$ and $\sigma-$. (b) Optical transitions obtained from the diagonalization of the spin + Zeeman + diamagnetism Hamiltonian in the subspace of the 24 heavy-hole exciton and Mn spin components; line thickness and colour scale for $\sigma+$, $\sigma-$ are proportional to absolute value of the projection of the exciton state on the $J_z = +1$, -1 exciton, respectively (green = low intensity, blue = high intensity). The two transitions which are forbidden at all magnetic fields ($|J_z = \pm 2, S_z = \mp 5/2\rangle$) are not plotted.

exchange and mixing of $|-1/2, 3/2, -3/2\rangle$ and $|1/2, -3/2, -3/2\rangle$ by the anisotropic e–h exchange; that is, the e–Mn exchange induces a mixing of states mediated by the anisotropy-induced coupling.

To summarize these geometrical effects, as for as this study is concerned an anisotropic stain distribution in the growth plane and a QD-shaped anisotropy induce a valence band mixing which strongly modifies the emission spectra of single Mn-doped QDs. The main observed effects are: (i) the appearance of anticrossings related to possible Mn–hole spin-flips, (ii) the separation of the six line structures into two sets of lines partially polarized, and (iii) the variation of the linear polarization directions which are no longer perpendicular to the exciton fine structure of a given QD. The last feature clearly evidences the competition between two types of anisotropy, one coming from the confinement potential (shape anisotropy) and one coming from the local strain distribution into the dot.

14.5   Carrier-controlled spin properties of a single magnetic atom

14.5.1   Carrier-induced spin splitting of a single Mn atom in a quantum dot

Investigating both the biexciton and the exciton transitions in the same Mn-doped QD, we analyse the impact of the Mn–exciton exchange interaction on the fine structure of the QD emission.

A single electron–hole pair is enough to induce a spontaneous splitting of the exciton–Mn system and a slight polarization of the Mn spin distribution. The injection of a second electron–hole pair cancels the exchange interaction with the Mn ion and the Mn spin degeneracy is almost completely restored.

In the following, we will compare the fine structure of the exciton and biexciton in a single Mn-doped QD. Figure 14.13 shows the emission spectra of one of these Mn-doped QDs for various excitation

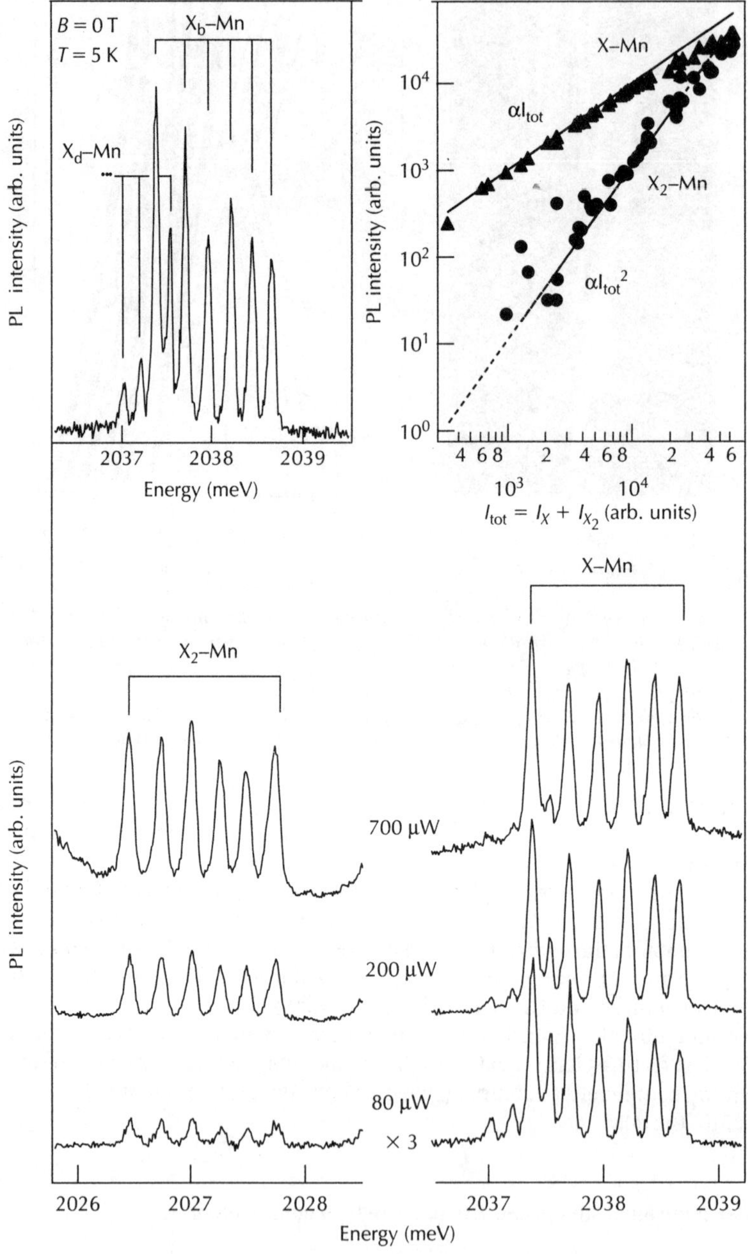

Figure 14.13   PL spectra of a single Mn-doped QD for different excitation intensity. X–Mn denotes the exciton–Mn emission while X_2–Mn corresponds to the recombination of the two-exciton–Mn (biexciton) states. The insets give a detail of the X–Mn emission spectra at low excitation density (X_b–Mn corresponds to the bright excitons and X_d–Mn corresponds to the dark excitons) and the evolution of the integrated intensities of X–Mn and X_2–Mn as a function of the total emission intensity.

densities. At low excitation, the spectrum is dominated by the six line fine structure (labelled X–Mn) which is attributed to a confined exciton interacting with a single Mn atom. The three additional lines, observed at low excitation intensity in the low energy side of the emission structure (enlarged view in the inset of Fig. 14.13), are due to the interaction of the dark exciton states ($J_z = \pm 2$) with the Mn^{2+} ion [69]. The PL contribution of the dark states is enhanced by the relaxation of the exciton–Mn system towards the low-energy levels. This relaxation is also responsible for the asymmetry in the emission intensity distribution of the bright states at low excitation intensity [4].

On increasing the excitation intensity, a new structure of six lines, labelled X_2–Mn, appears 10.9 meV below the exciton lines. Its intensity increases quadratically with the excitation and its structure reproduces the spectral pattern of X–Mn. Such intensity dependence has been observed in various QD systems and is regarded as a fingerprint of exciton and biexciton emission [62, 63]. The energy difference between the X–Mn and the X_2–Mn transitions (10.9 meV) corresponds to the typical binding energy of the biexciton measured in non-magnetic CdTe/ZnTe QDs [62]. We have also noticed that, as the excitation intensity increases, the contribution of the transitions associated with the dark exciton states progressively vanishes. This evolution, already observed in non-magnetic QDs [62], is due to the longer lifetime of the dark exciton states. Because of this difference in lifetime, for a given non-resonant excitation intensity, the probability to create a biexciton with two dark excitons is larger than with two bright excitons. The formation of the biexciton acts then as an efficient recombination channel for the dark exciton states and decreases their direct PL contribution.

Magneto-optic measurements confirm that the exciton and the biexciton transitions coming from the same Mn-doped QD are observed. The magnetic field dependence of the exciton–Mn and biexciton–Mn transition energies are presented as a contour plot in Figure 14.14. For X–Mn, six lines are observed in each circular polarization. Their energy follows the Zeeman and diamagnetic shift of the exciton in non-magnetic QDs [4]. The excitonic transitions present also a rich fine structure mainly characterized by a series of five anticrossings observed in $\sigma-$ polarization around 7 T. These anticrossings come from the mixing of the bright and dark exciton states [4] induced by a simultaneous spin flip of the electron and Mn^{2+} ion. The fine structure of the exciton–Mn and the biexciton–Mn systems under magnetic field present a perfect mirror symmetry. The anticrossings observed on the high energy lines of X–Mn in $\sigma-$ polarization are symmetrically observed on the low energy lines and in $\sigma+$ polarization for X_2–Mn (circles in Fig. 14.14) [70].

As it appears in the contour plot presented in Fig. 14.14, the relative intensities of the six emission lines observed in each circular polarization strongly depend on the applied magnetic field. The emission intensity, which is almost equally distributed over the six emission lines at zero field, is concentrated on the high-energy side of the $\sigma-$ emission and on the low-energy side of the $\sigma+$ emission at high magnetic field. The intensity distribution is similar for X–Mn and X_2–Mn.

In a QD, the biexciton ground state is a spin-singlet state ($J = 0$) and cannot be split by the magnetic field or the spin interaction part of the carriers–Mn Hamiltonian [71]. In this model, the creation of two excitons in the same QD cancels all the exchange interaction terms with the Mn^{2+} ion. Thus, the fine structure of the biexciton–Mn transitions is controlled by the final state of the recombination of the biexciton, i.e. the eigenstates of the exciton–Mn coupled system. The optical transitions directly reflect this mirror symmetry of the energy levels. Moreover, the intensity distribution of the X–Mn and X_2–Mn transitions are both controlled by the spin polarization of the Mn^{2+} ion. The spin of the Mn^{2+} ion is orientated by the applied magnetic field. At high field, when the biexciton recombines, the probability of leaving in the QD an exciton coupled with a Mn spin component $S_z = -5/2$ is enhanced. Therefore, in the two-photon cascade, occurring during the recombination of a biexciton, the polarization of the Mn spin will enhance the intensity of the high-energy biexcitonic transition in $\sigma-$ polarization and the one in the low-energy single-exciton transition in $\sigma+$ polarization. Symmetrically, the emission intensity is concentrated in the low-energy biexcitonic transition in $\sigma+$ polarization and in the high-energy single-exciton transition in $\sigma-$ polarization.

In summary, the analysis of the biexciton fine structure shows that the optical injection of a controlled number of carriers in an individual QD allows control of the spin splitting of a single magnetic ion. The exchange interaction with a single exciton acts as an effective local magnetic field

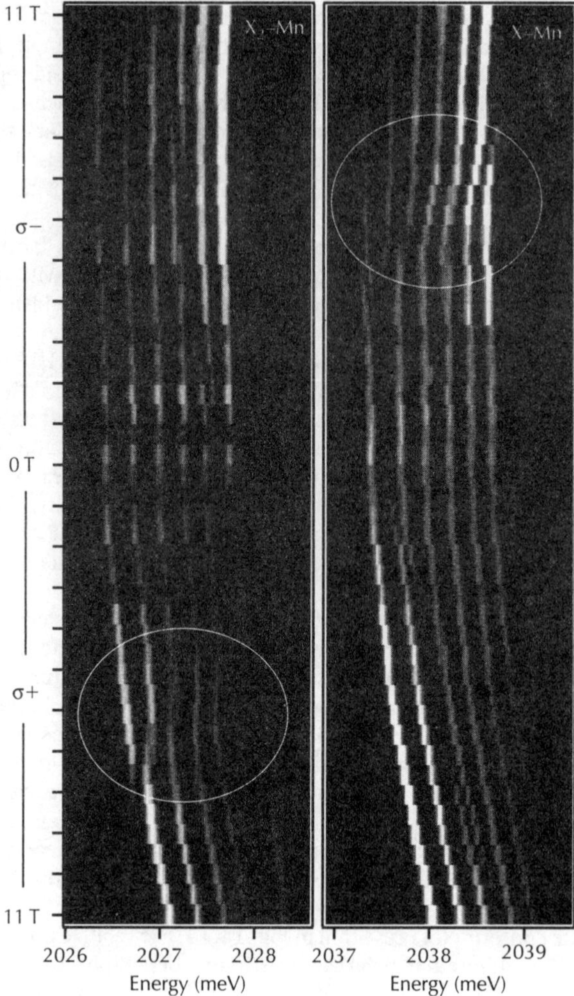

Figure 14.14   Intensity contour plot of the magnetic field dependence of the excitonic and biexcitonic transitions of a single Mn-doped QD. The bright areas correspond to the high PL intensity. The anticrossings observed for the exciton are symmetrically reproduced on the biexciton transitions.

which splits the Mn levels in zero applied magnetic field. The injection of a second exciton almost cancels the carriers – Mn exchange interaction and the Mn spin splitting is significantly reduced.

14.5.2   Electrical control of a single Mn atom in a quantum dot

CdTe/ZnTe QDs are p-type modulation doped by the transfer of holes from the p-doped ZnTe substrate and from surface states that act as acceptors [60, 61]. The occupation of the QDs by holes can be controlled by an external bias voltage V on an aluminium Schottky gate with respect to a back contact on the p-type substrate. The bias-dependent emission of a non-magnetic QD and a Mn-doped QD is presented in Fig. 14.15. For increasing V, the surface level states are shifted below the ground hole level in the QDs which results in the single hole charging of the dots. The optically generated excitons then form charged excitons with the bias-induced extra hole in the QD. At zero bias or negative bias, the Fermi level is above the ground state and the QDs are likely to be neutral. However, the separated capture of photo-created electron or holes can sometimes charge the dots so that weak contributions of X^+ or X^- are observed in the zero bias spectra.

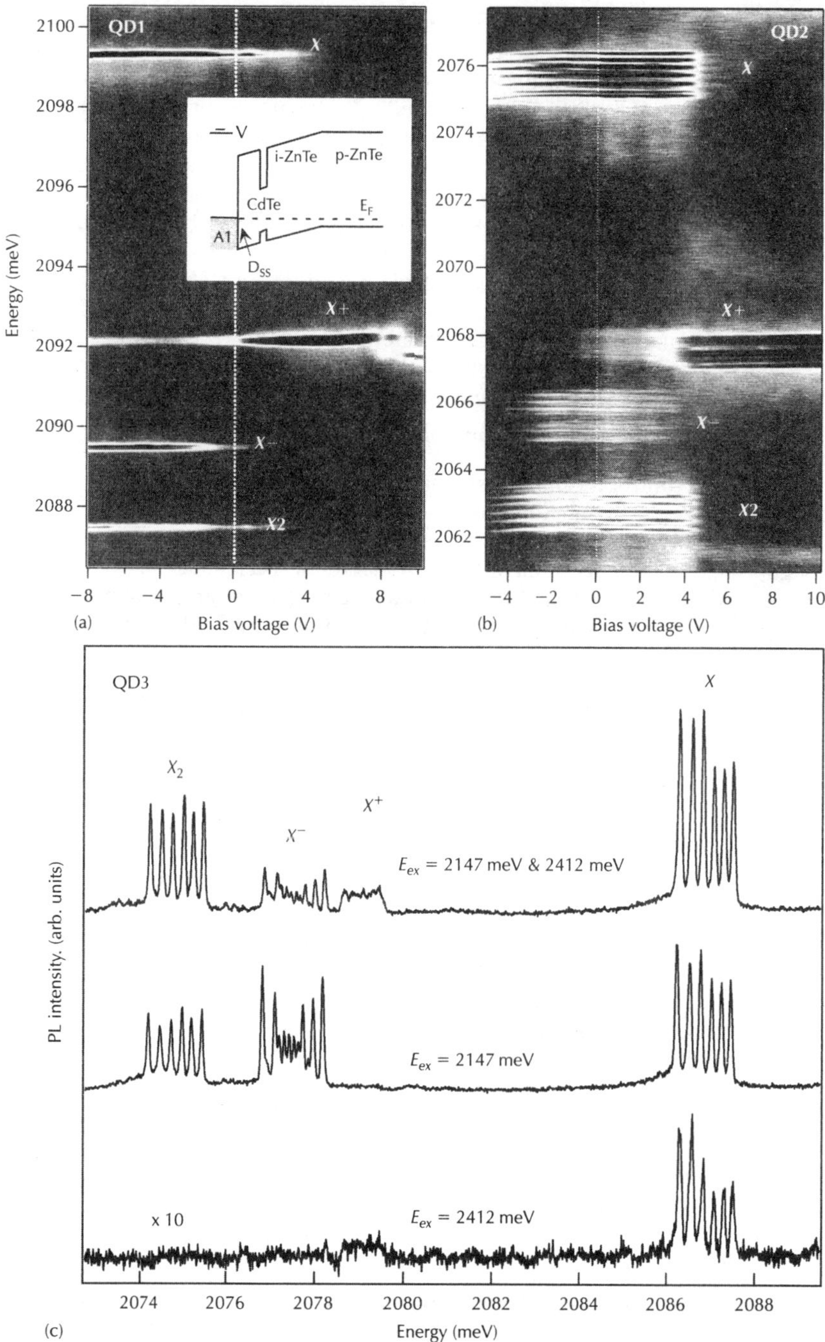

Figure 14.15   *Colour-scale plot of the photoluminescence intensity of a non-magnetic QD (a) and a single Mn-doped QD (b) in a Schottky structure as a function of emission energy and bias voltage. The series of emission lines can be assigned to QD s-shell transitions, namely the recombination of the neutral exciton (X), biexciton (X^2), positively charged exciton (X$^+$) and negatively charged exciton (X$^-$). (c) Details of the PL of a single Mn-doped QD under resonant excitation (E_{ex} = 2147 meV), non-resonant excitation (E_{ex} = 2412 meV) and both resonant and non-resonant excitation.*

At zero bias, excess electrons can also be injected in the QD using resonant optical excitation into the QD levels. Under resonant excitation (energy below the band gap of the barriers), optical transitions from delocalized valence band states to the confined electron levels will preferentially create electrons in the QD [64]: the probability of finding an excess electron in the QD is increased. As presented in Fig. 14.15c, the negatively charged exciton emission is then seen for some discrete excitation energies. After the recombination of the charged exciton X^-, a single hole is likely to be captured to neutralize the QD and create a neutral exciton. This neutralization process is responsible for the simultaneous observation of charged and neutral species under resonant excitation (Fig. 14.15c).

The charge state of the dot can also be optically tuned (Fig. 14.15c). By combining a weak non-resonant excitation with the resonant one, a few carriers are created in the ZnTe barrier. They do not significantly contribute to the luminescence (lower PL spectra in Fig. 14.15c) but reduce the PL contribution of X^- in favour of the neutral species. This evolution is characteristic of a photo-depletion mechanism in modulation-doped QDs [54]. High-energy photoexcited e–h pairs are dissociated in the space charge region surrounding the negatively charged QDs and neutralize the QDs.

These two charge control mechanisms (bias voltage and resonant excitation combined with photo-depletion) allow the interaction between individual carriers (electron or hole) and an individual magnetic atom to be independently probed. Let us first consider the negatively charged exciton. Figure 14.16a presents detail of the recombination spectrum for X^- coupled with a single Mn atom obtained at zero bias under resonant excitation. Eleven emission lines are clearly observed with intensity decreasing from the outer to the inner part of the emission structure.

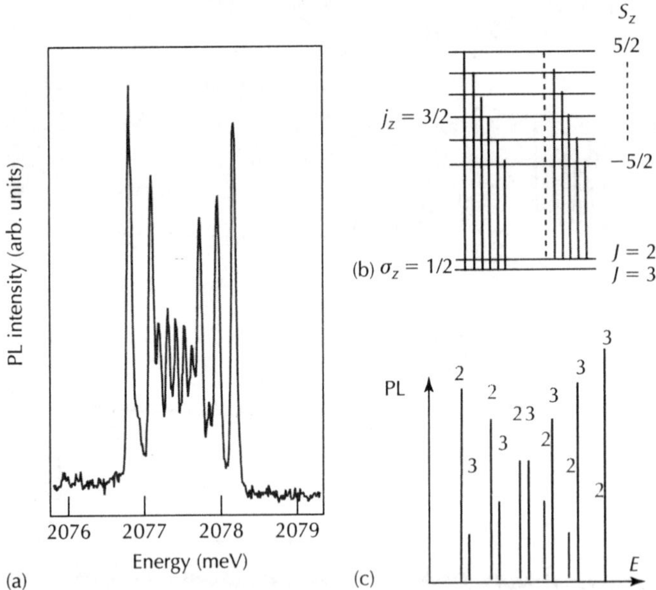

Figure 14.16   (a) Detail of the unpolarized emission spectrum of a negatively charged exciton (X^-) coupled with an Mn atom. (b) Scheme of the $\sigma+$ optical transitions of (X^-, Mn) and their respective PL intensity distribution (c).

A simple effective spin Hamiltonian quantitatively accounts for the emission spectrum shown in Fig. 14.16a. The emitting state in the X^- transition has two conduction band (CB) electrons and one hole coupled to the Mn. The effect of the two spin-paired electrons on the Mn is strictly zero. Thereby, the spin structure of the X^- state is governed by the interaction of the hole with the Mn. Based upon the results presented in section 14.3, we propose the following Hamiltonian:

$$\mathcal{H}_{h-Mn} = I_h \left(S_z j_z + \frac{\epsilon}{2} (j_- S_+ + j_+ S_-) \right) \tag{14.9}$$

where $\vec{S}$ is the Mn spin operator and $\vec{j}$ acts on the hole's lowest energy doublet. The first term is the spin conserving or Ising exchange whereas the second is only possible if there are some heavy–light hole mixing [49, 65]. From our measurements we find that ϵ is small and, in a first stage, we neglect it. Later we will show its influence. The 12 eigenstates of $\mathcal{H}_{h-Mn}$ with $\epsilon = 0$ are organized as six doublets (Fig. 14.16b) with well-defined S_z and j_z (Mn and hole spin along the z-axis). We label these states as $|S_z, j_z\rangle$. Recombination of one of the CB electrons with the hole of the X^- state leaves a final state with a single CB electron coupled to the Mn. The spin Hamiltonian of this system is the ferromagnetic Heisenberg model, $\mathcal{H}_{e-Mn} = -I_e \vec{S} \cdot \vec{\sigma}_e$. The 12 eigenstates of the Mn–electron complex are split into a ground state septuplet (total spin $J = 3$) and a five-fold degenerate manifold with $J = 2$. We label them all as $|J, J_z\rangle$.

Thereby, for each of the six doublets of X^- there are two possible final states after annihilation of an e–h pair, with either $J = 2$ or $J = 3$. From this consideration alone, we would expect 12 spectrally resolved lines. Their weight is given by both optical and spin conservation rules. Since electrons and holes reside in s and p bands, respectively, the $\Delta L = 1$ optical selection rule is immediately satisfied. The polarization of the photon imposes an additional selection rule on ΔM which is accounted for by the spin of the electron and hole. The Mn spin is not affected by the transition.

The weight of optical transitions between the initial state $|i\rangle = (\uparrow, \downarrow)_e \times |S_z, j_z\rangle$ and the final state $|f\rangle = |J, J_z\rangle$ is proportional to $|\langle f| \sum_\sigma P(\sigma, j_z) c_\sigma d_{j_z} |i\rangle|^2$, where c_σ annihilates a CB electron with spin σ, and d_{j_z} annihilates a VB hole with angular momentum j_z. Here $P(\sigma, j_z)$ is given by the polarization selection rule.

Let us consider, for instance, $\sigma+$ recombination transitions where the $(\downarrow_e, \Uparrow_h)$ e–h pair is annihilated. Each of the six doublets, characterized by their Mn spin projection S_z, can be an initial state. After the electron–hole annihilation, the resulting state is $|S_z, \uparrow_e\rangle$ which, in general, is not an eigenstate of $\mathcal{H}_{e,Mn}$. The intensity of the optical transition to a given final state $|J, Jz\rangle$ is proportional to the overlap $\langle J, J_z|S_z, \uparrow_e\rangle$, which is nothing but a Clebsh–Gordan coefficient. The highest energy transition, with $\sigma+$ polarization, would correspond to the initial state $(\uparrow, \downarrow)_e \times | +5/2, \Uparrow_h\rangle$ and a low energy final state $|J = 3, J_z\rangle$. After the photon emission, the state of the system is $|S_z = +5/2, \uparrow_e\rangle$ which is identical to $|J = 3, J_z = +3\rangle$ and thereby gives the highest optical weight (Fig. 14.16b). In contrast, emission from that initial state to $|J = 2, J_z\rangle$ is forbidden. The other five doublets have optical weights lying between $1/6$ and $5/6$ with both $|J = 2, J_z\rangle$ and $|J = 3, J_z\rangle$ final states. Thereby, the number of spectrally resolved lines in this model is 11.

The relative weight of the emission lines is accounted for by the model. According to the final state, the transitions belong to either the $J = 2$ or the $J = 3$ series. As the initial S_z decreases, the overlap of $|\uparrow_e S_z\rangle$ to the $J = 3$ ($J = 2$) states decreases (increases). As presented in Fig. 14.16c, the PL of X^- can be seen as a superposition of two substructures: six lines with intensities increasing with their energy position (transitions to $J = 3$ states) and five lines with intensities decreasing with increasing their energy position (transitions to $J = 2$ states).

Reversing the role of the initial and final states, and neglecting the small coupling of two holes to the Mn spin [74], this model should account for the emission from X^+ states. Actually, different energy splittings are observed for the different excitonic species in the same QD. For instance, in QD3 (Fig. 14.15c) one measured $\Delta E_X = 1.23\,\mathrm{meV}$, $\Delta E_{X-} = 1.36\,\mathrm{meV}$ and $\Delta E_{X+} = 0.95\,\mathrm{meV}$. The energy splitting is mostly due to the Mn–hole exchange coupling, which in turn is inversely proportional to the volume of the hole wave function. The difference between ΔE_{X+} and ΔE_{X-} indicates that a significant fraction of the confinement of the hole comes from the Coulomb attraction of the two electrons in the initial state of the X^- emission. In contrast, in the final state of the X^+ emission, there is no electron to attract the hole, resulting in a spread of the hole wave function and a smaller exchange energy. This difference appears directly in the emission structure shown in Fig. 14.15, where the peak structure of X^+ is not well resolved.

In Fig. 14.17a we show the intensity of X^- emission as a function of the direction of a linear analyser. It is apparent that the central lines are linearly polarized. This polarization can only be understood if we allow for some spin–flip interaction between the Mn and the hole (second term in Eq. 14.9). Provided that $\epsilon \ll 1$, the effect of this interaction is small both on the wave function and on the degeneracy of all the doublets except the third, which is split, as illustrated in the inset of Fig. 14.17. The split states are the bonding and antibonding combinations of $|S_z = -1/2, \Uparrow_h\rangle$

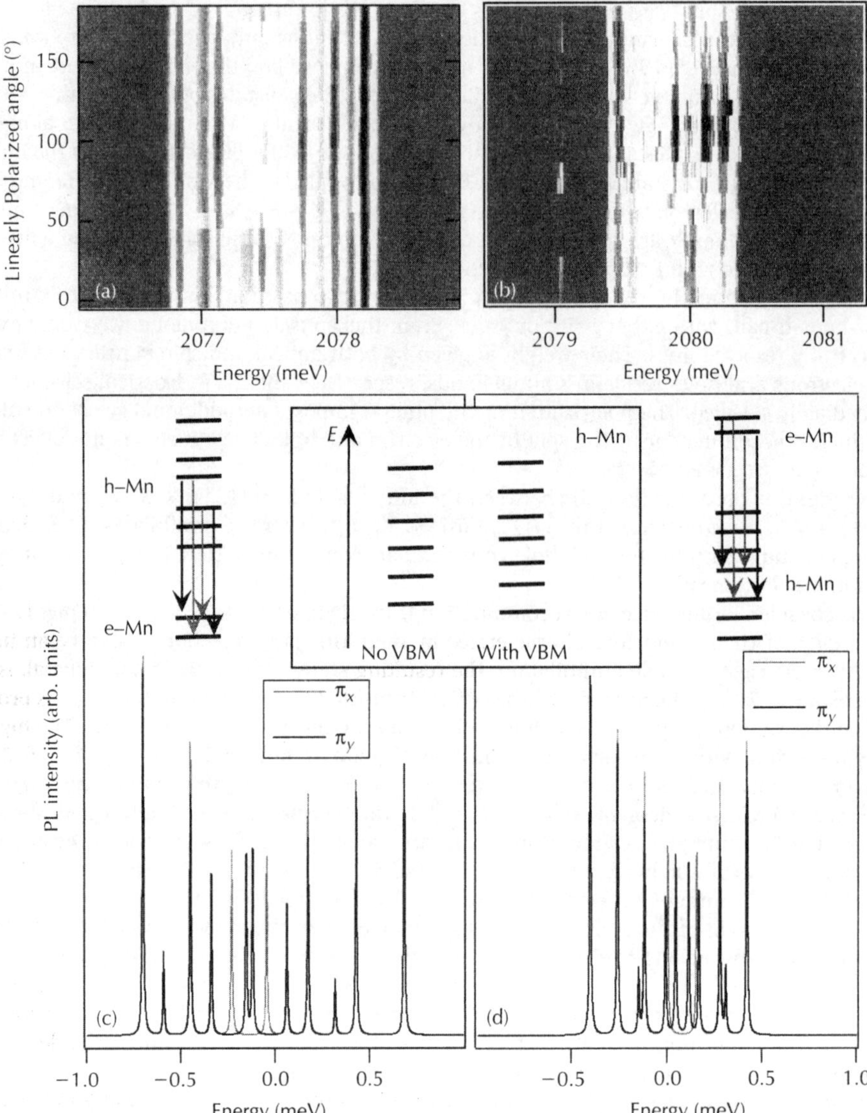

Figure 14.17 Colour scale plot of the dependence of the PL of (X^-, Mn) (a) and (X^+, Mn) (b) on the direction of a linear analyser, in the same QD. Three lines in the centre of the structure are linearly polarized. (c) and (d) Calculated linearly polarized PL spectra of (X^-, Mn) and (X^+, Mn) with exchange integrals I_e and I_h chosen to reproduce the overall splitting for X^+ and X^- presented in (a) and (b). Transitions are arbitrarily broadened by $10\,\mu eV$. The schemes in (c) and (d) show energy levels involved in (X^-, Mn) and (X^+, Mn) recombination with valence band mixing. The central inset presents the energy level scheme of the h–Mn system without and with valence band mixing (VBM).

and $|S_z = +1/2, \Downarrow_h\rangle$. These states are coupled, via linearly polarized photons, to the $|2,0\rangle$ and $|3,0\rangle$ e–Mn complex and four linearly polarized lines should be observed on the emission spectra as shown in the insets of Fig. 14.17c and d. Polarization directions are controlled by the distribution of strains through the Bir–Pikus Hamiltonian [65, 76].

We can obtain numerical values of I_h, I_e and ϵ for the charged excitons comparing the transition probabilities calculated with model (14.9) (Fig. 14.17c) to the experimental data. The electron–Mn exchange integral I_e is deduced from magneto-optics measurements on the neutral exciton [4]. I_h, the hole–Mn exchange integral, is then chosen to reproduce the overall splitting of the two charged species' emission structure. There is a good agreement between the experimental data (QD3 in Fig. 14.15c) and our model with $I_e = 40\,\mu eV$, $I_h(X) = 150\,\mu eV$, $I_h(X^+) = 95\,\mu eV$ and $I_h(X^-) = 170\,\mu eV$. The different values of I_h directly reflect the expected variation of the confinement of the hole. The main characteristics of the emission spectra are well reproduced, namely the number of emission lines, their intensity distribution and the linear polarization structure, with a slight valence band mixing coefficient $\epsilon = 0.05$. This small value of the valence band mixing coefficient shows that hole–Mn exchange interaction remains highly anisotropic.

14.6   Conclusion

Let's recall that the precise control of electronic spins in semiconductors is a very active field nowadays, the so-called spintronics, which needs, besides the potential applications, to have a deep insight of the effects which combine manipulation of charges and manipulation of spins. Among this large activity, the development of techniques for optical control of single spins in the solid state is motivated by the possibility of using single spins embedded in nanostructures for the implementation of quantum information processing devices. These devices will require the detection and manipulation of individual spins and ensure their interaction with the semiconductor carriers. However, in such a nanometre-size system, it is important to tune independently the quantum dot charge states and the magnetic doping in order to elucidate the interactions with the different charge carriers.

Quantum dots made of a II–VI semiconductor (such as CdTe) in which manganese atoms randomly substitute cadmium (Mn in II–VIs is an isoelectronic impurity, i.e. it is not electrically active) allow the various configurations to be properly tuned. In particular we can study the case of a single Mn atom interacting with an excess of one hole, which is the expected configuration in worldwide developed III–V semiconductor quantum dots.

We have shown that the injection of a controlled number of carriers in an individual II–VI QD permits spin splitting of a single magnetic ion to be controlled. The exchange interaction with a single carrier acts as an effective local magnetic field that splits the Mn levels in a zero-applied magnetic field. We have fabricated a device with electrically tunable magnetic properties, in analogy with two-dimensional DMS-based electrically active heterostructures [7–9], but scaling the number of controlled magnetic atoms down to one and the size of the active region down to a few nanometres. Micro-photoluminescence experiments permit the identification of three magnetically different ground states corresponding to three charge states ($\pm 1e$ and 0) and measurement of the exchange interaction of both a single electron and a single hole with a single magnetic atom. The final state of (X^+, Mn) transition (1h,1Mn) is non-degenerate in the absence of an external magnetic field. This splitting of the different spin configurations should efficiently increase the spin relaxation time of both Mn and holes.

The observation of an individual spin in a quantum dot opens new possibilities in information storage. The spin of an isolated Mn atom should present a relaxation time in the millisecond range. This property could be exploited to store digital information on a single atom. The device presented here is a first step towards the development of new memories in which an information digit would be stored on the spin state of an individual atom.

Acknowledgements

The authors like to acknowledge Prof. Joaquim Fernandez–Rossier (Univ. Alicante) for his theoretical support in the interpretation of the electrical control data.

References

1. H.C. Manoharan, C.P. Lutz, and D.M. Eigler, *Nature* **403**, 512 (2000).
2. P. Gambardella, S. Rusponi, M. Veronese, S.S. Dhesi, C. Grazioli, A. Dallmeyer, I. Cabria, R. Zeller, P.H. Dederichs, K. Kern, C. Carbone, and H. Brune, *Science* **300**, 1130 (2003).
3. A.M. Yakunin, A.Yu. Silov, P.M. Koenraad, J.-M. Tang, M.E. Flatte, W.V. Roy, J.D. Boeck, and J.H. Wolter, *Phys. Rev. Lett.* **92**, 216806 (2004).

4. L. Besombes, Y. Leger, L. Maingault, D. Ferrand, H. Mariette, and J. Cibert, *Phys. Rev. Lett.* **93**, 207403 (2004).

5. A.J. Heinrich, A. Gupta, C.P. Lutz, and D.M. Eigler, *Science* **306**, 466–469, (2004).

6. F. Jelezko, T. Gaebel, I. Popa, A. Gruber, and J. Wrachtrup, *Phys. Rev. Lett.* **92**, 076401 (2004).

7. H. Ohno, D. Chiba, F. Matsukura, T. Omiya, E. Abe, T. Dietl, Y. Ohno, and K. Ohtani, *Nature* **408**, 944 (2000).

8. H. Boukari, P. Kossacki, M. Bertolini, D. Ferrand, J. Cibert, S. Tatarenko, A. Wasiela, J.A. Gaj, and T. Dietl, *Phys. Rev. Lett.* **88**, 207204 (2002).

9. D. Chiba, M. Yamanouchi, F. Matsukura, and H. Ohno, *Science* **301**, 943 (2003).

10. Al.L. Efros, E.I. Rashba, and M. Rosen, *Phys. Rev. Lett.* **87**, 206601 (2001).

11. A.O. Govorov and A.V. Kalameitsev, *Phys. Rev. B* **71**, 035338 (2005). A.O. Govorov, *ibid.* **72**, 075359 (2005).

12. J.I. Climente, M. Korkusinski, P. Hawrylak, and J. Planelles, *Phys. Rev. B* **71**, 125321 (2005); F. Qu and P. Hawrylak, *Phys. Rev. Lett.* **95**, 217206 (2005); F. Qu and P. Hawrylak, *ibid.* **96**, 157201 (2006).

13. J. Fernandez-Rossier and L. Brey, *Phys. Rev. Lett.* **93**, 117201 (2004).

14. A.O. Govorov, *Phys. Rev. B* **72**, 075358 (2005).

15. F. Qu and P. Hawrylak, *Phys. Rev. Lett.* **96**, 157201 (2006).

16. A.V. Chernenko, P.S. Dorozhkin, V.D. Kulakovskii, A.S. Brichkin, S.V. Ivanov, and A.A. Toropov, *Phys. Rev. B* **72**, 045302 (2005).

17. A.A. Maksimov, G. Bacher, A. McDonald, V.D. Kulakovskii, A. Forchel, C.R. Becker, G. Landwehr, and L.W. Molenkamp, *Phys. Rev. B* **62**, R7767 (2000).

18. J. Seufert, G. Bacher, M. Scheibner, A. Forchel, S. Lee, M. Dobrowolska, and J.K. Furdyna, *Phys. Rev. Lett.* **88**, 027402 (2002).

19. G. Mackh, W. Ossau, D.R. Yakovlev, A. Waag, G. Landwehr, R. Hellmann, and E.O. Gobel, *Phys. Rev. B* **49**, 10248 (1994).

20. D.R. Yakovlev, K.V. Kavokin, I.A. Merkulov, G. Mackh, W. Ossau, R. Hellmann, E.O. Gobel, A. Waag, and G. Landwehr, *Phys. Rev. B* **56**, 9782 (1997).

21. P.S. Dorozhkin, A.V. Chernenko, V.D. Kulakovskii, A.S. Brichkin, A.A. Maksimov, H. Schoemig, G. Bacher, A. Forchel, S. Lee, M. Dobrowolska, and J.K. Furdyna, *Phys. Rev. B* **68**, 195313 (2003).

22. A. Hundt, J. Puls, and H. Henneberger, *Phys. Rev. B* **69**, 121309 (2004).

23. G. Bacher, A.A. Maksimov, H. Schomig, V.D. Kulakovskii, M.K. Welsh, A. Forchel, P.S. Dorozhkin, A.V. Chernenko, S. Lee, M. Dobrowolska, and J.K. Furdyna, *Phys. Rev. Lett.* **89**, 127201 (2002).

24. D.J. Norris, Nan Yao, F.T. Charnock, and T.A. Kennedy, *Nanolett.* **1**, 3 (2001).

25. S.C. Erwin, L. Zu, M.I. Haftel, A.L. Efros, T.A. Kennedy, and D.J. Norris, *Nature* **436**, 91 (2005).

26. A.K. Bhattacharjee and J. Perez-Conde, *Phys. Rev. B* **68**, 045303 (2003).

27. P.I. Archer, S.A. Santangelo, and D.R. Gamelin, *Nanolett.* **7**, 1037 (2007).

28. S. Mackowski, T. Gurung, H.E. Jackson, L.M. Smith, G. Karczewski, and J. Kossut, *Appl. Phys. Lett.* **87**, 072502 (2005).

29. J.K. Furdyna, *J. Appl. Phys.* **64**, R29 (1998).

30. *Semiconductors and Semimetals*, vol. 25, ed. by J.K. Furdyna and J. Kossut (Academic, Boston, 1988).

31. J. Schneider, U. Kaufmann, W. Wilkening, M. Baeumler, and F. Kohl, *Phys. Rev. Lett.* **59**, 240 (1987).

32. M. Linnarsson, E. Janzn, B. Monemar, M. Kleverman, and A. Thilderkvist, *Phys. Rev. B* **55**, 6938 (1997).

33. J.A. Gaj, W. Grieshaber, C. Bodin, J. Cibert, G. Feuillet, Y. Merle d'Aubigne, and A. Wasiela, *Phys. Rev. B* **50**, 5512 (1994).

34. Y. Shapira, S. Foner, D.H. Ridgley, K. Dwight, and A. Wold, *Phys. Rev. B* **30**, 4021 (1984).

35. J.M. Fatah, T. Piorek, P. Harrison, T. Stirner, and W.E. Hagston, *Phys. Rev. B* **49**, 10341 (1994).

36. W. Grieshaber, A. Haury, J. Cibert, Y. Merle d'Aubigne, A. Wasiela, and J.A. Gaj, *Phys. Rev. B* **53**, 4891 (1996).

37. J.A. Gaj, R. Planel, and G. Fishman, *Solid-State Commun.* **29**, 435 (1979).

38. A. Twardowski, P. Swiderski, M. von Ortenberg, and R. Pauthenet, *Solid-State Commun.* **50**, 509 (1984).

39. W. Pacuski, D. Ferrand, J. Cibert, C. Deparis, J.A. Gaj, P. Kossacki, and C. Morhain, *Phys. Rev. B* **73**, 035214 (2006).

40. J.M. Hartmann, J. Cibert, F. Kany, H. Mariette, M. Charleux, P. Alleyson, R. Langer, and G. Feuillet, *J. Appl. Phys.* **80**, 6257 (1996).
41. J. Cibert, Y. Gobil, Le Si Dang, S. Tatarenko, G. Feuillet, P.H. Jouneau, and K. Saminadayar, *Appl. Phys. Lett.* **56**, 292 (1990).
42. S.H. Xin, P.D. Wang, A. Yin, C. Kim, M. Dobrowolska, J.L. Merz, and J.K. Furdyna, *Appl. Phys. Lett.* **69**, 3884 (1996).
43. Y. Terai, S. Kuroda, K. Takita, T. Okuno, and Y. Masumoto, *Appl. Phys. Lett.* **73**, 3757 (1998).
44. F. Tinjod, I.C. Robin, R. Andre, K. Kheng, and H. Mariette, *J. Alloys Comp.* **371**, 63 (2004).
45. F. Tinjod, B. Gilles, S. Moehl, K. Kheng, and H. Mariette, *Appl. Phys. Lett.* **82**, 4340 (2003).
46. I.C. Robin, R. Andr, C. Bougerol, T. Aichele, and S. Tatarenko, *Appl. Phys. Lett.* **88**, 233103 (2006).
47. F. Tinjod and H. Mariette, *Phys. Status Solidi. B* **241**, 550 (2004).
48. R. Brazis and J. Kossut, *Solid-State Commun.* **122**, 73 (2002).
49. J. Fernandez-Rossier, *Phys. Rev. B* **73**, 045301 (2006).
50. G. Lentz, A. Ponchet, N. Magnea, and H. Mariette, *Appl. Phys. Lett.* **55**, 2733 (1989).
51. J.M. Moison, C. Guille, F. Houzay, F. Barthe, and M. Van Rompay, *Phys. Rev. B* **40**, 6149 (1989).
52. A.O. Govorov, *Phys. Rev. B* **70**, 035321 (2004).
53. J. Seufert, M. Rambach, G. Bacher, A. Forchel, T. Passow, and D. Hommel, *Appl. Phys. Lett.* **82**, 3946 (2003).
54. A. Hartmann, Y. Ducommun, E. Kapon, U. Hohenester, and E. Molinari, *Phys. Rev. Lett.* **84**, 5648 (2000).
55. L. Maingault, L. Besombes, Y. Leger, C. Bougerol, and H. Mariette, *Appl. Phys. Lett.* **89**, 193109 (2006).
56. L. Besombes, K. Kheng, and D. Martrou, *Phys. Rev. Lett.* **85**, 425 (2000); id. *J. Cryst. Growth* **214** 742 (2000).
57. I.A. Merkulov, D.R. Yakovlev, A. Keller, W. Ossau, J. Geurts, A. Waag, G. Landwehr, G. Karczewski, T. Wojtowicz, and J. Kossut, *Phys. Rev. Lett.* **83**, 1431 (1999).
58. D. Keller, D.R. Yakovlev, B. Knig, W. Ossau, Th. Gruber, A. Waag, L.W. Molenkamp, and A.V. Scherbakov, *Phys. Rev. B* **65**, 035313 (2001).
59. D. Scalbert, J. Cernogora, and C. Benoit la Guillaume, *Solid-State Commun.* **66**, 571 (1988).
60. W. Maslana, P. Kossacki, M. Bertolini, H. Boukari, D. Ferrand, S. Tatarenko, J. Cibert, and J.A. Gaj, *Appl. Phys. Lett.* **82**, 1875 (2003).
61. S. Bhunia and D.N. Bose, *J. Appl. Phys.* **87**, 2931 (2000).
62. L. Besombes, K. Kheng, L. Marsal, and H. Mariette, *Phys. Rev. B* **65**, 121314 (2002).
63. K. Brunner, G. Abstreiter, G. Bohm, G. Trankle, and G. Weimann, *Phys. Rev. Lett.* **73**, 1138 (1994).
64. A. Vasanelly, R. Ferreira, and G. Bastard, *Phys. Rev. Lett.* **89**, 216804 (2002).
65. A.V. Koudinov, I.A. Akimov, Yu.G. Kusrayev, and F. Henneberger, *Phys. Rev. B* **70**, 241305 (2004).
66. F.V. Kyrychenko and J. Kossut, *Phys. Rev. B* **70**, 205317 (2004).
67. Y. Léger, L. Besombes, L. Maingault, D. Ferrand, and H. Mariette, *Phys. Rev. B* **72**, 241309 (2005).
68. Y. Léger, L. Besombes, J. Fernández-Rossier, L. Maingault, and H. Mariette, *Phys. Rev. Lett.* **97**, 107401 (2006).
69. Y. Léger, L. Besombes, L. Maingault, and H. Mariette, *Phys. Rev. B* **76**, 045331 (2007).
70. Y. Léger, L. Besombes, L. Maingault, D. Ferrand, and H. Mariette, *Phys. Rev. Lett.* **95**, 047403 (2005).
71. V.D. Kulakovskii, G. Bacher, R. Weigand, T. Kummell, and A. Forchel, *Phys. Rev. Lett.* **82**, 1780 (1999); L. Besombes, K. Kheng, and D. Martrou, *Phys. Rev. Lett.* **85**, 425 (2000).
72. A. Kuther, M. Bayer, and A. Forchel, *Phys. Rev. B* **58**, R7508 (1998).
73. J. Allègre, B. Gil, J. Calatayud, and H. Mathieu, *J. Cryst. Growth* **101**, 603 (1990).
74. L. Besombes, Y. Leger, L. Maingault, D. Ferrand, H. Mariette, and J. Cibert, *Phys. Rev. B* **71**, 161307 (2005).
75. M.M. Glazov, E.L. Ivchenko, L. Besombes, Y. leger, L. Maingault, and H. Mariette, *Phys. Rev. B* **75**, 205313 (2007).
76. M. Tadic, F.M. Peeters, and K.L. Janssens, *Phys. Rev. B* **65**, 165333 (2002).
77. U. Bockelmann and G. Bastard, *Phys. Rev. B* **45**, 1688 (1992).
78. T. Tanaka, J. Singh, Y. Arakawa, and P. Bhattacharya, *Appl. Phys. Lett.* **62**, 756 (1993).
79. S. Cortez, O. Krebs, and P. Voisin, *J. Vac. Sci. Technol. B* **18**, 2232 (2000).

CHAPTER 15

Quantum Dot Charge and Spin Memory Devices

Jonathan Finley

*Walter Schottky Institut and Physik Department, Technische Universität München,
D-85748, Garching, Germany*

15.1 Introduction

Conventional non-volatile semiconductor memories represent binary information by switching the charge stored on a capacitor via a transistor. Each silicon-based capacitor–transistor subunit constitutes one memory cell within a large and independently addressable array. These fundamental concepts for dynamic random access memories (DRAM) were patented in 1967 [1] and introduced to the market a few years later by the Intel Corporation [2]. Over the 35 years since the first DRAM chips appeared, we have witnessed a seemingly relentless increase of the maximum integration density of Si-based memory and logic chips with minimum feature sizes reducing exponentially, from over $10\,\mu m$ in 1972 to less than $50\,nm$ in the latest prototype memory cells. This relentless trend towards miniaturization, and the seemingly self-fulfilling prophecy that has become known as Moore's law [3], shows little sign of abating despite minimum feature sizes rapidly approaching lengthscales where quantum mechanical effects begin to dominate the properties of the system. However, downscaling comes at the price of ever more stringent tolerances as lengthscales approach the limits of nanofabrication technologies. The ultimate non-volatile memory cell would represent one bit of information by storing a single charge carrier on a nanoscale capacitor [4]. Since conventional nanofabrication technologies are not yet able to manufacture conventional DRAM cells with sufficiently small dimensions, semiconductor quantum dots (QDs) may offer an elegant and direct method to create huge ensembles of electronic traps for single or very few carriers [5].

Quantum dots are quasi zero-dimensional nanostructures that can localize charge carriers in all three spatial directions to lengthscales comparable to their de Broglie wavelength. The effective three-dimensional confinement potential can be created using one of many approaches. Two of the most widely studied QD systems are formed either by modulating the chemical composition of the dot relative to its environment [6] or by electrostatically defining a potential minimum within a two-dimensional quantum film using metallic gates on the sample surface [7]. These two types of QDs differ fundamentally in their ability to confine carriers: chemically defined dots can simultaneously trap both electrons and holes and are, thus, *optically active*, while electrostatically defined dots tend to trap only one charge polarity (usually electrons) and are optically *inactive*. In this chapter we focus exclusively on optically active quantum dot nanostructures and investigate their suitability as optically addressable charge and spin memory structures. An inhomogeneously broadened ensemble of such QD-like absorbing centres has strong potential for application as a wavelength domain optical data storage medium [8, 9]. Optical data storage offers many advantages including the robustness associated with "non-contact" data access via a laser, high data storage densities, non-volatility and fast read and write access [10]. Like memory and logic chips, the technologies for optical data storage have downscaled rapidly from the first compact discs in the early 1980s to DVD discs, DVD-RW technologies that rely on phase change media and now *Blu-Ray* discs operating at shorter wavelengths (405 nm, *cf.* 640 nm for DVD-RW) [11]. However, the data storage densities of these conventional optical technologies are defined by the diffraction limited size of the laser focal volume, which sets the area occupied by a single bit of information. Moving to bit sizes below the diffraction limit will call for new concepts such as wavelength domain [8] or holographic optical data storage concepts [12].

In this chapter we review the use of self-assembled QD nanostructures for optical charge and spin storage devices. After discussing the major physical, electronic and optical properties we describe how *single charges* can be optically generated in QDs and illustrate some of the progress over the past ten years. As we will see, while other materials may eventually emerge to be more suitable for optical data storage, the methods developed provide a valuable tool to study fundamental properties of semiconductor QDs. We then continue to describe how the spin of optically generated charges can be orientated by controlling the polarization of the light used to generate them. This approach provides a direct method to probe electron and hole spin relaxation in self-assembled QDs – a topic with strong relevance for future spin-based quantum information processing [13].

15.1.1 Self-assembled quantum dots

Of the many types of QD materials for *optical applications*, nanoscale self-assembled islands formed by a strain-driven self-assembly are among the most widely investigated. This epitaxial growth mode is driven by the trade-off between strain and surface energies during heteroepitaxial growth of materials that have different lattice constants [14, 15]. Goldstein *et al.* were the first to report the observation of such island formation in a semiconductor system during the growth of InAs–GaAs superlattices [16]. While initially being considered to be an undesirable side effect of strained layer epitaxy, it was quickly realized that such nanoscale islands could be grown without defects and, thus, could exhibit excellent optical properties. From the earliest studies of self-assembled quantum dots, InAs grown on GaAs has remained one of the most important systems for optical applications.

Due to the 7% lattice mismatch, InAs initially grows on GaAs as a highly strained two-dimensional (2D) layer termed the wetting layer. This 2D growth mode continues until a thickness of approximately one atomic layer whereupon the growth transforms from 2D–3D clustering of islands, due to the reduction of the strain energy arising from island formation, which outweighs the additional surface energy of the multifaceted crystal surface. The process gives rise to a series of nano-sized islands that form on top of the underlying wetting layer as discussed in other chapters in this volume. Figure 15.1a compares a typical atomic force microscopy image recorded from uncapped self-assembled InAs QDs formed on GaAs with an atomic illustration of a single dot. For $In_xGa_{(1-x)}$ As self-assembled QDs grown on GaAs ($x \geq$), typical heights are ~ 2–8 nm with base widths of the order of ~ 5–25 nm. A real densities of the nanostructures can be tuned in the range $\sim 10^9$–10^{11} cm^{-2} (see Fig. 15.1a) by controlling the growth conditions. After formation the islands are overgrown by GaAs to form fully encapsulated QDs that are optically active.

15.1.2 Fundamental optical properties and single particle non-linearities

The confinement lengthscales of self-assembled QDs is such that the electron and hole single particle states are fully quantized with a discrete density of states. Since the vertical (z) height of the dots tends to be much larger than their lateral (x, y) dimensions as discussed above, the motion along the vertical and lateral directions is largely decoupled: the strong vertical confinement tends to determine the effective band gap of the dots together with the strain, while the spectrum of lowest lying single particle quantum states are representative of the effective lateral confinement potential [17]. Within the QD research community these orbital states are often denoted by the labels s, p, d, , , etc. in analogy to the orbital states of atoms, as depicted schematically

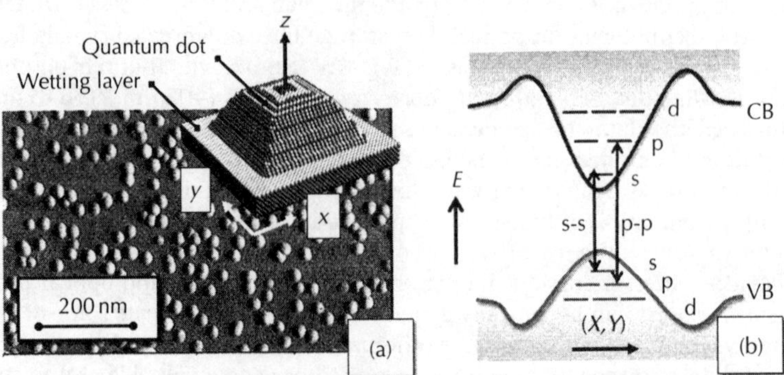

Figure 15.1   (a) Atomistic illustration and atomic force microscopy image of an uncapped ensemble of self-assembled InGaAs quantum dots. In the illustration, the Ga and As atoms are green and yellow, respectively, while In atoms are coded red. This image illustrates the typical random alloy composition of a self-assembled quantum dot – image reproduced with permission of M. Migliorato, University of Manchester, UK. (b) Schematic representation of the discrete orbital electronic structure in the conduction and valence bands of a self-assembled QD. Only the lateral confinement potential is depicted since the vertical confinement tends to be much stronger due to the smaller spatial extent of the QDs and the principal quantum number associated with vertical motion is identical for the lowest lying orbital states.

in Fig. 15.1b. However, one should appreciate that the real symmetry of the envelope wavefunctions is lower than real atoms due to the effects of inhomogeneous composition and shape of the dots and piezoelectric potentials. The energy separation of the orbital states lies in the range $\hbar\omega_{sp}^{e}$ ~10 −meV for electrons and $\hbar\omega_{sp}^{h}$ ~5 −meV for holes, depending on the microscopic details of the effective confinement potential. For electrons, these energy scales are up to ~3 $k_{B}T$ or greater at room temperature, as required for most device applications. The fundamental interband absorption energy of the QDs can be tuned over the range ~0.9eV to ~1.8eV either by varying the growth conditions (and hence size of the QDs), the dot composition or by growing the dots in a matrix other than GaAs. Many material combinations have been studied in the III–V semiconductor family including AlGaAs for increased emission energies [18] and InGaAs for reduced emission energies [19]. For an in-depth recent review of the fundamental optical properties of QDs and their numerous applications in optoelectronics the reader is referred to [20] and a number of other chapters in the present volume.

All types of semiconductor QDs have the common property that their interband optical response is highly sensitive to the charges they already contain. Thus, if one measures some property, such as the frequency of emitted photons, the result will depend on the number, orbital configuration and spin of charges already within the dot. In this respect, one can consider the optical response of a single QD to be non-linear even at the level of a single charge carrier. This extraordinary sensitivity to charge occupation already forms the basis of the operation of many QD devices within nanoelectronics and nanophotonics. Examples include the single electron transistor [7] and the deterministic generation of single photons for quantum cryptography [21] or entangled photon pairs generated by the cascaded biexciton–exciton decay [22]. Fundamentally, all these applications are enabled by strong Coulomb interactions between the charges trapped by the dots which are strongly enhanced by the complete spatial confinement. These ideas are illustrated in Fig. 15.2a which provides a schematic illustration of the interband optical emission spectrum of a single QD in the absence of Coulomb interactions (lower panel), and compares it with the situation with the strong Coulomb interactions switched on (upper panel) [23]. Without Coulomb interactions, the interband transitions in the dot mostly take place between single particle orbital states with the same quantum number (s, p, etc. – see Fig. 15.1b). The emission spectrum would then consist of a series of discrete transitions as depicted schematically in Fig. 15.2a. In reality, band mixing effects and reduction of the symmetry of the dot, due to an inhomogeneous composition profile and piezoelectric charges, would still give rise to weakly allowed transitions between other orbital states (s, s, etc.), although the spectrum would

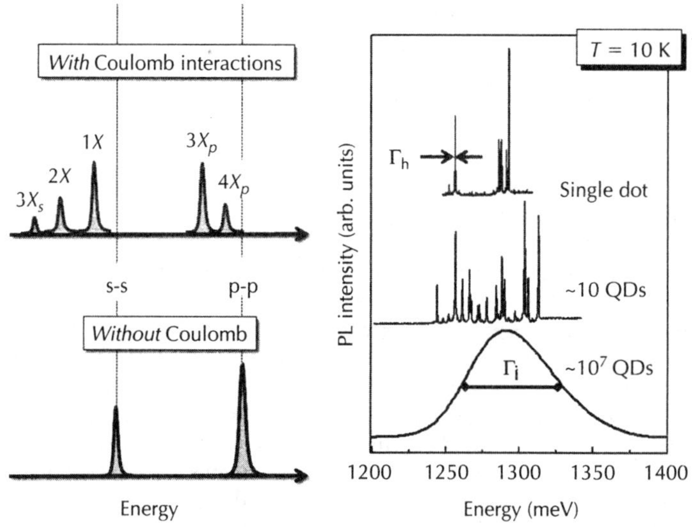

Figure 15.2   (a) Schematic representation of the form of the interband emission spectrum of a single QD without (lower) and with (upper panel) Coulomb interactions between confined carriers. As discussed in the text, Coulomb interactions lift the degeneracy of the orbital states allowing each occupancy state to be individually spectrally addressed. (b) Photoluminescence spectra recorded at low temperatures from an ensemble of InGaAs QDs, revealing a large inhomogeneous linewidth ($\Gamma_i \sim 70$ meV) and contrasting strongly with the homogeneous linewidth of a single dot ($\Gamma_h < 1 \mu eV$). The upper single dot spectrum also clearly shows the influence of Coulomb interactions on the emission spectrum.

remain rather simple [24]. The enhanced Coulomb interactions in QDs due to the strong confinement give rise to energy shifts between different occupancy states of the order of a few millielectronvolts. Each occupancy state of electrons and holes, such as the *single exciton* ($1X = 1e+$), *bi-exciton* ($2X = 2e+$), *charged exciton* ($X^- = 2e + 1h$, $X^+ = 1e+$), etc. then have a unique transition energy.* Furthermore, since the Coulomb energy shifts are much larger than the homogeneous linewidth of each transition ($\Gamma_{hom} < 1$), each occupancy state can be *optically selected* via the frequency of either the emitted photon or laser used to address the system. This property underpins the use of QDs for a diverse range of applications in the fields of classical and quantum information processing.

In order to experimentally observe the single particle non-linearities discussed above, one has to probe single dots in isolation [25]. Otherwise the minimum energy scale that can be probed is limited by the inhomogeneous linewidth of the ensemble ($\Gamma_{inhomog.} \sim 20 - 50$ nm). This necessity is nicely illustrated in Fig. 15.2b, which compares PL recorded from an ensemble of InGaAs QDs with data recorded from the same sample through shadow mask apertures [26]. In this experiment, single dots could be isolated by performing luminescence measurements through submicron diameter apertures in an opaque metal film. This measurement followed many others which, already at the beginning of the 1990s, provided the first direct evidence for the discrete electronic structure of QD nanostructures via spatially resolved optical studies of localized excitons in quantum well nanostructures [27, 28]. Shortly after these first studies, various groups reported similar results on different QD systems using confocal microscopy [28, 29, 30], near-field optics [31, 32] and cathodoluminescence [33] techniques. In the pioneering work of Brunner *et al.* [27, 28]

---

*As a further complication, transitions between quasi 0D orbital states in the dot and delocalized carriers in the 2D wetting layer continuum can also provide an intrinsic continuum absorption background for energies above the fundamental s–s transition [71]. This property is manifestly far away from the idea that such nanostructures behave as artificial atoms, while other properties of the dots, such as the observation of excitonic Rabi oscillations, firmly underscore the analogy between quantum dots and real atoms.

the discrete and highly non-linear optical properties of single QDs were highlighted, together with the possibility of controlling the spin of individual charge carriers via the optical polarization selection rules [27]. Studies of this type revealed that energy splittings between 1X and 2X transitions typically lie in the range of a few millielectronvolts, with 2X lying either to lower or higher energy than 1X, depending on the importance of direct Coulomb interactions and few particle correlation effects [34].

A further important development in the study and control of few particle states in QDs was establishing an electrical contact to single dots. This provides the unique possibility to control the QD charge occupancy using Schottky photodiode structures and has led to a number of advances such as photocurrent absorption spectroscopy [35], and the electrical detection of coherent optical control of excitons [36, 37]. The selective charging of self-assembled QDs was first demonstrated by Drexler *et al.* using capacitance spectroscopy of ensembles of InGaAs dots embedded in the depletion region of a Schottky diode [38]. This was ingeniously extended by Warburton *et al.* [39], followed by other groups [39, 40, 41], who combined this approach with near-field optical spectroscopy through metal shadow masks to directly study the interaction of excitons with additional electrons [39, 40] or holes [41]. These studies of charged excitons in QDs revealed a wealth of information on Coulomb exchange and correlation effects in single dots populated by a controlled number of charge carriers. Most recently, single dot *absorption* spectroscopy has been shown to be a particularly clean method to directly probe the fundamental quantum properties of excitonic transitions in QDs [42, 43] and even to facilitate measurement of the spin of a single electron via spin state selective experiments [44]. An up-to-date overview of the field of single dot spectroscopy can be found in the proceedings of the biannual International Conferences on Semiconductor Quantum Dots established in 2000, the most recent of which was held in May 2006 in Chamonix, France [45].

15.1.3   *Optical memory structures based on quantum dots*

For the purpose of the present chapter, the single particle non-linearities discussed above play an important role: they ensure that resonant optical excitation of QDs in the spectral range of their *s* interband transition results only in the generation of a single electron–hole pair, the single exciton, denoted by the nomenclature $1X = 1e +$. Upon exciting an inhomogeneously broadened ensemble with a narrowband source, one selects a subset of the dots from the ensemble via their exciton transition frequency. Other excitations, such as $2X = 2e + X^+ = 1e +, X^- = 2e +,$ etc. are detuned by many thousand linewidths from the exciton transition and are not excited via any linear optical processes. In 1995 Shunchui Muto suggested that such an inhomogeneously broadened ensemble of quasi zero-dimensional absorbing centres may be suitable for application as an ultra-dense, wavelength domain, optical data storage medium [46]. The use of both spatial and wavelength domains for data storage provide the potential to combine both high data storage capacities ($>1\,\mathrm{TB\,cm^{-2}}$) with very low switching energies. In this proposal, each bit of information is represented by a small number of e−h pairs, potentially even a single charge, stored within an ensemble of QD nanostructures. High data storage densities are attainable since data are stored in both spatial and frequency domains. One problem associated with the use of interband excitations for frequency domain optical data storage is that they are relatively short lived (typically ~1 ns) and electrically neutral, raising significant problems related to the efficient detection of such optically encoded information. This limitation can be overcome by implementing device concepts for *separation* of the optically generated e–h pair over timescales shorter than the recombination lifetime, and then sensing these charges to read out the optically encoded information.

Over the past ten years, these ideas have been explored by a number of groups with first investigations focused on optical charge generation and non-selective *electrical* sensing of the stored charge. In 1995 Yusa and Sakaki demonstrated an optically gated FET that incorporates a layer of InAs QDs in the vicinity of a two-dimensional electron channel [47]. The observed photoswitching operation of the channel resistivity in this device was attributed to the preferential trapping of non-resonantly photo-generated holes in the dot layer. Finley *et al.* extended this approach to conditions of *resonant* optical charge generation in the QD ground states and selective exciton ionization [48]. Later on, Shields *et al.* presented a similar scheme of electric readout in nanoscale FET structures that eventually led to the observation of discrete photo-charging events of individual QDs in the transport characteristics [49]. To date, a number of groups have shown that single charges can be electrically [50, 51,52, 53] or optically [54, 55, 56] injected into QDs. Alternatively, the quantum dots may be charged optically, an approach that, when combined with wavelength selective writing or local probe techniques, could approach the situation where each quantum dot stores a *single* bit of information as discussed in the introduction.

One of the crucial parameters for the use of InAs–GaAs QDs for optical memory applications is the thermal activation barrier for the trapped carriers – the larger it is, the longer the storage time becomes eliminating the need for frequent memory refresh cycles. Capacitance spectroscopy has been shown to be a powerful method to investigate charge retention times [57] together with deep-level transient spectroscopy and admittance spectroscopy for both InAs–GaAs [58, 59] and Ge–Si QDs [60]. GaSb–GaAs dots have similar structural properties to the InAs–GaAs system, but a significantly larger hole ground state localization energy is expected due to the type II band alignment attractive for holes [61]. The thermal activation barrier of hole-charged GaSb–GaAs QDs has been studied by admittance spectroscopy and shown to provide extremely long hole storage times, in excess of milliseconds even at room temperature [62].

In the following sections we discuss the operation of early prototype optical charge memory structures with *electrical* readout of the stored charge. These experiments are then contrasted with devices that enable all optical charging *and* readout later in the chapter. These "all optical" memory devices provide a very convenient method to directly orientate the spin of the optically generated charges via the optical polarization selection rules. We then continue to describe experiments in which we probe the electron and hole spin relaxation dynamics in self-assembled QD nanostructures. The spin orientation is shown to be remarkably stable, demonstrating that motional quantization dramatically suppresses spin relaxation via spin orbit interactions.

15.2   Optically induced charge storage

15.2.1   Electrical detection of stored charge

A number of groups have investigated the effects of a self-assembled InAs QD layer on the lateral transport properties of a two-dimensional electron gas (2DEG) in MODFET-type structures [51, 56, 63, 64]. These works have shown that the QDs can efficiently act as controllable charge traps and scattering centres which can be used to effectively tailor the transport properties of the 2DEG. Furthermore, several authors [51, 56] have demonstrated that such devices are strongly photosensitive forming the foundation for a basic QD optical memory element or highly sensitive QD phototransistor with single photon counting capabilities [65].

15.2.1.1   Charge storage device structure

Optically induced charging of self-assembled InAs QDs has been studied in [48, 66] and using an optically gated FET structure into which a layer of InAs quantum dots is embedded $\sim$50 nm from a 2D electron or hole conducting channel. The operating principles of the devices are similar to a MODFET, as illustrated by Fig. 15.3a, with the exception that the metallic gate electrode is replaced by a layer of self-assembled quantum dots that can be optically charged. By separating the 2D channel and QD layer using a wider band gap blocking barrier, photo-created excitons can be selectively ionized leaving electrons (or holes) stored preferentially within the dots.

Figure 15.3b illustrates schematically the fundamental operating principles of these structures. While the following discussion pertains to an electron storage device, the principles for holes, discussed below, are entirely analogous. The devices consist of a two-dimensional (2D) electron channel that was spatially separated by a blocking barrier from a layer of self-organized quantum dots (upper panel of Fig. 15.3b). The QD layer is embedded within the intrinsic region of a vertical *p-i-n* junction. By fabricating separate ohmic contacts to both the 2D *n*-channel and buried *p*-contact, the electric field in the intrinsic region and conductivity of the *n*-channel (σ) can be tuned by varying the DC bias applied to the *p-i-n* junction (V_{pn}). Following resonant optical excitation

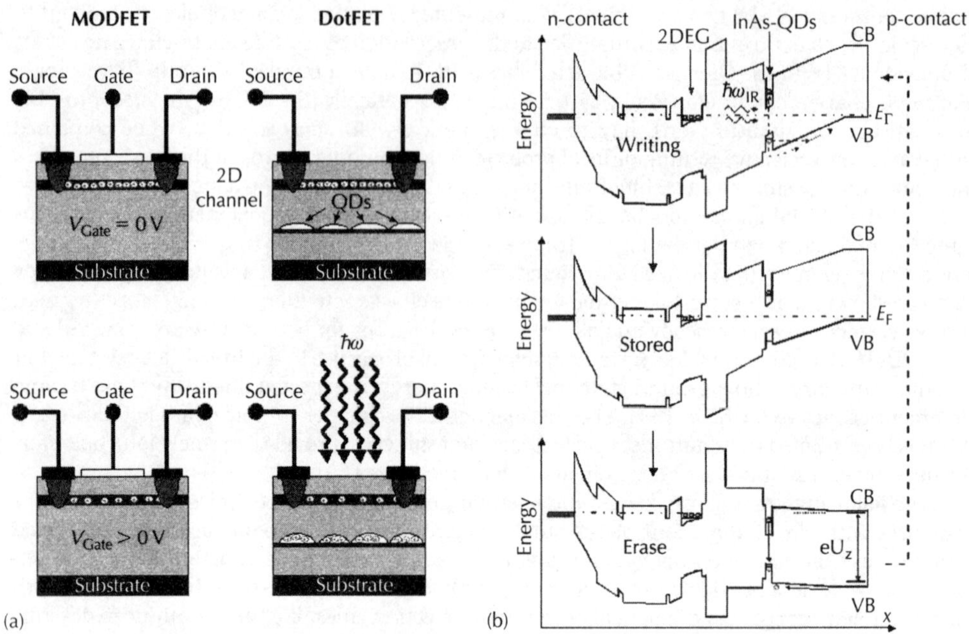

Figure 15.3 *(a) Comparison of a conventional MOSFET and an optically switchable DotFET realized by embedding a single layer of InGaAs quantum dots in the vicinity of a 2D conducting channel. (b) Schematic representation of the device bandstructure and operating principles as discussed in the text.*

of the quantum dots, the photo-created holes escape by thermal activation ($T > 100\,K$) or by tunnelling ($T < 100\,K$). By contrast, the electrons remain preferentially stored due to the presence of the wider band gap-blocking barrier (middle panel of Fig. 15.3b). The stored electrons within the QD layer then selectively deplete the 2DEG, resulting in a reduction of its in-plane conductivity which can be measured electrically. The magnitude of this conductivity change ($\Delta\sigma$) is expected to reflect the charge density stored in the QDs. The stored electrons are then removed from the QDs by forward biasing the vertical *p-i-n* junction (lower panel of Fig. 15.3b). This results in the injection of holes into the QD layer, where they become trapped by the dots, recombining radiatively with the stored electrons. The device then switches back to its initial state and another charge *write–store–reset* cycle can be performed.

Devices similar to those illustrated in Fig. 15.3b have been grown on a [100] orientated semi-insulated GaAs substrate and nominally consisted of the following epitaxial layers: 500 nm $p(Be) = 2.10^{18}\,cm^{-3}$ back contact followed by a 240 nm undoped (u.d.) GaAs spacer. After this, 2.25 ML of InAs was deposited at $530°C/0.04\,MLs^{-1}$ which forms the self-assembled QDs. This QD layer was capped with 5 nm GaAs before growth of the blocking barrier. This consisted of 30 nm undoped $Al_{0.3}Ga_{0.7}As$ and a five-period 2 nm(AlAs)/2 nm (GaAs) short period superlattice to inhibit impurity segregation in the 2D electron channel. The 10 nm thick $In_{0.1}Ga_{0.9}As$ electron channel was then grown followed by a 40 nm-wide modulation-doped $Al_{0.3}Ga_{0.7}As$ barrier region. The modulation doping was incorporated into two narrow (1.5 nm) δ-doped ($\sim2 \times 10^{12}\,cm^{-12}$) GaAs quantum wells in an effort to inhibit persistent photoconductivity associated with carrier excitation from DX centres in the $Al_{0.3}Ga_{0.7}As$ modulation-doped region [67]. Finally, the sample was capped with 10 nm of *n*-doped ($4 \times 10^{18}\,cm^{-3}$) GaAs. A reference sample was also grown which was nominally identical to the above but *without* the QD layer. After growth the wafers were processed into a Hall bar geometry using standard optical lithographic techniques and separate ohmic contacts were established to both the 2D-electron channel and the back *p*-contact to enable switching of the electric field as depicted in Fig. 15.3.

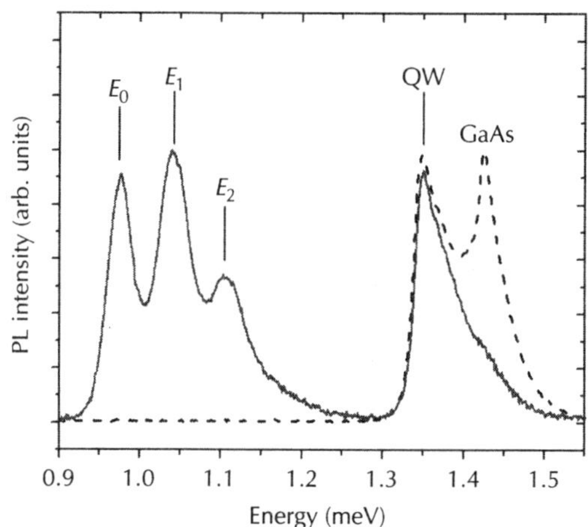

Figure 15.4   Photoluminescence spectra obtained at 293 K from the storage sample containing QDs (sample A – red full curve) and the reference sample without QDs (sample B – dashed blue curve).

15.2.1.2   Demonstration of selective electron and hole generation

Figure 15.4 compares high power ($\sim$5 Wcm^{-2}) ensemble PL spectra recorded from the sample with QDs (sample A – full line) and the reference sample without dots (sample B – dashed line). Over the energy range from 950 to 1200 meV a series of well-resolved peaks labelled E_n, $n = 0$, 1, 2 and 3 are observed which are separated by 64 $\pm$ 3 meV and exhibit a linewidth (FWHM) of $\sim$35 meV. The energy at which these features arise together with their characteristic dependence on the excitation power density identifies them as arising from radiative recombination of s–s and p–p transitions in the ensemble of QD ensemble studied. To higher energy, additional PL peaks are observed which arise from recombination in the In$_{0.1}$Ga$_{0.9}$As quantum well (QW) and bulk GaAs (GaAs) regions of the device. As expected, sample B (dashed line) generates no luminescence over the energy range 800–1200 meV confirming the absence of optically active centres.

In order to investigate optical charge storage effects a series of spectrally resolved photo-conductivity measurements was performed. This consisted of a 100 s illumination period after which the temporal dependence of the sample conductivity $-\sigma(t)$ was monitored before a reset electrical pulse was applied to the *p-i-n* junction [68]. The excitation energy was then changed and the cycle repeated. For these measurements the samples were held at $T \sim 145$ K in a stabilized environment designed to eliminate parasitic effects arising from ambient illumination and another persistent photoconductive effect that was found to be present for $T < 100$ K. Optical excitation was via a power-stabilized quasi monochromatic white light from a tungsten halogen lamp ($P_{max} \sim 1$ mWcm^{-2}, FWHM $\sim$ 5 meV) and the effect of exciting different quantum dots within the ensemble was studied by tuning the excitation energy (E_{ex}) through the QD ensemble absorption spectrum.

Figure 15.5 shows $\sigma(t)$ data obtained from sample A as a series of three illumination, recovery and reset phases were performed with $E_{ex} = 1016$ meV, 1024 meV and 1059 meV, respectively. The onset of illumination is denoted by the label **A**, with the label **B** showing the point at which the shutter is closed. The reset pulse is applied to the sample at **C**. The channel conductivity prior to illumination is denoted by σ_0, which for this sample was measured to be 1.5 $\pm$ 0.2 mS. As can be seen clearly in Fig. 15.5, for $E_{ex} = 1016$ meV the sample responds only weakly to illumination. By contrast, after increasing E_{ex} the photo-effect becomes much stronger until for $E_{ex} > 1020$ meV a rapid switching of σ is observed following illumination. We focus now on the illumination cycle at $E_{ex} = 1059$ meV [69] After closing the shutter (B) σ recovers weakly, saturating at a new level σ_2. Without resetting the sample, σ remained at σ_2 over timescales longer than *eight hours* at 145 K while recovering towards σ_0 by less than 10%. This directly reflects the potential for extremely long charge storage times in the deep trapping potential of the quantum dots, even at elevated temperatures. Sample B, which did not contain quantum dots, exhibited little or no photo-response, independent of E_{ex}, over the energy interval 900 meV $< E_{ex} < 1200$ meV [55].

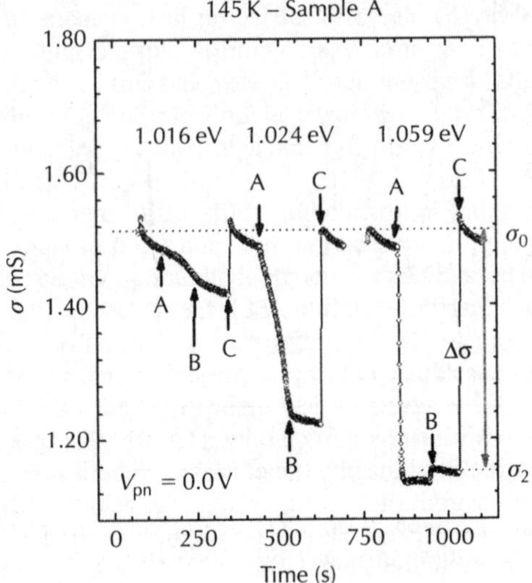

Figure 15.5   Temporal dependence of the n-channel conductivity for the QD containing sample (A) as a series of 100 s illumination, recovery and reset cycles are performed as described in the text. The measurements were taken with excitation energies of $E_{ex} = 1016\,meV$ (leftmost), $1024\,meV$ and $1059\,meV$ (rightmost), respectively. The arrows labelled A and B denote the points at which the illumination is applied and removed, respectively. The sample is "reset" by forward biasing the p-i-n junction to neutralize the stored charge at point C. The reference sample (B) showed a much weaker photo-response over the excitation range $900 < E_{ex} < 1200\,meV$ indicating that the measured response arises from optically stimulated charge storage in the QDs.

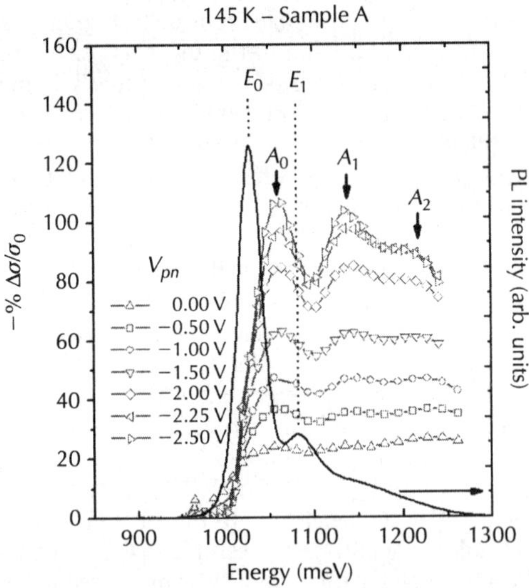

Figure 15.6   Spectral dependence of the photoconductive effect for sample A as a function of the reverse bias applied to the vertical p-i-n junction from $V_{pn} = 0$ to $-2.5\,V$. The thick line shows a PL spectrum for comparison.

Figure 15.6 shows the spectral dependence of the device photo-response for a number of one shot measurements over the energy range $950\,meV < E_{ex} < 1260\,meV$ and for different reverse biases applied to the vertical *p-i-n* junction (up to $V_{pn} = -2.5\,V$). The bold line shows the PL spectrum obtained at 145 K for comparison. The onset of the photo-response discussed above is very clearly observed for $E_{ex} \sim 1015\,meV$, firmly establishing the correspondence between the observed photo-effect and charging of the lowest orbital states of selected quantum dots with a single electron. Upon increasing V_{pn}, the magnitude of the photo-effect increases strongly (up to $-\Delta\sigma/\sigma_0 \sim 100\%$ for $V_{pn} = -2.5\,$V) reflecting the electrostatic depletion of the *n*-channel under reverse bias. The photosensitivity of the devices can be increased by depleting the 2D channel. Over the same energy range, only a very weak photo-effect ($\Delta\sigma/\sigma_0 < 5\%$) is observed from the reference sample. To higher energies, a series of peaks labelled A_{0-2} are observed. Before discussing these features, we note that the data plotted is the *saturation* value of the photo-response. Thus, the magnitude of the charge storage observed includes *all* possible QD excitation mechanisms regardless of the relative timescales over which they occur. The energy separation between A_0 and A_1 is 75 ± 5 meV, much larger than the splitting between ground and excited quantum dot peaks ($E_1 - E_0 = 60 \pm 5$ meV). Furthermore, these resonances are shifted from E_0 by 33 ± 5 meV, 115 ± 10 meV and 172 ± 20 meV for A_0, A_1 and A_2, respectively. This indicates that they do not simply reflect *direct* optical charging of the ground and excited states of the quantum dots. These observations were attributed to the participation of inelastic processes [55] over long timescales involved, a superposition confirmed by optical readout measurements discussed below. Following direct excitation of the quantum dots' excited states, stored electrons are expected to relax rapidly to the lowest orbital state [70] and multiple charging of the quantum dots may occur via the quasi continuum absorption band of the QDs associated with crossed 2D–0D transitions [71].

Similar experiments have also been performed on inverted devices in which *holes* are optically stored in the QDs and sensed using a 2D $In_{0.1}Ga_{0.9}As$ *p*-channel [72]. In all other respects the device operating principles and fabrication are similar to the electron storage sample discussed above. Besides being interesting from the fundamental physics viewpoint, hole storage may enable charge storage up to higher temperatures as a consequence of the larger effective mass, as mentioned above [62, 73]. As for the electron storage sample discussed above the modulation doped *p*-channel was separated by 50 nm from the InAs QD layer which was embedded within the intrinsic region of a vertical n-i-p junction. The growth conditions for the QD layer were nominally identical to the electron sample discussed above.

Figure 15.7 (inset) shows the photo-response of the hole storage sample following illumination at $E_{ex} = 1050$ meV, just above the peak of the ground state PL emission (Fig. 15.7 – bold line, top panel). As for the electron storage sample discussed above, a strong decrease ($\Delta\sigma$) in the channel

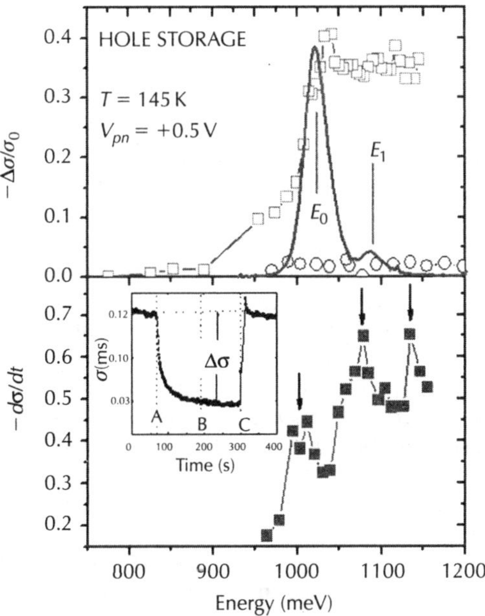

Figure 15.7   Inset: *Persistent photoconductivity arising from optically stimulated hole storage in the QD layer (top panel) QD charging spectrum for sample with (open squares) and without (open circles) the QD layer. (lower panel) Switching rate spectrum (dσ/dt) for the hole storage sample at $P_{ex} \sim 0.02\,mWcm^{-2}$.*

conductivity was observed following illumination (**A**), which remains persistent for many hours after closing the shutter (**B**). After resetting the device by forward biasing the n-i-p junction (**C**) the conductivity, once again, returns to the pre-illumination level (σ_0). A similar photo-response is not observed in a separate reference sample, which did not contain quantum dots. Fig. 15.7 (top panel) shows the spectral dependence of the induced photo-effect $\Delta\sigma/\sigma_0$ for both QD (open squares) and reference (open circles) samples, respectively. Since this data show the saturation level of the photo-response, care should be exercised while interpreting the detailed form of the charging spectrum. However, a clear onset is observed close to ~1030 meV confirming that the photo-effect arises from *hole* storage within the quantum dots. Furthermore, the peaked structure observed in the $\Delta\sigma/\sigma_0$ data provides evidence for *selective* storage of holes within the QDs. The filled squares in Fig. 15.7 (lower panel) show the spectral dependence of the device switching *rate* ($P_{ex} = 0.05\,\mathrm{mW/cm^{-2}}$) immediately after illumination for comparison. A series of peaks, marked by the bold arrows on the figure, are clearly observed in $d\sigma/dt$ arising close to the ground and excited state PL emission from the QDs. This observation indicates that the observed persistent photoconductivity arises from optically stimulated hole storage in the QDs.

The combination of the AlGaAs blocking barrier adjacent to the QD layer combined with the switchable electric field in the structure clearly enables charges to be selectively optically generated within a subset of self-assembled quantum dots. However, while the experiments discussed above clearly demonstrate the ability to selectively *generate* charges within the ensemble of dots, they cannot answer any the following questions:

- How many charges are generated per dot?
- What information does the spectral distribution of stored charge provide about the response of a QD ensemble to quasi resonant optical excitation?
- What are the mechanisms by which stored charges are lost from the ensemble?
- Can these ideas be extended from optical *charge generation* to *spin generation*?

To address these questions and reveal information about the optical charge storage process, further experiments were performed with modified devices that enable optical readout of the stored charge.

15.2.2   Optically detected charge storage

The first demonstration of full optical readout of optically induced charges was presented by Lundstrom *et al.*, employing electric field-induced charge transfer between a QD and a strain-induced potential minimum in an adjacent quantum well [74]. Optical readout is vital since spectral information about the stored charge distribution within the ensemble can be preserved by optical detection schemes; the selective transfer of one of the photo-generated charge carriers out of the QD does not destroy the energetic information encoded during the optical writing process provided that the second carrier continues to reside in the dot where it was generated. This is a vital ingredient of the optical memory scheme discussed in [46]. An exciton can then be re-formed in the ground state of the *same* QD by subsequent back transfer of charge, leading to photon emission from an identical quantum mechanical environment to that present during the initial absorption process. The selective optical generation and readout of a single charge carrier in a zero-dimensional electronic environment is therefore a key requirement with regards to preserving spatial and spectral information. Furthermore, as discussed later in the chapter it can also provide information about thermally induced charge redistribution [77] and the ideas can be extended from charge storage to spin storage [75] by utilizing the polarization degree of freedom of the excitation source.

We continue to discuss such a wavelength selective QD-based charge storage device with optically stimulated charge storage (electron or hole) and a demonstration of direct *optical readout* of the stored charge distribution in the spectral domain. We will see that similar device concepts are utilized to those discussed in the previous section, but spectral information is *retained* in the optical readout process. This provides further scope to investigate thermally induced charge redistribution and, crucially, electron and hole spin relaxation as discussed in the final sections of the chapter. Using these devices, wavelength domain "data storage" is demonstrated at $T = 10\,$K and charges are stored over timescales much longer than $25\,\mu$s. Analysis of the spectral distribution of the stored charge enables the study of resonant and phonon-assisted absorption and, furthermore, temperature dependent measurements allow direct investigations of thermal carrier loss from the QDs.

The operating principles and band diagram of a fully optical charge memory device are presented in Fig. 15.8. A single layer of InGaAs QDs is embedded within the intrinsic region of either an n- or p-type Schottky photodiode for hole or electron storage, respectively. In the following, we describe the operation of the hole storage device, the principles for the electron storage sample being entirely analogous. In the write mode (Fig. 15.8a), the device is reverse biased ($V = V_{store}$). Following resonant optical excitation of the QD sub-ensemble ($\hbar\omega_{aser} = \hbar\omega_{write}$), photo-generated excitons are rapidly ionized by the strong vertical electric field ($F \sim 150\,$kV/cm). While electrons readily tunnel out of the dot, the holes remain stored by virtue of an AlGaAs barrier immediately adjacent to the QD layer. Providing that V_{store} is chosen such that the exciton ionization rate far exceeds the radiative lifetime, holes are efficiently stored with a spectral distribution that reflects the response of the QD ensemble to resonant optical excitation. Optical readout of the stored charge distribution is achieved by application of a forward bias pulse ($V = V_{reset}$), resulting in non-resonant injection of majority charge carriers into the QD layer and subsequent neutralization of the stored charge by radiative recombination. The resulting charge storage spectrum directly reflects the spectral distribution of stored charge within the QD ensemble.

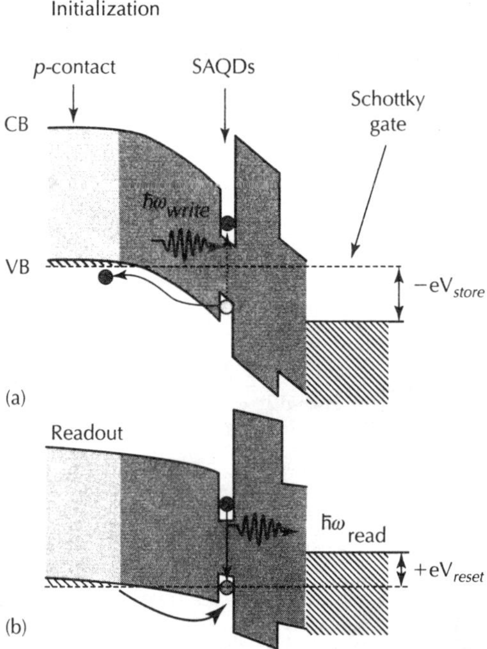

Figure 15.8   Schematic structure and operating principles of charge storage devices investigated. (a) In the charge storage mode ($V_{gate} = V_{store}$) photon to charge conversion is ensured by the negative gate potential and selective exciton ionization. (b) Conversion from charge to photon enabled by forward biasing the Schottky junction, hole reinjection and detection of the resulting storage EL at $\hbar\omega_{read} \sim \hbar\omega_{write}$ as discussed in the text.

As for the electrical memory devices discussed above, the optical memory structures were grown using molecular beam epitaxy on [100] semi-insulating GaAs substrates. After an undoped GaAs buffer, a back contact layer was deposited, consisting of $200\,$nm of heavily (p- or n-) doped GaAs ($N_{D/A} \sim 10^{18}\,$cm^{-3}). Following this, a $50\,$nm-undoped GaAs buffer was grown followed by the QD layer that consisted of 6 ML of $In_{0.5}Ga_{0.5}As$, deposited at 520°C and 0.1 ML/s. These growth conditions resulted in approximately lens-shaped QDs with a mean diameter (height) of 25 nm (4 nm). Moreover, the optical activity of these QDs is such that the emission from their fundamental transition is around ~950 nm and can be detected using high-sensitivity silicon-based detectors. The dot density was ~150 μm^{-2} measured by atomic force microscopy on a sequentially grown control sample. The QDs were capped by 1 nm GaAs, a 50 nm thick $Al_{0.4}Ga_{0.6}As$ blocking layer and completed with 10 nm GaAs. The wafers were then processed into photodiodes, with ohmic contacts established to the buried back contact and completed with a 10 nm thick, semi-transparent, top Schottky contact.

All optical experiments discussed below were performed using a standard micro-photoluminescence (PL) system and resonant optical excitation pulses were delivered using a mechanically modulated continuous-wave (CW) Ti–sapphire laser ($\hbar\omega_{write}$ = 1.25–1.38 eV) or the Ti–sapphire in combination with a laser diode ($\hbar\omega_{write}$ = 1.27 eV) for two colour experiments. An accurately timed sequence of optical (write) and electrical (readout) pulses were applied to the sample at a repetition rate of ~20 kHz. In order to ensure efficient suppression of scattered laser light during detection, an Si-based gated avalanche photodiode was used and only switched on during application of the readout voltage pulse. The charge storage time (Δt) is defined as the time delay between switching off the optical write pulse and the onset of the electrical readout.

Typical charge storage spectra for electron (dashed line) and hole (solid line) devices are presented in Fig. 15.9a for a storage time of Δt = 12 μs. The overall form of the charge storage spectra was found to be very similar for both samples, consisting of two principal components: a sharp peak, labelled **R** in Fig. 15.9a, very close to the excitation energy and a much broader band, labelled **S**, ~20–50 meV to lower energy. For both samples, $\hbar\omega_{write}$ was tuned to the high-energy side of the QD ensemble (~1.27 eV – dashed line in Fig. 15.9a) and the magnitude of the readout voltage was $|V_{reset}|$ = 0.8 V corresponding to the onset of majority carrier current flow.

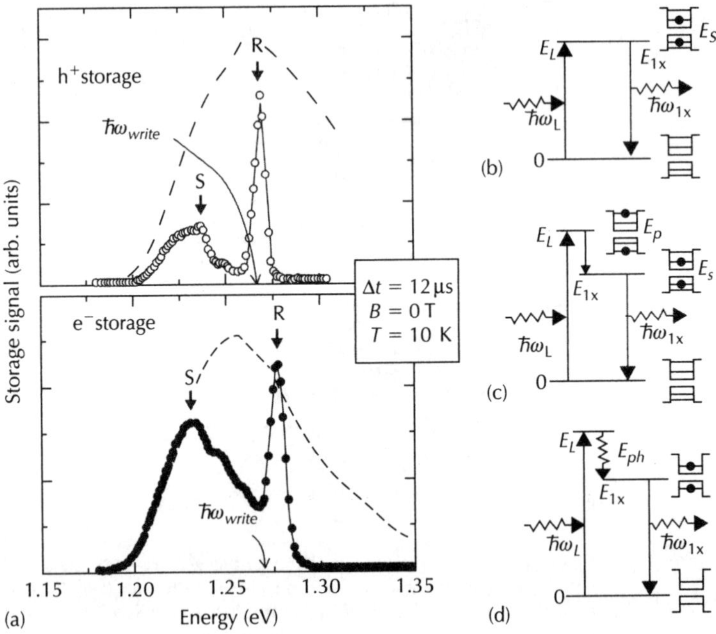

Figure 15.9   (a) Typical emission spectra after application of the reset bias pulse (Δt = 12 μs), for electron (lower panel) and hole (upper panel) storage devices, together with the non-resonantly excited QD ensemble luminescence spectrum of the hole storage device (dashed lines). The absorption processes resulting in the resonant (R) and low energy (S) emission features are also schematically depicted. (b) Ground state absorption giving rise to R. (c) Quasi-resonant excitation via excited state generating the broad emission band S. (d) Phonon-assisted absorption generating the peaks marked with S and arrows. Redrawn from [105] by permission of the American Physical Society.

The characteristic form of the charge storage spectra presented in Fig. 15.9a indicated the existence of two distinct absorption channels: peak **R** arises from resonant optical charging of QD ground states while **S** is due to inelastic charging involving phonons. Following resonant excitation energy relaxation cannot occur and consequently peak **R** is observed close to $\hbar\omega_{write}$ (see Fig. 15.9b). The small 5–7 meV shift observed between $\hbar\omega_{write}$ and **R** is due to the quantum confined Stark effect, since the electric field differs between charge generation and readout [76]. The broader, low energy emission band, labelled **S**, is due to QDs that are quasi-resonantly excited via their excited state. During this process, photo-generated excitons relax into their ground state before ionization occurs, resulting in a luminescence band shifted below $\hbar\omega_{write}$ by the energy separation between ground and excited states. In addition, phonon-assisted excitation may also contribute to **S**, due to enhanced exciton–LO-phonon coupling in QDs as evidenced by the barely resolved satellite features shown in Fig. 15.9, shifted by $\Delta E_{LO} \sim 32$ meV below $\hbar\omega_{write}$. The principal mechanisms contributing to **S** are indicated schematically in Fig. 15.9c and d.

Typical charge storage EL spectra recorded at $T = 10$ K and $\Delta t = 12\,\mu s$ are presented in Fig. 15.10 (lower panel) as the laser excitation energy is tuned from $\hbar\omega_{write} = 1.370$ eV to 1.305 eV. The inhomogeneously broadened PL recorded from the sample under non-resonant optical excitation is also presented for comparison by the dashed line, revealing a ground state maximum around ~ 1.340 eV. The positions of peaks **R** and **S** are plotted in the upper panel of the figure as a function of $\hbar\omega_{write}$, showing that **R** perfectly tracks $\hbar\omega_{write}$ while **S** saturates close to the maximum of the inhomogeneously broadened ensemble PL emission. These observations confirm that peak **R** arises from single electrons that are optically generated *directly* into QD ground states whereafter they do not undergo energy relaxation before readout occurs. In contrast, peak

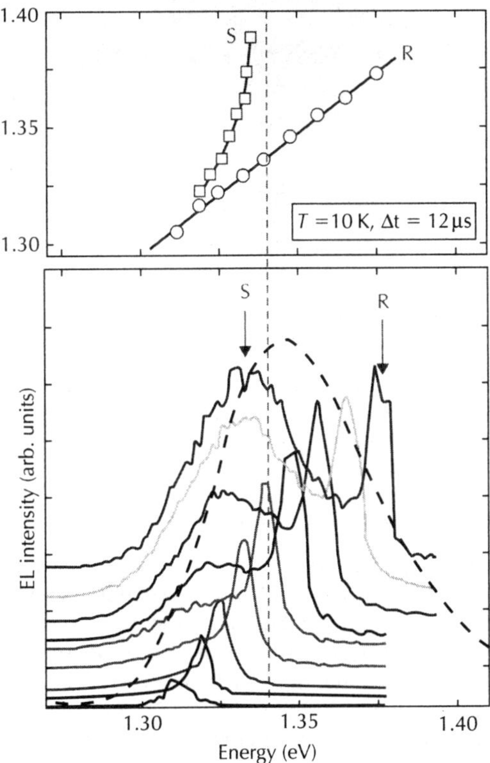

Figure 15.10   Typical electron charge storage spectra recorded at T = 10 K, $\Delta t = 12\,\mu s$ *as* $\hbar\omega_{write}$ *is tuned from* ~1.370 eV (red curve) to 1.305 eV (purple curve). The characteristically sharp peak R arising from resonant charge *generation and readout clearly tracks the laser excitation energy (top panel) while the energy of the broader satellite peak S saturates around* ~1.34 eV *confirming its origin as arising from quasi-resonant QD excitation via excited states. The dashed line shows the inhomogeneously broadened photoluminescence spectrum for comparison.*

S arises from electrons that are quasi-resonantly excited into QD excited states, from where they undergo energy relaxation into the ground state before readout occurs. The fundamental principle of wavelength selective optical QD charging proposed in [46] is, thus, demonstrated; the storage EL signal close to peak **R** corresponding to direct frequency selection of a sub-ensemble of QDs for electron storage and subsequent voltage triggered readout of the stored charge distribution.

The wavelength selective nature of the experiment can be confirmed by performing two-colour experiments, applying a sequence of optical pulses with different frequencies during each excitation cycle. The first pulse, delivered by a diode laser, was centred at ~1.27 eV while the second pulse from the Ti–sapphire, at $\hbar\omega_{write}$, was tuned over the ground state absorption spectrum. Typical charge storage spectra after $\Delta t = 12\,\mu s$ are presented in Fig. 15.11. For both excitation energies the emission spectrum has the form described above, consisting of a resonant peak **R** close to the excitation energies and a weak low energy side band **S**. In full agreement with the discussion above, the side band intensity increases as $\hbar\omega_{write}$ is shifted into the region of excited state absorption. These results demonstrate unambiguously that two separate and *distinct* sub-ensembles are charged by the two-excitation frequencies, reaffirming the conceptual basis for implementation of wavelength division multiplexing for information storage using inhomogeneously broadened media such as QDs. It is also remarkable that, for excitation intensities well below saturation, as is the case for the data presented in Fig. 15.11, crossing of these energies results in a superposition of the charge storage signal and no loss of "information".

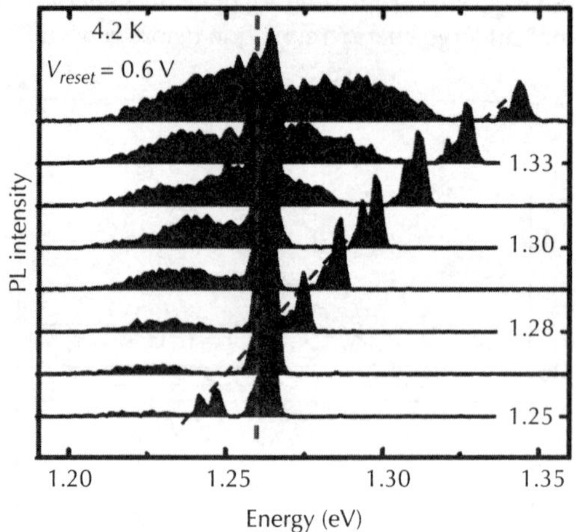

Figure 15.11 Demonstration of two colour charge storage experiment following simultaneous excitation with a laser diode at ~1.26 eV and a tunable pulse from a Ti–sapphire laser.

15.2.3 Thermal redistribution and loss of electrons and holes

Before continuing to discuss the potential to use the present devices to probe spin dynamics and deduce spin flip mechanisms for electrons and holes, we examine the temporal and thermal stability of the optically generated charge distribution. These measurements demonstrate that carriers optically generated directly into QD ground states (peak **R**) remain in the dot where they were generated over ultra-long timescales at low temperatures and are only thermally lost from the dots at elevated temperatures approaching ~100 K [77]. Analysis of the temperature dependence of the charge storage data yields the activation energy, showing that the wetting layer acts as the dominant redistribution channel.

Figure 15.12a shows the intensity of **R** for an electron storage sample as a function of Δt, recorded at temperatures of 10 K, 90 K and 100 K. At $T = 10$ K, no temporal evolution is observed indicating that resonantly generated electrons are not thermally activated out of the dots in which they were created over very long ($\gg 1$ ms) timescales. Indeed, evidence for charge dynamics over these timescales could only be observed at much higher temperatures. This behaviour is illustrated by the data presented in Fig. 15.12b which compares electron storage spectra recorded at $T \sim 110$ K for $\Delta t = 0.2\,\mu$s, $1.0\,\mu$s and $5\,\mu$s. As the storage time increases, a clear *decrease* in the intensity of **R** can be observed as electrons are thermally activated out of the dots in which they were resonantly generated. This behaviour is accompanied by an anticorrelated *increase* of the intensity of **S** showing that electrons may thermally redistribute throughout the QD ensemble (as depicted schematically by the inset on Fig. 15.12) at sufficiently high temperatures but, nevertheless, remain stored within the QD ensemble and are not lost. The results presented in Fig. 15.12 confirm that for $T < 50$ K, the resonant nature of the charge storage process is preserved over timescales $\gg 1$ ms, forming the basis for analysis of electron and hole spin relaxation dynamics presented later in the chapter.

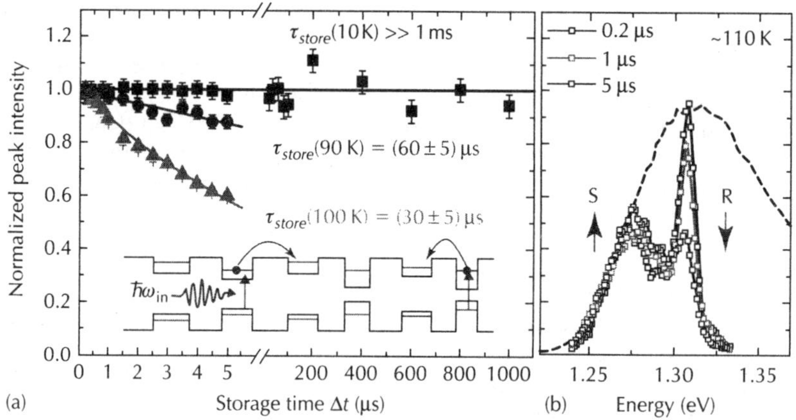

Figure 15.12   (a) Temporal stability of selectively generated electron distribution. Time decay transients of the peak R are presented at temperatures of 10 K, 90 K and 100 K demonstrating the complete absence of thermal charge redistribution via the wetting layer continuum at low lattice temperature. (b) Spectra recorded at ~110 K for $\Delta t = 0.2$, 1.0 and 5.0 μs showing that temporal charge redistribution dynamics are only important at high lattice temperature; the amplitude of peak R quenching with increasing storage time while peak S becomes stronger.

To identify the nature of the thermal redistribution channel, the intensity of **R** was measured as a function of time ($I_R(\Delta t)$) for different lattice temperatures and $\hbar\omega_{write}$. Typical data are presented in Fig. 15.13a for an electron storage sample and in Fig. 15.13b for a hole storage sample with $\hbar\omega_{write} = 1.272$ eV, reproduced from [77]. Both *electron* and *hole* storage devices exhibit similar behaviour but over a different temperature range. Over this time range $I_R(\Delta t)$ remains constant up to $T_{c,e} \sim 90$ K for the electron storage device, while for the hole storage device it only

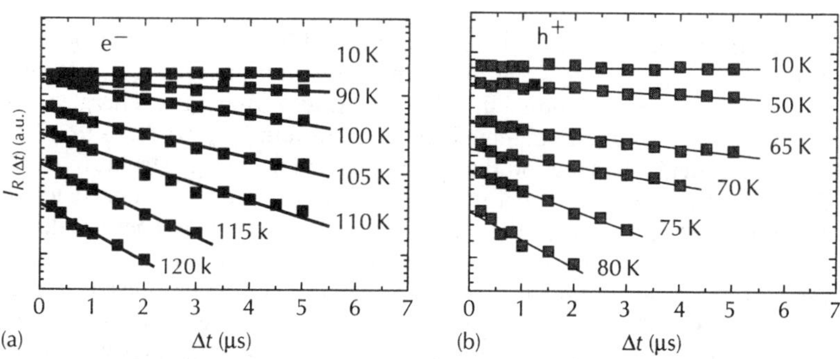

Figure 15.13   Temporal evolution $I_R(\Delta t)$ of the peak R intensity as a function of storage time Δt, at increasing temperatures, for the electron (a) T = 10 to 110 K and hole (b) T = 10 to 80 K storage devices. The data are represented on a logarithmic scale. The excitation energy is $\hbar\omega_w = 1.272$ eV. Grey curves: Single exponential decay fits of $I_R(\Delta t)$. Data replotted from [77] by permission of the American Physical Society.

remains constant up to $T_{c,h} = 50$ K. For $T > T_{c,e(h)}$ the intensity of peak **R** decays exponentially with Δt. Near the critical temperature, the decay time constant τ, as determined by a single exponential decay fit (grey curves in Fig. 15.13a and b) is $\tau \sim 10\,\mu$s, and decreases continuously as the thermal redistribution channel becomes accessible [77]. These results demonstrate that for temperatures up to $T_{c,e(h)}$ no charge redistribution occurs and spectral information is preserved over microsecond timescales. For higher temperatures, the temporal behaviour of the peak **R** intensity as well as the observed excitation energy dependence of the decay times suggests a thermally activated charge redistribution process among the QD ensemble via a high energy channel, which introduces an important time-dependent mechanism for loss of spectral selectivity in the QD charging process.

Figure 15.14a and b show an Arrhenius analysis of the fitted temperature-dependent carrier redistribution rate for both electron (a) and hole (b) storage devices for the three different excitation energies investigated ($\hbar\omega_{write} = 1.272$, 1.298 and 1.342 eV). This analysis is based on the detailed balance between the QD capture and emission rates at thermal equilibrium, under the assumption that the QD capture cross-section is temperature independent. In this case, the QD thermal emission rate $\tau^{-1}(T)$ can be expressed as $\tau(T)^{-1} = AT^{3/2}$ exp (where is a temperature-independent constant, the Boltzmann constant and the activation energy for the thermally activated redistribution process [57]. Thus, a plot of $ln(\tau$ versus e/i should yield a straight line with a slope equal to the activation energy corresponding to the mechanism driving the thermal redistribution.

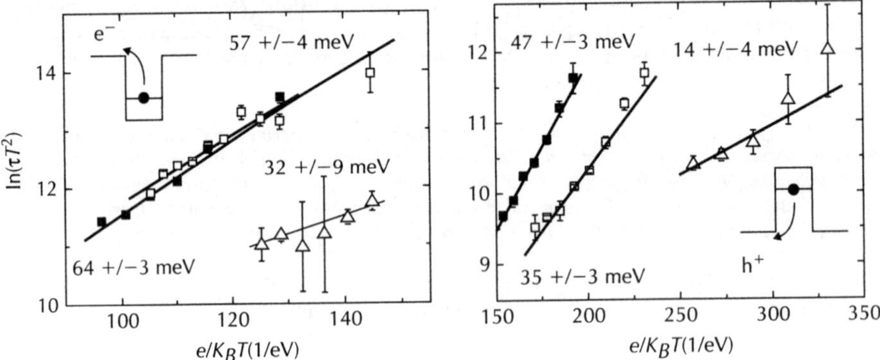

Figure 15.14   Arrhenius analysis of the fitted charge redistribution rates for electrons (left panel) and holes (right panel). As discussed in the text, analysis of the measured charge redistribution rates yield activation energies for the thermally driven redistribution processes.

The results obtained from our analysis are summarized in Table 15.1. The first column displays the excitation energy whereas the measured activation energies $E_{A,e(h)}$ for electrons (e) and holes (h), respectively, are presented in columns 2 and 3. Activation energies ranging from $E_{A,e} = 32 \pm 9$ up to 64 ± 3 meV are obtained for electrons and $E_{A,h} = 14 \pm 4$ to 47 ± 3 meV for holes. These values provide information on the single particle energy structure of the conduction and valence bands, namely the energy difference between the stored particle ground state and the redistribution channel band edge. As expected, the activation energies decrease as the excitation energy is increased; smaller activation energies are obtained for holes. The interband energy of the recombination channnel E_{ch} can be estimated from these single particle results since the electron and hole storage devices' electronic structures are similar, a concept which is supported by the similar interband optical properties in emission (PL) and absorption (photocurrent) spectroscopy. The fourth column of Table 15.1 shows the calculated values of E_{ch}, obtained by summing the single particle activation energies $E_{A,e(h)}$ and the excitation energy, $\hbar\omega_{write}$, while neglecting excitonic effects. Remarkably, E_{ch} is found to be almost constant at $\sim$1.390 eV for all three excitation energies investigated, confirming that redistribution occurs via the wetting layer.

Table 15.1   Summary of the activation energy analysis. *First column*: Excitation energy $\hbar\omega_{write}$. *Second and third columns*: Activation energies $E_{A,e}$ and $E_{A,h}$ for electrons and holes, respectively, obtained using an Arrhenius analysis of the fitted decay time. *Fourth column*: Estimated energy E_{ch} of the redistribution channel obtained by summing $\hbar\omega_{write}$, $E_{A,e}$ and $E_{A,h}$.

1272	64 ± 3	47 ± 3	1393 ± 6
1298	57 ± 4	35 ± 3	1390 ± 7
1342	32 ± 9	14 ± 4	1388 ± 13

Further increase of the sample temperature to values much higher than the critical temperature results in loss of stored charge from the QD layer over the $Al_{0.4}Ga_{0.6}As$ blocking barrier. This effect manifests itself as a temporal decay of the total integrated intensity $I_{tot}(\Delta t)$ of the storage spectra with storage time unlike the temporal redistribution discussed above. For the electron storage device, $I_{tot}(\Delta t)$ was found to be virtually time independent up to $T = 130\,K$ signifying that electrons do not escape from the QD layer. At higher temperatures, the escape time constant was found to be $\tau_{esc} \sim 10\,\mu s$ at $150\,K$. Similar results were obtained for the hole storage sample.

To summarize this section, the results discussed have demonstrated efficient wavelength selective storage of electrons and holes in InGaAs-GaAs self-assembled QD-based devices and their optical readout. Furthermore, they show that wavelength domain data storage can be achieved, with ultra-long charge retention times in excess of milliseconds at $T = 10\,K$. Analysis of the form of the storage spectra revealed the contribution of several absorption mechanisms, such as purely resonant absorption in the QD ground states, quasi-resonant absorption in the QD excited states as well as LO-phonon-assisted absorption. The potential for wavelength division multiplexing has been demonstrated using two-colour experiments, where independent charging of two distinct sub-ensembles of QDs has been clearly identified. Finally, time resolved measurements were used to probe post-absorption redistribution of resonantly generated charge among the QD ensemble. Thermal charge redistribution was shown to be negligible for low temperatures ($T < 60\,K$) and occur over microsecond timescales at elevated temperatures ($T \sim 100\,K$). An activation energy analysis demonstrated that the 2D wetting layer beneath the QDs is the dominant redistribution channel. We continue in the next section to discuss how these concepts for optical *charge* generation can be extended to achieve optical *spin* generation. The high stability of optically generated charges allows investigation of spin relaxation dynamics in self-assembled QDs.

15.3   Optical spin orientation

15.3.1   Selection rules for the neutral exciton transition in QDs

To extend the concepts introduced above from optical charge storage to "*spin* storage" a key consideration is the spin structure of the QD ground state exciton transition. The exciton states are constructed from electron and heavy-hole* single particle basis states with spin projections of $J_{e,z} = +1/2\hbar$, or $-1/2\hbar$ (e↑ or e↓) and $J_{h,z} = -3/2\hbar$, $+3/2\hbar$ (h↓ and h↑), respectively [78]. As a result, four exciton states can be formed with total angular momentum projections of $J_{X,z} = S_z^e + J_z^h = \pm 1\hbar$ (e↓h↑ and e↑h↓) and $\pm 2\hbar$ (e↑h↑ and e↓h↓), respectively. Since a circularly polarized photon conveys a single unit of angular momentum ($+1\hbar$ for σ^+ and $-1\hbar$ for σ^-) and the optical transition takes place to the crystal ground state, only the $J_{X,z} = \pm 1\hbar$ (e↓h↑ and e↑h↓) states are optically active, the $J_{X,z} = \pm 2\hbar$ states remaining dark. As depicted schematically in Fig. 15.15, the isotropic electron–hole exchange interaction lifts the degeneracy of these four states forming a $J_{X,z} = \pm 1$ bright doublet at higher energy, with the $J_{X,z} = \pm 2$ dark states shifted by $\delta_0 \sim 100\,\mu eV$ to lower energy. These ideas have been broadly confirmed by single dot spectroscopy experiments [78, 79].

---

*The predominant heavy-hole character of the ground state arises from the large bi-axial compressive strain present in InGaAs–GaAs QDs.

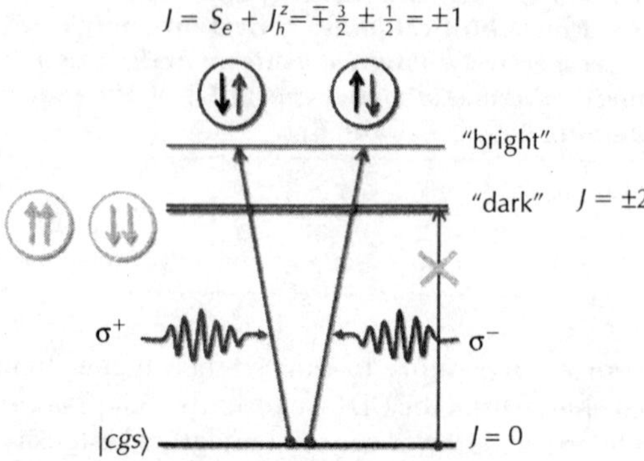

Figure 15.15   *Schematic representation of the spin fine structure of the heavy-hole exciton in a self-assembled QD. The optically active bright exciton states* J = ± 1 *can be addressed using circularly polarized light and are spectrally split from the dark* J = ±2 *states by the isotropic exchange interaction energy. An anisotropic component of the exchange interaction energy mixes the nominally pure* J = +1 *and* −1 *eigenstates to produce a pair of linearly polarized transitions separated by the anisotropic exchange interaction energy.*

The nominally degenerate bright states (e↑h↓ and e↓h↑) are generally mixed in dots due to a reduction of their symmetry. This symmetry reduction is related to the lack of inversion symmetry of the underlying zinc-blende crystal structure and possible asymmetries in the effective confinement potential due to piezocharges and dot morphology. This results in the formation of two linearly polarized eigenstates $| L\pm >$ separated by a few $\sim 10\,\mu eV$ (δ_1) in the absence of a static magnetic field [78, 80]. The general character of these bright exciton eigenstates in a QD subject to a static magnetic field applied parallel to the QD growth direction can be represented by:

$$| L\pm \rangle = \frac{\left|+1\right\rangle + \left(r \pm \sqrt{1+r^2}\right)\left|-1\right\rangle}{\sqrt{2\left(1+r^2 \pm r\sqrt{1+r^2}\right)}} \tag{15.1}$$

where the parameter r is the ratio of the exciton Zeeman energy to the anisotropic exchange splitting ($r = g_{ex}\mu_B B$). In order to enable optical selection of pure $J_{X.z} = \pm 1$ spin states all the measurements described below were performed with static magnetic fields $B > 4\,T$ applied parallel to the growth axis of the dots. This results in the formation of pure spin eigenstates provided that $g_{ex}\mu_B B >>$ (i.e. $r>$) [78]. In this case, excitation with circularly polarized light with σ^- or σ^+ helicity then selectively generates electrons with up (e↑) or down (e↓) spin orientation, respectively, after the charge storage mechanism is completed.* All the measurements presented in this chapter were recorded in the regime where $r>$ and the excitonic states involved are well represented by Fig. 15.16. When the charge storage signal is read out after a time delay Δt, holes with random spin orientation are injected into the dots from the back contact forming equal populations of bright and dark states. Only the bright excitons recombine during the gating time of the single photon detector and the degree of circular polarization of the emitted EL (DoP = $(I_{\sigma+} - I_{\sigma-})/(I_{\sigma+} + I_{\sigma-})$) provides a direct optical probe of the electron spin orientation a time Δt after optical initialization.

In order to test these expectations we performed charge storage measurements with circular polarization discrimination in both excitation and detection channels and large static magnetic fields applied parallel to the QD growth axis. Selected examples of the results of these measurements recorded at $T = 1\,K$, $B = 8\,T$ and a storage time of $\Delta t = 2\,\mu s$ are presented in the upper two panels of Fig. 15.16a. Storage spectra following *excitation* with σ^+ (Fig. 15.16a – upper panel) and σ^- (Fig. 15.16a – middle panel) polarized light were recorded and analysed with σ^-

---

* This statement assumes that the electron component of the exciton *g*-factor is negative as is normally the case for self-assembled quantum dots.

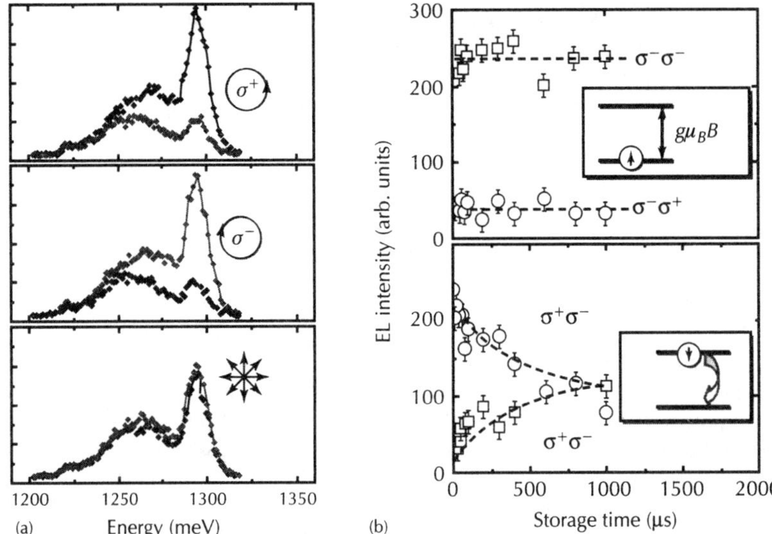

Figure 15.16   *(a) Charge storage spectra recorded with circular polarization discrimination in the excitation and detection channels at B = 8 T and T = 1 K (upper panel = σ^+ excitation, middle panel = σ^- excitation, lower panel = linearly polarized excitation). (b) Intensity of storage luminescence recorded with circular polarization discrimination of both the excitation and detection channel. The upper panel shows the result of optically pumping spins into the lowest Zeeman level. No time evolution is observed since the system is prepared in a state close to thermal equilibrium. In contrast, pumping spins into the upper Zeeman level reveals a clear time evolution (lower panel), from which the spin–flip lifetime is measured. Data redrawn with permission from [75].*

(σ^+) discrimination in the *detection* channel as shown by the red (black) curves on Fig. 15.16. Under these experimental conditions, after spin initialization with circularly polarized light the storage EL in the vicinity of the resonant peak (**R**) is found to be predominantly co-polarized with a DoP $\sim$ 65%. This indicates that spin orientation of the optically generated single *electrons* is preserved over the $\Delta t = 2\,\mu$s storage time. In contrast, following excitation with randomly linearly polarized light (Fig. 15.16a – lower panel) we observe zero degree of circular polarization demonstrating that spin alignment due to, e.g., inter-Zeeman level thermalization does not occur over such short storage times. This observation already suggests that the longitudinal spin relaxation time (T_1) is much longer than the $\sim$2 μs storage time investigated in Fig. 15.16. Here, we note that the population of the two Zeeman levels can be described by Boltzmann statistics, i.e. $N_\downarrow/N_\uparrow \approx \exp(-\,|\,g_{ex}\mu_B B\,|\,/k_B T)$. For $T = 1$ K, we estimate that $N_\downarrow/N_\uparrow \sim 10^{-5}$ demonstrating that the single electron spins probed should occupy the lower Zeeman level in the system at thermal equilibrium if thermalization were fast compared with Δt.

The results presented in Fig. 15.16a demonstrate clearly that the storage signal exhibits a pronounced *polarization memory* effect for both spin orientations (excitation helicities). This arises from the reversible transfer from optical polarization to electron spin orientation followed by electron spin storage for a time $\Delta t = 2\,\mu$s and back-transfer from electron spin orientation into optical polarization. Analysis of the dynamics of the optical polarization provides direct information about the timescales for spin relaxation in self-assembled QDs as discussed later on. However, before discussing investigations of spin relaxation dynamics we review some of the fundamental mechanisms for spin relaxation in semiconductor nanostructures.

15.3.2   Spin relaxation in semiconductor nanostructures

Spin flip relaxation in bulk III–V semiconductor materials is primarily due to scattering processes that can couple to the spin of the electron via the spin orbit interaction (SOI). Coupling of the spin to the orbital motion is well known in atoms to have relativistic origins, arising from the interaction of the electron spin with an effective magnetic field experienced by the electron as it moves in the electric field of the nucleus. Quite generally, an electron moving with momentum in a vacuum experiences an effective magnetic field $\vec{B}_{eff} = (\vec{E} \times \vec{p}) / 2rr$, the origin of which can easily be appreciated by visualizing the electron in its rest frame whereupon the positively charged nucleus executes orbital motion around it, generating a magnetic field at the position of the electron. In semiconductors, SOI will result in the electron spin precessing in $\vec{B}$ as as it propagates with a momentum $\vec{p} =$ through the material. The direction of and $\vec{E}$ will be constant during uninterrupted ballistic motion but change upon scattering from phonons, impurities or other charge carriers. As a result, the randomization of the electron's momentum by scattering will be accompanied by a scrambling of its spin orientation. Rather surprisingly the angle through its spin as it is rotated while it executes ballistic motion turns out to be independent of the velocity $\vec{v} = \vec{p}$. This rather surprising result is a consequence of the fact that the precession frequency of the spin depends on $| \vec{B}$ that itself is determined by $\vec{v} = \vec{p}$. Thus, if an electron moves with a larger velocity it also precesses faster and the net rotation over a given path length is the same. As a result, the strength of the SOI in any material can generally be characterized by the distance over which an electron must travel before its spin rotates by an angle π. This length is termed the *spin–orbit length* and is typically of the order of $\ell_{SO} \sim 1$–$10\,\mu m$ in Ga(In)As semiconductor heterostructures, defining the distance over which an electron should propagate before the spin can flip its orientation. As shown in Fig. 15.1, the typical dimensions of InGaAs QDs are far smaller than ℓ_{SO} and, thus, one would expect that SOI becomes increasingly ineffective as the confinement lengthscales increase. Electron spin relaxation in bulk semiconductors is an ultra-fast process, typically proceeding over timescales faster than $\sim 10\,ps$, primarily due to SOI mediated by phonon or impurity scattering through the continuum of electronic states [81, 82]. However, spin relaxation in QDs is expected to be slowed due to the weaker SOI, as discussed above, and their discrete electronic structure that makes it much more difficult to fulfil energy conservation requirements during spin–flip [83, 84, 85]. While the precise mechanism driving spin relaxation in QDs remains the subject of discussion, a number of workers have investigated theoretically the comparative roles of linear spin–orbit interaction [83, 84, 85, 86] and hyperfine coupling to the nuclear spin system [87, 88]. The scattering rate due to SOI has been shown to be suppressed by several orders of magnitude in QDs due to their discrete electronic structure and the need, therefore, for the participation of a limited spectrum of phonons to satisfy energy conservation requirements [83, 84, 85, 88]. Under conditions of large static B-fields, the nuclear hyperfine coupling perturbs only the precession frequency of the electron spin about the static field direction, causing dephasing, but is not expected to flip the orientation of the electron spin and impact upon T_1 [87, 88].

These ideas from theory have been broadly supported by measurements of long spin lifetimes for *excitons* in quantum dots, which essentially reveal no evolution of the exciton spin over its radiative lifetime (~ 1 ns) [89, 90]. However, such measurements are incapable of directly probing spin dynamics, since they are so slow, calling for investigations of single charges (electrons and holes) in QDs for which radiative recombination cannot occur. The spin memory devices investigated here provide a direct route towards the investigation of the electron and hole longitudinal spin lifetimes.

15.3.3   *Polarization dynamics as a probe of electron spin relaxation dynamics*

Spin relaxation can be directly measured using spin memory devices: electrons are optically initialized in the higher energy Zeeman level by exciting the system with polarized light. The spin relaxation time is then directly measured by monitoring the intensity of the emission recorded with helicity as a function of the storage time ($I_{+, +}()$. In the absence of spin relaxation over the time one should observe a strong *polarization memory* in the storage signal as discussed above. In contrast, complete spin relaxation would result in steady-state spin populations of the two spin states according to Boltzmann statistics, independent of the helicity of the optical excitation. For the low temperatures and high magnetic fields studied here, this would correspond to all electron spins occupying the lowest Zeeman level, or complete polarization of the storage EL. Reference to the data presented in Fig. 15.16a confirms that spin relaxation is slower than ~2 μs at $B = 8$ T and $T = 1$ K.

Examples of such time resolved spin storage measurements, recorded at $T = 1$ K and $B = 8$ T, are presented in Fig. 15.16b. The data shows the temporal evolution of the storage luminescence intensity following σ excitation over the time range 0.001 ms << 1 ms. Following excitation with polarized light to pump electrons into the lower energy Zeeman level the storage EL is found to be predominantly polarized as expected. Furthermore, it exhibits no detectable evolution up to ~1 ms (Fig. 16b – upper panel) since the system is initialized close to thermal equilibrium. In contrast, following excitation using polarized light, to generate electrons into the upper Zeeman level, a very marked time dynamics of the luminescence polarization is observed (Fig. 15.16b – lower panel). For $\Delta t = 0.001$ the measurement reveals a maximum degree of circular polarization of $DoP \sim +61$, decaying over time to a few per cent at $\Delta t \sim 1$ as electrons flip their spin as depicted schematically in the figure. A very long spin lifetime of $T_1{}^g = 1.1 \pm 0.2$ is obtained, more than four orders of magnitude longer than spin–flip times for quantum wells and seven orders of magnitude longer than the corresponding time in bulk III–V semiconductors.

In order to establish the mechanism driving spin–flip events we investigated the dependence of T_1 on magnetic field and the lattice temperature. The data recorded at $T = 1$ K is plotted in Fig. 15.17a on a double logarithmic representation. The decay time constants extracted using the above method are found to be strongly dependent on the magnetic field, reducing dramatically from $T_1 = 20 \pm 6$ ms at $B = 4$ T to 0.1 ± 0.01 ms at 12 T. These data suggest a clear power-law dependence $T_1 \propto B^m$. A least squares fit to the data shown in Fig. 15.17a yields $m = -4.5 \pm 0.2$. We continue by discussing the origin of this very strong magnetic field dependence. As discussed in [83, 84] and [85], one of the most significant mechanisms driving spin–flip transitions is due to spin–orbit coupling of the Zeeman levels via one-phonon scattering processes. In this case, only vibrational modes with an energy $E_{ph} = g_e\mu_B B$ couple the two spin states.

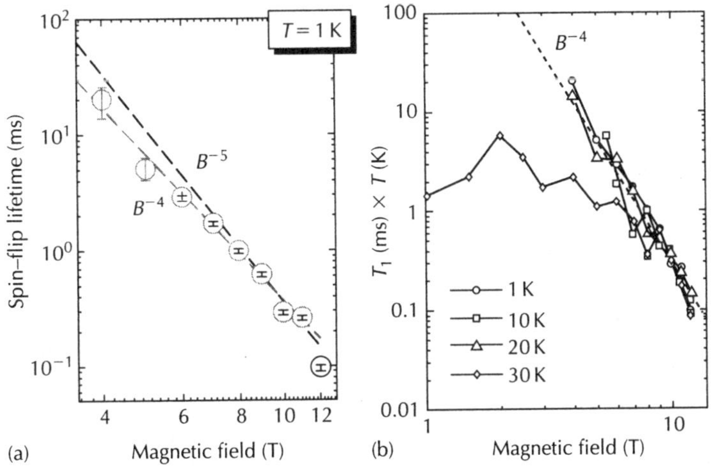

Figure 15.17 (a) The evolution of the electron spin lifetime at T = 1 K as a function of magnetic field reproduced with permission from [75]. (b) Measured electron spin lifetimes scaled with the lattice temperature at 1, 10, 20 and 30 K. The strong exponent and linear temperature dependence can clearly be seen.

The spin relaxation time can be calculated using perturbation theory and has the form [83]

$$T_1 = A\frac{\hbar(\hbar\omega_{sp})^4}{(g_e\mu_B B)^5}\left(\frac{1}{N_{ph}(E_{ph},T) + 1}\right) \tag{15.2}$$

where A is a dimensionless constant that reflects the effective spin piezoelectric phonon coupling strength in the material,* $\hbar\omega_{sp}$ is the single particle s–p level spacing and $N_{ph}(E_{ph},T)$ is the occupation number of the phonon mode. At temperatures such that $k_BT \gg g_e\mu_BB$, N_{ph} reduces to $N_{ph} \sim k_BT/g_e\mu_BB$ and the spin lifetime becomes:

$$T_1 \approx A\frac{\hbar(\hbar\omega_{sp})^4}{(g_e\mu_BB)^4}\cdot\left(\frac{1}{k_BT}\right) \tag{15.3}$$

Thus, the electron spin lifetime due to linear spin–orbit interaction mediated by one-phonon scattering events should exhibit the following characteristics:

- Varies with magnetic field according to $T_1 \propto B^{-4}$.
- Varies with temperature according to $T_1 \propto (1/T)$.
- Exhibits a very strong dependence on the dot size via $\hbar\omega_{sp}$.

Temperature dependent data T_1^e is presented in Fig. 15.17b, and plotted in the form $T_1 \cdot T$ vs. -field for temperatures of 1 K, 10 K, 20 K and 30 K. For T$\leq$, the experimental data plotted in this representation is very similar, confirming the temperature dependence predicted by Eq. 15.3 and a magnetic field exponent of $m = -4.5 \pm 0.2$, close to the characteristic dependence predicted by Eqs 15.2 and 15.3. When combined, these observations provide strong experimental evidence that inter-Zeeman level spin–flip transitions in QDs are indeed dominated by SOI involving one-phonon scattering processes at low temperature. The longest T_1 times measured of $\sim$20 ms at $T = 1$ K and $B = 4$ T are limited by the sensitivity of our optical detection scheme and represent a lower limit. The absence of any observable saturation of T_1 in Fig. 15.17 suggests that the relaxation time could be much longer at lower fields. For example, extrapolating the observed $T_1 \propto B^{-4}$ dependency to lower fields would indicate $T_1 = 80$ ms at $\sim$3 T, reaching $\sim$1 s at $B \sim 1.6$ T. More recent work on electrostatically defined QDs has revealed spin lifetimes in excess of seconds at low temperatures [91]. Theoretical work has suggested that if spin orbit coupling mediated by single phonons is the *dominant* spin scattering mechanism, then the spin coherence time should approach the limit $T_2 = 2 T_1$ since fluctuations of the effective magnetic field are always perpendicular to the applied magnetic field [92]. However, recent studies have shown that this is not the case since hyperfine coupling with the nuclear spin system dominates electron spin relaxation and dephasing at lower magnetic fields [93].

15.3.4   Hole spin relaxation in QDs

Unlike electrons, holes couple more weakly to the nuclear spins via the hyperfine contact interaction since they have p-like central cell symmetry [94]. This may provide an attractive route towards *hole–spin*-based applications free from the complications caused by the fluctuating nuclear spin system. However, the hole–spin lifetime in III–V semiconductor nanostructures is generally much shorter than discussed above due to strong SO mixing of heavy (HH) and light hole (LH) valence bands [95, 96]. This mixing is inhibited by motional quantization effects and enhanced hole–spin lifetimes have been reported for quantum wells ($\sim$100 ps [97]), extending beyond $\sim$1 ns when optically driven spin heating effects are avoided [98]. For QD nanostructures, is expected to become even longer due to the combined effects of bi-axial compressive strain and motional quantization. These expectations have recently been supported by studies of negatively charged trions in InAs [99] and CdSe [100] QDs, which indicate that $T_1^h \geq$ ns, limited by the timescale for radiative recombination of the trions. Recent calculations have indicated that can become much longer for isolated holes [101], even exceeding in the limit when the energy separation between HH and LH bands far exceeds the orbital quantization energy in the valence band [95].

---

*Using typical material properties applicable to GaAs gives $A = 71$, which together with reasonable values for $|g_e| = 0.8$ and $\hbar\omega_{sp} = 30$ meV provide good quantitative agreement with our experimental data as indicated by the dotted line in Figure 15.17.

In the last few months we applied our optical charge storage techniques to probe single *hole–spin* relaxation in small ensembles of self-assembled InGaAs QD [102]. Time and polarization resolved magneto-optical spectroscopy were performed to obtain as a function of static magnetic field and lattice temperature, as for the electron storage samples discussed above. Most remarkably, our results demonstrated that single hole–spin relaxation can proceed over timescales which are *comparable* to electrons in nominally identical QDs. This shows that the valence band SO coupling does not necessarily set the timescale for hole–spin relaxation as is the case in higher-dimensional systems. Explicit expressions for due to SO-mediated single-phonon scattering produce very good quantitative agreement, demonstrating that hole-spin relaxation in self-assembled QDs is mediated by the same mechanism as for electrons.

Figure 15.18 shows the temporal evolution of the peak intensity of the resonant peak at $B =$ for storage times in the range $0.3\,\mu s \ll 60\,\mu s$, recorded with circularly polarized excitation and detection. Here, denotes the intensity of peak **R** recorded with polarization in both excitation and detection channels, the other three curves completing the four polarization permutations. The full dark grey lines in Fig. 15.18 show the storage EL intensities and for the corresponding excitation polarization averaged over the two detection polarizations. Within experimental error and are independent of, confirming that holes are not thermally redistributed between the dots via wetting layer states during the range of storage times investigated. Careful examination of the data presented in Fig. 15.18 indicates the presence of a significant polarization memory effect; the storage EL is co-polarized with the excitation laser for short storage times () with a degree of circular polarization up to 10% as shown in the upper panel of Fig. 15.18. As Δt increases, the DoP decreases exponentially for both and excitation, reaching an equilibrium value close to zero for long storage times ($\Delta t > 40\,\mu s$).

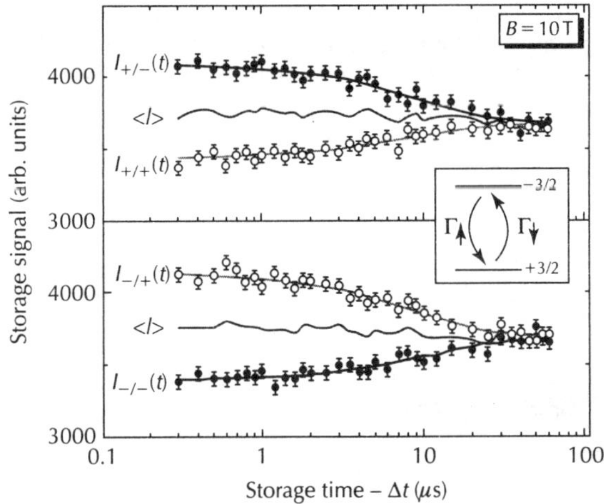

Figure 15.18 Polarization resolved intensity of resonant hole storage signal as a function of storage time. The four curves show data for the four combinations of excitation and detection polarizations. Inset: Two-level model used to represent hole spin states.

Quantitative information can be extracted from these data using rate equations applicable to a non-interacting ensemble of two-level systems to represent the total population of $+3/2$ and $-3/2$ holes in the QDs probed by the experiment (Fig. 15.18 (inset)). The rates for up-scattering ($-3/2 \rightarrow +3/2$) and down-scattering ($+3/2 \rightarrow -3/2$) are represented by and, respectively. Using this model the population of a specific spin species, a time t after generation, is given by:

$$N_{\uparrow(\downarrow)}(t) = (N_{\uparrow(\downarrow)}(0) - N_{\uparrow(\downarrow)}^{equ})\exp\left(-\frac{t}{T_2^h}\right) + N \tag{15.4}$$

where $N_{\uparrow(\downarrow)}(t) = (\Gamma_\uparrow \omega / \Gamma_\uparrow + \Gamma_\downarrow)$ is the Boltzmann distribution at thermal equilibrium, $T_1^h = 1/(\Gamma_\uparrow +$ is the hole–spin lifetime and $N_{\uparrow(\downarrow)}$ the initial population of spin up (down) holes at

$\Delta t =$. Since the intensity of the storage luminescence a time after generation is proportional to the curves presented in Fig. 15.18b can each be fitted by $I_{n/m}(\Delta t) = (I_{n/m}(0) - \langle I^n \rangle)\exp(-\Delta t/T_2^h)$, where $I_{n/m}$ is the corresponding intensity at Δt and $\{n, m\} = \{+, +\}, \{+, -\}\{-, +\}$, and $\langle I^- \rangle = \langle I^+ \rangle =$ is the saturation value for long storage times. The best fits to all four decay curves are presented as full lines in Fig. 15.18b. Most importantly, all four traces are well described using a *single* value for the hole–spin lifetime of $T_1^h = 11 \pm$ using this method.

The observation of such remarkably slow hole–spin relaxation dynamics in QDs contrasts strongly with bulk III–V materials where spin–flip scattering occurs over subpicosecond timescales due to SO mixing of the valence band spin states close to the Γ-point in the bandstructure [96]. This mixing is partially inhibited by motional quantization effects in quantum wells, leading to enhanced hole–spin lifetimes longer than ~ 1 ns at low temperatures [98]. In contrast, the fully quantized electronic structure in QDs restricts the phase space for such quasi-elastic scattering processes and results in much longer hole–spin lifetimes in the present experiment [101].

Finally, we discuss the dependence of on external magnetic field B along the growth direction and lattice temperature T, as for the electron sample discussed in the previous section. Fig. 15.19a shows the excess degree of circular polarization following excitation and detection plotted as a function of Δt for $B = 5 - 11\,T$ at $T = 8\,K$. In all cases the time dependence of the polarization is well described by a monoexponential decay (Eq. 15.4), with a time constant that becomes longer as the B-field is reduced. The values of obtained are presented in Fig. 15.19b and directly compared with equivalent data recorded from an electron spin memory device measured under the same experimental conditions. As the B-field is reduced from 12 to 1.5 T the hole–spin lifetime is found to increase from $8 \pm 3\,\mu s$ to $270 \pm 180\,\mu s$. The spin relaxation time measured at $B = 1.5\,T$ is more than four orders of magnitude longer than recent reports for single CdSe dots where a lower limit of 10 ns was measured [100]. Furthermore, the ratio T_1^h lies in the range ~ 5–10 over the whole range of the B-field investigated, in strong contrast with quantum wells where T_1^h is smaller than $\sim 10^{-3}$ due to SO mixing of HH and LH valence bands. In the absence of such SO mixing we expect hole–spin relaxation to be dominated by SO-mediated spin–lattice coupling, as is the case for electron spin relaxation in self-assembled QDs. To test this hypothesis we studied the temperature dependence of. Fig. 15.19c shows the temperature dependence of for

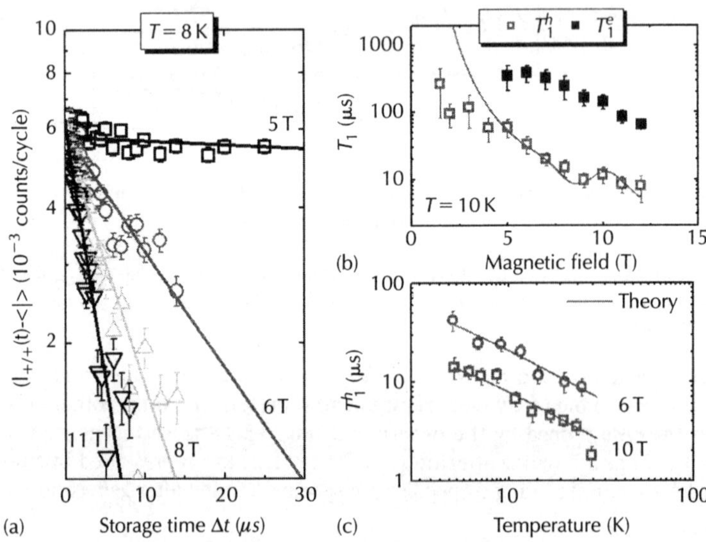

Figure 15.19   (a) Temporal dependence of polarization intensity as a function of Δt for magnetic field ranging from 5 to 11 T. (b) Comparison of the B-field dependence of and (filled symbols) for identical QD material and experimental conditions. The filled line shows the calculated hole spin relaxation time using the model of [101] and the following parameters: $g_h =$, $\hbar = 6\,meV$, $= 100\,meV$, $m^*_h = 0.03\,m_0$ and $T = 8\,K$. (c) Comparison of measured and calculated dependence of on lattice temperature.

magnetic fields of $B = 6$ T and 10 T. For both magnetic fields a very clear $T_1^h \propto$ dependence is observed, exactly as was found previously for electron spin relaxation in the previous section. The observation of mono-exponential decays and comparable spin lifetimes for electrons and holes combined with temperature and strong B-field dependencies strongly indicates that the dominant relaxation mechanism for hole spins is, indeed, due to SO-mediated spin–lattice coupling as has been shown to be the case for electrons.

15.4 Outlook

One of the major challenges for investigations in the immediate future is the determination of coherence (T_2) and inhomogeneous spin dephasing (T_2^*) times for single electrons and holes in self-assembled QD nanostructures. These quantities may be accessible using time resolved coherent optical measurements or by combining the spin storage experiments with electron spin resonance techniques. Another important aspect is the efficient readout of single electron spins. This may be achieved optically by incorporating the dots within microcavities with enhanced collection efficiency and, thus, enhanced sensitivity. In addition, high-efficiency electrical readout of such optically prepared spin systems may also be feasible [103]. The use of the spin degree of freedom in quantum dots for more complex quantum operations will also require extension of the present techniques to coupled dots or arrays of dots. Here, frequency selective optical addressing provides maximum flexibility when compared to spin to charge conversion. Furthermore, we have very recently demonstrated that coupling of self-assembled InGaAs quantum dots is indeed possible by electric field tuning [104]. This is a first step towards up-scaling of the spin control from single quantum dots to more complex coupled quantum systems for spin-based quantum information processing.

Acknowledgements

This work was financed by the Deutsche Forschungsgemeinschaft via SFB631 (*Festkörperbasierte Quanteninformationsverarbeitung: Physikalische Konzepte und Materialaspekte*) and the Nanosystems Initiative Munich (NIM) Cluster of excellence. I would like to acknowledge the important contributions from many of my colleagues and, above all, the hard work and dedication of many PhD, Diploma and Masters students over the past few years researching the topics presented in this chapter.

References

1. R.H. Dennard, United States Patent 3387286 (1967).
2. For a history of Intel and the development of DRAM technologies see http://www.intel.com/museum/corporatetimeline/index.htm
3. Term coined around 1970 by the Caltech professor, VLSI pioneer, and entrepreneur Carver Mead – see http://en.wikipedia.org/wiki/Moore%27s_Law
4. S.M. Sze, in: *Future Trends in Microelectronics*, ed. by S. Luryi, J. Xu, and , A. Zaslavsky. (John Wiley & Sons, Inc.1999)291 pp.
5. D. Bimberg, M. Grundmann, and N.N. Ledentsov. *Quantum Dot Heterostructures*, (John Wiley & Sons, Chicester, 1998).
6. D. Bimberg, Der Zoo der Quantenpunkte, *Physik Journal*, **5**, 43 (2006).
7. L.P. Kouwenhoven, C.M. Marcus, P.L. McEuen, S. Tarucha, R.M. Westervelt, and N.S. Wingreen, in: *Mesoscopic Electron Transport*, ed. by L.L. Sohn, L.P. Kouwenhoven, and G. Schön, Series E: Applied Sciences (Kluwer Academic Dordrecht), Vol. 345, p. 105 (1997).
8. F.M. Schellenberg, W. Lenth, and G.C. Bjorklund. *Appl. Optics*. **25**, 3207 (1986).
9. W.E. Moerner, ed., *Persistent Spectral Hole Burning: Science and Applications* (Springer Verlag, New York, 1988).

10. *Optical data storage* by H. Coufal, and G. W. Burr, in: International Trends in Applied Optics, ed. A.H. Guenther, SPIE (2002) – ISBN 081944510X.

11. For a comparison of the specifications of many optical data storage technologies see www.blu-ray.com.

12. J. Ashley, M.-P. Bernal, G.W. Burr, H. Coufal, H. Guenther, J.A. Hoffnagle, C.M. Jefferson, B. Marcus, R.M. Macfarlane, R.M. Shelby, and G.T. Sincerbox, *IBM Journal of Research and Development*, **44**, 3 (2000).

13. V. Cerletti, W.A. Coish, O. Gywat, and D. Loss, *Nanotechnology*, **16**, R27–rR49 (2005).

14. I.N. Stranski and L. Krastanow, *Sitzungsber. Akad. Wiss. Wien, Math.-Naturwiss. Kl., Abt. 2B*, **146**, 797 (1938).

15. For a recent review of strain-driven self-assembly with a focus on growth and structural properties, see J. Stangl, V. Holy and G. Bauer, Rev. Mod. Phys. **76**, 725 (2004)

16. L. Goldstein, F. Glas, J.Y. Marzin, M.N. Charasse, and G. LeRoux, *Appl. Phys. Lett.*, **47**, 1099 (1987).

17. C. Pryor, *Phys. Rev. B*, **60**, 2869–2874 (1999).

18. D.S. Sizov, Yu.B. Samsonenko, G.E. Tsyrlin, N.K. Polyakov, V.A. Egorov, A.A. Tonkikh, A.E. Zhukov, S.S. Mikhrin, A.P. Vasil'ev, Yu.G. Musikhin, A.F. Tsatsul'nikov, V.M. Ustinov, and N.N. Ledentsov, *Semiconductors*, **37**, 5 (2003).

19. J.G. Lim, Y.J. Park, Y.M. Park, J.D. Song, W.J. Choi, I.K. Han, W.J. Cho, J.I. Lee, T.W. Kim, H.S. Kim, and C.G. Park, *J. Cryst. Growth*, **275**, 415–421 (2005).

20. D.J. Mowbray and M.S. Skolnick, *J. Phys. D. Appl. Phys.*, **38**, 2059–2076 (2005).

21. A. Shields and Z.L. Yuan, *Physics World*, **20**, 3 (2007).

22. R.M. Stevenson, R.J. Young, P. Atkinson, K. Cooper, D.A. Ritchie, and A.J. Shields, *Nature*, **439**, 179 (2006).

23. P. Hawrylak, *Phys. Rev. B*, **60**, 5597 (1999).

24. Oliver Stier, *Electronic and Optical Properties of Quantum Dots and Wires*, ISBN 3-89685-366-X, (2002) PhD thesis, TU-Berlin.

25. A. Zrenner, *J. Chem. Phys.*, **112**, 7790 (2000).

26. J.J. Finley, P.W. Fry, A.D. Ashmore, A. Lemaître, A.I. Tartakovskii, R. Oulton, D.J. Mowbray, M.S. Skolnick, M. Hopkinson, P.D. Buckle, and P.A. Maksym, *Phys. Rev. B*, **63**, 161305 (2001).

27. K. Brunner, U. Bockelmann, G. Abstreiter, M. Walther, G. Böhm, G. Tränkle, and G. Weimann, *Phys. Rev. Lett.*, **69**, 3216 (1992).

28. K. Brunner, G. Abstreiter, G. Böhm, G. Tränkle, and G. Weimann, *Appl. Phys. Lett.*, **64**, 3320 (1994).

29. A. Zrenner, L.V. Butov, M. Hagn, G. Abstreiter, G. Böhm, and G. Weimann, *Phys. Rev. Lett.*, **72**, 3382 (1994).

30. K. Brunner, G. Abstreiter, G. Böhm, G. Tränkle, and G. Weimann, *Phys. Rev. Lett.*, **73**, 1138 (1994).

31. H.F. Hess, E. Betzig, T.D. Harris, L.N. Pfeiffer, and K.W. West, *Science*, **264**, 1740 (1994).

32. D. Gammon, E.S. Snow, B.V. Shanabrook, D.S. Katzer, and D. Park, *Science*, **273**, 87 (1996).

33. M. Grundmann, J. Christen, N.N. Ledentsov, J. Böhrer, D. Bimberg *et al.*, *Phys. Rev. Lett.*, **74**, 4043 (1995).

34. S. Rodt, A. Schliwa, K. Pötschke, F. Guffarth, and D. Bimberg, *Phys. Rev. B*, **71**, 155325 (2005).

35. E. Beham and A. Zrenner, in: *Adv. in Solid State Phys.* 43, ed. By B. Kramer, 287–300 (2003).

36. A. Zrenner, E. Beham, S. Stufler, F. Findeis, M. Bichler, and G. Abstreiter, *Nature*, **418**, 612 (2002).

37. A.J. Ramsay, R.S. Kolodka, F. Bello, P.W. Fry, W.K. Ng, A. Tahraoui, H.Y. Liu, M. Hopkinson, D.M. Whittaker, A.M. Fox, and M.S. Skolnick, *Phys. Rev. B*, **75**, 113302 (2007).

38. D.L. Drexler, D. Leonard, W. Hansen, J.P. Kotthaus, and P.M. Petroff, *Phys. Rev. Lett.*, **73**, 2252 (1994).

39. R.J. Warburton, C. Schäflein, D. Haft, F. Bickel, A. Lorke, K. Karrai, M.M. Garcia, W. Schoenfeld, and P.M. Petroff, *Nature*, **405**, 926 (2000).

40. F. Findeis, M. Baier, A. Zrenner, M. Bichler, G. Abstreiter, U. Hohenester, and E. Molinari, *Phys. Rev. B*, **63**, 121309 (2001).

41. J.J. Finley, M. Sabathil, P. Vogl, G. Abstreiter, R. Oulton, A.I. Tartakovskii, D.J. Mowbray, M.S. Skolnick, S.L. Liew, A.G. Cullis, and M. Hopkinson, *Phys. Rev. B*, **70**, 201308 (2004).

42. B. Alén, F. Bickel, K. Karrai, R.J. Warburton, and P.M. Petroff, *Appl. Phys. Lett.*, **83**, 2235 (2003).

43. A. Högele, S. Seidl, M. Kroner, K. Karrai, R.J. Warburton, B.D. Gerardot, and P.M. Petroff, *Phys. Rev. Lett.*, **93**, 217401 (2004).
44. M. Atatüre, J. Dreiser, A. Badolato, and A. Imamoglu, *Nature Phys*, **3**, 101 (2007).
45. Proceedings of the International Conference on Semiconductor Quantum Dots, Chamonix, (2006), edited by *Phys. Stat. Sol. (b)* 243, No. 15, 3855–3998 (2006) and *Phys. Stat. Sol. (c) 3*, No. 11, 3641–4044 (2006).
46. S. Muto, *Jpn. J. Appl. Phys. Part 2*, **34**, L210 (1995).
47. G. Yusa and H. Sakaki, *Appl. Phys. Lett.*, **70**, 345 (1997).
48. J.J. Finley, M. Skalitz, M. Arzberger, A. Zrenner, G. Böhm, and G. Abstreiter, *Appl. Phys. Lett.*, **73**, 2618 (1998).
49. A.J. Shields, M.P. O'Sullivan, I. Farrer, D.A. Ritchie, R.A. Hogg, M.L. Leadbeater, C.E. Norman, and M. Pepper, *Appl. Phys. Lett.*, **76**, 3673 (2000).
50. L. Zhuang, L. Guo, and S.Y. Chou, *Appl. Phys. Lett.*, **72**, 1205 (1998).
51. H. Yusa and H. Sakaki, *Appl. Phys. Lett.*, **70**, 345 (1997).
52. M. Geller, C. Kapteyn, L. Müller-Kirsch, R. Heitz, and D. Bimberg, *Phys. Stat. Sol. (b)*, **238**, 258–261 (2003).
53. A. Marent, M. Geller, D. Bimberg, A.P. Vasiev, E.S. Semenova, A.E. Zhukov, and V.M. Ustinov, *Appl. Phys. Lett.*, **89**, 072103 (2006).
54. K. Imamura, Y. Sugiyama, Y. Nakata, S. Muto, and N. Yokoyama, *Jpn. J. Appl. Phys.*, **34**, L1445 (1995).
55. J.J. Finley, M. Skalitz, M. Arzberger, A. Zrenner, G. Böhm, and G. Abstreiter, *Appl. Phys. Lett.*, **73**, 18 (1998).
56. W. Wieczorek, T. Warming, M. Geller, D. Bimberg, G.E. Cirlin, A.E. Zhukov, and V.M. Ustinov, *Appl. Phys. Lett.*, **88**, 182107 (2007).
57. C. Kapteyn, PhD thesis, *Carrier Emission and Electronic Properties of Self-Organized Semiconductor Quantum Dots* (Mensch & Buch Verlag, Berlin, 2001).
58. C.M.A. Kapteyn, F. Heinrichsdorff, O. Stier *et al.*, *Phys. Rev. B*, **60**, 14265 (1999).
59. C.M.A. Kapteyn, M. Lion, R. Heitz *et al.*, *Appl. Phys. Lett.*, **76**, 1573 (2000).
60. C. Miesner, T. Asperger, K. Brunner, and G. Abstreiter, *Appl. Phys. Lett.*, **77**, 2704 (2000). and C.M.A. Kapteyn, M. Lion, R. Heitz, D. Bimberg, C. Miesner, T. Asperger, K. Brunner, and G. Abstreiter, Appl. Phys. Lett. 77, 4169 (2000).
61. F. Hatami, N.N. Ledentsov, M. Grundmann *et al.*, *Appl. Phys. Lett.*, **67**, 656 (1995).
62. A. Marent, M. Geller, D. Bimberg, A.P. Vasi'ev, E.S. Semenova, A.E. Zhukov, and V.M. Ustinov, *Appl. Phys. Lett.*, **89**, 072103 (2006).
63. H. Sakaki, G. Yusa, T. Someya, Y. Ohno, T. Noda, H. Akiyama, Y. Kadoya, and H. Noge, *Appl. Phys. Lett.*, **67**, 3444 (1995).
64. E. Ribeiro, E. Miller, T. Heinzel, H. Auderset, K. Ensslin, G. Medeiros-Ribeiro, and P.M. Petroff, *Phys. Rev. B*, **58**, 1506 (1998).
65. B.E. Kardynal, A.J. Shields, M.P. O'Sullivan, N.S. Beattie, I. Farrer, D.A. Ritchie, and K. Cooper, *Meas. Sci. Technol.*, **13**, 1721–1726 (2002).
66. D. Heinrich, J. Hoffmann, J.J. Finley *et al.*, *Thin Solid Films*, **380**, 192–194 (2000).
67. D. Heinrich, J. Hoffmann, J.J. Finley *et al.*, *Phys. Stat. Sol. (b)*, **224**, 357–360 (2001).
68. D.J. Chadi and K.J. Chang, *Phys. Rev. Lett.*, **61**, 7873 (1988).
69. The samples were typically reset by passing between Ipn = $1 \sim 10\,\mu$A for times between 100 and 300 ms.
70. We note that the form of the excitation cycle for Eex = 1059 meV is representative of the measured response to higher Eex, up to 1250 meV while the Eex = 1016 meV cycle typifies the lower excitation energy response.
71. S. Raymond, P.J. Poole, S. Fafard, A. Wojs, P. Hawrylak, S. Charbonneau, D. Leonard, P.M. Petroff, and J.L. Merz, *Phys. Rev. B*, **54**, 11548 (1996).
72. A. Vasinelli *et al.*, *Phys. Rev. Lett.*, **89**, 216804 (2002). and R. Oulton et al., *Phys. Rev. B* **68**, 235301 (2003).
73. D. Heinrich, J. Hoffmann, J.J. Finley *et al.*, *Physica E*, **7**(3–4), 484–488 (2000).
74. For the electron storage sample, the strength of the observed photo effect reduces for T > 170 K, and is not observed for T > 230 K.

75. T. Lundstrom, W. Schoenfeld, H. Lee, and P.M. Petroff, *Science*, **286**, 2312 (1999).
76. M. Kroutvar, Y. Ducommun, D. Heiss, M. Bichler, D. Schuh, G. Abstreiter, and J.J. Finley, *Nature*, **432**, 81 (2004).
77. During writing, $|F| \sim 10^4$ V/cm whereas the device is close to flat band conditions (F $\sim$ 0 V/cm) during the read cycle.
78. Y. Ducommun, M. Kroutvar, M. Reimer, M. Bichler, D. Schuh, G. Abstreiter, and J.J. Finley, *Appl. Phys. Lett.*, **85**, 2592 (2004).
79. M. Bayer, G. Ortner, O. Stern, A. Kuther, A.A. Gorbunov, A. Forchel, P. Hawrylak, S. Fafard, K. Hinzer, T.L. Reinecke, S.N. Walck, J.P. Reithmaier, F. Klopf, and F. Schäfer, *Phys. Rev. B*, **65**, 195315 (2002).
80. J.J. Finley, D.J. Mowbray, M.S. Skolnick, A.D. Ashmore, C. Baker, A.F. Monte, and M. Hopkinson, *Phys. Rev. B*, **66**, 153316 (2002).
81. W. Langbein, P. Borri, U. Woggon, V. Stavarache, D. Reuter *et al.*, *Phys. Rev. B*, **69**, 161301 (2004).
82. *Optical Orientation* ed. by F. Meier, and B.P. Zakharchenya, Chap. 3 (North-Holland, Amsterdam, 1984). See also *Superlattices and Other Microstructures*, ed. by E.L. Ivchenko, and G. Pikus (Springer, Berlin, 1995).
83. L. Vina, *J. Phys. Condens. Matter*, **11**, 5929 (1999).
84. A.V. Khaetskii and Y. Nazarov, *Phys. Rev. B*, **64**, 125316–1 (2001).
85. A.V. Khaetskii and Y. Nazarov, *Phys. Rev. B*, **61**, 12639 (2000).
86. L.M. Woods, T.L. Reinecke, and Y. Lyanda-Geller, *Phys. Rev. B*, **66**, 161318 (2002).
87. J.L. Cheng, M.W. Wu, and C. Lü, *Phys. Rev. B*, **69**, 115318 (2004).
88. A. Khaetskii, D. Loss, and L. Glazman, *Phys. Rev. B*, **67**, 195329 (2003).
89. I.A. Merkulov, Al.L. Efros, and M. Rosen, *Phys. Rev. B*, **65**, 205309 (2002).
90. M. Paillard, X. Marie, P. Renucci, T. Amand, A. Jbeli, and J.M. Gerard, *Phys. Rev. Lett.*, **86**, 1634 (2001).
91. R.J. Epstein, D.T. Fuchs, W.V. Schoenfeld, P.M. Petroff, and D.D. Awschalom, *Appl. Phys. Lett.*, **78**, 733 (2001).
92. S. Amasha, K. MacLean, I.P. Radu, D.M. Zumbuhl, M.A. Kastner, M.P. Hanson, and A.C. Gossard. Preprint at cond-mat arXiv:0707.1656.
93. V.N. Golovach, A. Khaetskii, and D. Loss, *Phys. Rev. Lett.*, **93**, 016601 (2004).
94. R. Hanson, L.P. Kouwenhoven, J.R. Petta, S. Tarucha, L.M.K. Vandersypen. To be published in *Rev. Mod. Phys.* Preprint available at arXiv:cond-mat/0610433.
95. D. Klauser, W.A. Coish, and D. Loss, *Phys. Rev. B*, **73**, 205302 (2006).
96. C. Lü, J.L. Cheng, and M.W. Wu, *Phys. Rev. B*, **71**, 075308 (2005).
97. D.J. Hilton and C.L. Tang, *Phys. Rev. Lett.*, **89**, 146601 (2002).
98. T.C. Damen, L. Vina, J.E. Cunningham, J. Shah, and L.J. Sham, *Phys. Rev. Lett.*, **67**, 3432 (1991).
99. B. Baylac, T. Amand, X. Marie, B. Dareys, M. Brousseau, G. Bacquet, and V. Thierry-Mieg, *Solid. State Commun.*, **93** 1, 57 (1995).
100. S. Laurent, B. Eble, O. Krebs, A. Lemaitre, B. Urbaszek, X. Marie, T. Amand, and P. Voisin, *Phys. Rev. Lett.*, **94**, 147401 (2005).
101. T. Flissikowski, I.A. Akimov, A. Hundt, and F. Henneberger, *Phys. Rev. B*, **68**, 161309(R) (2003).
102. D.V. Bulaev and D. Loss, *Phys. Rev. Lett.*, **95**, 076805 (2005).
103. D. Heiss, S. Schaeck, H. Huebl, M. Bichler, G. Abstreiter, J.J. Finley, D.V. Bulaev, and Daniel Loss, arXiv:0705.1466 (2007).
104. O. Gywat, H.A. Engel, D. Loss, R.J. Epstein, F.M. Mendoza, and D.D. Awschalom, *Phys. Rev. B*, **69**, 205303 (2004).
105. H. Krenner, M. Sabathil, E.C. Clark, A. Kress, D. Schuh, M. Bichler, G. Abstreiter, and J.J. Finley, *Phys. Rev. Lett.*, **94**, 057402 (2005).
106. M. Kroutvar, Y. Ducommun, J.J. Finley, M. Bichler, G. Abstreiter, and A. Zrenner, *Appl. Phys. Lett.*, **83**, 443 (2003).

CHAPTER 16

Engineering of Quantum Dot Nanostructures for Photonic Devices

L. Seravalli, G. Trevisi, M. Minelli, P. Frigeri, and S. Franchi

CNR-IMEM Institute, Parco delle Scienze, 37a, I-43100 Parma, Italy

In order to take full advantage of the peculiar optical properties of quantum dot (QD) nano-structures, their band structure must be engineered to optimize a few relevant parameters. In this chapter we review some approaches used to design and prepare structures for both the 1.3–1.6 μm spectral window of low hydroxyl-content optical fibres, and the 0.98–1.04 μm one for telecom and medical applications.

As for long wavelength applications, we discuss first the approach termed as *QD strain engineering*: in InAs/InGaAs structures on GaAs substrates the composition of InGaAs confining layers and the thickness of the metamorphic lower one determine both the band discontinuities between QDs and confining layers and the energy gap of the QD material, through its strain; the availability of the two degrees of freedom make it possible to tune both the emission energy and the activation energy for thermal quenching of emission, a parameter that determines the room temperature (RT) emission efficiency. By using QD strain engineering we obtained from structures grown by molecular beam epitaxy photoluminescence emission at RT up to 1.44 μm under excitation power densities as low as 5 W/cm^2. A simple effective-mass model, validated by our experimental results, offers a rationale for the achievement of efficient RT emission at long wavelength; the results also show that in long wavelength structures the inevitably low-band discontinuities hamper the achievement of room temperature operation.

We review our results on the insertion of InAlAs additional barriers embedding QDs and set amid the InGaAs confining layers; it is shown that the blue shift of emission wavelength due to the additional barriers can be effectively counterbalanced by the red shift induced by QD strain engineering; as a consequence, in QD strain engineered structures with enhanced barriers the activation energies can be significantly increased so that RT emission wavelengths in excess of 1.5 μm are experimentally obtained.

AlGaAs confining layers and InGaAs QDs have been successfully used in order to respectively increase the band discontinuities and the QD energy gap for 0.98–1.04 μm emitting structures. Experimental results and model calculations allow us to discuss the effect of QD and confining layer composition on the QD morphology. In particular, we show how the CL composition, besides band discontinuities, affects also the QD dimensions and other details of the QD band structure. Furthermore, by studying the effect of the change in composition of InGaAs QDs, we identify different mechanisms contributing to the blue shift of the emission; 0.98 μm RT emission was achieved from structures consisting of InGaAs QDs embedded in Al$_{0.30}$Ga$_{0.70}$As CLs.

16.1 Introduction

Self-assembled semiconductor quantum dots (QDs) have unique properties that make them extremely interesting for photonic devices with improved performances such as QD lasers [1–4].

Such properties are intrinsic and are related to the zero-dimensional confinement of carriers in the structures.

Moreover, by using QD structures advantage may be taken of the lattice mismatch between components of the structure (about 7% when growing InAs QDs on GaAs substrates) that – in the case of quantum well (QW) structures – would be intolerably high, whereas in QD structures can be profitably used as the driving force for the self-assembly of coherent nanostructures; these structures grown on GaAs may have a sufficiently high quality to show room temperature (RT) emission at 1.3 μm, a very important advantage with respect to the current technology based on QW structures on InP substrates [2].

These considerations explain why a considerable amount of research is devoted worldwide to the study of QD structures with efficient RT emission in the spectral windows for lightwave communications (0.98 μm and 1.3–1.6 μm) and medical applications (1.04 μm). However, the extension of RT emission to wavelengths up to 1.6 μm is still problematic for QD structures grown on GaAs.

In order to tune the QD emission in regions of interest, it is necessary to control the band structure details that determine the optical properties of QD structures. Therefore, it is crucial to reach a firm understanding of the complex dependence of QD band structure on material properties, growth conditions and design parameters. Ultimately, this knowledge will not only allow the QD band structure to be effectively engineered, but in some cases will also hint at new methods to extend the spectral range of QD operation.

In this contribution, we thoroughly discuss the engineering of QD emission in InAs/InGaAs and InGaAs/AlGaAs QD structures intended for 1.3–1.6 μm and 0.98–1.04 μm operation, respectively. In section 16.2, we review different approaches proposed to red shift InAs QD emission towards the long wavelength spectral region and discuss the various parameters considered. Then, in section 16.3, we present the *quantum dot strain engineering* approach, which relies on the control of the band structure of InAs/InGaAs QDs by means of both the composition of InGaAs confining layers (CLs) and the strain of QDs, determined by the thickness of a metamorphic lower CL (LCL) on which QDs are deposited. By using a single-band effective-mass model validated by the experimental results obtained on QD structures with different CL composition and QD strain we show how QD strain engineering allows the separate control of QD–CL band discontinuities and energy gap of the QD material. These band structure parameters can be considered as two degrees of freedom that can be used simultaneously to tune the emission energy and to optimize the activation energy for emission quenching at RT, a parameter that determines the emission efficiency; this approach results in structures grown on GaAs with RT emission up to 1.44 μm under excitation power densities as low as 5 W/cm^2.

Moreover, a deeper analysis of model calculations leads us to discuss a rationale for the achievement of efficient RT emission at long wavelength; in particular, we highlight that the inevitable requisite for red shifting the emission towards 1.6 μm is the decrease of the potential barriers that confine carriers which results in the decrease of the emission efficiency at RT; then, we show how the use of additional InAlAs barriers embedding QDs increases the emission efficiency of such structures so that 1.51 μm RT emission is obtained under low excitation power densities, an interesting step in the quest for 1.6 μm operation.

In section 16.4 we show how In(Ga)As/AlGaAs structures allow the achievement of blue shifted QD emission necessary to match the 0.98–1.04 μm window of operation. The use of compositions of both CLs and QDs as tuning parameters is thoroughly discussed. In the former case, an analysis of the physical mechanisms responsible for the blue shift is carried out, thanks to the comparison between experimental emission energies and model calculations in InAs/AlGaAs structures; in the latter case the effects of the change of QD composition on QD morphology and band structure are discussed and the causes of the increase of the emission energies are identified. Finally, in section 16.5 we draw some conclusions and envision possible paths for future work.

16.2   QD nanostructures for long wavelength emission

The emission energy of InAs QDs embedded in GaAs confining layers depends on a number of parameters of structures, such as InAs coverages, QD shapes and dimensions, as well as on growth conditions, as it is shown, for instance in [5], where it was reported that the emission energy at 10 K of structures may range between 1.05 and 1.31 eV, which would correspond to 1.25–1.00 μm at RT.

In order to red shift the InAs QD emission towards longer wavelengths, a first proposed approach has been the increase of QD sizes to reduce the confinement energies for carriers. This can be obtained by an increase of the InAs coverage; anyway the enlargement of QDs may result in non-coherent structures, leading to the nucleation of dislocations with deleterious consequences on the optoelectronic properties. Therefore, various methods to increase QD sizes have been used, including: (i) the use of the ALMBE technique [6–8], (ii) other variants of MBE growth based on the concept of alternating molecular beams [9–11], (iii) very low growth rates for InAs deposition [12–15] and (iv) the use of a seed layer of dots in order to control shape, size and density of QDs grown on a second layer [14, 16–18].

The second approach to red shift the emission, which has been more generally used in the last decade, consists in embedding InAs QDs in InGaAs confining layers (CLs) instead of GaAs ones [19–31]. The interpretation of this effect, however, is still debated due to the concomitant presence of different mechanisms all contributing to the red shift of QD emission energy: (i) the reduction of QD–CL band discontinuities, which we denote here by the sum ΔE of the conduction and valence band offsets, (ii) the decrease of the energy gap E_{QD} of the QD material, due to the reduced QD strain ensuing from the reduction of the lattice mismatch between InAs QDs and InGaAs CLs, and (iii) the change in QD sizes due to intermixing effects between CLs and QDs.

Such contributions cannot be easily singled out, therefore for some authors the most relevant effect is the reduction of E_{QD} [14, 20, 21, 25, 31], while others have identified the most important mechanism in the substantial changes in QD composition and sizes stemming from the strain-driven In migration from InGaAs CLs to InAs QDs (activated alloy phase separation) [19, 26, 29, 32, 33].

A comprehensive understanding of the complex interplay of design and growth parameters is therefore a crucial need in order to be able to reliably engineer InAs/InGaAs QD nanostructures for photonics devices operating in the long wavelength range.

16.3 Quantum dot strain engineering

When the emission energy is tuned by using the CL composition two effects are to be considered that are changes of: (i) E_{QD} related to the QD strain variation and (ii) QD–CL band discontinuities. In order to disentangle these effects the *quantum dot strain engineering* [34] approach has been proposed which is based on the control of: (i) CL composition and (ii) QD–CL mismatch and, then, the QD strain by using the thickness of the InGaAs lower confining layer (LCL). Such a LCL acts as a metamorphic buffer [34–38], the lattice parameter of which depends on its thickness through the mechanism of strain relaxation that takes place to minimize the elastic energy stored in the lattice-mismatched LCL [39].

InGaAs layers grow on GaAs in two different regimes, depending on their thickness: (i) the pseudomorphic regime, for thicknesses t smaller than the so-called critical thickness t_c for plastic relaxation, where the lattice parameter in the plane of the heterointerface equals that of GaAs, and (ii) the metamorphic regime, for thicknesses larger than t_c, where for increasing t the lattice parameter increases up to that of free-standing (unstrained) InGaAs (a_{InGaAs}).

The lattice constant a_{InGaAs} is calculated according to Vegard's law:

$$a_{InGaAs} = a_{GaAs} + (a_{InAs} - a_{GaAs})x \tag{16.1}$$

where a_{InAs} and a_{GaAs} are the free-standing lattice parameters of InAs and GaAs, respectively, and x is the In composition; the in-plane lattice parameter a_{LCL} of the InGaAs metamorphic buffers is given by:

$$a_{LCL} = \frac{a_{InGaAs}}{1 + \varepsilon_{res}} \tag{16.2}$$

where ε_{res} is the in-plane residual strain of the metamorphic buffer that, independently of the considered strain relaxation model, does depend on the thickness t of the buffer layer, as it will be discussed below. Then, the lattice mismatch f between InAs QDs and InGaAs LCL is:

$$f = (a_{InAs} - a_{LCL})/a_{LCL} \qquad (16.3)$$

Photoreflectance (PR) experiments [38] have shown that the upper confining layers (UCL) grow pseudomorphically on the lower ones with the same composition x. Therefore, both upper and lower CLs have the same lattice parameter a_{CL} and the same mismatch f with respect to the QDs.

Two classes of models have been put forward to describe the residual strain due to plastic relaxation; they are the so-called non-equilibrium models of strain relaxation (energy-balance [40] and nucleation models [41]) and the equilibrium ones that take into account the thermo-dynamic equilibrium between a rectangular grid of misfit dislocations and the strained epitaxial layer [42, 43]. The former and the latter models predict a $\varepsilon_{res} \propto t^{-n}$ dependence, with $n = 1/2$ [41] and $n = 1$ [44], respectively.

By PR measurements [45], as well as by X-ray diffraction and Raman scattering experiments [46] performed on structures similar to those considered here, it was shown that the measured values of the residual strain compare quite favourably with the Marée *et al.* [41] ones, while they definitely do not support predictions of equilibrium models [44].

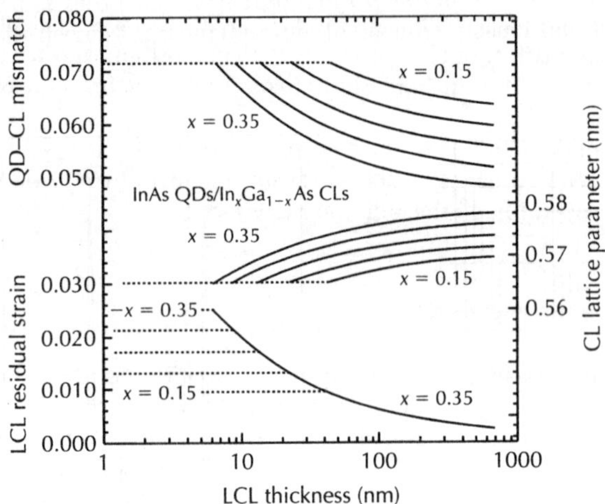

Figure 16.1   Dependencies of LCL residual strain, CL lattice parameter and QD–CL lattice mismatch on LCL thickness in In$_x$Ga$_{1-x}$As structures in the pseudomorphic (dashed lines) and metamorphic (continuous lines) regimes. The strain relaxation in LCLs has been calculated according to [41]. Reprinted with permission from [38]. Copyright 2007, American Institute of Physics.

The LCL in-plane residual strain ε_{res} following the $t^{-1/2}$ dependence, the CL lattice parameter a_{CL} and the QD–CL mismatch f are plotted in Fig. 16.1 as functions of t for different values of x, highlighting that by growing QDs on metamorphic InGaAs buffers the QD–CL mismatch can be controlled in a wide range by acting on both x and t. As discussed in [47] both the hydrostatic (ε_h) and biaxial (ε_b) components of the QD strain defined as $\varepsilon_h = \varepsilon_{xx} + \varepsilon_{yy} + \varepsilon_{zz}$ and $\varepsilon_b = 2\varepsilon_{zz} - \varepsilon_{xx} - \varepsilon_{yy}$ (where ε_{ij} are the components of the strain tensor, z being the direction parallel to the QD axes) are proportional to the mismatch f, with proportionality constants that depend on the QD sizes and shapes. The heavy-hole energy gap E_{QD} of the QD material (the smallest one in compressively strained semiconductors) is given by [48, 49]:

$$E_{QD} = E_{QD}^0 + \delta E_H + \delta E_c - \frac{\delta E_S}{2} \qquad (16.4)$$

where E_{QD}^0 is the energy gap of unstrained material and:

$$\delta E_c = a_c \varepsilon_h$$
$$\delta E_H = a_v \varepsilon_h \qquad (16.5)$$
$$\delta E_S = b \varepsilon_b$$

where a_c, a_v and b are the hydrostatic deformation potentials of conduction band and valence band and shear one, respectively.

Therefore, we can conclude that the CL composition x and the LCL thickness t are available to change the QD strain and then the energy gap of the QD material; E_{QD} as well as the x-dependent band discontinuities can be used as two design parameters that determine the QD emission energy. The relevant consequences of the QD strain engineering [34] approach in designing QD structures will be discussed in section 16.3.2.

16.3.1 The quantum dot strain engineering approach: experimental

Experimental confirmations of the effect of the InGaAs LCL thickness on the emission energy of overlying InAs QDs, provided by photoluminescence (PL) studies of MBE-grown metamorphic InAs/InGaAs nanostructures [34, 38], are presented in Fig. 16.2a. The symbols correspond to 10 K peak emissions for 3 ML InAs QDs embedded in InGaAs CLs, plotted as a function of the QD–CL mismatch f (depending on the LCL thickness t) for different CL compositions x. The results show that the emission is red shifted by the reduction of f (i.e. by the decrease of the QD strain), alongside with the increase of x [34].

From atomic force microscopy (AFM) characterization of uncapped structures it was found that the QD height distribution (peaked at 4 ± 1 nm) is scarcely dependent on x and f, while the value of the most frequent QD diameter (21 ± 4 nm) is almost constant for $x \leq 0.33$, but dramatically increases up to 29 ± 4 nm for $x = 0.35$; on the contrary, diameters are basically unaffected by the QD–CL mismatch. As for capped structures, we do not expect intermixing and segregation effects [19, 26, 32] that could modify QD sizes, since our QDs and UCLs are purposely grown by ALMBE at relatively low temperatures (460°C and 360°C, respectively). All these observations allow us to conclude that the red shift of QD emission energy due to the reduction of the mismatch is caused only by the decrease of QD strain.

The right-hand side axis in Fig. 16.2a refers to the expected wavelength of radiation that would be generated at RT from structures emitting at 10 K at the energies given on the left-hand side scale;

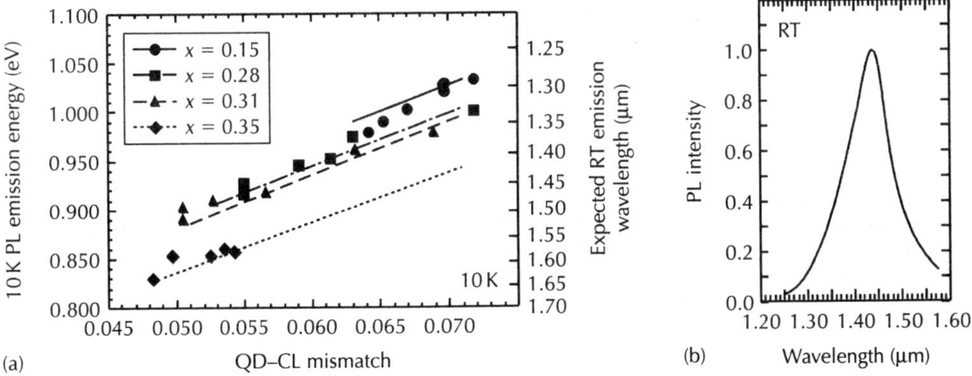

Figure 16.2   (a) 10 K PL emission energy (left vertical scale) and expected RT emission wavelength (right scale) as functions of QD–CL mismatch in InAs/In$_x$Ga$_{1-x}$As QD strain engineered structures. The wavelength scale is obtained from the energy one by taking into account the Varshni shift from 10 K to RT. The lines represent the results of the model calculations. (b) RT emission spectrum peaked at 1.44 μm, from a structure with x = 0.24 and t = 145 nm (f = 0.0592), not shown in panel (a). The excitation power density is 5 W/cm². Reprinted with permission from [38]. Copyright 2007, American Institute of Physics.

the expected RT wavelength scale has been calculated from the measured 10 K emission energy one considering a Varshni [50] shift of 70 meV when increasing the temperature from 10 K to RT. It is worth noting that the QD emission energy could be shifted down to values that would correspond to wavelengths beyond the 1.55 μm region at RT, if the emission were not quenched. However, due to the significant thermal quenching of PL emission (to be discussed below) for In-rich CLs, the longest emission wavelength observed at RT under low excitation power density (5 W/cm^2) was 1.44 μm, for a structure with $x = 0.24$ and $f = 0.0592$ (Fig. 16.2b).

Lines in Fig. 16.2a represent calculations of QD ground state emissions provided by a model that will be presented and discussed in depth in section 16.3.2. It is noticeable how these calculations and experimental results agree quite satisfactorily with discrepancies not larger than 25 meV, which can be accounted for by uncertainties in QD sizes as measured by AFM and by approximations in the model and in its input parameters. This agreement confirms the conclusions made on the relevance of QD strain reduction to red shift the emission.

The model can be used to design structures emitting at predetermined wavelengths by selecting the appropriate pairs of x and t; indeed, Fig. 16.3 [36] presents the experimental PL emission energies at 10 K (symbols) of structures with values of x and t calculated by the model in order to have RT emission at the wavelengths of 1.3, 1.4, 1.5 and 1.55 μm; the dashed lines are drawn at the 10 K emission energies corresponding to the indicated RT wavelengths. The horizontal error bar represents the $\pm 3\%$ error in mismatch that may stem from the uncertainty in the RHEED calibrations of In and Ga fluxes, while the vertical bar on model calculations refers to errors of ± 4 nm and ± 1 nm for the most frequent values of the QD diameter and height, respectively. Once again, the agreement between experimental results and model calculations is satisfactory, thus showing how our model can be used to predict the experimental values of QD emission at low temperatures, where quenching effects are irrelevant.

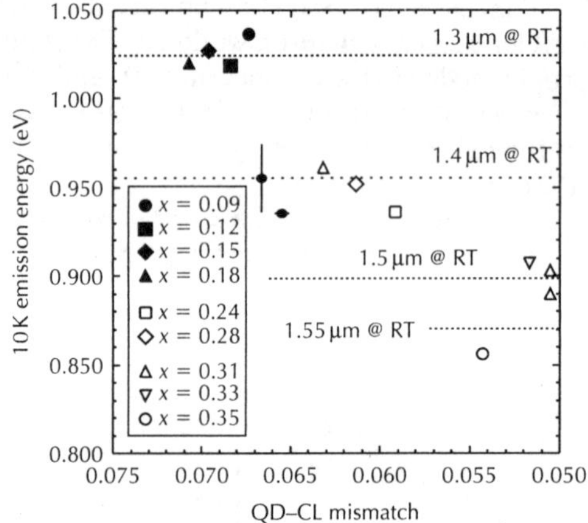

Figure 16.3   10 K PL emission energies (symbols) as functions of the QD–CL mismatch in InAs/In$_x$Ga$_{1-x}$As structures with (x, f) pairs selected for expected 1.3, 1.4, 1.5 and 1.55 μm RT emission. The dashed lines are drawn at 10 K emission energies corresponding to the RT wavelengths shown in the figure. Reprinted with permission from [36]. Copyright 2005, American Institute of Physics.

Thermal quenching of the QD emission intensity is caused by the increased effectiveness of non-radiative mechanisms for confined carrier loss, in competition with radiative recombination. By fitting the equation:

$$I = \frac{I_0}{1 + a\,\exp(-E_1/kT) + b\,\exp(-E_2/kT)} \tag{16.6}$$

to the temperature (T) dependence of the photoluminescence integrated intensity (I) it is possible to derive the values of the activation energies E_1 and E_2 of two different quenching processes. As discussed in [51, 52], the non-radiative process related to the larger activation energy E_1 is the confined carrier thermal escape from QD levels, whose activation energy is given by the sum of the energy differences between the final levels of the escape process (such as CL ones) and that of QDs for both electrons and holes. The smaller activation energy E_2 may be related to extrinsic processes likely caused by defects in the structures [53, 54].

To investigate channels where confined carriers escape to, the experimental values of E_1 can be compared with the sum of the energy differences of QD and CL states for electrons and heavy holes, as calculated by our model. In Fig. 16.4 [36] values of activation energies E_1 for structures of Fig. 16.3 obtained both by model calculations and by fitting Eq. 16.6 to the experimental data are plotted as functions of CL composition as lines and symbols, respectively. Both experimental and model values decrease as x increases; an effect that can be easily understood in terms of shrinking band discontinuities between QDs and CLs and that will be discussed in more depth in section 16.3.2.

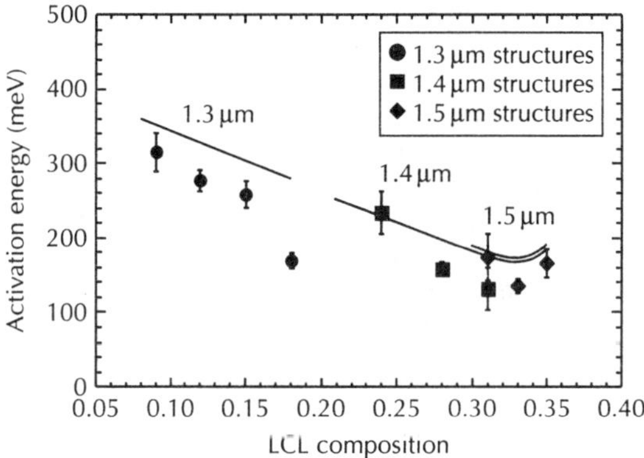

Figure 16.4   Experimental (symbols) activation energies for PL thermal quenching as a function of the CL composition x of InAs/In$_x$Ga$_{1-x}$As structures designed for expected 1.3, 1.4 and 1.5 µm operation at RT. Lines represent results of model calculations. Reprinted with permission from [36]. Copyright 2005, American Institute of Physics.

A second important observation is that the measured E_1 values fall consistently below the model ones; this has been explained by assuming that the final states of the escape process are wetting layer (WL) rather than CL ones, interpretation that was supported by the photoreflectance study of the same structures that evidenced the occurrence of transitions to WL states [38].

The smaller activation energy E_2 has values in the 30–70 meV range and is caused by extrinsic processes such as non-radiative recombinations due to defects [36, 54] or losses of photogenerated carriers in confining layers [53, 55].

The dependence of activation energies E_1 on the QD–CL mismatch f deduced from experiments (symbols) and model calculations (lines) for some structures of Fig. 16.2 is presented in Fig. 16.5. As in Fig. 16.4, model calculations considering CL states as escape channels predict values 50–90 meV above those derived from the fitting of PL integrated intensity data.

16.3.2   The quantum dot strain engineering approach: model and discussion

As mentioned in the previous section, a simple and yet effective model was developed to calculate QD ground state emission in strain engineered structures [34, 38]. It is based on a single-band, effective-mass approach for QDs with cylindrical symmetry, following the work of Marzin and Bastard [56]. Although a large number of more refined and complex models have been presented in the literature, including four-band [57] and eight-band [58] $\mathbf{k} \cdot \mathbf{p}$ treatments, pseudopotential [59] and Green function deterministic numerical methods [60], it has been pointed out in

[58, 60] that such a simpler approach is more suitable for the description of fundamental transitions in QDs, as those under consideration here. For a comprehensive review of theoretical work on QD modelling, we refer the reader to [61].

We considered QDs with truncated conical shapes and with base diameters and heights given by the most frequent values in the respective distributions derived by AFM characterization of uncapped structures; values of the ratio between base and top diameters were taken similar to those reported in the literature (2.5–3.0) [62, 63]. The composition of QDs was considered uniform and equal to the nominal one; this assumption is justified since, as discussed in section 16.3.1, the relatively low growth temperatures of QDs and UCLs are expected to limit segregation and interdiffusion of elements.

The band discontinuities between QDs and CLs were calculated according to the model solid theory (MST) [64], inclusive of strain effects. The MST parameters of InGaAs were deduced from those of the binary constituents as proposed by Cardona and Christensen [65], while the effective-mass values (in units of free electron mass) for InAs QDs were taken as 0.04 for electrons, to include the effect of strain, and 0.59 for heavy holes [57]. The values of other parameters such as elastic constants and electronic band structure ones were taken from the recent literature [66]. The binding energy of excitons was assumed to be 20 meV [67].

The calculation of the hydrostatic and biaxial components of the QD strain, determining the band-edge profiles of QDs and CLs, were carried over following the analytical approach of Andreev *et al.* [47], where values for InAs QDs embedded in GaAs are given as linear functions of the QD–CL mismatch. While the hydrostatic strain is constant inside the QDs, the biaxial one is position dependent, therefore we have taken an average value derived from the results of [47, 68]; in particular, we assumed the value of 0.05 for InAs QDs in GaAs confining layers and scaled the value for InAs QDs in InGaAs CLs. Finally, the biaxial component of strain in the CLs induced by the presence of QDs was disregarded.

The Schrödinger equation for electrons and heavy holes was solved considering both the QD and a 1.6 ML thick WL as included in a cylindrical box larger than the QD; the envelope functions of carriers were written in terms of a complete orthonormal set of eigenfunctions of the box [56]. Then, the Schrödinger equation was transformed in a system of linear algebric equations, which were solved with well-known computational techniques.

The experimental data of Figs 16.2–16.5 proved that model calculations are indeed accurate in predicting the dependence of emission energy and activation energy of carrier thermal escape on the two design parameters, the CL composition x and the QD–CL mismatch f. Thanks to this experimental validation, a deeper analysis of model calculations can provide a more quantitative insight on the fundamental features of QD strain engineering and may allow for a rationalization of our and literature experimental data.

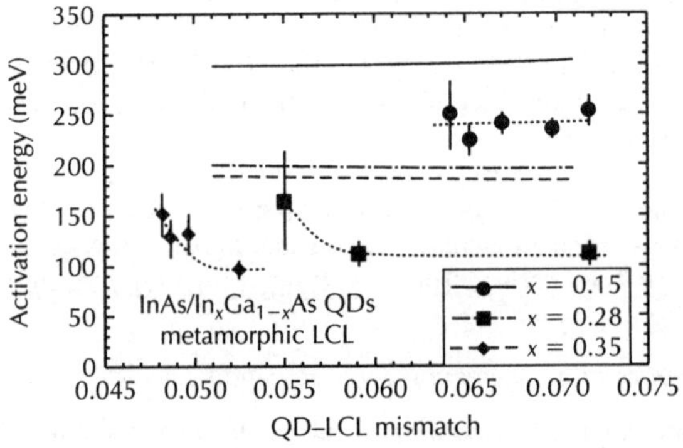

Figure 16.5   Activation energies for PL thermal quenching as a function of the QD–CL mismatch. The dotted lines are guides for the eye, while continuous, dashed and dotted–dashed lines represent the results of model calculations. Reprinted with permission from [38]. Copyright 2007, American Institute of Physics.

The upper surface in Fig. 16.6 represents the calculated 10 K emission energy of QDs as a function of x and f [38], while the lower one represents the sum of the energy differences between

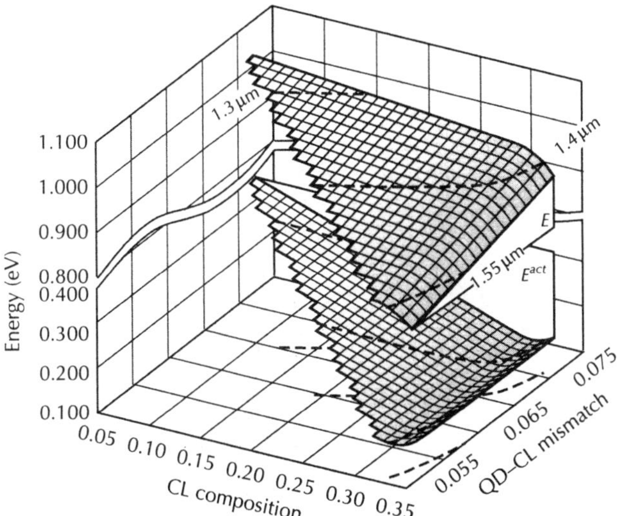

Figure 16.6   Calculated PL emission energy at 10 K (upper surface) and activation energy E_1 for thermal escape of carriers (E^{act} lower surface) as functions of the QD–CL mismatch f and the CL composition x. The dashed lines represent the (x, f) pairs that are expected to result in RT emission at the indicated wavelengths. Reprinted with permission from [38]. Copyright 2007, American Institute of Physics.

QD and CL states for electron and heavy holes; this quantity corresponds to the activation energy for thermal escape of confined carriers from QD to CL states. The surfaces have been drawn only for values of (x, f) pairs that correspond to in-plane LCL residual strain ε_{res} larger than 0.002, the minimum value for which the Marée *et al.* model is considered reliable [41]. The dashed lines on the (x, f) plane indicate the (x, f) pairs of structures expected to have RT emission at 1.3, 1.4 and 1.55 μm; these lines are also projected on the upper surfaces. As already discussed when presenting Fig. 16.3, it was experimentally proved [36] that structures designed with such values of x and f showed a satisfactory agreement between predetermined and measured values of emission wavelengths. The sharp decrease of emission energy when x increases beyond 0.33 is due to the abrupt change of QD diameters that was discussed in section 16.3.1; due to the sudden widening of the QD potential well, electron and hole confinement energies decrease, resulting in an additional red shift of the QD emission. The same argument also explains the slight increase of the activation energies for $x \leq 0.33$, which is noticeable in Fig. 16.6 and in the experimental data of Fig. 16.4.

Looking at Fig. 16.6 it can be clearly seen that in the QD strain engineering approach the QD–CL mismatch and the CL composition can be considered as two degrees-of-freedom that allow the engineering of both the emission energy and the activation energy E_1 for thermal escape of carriers from QDs. Let us consider, for example, the (x, f) pairs corresponding to RT emission at 1.4 μm: by choosing pairs at the high- or at the low-x side of the dashed line relatively low or high values of E_1 are obtained, respectively, which are about 170 and 250 meV, while the emission wavelength at RT in both cases is 1.4 μm. It is worth noting that the longest emission wavelength obtained so far (1.44 μm, section 16.3.1) is from structures with (x, f) pairs lying in the low-x side of the (x, f) plane.

Very interestingly, Figs 16.6 and 16.4 also show that structures expected to emit at 1.55 μm at RT have an inevitably small E_1 (~180 meV) which hampers the achievement of RT operation, unless different design criteria of structures are introduced (section 16.3.3).

Figure 16.7 shows the 10 K emission energy from strain-engineered InAs/InGaAs structures as a function of x and f; the dashed lines represent (x, f) pairs of structures that would emit at RT at the wavelengths indicated, while the thick continuous lines both on the (x, f) plane and on the surface show the relationship that exists between x and f in structures with given values of the thickness t of the lower CL. The Figure can be useful to illustrate the difference between

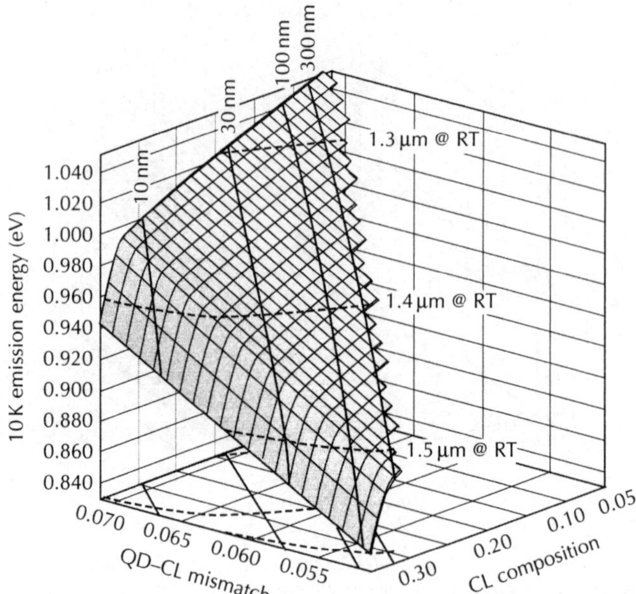

Figure 16.7   Calculated PL emission energy (10 K) as a function of the QD–CL mismatch f and the CL composition x. The dashed lines represent the (x, f) pairs that are expected to result in RT emission wavelengths at 1.3, 1.4 and 1.5 µm. The continuous lines show the (x, f) relationship for structures with given values of LCL thickness.

the usual design criteria followed in the literature and that proposed by QD strain engineering. In the former case, only x is used to tune the emission wavelength, while t is not considered as a design parameter. In this case, at the value of x determined in such a way only a definite value of the activation energy for thermal escape of carriers results. Instead, in the case of QD strain engineering x and the t-dependent f can be used to engineer two different parameters that are emission and activation energies, thus allowing the maximization of the emission efficiency at predetermined operation wavelengths.

A more general understanding of the dependence of significant parameters of the structure on x and f can be reached by studying how the energy gap of the QD material E_{QD} and the sum ΔE of band discontinuities for electrons and heavy holes depend on the two design parameters under consideration. Such quantities have been calculated by our model and are plotted in Fig. 16.8 [38], where the upper and the lower surfaces represent E_{QD} and ΔE. As in Fig. 16.6, dashed lines indicate the couples of values (x, f) of structures that should emit at 1.3, 1.4 and 1.55 µm at RT.

The slight dip in E_{QD} for $x \geq 0.33$ is due to the increase in QD lateral sizes which modifies the dependence of the QD strain on QD–CL mismatch, as discussed in [47]. This causes also a reduction in the slope of ΔE.

It can be seen that the effect of the QD–CL mismatch on the energy gap is quite relevant, while the CL composition for $x < 0.33$ has hardly any effect on it. On the other hand, the sum of the band discontinuities depends strongly on the CL composition x and is basically unaffected by the mismatch f.

From all above considerations, a qualitative understanding of this approach is possible by considering the following simple arguments: (i) the energy gap of strained InAs depends only on the QD–CL mismatch and is basically unaffected by the CL composition (unless this parameter causes changes in QD morphology), (ii) the sum ΔE of the band discontinuities, given by the difference between the energy gap of CL and QD material, depends on x as only the energy gap of InGaAs CLs has such a dependence and (iii) ΔE is essentially independent of f as both the CL and QD energy gaps have a quite similar dependence on mismatch.

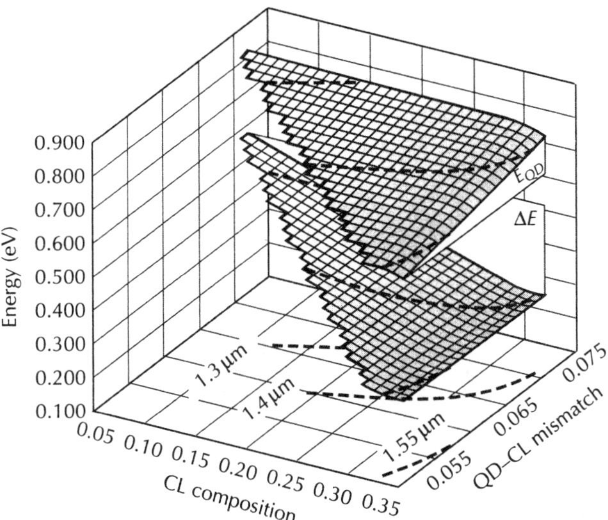

Figure 16.8   Calculated energy gap of the QD material (strained InAs) (upper surface) E_{QD} *and sum of band discontinuities for electrons and holes* ΔE *(lower surface) as functions of the QD–CL mismatch f and the CL composition x in InAs/In$_x$Ga$_{1-x}$As QD strain engineered nanostructures. The dashed lines represent the (x, f) pairs that are expected to result in RT emission at the indicated wavelengths. Reprinted with permission from [38]. Copyright 2007, American Institute of Physics.*

A different perspective on the QD strain engineering approach can be gained by considering the QD material (strained σ-InAs) as a "virtual" semiconductor material with its own lattice parameter and energy gap represented in Fig. 16.9a by points on the dashed line labelled σ-InAs. Instead, the closed circles and the continuous lines represent free-standing III–V binaries and ternaries, respectively. From the Figure it can be noted that by increasing the lattice parameter of strained InAs (σ-InAs) from that of GaAs towards the InAs one E_{QD} is red shifted according to Eqs 16.4 and 16.5.

Figure 16.9b, which is a blow-up of Fig. 16.9a, shows that E_{QD} slightly depends also on QD morphology, as mentioned above when discussing features of Fig. 16.8: for $x = 0.05$–0.30 the lines representing the energy gap of σ-InAs lean one on the other, while for $x = 0.35$ the line sets at a smaller energy since for larger dots the material is slightly less compressed as the proportionality constant between QD strain and QD–CL mismatch changes [47].

The shaded area denoted by CL in Fig. 16.9a and the lines labelled σ-InGaAs and CL in Fig. 16.9b represent the metamorphic CLs with different composition in the 0.05–0.35 range. From Fig. 16.9a it can be seen that the use of strain to shift the energy gap of the QD material can be considered as an alternative approach as compared to the change of QD composition. In the figure the arrows (labelled InNAs, InAsSb, InGaAs and σ) diverging from the point representing σ-InAs pseudomorphic on GaAs show, respectively, the effects of the introduction of N, Sb and Ga in the InAs lattice and that of the reduction of strain in σ-InAs.

As a matter of fact, the use of In(Ga)AsN QDs has already proved to be a possible path towards the extension of QD emission in the $1.55\,\mu$m window [69–75]. On the other hand, InAsSb QDs have been attracting relatively less attention, despite the fact that incorporation of antimony in InAs QDs has shown promising properties both in terms of QD emission red shifting [76–78] and improved QD growth [78–80]. On the other hand, the use of InGaAs to blue shift the QD emission energy is a well-established method, allowing for $0.98\,\mu$m emission from QD structures, as will be thoroughly discussed in section 16.4. The resort to ternary or quaternary alloys instead of InAs for QDs, however, may present some elements of concern, such as the possible lower material quality and the critical control of compositions; therefore further research work could help clarify advantages and drawbacks of different QD materials. In any case, QD strain engineering

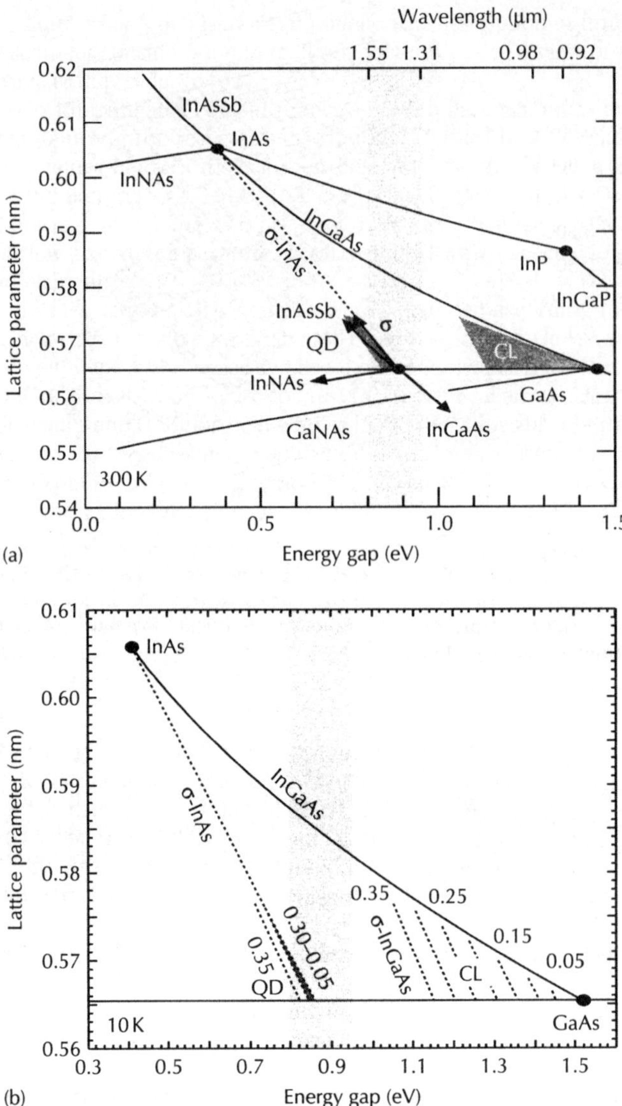

Figure 16.9   *(a) Lattice parameters as functions of energy gaps for III–V binaries (closed circles) and ternaries (continuous lines). The dashed line labelled σ-InAs represents the strained InAs (QD material). The shaded area labelled CL indicates metamorphic CLs in QD strain engineered nanostructures. Arrows represent the effect of change in composition of QDs from InAs to ternary materials (InAsSb, InNAs and InGaAs), and the effect of reduction of strain for InAs QDs (labelled σ). (b) Blow-up of (a) showing lattice parameters as functions of energy gaps for InAs QDs and $In_xGa_{1-x}As$ CLs under metamorphic regime (dashed lines) in structures with different values (0.05–0.35) of the CL composition x.*

and the use of multinary QDs are not mutually exclusive approaches and advantage can be taken of the former solution to limit the incorporation of Sb and N in QDs that would be necessary to obtain a predetermined red shift.

Figure 16.9 shows once more the quintessential features of QD strain engineering already evidenced by Figs 16.6 and 16.8: (i) the QD emission can be red shifted by decreasing the strain of QDs in metamorphic structures; (ii) the increase in x composition of CLs can also contribute to red shift the emission through the reduction of band discontinuities and then of confinement energies, but – concomitantly – it reduces the activation energy for carrier thermal escape and, then, the RT emission efficiency. A further comment, which only marginally emerges from Fig. 16.9, is: (iii) the increase of QD sizes can be instrumental for the red shift of emission, as it was done in the literature [8, 32], since confinement energies decrease while the activation energy and, hence, the emission efficiency increase.

16.3.3   *Further steps towards QD emission at 1.55 µm at room temperature*

As discussed in the comments to Figs 16.4–16.8, QD strain engineered structures designed for long wavelength operation suffer a consistent reduction of the activation energy for thermal escape of carriers ($E_1 \sim 180\,\text{meV}$), which inevitably implies a smaller emission efficiency at RT. Indeed, to red shift the QD light emission in the 1.55 µm region it is necessary: (i) to increase the values of the CL composition in order to decrease the carrier confinement energies and/or (ii) to decrease the energy gap of the QD material by decreasing the QD strain. However, point (i) implies the deleterious consequence of the reduction of QD–CL band discontinuities and, hence, the decrease of the activation energy, while point (ii) can be pursued only to a limited extent since the decrease of QD–CL mismatch (Fig. 16.1) and that of QD strain saturates with increasing t. These considerations explain why, as it has been already pointed out, the longest wavelength experimentally observed at RT under low excitation conditions ($5\,\text{W/cm}^2$) is 1.44 µm [38], in spite of the fact that emission at much longer wavelength could be obtained if transitions were not quenched (Figs 16.2 and 16.3). A modification of the structure is hence necessary to get QD emission at RT in the 1.55 µm region.

An interesting option is provided by embedding QDs in additional potential barriers made of materials, such as InAlAs, with band discontinuities with respect to InAs higher than those given by InGaAs CLs; this method has been proved to be effective in reducing PL thermal quenching of QD emission in structures for 1.3 µm emission [28, 32, 81–83].

In [84], we reported on a detailed study of the effect of barrier parameters on the properties of strain engineered metamorphic QD nanostructures, with InAs QDs embedded in $In_{0.15}Al_{0.85}As$ or GaAs additional barriers set amid relaxed $In_{0.15}Ga_{0.85}As$ CLs (insets in Fig. 16.10). As Al-containing

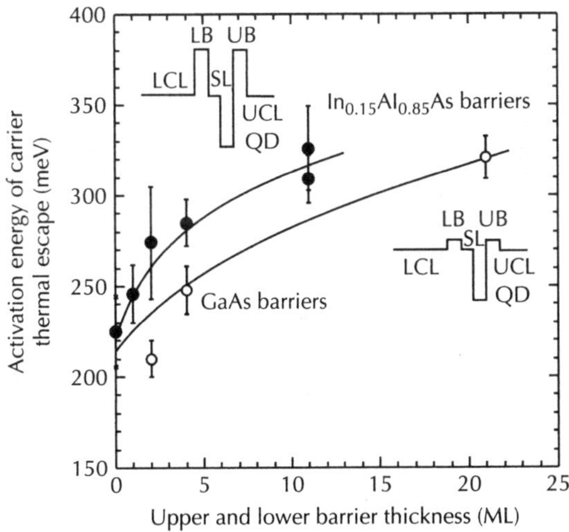

Figure 16.10   Measured activation energy for thermal escape of carriers from QD structures with InAlAs and GaAs barriers (closed and open circles, respectively) as a function of the barrier thickness. Insets show schematics of the conduction band profile for the structures, consisting of lower and upper InGaAs confining layers (LCL and UCL), InGaAs spacer layers (SL), and lower and upper InAlAs or GaAs barriers (LB and UB). Lines are guides for the eye. Reprinted from [84]. Copyright 2006, with permission from Elsevier.

alloys cause a reduction of sizes of overgrown QDs [28], a 4 ML thick $In_{0.15}Ga_{0.85}As$ spacer layer (SL) was grown atop of the InAlAs lower barrier, in order to not affect QD dimensions.

The additional barriers cause a relevant reduction of the PL quenching at high temperatures, resulting in an increase of the PL intensity at RT (not shown here) by ~ 10 and ~ 500 (extrapolated values) when using 21 ML thick GaAs and $In_{0.15}Al_{0.85}As$ barriers, respectively. In order to have a more quantitative insight on the barrier effect, activation energies E_1 were derived for structures with different barrier thicknesses, as plotted in Fig. 16.10. It can be noticed how GaAs barriers result in the increase of E_1 by about 100 meV, while the InAlAs ones cause a larger increase. In this last case the PL quenching was so reduced that, for the structure with 21 ML thick $In_{0.15}Al_{0.85}As$ barriers, Eq. 16.6 could not be reliably fitted to the experimental data, taken only up to RT.

However, the use of higher barriers has the drawback of a blue shift of the QD emission wavelength, as the carrier confinement energies are also affected by the modified band-structure. This is shown in Fig. 16.11 where QD emission energy as measured by PL at 10 K is plotted as a function of the barrier thickness, for the same structures of Fig. 16.10. The emission is blue shifted as the barrier thickness increases: 21 ML thick InAlAs barriers result in a blue shift of ~ 80 meV, while GaAs counterparts cause a ~ 60 meV one. Some reports [32, 85, 86] indicate that the QD emission energy presents an initial red shift and then a blue shift when increasing the thickness of the InAlAs barrier. This effect has been interpreted as being due to In enrichment of QDs when they are capped with InAlAs and reduced In segregation and intermixing [86]. However, other works [87, 88] evidenced how the presence of an initial red shift depends on the growth temperature of QDs, being absent for QDs grown at 480°C. As in our case QDs are grown at 460°C [84], we can conclude that intermixing and segregation effects are negligible and, therefore, only the blue shift due to increased confinement potential is present. Since it was observed from AFM characterization [84] on structures of Figs 16.10 and 16.11 that QD sizes do not change when adding InAlAs barriers (provided an InGaAs SL is grown before QD deposition), it can be concluded that the blue shift of the QD emission is entirely related to the effectiveness of barriers in increasing the confinement of carriers.

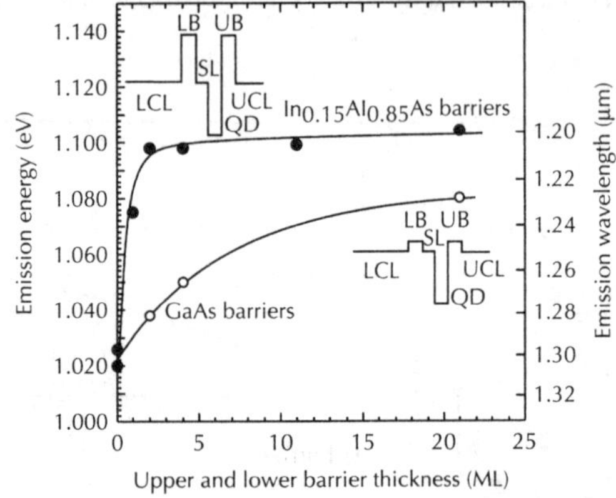

Figure 16.11   *Measured 10 K emission energy (left scale) and corresponding RT emission wavelength (right scale) of QD structures with InAlAs and GaAs barriers (closed and open circles, respectively) as functions of the barrier thickness. Insets show schematics of the conduction band profile for such structures. Lines are guides for the eye. Reprinted from [84]. Copyright 2006, with permission from Elsevier.*

This detrimental effect of the enhanced barriers raises concerns about their usefulness in structures aiming at long wavelength operation. However, the blue shift can be compensated for by a suitable red shift determined by the intentional reduction of QD strain obtained by QD strain engineering [84].

Figure 16.12a provides an example, where a structure with a 190 nm thick $In_xGa_{1-x}As$ LCL ($x = 0.31$, $f = 0.063$), which should emit at RT at about 1.4 µm, has an activation energy so

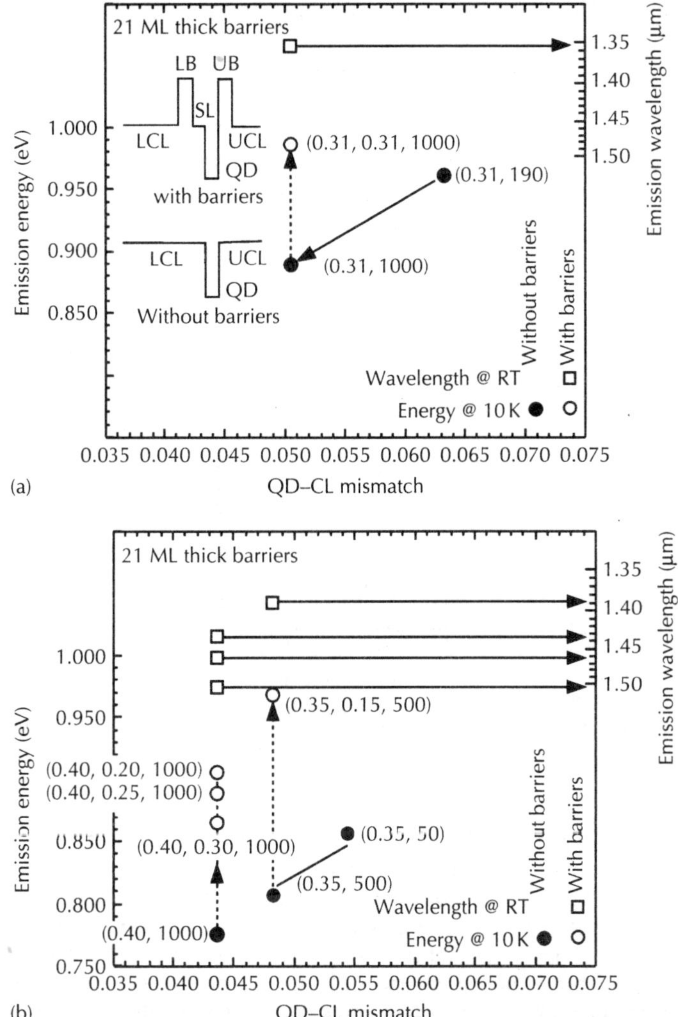

Figure 16.12   *(a) Measured 10 K emission energy (open and closed circles, left scale) and RT emission wavelength (squares, right scale) of structures without and with barriers (closed and open symbols, respectively) as functions of QD–CL mismatch for structures with In$_{0.31}$Ga$_{0.69}$As CLs. The full and dashed arrows show the effects of QD strain reduction and of the additional barriers, respectively. The (x, z, t) terns and the (x, t) pairs give the composition of In$_x$Ga$_{1-x}$As CL (x) and of In$_z$Al$_{1-z}$As barriers (z) and the thicknesses of LCL (t) in the case of structures with and without 21 ML thick barriers, respectively. The inset schematically shows the conduction band profile in structures without and with lower and upper barriers (LB and UB). (b) Same as panel (a) but for structures with x = 0.35 and x = 0.40, aiming at long wavelength operation.*

low (129 ± 29 meV) that RT emission is inhibited. The QD emission is first red shifted by using a 1000 nm thick metamorphic LCL and then blue shifted when 21 ML thick In$_{0.31}$Al$_{0.69}$As barriers are added: the new design results in an activation energy of 191 ± 10 meV and a RT light emission at 1.35 μm.

On the basis of the combined approach of QD strain engineering and enhanced barriers, we worked to extend the RT range of operating wavelengths of metamorphic QD nanostructures towards the 1.55 μm region. For this purpose, we reduced further the QD strain by using: (i) $x = 0.35$ and $f = 0.048$ and (ii) $x = 0.40$ and $f = 0.044$, as shown in Fig. 16.12b. For the first structure we obtained room temperature emission at $1.39\,\mu m$ by adding $In_{0.15}Al_{0.85}As$ barriers; in the structures of the second type, upon the insertion of $In_zAl_{1-z}As$ barriers RT emissions were achieved at $1.43\,\mu m$ for $z = 0.20$, at $1.46\,\mu m$ for $z = 0.25$ and at $1.51\,\mu m$ for $z = 0.30$, a value rarely reported for InAs/InGaAs structures grown on GaAs substrates. The PL spectrum at RT of this last structure is reported in Fig. 16.13. It should be mentioned that in the structure with $z = 0.20$ the 21 ML thick ($\sim$6 nm) barriers are very close to lattice relaxation, as the critical thickness (depending on the mismatch between CLs and barriers) is 12 nm [41]. Nevertheless, PL spectrum for this structure shows no degradation, indicating that strain relaxation has not yet occurred.

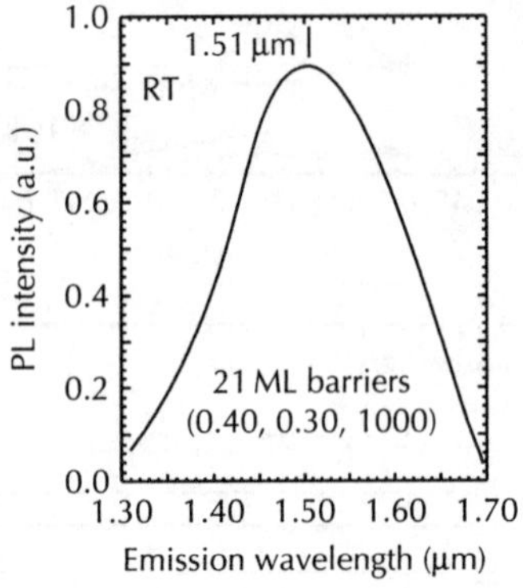

Figure 16.13   *PL emission spectrum at RT of the structure with 1000 nm thick $In_{0.40}Ga_{0.60}As$ LCL and 21 ML thick $In_{0.30}Al_{0.70}As$ barriers, peaked at 1.51 µm. The excitation power density is 5 W/cm².*

It is also interesting to notice how the In molar fraction z of InAlAs barriers can be regarded as an additional parameter to tune the emission properties of QD structures, since the band discontinuities due to additional barriers depend on z and determine the increase of the carrier confinement that influences the QD emission energies and the activation energies for thermal emission quenching, as discussed above. Indeed, it is possible to identify optimized barriers that minimize the blue shift, while assuring an activation energy large enough to obtain QD emission at RT.

This combined approach looks to be a very promising way towards InAs QDs emitting in the $1.55\,\mu m$ window, widening the spectral window of operation of such nanostructures. As a matter of fact, the usefulness of this approach is confirmed by very recent results obtained on long wavelength QD lasers based on metamorphic structures that were obtained in a very much device-orientated approach to structure design [37, 89, 90].

16.4   QD nanostructures for 0.98–1.04 µm emission

The preparation of QD structures based on GaAs substrates emitting in the 0.98–$1.04\,\mu m$ range for telecom and medical application requires the blue shift of emission as compared to that of InAs/(In)GaAs QD structures; since the earliest results reported in the literature [91, 92], this is usually achieved by using InGaAs QDs embedded in AlGaAs CLs. This result is attributed to the increase in both the energy gap of QDs and the QD–CL band discontinuities (and, then, in the confinement energies of carriers in the QDs) due to the introduction of Ga and Al in QDs and CLs, respectively.

In these structures the lattice mismatch between the GaAs substrates and the AlGaAs confining layers is negligibly small, so that the QD strain engineering approach cannot be followed, as it is done in the case of metamorphic InAs/InGaAs QD structures. Instead, the main parameters that can be used to engineer both the emission wavelength and efficiency are the compositions of

QDs and CLs and QD dimensions that can be controlled, for instance, by means of the In(Ga)As coverage.

As it will be shown below, the emission properties are also determined by QD shapes and composition profiles that are related in a subtle way to nominal compositions and growth conditions. The last considerations explain why simple models such as the one described in section 16.3.2 cannot generally be used for InGaAs/AlGaAs QD nanostructures.

16.4.1   InGaAs/AlGaAs QDs: the effect of CL composition

The effect of the CL composition on the emission energy of InAs/AlGaAs and $In_{0.50}Ga_{0.50}As$/ AlGaAs QD structures is presented in Fig. 16.14. Unlike the LCLs that were deposited at 600°C, the QDs and UCLs were grown by ALMBE at relatively low temperatures (460 and 360°C, respectively), with the aim of reducing interdiffusion and segregation of constituent elements. In the Figure, the blue shift of emission with increasing Al composition in CLs is clearly evidenced for both the $In_{0.50}Ga_{0.50}As$ (circles) and InAs (squares) QD systems. Moreover, we observe that, by using $In_{0.50}Ga_{0.50}As$ QDs and $Al_{0.30}Ga_{0.70}As$ CLs, we obtained efficient RT emission (empty circles) at the 0.98 μm wavelength of optoelectronic interest.

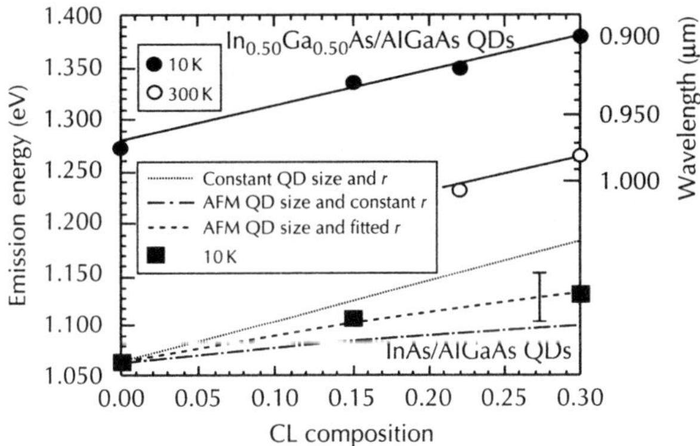

Figure 16.14   *10 K (full symbols) and RT (empty symbols) PL emission energies (left vertical scale) and respective wavelengths (right vertical scale) as functions of the CL composition in $In_{0.50}Ga_{0.50}As$ (circles) and InAs (squares) QDs embedded in $Al_xGa_{1-x}As$. The solid lines are guides for the eye, while the dotted, dashed and chained lines represent the results of model calculations.*

In order to gain a deeper insight on the effect of CL composition, we compare the emission energy of InAs/AlGaAs QD structures (squares in Fig. 16.14) with the results of the single-band effective-mass model calculation discussed in section 16.3.2 (lines). QD heights and base diameters were evaluated by means of the statistical analysis of AFM measurements performed on uncapped structures grown under the same conditions. The morphology investigations revealed that, by increasing the Al composition in CLs from 0 to 0.30, the most frequent value of QD heights decreases from 4.9 to 2.8 nm, while that of base diameters remains almost constant (24 nm); the dependence of QD sizes on the composition of the underlying AlGaAs surfaces is usually attributed to the larger reactivity of Al as compared to that of Ga [93].

By analysing Fig. 16.14, we observe that the composition-dependent increase in the QD–CL band discontinuities alone (chained line) cannot justify the observed blue shift of PL emission. In order to reproduce the experimental data, the model calculations have to take into account not only the decrease in QD sizes as observed by AFM (dotted line) but also the change (from 4 to 2) in the major-to-minor base–diameter ratio (r) of the assumed truncated-conical QD shape (dashed line). We note that r may have a more general meaning than that of the ratio between diameters: indeed, it can be considered as a phenomenological parameter that accounts for non-uniform composition profiles induced by segregation and mixing effects at the QD–CL interfaces and for other effects not explicitly considered by the model. These considerations suggest that the CL composition in the InAs/AlGaAs QD system affects not only band discontinuities and QD sizes but also other details of the QD band structure; these details are related to QD shapes and composition profiles that: (i) cannot be directly controlled even under our low-growth temperature conditions and (ii) are not easily modelled.

Our considerations are supported by a few experimental observations: by means of scanning tunnelling microscopy and Raman scattering measurements performed on nominal InAs QDs embedded in AlGaAs CLs it has been shown that the Al composition in CLs affects both the sizes [93] and composition profiles [94, 95] of QDs.

The CL composition is known to affect not only the QD emission energy but also the PL emission efficiency [96]. By analysing with Eq. 16.6 the temperature dependence of the integrated PL intensity of $In_{0.50}Ga_{0.50}As$ QDs embedded in $Al_xGa_{1-x}As$ CLs, we derived the high-T activation energy E_1 for thermal quenching of photoluminescence (full circles in Fig. 16.15), which is generally attributed to the thermal escape of carriers from QDs [51, 52]. The interesting feature of the plot is that the activation energy increases with the Al composition in CLs, which linearly determines the QD–CL band discontinuities.

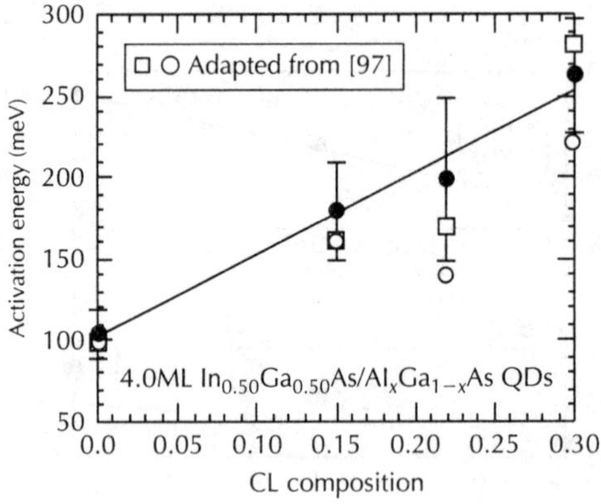

Figure 16.15   Measured activation energy for carrier thermal escape from $In_{0.50}Ga_{0.50}As/Al_xGa_{1-x}As$ QD structures (full circles, present work) and energy separations between WL and QD confined levels (empty symbols), as derived from [97], represented as functions of the CL composition. The solid line is a guide for the eye.

Structures similar to the ones considered here were previously studied by means of PL and resonant PL techniques [53, 97]; the authors were able to deduce the energy of the WL levels (defined as the difference between the energy of electron and hole confined levels in WLs) by: (i) fitting the temperature dependence of some PL features with the model developed in [51] and (ii) analysing differences in resonant and non-resonant PL spectra of the structures. The results of the two methods are very close to each other and compare favourably with the values calculated by describing WLs in terms of quantum wells of suitable thicknesses [53, 97]. The empty symbols in Fig. 16.15 represent the energy separations ΔE_{QD-WL} between QD (obtained from PL measurements [97]) and WL levels (deduced by methods (i) (empty squares) and (ii) (empty circles)).

We observe that ΔE_{QD-WL} agrees satisfactorily with the activation energy E_1 for thermal escape of carriers from QDs (full circles), thus showing that our results are consistent with those of [53, 97], and that the final states of the carrier escape process are the WL levels.

We note that E_1 and ΔE_{QD-WL} increase by increasing the Al composition in the CLs since the less confined and, therefore, shallower WL levels are more sensitive to the increase in band discontinuities with respect to more confined and, therefore, deeper QD levels.

16.4.2   InGaAs/AlGaAs QDs: the effect of QD composition

As for the dependence of QD emission energy on QD composition, we analysed the emission of MBE $In_yGa_{1-y}As$ QDs embedded in $Al_{0.30}Ga_{0.70}As$ CLs with y ranging from 0.40 to 0.70.

The structures were prepared by using: (i) a relatively high Al composition in the CLs in order to enhance the QD-to-WL energy separation, as a result of the previous discussion, and (ii) growth temperatures of 470°C for QDs (by MBE) and the lower part of UCLs (by ALMBE) and of 600°C for the uppermost part of UCLs (by MBE), being the relatively high values of the last temperature a requirement to improve the material quality so as to allow the preparation of optimized QD laser structures [98].

The results of the 10 K emission energy as a function of the InGaAs coverage θ for different values of y are shown in Fig. 16.16, where we observe that the blue shift of emission energy is achieved by: (i) decreasing θ at constant In composition in QDs or (ii) decreasing y at constant InGaAs coverage.

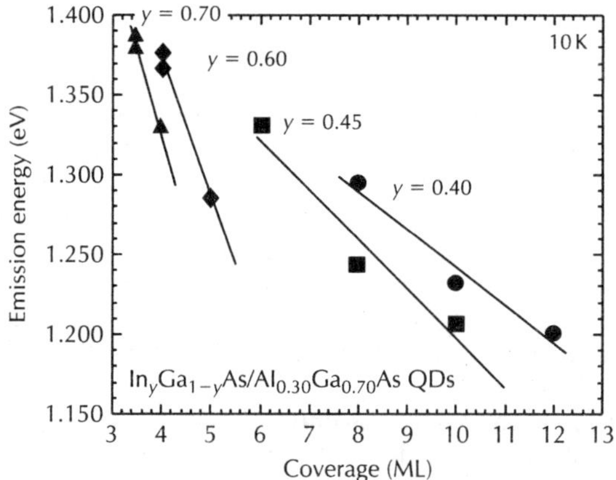

Figure 16.16   10 K PL emission energies of In$_y$Ga$_{1-y}$As/Al$_{0.30}$Ga$_{0.70}$As QD structures as functions of the QD coverage for different QD compositions. The lines are guides for the eye.

As for the effect of coverage, we note that, for constant In composition in QDs, the decrease in θ yields a reduction of QD sizes that results in higher carrier confinement energies and, then, in the blue shift of the emission energy. It should be noted that the lower limit for θ at any given y is represented by the critical coverage θ_c for the two-dimensional to three-dimensional growth transition for self-assembled QDs; as a consequence, the emission energy cannot be indefinitely blue shifted by decreasing θ at constant y, as shown in Fig. 16.16.

As for the change in QD composition at constant θ, four effects are to be considered that act in different ways on the band structure and, therefore, on the emission energy: by decreasing the In composition in QDs at constant θ: (1) the energy gap of the QD material E_{QD} widens and, therefore, since the CL band edges are unaffected by the change in y (2) the band discontinuities ΔE diminish, (3) owing to the dependence of θ_c on y [99–102], the effective amount of InGaAs that self-assembles into QDs is reduced and, hence, the QD sizes decrease, and (4) the QD–CL lattice mismatch decreases thus leading to the formation of larger QDs, as observed by Orr [103] and Li [101] by means of kinetic Monte Carlo simulations and experiments, respectively. As it can be seen in Fig. 16.16, the combined effect of the four mechanisms is the blue shift of the emission energy.

It remains to be explained the apparently surprising observation that for higher In composition y it is possible to have higher emission energies: in this connection, it should be recalled that θ_c decreases for increasing y [99–102]; therefore, ranges of smaller coverages can be accessed, thus resulting in blue shifted emission (Fig. 16.16).

As a result of these mechanisms, 0.98 μm RT emission has been obtained from the low-coverage MBE structures with In$_{0.70}$Ga$_{0.30}$As QDs embedded in Al$_{0.30}$Ga$_{0.70}$As CLs. We observe that structures characterized by both blue shifted emission energy and high emission efficiency at RT may be obtained by means of the simultaneous increase in: (i) carrier confinement energies and/or QD material energy gap and (ii) energy separation between WL and QD states. Both conditions can be fulfilled when the Al composition in CLs is increased (section 16.4.1); instead, we note that the approach of intentionally decreasing the QD sizes and/or enhancing E_{QD} in order to blue shift the emission (section 16.4.2) should not be followed since the increase in the transition energy would be obtained concomitantly to the decrease in the energy separation between WL and QD states, which would give rise to enhanced thermal escape of carriers from QDs.

16.5 Conclusions

In this work we reviewed methods allowing the extension of the spectral window of QD emission, through the engineering of the InAs/GaAs QD structure, taking advantage of predictions provided by a simple effective-mass model. In order to match the 1.3–1.6 µm window we considered InAs/InGaAs metamorphic QD nanostructures, where band discontinuities and energy gap of the QD material are controlled by means of two design parameters that are the composition of the InGaAs CLs and the QD–CL mismatch (which determines the QD strain). We demonstrated how by this approach, termed *QD strain engineering*, we can substantially red shift the emission wavelength of InAs QDs embedded in InGaAs CLs and optimize the activation energy for thermal escape of confined carriers, reaching 1.44 µm emission at RT, under excitation power densities as low as 5 W/cm^2. Moreover, the careful analysis of model calculations led us to conclude that the proposed approach yields two degrees of freedom, the CL composition and the QD–CL mismatch, which allow us to control the QD–CL band discontinuities and QD energy gap, respectively; the availability of such parameters permits us not only to tune the QD emission wavelength at predetermined values, but also to maximize the activation energy for thermal quenching of emission. To overcome the reduction of the barrier discontinuities and, therefore, of the activation energy, which is a consequence of red shifting the emission, we combined the use of QD strain engineering with enhanced additional barriers, achieving RT emission up to 1.51 µm, thus proving that this approach is of great interest to extend the QD emission wavelengths to 1.55 µm and beyond.

To match the 0.98–1.04 µm window, AlGaAs confining layers and InGaAs QDs have been successfully used in order to increase the confinement energies and the QD energy gap, respectively. Experimental results and model calculations allowed us to discuss the effect of QD and CL composition on the QD morphology, highlighting its impact on the QD emission blue shift and on the activation energy for confined carrier thermal escape. In particular, we showed how the CL composition, besides band discontinuities, affects also the QD dimensions and other details of the QD band structure, resulting in a complex interplay of different effects. Furthermore, by studying the effect of the change in composition of InGaAs QDs, we identified different mechanisms contributing to the blue shift of the emission and achieved 0.98 µm RT emission from different structures consisting of InGaAs QDs embedded in $Al_{0.30}Ga_{0.70}As$ CLs.

By comparing the results of the model in the case of InAs/InGaAs and of In(Ga)As/AlGaAs structures we can conclude that the very satisfactory predictions of the model (on emission energy, activation energy for thermal quenching of emission, as well as on approaches that result in the substantial improvement of the emission efficiency) rely not only on the soundness of the assumptions but also on the circumstances that: (i) important quantities such as compositions of QDs and CLs are very close to the nominal ones and (ii) the QD sizes evaluated by AFM are conceivably very little modified during the subsequent steps of growth, such as the deposition of UCLs. These results are related to the low-temperature growth conditions that have been used throughout our experiments and that limit activated alloy phase separation [19, 26, 29, 32, 33], intermixing and segregation phenomena. On the other hand, the relatively high reactivity of Al gives rise to important changes in QD size, shape and composition in In(Ga)As/AlGaAs structures, which would require the use of specific techniques to measure these quantities in the complete structures to be modelled.

In conclusion, we demonstrated how by a careful design of structures grown under well-defined conditions it is possible to cover the 0.98–1.04 µm and 1.3–1.6 µm windows with QD nanostructures based on GaAs substrates, a result that could open the path to the realization of QD photonic and optoelectronic devices for a wide range of applications.

Acknowledgements

This work has been supported by the MIUR-FIRB project "Nanotecnologie e Nanodispositivi per la Società dell'Informazione" and by the SANDiE Network of Excellence of EU, contract no. NMP4-CT-2004-500101.

Note added in proof

Very recently, emissions upto $1.59\,\mu m$ at room temperature were obtained under excitation power densities as low as $5\,W/cm^2$ from metamorphic QD nanostructures on GaAs substrates; the structures consist of InAs QDs, $In_{0.45}Ga_{0.55}As$ CLs and $In_{0.35}Al_{0.65}As$ additional barriers. These results demonstrate how by using the combined approach of QD strain engineering and barrier enhancing it is possible to: (i) tune the emission wavelength in the whole 1.3–$1.6\,\mu m$ range, (ii) increase the activation energy of PL thermal quenching with the consequent improvement of RT emission efficiency [104].

References

1. D. Bimberg, M. Grundmann, and N.N. Ledentsov. *Quantum Dot Heterostructures*, (Wiley, Chichester, 1999).
2. M.S. Skolnick and D.J. Mowbray, *Annu. Rev. Mater. Sci.* **34**, 181 (2004).
3. V.M. Ustinov, A.E. Zhukov, A.Yu. Egorov, and N.A. Maleev. *Quantum Dot Lasers*, (Oxford University Press, Oxford, 2003).
4. P. Bhattacharya, S. Ghosh, and A.D. Stiff-Roberts, *Annu. Rev. Mater. Sci.* **34**, 1 (2004).
5. M. Grassi Alessi, M. Capizzi, A.S. Bhatti, A. Frova, F. Martelli, P. Frigeri, A. Bosacchi, and S. Franchi, *Phys. Rev. B* **59**, 7620 (1999).
6. Atomic layer molecular beam epitaxy (ALMBE) is a variant of molecular beam epitaxy (MBE) where group III and group V species impinge on the substrate alternatively in monolayer or sub-monolayer amounts per cycle; F. Briones and A. Ruiz, *J. Cryst. Growth* **111**, 194 (1991).
7. R.P. Mirin, J.P. Ibbetson, J.E. Bowers, and A.C. Gossard, *J. Cryst. Growth* **175–176**, 696 (1997).
8. A. Bosacchi, P. Frigeri, S. Franchi, P. Allegri, and V. Avanzini, *J. Cryst. Growth* **175–176**, 771 (1997).
9. R.P. Mirin, J.P. Ibbetson, K. Nishi, A.C. Gossard, and J.E. Bowers, *Appl. Phys. Lett.* **67**, 3795 (1995).
10. K. Mukai, N. Ohtsuka, M. Sugawara, and S. Yamazaki, *Jpn. J. Appl. Phys.* **33**, L1710 (1994).
11. S. Krishna, D. Zhu, J. Xu, K.K. Linder, O. Qasaimeh, P. Bhattacharya, and D.L. Huffaker, *J. Appl. Phys.* **86**, 6135 (1999).
12. Y. Nakata, K. Mukai, N. Ohtsuka, M. Sugawara, H. Ishikawa, and N. Yokoyama, *J. Cryst. Growth* **93**, 208 (2000).
13. R. Murray, D. Childs, S. Malik, P. Siverns, C. Roberts, J. Hartmann, and P. Stavrinou, *Jpn. J. Appl. Phys.* **38**, 528 (1999).
14. E.C. Le Ru, P. Howe, T.S. Jones, and R. Murray. *Phys. Rev. B* **67**, 165303 (2003).
15. M.J. da Silva, A.A. Quivy, S. Martini, T.E. Lamas, E.C.F. da Silva, and J.R. Leite, *Appl. Phys. Lett.* **82**, 2646 (2003).
16. Q. Xie, A. Madhukar, P. Chen, and N.P. Kobayashi, *Phys. Rev. Lett.* **75**, 2542 (1995).
17. I. Mukhametzhanov, R. Heitz, J. Zheng, P. Chen, and A. Madhukar, *Appl. Phys. Lett.* **73**, 1841 (2003).
18. Z. Mi and P. Bhattacharya, *J. Appl. Phys.* **98**, 023510 (2005).
19. V.M. Ustinov, N.A. Maleev, A.E. Zhukov, A.R. Kovsh, A. Yu. Egorov, A.V. Lunev, B.V. Volovik, I.L. Krestnikov, Y.G. Musikhin, N.A. Bert, P.S. Kop'ev, Zh.I. Alferov, N.N. Ledentsov, and D. Bimberg, *Appl. Phys. Lett.* **74**, 2815 (1999).
20. K. Nishi, H. Saito, S. Sugou, and J. Lee, *Appl. Phys. Lett.* **74**, 1111 (1999).
21. N.T. Yeh, T.E. Nee, J.I. Chyi, T.M. Hsu, and C.C. Huang, *Appl. Phys. Lett.* **76**, 1567 (2000).
22. A. Stintz, G.T. Liu, A.L. Gray, R. Spillers, S.M. Delgado, and K.J. Malloy, *J. Vac. Sci. Technol. B* **18**, 1496 (2000).

23. F. Ferdos, M. Sadeghi, Q.X. Zhao, S.M. Wang, and A. Larsson, *J. Cryst. Growth* **227–228**, 1140 (2001).
24. H.Y. Liu, X.D. Wang, J. Wu, B. Xu, Y.Q. Wei, W.H. Jiang, D. Ding, X.L. Ye, F. Lin, J.F. Zhang, J.B. Liang, and Z.G. Wang, *J. Appl. Phys.* **88**, 3392 (2000).
25. J. Tatebayashi, M. Nishioka, and Y. Arakawa, *Appl. Phys. Lett.* **78**, 3469 (2001).
26. F. Guffarth, R. Heitz, A. Schliwa, O. Stier, N.N. Ledentsov, A.R. Kovsh, V.M. Ustinov, and D. Bimberg, *Phys. Rev. B* **64**, 085305 (2001).
27. J.X. Chen, A. Markus, U. Oesterle, A. Fiore, R.P. Stanley, M. Ilegems, J.F. Carlin, R. Houdrè, L. Lazzarini, L. Nasi, M.T. Todaro, E. Piscopiello, R. Cingolani, M. Catalano, J. Katcki, and J. Ratajczak, *J. Appl. Phys.* **91**, 6710 (2002).
28. H.Y. Liu, I.R. Sellers, M. Hopkinson, C.N. Harrison, D.J. Mowbray, and M.S. Skolnick, *Appl. Phys. Lett.* **83**, 3716 (2003).
29. J.S. Kim, J.H. Lee, S.U. Hoing, W.S. Han, H.S. Kwack, C.W. Lee, and D.K. Oh, *J. Appl. Phys.* **94**, 6603 (2003).
30. T.P. Hsieh, N.T. Yeh, P.C. Chiu, W.H. Chang, T.M. Hsu, W.J. Ho, and L.I. Chyi, *Appl. Phys. Lett.* **87**, 151903 (2005).
31. Z.Z. Sun, S.F. Yoon, W.K. Loke, and C.Y. Liu, *Appl. Phys. Lett.* **88**, 203114 (2006).
32. M.V. Maximov, A.F. Tsatsul'nikov, B.V. Volovik, D.S. Sizov, Y.U.M. Shernyakov, I.N. Kaiander, A.E. Zhukov, A.R. Kovsh, S.S. Mikhrin, V.M. Ustinov, Zh.I. Alferov, R. Heitz, V.A. Shchukin, N.N. Ledentsov, D. Bimberg, Yu.G. Musikhin, and W. Neumann, *Phys. Rev. B* **62**, 16671 (2000).
33. W.H. Chang, H.Y. Chen, H.S. Chang, W.Y. Chen, T.M. Hsu, T.P. Hsieh, J.I. Chyi, and N.T. Yeh, *Appl. Phys. Lett.* **86**, 131917 (2005).
34. L. Seravalli, M. Minelli, P. Frigeri, P. Allegri, V. Avanzini, and S. Franchi, *Appl. Phys. Lett.* **82**, 2341 (2003).
35. A.E. Zhukov, A.P. Vasil'yev, A.R. Kovsh, S.S. Mikhrin, E.S. Semenova, A.Yu. Egorov, V.A. Odnoblyudov, N.A. Maleev, E.V. Nikitina, N.V. Kryjanovskaya, A.G. Gladyshev, Y.M. Shernyakov, M.V. Maximov, N.N. Ledentsov, V.M. Ustinov, and Zh.I. Alferov, *Semiconductors* **37**, 1411 (2003).
36. L. Seravalli, M. Minelli, P. Frigeri, P. Allegri, V. Avanzini, and S. Franchi, *Appl. Phys. Lett.* **87**, 063101 (2005).
37. Z. Mi, P. Bhattacharya, and J. Yang, *Appl. Phys. Lett.* **89**, 153109 (2006).
38. L. Seravalli, M. Minelli, P. Frigeri, S. Franchi, G. Guizzetti, M. Patrini, T. Ciabattoni, and M. Geddo, *J. Appl. Phys.* **101**, 024313 (2007).
39. A.V. Drigo, A. Aydinii, A. Carnera, F. Genova, C. Rigo, C. Ferrari, P. Franzosi, and G. Salviati, *J. Appl. Phys.* **66**, 1975 (1989).
40. R. People and J.C. Bean, *Appl. Phys. Lett.* **49**, 229 (1986).
41. P.M.J. Marée, J.C. Barbour, J.F. Van der Veen, K.L. Kavanagh, C.W.T. Bulle-Lieuwma, and M.P.A. Viegers, *J. Appl. Phys.* **62**, 4413 (1987).
42. J.W. Matthews and A.E. Blakeslee, *J. Cryst. Growth* **27**, 118 (1974).
43. C.A.B. Ball and J.H. van der Merwe, in: *Dislocation in Solids*, ed. by, F.R.N. Nabarro. (North-Holland, Amsterdam, 1983).
44. D.J. Dunstan, *Philos. Mag. A* **73**, 1323 (1996).
45. M. Geddo, G. Guizzetti, M. Patrini, T. Ciabattoni, L. Seravalli, P. Frigeri, and S. Franchi, *Appl. Phys. Lett.* **87**, 263120 (2005).
46. V. Bellani, C. Bocchi, T. Ciabattoni, S. Franchi, P. Frigeri, P. Galinetto, M. Geddo, F. Germini, G. Guizzetti, L. Nasi, M. Patrini, L. Seravalli, and G. Trevisi, *Europ. Phys. J. B.* **56**, 217 (2007).
47. A.D. Andreev, J.R. Downes, D.A. Faux, and E.P. O'Reilly, *J. Appl. Phys.* **86**, 297 (1999).
48. F.H. Pollak, Strained layer superlattices: Physics, in: *Semiconductors and Semimetals*, Vol. 32, ed. by T.P. Pearsall (Academic, Boston, 1990).
49. L.C. Andreani, D. De Nova, S. Di Lernia, M. Geddo, G. Guizzetti, M. Patrini, C. Bocchi, A. Bosacchi, C. Ferrari, and S. Franchi, *J. Appl. Phys.* **78**, 6745 (1995).
50. Y.P. Varshni, *Physica (Utrecht)* **34**, 149 (1967).
51. S. Sanguinetti, M. Henini, M. Grassi Alessi, M. Capizzi, P. Frigeri, and S. Franchi, *Phys. Rev. B* **60**, 8276 (1999).
52. E.C. Le Ru, J. Fack, and R. Murray, *Phys. Rev. B* **67**, 245318 (2003).

53. D. Colombo, S. Sanguinetti, E. Grilli, M. Guzzi, L. Martinelli, M. Gurioli, P. Frigeri, G. Trevisi, and S. Franchi, *J. Appl. Phys.* **94**, 6513 (2003).
54. S. Mazzucato, D. Nardin, M. Capizzi, A. Polimeni, A. Frova, P. Frigeri, L. Seravalli, and S. Franchi, *Mat. Sci. Eng. C* **25**, 830 (2005).
55. K. Mukai, N. Ohtsuka, and M. Sugawara, *Appl. Phys. Lett.* **70**, 2416 (1997).
56. J.Y. Marzin and G. Bastard, *Sol. Stat. Commuu.* **92**, 437 (1994).
57. M.A. Cusack, P.R. Briddon, and M. Jaros, *Phys. Rev. B* **54**, R2300 (1996).
58. H. Jiang and J. Singh, *Phys. Rev. B* **56**, 4696 (1997).
59. A.J. Williamson, L.W. Wang, and A. Zunger, *Phys. Rev. B* **62**, 12963 (2000).
60. G. Cipriani, M. Rosa-Clot, and S. Taddei, *Phys. Rev. B* **62**, 4413 (2000).
61. A.D. Yoffe, *Adv. in Phys.* **50**, 1 (2001).
62. D.M. Bruls, J.W.A.M. Vugs, P.M. Koenraad, H.W.M. Salemink, J.H. Wolter, M. Hopkinson, M.S. Skolnick, F. Long, and S.P.A. Gill, *Appl. Phys. Lett.* **81**, 1708 (2002).
63. Q. Gong, P. Offermans, R. Notzel, P.M. Koenraad, and J.H. Wolter, *Appl. Phys. Lett.* **85**, 5697 (2004).
64. C.G. Van de Walle, *Phys. Rev. B* **39**, 1871 (1989).
65. M. Cardona and N.E. Christensen, *Phys. Rev. B* **37**, 1011 (1988).
66. I. Vurgaftman, J.R. Meyer, and L.R. Ram-Mohan, *J. Appl. Phys.* **89**, 5815 (2001).
67. M. Grundmann, O. Stier, and D. Bimberg, *Phys. Rev. B* **52**, 11969 (1995).
68. G.S. Pearson and D.A. Faux, *J. Appl. Phys.* **88**, 730 (2000).
69. M. Sopanen, H.P. Xin, and C.W. Tu, *Appl. Phys. Lett.* **76**, 994 (2000).
70. K.C. Yew, S.F. Yoon, Z.Z. Sun, and S.Z. Wang, *J. Cryst. Growth* **247**, 279 (2003).
71. S. Tomic, *Phys. Rev. B* **73**, 125348 (2006).
72. T. Kita, T. Mori, H. Seki, M. Matsushita, M. Kikuno, O. Wada, H. Ebe, M. Sugawara, Y. Arakawa, and Y. Nakata, *J. Appl. Phys.* **97**, 024306 (2005).
73. O. Schumann, L. Geelhaar, H. Reichert, H. Cerva, and G. Abstreiter, *J. Appl. Phys.* **96**, 2832 (2004).
74. G. Bais, A. Cristofoli, F. Jabeen, M. Piccin, E. Carlino, S. Rubini, F. Martelli, and A. Franciosi, *Appl. Phys. Lett.* **86**, 233107 (2005).
75. V. Sallet, G. Patriarche, M.N. Merat-Combes, L. Largeau, O. Mauguin, and L. Travers, *J. Cryst. Growth* **290**, 80 (2006).
76. K. Suzuki and Y. Arakawa, *Phys. Status Solidi B* **224**, 139 (2001).
77. Y.M. Qiu, D. Uhl, and S. Keo, *Appl. Phys. Lett.* **84**, 263 (2004).
78. M. Kudo, T. Nakaoka, S. Iwamoto, and Y. Arakawa, *Jpn. J. Appl. Phys. Part 2* **44**, L45 (2005).
79. Y.M. Qiu and D. Uhl, *Appl. Phys. Lett.* **84**, 1510 (2004).
80. Y. Sun, S.F. Cheng, G. Chen, R.F. Hicks, J.G. Cederberg, and R.M. Biefeld, *J. Appl. Phys.* **97**, 053503 (2005).
81. J. He, B. Xu, and Z.G. Wang, *Appl. Phys. Lett.* **84**, 5237 (2004).
82. Z.Y. Zhang, B. Xu, P. Jin, X.Q. Meng, Ch.M. Li, X.L. Ye, and Z.G. Wang, *J. Appl. Phys.* **92**, 511 (2002).
83. Y.Q. Wei, S.M. Wang, F. Ferdos, J. Vukusic, A. Larsson, Q.X. Zhao, and M. Sadeghi, *Appl. Phys. Lett.* **81**, 1621 (2002).
84. L. Seravalli, P. Frigeri, P. Allegri, V. Avanzini, and S. Franchi, *Mat. Sci. Eng. C* **27**, 1046 (2007).
85. R. Jia, D.S. Jiang, H.Y. Liu, Y.Q. Wei, B. Xu, and Z.G. Wang, *J. Cryst. Growth* **234**, 354 (2002).
86. I.R. Sellers, H.Y. Liu, M. Hopkinson, D.J. Mowbray, and M.S. Skolnick, *Appl. Phys. Lett.* **83**, 4710 (2003).
87. M. Arzberger, U. Kasberger, G. Bohm, and G. Abstreiter, *Appl. Phys. Lett.* **75**, 3968 (1999).
88. K.P. Chang, S.L. Yang, D.S. Chuu, R.S. Hsiao, J.F. Chen, L. Wei, J.S. Wang, and J.Y. Chi, *J. Appl. Phys.* **97**, 083511 (2005).
89. A.E. Zhukov, A.R. Kovsh, V.M. Ustinov, N.N. Ledentsov, and Zh.I. Alferov, *Microelectron. Eng.* **81**, 229 (2005).
90. T. Kettler, L.Ya. Karachinsky, N.N. Ledentsov, V.A. Shchukin, G. Fiol, M. Kunts, A. Lochmann, O. Schulz, L. Reissmann, K. Posilovic, D. Bimberg, I.I. Novikov, Y.u.M. Shernyakov, N.Y.u. Gordeev, M.V. Maximov, N.V. Kryjanovskaya, A.E. Zhukov, E.S. Semenova, A.P. Vasil'yev, V.M. Ustinov, and A.R. Kovsh, *Appl. Phys. Lett.* **89**, 041113 (2006).

91. A.E. Zhukov, A.Yu. Egorov, A.R. Kovsh, V.M. Ustinov, N.N. Ledentsov, M.V. Maksimov, A.F. Tsatsulnikov, S.V. Zaitsev, N.Yu. Gordeev, P.S. Kop'ev, Zh.I. Alferov, and D. Bimberg, *Semiconductors* **31**, 411 (1997).

92. F. Klopf, J.P. Reithmaier, and A. Forchel, *J. Cryst. Growth.* **227–228**, 1151 (2001).

93. P. Ballet, J.B. Smathers, H. Yang, C.L. Workman, and G.J. Salamo, *J. Appl. Phys.* **90**, 481 (2001).

94. P. Offermans, P.M. Koenraad, J.H. Wolter, K. Pierz, M. Roy, and P.A. Maksym, *Physica E* **26**, 236 (2005).

95. J. Ibáñez, R. Cuscó, L. Artús, M. Henini, A. Patanè, and L. Eaves, *Appl. Phys. Lett.* **88**, 141905 (2006).

96. A. Polimeni, A. Patanè, M. Henini, L. Eaves, and P.C. Main, *Phys. Rev. B* **59**, 5064 (1999).

97. M. Gurioli, S. Testa, P. Altieri, S. Sanguinetti, E. Grilli, M. Guzzi, G. Trevisi, P. Frigeri, and S. Franchi, *Physica E* **17**, 19 (2003).

98. H.Y. Liu, I.R. Sellers, T.J. Badcock, D.J. Mowbray, M.S. Skolnick, K.M. Groom, M. Gutierrez, M. Hopkinson, J.S. Ng, J.P.R. David, and R. Beanland, *Appl. Phys. Lett.* **85**, 704 (2004).

99. D. Leonard, M. Krishnamurthy, C.M. Reaves, S.P. Denbaars, and P.M. Petroff, *Appl. Phys. Lett.* **63**, 3203 (1993).

100. K. Kamath, P. Bhattacharya, and J. Phillips, *J. Cryst. Growth.* **175–176**, 720 (1997).

101. H. Li, Q. Zhuang, Z. Wang, and T. Daniels-Race, *J. Appl. Phys.* **87**, 188 (2000).

102. A.G. Cullis, D.J. Norris, M.A. Migliorato, and M. Hopkinson, *Appl. Surf. Sci.* **244**, 65 (2005).

103. B.G. Orr, D. Kessler, C.W. Snyder, and L. Sander, *Europhys. Lett.* **19**, 33 (1992).

104. L. Seravalli, P. Frigeri, G. Trevisi, and S. Franchi, *Appl. Phys. Lett.* **92**, 213104 (2008).

CHAPTER 17

Advanced Growth Techniques of InAs-system Quantum Dots for Integrated Nanophotonic Circuits

Kiyoshi Asakawa,[1] Nobuhiko Ozaki,[1] Shunsuke Ohkouchi,[2] Yoshimasa Sugimoto,[1,3] and Naoki Ikeda[3]

[1]*Center for Tsukuba Advanced Research Alliance (TARA), University of Tsukuba, 1-1-1 Tennoudai, Tsukuba, Ibaraki 305-8577, Japan;*
[2]*Nano Electronics Research Laboratories, NEC Corporation, 34 Miyukigaoka, Tsukuba, Ibaraki 305-8501, Japan;*
[3]*National Institute for Materials Science (NIMS), 1-2-1 Sengen, Tsukuba, Ibaraki 305-0047, Japan*

17.1 Introduction

For the last two decades, semiconductor quantum dots (QDs) have been intensively investigated from the viewpoints of both band engineering and growth technology. As a result, they have provided a great deal of attractive electronic/optoelectronic devices thanks to their high density-of-states specific to the low-dimensional structures. Some of the results involve QD-based laser diodes [1] and optical amplifiers [2] in the commercial base, while others have been exploited intensively for advanced electronic/optoelectronic devices.

As a topic of growth technology, site-controlled quantum dots (SCQDs) by means of patterned substrates or nano-probe-assisted technologies have attracted much attention for creation of functional QD structures such as single QD and arrayed QDs, as well as achievement of QDs with high uniformity and high density. At the first stage for creating such an SCQD, we proposed an idea of an *in situ* STM (scanning tunnelling microscope)-probe-assisted MBE technique and demonstrated two-dimensionally arrayed InAs SCQDs with $50 \sim 100\,nm$ pitches [3]. At the second stage, we have developed an *in situ* AFM (atomic force microscope) probe with a specially designed cantilever capable of a practical throughput ($1 \sim 10$ msec/dot, for example). By using the new probe, we have reproducibly fabricated uniform In (indium) nanodots in the selected area first. Since the AFM chamber is connected *in vacuum* to an MBE chamber, In dots could be directly converted to InAs QDs by subsequent arsenic-flux irradiation [4]. For implementation of a stacked array of uniform and high-density QDs, combination of the SCQD and subsequent S–K mode QD growth was successfully demonstrated.

As a new topic of application, on the other hand, we proposed a unique research scheme by combining QDs with another nanostructure, i.e. a photonic crystal (PC), to provide key photonic devices for future advanced telecommunication systems, as shown in Fig. 17.1 [5]. In the left-hand wing is an ultra-fast digital photonic network of the future, where an ultra-small and ultra-fast symmetrical Mach–Zehnder (SMZ)-type [6] all-optical switch (PC-SMZ) has been developed so far in the first phase by using GaAs-based two-dimensional PC slab waveguides embedded with InAs-based QDs (Fig. 17.1) [7]. In the second phase, the PC-SMZ is now evolving into a new functional key device, i.e. an ultra-fast all-optical flip-flop (PC-FF) device that is essential for the digital photonic network [5].

As long as QD technology is concerned all through these two phases, a selective-area-growth (SAG) of QDs has been shown to be an important subject for exhibiting an optical non-linear effect in the selective optical non-linear arms particularly for the PC-SMZ and PC-FF. For this purpose, a metal-mask (MM)/molecular beam epitaxy (MBE) method of InAs QDs has been developed [8]. In the right-hand wing in Fig. 17.1, on the other hand, another category of a future quantum information system is described, where a single photon qubit composed of a single QD embedded in a PC-based high-Q cavity is situated as another important product of the PC/QD-combined nano-photonic structure [9–11]. The AFM-probe-assisted SCQD growth method mentioned above, renamed recently as a nano-jet probe (NJP) method, has been developed for this purpose because of its potential to position a single QD at the centre of the point defect PC high-Q cavity [12–14].

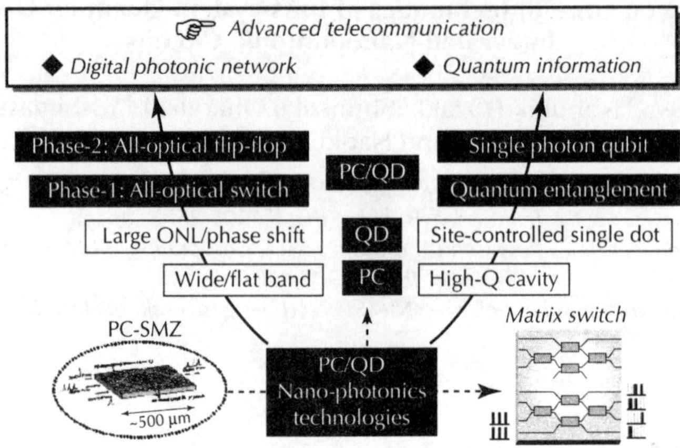

Figure 17.1   Schematic research scenario of PC/QD combined nano-photonic structures for advance telecommunication systems.

In this chapter, two types of unique QD growth techniques have been reviewed. The MM method for InAs QDs in the PC-based all-optical switching devices is shown in section 17.2 and the advanced techniques for the growth of InAs SCQDs by using the NJP method is described in section 17.3.

17.2   Selective-area-growth of InAs quantum dots using the metal-mask/MBE method

17.2.1   Introduction

A selective-area-growth (SAG) technique for self-assembled QDs is useful for various applications of the QDs. In the case of our proposed all-optical devices, PC-SMZ [7] and PC-FF [5], the SAG technique is intensively required for their effective operations. For instance, although the embedded QDs in the PC waveguides (WGs) of PC-SMZ act as a phase shifter due to their optical non-linearity, the QDs simultaneously absorb the propagation light and result in a low transmittance in the PC-WG. Therefore, the QD growth in limited regions in the WGs is desirable rather than an entire growth on a substrate. Furthermore, the PC-FF, which composes two PC-SMZs, requires SAG of two kinds of QDs having different absorption wavelengths for embedding the QDs in each optical non-linearity arm in the PC-SMZs.

In this section, we describe a metal mask (MM)/MBE method [8] for SAG of self-assembled InAs QDs. First, we introduce the MM/MBE method and characterizations of the SA-grown QDs by atomic-force-microscopy (AFM) and photoluminescence (PL) measurements, including two-dimensional (2D) PL intensity mapping. Successful SAG of InAs QDs was clearly demonstrated, while the QD density and homogeneity were almost equal to those of the QDs grown without the MM. Second, we describe technologies for controlling the PL wavelength of InAs QDs by using a strain-reducing-layer (SRL) [15] inserted on the QDs. Sequential variation of the PL peak wavelength as a function of SRL thickness was attained by the SRL insertion method. Finally, we report on fabrications of a PC-WG with SA-grown QDs. For avoiding reflection due to a step at a boundary of the selective growth area, we reduce the step height with a developed growth sequence. Clear transmission spectra and pulse intensities/phase modulations in the PC-WG due to the optical non-linearity of the QD are confirmed.

17.2.2   Experimental apparatus and procedures for MM/MBE method

Figure 17.2a shows an image of the prepared MM attached on a holder. The tantalum (Ta) MM is employed to cover undesired growth areas during conventional MBE growth. The MM holder can be mounted on and off in an ultra-high vacuum (UHV) chamber without exposing the sample to the atmosphere.

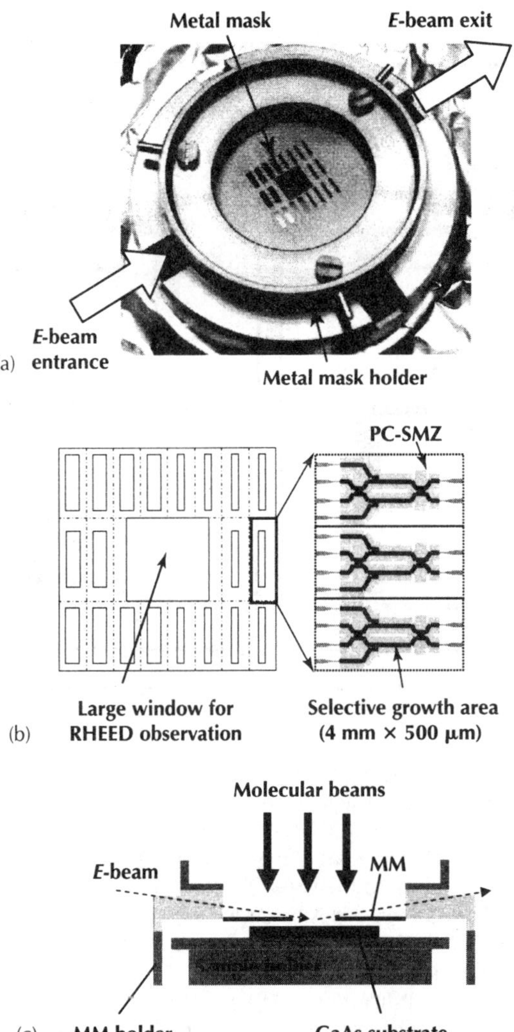

Figure 17.2   (a) Photograph of a metal-mask (MM) holder. (b) Schematic of patterns on the MM designed for PC-SMZ fabrications. (c) Cross-sectional illustration of the MM and e-beam incident for RHEED observations during the MBE growth.

As illustrated in Fig. 17.2b, the MM has a large window (6 mm × 6 mm) at the centre and several small windows of 4 mm × 0.5–1.0 mm around it. The small windows are prepared for SAG of QDs, suited for the PC-SMZ. On the other hand, the large window is used for RHEED (reflection high-energy electron-diffraction) observations and temperature measurements of the sample surface by using a pyrometer during the MBE growth. An electron beam for RHEED observations goes through the tunnels formed on the frame of the MM holder and is diffracted at the surface where molecular beams are irradiated through the large open window during MBE growth, as shown in Fig. 17.2c. These *in situ* observations enable us to optimize growth conditions for QDs.

We have performed the SAG of the QD using two methods, i.e. "mask" and "cap/anneal" methods [16], as schematically shown in Fig. 17.3. In the mask method, QDs are grown when the MM covers the substrate, namely the SAG of QDs is directly operated with the MM. In the cap/anneal method, on the other hand, the MM is used for selective growth of capping layers of QDs grown on the entire substrate. After the entire QDs' growth, a selective area GaAs layer is grown by using the MM for capping selective area QDs, followed by an annealing to evaporate the rest of the QDs.

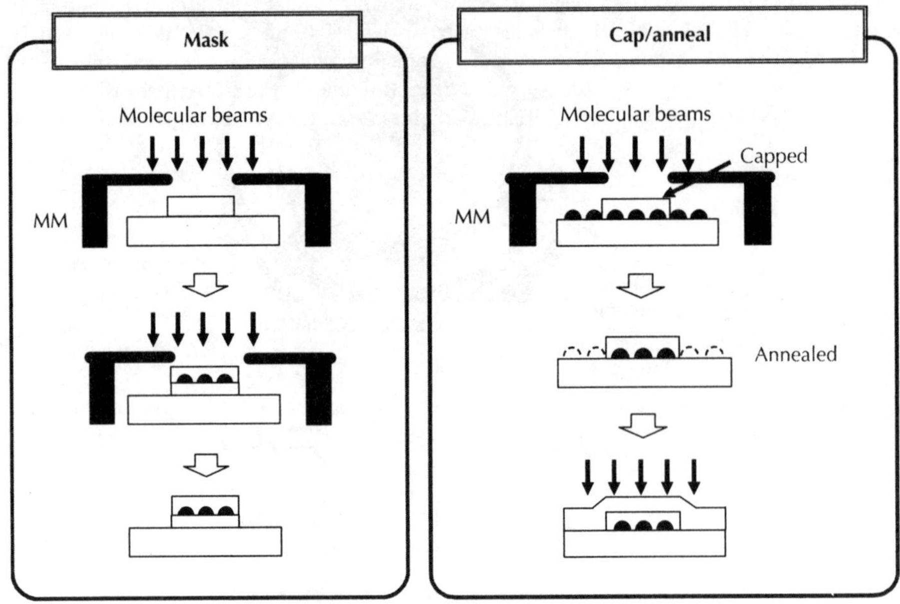

Figure 17.3   Two SAG sequences: "mask" and "cap/anneal" methods.

In both methods, sequences before QD growth are identical. At first, a multilayer structure of a 50 nm thick GaAs layer/30 nm thick $Al_{0.3}Ga_{0.7}As$ layer/GaAs buffer layer was grown on the entire GaAs substrate without the MM at approximately 540°C. In the mask method, the MM holder was then mounted after the sample was cooled down below 200°C. Following this, the sample was gradually heated up to the InAs QD growth temperature, typically 470–490°C, and 2.6 monolayers (ML) of InAs were deposited at a rate of 0.2 ML/s for the QD growth. During this process, we confirmed that the RHEED patterns varied from streaky to spotty patterns, as shown in the photographs in Fig. 17.4a and b, respectively. A chevron pattern originated from the QD facet, as shown in Fig. 17.4b, indicating 3D QD growth.

After the sample was slightly cooled down by approximately 40°C to avoid the degradation of the QD, a 3 nm thick $In_{0.2}Ga_{0.8}As$ layer/3 nm thick GaAs cap layer was sequentially grown. In the cap/anneal method, on the other hand, the InAs QDs were grown entirely on the GaAs layer with a 2.6 monolayer (ML) InAs deposition at a rate of 0.2 ML/s without the MM. Then, the MM was mounted on the sample after cooling down below 200°C and the sample temperature was

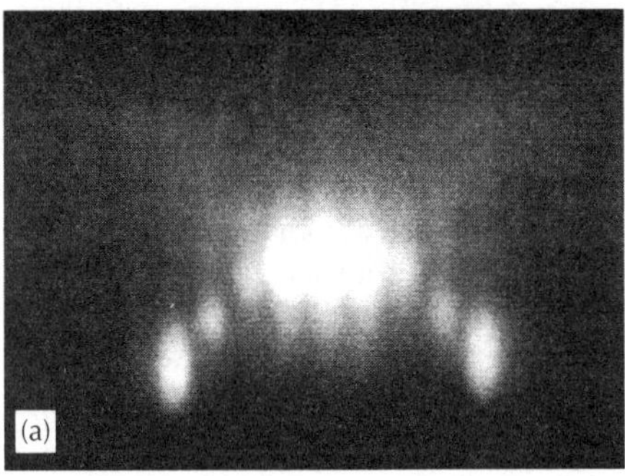
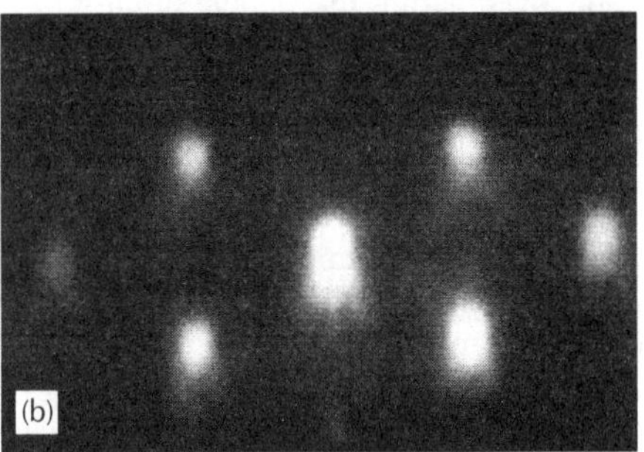

Figure 17.4   RHEED patterns obtained (a) before and (b) after growth of QDs with the MM.

gradually raised up to typically 430–450°C. Selective capping layers of 3 nm thick $In_{0.2}Ga_{0.8}As$ and 20 nm thick GaAs were grown at the temperature via the MM. After removing the MM, the temperature of the sample was gradually raised up to the QD growth temperature, for instance 480°C, and was kept for 3 min to evaporate the QDs on regions uncovered by the capping layer. SAG of QDs was completed in this way. After SAG of QDs using both methods, a multilayer structure of 50 nm thick GaAs capping layer/30 nm thick $Al_{0.3}Ga_{0.7}As$ layer/50 nm thick GaAs layer was grown on the entire substrate. The $Al_{0.3}Ga_{0.7}As$ layers were inserted below and above the QD layer as an energy barrier against free carriers in the PL measurement of the QD. As mentioned later, the capping layer of 3 nm thick $In_{0.2}Ga_{0.8}As$ on the QD is employed for reducing the strain in grown QDs and a red shift of the PL from QDs to around 1300 nm in wavelength.

17.2.3   Structural and optical properties of SA-grown Qds

We characterized the SA-grown QDs by AFM observations and PL measurements. Afterwards, we showed results from samples grown by the mask method on behalf of the two SAG methods.

Figure 17.5 shows typical AFM images of regions masked and unmasked by the MM. The image of an unmasked region shows formation of InAs QDs whose density is around 4×10^{10} cm^{-2}, whereas no QD was formed on the masked regions of the surface. Such high density is comparable to that of QDs conventionally grown without the MM. The mean lateral size and height of the QD was approximately 40 nm and 5 nm, respectively.

Masked region   Unmasked region

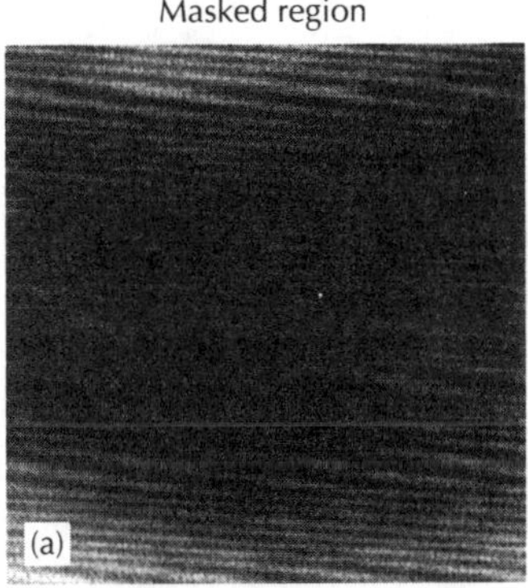
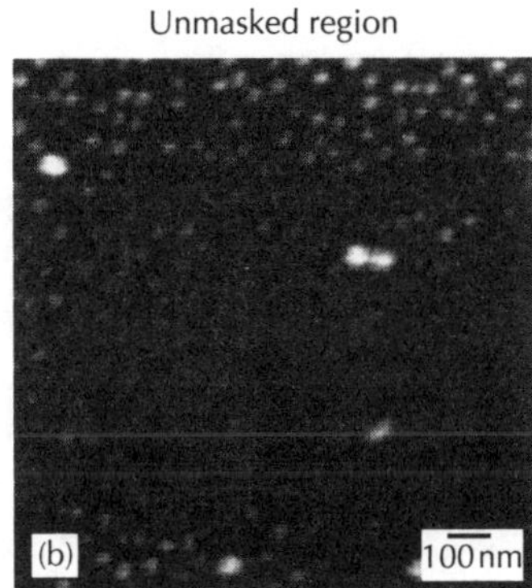

Figure 17.5   AFM images of masked and unmasked regions with the MM. SA-grown QDs with a density of $4 \times 10^{10} cm^{-2}$ were observed on the unmasked regions.

For investigating the optical quality of SA-grown QDs, PL measurements were obtained at room temperature (RT). An He–Ne laser ($\lambda = 633$ nm), focused to a diameter of several microns, was used for QD excitation with a power of 80 μW. A stepping-motor attached to the sample stage was utilized for positioning the excitation area and 2D mapping of the PL intensity.

PL spectra and 2D intensity mapping from the unmasked and masked regions are shown in Fig. 17.6. The spectra shown with black and grey lines are measured from the unmasked regions which are indicated in the PL intensity mapping by arrows. High contrast in the mapping shows high PL intensity attributed to the SA-grown QDs. The PL peak and the FWHM are 1295 nm (0.957 eV) and approximately 32 meV, respectively, which are almost the same as those of the QDs grown without the MM. This small FWHM value obtained even at RT indicates the high homogeneity of the SA-grown QDs comparable to that of QDs grown without the MM.

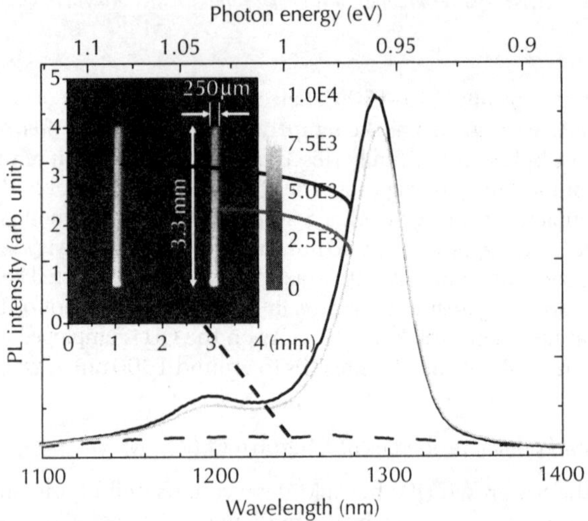

Figure 17.6   PL spectra and intensity mapping from SA-grown sample at room temperature.

Moreover, these PL spectra measured at two different regions show almost identical intensity and linewidth, suggesting that such high density and homogeneous QDs were grown on the selective areas of the substrate independent of the position. On the other hand, no peak was found in the PL spectrum from the masked region, as shown by a dashed line. Figure 17.7 shows a PL intensity mapping of the entire masked and unmasked areas. Through these PL intensity mapping images, the successful SAG of the entire region underneath the MM was clearly exhibited.

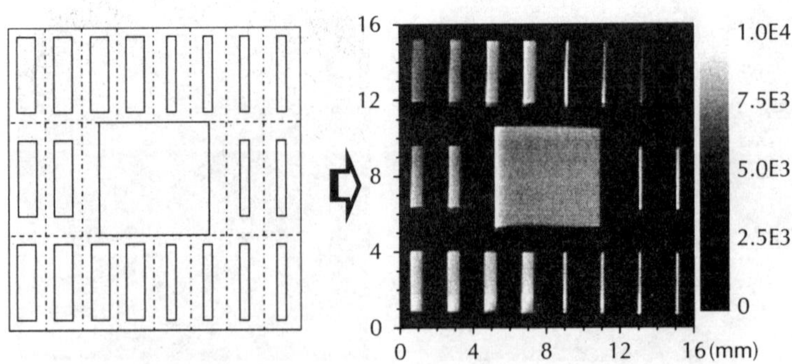

Figure 17.7   PL intensity mapping image from the entire region under the MM. Successful SAG corresponding to the MM pattern was confirmed.

Each QD-grown area decreased to approximately 0.25 mm × 3.3 mm SAG area obtained from a window size of 0.5 mm × 4 mm. This shrinkage of the SAG area is considered to be due to thermal evaporation of QDs near the edge of the windows by the heated MM.

17.2.4   *PL wavelength control of InAs QDs with a strain-reducing-layer*

As previously mentioned, successful SAG of QDs was attained by the MM/MBE method. However, when applying the MM/MBE method to realize the PC-FF [5], controlling absorption wavelength of SA-grown QDs is also required. For that purpose, we inserted a strain-reducing-layer (SRL) between QDs and spacers. This method is operated by covering grown QDs with an $In_{0.2}Ga_{0.8}As$ layer to reduce strain in the QDs, and it results in a red shift of the PL emission peak of QDs [15].

We have investigated the effectiveness of the SRL for controlling the absorption wavelength of QDs. Figure 17.8a summarizes the PL spectra from strain-reduced QDs with different SRL thicknesses ranging from 0 to 6 nm. The PL spectra exhibit the PL peak wavelength that shifted from 1240 nm to 1320 nm and with the FWHM that was almost maintained at approximately 30 meV, as shown in Fig. 17.8b.

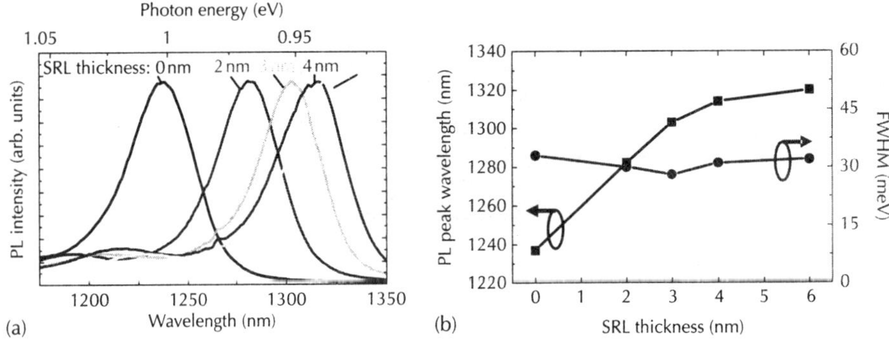

Figure 17.8   Effectiveness of a strain-reducing-layer (SRL) for tuning the PL peak wavelength. (a) Summary of PL spectra from QDs as a function of SRL thickness. (b) PL peak wavelength and FWHM variations against the SRL thickness.

This result clearly indicates that the SRL can control the PL wavelength of QDs without degradation of the QD optical quality. Thus, combination of the MM method and the insertion of the SRL has a potential to develop the PC-FF, which requires SAG of QDs with different absorption wavelengths at different areas.

17.2.5   *Selective area-grown QD embedded in PC-WG*

Here, we report on an application of the MM/MBE method to fabricate the SA grown QDs embedded in PC-WGs. As previously mentioned, we developed the SAG technique for our proposed PC-based devices. Considering an application of the SAG to PC-WGs, the step height of the SAG area should be suppressed as much as possible since the step causes undesired reflections of the propagating light and narrowing the bandwidth of the transmitted light. Thus, we attempted to minimize a step height of the SAG with optimizing the growth sequences. As a result, the SAG method enables us to minimize the step height within 6 nm at the SAG region against the core thickness 250 nm, as schematically shown in Fig. 17.9. Also, we succeeded in characterizations of the PC-WG with QDs partially embedded, i.e. the transmittance of the continuous wave (CW) light and the intensity and phase modulations of light pulses due to the optical non-linear phenomenon induced by the embedded QDs.

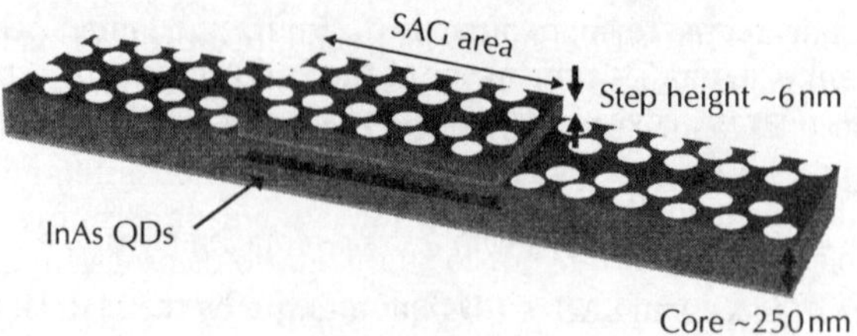

Figure 17.9　Schematic of PC-WG embedding SA-grown QDs. The ratio of a step height at SAG area to a core thickness is below 2.4%, which should be negligible for pulse propagations.

We prepared a GaAs-based straight PC-WG composed of an air bridge-type 2DPC slab and single InAs QDs layer embedded in a selective area of the WG. Figure 17.10 shows a structure of the MBE-grown sample for fabricating the PC-WG with SA-grown QDs. A 250 nm thickness GaAs core layer with QDs on a 2 μm thickness sacrificial clad layer was grown on a GaAs substrate. To minimize the step height of the selective area, the MM was used for only QD layers and capping layers of 6 nm in thickness.

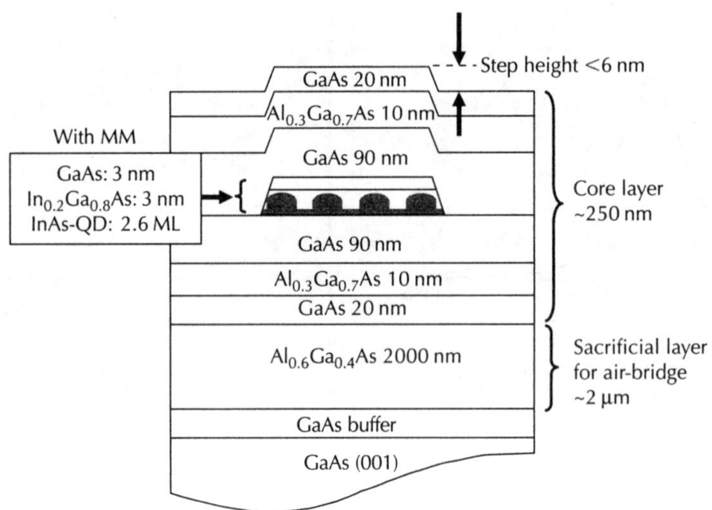

Figure 17.10　MBE-grown sample structure for PC-WG with SA-grown QDs.

The MBE-grown sample was then transferred to electron beam lithography and dry etching processes for fabricating the air bridge-type 2DPC slab [17]. The PC is constructed with holes of hexagonal arrays. In the PC, a missing line of holes is prepared as a straight WG whose input and output ports involve solid immersion lens and taper structures [18] resulting in a high coupling efficiency. By marking the position of SAG in advance, one can set the SA-grown QDs within the WG. The lattice constant and radius of air holes are designed to include the centre wavelength of the SA-grown QDs, around 1300 nm, within the transmission band of the WG.

Figure 17.11a shows an optical microscope image of a fabricated sample. The area surrounded by the dashed lines is embedded with the QDs. A PL intensity mapping obtained from the same area in Fig. 17.11a exhibits clear PL emission of 1.29 μm in wavelength from QDs embedded in a selective area, as shown in Fig. 17.11b. These images indicate successful fabrication of the PC-WG, embedding the SA grown QDs.

We then characterized the fabricated PC-WGs by CW transmission and pump/probe pulse measurements. The transmission spectra were observed with CW white light as a light source

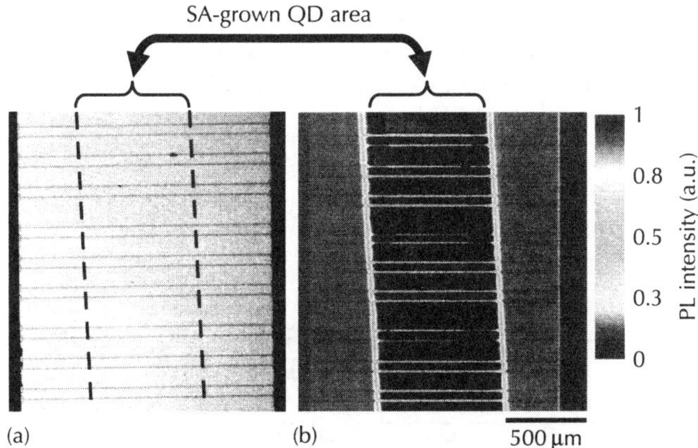

Figure 17.11   (a) *Optical microscope image of a fabricated sample with PC/QD straight WGs. (b) PL intensity mapping from the sample. PL intensity of 1290 nm in peak wavelength due to the SA-grown QDs is clearly shown.*

and a conventional measurement set-up. Figure 17.12b shows the transmission spectra obtained from the PC-WG. The transmission band through the WG mode is seen from 1270 nm to 1330 nm indicated by a black arrow. This value almost corresponds to the transmission band of the straight WG without QDs, therefore, the step height of the SAG area, below 6 nm, should not influence the transmission. An absorption attributed to the QDs is observed around 1290 nm, as depicted by a dashed line, corresponding to the inverse of the PL spectrum shown in Fig. 17.12a. From the above results, we confirmed a successful fabrication of the PC-WG embedding the SA-grown QDs; the transmission band was set to include the QD absorption centre wavelength, while no influence of the step height was observed.

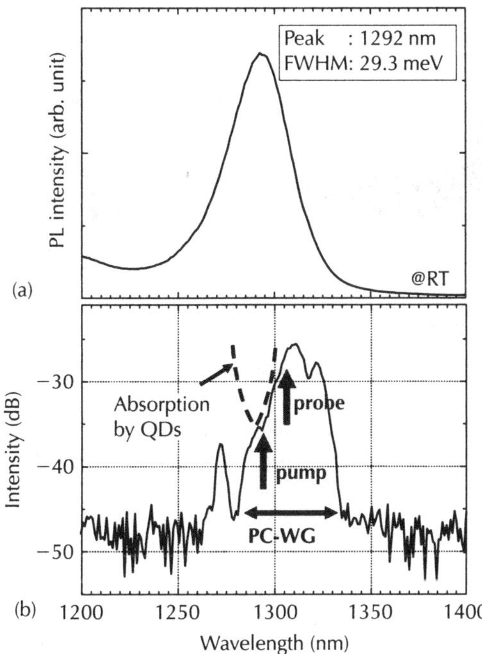

Figure 17.12   (a) *PL spectrum from the SA-grown QDs at RT. (b) The transmission spectrum obtained from the PC/QD WG.*

Then, we measured the non-linear phenomena by the QD, amplitude and phase shift of signal pulses (SPs) induced by control pulses (CPs), which are essential for the operation of PC-SMZ. The optical non-linearity-induced modulations were measured by utilizing a two-colour pump and probe set-up, as shown in Fig. 17.13 [7].

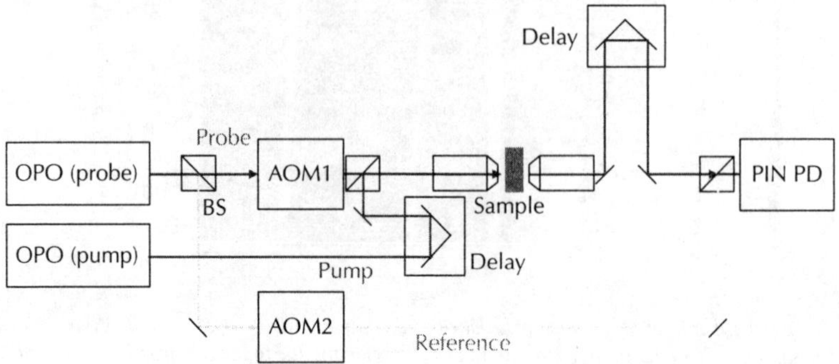

Figure 17.13   Schematic of the experimental apparatus for two-colour pump/probe measurements.

The SPs and CPs were collinearly coupled into the PC-WG. By considering the transmission spectrum shown in Fig. 17.12b, we set the CPs' centre wavelength to the centre of the QDs' absorption wavelength, 1290 nm, and set the SPs' centre wavelength to a longer wavelength, 1310 nm. The ONL-induced amplitude and phase shifts of the SPs were measured by lock-in-based heterodyne detection. Figure 17.14 shows typical non-linearity-induced amplitude and phase modulations of SP against time delay between the CP and the SP. The shift of the amplitude and the phase of SPs due to the CPs were clearly observed. The energies of the SP and the CP in the WG were estimated to be 33.8 fJ/pulse and 338 fJ/pulse, respectively. The net energy is estimated from the coupling efficiency into the PC-WG.

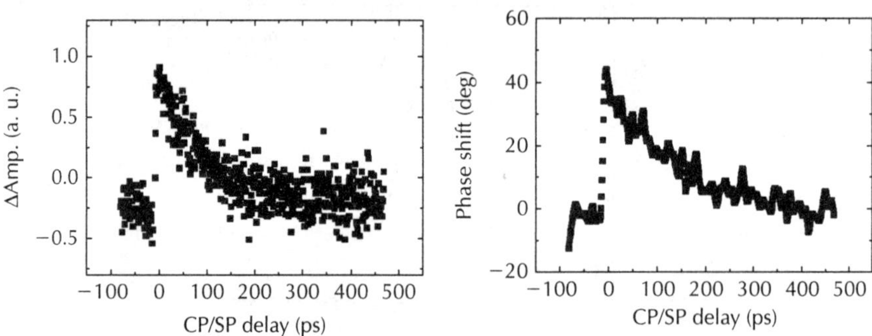

Figure 17.14   The ONL-induced amplitude and phase shifts of SPs against time delay between CPs and SPs.

In this case, we have obtained the phase shift value around 45°. This value is consistent with that previously obtained from the sample with entire-grown three-layer-stacked QDs, e.g. 130° [7], considering that this sample includes a single layer of QDs. However, for PC-SMZ operations, which requires the SP phase shift of at least 180°, further improvement of the non-linearity efficiency is necessary, e.g. by QD stacking growth or utilizing the slow light in the hetero V_g (group velocity) PC-WG [19].

17.2.6   Summary

We have succeeded in the SAG of InAs QDs by the developed MM/MBE method. The SA-grown QDs, which exhibit comparable structural and optical qualities with conventional QDs grown without the MM, can be utilized in PC-SMZ. The PL peak wavelength was red shifted with a strain-reducing-layer from 1240 nm to 1320 nm with the SRL thickness. The combination of these techniques is promising for realizing an all-optical digital flip-flop device: PC-FF.

In addition, we have fabricated the PC-WG embedding SA-grown QDs. The non-linearity-induced amplitude and phase shifts of signal pulses due to the control pulses were confirmed. Although the embedded QD layer was single, around a 45° phase shift was observed. More non-linearity efficiency improvements, by such as multilayer QD stacking and utilization of slow light, enable us to obtain sufficient phase shift value for the switch operation in PC-SMZ. These results indicate that the SAG technique can be applied to fabricate the PC-WG with QDs partially embedded for PC-SMZ and PC-FF.

17.3 Site control of InAs QDs using the nano-jet probe (NJP)

17.3.1 Introduction

In the previous section, we reported on a selective-area-growth (SAG) technique involving self-assembled InAs QDs that uses an *in situ* metal mask (MM). This MM method is a simple technique to obtain high-quality InAs QDs in selected areas for PC-SMZ and PC-FF by merely positioning shadow masks on the sample surface. However, the QDs fabricated by the MM method are randomly distributed on the sample surface, since their growth proceeds in the Stranski–Krastanov (SK) mode. In other words, the MM method cannot control the nucleation sites of the QDs precisely. If the nucleation sites can be intentionally controlled on a nanometre scale, then the QDs can be arranged hexagonally on the sample surface. This is the best way to obtain a high density of QDs. Additionally, Lee *et al.* pointed out in their calculation that the uniformity of QDs can be improved by growth on sample surfaces containing regularly arranged nucleation sites [20]. Furthermore, considering the application of QDs to cavity quantum electrodynamics (QED) devices such as single-photon emitters, which comprise QDs embedded in a photonic crystal (PC)-based cavity [10, 11], controlling the nucleation sites of QDs is one of the key issues since the position of the QDs determines the performance of the devices.

In this section, we report on an advanced technique for obtaining site-controlled QDs that enables the formation of the required number of QDs at the desired locations. In previous papers, we reported a scanning-tunnelling-microscope (STM)-probe-assisted site-control technique for InAs QDs and demonstrated the formation of two-dimensionally (2D) arrayed QDs with different constant (50 ∼ 100 nm) pitches [3]. However, when we consider applying this method to our recently proposed PC-based all-optical switches [5], which require many uniform QDs in the selected areas, the capability of the reported method for selective QD formation using the STM probe is not sufficient for practical nanofabrication use since the throughput of this technique is 0.5–1 s/dot. Furthermore, a change in the shape of the apex of the STM tip during the fabrication process is an inevitable problem since this technique utilizes a part of the tip itself as a depositing material for creating the nucleation sites of QDs [3].

To solve these problems, we have recently proposed a new technique that employs an *in situ* AFM probe with a specially designed cantilever, which is referred to as the nano-jet probe (NJP) [12, 21]. The NJP provides the capability to fabricate high-density 2D indium (In) nano-dot arrays on GaAs substrates within a selected area; these arrays can be directly converted into InAs arrays by subsequent annealing with arsenic flux irradiation [4]. Using this method, we have achieved site-controlled QD formation at desired locations on the sample surface [22].

17.3.2 Experimental apparatus and procedures

Figure 17.15 shows a schematic illustration of the micro-fabricated cantilever (NJP) and the procedure for nano-dot formation developed in this study. The nano-dot formation was realized using a UHV-AFM probe with a specially designed cantilever. The probe had a hollow pyramidal tip with a submicron-sized aperture on the apex and an In-reservoir tank within the stylus. This cantilever belongs to the piezoelectric type, and it is used for nano-dot fabrication as well as for sensing the force in AFM observations. The nano-dot formation was performed in a non-contact mode. By applying a voltage pulse between the pyramidal tip and the sample, In clusters were extracted from the reservoir tank within the stylus through the aperture, resulting in In nano-dot formation.

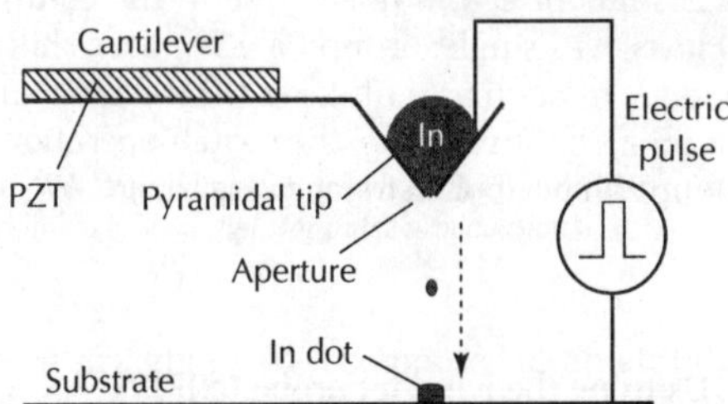

Figure 17.15   Schematic illustration of the NJP and the procedures involved in nano-dot formation.

Figure 17.16 shows the fabrication process of the cantilever developed in this study. We used a commercially available batch-fabricated silicon cantilever with a hollow pyramidal stylus. The developed cantilever was embedded with a lead zirconate titanate (PZT) piezoelectric thin film on its beam, sensing the force during AFM observations as well as actuating. The cantilever for nano-dot formation was prepared as follows. First, a submicron-sized aperture was formed on the apex of the stylus using a focused ion beam (FIB) system (SIM 9200: Seiko Instruments Inc.) with a Ga^+ ion beam operating at 30 kV. The beam diameter was approximately 50 nm at 300 pA. The typical diameter of the aperture was approximately 500 nm and this diameter can be reduced to a few tens of nanometres by focusing the ion beam on the irradiated region. The measured mechanical resonance frequency of the cantilever is 80–100 kHz and the calculated spring constant is approximately 150 N/m. Next, In was deposited into the hollow stylus from the opposite side by an evaporator operated in high vacuum. The average thickness of the deposited In layer was of the order of a few microns. The amount of charged In is sufficient to form at least one million In nano-dots.

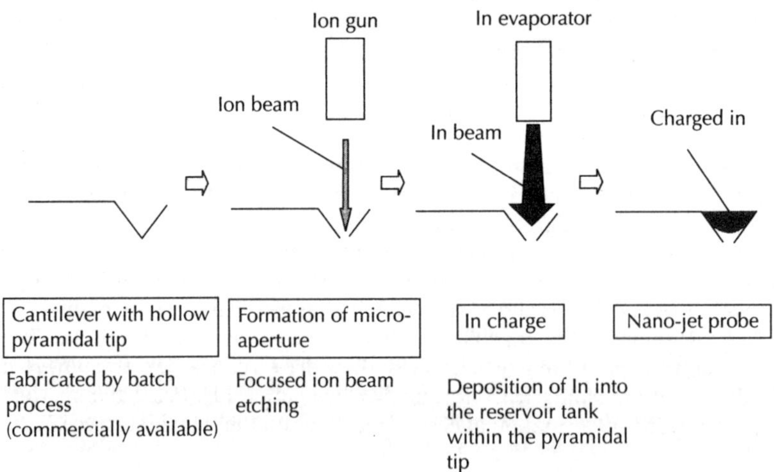

Figure 17.16   Schematic illustration of the fabrication processes of the NJP.

Figure 17.17 shows the scanning electron microscope (SEM) images of the fabricated cantilever and a long-distance optical microscope image of the NJP, which is located above the sample surface. The length and width of the cantilever beam are approximately 500 µm and 150 µm, respectively. The pyramidal tip integrated at the end of the beam has a base area of approximately $50\,\mu m^2$ and is approximately $30\,\mu m$ in height; a submicron-sized aperture is situated at the apex of the stylus. In the SEM image of the close-up view, as shown in Fig. 17.17, a submicron-sized aperture can be observed at the apex of the stylus.

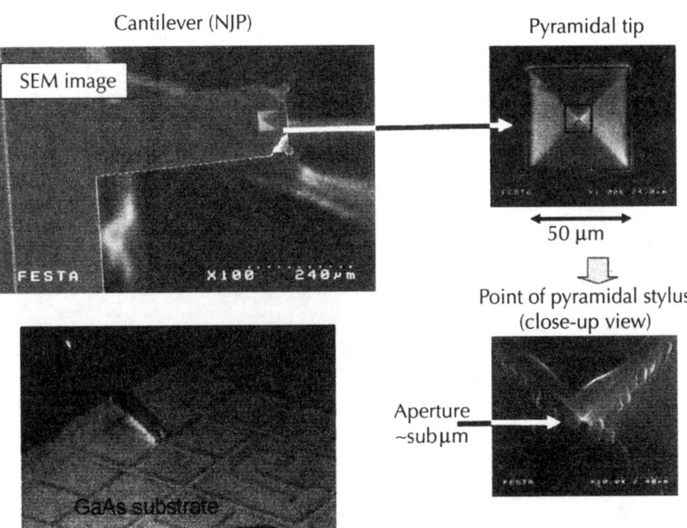

Figure 17.17   Scanning electron microscope images and long-distance optical microscope image of the NJP. (a) The entire image of the cantilever. (b) A high magnification image of the pyramidal tip. (c) A close-up view of the apex of the pyramidal stylus (a submicron-sized aperture can be observed at the apex in the image). (d) A long-distance microscope image of the NJP, which is located above the patterned GaAs substrate.

The experiments of In nano-dot formation were performed with a conventional UHV-AFM (made by Unisoku Co. Ltd). Considering the application for mass production use which requires the deposition of In dots on a large area, our AFM system has a linear motion scanning stage with a range of $100\,\mu m \times 100\,\mu m$ and a capacitive feedback sensor for high accuracy and repeatability in the sub-nanometre range. This stage enables the precise positioning of the nano-dot formation as well as large-area scanning in the AFM observation after the deposition of nano-dots. Further, a pulsed voltage of either polarity under $150\,V$ for $0.1\,ms$–$1000\,ms$ can be applied to the stylus at any position on the sample within the large scanning range by employing computer control.

Our apparatus has a specially designed tip-repositioning function. This function enables us to repeatedly observe AFM images of identical nanostructures with ease, even after demounting the sample holder from the AFM unit and remounting it. For this purpose, a long-distance optical microscope was used for monitoring the approach of the probe on the sample surface through a viewport of the AFM chamber. Additionally, mesa-array-patterned substrates (typically $400\,\mu m^2$ mesa, as shown in Fig. 17.17), which were prepared by the conventional photolithography technique, were used for the coarse positioning of the AFM probe. Since the optical microscope has a high resolution, it allowed the probe to be accurately positioned at the centre of the specified mesa top surface. Subsequently, the large scanning area of the piezoelectric scanner ($100\,\mu m^2$) enabled us to easily identify the target nanostructures such as the QDs, which had been formed around the centre of the mesa. Using this function, a step-by-step evaluation of SC QDs is available at each point during the MBE growth, AFM observation, and In nano-dot fabrication.

17.3.3   Fabrication of site-controlled QDs

17.3.3.1   In nano-dot formation

Figure 17.18 shows a series of the illustrations for the experimental procedures involved. The first step is the In nano-dot deposition and the second is the crystallization process of the In nano-dots into the InAs QDs using the droplet epitaxy technique. In order to deposit nano-dots, a single voltage pulse was applied between the tip and the GaAs substrate using an AFM operated in the non-contact mode at room temperature. Here, the magnitude of the pulsed voltage was defined as the value applied to the substrate (with the tip grounded). The value of the applied pulsed voltage for the nano-dot formation varied in the range of 70–140 V, although it was somewhat dependent on the cantilever. During the deposition, the feedback loop was turned off and no additional bias voltage was applied between the tip and the sample. After the deposition, the AFM images were observed using the same tip that was used for the deposition of the nano-dots.

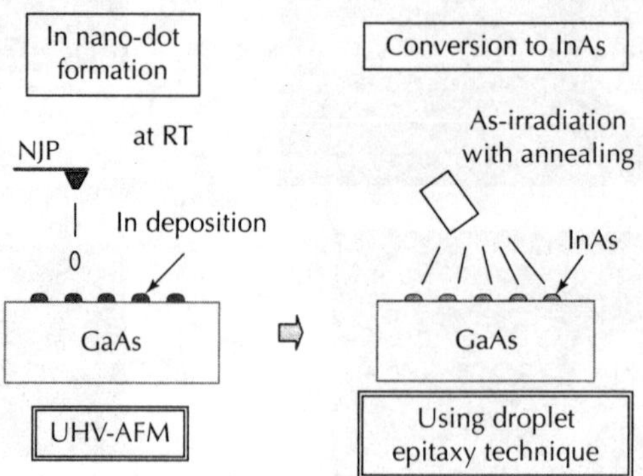

Figure 17.18   Schematic illustrations of the site-controlled formation processes for InAs QD structures using the NJP method.

Figure 17.19 shows the AFM image of an In nano-dot pattern deposited on an Si surface for a preliminary study. Each dot was formed by applying an electric pulse of 75 V for 7 ms between the tip and the sample. The typical diameter of the formed dots was 30–40 nm in diameter. Although the dots are rough in pitch and disordered, an intentionally patterned word "FESTA" can be recognized in the AFM image. The deformation in the pattern was caused due to the positioning of each dot by hand, and it can be easily improved with regard to the order and pitch by moving the cantilever using computer control.

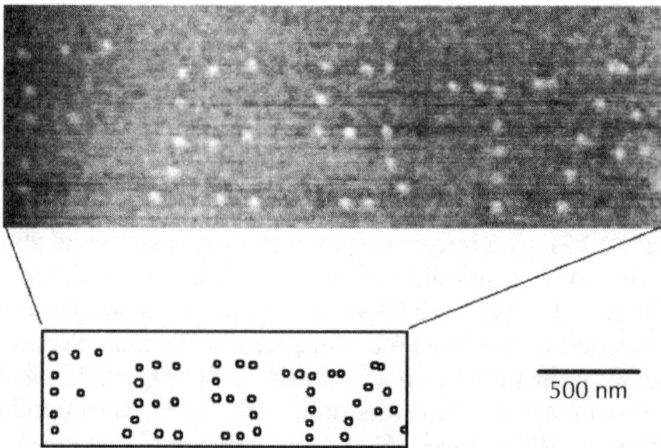

Figure 17.19   AFM image of an In nano-dot pattern ("FESTA") deposited on an Si substrate. The deformed pattern of the "FESTA" with a rather large pitch (~100 nm) is caused by hand patterning for a preliminary study.

The success of the In nano-dot formation depends on parameters such as the voltage, the width and shape of the applied electric pulse, and the distance between the tip and the sample during the deposition. Further, the size of the In nano-dots increases with the applied voltage pulse.

There is a likelihood that In nano-dot formation has a threshold for the applied pulsed voltage. The threshold value for forming the nano-dots, which is somewhat dependent on the cantilever, usually varied between 70 and 120 V. This threshold value is considerably larger when compared to the results reported previously by other researchers [23, 24].

Figure 17.20 shows one possible mechanism of the In nano-dot formation. The top of the In mount is actually positioned approximately 1 μm inside the aperture since the thickness of the walls of the pyramidal tip is around 1 μm. Additionally, the distance between the top of the pyramidal tip and the sample is around 10 nm since the AFM is operated in the non-contact mode. After moving the tip closer to the sample, we applied a pulsed voltage between the tip and the sample. In the early stage of deposition, local melting of In due to field evaporation caused by high applied voltage [25] may occur on the top of the In mount (Fig. 17.20a). A small protrusion of In may then be formed in the next stage of deposition (Fig. 17.20b). In the subsequent stage of deposition, this protrusion is elongated by the strong electric field, and the melted In forms a so-called Taylor cone due to the equilibrium between the electrical and the surface tension forces [26]. Finally, an In droplet is separated from the bridge and an In nano-dot is formed on the substrate. Although the details of the mechanism involved in the In nano-dot formation are unclear, the mechanism of the In nano-dot formation is thought to be similar to that of an electric droplet ejection in inkjet patterning [27].

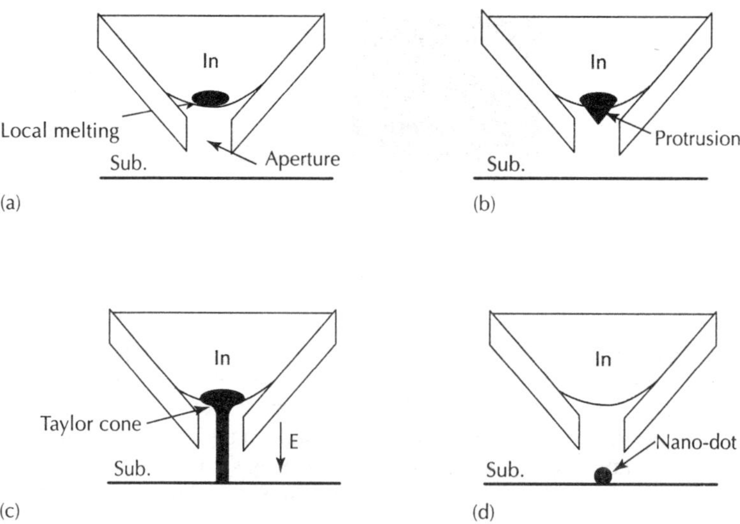

Figure 17.20 Schematic illustrations of the possible mechanism of nano-dot formation.

The size of the In nano-dots does not depend on the aperture size, that is, the In nano-dots are considerably smaller when compared to the aperture, the diameter of which is typically 30–50 nm. In nano-dots of nearly constant size were formed for aperture sizes of 1 μm, 500 nm and 200 nm. These results can be explained by the limitations on the extent of the melting of In on the top of the In mount since the melting area is determined by the applied voltage rather than the aperture size.

Figure 17.21a shows the AFM image of a fabricated 2D In nano-dot array with periods of 1 μm deposited on a GaAs substrate. Each nano-dot was generated by the application of a single voltage pulse of 120 V for 10 ms. In the AFM image, each observed nano-dot does not possess a clear shape. This is likely to be a tip-induced artifact. In the case of the NJP, the point of the top apex of the pyramidal tip was removed in order to form the micro-aperture. As a result, the blunt tip sometimes causes the artifact observed in the AFM image. However, in certain cases, the resolution of the AFM image improves despite the use of the same tip. Figure 17.21b shows an AFM image of the same sample at higher magnification. In this image, a clear nano-dot shape is observed. This image is considered to be produced by a coincidental protrusion formed around the apex of the tip during scanning. These results suggest that the AFM image observed using the NJP does not always reflect the real shape of the nano-dots. Therefore, STM or AFM observations employing a conventional tip should be used for the precise characterization of the nano-dots. Nevertheless, the NJP is sufficiently capable of recognizing whether or not the nano-dot has been successfully formed.

Figure 17.21   AFM images of an In nano-dot array deposited on a GaAs substrate. The spacing between each nano-dot is 1 μm. Image (b) shows a higher magnification of the upper-left region of (a).

Figure 17.22 shows the AFM images of 2D In nano-dot arrays with periods of 100 nm and 50 nm deposited on GaAs substrates. Each nano-dot was formed on an MBE-grown GaAs surface by applying a single voltage pulse of 140 V for 10 ms. In Fig. 17.22a, GaAs surface steps produced during MBE growth were observed on the sample in addition to In nano-dots. The surface density of the In nano-dots shown in Fig. 17.22b correspond to 4×10^{10} cm^{-2}, which satisfies the requirements of the device fabrication from the viewpoint of the QD surface density. Although, in this case, it took approximately 2 min to form 20×20 In nano-dots (corresponding to approximately 3 dots/s), a throughput of 20 dots/s was achieved by optimizing the fabrication conditions. This throughput (corresponding to 10 000 In nano-dots formed in approximately 10 min) is considered to be sufficient for use in mass production.

In the AFM images shown in Figs. 17.21 and 17.22, there seems to be some non-uniformity in the dots. However, the precise determination of this fluctuation is difficult at present because of the poor resolution of the NJP for inspection, as mentioned in the previous section. The In nano-dot array formation was demonstrated for a preliminary study. Therefore, fluctuations in the position, size, and shape of the In nano-dot arrays can be decreased by further optimization of the conditions for nano-dot formation.

17.3.3.2   InAs conversion process

The ordered In nano-dots are directly converted into InAs QD arrays by subsequent irradiation with arsenic flux in the MBE chamber which is connected to the AFM chamber through a UHV tunnel. The droplet epitaxy technique makes it possible to crystallize In nano-dots into InAs dots. This technique was proposed by Chikyow and Koguchi, who deposited Ga droplets on an AlGaAs surface and then converted them to GaAs dots by annealing with As flux [28]. We applied

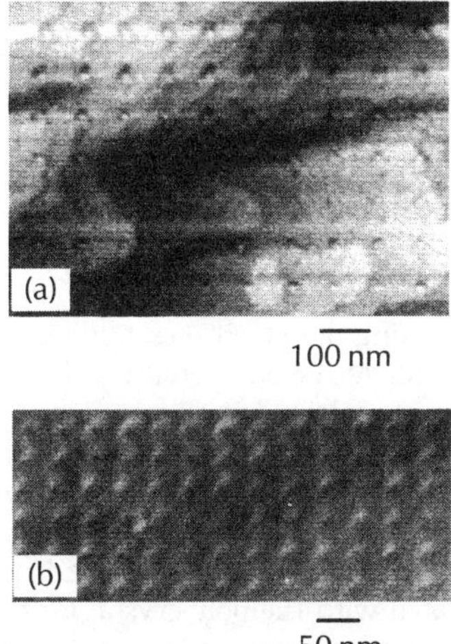

Figure 17.22   AFM images of high-density In nano-dot arrays deposited on an MBE-grown GaAs surface. The spacing between the In nano-dots is (a) 100 nm and (b) 50 nm. The surface density of the In nano-dots for each sample corresponds to $1 \times 10^{10}\,cm^{-2}$ and $4 \times 10^{10}\,cm^{-2}$, respectively.

this technique to the process of converting In nano-dots into InAs dots. The crystallization of In nano-dots into InAs dots can be attributed to a vapour–liquid–solid (VLS) mechanism [29, 30]. The annealing process with As flux incorporates As atoms into melted In, resulting in the crystallization of InAs. Figure 17.23 shows schematic illustrations of the conversion mechanism of In droplets into InAs dots.

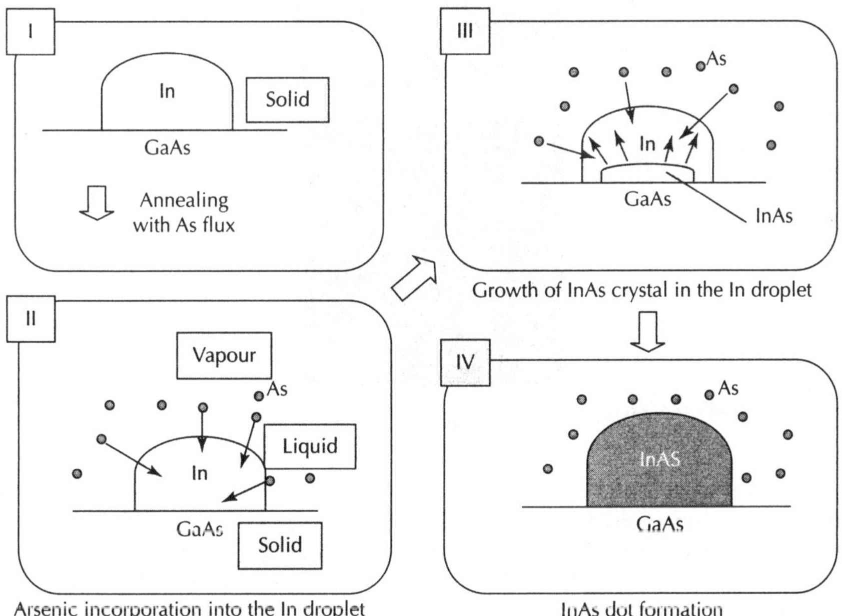

Figure 17.23   Schematic illustrations of the conversion mechanism of In droplets into InAs dots.

Before applying this technique to the process of converting In nano-dots into InAs dots, the crystallization process was investigated by using In droplets, which were formed by the conventional MBE technique. The sample was prepared by the deposition of 3 ML of In at 200°C on a GaAs (2 × 4) reconstructed surface, which was terminated by the deposition of 0.75 ML of Ga atoms. An AFM image of the sample prepared in this manner revealed the In droplet formation (Fig. 17.24a). The annealing condition was optimized using the In droplets formed by this procedure. The temperature of the sample was gradually increased (10°C/min) up to 400°C under an arsenic flux pressure of 5×10^{-5} Torr, after which the sample was annealed for 20 min. Figure 17.25 shows the typical RHEED patterns along the [110] direction during the crystallization process of the In droplets. Arsenic flux was irradiated before increasing the substrate temperature. A diffused spotty pattern was observed before the annealing started (Fig. 17.25a). This means that a part of the In droplets was crystallized at room temperature (below the melting point). This diffused spotty pattern gradually changed to a clear spotty one as the substrate temperature increased. Additionally, a clear ring pattern appeared when the substrate temperature increased above 200°C (Fig. 17.25c), indicating poly-crystal formation on the sample. This means that the crystallization process of In droplets is incorporated with the poly-crystal formation. Spotty and ring patterns continued to coexist until the substrate temperature reached 400°C. Finally, when the substrate temperature exceeded 400°C, the ring pattern disappeared and the RHEED patterns showed only clear spotty patterns with extended streaks, indicating that the crystallization process was completed. In addition, facets were formed on the converted InAs dots (Fig. 17.25f). The AFM image of the converted In droplets is shown in Fig. 17.24b. The crystallization of In nano-dots formed by the NJP was performed under the same conditions as those for the optimized one investigated by using In droplets.

Figure 17.24   AFM images of In droplets (a) before and (b) after the crystallization process.

The actual conversion process was performed as follows. After the In nano-dots were deposited, the sample was transferred to the MBE chamber to crystallize them. Figure 17.26 shows the time sequence for the crystallization process. The temperature of the sample was gradually raised to 420°C under an arsenic flux of 5×10^{-5} torr and the sample was annealed for 40 min. Subsequently, it was transferred to the AFM chamber and its surface structure was observed by AFM. In order to identify the region where the In nano-dots were deposited, the tip-repositioning system, which was mentioned in the previous section, was used. The AFM image revealed the

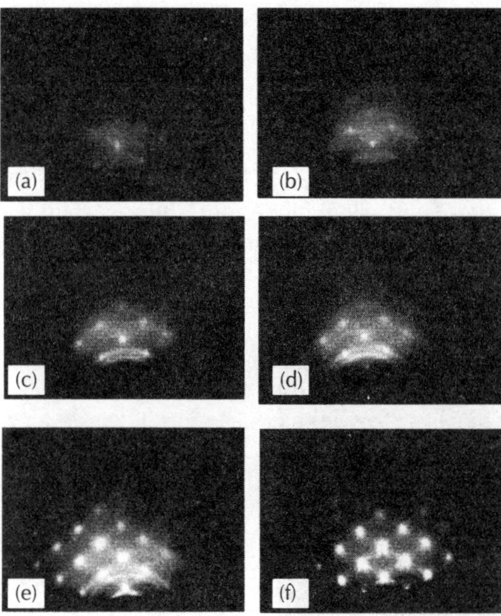

Figure 17.25   RHEED images during the InAs conversion process using In droplets formed by the conventional method. The substrate temperatures are (a) room temperature before annealing, (b) 85°C, (c) 210°C, (d) 265°C, (e) 360°C, and (f) 400°C.

formation of InAs nano-dot arrays at the same position where the In nano-dots had been deposited. Figure 17.27 shows the AFM images and schematic illustrations of the nano-dots before and after the conversion. Before the conversion, the In nano-dots appeared cone shaped, as shown in Fig. 17.27a. After the conversion, however, they assumed anisotropic shapes, as shown in Fig. 17.27b. To be precise, the nano-dots were elongated along the [$\bar{1}$10] direction after the conversion. Additionally, high-index facets were formed on the side walls of the converted nano-dots (Fig. 17.27c). These features of the converted nano-dots correspond to those of the InAs self-assembled QDs. Therefore, based on these observed changes in the features of the nano-dots, we concluded that the In nano-dots had been converted into InAs dots.

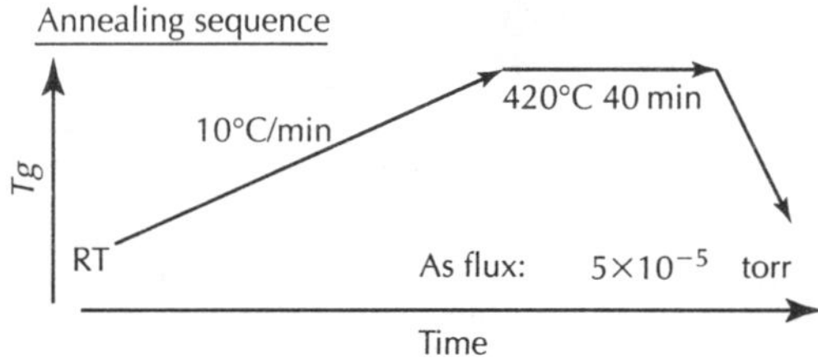

Figure 17.26   Time sequence for the crystallization process.

17.3.3.3   Stacking of ordered InAs QDs

As the next step, we stacked InAs QDs using the converted InAs dots as the template. Figure 17.28 shows schematic illustrations of the experimental procedures. First, by using the NJP, we fabricated 2D In nano-dot arrays (Fig. 17.28a). Then, in order to convert these ordered In nano-dot arrays to InAs QDs, the sample was transferred to the MBE chamber through the UHV tunnel and was subsequently annealed by irradiation with arsenic flux (Fig. 17.28b).

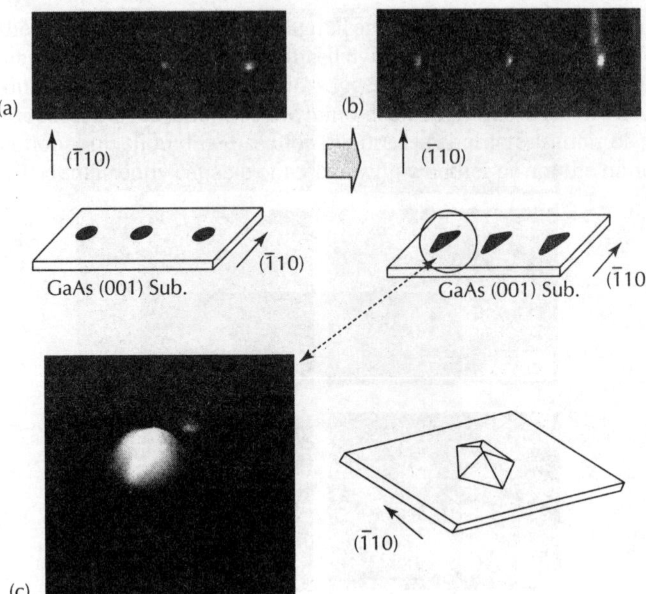

Figure 17.27   AFM images of the In nano-dots (a) before and (b) after the crystallization process. Image (c) is a 3D close-up view of the converted InAs dot.

Subsequently, self-assembled InAs QDs were grown on the converted InAs 2D arrays. We grew a 10 nm thick GaAs layer and deposited 1.4 ML of InAs on the converted InAs QD arrays; the surfaces were then observed by AFM (Fig. 17.28c). Figure 17.29 shows a typical AFM image of a 2D In nano-dot array fabricated by the NJP method. The typical size of the In nano-dots is approximately 40 nm and their height was approximately 3–4 nm. In nano-dots were deposited in the area of approximately 4 μm × 4 μm with periods of 200 nm, as shown in the inset of Fig 17.29.

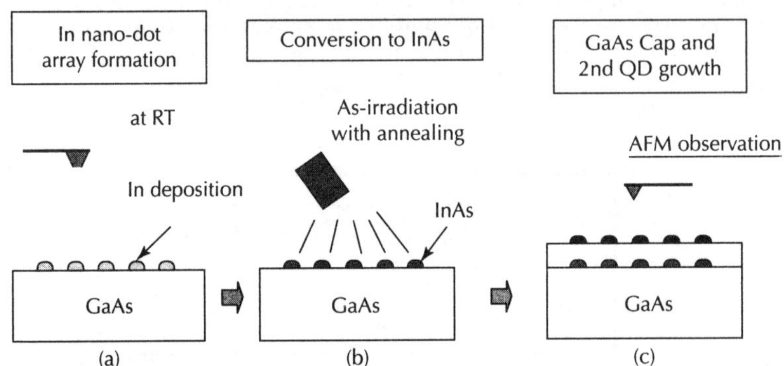

Figure 17.28   Fabrication process of stacked InAs QDs using In(As) dot arrays formed by the NJP method as templates.

Figure 17.30 shows an AFM image of the InAs-deposited GaAs surface, which is stacked on the converted InAs dots. The AFM image revealed the selective formation of high-density InAs dots only in the square region where the In nano-dots were formed by the NJP method. Despite taking into account the deformation effect due to the blunt shape of the NJP, the size of the InAs dots thus formed is large compared to the InAs QDs grown by the conventional SK mode. The increase in the density of the InAs dots and their large size were probably due to the thin GaAs covered layer and a large amount of InAs deposition, that is, a QD on the first layer generates a

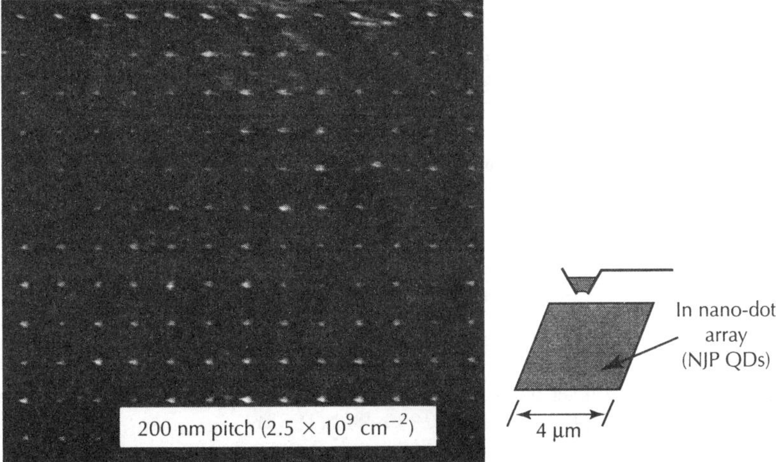

Figure 17.29   AFM image of a fabricated In nano-dot array.

large potential strain field on the GaAs covered layer and creates a few nucleation sites. This configuration can be easily changed by varying the thickness of the covered layer and the amount of InAs for the second layer QDs; in this way, the density of the second layer QDs can be controlled. These results indicate that selective growth of InAs QDs can be achieved by using the In nano-dots produced by the NJP method.

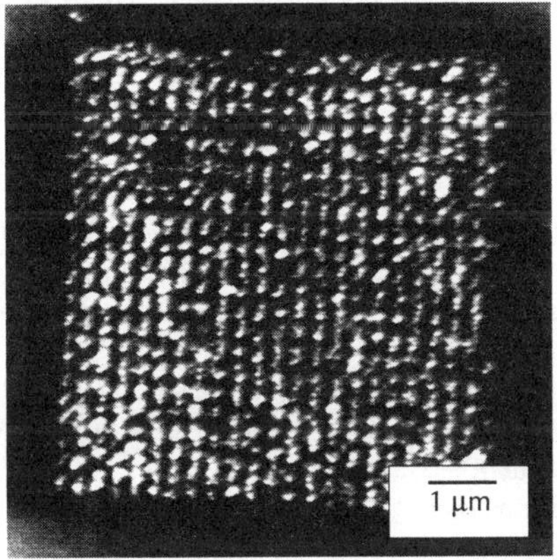

Figure 17.30   AFM image of the stacked high-density QD array.

17.3.4   Summary

We have demonstrated the use of a new nanoprobe-assisted technology that forms InAs QDs on a GaAs substrate by a microfabrication technique (NJP) employing a specially designed cantilever. Using the NJP, uniform In nano-dot arrays were formed in a reproducible manner on an MBE-grown surface by applying electric pulses between the tip and the sample. These In nano-dot arrays were converted directly into InAs QD arrays by subsequent annealing with irradiation of arsenic flux. Furthermore, we have demonstrated the selective area growth of high-density InAs QDs by using site-controlled InAs QDs that were formed in the desired regions as templates.

The NJP method enables the formation of the required number of InAs QDs in a desired region. In other words, the NJP method offers perfect selectivity in forming InAs QDs. An example of the integration of QDs into photonic devices is a photonic crystal-based ultra-fast all-optical switch which we are currently developing [5]. These devices use QDs as a non-linear optical material to switch the optical signal pulses. For these devices, QDs should be formed only in the photonic crystal straight waveguide area. The NJP method will enable the formation of the required number of QDs in a desired region with both high uniformity and high density. Furthermore, we can achieve the required volume of QDs for optical switching by combining this site-controlled QD formation technique with the subsequent stacking of spatially ordered InAs QD arrays using MBE growth [31]. In addition to our proposed PC-based devices [5], the NJP method has applications to other optical switching devices, including future high-performance functional devices such as regular arrays of quantum bits and single-photon emitters for quantum computers and quantum communications [10, 11].

References

1. Y. Arakawa and H. Sakaki, Multidimensional quantum well laser and temperature dependence of its threshold current, *Appl. Phys. Lett.* **40**, 939–941 (1982).
2. M. Sugawara, N. Hatori, T. Akiyama, Y. Nakata, and G. Ishikawa, Quantum-dot semiconductor optical amplifiers for high bit-rate signal processing over 40 Gbit/s, *Jpn. J. Appl. Phys.* **40**, L488–L491 (2001).
3. S. Kohmoto, H. Nakamura, T. Ishikawa, and K. Asakawa, Site-controlled self-organization of individual InAs quantum dots by scanning tunneling probe-assisted nano-lithograph, *Appl. Phys. Lett.* **75**, 3488–3490 (1999).
4. S. Ohkouchi, Y. Nakamura, H. Nakamura, and K. Asakawa, InAs nano-dot array formation using nano-jet probe for photonics applications, *Jpn. J. Appl. Phys.* **44**, 5777–5780 (2005).
5. K. Asakawa, Y. Sugimoto, Y. Watanabe, N. Ozaki, A. Mizutani, Y. Takata, Y. Kitagawa, H. Ishikawa, N. Ikeda, K. Awazu, X. Wang, A. Watanabe, S. Nakamura, S. Ohkouchi, K. Inoue, M. Kristensen, O. Sigmund, P.I. Borel, and R. Baets, Photonic crystal and quantum dot technologies for all-optical switch and logic device, *New J. Phys.* **8**, 208 (2006).
6. K. Tajima, All-optical switch with switch-off time unrestricted by carrier lifetime, *Jpn. J. Appl. Phys.* **32**, L1746–L1749 (1993).
7. H. Nakamura, Y. Sugimoto, K. Kanamoto, N. Ikeda, Y. Tanaka, Y. Nakamura, S. Ohkouchi, Y. Watanabe, K. Inoue, H. Ishikawa, and K. Asakawa, Ultra-fast photonic crystal/quantum dot all-optical switch for future photonic network, *Opt. Express* **12**, 6606–6614 (2004).
8. N. Ozaki, Y. Takata, S. Ohkouchi, Y. Sugimoto, Y. Nakamura, N. Ikeda, and K. Asakawa, Selective area growth of InAs quantum dots with a metal mask towards optical integrated circuit devices, *J. Cryst. Growth* **301–302**, 771–775 (2007).
9. T. Yoshie, A. Scherer, J. Hendrickson, G. Khitrova, H.M. Gibbs, G. Rupper, C. Ell, O.B. Schchekin, and D.G. Deppe, Vacuum Rabi splitting with a single quantum dot in a photonic crystal nanocavity, *Nature* **432**, 200–203 (2004).
10. E. Waks, K. Inoue, C. Santori, D. Fattal, J. Vuckovic, G.S. Solomon, and Y. Yamamoto, Secure communication: quantum cryptography with a photon turnstile, *Nature* **420**, 762 (2002).
11. A. Imamoglu, D.D. Awschalom, G. Burkard, D.P. DiVincenzo, D. Loss, M. Sherwin, and A. Small, Quantum information processing using quantum dot spins and cavity QED, *Phys. Rev. Lett.* **83**, 4204–4207 (1999).
12. S. Ohkouchi, Y. Nakamura, H. Nakamura, and K. Asakawa, Indium nano-dot arrays formed by field-induced deposition with a nano-jet probe for site-controlled InAs/GaAs quantum dots, *Thin Solid Films* **464–465**, 233–236 (2004).
13. Y. Nakamura, N. Ikeda, S. Ohkouchi, Y. Sugimoto, H. Nakamura, and K. Asakawa, Regular array of InGaAs quantum dots with 100-nm-periodicity formed on patterned GaAs substrates, *Physica E* **21**, 551–554 (2004).
14. Y. Nakamura, N. Ikeda, S. Ohkouchi, Y. Sugimoto, H. Nakamura, and K. Asakawa, Two-dimensional InGaAs quantum-dot arrays with periods of 70–100 nm on artificially prepared nanoholes, *Jpn. J. Appl. Phys.* **43**, L362–L364 (2004).
15. K. Nishi, H. Saito, S. Sugou, and J.S. Lee, A narrow photoluminescence linewidth of 21 meV at 1.35 μm from strain-reduced InAs quantum dots covered by $In_{0.2}Ga_{0.8}As$ grown on GaAs substrates, *Appl. Phys. Lett.* **74**, 1111–1113 (1999).

16. S. Ohkouchi, Y. Nakamura, H. Nakamura, N. Ikeda, Y. Sugimoto, and K. Asakawa, Selective growth of high quality InAs quantum dots in narrow regions using in situ mask, *J. Cryst. Growth* **293**, 57–61 (2006).

17. Y. Sugimoto, N. Ikeda, N. Carlsson, K. Asakawa, N. Kawai, and K. Inoue, Fabrication and characterization of different types of two-dimensional AlGaAs photonic crystal slabs, *J. Appl. Phys.* **91**, 922–929 (2002).

18. N. Ikeda, H. Kawashima, Y. Sugimoto, T. Hasama, K. Asakawa, and H. Ishikawa, Coupling characteristic of micro planar lens for 2D photonic crystal waveguides, 19th International Conference on InP and related materials (IPRM) (2007).

19. N. Ozaki, Y. Kitagawa, Y. Takata, N. Ikeda, Y. Watanabe, A. Mizutani, Y. Sugimoto, and K. Asakawa, High transmission recovery of slow light in a photonic crystal waveguide using a hetero group-velocity waveguide, *Opt. Exp.* **15**, 7974–7983 (2007).

20. C. Lee and A.L. Barabasi, Spatial ordering of islands grown on patterned surfaces, *Appl. Phys. Lett.* **73**, 2651–2653 (1998).

21. S. Ohkouchi, Y. Nakamura, H. Nakamura, and K. Asakawa, Nano-probe-assisted technology of indium-nano-dot formation for site-controlled InAs/GaAs quantum dots, *Physica E* **21**, 597–600 (2004).

22. S. Ohkouchi, Y. Sugimoto, N. Ozaki, H. Ishikawa, and K. Asakawa, Selective growth of InAs quantum dots using In nano-dot arrays formed by nano-jet probe method, *J. Cryst. Growth* **301–302**, 726–730 (2007).

23. H.J. Mamin, P.H. Guethner, and D. Ruger, Atomic emission from a gold scanning-tunneling-microscope tip, *Phys. Rev. Lett.* **65**, 2418–2421 (1990).

24. N. Aoki, K. Fukuhara, T. Kikutani, A. Oki, H. Hori, and S. Yamada, Fabrication of buried metal dot structure in split-gate wire by scanning tunneling microscope, *Jpn. J. Appl. Phys.* **35**, 3738–3742 (1996).

25. K. Higa and T. Asano, Joule heating of field emitter tip fabricated on glass substrate, *Jpn. J. Appl. Phys.* **43**, 2749–2750 (2004).

26. G. Taylor, Electrically driven jets, *Proc. Roy. Soc. Lond. A* **313**, 453–475 (1969).

27. R.P.A. Hartman, D.J. Brunner, D.M.A. Camelot, J.C.M. Marijnissen, and B. Scarlett, Electrohydrodynamic atomization in the cone-jet mode physical modeling of the liquid cone and jet, *J. Aerosol Sci.* **30**, 823–849 (1999).

28. T. Chikyow and N. Koguchi, MBE growth method for pyramid-shaped GaAs micro crystals on ZnSe(001) surface using Ga droplets, *Jpn. J. Appl. Phys.* **29**, L2093–L2095 (1990).

29. R.S. Wagner and W.C. Ellis, Vapor-liquid-solid mechanism of single crystal growth, *Appl. Phys. Lett.* **4**, 89–90 (1964).

30. N. Koguchi, K. Ishige, and S. Takahashi, New selective molecular-beam epitaxial growth method for direct formation of GaAs quantum dots, *J. Vac. Sci & Technol. B* **11**, 787–790 (1993).

31. Y. Nakamura, N. Ikeda, S. Ohkouchi, Y. Sugimoto, H. Nakamura, and K. Asakawa, Two-dimensional InGaAs quantum-dot arrays with periods of 70–100 nm on artificially prepared nanoholes, *Jpn. J. Appl. Phys.* **43**, L362–L364 (2004).

CHAPTER 18

Nanostructured Solar Cells

S.G. Bailey,[1] Seth Hubbard,[2] and R.P. Raffaelle[2]

*[1]Space Environment and Experiments Branch, NASA Glenn Research Center,
MS 302-1, Cleveland, OH 44135;
[2]Rochester Institute of Technology, Rochester, NY 14623*

18.1 Introduction

There has been considerable investigation regarding the potential for the use of nanomaterials and nanostructures to increase the efficiency of photovoltaic devices [1]. Examples of such investigation include the use of materials and/or combinations of materials that provide either one-dimensional, two-dimensional, or even three-dimensional quantum confinement (i.e. quantum wells [2]; quantum wires [3]; or quantum dots [4,5], respectively). Quantum confinement refers to the fact that a dimension or dimensions associated with a given material are smaller than the Bohr exciton radius for that material. This results in the possible electronic states associated within the material to deviate from the ordinary band theory of solids and begin to show discrete-like optoelectronic behaviour more familiar with atoms or molecules. The term nanomaterial would generally be reserved for isolated materials that have at least some degree of quantum confinement (e.g. colloidal suspensions of semiconducting nanocrystals, fullerenes, single- or multi-wall carbon nanotubes). A nanostructure would commonly refer to a composite material in which the boundaries of one material by another results in the dimensions of the first material to be on the nanoscale. A prime example of this would be an extremely thin (i.e. a few nanometres) epitaxially grown layer of a narrow band semiconductor within a wider band gap host material. This resulting "nanostructured" material would be a quantum well if the thickness of this narrow band gap layer is of a sufficiently small dimension, as mentioned above.

Nanostructures can also be comprised of arrays of individual nanomaterials. Semiconducting quantum dots (QDs) can be combined in a three-dimensional array, often through the use of self-ordering. The discrete-like energy levels of the quantum dots will combine and form bands of allowed energy states in an analogous way in which atomic energy levels combine to produce the energy bands in conventional solids. In this case, it is easy to see why quantum dots are sometimes referred to as "super atoms". By changing the quantum dot size and the spacing between the quantum dots, the optoelectronic bandstructure will change in ways similar to the way a semiconductor bandstructure changes with compositional doping. The shift in the effective band gap will vary inversely with the particle size of the dot [6].

Quantum wells or quantum wires can also be arrayed. The use of many closely spaced quantum wells, or a multiple quantum well, is a very well-known structure within the optoelectronic industry. This has been a prime means of "band gap engineering" semiconducting materials. By changing the quantum well width, spacing of the wells (i.e. the thickness of the barriers between the wells), and the composition well and/or barrier material, one can effectively "tune" the band gap to values that cannot be achieved using merely compositional changes alone.

The role of a nanomaterial or nanostructure in a given photovoltaic solar cell design can vary dramatically. In some cases, the goal may be simply to provide the means to disassociate excitons throughout a bulk material as in the use of colloidal quantum dots in organic or polymeric solar cells. In other cases, the use of a nanostructure is motivated by the desire to band gap engineer a device in a similar approach to what is done more conventionally through compositional changes in a device, albeit in regimes where compositional solutions may not be readily available. A more ambitious use of nanostructures for photovoltaics lies in the desire to exploit underlying quantum confinement concepts to fundamentally change the way we can exploit photoconversion for electrical energy production. Examples of quantum mechanical phenomena offered by nanostructures that could be exploited to exceed the standard Shockley–Quiesser limit for a conventional solar cell [7] (i.e. the maximum thermodynamic efficiency for the conversion by a single band gap cell of unconcentrated solar irradiance into electrical free energy of approximately 31%) are hot carrier effects, multiple exciton generation, phonon bottlenecking, and up- and down-conversion. Hot-carrier solar cells [8] attempt to use nanostructures to obviate the normal thermalization processes associated with conventional solar cells, or in other words, extract the carriers before they relax down to the conduction band edge of the host semiconductor. Solar cells attempting to exploit the multiple exciton generation processes demonstrated by colloidal by quantum dots hope to produce multiple electron–hole pairs per photon through this inverse Auger impact ionization [9]. Finally, the use of quantum dot arrays may allow approaches that expand the spectral response of conventional solar cells by shifting incoming photon energy from inaccessible regions or regions of the spectrum that are inefficiently converted to more favourable energy ranges, and potentially provide the means for multiband intermediate band cells [10]. This chapter discusses a few of the recent approaches to produce nanostructured solar cells.

18.2 Quantum dot solar cells

The term quantum dot solar cells is somewhat ambiguous as it is colloquially used to refer to two general types of solar cells in which both the role of the quantum dots, the materials and synthesis methods used to create them, and the types of cells in which they are used, are drastically different. These two classes of cells are those which use colloidal quantum dots in a polymer matrix and their crystalline counterparts which utilize epitaxially grown, normally Stranski–Krastanov (SK) mode, quantum dots (QDs). The motivation for the former class is due to the rapid increase in device efficiencies in polymer solar cells with the addition of nanomaterials, such as CdSe quantum dots and more recently fullerenes. This work has led to the type of solar cell called a distributed heterojunction solar cell. The latter class of quantum dot solar cells are those involving the epitaxially grown quantum dots, usually comprised of III–V materials in the form of quantum dot arrays (QDAs). Much of the motivation for the research in this area was provided by the theoretical work on intermediate band solar cells (IBSCs). It has been proposed that an IBSC could be achieved via a quantum dot array embedded into a suitable host solar cell structure. It is still somewhat of an open debate whether or not the IBSC band solar cell as proposed in the aforementioned theoretical treatments is theoretically achievable using a QDA.

18.2.1 Intermediated band solar cells

Theoretical results of Luque and Marti have shown that a photovoltaic device with a single intermediate band of states resulting from the introduction of quantum dots offers a potential efficiency of 63.2% [11]. This was extended to two intermediate bands and a limiting efficiency of 71.7% was calculated [9]. The enhanced efficiency results from converting photons of energy less than the band gap of the cell by an intermediate band. The intermediate band provides a mechanism for low-energy photons to excite carriers across the energy gap by a two-photon process. In principle, the use of a QDA to achieve this intermediate band may have some distinct advantages to that of a multiple quantum well (MQW), such as in directionality of the illumination that can result in the ability to optically excite carriers from the intermediate band into the host semiconductor conduction band and in terms of the practical range of tunability offered by a QDA over that of an MQW.

Quantum dots offer the potential to control the intermediate band energies since the individual quantum energy levels associated with isolated quantum dots is a function of their size, separation from one another, and material composition. Placing the appropriate quantum dot material of the necessary size into an organized matrix within an ordinary p-i-n structure solar cell should result in the formation of accessible energy levels within what would normally be the forbidden band of the device. A simple schematic of one such proposed intermediate band solar cell structure is shown in Fig. 18.1.

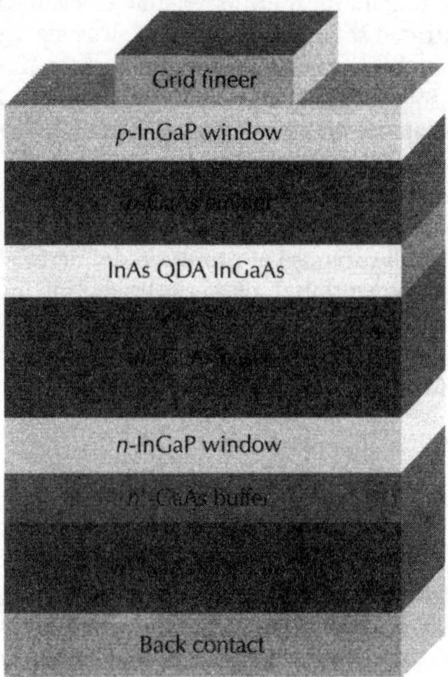

Figure 18.1   Intermediate band cell structure.

The electronic wavefunctions associated with the discrete electronic states of the individual quantum dots in the QDA array will overlap creating "mini-bands" in the host semiconductors' band gap. The composition and related properties (i.e. bulk band gap, electron affinity, effective masses, etc.), size, symmetry, and spacing of the quantum dots need to be chosen to produce mini-bands which are appropriately spaced within the band gap of the host material (see Fig. 18.2). The majority of the work to date on this type of QDA for use in a solar cell has been on the development of InAs QDAs in GaAs. Fortunately, the strain associated with the SK growth of QDs acts to self-align the QD into vertical columns to produce the desired arrays as each additional layer is added in this system. More recently, it has been proposed that QDAs of Si could be used in a similar fashion in an amorphous silicon solar cell [12].

A major challenge in designing solar cells which can take maximum advantage of a QDA is the balance between balancing absorption and transport. A QDA with many layers could improve the subhost band gap absorption at the cost of the efficient extraction photo-generated carriers from the quantum-confined region [13] due to the diminished electric field across the QDA within the intrinsic region of the device and increased probability of recombination and radiative losses. Transport of carriers generated in bulk would presumably also suffer as the QDA is widened. On the other hand, a QDA that is too thin would provide little benefit from photoconversion of the subhost band gap illumination to offset the increased defect density that would be likely to result from its inclusion. Recent modeling of an IBSC based on quantum dot supracrystals was undertaken to engineer optimum array dimensions necessary for maximum solar cell efficiency [14].

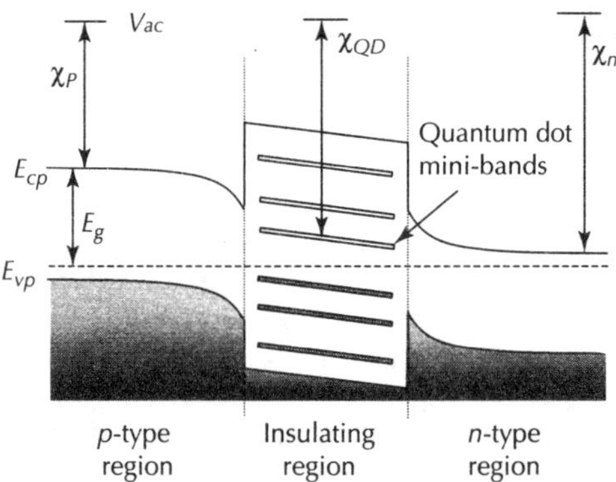

Figure 18.2   Idealized energy band diagram of an intermediate band solar cell.

Using the material system of an InAsN/GaAsSb quantum dot supralattice, it was shown that the power conversion efficiency of the IBSC was maximized by a quantum dot size of approximately 4.7 nm. It was noted that solar cells based on these quantum dots may also show enhanced radiation hardness [15] and improved collection efficiency [16, 17].

Transport through the QDA can be enhanced by reducing the spacing between the QD layers. This, of course, will change the absorption spectrum and reduce the 3D quantum confinement. Taken to the extreme, this type of a cell would approach what would be in a sense a quantum wire solar cell. The closely spaced QDs would form nanoscale wires across the intrinsic region of the device (see Fig. 18.3). In such a device, the presence of a one-dimensional electron gas across the junction region will result in a fast collection of the carriers photo-generated in the quantum wires region, and the solar cell output will thus fully benefit from improved absorption properties associated with the presence of quantum confined states. In practice, arrays of vertical quantum wires can be formed by a self-organization process occurring during the deposition of ultra short period stacks of highly strained/strain balanced films of III–V semiconductors (e.g InAs/AlAs on InP, InGaAs/InGaP on GaAs, etc.). One could also envision potentially producing solar cells with such structures utilizing other materials through electron lithography (wire widths in excess of 20–30 nm).

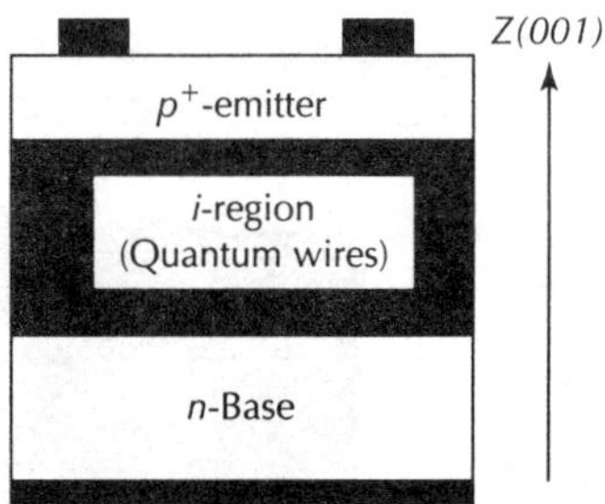

Figure 18.3   Schematic representation of the proposed vertical quantum wire device.

Modeling of a quantum wire structure was undertaken in which the dots were assumed to stack vertically in the cell [18]. For an indium phosphide emitter and base with indium arsinide wires, there was a peak efficiency with quantum wire length. It has been suggested that the quantum wire solar cell, while circumventing carrier collection issues (when compared to quantum dot and quantum well systems), may also provide a significantly larger density of states and consequently enhanced photo-absorption.

18.3 Quantum dot growth

18.3.1 *Stranksi–Kranstanow quantum dots*

The ideal QD behaves much like the idealized box in the famous quantum mechanical of the particle-in-a-box problem. This QD is a structure capable of confining carriers (electrons and holes) in three-dimensions, allowing zero dimensions (0D) in their degrees of freedom, and creating atom-like levels with discrete (Dirac delta) densities of states. This is why quantum dots can also be visualized and described as "artificial atoms". In SK grown QDs, this structure is achieved by having a very small volume of a low band gap material embedded in larger band gap material(s), thus creating the potential barrier for quantum confinement of the carriers. Until recently, such three-dimensional confinement in semiconductor structures existed mostly in theoretical treatments. In addition to SK growth, other approaches to manufacture quantum dots made use of photolithography and etching. While providing good control of the order and offering design flexibility; these types of processing affected the interfaces, introducing impurities and surface damage that limited the optical performance of these quantum structures.

The SK growth mode for III–V QDs takes advantage of a strain-induced transformation that happens naturally in the initial stages of growth for lattice-mismatched materials. The growth usually starts layer by layer, and after a certain critical thickness is reached, the structure spontaneously forms nanometre-size islands (SK mode [19]) that show good size uniformity and large surface densities. In this process, the growth is interrupted [20] immediately after formation of the islands, but before the islands reach a size for which strain relaxation and misfit dislocations occur. This spontaneous island formation during growth precludes the interface quality problems often associated with *ex situ* processed quantum structures of low dimensionality. The mechanisms of SK strain-induced QD growth in molecular beam epitaxy (MBE) are fairly straightforward. However, in the case of organometallic vapour phase epitaxy (OMVPE) to initiate the SK growth mode, growth temperatures must be reduced to well below the values typically associated with mass-transport limits [21]. The result is that QD nucleation, growth, and uniformity becomes much more sensitive to OMVPE growth parameters, such as growth temperature, growth rate and V/III ratio. However, for commercial production of solar cells, the OMVPE has definite advantages in terms of production rate, scalability, and cost to its MBE counterpart.

The growth of SK mode InAs QDs on GaAs by organometallic vapour phase epitaxy (OMVPE) or metal organic chemical vapour deposition (MOCVD) commonly use metallorganics, such as trimethyl gallim (TMGa) and trimethyl indium (TMIn) as precursor materials, along with hydrides phosphine (PH_3), arsine (AsH_3), and 1% AsH_3 in hydrogen. The typical growth temperatures are between 500 and 700°C and a pressure of around 600 torr. An example of how OMVPE growth conditions effect InAs quantum dots on GaAs can be seen in Fig. 18.4.

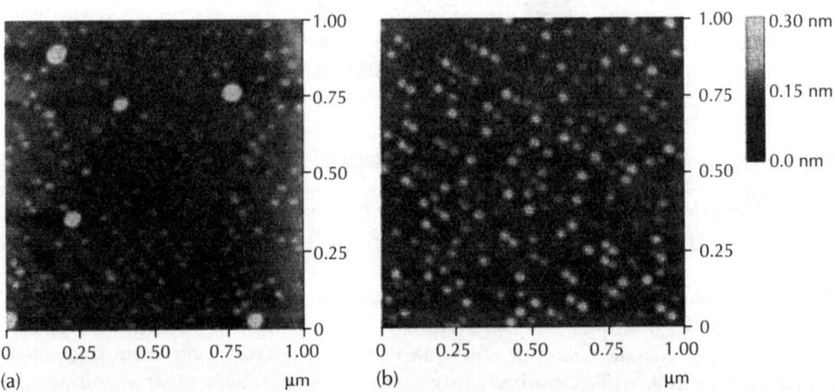

Figure 18.4 AFM micrographs near wafer center for OMVPE-grown SK mode InAs QDs grown on GaAs with V–III ratios of (a) 58 and (b) 12 [20].

There are other examples of quantum dot formation by OMVPE. A recent study of InN dots synthesized on GaN templates has been reported [22]. Using sapphire substrates, a 1.5 micron thick GaN template was grown using a standard two-step process prior to the deposition of InN. The lattice mismatch between InN and GaN is quite large (>10%), causing to proceed in a three-dimensional mode. Indeed, AFM images revealed the formation of discrete InN islands on the GaN surface, as illustrated in Fig. 18.5 and similar to reports from other groups [23, 24]. X-ray diffraction measurements confirmed the deposition of InN in the sample set grown on GaN templates (see Fig. 18.5 inset). The strong peak near 17.3 ° seen in the Fig. 18.5 inset corresponds to diffraction from the GaN lattice in the buffer layer, while the weaker peak near 15.8 ° is in the range expected for epitaxial InN. The intensity of the InN-related peak generally increases with growth time, and both the position and full width-at-half maximum exhibit a slight dependence on the growth conditions.

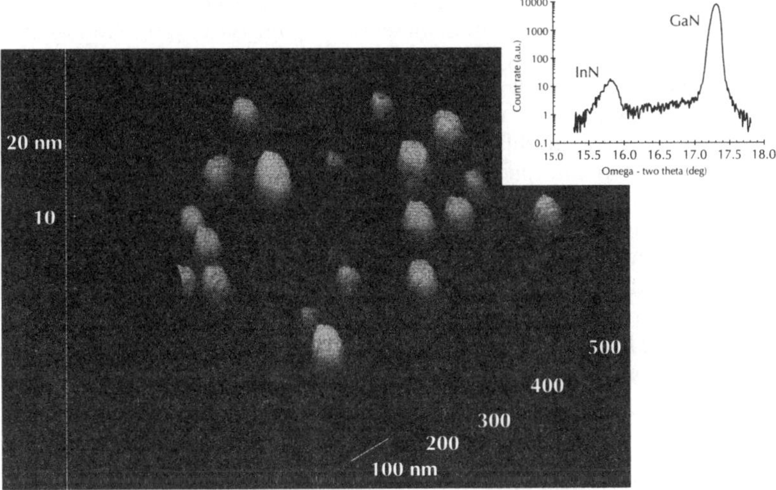

Figure 18.5   AFM image of InN quantum dots grown on a GaN buffer layer. The presence of InN material has been verified by X-ray diffraction spectrum shown on the inset [21].

Figure 18.6 demonstrates the evolution in the height of the InN islands with the growth time under fixed deposition conditions. The InN islands do not form until approximately 15 to 20 seconds of growth time, and they are still discrete after 900 seconds of growth. The delayed

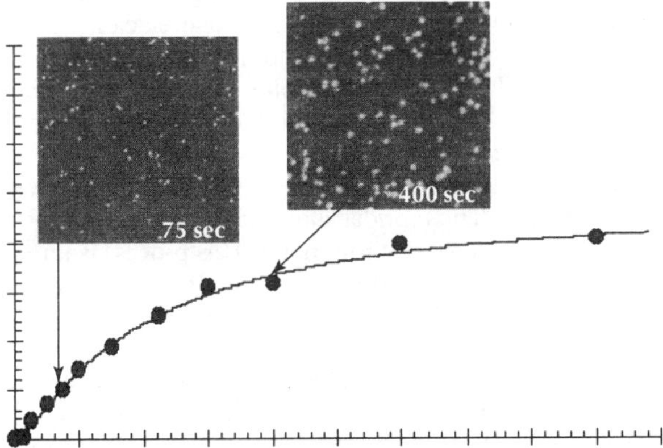

Figure 18.6   Height of the InN islands as a function of growth time for a series of samples deposited at a temperature set point of 720 °C. Insets show the AFM image from 5 mm × 5 mm scans of samples grown for 75 seconds and 400 seconds. The lines through the data points are provided as a guide to the eye [21].

onset of three-dimensional islands is consistent with a Stranski–Krastanov growth mode for the deposition of InN on a GaN template. The height of the InN islands increases rapidly from 7 nm to 50 nm as the growth time increases from 25 to 225 seconds, and then begins to saturate, approaching 80 nm as the growth time increases further from 300 to 900 seconds. At first, the density of the InN islands remains constant near $5 \times 10^8 \mathrm{cm}^{-2}$, and the distribution of island sizes is relatively narrow, as illustrated by the AFM image of the sample grown for 75 seconds (see inset in Fig. 18.5). However, a second group of much smaller islands appears in structures with a growth time of 300 seconds or greater, resulting in a total InN island density approaching $1 \times 10^9 \mathrm{cm}^{-2}$. The bimodal distribution of InN islands is evident in the AFM image of the 400 second sample also seen in the inset in Fig. 18.5. The appearance of the second group of InN islands near 300 seconds in growth time corresponds to the saturation in the height of the initial set of InN islands.

Photoluminescence emissions from both dots and wetting layer are observed in other lattice-mismatched III–V material systems, most notably InAs on GaAs [25]. Non-radiative recombination at surface states is well known to degrade the photoluminescence of many III–V semiconductors, and most PL studies of InAs dots and wetting layers on GaAs employ capped structure and/or low-temperature measurements to enhance emission intensities. However, nitride materials can behave quite differently. For example, room temperature photoluminescence emission is observed in thin, uncapped GaN films grown on wider energy gap AlGaN layers [26]. While surface states may be less severe in nitride materials, it has been observed that the buffer layer quality has a strong impact on the intensity of PL emissions from InN dots.

Quantum dots of Si have also been formed by alternating layers of silicon dioxide followed by a silicon-rich layer of the same material and then heat treating [27]. On heating, the surface energy minimization favoured the precipitation of silicon into spherical quantum dots. These quantum dots could be inserted into an amorphous silicon solar cell. A proposed all-silicon tandem cell has been proposed.

18.4 Organic quantum dot solar cells

Improvement in polymer solar cell performance has long suffered from several shortcomings (i.e. low electron mobility, absorption characteristics that were not well suited to the solar spectrum, environmental degradation, etc.). However, the lure of potential low cost and large area manufacturing of thin film flexible solar cells continued to attract researchers [28]. A main fundamental limitation associated with the use of organic thin films for PV is that photonic absorption in these materials produces bound state excitons. Dissociation of these charge pairs requires a substantial potential difference across a polymer–metal or polymer–semiconductor junction, provided the excitons are near the interface. Thus, polymer solar efficiencies improved dramatically with the introduction of the use of nanomaterial/conjugated polymer complexes or what are now commonly referred to as bulk heterojunction solar cells [29]. This refers to the fact that the materials that are responsible for both the donors and acceptors are intimately mixed throughout the absorber layer of the device.

A suitable choice of QD, or alternatively a fullerene, can provide the necessary electron-accepting impurity in a polymer matrix. Thus, under illumination a preferential transfer of electrons from the excitons to the acceptors leaves behind holes. This process is known as photo-induced charge transfer. The carriers, once liberated from one another, are now free to be transported through the conjugated polymer. All that remains is to incorporate these composites or blends into a suitable device structure.

A variety of acceptor materials have been introduced into conjugated polymers to produce photovoltaic devices (i.e. buckminster fullerenes [30], CdSe quantum dots, CdSe nano rods [31], and single-wall carbon nanotubes [32] (SWCNT)) [33]. The devices are produced by placing the doped polymeric films between a transparent conductive oxide, typically indium tin oxide (ITO), top contact and a metallic back contact. There has also been recent work on thin film polymeric solar cells which incorporate quantum dots as well as quantum dot/SWCNT complexes [34].

Typically, very high loading levels of semiconductor nanoparticles are required to overcome the poor electron transport properties of the polymer. Inclusion of SWCNTs to the polymer/quantum dot mixture can improve the transport problem.

The most commonly used colloidal quantum dot system for organic solar cell applications has been CdSe. However, new reports on other II–VIs, III–Vs, and ternary compounds such as $CuInS_2$ and $CuInSe_2$ are increasing in frequency. CdSe quantum dots are conventionally synthesized from CdO as the precursor, following the established protocol. In short, CdO, stearic acid, and 1-octadecene are heated to 200 °C under $Ar_{(g)}$ and then cooled to room temperature. Following the addition of trioctylphosphine oxide (TOPO) and octadeclyamine (ODA), the reaction mixture is heated and a selenium–trioctylphosphine solution is injected. A combination of time and temperature is then used to control the QD size. Extraction of the CdSe–TOPO quantum dots requires sequential washes using a methanol:hexanes mixture until a distinct interfacial separation is achieved. The CdSe–TOPO quantum dots are normally precipitated with acetone and can be resuspended in chloroform for additional characterization. Figure 18.7 shows two samples of colloidal CdSe QDs of different sizes under UV illumination. The different colors are a result of the varying degrees of quantum confinement related to the difference in particle size. The green sample have the higher effective band gap and thus the smaller nanocrystallites.

(a)   (b)

Figure 18.7 Digital photographs of colloidal InAs quantum dot suspensions showing the changes in the photoluminescence, and thus effective band gap, with a variation in the size (green to red, smaller to larger) of these semiconducting nanocrystals under (a) ambient white light and (b) UV illumination.

18.5 Quantum dot solar cell behaviour

18.5.1 III–V quantum dot solar cell behaviour

The key operating principle of the quantum dot enhanced solar cell of the type shown in Fig. 18.1 is absorption of long wavelength (infrared) photons by the QDs. This allows collection of some portion of the light that would normally be lost in non-QD enhanced solar cells. This enhanced absorption of infrared light leads to increased photocurrent production and thus, increased solar cell short circuit current (I_{SC}). In the case of ultra-high-efficiency InGaP/GaAs/Ge triple-junction solar cells (TJSC), the GaAs junctions are typically the current limiting junction, thus increasing the I_{SC} of the stack using QD enhancement could potentially lead to improved global conversion efficiency for what currently is the highest efficiency solar cell design being produced today. It has been predicted that quantum dot enhanced TJSCs have an efficiency ceiling of 47% under one-sun 6000 K black body illumination spectrum [35]. Additionally, quantum dot array enhanced GaAs cells have the added benefit of possible intermediate band effects [36], anisotropic absorption [37] and enhanced radiation tolerance [38].

Figure 18.8 shows a band diagram for the quantum dot enhanced GaAs solar cell. As can be seen, photons below the GaAs band gap are able to excite carriers in the InAs quantum dots. These carriers may escape by either of two means: photon-assisted extraction from deep bound QD levels or thermal-assisted extraction from higher energy bound QD levels. Carriers are then swept across the junctions due to the strong built-in electric field of the p-i-n device.

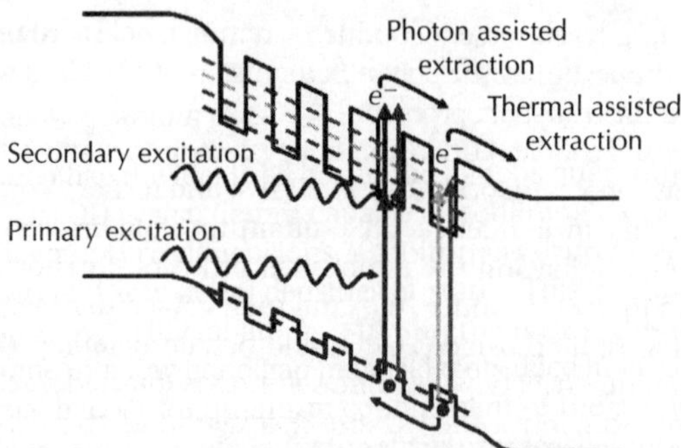

Figure 18.8   Idealized band diagram for a p-i-n solar cell with an embedded quantum dot array in the intrinsic region showing potential photoconversion mechanisms.

Measurements of the cell quantum efficiency show the QD absorption effects seen in the above Figure. As seen in Fig. 18.9, the quantum efficiency (QE) of the QD enhanced solar cells directly show the effects of QD absorption [39]. The cell quantum efficiency measures the photocurrent produced by the cell per incident photon as a function of the photon wavelength. For a standard GaAs cell, the quantum efficiency will be more or less constant over the visible region of the spectrum and rapidly fall off at the GaAs band edge near 870 nm (below band gap photons produce no photocurrent in the cell). As can be seen in the baseline p-i-n cell (without QDs), the quantum efficiency is relatively high over the 500–870 nm region.

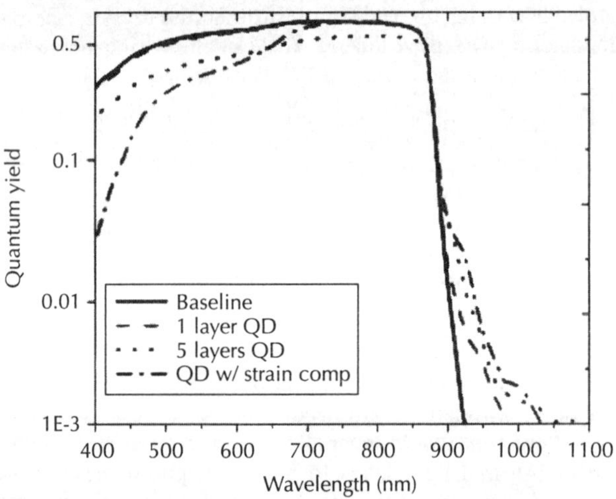

Figure 18.9   Quantum efficiency of the p-i-n cells with and without QD and strain compensation.

However, with five layers of QDs, the cells show an extended QD spectrum at wavelengths below 870 nm. This indicates that a portion of the short circuit current is being generated from below band gap absorption in the QD region, one of the requirements of QD cell enhancement. The five-layer QD cell shows increased below-band gap QE, due to the increased absorption cross-section of the five-layer QD cell. The QD cells show 1–2% QE in the 870–950 nm range. The QE would be expected to increase even more with additional layers of QDs. The quantum confinement in the QDs should lead to absorption peaks corresponding to transitions from the valence band to the discrete QD energy levels. These discrete transitions are evident in the five-layer QE spectrum, as evidenced in peaks appearing in the below-gap tail region.

18.5.2   Organic bulk heterojunction solar cells

The efficiencies of bulk heterojunction (BHJ) solar cells based on conventional sandwich structure ITO/PEDOT:PSS/active layer/Al, where the active layer is conjugated polymer, such as MEH-PPV or P3HT and a quantum dot, fullerene, or derivatized fullerene (i.e. PCBM) blend have reached approximately the 5% level under a 1 sun AM 1.5 illumination [www.nrel.gov/pv/thin_film/does/kaz_best_research_cells.ppt]. At this point, these cells are starting to exhibit what amounts to a much more conventional current versus voltage behaviour, which is reminiscent of their crystalline counterparts of a decade or so ago. A thermal treatment of active layer in these types of solar cells, both before and after contacting, has been shown to be critical to their improvement as shown in Fig. 18.10.

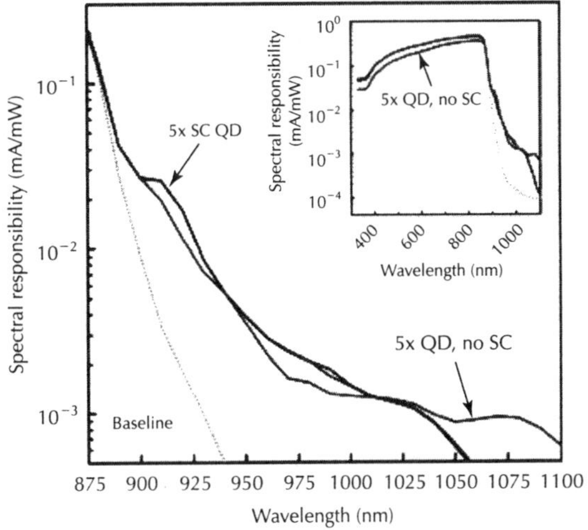

Figure 18.10   *The spectral response for a p-i-n GaAs solar cell without quantum dots and with both non-strain-compensated and strain-compensated quantum dot arrays (QDAs) in the intrinsic region demonstrating the sub-GaAs band gap photoconversion that is achievable with the QDA introduction. Inset: Spectral response in the above GaAs band gap region showing how the use of strain compensation can be used to eliminate the loss of above GaAs band gap photoconversion associated with ordinary QDA inclusion.*

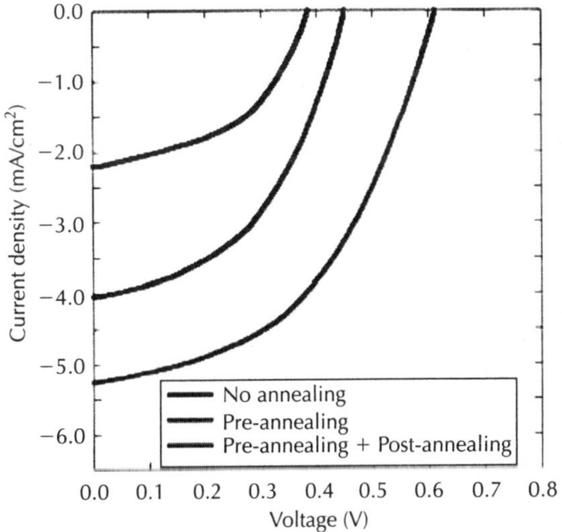

Figure 18.11   *Current versus voltage characteristics under 1 sun AM 1.5 illumination for a P3HT/PCBM bulk heterojunction solar cell with and without thermal annealing.*

The band offsets between a given conjugate polymer host and the various types of quantum dots or nanomaterials used are critical to the performance of BHJ devices. The QD composition and size, and thus their effective band gap, electron affinity, and ionization potentials, need to be appropriately matched to the polymer host or they will act as either electron traps, hole traps, or both. Even when band offsets are favourably aligned for electron and hole transport, small changes in the effective optoelectronic structure will affect the photovoltaic performance of the BHJ devices. Figure 18.12 shows the effect of changing the fullerene type in a typical BHJ device. Unfortunately, the effect band offsets for the materials used based upon optical spectroscopy or electrochemical measurements made on the polymers or nanomaterials separately can be misleading (see Fig. 18.13). The actual energy levels in the blends are known to shift due to the basic electronic interaction of materials once placed in intimate contact.

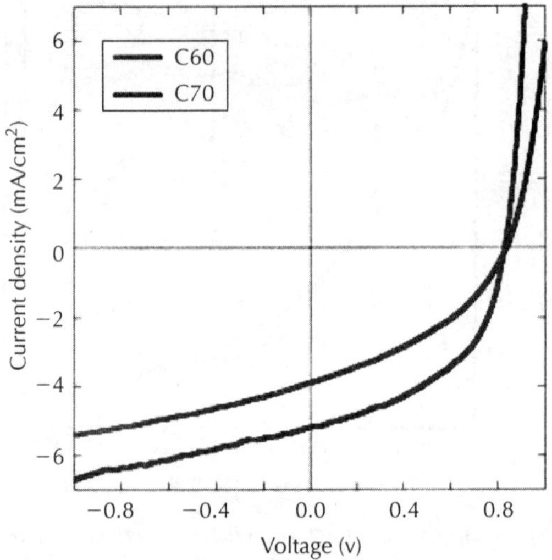

Figure 18.12   Current versus voltage behaviour under a 1 sun AM 1.5 illumination of an MEH-PPV bulk heterojunction solar cell as a function of the fullerene used in the active region [40].

The advantageous affects of thermal annealing of polymer solar cells, as demonstrated above, are well known in the field. However, it has also been recently shown that these types of solar cells exhibit a positive temperature coefficient as well (see Fig. 18.14). The short circuit current density is shown to improve dramatically with increasing temperature with only a slight drop in the open circuit voltage. The net result is almost a factor of 3 improvement in 1 sun AM 1.5 photovoltaic efficiency over a temperature range of 77 K to 330 K for a conventional bulk heterojunction solar cell.

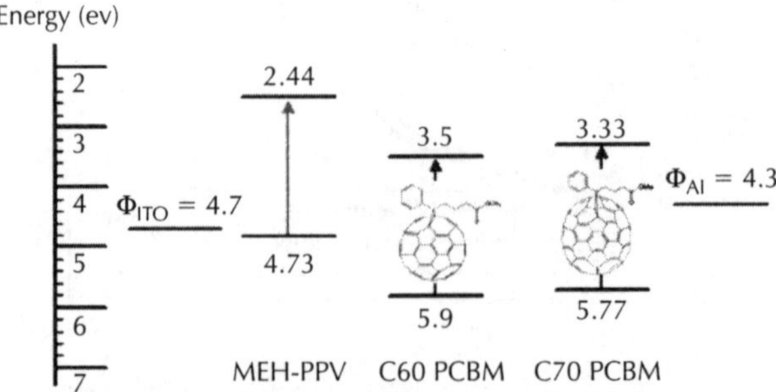

Figure 18.13   Experimentally determined values for the work functions, ionization potentials, and electron affinities associated with the materials used in the devices measured in Fig. 18.11 [40].

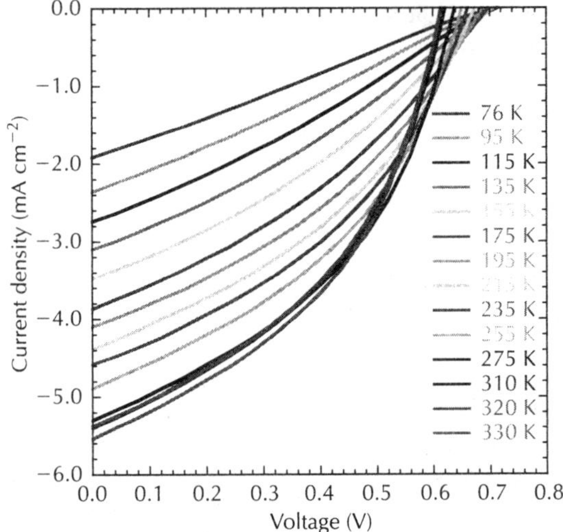

Figure 18.14   *Temperature dependent current versus voltage characteristics for a conventional P3HT/PCBM bulk heterojunction solar cell showing a positive photovoltaic efficiency temperature coefficient [40].*

18.6   Conclusions

Nanostructured solar cells will play an important role in enhancing the efficiency of future generations of solar cells, whether they are III–V, II–VI, or hybrid organic–inorganic cells. There is a great deal of potential in multiple approaches for these nanostructures. It is also clear that further work will be necessary in many material areas; however, substantial progress has been made in the last five years. The future looks bright for nanostructured photovoltaics.

References

1. M.A. Green. *Third Generation Photovoltaics: Advanced Solar Energy Conversion*, (Springer-Verlag 2006) pp. 1–125.
2. D.B. Bushnell, K.W.J. Barnham, J.P. Connolly, N.J. Ekins-Daukes, R. Airey, G. Hill, and J.S. Roberts, Light-trapping structures for multi-quantum well solar cells, *Proceedings of the 29th Photovoltaic Specialists Conference* (IEEE, New York, 1035–103 (2002).
3. A. Freundlich, A. Alemu, and S. Bailey, Quantum wire solar cell, *Proceedings of the 31st IEEE Photovoltaic Specialist Conference* (2005).
4. R. Raffaelle, S. Castro, A. Hepp, and S. Bailey, Quantum dot solar cells, *Progress in Photovoltaics* **10** Number 6, 433–439 (2002).
5. A.J. Nozik, *Annu. Rev. Phys. Chem.* **52**, 193 (2001).
6. D. Halliday, R. Resnick, and J. Walker. *Fundamentals of Physics*, 7th ed. (John Wiley & Sons, 1094, 2005).
7. W. Shockley and H.J. Queisser, *J. Appl. Phys.* **32**, 510 (1961).
8. A.J. Nozik, *Annu. Rev. Phys. Chem.* **52**, 193 (2001).
9. P.T. Landsberg, H. Nussbaumer, and G. Willeke, *J. Appl. Phys.* **74**, 1451 (1993).
10. A. Brown, M. Green, and R. Corkish, Limiting efficiency for a multi-band solar cell containing three and four bands, *Physica E* **14**, 121–125 (2002).
11. A. Luque and A. Marti, Increasing the efficiency of ideal solar cells by photon induced transitions at intermediate levels, *Phys. Rev. Lett.* **78**, 5014–5017 (1997).
12. M. Green, Proceedings of the *IEE 4th World Conference on Photovoltaic Energy Conversion*, **15** (2006).

13. I. Serdiuokova, C. Monier, M. Vilela, and A. Freundlich, *Appl. Phys. Lett.* **72**, 2812 (1999).
14. Q. Shao, A. Balandin, A. Fedoseyev, and M. Turowski, *Appl. Phys. Lett.* **91**, 1 (2007).
15. F. Guffarth, R. Heitz, M. Geller, C. Kapteyn, H. Born, R. Sellin, A. Hofffmann, D. Bimberg, N. Soholev, and M. Carno, *Appl. Phys. Lett.* **82**, 1941 (2003).
16. M. Green, *Handbook of Semiconductor Nanostructures and Nanodevices*, Photovoltaic Applications of Nanostructures, Vol. 4, ed. by A.A. Banlandin and K.L. Wang, 219–237 (2006).
17. O. Yilmaz, S. Chaudhary, and M. Ozkan, *Nanotechnology* **17**, 3662 (2006).
18. A. Freundlich, A. Alemu, and S. Bailey, Quantum wire solar cell, *Proceedings of the 31st IEEE Photovoltaic Specialist Conference* (2005).
19. N. Stranski and L. Von Krastanov, *Akad. Wiss. Lit. Mainz Math. Naturwiss. Kl* **146**, 797 (1939).
20. D. Leonard, M. Krishnamurthy, C.M. Reaves, S.P. Denbaars, and P.M. Petroff, *Appl. Phys. Lett.* **63**, 3203 (1993).
21. S.M. Hubbard, D. Wilt, S. Bailey, D. Byrnes, and R. Raffaelle, *Proceedings of the IEEE 4th World Conference on Photovoltaic Energy Conversia* **1**, 118–121 (2006).
22. O. Laboutin, R. Welser, L. Guido, A. Sood, and S. Bailey, *Proceedings of the 20th Space Photovoltaic Research and Technology Conference* (2007).
23. S. Ruffenach, O. Briot, M. Moret, and B. Gill, Control on InN quantum dot density using rate gases in metal organic vapor phase epitaxy, *Appl. Phys. Lett.* **90**, 153102 (2007).
24. W.C. Ke *et al.*, Photoluminescence properties of self-assembled InN dots embedded in GaN grown by metal organic vapor phase epitaxy, *Appl. Phys. Lett.* **88**, 191913 (2006).
25. C.A. Duarte *et al.*, Influence of the temperature on the carrier capture in self-assembled InAs/GaAs quantum dots, *J. Appl. Phys.* **93**, 6279–6283 (2003).
26. J.F. Muth, X. Zhang, A. Cai, D. Fothergill, J.C. Roberts, P. Rajagopal, J.W. Cook, Jr, E.L. Piner, and K.J. Linthicu, Gallium nitride surface quantum wells, *Appl. Phys. Lett.* **87**, 192117 (2005).
27. M. Zacharias *et al.*, *Appl. Phys. Lett.* **80**, 661 (2002).
28. G. Yu, J. Gao, J. Hummelen, F. Wudl, and A.J. Heeger, *Science* **270**, 1789–1791 (1995).
29. M. Granstrom, K. Petritsch, A.C. Arias, A. Lux, M.R. Andersson, and R.H. Friend, *Nature* **395**, 257–360 (1998).
30. N. Camaioni, G. Ridolfi, G. Casalbore-Miceli, G. Possamai, L. Garlaschelli, and M. Maggini, *Solar Energy Materials & Solar Cells* **76**, 107–113 (2003).
31. W.U. Huynh, J.J. Dittmer, and A.P. Alivisatos, *Science* **295**, 2425–2427 (2002).
32. E. Kymakis and G.A.J. Amaratunga, *Appl. Phys. Lett.* **80**, 112–114 (2002).
33. N. Hoppe and N.S. Saracifti, Nanostructure and nanomorphology engineering in polymer solar cells in: *Nanostructured Materials for Solar Energy Conversion*, Ed. by Tetsuo Soga (Elsevier, pp. 277–318, 2006).
34. K. Kim, J. Liu, M.A.G. Namboothiry, and D.L. Carroll, *Appl. Phys. Lett.* **90**, 163511/163511-163511/163513 (2007).
35. R.P. Raffaelle, S. Sinharoy, J. Andersen, D. Wilt, and S.G. Bailey, *Proceedings of the IEEE World Conference on Photovoltaic Energy Conversion* **1**, 162–166 (2006).
36. A. Luque and A. Marti, *Phys. Rev. Let.* **78**, 5014–5017 (1997).
37. A. Marti, N. Lopez, E. Antolin, E. Canovas, C. Stanley, C. Farmer, L. Cuadra, and A. Luque, *Thin Solid Films* **511–512**, 638 (2006).
38. C.D. Cress, S.M. Hubbard, B.J. Landi, D.M. Wilt, and R.P. Raffaelle, *Appl. Phys. Lett.* **91**, 183108 (2007).
39. S. Hubbard, R. Raffaelle, R. Robinson, C. Bailey, D. Wilt, D. Wolford, W. Maurer, and S. Bailey, *Proceedings of the 20th Space Photovoltaic and Research and Technology Conference*, (2007).
40. A. Anctil, A. Merrill, C. Cress, B.J. Landi, and R.P. Raffaelle, *Proceedings of the Materials Research Society Symposium H; Nanostructured Solar Cells*, Nov. 26–29 (2007).

CHAPTER 19

Quantum Dot Superluminescent Diodes

M. Rossetti,[1] L.H. Li,[1] A. Fiore,[1] L. Occhi,[2] and C. Vélez[2]

[1]*Ecole Polytechnique Fédérale de Lausanne (EPFL), Station 3, Ch-1015 Lausanne (Switzerland)*
[2]*EXALOS AG, Wagistrasse 21, CH-8952 Schlieren (Switzerland)*

19.1 Introduction

19.1.1 Superluminescent diodes and their applications

Superluminescent light-emitting diodes (SLEDs) are semiconductor light sources that show a broadband output optical spectrum (like LEDs) at high power levels (like semiconductor lasers). Sometimes they are referred as "high-power LEDs" or as "laser diodes operating below threshold". SLEDs have features typical of LEDs like low coherence, as well as features of semiconductor lasers like high optical power. This combination is extremely useful for a wide range of applications.

SLEDs are based on a pn-junction embedded in an optical waveguide. When electrically biased in the forward direction they show optical gain and generate amplified spontaneous emission over a wide range of wavelengths. SLEDs are designed to have high single-pass amplification for the spontaneous emission generated along the waveguide but unlike laser diodes, insufficient feedback to achieve lasing action. The suppression of the cavity modes is typically achieved by tilting the waveguide with respect to the end facets and by providing the facets with an anti-reflection coating. With the suppression of the cavity modes a low ripple or smooth spectra is obtained. The main technological challenge is to achieve high optical output powers with a smooth spectrum.

In other words, the SLED is a light source, which combines the spatial coherence of a laser diode with the temporal incoherence of an LED. The spatial coherence translates into small beam divergence, which greatly facilitates the light coupling into a single mode fibre with similar efficiency as for laser diodes. Typically, more than 50% of the chip facet power is coupled to the single mode fibre. The large optical bandwidth or the low temporal coherence is of great advantage for applications where interferences cause problems like speckle or ghost signals. SLEDs are much more powerful than standard LEDs and are particularly advantageous for applications requiring high power density. SLEDs typically have single-mode output powers of the same order of magnitude as single-mode laser diodes of several tens of milliwatts when biased with several hundreds of milliamperes.

Although SLEDs have been known in the literature for many years, it has been only recently that they have become more popular or even standard devices for various applications. The main reason is that only recently devices showing high output power, smooth spectra and large bandwidth have been demonstrated and made available in larger quantities commercially [1].

The availability of large optical spectral bandwidth at high power levels makes SLEDs the most cost-effective light source solution for a wide range of applications, as summarized in Table 19.1.

Table 19.1   SLED are versatile optical sources. Overview of the main SLED application fields.

Application field	SLED advantage	Examples
Optical coherence tomography (OCT)	high resolution, fast scan time	endoscopes, dermatology, cornea and retina diagnostics cardiovascular and gastrointestinal imaging
Fibre optic gyroscopes (FOG)	reliability, stability, cost effective	navigation for satellites, airplanes, ships, land vehicle
Fibre optic test equipment	high power density over a large spectral range, telecom reliability standards, cost effective	chromatic and polarization mode dispersion, optical channel monitoring, passive components characterization
Fibre optic sensors	cost effective, robust	temperature and strain measurements in bridges or oil pipelines

19.1.2   Application requirements and markets

Optical coherence tomography (OCT) is an emerging technology for producing high-resolution cross-sectional imaging. Tissue structure can be imaged on the micron scale in real time. The principle of OCT is analogous to that of ultrasound imaging, except that it uses light instead of sound (see Fig. 19.1). The light is provided by a broadband source like the SLED. Although sound waves are able to pass through tissue, light, which has a much shorter wavelength, is unable to penetrate beyond 2 mm in most non-transparent tissue. Thus, the use of OCT is limited to optically transparent tissues, such as retina, or endoscopic examinations. The large optical bandwidth together with the good suppression of the cavity modes allows short coherence lengths from some tens of micrometres down to few micrometres. Moreover, the high power level of the SLEDs allows fast scan times required in ophthalmologic, cardiovascular, and gastrointestinal imaging applications. Quantum dot-based SLEDs are leading the development of new SLEDs with increased spectral bandwidth and large optical output power, in order to achieve higher image resolution to perform tissue characterization on a scale never before possible within the human body. This technology has the potential to dramatically change the way that physicians, researchers and scientists see and understand the human body in order to better diagnose and treat diseases.

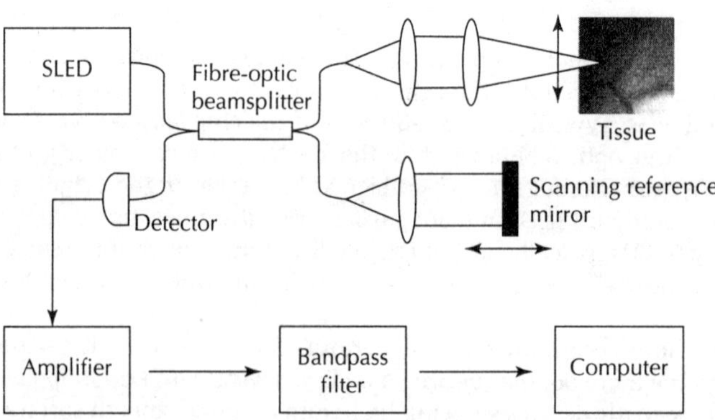

Figure 19.1   Basic block diagram of an optical coherence tomography set-up.

Another main SLED application is in navigation systems employing fibre optic gyroscopes (FOGs). Fibre optic gyroscopes are mainly used in avionics and aerospace. They are at the heart of very precise rotation measurement systems. FOGs are based on the measurement of the Sagnac phase shift occurring to the radiation propagating along a fibre optic coil when it is subject to a rotation around the winding axis. When an FOG is mounted within a navigation system it allows the tracking of change of orientation. The basic elements within an FOG are a light source, a single-mode polarization maintaining fibre coil, a coupler and a detector (see Fig. 19.2). By means of the optical coupler, the light emitted by the source is injected into the fibre coil in counter-propagating directions. When the fibre coil is at rest the two light waves interfere constructively at the detector and a maximum signal is detected. During coil rotation the two optical waves experience different optical paths that depend on the rotation rate; the phase difference between the two fields causes variations of the intensity detected by the photodiode and provides information on the rotation rate. SLED-based gyroscopes rely on the large bandwidth of the source to reduce the undesired effect of scattering along the fibre and of reflections at the facets of the internal optical components that could decrease the sensitivity for very low rotation rates.

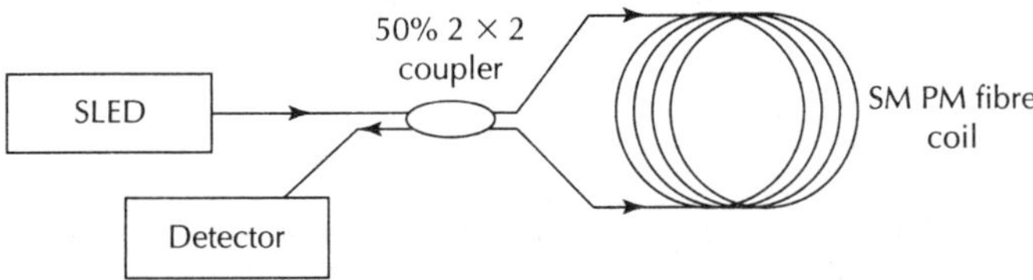

Figure 19.2   Basic block diagram of a fibre optic gyroscope.

The requirements on FOGs and on the embedded optical sources are low cost, reduced form factor size, low power consumption, stability and reliability. Quantum dots (QDs) have the potential to successfully combine all these properties since they are potentially less sensitive to temperature.

Broadband optical sources are also used in fibre optics for component test or network management purposes. An example of the first is the characterization of optical components suitable for coarse wavelength division multiplexing (CWDM) application, which requires multiple wavelength sources. Thanks to their large optical bandwidth, QD SLEDs allow the integration of fewer sources, with wavelengths spanning the whole CWDM band (i.e. 1200–1700 nm), in single test equipment. In optical networks SLEDs are used to measure the polarization-mode dispersion of optical fibres.

SLEDs are preferred sources for these purposes, because of many aspects. First, SLEDs are fully compatible with the wavelengths used in the optical communication systems. Second, the SLED's optical waveguide allows an efficient coupling of the light into single-mode optical fibres. Third, SLEDs answer well to the pressures in today's telecommunications carrier market, which requires test solutions that are cost effective, reliable, provide better productivity and reduce the number of testers in the field.

The final main SLED application field is in fibre optic sensors for strain and temperature measurements in civil engineering, structural analysis and composite material manufacturing applications. Fibre optic sensors have a number of advantages over conventional sensors. Their immunity to electromagnetic fields, their ability to measure at many points along a single fibre, and their ability to be embedded within, or bonded to, structures makes them a highly flexible solution. Over the past 25 years, fibre optic sensors have seen a continuous improvement in quality and performance. SLEDs are used for these applications because of their large optical bandwidth and robustness.

19.1.3   Quantum dots for superluminescent diodes

SLEDs are based on amplified spontaneous emission in an active semiconductor waveguide, which can simultaneously provide large bandwidth and high output power. In the most simplified picture, assuming a uniform gain profile along the propagation direction, the spatial dependence of light power $P(z)$ along the waveguide is described by the equation:

$$\frac{dP}{dz} = \beta P_{sp} + gP \tag{19.1}$$

where P_{sp} is the power spontaneously emitted per unit length, β the spontaneous emission coupling factor, and g the net modal gain. The solution, assuming a perfect antireflection coating at $z = 0$, is:

$$P(z) = \frac{\beta P_{sp}}{g}\left(e^{gz} - 1\right). \tag{19.2}$$

The exponential growth of the power with distance is at the basis of the high efficiencies achievable in SLEDs. However, the exponential factor in Eq. 19.2 also implies a strong wavelength dependence, as $g = g(\lambda)$. Neglecting the other spectral dependences, the output spectral density (at $z = L$) drops to half of the peak value at the wavelength λ^* where the gain drops by Δg:

$$\frac{\Delta g}{g} = \frac{\left|g\left(\lambda^*\right) - g\left(\lambda_0\right)\right|}{g\left(\lambda_0\right)} = \frac{\ln 2}{gL}. \tag{19.3}$$

At typical working conditions, for output powers $>1\,\text{mW}$ a $gL \approx 6\text{--}10$ is needed, which gives $\Delta g/g \approx 10\%$. In order to achieve wide output spectra, an extremely flat gain spectrum must thus be provided.

Quantum wells (QWs) are in principle well suited for wide gain spectrum, as their density of states is flat – at least ideally – within each subband. However, in practice very large bandwidths ($\gg 50\,\text{nm}$) are hard to achieve with QWs. This is due to the large density of states corresponding to the first QW subband – in order to completely fill one subband, an extremely large current density ($\approx 10\,\text{kA/cm}^2$) would be needed. Additionally, the modal gain corresponding to the first subband is already very high, which typically drives the device into lasing unless special (technologically demanding) antireflection structures are used.

Semiconductor quantum dots (QDs) – semiconductor nanostructures where electrons and holes are confined in three dimensions, resulting in quantization of the energy levels – have several advantages over bulk and QW active regions for SLEDs:

1. Inhomogeneously broadened gain spectrum: In self-assembled QDs grown in the Straski–Krastanov regime (see section 19.2 and Fig. 19.3), the natural size dispersion of the QDs in the ensemble results in a dispersion of the energy states and thus of the interband transition energies – a prototypical example of inhomogeneous broadening. More precisely, at low temperature (10 K) and low excitation level, each QD contributes few

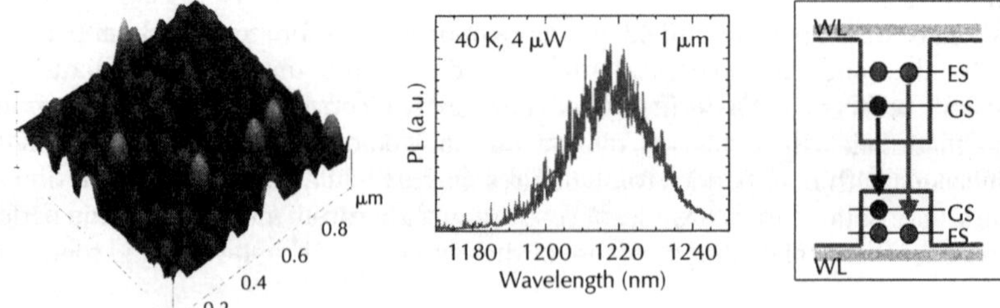

Figure 19.3   Left: Atomic force microscopy image of self-assembled QDs grown by molecular beam epitaxy of InAs on GaAs. Centre: Low-temperature photoluminescence from an ensemble of about 100 QDs, showing how single QD lines merge in a wide Gaussian emission. Right: Simplified schematics of energy levels and transitions in a QD (GS: ground state. ES: excited state. WL: wetting layer).

ultra-narrow (<1 Å) emission lines corresponding to the recombination of excitons, biexcitons and other multi-excitonic states. The lines from a large ensemble of QDs (typically $\approx 10^6$–10^7 QDs are embedded in the active region of a waveguide) then merge in a broad (30–100 nm) Gaussian-shaped emission line. At room temperature and/or high excitation levels typical of device operation, a significant homogeneous broadening (10–20 nm) of the emission line of each QD also contributes to the total spectral width. While inhomogeneous broadening has always been perceived as the major obstacle to achieving high performance in lasers, it represents a clear advantage for SLED applications. As discussed thoroughly in section 19.2, the growth conditions can be tuned to increase the QD size dispersion, and thus the gain spectral width. Additionally, several QD layers can be stacked with different centre wavelengths ("chirped multilayers"), resulting in an even wider combined gain spectrum.

2. Limited density of states: The number of electronic states in a QD ensemble is directly related to their areal density (there are two spin-degenerate states on the lowest ("ground") state for each QD). The areal density depends on the growth conditions and is typically limited to few 10^{10} dots/cm^2, particularly due to a limitation on the average strain in the system where In-rich QDs are needed to shift the wavelength towards the 1300 nm region. By stacking multiple layers, this can be increased to the few 10^{11} dots/cm^2 range. This density of states (DOS) is roughly a factor of 10 lower than the typical DOS in the first subband of a single QW in a comparable energy interval. This in turn results in lower maximum modal gain corresponding to full inversion of the states (in the few 10s cm^{-1} range for QDs, as compared to 100–200 cm^{-1} for a single QW). This lower gain is a major limitation for lasers, producing gain saturation and consequently poor frequency response. On the contrary, the relatively low density of states can be an advantage for SLEDs. In fact, to achieve population inversion over a large energy range at reasonable current levels, a low density of states per unit energy is preferable. The ground state (GS) of a single InAs/GaAs QD layer emitting around 1300 nm is typically filled with ≈ 100–200 A/cm^2, and a gain spectrum >100 nm wide, combining ground and excited states (ES, see right part of Fig. 19.3), is obtained with less than 1 kA/cm^2. In QWs, as mentioned above, filling all states in this energy range would require 10 kA/cm^2 or more. Ultimately, the wide emission spectrum of QD SLEDs is the consequence of the relatively small density of states per unit energy. Achieving high-output powers obviously requires a certain level of chip gain ($gL \approx 6$–10), which is nowadays achievable by stacking 10–20 layers of QDs.

19.2   QD growth for SLEDs

Large spectral width is one of the most important features of SLEDs [2, 3], as the correspondingly short coherence length can significantly improve the spatial resolution in coherence-based systems. The large spectral width can be realized by using the techniques such as multiple-quantum well (QW) engineering and QW intermixing [4–7]. However, increasing the SLED spectral width over 100 nm in the 1300 nm wavelength region is still a challenge. As mentioned in section 19.1, and originally proposed in [8], the naturally occurring size dispersion in self-assembled growth of quantum dots (QDs) can be beneficial for SLEDs requiring larger spectral width. Generally, inhomogeneous size distribution of QDs in the active region is disadvantageous for laser applications. However, for the wide spectra device applications such as SLEDs, it becomes an effective and intrinsic merit due to the broadened gain spectrum. Size dispersion in the tens of per cent range results in a spectral width over 100 nm from the QD active region as predicted theoretically. Therefore, considerable efforts have been invested in developing QD SLEDs [9–22]. Initial reports employed the uniform stacked In(Ga)As/GaAs QDs which are routinely used in laser devices, showing a narrow spectral width and/or a short emission wavelength [10–13, 15]. To increase the spectral width, deliberate increase of the dot size dispersion is a straightforward method [8, 17, 21]. In addition, employing multiple QD layers with different amounts of InAs deposition in the QDs is another option [9, 17]. Large photoluminescence (PL) spectral width over 100 nm can be obtained by using both methods [9, 21]. However, for SLED applications, a large PL linewidth is not sufficient, high radiative efficiency and uniform optical gain must be obtained across the entire spectral range. For example, varying the InAs coverage also affects the QD density, producing a non-uniform gain spectrum. As an alternative, stacking InAs-chirped multiple QD (CMQD) structures with constant InAs coverage and varying In composition in the InGaAs capping layer was proposed by us [4, 17] and also used by other groups [18–20]. In fact, when the QDs are covered/surrounded by an InGaAs layer, a red shift of the emission wavelength is observed, depending on the composition and thickness of the InGaAs layer [23, 24]. Because the emission wavelength of each QD layer in the chirped structure can be individually controlled by a change in the InGaAs matrix, the composite emission of the chirped structure will yield a broad gain spectrum. In such CMQD structures, the QD areal density is not significantly affected by the InGaAs layer, so that the available gain and the wavelength are effectively decoupled and a flat gain spectrum can be obtained. The combined spectral broadening depends on the spectral width of the individual QD layer, the spectral separation among each QD layer and the number of the stacked layers. Using the CMQD structure, a composite PL spectral width of 78 nm was obtained from the ground state (GS) of uniform QDs. Incorporating such a CMQD structure into a SLED device structure and combining the emission from GS and excited state (ES) of QDs, SLEDs with spectral width of 121 nm, covering from 1165 to 1286 nm, were demonstrated [14]. In such structures a dip may appear in the spectrum between GS and ES, due to insufficient broadening of the GS and ES lines. The dip can be effectively eliminated by optimizing the chirped stacking structure and QD growth conditions. In particular, replacing uniform QDs by QDs with increased size dispersion in the CMQD structure, smooth and broad output spectra up to 115 nm without dip were obtained from SLEDs [17]. In the following paragraphs, the growth optimization of the CMQD structure and single-layer InAs QDs with increased size dispersion are described.

19.2.1   Chirped multiple QD structure

High uniform InAs QDs used for laser devices have high peak gain but provide relative small gain spectral width which limits their applicability in the SLEDs. The CMQD structure was proposed to take advantage of the optimized gain characteristics of these QDs while increasing the total gain linewidth in a controllable way [14]. In this part, we describe the growth of the CMQD structure. The samples are grown on (100) GaAs substrates by using solid-source molecular beam epitaxy (MBE). InAs QDs form by continuous deposition of InAs material (nominal thickness 3 monolayers) at the growth temperature of 530°C and growth rate of 0.163 μm/h, which are then covered by a few nm of thin InGaAs capping layer. In the CMQDs structure, the InAs/InGaAs QD layers are separated by 40 nm GaAs barriers to avoid as much as possible strain-driven growth coupling between different dot layers. Typical PL test structures are finally completed capping the dots with a 100 nm thick GaAs barrier, and are then characterized at room temperature (RT) exciting the dots with a 632.8 nm He–Ne laser and detecting the photoluminescence with an uncooled InGaAs detector.

Varying indium content and thickness of the InGaAs capping allows a fine tuning of the PL characteristics of each QD layer, which is very useful for the optimization of CMQDs. The dependences of the RTPL peak wavelength of the samples as a function of the In composition and the thickness of the InGaAs layer are shown in Fig. 19.4. Insets show the normalized PL spectra of the samples with the different In composition or the thickness of the InGaAs layer. Between 0 and 5 nm, the PL peak wavelength is proportional to the thickness of the InGaAs layer as shown in Fig. 19.4a. The PL peak wavelength increases from 1140 to 1300 nm upon increasing the InGaAs layer thickness from 0 to 6 nm. Similar behaviours are also observed when the In composition of the InGaAs layer is changed, which is shown in Fig. 19.4b. The significant PL peak wavelength variation mainly appears in the range of In composition between 0 and 20%. A further increase of In composition does not contribute to the further PL peak wavelength shift. These findings are in good agreement with previous reports [23, 24]. The red shift of the QD emission with increasing In composition and thickness of the InGaAs layer can be attributed to a reduction in the strain in the QDs [24], to the suppressed In segregation from the QDs into the cap layer, and to the activated spinoidal decomposition of the cap layer [25].

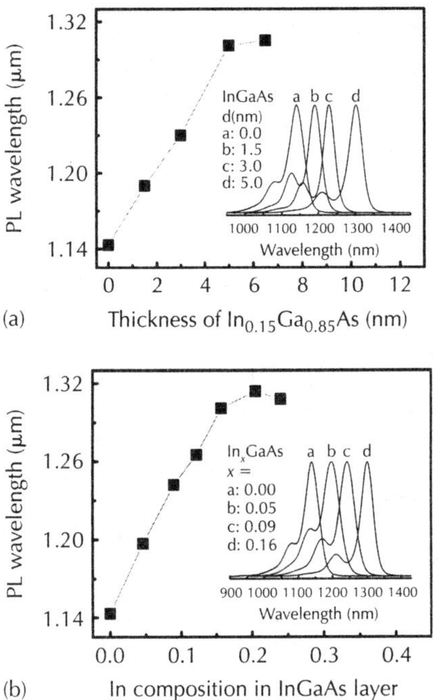

Figure 19.4 (a) *RTPL peak wavelength as a function of the thickness of the In$_{0.15}$Ga$_{0.85}$As layer. (b) RTPL peak wavelength as a function of the In composition of the 5 nm thick InGaAs layer. Insets show the normalized PL spectra of the samples with different In composition or thickness of the InGaAs layer.*

By appropriately choosing In composition or thickness of the InGaAs capping layers, the emission of CMQDs should cover a wide spectral range. This idea can be easily validated by implementing several QD layers with different capping properties in the same structure. Figure 19.5 (continuous line) shows the normalized RTPL spectra of a CMQD structure containing five QD layers. The In composition in the InGaAs layer is varied from 0.09 to 0.15 in steps of 0.015, which results in a wide PL spectral width of 78 nm. The PL emission is dominated by GS transition at around 1270 nm, while the high energy tail merging with the main peak comes from ES transition. The achieved spectral width is much larger than the typical spectral width of 44 nm from the single-layer QD samples (dashed–dotted line), which verifies the validity of using the CMQD structure for spectral broadening.

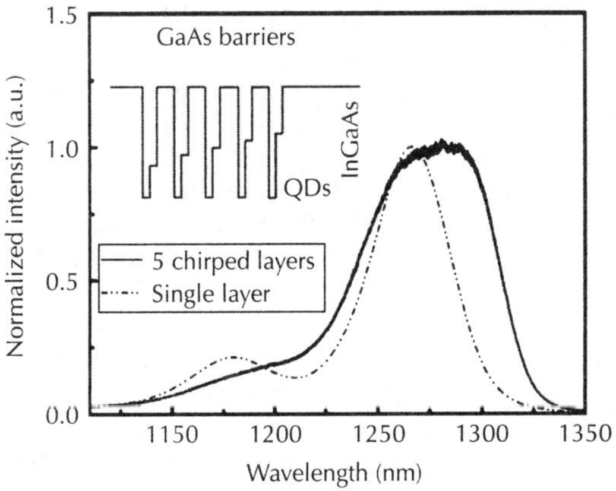

Figure 19.5 *Normalized RTPL spectra of a single layer of InAs QDs and a chirped multiple InAs QDs sample. Inset: Schematic band diagram of the chirped QD structure.*

19.2.2   *Increasing the inhomogeneous spectral width of a single layer of QDs*

Although using the CMQD structure is effective to increase the spectral width, the number of QD layers needed for a given total linewidth without a spectral dip, and the required accuracy in the In content, depends on the linewidth of each layer. Increasing the inhomogeneous linewidth of each layer helps to obtain a broad and uniform spectrum. In this paragraph, we show the effect of growth conditions on the linewidth, and find optimized conditions for large linewidth and high radiative efficiency.

Growth temperature and amount of InAs deposited to form the dots have a strong impact on the QD size dispersion over the ensemble and on the material quality. Figure 19.6 shows the RTPL spectra of 2.2 ML InAs QDs deposited at various growth temperatures from 485 to 530°C and at a growth rate of 0.1 ML/s (monolayer per second). QDs are capped with a 5 nm thick InGaAs layer and covered by a 100 nm thick barrier. The spectral width of the QDs increases and the PL efficiency deteriorates when lowering the growth temperature. This can be attributed to a reduced diffusion length of adatoms at lower growth temperatures [26]. Nucleation centres on the epitaxial surface increase, giving rise to a high density of small dots. The size distribution of these dots is inhomogeneous which results in the broad emission spectrum. There is an optimized growth temperature around 510°C which gives a broad spectrum emitting near 1300 nm. Its PL integrated intensity is comparable to that of the samples grown at higher temperature.

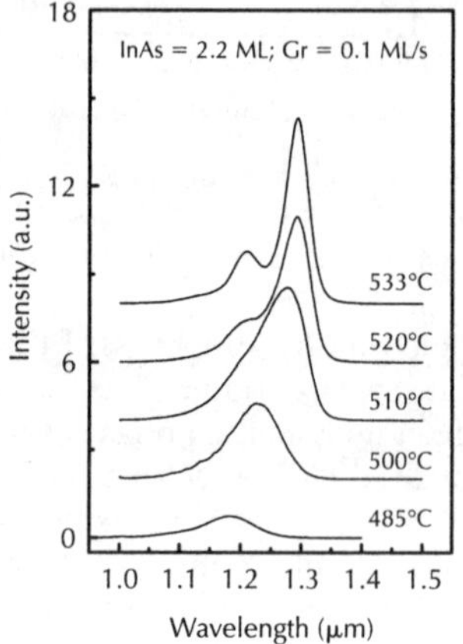

Figure 19.6   RTPL spectra of the 2.2 ML InAs QDs deposited at various growth temperatures for a growth rate of 0.1 ML/s.

Figure 19.7 shows the RTPL spectra of the QDs with different InAs thicknesses deposited at growth temperatures of 485 and 510°C. With increasing InAs thickness, the PL peaks of the samples grown at both temperatures shift systematically to the long wavelength side due to an increased dot size. For the QDs deposited at 485°C, the spectra are broad. However, the QDs show very low PL efficiency and the PL emission peak is blue shifted as compared to the QDs grown at 510°C, which prevents accessing the 1300 nm range. Although the spectral widths are large, the QDs grown at 485°C may not be suitable for device applications. For the QDs deposited at 510°C, the spectral width of the QDs becomes small with increasing InAs thickness due to the improved dot size uniformity [27]. A broad spectral width up to 96 nm is only obtained when the InAs thickness is ≤2.2 ML due to the dot size dispersion. There might exist several groups of QDs with different sizes, each of them emitting at different wavelengths, thereby giving rise to a broad spectrum. The PL integrated intensities of these QDs are comparable to that of the 2.4 ML thick InAs QD layer, which has the strongest PL intensity. In fact, 2.4 ML thick InAs QDs have been successfully used in low threshold current density lasers, implying high radiative recombination efficiency. It should be noticed that the PL intensity of QDs with thicknesses larger than 2.4 ML degrades, probably due to the crystal defects such as dislocations which may have been generated in larger islands, so that they grow faster or even at the expense of the coherent ones

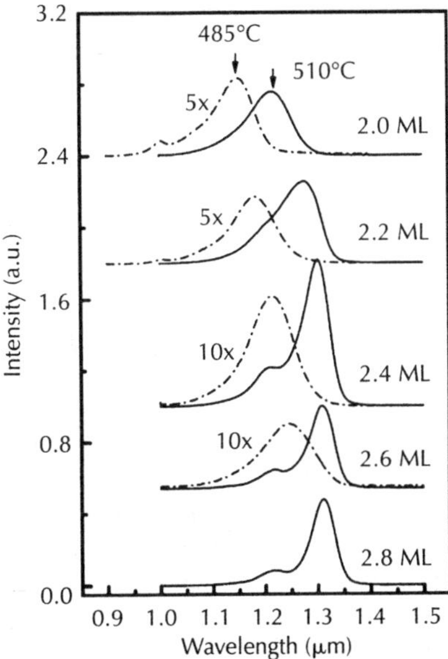

Figure 19.7 RTPL spectra of the QDs with different InAs thicknesses deposited at growth temperatures of 485°C and 510°C.

[28, 29]. In contrast, there is no dislocation in QDs with InAs thicknesses less than 2.4 ML, implying efficient radiative recombination.

By incorporating such dispersed InAs QDs into a CMQD structure, even larger spectral width and higher gain can be expected. Two approaches were developed to stack QD layers having chirped emission wavelength, by varying (i) the In composition in the InGaAs layer and (ii) the InAs thickness. The RTPL spectra from the two types of structure are depicted in Fig. 19.8. A spectral width of 91 nm (solid line) was achieved in a seven-layer stack with identical InAs QDs (2.2 ML thick) capped by three different In compositions of an InGaAs layer (12%, 15% and 17%; structure A). The value is larger than the value of 78 nm achieved by stacking

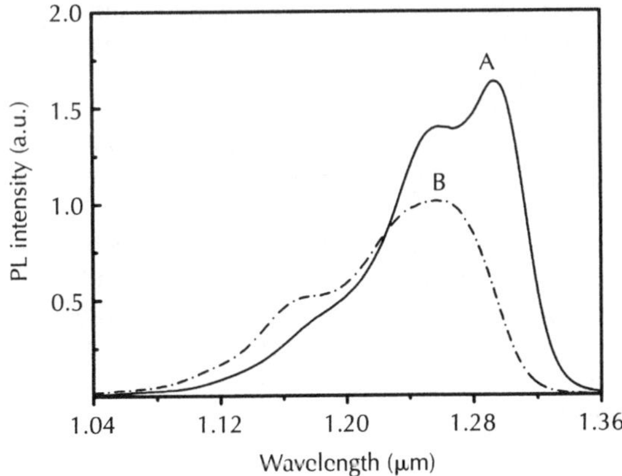

Figure 19.8 RTPL spectra from the two types of chirped QD structures. Structure A: seven-layer stack with identical thick InAs QDs (2.2 ML) capped by InGaAs layers with three different In compositions (12%, 15% and 17%); structure B: seven-layer stack with two different InAs thicknesses (2.1 and 2.2 ML) capped by InGaAs layers with the same In compositions.

chirped high uniform InAs QD layers. The largest spectral width of 124 nm was achieved in a seven-layer stack with two different InAs thicknesses (2.1 and 2.2 ML) capped by InGaAs layers with the same In composition (dash–dot line; structure B). These spectral widths correspond to the combined GS+ES emission. For the sample with a different InAs thickness, the deconvolved GS spectral width is still larger than 100 nm. The optimized CMQDs with 124 nm PL linewidth (structure B), once incorporated in an SLED device structure, result in a smooth spectral width (GS+ES) of 115 nm [17] as will be detailed in section 19.4.2.

19.2.3   Conclusions and perspectives

In this section we investigated the growth of size inhomogeneous InAs QDs and CMQD structures, in order to broaden the gain spectrum of the QD active region in the 1200–1300 nm wavelength range. By intentionally increasing the size dispersion of InAs QDs, a broad PL spectral width is achieved from a single layer of QDs. No evidence of degradation of optical properties was observed. The results confirm that the CMQD structure is effective for devices which need a broad-gain spectrum such as SLEDs, semiconductor optical amplifiers and tunable lasers. Although this discussion has been conducted on the InAs/GaAs material system for applications in the 1300 nm region, the concepts and approaches can be easily extended to the other material systems. In particular, it becomes possible to realize QD or quantum dash SLEDs covering various wavelength ranges based on the material systems such as InAs QDs/quantum dashes on InP substrate (1550 nm telecom wavelength) [30, 31], InP/InGaP QDs on GaAs substrate (700–800 nm red emission) [32, 33], and InGaN QDs on GaN substrate (400–500 nm blue emission) [34, 35].

19.3   Wide-spectrum InAs/GaAs QD SLEDs

19.3.1   Gain and length requirements

The typical structure of QD-SLEDs emitting in the 1.2–1.3 μm wavelength region is made of a certain number of QD layers separated by 35–40 nm GaAs spacers to avoid strain and coupling effects, and embedded in the GaAs intrinsic region of a p-i-n junction. The p- and n-regions are $Al_xGa_{1-x}As$ layers, which provide both optical and electrical confinement. Starting from these structures, devices are realized by dry-etching tilted or bent waveguides (see inset of Fig. 19.11) which are then passivated by BCB planarization or by steam oxidation of the exposed AlGaAs cladding depending on the Al content in the alloy. Even though high Al content results in a high optical confinement, a low Al concentration can be desirable for the applications, to minimize the reliability issues related to the aluminium oxidation and to improve the far-field aspect ratio of the emission. Ridges are a few μm in lateral size to ensure a single lateral mode operation and therefore an efficient coupling into tapered single-mode fibres (the typical coupling efficiency is 40–50% when the SLED operates in the amplification regime), which is highly desirable for applications. The last fabrication steps are p-contact (Ti/Pt/Au) and n-contact (Ni/Ge/Au/Ni/Au) deposition, on top and on the substrate side, respectively, and finally cleaving chips of the proper length.

   The most important parameters related to the superluminescent diode output power are the chip length L and net gain g, which relates to modal gain g_{mod} and internal loss α_i through the relation $g = g_{mod} - \alpha_i$. As mentioned in section 19.1.3, in a simplified picture where the output facet reflectivities are assumed be negligible and the gain and spontaneous emission rate are assumed to be independent from positions inside the waveguide, an exponential increase in output power is expected when increasing g or L:

$$P_{out} = \frac{\beta P_{sp}}{g}\left[e^{gL} - 1\right] \tag{19.4}$$

In Fig. 19.9 a the output power measured over a series of QD SLEDs with similar geometrical characteristics (tilted waveguides with ridge widths of 3–4 μm, 4 mm long) are reported in the

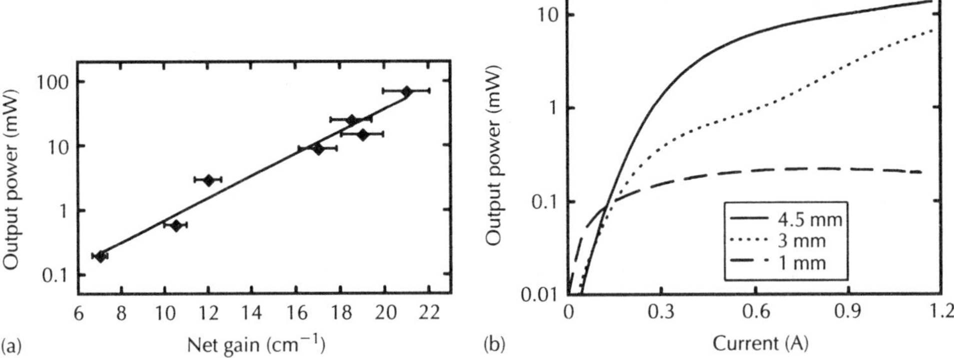

Figure 19.9   (a) Output power measured over a series of 4 mm long, tilted QD-SLEDs in the regime of GS = ES emission, versus the corresponding maximum net gain. The data are fitted using Eq. 19.4. (b) L–I characteristics of tilted-waveguide SLEDs for varying device length (fabricated from the same wafer).

regime of emission where GS = ES versus the measured maximum net gain g_{max}. In spite of the many differences between the samples (number of QD layers, cladding compositions, bandwidths, etc.) the data points show a clear exponential increase with increasing gain, in agreement with Eq. 19.4, and provide a path to the achievement of high optical powers. Outputs of 10 mW or higher can be achieved only when g_{max} is of the order of 18–20 cm^{-1} (i.e. chip gain $g_{max}L \simeq 8$). A fit is reported in the plot, which was made using Eq. 19.4, where $\beta P_{sp}/g$ was assumed constant and used as a fitting parameter and L was fixed to 4 mm. This corresponds to a spontaneously emitted output power per unit length coupled into the guided mode at a gain of 20 cm^{-1} of 25 µW mm^{-1}.

The effect of the device length on the *L–I* characteristics of tilted waveguide superluminescent diodes can be observed in Fig. 19.9b. The Figure shows the *L–I* characteristics of three devices fabricated from the same wafer and cleaved to 1 mm, 3 mm and 4.5 mm length. These devices contained five identical QD layers with a relatively narrow gain spectrum (20–40 nm around 1.3 µm, depending on the injection and amplification regime) embedded into a GaAs/ $Al_{0.8}Ga_{0.2}As$ waveguide. Measurements were performed in pulsed regime, at room temperature. The output power of the 1 mm long devices saturates around 300 µW and does not show any evident exponential characteristic in the *L–I* curve, which would be typical of the superluminescent emission. The 3 mm long device in contrast reaches output powers in the mW range. For this device two different regimes of amplification can be observed, one at low and one at higher injection, which are related to recombination from different bound states in the QDs. The double exponential increase is an intrinsic phenomenon in QD superluminescent diodes where the relatively low density of states of the lower energy levels leads to state filling and subsequent population of the higher energy levels. This phenomenon, which can be exploited in quantum dots for the achievement of very large bandwidths combining the emission from different states, will be the subject of a deeper analysis in the next paragraphs. The *L–I* characteristics of the 4.5 mm long SLED reported in Fig. 19.9b show even higher amplification with output powers exceeding 10 mW mainly composed of ground state emission.

The two parameters having an important effect on the maximum modal gain g_{mod}^{max} are the number of QDs contributing to the emission and the gain spectral bandwidth. The number of dots introduced in the device active region, which are related to the DOS of the system, is directly proportional to the modal gain. Figure 19.10a shows g_{mod}^{max} measured over three samples containing five, ten and 18 QD layers, at the peak wavelength, all of them having similar PL linewidths of about 80 nm. These devices show a typical g_{mod}^{max} of 1–2 cm^{-1} per dot layer at the peak position, which becomes slightly higher for samples with a narrower inhomogeneous broadening. The line in the plot is a guide to the eye and has been intentionally considered sub-linear. This is because the three waveguiding structures are not identical and the width t_{opt} of the optical

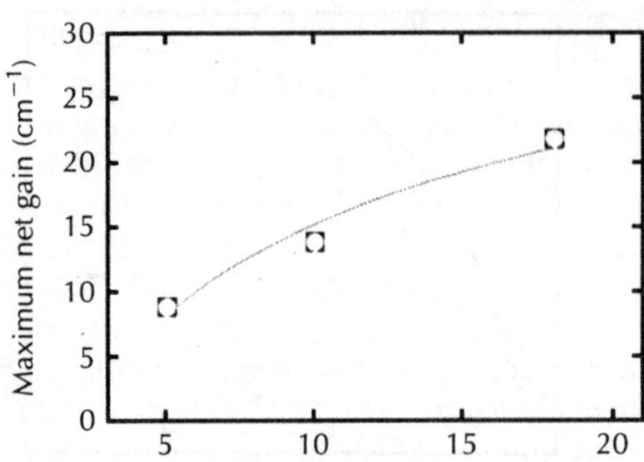
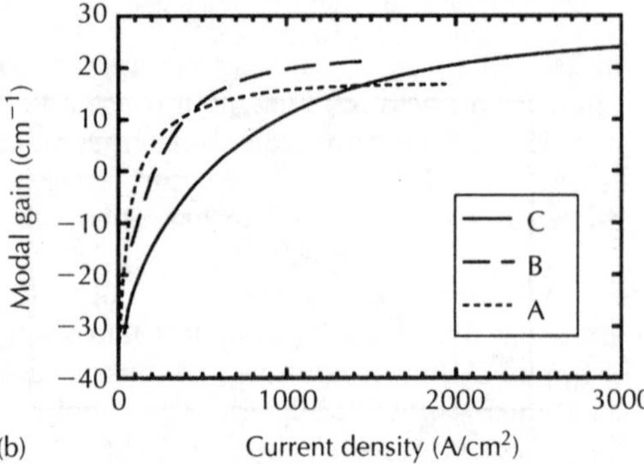

Figure 19.10   (a) Maximum modal gain versus number of QD layers for samples showing a comparable PL linewidth. (b) Modal gain of 2 mm long and 3 µm wide tilted ridge-waveguide devices with identical epitaxial structure but different doping levels in the active region.

mode in the growth direction is expected to be larger for the five and ten QD layer samples, leading to a smaller confinement factor (g_{mod}^{max} is inversely proportional to t_{opt}). t_{opt} indeed depends on the device geometry and in particular varies proportionally to the GaAs core region thickness, and depends on the refractive index contrast between active region and claddings (higher Al content in the claddings means higher refractive index contrast and higher confinement and therefore smaller t_{opt}). The five QD layer sample has 80% Al concentration against 35% for the other two samples. Also, the GaAs core region is thicker in the 18 QD layer sample than in the others. As the spacing between different dot layers must be kept large (35–40 nm) to reduce strain and coupling effects, the higher number of QD layers requires the growth of a thicker active region, thus resulting in a larger t_{opt}.

Once the number of QD layers is fixed (each layer with approximately fixed QD density of 3.10^{10} cm^{-2}), the material gain at the peak position depends only on the shape of the broadened density of states, whose integrated value is proportional to the total number of dots. If we consider Gaussian broadenings for the GS DOS, which is usually the case for single QD layers as confirmed by the shape of the PL emission, the peak gain is reduced proportionally to the increase in width of the broadening. The decrease in gain was observed measuring the peak gain in samples with ten QD layers and a different PL linewidth. g_{mod}^{max} decreased from 20 to 15 cm^{-1} for a low-excitation PL linewidth increase from 50 to 80 nm.

Number of QD layers and inhomogeneous broadening are not the only two parameters affecting the gain characteristics of QD-SLEDs. The thermal redistribution of carriers over the energy levels of a semiconductor is one of the most important parameters limiting the gain performance of devices operating at room temperature. This is true for bulk materials, where electrons and holes occupy the continuous energy bands with quasi-Fermi distributions, but also for QDs where the presence of close excited state energy levels may reduce the occupation and therefore the gain of the lower energy levels. In [36], for example, is shown how the proximity of the WL levels (especially in the valence band) of QDs emitting around 1000 nm generates a carrier redistribution reducing the maximum modal gain. Furthermore, the presence of many closely spaced energy levels for the holes in the valence band could produce a similar effect. A few years ago the introduction of *p*-doping in the QD active region was proposed as one possible solution to overcome this problem by forcing a certain number of extra holes to be in the dots [37]. Since this moment, many groups all over the world have been studying the effect of p-doping on the temperature, gain and modulation characteristics of quantum dot lasers. Here we report the effect of p-doping on the gain characteristics of the QD material. Later on, the effect of p-doping on the *L–I* and temperature characteristics of SLEDs will be discussed. Three samples with identical epitaxial structure, grown under the same conditions, but with different doping levels in the active region were processed into SLEDs. Each of them contains ten InAs QD layers in a GaAs active region embedded between 1.5 µm thick Al$_{0.35}$Ga$_{0.65}$As cladding layers. In sample A the dot layers are separated by undoped GaAs spacers, while in samples B and C carbon doping is introduced inside a thin region within each GaAs spacer (10 nm thick regions, 10 nm above each QD layer). The doping level was estimated to correspond to the introduction of about eight and 15 extra holes per QD, for samples B and C, respectively. The spectral characteristics showed very similar behaviour for the three samples, except for small variations in the spectral centre wavelength (1267, 1270 and 1300 nm for samples A, B and C, respectively), demonstrating that for the considered doping levels the dispersion of the QD density of states is not substantially modified. In Fig. 19.10b the modal gain curves for TE polarization (electric field perpendicular to the growth direction) at the GS peak position versus current density are reported for the three samples. Each line interpolates a series of 80 data points obtained through the analysis of the amplification undertaken by a tunable laser injected in the tilted waveguides and tuned at the GS peak wavelength. The measurements were verified to be reproducible over several nominally identical devices. The plot shows an increasing maximum modal gain with increasing doping level. The maximum value of g_{mod} for samples A, B and C is 18, 22 and 25 cm^{-1}, respectively. The increase can be attributed to the modified carrier distribution in the valence band due to the introduction of several acceptors per QD, which has the effect of pushing the quasi-Fermi level deeply inside the band, and thus increasing the hole population contributing to the gain. We note that an increased injection current is needed in p-doped devices to achieve the same gain level. P-doping is expected to increase monomolecular (through dopant-related defects), radiative and Auger (through increased hole population) recombination rates. A combination of these effects is likely to produce the shift of the gain curves towards higher current. P-doping has an impact on the optical losses as well, through increased free carrier absorption and photon scattering due to the introduction in the active region of doping-related crystal defects. From the measured Fabry–Perot fringe visibility and laser thresholds in untilted devices we estimate the optical losses as $\alpha_i = 1.8$ cm^{-1}, 3.5 and 5 cm^{-1} for samples A, B and C, respectively. The increase in modal gain in p-doped structures B and C exceeds the increase in optical losses, resulting in a larger net gain which, as we will see in the next section, results in higher output power for p-doped SLEDs.

As already discussed, the SLED output power does not depend simply on the material gain but also on the distance travelled by photons inside the waveguide before output. In tilted devices this can be assumed to be equal to the device length (single-pass amplification). In contrast, combining the low reflectivity of one facet with the high reflectivity of the opposite, in bent devices the photon may cover a larger length than the cavity length before output. This leads to higher light amplification and therefore higher output power, provided that the waveguide curvature does not introduce significantly higher internal loss. With this purpose bent stripes were realized by using an arc of circumference whose tangents at the two end points are 7° tilted and perpendicular to the opposite facets of a 2 mm long cavity, respectively. Longer cavities were also realized by adding a straight piece of waveguide to the perpendicular facet side. This combines the low reflectivity of the tilted facet with the high reflectivity of the perpendicular facet so that the light travelling inside the waveguide may experience double-pass amplification. A comparison between the L–I characteristics of a 2 mm long bent and 2 mm long tilted waveguide are reported in Fig. 19.11. Both devices were fabricated with the same process and from the same wafer, thus demonstrating that the bent structure is effective in increasing the light amplification if compared to a single-pass guide. While the output power of the tilted waveguide is limited to few mW at high injection, bent waveguides achieve 30 mW at the same injection.

In the following paragraphs, we will first describe the characteristics of SLEDs fabricated from samples with a narrow gain, optimized for emission at 1.3 μm, as they are the standard gain material for high-performance QD lasers. This results in devices with narrow optical bandwidths but at the same time provides useful information for the understanding of the device characteristics. Then, in section 19.3.3 we will discuss the properties of SLEDs with chirped active regions showing that through the optimization of growth conditions – as discussed in section 19.2 – the device output may exhibit bandwidths larger than 100 nm. The large bandwidth is obtained in a regime of two-state emission (GS + ES) that, for optimized chirping, results in a smooth spectral superposition (flat top). The coherence properties and temperature characteristics of both types of devices are addressed in sections 19.3.4 and 19.3.5.

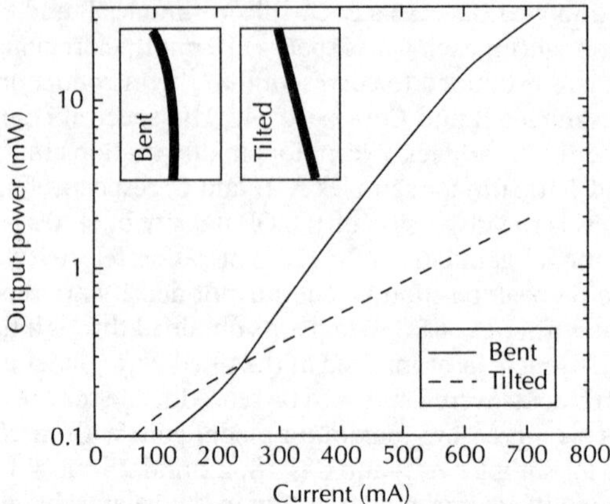

Figure 19.11   Bent versus tilted SLED structures (2 mm long chips): bent waveguides allow the achievement of higher output powers at the same injection levels. A schematic drawing of the two structures is reported in the inset.

19.3.2   Narrow-gain devices

In this paragraph we report the spectral and L–I characteristics of SLEDs realized from samples containing five and ten identical QD layers. More details about these devices can be found in [13, 15]. Tilted-ridge SLEDs fabricated from these samples were mounted p-side up on a temperature controlled heat-sink, and tested under pulsed operation. In Fig. 19.12a the L–I characteristics of a 3 mm long SLED are reported for the heat-sink temperature of 10°C. Two separated superlinear behaviours can be identified in the curve: one at low injection (0–200 mA) and one at higher injection (600–800 mA) which are related to the recombination from GS and ES energy levels of the dot. The superlinear increase in output power together with the high power (of the order of 10 mW at high injection) are the signature of operation in the gain regime.

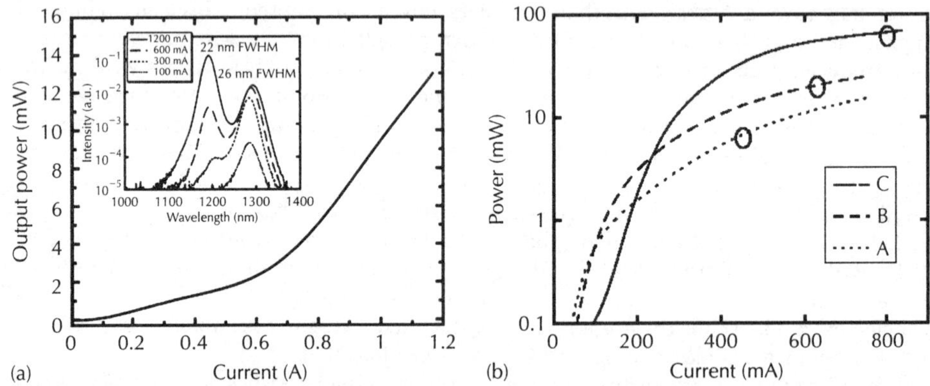

Figure 19.12   (a) L–I characteristics of a 3 mm long tilted ridge-waveguide SLED containing five QD layers, measured at 10°C in pulsed regime. The corresponding spectra are reported in the inset. (b) L–I characteristics of 4 mm long tilted ridge-waveguide SLEDs containing ten QD layers and different p-doping levels in the active region. The circles identify the regime of emission where GS and ES contributions to the spectra are comparable.

A better understanding of the L–I characteristics can be achieved observing the corresponding spectra reported in the inset of Fig. 19.12a. At low injection the GS provides the main contribution to the emission with 26 nm full width at half maximum (FWHM) and up to 2 mW output power. At higher currents this emission saturates because of the GS filling and a second line due to recombination from the first ES appears. This line is eventually dominant (exceeding 10 mW) at high injection currents due to the higher degeneracy of ES energy levels and shows 22 nm FWHM. The ES level can contribute to gain with twice as many carriers as the corresponding GS level due to higher angular momentum degeneracy [38]. The multistate emission, peculiar in QD systems, is promising for the achievement of very broad spectral characteristics, provided that the emission from each line can be broadened enough to achieve a smooth spectral superposition at the injection level when the intensity of the two lines is comparable.

In the previous paragraph we discussed the influence of p-doping on the modal gain, reporting an increase of g_{mod}^{max} for increasing doping level. Figure 19.12b reports the L–I characteristics of three identical 4 mm long SLEDs realized from the same samples whose gain curves are reported in Fig. 19.10. The effect of p-doping is confirmed by the analysis of the L–I characteristics shown in Fig. 19.12b. The output powers in the regime of GS = ES emissions (circles in figure) go up to 60–70 mW for the highly p-doped device while stay limited to less than 10 mW for the undoped one. These measurements demonstrate that the use of p-doping in QD superluminescent diodes results in increased output powers and confirm the estimation of a modal gain increase larger than the internal loss increase for the considered doping levels.

These samples, as the five QD layers sample previously presented, were not optimized for achieving large bandwidths and the spectral characteristics were very similar to the ones reported in the inset of Fig. 19.12a, with well-separated GS and ES emissions. In the next paragraph we report the characteristics of QD-SLEDs with improved spectral characteristics.

19.3.3 Wide-gain devices

As discussed in sections 19.2.1 and 19.2.2 the spectral characteristics of QD-SLEDs can be improved through the use of chirped multilayers and/or increasing the inhomogeneous broadening of each QD layer by optimizing the growth conditions. Some feedback about the optical properties of such structures can be obtained through the analysis of the spectral and L–I characteristics of ridge-waveguide lasers. As showed in Fig. 19.13a, 2 mm long lasers realized from a five-QD layer chirped structure (PL emission reported in Fig. 19.5) show low threshold current densities of about 200 A/cm^2 and high slope efficiencies of 0.22 W/A per facet, signatures of high recombination efficiency and therefore good crystal quality. Figure 19.13b displays the corresponding spectra. The device starts lasing through GS recombination around 1265 nm. The emission line progressively broadens with increasing current up to a linewidth of 35 nm, demonstrating that chirping is effective in broadening the gain spectrum. At very high injection lasing takes place also through ES recombinations and a second lasing line appears also on the high-energy side of the spectrum, as already observed on standard 1.3 μm lasers [39, 40]. The very large lasing line, stemming from a uniformly broadened spectral gain, is very promising for the

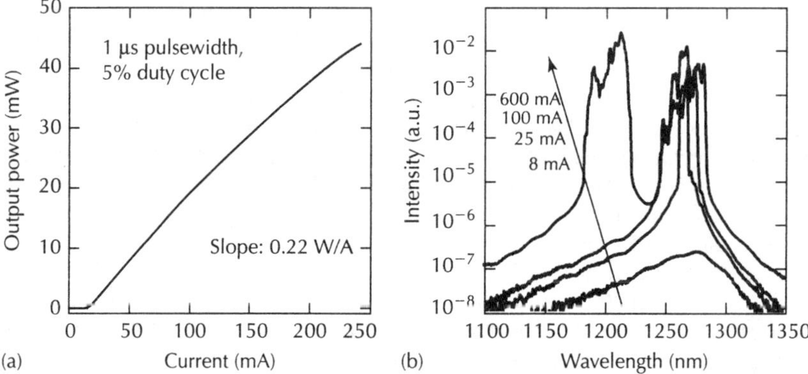

Figure 19.13 (a) L–I *characteristics of a 2 mm long and 5 μm wide Fabry–Perot ridge-waveguide laser realized using five chirped QD layers.* (b) *Corresponding spectra.*

realization of tunable lasers with large tunability ranges. In the framework of the project EU FP6 integrated project "ZODIAC", tunable distributed feedback lasers were realized from this sample with a tunability range of about 70 nm.

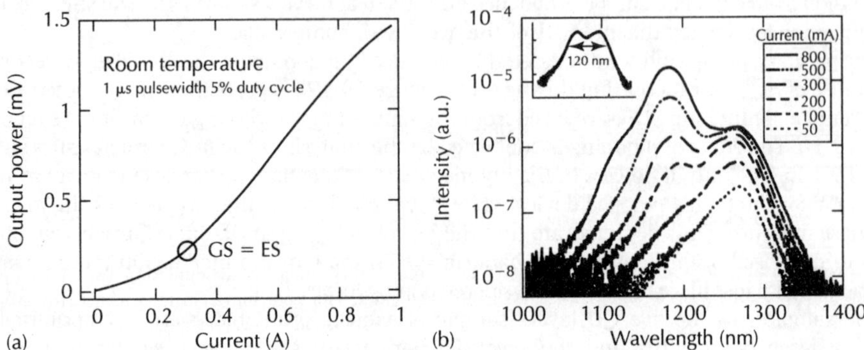

Figure 19.14   (a) L–I curve of a 4 mm long SLED realized using five chirped QD layers, measured at room temperature in pulsed regime. The circle identifies the regime of comparable GS and ES emissions. (b) Corresponding spectra. The 300 mA spectrum is reported in the inset, showing a bandwidth of 120 nm.

Tilted ridge-waveguide SLEDs fabricated from the same wafer show a behaviour similar to the one reported for the narrow gain sample but with smaller saturated output powers, which results from the decreased maximum modal gain produced by the bandwidth increase [14, 17]. In Fig. 19.14a the room-temperature L–I characteristics of a 4 mm long and 5 μm wide SLED are reported. The saturated output power in this case goes up to a few hundred μW for the GS emission and up to a few mW when the SLED operates in the ES emission regime. Figure 19.14b displays the spectral characteristics of the same device. At low injection current the emission from the GS dominates, with a spectral width of 55 nm (FWHM) centred around 1265 nm. The ground state then saturates with increasing current and the carriers begin to fill the excited states, resulting in a broadening of the spectrum on the short wavelength side as already observed on a narrow gain sample. At currents >400 mA the ES emission eventually dominates the spectrum, due to the higher maximum gain on the ES. At 300 mA, due to simultaneous contribution of the two states, a combined spectral width of 120 nm is obtained, with a 3 dB dip between GS and ES peaks (see spectrum in inset). This bandwidth is very promising for the application in high-resolution OCT systems; however, the output power in this regime is limited to a few hundred μW and does not meet the mW-range output powers required by the application. A way to achieve higher output powers, as detailed in section 19.3.1, is the use of a higher number of QD layers in the active region, p-doping the active region and using bent waveguides. Tilted-ridge SLEDs, 4 mm long, with similar linewidth characteristics but a higher number of QD layers in the active region result in improved output power as shown in the inset of Fig. 19.15a. The power in the regime of large bandwidth ($\geq$100 nm) increases from the 300 μW of the five QD layer SLED up to 2–3 mW for a ten QD layer SLED and up to 10–15 mW for an 18 QD layer SLED. Both the ten and the 18 QD layer samples maintain a spectral characteristic similar to the one of the five QD layer sample, with a 3 dB peak to valley ratio between GS and ES. Best SLED performance in a chirped structure can be achieved using 18 QD layers with p-doped GaAs spacers (10 nm thick regions, 10 nm above each QD layer, doping level 5e17 cm^{-3}) and $Al_{0.35}Ga_{0.65}As$ claddings. Three repetitions of six chirped QD layers were used, where chirping was performed using different In content in the capping layers of five QD layers and also adjusting the amount of InAs deposed to grow the QDs of the sixth layer.

In Fig. 19.15a we report the L–I characteristics of a 4 mm long tilted SLED realized from this structure. The measurements were taken in pulsed regime (50 kHz, 4% duty cycle, at a heat-sink temperature of 20°C. The output power reaches several tens of mW at the highest current levels. A similar contrbution from GS and ES (broad spectral regime indicated by a circle) is achieved at a current of 1200 mA, corresponding to a high output power of $\approx$ 30 mW. This could be further increased by applying anti-reflection (AR) coatings.

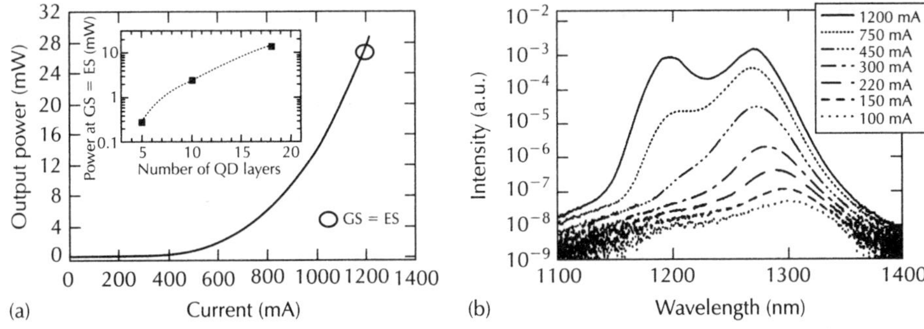

Figure 19.15 *(a) L–I characteristics of a 4 mm long and 3 μm wide tilted SLED containing 18 chirped QD layers and p-doping. In the inset are reported the output powers of SLEDs containing five, ten and 18 undoped QD layers (squares) emitting in the GS = ES regime. The dashed line is a guide to the eye. (b) Spectra corresponding to the L–I characteristics shown in (a).*

Figure 19.15b shows the corresponding spectral characteristics. The device shows the usual GS–ES spectral behaviour, with a large GS + ES combined bandwidth at high injection. A bandwidth of 100 nm (FWHM) centred at 1235 nm, with a relatively small 5 dB valley between GS and ES emissions is reached around 1200 mA. The pulsed operation provides an estimate of the best performance that could be achieved with these devices; however, a CW operation is necessary for the applications, which could be limited by heating at the high currents where the best bandwidth/power performance is achieved. In this regard the use of bent-stripe devices may be helpful, because it allows the achievement of higher output powers at lower currents. The application of bent structures also opens the way to the use of shorter cavities which may be desirable for mass production, and which could be further pursued by increasing the reflectivity of the straight facet through high-reflection (HR) coatings. This kind of structure can compensate for the low gain showed by QDs emitting in the 1.3 μm region as compared to quantum wells (QWs). A comparison between the pulsed and CW L–I characteristics measured on a 2 mm bent SLED are reported in Fig. 19.16. CW and pulsed characteristics are superposed up to 300–400 mA (corresponding to 5–7 kA/cm²). At higher bias the CW characteristics become dominated by heating effects and show a roll-over around 700 mA. The CW output power in the regime of large bandwidth is about 2–3 mW corresponding to 100 nm spectral bandwidth and a 3 dB dip between GS and ES peaks. The spectral characteristics are shown in the inset. Higher output power could be achieved by applying HR coatings on the straight facet and AR coatings on the tilted facet.

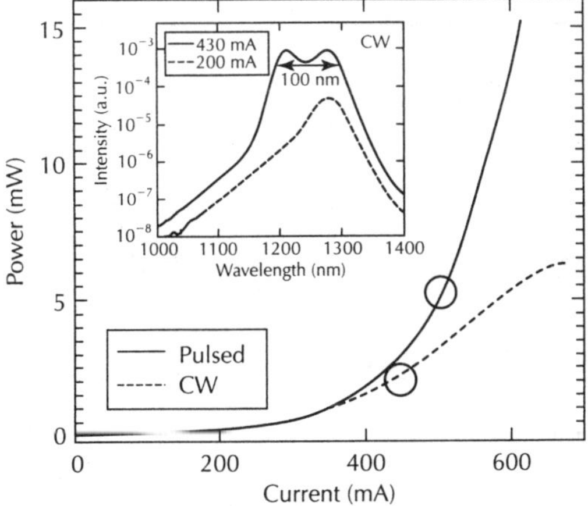

Figure 19.16 *Pulsed (continuous line) and CW (dashed line) L–I characteristics of a 2 mm long bent QD-SLED containing 18 p-doped QD layers. The corresponding CW spectra are reported in the inset.*

The modulation between GS and ES emissions affects the coherence properties of the emission resulting in a coherence function with non-negligible side lobes, which can reduce the effective resolution achieved once the SLEDs are applied to OCT. A detailed description of the coherence properties of QD-SLEDs is reported in the next section.

19.3.4　Coherence properties

The coherence length l_c is a quantity frequently used to characterize the coherence properties of the light source, as it is directly related to the resolution achievable in the OCT technique. l_c is related to the maximum path difference between the two arms of an optical interferometer for which the light wave is still capable of generating an interference pattern and it is often defined as the FWHM of the first order degree of coherence. To a first approximation it can also be expressed in terms of the spectral width as:

$$l_c = k \frac{\lambda^2}{\Delta\lambda} \tag{19.5}$$

where λ is the central wavelength, $\Delta\lambda$ the spectral broadening and k is a spectrum form factor corresponding to 0.32 for Lorentzian-, to 0.66 for Gaussian- and to 1 for rectangular-shaped spectra (values derived using the power-equivalent width of the corresponding coherence functions [41]).

The first-order degree of coherence is easily measured with an optical interferometer. Besides the FWHM, its shape is also of interest for OCT. The spectral modulations (as the valley between GS and ES emissions) result in short-range side lobes of the coherence function which can reduce the effective OCT resolution, while the presence of a short-range spectral modulation (as the residual FP modulation) results in long-range secondary subpeaks. The residual Fabry–Perot modulation (commonly called spectral ripple) results in parasitic subpeaks of the coherence function at an optical path difference equal to $2n_{eff}L$ were L is the SLED length and n_{eff} the refractive index for the guided mode. A high intensity of such secondary subpeaks, which is determined by the spectral modulation depth, is undesirable as it may produce distortion effects when using the SLED for imaging. As QD-SLEDs usually present low gain values, ripple is not expected to be a concern and its analysis will be omitted here. Also, the devices were not optimized to minimize the optical feedback through the application of AR coatings, which would be desirable for the analysis of the secondary subpeaks of the coherence function.

Figure 19.17b displays the short-range fringe visibility measured with a Michelson interferometer-based optical spectrum analyser on two different 2 mm long tilted SLEDs, one

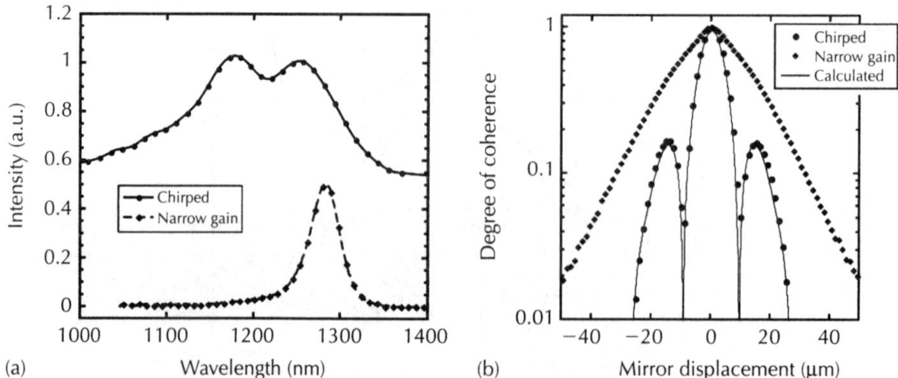

Figure 19.17　(a) CW spectra measured over a narrow gain and a chirped 2 mm long SLED. The spectra have been shifted vertically for clarity. (b) Corresponding degree of coherence measured observing the fringes' visibility in a Michelson interferometer (dots and rhombs). The degree of coherence calculated through inverse Fourier transform of the chirped spectrum is also reported (continuous line).

with a narrow gain spectrum and the other chirped. The measurements are reported for the narrow gain device emitting in a pure GS regime (rhombs) and for the chirped device emitting in a GS + ES regime (dots). The corresponding spectra are displayed in Fig. 19.17a in linear scale, and show an FWHM of 45 and 180 nm, corresponding to an FWHM of the measured degree of coherence of 28 and 9 μm, respectively. The chirped spectrum has been shifted vertically for clarity. While the single peak spectra correspond to a single peak coherence function (in the ideal case of a Gaussian spectrum the coherence function is also Gaussian), the chirped GS + ES emission generates side lobes that could result in a reduced OCT resolution. The side lobes in the degree of coherence of the chirped spectrum show an 8 dB suppression ratio. During these measurements the chirped sample operated on a very low gain regime and low output power of a few hundred μW (five QD layers, 2 mm tilted cavity) which resulted in the extremely broad bandwidth of 180 nm and $l_c \approx 10\,\mu m$. On a high-gain SLED the bandwidth is rather expected to be about 100 nm, which would result in $l_c \approx 15\,\mu m$.

Spectral emission and coherence function are connected through a Fourier transform. It can be shown that the spectral density $S(v)$ and coherence function $G(\tau)$ of the radiation are related through the Wiener–Khinchin theorem [41]. The theorem is a useful tool as it allows the calculation of the degree of coherence starting from any arbitrary spectral shape of the emission, thus avoiding repetitive and time-consuming experiments. We checked the validity of calculations through a comparison with the coherence measurements. Together with measurements, the degree of coherence calculated through inverse Fourier transform of the chirped spectrum is reported in Fig. 19.17b (continuous line). We find a very good agreement, with almost negligible errors over more than two orders of magnitude. The calculations can then be applied to evaluate the coherence function for SLEDs emitting in the regime of high power and GS + ES emission, as for the case of the 18 QD layer, p-doped SLEDs described in the previous paragraph. The degree of coherence calculated from a spectrum with 100 nm bandwidth and a 3 dB dip between GS and ES emissions (spectrum in inset of Fig. 19.16) shows 15 μm FWHM and a 5 dB suppression ratio of side lobes. As we will see in section 19.4, a much better side lobe suppression can be obtained with a flat-top spectral emission.

19.3.5   Temperature characteristics

Since their theoretical proposal QD devices have been predicted to show better temperature stability when compared to their QW or bulk counterpart [42, 43]. Due to the discrete nature of QD energy levels, carriers are supposed to suffer a smaller thermal dispersion resulting in a lower thermal sensitivity of devices. In spite of these predictions most of the devices realized so far showed a temperature dependence similar to the one obtained on InP-based devices for 1.3 μm emission, with laser characteristic temperature T_0 smaller than 100 K in the 20–80°C interval [44–47]. To explain the temperature sensitivity of the QD lasers, several different mechanisms have been proposed in the literature. Re-emission of carriers towards the barrier and wetting layer activated by increasing temperature and the subsequent radiative and/or non-radiative recombinations outside the dots is one of these [48, 49]. Indeed Matthews *et al.* showed that the temperature sensitivity of short wavelength (1 μm) QD lasers is caused by gain saturation due to the presence of a high density of states in the WL [36]. Other authors, in contrast, have attributed the temperature characteristics of QD lasers to the thermal sensitivity of non-radiative recombinations [50] or to the spreading of holes over the closely spaced energy levels of the valence band [37]. The situation is different in p-doped QD devices, for which lasers with improved temperature characteristics have been recently demonstrated [56–58]. The idea of using p-doping to reduce the temperature dependence, which was first proposed in QDs by Shchekin and Deppe [37], derives from the hypothesis of a thermal dependence governed by the spreading of holes over the closely spaced energy levels in the valence band. Forcing a high number of holes to be confined in the WL and QD valence band through p-doping should therefore reduce the influence of thermal spreading. In contrast to this picture, two independent authors have recently reported that the temperature dependence of the current threshold in p-doped QD lasers at room temperature stems from a redistribution of carriers over the QD ensemble [51, 52]. This is likely to happen around room temperature (in undoped devices a similar phenomenon takes place around 200 K) due to the strong Coulomb attraction in p-doped devices, which results in an increased potential barrier. As we will see in the following paragraph, p-doping is not only beneficial to improve the gain characteristics of QD-SLEDs, but also results in a higher temperature stability of the SLED L–I curves. While the temperature dependence in undoped devices can be explained in terms of a temperature increasing non-radiative recombination, a more complex dynamics is at the origin of the higher stability in p-doped structures.

Figure 19.18 shows the typical temperature characteristics of undoped QD-SLEDs. Temperature controlled measurements were performed mounting the devices p-side up on a copper sub-mount with the double function of n-electrode and heat-sink. The temperature of the heat-sink was controlled with a Peltier thermoelectric cooler. In the left panel the L–I characteristics of 4 mm long and 3 μm wide narrow gain device (ten QD layers) are reported for increasing temperatures in the interval 20–85°C. At high injection the output power exhibits a thermal degradation of 15 dB, which is similar to the decrease observed in SLEDs realized with the competing InP-based technology. The reduction in output power is associated with a reshaping of the emission spectra, which vary their bandwidth and GS/ES intensity ratio. The typical spectral behaviour is shown in the right panel of Fig. 19.18 where the spectra of the chirped five QD layer SLED described in section 19.3.3 are displayed for increasing temperature (20–60°C), at the fixed current of 300 mA. The corresponding L–I caracteristics are reported in the inset, with a decrease similar to the narrow gain sample. The current was fixed at 300 mA so as to have a broad spectral emission (GS = ES) at 20°C. When the temperature is increased the spectral intensity is decreased with a higher suppression for the ES emission than for the GS, which must be related to a smaller steady-state population of carriers in the dots. One possible explanation for the reduced carrier density versus temperature might be found in an increase in non-radiative recombinations. The overall spectrum moves to the red linearly with a rate of about 0.5 nm/K due to the usual T-dependence of the band gap in InAs and GaAs.

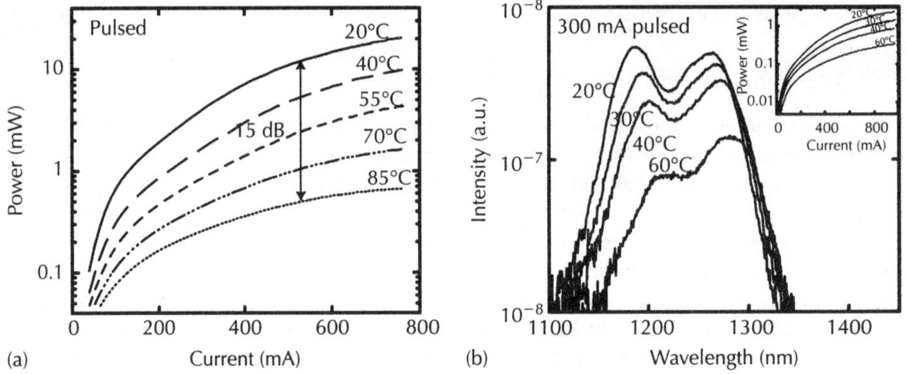

Figure 19.18   (a) L–I characteristics for varying temperatures of a 4 mm long, 3 μm wide narrow-gain SLED with tilted ridge-waveguides. (b) Spectral characteristics of a 4 mm long, 5 μm wide chirped SLED with tilted ridge-waveguides for varying temperatures. The corresponding L–I curves are shown in the inset.

As we have described in section 19.3.1 the L–I characteristics of a superluminescent diode are determined by modal gain and internal losses. Both these quantities were therefore analysed versus temperature to identify the origin of the temperature degradation. Figure 19.19a shows the modal gain curves measured on a tilted ridge-waveguide at 20, 40, 60 and 80°C at the GS peak position (narrow gain, same sample as Fig. 19.18a). The gain was measured with the technique described in section 19.3.1, studying the amplification of a tunable laser injected and detected from the opposite facets of the waveguide. The laser wavelength was adjusted to the GS peak position for each temperature. The curves show a stronger saturation of gain for increasing temperature. This behaviour was confirmed over a series of different samples with undoped active regions and is therefore a general finding. The internal losses, in contrast, show a completely temperature independent characteristic. They were measured through the analysis of the transmission of the tunable laser tuned out of resonance with the QD spectrum. A plot of the experimental points together with a linear fit are displayed in the inset of Fig. 19.19a. SLED characteristics are therefore completely determined by the temperature variation in modal gain.

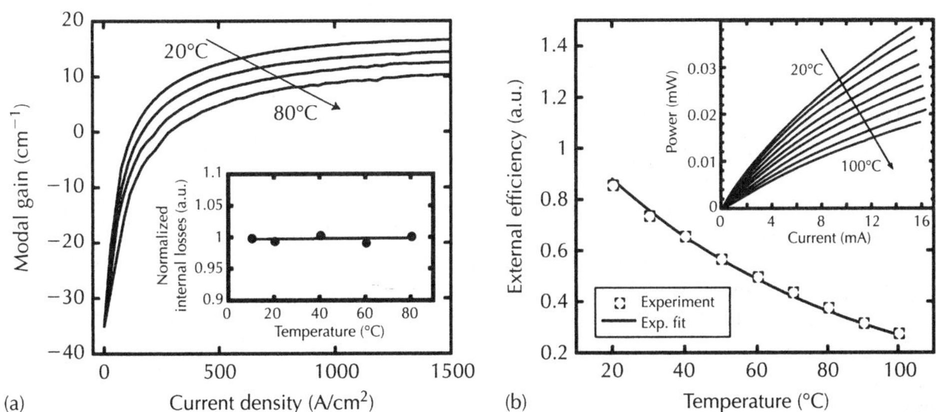

Figure 19.19   (a) Modal gain curves for different temperatures, measured on a narrow-gain 2 mm long tilted ridge-waveguide (realized from the same wafer of the SLEDs whose L–I characteristics are reported in Fig. 19.18). The normalized internal loss versus temperature is reported in the inset (circles) together with a linear fit. (b) Single facet external efficiencies of short tilted waveguides (500 μm) realized from the same wafer. The points are fitted assuming an exponential decrease of the ratio τ_{sp}/τ_{nr}. The corresponding L–I characteristics are displayed in inset.

Two physical processes are seen as possibly responsible for the temperature dependence of gain in these QDs. The first is the dispersion of holes over the closely spaced energy levels in the valence band. This would result in an increase in the spontaneous emission decay time τ_{sp} due to the decreased probability of electron–hole recombination. The second is the increase in non-radiative recombination which can derive from Auger [53] and/or defect assisted processes [54]. This would result in a decrease in the non-radiative lifetime τ_{nr}, thus subtracting carriers to the radiative recombinations. Additional insight about the different processes is obtained from the L–I characteristics of LEDs versus temperature. Indeed, in an LED the effects related to stimulated emission and absorption can be neglected resulting in a simple dependence of the emitted output power on the carrier populations. This can be done measuring the low-injection, edge-emitted electroluminescence from short tilted waveguides $L = 500$ μm (where gain and losses can be neglected), fabricated on the same wafer as the SLED. The assumption of low gain and losses is experimentally confirmed by the absence of any evident spectral narrowing of the edge-emitted electroluminescence for increasing current injections. The inset of Fig. 19.19b shows the L–I characteristics of such a structure in the 20–100°C temperature interval. The efficiency decrease versus temperature is exponential at all the current levels, and a characteristic temperature $T_0 = 70$ K is found by exponential fitting of the low injection efficiency. The experimental points together with the fit are displayed in Fig. 19.19b. A similar T_0 is found from fitting of the threshold characteristics of Fabry–Perot lasers realized from the same sample, suggesting that the same processes are at the origin of the temperature characteristics of lasers around room temperature. However, as the internal efficiency at fixed injection in an LED structure is expected to vary proportionally to the ratio τ_{nr}/τ_{sp}, the intensity decrease of the curves versus temperature does not allow one to distinguish between an increase in the spontaneous emission decay time τ_{sp} and a decrease in the non-radiative lifetime τ_{nr}.

Further information about the temperature behaviour is provided by the analysis of the temperature-dependent decay time of the photoluminescence signal. Measurements were performed on a test sample containing one QD layer grown under the same conditions used for the laser structure. At low temperature (0–200 K) the decay time slightly increases which can be attributed to an increase of τ_{sp} determined by the hole spreading over the closely spaced energy levels in the valence band. In contrast, around room temperature the decay time undergoes an important decrease (associated with the efficiency decrease) which can be associated with a reduction of τ_{nr} (increased non- radiative recombinations). We attribute to this effect the temperature characteristics of undoped QD devices around room temperature. A similar temperature dependence has been recently reported in [55], where the authors have measured the decay time studying the time-resolved differential photoluminescence induced by pump and probe laser pulses. Increasing non-radiative recombinations could be due to a temperature-activated non-radiative channel in the dots, or to thermal evaporation and subsequent non-radiative recombination in the WL. Detailed calculations show that the T-dependence of gain, LED efficiency, laser thresholds and lifetime can all be explained by hole evaporation to the WL and defect-related non-radiative recombination in the WL.

The situation is different in p-doped QD devices, for which lasers with improved temperature characteristics have been recently demonstrated [56–58]. Figure 19.20a shows a comparison between the temperature degradation of output power at fixed current for SLEDs with and without p-doped actve regions (same devices as described in section 19.3.2). The output power shows good temperature stability in the temperature interval 20–60°C and an important decrease at higher temperatures for the highly doped device (C). In contrast, the power decrease in the undoped SLED (A) is exponential over the full temparature range 20–85°C, while an intermediate behaviour is observed for the slightly doped device (B). Similar characteristics have been reported in edge emitting lasers, with temperature insensitive threshold characteristics in the interval 20–70°C for p-doped structures [57, 58]. As in the case of the intrinsic sample we measured gain characteristics and internal losses over 2 mm long tilted waveguides. Also the p-doped samples show temperature independent internal losses and strongly T-dependent gain curves, confirming that gain is the main driving force for the temperature dependence of device characteristics. The gain curves measured at the GS peak position of sample C are displayed in Fig. 19.20b. At high injection the curves show a gain saturation versus temperature similar to the undoped device. In this regime the carrier concentration is very high and the dots are expected to be strongly filled (GS, ES and part of the higher excited states) so that the gain decrease is likely produced by a similar decrease in the non-radiative lifetime for p-doped and undoped devices. On the other hand, the behaviour is different at smaller injection, where the doped sample inverts the trend showing higher gain values at higher temperatures. The inset of Fig. 19.20b reports the modal gain extracted at a fixed current corresponding to a gain of about 8 cm^{-1} at 20°C versus temperature, for the undoped and p-doped samples. The gain decrease for the intrinsic sample is linear while

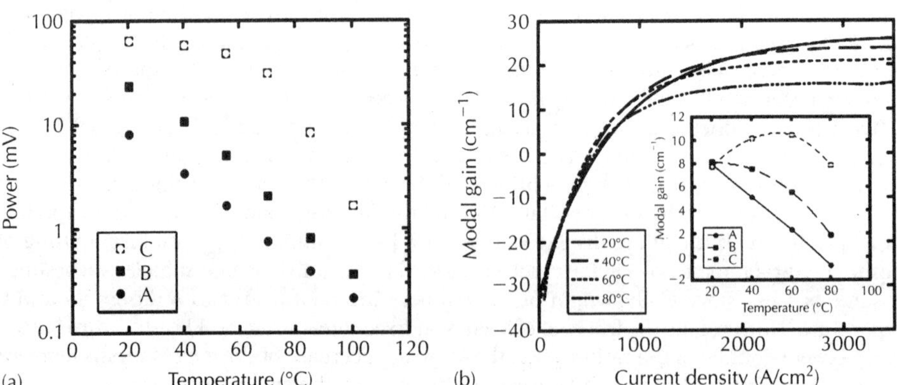

Figure 19.20   *(a) Output power versus temperature measured at fixed current for QD-SLEDs with identical epitaxial structure but different doping levels of the active region. A = undoped, B = slightly p-doped, C = highly p-doped. The currents are chosen so that GS = ES emission at room temperature. (b) Modal gain curves for different temperatures, measured on a narrow-gain 2 mm long tilted ridge-waveguide with p-doped active region (sample C). Inset: Modal gain versus temperature measured at fixed injection for samples A, B and C. The lines are guides to the eye.*

the highly *p*-doped sample (C) shows a gain increase up to 60°C, and a 80°C value as high as 20°C (the situation is again intermediate for the slightly *p*-doped sample). This effect might be produced by thermal escape of carriers from the dots and transfer into the deeper levels (larger dots) where they can take part more effectively to the gain at the peak position. The decrease in current threshold exhibited by lasers with undoped QDs at lower temperatures (around 200 K) has also been attributed to this effect [50]. The likely reason why thermalization does not start to become significant until higher temperatures in *p*-doped devices is the more difficult carrier evaporation due to the electrostatic attraction and band bending. The presence of a more efficient capture in *p*-doped QDs has been demonstrated through the analysis of time-resolved PL decay times [59], while the thermal redistribution of carriers has been reported observing the narrowing of the spontaneously emitted electroluminescence spectra versus temperature [51, 52].

To summarize, an important decrease (15 dB) in the output powers of undoped SLEDs is observed in the 20–85°C temperature interval, which is comparable to the typical thermal degradation observed in InP-based SLEDs. This is produced by thermally activated non-radiative recombinations, which can be deduced through the analysis of the *L–I* curves and extraction efficiencies of short tilted cavity devices and through the analysis of the temperature-dependent PL lifetime measured in time-resolved experiments. Detailed calculations show that the main process at the origin of such degradation is thermal evaporation of holes towards the wetting layer and subsequent non-radiative recombinations. In contrast, a better temperature stability can be achieved through the use of *p*-doping in the 20–60°C temperature interval. The temperature characteristics of devices with *p*-doped active regions present a more complex behaviour. Better stability can be achieved for the SLED emission and for the laser thresholds but usually at the expense of a higher threshold current. Here the temperature dependence is driven by a different interplay versus temperature of non-radiative recombinations, thermalization of carriers over the QD ensemble and variation of the radiative decay times.

19.4 Modelling QD SLEDs

Rate-equation models (REMs) are widely used in the literature to model a QD system [60–64], inspired by the description of average carrier densities in bulk materials, quantum wells or quantum wires. In this approach the temporal evolution of carrier populations and photon numbers inside an optical cavity may be modelled through a system of coupled differential equations. For the simulation of laser diodes a mean field model considering all the quantities averaged over the cavity position may be fully satisfactory as the finite reflectivities of the cavity facets make the photon distribution rather uniform across the device length. In contrast, in devices operating on a single pass like superluminescent diodes, photon numbers and carrier populations can be strongly position dependent and a more sophisticated model developed with a travelling-wave (TW) approach is necessary. A TW-REM model for the calculation of QD-SLED characteristics is introduced in the next two sections, first in a single-mode approximation that allows the calculation of the *L–I* curves of narrow gain devices, and then introducing homogeneous and inhomogeneous broadenings to reproduce the spectral characteristics.

19.4.1 Travelling-wave rate-equation models

The fact that the spatial dependence of gain saturation cannot be neglected in low *Q* cavities was pointed out in some early papers at the beginning of the 1970s [65, 66]. This has led to a revision of the rate equations which incorporates a spatial dependence to study the amplification of light pulses in amplifiers [67], to analyse mode locking in lasers [68], or to model the *L–I* characteristics of superluminescent diodes [69]. From the model presented in [69], here we develop a travelling-wave rate-equation model (TW-REM) that can be used to model the *L–I* characteristics of QD-SLEDs. The TW-REM will be further developed in the next section, to account for the spectral broadening of the QD emission, which will allow the calculation of SLED spectra.

A starting point is to express the photon numbers into the guided mode as position dependent. This can be done introducing the linear photon densities $n_{\varphi\pm}$ (number of photons in the optical mode per unit length) at the spatial point x, which are directly related to the field amplitudes $E^{\pm}$ of a forward $(+)$ and backward $(-)$ travelling wave:

$$n_{\varphi} = n_{\varphi+} + n_{\varphi-} \tag{19.6}$$

with the photon densities $n_{\varphi\pm} \propto |E^{\pm}|^2$ in a slowly varying envelope approximation. As for photons, we consider linear densities for carriers rather than carrier numbers, dependent on time and x position along the cavity.

The QD system is modelled (see sketch in inset of Fig. 19.21a) considering two bound states (GS and ES) and a reservoir level with much higher degeneracy where the current is directly injected (corresponding to the wetting layer (WL) in the real system). We neglect the presence of the highest excited states (QDs emitting at $1.3\,\mu$m usually show 4–5 distinct emissions stemming from recombinations between different bound states [70]) and consider a direct capture from the wetting layer to the first excited state, thus reducing the number of equations to be solved. Electrons and holes are considered as bound excitons, and are described through a single equation for each energy level (GS, ES and WL). Carrier transitions between different energy levels are described in terms of a constant relaxation time from the ES to the GS, and capture time from the WL to the ES. Capture and relaxations are the ones observed in time-resolved PL experiments, and the energy separation between GS, ES and WL corresponds to the separation between the energies associated with the transitions. A drawback of this model is the absence of closely spaced energy levels for the holes. This results in almost temperature-independent calculations in the range 0–350 K, as the spacing between energy levels is larger than $k_{\mathrm{B}}T$. This does not represent a limitation if the model is applied for calculations at a fixed temperature. In general, the dynamics of electron–hole populations in QD ensembles is still the object of discussions and it is not completely clear whether they can be considered excitons or free carriers. This is because of the complex correlation existing between electron and hole capture processes and likely mediated by Auger mechanisms.

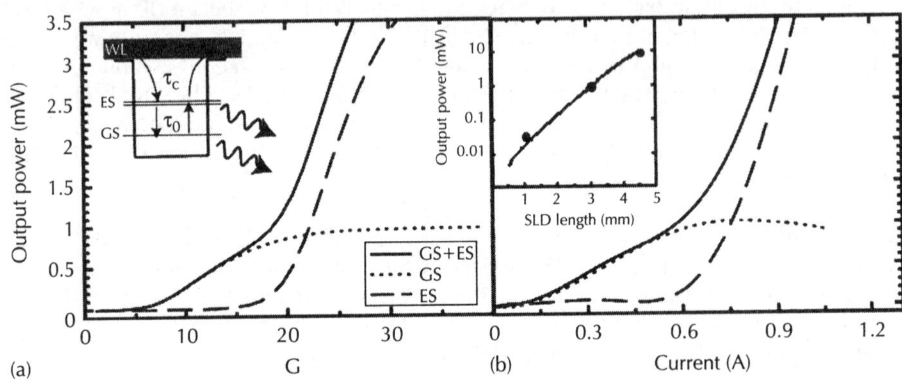

Figure 19.21   (a) Calculated L–I characteristics of a 3 mm long SLED containing five QD layers (continuous line). The GS (dotted) and ES (dashed) contributions to the full L–I characteristics are also reported. A sketch of the rate equation model used for the L–I characteristics calculation of QD-SLEDs is displayed in the inset. (b) L–I characteristics measured on a 3 mm long narrow-gain SLED containing five QD layers. GS and ES contributions to the L–I were separated by spectral filtering. Inset: Measured (dots) and calculated (line) saturated GS output power versus cavity length for tilted SLEDs.

Figure 19.21 displays a schematic representation of the rate-equation model. Carriers directly injected in the WL are then captured by the QD excited state where they can relax towards the ground state. The thermal escape towards higher energy levels is also considered, and radiative recombination can take place from both GS and ES. The equations for the time-and space-dependent carrier densities $n_{GS}(x, t)$, $n_{ES}(x, t)$, $n_{WL}(x, t)$, and for the photon numbers $n_{\varphi\pm}(x, t)$ can be summarized as follows:

$$\frac{\partial n_{WL}}{dt} = \frac{GN_D}{\tau_r} - \frac{n_{WL}}{\tau_b} - \frac{n_{WL}\left(1 - f_{ES}\right)}{\tau_c} + \frac{n_{ES}}{\tau_{esc}^{ES}} \tag{19.7}$$

$$\frac{\partial n_{ES}}{dt} = -\frac{n_{ES}}{\tau_r} + \frac{n_{WL}\left(1 - f_{ES}\right)}{\tau_c} - \frac{n_{ES}\left(1 - f_{GS}\right)}{\tau_0} + \frac{n_{GS}\left(1 - f_{ES}\right)}{\tau_{esc}^{GS}}$$
$$- \frac{n_{ES}}{\tau_{esc}^{ES}} - 2B_{GS}\left(n_{\varphi+}^{ES} + n_{\varphi-}^{ES}\right)\left(n_{ES} - 2N_D\right) \tag{19.8}$$

$$\frac{\partial n_{GS}}{dt} = -\frac{n_{GS}}{\tau_r} + \frac{n_{ES}\left(1 - f_{GS}\right)}{\tau_0} - \frac{n_{GS}\left(1 - f_{ES}\right)}{\tau_{esc}^{GS}} - 2B_{GS}\left(n_{\varphi+} + n_{\varphi-}\right)\left(n_{GS} - N_D\right) \tag{19.9}$$

$$\frac{dn_{\varphi\pm}^{ES}}{dt} \pm \frac{c}{n_{eff}}\frac{dn_{\varphi\pm}^{ES}}{dx} = 2B_{ES}n_{\varphi\pm}^{ES}\left(n_{ES} - 2N_D\right) - \frac{n_{\varphi\pm}^{ES}}{\tau_\varphi} + \beta\frac{n_{ES}}{\tau_r} \tag{19.10}$$

$$\frac{dn_{\varphi\pm}^{GS}}{dt} \pm \frac{c}{n_{eff}}\frac{dn_{\varphi\pm}^{GS}}{dx} = 2B_{GS}n_{\varphi\pm}^{GS}\left(n_{GS} - N_D\right) - \frac{n_{\varphi\pm}^{GS}}{\tau_\varphi} + \beta\frac{n_{GS}}{\tau_r} \tag{19.11}$$

where a degeneracy 2 (4) was considered for the GS (ES) energy level, the radiative recombination is described by a lifetime $\tau_r = 1$ ns identical for GS and ES (corresponding to the one determined by time-resolved photoluminescence experiments), and a lifetime $\tau_b = 1$ ns is assumed for the WL population. N_D is the number of dots per unit length, which is fixed by the number of QD layers, the QD areal density and the device lateral width. The current injection in the wetting layer is expressed for convenience in terms of the adimensional coefficient G representing the number of electron–hole pairs injected in the WL per QD and per unit of radiative lifetime. The carrier exchange rate from one energy level to another always depends on a Pauli-blocking term of the form $(1 - f)$, f being the occupation function of the final state, which prevents the carriers from being transferred when the final state is close to saturation. This term was neglected in the escape term from ES to WL, as the degeneracy of the WL is much higher than the one of the bound states of the dot. The escape times are derived by assuming that the system reaches quasi-Fermi equilibrium in the absence of external excitation and considering the relative degeneracies of GS and ES:

$$\tau_{esc}^{GS} = \frac{\tau_0}{2}e^{\frac{E_{ES} - E_{GS}}{k_BT}} \tag{19.12}$$

$$\tau_{esc}^{ES} = \tau_c\frac{4N_D\pi h^2 e^{\frac{E_{WL} - E_{ES}}{k_BT}}}{m_r k_BT} \tag{19.13}$$

where E_{GS}, E_{ES} and E_{WL} are the respective transition energies, which can be determined by high excitation photoluminescence experiments, and τ_{esc}^{ES} is calculated integrating the escape rate over the lowest WL subband, and using the WL-reduced mass m_r in the excitonic approximation.

In the photon equations β is the spontaneous emission coupling factor, characterizing the fraction of spontaneously emitted photons that go into the optical mode, and the photon lifetime can be expressed as $\tau_\varphi = n_{eff}/c\alpha_i$, in terms of the internal loss α_i and refractive index of the guided mode n_{eff}. The mirror losses are taken into account as boundary conditions for the spatial-dependent photon equations at the facet position ($x = 0$ and $x = L$), $n_{\varphi+}(0) = R_1 n_{\varphi-}(0)$, $n_{\varphi-}(L) = R_2 n_{\varphi+}(L)$, where R_1 and R_2 are the facet reflectivities. The Einstein factor $B_{GS, ES}$ for absorption/stimulated emission appearing in Eqs 19.8–19.11 can be expressed as:

$$B = \frac{\lambda^4}{n_{eff}^3 V 8\pi\tau_r\Delta\lambda} \tag{19.14}$$

with the laser mode wavelength λ, the gain spectral linewidth $\Delta\lambda$, and the optical mode volume V.

Figure 19.21 shows a comparison between the measured and calculated $L–I$ characteristics of a 3 mm long narrow-gain QD-SLED containing five QD layers. The spectrally resolved GS/ES $L–I$ characteristics used high-pass/low-pass filters with a cut-off at 1250 nm. The results of measurements performed at room temperature are shown in Fig. 19.21b (dotted line: GS emission, dashed line: ES emission, continuous line: total emission). The saturated GS output power (points in the inset in Fig. 19.21b) presents an exponential increase with device length, reaching 10 mW for the 4.5 mm long device. The corresponding calculated $L–I$ characteristics displayed in Fig. 19.21a and the calculated GS output power at saturation for increasing chip length (continuous line in inset of Fig. 19.21b) agree very well with the measured characteristics, proving that the model can be successfully applied for calculating the $L–I$ characteristics of QD-SLEDs.

The carrier relaxation from ES to GS plays an important role in the description of the GS–ES competitive behaviour, and τ_0 is set to 7 ps, consistent with QD laser calculations [62]. The radiative lifetime τ_r was evaluated observing the low-temperature decay time of the GS emission in time-resolved phtoluminescence (TRPL) experiments and was assumed to be the same for the ES. Capture time τ_c does not play an important role in a single mode model and was fixed to 1 ps which is in the range of values reported by other authors [71, 72]. The wetting layer lifetime τ_b was arbitrarily fixed at 1 ns. The theoretical values of the Einstein coefficient B are corrected through a factor B_{fit} (of the order of 0.3) in the model which is needed to reproduce the gain saturation characteristics of the devices. This reduced gain value must be likely related to thermal spreading of holes in closely spaced valence-band states and evaporation towards the WL. B_{fit} has been fitted separately for GS and ES and resulted in a $B(ES) = 0.7B(GS)$. This difference corresponds to a gain reduction for the ES, which can be attributed to the increased ES inhomogeneous broadening. Indeed, the factor 0.7 roughly corresponds to the ratio between the FWHM of GS and ES emissions as it has been determined by a multi-Gaussian fit of the high-excitation PL spectra measured on a sample containing similar QDs. Other parameters entering in the model are the facet reflectivities (assumed negligible for an SLED: $R_1 = R_2 = 0$) and the spontaneous emission coupling coefficient β which has been set to $5 \cdot 10^{-3}$ in agreement with [73].

The TW-REM described here can provide useful information about the $L–I$ characteristics and the GS/ES competitive behaviour in QD-SLEDs; however, it does not take into account any spectral effect, which is important in SLEDs. The model is further developed in the next section to account for the spectral broadening.

19.4.2 Modelling of spectral characteristics – towards flat-top spectra

To account for the shape of the spectral emission both the inhomogeneous- and homogeneous-type broadenings must be considered. The first one, being mainly related to the size dispersion over the QD ensemble, is assumed Gaussian (or a linear combination of Gaussians when applied to calculation of chirped samples). The DOS is discretized into a number of identical energy intervals of width ΔE. Given the linear density of dots N_D as defined in the previous section, the number of dots N_D^{iGS} whose GS levels lie inside the energy interval ΔE centred around E_i can be written as:

$$N_D^{iGS} = N_D \frac{\displaystyle\int_{E_i-\Delta E/2}^{E_i+\Delta E/2} F(E)dE}{\displaystyle\int_{-\infty}^{\infty} F(E)dE} \tag{19.15}$$

where $F(E)$ defines the inhomogeneous broadening shape for the GS transition. For the ES energy levels we assume an identical broadening, with a peak value centred at shorter wavelength (fixed according to PL spectra). This corresponds to the assumption of a spacing $E_{i+\theta} - E_i$ between GS and ES energy levels of the same dot not dependent on the QD size. Under this assumption the number of QDs with GS energies lying in the interval centred in E_i is identical to the number of QDs with ES energies lying in the interval $E_{i+\theta}$ so that rate equations coupling GS levels in the interval E_i with ES levels in the interval $E_{i+\theta}$ may be safely handled. The WL is still considered as a discrete reservoir energy level as in the case of a single-mode model. A scheme of the energy discretization of the density of states is reported in Fig. 19.22a. For the calculation we assumed 75 meV spacing between GS and ES transitions, and a Gaussian dispersion of the GS and ES energy levels with 40 meV FWHM. The WL is not reported in the picture (1.25 eV: out of scale).

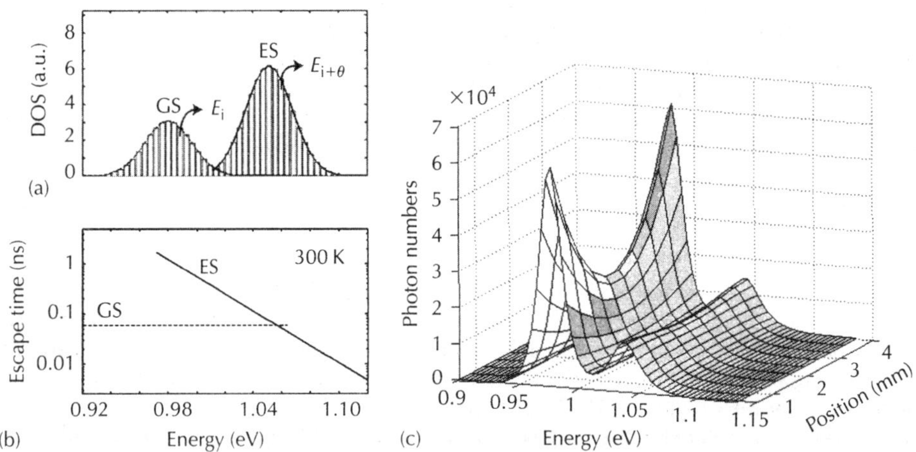

Figure 19.22 (a) *Discretized density of states of a dispersed QD ensemble. Identical Gaussian dispersions (40 meV FWHM) were assumed for the GS and ES transitions. (b) GS and ES escape times calculated for $\tau_c = 1$ ps and $\tau_0 = 7$ ps. The constant value of τ_{esc}^{GS} reflects the assumption of a constant GS–ES spacing for all the QD families. In contrast, τ_{esc}^{ES} decreases exponentially moving towards high energies (shorter distance from WL.) (c) Photon densities versus energy and cavity position calculated for a 4 mm long SLED, containing five QD layers. Model parameters: $B_{fit} = 0.25$, $T = 300$ K, DOS broadening: $FWHM_{GS} = FWHM_{ES} = 40$ meV, $\Gamma = 10$ meV.*

The GS, ES and WL populations (linear densities) n_{GS}, n_{ES} and n_{WL} may be described at a fixed cavity position according to three equations similar to Eqs 19.7–19.9, except that now the spectral position (described by the index i) must also be considered. In the absence of photons the equations can be expressed as follows:

$$\frac{\partial n_{WL}}{\partial t} = \frac{GN_D}{\tau_r} - \frac{n_{WL}}{\tau_b} - \frac{n_{WL}\sum_i N_D^{iES}\left(1-f_{ES}^i\right)}{\tau_c N_D} + \sum_i \frac{n_{ES}^i}{\tau_{esc}^{iES}} \tag{19.16}$$

$$\frac{\partial n_{ES}^i}{\partial t} = -\frac{n_{ES}^i}{\tau_r} - \frac{n_{ES}^i}{\tau_{esc}^{iES}} + \frac{n_{WL}N_D^{iES}\left(1-f_{ES}^i\right)}{\tau_c N_D} - \frac{n_{ES}^i\left(1-f_{GS}^{i-\theta}\right)}{\tau_0} + \frac{n_{GS}^{i-\theta}\left(1-f_{ES}^i\right)}{\tau_{esc}^{i-\theta GS}} \tag{19.17}$$

$$\frac{\partial n_{GS}^i}{\partial t} = -\frac{n_{GS}^i}{\tau_r} + \frac{n_{ES}^{i+\theta}\left(1-f_{GS}^i\right)}{\tau_0} - \frac{n_{GS}^i\left(1-f_{ES}^{i+\theta}\right)}{\tau_{esc}^{iGS}} \tag{19.18}$$

where the carrier exchange between WL and ES is described through a sum of contributions deriving from QDs with different energies. Due to the assumption of a constant spacing between GS and ES energy transitions, the GS escape time will not depend on the index i and can be derived as in Eq. 19.12. In contrast, the broadening of the ES transition results in a different spacing from the WL transition for QDs belonging to different energy intervals, thus decreasing the escape times of the QDs lying in energy intervals closer to the WL:

$$\tau_{\text{esc}}^{i\text{ES}} = \tau_{\text{c}} \frac{4N_{\text{D}}^{i\text{ES}} \, \pi h^2 e^{\frac{E_{\text{WL}} - E_i}{k_{\text{B}}T}}}{m_{\text{r}}k_{\text{B}}T}. \tag{19.19}$$

The GS and ES escape times calculated at $300\,\text{K}$ are reported in Fig. 19.22b, using the reduced mass of an $\text{In}_{0.15}\text{Ga}_{0.85}\text{As}$ quantum well (corresponding to the WL).

To account for the effects of absorption/stimulated emission, Eqs 19.16–19.18 must be coupled to the equations describing the energy-dependent photon densities $n_{\varphi+}^i$ and $n_{\varphi-}^i$. These are defined over the same energy intervals introduced for GS and ES, and can interact with carriers at different energies inside the homogeneous broadening, which we assume to be Lorentzian. This will be described as a matrix whose elements ij correspond to the values of the Lorentzian function centred in the energy interval E_i, calculated at the position E_j:

$$L_i^j = \frac{\Gamma}{\pi} \frac{1}{(E_j - E_i)^2 + \Gamma^2} \tag{19.20}$$

where Γ represents the excitonic linewidth. At low temperature Γ is limited by the transition lifetime and is in the μeV range for typical radiative lifetimes of $1\,\text{ns}$. Increasing temperature Γ becomes dominated by other processes such as phonon scattering or carrier–carrier scattering, which results in broadenings of 10–$20\,\text{meV}$ at room temperature [74]. As our interest is limited to the high temperature regime, we fix Γ to $10\,\text{meV}$.

An absorption/stimulated emission term can then be introduced in Eqs 19.17 and 19.18, by summing all the contributions due to photon densities lying inside the homogeneous broadening, weighted by the homogeneous broadening matrix:

$$-2B_i \left(n_{\text{ES}}^i - 2N_{\text{D}}^{i\text{ES}} \right) \sum_j L_j^i \left(n_{\varphi+}^j + n_{\varphi-}^j \right) \tag{19.21}$$

$$-2B_i \left(n_{\text{GS}}^i - N_{\text{D}}^{i\text{GS}} \right) \sum_j L_j^i \left(n_{\varphi+}^j + n_{\varphi-}^j \right) \tag{19.22}$$

where the expression of the coefficient B_i depends on the wavelength λ_i associated to every energy interval:

$$B_i = \frac{\lambda_i^4}{n_{\text{eff}}^3 8V\pi\tau_{\text{r}}\Gamma}. \tag{19.23}$$

Unlike the single-mode model, where the inhomogeneous broadening was taken into account in the coefficient B, this is not necessary here as it is explicitly treated in the model. Finally, equations for the forward travelling and backward travelling photon densities must be coupled to the carrier equations. These can be written as:

$$\frac{\partial n_{\varphi\pm}^i}{\partial t} \pm v_{\text{g}} \frac{\partial n_{\varphi\pm}^i}{\partial x} = \frac{\beta}{\tau_{\text{r}}} \sum_j L_j^i \left(n_{\text{GS}}^j + n_{\text{ES}}^j \right) - \frac{n_{\varphi\pm}^i}{\tau_\varphi}$$
$$+ 2B_i n_{\varphi\pm}^i \sum_j L_j^i \left(n_{\text{GS}}^j - N_{\text{D}}^{j\text{GS}} + n_{\text{ES}}^j - 2N_{\text{D}}^{j\text{ES}} \right). \tag{19.24}$$

We note that the population inversion at a fixed energy position E_i:

$$n_{\text{GS}}^i - N_{\text{D}}^{i\text{GS}} + n_{\text{ES}}^i - 2N_{\text{D}}^{i\text{ES}} \tag{19.25}$$

depends on the GS and ES populations in the two different QD families having a GS or ES transition energy around E_i. Considering both GS and ES gain is important to reproduce the experimental spectra particularly in the energy region in between GS and ES peak position, where the GS and ES DOS may be comparable for QD ensembles with large broadenings. Once the full system of equations is integrated, the results provide the values of the populations and of the photon numbers versus time, energy and position in the cavity: $n_{\mathrm{GS,ES,WL}}(x,E,t)$ and $n_{\varphi\pm}(x,E,t)$. The steady-state spectral characteristics are then reproduced, waiting some time to reach a steady-state solution of the system (typically few times τ_r using the initial conditions $n_{\mathrm{GS,ES,WL}}(t=0)=0$ and $n_{\varphi\pm}(t=0)=0$.

Typical SLED calculations ($R_1 = R_2 = 0$) performed with this model are reported in Fig. 19.22c. The picture displays the total photon density ($n_{\varphi+} + n_{\varphi-}$) versus cavity position and energy, for a 4 mm long device operating in the gain regime. The device operates in a regime of important GS amplification (slightly lower for the ES) that is evident from the higher photon densities at the device facets. The shape of the density of states (energy dispersion for the GS energy levels) can be chosen arbitrarily to reproduce any type of chirped structure. Model parameters were fixed, as above, according to the experimental characteristics (gain, internal losses, number of QDs, device geometry) and to the literature or fitting of the laser characteristics ($\tau_c = 1\,\mathrm{ps}$, $\tau_0 = 7\,\mathrm{ps}$, $\Gamma = 10\,\mathrm{meV}$). In contrast to real SLEDs where QD layers with different properties are used to broaden the spectrum, the model is developed assuming a single QD distribution. This neglects the fact that in real devices the carriers captured into a given QD layer cannot escape to a different QD layer through the WL (the GaAs barrier suppresses the exchanges), which in contrast is possible in the model where we have considered a single WL energy level.

In Fig. 19.23 we report a comparison between measured and calculated spectra for two tilted QD-SLEDs, one containing five identical QD layers with a narrow spectral gain (top) and the second containing five chirped QD layers with a broader emission (bottom). The calculations were

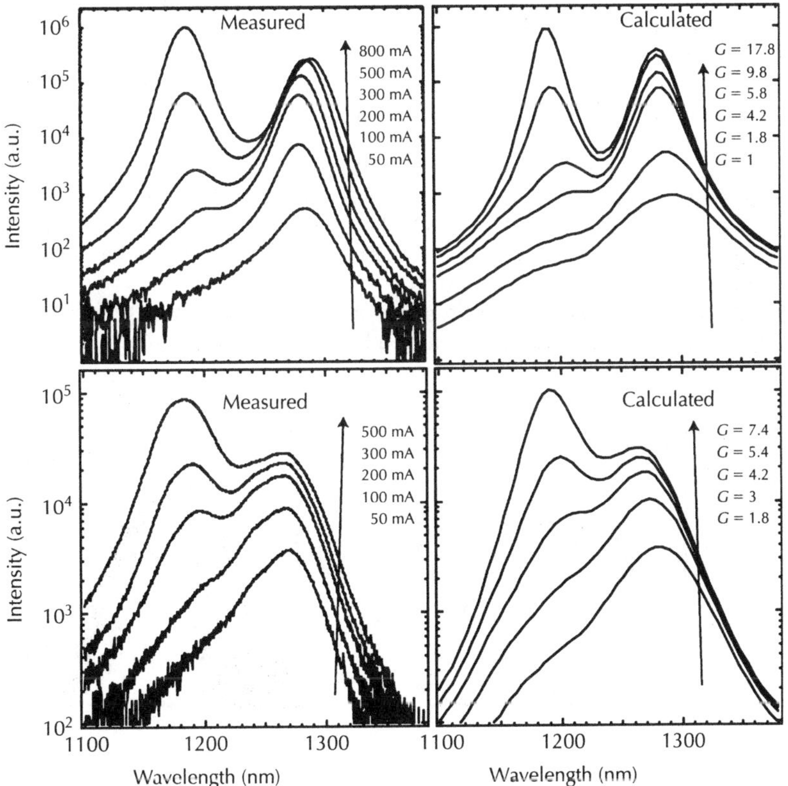

Figure 19.23   Calculated (right) and measured (left) spectral characteristics for a 4 mm long narrow gain SLED (top) and a 4 mm long chirped SLED (bottom), for different currents.

performed assuming negligible facet reflectivities and fixing the density of states shape in agreement with the spectral shape extracted from PL experiments or from the electroluminescence of short tilted cavities where the effect of optical gain and absorption can be neglected. For the narrow gain sample it corresponds to a Gaussian dispersion with FWHM of $\approx 40\,meV$ ($\approx 40\,nm$), with a GS transition centred at 0.97 eV (1280 nm) and a GS–ES spacing of $\approx 70\,meV$ (≈ 95 nm). The chirped sample contains five layers of identical QDs capped by a 5 nm thick $In_xGa_{1-x}As$ layer with x varying linearly between 9% and 15%. The density of states shape in this case was then fixed as a sum of five equally spaced Gaussians, each one with an FWHM similar to the one measured by PL measurements on test structures containing one single QD layer, and with a spacing such that the DOS reproduces the PL spectra measured on the chirped structure. The GS–ES energy level separation here was fixed similarly to the previous case at $\approx 70\,meV$. The gain was fixed from the analysis of laser characteristics. All the other parameters were fixed as in the previous paragraph. For both samples a good agreement is found between experiment and calculations without the use of fitting parameters, proving the validity of our model. The shift of the GS emission for the narrow-gain sample at the highest injections is due to heating and possibly many-body interactions taking place at high injection, which are not considered in the model. A slight disagreement is also found concerning the spectral width of the ES emission at high injection, which is smaller in calculations. The effect may be due to the assumption of identical inhomogeneous broadenings for GS and ES in the model, which is not necessarily the case in the real system. Moreover, on the chirped sample, where the effect is more evident, the model does not account for the injection through separate wetting layers for each dot family and a stronger thermal redistribution narrowing the spectrum towards the red can take place. In the real sample this is not allowed because "red-emitting" and "blue-emitting" dots are separated by GaAs barriers (different dot layers) and are not thermally coupled.

Using the same parameters as the ones used for the chirped sample, the shape of the density of states was then varied to calculate the inhomogeneous broadening required to achieve a flat-top SLED emission. Assuming Gaussian dispersions centred at 1265 nm for the GS and 1180 nm for the ES, a flat top spectral superposition can be achieved in a device only with a minimum FWHM of the GS DOS of about 110 nm. This represents a target for the PL spectra of samples with dot layers optimized for the achievement of large size dispersions. Assuming a chirped spectral distribution constituting five equally spaced Gaussian contributions (five chirped QD layers), the minimum spacing required was found to be 15 meV (about 18 nm spacing between the GS peak emission of different QD families). This spacing can be easily translated into In content of the InGaAs capping covering the dots which tunes the emission to the targeted wavelength. Figure 19.24a displays the calculated spectra of a five QD layer chirped SLED, 4 mm long. The spectra are reported for varying Gaussian spacings.

Growth optimization resulted in a flat-top spectra emission in seven QD layer SLEDs as can be seen in Fig. 19.24b. SLEDs 2 mm long, show a bandwidth of 115 nm at a current density of

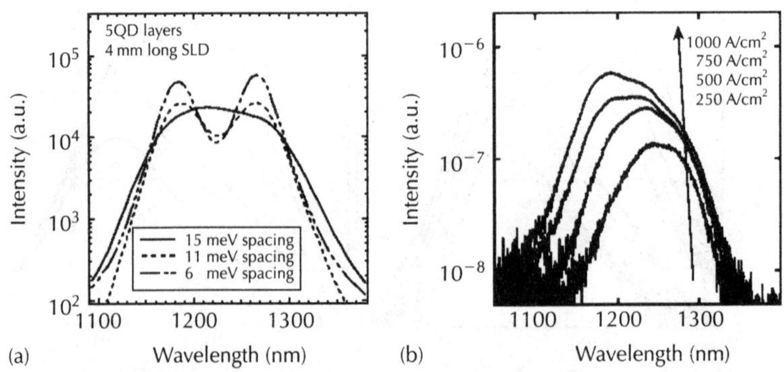

Figure 19.24   (a) Calculated spectra in the regime GS = ES. DOS constituted by five Gaussians equally spaced of 6 meV, 11 meV and 15 meV. (b) Spectra versus injection of a flat-top QD-SLED with optimized emission linewidth.

750 A/cm^2, with a dip-free GS + ES superposition. The SLEDs were obtained by properly choosing the spacing between the GS emissions of two different QD families optimized for broader inhomogeneous broadening (as described in section 19.2). Typical output powers for these devices, similar to the five QD layer chirped SLEDs presented in section 19.3, were of the order of a few hundred μW in the large-bandwidth regime. The maximum modal gain (at GS peak position) for these structures does not overcome 10 cm^{-1}, which is the main limiting factor for the SLED output power. Higher powers, as already discussed in previous paragraphs, can be achieved by increasing the number of QD layers, introducing p-doping in the active region and fabricating bent structures.

The improvement in the coherence properties for a flat-top emission can be observed in Fig. 19.25a, where the coherence functions (obtained by inverse Fourier transform) corresponding to the spectra of Fig. 19.24a are displayed. We observe an important reduction of the side lobes' intensity (13 dB suppression ratio) for the flat-top spectrum, while the side-lobes are only 3 dB suppressed when a 5 dB dip is present between GS and ES. In Fig. 19.25b we report also the calculated degree of coherence associated to the flat-top SLED spectra of Fig. 19.24b. The curves correspond to spectra taken at different injection levels. At low injection the emission is dominated by the only GS (FWHM 85 nm) which results in an FWHM of the coherence function of 17 μm. In the broad emission regime (FWHM 115 nm) the FWHM of the degree of coherence is reduced to 12 μm, with a high suppression of side lobes (16 dB). At high injection the spectrum becomes dominated by the ES emission (FWHM 80 nm) with a coherence function of the same width (coherence length proportional to $\lambda^2/\Delta\lambda$) but increasing intensity tails on the two sides.

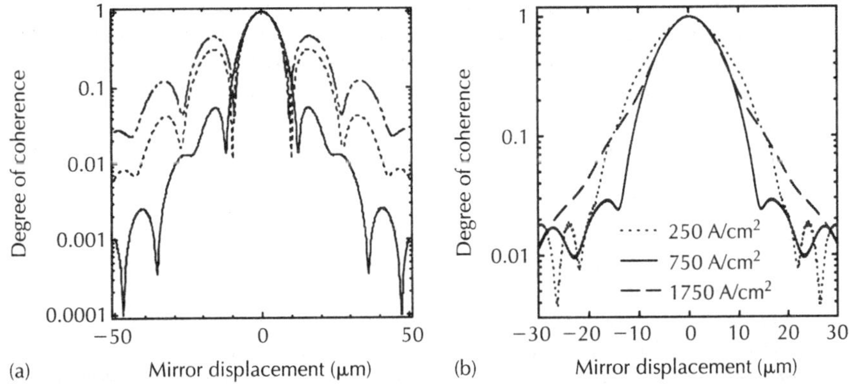

Figure 19.25 (a) *Calculated degree of coherence for the SLEDs simulated in Fig. 3.16 and (b) for the 2 mm long chirped SLED (flat-top emission) whose spectral characteristics are reported in Fig. 3.16b.*

19.5 Conclusion and perspectives

19.5.1 Comparison of present performance with commercial devices and application needs

In order to compare commercially available devices with today's state of the art quantum dot SLEDs we will consider the following main parameters: wavelength, optical bandwidth, spectral shape, and output power.

- *Wavelength:* The QD SLEDs reported here emit mainly in the 1200–1300 nm range. This wavelength range is compatible with most SLED application fields, as in optical coherence tomography (OCT), fibre-based sensors, and test equipment applications. Further work would, however, be required to shift the centre wavelength more towards the 1300–1400 nm range. This will have the advantage of avoiding multi-mode operation in the widely available standard single-mode fibres (e.g. SMF-28) used in fibreoptic telecommunications, because these fibres have a cut-off wavelength between 1260 and 1280 nm. This is especially critical when building fibre-based interferometers.

- *Optical bandwidth and spectral shape:* The biggest advantage of using QD-SLED is certainly the large optical bandwidth that can be obtained. The values reported in this work are 50% larger than today's commercially available single device SLEDs. Although for many applications only a large optical bandwidth is required, for some main applications like OCT or FOG the spectral shape is also very important. Thus, further work would be required in order to achieve a Gaussian-like spectral shape.
- *Output power:* Commercially available SLEDs can emit up to 30 mW from a single-mode fibre. These power levels are required for OCT applications where the high power enables the high scan rates. The QD-SLEDs reported in this work are already in this order of magnitude; some development in order to further increase the output power will be beneficial for OCT applications.

Other important parameters such as spectral ripple and lifetime of the device are expected to be similar to today's commercially available SLEDs, since the used substrates and production techniques are the same.

19.5.2   Potential improvements

As indicated above, QD-SLEDs emitting in the 1200–1300 nm wavelength region have a performance (>100 nm bandwidth with >10 mW output power) already exceeding QW and bulk devices. Further improvements include the engineering of the spectral shape and higher cw output power. The output spectrum can be optimized depending on the application: in particular the dip between the GS and ES emission can be reduced to the level required by the specification on the SMSR, by optimizing the design of the multilayer. Also the cw output power can be maximized by an optimized technological approach (anti-reflective/high-reflection coatings, mounting, cooling). On the research side, the more fundamental question is whether a much wider spectrum can be obtained, along with a high-power spectral density, and whether the QD approach can be extended to other material systems for other spectral regions (near-IR, red, blue).

At present, the maximum spectral width is limited by the fact that only two bound states in the QDs are contributing to the emission. In fact, when the current is increased beyond the point GS = ES, the ES emission dominates due to higher ES degeneracy and thus higher maximum gain. A combined emission from the GS, ES and second excited state (ES2), with comparable contributions, cannot be achieved in these simple devices. However, more complicated structures could be used for this purpose. In particular, multi-section devices have been proposed to achieve an ultra-wide emission spectrum >150 nm [75]. One section is biased at high current levels to achieve inversion of all three states. A second section has a lower bias, thus favouring the GS emission. By a careful choice of section lengths and bias currents, the intensities from the three states can be made comparable, giving a combined bandwidth of 160 nm [75]. Note that by varying the section currents the spectrum can be changed significantly (e.g. from 1200 nm to 1300 nm centre wavelength), resulting in a tunable broadband source. An additional degree of freedom is the choice of the epitaxial structure: QD layers with a GS emission spanning a large wavelength range (e.g. between 1000 and 1300 nm) can be easily integrated within the same active region. Careful engineering of the coupling of the different states with the optical spectrum, together with the use of multi-section structures, should allow the bandwidth to be extended well above 200 nm. A common problem in these approaches is the current needed to bring to full inversion (i.e. including ES states) all QDs in an already broad ensemble. A large chip gain ($gL \approx 6$–10, as compared to the value $gL \approx 1.2$ needed to reach threshold in a cleaved-facet laser) is needed across the entire spectrum, requiring long devices and very high currents. Also, devices with lengths >2 mm become impractical for packaging. In order to reduce physical device length and the number of QDs to be inverted, increasing the effective length with bent waveguides and high-reflection coatings on one facet is very useful – see the results in section 19.3. More clever structures could also be devised to increase the effective length of light amplification without creating a cavity effect (leading to lasing) and without increasing the number of QDs.

Another direction of research consists of extending the concepts presented here to other materials and wavelength regions. The InAs/GaAs system can cover the 800–1350 nm region, and high-performance QD SLEDs were demonstrated down to 1000 nm [76, 77], a region of interest for OCT. In the $\approx$ 1550 nm region, wide-gain QD semiconductor optical amplifiers have been demonstrated [78] in the InAs/InP material system, which shows a significant potential for SLED applications. For wavelengths between 700 and 800 nm, the InP/InGaP and InAlAs/AlGaAs QD systems have been investigated for laser applications [79–81]. Laser threshold current densities are in the kA/cm^2 range, much higher than with best InAs/GaAs devices, possibly due to lower material quality – significant progress on the material side will be needed to produce high-efficiency and broad-gain SLEDs. On the blue side of the spectrum, the InGaN/AlGaN QD material system has the potential to cover a large wavelength region from 350 to 500 nm (see, e.g., [82, 83]), with important potential applications in ultra-high-resolution OCT. However, little or no work has been realized on QD or even QW SLEDs, presumably due to the technological challenges related to this material. Significant advances are therefore expected in this area in the next few years.

References

1. See www.exalos.com
2. S. Kondo, H. Yasaka, Y. Noguchi, K. Magri, M. Sugo, and O. Mikami, *Electron. Lett.* **36**(25), 2093 (2000).
3. W. Burns, C.L. Chen, and R. Moeller, *J. Lightwave Technol.* **1**, 98 (1983).
4. S. Kondo, H. Yasaka, Y. Noguchi, K. Magari, M. Sugo, and O. Mikami, *Electron. Lett.* **28**, 132 (1992).
5. A.T. Semenov, V.R. Shidlovski, and S.A. Safin, *Electron. Lett.* **29**, 854 (1993).
6. O. Mikami, H. Yasaka, and Y. Noguchi, *Appl. Phys. Lett.* **56**, 987 (1990).
7. C.F. Lin and B.L. Lee, *Appl. Phys. Lett.* **71**, 1598 (1997).
8. Z.Z. Sun, D. Ding, Q. Gong, W. Zhou, B. Xu, and Z.G. Wang, *Optical Quant. Electron.* **31**, 1235 (1999).
9. Z.Y. Zhang, X.Q. Meng, P. Jin, Ch.M. Li, S.C. Qu, B. Xu, X.L. Ye, and Z.G. Wang, *J. Cryst. Growth* **243**, 25 (2002).
10. D.C. Heo, J.D. Song, W.J. Choi, J.I. Lee, J.C. Jung, and I.K. Han, *Electron. Lett.* **39**, (11), 863 (2003).
11. Z.Y. Zhang, Z.G. Wang, B. Xu, P. Jin, Z.Z. Sun, and F.Q. Liu, *IEEE Photonics Technol. Lett.* **16**(1), 27 (2004).
12. N. Liu, P. Jin, and Z.G. Wang, *Electron. Lett.* **41**(25), 55 (2005).
13. M. Rossetti, A. Markus, A. Fiore, L. Occhi, and C. Velez, *IEEE Photonics Technol. Lett.* **17**(3), 540 (2005).
14. L.H. Li, M. Rossetti, A. Fiore, L. Occhi, and C. Velez, *Electron. Lett.* **41**(1), 41 (2005).
15. M. Rossetti, L.H. Li, A. Fiore, L. Occhi, C. Velez, S. Mikhrin, and A. Kovsh, *IEEE Photonics Technol. Lett.* **18**, 1946 (2006).
16. C.E. Dimas, H.S. Djie, and B.S. Ooi, *J. Cryst. Growth* **288**, 153 (2006).
17. L.H. Li, M. Rossetti, A. Fiore, L. Occhi, and C. Velez, *Phys. Stat. Sol. (b)* **243**(15), 3988 (2006).
18. S.K. Ray, K.M. Groom, R. Alexander, K. Kennedy, H.Y. Liu, M. Hopkinson, and R.A. Hogg, *J. Appl. Phys.* **100**, 103105 (2006).
19. S.K. Ray, K.M. Groom, H.Y. Liu, M. Hopkinson, and R.A. Hogg, *Jpn. J. Appl. Phys.* **45**, 2542 (2006).
20. S.K. Ray, K.M. Groom, M.D. beattie, H.Y. Liu, M. Hopkinson, and R.A. Hogg, *IEEE Photonics Technol. Lett.* **18**, 58 (2006).
21. C.Y. Ngo, S.F. Yoon, W.J. Fan, and S.J. Chua, *Appl. Phys. Lett.* **90**, 11103 (2007).
22. Y.C. Xin, A. Martinez, T. Saiz, A.J. Moscho, Y. Li, T.A. Nilsen, A.L. Gray, and L.F. Lester, *IEEE Photonics Technol. Lett.* **19**(7), 501 (2007).
23. J.X. Chen, A. Markus, A. Fiore, U. Oesterle, R.P. Stanley, J.F. Carlin, R. Houdre, L. Lazzarini, L. Nasi, M.T. Todaro, E. Piscopiello, R. Cingolani, M. Catalano, J. Katchi, and J. Ratajczak, *J. Appl. Phys.* **91**(10), 6710 (2002).

24. K. Nishi, H. Saito, S. Sugou, and J. -S. Lee, *Appl. Phys. Lett.* **74**, 1111 (1999).
25. V.M. Ustinov, N.A. Maleev, A.E. Zhukov, A.R. Kovsh, A.Y. Egorov, A.V. Lunev, B.V. Volovik, I.L. Krestnikov, Y.G. Musikhin, N.A. Bert, P.S. Kop'ev, Z.I. Alferov, N.N. Ledentsov, and D. Bimberg, *Appl. Phys. Lett.* **74**, 2815 (1999).
26. T. Ngo, P.M. Petroff, H. Sakaki, and J.L. Merz, *Phys. Rev. B* **53**, 9618 (1996).
27. L. Chu, M. Arzberger, G. Böhm, and G. Abstreiter, *J. Appl. Phys.* **85**(4), 2355 (1999).
28. J. Drucker, *Phys. Rev. B* **48**, 18203 (1993).
29. D. Leonard, K. Pond, and P.M. Petroff, *Phys. Rev. B* **50**, 11687 (1994).
30. H.S. Djie, C.E. Dimas, and B.S. Ooi, *IEEE Photonics Technol. Lett.* **18**, 1747 (2006).
31. S. Ruffenach, O. Briot, M. Moret, and B. Gil, *Appl. Phys. Lett.* **90**, 153102 (2007).
32. J. Lutti, P.M. Smowton, G.M. Lewis, P. Blood, A.B. Krysa, J.C. Lin, P.A. Houston, A.J. Ramsay, and D.J. Mowbray, *Appl. Phys. Lett.* **86**, 011111 (2005).
33. S. Fafard, K. Hinzer, S. Raymond, M. Dion, J. McCaffrey, Y. Feng, and S. Charbonneau, *Science* **22**, 1350 (1996).
34. Y. Narukawa, Y. Kawakami, M. Funato, S. Fujita, S. Fujita, and S. Nakamura, *Appl. Phys. Lett.* **70**, 981 (1997).
35. K.P. O'Donnell, R.W. Martin, and P.G. Middleton, *Phys. Rev. Lett.* **82**, 237 (1999).
36. D.R. Matthews, H.D. Summers, P.M. Smowton, and M. Hopkinson, *Appl. Phys. Lett.* **81**(26), 4904 (2002).
37. O.B. Shchekin and D.G. Deppe, *Appl. Phys. Lett.* **80**(15), 2758 (2002).
38. A.J. Williamson, L.W. Wang, and Alex. Zunger, *Phys. Rev.* **62**(19), 12963 (2000).
39. A. Markus, J.X. Chen, C. Paranthoen, A. Fiore, C. Platz, and O. Gauthier-Lafaye, *Appl. Phys. Lett.* **82**(12), 1818 (2003).
40. E.A. Viktorov, P. Mandel, Y. Tanguy, J. Houlihan, and G. Huyet, *Appl. Phys. Lett.* **87**(5), 3113 (2005).
41. B.E.A. Saleh and M.C. Teich, *Fundamentals of Photonics*, (Wiley-Interscience, New York, 1991).
42. Y. Arakawa and H. Sakaki, *Appl. Phys. Lett.* **40**, 939 (1982).
43. R. Dingle and C.H. Henry, U.S. Pat. 3982207 (1976).
44. G. Park, O.B. Shchekin, D.L. Huffaker, and D.G. Deppe, *Electron. Lett.* **36**(15), 1283 (2000).
45. A.R. Kovsh, N.A. Maleev, A.E. Zhukov, S.S. Mikhrin, A.P. Vasil'ev, Yu.M. Shernyakov, M.V. Maximov, D.A. Livshits, V.M. Ustinov, Zh.I. Alferov, N.N. Ledentsov, and D. Bimberg, *Electron. Lett.* **38**(19), 1104 (2002).
46. R. Krebs, F. Klopf, J.P. Reithmaier, and A. Forchel, *Jpn. J. Appl. Phys.* **41**(2B), 1158 (2002).
47. G.T. Liu, A. Stintz, H. Li, T.C. Newell, A.L. Gray, P.M. Varangis, K.J. Malloy, and L.F. Lester, *IEEE J. Quantum Electron.* **36**(11), 1272 (2000).
48. L.V. Asryan and R.A. Suris, *IEEE J. Quantum Electron.* **34**(5), 841 (1998).
49. M. Grundmann, O. Stier, S. Bognar, C. Ribbat, F. Heinrichsdorff, and D. Bimberg, *Phys. Stat. Sol. (A)* **178**(1), 255 (2000).
50. I.P. Marko, A.R. Adams, S.J. Sweeney, D.J. Mowbray, M.S. Skolnick, H.Y. Liu, and K.M. Groom, *IEEE J. Sel. Topics Quantum Electron.* **11**(5), 1041 (2005).
51. I.P. Marko, N.F. Mass, S.J. Sweeney, A.D. Andreev, A.R. Adams, N. Hatori, and M. Sugawara, *Appl. Phys. Lett.* **87**(21), 1 (2005).
52. I.C. Sandall, P.M. Smowton, J.D. Thomson, T. Badcock, D.J. Mowbray, H.Y. Liu, and M. Hopkinson, *Appl. Phys. Lett.* **89**(15), 1118 (2006).
53. I.P. Marko, A.D. Andreev, A.R. Adams, R. Krebs, J.P. Reithmaier, and A. Forchel, *Electron. Lett.* **39**(1), 58 (2003).
54. M. Sugawara, K. Mukai, and Y. Nakata, *Appl. Phys. Lett.* **75**(5), 656 (1999).
55. X. Mu, Y.J. Ding, B.S. Ooi, and M. Hopkinson, *Appl. Phys. Lett.* **89**(18), 181921–1 (2006).
56. O.B. Shchekin, J. Ahn, and D.G. Deppe, *Electron. Lett.* **38**(14), 712 (2002).
57. S. Fathpour, Z. Mi, P. Bhattacharya, A.R. Kovsh, S.S. Mikhrin, I.L. Krestnikov, A.V. Kozhukhov, and N.N. Ledentsov, *Appl. Phys. Lett.* **85**(22), 5164 (2004).
58. I.I. Novikov, N.Y. Gordeev, L.Y. Karachinskii, M.V. Maksimov, Y.M. Shernyakov, A.R. Kovsh, I.L. Krestnikov, A.V. Kozhukhov, S.S. Mikhrin, and N.N. Ledentsov, *Semiconductors* **39**(4), 477 (2005).

59. K. Gündogdu, K.C. Hall, T.F. Boggess, D.G. Deppe, and O.B. Shchekin, *Appl. Phys. Lett.* **85**(20), 4570 (2004).
60. M. Sugawara, K. Mukai, and H. Shoji, *Appl. Phys. Lett.* **71**(19), 2791 (1997).
61. S. Sanguinetti, M. Henini, M. Grassi Alessi, M. Capizzi, P. Frigeri, and S. Franchi, *Phys. Rev.* **60**(11), 8276 (1999).
62. A. Markus, J.X. Chen, O. Gauthier-Lafaye, J.G. Provost, C. Paranthoen, and A. Fiore, *IEEE J. Sel. Topics Quantum Electron.* **9**(5), 1308 (2003).
63. A.A. Dikshit and J.M. Pikal, *IEEE J. Quantum Electron.* **40**(2), 105 (2004).
64. P. Dawson, O. Rubel, S.D. Baranovskii, K. Pierz, P. Thomas, and E.O. Göbel, *Phys. Rev.* **72**(23), 1 (2005).
65. R. Ulbrich and M.H. Pilkhun, *Appl. Phys. Lett.* **16**(12), 516 (1970).
66. S. Hasuo and T. Ohmi, *J. Appl. Phys.* **13**(9), 1429 (1974).
67. E. Schöll, *IEEE J. Quantum Electron.* **24**(2), 435 (1988).
68. J.E. Bowers, P.A. Morton, A. Mar, and S.W. Corzine, *IEEE J. Quantum Electron.* **25**(6), 1426 (1989).
69. N.S.K. Kwong, K.Y. Lau, and N. Bar-Chaim, *IEEE J. Quantum Electron.* **25**(4), 696 (1989).
70. I.E. Itskevich, M.S. Skolnick, D.J. Mowbray, I.A. Trojan, S.G. Lyapin, L.R. Wilson, M.J. Steer, M. Hopkinson, L. Eaves, and P.C. Main, *Phys. Rev.* **60**(4), R2185 (1999).
71. T. Müller, W. Parz, F.F. Schrey, G. Strasser, and K. Unterrainer, *Semiconductor Science and Technology* **19**(4 SPEC. ISS:), S287 (2004).
72. J. Siegert, S. Marcinkevicius, and Q.X. Zhao, *Phys. Rev.* **72**(8), 5316 (2005).
73. Y. Zhao, W. Han, J. Song, X. Li, Y. Liu, D. Gao, G. Du, H. Cao, and R.P.H. Chang, *J. Appl. Phys.* **85**, 3945 (1999).
74. M. Bayer and A. Forchel, *Phys. Rev. B* **65**(4), 413081 (2002).
75. Y.-C. Xin, A. Martinez, T. Saiz, A.J. Moscho, Y. Li, A. Nilsen, A.L. Gray, and L.F. Lester, *IEEE Photon. Technol. Lett.* **19**, 501 (2007).
76. D.C. Heo, J.D. Song, W.J. Choi, J.I. Lee, J.C. Jung, and I.K. Han, *Electron. Lett.* **39**, 863 (2003).
77. Z.Y. Zhang, Z.G. Wang, B. Xu, P. Jin, Z.Z. Sun, and F.Q. Liu, *IEEE Photonics Techn. Lett.* **16**, 27 (2004).
78. T. Akiyama, M. Ekawa, M. Sugawara, K. Kawaguchi, H. Sudo, A. Kuramata, H. Ebe, and Y. Arakawa, *IEEE Photonics Techn. Lett.* **17**, 1614 (2005).
79. S. Fafard, K. Hinzer, S. Raymond, M. Dion, J. McCaffrey, Y. Feng, and S. Charbonneau, *Science* **274**, 1350 (1996).
80. H.Y. Liu *et al.*, *Appl. Phys. Lett.* **80**, 3769 (2002).
81. Y.M. Manz, O.G. Schmidt, and K. Eberl, *Appl. Phys. Lett.* **76**, 3343 (2000).
82. Y. Arakawa and S. Kako, *Phys. Stat. Sol. a Applications and Materials Science* **203**, 3512 (2006).
83. T. Taliercio, P. Lefebvre, A. Morel, M. Gallart, J. Allegre, B. Gil, H. Mathieu, N. Grandjean, and J. Massies, *Materials Science and Engineering B – Solid State Materials for Advanced Technology* **82**, 151 (2001).

CHAPTER 20

Quantum Dot-based Mode-locked Lasers and Applications

A. Martinez,[1] C. Gosset,[1] K. Merghem,[1] G. Moreau,[1] F. Lelarge,[2] and A. Ramdane[1]

[1]CNRS Laboratory for Photonics and Nanostructures (CNRS LPN) Route de Nozay,
91460 Marcoussis, France;
[2]ALCATEL THALES III–V Lab, Route de Nozay, 91460 Marcoussis, France

20.1 Introduction

Self-assembled semiconductor quantum dot (QD)-based lasers are attracting considerable interest owing to their remarkable optoelectronic properties that result from the three-dimensional carrier confinement. They are indeed expected to exhibit much improved performance over that of quantum well devices [1]. Extremely low threshold currents [2] as well as high temperature stability [3–4] have readily been demonstrated in the InAs/GaAs material system with temperature-independent transmission experiments already carried out at 10 Gbit/s [5].

Progress in the growth of self-assembled nanostructures has allowed the stacking of several layers of QDs into the active region, thus enabling the achievement of high modal gain. Fast carrier dynamics such as sub-picosecond ground state (GS) gain recovery time of QD semiconductor optical amplifiers has been demonstrated as well as sub-picosecond absorption recovery time for QD saturable absorbers. Low relative intensity noise (RIN) is yet another interesting property for QD based lasers.

QD-based active layers have hence recently been used in the fabrication of mode-locked lasers. Inhomogeneous broadening of the gain spectrum due QD size distribution is indeed an asset for MLLs as this should result in pulse shortening. Indeed, the sub-picosecond barrier has recently been crossed with the demonstration of 400 fs pulsewidth at a 20 GHz repetition rate based on an InAs/GaAs QD material system [6], and 500 fs pulsewidth at 53 GHz generated by InAs/InP QDash lasers [7]. Ultra-fast carrier dynamics [8, 9] is also a major advantage in the short pulse generation process. Interestingly, the low optical confinement factor, a high degree of population inversion and low loss [10] should lead to low phase noise and results in low timing jitter.

Small jitter of the generated pulses is further expected owing to the theoretically predicted low linewidth enhancement factor (Henry factor) of low-dimensional active layer-based lasers.

Section 20.2 discusses short pulse generation and low timing jitter of In(Ga)As quantum dot mode-locked lasers emitting in the 1.1–1.3 μm range. Section 20.3 describes the performances of InAs/InP quantum dash mode–locked lasers emitting in the 1.55 μm window. Section 20.4 highlights the potential of In(Ga)As/GaAs QD MLLs for optically interconnected systems. Finally, 40 Gb/s all optical clock recovery using InAs/InP QDash MLLs will be reported.

20.2 InAs/GaAs quantum dot mode-locked lasers

The first demonstration of monolithic passively mode-locked lasers using the QD-based active region was reported by Huang *et al.* [11]. The authors showed stable mode locking at 1278 nm at a repetition rate of 7.4 GHz in a two-section device (a gain section electrically isolated from an absorber section) at room temperature in a two dot-in-a-well (DWELL) layer. This first paper evidenced some of the fundamental trends observed in QD MLLs: shortest pulses are obtained just above threshold while pulsewidth quickly broadens with the injection current. They demonstrated 17 ps pulsewidth with a photocurrent radio frequency spectrum of 370 kHz width, which is an indication of phase noise level. The authors pointed out that higher ground state gain (corresponding to ~1.3 μm laser emission) should lead to improvement in the MLL performances and that QD MLLs are potentially attractive for ultra-short pulse generation.

In this chapter we will restrict our discussion to monolithic two-section configurations made of a gain and absorber part as illustrated in Fig. 20.1. However, some of the properties have also been reported for external cavity MLLs implemented from a two-section laser placed into an external cavity. Our discussion will try to emphasize device properties arising from the use of QD material.

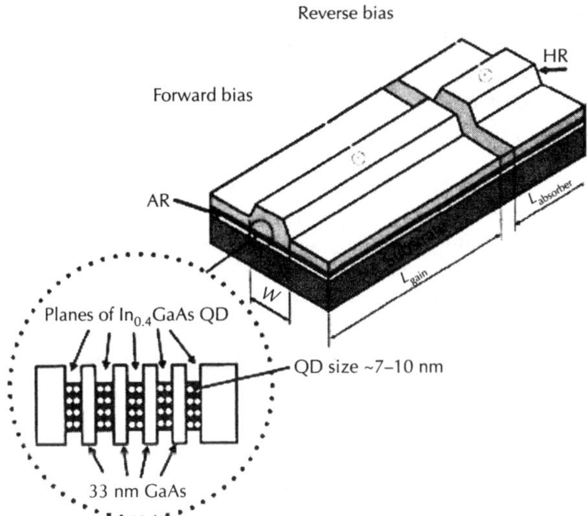

*Figure 20.1 Schematic view of a two-section quantum dot mode-locked laser. The gain section is forward biased and AR coating is applied to the facet, the saturable absorber is reverse biased and the facet is HR coated [6]. (Reused with permission from E.U. Rafailov, Appl. Phys. Lett. **87**, 081107 (2005). Copyright 2005, American Institute of Physics.)*

While In(Ga)As/GaAs QD MLLs always rely on a two-section configuration, we will show that mode locking in an InAs/InP QDash material system (section 20.3) can occur in a single-section FP laser without resorting to a saturable absorber. Moreover, in our discussion, we are reminded here that passive mode locking means pulses are generated by driving the gain and absorber sections in continuous wave mode; while hybrid mode locking corresponds to the application of an external RF signal for loss modulation of the absorber.

20.2.1 Quantum dot active medium for ultra-short pulse generation

Some of the main advantages of QD-based active medium can easily be understood by analysing pulse duration versus different parameters of MLLs. Using a model based on analytic theory, it was demonstrated that for two-section configuration (gain and absorber) the full width half maximum (FWHM) of the pulse width can be described by [12–15]:

$$\Delta\tau_{FWHM} = 1.76 \sqrt{\left(\frac{2gI_s^a}{\alpha_0 \Delta\omega_g^2 |A_0|^2} \right)} \qquad (20.1)$$

where g represents the gain of the laser, I_s^a is the saturation energy of the absorber, α_0 is a constant loss of the absorber, $\Delta\omega_g$ the width of the gain spectrum and A_0 the envelope of the pulse at the initial time and position.

The modal gain of QD lasers is lower than that of QW-based lasers due to a reduced optical confinement factor Γ. It will be shown later that low Γ may prove to be an advantage as long as GS gain saturation can be avoided through stacking of multiple layers of QDs up to 10 or 15. The amount of gain also determines the minimum cavity length which allows GS emission and hence sets a limit on the highest achievable repetition rate.

Relation (20.1) shows that an absorber with low saturation energy I_s^a should yield reduced pulsewidth. This essential property will be discussed in section 20.2.3.2 through the carrier dynamics in the absorber section.

Inhomogeneous broadening of the gain spectrum due QD size distribution (Gaussian-like distribution of the QD size) is indeed an asset for mode-mocked lasers as this should result in a wider effective gain bandwidth and thus pulse shortening.

In conclusion, the section devoted to monolithic QD MLLs grown on GaAs substrates is organized as follows. The first paragraph (section 20.2.2) will present the properties related to the modal gain of InAs/GaAs QDs and the carrier dynamics in the gain regime. We will consider their consequences on the pulse duration as well as the role of the Henry factor in the pulse shortening/broadening.

In a second part (section 20.2.3), the possibility of using QD active material as a saturable absorber will be highlighted. It will be shown that increasing the reverse voltage in the absorber section reduces by one order of magnitude the absorber recovery time, allowing sub-picosecond pulse generation.

Moreover, section 20.2.4 will present the noise properties of the QD material system and demonstration of sub-picosecond timing jitter will be reported.

Finally, the performances of QD MLLs will be summarized to compare the results reported so far and discuss perspectives.

20.2.2   *Gain properties of quantum dot mode-locked lasers*

20.2.2.1   *The role of inhomogeneous broadening*

Owing to their wider inhomogeneous broadened gain spectrum, QD MLLs are expected to produce shorter pulses compared to their QW-based counterparts. The first demonstration of passive QD MLLs showed pulsewidths of 17 ps [11] at a 7.4 GHz repetition rate and the average power did not exceed 2 mW. Due to GS gain saturation in a two dot-in-a-well (DWELL) layer, the authors could not reach a broad gain regime and pulses were not obtained for injection currents greater than 45 mA.

While the 3 dB optical bandwidth (~ 1 nm) does not exceed that of QW MLL lasers, one solution to reduce the pulsewidth would be to exploit the potentially wider effective gain bandwidth, achievable at high injection current in high modal gain structures. This could be obtained by stacking a much larger number of QD layers, as pointed out by the authors.

Owing to the progress in growth and quality of this material system, the possibility of stacking more QD layers into the active region opens the door to shorter cavity lasers emitting on the GS. Kuntz *et al.* demonstrated improved performances of QD MLLs using higher GS modal gain by stacking five QD layers [16]. Passive MLL at 20 GHz was evidenced by using a 2 mm long cavity (1500 μm long gain/500 μm long absorber). Careful adjustment of the bias/voltage can provide transform-limited pulses [16]. Indeed, assuming Gaussian pulse shape, Kuntz *et al.* measured an FWHM of 12 ps which is in good agreement with the Fourier limit of 13 ps corresponding to a spectral FWHM of 0.180 nm. Based on the same active layer [16], the authors report GS emission in a two-section configuration for cavities as short as 1130 μm (980 μm and 150 μm, respectively, for the gain and absorber lengths) leading to passive ML at a repetition rate of 35 GHz. Transform-limited pulses were generated with a minimum pulse duration of 7 ps, which is half of the first value reported in a two QD layer structure. These results show that QD MLLs at high repetition rate can be obtained together with pulse duration of few ps and Fourier-limited pulses.

Kuntz *et al.* remarked that, in the case of 7 ps pulses, they were using only 2% of the intrinsic spectral width of the QD gain medium. According to formula (20.1), the gain bandwidth can further be increased to achieve sub-picosecond pulses [16].

It is worth noticing that these devices exhibited mode-locking in regions where the gain current did not exceed 70 mA (for a threshold current of 30 mA) thus limiting the width of the optical spectrum. Finally, for the 7 ps transform-limited pulze, the peak power from one facet amounts to 6 mW. This is rather low as some applications such as optically interconnected systems require peak powers up to a few watts [16].

Optimized design of the MLL cavity includes careful adjustment of the gain/absorber length to avoid GS gain saturation at high current and simultaneous enhanced absorber saturation owing to reduced optical losses, using HR coating [6]. This allowed the demonstration of picosecond as well as sub-picosecond pulse duration with an average power of 45 mW ($\sim$1.1 W peak power) at 1260 nm at 21 GHz repetition rate. Through a careful adjustment of the gain current and reverse voltage (185 mA, -7.6 V), an FWHM of 14 nm was achieved (Fig. 20.2) leading to pulses as short as 390 fs, the lowest pulsewidth ever reported in the InAs/GaAs QD material system [6]. However, the time bandwidth product of $\sim$1 indicated substantial residual chirping. The optical spectrum is indeed two orders of magnitude larger than previous results based on a similar five QD layer structure [16]. The average power in this regime amounts to 25 mW, i.e. $\sim$3 W peak power. These results highlight the potential of QD MLLs for ultra-short pulse generation by exploiting the intrinsic wide gain spectrum of the active medium. This reflects the impact of the gain spectrum on the shortening of the pulses and also shows that to benefit from the potentially wide gain spectrum, one has to ensure high GS modal gain to be able to drive the section at high currents (Figs 20.2 and 20.3). It relies in part on a combination of three criteria in this case: high modal gain (five QD layers), long gain section (1.8 mm), and HR coating which reduces the mirror loss and provides lower threshold net gain into the cavity.

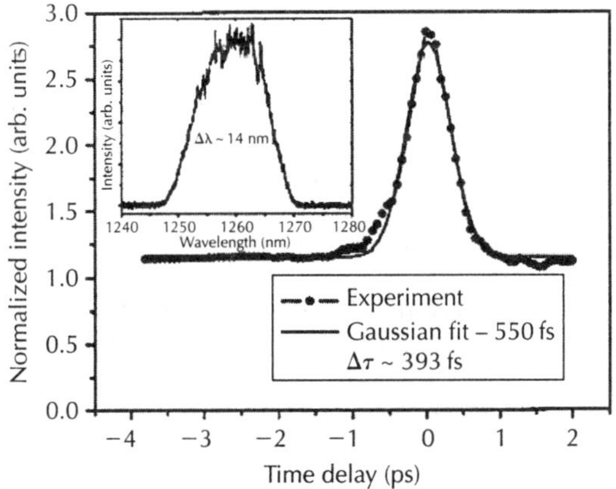

*Figure 20.2   Measured intensity autocorrelation and the corresponding optical spectrum of the QD MLL (inset). Note the spectrum showing a large bandwidth [6]. (Reused with permission from E.U. Rafailov et al., Appl. Phys. Lett. **87**, 081107 (2005). Copyright 2005, American Institute of Physics.)*

20.2.2.2   The role of Henry factor in the pulsewidth broadening

As previously mentioned, the pulsewidth is limited by the gain bandwidth of the active medium. However, there may be deviation from this trend due to chirping of the pulses [17], where the chirp illustrates the wavelength shift of a longitudinal mode when the gain is changed. Indeed, the role the linewidth enhancement factor (LEF) in QW-based mode locked lasers was theoretically investigated [17]. Simulations showed that for QW MLLs the time–bandwidth product scales linearly with the LEF, assuming that this parameter is the same for gain and absorber sections.

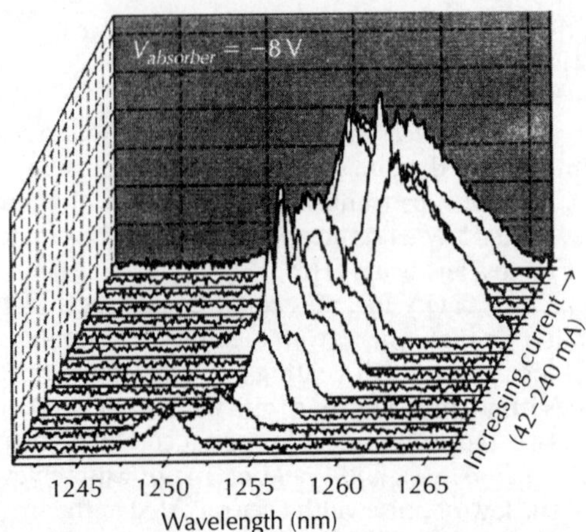

*Figure 20.3   Evolution of the output spectra of the QD MLL versus the injection current in the gain section for a constant reverse voltage [6]. (Reused with permission from E.U. Rafailov et al., Appl. Phys. Lett. **87**, 081107 (2005). Copyright 2005, American Institute of Physics.)*

Several theoretical and experimental investigations showed a near-zero Henry factor of QD lasers *below* threshold. But above threshold, excited state band filling and carrier injection into the wetting layer breaks the symmetry of the gain spectrum which results in the divergence of Henry factor with current. For some specific applications, Fourier-limited very short pulses with high peak power are desirable. This implies working within the range of high injection current to generate output peak power of typically a few W. However, it was shown that the Henry factor of directly modulated QD lasers strongly depends on the injection rate. When GS gain saturation occurs in QD lasers, the Henry factor starts diverging [18, 19]. Even giant linewidth enhancement factors (LEF) of 50 for GS lasing were measured at high currents in low modal gain structures (three QD layers) [20]. This behaviour would induce chirping of the pulses in QD MLLs with significant pulse broadening. A solution to overcome the divergence of the Henry factor α_H relies on increasing the optical confinement factor Γ by stacking several QD layers, which in turn increases the GS modal gain. It was indeed shown that stacking up to ten layers into the active region prevents the Henry factor from diverging and yields α_H similar to the best values obtained in QW lasers [21].

The first evidence of the impact of the Henry factor in the broadening of the pulses in QD MLLs was reported by Choi *et al.* [22]. The authors showed that the LEF of pulses generated by external cavity two-section QD MLL increases with bias current while the opposite trend is observed for the absorber section when the bias voltage increases. It was pointed out that the total chirp can be controlled through adjustment of the current and reverse voltage of the gain/absorber section.

More recently, measurement of the pulse duration in monolithic QD MLLs versus injection current showed an increase in the pulse width for a constant reverse bias in the absorber. The pulses broaden from 4 ps to ~14 ps when the current increases from 50 mA to ~270 mA [23, 24]. This trend is explained by the increase of the Henry factor due to GS gain saturation as long as the laser emits on the GS. The time bandwidth product of pulses generated in diode lasers is given by $\Delta\tau \cdot \Delta\nu = K(1 + \alpha_H^2)^{1/2}$, where K is a constant depending on the pulse shape [12]. While the authors did not give values of the increase of the optical bandwidth $\Delta\nu$ with current, these results seem to be compatible with Henry factors measured in similar layer structures [19]. Nevertheless, a systematic investigation of the Henry factor while the absorber is biased would be required to evaluate the impact of this latter section on the chirp of pulses. The impact of the Henry factor on the pulsewidth in self-pulsating lasers based on a single section FP laser in the InAs/InP QDash material system has not yet been investigated (section 20.3).

QD active medium provides a means to achieve both very short pulses (few ps or less) and high peak power at high repetition rate (>20 GHz). The exploitation of the inherent large gain bandwidth ultimately relies on stacking several QD layers to avoid GS gain saturation, responsible for pulse broadening through an increase of the Henry factor.

However, applications such all optical clock recovery may require higher repetition rates in excess of 40 GHz.

20.2.2.3 High modal gain for high repetition rate (40–80 GHz)

As the repetition rate is given by the inverse of the round trip cavity time, the cavity length needs to be in the range 0.5–1 mm to match the 40–80 GHz range. Increasing the modal gain is mandatory to obtain GS emission on such short cavities. The solution obviously relies on stacking a high number of QD sheets compatible with single-mode operation and/or HR coating, which decreases the threshold gain.

Fourier-limited pulses were, for instance, generated at high repetition rate (40–50 GHz) using the same five QD layer structure as in [16] owing to an HR coating on the absorber facet [25]. Passive mode locking was hence demonstrated at 40 GHz and 50 GHz with a typical pulse duration of 3 ps. Taking into account the spectral FWHM, the time bandwidth product showed Fourier-limited pulses for both repetition rates.

Further improvement in the modal gain was achieved through stacking ten layers of QDs. This resulted in the demonstration of passive 40 GHz mode locking with typical pulse duration of 2.5 ps and time bandwidth products ranging from 0.5 to 0.9 [26]. In this case, the device consisted of a 940 μm long gain and a 100 μm long HR-coated absorber section.

Enhancement of the optical gain using 15 QD layers allowed a record repetition rate of 80 GHz to be achieved associated with a 1.9 ps pulsewidth obtained from a 0.5 mm long passive mode-locked laser based on InAs/GaAs [27]. The time bandwidth product amounts to 1.5, leaving room for further pulsewidth shortening.

As a conclusion, most of the effects that take place in the gain region of QD MLLs and affect the pulse duration and the time bandwidth product have been presented. Another advantage of QD material system arises from the low saturation energy and fast recovery time (down to a few hundred fs) of the absorber, which makes QD saturable absorbers ideal candidates for monolithic MLLs [28].

20.2.3 Quantum dot saturable absorber

20.2.3.1 Saturable absorber for short pulse generation

When a pulse enters the absorber, the leading and trailing edges of the pulse experience high losses (unsaturated loss), while the peak pulse faces lower losses because the absorber exhibits a sudden decrease of the absorption losses (saturation of the loss) [17]. The time during which the absorber loss is saturated corresponds to a temporal window for which net gain is obtained. The temporal width of this saturated loss is directly determined by the carrier lifetime in the absorber. The shorter the carrier lifetime is, the narrower the window will be and hence the pulsewidth.

Two conditions are required to obtain mode locking with a two-section configuration. As previously discussed, shorter pulsewidths are obtained with an absorber that can be more easily saturated [12] because the pulse duration scales with the squared root of saturation energy of the absorber I_s^a. Moreover, short pulses imply absorber recovery time faster than gain recovery time [17]. In other words, the carrier lifetime in the absorber has to be shorter than ~ 1 picosecond to ensure the generation of sub-ps pulses. Broadband absorption spectrum related to inhomogeneous broadening of the QD ensemble is also a key parameter for short pulse generation [29].

Lester and co-workers recently measured the saturation energy of a six-DWELL layer structure emitting pulses at ~ 1210 nm at 5 GHz repetition rate [30]. The absorption saturation powers were calculated for different reverse bias voltages: they vary from ~ 0.5 W to 2.2 W for a -5 V applied voltage.

In the following, we will briefly mention the first works based on the use of the QD medium as a saturable absorber for pulse generation and we will focus on the absorber recovery time. Then the impact of the fast carrier dynamics of the absorber will be discussed regarding the performances of QD MLLs.

20.2.3.2 *Ultra-fast QD saturable absorber dynamics*

While the first use of a QD saturable absorber was reported by Lester and co-workers in a monolithic configuration, the fast character of the quantum dot saturable absorber was further evidenced in stable passive mode locking of a solid-state laser [29, ref. 5 therein]. The authors demonstrated that pulses can be shortened by applying a reverse voltage to the absorber. Recent work also demonstrated that the careful adjustment of the reverse voltage of a QD saturable absorber allows the control of the pulsewidth in a monolithic two-section device [16, 25]. These last papers evidenced standard passive mode locking, where the pulsewidth can be reduced from 18 to 8 ps by increasing the reverse bias from 0 V to -2 V for a constant gain current at a 20 GHz repetition rate.

The electric field dependence of carrier dynamics of a QD saturable absorber waveguide has been investigated by means of sub-picosecond pump–probe experiments at room temperature [9]. Differential time spectroscopy aimed at measuring the absorber recovery in a degenerate scheme at the 1.28 µm GS transition. The investigated waveguides were fabricated from a five-QD layer structure, comparable to the p-i-n diode used for MLL in [16].

The authors demonstrated a decrease in the carrier recovery time from 62 ps to 700 fs, nearly two orders of magnitude, with an increasing reverse voltage from 0 V to 10 V (Fig. 20.4). The absorption recovery time scales as a simple exponential function of the reverse voltage. The experiments showed that at low bias, thermionic emission is the dominant carrier escape mechanism while above $\sim$4 V, the tunnelling process becomes dominant.

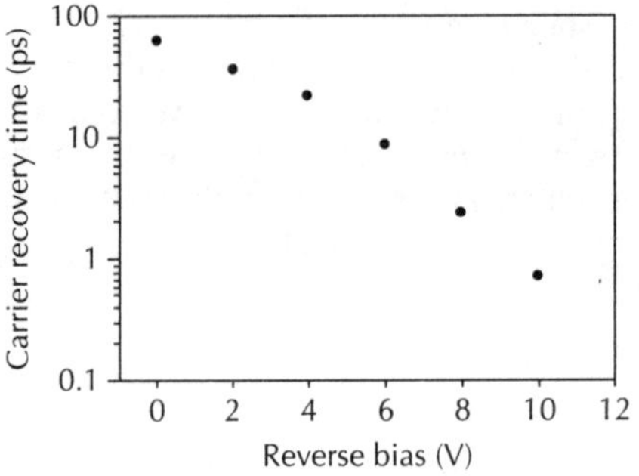

Figure 20.4 *Absorption recovery time of QD saturable absorber as a function of the reverse voltage [9]. (Reused with permission from D.B. Malins et al., Appl. Phys. Lett. **89**, 171111 (2006). Copyright 2006, American Institute of Physics.)*

In [9], to saturate the QD absorber, pump pulse energy of 580 fJ was employed. Previous work performed on InGaAs/InGaAlAs QW electro-absorption modulators demonstrated similar saturation energy of 0.4 pJ and absorption recovery time of 5 ps at 4 V for an optimized band gap engineered structure [31].

These very interesting results highlight the potential of ultra-fast QD saturable absorbers for sub-ps pulse generation. This also gives a general trend for the adjustment of the pulsewidth because as the reverse voltage is increased, the average output power falls off approximately exponentially. The next section is dedicated to the implementation of QD saturable absorber and the adjustment of the reverse voltage to achieve ultra-short pulses.

20.2.3.3 *QD saturable absorber for sub-picosecond pulse generation*

The first demonstration of the impact of absorber carrier lifetime on the pulsewidth in monolithic mode-locked lasers was reported by Kuntz *et al.* [16]. The authors performed a systematic investigation of the pulse duration versus applied voltage in the absorber section for a range of constant gain currents (Fig. 20.5). Passive mode locking was investigated using a five QD layer structure in a two-section configuration (1500 μm long gain and 50 μm long absorber), which allowed mode locking at 20 GHz [16]. They demonstrated a reduction of the pulsewidth from 18 to 8 ps when the reverse voltage is decreased from 0 V to 3 V. These lasers exhibit a common behaviour already observed in QW-based MLLs, implying that InAs/GaAs QD MLLs are governed by similar physical mechanisms.

Moreover, Kuntz *et al.* showed that for a constant current, decreasing the reverse voltage from 0 V to ~3 V forces the laser to operate in three different regimes: incomplete mode locking (low V), mode locking (intermediate V) and no lasing (high V). These three different regimes were also reported by Todaro *et al.* [32]. In this latter regime (Fig. 20.5, [16]), increasing the reverse bias in the range 2.5–3 V induces an increased absorption which prevents lasing when the gain current is not large enough to sustain net gain into the cavity. Providing a higher modal gain through stacking more QD layers is a solution to further reduce the pulse duration, as discussed in section 20.2.2.2 through the use of higher reverse voltages.

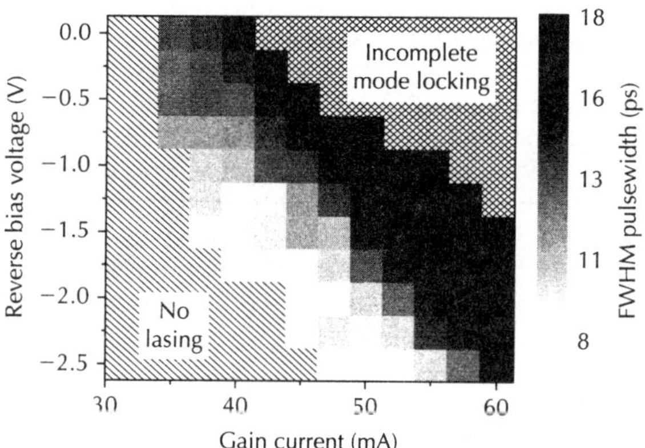

*Figure 20.5   Plot of pulse width dependence on reverse bias voltage and gain current [16]. (Reused with permission from M. Kuntz et al., Appl. Phys. Lett. **85**, 843 (2004). Copyright 2004, American Institute of Physics.)*

To obtain much shorter pulsewidths by increasing the reverse voltage implies increasing the available gain into the cavity, to compensate for the extra losses. This can be achieved by lengthening the gain section and adding an HR coating on the absorber facet [6] as previously discussed in section 20.2.2.1.

Zhang *et al.* also reported a decrease in the pulsewidth from 8 ps to 5.5 ps with increasing the reverse bias [33] from 6 V to 7.5 V, for a constant gain current of 110 mA in passive QD MLLs.

Further experiments on QD MLL lasers showed an exponential-like dependence of the pulsewidth with the reverse bias [28, 34] (Fig. 20.6). While the authors demonstrated nearly transform-limited 780 fs pulses at −7 V and 250 mA injection current [28], deviation from this point shows an increase of the pulsewidth when the reverse voltage is increased. The authors suggested an exponential dependence of the absorber recovery time on the applied reverse bias, which was indeed confirmed later in Malins *et al.* [9]. On the contrary, increasing the bias current in the gain section induces a linear increase in pulsewidth, which could be attributed to self phase modulation and/or α_H increase [9, 17].

Finally, reduction in the FWHM autocorrelation traces when increasing the reverse voltage of the absorber was also reported in an external cavity mode locking of a quantum dot two-section diode laser [35, 36]. The external cavity mode-locked laser is based on a two-section device (3 mm long gain and 300 μm long absorber) and a semiconductor optical amplifier (SOA) to amplify the pulse energy. Both devices are fabricated from a ten QD layer structure. These investigations focused on a fundamental cavity frequency of 4.9 GHz. For a fixed gain current of

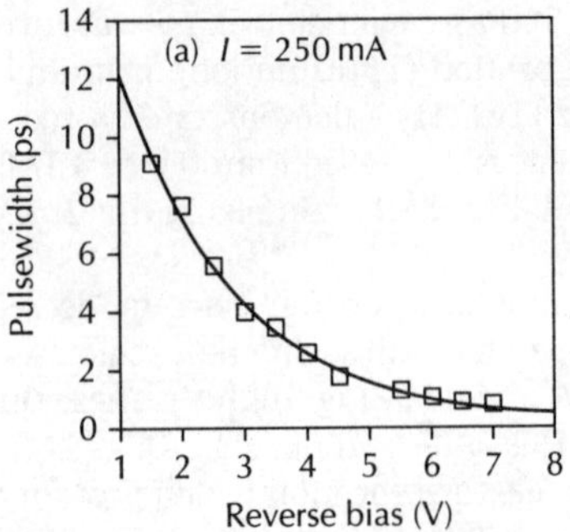

*Figure 20.6   Pulsewidth for constant gain current versus reverse voltage [28]. (Reused with permission from M.G. Thompson et al., Appl. Phys. Lett. **88**, 133119 (2006). Copyright 2006, American Institute of Physics.)*

115 mA (Fig. 20.7, [35]), the FWHM of the autocorrelation trace decreases from 38 ps to 3.5 ps when the reverse voltage increases from 2.5 V to 8 V. Nearly transform-limited pulses of 3.5 ps with a 1.3 nm spectral bandwidth are reported.

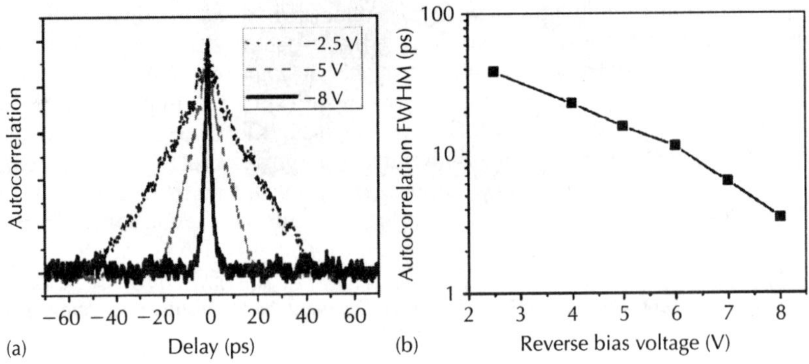

*Figure 20.7   Full width half maximum of autocorrelation signal of pulses generated from an external cavity QD mode-locked laser versus reverse voltage [35]. (Reused with permission from Myoung-Taek Choi et al., Appl. Phys. Lett. **87**, 221107 (2005). Copyright 2005, American Institute of Physics.)*

The demonstration of sub-picosecond pulsewidth using high reverse bias is, however, similar to simulated trends observed in QW-based material systems reported earlier [17]. The authors further showed that in QW MLLs the pulse-width scales exponentially with the ratio of the non-linear absorber saturation coefficient to the non-linear gain saturation coefficient (each coefficient is proportional to the inverse of the associated saturation energy), and that pulse duration scales linearly with the absorber lifetime. This proves that the absorber saturation needs to be stronger than the gain saturation to ensure a pulse shortening mechanism in QW MLLs [17].

The experimental results identified from these QD MLLs [16, 26, 28, 33, 34] confirm that monolithic two-section QD MLLs are governed by similar physical mechanisms as QW MLLs in terms of carrier dynamics in the gain and absorber sections. But the achievable faster carrier dynamics of QD saturable absorbers allows one order of magnitude improvement in the pulsewidth reduction compared to standard QW MLLs emitting at 1.5 μm [17].

A key parameter of MLLs is the small fluctuation in the arrival time of pulses, illustrated by the timing jitter. The next section will present a recent demonstration of reduced timing jitter in QD MLLs compared to standard QW MLLs.

20.2.4   Low timing jitter of quantum dot mode-locked lasers

For many applications such as electro-optic sampling, high bit rate Optical Time Division Multiplexing (OTDM), and optical interconnect, a low timing jitter of ultra-short pulses is required because the fluctuation in the time arrival degrades the bit error rate (BER) and time resolution [33]. This parameter represents the deviation from a perfect train of identical pulses [12]. Here we rapidly stress the origin of low timing jitter of QD MLL and present the main results.

Theoretical investigations showed that QD amplifiers exhibit improved performance in terms of noise [10, 37]. Noise figures better than 4.5 dB are expected owing to three characteristics [10]: low modal gain due to a lower confinement factor Γ (i.e. $\leq 1\%$), a lower differential gain owing to a reduced Γ, and a large energy separation between GS, ES and WL [10]. The energy diagram of this material system is thus equivalent to a three-level system allowing a high degree of population inversion of the GS which implies a population inversion factor n_{sp} close to 1. These features are also present in QD MLLs.

The low confinement factor leads to a reduced coupling of the amplified spontaneous emission to optical mode, thereby reducing the noise level of QD lasers compared to QW lasers. The low differential gain implies a large saturation energy of the gain section, which is useful as the available high optical power will reduce the timing jitter [33]. Timing jitter as well as intensity noise in MLLs arises from spontaneous emission, random fluctuations of driving current and external circuitry [12, 38]. While the absolute timing jitter refers to all sources of noise, residual timing jitter represents the noise contribution of the laser alone [12]. Here we will focus on the residual timing jitter to underline the potential of QD MLLs over QW MLLs.

Gain or photon density variation can also impact on the timing jitter through the phase-amplitude coupling factor or so-called linewidth enhancement factor (LEF). This parameter is expected to be lower than in QW-based counterparts as discussed in section 20.2.2.2. Experimentally, we have already shown that the Henry factor can be similar to that of the best QW lasers at high currents (i.e. ~ 3) using high modal gain structures to prevent GS gain saturation [24]. The demonstrated reduced timing jitter of QD MLLs cannot hence solely be attributed to a low LEF.

Early measurements of timing jitter in passive QD MLLs emitting at $\sim 1.3\,\mu$m have yielded values lower than 1 ps [33, 34, 39] and lower than 2 ps at a repetition rate of 35 GHz [16]. This is smaller than that of passive QW MLLs, integrated over a frequency range from 150 kHz to 50 MHz [40]. Thompson *et al.* measured the timing jitter of 18 GHz monolithic passive QD mode-locked lasers emitting at 1.3 μm [39]. They estimated an upper limit of 600 fs for the RMS timing jitter from the single-sideband phase noise (SSBPN) spectrum over a frequency range of 2.5–50 MHz. In this regime, near transform-limited optical pulses of ~ 10 ps (time bandwidth product of 0.315) were demonstrated in this passive monolithic QD ML laser with ten QD layers. The 24 μm wide waveguide supported several tranverse modes and limited the peak power to ~ 3 mW.

Timing and amplitude jitter of pulses from narrow ridge devices were extracted by comparison of autocorrelation and cross-correlation traces [16]. The uncorrelated jitter was found to be less than 1 ps in active ML at 20 GHz and less than 2 ps in passive ML at 35 GHz. The authors did not indicate the integration range of the SSB PN spectrum.

The first unambiguous demonstration of sub-picosecond timing jitter in passive QD MLLs was reported in 2005 by Zhang *et al.* at a 5 GHz repetition rate [33]. The integration of the SSBPN spectral density $L(f)$ (fourth harmonic) over the offset frequency range of 30 kHz to 30 MHz led to a root mean squared timing jitter of 0.91 ps. Compared to the standard timing jitter of QW passively MLLs (12.5 ps over [150 kHz; 50 MHz]), these QD ML lasers exhibit more than one order of magnitude improvement [33]. The results are explained by the Dirac function-like density of states in QDs as well as low transparency current implying that only a reduced part of carriers is involved in non-stimulated emission [33]. These arguments confirm the origin of the low noise figure evoked in QD SOAs by Berg *et al.* [10].

More recently, very impressive results were reported with two-section lasers at a 40 GHz repetition frequency by Thompson *et al.* [26]. Pulsewidths of 2.5 ps are measured with a time bandwidth product of 0.5 for passive and hybrid QD MLLs. Timing jitter as low as 219 fs was demonstrated in the [16 MHz; 320 MHz] range. Its value amounts to 843 fs in the [150 kHz; 320 MHz] range (integration range specified by ITU-T for a passive MLL) [26]. This is significantly lower than any other reported result in passive QW MLLs at 40 GHz where timing jitter is in the range of 12 ps. The timing jitter for hybrid ML at 40 GHz (124 fs in [20 kHz; 320 MHz] range) was not as good as that of hybrid QW MLLs (73 fs). Further optimization of the fabrication of the absorber should lead to higher absorber modulation and improved timing jitter in the hybrid regime.

An interesting study performed by Thompson *et al.* [34] showed that the timing jitter of passive QD MLLs can be optimized through a careful adjustment of the absorber length at the expense of the pulsewidth. Indeed, a trade-off exists between the pulse duration and the timing jitter: while the pulsewidth decreases with an increased absorber length for a given reverse voltage, the timing jitter is found to decrease when the absorber length is reduced. This implies that the lowest timing jitter is obtained for longer pulsewidths. Typical timing jitter integrated over [20 kHz; 50 MHz] of 100 ps are extracted for 800 fs pulses generated with a 520 μm long absorber, while timing jitter below 400 fs is obtained for 1.8 ps pulses for a 260 μm long absorber [34]. This trend is attributed to the decrease in overall cavity loss for shorter absorber devices.

Finally, one must mention extraction of the residual timing jitter of an external cavity 12.8 GHz actively mode-locked laser performed by Delfyett and co-workers [41]: it amounts to 7.5 fs in the 1 Hz–10 MHz frequency range. All these results show that QD MLLs are ideal sources for both short pulses and low noise optical pulse trains.

As discussed above, the evaluation of the timing jitter using the spectral technique is particulary suited for repetition rates lower than the typical bandwidth of high-speed photodiodes (50 GHz). Based on a cross-correlation technique, Tourrenc *et al.* measured the timing jitter of passive ML QD lasers [42].

Interestingly, the two techniques lead to similar conclusions regarding the evolution of the timing jitter versus the gain current for a given absorber bias [33, 42]. The jitter appears to improve with an increase of the gain leading to typical values as low as 1 ps for a 5 GHz repetition rate [33] and as low as 20 fs/pulse cycle at an 18 GHz repetition rate [42].

In summary, three main properties arise from the application of the reverse voltage to the QD saturable absorber. First, the low differential gain results in high saturation energy of the gain section. Moreover, the absorber recovery time decreases exponentially with the reverse voltage, which implies an exponential decrease of the pulsewidth with voltage. Finally, reduction of the optical confinement factor implies a low phase noise with direct consequence on the timing jitter. All these trends prove that the carrier dynamics of QD saturable absorbers is a key element for the generation of both sub-ps pulses and sub-ps timing jitter, provided high GS modal gain is again available.

20.2.5 Summary

Table 20.1 summarizes the performances discussed above for the sake of comparison. To match the requirement of desired applications (optically interconnected systems, optical time division multiplexing, etc.), the active layer structure, the regime of mode locking, the pulsewidth, the repetition rate, the time bandwidth product and timing jitter are reported.

Different applications require a combination of pulse characteristics, and some applications may require low timing jitter, high peak power or ultra short-pulse duration, and sometimes a combination of all these three parameters [34].

20.3 InAs/InP quantum dash mode-locked lasers emitting at 1.55 μm

20.3.1 Sub-picosecond pulse generation at very high repetition rate

Growth of InAs quantum dots on InP (100) substrates generally leads to the formation of quantum "dash" nanostructures (elongated dots) and the litterature has been less fertile in device results related to 3D confinement of the carriers [43–47].

These nanostructures however, exhibit very interesting properties pertaining to short pulse generation using mode-locked lasers.

Table 20.1 Summary of the performances of the QD MLLs emitting at 1.3 μm.

Reference	Active layer	Regime of mode locking	Pulsewidth	Repetition rate	Peak power	Time bandwidth product	Timing jitter
[11]	Two DWELL layers	Passive	17 ps	7.4 GHz	~20 mW	3.1 (1 nm × 17 ps)	n/a
[16]	Five QD layers	Passive	8–18 ps	35 GHz	6 mW	~0.44 (180 pm × 13 ps)	<2 ps
[16]		Hybrid	7 ps	20 GHz			<1 ps
[6]	Five QD layers	Passive	390 fs	21 GHz	3 W	1 (14 nm × 2 ps)	n/a
[33]	Six DWELL layers	Passive	5.7 ps	5.1 GHz	290 mW		0.91 ps [30 kHz; 30 MHz]
[33]		Passive	4 ps	~5 GHz			<3 ps
[28]	Five QD layers	Passive	800 fs	24 GHz	500 mW	0.5	n/a
[34]	Five QD layers	Passive	5.5 ps	20.5 GHz	45 mW		390 fs [20 kHz; 50 MHz]
[26]	Ten QD layers	Passive	~2.5 ps	40 GHz		0.5–0.9	219 fs [16 MHz; 320 MHz] 843 fs [150 kHz; 320 MHz]
[26]	Ten QD layers	Hybrid					124 fs [20 kHz; 320 MHz]
[26]	Ten QD layers	Passive	2.2 ps	80 GHz	11 mW	1.5	n/a

The "elongated dots" provide a higher optical confinement factor compared to that of QDs grown on GaAs susbstrates. This results in high modal gain ($48\,\mathrm{cm}^{-1}$ for a nine dash-in-a-well layer structure), leading to static performances (threshold current and quantum efficiency) similar to those of QW-based lasers [48]. An optimized laser structure consisting of a stack of six QDash layers has in particular allowed laser emission for very short cavity lengths of $\sim 200\,\mu\mathrm{m}$ owing to the available high modal gain (typically $36\,\mathrm{cm}^{-1}$) [48]. This is an asset for very high achievable repetition rates of MLLs, in excess of 80 GHz, compared to the lower modal gain obtained for QDs grown on GaAs.

QDash lasers further exhibit an inherently wider inhomogeneous broadening due to overlapping states in the high energy portion of the spectrum [49]. The expected wider gain spectrum of the QDash lasers should be an advantage compared to QD lasers for the generation of a much shorter pulsewidth.

The first demonstration of passive QDash-based MLLs at 1.56 μm was reported by Martinez *et al.* in a two-section configuration based on gain and saturable absorber sections [7]. Narrow ridge single-mode waveguides yielded sub-picosecond pulses of ~ 500 fs at a 53 GHz repetition rate without any compression scheme. The reported full width at half maximum of the emission spectrum is ~ 11 nm for this pulse duration highlighting the potential of QDash material for short pulse generation.

At the same time, based on a similar material system, Renaudier *et al.* [50] showed self-pulsation at 45 GHz in a single section Fabry–Perot laser, i.e. without resorting to an absorber section. In this latter configuration, enhanced four-wave mixing (FWM) is invoked as the driving mechanism of the passive mode locking which induces a strong correlation between the longitudinal modes (section 20.3.3).

Indeed, InAs/InP quantum dash-based lasers have recently demonstrated sub-picosecond pulse generation (800 fs) at a very high repetition rate of 134 GHz with an extinction ratio better than 20 dB [51, 52]. This very short pulse duration was achieved using a *one-section* Fabry–Perot laser without resorting to any compression scheme. The time bandwidth product of 0.46 indicates that the pulses are close to being Fourier transform limited. These high performances were not obtained at the expense of output power which amounts to about 10 mW (implying a peak power of ~0.1 W).

In this material system, all that remains is to conduct systematic investigations in two-section devices on how the bias current and the absorber reverse voltage modify the temporal properties of the pulses, as carrier dynamics of the absorber will play an important role.

20.3.2   Extremely narrow radio frequency spectrum

Analysis of the radio frequency photocurrent spectrum has shown a record low linewidth of 47 kHz for a Fabry–Perot laser emitting pulses at 43 GHz repetition rate [7, 52] (Fig. 20.8). This was the lowest linewidth ever reported for any passively mode-locked laser. Efficient phase correlation between longitudinal modes involved in the mode-locking process is found responsible for such performances, as it will be discussed in the next part. With such a low linewidth, the high-frequency timing jitter can be estimated to be less than ~200 fs [38].

The investigation of mode locking in a two-section laser including an unbiased short absorber reveals similar performances with a spectral linewidth of 47 kHz for a 43.6 GHz pulsation frequency [52]. Passive mode locking was confirmed through the measurement of the linewidth of the longitudinal mode which amounts to 17 MHz, approximately two orders of magnitude higher than the FWHM of the RF photocurrent spectrum.

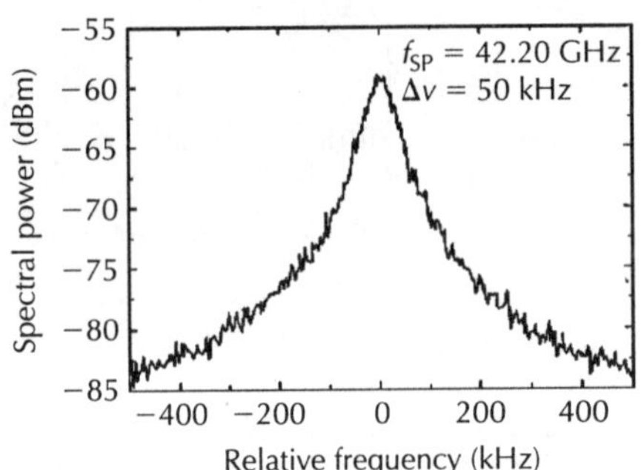

*Figure 20.8   Photocurrent spectrum of passively mode-locked laser around the fundamental frequency (resolution bandwidth = 10 kHz) [52] (Reused with permission from C. Gosset et al., Appl. Phys. Lett. **88**, 241105 (2006). Copyright 2006, American Institute of Physics.)*

Similar experiments performed by Renaudier *et al.* [50] demonstrated linewidths of the mode-beating spectrum as low as 70 kHz using a single-section laser generating pulses at a 45 GHz frequency. Even lower linewidths of ~15 kHz have been evidenced in lower confinement QDash laser structures [47]. This is attributed to a reduced coupling of the spontaneous emission to the lasing mode owing to the low optical confinement factor of QDash-based MLLs.

As a conclusion, at this stage of the investigation of QDash MLLs, it appears that the absorber section does not play a major role in the reported mode locking with narrow FWHM of the RF spectrum. The dominant mechanism arises from the gain section. Indeed, reduction of the dimension of active medium from two dimensions to one dimension should be responsible for the enhancement of four-wave mixing wavelength conversion efficiency in discrete level systems [52, ref. 16 therein].

Besides passive mode locking, active mode locking performed in a six QDash layer FP laser [50] indicates a high-frequency timing jitter as low as 82 fs around the fundamental frequency of 45 GHz, thus demonstrating the reduced intrinsic noise owing to the narrow linewidth of the RF spectrum.

The record low linewidth of the RF photocurrent highlights the potential of the QDash material system for extremely low jitter applications. In particular, we will present in the next section how jitter filtering effects can be useful to realize all optical clock recovery at 40 Gb/s.

20.3.3   Phase amplitude characterization

Enhanced four-wave mixing which favours passive phase-locking between longitudinal modes is invoked to be the main mechanism responsible for QDash mode locking. Recent experiments demonstrated FWM efficiency in QDash semiconductor optical amplifiers (QDash SOA) similar to that of bulk counterpart, although the optical confinement factor is 30 times lower [47].

Evidence of phase correlation in QDash MLLs was further brought by the measurement of the optical spectral phase of longitudinal modes. This relies on the successive analysis of the auto-correlation intensity signal of groups of three adjacent modes [53]. The investigated laser was a one-section passive MLL at 42 GHz pulsation frequency. The evolution of the spectral phase shows a linear dependence versus wavelength in the central part of the spectrum, leading to a transform limited pulse reconstructed through modelling. A non-linear dependence of the spectral phase is evidenced in the lower and higher portion of the spectrum. Simulations also confirm that these uncorrelated modes give rise to a pedestal in the intensity autocorrelation trace and result in non-transform-limited pulses. The phase variation is attributed to inhomogeneous broadening, typical of the QDash active medium [53].

It can be argued that filtering out the uncorrelated longitudinal modes should improve the quality of the pulses as well as the performances of the QDash MLLs. Indeed, very recent experiments demonstrated greatly improved extinction ratios as well as time bandwidth product by an adequate spectrum filtering of undesirable groups of longitudinal modes [54]. The tested device was a 1 mm long Fabry-Perot laser processed from a six DWELL layer structure. Careful selection of the centre frequency and spectral width of the filter results in an improved extinction ratio of the intensity autocorrelation signal from 2 dB to 17 dB, with a time bandwidth product of 0.44. Spectrum filtering also warrants intensity autocorrelation traces without pedestal, thus confirming the above arguments, and leads to an RF photocurrent linewidth as narrow as 30 kHz.

In summary, QDash-based MLLs have demonstrated very promising performances such as subpicosecond widths of transform limited pulses, very low RF linewidths implying extremely low high-frequency timing jitter and over 100 GHz repetition rates. Systematic investigation of the mode-locked mechanisms is still under way to fully exploit all the potential of this novel material system.

20.4   Applications

Practical applications of mode-locked lasers rely on short pulse duration and low timing jitter in such diverse areas as high bit rate optical communications (optical time division multiplexing, optical sources, clock recovery), microwave signal generation, optical sampling, optical interconnects, etc.

Two specific applications will be highlighted, namely pulse generation for optical interconnect and clock distribution in integrated circuits (ICs) [55] and clock recovery at 40 Gbit/s.

In(Ga)As/GaAs quantum dot mode-locked lasers operating up to 80°C have recently been demonstrated, using an undoped active layer [56]. Further improvement in the temperature behaviour is, however, still needed. A main requirement for this former application is the capability to operate the devices above about 100°C.

The second part of this section will present recent exploitation of jitter filtering effect of QDash MLLs for clock recovery at 40 Gbit/s in fibre optic communications.

20.4.1   Optical interconnects applications

In the context of optical interconnects [55] an important requirement for integrated circuits manufacturers regarding clock generation and distribution is the ability to operate at least at

100°C. Another important issue is related to the value of the IC' frequency, ~5 GHz for the next generation processors. This implies cavity lengths of the order of 8 mm which requires specific processing and mounting of the devices. Indeed, MLLs are ideal candidates for interconnects owing to their compactness, low power consumption and DC-biased operation [33]. Moreover, monolithic MLLs based on a two-section configuration offer competitive advantage in terms of reduced cost for volume production.

As CMOS transistor technology improves, problems such as frequency and distance-dependent loss and distortion appear to be a major problem for operations at frequencies in the GHz range [55]. Optical interconnected systems offer potential benefits because optical signal quality is preserved with distance. QD MLLs are very attractive for such applications since they can provide sub-picosecond pulsewidth at repetition rates from 5 GHz up to 80 GHz, and peak power up to a few watts. The ultra-short generated pulses offer advantages over QW MLLs for "optical signalling": receiver sensitivity, signal retiming and time division multiplexing are important benefits. Moreover, the sub-picosecond timing jitter of these lasers is interesting for optical clock distribution and precise non-contact chip testing.

The first report of MLLs for optical interconnects was made in 2005 [33], where sub-picosecond timing jitter and pulse duration shorter than 10 ps at 5 GHz were demonstrated. Growth conditions of the InAs/GaAs six DWELL layers were adjusted to target two different applications: intra-chip/inter-chip clocking and signalling applications at 1250 nm (transparent Si waveguide and SiGe photodetector) and signal generation and processing for data- and telecommunication.

As discussed in the introduction to this chapter, owing to the large energy separation between the ground state (GS) and the excited state (ES), quantum dot directly modulated lasers and QD mode-locked lasers are expected to demonstrate reduced temperature sensitivity compared to their QW counterparts. This potentially opens the way to fabrication of these devices without a thermoelectric cooler, thereby reducing the complexity and the cost of the next generation of optical sources.

However, until recently QD DMLs exhibited a standard characteristic temperature of ~50 K [56]. Deppe *et al.* showed theoretically that the temperature sensitivity of InAs/GaAs QD lasers was largely determined by the closely spaced energy levels of holes in the valence band [57]. Hence, thermal hole spill-out is the main limiting physical mechanism which reduces the gain and the differential gain of InAs/GaAs QD lasers when the temperature increases [57]. Moreover, as GS gain saturation was evidenced to be responsible for pulse broadening in MLLs, one may expect degradation of the performances of the QD MLLs with temperature. Improvement of the temperature stability of QD lasers may be achieved by p-type doping of the active layer [3, 4, 57–60].

However, the demonstration of greater resilience of QD MLLs to temperature up to 80°C and investigation of the temperature dependence of pulse duration was recently reported in undoped layers.

Mode-locking operation using a two-section QD laser in the temperature range 20°C to 80°C was demonstrated [56]. The active region is based on an undoped five QD layer structure and the two-section configuration is similar to that in previous QD MLLs which showed high power sub-ps pulse generation at 21 GHz [6]. The characteristic temperature T_0 amounts to 41 K and does not change significantly with the reverse bias in the absorber section. This work led to the observation of stable mode locking from 20°C to 70°C with signal to noise ratios greater than 20 dB and FWHM of the RF spectra smaller than 80 kHz. At 80°C, the mode locking became less stable with an SNR of 15 dB and an FWHM of the beating spectrum of 700 kHz [56]. To maintain mode locking with temperature above 70°C for a given constant current, the reverse bias voltage has to be reduced to avoid excess cavity losses which prevents mode locking.

More recently, experimental characterizations of the temporal width of the preceding pulses demonstrated that the pulse duration in this QD MLL decreases with temperature [61]. The current is kept at a constant value of 190 mA while the reverse voltage is adjusted to its optimum value in terms of pulse duration for each temperature. The minimum pulse width decreases from ~10 ps to ~6.5 ps from 25°C to 70°C, resulting in a seven-fold decrease of the time bandwidth product from ~20 to ~3.5. The spectra were found to become narrower with the temperature (Fig. 20.9). Hence, the reduction in the pulse duration may be attributed to an increase in the homogeneous

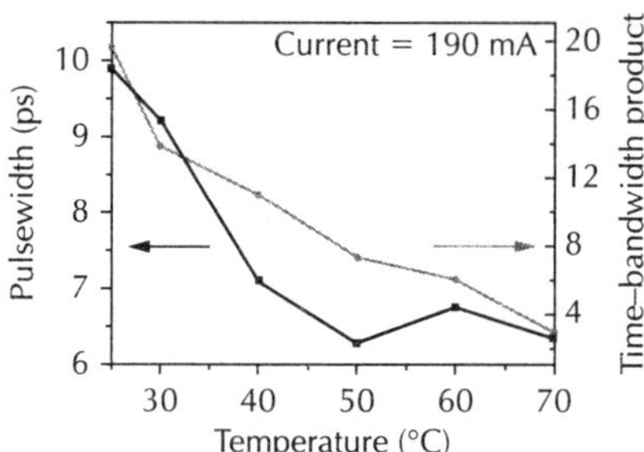

*Figure 20.9   Pulse duration and time bandwidth product versus temperature [61]. (Reused with permission from M.A. Cataluna et al., Appl. Phys. Lett. **90**, 101102 (2007). Copyright 2007, American Institute of Physics.)*

linewidth, or to thermal coupling of carriers with the wetting layer which impacts on the number of modes experiencing gain at threshold [61].

The *p*-type doping approach was proven to be successful to improve the temperature stability of QD lasers with the demonstration of very high characteristic temperature [3, 4] as well as temperature-independent 10 Gb/s direct modulation up to 85°C [58, 59]. While this technique has been applied to DML lasers, there are no reports to date of QD MLLs based on *p*-type doping of the active region. One has to point out that this technique is accompanied by substantial internal losses [3, 4] which may prevent mode locking. Thus a trade-off must be found in the future between improved characteristic temperature and reasonable losses to achieve MLL using *p*-type-doped QD active medium.

20.4.2   40 Gb/s all-optical clock recovery at 1.55 µm

Of particular significance is the recent demonstration of 40 Gbit/s all-optical clock recovery [62, 63], compliant with ITU-T recommendation. This used a one-section self-pulsating QDash laser which realizes high-frequency jitter suppression owing to its inherently narrow free-running spectral linewidth. This result paves the way to all-optical clock recovery up to 160 Gbit/s.

The potential of jitter suppression for clock recovery at 40 Gb/s was demonstrated through the phase noise measurement of a six QDash layer self-pulsating Fabry–Perot laser. The lasing wavelength is centred at 1.5 µm and the laser exhibits a free-running spectral linewidth of ~20 kHz for a fundamental frequency of 40 GHz [62]. The experiment relies on the generation of a reference clock at 40 GHz whose jitter can be monitored on the incoming data and then measured on the recovered clock using a jitter analyser in the range [100 Hz; 500 MHz].

Preliminary calibrations require the evaluation of the recovered clock through the phase noise measurement of the unjittered reference clock and of the self-pulsating QDash laser. The phase noise curves of the MLL indicate three different regimes [62] as evidenced in an earlier paper [63]. In the low-frequency region, i.e. below 60 kHz, the phase noise is dominated by the noise of the incoming signal. The high-frequency region (>4 MHz), where the noise of the laser itself dominates, shows a jitter suppression defined by a $1/f^2$ slope, attributed to the narrow linewidth of the RF spectrum. A previous work indeed measured a cut-off frequency of 9 MHz for this laser, much better than the 100 MHz typical cut-off frequency of bulk DBR lasers for the same repetition rate [63]. In the intermediate region, the noise originates from both contributions of clock and self-pulsating laser. The RMS jitter amounts to ~0.16 ps and is independent of PRBS length of incoming OTDM signal at 40 Gbit/s.

Experiments performed using jittered incoming signal evidenced a drastic decrease of the timing jitter attributed to the filtering effect of the QDash laser. The RMS jitter is reduced from 1.37 ps for the incoming data to 0.31 ps for the recovered clock from the analysis of phase noise curves. Furthermore, the authors report for the first time the jitter transfer function which complies with the ITU standard for clock recovery at 40 Gb/s.

These results highlight the real potential of QDash-based MLLs for all-optical clock recovery at 40 GHz based on the narrow linewidth of the RF spectrum.

20.5 Conclusion and perspectives

Quantum dot-based monolithic mode-locked lasers have recently been the subject of intense investigations owing to the unique properties expected from 3D confinement of carriers. Significant advances have readily been achieved, opening the way to a large range of photonic applications.

Both the InAs/GaAs and InAs/InP QD material systems have been investigated with the demonstration of sub-picosecond pulse generation, multi-GHz repetition rates up to 134 GHz for fundamental mode locking, and more interestingly record timing jitter which is of prime importance for many applications. Specific configurations have also led to improved output powers [6].

Based on these achievements, the first applications such as clock recovery at 40 Gbit/s in fibre optic communication systems, and very short pulse emitters for optical interconnects, have already been reported.

Further improvements are still expected in device characteristics such as temperature stability, output power, control of pulse characteristics, etc. to take full advantage of this new generation of compact, ultra-short pulse sources based on QD material.

Acknowledgements

The authors acknowledge the EC SANDiE Network of Excellence for partial support of this work.

References

1. D. Bimberg, N. Kirstaedter, N.N. Ledentsov, Zh.I. Alferov, P.S. Kop'ev, and V.M. Ustinov, InGaAs-GaAs quantum-dot lasers, *IEEE J. Sel. Topics in Quantum Electron.* **3**(2), 196–205 (1997).
2. G.T. Liu, A. Stintz, H. Li, K.J. Malloy, and L.F. Lester, Extremely low room-temperature threshold current density diode lasers using InAs dots in $In_{0.15}Ga_{0.85}As$ quantum well, *IEE Electron. Lett.* **35**(14), 1163–1165 (1999).
3. O.B. Shchekin and D.G. Deppe, Low-threshold high-T_0 1.3-μm InAs quantum-dot lasers due to p-type modulation doping of the active region, *IEEE Photon. Technol. Lett.* **14**(19), 1231–1233 (2002).
4. I.I. Novikov, N.Y.u. Gordeev, L.Ya. Karachinski, M.V. Maksimov, Yu.M. Shernyakov, A.R. Kovsh, I.l. Krestnikov, A.V. Kozhukhov, S.S. Mikhrin, and N.N. Ledentsov, Effect of p-type doping of the active region on the temperature stability of InAs/GaAs QD lasers", *Semiconductors* **39**, 477–480 (2005).
5. M. Ishida, N. Hatori, K. Otsubo, T. Yamamoto, Y. Nakata, H. Ebe, M. Sugawara, and Y. Arakawa, Low-driving-current temperature-stable 10 Gbit/s operation of p-doped 1.3 μm quantum dot lasers between 20 and 90°C, *IEE Electron. Lett.* **43**(4), 219–221 (2007).
6. E.U. Rafailov, M.A. Cataluna, W. Sibbett, N.D. Il'inskaya, Yu.M. Zadiranov, A.E. Zhukov, V.M. Ustinov, D.A. Livshits, A.R. Kovsh, and N.N. Ledentsov, High-power picosecond and femtosecond pulse generation from a two-section mode-locked quantum-dot laser, *Appl. Phys. Lett.* **87**, 081107 (2005).
7. A. Martinez, C. Gosset, K. Merghem, F. Lelarge, J. Landreau, G. Aubin, and A. Ramdane, Sub-picosecond pulse generation at 1.56 μm using a mode-locked quantum dot laser, *Techni. Digest of European Conference on Optical Communications 2005*, paper Tu1.5.4 (2005).
8. T. Akiyama, H. Kuwatsuka, T. Simoyama, Y. Nakata, K. Mukai, M. Sugawara, O. Wada, and H. Ishikawa, Linear and non linear properties of QD material, *IEEE J. Quantum. Electron.* **37**(8), 1059–1065 (2001).
9. D.B. Malins, A. Gomez-Iglesias, S.J. White, W. Sibbett, A. Miller, and E.U. Rafailov, Ultrafast electro-absorption dynamics in an InAs quantum dot saturable absorber at 1.3 μm, *Appl. Phys. Lett.* **89**, 171111 (2006).
10. T.W. Berg and J. Mørk, Quantum dot amplifiers with high output power and low noise, *Appl. Phys. Lett.* **82**(18), 3083–3085 (2003).

11. X. Huang, A. Stintz, H. Li, L.F. Lester, J. Cheng, and K.J. Malloy, Passive mode-locking in $1.3\,\mu m$ two-section InAs quantum dot lasers, *Appl. Phys. Lett.* **78**(19), 2825–2827 (2001).

12. P. Vasil'ev. *Ultrafast Diode Lasers: Fundamentals and Applications*, (Artech House Publishers, Chapters 3 and 4, 1995).

13. A.E. Siegman and D.J. Kuizenga, Simple analytical expression for AM and FM mode-locked pulses in homogenous lasers, *Appl. Phys. Lett.* **14**(6), 181–182 (1969).

14. D.J. Kuizenga and A.E. Siegman, FM and AM mode-locking of the homogeneous laser – Part I: Theory, *IEEE J. Quantum Electron.* **6**, 694–708 (1970).

15. H.A. Haus, A theory of forced mode-locking, *IEEE J. Quantum Electron.* **11**, 323–330 (1975).

16. M. Kuntz, G. Fiol, M. Lämmlin, D. Bimberg, M.G. Thompson, K.T. Tan, C. Marinelli, R.V. Penty, I.H. White, V.M. Ustinov, A.E. Zhukov, Y.M. Shernyakov, and A.R. Kovsh, 35 GHz mode-locking of $1.3\,\mu m$ quantum dot lasers, *Appl. Phys. Lett.* **85**(5), 843–845 (2004).

17. K.A. Williams, M.G. Thompson, and I.H. White, Long wavelength monolithic mode-locked diode lasers, *New Journal Physics*, **6**(70), 1–30 (2004).

18. A. Markus, J.X. Chen, O. Gauthier-Lafaye, J.G. Provost, C. Paranthoën, and A. Fiore, Impact of intraband relaxation on the performance of a quantum-dot laser, *IEEE J. Sel. Topics in Quantum Electron.* **9**(5), 1308–1314 (2003).

19. A. Martinez, A. Lemaître, K. Merghem, L. Ferlazzo, C. Dupuis, A. Ramdane, J.-G. Provost, B. Dagens, O. Le Gouezigou, and O. Gauthier-Lafaye, Static and dynamic measurements of the α-factor of 5-quantum-dot-layer single-mode lasers emitting at $1.3\,\mu m$ on GaAs, *Appl. Phys. Lett.* **86**, 211115 (2005).

20. B. Dagens, A. Markus, J.X. Chen, J.-G. Provost, D. Make, O. Le Gouezigou, J. Landreau, A. Fiore, and B. Thedrez, Giant linewidth enhancement factor and purely frequency modulated emission from quantum dot laser, *IEE Electron. Lett.* **41**(6), 323–324 (2005).

21. D.-Y. Cong, A. Martinez, K. Merghem, G. Moreau, A. Lemaître, J.-G. Provost, O. Le Gouezigou, M. Fischer, I. Krestnikov, A.R. Kovsh, and A. Ramdane, Optimisation of a- factor for quantum dot InAs/GaAs Fabry–Pérot lasers emitting at $1.3\,\mu m$, *IEE Electron. Lett.* **43**(4), 222–224 (2007).

22. M.-T. Choi, W. Lee, J.-M. Kim, and P.J. Delfyett, Ultrashort, high-power pulse generation from a master oscillator power amplifier based on external cavity mode locking of a quantum-dot two-section diode laser, *Appl. Phys. Lett.* **87**, 221107 (2005).

23. M.A. Cataluna, W. Sibbett, D.A. Livshits, J. Weimert, A.R. Kovsh, and E.U. Rafailov, Stable mode locking via ground or excited-state transitions in a two-section QD laser, *Appl. Phys. Lett.* **89**, 081124 (2006).

24. M.A. Cataluna, A.D. McRobbie, W. Sibbett, D.A. Livshits, A.R. Kovsh, and E.U. Rafailov, New mode locking regime in quantum-dot laser: enhancement by simultaneous cw excited-state emission, *Techni. Digest Conference on Laser and Electro-Optic 2006*, paper CThH3 (2006).

25. M. Kuntz, G. Fiol, M. Lämmlin, D. Bimberg, M.G. Thompson, K.T. Tan, C. Marinelli, A. Wonfor, R. Sellin, R.V. Penty, I.H. White, V.M. Ustinov, A.E. Zhukov, Yu.M. Shernyakov, A.R. Kovsh, N.N. Ledentsov, C. Shubert, and V. Marembert, Direct modulation and mode locking of $1.3\,\mu m$ quantum dot lasers, *New J. Phys.* **6**(181), 1–11 (2004).

26. M.G. Thompson, D. Larson, A.R. Rae, K. Yvind, R.V. Penty, I.H. White, J. Hvam, A.R. Kovsh, S.S. Mikhrin, D.A. Livshits, and I.L. Krestnikov, Monolithic hybrid and passive mode-locked 40 GHz, *Techni. Digest of European Conference on Optical Communications 2006*, paper We4.6.3 (2006).

27. M. Laemmlin, G. Fiol, C. Meuer, M. Kuntz, F. Hopfer, A.R. Kovsh, N.N. Ledentsov, and D. Bimberg, Distortion-free optical amplification of 20–80 GHz mode-locked laser pulses at $1.3\,\mu m$ using quantum dots, *IEE Electron. Lett.* **42**(12), 697–699 (2006).

28. M.G. Thompson, A. Rae, R.L. Sellin, C. Marinelli, R.V. Penty, I.H. White, A.R. Kovsh, S.S. Mikhrin, D.A. Livshits, and I.L. Krestnikov, Subpicosecond high-power mode locking using flared waveguide monolithic quantum-dot lasers, *Appl. Phys. Lett.* **88**, 133119 (2006).

29. A.A. Lagatsky, E.U. Rafailov, W. Sibbett, D.A. Livshits, A.E. Zhukov, and V.M. Ustinov, Quantum-dot-based saturable absorber with p-n junction for mode-locking of solid-state lasers, *IEEE Photon. Technol. Lett.* **17**(2), 294–296 (2005).

30. Y.-C. Xin, L.F. Lester, A.L. Gray, and L. Zhang, Characterization of the static and dynamic parameters in a $1.3\,\mu m$ quantum dot mode-locked laser, *Techni. Digest Conference on Laser and Electro-Optic 2007*, paper CMM3 (2007).

31. N. El Dahdah, G. Aubin, J.-C. Harmand, A. Ramdane, A. Shen, F. Devaux, A. Garreau, and B.-E. Benkelfat, Ultrafast InGaAs/InGaAlAs multiple-quantum-well electro-absorption modulator for wavelength conversion at high bit rates, *Appl. Phys. Lett.* **84**(21), 4268–4270 (2004).

32. M.T. Todaro, J.-P. Tourrenc, S.P. Hegarty, C. Kelleher, B. Corbett, G. Huyet, and J.G. McInerney, Simultaneous achievement of narrow pulse width and low pulse-to-pulse timing jitter in 1.3 μm passively mode-locked quantum-dot lasers, *OSA Optics Lett.* **31**(21), 3107–3109 (2006).

33. L. Zhang, L. Cheng, A.L. Gray, S. Luong, J. Nagyvary, F. Nabulsi, L. Olona, K. Sun, T. Tumolillo, R. Wang, C. Wiggins, J. Zilko, Z. ZOu, and P.M. Varangis, Low timing jitter, 5 GHz optical pulses from monolithic two-section passively mode-locked 1250/1310 nm quantum dot lasers for high speed optical interconnects, *Techn. Digest of Optical Fiber Communications 2005*, paper OWM4 (2005).

34. M.G. Thompson, A. Rae, R.V. Penty, I.H. White, A.R. Kovsh, S.S. Mikhrin, D.A. Livshits, and I.L. Krestnikov, Absorber length optimisation for sub-picosecond pulse generation and ultra-low jitter performance in passively mode-locked 1.3 μm quantum-dot laser diodes, *Techn. Digest of Optical Fiber Communications 2006*, paper OThG3, (2006).

35. M.-T. Choi, W. Lee, J.-M. Kim, and P.J. Delfyett, Ultrashort, high-power pulse generation from a master oscillator power amplifier based on external cavity mode locking of a quantum-dot two-section dioed laser, *Appl. Phys. Lett.* **87**, 221107 (2005).

36. J. Kim, M.-T. Choi, and P.J. Delfyett, Pulse generation and compression via ground and excited states from a grating coupled passively mode-locked quantum dot two-section diode laser, *Appl. Phys. Lett.* **89**, 261106 (2006).

37. T. Akiyama, M. Ekawa, M. Sugawara, K. Kawaguchi, H. Sudo, A. Kuramata, H. Ebe, and Y. Arakawa, An ultrawide-band semiconductor optical amplifier having an extremely high penalty-free output power of 23 dBm achieved with quantum dots, *IEEE Photon. Technol. Lett.* **17**(8), 1614–1616 (2005).

38. D. Von der Linde, Characterization of the noise in continuously operating mode-locked lasers, *Appl. Phys. B*, **39**, 201–217 (1986).

39. M.G. Thompson, K.T. Tan, C. Marinelli, K.A. Williams, R.V. Penty, I.H. White, M. Kuntz, D. Ouyang, D. Bimberg, V.M. Ustinov, A.E. Zhukov, A.R. Kovsh, N.N. Ledentsov, D.-J. Kang, and M.G. Blamire, Transform-limited optical pulses from 18 GHz monolithic modelocked quantum dot lasers operating at ~1.3 μm, *IEE Electron. Lett.* **40**(5), 346–347 (2004).

40. D.J. Derickson, P.A. Morton, J.E. Bowers, and L.R. Thornton, Comparison of timing jitter in external and monolithic cavity mode-locked semiconductor-lasers, *Appl. Phys. Lett.* **59**(26), 3372–3374 (1991).

41. M.-T. Choi, J.-M. Kim, W. Lee, and P.J. Delfyett, Ultralow noise optical pulse generation in an actively mode-locked quantum-dot semiconductor laser, *Appl. Phys. Lett.* **88**, 131106 (2006).

42. J.-P. Tourrenc, S. O'Donoghue, M.T. Todaro, S.P. Hegarty, M.B. Flynn, G. Huyet, J.G. McInerney, L. O'Faolain, and T.F. Krauss, Cross-correlation timing jitter measurement of high power passively mode-locked two-section quantum-dot lasers, *IEEE Photon. Technol. Lett.* **18**(31), 2317–2319 (2006).

43. R.H. Wang, A. Stintz, P.M. Varangis, T.C. Newell, H. Li, K.J. Malloy, and L.F. Lester, *IEEE Photon. Technol. Lett.* **13**(8), 767–769 (2001).

44. S. Deubert, A. Somers, W. Kaiser, R. Schwertberger, J.P. Reithmaier, and A. Forchel, *J. Cryst. Growth* **278**, 346 (2005).

45. F. Lelarge, B. Rousseau, B. Dagens, F. Poingt, F. Pommereau, and A. Accard, Room temperature continuous-wave operation of buried ridge stripe lasers using InAs-InP (100) quantum dots as active core, *IEEE Photon. Technol. Lett.* **17**(7), 1369–1371 (2005).

46. J.P. Reithmaier, A. Somers, S. Deubert, R. Schwertberger, W. Kaiser, A. Forchel, M. Calligaro, P. Resneau, O. Parillaud, S. Bansropun, M. Krakowski, R. Alizon, D. Hadass, A. Bilenca, H. Dery, V. Mikhelashvili, G. Eisenstein, M. Gioannini, I. Montrosset, T.W. Berg, M. van der Poel, J. Mørk, and B. Tromborg, InP based lasers and optical amplifiers with wire-/dot-like active regions, *J. Phys. D: Appl. Phys.* **38**, 2088–2102 (2005).

47. F. Lelarge, B. Dagens, J. Renaudier, R. Brenot, A. Accard, F. van Dijk, D. Make, O. Le Gouezigou, J.-G. Provost, F. Poingt, J. Landreau, O. Drisse, E. Derouin, B. Rousseau, F. Pommereau, and G.-H. Duan, Recent advances on InAs/InP quantum dash based semiconductor lasers and optical amplifiers operating at 1.55 μm, *IEEE J. Sel. Topics in Quantum Electron.* **13**(1), 111–124 (2007).

48. G. Moreau, S. Azouigui, D.-Y. Cong, K. Merghem, A. Martinez, G. Patriarche, A. Ramdane, F. Lelarge, B. Rousseau, B. Dagens, F. Poingt, A. Accard, and F. Pommereau, Effect of layer stacking and p-type doping on the performance of InAs/InP quantum-dash-in-a-well lasers emitting at 1.55 μm, *Appl. Phys. Lett.* **89**, 241123 (2006).

49. H. Dery, E. Benisty, A. Epstein, R. Alizon, V. Mikhelashvili, G. Eisenstein, R. Schwertberger, D. Gold, J.P. Reithmaier, and A. Forchel, On the nature of quantum dashes, *J. Appl. Phys.* **95**(11), 6103–6111 (2004).

50. J. Renaudier, R. Brenot, B. Dagens, F. Lelarge, B. Rousseau, F. Poingt, O. Legouezigou, F. Pommereau, A. Accard, P. Gallion, and G.-H. Duan, 45 GHz self-pulsation with narrow linewidth in quantum dot Fabry–Perot semiconductor lasers at 1.5 μm, *IEE Electron. Lett.* **41**(18), 1007–1008 (2005).

51. C. Gosset, K. Merghem, A. Martinez, G. Moreau, G. Patriarche, G. Aubin, J. Landreau, F. Lelarge, and A. Ramdane, Subpicosecond pulse generation at 134 GHz and low radiofrequency spectral linewidth in quantum dash-based Fabry–Perot lasers emitting at 1.5 μm, *IEE Electron. Lett.* **42**(2), 91–92 (2006).

52. C. Gosset, K. Merghem, A. Martinez, G. Moreau, G. Patriarche, G. Aubin, and A. Ramdane, Subpicosecond pulse generation at 134 GHz using quantum dash-based Fabry–Pérot laser emitting at 1.56 μm, *Appl. Phys. Lett.* **88**, 241105 (2006).

53. C. Gosset, K. Merghem, G. Moreau, A. Martinez, G. Aubin, J.-L. Oudar, A. Ramdane and F. Lelarge, Phase-amplitude Characterization of a high repetition rate Quantum Dash Passively Mode-Locked Laser, *Optics Letters.* **31**(12), 1848 (2006).

54. K. Merghem, C. Gosset, A. Martinez, G. Moreau, F. Lelarge, G. Aubin, and A. Ramdane, Effect of spectrum filtering on the performances of quantum-dash mode-locked lasers emitting at 1.55 micrometer, *Techni. Digest Conference on Lasers and Electro-Optics/Europe-IQEC Conference*, paper CB13-5-THU, (2007).

55. G.A. Keeler, B.E. Nelson, D. Agarwal, C. Debaes, N.C. Helman, A. Bhatnagar, and D.A.B. Miller, The benefits of ultrashort optical pulses in optically interconnected systems, *IEEE J. Sel. Topics in Quantum Electron.* **9**(2), 477–485 (2003).

56. E.U. Cataluna, A.D. Rafailov, W. McRobbie, D.A. Sibbett, Livshits, and A.R. Kovsh, Stable mode-locked operation up to 80°C from an InGaAs quantum-dot laser, *IEEE Photon. Technol. Lett.* **18**(14), 1500–1502 (2006).

57. D.G. Deppe, H. Huang, and O.B. Shchekin, Modulation characteristics of quantum-dot lasers: the influence of p-type doping and the electronic density of states on obtaining high speed, *IEEE J. Quantum Electron.* **38**(12), 1587–1593 (2002).

58. M. Ishida, N. Hatori, K. Otsubo, T. Yamamoto, Y. Nakata, H. Ebe, M. Sugawara, and Y. Arakawa, Low-driving-current temperature-stable 10 Gbit/s operation of p-doped 1.3 μm quantum dot lasers between 20 and 90°C, *IEE Electron. Lett.* **43**(4), 219–221 (2007).

59. B. Dagens, O. Bertran-Pardo, M. Fischer, F. Gerschütz, J. Koeth, I. Krestnikov, A. Kovsh, O. Le Gouezigou, and D. Make, Uncooled directly modulated quantum dot laser 10Gb/s transmission at 1.3 μm, with constant operation parameters, *Techni. Digest of European Conference on Optical Communications 2006*, PD Paper Th4.5.7, (2006).

60. T.J. Badcock, R.J. Royce, D.J. Mowbray, M.S. Skolnick, H.Y. Liu, M. Hopkinson, K.M. Groom, and Q. Jiang, Low threshold current density and negative characteristic temperature 1.3 μm InAs self-assembled quantum dot lasers, *Appl. Phys. Lett.* **90**, 111102 (2007).

61. E.A. Cataluna, P. Viktorov, W. Mandel, D.A. Sibbett, J. Livshits, A.R. Weimert, J. Kovsh, and E.U. Rafailov, Temperature dependence of pulse duration in mode-locked quantum-dot laser, *Appl. Phys. Lett.* **90**, 101102 (2007).

62. J. Renaudier, B. Lavigne, F. Lelarge, M. Jourdan, B. Dagens, O. Legouezigou, P. Gallion, and G.-H. Duan, Standard-compliant jitter transfer function of all-optical clock recovery at 40 GHz based on a quantum-dot self-pulsating semiconductor laser, *IEEE Photon. Tech. Lett.* **18**(11), 1249–1250 (2006).

63. G.-H. Duan, B. Lavigne, J. Renaudier, F. Lelarge, O. Legouezigou, B. Dagens, and A. Accard, Importance of the beating spectral linewidth on the perfomance of the recovered clock at 40 Gb/s using self-pulsating lasers, *Proc. Conference on Lasers and Electro-Optics 2006*, paper CMN3 (2006).

CHAPTER 21

Quantum Dot Infrared Photodetectors by Metal-Organic Chemical Vapour Deposition

Manijeh Razeghi, Wei Zhang, Hochul Lim, and Stanley Tsao

Center for Quantum Devices, Department of Electrical Engineering and Computer Science, Northwestern University, Evanston, IL 60208

21.1 Introduction

The strong interest in low-dimensional semiconductor structures originates from their exciting electronic properties, which have an important impact on the performance of high-speed electronic and photonic devices. The quantum dots (QDs), also known as quantum boxes, are nanometre scale islands in which electrons and holes are confined in three-dimensional potential boxes. They are expected to show a zero-dimensional, δ-function density of states and are able to quantize an electron's free motion by trapping it in a quasi-zero-dimensional potential confinement. As a result of the strong confinement imposed in all three spatial dimensions, quantum dots are similar to atoms. They are frequently referred to as "artificial atoms". Due to this confinement, novel physical properties will emerge, which can lead to new semiconductor devices as well as drastically improved device performance.

As the particles are confined in all three dimensions, there is no dispersion curve and the density of states is just dependent on the number of confined levels. For one single dot, only two (spin-degenerate) states exist at each energy level and the plot of the density of states versus energy will be a series of δ-functions. Figure 21.1 shows the change of the density states from a bulk system to the low-dimensional systems of quantum well (QWL), quantum wire (QWR) and QD.

In QDs, the width of the electron energy distribution is zero in an ideal case. This means that electrons in those structures are distributed in certain discrete energy levels and the energy distribution width is fundamentally independent of temperature. In real semiconductor structures, due to many interaction processes such as electron–electron and electron–phonon scattering (which can also be reduced by QDs due to the lack of phonons to satisfy the energy conservation, which is the so-called phonon bottleneck [1]), a certain width in the electron energy distribution exists; However, it is expected to be much smaller compared to bulk and QW.

QDs have generated great interest as a new material and structure that are expected to lead to novel semiconductor devices or to improve current existing devices' performance. One such example is QD lasers. The main advantages of QD lasers over the conventional QW lasers are lower threshold current density, high gain, weak temperature dependence (high characteristic temperature T_0) and low chirp [2, 3].

Another application of QDs is quantum dot infrared detectors (QDIPs), which is the main focus of this chapter. Many applications require thermal imaging that is collected by infrared focal plane arrays (FPAs). So far, most mid-wavelength infrared (MWIR) and long-wavelength infrared (LWIR) photodetector FPAs are based either on HgCdTe (MCT) [4] or quantum well infrared photodetectors

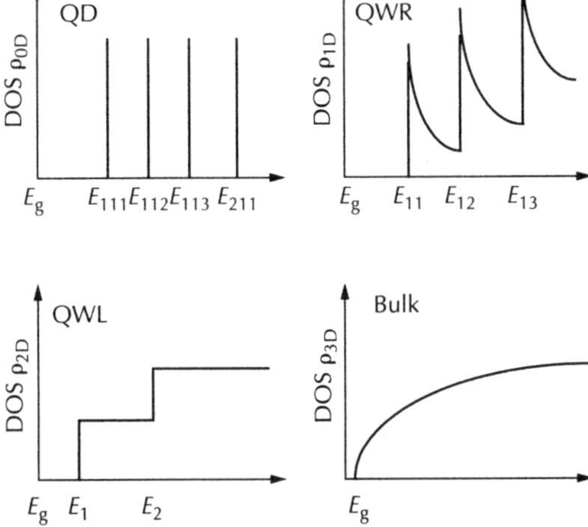

Figure 21.1 Density of state of 200-dimensional (upper-left), one-dimensional (upper-right), two-dimensional (lower-left), and bulk (lower-right) systems.

(QWIPs) [5]. MCT FPAs suffer from difficult material growth, device instability, high array non-uniformity, and very high cost. QWIPs have been used successfully for FPAs. However, a limitation of QWIPs is that, due to the transition selection rules, the most widely used n-type QWIPs are not sensitive to normally incident light and typically have a narrow response range in the infrared. P-type QWIPs are able to detect normal incident light due to the band mixture; however, they have not found many practical applications due to their low detectivity.

An extension of QWIP is QDIP, which utilizes intersubband absorption between bound states in the conduction/valence band in QDs. Given high uniformity and high density quantum dot layer, QDIPs are predicted to outperform QWIPs due to their inherent sensitivity to normal incidence radiation and reduced phonon scattering. Higher temperature operation and lower dark current are also expected for this type of device [6]. So far, most QDIPs reported have showed inferior performance compared to that of QWIPs with similar parameters. The major challenges facing QDIPs are QD fabrication. To justify its potential advantages, QDIPs need high-uniform and high-density QD layers. New device designs for QDIPs are also required to further improve their performance as an infrared photodetector.

The condition for new electronic properties to occur in a QD device is that the lateral size of their active region must be smaller than the coherence length and the elastic scattering length of the carriers. Additional quantum-size effects require the structural features to be reduced to the range of the de Broglie wavelength. The advantages in operation depend not only on the absolute size of the nanostructures in the active region, but also on the uniformity of size. A large distribution of sizes would "smear" the density of states of QDs thus making it more like that of bulk material. Therefore, the repeatable fabrication of these nanometre three-dimensional quantum structures requires methods with atomic scale accuracy, which is a major challenge for current micro- or nanostructure material technologies.

The fabrication technique of quantum dots can be categorized into a "top-down" method using lithography and etching, and a "bottom-up" method using self-assembly. The "top-down" method usually includes electron beam lithography, dry etching, and sometimes regrowth on patterned substrates. QDs can be etched from QW structures via low energy electron-beam lithography. Another method of creating quantum dots is realized by applying voltage to nano-electrodes. The spatially modulated electric field created by the voltage localizes the electrons in a small area. Quantum dots can also be created through the selective growth of a narrow gap semiconductor material on a patterned wide gap substrate. The problem of such a "top-down" method is the low optical efficiency of such dots: high surface-to-volume ratios of these nanostructures and associated high surface recombination rates, plus the damage introduced during the fabrication, prevent the successful formation of high-quality QD devices.

A recent breakthrough in QD fabrication techniques is self-assembly based on the Stranski–Krastanov growth mode [7]. In this method, the lattice constants of the substrate and the crystallized material differ greatly, so only the first few deposited monolayers crystallize in the form of epitaxial strained layers where the lattice constant is equal to that of the substrate. When the critical thickness of the epitaxial layer is exceeded, the significant strain that occurs in the epitaxial layer leads to the break down of this ordered structure and to the spontaneous creation of randomly distributed islets (e.g. quantum dots) of regular shapes and similar sizes. The small sizes of the self-assembled quantum dots, the homogeneity of their shapes and sizes in a macroscopic scale, the perfect crystal structure, and the fairly convenient growth process, without the necessity to precisely deposit electrodes or etching, are among this method's greatest advantages.

This rest of the chapter is organized as follows: in section 21.2 we discuss the theoretical calculations and modelling of QDIPs; in section 21.3, the MOCVD growth and characterization of the materials for QDIP device are presented. We show that the size, shape, and density of the InAs QDs are strongly affected by growth conditions such as temperature, V/III ratio, growth rate, ripening time and the matrix materials. Section 21.4 focuses on the fabrication and measurement of QDIP devices. Next, FPAs based on our QDIPs are presented in section 21.5. Finally, we will summarize and discuss briefly how to further improve the QDIP device performance.

21.2   Theoretical modelling of quantum dot infrared detectors

It is important to understand the correlations between the design parameters and the device characteristics for QDIPs. However, in order to make correlations between them, one needs to know what is going on inside the device and should have certain pictures about the physics of the device. In this section, the modelling of QDIPs will be presented based on the semi-phenomenological theory. The modelling includes the calculation of energy levels and oscillator strengths, responsivity, dark current, gain and detectivity.

21.2.1   *Single-band effective mass envelope function method*

The simplest model for the quantum dot energy level calculation is when a quantum dot is surrounded by an infinite potential barrier. Even though it is not realistic, a general idea about the discrete energy levels and other properties can be obtained. The wavefunctions of a quantum dot surrounded by an infinite potential barrier are used for the basic functions of the single-band finite potential problem. In the case where the quantum dot has cylindrical symmetry, the basis wavefunctions can be obtained by solving Schrödinger's equation in the cylindrical infinite potential wall.

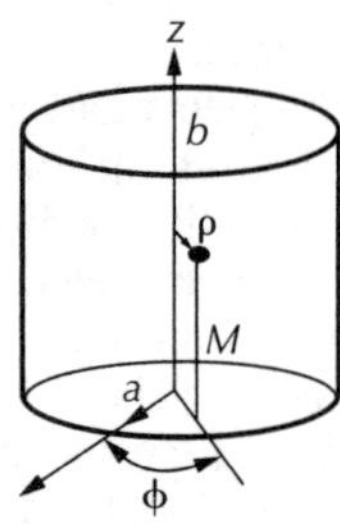

Figure 21.2   One particle inside a cylindrical infinite potential box.

The cylinder has radius a and height b. The Hamiltonian has the cylindrical symmetry which is azimuthal around the z axis and is given as:

$$\hat{H} = \frac{1}{2M}\left(\hat{p}_\rho^2 + \hat{p}_z^2 + \frac{L_z^2}{\rho^2}\right) \tag{21.1}$$

where $\hat{p}_\rho^2 = -\hbar^2 \frac{1}{\rho}\frac{\partial}{\partial\rho}\left(\partial\frac{\partial}{\partial\rho}\right)$ and $\hat{L}_z^2 = -\hbar^2 \frac{\partial^2}{\partial\phi^2}$.

The boundary condition for this problem is:

$$\sin k_q b = J_m(K_{mn}a) = 0 \tag{21.2}$$

where J_m is the m-th order of Bessel function and K_{mn} is the n-th zero of J_m.

After solving the Hamiltonian with the boundary condition above, the eigenfunctions and eigenenergies can be calculated as follows:

$$\psi_{qmn}(x, y, z) = \sqrt{\frac{2}{\pi b[aJ'_m(K_{mn}a)]^2}}\, J_m(K_{mn}\rho)\sin k_q z\, e^{im\phi} \tag{21.3}$$

$$E_{qmn} = \frac{\hbar^2}{2M}(K_{mn}^2 + k_q^2) \tag{21.4}$$

For the design of a real QDIP device, it is desirable to have such a tool that the energy levels and their transitions can be easily calculated and can be applied to the design of the device structure. The process between the design and the growth should be quick. As a further approximation to multi-band effective mass theory, the single-band effective mass envelope function method can be used for the energy levels of quantum dots.

Gershoni *et al.* [8] developed a numerical method in which they expand the envelope function of a rectangular quantum wire which is a 2D confined system using a complete orthonormal set (COS) of periodic functions, which are solutions for a rectangular wire with an infinite barrier height and suitably chosen dimensions. The advantage of this method is that it can be applied to structures of arbitrary shape. Moreover, all the matrix elements can be calculated analytically. Gangopadhyay and Nag [9] extended this method to study 3D confined structures such as pararellelipeds and cylinders.

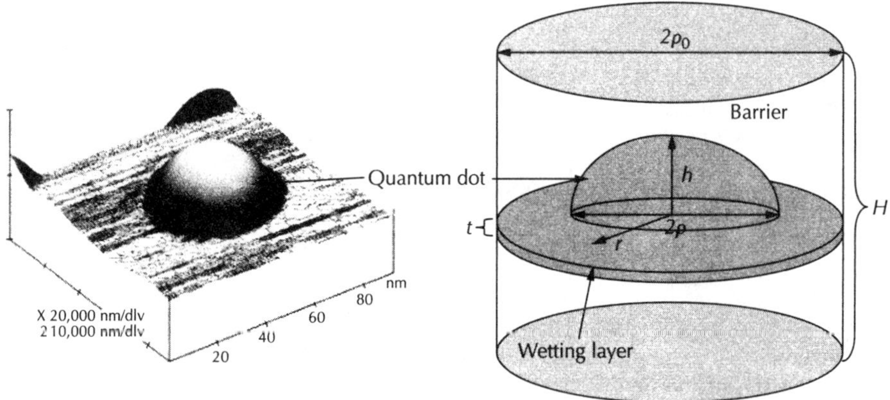

Figure 21.3 Left: *AFM image of uncapped InAs quantum dot on InP.* Right: *The calculation model of a capped quantum dot.*

The aim of this section is to extend Gershoni *et al.*'s method to determine the energy levels of current quantum dots under discussion.

Our InAs QDs on InP grown by MOCVD have lens-like shape as shown in Fig. 21.3. In order to model the quantum dot energy levels, it is necessary to simplify the geometry of quantum dots into perfect lens shape as shown on the right of Fig. 21.3. The quantum dot is varied inside the barrier cylinder which has infinite potential wall. The size of the barrier cylinder should be independent of the calculation results when it is bigger than a certain size. Usually the size of the barrier cylinder is chosen so that four times QD height is the height of the cylinder and four times QD radius is the radius of the cylinder. The dimensions of the lens-shaped quantum dot are usually taken from AFM (atomic force microscope) measurement. For example, one of the InAs QDs on InP is shown on the left of Fig. 21.3.

Schrödinger's equation for the envelope function in the effective mass approximation can be written as:

$$-\left(\nabla \frac{1}{m^*(\vec{r})} \nabla\right)\Psi(\vec{r}) + V(\vec{r})\Psi(\vec{r}) = E\Psi(\vec{r}) \tag{21.5}$$

The unit length is the Bohr radius $a_0 (= \hbar^2/m_e^2)$ 0.529 Å and the unit energy is the Rydberg constant Ry $(=M_e e^4/2\hbar^2)$ 13.6 eV. This Schrödinger's equation is Hermitian and its wavefunctions are orthogonal and the probability current is conserved at the interface of the heterojunction. The envelope function of the quantum dot with a lens shape $\Psi(x, y, z)$ is then expanded in terms of a complete orthonormal set of solutions $\psi_{lmn}(x, y, z)$ of the cylindrical problem with infinite barrier height (see Eq. 21.3):

$$\Psi(x, y, z) = \sum_{lmn} a_{lmn}\psi_{lmn}(x, y, z) \tag{21.6}$$

$$\psi_{lmn}(x, y, z) = \sqrt{\frac{2}{H\pi[\rho_0 J_{m+1}(k_{mn}\rho_0)]^2}} J_m(k_{mn}r) \sin\left[l\pi\left(\frac{1}{2} - \frac{z}{H}\right)\right]e^{im\phi} \tag{21.7}$$

The boundary condition is $J_m(k_{mn}\rho_0) = 0$. We have chosen the domains $[-H/2, H/2]$ and $[0, \rho_0]$ for the variation z and r. This approach does not need explicit matching wavefunctions across the boundary between the barrier and dot materials. This method is easily applicable to an arbitrary confining potential. Substituting Eq. 21.6 into Eq. 21.7, multiplying on the left by $\psi_{l'm'n'}^*$, and finally integrating over the cylinder, yields the matrix equation:

$$(A_{lmnl'm'n'} - E\delta_{mm'}\delta_{nn'}\delta_{ll'})a_{lmn} = 0 \tag{21.8}$$

The matrix element $A_{lmnl'm'n'}$ is given by:

$$\begin{aligned}
A_{lmnl'm'n'} = &\int_{\text{cylinder}} \nabla\psi_{l'm'n'}^*(\vec{r})\frac{1}{m^*(\vec{r})}\nabla\psi_{lmn}(\vec{r})dv \\
&+ \int_{\text{cylinder}} \psi_{l'm'n'}^*(\vec{r})V(\vec{r})\psi_{lmn}(\vec{r})dv
\end{aligned} \tag{21.9}$$

The problem in above equation is the discontinuity of effective mass in passing from the well region into barrier region. In order to overcome this problem, the integral can be split into three parts, within each of which the effective mass is constant [10]. First, we take an integral with $m^* = m_B$ over the whole cylinder which also includes the quantum dot and wetting layer (the well region). Second, we subtract the integral with $m^* = m_B$ over the well region and third, we add the integral with $m^* = m_W$ over the well region. The same procedure has been adopted for the integral containing the potential. The final expression for the matrix element is:

$$A_{lmnl'm'n'} = \left[\frac{1}{m_B}\left[k_{mn}^2 + \left(\frac{\pi l}{H}\right)^2\right] + V_0\right]\delta_{ll'}\delta_{mm'}\delta_{nn'}$$
$$+ \left(\frac{1}{m_W} - \frac{1}{m_B}\right)\int\limits_{QD+WI} \nabla\psi_{l'm'n'}^*(\vec{r})\nabla\psi_{lmn}(\vec{r})dv$$
$$- V_0\int\limits_{QD+WI} \psi_{l'm'n'}^*(\vec{r})\psi_{lmn}(\vec{r})dv \qquad (21.10)$$

where the subscript QD + WL in the integrals means that the integration is over the quantum dot and wetting layer inside a cylinder. The first term of Eq. 21.10 is simply the free particle energy inside a cylinder.

In order to calculate the matrix element $A_{lmnl'm'n'}$, the integrals need to be calculated over the quantum dot and wetting layer. But these volume integrations can be done analytically for the lens shape geometry.

In this work, we have used 70 sine functions and 10 Bessel functions as basis functions for expanding the envelope functions. Equation 21.10 is a 700×700 matrix and can be solved numerically. This method which we have discussed is not the most accurate of the available methods [11, 12], but it is simple to use, versatile, and good enough for our modelling for QDIPs.

21.2.2 Oscillator strength

In order to calculate the optical absorption spectrum or analyse the photocurrent spectrum in the QDIP, one requires the energy levels and the oscillator strengths for transitions between the various states. The oscillator strength is the measure of the interaction between the light and electrons (or holes). When the incoming light enters the quantum dots, the electrons in the discrete energy levels gain the energy and experience dipole transitions. The rate of the dipole transition can be obtained from Fermi's golden rule. The oscillator strength for a transition from a level i to a level j is given by:

$$f_{ij} = 2|\langle i|\,\vec{\eta}\cdot\vec{p}\,|j\rangle|^2/(m^*\hbar\omega_{ij}) \qquad (21.11)$$

where $|i\rangle$ and $|j\rangle$ are wavefunctions of the quantum dot, $\vec{\eta}$ is the photon polarization, $\vec{p}$ is the electron momentum operator, and ω_{ij} is the transition frequency. It may be noted that a spherical QD of cubic material is optically isotropic, and the oscillator strength is independent of the polarization of the incoming light. In reality, the quantum dot has asymmetric shape and therefore the oscillator strength of the quantum dot has strong dependence on polarization. For the normal incidence of light, which is perpendicular to the growth plane, the incoming light has in-plane polarization (or s-polarization) which is parallel to the growth plane. If the quantum dot has rotational symmetry, the wavefunctions of the quantum dot also has rotation symmetry. In such a case, the in-plane coordinates such as x and y do not make any difference. For the x-polarization (or s-polarization), $\vec{\eta}\cdot\vec{p}$ is $-\hbar\partial/\partial x$ and for the z-polarization, it is $-i\hbar\partial/\partial z$. In order to calculate the oscillator strength numerically, it is necessary to know the wavefunctions and their derivatives. In a later section, the results of the calculation of the oscillator strength for the quantum dots in our QDIPs will be discussed.

21.2.3 Absorption

The absorption of the light in the QDIPs mostly happens in the quantum dots. The absorption coefficient $\alpha(\omega)$ can be written as

$$\alpha = \frac{\pi\hbar N_d n_{op}e^2}{m^*\varepsilon\varepsilon_0 c}\left[\frac{\Gamma}{(\hbar\omega - E_{eg})^2 + \Gamma^2}\right]n_g(1 - n_e)f_{ge} \qquad (21.12)$$

where n_{op} is the refractive index, c is the velocity of light, and Γ is the total level width. The absorption coefficient also involves the following quantities: (i) the dot density N_d, (ii) the oscillator strength f_{ge}, and (iii) the probability n_g that the carriers remain in the initial state and the probability n_e that the carriers stay in the excited state. As you can see, the absorption can be increased by increasing the dot density and the oscillator strength and maximizing $n_g(1 - n_e)$. In order to increase the dot density, the growth condition of the quantum dot should be optimized. Increasing the oscillator strength is not a simple problem and will be discussed later. The occupation probabilities n_g and n_e can be calculated if the energy levels of the quantum dot are known. Assuming Boltzmann statistics for convenience, n_g can be written:

$$n_g = \frac{e^{-E_g/kT}}{\sum_s d_s e^{-E_s/kT} + \sum_t e^{-E_t} + \int_{\varepsilon_c} d\varepsilon\, \rho(\varepsilon) f(\varepsilon) / N_d} \tag{21.13}$$

where the E_s are the quantum dot energy levels, d_s the degeneracy, the "t" sum is over traps including the new eigenstates formed by electron–phonon resonance; $\rho(\varepsilon)$ is the band density of states and $f(\varepsilon)$ the Fermi function. As we can see, the absorption of the quantum dot depends on the occupation probabilities of the levels. Experimentally these occupation probabilities can be controlled through the doping of the quantum dot.

21.2.4   Modelling of responsivity and photocurrent

When the incoming infrared light is absorbed in the QDIP, the electrons inside quantum dots are photoexcited into the continuum state under the bias, and those electrons can be collected as a photocurrent. The photoexcited electrons should escape from quantum dots to be detected. This escape process involves tunnelling and thermal activation. If R is the responsivity as a function of temperature T and applied bias V, then the photocurrent I_P flowing is given by:

$$I_P = AR(T,V)P_L \tag{21.14}$$

where P_L is the optical power per unit area, and A is the illuminated area of the device. Collecting together the terms, we can write the peak responsivity in terms of the quantum efficiency η and the gain g:

$$R = \frac{eg\eta}{\hbar\omega} \tag{21.15}$$

We can define the quantum efficiency η

$$\eta = \alpha L \left[\frac{\nu_{ec} e^{-E_{eff}/kT}}{\nu_0 + \nu_{ec} e^{-E_{eff}/kT} + \nu_t e^{-\Delta/kT}} \right] \tag{21.16}$$

where the first factor αL is the absorbance, L is the device length, and $\alpha(\omega)$ the absorption coefficient. The factor "g" in Eq. 21.15 is the gain defined as the ratio of the recombination time over the transit time:

$$g = \frac{\mu F}{LC_{be}} \tag{21.17}$$

where $\mu F/L$ is the transit time, and C_{be} the capture rate of electrons from the continuum band into the bound excited state of the quantum dot. The third factor in Eq. 21.16 is the ratio of the escape rate $W_{ec}(=\nu_{ec}\exp[-E_{eff}/kT])$ out of the excited state to the continuum and the inverse lifetime of the excited state (ν_0). In this section, we will concentrate on this factor which is related to the escape rate, and the quantum efficiency and gain will be discussed in a later section.

21.2.5 Escape rate

In the dot, the electrons can tunnel out back to the extended state outside the quantum dot if the electric field is applied. Transmission out of the dot through a triangular barrier which is created by an electric field, known as Fowler–Nordheim tunnelling, is given as follows:

$$T = \exp\left[-\frac{2}{\hbar}\int_0^x \sqrt{2m(E_{ec} - eFx')}dx'\right].$$

(21.18)

Another process involved for the escape process is thermal activation. In Fig. 21.4, the physical picture is that (1) the electron in the excited state can thermally activate to the continuum state or (2) directly tunnel out or (3) thermally activate and tunnel to continuum state. In order to estimate the escape rate, the sum of all the possible paths should be required.

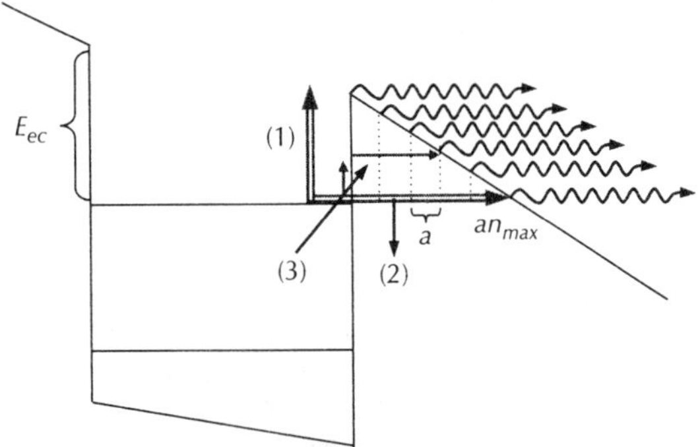

Figure 21.4 *Escape paths out of the excited state to the continuum state.*

If the longest tunnelling path is divided into a lattice constant a, the possible escape occurs at the point whose distance from the well is multiple lattice constants (na) (Fig. 21.4).

The escape rate through one path is given by:

$$W_{0n} = \nu_{ec}\exp\left[-2\gamma\int_0^{na}\sqrt{(E_{ec} - eFx')}\,dx'\right]\exp\left(-\frac{E_{ec} - eFna}{kT}\right)$$

(21.19)

where ν_{ec} is the attempt frequency, F is the applied electric field, $\gamma = \sqrt{2m^*}/\hbar$ and E_{ec} is the height of the escape barrier. The total escape rate is the sum of the all the escape rates through each path. With appropriate approximation, the escape rate is finally given by:

$$W_{ec} = \nu_{ec}\,\exp[-E_{\text{eff}}(F)/kT]$$

$$= \nu_{ec}g_c\,\frac{\exp[-E_{ec}/kT] - \exp[-2\gamma aE_{ec}^{1/2}]\exp[-4\gamma E_{ec}^{3/2}/3eF]\exp[eFa/kT]}{1 - \exp[-2\gamma aE_{ec}^{1/2}]\exp[eFa/kT]}$$

(21.20)

where g_c is the density of final states of escaping charge at the band edge which is reachable within kT. ν_{ec} varies between (a) the value of a phonon frequency multiplied by the probability of finding the charge at a given site in the quantum dot localized state, giving $\nu_{ec} \sim 10^9$ to 10^{10} Hz, and (b) the excited state pure tunnelling attempt frequency $\sim E_{ec}/h$. g_c is given by $g_c \sim 10^4(kT/e)^{3/2} \sim 10$, assuming three-dimensional plane wave-like states. Even though clearly ν_{ec} is not a constant but depends on the path chosen, we shall assume that the product $\nu_{ec}g_c$ is a fit parameter varying between 10^{10} and 10^{13} Hz.

21.2.6 Gain

There are two different definitions of gain in the QWIP which can also be applied to the QDIP. One is the noise gain and the other is photoconductive gain. The gain factor g in Eq. 21.15, which is the photoconductive gain (g_e) cannot be deduced easily from the experiment. Later, in order to fit the experimental responsivity to a theoretical one, we should use the noise gain instead of the photoconductive gain. The assumption that the photoconductive gain is equal to the noise gain should be checked before using the approximation. Borrowing from the QWIP case, one uses the generation–recombination (G–R) noise formula which says that the noise power spectrum or the square of the noise current I_n^2 is proportional to the dark current I_{dark} at given bandwidth Δf[13]. But this equation can be also expressed with the correctional factor F the photoconductive gain [14]:

$$I_n^2 = 4eg_n I_{dark}\Delta f = 4eg_e F I_{dark}\Delta f. \tag{21.21}$$

Here $F = r - (1 - r)1 - p_c/p_c \ln(1 - p_c)$ where p_c is the capture probability and $r = t_w/t_p$ where t_w and t_p are transit times across one well and one period, respectively. For the case when r is close to 1, the correction is small for all p_c. For the small capture probability ($p_c < 0.2$), the correction factor F is greater than 0.9 and relatively independent of r (see Fig. 21.5).For the small p_c, the photoconductive gain can be equal to the noise gain to a good approximation. The photoconductive gain in the QDIP can be expressed in terms of the capture probability p_c, the number of quantum dot layers N, and the fill factor of the quantum dots Q.

$$g_e = \frac{1 - p_c/2}{QNp_c}. \tag{21.22}$$

with Eq. 21.22, the capture probability of the electrons in the QDIP can be extracted.

Identifying the proportionality factor "g_n" with gain as defined by Eq. 21.21 implies that the current really is one of generating and recombining carriers in a band. Note also that some authors, in analogy to QWIPs, assume that the recombination in QDIPs is drift limited [15]. From this it would follow that the drift velocity dependence in the gain drops out and the gain can be a constant. This is, however, not justified in our devices which we are going to discuss later, where the wetting layer scatters and reduces the band mobility, but is too thin to capture charge.

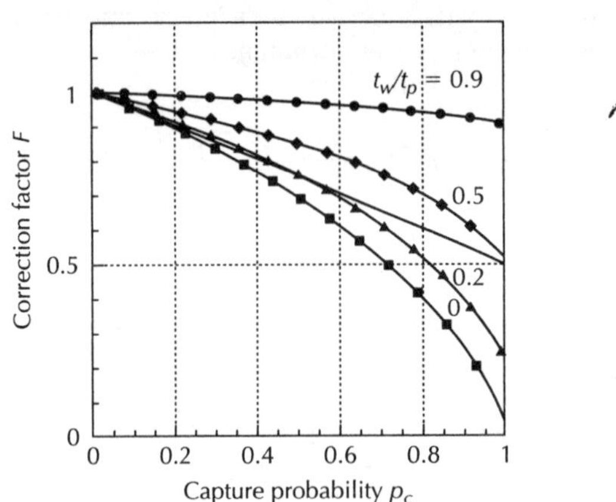

Figure 21.5   The correction factor F as a function of capture probability p_c for $r = 0, 0.2, 0.5,$ and 0.9. The straight solid line is predicted by the theory presented by Beck ([16], Ref. 14).

21.2.7   Dark current

Let's consider now the transport of the electrons in the dark situation [17–19]. This can be understood as follows: electrons are emitted from a dot by field assisted tunnelling and thermal activation on a timescale which depends on the local temperature and bias, and on the eigenstates. A charge already in the top most excited state will, for example, be emitted with an escape rate with Eq. 21.20. In the dark current, this time has to be lengthened by dividing it with the probability that the level is occupied.

In the meantime, while charges are being emitted from the dot, other charges are being injected from the electrode. In the steady state, as many are coming in as are going out. The electrode injection rate I_{inj}, assuming injection into a drifting state, rather than a pure eigenstate, is given by

$$I = eA \int_0^{E_b} dE f(E)\rho(E)\mu F \, \exp\left[-\left(\frac{2m^*}{\hbar^2}\right)^{1/2} \frac{4}{3eF}\{[E_b - E]^{3/2}\right] + eA \int_{E_B} dE\rho(E)f(E)\mu F \qquad (21.23)$$

where the first term is Fowler–Nordheim tunnelling through the injection barrier, A is the electrode area, and $f(E)$ the Fermi function. The second term is the band contribution. The equality of injection current to bulk current establishes the Fermi level in the bulk.

Assuming the system is roughly neutral, it follows that quantum dots which have just emitted will eventually be replenished by charges which are flowing about in the band. The typical timescale $1/C_{be}$ for reoccupation is 10^{-9} to 10^{-10} s, which is much longer than the typical transit time $L/\mu F \sim 10^{-12}$ s. So it follows that photoexcited charges will be flowing around the circuit at very high speeds before they get captured again.

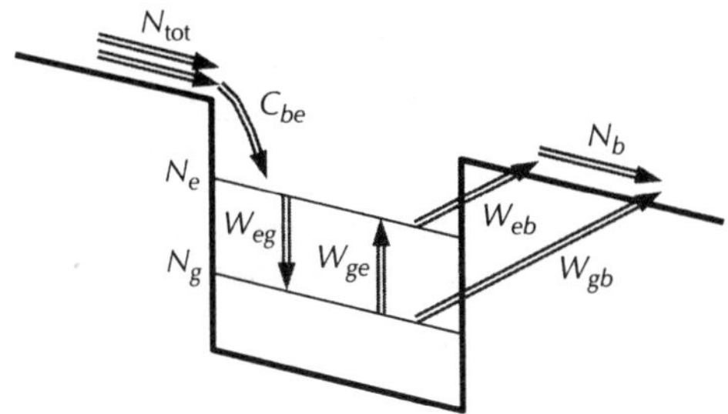

Figure 21.6   Schematic diagram for the processes of the electrons that escape from, capture into, excite and recombine in the dot.

The theory of the dark current is developed using a rate equation approach. In Fig. 21.6, the schematic diagram of the processes involved in the dark situation is shown in a two-level system. But this can be easily expanded into a multi-level system. N_{tot} is the number of electrons which are provided to a quantum dot through the current injection from the electrode and proportional to the density of the quantum dots in one layer. N_e and N_g are the numbers of the electrons staying in the excited states and the ground state, respectively. N_b is the number of electrons which are emitted from a dot to the barrier continuum. C_{be} is the capture rate from the band continuum to the excited state. W_{eb} and W_{gb} are the escape rates from each quantum level (e: excited state and g: ground state) to the band continuum. These escape rates can be obtained with the same method as the escape rate for the responsivity, Eq. 21.20, because the photoexcited electrons and electrons under dark condition in a dot follow the same routes such as thermal activation and tunnelling. W_{eg} is the transition rate from the excited state to the ground state. W_{ge} is the transition rate from the ground state to the excited state. Because of the charge conservation:

$$N_{tot} = N_b + N_g + N_e \qquad (21.24)$$

should be satisfied at any time. In Eq. 21.24, N_{tot} is a constant which is determined as

$$\dot{N}_{tot} = Ae\mu(F)FN_d. \qquad (21.25)$$

But the other numbers N_b, N_g, and N_e change under circumstances. For the steady state, detailed balance rate equations can be obtained for each N.

If we solve the above equations to get the relation between N_b and N_{tot}, we can get:

$$N_b = \frac{N_{tot}(W_{eb}W_{gb} + W_{eb}W_{ge} + W_{eg}W_{gb})}{(W_{eb}W_{gb} + W_{eb}W_{ge} + W_{eg}W_{gb}) + C_{be}(W_{eg} + W_{gb} + W_{ge})}$$

$$\approx \frac{N_{tot}W_{eb}}{W_{eb} + C_{be}(1 + W_{eg}/W_{ge})} \qquad (21.26)$$

Here we use the approximation that W_{gb} is very small and can be neglected in most cases. Because W_{eg}/W_{ge} is the very occupation probability in the quantum dot level "e" which obeys a Fermi distribution with a self-consistent Fermi level and temperature, Eq. 21.26 becomes:

$$N_b = \frac{N_{tot}W_{eb}f_e}{C_{be}[1 + f_e + (f_eW_{eb}/C_{be})]}. \qquad (21.27)$$

In general, the quantum dot has more than two levels. In the multi-level case, the number of electrons emitted from each level and therefore the dot emission currents from each level can add up. The general expression for the dot emission current can be written:

$$I_{dot} = \frac{Ae\mu(F)Fn_d}{C_{be}}\left\{\sum_s \frac{f_sW_{sb}}{(1 + f_s + W_{sb}/C_{be})}\right\}$$

$$W_{sb} = \nu_{sb}g_s \frac{\exp[-E_{sb}/kT] - \exp[-2\gamma aE_{sb}^{1/2}]\exp[-4\gamma E_{sb}^{3/2}/3eF]\exp[-E_{sb}/kT]}{1 - \exp[-2\gamma aE_{sb}^{1/2}]\exp[eFa/kT]} \qquad (21.28)$$

where A is area, e is the electronic charge, F is the electric field, $\mu(F)$ is band mobility and n_d is the density of quantum dots. f_s is the Fermi function at level "s" and the Fermi level has to be determined self-consistently for each bias V and temperature T by matching injection and quantum dot escape current. W_{sb} is the escape rate from each quantum level s to the band continuum. For simplicity we define activation energy, also $\sum_s 2f_s = \langle n_{dot}\rangle$ which is the mean number of electrons in a dot:

$$I_{dot} = \frac{Ae\mu(F)FN_d}{(1 - n_e)C_{be}}\langle g_s\nu_{sc}\rangle \exp[-E_D(F)/kT]. \qquad (21.29)$$

The total current across the device area includes also the uniform band contribution as in Eq. 21.23. The prefactor $g_s\nu_{sc}$ is the product of the density of final states and the sum over all paths from a level "s" to the continuum "c".

21.2.8   Modelling of detectivity

One of the measures for device performances is the specific detectivity D^*, defined as the ratio of responsivity over the square root of the dark current density at a given bandwidth multiplied by the gain [13].

This can be written in the elegant form (unit band width, D^* in $\mathrm{cmHz}^{1/2}/\mathrm{W}$):

$$D^* = \frac{R^*}{[eg(I_D)]^{1/2}}. \qquad (21.30)$$

Theoretical modelling of peak detectivity Eq. 21.30 can be rewritten combining Eq. 21.15 and Eq. 21.29 as follows:

$$D^* = \left[\frac{\langle Q\nu_{ec}\rangle^{1/2}}{\nu_0 + \nu_t e^{-\Delta/kT} + \langle\nu_{ec}\rangle e^{-E_{ec}/kT}}\right]\frac{\langle\alpha(\omega,T)\rangle}{A^{1/2}N_d\hbar\omega}e^{(E_D/2kT - E_{ec}/kT)} \qquad (21.31)$$

where $Q = N_dAL$ is the total number of quantum dots in the device.

21.2.9   Summary

The origins of novel properties and predicted device improvement of quantum dots devices are discussed. It has been shown that the steep δ-function density of states of quantum dots is responsible for most of the improvements compared with QWIP, including high-temperature operation, high responsivity, high selectivity and narrow spectrum. Another advantage is normal incidence detection due to 3D confinement. It has been predicted that with improved quantum dots fabrication QDIP should outperform QWIP.

21.3   InP-based QDIP materials growth and characterizations

The material growth is the most challenging step in QDIP device fabrication. In this section, the details regarding the MOCVD growth and characterization of InP, GaInAs, AlInAs materials and, most importantly, InAs quantum dots are described. The next chapter will present the detector characterization for the structures fabricated from these materials.

Table 21.1   Physical properties of InAs and InP.

	T (K)	InAs	InP
Crystal structure		Cub (ZnS)	Cub (ZnS)
Lattice constant (Å)	300	6.0584	5.8688
Energy gap (eV)	300	0.36	1.35
	0	0.41	1.42
Melting point (K)		1215	1335
Coefficient of thermal expansion $(10^{-6}\,\mathrm{K}^{-1})$	300	4.52	5.0
$m_e{}^*/m_0$	4	0.024	0.077
$m_{lh}{}^*/m_0$	4	0.025	0.12
$m_{hh}{}^*/m_0$	4	0.37	0.55
$m_{sos}{}^*/m_0$	4	0.14	0.12
Electron mobility (cm^2/V s)	300	3×10^4	5×10^3
	77	8×10^4	4×10^4
Hole mobility (cm^2/V s)	300	200	150
	77	500	1200
Intrinsic carrier Conculation (cm^{-3})	300	9×10^{14}	1×10^8
Static dielectric constant		15.15	12.5
High-frequency dielectric constant		12.25	9.61
Elastic moduli (10^{12} dyn cm^{-2})	300		
C_{11}		8.329	10.11
C_{12}		4.526	6.61
C_{44}		3.959	4.56
Refractive index		3.44	3.35
Optical phonon (cm^{-1})		242	320
LO		220	288
TO			

21.3.1   Growth and characterization of matrix material

"Epi-ready" semi-insulating (SI) InP substrates are the templates for all the growths in this work. All of the growths in this work are done with an Emcore D125 low-pressure (LP) MOCVD. First of all, homoepitaxial growth of InP is optimized. The growth of high-quality InP is important since it serves as the nucleation layer for the entire growth as well as the barrier in the active region. InP has a wide growth window. Usually within one growth of a device structure, different growth temperatures are needed. For example, the bottom and top InP contact layers are usually grown at a higher, optimum temperature to ensure the excellent quality; while a much lower temperature will be used for the InP barrier in the active region to accommodate the growth of InAs QDs at lower temperature.

The as-grown InP sample shows a mirror-like surface. The $1\,\mu m \times 1\,\mu m$ atomic force microscope (AFM) scan image shows a very smooth surface with clear atomic steps and a roughness (RMS) of $1.16\,\text{Å}$, indicating excellent structural quality.

GaInAs is an important alloy because $Ga_{0.47}In_{0.53}As$ is lattice matched to InP and has a lower band gap energy. By using $Ga_{0.47}In_{0.53}As$ as the matrix, the confinement, and thus the intersubband transition energy, of InAs QD can be tuned. For example, GaInAs QWs can be inserted between InAs QD and InP barrier to form a "dot in a well" structure. By changing the thickness or the composition of this GaInAs layer, the detecting wavelength may be tuned. Besides, it is also found experimentally that by using GaInAs the form and size of the QD can be changed, as will be shown later.

As part of the active region, the GaInAs layer has to be grown at the same temperature for QDs. For some of the work here, a growth temperature of 500°C was chosen. In order to optimize the growth of GaInAs, a $0.5\,\mu m$ InP buffer layer was first grown at optimum condition. The temperature was decreased to 500°C and bulk GaInAs was grown. The lattice match condition was obtained by adjusting the flow rates of trimethylindium (TMIn) and triethylgallium (TEGa).

Under the optimized condition, the as-grown sample shows excellent morphology with very few defects. X-ray diffraction shows an FWHM as small as 32.3 arcsec. AFM shows atomic steps and an RMS of $1.67\,\text{Å}$ for a $1\,\mu m \times 1\,\mu m$ scan, as shown in Fig. 21.7.

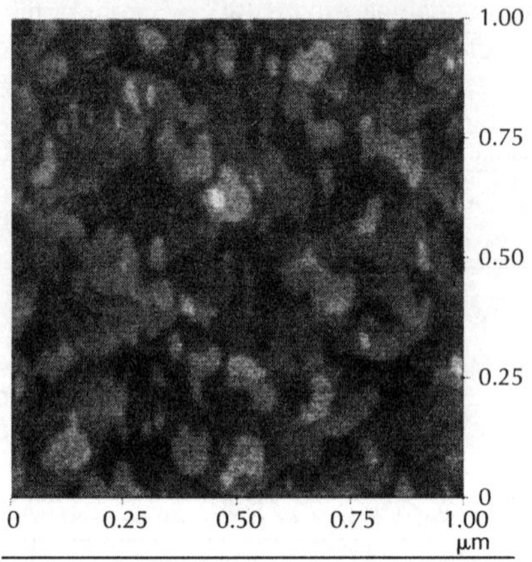

Figure 21.7 *Surface morphology of GaInAs grown at T = 500°C ($1\,\mu m \times 1\,\mu m$ AFM).*

21.3.2 Growth and characterization of InAs QDs

Growth of InAs quantum dots is not only the most important to the success of the QDIP device operation but also the most challenging task. So far, most quantum dot research has been done with In(Ga)As quantum dots on GaAs substrate grown by MBE. Very limited information is available for InAs quantum dots growth on an InP material system by MOCVD. Various growth parameters need to be optimized to first enable the growth of quantum dots then change the form, size and density of QD in a controlled manner. In this section, the growth and characterizations of single-layer InAs quantum dots on both InP and GaInAs matrix are presented.

21.3.3 Single-layer InAs quantum dots on InP

Self-assembly growth of InAs quantum dots on an InP matrix are first realized. The relationship between growth parameters and characteristics of quantum dots, such as dot size, dot density, and dot size uniformity, are fully studied. Some important parameters being optimized for quantum dot growth are growth temperature, growth rate, and V/III ratio. The lower growth rate is necessary to obtain more uniform-sized dots and prevent the forming the large islands. For MOCVD, lowering the growth rate is limited technically by the minimum flow rate allowed by mass flow control in most cases. The substrate temperature is a dominant factor for the diffusion process during quantum dot growth. After optimization, the optimum growth parameters will be used for final QDIP structure growth.

21.3.3.1 Growth temperature

First, the growth temperature was optimized. The structure consists of $0.5\,\mu$m InP buffer grown at 590°C, then 5 ML InAs grown at 490°C, 500°C and 510°C for three samples. During the QD growth, the flow rate is 50 sccm for both TMIn and arsine. The AFM of these three samples are shown and compared in Fig. 21.8. It can clearly seen that at 490°C, the dot density is low while at 510°C although dot density increases the large defect density increases too. As a result, 500°C is chosen as the optimium growth temperature. Too low temperature causes low mobility of the adsorbed atoms on the substrate surface and too high temperature causes high coalescence rate between formed quantum dots.

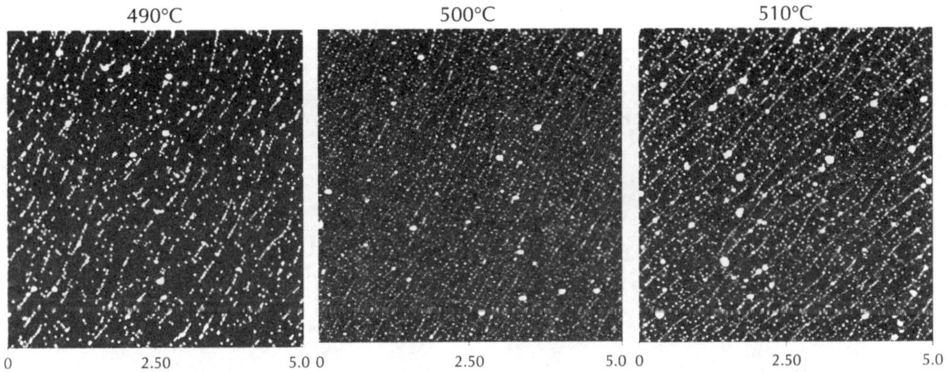

Figure 21.8   Comparison of growth of InAs QD on InP at different temperatures.

21.3.3.2 V/III ratio

The effect of V/III ratio on the formation of QD was studied by changing the flow rate of arsine. The growth temperature was fixed at 500°C. The growth structures and conditions were identical as above except that different flow rates of arsine were used during growth of InAs QDs. As shown in Fig. 21.9, dot density is increased when the V/III ratio is decreased due to increased movement of the adatoms.

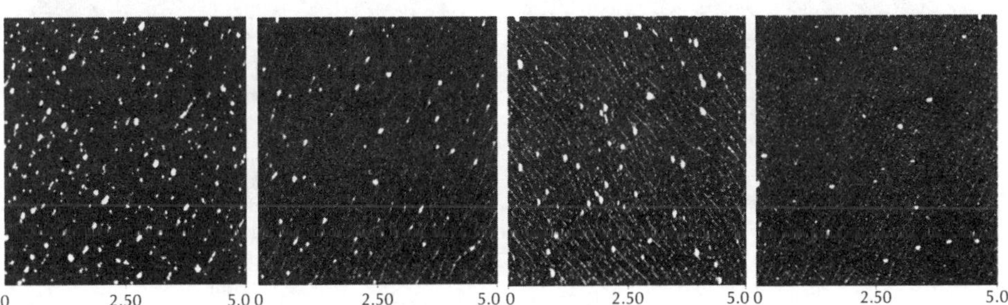

Figure 21.9   Comparison of growth of InAs QD on InP with different arsine flow rates: (from left to right) 50 sccm, 25 sccm, 12 scmm and 8 sccm.

21.3.4   Single-layer InAs quantum dots on GaInAs

InAs QDs were also grown on a GaInAs matrix and compared with those grown on an InP matrix. The structure consists of $0.5\,\mu m$ buffer grown at 590°C, then 3 ML (10 Å) $Ga_{0.47}In_{0.53}As$ quantum well and 5 ML InAs QDs at the optimized conditions mentioned above.

The dot size, density and distribution are compared with the results of the InP matrix in Fig. 21.10. It can be found that even though the GaInAs layer is quite thin, it affects the growth of InAs QDs. The dot size and distribution are more uniform and the defect size decreases. To explain such difference, similar structures were grown without the QDs on top. By comparing the results with and without QD for both GaInAs and InP matrix materials, it can be found that the cause might be the different surface condition of the matrix materials. For the InP matrix, atomic steps can be clearly seen and the InAs QDs seem to follow these steps. This argument is reasonable because most adatoms will incorporate at an atomic step resulting from a miscut substrate. Although the surface of GaInAs is smooth too, the atomic steps are much smaller and more random, which may help form more randomly distributed but uniform QDs.

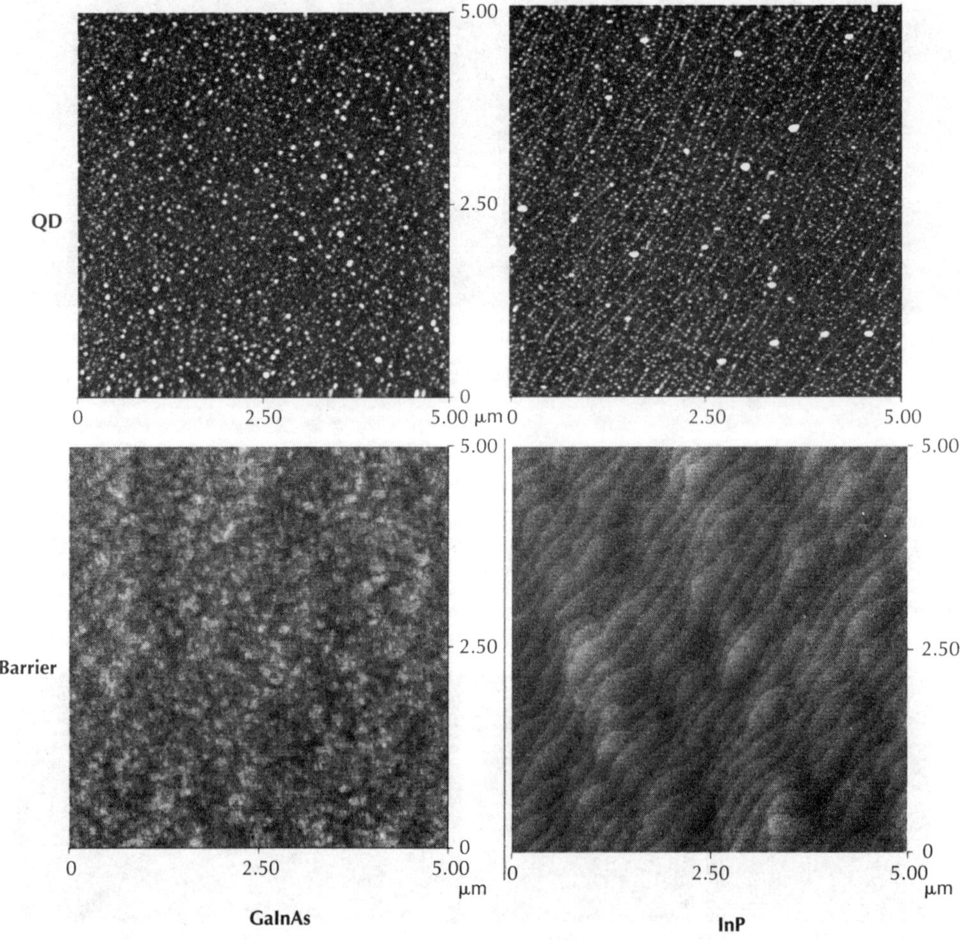

Figure 21.10   Comparison of InAs QD grown on GaInAs (top left) and on InP (top right). Also shown here are the matrix surfaces before the deposition of InAs QDs.

21.3.4.1   Single-layer QDs growth with dilute arsine

As discussed previously, dot density increases with the decrease of arsine due to increased movement of the adatoms. To follow this trend, a dilute arsine (5% arsine/hydrogen mixture) was used for the growth of InAs QDs. All the other conditions remained identical. Dilute arsine flow rates of 100, 75, 50 and 25 sccm were used. The AFM images of the QDs are shown in Fig. 21.11. It seems that the dot density remains the same when the dilute arsine flow changes. However, the dot form changes from tall and thin to short and fat from 100 sccm to 25 sccm.

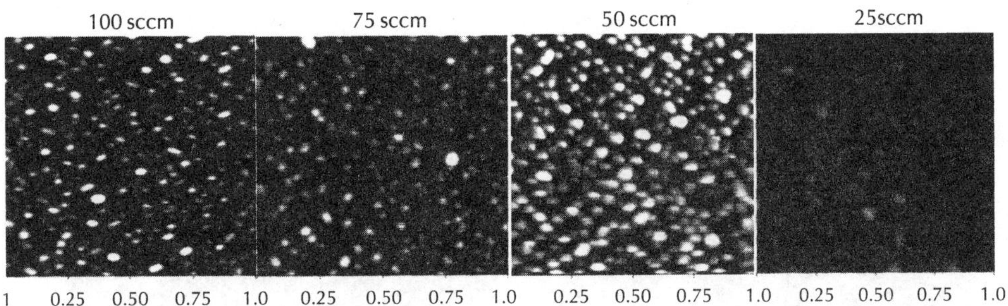

Figure 21.11   Comparison of growth of InAs QD on GaInAs matrix with different 5% dilute arsine flow rates: (from left to right) 100, 75, 50 and 25 sccm.

21.3.5   Doping of InAs quantum dots

As discussed in section 21.2, QDs need to be doped for the proper QDIP operation. Doping of QDs is much different from the doping of bulk semiconductors or quantum wells. Doping may also affect the quantum dot size and density. Unfortunately, there is no easy way to directly estimate the doping level in the QDs. However, it is natural to assume that the device that gives the best result shall have the proper doping. In this work, doping of QDs is optimized by comparing the results from a batch of samples with the same structure but different doping (silane flow rate).

21.3.6   Conclusions

The optimized growth conditions for InP and GaInAs bulk materials and InAs quantum dots are presented. The growth conditions such as growth temperature, V/III ratio, and matrix material are optimized. After InP and GaInAs matrix materials are optimized, the growth of InAs quantum dots on both InP and GaInAs matrices are done and compared. The dependence of dot size, density and distribution upon growth parameters such as growth temperature, V/III ratio, as well as different matrix material is presented. InAs quantum dots with uniform size, high density and low defect density are realized under the optimized conditions. Finally, the importance and method for doping of InAs quantum dots are discussed.

21.4   InP-based QDIP device results

The realization of the QDIPs in this work is performed by the careful optimization of growth and doping of the InP buffer and barrier, GaInAs matrix and InAs quantum dots. Then the layers are grown together and optimized again if necessary. Finally, with the methods previously described, the device structures are grown, fabricated, and tested. This order of design, growth optimization, and QDIP operating characteristics is presented in this section.

21.4.1   InAs/GaInAs/InP QDIP

As discussed in the previous section, QDs grown on different matrix materials can result in different dot size, density and uniformity. Previously, InAs QDs were grown after 10 Å GaInAs QW on top of an InP barrier rather than directly on InP in order to stop As and P exchange and have better density and uniformity. Here we replace the 10 Å lattice-matched GaInAs layer with a 10 Å strained GaAs layer and study its effect on the formation of quantum dots.

As shown in Fig. 21.12, under the same growth conditions, InAs QDs grown on GaAs (right) show smaller dot size, higher dot density and improved uniformity compared with those grown on GaInAs (left). Dot density and uniformity are important to device performance while smaller dots would enable further increase of the dot density.

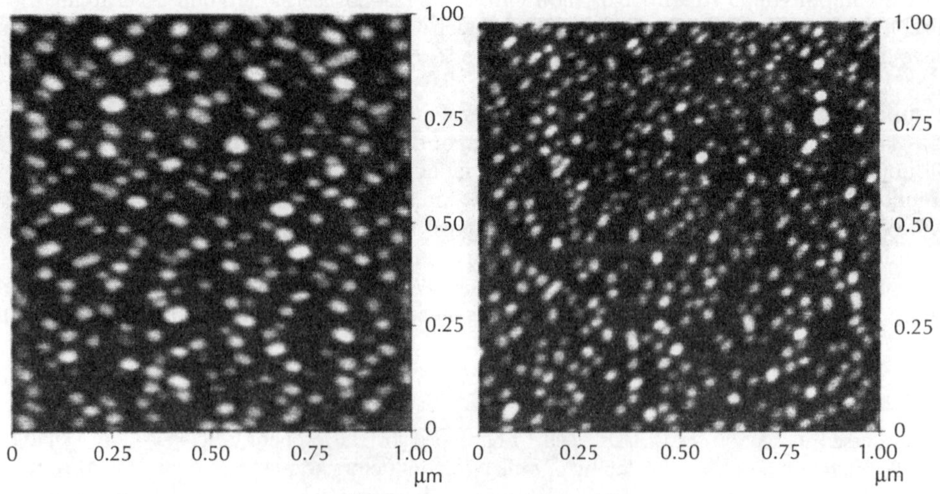

Figure 21.12   Comparison of InAs QD grown on different matrices. Left: 5 ML InAs QD grown on 10 Å GaInAs/ InP. Right: 5 ML InAs QD grown on 10 Å GaAs/InP.

The QDIP device has the following structure. First, a 0.5 μm undoped InP buffer was grown at 590°C followed by a 0.5 μm bottom InP contact layer doped with dilute SiH_4 to $n = 1 \times 10^{18}\,cm^{-3}$. Next the active region was grown at 500°C. The active region consisted of ten periods of the following structure: 400 Å InP barrier, 10 Å GaAs, InAs QDs and 10 Å $In_{0.53}Ga_{0.47}As$. For InAs QDs, the nominal growth rate was 0.42 ML/s and the growth time was 12 seconds. After the QD layer was deposited, 60 seconds of ripening time was given with dilute AsH_3 flowing. The InAs QDs were doped with dilute SiH_4 (200 ppm). Finally, a 0.2 μm n-type $In_{0.53}Ga_{0.47}As$ ($n = 1 \times 10^{18}\,cm^{-3}$) top contact layer was grown at 590°C.

To test the QDIP's performance, 400 μm × 400 μm detector test mesas were fabricated with lithography and dry etching with electron–cyclotron resonance reactive ion etching. Ti/Pt/Au bottom and top metal contacts were made via e-beam metallization, lift-off technique and alloying at 400°C for 2 minutes. The sample was then mounted to a copper heat-sink and attached to the cold finger of a liquid nitrogen cryostat equipped with a temperature controller.

21.4.1.1   Spectral response

The spectral response of the QDIP was tested on a Fourier transform infrared (FTIR) spectrometer. Photoresponse peaked at 6.4 μm and had a cut-off at 6.72 μm (see Fig. 21.14). The spectral width ($\Delta\lambda/\lambda_{peak}$) was 20%, indicating bound-to-bound intersubband absorption. The shape, peak and cut-off of this QDIP showed negligible change with varying temperature (from $T = 77\,K$ to $T = 120\,K$) and bias (from $V_b = -1\,V$ to $V_b = 1\,V$).

21.4.1.2   Dark current

The dark current of the QDIP was measured at 77 K using a closed cycle cryostat and the result is shown in Fig. 21.15. The dark current was still very high compared with QWIP and far from

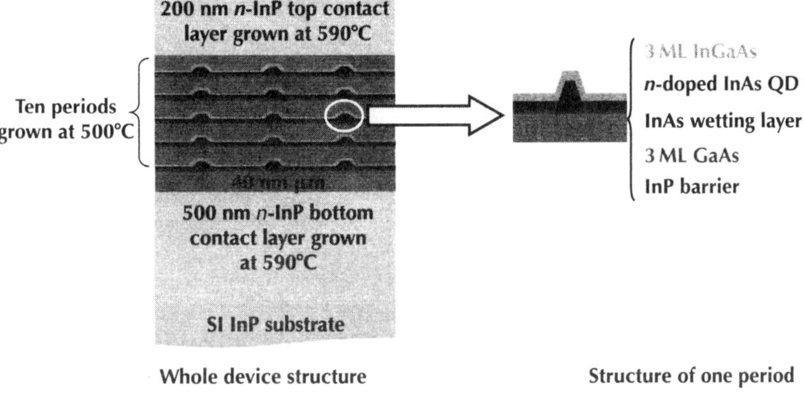

Figure 21.13   InAs/GaInAs/InP QDIP structure.

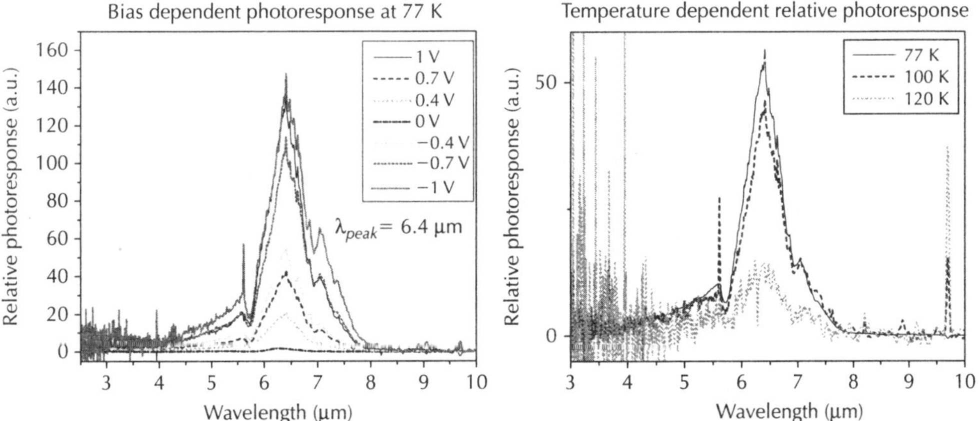

Figure 21.14   Normalized relative spectral response curve for the ten stack InAs/GaInAs/InP QDIP.

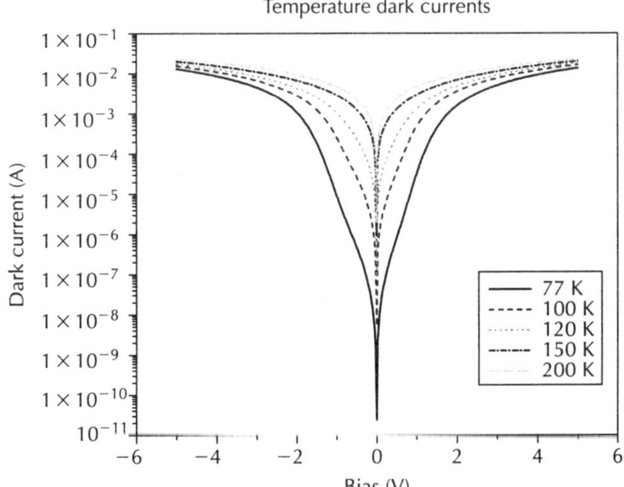

Figure 21.15   Dark current measured as a function of bias for the ten stack InAs/GaInAs/InP QDIP at different temperatures.

the predicted value. As will be shown next, this problem can be partly solved by using a current blocking layer.

21.4.1.3 Responsivity

The absolute magnitude of the blackbody responsivity (R_{bb}) was determined by measuring the photocurrent I_p with a calibrated blackbody source which was set at 800°C. Test mesa was front illuminated from the top of the mesa with normal incident infrared radiation. Peak responsivity (R_p), a more general merit for infrared detector, can be calculated as:

$$R_p = R_{bb} \frac{\int_0^\infty M_{det}(\lambda)d\lambda}{\int_0^\infty M_{det}(\lambda)R(\lambda)d\lambda} \tag{21.32}$$

where $R(\lambda)$ is the normalized photoresponse spectrum, and $M_{det}(\lambda)$ is the blackbody exitance spectrum, and the results at both 77 K and 100 K are shown in Fig. 21.16. At $T = 77$ K and a bias of -1.7 V, a peak responsivity of 1.5 A/W was observed for our QDIP. The peak responsivity at 100 K did not change around from -1 V to 0.8 V, but at higher biases, the responsivity decreased compared to that of 77 K.

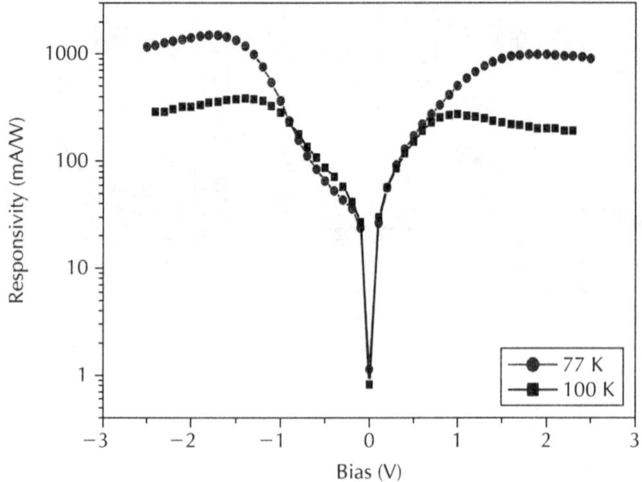

Figure 21.16 Responsivity measured as a function of bias for the ten stack InAs/GaInAs/InP QDIP at 77 K and 100 K.

21.4.1.4 Noise current

Noise current was measured and the result is shown in Fig. 21.17. The noise current (i_n) was measured at both $T = 77$ K and $T = 100$ K using a fast Fourier transform spectrum analyser. As shown in Fig. 21.17, very low noise was observed around 0 V at 77 K. Beyond these ranges, the noise was dominated by generation–recombination noise of the dark current [20]. The noise current increases very fast with increasing temperatures.

21.4.1.5 Detectivity

The specific detectivity was calculated based on the peak responsivity and noise current. The peak detectivities (D^*) can be calculated from $D^* = R_p(A \cdot \Delta f)^{1/2}/i_n$, where $A = 1.375 \times 10^{-3}$ cm^2 is the illuminated detector area and $\Delta f = 1$ Hz is the bandwidth. The detectivities of our QDIP as a function of bias at both $T = 77$ K and $T = 100$ K are shown in Fig. 21.18. The highest detectivities

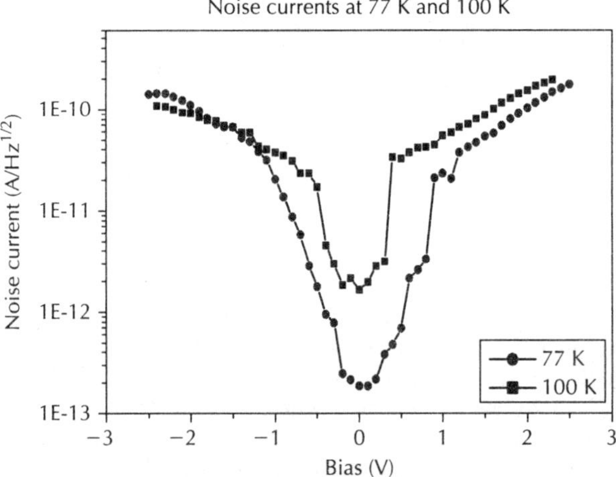

Figure 21.17   Noise current measured as a function of bias for the ten stack InAs/GaInAs/InP QDIP at 77 K and 100 K.

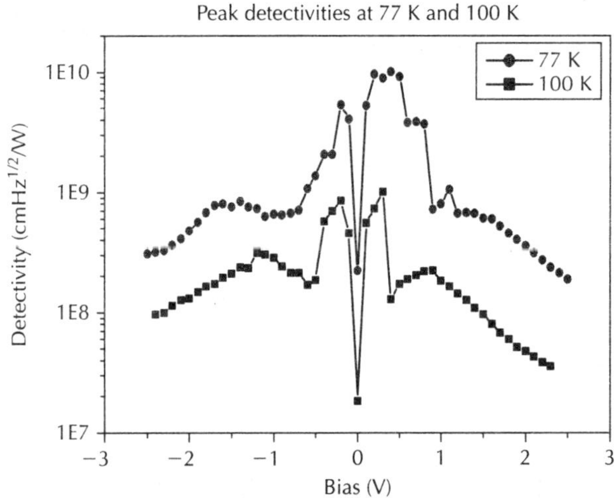

Figure 21.18   Specific detectivity vs bias for the ten stack InAs/GaInAs/InP QDIP at 77 K and 100 K.

of our QDIP were $1.0 \times 10^{10}\,\mathrm{cmHz}^{1/2}/\mathrm{W}$ at a bias of $0.4\,\mathrm{V}$ at $T = 77\,\mathrm{K}$ and $1.0 \times 10^{9}\,\mathrm{cmHz}^{1/2}/\mathrm{W}$ at a bias of $0.3\,\mathrm{V}$ at $T = 100\,\mathrm{K}$.

21.4.1.6   Quantum efficiency

The internal quantum efficiency of our QDIP was extracted using Eq. 21.15 in section 21.2 and was found to be less than 0.1%.

21.4.2   InAs/AlInAs/InP QDIP

As shown above, the dark current of our InAs/InGaAs/InP QDIP is still very high. A high band gap material such as AlInAs can act as a current blocking layer (CBL) and reduce the dark current and improve the performance in the In(Ga)As/GaAs QDIP system. In this section, a device structure with AlInAs CBL is used to reduce the dark current [21].

First, a $0.5\,\mu$m undoped InP buffer was grown at 590°C followed by a $0.5\,\mu$m bottom InP contact layer doped with dilute SiH_4 to $n = 1 \times 10^{18}$ cm^{-3}. Next, the active region was grown at 500°C. The active region consisted of 10 periods of the following structure: 400 Å InP barrier, 10 Å GaAs, InAs QDs, and 30 Å $Al_{0.48}In_{0.52}As$ CBL. For InAs QDs, the nominal growth rate was 0.42 ML/s and the growth time was 12 seconds. After the QD layer was deposited, 60 seconds of ripening time was given with dilute AsH_3 flowing. The InAs QDs were doped with dilute SiH_4 (200 ppm) with a flow rate of 35 sccm. Finally, a $0.2\,\mu$m n-type InP ($n = 1 \times 10^{18}$ cm^{-3}) top contact layer was grown at 590°C.

21.4.2.1 Dark current

The dark current (I_d) of a QDIP mesa was measured as a function of bias (V_b) at different temperatures, as shown in Fig. 21.19. Also shown is the 300 K background photocurrent with a 45° field of view (FOV). The background-limited performance (BLIP) was obtained at 100 K for the $-1.9\,V < V_b < 3.7\,V$ range. A very low dark current was observed for this QDIP due to the $Al_{0.48}In_{0.52}As$ CBL. At $T = 77$ K, a dark current below the pA range was observed between -0.8 V and 1.8 V. An asymmetric I–V relationship was also observed in this QDIP, especially at lower temperatures. The asymmetry of the dark current for our QDIP can be attributed to several factors. First, it could be due to the asymmetry of the device structure as every GaAs/InAs/ AlInAs/InP period of the active region is asymmetric. Second, the InAs QD has a lens shape which is not symmetric in the growth direction. Third, dopant diffusion into the quantum dot layer during the material growth might create a small built-in electric field. A complete understanding of the dark current characteristics of our QDIP needs further theoretical and experimental investigation.

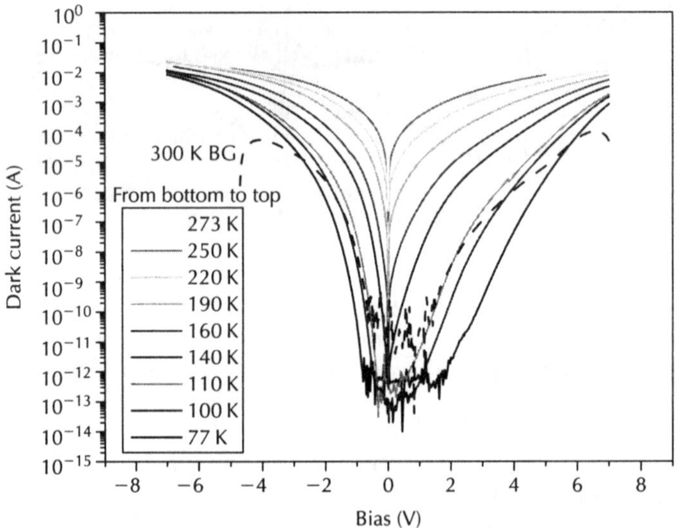

Figure 21.19   *Dark current measured as a function of bias for InAs/GaAs/AlInAs/InP QDIP at different temperatures. Also shown is the 300 K background photocurrent with a 150° field of view (dashed line).*

21.4.2.2 Spectral response

The spectral response of the QDIP was tested on an FTIR spectrometer. Photoresponse peaked at $6.4\,\mu$m with a cut-off at $6.6\,\mu$m (see Fig. 21.20). The spectral width ($\Delta\lambda/\lambda_{peak}$) was 12%, which indicated bound-to-bound intersubband absorption. The shape, peak and cut-off of this QDIP showed negligible change with varying temperature (from $T = 77$ K to $T = 160$ K) and bias (from $V_b = 1$ V to $V_b = 3$ V).

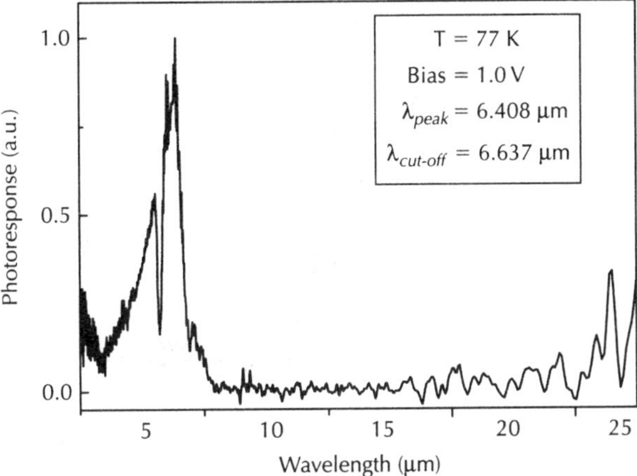

Figure 21.20   The relative spectral response measured at bias of 1 V and temperature of 77 K.

21.4.2.3 Responsivity

Peak responsivity (R_p) at both 77 K and 100 K are shown in Fig. 21.21. An asymmetry was also observed for the peak responsivity, both at 77 K and 100 K. The asymmetry of the responsivity is caused by the asymmetry of the potential in the QD itself. At $T = 77$ K and bias of -5 V, a peak responsivity of 1.0 A/W was observed for our QDIP.

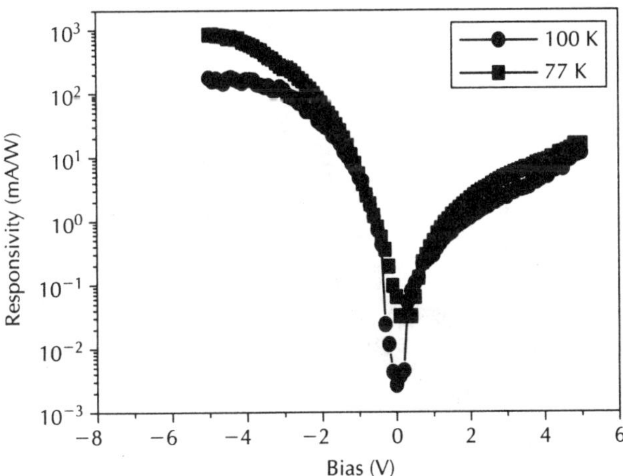

Figure 21.21   Peak responsivity as a function of bias for the QDIP at 77 K and 100 K.

21.4.2.4 Noise current

As shown in Fig. 21.22, very low noise was observed and was almost constant for biases -1.0 V $< V_b <$ 2.5 V at 77 K and -0.8 V $< V_b <$ 2.3 V at 100 K. Beyond these ranges, the noise was dominated by generation–recombination noise of the dark current. Based on the measured noise current and dark current (i_d), we extracted the gain g of the device using $I_n^2 = 4egI_d$ for unit bandwidth. The gain of the device depends strongly on the bias and increases from 5 to 5000 when the bias changes from -1.2 V to -3.5 V. The high value of the gain in a given voltage range is a unique feature of this QDIP technology. The gain in our QDIP devices is much higher than the gain measured in typical quantum well infrared detectors.

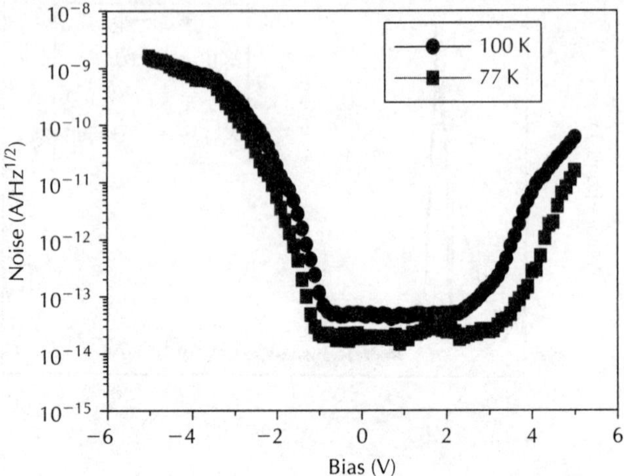

Figure 21.22   *Noise current as a function of bias at 77 K and 100 K.*

21.4.2.5   Detectivity

The detectivities of our QDIP as a function of bias at both $T = 77\,$K and $T = 100\,$K are shown in Fig. 21.23. The highest detectivities of our QDIP were $1.0 \times 10^{10}\,$cmHz$^{1/2}$/W at a bias of $-1.1\,$V and $2.3 \times 10^9\,$cmHz$^{1/2}$/W at a bias of $-0.9\,$V for $T = 77\,$K and $T = 100\,$K, respectively.

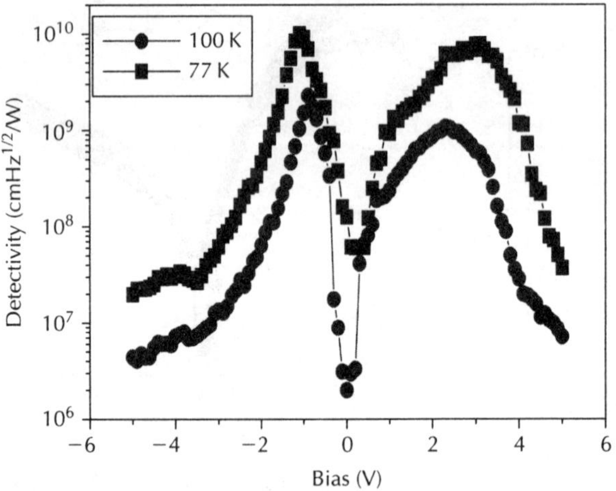

Figure 21.23   *Peak detectivity as a function of bias at 77 K and 100 K.*

21.4.2.6   Quantum efficiency

The internal quantum efficiency (η) of our QDIP was calculated and was found to be less than 0.1%. The very low quantum efficiency so far is due to the (i) low oscillator strength for s-polarized light, (ii) less-than-unity fill factor of the quantum dot layer and (iii) a thin active region of ten layers of QDs. A low oscillator strength must be expected for s-polarized light in a flat (50 nm by 5 nm) lens-like QD at this wavelength. The quantum efficiency can be improved by increasing dot density, optimizing the dot shape, size and uniformity [22].

21.4.3   MWIR InP-based QDIP

Our world first InP-based InAs QDIPs with high detectivity have a peak detection wavelength of 6.4 μm. The wavelength 6.4 μm is outside the MWIR(3 ~ 5 μm) or LWIR (8 ~ 12) μm window where there is low absorption in the atmosphere. As has been shown in the previous section, the size and shape of the quantum dot can be changed by varying the growth conditions. In this section, the peak detection wavelength of QDIP is tailored by dot engineering in order to achieve an MWIR QDIP.

21.4.3.1   Optimization of the single-layer QD growth

In order to have QDs that will detect shorter wavelength in the MWIR range, the QD growth conditions were thoroughly optimized again as in section 21.3, but specifically for the goal to be smaller size, high density, and high uniformity QDs. Growth parameters including growth temperature, growth rate, V/III ratio and ripening time have been optimized.

First, growth temperature was studied again but in a broader range between 440°C and 520°C with an interval of 20°C. The AFMs of a single-layer QD grown at these temperatures are compared in Fig. 21.24. It is clearly seen that the dot size decreases and the dot density increases with the decrease of the growth temperature. The growth temperature of 440°C was chosen.

Next, the growth rate of the QD was studied. The growth rate was varied by varying the flow rate of the group III precursor, in this case TMIn. Three flow rates were used: 50, 100 and 150 sccm. The corresponding growth time was adjusted to make the total amount of material deposited (the product of the TMIn flow rate and growth time) the same. The rest of the growth conditions were identical. The dot density was checked with AFM and Fig. 21.25 shows the dot density as a function of the flow rate of TMIn. It is shown that higher growth rate gives higher dot density. Although 150 sccm gives a slightly higher dot density, PL shows that 100 sccm has better quality. Thus 100 sccm was used for optimization after.

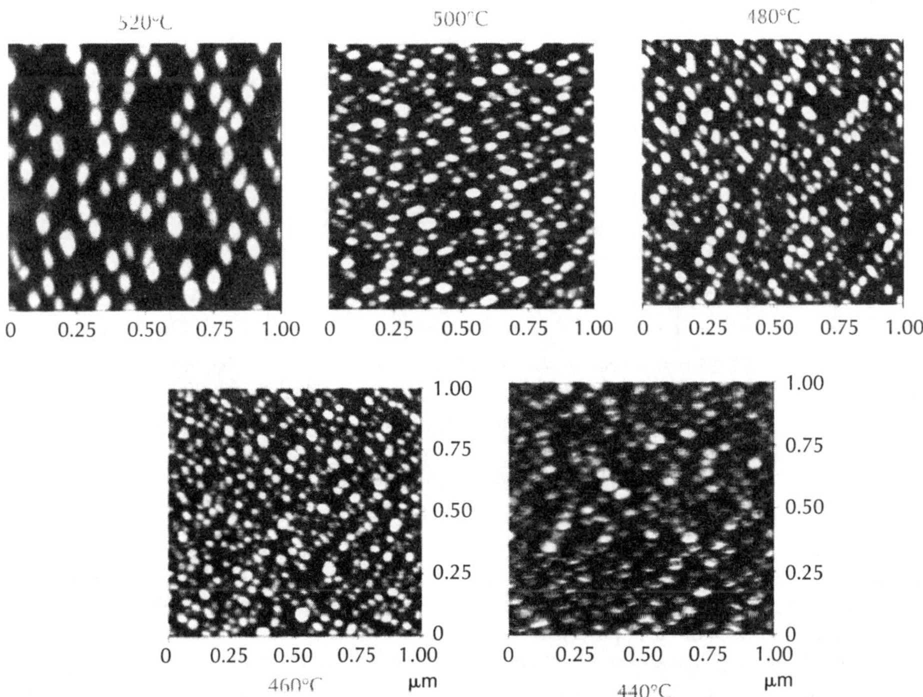

Figure 21.24   AFM images of single-layer InAs QDs grown at different temperatures.

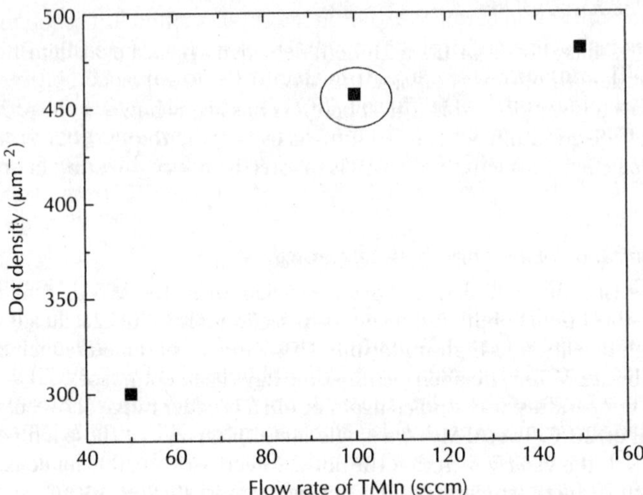

Figure 21.25   Dot density as a function of the trimethylindium flow rate.

The V/III ratio was optimized by varying the flow rate of the group V precursor: 5% dilute arsine. The flow rates of 25, 50, 75, 100 and 200 sccm were used. The flow rate of TMIn was 100 sccm. Dilute arsine gives the highest dot density of 50 sccm as shown in Fig. 21.26 and was chosen for the growth after.

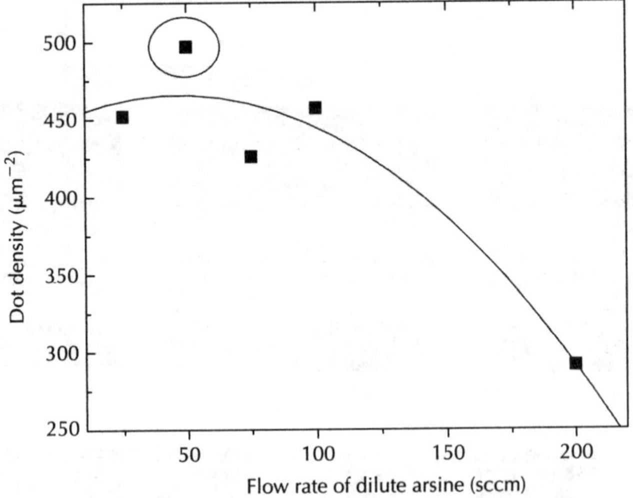

Figure 21.26   Dot density as a function of V/III ratio.

Next, the ripening time is optimized. The ripening time is the time that needs to elapse after the dot material was deposited and before it is covered by the next layer. Usually it is used to allow the dot to form, or to "ripen". Ripening times of 60, 30, 10 and 0 s were used to study its effect on the dot formation. For this optimization, the method used so far, of single-layer uncapped QDs scanned by AFM, is *no longer* a good characterization technique because during the cooling down after growth, the QDs on the surface may still change when it is hot. Instead, a cap layer after QD growth (as in the device growth) was to be grown after ripening, then the samples were studied with photoluminescence (PL).

Figure 21.27 shows the room temperature PL of QDs grown at 440°C with different ripening times. Several trends can be seen here with the decrease of the ripening time from 60 s to 0 s. First, there is clearly a blue shift, which indicates smaller dot size. Second, the full width half maximum (FWHM) decreases, indicating better uniformity. Finally, the intensity is higher which indicates higher dot density and quality. The same study was also done for the QDs grown at a higher temperature of 500°C and similar trends were observed but with a shift of the PL peak toward a longer wavelength (Fig. 21.28). This is clearly the result of the larger dot size at higher growth temperature, as shown by Fig. 21.24.

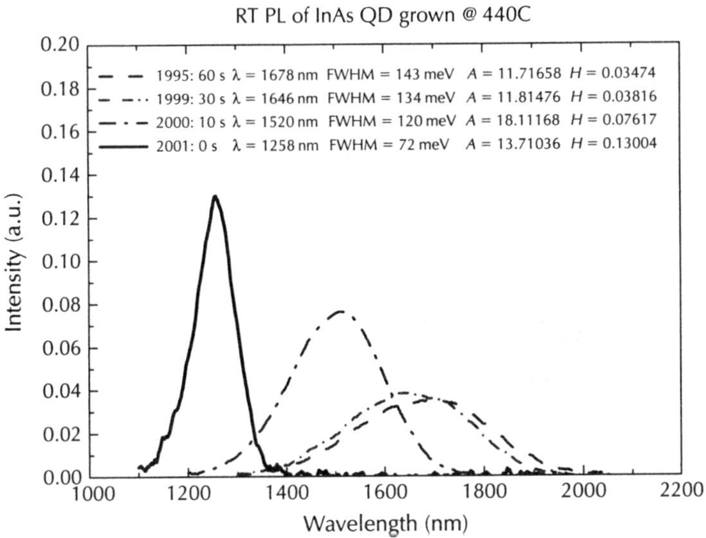

Figure 21.27   Room temperature PL of four InAs QD samples (#1995, 1999, 2000 and 2001) grown at 440°C with ripening times of 60, 30, 10 and 0 s, respectively. Note the change of the wavelength λ, FWHM, integrated intensity A (for area) and peak intensity H (for height) with the decrease in the ripening time.

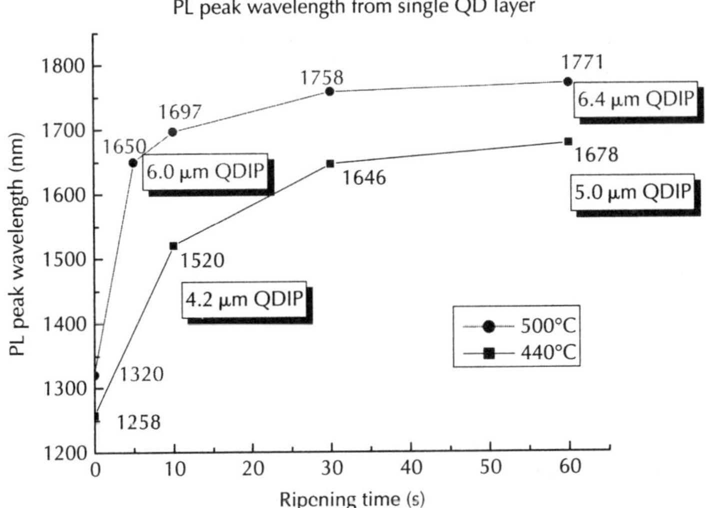

Figure 21.28   PL peak wavelength as a function of the ripening time for QDs grown at 440°C and 500°C.

As can be seen in Fig. 21.28, with the combination of the growth temperature and ripening time, the PL peak wavelength can be tuned in a wide range. However, detection in QDIPs is caused by the intersubband transition rather than the interband transition in PL. Thus it is necessary to check the tunability of dot detection wavelength with FTIR. Figure 21.29 shows that peak detection wavelengths from 4 μm to 6.4 μm have been obtained by changing the growth temperature and ripening time. At 500°C, the minimum wavelength that can be reached is around 6 μm with even 0 s ripening time. At 440°C, the 4 μm and 5 μm were obtained with 10 and 60 s ripening times.

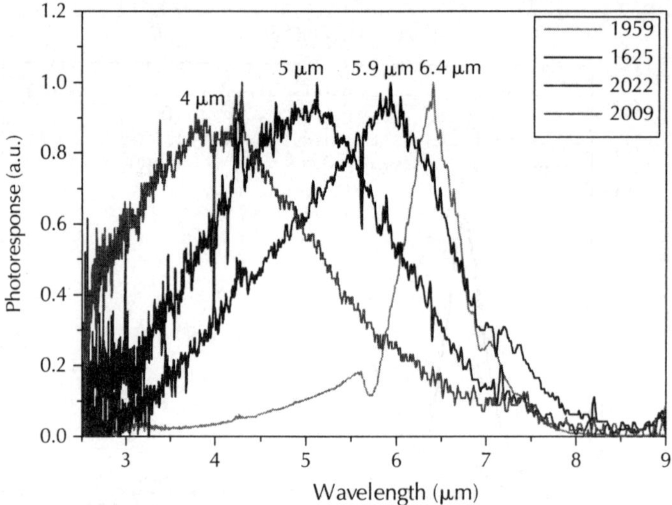

Figure 21.29   Photocurrent spectra of QDIPs with different detection wavelengths.

21.4.3.2   Optimization of the multi-stack QD growth

As shown above, in order to have an MWIR QDIP, at least the QD and the adjacent layer have to be grown at 440°C. One concern is that if the InP barrier is grown at such a low temperature (note the optimum growth temperature for InP in this work is around 590°C, as shown in the previous section) the barrier quality would be poor. This will have two consequences: first, the surface after the barrier growth will be rough (as proven by the AFM image of the surface of a 40 nm InP barrier grown at the 440°C on the left in Fig. 21.30). As a result, the next QD layers will be grown on a much different surface compared with the first QD layer. This will cause the non-uniformity of QDs at different layers and the degradation of quality with the increase of the QD layers. Second, the barrier grown at low temperature may have poor electronic and optical quality. Hall measurements of the InP grown at 500°C and 440°C have shown orders of magnitude in higher background concentration compared with that grown at 590°C. This may lead to high dark current. In order to solve this problem, a two-temperature barrier growth technique is used. As shown in Fig. 21.30, after the QD growth, a thin InP layer with a thickness of 10 nm is grown at the same temperature as the QD and then the temperature is ramped up and the rest of the barrier is grown at a higher temperature, 500°C and 590°C, respectively. The AFM scans show clearly that with this technique, the morphology improves: the roughness (RMS) of the three samples is 9 Å, 3.9 Å and 1.24 Å, respectively. Atomic steps are seen for the 500°C barrier (in the middle) and are very clear at 590°C (on the right). The surface for the InP barrier grown with 590°C is almost the same as the InP before the first QD layer, which indicates the reproducibility of QD at the next layer.

With this two-temperature technique, another concern arises: will the high temperature affect the QD by annealing? In order to check this effect, PLs of the above three samples were measured and the result is shown in Fig. 21.31. The two-temperature growth causes an almost negligible

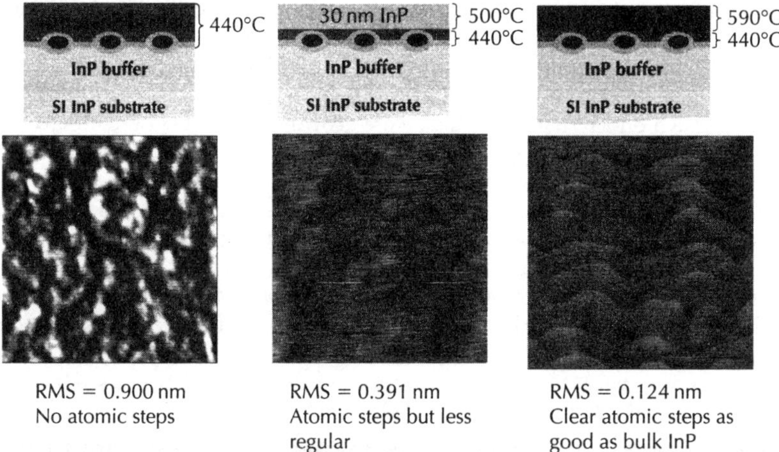

Figure 21.30   Comparison of AFM for the InP barrier after InAs QD. Left: 40 nm InP barrier grown at the same temperature of 440°C as QDs. Middle: 10 nm InP grown at 440°C plus 30 nm InP grown at 500°C. Right: 10 nm InP grown at 440°C plus 30 nm InP grown at 590°C.

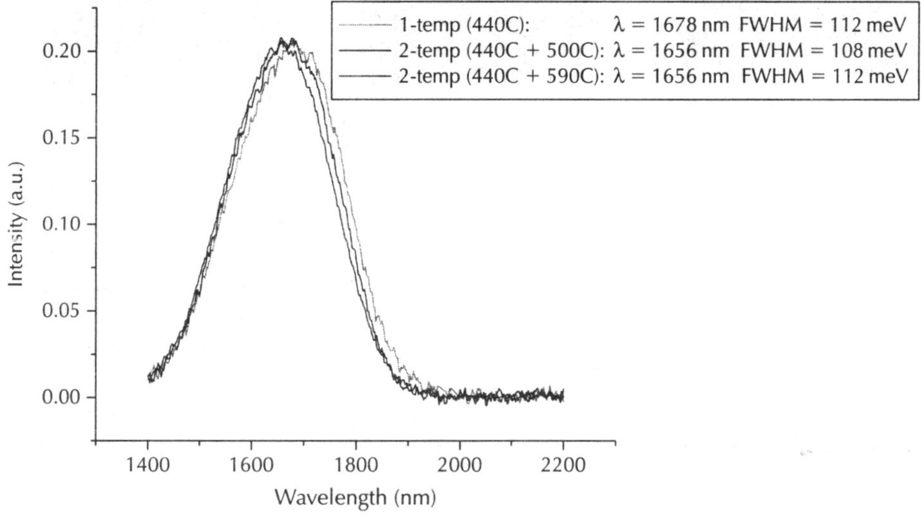

Figure 21.31   Room temperature PL single-layer InAs QD covered with 40 nm InP barrier grown at different conditions.

blue shift and almost the same FWHM, indicating there is almost no effect of the high temperature on the dot properties such as size, shape and composition.

21.4.3.3   Device results

With the QD growth conditions optimized, three devices with different QD growth conditions were grown. An Emcore LP-MOCVD reactor was used to grow QDIP device structures. Trimethylindium and triethylgallium were used as group III precursors while pure phosphine, pure arsine and 5% dilute arsine were used as group V precursors. The typical device structures are as follows. First, a 0.5 μm undoped InP buffer layer was grown at 590°C followed by a 0.5 μm bottom InP contact layer doped with dilute SiH_4 to $n = 1 \times 10^{18}\,cm^{-3}$, then the active region which is a ten-period quantum dot layer separated by InP barriers, and a 0.2 μm top $Ga_{0.47}In_{0.53}$ As contact layer doped with dilute SiH_4 to $n = 5 \times 10^{17}\,cm^{-3}$. The structure of device A is similar to that of the InAs/AlInAs/InP QDIP discussed above in section 21.4.2. Its active region was grown at 500°C. The active regions of the other two devices (device B and device C) were grown using the two-temperature technique. The InAs QDs of device B and C were grown at 440°C. The nominal growth rate was 0.84 ML/s and the growth time was 6 s. The InAs QDs of device B were grown on InP barriers which were grown at 590°C. On the other hand, the QDs of device C were grown on 3 nm $Al_{0.48}In_{0.52}As/InP$ barriers which were grown at 590°C. Before the deposition of InAs layers in device B and C, thin 1 nm GaAs layers were inserted on the InP or $Al_{0.48}In_{0.52}$ As in order to produce more uniform QDs as in device A [23]. The growth interruption time after the QD growth was 30 s under dilute AsH_3 flowing. After the 30 s interruption, the first 10 nm InP capping barriers were grown at 440°C and the remaining 30 nm InP barriers were grown at 590°C. The reason why we used two-temperature barrier growth is because we were able to improve the InP barrier quality which in turn leads to a decrease in the dark currents.

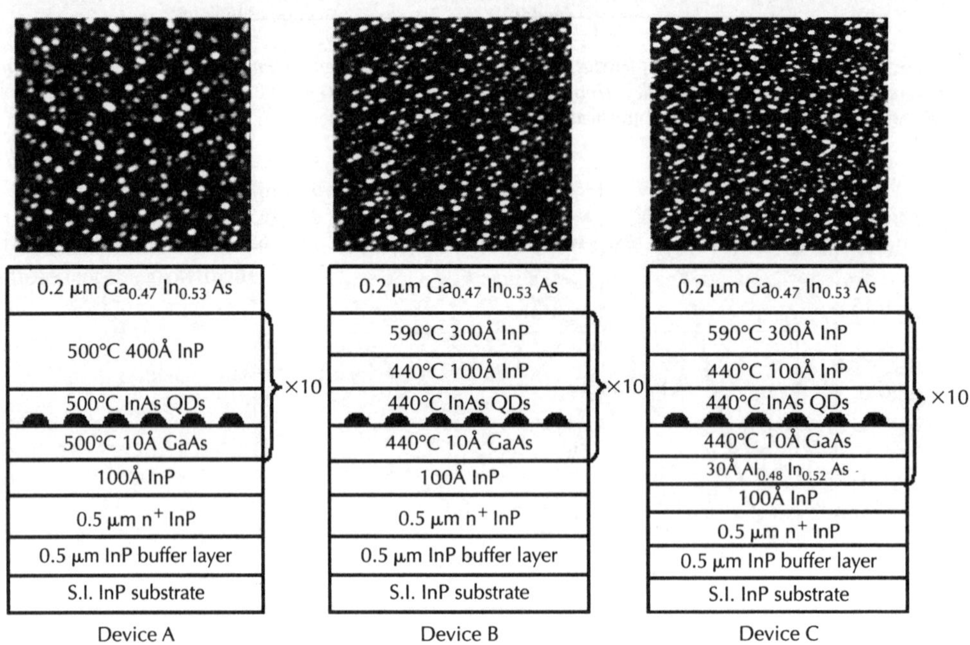

Figure 21.32   The 1 μm × 1 μm AFM images of the single layer of uncapped InAs QDs corresponding to the structures of device A, B and C. In the AFM images, the data scale of the height is 10 nm. In the device structures, the shown temperature is the growth tem1perature and if the temperature is not shown, the growth temperature is 590°C.

The schematic diagrams of device A, B, and C and 1 μm × 1 μm AFM images of QDs corresponding to each device are shown in Fig. 21.32. The density of InAs QDs is around $3\sim5 \times 10^{10}\,cm^2$. The InAs QDs of device A have a lens-like shape with a typical base diameter of 50 nm and a height of 5 nm, and those of device B and C have a similar shape with a typical base diameter of 50 nm and a height of 4 nm which is smaller than the height of QDs of device A. The decrease of the height is caused by the different growth temperature and growth rate of QDs.

To test the performance of the QDIPs, 400 μm × 400 μm detector test mesas were fabricated as before. The samples were then mounted to a copper heat-sink and attached to the cold finger of a liquid nitrogen cryostat equipped with a temperature controller.

The spectral response of the QDIP was tested by FTIR. The source light was normally incident. The bias is applied to the top contact and the bottom contact is always grounded. In Fig. 21.33,

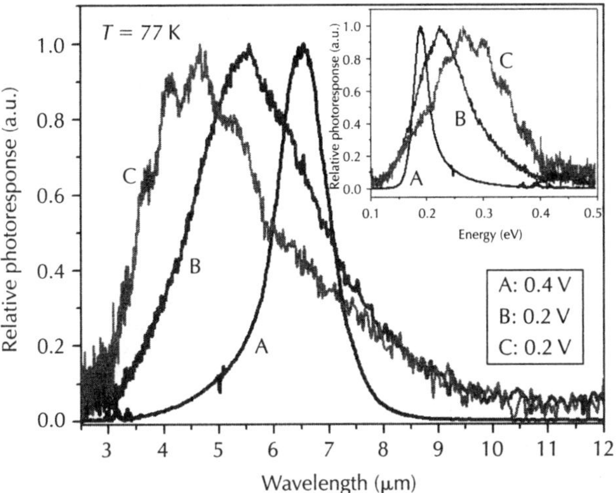

Figure 21.33   Normalized photoresponses of device A, B and C measured by a Fourier transform infrared spectrometer at 77 K under the applied biases. The inset shows the photoresponses as a function of photon energy.

the photoresponse of device A peaked at $6.5\,\mu$m ($190\,$meV) at $77\,$K and at the bias of $0.4\,$V and the half width ($\Delta\lambda/\lambda_{peak}$) is only 18%. The narrow photoresponse is a strong indication of a bound-to-bound transition in the QD energy levels [24]. On the other hand, the photoresponses from device B and C have peaks at $5.5\,\mu$m ($224\,$meV) and $4.7\,\mu$m ($266\,$meV), respectively. The half width of device B and C is 46% and 57%, respectively. These broad responses are very different from those of device A. They are due to bound-to-continuum transitions [25]. In particular, looking at device C, we see that the multipeaks appear on the high-energy side above $0.23\,$eV.

In order to explain this change of the optical transition scheme from device A to device C, we have used the effective mass embedding method to calculate QD energy levels [26]. The single-electron Hamiltonian for QDs can be written as:

$$H(\vec{r}) = -\hbar^2\nabla\cdot\frac{1}{m^*(\vec{r})}\nabla + V(\vec{r})$$

where the effective mass $m^*(\vec{r})$ is locally varying and $V(\vec{r})$ is the potential of the electrons determined by the conduction band offset. The strain effect of quantum dots has been taken into account in the effective masses and the confining potential [27]. We ignored the effect of the electric field on the energy levels because the electric field is of order $10^4\,$V/cm and this gives an energy change of order $5\,$meV across the height of the QD. The effective masses of InAs and InP are $0.04\,m_0$ and $0.077\,m_0$ respectively, and the electron confining potential is $400\,$meV [28, 29]. The lens shape and the geometry parameters of the QDs as measured with AFM were inserted into the calculation. The envelope functions $\psi_n(\vec{r})$ are the solutions of the Hamiltonian. They can be obtained by taking a linear superposition of the Bessel and sine functions generated by the near-cylindrical symmetry of the lens. The number of the basis functions used is 700 [30].

In order to match the observed energy and satisfy the selection rule we identify the main photo-transition of device A as the one which links the first excited state to the bound state which is below the top of the barrier by $24\,$meV and is expected to lead to a fairly sharp response.

When the height of quantum dots decreases, from the QDs of device A to those of device B, the confinement becomes stronger and this pushes the ground state more toward the band edge of the barrier. According to our calculations, the actual energy level configuration changes, so that the most probable photo-transition can change from the bound-to-bound transition to the bound-to-continuum transition. In device B, the main transition is from the ground state to the continuum state of the barrier. For device C, the $Al_{0.48}In_{0.52}As$ layer below the QDs increases

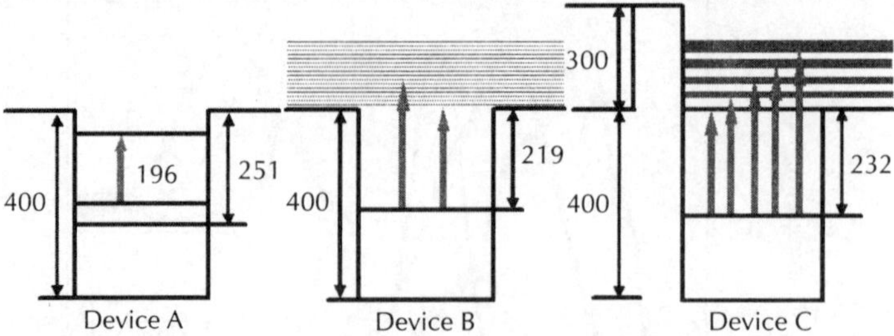

Figure 21.34   The schematic diagram of calculated energy levels of the quantum dots for device A, B and C. The arrows show the relevant intersubband transitions corresponding to main peaks of each device.

the electron confining potential by 300 meV in our calculation [31], therefore the ground state moves down slightly compared to device B. In device C, the multiple peaks above 230 meV are observed at 270, 300, and 340 meV. We believe that those peaks are transitions which link the ground state of the QDs to the minibands formed by the thin $Al_{0.48}In_{0.52}As$ layers and InP barriers. Although the spacing is wider compared to a conventional superlattice [32], the simple calculation of the energy levels of ten-period 3 nm $Al_{0.48}In_{0.52}As$/40 nm InP produced narrow minibands located above the InP band edge by 45, 70 and 100 meV. The calculations reproduce the transition energy from the ground state to the band edge which is 232 meV; furthermore, the multiple peaks in the data agree with the calculated energies linking the ground state to the minibands. But all the transitions are broadened by the inhomogeneous size distribution of the quantum dot. We were able to shift the peak detection wavelengths of QDIPs from 6.5 μm to 4.7 μm using a combination of the change of the quantum dot growth conditions and the addition of the heterostructure.

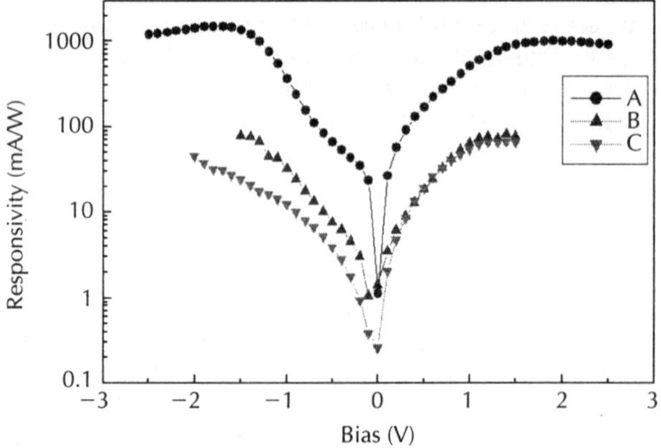

Figure 21.35   Peak responsivities as a function of bias at 77 K.

The blackbody responsivities (R_{bb}) for device A, B, and C were calculated by measuring the photocurrent (I_p) with a calibrated blackbody source at 800°C. At $T = 77$ K and a bias of 1.5 V, we obtained peak responsivities (R_p) of 77 and 65 mA/W, respectively, in device B and device C as shown in Fig. 21.35. The reason for the decrease in the responsivity of B and C when compared to device A is, we believe, because the transition is changing from what is a relatively sharp high oscillator strength bound to bound, to a much more diffuse bound-to-continuum transition.

After the noise currents (i_n) were measured at $T = 77$ K using a fast Fourier transform spectrum analyser, the peak detectivities (D^*) were calculated from $D^* = R_p(A \cdot \Delta f)^{1/2}/i_n$, where $A = 1.375 \times 10^{-3}$ cm^2 is the illuminated detector area and $\Delta f = 1$ Hz is the bandwidth. The detectivities of QDIPs as a function of bias at $T = 77$ K are shown in Fig. 21.36. The detectivities

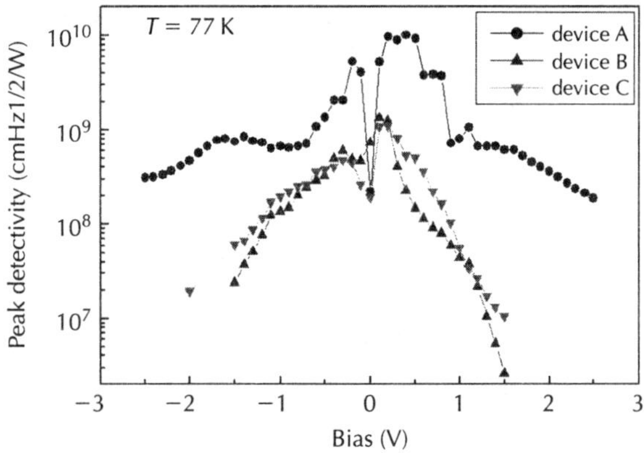

Figure 21.36   Peak detectivities as a function of bias at 77 K.

of device B and C, which are MWIR QDIPs, were $1.0 \times 10^9\,\mathrm{cmHz^{1/2}/W}$ at a bias of 0.2 V. The lower detectivities of both device B and C are due to lower peak responsivity when compared to the responsivity of device A.

21.4.4   InAs QD/InGaAs QW/AlInAs barrier MWIR QDIP

The device structure is shown in Fig. 21.37 and grown by LP-MOCVD [33]. Different from previous devices, the growth temperature of the whole device structure is 590°C. First, a $0.5\,\mu\mathrm{m}$ thick undoped InP buffer layer followed by a $1.0\,\mu\mathrm{m}$ thick bottom InP contact layer n-type doped to $n = 1.5 \times 10^{18}\,\mathrm{cm^{-3}}$ was grown. Then the active region was grown, consisting of 25 stacks of InAs QD/GaInAs QW layers with 29 nm AlInAs barrier layers. The 3.5 nm GaInAs QW layer on top of each QD layer had a doping level of $n = 1 \times 10^{18}\,\mathrm{cm^{-3}}$. Finally, we grew a $0.5\,\mu\mathrm{m}$ thick top InP contact layer doped to $n = 1.5 \times 10^{18}\,\mathrm{cm^{-3}}$. The InAs QDs were grown on the AlInAs barrier layers via self-assembly with the nominal QD growth rate of 0.5 ML/s and the growth time of 3.6 s. An array of $400 \times 400\,\mu\mathrm{m^2}$ detector mesas was fabricated.

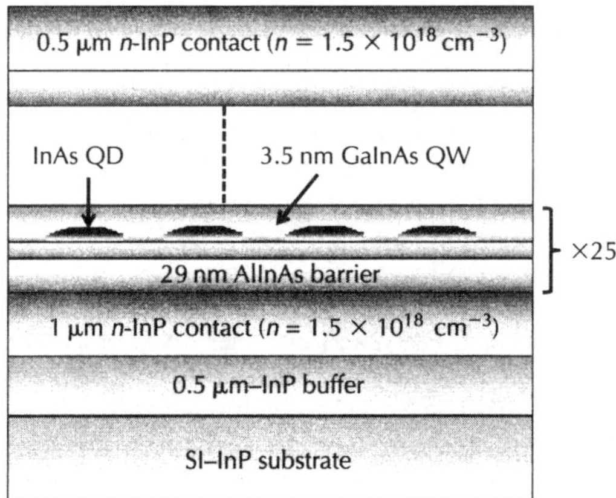

Figure 21.37   Schematic illustration of the device structure grown with low-pressure metal-organic chemical vapour deposition.

Spectral responses at several temperatures and applied biases were observed by FTIR in the normal incidence configuration without any optical coupling structures (Fig. 21.38). In this device structure, both the InAs QD layers and GaInAs QW layers are involved in the infrared absorption

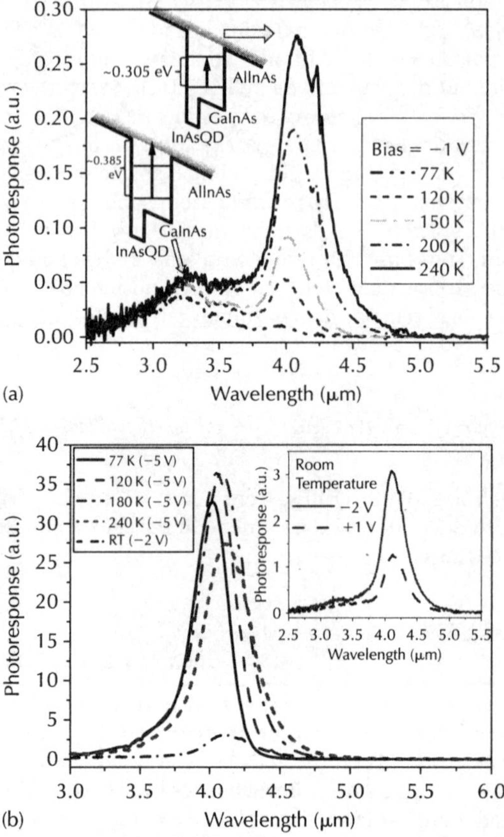

Figure 21.38   (a) Photoresponses at different temperatures for −1 V bias; (b) photoresponses at different temperatures for −5 V applied bias and −2 V for room temperature (RT). The inset shows the photoresponses measured at RT for various biases.

process. The coupling of QDs and QWs has been used in other QDIP device structures, such as dot-in-a-well (DWELL) [34], where the intersubband transition occurs between the hybrid states of the quantum dot and the quantum well. In our device structure, we believe the initial state is not necessarily from a localized "pure" quantum dot state but from a delocalized "mixed" state of the quantum well and the quantum dot as shown in the inset of Fig. 21.38a. At an applied bias of −1 V, there are two peaks, around 3.2 μm and 4.1 μm, as shown in Fig. 21.38a. The intensity of the peak around 3.2 μm does not increase significantly as the temperature increases. The peak around 3.2 μm comes from a bound-to-continuum transition where the electrons are photo-excited from the ground state to a continuum state as depicted in the inset of Fig. 21.38a. This is the reason why the increase of the temperature does not improve the photoresponse around 3.2 μm. On the other hand, the photoresponse around 4.1 μm increases significantly with the temperature because it comes from a bound-to-bound transition in the InAs QD/GaInAs QW hybrid states and thus the temperature can help the photo-excited electrons escape to the continuum as depicted in the inset of Fig. 21.38a. For all temperatures except room temperature, at an applied bias of −5 V (−2 V for room temperature), the peak around 4.1 μm was dominant in the spectral response, as shown in Fig. 21.38b. The strong sensitivity to the applied bias is another indicator that the transition of the photo-excited electrons takes place between bound states of the QD/QW hybrid.

The peak responsivity (R_p) was measured as a function of bias and temperature as shown in Fig. 21.39(a). The responsivity increased with temperature from 120 K to 200 K and started decreasing above 200 K. The peak responsivity was measured to be 822 mA/W at 150 K and

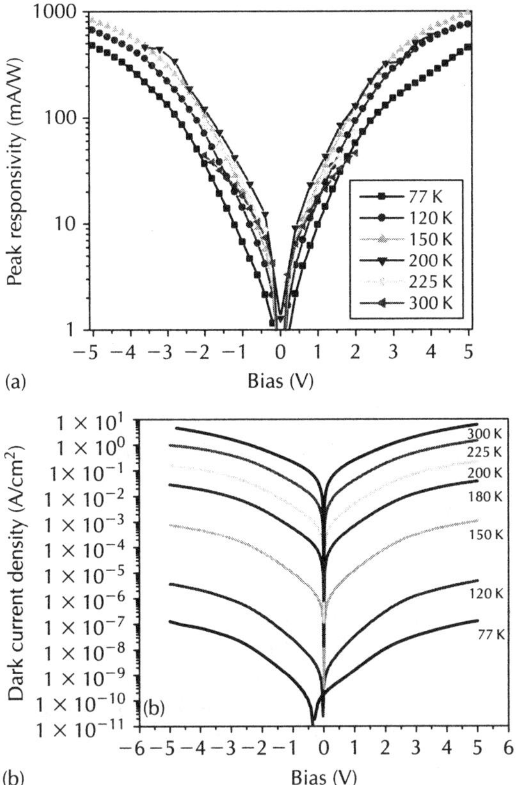

(a)

(b)

Figure 21.39   (a) Peak responsivity at different temperatures as a function of applied bias; (b) Dark current density at different temperatures as a function of applied bias.

-5 V. In QDIPs or QWIPs, the photocurrent can increase or decrease with the temperature depending on whether the relaxation to the lower state or the escape to the continuum state is favourable. Above a certain temperature, the adverse thermal increase of the relaxation of the photo-excited electrons back to the lower state dominates any improvement in escape [35]. In our system, that turnover is believed to take place at around 200 K, above which the responsivity starts decreasing with increasing temperature.

The dark current density of this device was measured as functions of bias and temperature (Fig. 21.39b). A remarkably low dark current density was obtained in this device. At 200 K and -5 V, the dark current density was measured at 163 mA/cm^2. High dark current usually limits the capability for high temperature operation in photoconductors. Therefore, it is crucial to achieve a low dark current with a reasonable photocurrent at high temperature. In QDIPs, low dark currents can be engineered by introducing a current blocking layer [36]. But this current blocking layer will also decrease the photocurrent because the dark current and photocurrent follow the same transport path, as shown previously in section 21.4.2. In our device, the QD layers decrease the dark current without significantly compromising the photocurrent. We think the InAs QD layers act as mobility traps for the dark carriers, but do not seriously affect the escape of the photo-excited carriers.

The specific detectivity (D^*) was calculated and is shown in Fig. 21.40. The maximum D^* of 2.8×10^{11} cmHz$^{1/2}$/W was measured at 120 K. The room temperature detectivity was 6×10^7 cmHz$^{1/2}$/W.

The quantum efficiency η can be obtained from the relation $\eta = R_p h\nu/qg$ where $h\nu$ is the incoming photon energy, q is the charge of the carrier, and g is the photoconductive gain. As a good approximation, the noise gain can be used instead of the photoconductive gain [37]. A very high quantum efficiency of 35% was obtained in this device for normal incidence. This high quantum

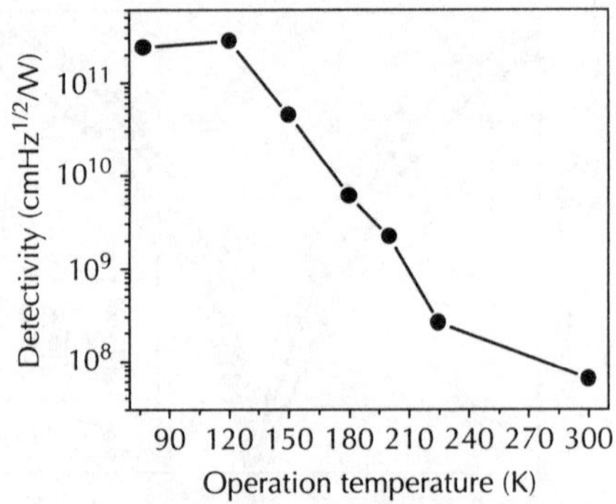

Figure 21.40   Maximum detectivity at different temperatures.

efficiency might be due to the high oscillator strength for the normal incident light and a higher number of carriers available for the absorption compared to conventional QDIPs where the number of photoactive carriers is limited by the number of QDs.

21.4.5   Summary

In this section, we presented several QDIP devices based on our InAs QD grown by MOCVD. Our first device shows a detection wavelength of 6.4 µm. Its detectivity is 1.0×10^{10} cmHz$^{1/2}$/W at a bias of 0.4 V at $T = 77$ K. This was the first report of D^* of an InP-based QDIP by any growth method. Next, a device structure with an AlInAs current blocking layer is shown. The dark current and noise current is substantially reduced although D^* is almost the same. The detection wavelength of our QDIP is tuned by dot engineering. A series of devices with detection wavelengths of 6.5, 5.5 and 4.7 µm are demonstrated. The phototransition scheme is explained by our theoretical model. Finally, we show a high-performance InAs mid-infrared QDIP, which operates up to room temperature. The peak detection wavelength is 4.1 µm. The peak responsivity and the specific detectivity at 120 K were 667 mA/W and 2.8×10^{11} cmHz$^{1/2}$/W, respectively. Low dark current density and a high quantum efficiency of 35% are obtained in this device.

21.5   InP-based QDIP focal plane arrays

Infrared imaging systems with high operating temperature will reduce imaging system cost and complexity by reducing the cooling requirements normally associated with running at cryogenic temperatures. The ability to maintain a photocurrent up to high temperatures with a low dark current makes QDIPs especially suitable for FPA applications where large dark currents, like those associated with QWIPs operating at high temperatures, would prevent high temperature operation due to saturation of the signal collecting capacitor on the readout integrated circuit (ROIC). Other detector technologies with potential for high operating temperature in the MWIR window include HgCdTe [4] and type II superlattices [38].

Here we present an infrared focal plane array with a very low dark current operating at temperatures up to 200 K. The FPA is based on our high temperature high performance QDIP shown in 21.4.4. The responsivity of this detector increases with increasing temperature up to 180 K, then remains fairly constant up to 200 K and begins to drop above 200 K. The photocurrent spectrum and signal are both measurable up to room temperature for this device. The persistence of the photocurrent up to room temperature gives this device the potential for high temperature imaging provided that the noise and dark currents are suitably low. The dark current density of the detector was very low even at high temperatures. This makes the detector a candidate for high-temperature FPA applications because the ROIC readout capacitor will not be saturated too quickly.

A 320 × 256 focal plane array with 30 µm pitch and 25 µm × 25 µm detectors was fabricated [39]. The FPA pixel definition and metallization were essentially the same as for the test mesa fabrication [40]. After array fabrication, the Indigo ISC9705 ROIC is hybridized to the FPA by indium bump bonding. Then, the FPA substrate was thinned using mechanical lapping and polishing. Finally, the hybridized die was then mounted in and wire bonded to a leadless ceramic chip carrier.

The FPA was tested using a CamIRa system from SE-IR Corp. The imaging system cryostat was equipped with a Ge window with 94% transmission and a midwave, f/2 ASIO series lens from Janos Technology with 90% transmission. All imaging and measurements were taken with a 300 K background.

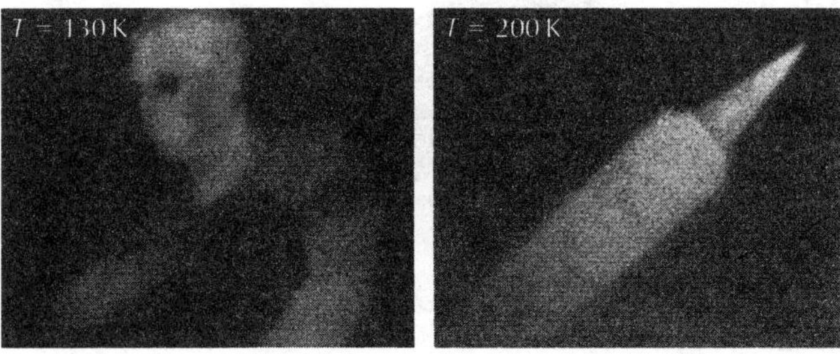

Figure 21.41   Focal plane array imaging taken at 130 K and 200 K, which was also the maximum operating temperature of the array.

Images obtained from the focal plane array are shown in Fig. 21.41. During testing, imaging was achieved at operating temperatures from 77 K to 200 K. The imaging tests were carried out at a fixed frame rate of 32.64 Hz and, depending on the operating temperature, with applied biases from -1 V to around -3 V and integration times from 0.34 ms up to 30.41 ms. Two-point non-uniformity correction was used for the imaging. Imaging of human targets was possible up to around 150 K, and a hot soldering iron could be imaged up to 200 K. Above 200 K the dark current of the detector became too high for imaging.

The FPA photocurrent was measured using an extended area blackbody source from CI Systems. The photocurrent of the FPA was measured with an extended area blackbody set at 35°C. The responsivity and conversion efficiency could then be calculated by dividing the measured photocurrent by the appropriate optical input. The mean responsivity and mean conversion efficiency of the FPA were 34 mA/W and 1.1%, respectively. The temporal noise of the focal plane array was measured by taking the standard deviation of the FPA signal. A common metric for focal plane array performance is the noise equivalent temperature difference (NEDT). In order to measure the NEDT, the FPA signal is measured for two different temperature targets provided by the extended area blackbody (25°C and 35°C). The differential signal for the two temperatures divided by the temporal noise gives the signal to noise ratio (SNR), and the target temperature difference divided by the SNR gives the NEDT. At 120 K the average NEDT was 344 mK and the percentage of functional pixels was 99%. The histogram of the NEDT is shown in Fig. 21.42. It should be noted that the NEDT measurements were taken without two-point non-uniformity correction. The current injection efficiency, which is the ratio of the integrated current to the photocurrent, of the array was calculated using:

$$\eta_{inj} = \frac{I_{int}}{I_{ph}} = \frac{R_{det}g_m}{1+R_{det}g_m} \left| \frac{1}{1 + \dfrac{j\omega C_{det}R_{det}}{1+R_{det}g_m}} \right| \tag{21.33}$$

where g_m is the ROIC transconductance, R_{det} is the detector differential resistance, and C_{det} is the detector capacitance of 8.3×10^{-14} F. The injection efficiency was 99% at 120 K. At 200 K the injection efficiency was still 98% due to the high differential resistance even at high temperature.

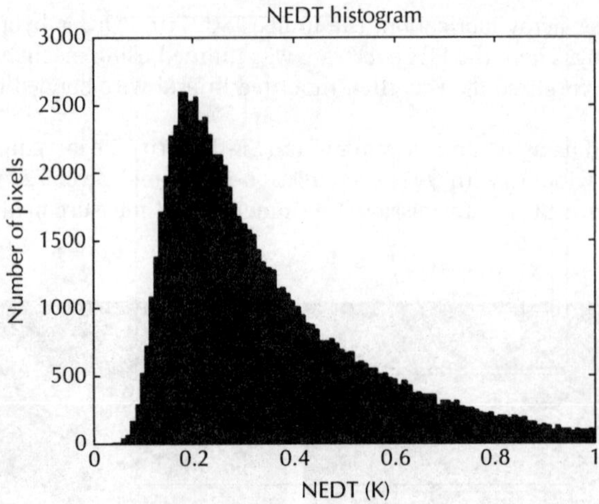

Figure 21.42   Histogram of the focal plane array noise equivalent temperature difference (NEDT) at 120 K. The mean NEDT was 344 mK.

In conclusion, focal plane array imaging based on an InP-based, InAs QD/InGaAs QW/InAlAs barrier detector with imaging at temperatures up to 200 K has been demonstrated. The very low dark current of the device enabled the high temperature imaging capability. The improvement of performance for this FPA will need to come primarily in two areas. The first is achieving a further reduction of the dark current at high temperatures, and the second is achieving a higher responsivity at low bias.

21.6   Summary

In this chapter, InAs QDIPs based on the GaInAs/AlInAs/InP system were demonstrated by MOCVD. The dependence of the dot formation on the MOCVD growth conditions were systematically studied and characterized. High density, uniform InAs QDs were obtained at the optimum conditions. Based on such QDs, several QDIP device structure were fabricated and characterized. The wavelength tunability of the QDIPs was realized within a certain range. With the improvement of structure design and material quality, some of the devices' have outperformed the corresponding QWIP. A 320 × 256 MWIR FPA based on our high performance QDIP was demonstrated. Imaging at temperatures up to 200 K was demonstrated.

Despite the exciting progress that has been made in this field, the performance of QDIP devices is still from desired. In order to further improve the device performance, QDIPs can be improved at several levels:

- QDs are the key element of the device. The dot size directly affects the detection wavelength. The shape of the dot may affect the oscillator strength and thus quantum efficiency. The density and uniformity affect device performance such as responsivity and detectivity. Thus it is crucial to have high density and uniformity QDs and still maintain the right size.
- Proper doping of quantum dots is necessary to maintain the balance between high responsivity and low dark current, which is the key to high-temperature operation.
- The choice and quality of barrier material are important. A barrier provides confinement of the QDs and so affects the detections' wavelength as well as transport properties.
- The stack number of QD layers in the QDIP active region can be increased to improve the absorption. Ideally, each layer of QD stacks would be the same, thus QDIP would be like QWIP, in which detectvity D^* scales with the square root of the number of stacks $\sqrt{N}$

within a certain range [41]. In reality, however, due to the strain, the quality of stacks might degrade with the increase in stacks, which would limit the success of this approach.

- An external scheme may also used to improve the device performance. Resonant cavity enhanced (RCE) detectors have been successfully used in near-infrared telecommunications. The internal quantum efficiency of QDIPs may also be enhanced by RCE structure. This may be realized by placing a pair of distributed Bragg reflectors (DBR). One limit to this approach is that it requires the growth of many quarter wavelength layers in order to achieve the desired reflectivity. Such growth will be tedious and more and more difficult with the increase in the wavelength. An alternative is to make the DBR with higher index contrast material, for example with dielectric materials. This will have to be done after the growth and may require the removal of the substrate. The normal trade-off of the DBR scheme is a reduction in the bandwidth of the photoresponse. However, a QDIP has a natural narrow bandpass detection scheme that is compatible with resonant cavity technology.

- Besides the improvement of QDIP itself, the QDIP FPA requires extra consideration such as the matching of the device operation parameters with characteristics of the ROIC.

References

1. K. Mukai and M. Sugawara, *Semiconductor and Semimentals.* Volume **60**, Chapter 5 (Acadamic Press, San Diego, 1999).
2. Y. Arakawa and H. Sakaki, *Appl. Phys. Lett.* **40**, 939 (1982).
3. M. Asada, Y. Miyamoto, and Y. Suematsu, *IEEE. J. Quantum Electron.* **QE-22**, 1887 (1986).
4. A. Piotrowski, P. Madejczyk, W. Gawron, K. Klos, J. Pawluczyk, M. Grudzien, J. Piotrowski, and A. Rogalski, *Proc. of SPIE* **5732**, 273 (2005).
5. B.F. Levine, *J. Appl. Phys.* **74**, R1–R81 (1993).
6. V. Ryzhii, *Semicond. Sci. Technol.* **11**, 759–765 (1996).
7. L. Goldstein, F. Glas, J. Marzin, M. Charasse, and G. LeRoux, *Appl. Phys. Lett.* **47**, 1099 (1985).
8. D. Gershoni, H. Temkin, G.J. Dolan, J. Dunsmuir, S.N.G. Chu, and M.B. Panish, *Appl. Phys. Lett.* **53**, 995 (1998).
9. S. Gangopadhyay and B.R. Nag, *Nanotechnology* **8**, 14 (1997).
10. M. Califano and P. Harrison, *Phys. Rev. B* **61**, 10959 (2000).
11. E.P. Pokatilov, V.A. Fonoberov, V.M. Fomin, and J.T. Devreese, *Phys. Rev. B* **64**, 245328 (2001).
12. J. Williamson, L. Wang, and A. Zunger, *Phys. Rev. B* **62**, 12963 (2000).
13. E. Rosencher and B. Vinter. *Optoelectronics*, (Cambridge University Press, Cambridge, 2002).
14. K.K. Choi, *Appl. Phys. Lett.* **65**, 1268 (1994).
15. Z. Ye, J.C. Campbell, Z. Chen, E.-T. Kim, and A. Madhukar, *Appl. Phys. Lett.* **83**, 1234 (2003).
16. W.A. Beck, *Appl. Phys. Lett.* **63**, 3589 (1993).
17. B.F. Levine, A. Zusman, S.D. Gunapala, M.T. Asom, J.M. Kuo, and W.S.J. Hobson, *Appl Phys.* **72**, 4429 (1992).
18. V. Ryzhii, *Jap J. Appl. Phys.* **40**, L148 (2001).
19. V. Ryzhii, I. Khmyrova, and V. Mitrin, *Semic. Science and Technology* **19**, 8 (2004).
20. Z.M. Ye, J.C. Campbell, Z.H. Chen, E.T. Kim, and A. Madhukar Noise and photoconductive gain in InAs quantum-dot infrared photodetectors, *Appl. Phys. Lett.* **83**, 1234 (2003).
21. W. Zhang, H. Lim, M. Taguchi, S. Tsao, B. Movaghar, and M. Razeghi, *Appl. Phys. Lett.* **86**, 191103 (2005).
22. J. Phillips, *J. Appl. Phys.* **91**, 4590 (2002).
23. Y.M. Qiu and D. Uhl, *J. Cryst. Growth* **257**, 225 (2003).
24. E.T. Kim, A. Madhukar, Z. Ye, and J.C. Campbell, *Appl. Phys. Lett.* **84**, 3277 (2004).
25. D. Pan, E. Towe, and S. Kennerly, *Appl. Phys. Lett.* **73**, 1937 (1998).
26. D. Gershoni, H. Temkin, G.J. Dolan, J. Duinsmuir, S.N.G. Chu, and M.B. Panish, *Appl. Phys. Lett.* **53**, 995 (1988).
27. M.A. Cusack, P.R. Briddon, and M. Jaros, *Phys. Rev. B* **54**, R2300 (1996).
28. R. People, *J. Appl. Phys.* **62**, 2551 (1987).

29. M. Holm, M.-E. Pistol, and C. Pryor, *J. Appl. Phys.* **92**, 932 (2002).
30. H. Lim, W. Zhang, S. Tsao, T. Sills, J. Szafraniec, K. Mi, B. Movaghar, and M. Razeghi, *Phys. Rev. B* **72**, 085332 (2005).
31. M. Allovon and M. Quillec, *IEE Pro.* **139**, 148 (1992).
32. M. Razeghi, *The MOCVD Challenge*, Vol. 1 (Adam Hilger, Bristol, 1989).
33. H. Lim, S. Tsao, W. Zhang, and M. Razeghi, *Appl. Phys. Lett.* **90**, 131112 (2007).
34. S. Krishna, *J. Phys. D: Appl. Phys.* **38**, 2142–2150 (2005).
35. H. Lim, W. Zhang, S. Tsao, T. Sills, J. Szafraniec, K. Mi, B. Movaghar, and M. Razeghi, *Phys. Rev. B* **72**, 085332 (2005).
36. S.Y. Wang, S.D. Lin, H.W. Wu, and C.P. Lee, *Appl. Phys. Lett.* **78**, 1023 (2000).
37. Z. Ye, J.C. Campbell, Z. Chen, E.-T. Kim, and A. Madhukar, *Appl. Phys. Lett.* **83**, 1234 (2003).
38. M. Razeghi, Y. Wei, A. Hood, D. Hoffman, B.M. Nguyen, P.Y. Delaunay, E. Michel, and R. McClintock, *Proc of SPIE* **6206**, 62060N–1 (2006).
39. S. Tsao, H. Lim, W. Zhang, and M. Razeghi, *Appl. Phys. Lett.* **90**, 201109 (2007).
40. J. Jiang, K. Mi, S. Tsao, W. Zhang, H. Lim, T. O'Sullivan, T. Sills, M. Razeghi, G.J. Brown, and M.Z. Tidrow, *Appl. Phys. Lett.* **84**, 2232 (2004).
41. K.K. Choi. *The Physics of Quantum Well Infrared Photodetectors*, (World Scientific Publishing Co., Singapore, 1997).

CHAPTER 22

Quantum Dot Structures for Multi-band Infrared and Terahertz Radiation Detection

G. Ariyawansa and A.G.U. Perera

Department of Physics and Astronomy, Georgia State University, Atlanta, GA 30303

22.1 Introduction

In the field of infrared (IR) detector technology, quantum dot infrared photodetectors (QDIPs) have attracted the attention of researchers aiming to develop IR optoelectronic devices with improved performance. Compared to quantum well infrared photodetectors (QWIPs), QDIPs have additional degrees of confinement, leading to three major advantages [1]: (i) QDIPs are sensitive to normal-incidence IR radiation, which is forbidden in n-type QWIPs due to polarization selection rules, (ii) QDIPs exhibit comparatively long effective carrier lifetimes, ~ 100s of picoseconds, which has been confirmed by both theory [2] and experiment [3], and (iii) QDIPs exhibit low dark current. Ideally, QDIPs should show improved performance characteristics such as high responsivity, high detectivity, and high operating temperatures. However, such dramatic improvements have not yet been demonstrated due to the fact that detection in most QDIPs is based on bound-to-continuum transitions, as opposed to transitions between the ground state and the first excited state in the QD.

QDIPs ranging from single-element detectors [4–9] to focal plane arrays (FPAs) [10–12] have been demonstrated. In addition to the aforementioned advantages, QDIPs show improved radiation hardness [13, 14] and polarization-sensitive spectral responses [15, 16]. QDIPs operating at temperatures above 77 K [5, 17–22] indicate the possibility of developing uncooled IR imaging systems. In a recent publication, Matthews *et al.* [23] reported an extremely long carrier lifetime of 3–600 ns for a dots-in-a-well (DWELL) detector, which also exhibits a photoconductive gain of 10^4–10^5 in the 20–100 K temperature range. In addition to InAs/GaAs (or InGaAs/GaAs) material system, QDIPs are being developed using SiGe/Si [24–26] and GaN/AlN material systems [27]. The behaviour of QDs under an applied magnetic field [28, 29] has recently become a point of interest in order to understand physical mechanisms of QDs as well as future spintronic devices. The spin of an electron in a QD can be used as a qubit [30–33] for quantum information processing.

Multi-band IR detection has become an important tool in the field of IR technology due to a number of applications. Detecting an object's IR emission at multiple wavelengths can be used to eliminate background effects and reconstruct the object's absolute temperature. This also plays an important role in differentiating the object from its background. Identification of a missile against its plume is an example. However, measuring multiple wavelength bands typically requires either multiple detectors or a single broad-band detector coupled to a filter wheel in order to select different wavelength regions of the incident radiation. Both of these techniques are associated with complicated detector assemblies, separate cooling systems, electronic components, and optical elements such as lenses, filters, and beam splitters. Consequently, such sensor systems (or imaging systems) require complex control mechanism hardware to achieve the necessary fine optical alignment. These requirements generally give rise to increased costs, which can be avoided by a single detector responding in multiple bands. The multi-spectral features are processed using colour fusion algorithms in order to extract signatures of the object with a maximum contrast. The development of multi-band detector systems has led to an increased effort by researchers [34, 35] to develop image fusion techniques. Fay *et al.* [34] have reported a colour-fusion technique using a multi-sensor imagery system consisting of four separate detectors, which operate in different wavelength regions.

There are many applications which will benefit from multi-band detectors. In landmine detection [36], the number of false positives can be reduced using a multi-spectral approach, allowing the locations of the land-mines to be identified. Military applications include the use of multi-band detectors to detect muzzle flashes, which emit radiation in different wavelength regions [37] to locate the position of enemy troops and operating vehicles. Multi-band FPAs responding in very long wavelength infrared $(14–30\,\mu m)$ can be used for space surveillance and space situational awareness [38], where the observation of extremely faint objects against a dark background is required. Present missile-warning sensors are built specifically to detect the ultraviolet (UV) emission from missile plumes. However, the UV emission from the plume of modern missiles is low, thus, making UV-based missile detection impractical. As a solution, IR emission [39] from the plume can be used. The detector system should be able to distinguish the missile plume from its complex background in order to avoid any possible false alarms. Thus, a single-band detector would not be suitable. Using a two-colour (or multi-colour) detector, which operates in two IR bands, which the IR emission of the missile plume falls in, the contrast between the missile plume and the background can be maximized. Additionally, a multi-band detector can be used as a remote thermometer [40], where the object's emission in two wavelength bands is detected and the resulting two components of the photocurrent can be solved to extract the object's temperature.

The predominant drawback of multi-band detectors is the inability to separate the photocurrent components generated by the detection mechanisms without using external optical filters. In order to overcome this, several approaches have been reported. Detector structures with multi-stack active regions [41] use separate electrical contacts to collect the photocurrent components, generated in each active region separately. Detectors with bias-selectable response peaks [10, 18] allow the selection of one peak at a time using the applied bias. In this chapter, DWELL and tunnelling quantum dot infrared photodetector (T-QDIP) structures reported by Ariyawansa *et al.* [42] and Bhattacharya *et al.* [43], respectively, are discussed in detail. Device design concepts, growth, and detection mechanisms in each of these structures along with theoretical and experimental results are discussed in sections 22.2 and 22.3. Some of the other multi-band detectors reported include QWIPs [44–52], MCT detectors [53–58], and homojunction- [59, 60]/heterojunction- [61] interfacial workfunction internal photoemission (HIWIP/HEIWIP) detectors, which are not discussed here as the focus is on multi-band QDIP structures.

22.2   Multi-band quantum dots-in-a-well (DWELL) infrared photodetectors

During the past few years, there has been extensive research in developing DWELL [8, 9, 18, 42, 62, 63, 64, 65] IR detectors. In a typical DWELL structure, InAs QDs are placed in a thin InGaAs quantum well (QW), which in turn is positioned in a GaAs matrix. The DWELL heterostructure provides strong confinement for carriers trapped in the QD by lowering the ground state of the QD with respect to the GaAs band edge, resulting in low thermionic emission. There can be one or more confined energy states in the QD, with the position and separation of energy states dependent on the size of the QD as well as the confinement potential. The detection mechanism of a DWELL detector involves the transitions of electrons from the QD ground state to an excited state in either the QD or QW. Energy states associated with the QW can be bound, quasi-bound, or part of the continuum. These different possible transitions lead to multi-colour characteristics. A schematic diagram of the conduction band profile of a DWELL structure is shown in

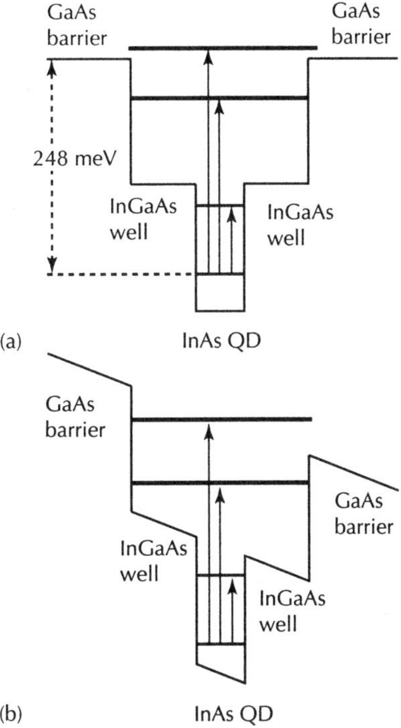

Figure 22.1   Conduction band profile of a DWELL structure (a) under zero bias and (b) under a negative bias. The energy states corresponding to possible transitions leading to spectral response peaks are indicated by arrows.

Fig. 22.1, along with different transitions between energy states as indicated by the arrows. The photocurrent, a result of the photoexcitation of carriers, is proportional to the product of the oscillator strength and the carrier escape probability. A response peak resulting from a bound-to-bound transition has a stronger oscillator strength and a smaller escape probability than a response peak resulting from a bound-to-continuum transition. However, the escape probability can be increased by applying an external electric field. Hence, a bound-to-continuum peak can be observed even at low biases, whereas a bound-to-bound peak dominates at high applied fields due to enhanced escape probability by field-assisted tunnelling. The energy states in the QD and the QW can be adjusted independently by changing the parameters associated with each. As a result, DWELL structures open up a variety of possible designs, leading to multi-band IR detectors.

In this section, three-colour InAs/InGaAs DWELL detectors, with different well widths, are discussed as previously reported by Ariyawansa *et al.* [42]. Three DWELL detectors (labelled as 1388, 1373 and 1299) with well widths (120 Å, 110 Å and 90 Å, respectively) were characterized. The detectors showed response peaks at three distinct wavelengths in the mid, long, and very long wavelength/far infrared (MWIR, LWIR, and VLWIR/FIR) regions. The 1388 detector had peak wavelengths at $\sim$6.25 μm, $\sim$10.5 μm and $\sim$23.3 μm. All observed peaks correspond to the energy difference of intersubband transitions in the DWELL heterostructure. The two peaks at 6.25 μm and 10.5 μm are a result of bound-to-bound transitions from the ground state in the QD to a state in the QW, whereas the longer wavelength peak ($\sim$23.3 μm) is due to an intersubband transition between QD levels. The 23.3 μm peak has a detectivity of 7.9×10^{10} cm $\sqrt{\text{Hz}}$/W at 4.6 K under a -2.2 V bias. Spectral response in the 1388 detector has been observed up to 80 K. The operating wavelength of these detectors in the MWIR and LWIR regions can be tailored by changing the width of the QW. When the width of the QW is increased, the two peaks in the MWIR and LWIR regions show a red shift, while the VLWIR peak remains around $\sim$23.3 μm. This observation confirms that the LWIR peaks are associated with dot-to-well transitions and the VLWIR peak originates from transitions between energy states within the well.

22.2.1   DWELL device structure

The DWELL detector structures reported by Ariyawansa *et al.* [42] (1388, 1373, and 1299) were grown [66] in a VG-80 solid source molecular beam epitaxy (MBE) system with a cracked As_2 source at the University of New Mexico. The GaAs layers were grown at a substrate temperature $T_{sub} = 580°C$, whereas the $In_{0.15}Ga_{0.85}As$ QW and the InAs QDs were grown at $T_{sub} = 480°C$. The temperature was measured using an optical pyrometer. Using standard lithography, metal evaporation and wet etching techniques, *n-i-n* detector mesas were fabricated for top-side illumination. Mesas with various circular optically active areas (diameters ranging from 25 to 300 μm) were fabricated to test for leakage current and uniformity of structures. A more detailed discussion on the growth process can be found in [66].

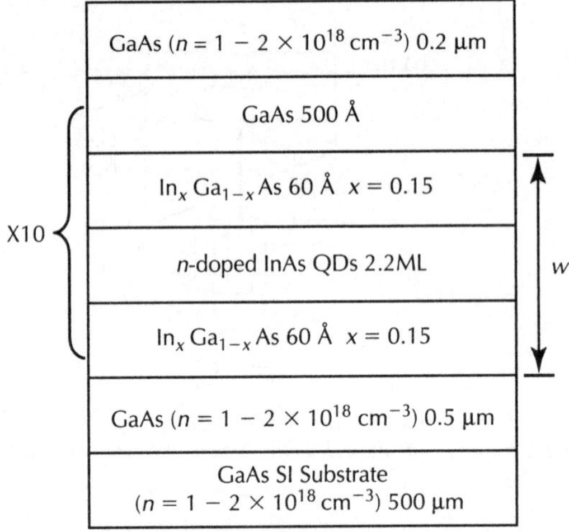

Figure 22.2   A schematic diagram of a DWELL structure. The width of the QW, i.e. the combined thickness of $In_{0.15}Ga_{0.85}$ As layers (indicated as w in the figure), is different for each detector. Structures 1299, 1373 and 1388 have well widths of 90, 110, and 120 Å, respectively, while all other parameters are the same. After [42].

The structure of the 1388 detector is shown in Fig. 22.2. The QDs were doped *n*-type with silicon to a sheet density of 5×10^{11} cm^{-2}, which corresponds to about one electron per QD. The QW was not intentionally doped. It has been found [65] that the optimal doping concentration for DWELL detectors corresponds to about one electron per QD. The size of the QDs is a critical parameter in the detector design and is controlled by growth conditions, especially the temperature and growth rate. Any inhomogeneous QD size fluctuation will lead to a broadening of spectral response peaks. There is a 2D distribution of QDs on a plane perpendicular to the growth direction; the formation of QDs above another QD is not possible since there is no barrier to separate the two QD layers. The width of the QW, i.e. the combined thickness of $In_{0.15}Ga_{0.85}As$ layers, is denoted by ω. The other two detectors (1373 and 1299) are identical to the 1388 sample except for the width of the QW. In 1373 and 1299 detector structures, the thickness of the bottom $In_{0.15}Ga_{0.85}As$ layer is 50 Å and 30 Å, respectively, while the top $In_{0.15}Ga_{0.85}As$ layer thickness is the same (60 Å) for both structures; thus, providing a 110 and 90 Å well width, respectively. There are ten periods of $In_{0.15}Ga_{0.85}As/n\text{-}InAs/In_{0.15}Ga_{0.85}As$ regions in each of the three detector structures. Square mesa devices with 400×400 μm dimensions were processed, and a 300 μm diameter opening was left in the top contact for front-side illumination. Photoluminescence measurements of the ground-state transition of the QD (1.25 μm at $T = 300$ K) with a 60:40 conduction to valence band ratio, were used to estimate the ground state of the QD, which is approximately 250 meV below the GaAs band edge. There can be at least two bound states in the QD and one confined state in the QW [9], as shown in Fig. 22.1. Possible carrier transitions, leading to the spectral response peaks, are indicated by arrows.

An energy level calculation model for a DWELL system was proposed by Amtout *et al.* [62]. DWELL structures with different QDs have been tested experimentally, and electronic spectra obtained by the model are in good agreement with the experimental results. The shape of the QDs is pyramidal as confirmed by transmission electron microscopy. The Hamiltonian of a system with a quasi-zero-dimensional QD placed in a two-dimensional QW is defined with a potential energy consisting of four terms: the potential in the QD region, the potential in the QW region, the potential in the barrier region, and the potential from the applied electric field. The eigenfunctions of the Hamiltonian are derived using a Bessel function expansion. The DWELL detectors tested by Amtout have QDs with base dimensions of 110 and 140 Å and heights of 65 and 50 Å, respectively. The energy spacing between the first and second energy levels obtained from the model for the two samples are 132 and 150 meV, whereas experimental analysis showed energy spacing of 123 and 140 meV, respectively. Although the energy states are shifted by the electrostatic potential from the bias field, the energy spacing between the first two energy states was not changed by the applied electric field. This can be observed in the spectral response of many DWELL detectors.

22.2.2 Device characterization techniques

The spectral response of the device under test and an Si composite bolometer with a known sensitivity were measured with a Fourier transform infrared (FTIR) spectrometer. The two spectra were obtained concurrently and used the same combination of optical windows, beam splitter and filters, so that the optical path was identical. The device spectrum (I_d) was divided by the bolometer spectrum (I_b) and then multiplied by the bolometer sensitivity (S_0) in order to obtain the voltage responsivity of the device:

$$R(V/W) = \frac{GS_0I_d}{I_b}. \tag{22.1}$$

here G is a geometrical factor which corrects for differences in the radiation incident area of the detector and the bolometer. To obtain the current responsivity, the voltage responsivity was divided by the effective resistance. Since the detector and load resistor act as a voltage divider for the photocurrent, the effective resistance R_e is the parallel combination of the load R_l and the detector dynamic resistance $R_d = dV/dI$, yielding $R_e = R_lR_d/(R_l + R_d)$. The final current responsivity is given by

$$R(A/W) = \frac{GS_0I_d(R_l + R_d)}{(I_bR_lR_d)}. \tag{22.2}$$

The specific detectivity (D^*) of the devices at different temperatures and applied biases was obtained from the measured peak responsivity (R_p), noise current density (S_i), and the illuminated area of the detector (A). The noise current density was measured with a dual channel fast Fourier transform (FFT) signal analyser and a low-noise preamplifier. A thick copper plate, which is at the device temperature, was used to block radiation, thus providing the dark conditions necessary for the measurements. The value of D^* is calculated from:

$$D^*(\text{Jones}) = \frac{R_pA^{1/2}}{S_i^{1/2}} \tag{22.3}$$

with $S_i = i_n^2/\Delta f$, i_n is the noise current, and Δf is the noise bandwidth.

The quantum efficiency is given by:

$$\eta = \frac{hcR_p}{gq\lambda} \tag{22.4}$$

where h is the Planck constant, c is the speed of light, g is the photoconductive gain, q is the electron charge, and λ is the wavelength.

The photoconductive gain (g) can be defined as the ratio of the total collected carriers to the total excited carriers generated by both thermal and photoexcitations. Assuming carrier emission and capture are due to generation–recombination (g–r) processes, g can be derived experimentally from [5]:

$$g = \frac{S_i}{4qI_d} + \frac{1}{2N} \tag{22.5}$$

where I_d is the measured dark current and N is the number of QD layers.

22.2.3   Experimental results and effects of the well width on response peaks

In order to explain the transition between energy states leading to response peaks, the experimental results of the three DWELL detectors reported by Ariyawansa *et al.* [42] will be discussed. A comparison of dark current–voltage (I–V) characteristics for all three structures is shown in Fig. 22.3. The sample 1388 showed the lowest dark current, and it increased when the width of the QW was increased. All three detectors exhibited three distinct peaks (three colours) in the MWIR, LWIR, and VLWIR regions (labelled as first, second, and third peak, respectively). The results for the 1388 detector under different bias voltages are shown in Fig. 22.4. The band diagram, with corresponding transitions indicated by arrows, is shown in the inset to Fig. 22.4. The origin of each peak will be explained in detail in the following sections.

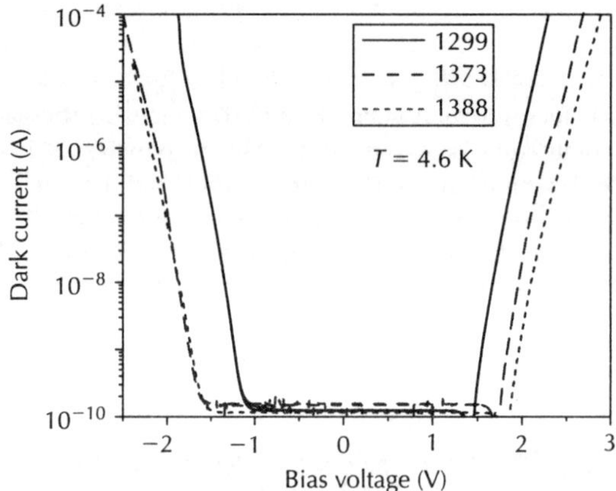

Figure 22.3   A comparison of dark I–V characteristics of three DWELL structures (1299, 1373, and 1388) at 4.6 K. The mesas tested have the same electrical area. The sample 1388 showed the lowest dark current, and a decrease of dark current is observed as the width of the QW increases. Modified after [42].

The spectral response of the three detectors in the MWIR/LWIR regions for $-0.5\,$V and $-1.4\,$V bias voltages is shown in Fig. 22.5. All three detectors showed two distinct peaks in this wavelength range. The 1299 detector exhibited the first peak at $\sim 4.2\,\mu$m and the second peak at $\sim 8.1\,\mu$m. A semi-empirical estimate based on the photoluminescence spectra with a 60:40 split of the band gap difference gives a 225–$250\,$meV ($\sim 4.9 - 5.5\,\mu$m) energy separation between the ground state of the QD and the conduction band edge of GaAs. Hence, the $4.2\,\mu$m peak is probably due to transitions from the ground state of the QD to the continuum state of the QW, and the second peak is due to transitions from the ground state of the QD to a bound state in the QW, as shown in Fig. 22.1. Moreover, it has been shown [67] that the line width ($\Delta\lambda/\lambda$) of a peak due to transitions from bound-to-bound states is narrower than that of transitions from

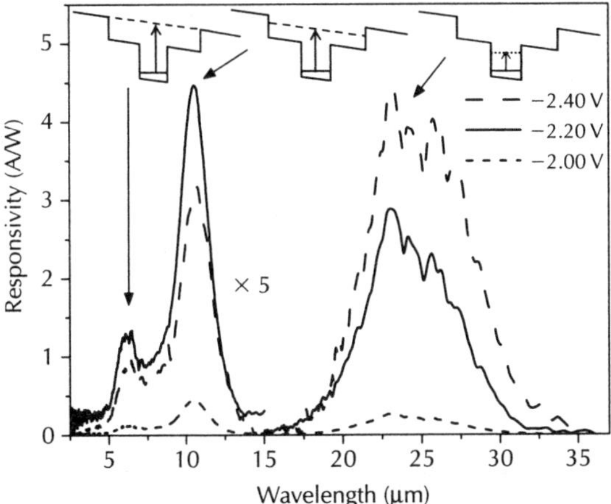

Figure 22.4   *Three-colour response of 1388 detector under different bias voltages at 4.6 K. The band diagram with the corresponding transitions indicated by arrows are shown in the inset.*

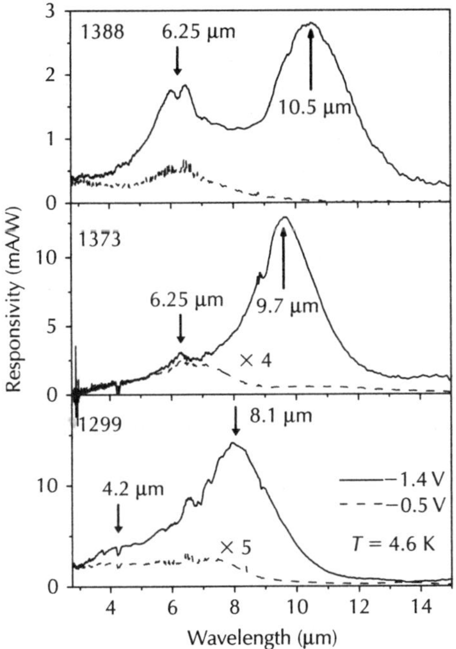

Figure 22.5   *The first two peaks of the three detectors biased with −1.4 V and −0.5 V at 4.6 K. Arrows indicate the peak positions and the × sign implies that the curve has been multiplied by the specified number. Modified after [42].*

bound-to-continuum states. The linewidth of the 4.2 μm peak is approximately 42%, whereas the linewidth of the 8.1 μm peak is approximately 28%; this observation is consistent with the aforementioned description. The escape probability of carriers excited to a bound state increases with increasing bias due to field-assisted tunnelling. Thus, the bound-to-bound peak (at 8.1 μm) shows a stronger dependency on the applied bias than the bound-to-continuum peak (at 4.2 μm), as evident from Fig. 22.5.

When the width of the QW is increased the energy spacing between the states in the QW decreases. At the same time, the energy spacing between QD ground state and states in QW also decreases. Thus, a red shift of the first and second peaks is observed. The results for the 1299 and 1388 detectors confirm this notion. In addition, the 1388 detector exhibits a quasi-bound state close to the band edge of the GaAs barrier, and hence the first peak of the 1388 detector is due to transitions from the ground state of the QD to the quasi-bound state in the QW. This can be confirmed by the red shift and the reduced linewidth for the first peak of the 1388 detector as compared to the first peak of the 1299 detector. Based on width of the QW in detector 1373, its peaks are expected to be located in between the peaks of the 1299 and 1388 detectors. However, the experiment showed a longer red shift than expected in both first and second peaks of the 1373 detector with respect to the 1299 detector. This discrepancy in the result for the 1373 detector could be explained by an unintentional change in the QD size during the growth process. Further discussion on this will be given in following sections. Moreover, several unexplained features in the responsivity spectra, such as line splitting, were also observed. Based on doping concentration and sheet density of QDs, it has been found [8] that a single QD consists of one electron. Multiple electrons within a QD could lead to a splitting of photoresponse peaks due to intralevel and interlevel Coulomb interactions [68]. Therefore, the secondary peaks superimposed on the primary peaks may result from either different QD sizes in the same DWELL structure or Coulomb interactions between multiple electrons in the QD. The expected red shift due to Coulomb interaction with an applied electric field could be compensated by the blue shift due to the Stark effect [68]. Splitting of absorption peaks is also possible through interdot coupling [69], which depends on the random distribution of QDs.

The spectral responsivity of the third peak of the 1388 detector in the VLWIR region, under different applied electric fields, is shown in Fig. 22.6. For QDs with a base diameter of 20 nm and a height of 7–8 nm, the energy separation between the ground and first excited state calculated from 8-band $\mathbf{k} \cdot \mathbf{p}$ model [70], is about 50–60 meV. Thus, it is believed that the VLWIR peak at ~23.3 μm is due to transitions between two bound states within the QD. The energy level diagram corresponding to this transition is shown in the inset to Fig. 22.6. The variation of the peak responsivity of the 23.3 μm peak with applied bias is shown in Fig. 22.7. As shown in the energy band diagrams in the inset to Fig. 22.7, carriers excited from the ground state to the first excited state in the QD have to tunnel through the barrier into the continuum, in order to be collected by the external circuit. With increased field strength, the barrier is pulled down strongly, allowing the excited carriers to tunnel through a thinner barrier. This leads to field-assisted tunnelling, where the applied field increases the escape probability of excited carriers. As a result, the experimental response curves showed a drastic increase in response when the reverse bias was increased from -1.0 V to -2.4 V. Moreover, the peak at 23.3 μm was broader than that would be expected for a bound-to-bound transition in the QD. This is attributed to an ~10% size fluctuation of the QDs during the self-assemble growth process.

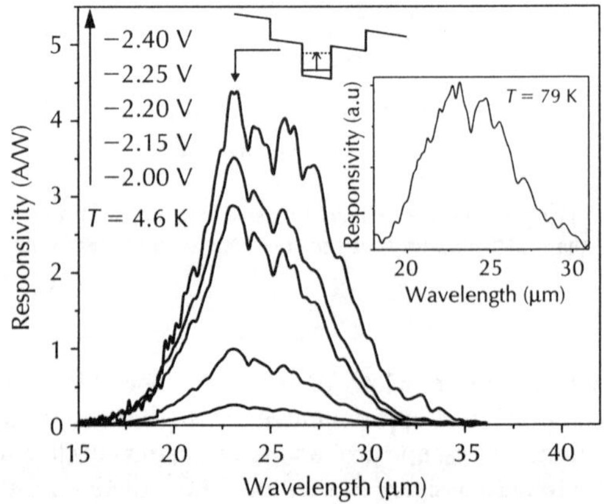

Figure 22.6   The VLWIR response of 1388 at different bias values (negative indicates that the top contact is negative). The band diagram represents the transition leading to the response. The inset shows the responsivity at 79 K. After [42].

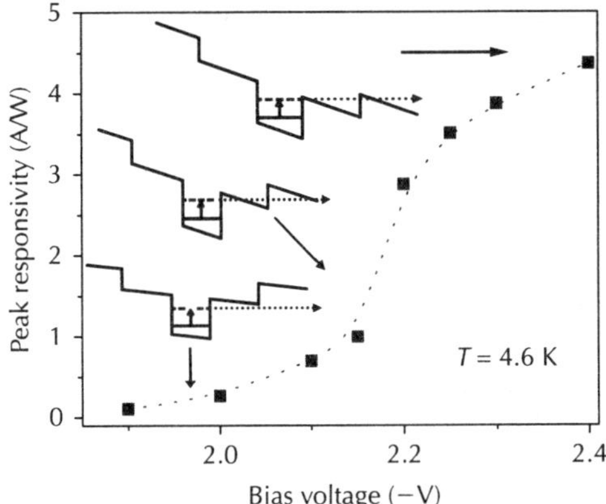

Figure 22.7   Variation of the peak responsivity of the 1388 detector with applied bias. Variation of the conduction band profile with applied bias (electric field) shows that excited electrons escape through the barrier by field-assistant tunnelling. After [42].

Device parameters and the figures of merit for the three detectors (1299, 1373, and 1388) under the optimum operating conditions are given in Table 22.1. The VLWIR peak of the 1373 detector could be obtained up to 60 K, while the VLWIR peaks of the 1299 and 1388 detectors were observed up to 80 K. The highest observed detectivity for the 1388 detector at 23.3 µm under a −2.2 V bias at 4.6 K was reported as $\sim 7.9 \times 10^{10}$ cm $\sqrt{\text{Hz}}$/W . At 80 K a maximum detectivity of 3.2×10^{8} cm $\sqrt{\text{Hz}}$/W was reported under a −1.4 V bias for the 1388 detector. At 23.3 µm, the 1388 detector exhibited lower responsivity and a lower noise current than those of the 1299 detector (a comparison of dark current is given in Fig. 22.3), resulting in a higher detectivity for the 1388 detector than for the 1299 detector. The improvement in operating temperature of the VLWIR response (up to 80 K), compared with a typical VLWIR QWIP [71] operating at $\sim$10–20 K, demonstrates the benefit of a quasi-zero dimensional confinement.

Table 22.1   Responsivity and detectivity of the three response bands of 1299, 1373, and 1388 DWELL detectors reported by Ariyawansa *et al*. [42] at 4.6 K with −1.4 V bias ($\sim$−22.5 kVcm^{-1} field). All three structures have InAs QDs placed in an InGaAs/GaAs QW. The QDs were doped *n*-type with silicon to a sheet density of 5×10^{11} cm^{-2}. The QDs in all three structures were grown to the same in size. After Reference [42].

Sample number	Well width (Å)	λ_{peak} (µm)	Responsivity (mA/W)	Detectivity (cm $\sqrt{\text{Hz}}$/W)
1299	90	4.2	3.9	1.1×10^{9}
		8.1	14.0	3.9×10^{9}
		23.3	60.2	1.9×10^{10}
1373	110	6.25	3.0	5.4×10^{9}
		9.7	12.9	2.3×10^{10}
		25.5	3.8	6.9×10^{9}
1388	120	6.25	1.8	6.2×10^{9}
		10.5	2.8	1.7×10^{10}
		23.3	25.6	6.6×10^{10}

As shown in Fig. 22.8, the VLWIR peak of both the 1299 and 1388 detectors occurs at approximately the same wavelength (23.3 µm). Changing the width of the QW does not affect the energy states in the QD, thus confirming that the VLWIR peak is due to transitions between

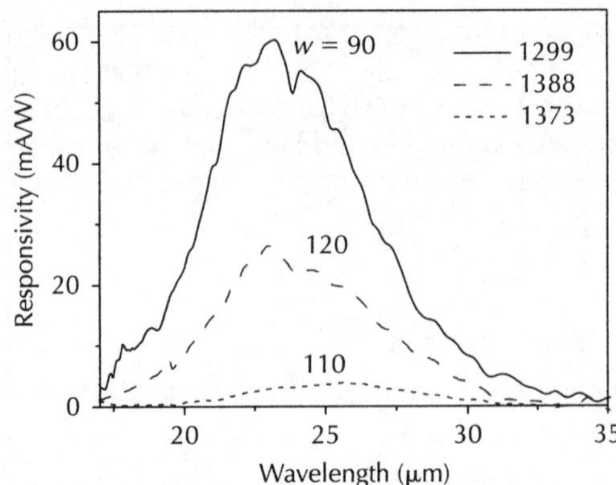

Figure 22.8   Spectral response of the VLWIR peak for all three detectors at 4.6 K under a −1.4 V bias (−23.7, −22.9, and −22.5 kVcm^{-1} field for 1299, 1373, and 1388, respectively). After [42].

QD states. However, the VLWIR peak of the 1373 detector appeared at 25.5 μm, as shown in Fig. 22.8, and red shifted with respect to the VLWIR peaks of the 1299 and 1388 detectors. This observation was attributed to the unintentional increase of QD size in the 1373 detector during the growth process, resulting in a decreased energy spacing between the ground and the first excited states in the QD. In addition, this would also decrease the energy spacing between the ground state in the QD and the bound state in the QW. As a result, the first two peaks will appear at longer wavelengths than expected. This shift was observed in the spectral response curves of the 1373 detector (see Fig. 22.5).

22.3   Tunnelling quantum dot infrared photodetectors (T-QDIPs)

At present, commercially available IR detectors only work at cryogenic temperatures, thus single-element devices and FPAs made of these detectors require cryogenic coolers and assemblies. These detector systems are not only complicated and bulky, but also very expensive. The most effective way to overcome these drawbacks would be the development of IR detectors capable of operating at room temperature. However, the development of room temperature IR detectors is a challenge as the rate of thermal excitations leading to the dark current increases exponentially with temperature. While a QD-based structure is a potential choice, conventional QDIP structures have not yet shown adequate performance above 150 K. At temperatures above 150 K electron occupation is dominated by the excited states in the QD; and, as a result, the reduction in the dark current is not significant. As a solution, Bhattacharya *et al.* [43] have explored a new resonant tunnel-based QD device architecture, demonstrating room temperature IR detection at 6 and 17 μm.

Aslan *et al.* [72] have recently observed resonant tunnelling through a QD layer. In general, any device structure designed to reduce the dark current will also reduce the photocurrent. Recently reported T-QDIPs [43, 70, 73, 74, 75] used resonant tunnelling to selectively collect the photocurrent generated within the QDs, while the tunnelling barriers (double-barrier system) block the majority of carriers contributing to the dark current. The maximum operating temperature of detectors is associated with the dark current, which in turn related to the detector response wavelength region. Ideally, the resonant tunnelling approach can be used to develop IR detectors operating at high temperatures irrespective of the response wavelength region. The characteristics of the room temperature T-QDIP reported by Bhattacharya *et al.*, showing two-colour response at wavelengths of ∼6 and ∼17 μm are discussed in this section.

A T-QDIP structure can be considered as an extended DWELL structure. That is, a DWELL structure coupled with a double-barrier system transforms into a T-QDIP structure, which has several advantages over DWELL structures. Conventional QDIPs have inherently low dark current, which can be further reduced using a DWELL structure. Compared to DWELL detectors, T-QDIPs exhibit lower dark current due to dark current blocking by the double-barrier system. As a result, T-QDIPs have the potential to achieve the highest possible operating temperature. Additionally, photocurrent filtering by means of resonant tunnelling in T-QDIPs offers a solution for low quantum efficiency, which has been observed in other QD-based devices. Quantum efficiency can be increased further through resonant cavity enhancement [76–78], which increases the absorption in the active region without increasing the dark current. In addition, several other important properties of T-QDIPs include the tunability of the operating wavelength and the multicolour (band) nature of the photoresponse based on different transitions in the structure. The operating wavelength can be tailored by changing the device parameters of the QW, QD, and double-barrier system. Utilizing transitions between the energy levels of the QD and the energy levels of the QW, it is possible to obtain detectors with multiple distinct response peaks.

22.3.1 Theoretical background and T-QDIP structure

Incorporating resonant tunnelling into QDIP structure design allows for the reduction of the dark current without reducing the photocurrent, thus leading to high performance. A typical T-QDIP consists of InGaAs QDs embedded in an AlGaAs/GaAs QW, which is then coupled to an AlGaAs/InGaAs/AlGaAs double-barrier system. The conduction band profile of a T-QDIP structure under an applied reverse bias is shown in Fig. 22.9. Pulizzi *et al.* [79] have reported resonant tunnelling phenomena for a similar QD-based structure coupled with a double barrier. The photocurrent generated by a transition from a state in the QD (E_1, E_2 or E_3) to a state in the QW, which is coupled with a state in the double-barrier system, can be collected by resonant tunnelling. In this discussion, the energy state in the QW is denoted as the resonant state, E_r, since it is associated with resonance tunnelling. The double-barrier system blocks the majority of carriers contributing to the dark current (carriers excited to any state other than the resonant state in the QW). It can be shown that the tunnelling probability is near unity for carriers excited by radiation with energy equal to the energy difference between the QD ground state and the resonant state.

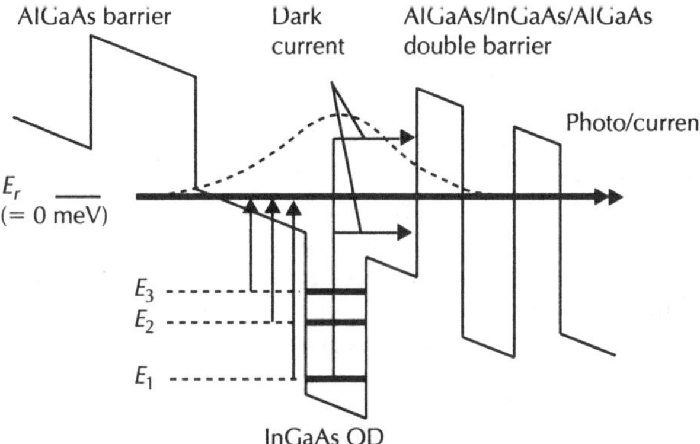

Figure 22.9   Schematic diagram of the conduction band profile of a T-QDIP structure under a bias. E_1, E_2 and E_3 are the energy level positions in the QD with respect to the resonant-state E_r. Only the carriers excited to the resonant state contribute to the photocurrent. Modified after [43].

The first step in designing a T-QDIP is the calculation of the QD energy levels using the 8-band **k · p** model [70]. This model uses the strain in the QD, which is calculated using the valence force field (VFF) model. VFF model has proven successful in calculating the strain tensor in self-assembled QDs. The size of the QD and the confinement potential should be designed such that the required spacing between energy levels can be obtained. For example, in order to design a two-colour T-QDIP with response peaks at 5 and 10 μm, the ground and first excited states in the QD (labelled as E_1 and E_2 in Fig. 22.9) should be 248 and 124 meV below the resonant-state energy. The energy spacing between states E_2 and E_1 is equal to 124 meV. Theoretical calculations indicate that the energy difference between the ground and first excited state in small InAs/GaAs QDs, with a height of $60\,\text{Å}$ and a width of $110\,\text{Å}$, is approximately $124\,\text{meV}$. The width and the confinement potential of the QW are adjusted to obtain the resonant state at $248\,\text{meV}$ above E_1 state (at $124\,\text{meV}$ above E_2 state). The doping concentration in the QDs should be sufficiently high so that both states in the QD are filled with ground-state electrons. The energy states in the QW, including the presence of the wetting layer and the double-barrier system, are calculated by solving the one-dimensional Schrödinger equation. The transmission probability for the double-barrier structure is calculated using the transfer matrix method [80].

The double-barrier system (AlGaAs/InGaAs/AlGaAs) is integrated with each QD layer of the QDIP, and is designed such that the resonant state coincides with a bound state in the double-barrier system under certain bias conditions. In this way, a higher potential barrier for thermal excitations can be introduced, while the photoexcitation energy remains very low. Due to the energy-dependent tunnelling rate of the double-barrier system, the dark current resulting from carriers with a broad energy distribution is suppressed. Thus the dark current can be significantly reduced, particularly at high temperatures.

The intersubband absorption coefficient of a photon with energy $\hbar\omega$ in a QD layer can be expressed as [5]:

$$\alpha(\hbar\omega) = \frac{\pi e^2 \hbar}{\varepsilon_0 n_0 c m_0^2 V_{av}} \frac{1}{\hbar\omega} \sum_{fi} |\,a \cdot p_{fi}\,|^2 N(\hbar\omega) \tag{22.6}$$

where V_{av} is the average QD volume, a is the polarization of the incident light, p_{fi} is the momentum matrix element between energy states, and $N(\hbar\omega)$ is the electron density of states. The Gaussian inhomogeneous broadening due to the fluctuations in QD size, $N(\hbar\omega)$, is:

$$N(\hbar\omega) = \frac{1}{\sqrt{2\pi}\sigma} \exp\left\{ \frac{-(E_{fi} - \hbar\omega)^2}{2\sigma^2} \right\} \tag{22.7}$$

where E_{fi} is the energy separation between states and σ is the linewidth of the transition. The momentum matrix element is calculated from the QD wavefunctions, which can be obtained from the 8-band $\mathbf{k} \cdot \mathbf{p}$ model. The spectral response of the detector is characterized by peak wavelength (λ_p), peak responsivity (R_p), and the peak quantum efficiency (η_p). Responsivity is given by $R = q\eta\lambda/hc$, where q is the electron charge, λ is the wavelength, h is the Planck constant and c is the speed of light. Quantum efficiency can be calculated from the absorption coefficient (from Eq. 22.6) and the thickness of the absorption region.

In order to achieve background-limited infrared performance (BLIP) conditions at high temperatures, the detector should exhibit an extremely low dark current density. A T-QDIP designed to have strong resonant tunnelling is capable of achieving high BLIP temperatures. The dark current, I_d, of a T-QDIP structure at a bias, V, is given [74] by:

$$I_d(V) = ev(V)n_{em}(V)A \tag{22.8}$$

where A is the device area, v and n_{em} (given by Eqs 22.9 and 22.10) are the average electron drift velocity in the barrier material and the concentration of electrons excited out of the QD, respectively. The electron drift velocity is given by:

$$v(V) = \frac{\mu F(V)}{\sqrt{1 + \left(\dfrac{\mu F(V)}{v_s}\right)^2}} \tag{22.9}$$

where μ is the electron mobility, F is the electric field, and v_s is the electron saturation velocity. The excited electron density from the QD is given by:

$$n_{em}(V) = \int_{-\infty}^{\infty} N(E)f(E)T(E, V)dE \tag{22.10}$$

where $f(E)$ is the Fermi–Dirac distribution function, $T(E, V)$ is the tunnelling probability calculated by the transfer matrix method [80], and $N(E)$ is the density of states, which is given by the following equation:

$$N(E) = \sum_i \frac{2N_d}{L_p} \frac{1}{\sqrt{2\pi}\sigma} \exp\left\{\frac{-(E_{fi} - E)^2}{2\sigma^2}\right\}$$
$$+ \frac{4\pi m^*}{L_p h^2} H(E - E_W) + \frac{8\pi\sqrt{2}}{h^3} (m^*)^{(3/2)} \sqrt{E - E_C} H(E - E_C) \tag{22.11}$$

where the first term is the density of states of the QD state and N_d is the surface density of QDs. The second term is the density of the wetting layer states, where E_W is the wetting layer state, and $H(x)$ is a step function with $H(x) = 1$ for $x \geq 0$ and $H(x) = 0$ for $x < 0$. The third term describes the density of states in the barrier material, where E_C is the conduction band edge of the barrier material.

As shown in Fig. 22.9, the carriers excited to any state other than the resonant state are blocked by the tunnel barriers. However, for efficient dark current blocking, the broadening of the resonant state has to be at a minimum. That is, the resonant state should be strongly bound. Basic parameters should be adjusted so that the tunnelling probability remains close to unity and the carrier escape lifetime is smaller than the carrier recombination lifetime. The Fermi level in the QD (hence QD ground state) should be below the band edge of the QW; however, adjusting the ground state will change the energy difference between the QD ground state and the resonant state, which will affect the peak response wavelength. Thus, all these factors need to be taken into account to design an optimized detector exhibiting low dark current.

22.3.2 Two-colour room temperature T-QDIPs

As reported by Bhattacharya *et al.* [43] a T-QDIP structure (MG386) was grown by MBEs and then characterized using *I–V*, spectral response, and noise measurements. The results demonstrated the detector's ability to operate at room temperature due to resonant tunnelling phenomenon present in the structure. The detector showed a two-colour response at wavelengths of ~ 6 and $\sim 17\,\mu m$ up to room temperature. A detailed explanation of the device structure, spectral response and device performance are given in following sections.

22.3.2.1 T-QDIP structure and experiment

The T-QDIP structure MG386 (reported by Bhattacharya *et al.*) is schematically shown in Fig. 22.10. Self-organized $In_{0.1}Ga_{0.9}As$ QDs were grown on a GaAs substrate. A stack of $Al_{0.3}Ga_{0.7}As/In_{0.1}Ga_{0.9}As/Al_{0.3}Ga_{0.7}As$ layers serve as the double-barrier system. The conduction band profile of this T-QDIP structure under an applied reverse bias is shown in Fig. 22.9. The GaAs and AlGaAs layers were grown at 610°C and the InGaAs or InAlAs QD layers were grown at 500°C. Vertical circular mesas for top illumination were fabricated by standard photolithography, wet chemical etching and contact metallization techniques. The *n*-type top ring contact and the *n*-type bottom contact were formed by evaporated Ni/Ge/Au/Ti/Au with a thickness of 250/325/650/200/2000 Å. The radius of the optically active area is 300 μm. Samples with devices to test were mounted on chip carriers using silver epoxy. Then electrical contacts were made by bonding gold wires from devices to the chip carrier leads. Characterization techniques used are explained in section 22.2.2.

22.3.2.2 Dark current measurements

Dark *I–V* measurements for MG386, at different temperatures ranging from 80 to 300 K, are shown in Fig. 22.11a. Positive (or negative) bias denotes positive (or negative) polarity on the top contact. A comparison of the dark current density between DWELL (1299) [9] and T-QDIP (MG386) detectors at 80 K is shown in Fig. 22.11b. Dark current densities at a bias of $-2\,V$ are 3×10^{-1} and $1.8 \times 10^{-5}\,A/cm^2$ for DWELL and T-QDIP, respectively. The reduction in dark

GaAs n^+ contact
40 Å Al$_{0.1}$Ga$_{0.9}$As (i)
40 Å GaAs (i)
In$_{0.4}$Ga$_{0.6}$As QDs (i)
10 Å GaAs (i)
30 Å Al$_{0.3}$Ga$_{0.7}$As (i)
40 Å In$_{0.1}$Ga$_{0.9}$As (i)
30 Å Al$_{0.3}$Ga$_{0.7}$As (i)
400 Å GaAs (i)
GaAs n^+ contact
S.I. GaAs substrate

10 ×

Figure 22.10   A schematic heterostructure of a T-QDIP grown by MBE. InGaAs QDs are placed in a GaAs QW. The AlGaAs/InGaAs/AlGaAs layers serve as a double-barrier system to decouple the dark and photo currents. The letter "i" stands for intrinsic. After [43].

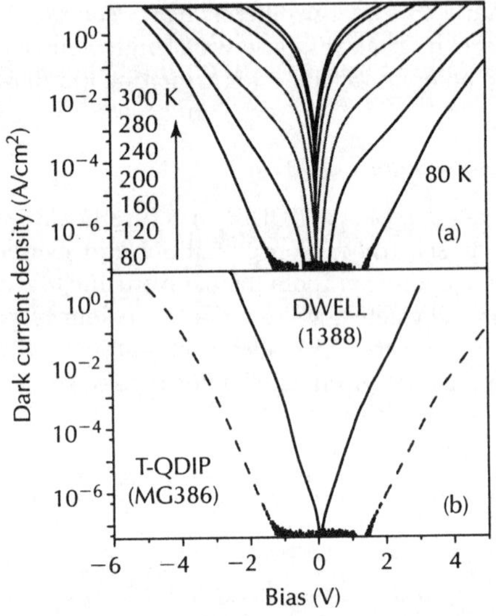

Figure 22.11   (a) Dark current density of the T-QDIP sample (MG386) as a function of bias in the temperature range 80–300 K. (b) A comparison of the experimental dark current density between a DWELL detector (1299) and a T-QDIP (MG386) at 80 K. The reduction of the dark current in T-QDIP is attributed to dark current blocking by the double-barrier system. Modified after [43].

current of the T-QDIP is associated with the double-barrier system in the structure. Moreover, dark current densities for MG386 at a bias of 1 V were 0.21, 0.96, and 1.55 A/cm^2 at 240, 280 and 300 K, respectively. These dark current density values are lower than the dark current values of other IR detectors operating in comparable wavelength regions at the same temperature. Based on the dark current variation as a function of bias, negative conductance peaks were not visible even though resonant tunnelling takes place in the structure. This observation can be expected for T-QDIPs since sequential resonant tunnelling through ground state is not possible. In T-QDIP structures, there is no coupling between the QD ground state and states in the double barrier (unlike in superlattice structures [81]). Also, each active region of the T-QDIP is separated by a thick spacer layer (400 Å GaAs), which does not allow any significant coupling between two active regions (two periods). Thus, *I–V* curves are not expected to display negative conductance regimes. Furthermore, it is important to underline the thin AlGaAs barriers (30 or 40 Å) on both sides of the QW. Even though the double barrier is placed only on one side of the QW (on the right side according to Fig. 22.9), tunnelling through the single barrier on the opposite side (on the left side of the QW) is also possible. However, the transmission through this barrier is lower compared to that through the double barrier. Thus, an asymmetric *I–V* characteristic was observed.

22.3.2.3 *Spectral responsivity*

The spectral response of MG386 at 80 K under different bias values is shown in Fig. 22.12a, and the variation of the peak responsivity at 6.2 μm is shown in Fig. 22.12b. Based on calculations, the allowed confined energy states in the QD E_1, E_2, and E_3 are located at -161, -103, and -73 meV with respect to the resonant state (see Fig. 22.9). Thus, the peak at ∼6 μm is due to transitions from the ground state of the QD to the resonant state in the structure, which is consistent with the calculated energy spacing between corresponding states ($\Delta E = 161$ meV). The peak responsivity and the conversion efficiency (the product of quantum efficiency and the photoconductive gain) of the 6 μm peak at 80 K and -4.5 V are ∼0.75 A/W and 16%, respectively. Under reverse bias (top contact is negative), the photoexcited electrons tunnel through the double barrier by resonant tunnelling. Similarly, under forward bias photoexcited electrons tunnel through the single barrier (on the opposite side of the double-barrier system). Due to the variations in transmission through the single and double barriers, the response under reverse bias is significantly higher than the response under forward bias, as evident from Fig. 22.9. However, the responsivity shows a strong dependence on the applied bias in both positive and negative directions. This behaviour is attributed to resonant tunnelling similar to that of double-barrier superlattice structures [72, 82]. Applying a bias across the structure can fine-tune the alignment of the bound state in the QW (resonant state) and the bound state in the double-barrier system, allowing for resonant tunnelling conditions. The observed bias dependence of the responsivity indicates that resonant conditions are satisfied over a considerable range of applied bias voltages. This behaviour could be associated with thin barriers and the broadening of the energy states (δE) in the system.

The spectral response of the MG386 detector under a -2 V bias at different temperatures in the range of 80–300 K is shown in Fig. 22.13a and b. Two distinct peaks centred around ∼6 and ∼17 μm were observed at high temperatures, and a weak response around 11 μm was also present. The peak at 17 μm results from transitions between the second excited state of the QD and the resonant state ($\Delta E = 73$ meV). The linewidth is ∼26 meV, which corresponds

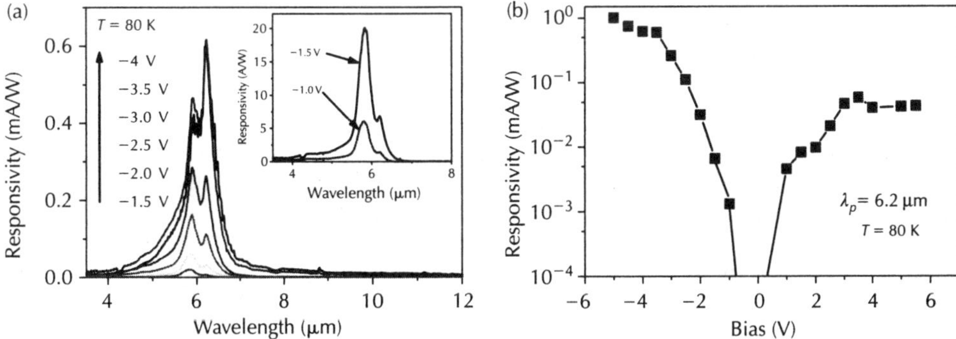

Figure 22.12 (a) *Bias dependence of the spectral responsivity of T-QDIP (MG386) at 80 K from -1.5 to -4 V. The inset shows the responsivity at low bias values (from -0.25 to -1.5 V). (b) Variation of the peak responsivity with applied bias at 80 K.*

to the inhomogeneous broadening of QD states at 300 K. Due to the symmetry of QD geome-try, excited states have a higher degeneracy (8) than the ground state (2). The carrier density in excited states increases with increasing temperature, as compared to that in the ground state. As a result, the 17 μm peak was dominant above 200 K, as evident from Fig. 22.13. The weak response at ∼11 μm corresponds to the energy separation between the first excited QD state and the resonant state ($\Delta E = 102\,\text{meV}$).

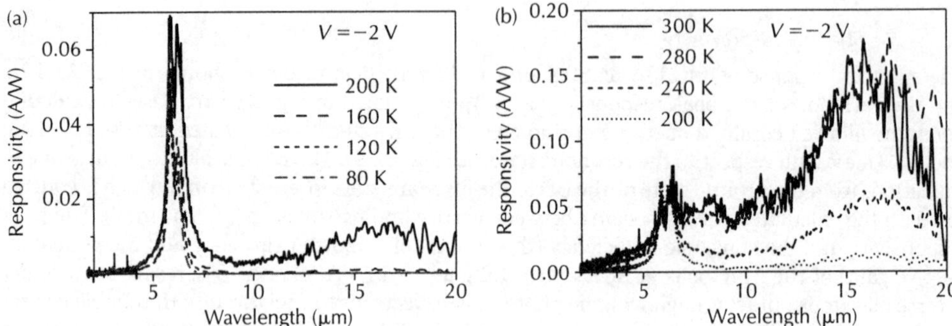

Figure 22.13   *Spectral responsivity of T-QDIP (MG386) in the temperature range (a) 80–200 K, and (b) 200–300 K under a −2 V bias. Two distinct peaks centred around ∼6 and ∼17 μm can be observed at high temperatures, and a weak response around 11 μm is also visible. Modified after [43].*

22.3.2.4   Noise measurements and detectivity

The noise current spectra of QWIPs [83, 84, 85] have been reported in the literature. In general, the total noise current is a contribution of Johnson noise, $1/f$ noise, and generation–recombination noise (*g–r*) components. However, QWIPs [83] and QDIPs [86] are expected to show very low $1/f$ noise characteristics. The frequency dependent noise spectrum, which is due to $1/f$ and generation–recombination noise components, has the form [87, 88]:

$$S(f) = C + \frac{B}{f} + \frac{A}{1 + \left(\dfrac{f}{f_c}\right)^2} \tag{22.12}$$

where *A*, *B*, and *C* are constants. The cut-off frequency, f_c, is given by $f_c = 1/2\pi\tau$ where τ is the electron lifetime, which is given by:

$$\tau \propto T^{-2} \exp\left(\frac{E_A}{kT}\right) \tag{22.13}$$

where *T* is the temperature, *k* is the Boltzmann constant and E_A is the activation energy of the thermally activated trap level. Noise current density spectra are used to determine the varia-tion of τ with temperature. Based on Eq. 22.13, the plot of $\log(\tau/T^2)$ against $1/T$ would result in a straight line with a slope of E_A/k, which can be used to calculate the activation energy. Furthermore, the electronic states, carrier capture, and carrier transport properties in a QD-based structure have been studied [89] using capacitance–voltage spectroscopy [90, 91] and deep-level transient spectroscopy [92, 93]. A strong negative capacitance phenomenon, which is originated from carrier capture and emission at interface states, has been observed in HIWIP [94] and QWIP [95] detector structures.

The specific detectivity (D^*) of the MG386 detector (reported by Bhattacharya *et al.* [43]) at different temperatures and applied biases was obtained from the measured peak responsivity and noise density spectra, as described in section 22.2.2. At 80 K and under a bias of −2 V, the maximum D^* was found to be $1.2 \times 10^{10}\,\mathrm{cmHz}^{1/2}/\mathrm{W}$. The variation in D^* at $6.2\,\mu\mathrm{m}$ with changing bias at $80\,\mathrm{K}$ is shown in Fig. 22.14. The rate of increasing noise current with increasing bias was much higher than the rate of increasing responsivity with increasing bias, resulting in lower D^* at higher bias voltages (beyond $\pm 2\,\mathrm{V}$). This variation in D^* as a function of bias is expected for a typical photodetector. The value of D^* at $17\,\mu\mathrm{m}$ and $300\,\mathrm{K}$ was of the order of $10^7\,\mathrm{cmHz}^{1/2}/\mathrm{W}$, and with some redesigning of the device heterostructure, a higher D^* can be obtained for the same conditions.

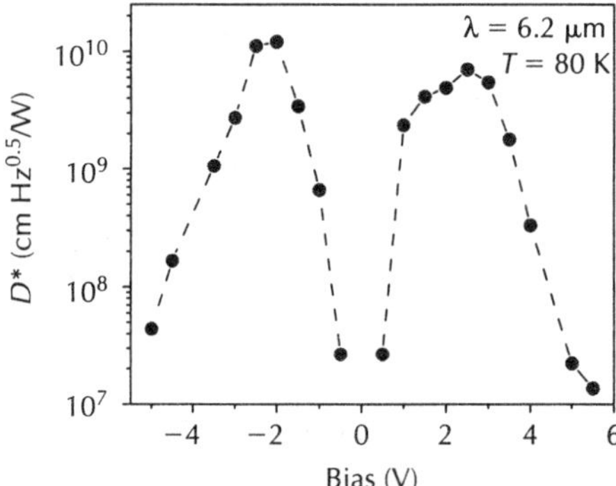

Figure 22.14   Variation of detectivity of the T-QDIP (MG386) at 6.2 μm as a function of bias at 80 K. The rate of increasing noise current with bias is much higher than the rate of increasing responsivity with bias, resulting in lower D at higher bias voltages (above ±2 V).*

22.3.3   T-QDIPs for terahertz radiation detection

With an increasing interest in the terahertz region of the spectrum (0.1–$3.0\,\mathrm{THz}$), there is a need for terahertz detectors exhibiting low dark current and high operating temperatures for applications in imaging, communication, security and defence. The primary challenge in developing terahertz detectors is the reduction of dark current (due to thermal excitations) associated with terahertz detection mechanisms in the device structure. At the present time, terahertz detectors such as Ge BIB detectors [96], photoconductors triggered by femtosecond laser pulses [97], QW detectors [98], HIWIP detectors [99], HEIWIP detectors [100], and thermal detectors, such as bolometers and pyroelectric detectors, are being studied. However, all of these detectors operate at low temperatures. A typical detector structure, in which the transitions leading to terahertz detection occur between two electronic states with an energy difference of ΔE ($\sim 4.1\,\mathrm{meV}$ for $1\,\mathrm{THz}$), would not be suitable for high-temperature terahertz detection since the dark current due to thermal excitations become dominant even at $77\,\mathrm{K}$ due to the small ΔE. A T-QDIP structure [73], in which the photocurrent is selectively collected while the dark current is blocked, can be adjusted to obtain terahertz response, thus offering a suitable platform for high operating temperature terahertz detectors.

A schematic diagram of a T-QDIP terahertz detector (MG764), reported by Su *et al.* [73], is shown in Fig. 22.15. The heterostructure was grown by MBE on (001)-oriented semi-insulating GaAs substrate. The GaAs and AlGaAs layers were grown at $610\,^{\circ}\mathrm{C}$ and the rest of the structure was grown at $500\,^{\circ}\mathrm{C}$. The top and bottom GaAs contact layers were n-doped with Si to a level of $2 \times 10^{18}\,\mathrm{cm}^{-3}$. Mesa-shaped vertical n-i-n devices for top illumination were fabricated by standard photolithography, wet chemical etching, and contact metallization techniques. The top and bottom n-type contacts were formed by evaporated Ni/Ge/Au/Ti/Au (thickness = $250/325/650/200/2000\,\text{Å}$, respectively) followed by annealing. In order to obtain a transition leading to a response in terahertz region, the excited states in the QD were pushed towards the resonant state

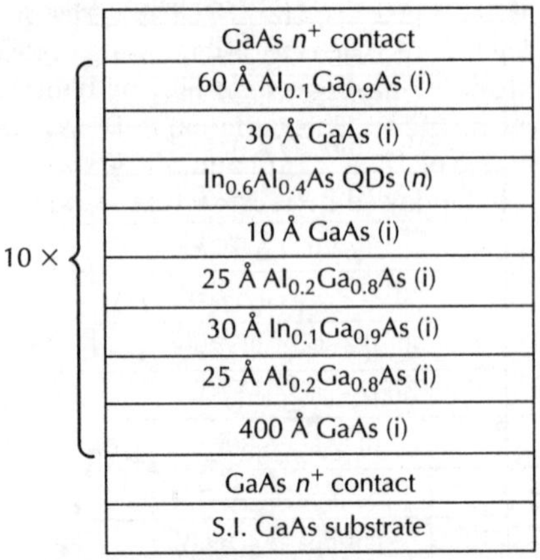

GaAs n^+ contact
60 Å $Al_{0.1}Ga_{0.9}As$ (i)
30 Å GaAs (i)
$In_{0.6}Al_{0.4}As$ QDs (n)
10 Å GaAs (i)
25 Å $Al_{0.2}Ga_{0.8}As$ (i)
30 Å $In_{0.1}Ga_{0.9}As$ (i)
25 Å $Al_{0.2}Ga_{0.8}As$ (i)
400 Å GaAs (i)
GaAs n^+ contact
S.I. GaAs substrate

(10 × for the bracketed layers)

Figure 22.15   A schematic heterostructure of a terahertz T-QDIP (MG764) with n-doped $In_{0.6}Al_{0.4}As$ QDs. The growth of smaller QDs compared to InAs or InGaAs QDs was achieved using InAlAs material. The QD size has been considerably reduced to 40 Å (height) and 130 Å (width). After [73].

by forming smaller QDs. The QDs were doped to raise the Fermi level so that photoexcitations take place from an upper state in the QD to the resonant state. In order to reduce QD size, $In_{0.6}Al_{0.4}As$ was used instead of InGaAs because the Al-containing islands (QDs) are smaller in size compared to InAs islands due to the smaller migration rate of Al adatoms on the growing surface during epitaxy.

22.3.3.1   Growth of "small" quantum dots by MBE

The typical size of the near-pyramidal InAs/GaAs self-organized QDs [9, 19] is $\sim$60–70 Å (height)/$\sim$200–250 Å (base), and QD density varying between the range 5 and $10 \times 10^{10} cm^{-2}$. Typical electron intersublevel energy separation in such QDs ranges between 40 and 80 meV. Based on calculated results, a large energy spacing ($\sim$124 meV) between the QD ground and first excited states can be obtained using smaller QDs. Smaller QDs also provide a large QD density for the same amount of adatom change, which increases the absorption of radiation. QD size is reduced by inhibiting growth kinetics on the surface. This can be done either by growing QDs on an Al-containing material or by incorporating a small amount of Al into the QD material. The presence of Al reduces the adatom migration lengths on the growing surface (insufficient kinetics), resulting in smaller QDs. During the investigation of terahertz T-QDIP operating at high temperatures reported by Su *et al.* [73], the QD size was considerably reduced to 40 Å (height) and 130 Å (width), consequently the QD density increased by an order of magnitude.

22.3.3.2   Dark current and responsivity

The dark current density of the terahertz T-QDIP (MG764) at different temperatures is shown in Fig. 22.16. The T-QDIP showed a lower dark current density compared to other terahertz detectors [99] operating in the $\sim$20–60 μm range. The spectral response of the detector at 80 and 150 K is shown in Fig. 22.17. The conduction band profile of the terahertz T-QDIP is similar to the diagram shown in Fig. 22.9. In this structure, the location of the E_2 state is -24.6 meV (54.6 μm) with respect to the E_r state, and the terahertz response originates from transitions between the E_2 and E_r states. Responsivity values at 80 and 150 K were 6 and 0.6 mA/W, respectively. The sharp dip around 37 μm is due to the reststrahlen band of GaAs, which is present in all GaAs-based photon detectors [100, 101]. The observed full width at half maximum (FWHM) of the spectral response is $\sim$35 meV. Spectral broadening arises from the inhomogeneous size distribution of self-organized QDs. The measured detectivity for MG764 was $\sim$5 × $10^7 cmHz^{1/2}$/W at

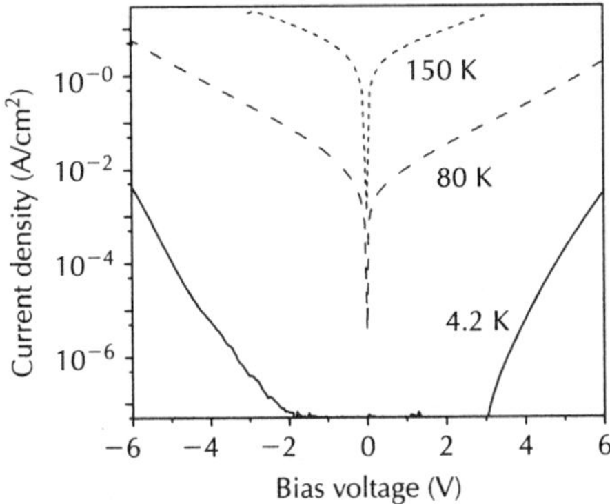

Figure 22.16   The dark current density of terahertz T-QDIP (MG764) as a function of bias in the temperature range 4.2–150 K. In the reported response range, the T-QDIP shows a lower dark current density compared to other terahertz detectors. After [73].

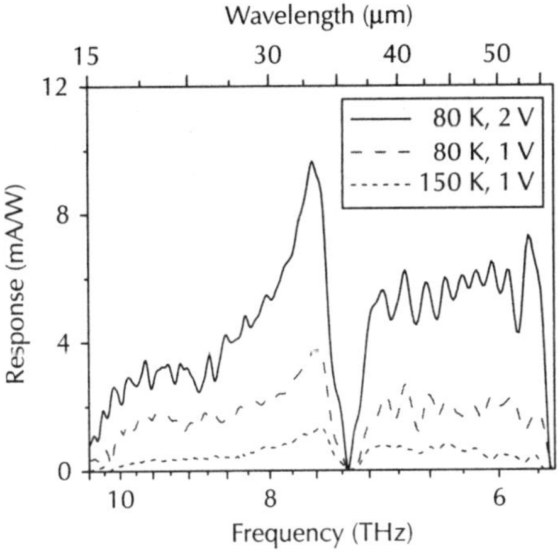

Figure 22.17   Spectral responsivity of terahertz T-QDIP (MG764) in the temperature range 80–150 K. The dip at 37 μm is the reststrahlen region of GaAs. Terahertz operation at high temperature (150 K) is made possible by the incorporation of resonant tunnelling phenomena into the device structure. After [73].

80 K under a bias of 1 V. Therefore, it can be concluded that terahertz operation at high temperatures (up to 150 K) is possible by incorporating resonant tunnelling phenomena into the device design. However, in order to achieve terahertz detection in the 1–3 THz region at high temperatures, several issues need to be resolved such as the growth of small QDs with reduced size fluctuation, optimization of structure parameters, and tight resonant conditions to maintain a low dark current.

22.4   Improvement of QDIP performance

Potential approaches to improve QDIP performance include using multi-periods of QD active layers [75], resonant cavity (RC) enhancement [78], and photonic crystal cavity (PhC) [102] enhancement. In order to increase IR absorption, Chakrabarti *et al.* [75] have reported an MBE grown QDIP consisting of 70 periods of QD active layers. A 400 Å $Al_{0.3}Ga_{0.7}As$ current-blocking barrier was incorporated underneath the GaAs top contact layer. A low dark current density (1.83×10^{-2} Acm^{-2} at 175 K under a -2.0 V bias) was observed at high operating temperatures. The response in the 2–6 μm range, peaking at 3.9 μm, was due to transitions of electrons from the QD ground state to the continuum state. At 150 K, a peak responsivity of 0.12 A/W was observed. A maximum detectivity of 10^{11} cmHz$^{1/2}$/W at 100 K for -2 V, which is reasonably high at this temperature, was also reported. The improvement in detector performance was due to the increased number of QD active layers.

Using photonic crystals as a microcavity resonator, Posani *et al.* [102] has demonstrated a significant improvement in the responsivity, conversion efficiency, and detectivity for a DWELL detector. A two-dimensional array of holes (PhC) with a 2–3 μm diameter and a lattice spacing of 2.4 μm was fabricated on the DWELL structure using e-beam lithography. A schematic diagram and an image of a hexagonal PhC are shown in Fig. 22.18a and b. PhCs provide a photonic band gap for normal incidence radiation. By introducing a defect into the crystal, a cavity can be built, which operates in the wavelength range corresponding to the photonic band gap. The above parameters were chosen [102] to obtain resonant cavity effects at 8.1 μm. Two structures, a DWELL detector and a PhC-DWELL detector [102], were tested and performance was compared to determine the effects of PhC. The DWELL detector exhibited two-colour response characteristics with peaks at 6 and 10 μm. The 6 μm (dominant at low bias) was due to transitions from the QD ground state to the continuum, while the 10 μm peak (dominant at high bias) originated from transitions from the QD ground state to a bound state in the QW. These electronic transitions and corresponding response peaks are similar to those of the DWELL detectors reported earlier in this chapter. An increase in photocurrent by an order of magnitude was reported for PhC-DWELL, while both DWELL and PhC-DWELL had the same dark current. A comparison of the responsivity of a PhC-DWELL and a standard DWELL is shown in Fig. 22.18c. At 9 μm and 78 K under a -2.6 V bias, the PhC-DWELL detector had a conversion efficiency of 95%, as opposed to the conversion efficiency of 7.5% for the DWELL detector. Using conventional optical lithography, PhCs with 2–3 μm hole diameter can be easily incorporated into FPA fabrication.

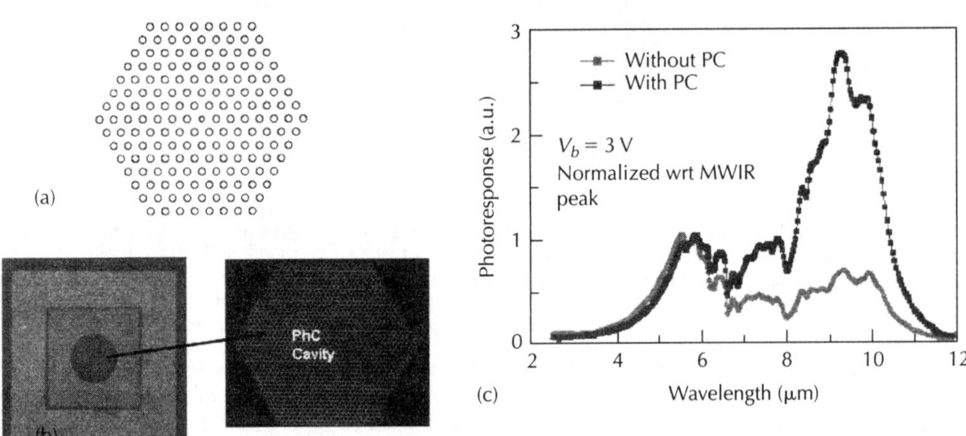

Figure 22.18   *(a) Photonic crystal resonant cavity comprising a hexagonal pattern of air holes. (b) Image of a PhC defined in a DWELL detector. (c) A comparison of the spectral responsivity of PhC-DWELL and standard DWELL detectors at 50 K under a -3 V bias. After [102].*

Similar to PhC-DWELL, an RC-DWELL, in which the active QD region was placed in a Fabry–Pérot resonant cavity, was reported by Attaluri *et al.* [78]. This allows radiation to pass multiple times through the active region within the cavity. The RC was designed to increase the optical field strength at the peak wavelength of a standard DWELL detector. The RC-DWELL consists of a regular DWELL (including top and bottom contact layers) grown on a distributed Bragg reflector (bottom reflector), as shown in Fig. 22.19a. The top reflector was the semiconductor–air interface (top contact layer), which has about 30% reflectivity. The DWELL detector showed two response peaks at 6 and $10\,\mu m$. The resonant optical cavity was active at $9.5\,\mu m$; consequently, the RC-DWELL showed an enhance response at $9.5\,\mu m$ (see Fig. 22.19b). A peak responsivity of 0.76 A/W, detectivity of $1.4 \times 10^{10}\,cmHz^{1/2}/W$ and 10% conversion efficiency at 1.4 V bias and 77 K were reported for RC-DWELL, while DWELL showed only 1.25% conversion efficiency at $10\,\mu m$ under the same conditions. The detectivity of the RC-DWELL was $3 \times 10^{9}\,cmHz^{1/2}/W$ at 77 K under a 1.2 V bias, showing a factor of 3 improvement compared to the conventional DWELL detector. Another important feature of incorporating an RC into a DWELL structure is the increase in photocurrent without increasing the dark current.

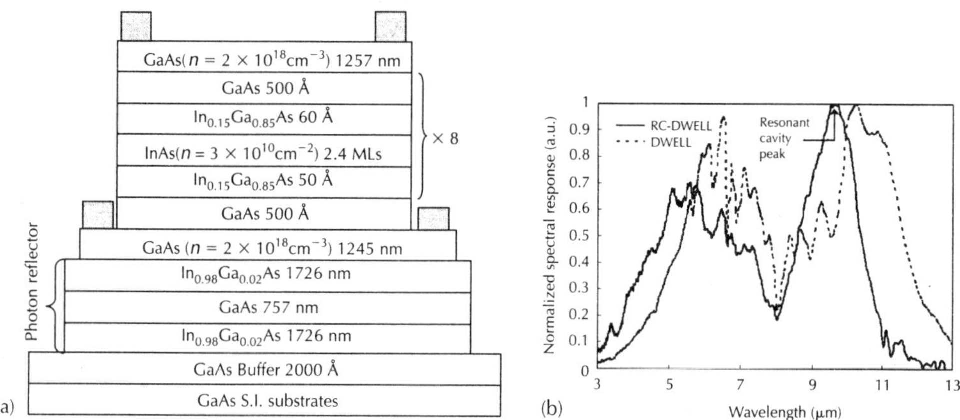

Figure 22.19   (a) Schematic structure of an RC-DWELL detector with eight-period QD active layers (InAs/ In$_{0.15}$Ga$_{0.85}$As). (b) A comparison of the spectral responsivity of an RC-DWELL detector and a standard DWELL

22.5   Present performance capabilities of QDIPs

A summary of specifications for several recently reported QDIPs is given in Table 22.2. Pal *et al.* [103] reported a super-lattice QDIP with 50-period InAs/GaAs QD active layers and an AlGaAs current-blocking layer to suppress the dark current. The spectral responsivity and detectivity of this detector is shown in Fig. 22.20a and b. Gunapala *et al.* [11] have demonstrated up to 1 A/W response at 77 K for a DWELL detector coupled with a grating. A two-colour DWELL detector with response peaks at 3.2 and $4.1\,\mu m$ operating up to room temperature was demonstrated by Lim *et al.* [22]. Several other QDIPs [75, 104] operating at high temperatures are also listed in the table. A historical analysis of the QDIP development shows rapid progression towards high-performance detectors capable of room temperature operation.

In a recent investigation by Jayaweera *et al.* [105], a new concept for developing photon detectors, based on the displacement currents in a QD-embedded dielectric media, operating at room temperature, has been reported. A preliminary detector structure along with its photoresponse results were also reported. The detector structure consists of PbS QDs embedded in a dielectric thin film, which is sandwiched between two conducting glass contacts. The dielectric thin film was made of paraffin wax, which has a high dielectric constant. The incident photons are absorbed in QDs, generating an electron–hole pair. When an electric field is applied across the contacts, the electrons drift towards the positive contact, while holes drift towards the negative contact. Since carrier transport through the dielectric medium is not possible, there will be a charge separation in the medium. This polarizability with photon absorption, which changes the capacitance of the device, is the key detection mechanism in the structure proposed by Jayaweera *et al.* QDs in a dielectric medium have shown several orders of magnitude higher polarizability than atoms and molecules. The incident radiation is modulated and the photogenerated displacement current is measured under an applied bias across the contacts. The displacement current

Table 22.2   A summary of recently reported QDIPs along with their figures of merit. λ_P, λ_0, R_p, T, η, g, D^*, and T_{max} stand for peak wavelength, zero response wavelength threshold, peak responsivity, operating temperature, quantum efficiency, photoconductive gain, detectivity, and maximum reported operating temperature, respectively.

Ref.	Description	λ_p (μm)	λ_0 (μm)	R_p (A/W) @ T (K)	$\eta \times g$ (%)	D^* (Jones) @ T (K)	T_{max} (K)
[10]	DWELL	~10	~12	~0.67 @ 120	8.3	2.6×10^{10} @ 78	240
[11]	DWELL	8.1	~10	~1 @ 77	15	1×10^{10} @ 77	–
[22]	DWELL	4.1	~4.75	~0.67 @ 120	20	2.8×10^{11} @ 120	300
[42]	DWELL	10.5	~13	0.125 @ 80	1.5	1.5×10^{9} @ 80	80
[43]	T-QDIP	5.7	~8	0.75 @ 80	16	2.4×10^{10} @ 80	300
[75]	70 InAs/GaAs QDs	~3.5	~6	~0.12 @ 100	4	1×10^{11} @ 100	150
[103]	50 InAs/GaAs QDs	4.9,5.8	~8	4.3 @ 100	91	2×10^{9} @ 100	220
[104]	DWELL	9.3	~10	0.7 @ 78	9	3×10^{11} @ 78	100

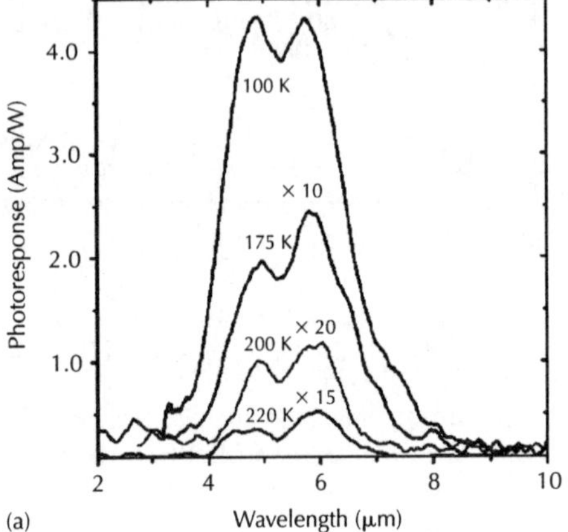

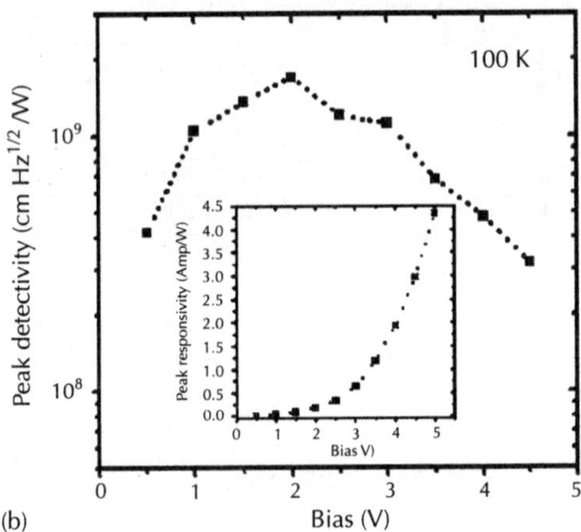

Figure 22.20   (a) Spectral responsivity of a QDIP with a current-blocking layer composed of an $Al_{0.3}Ga_{0.7}As$ layer. This enabled operation up to 220 K. (b) Variation of the peak detectivity with temperature. The photoconductive gain as a function of bias at 100 K is shown in the inset. After [103].

is a direct measurement of the incident radiation intensity and does not contain any contribution from the dark current. Hence, this approach is suitable for developing room temperature detectors. At 300 K and 40 V bias, the responsitivity and the detectivity at the peak absorption ($\lambda_P = 540$ nm) were found to be 195 V/W and 3×10^8 cmHz$^{1/2}$/W, respectively. As an advantage, a sophisticated growth and fabrication techniques are not required for this type of detector development. However, in this approach the incident light has to be modulated in order to obtained a photoresponse. A closely related device structure, which operates in photocurrent mode, has been reported by Konstantatos *et al.* [106] The device structure consists of spin-coating colloidal PbS QDs fabricated onto interdigitated gold electrodes. A detectivity in the order of 10^{13} cmHz$^{1/2}$/W has been reported at 1.3 μm at room temperature. A high photoconductive gain (~10 000) and consequently a high responsivity of 10^3 A/W were observed at room temperature.

22.6   Quantum dot focal plane arrays (FPAs)

The development of FPAs using successful single-element detector structures has remarkably increased potential applications in IR imaging. Multi-colour FPAs would lead to additional detection capabilities such as the construction of a true thermal map of a target. FPAs operating in atmospheric windows (3–5 and 8–14 μm) are required for a number of applications including night vision cameras and missile tracking, whereas FPAs operating in the FIR region are particularly suited for applications in astronomy and space situational awareness [38]. Most high-performance FPAs developed so far have been based on InSb, MCT and QWIPs; however, QD-based FPAs have also attracted the attention of the infrared community due to the development of successful single element QDIPs [22, 103, 107]. Krishna *et al.* [10] have reported a 320×256 pixel two-colour FPA using an InAs/InGaAs DWELL detector. According to Krishna *et al.* one of the major drawbacks in developing QD-based FPAs is the growth of thick active layers, providing sufficient absorption of radiation, without causing misfit dislocations. The growth of self-assembled QDs needs sufficient strain in the QD regions, reducing the possibility of growing thick QD layers. The two-colour DWELL detector, which was used for the FPA, showed wavelength selectivity with applied bias, where a $5.5\,\mu$m peak was enabled at low bias and 8–$10\,\mu$m peak was activated by high biases. This has a peak responsivity of 1 A/W and D* of 2–7×10^{10}cmHz$^{1/2}$/W. Thermal imaging using the FPA was demonstrated at 80 K with different optical filters (3–5 and 8–$12\,\mu$m). Moreover, the operability of the FPA was reported to be greater than 99%, and the noise equivalent temperature difference (NEΔT) was estimated to be less than 100 mK for $f/1$ (for 3–$5\,\mu$m) and $f/2$ (5–$9\,\mu$m) optics.

Gunapala *et al.* [11] have demonstrated a 640×512 pixel QD-based IR imaging FPA operating in the LWIR region. The detector was modelled on a DWELL structure, where InGaAs QDs were placed in a GaAs/AlGaAs QW. The 30-stack DWELL detector has shown a peak absorption quantum efficiency of 2.7%. Under normal incidence configuration, the detector had a much stronger (almost one order of magnitude) responsivity compared to the responsivity of a typical QWIP under $45°$ incidence configuration. In addition, it was found that the $45°$ incidence responsivity of the detector (~ 1 A/W) was 4–5 times stronger than the normal incidence responsivity, as shown in Fig. 22.21a. DWELL detectors can be fabricated with a reflection grating to couple normal incidence light to the z-polarization-sensitive absorption mechanism in the DWELL structure. A DWELL detector (reported by Gunapala *et al.* [11]) fabricated with a reflecting grating showed a factor of 3–4 enhancement in the responsivity, as shown in Fig. 22.21b. Peak detectivity of 1×10^{10}cmHz$^{1/2}$/W at $8.1\,\mu$m and 77 K was reported. Using this detector structure, a 640×512 pixel FPA has been fabricated. Additionally, an LWIR imagery system with a noise-equivalent temperature difference of 40 mK at 60 K showing a high uniformity was constructed. A 3-inch GaAs wafer with 12 640×512 pixels QDIP FPAs is shown in Fig. 22.22a, while an image of a human target taken with the 640×512 pixels FPA camera is shown in Fig. 22.22b.

In a recent publication, Razeghi group demonstrated [12] a 320×256 pixel MIR focal plane array operating at temperatures up to 200 K. The FPA was based on a DWELL structure [22], which consists of InAs QDs imbedded in a InGaAs/InAlAs QW. It was reported that the detector has two-colour characteristics up to room temperature with response peaks at 3.2 and $4.1\,\mu$m, as shown in Fig. 22.23. The photoresponse peak at $4.1\,\mu$m originated from bound-to-bound intersubband transitions between delocalized mixed states in the QD and QW unlike

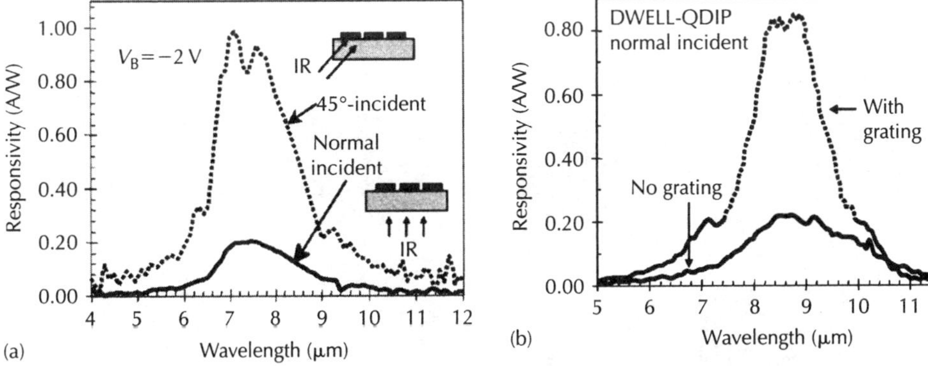

Figure 22.21 (a) A comparison between the spectral responsivity of a DWELL detector measured under normal incidence and $45°$ incidence configurations. (b) Normal incidence spectral responsivity with and without reflection gratings. After [11].

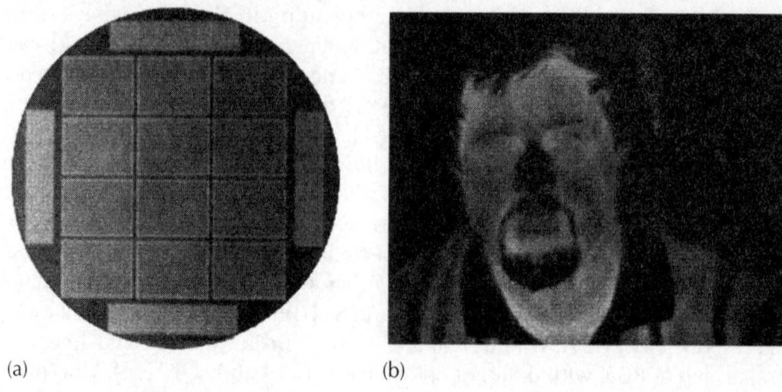

Figure 22.22   (a) A 3-inch. GaAs wafer with 12 640 × 512 pixels QDIP FPAs. (b) An image of a human target taken with the 640 × 512 pixels QDIP FPA camera. After [11].

the bound-to-bound transitions between pure QD states and QW states for DWELL structures reported in section 22.2. The detector had a peak responsivity of 34 mA/W and maximum D^* of $2.8 \times 10^{11}\,\text{cmHz}^{1/2}/\text{W}$ at 120 K, while a D^* of $6 \times 10^7\,\text{cmHz}^{1/2}/\text{W}$ was obtained at 300 K. The DWELL FPA showed imaging capabilities at 120 and 200 K. A conversion efficiency of 1.1% and a noise equivalent temperature difference of 344 mK at 120 K were observed. The FPA camera is capable of imaging human targets up to 150 K. Images of a human target and a hot soldering iron taken with the FPA camera at 130 and 200 K, respectively, are shown in Fig. 22.24. Based

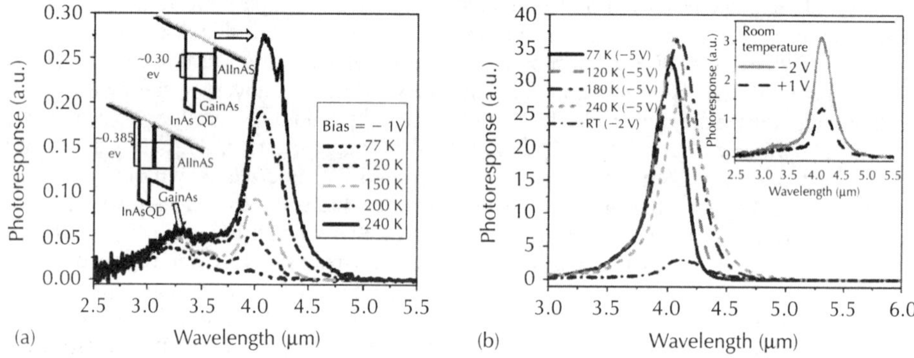

(a)   Wavelength (μm)   (b)   Wavelength (μm)

Figure 22.23   (a) Spectral responsivity of a DWELL detector at different temperatures under a (a) − 1 V bias, and (b) − 5 V bias (response at room temperature (RT) was obtained under a − 2 V bias). The variation of the response at room temperature is shown in the inset. After [22].

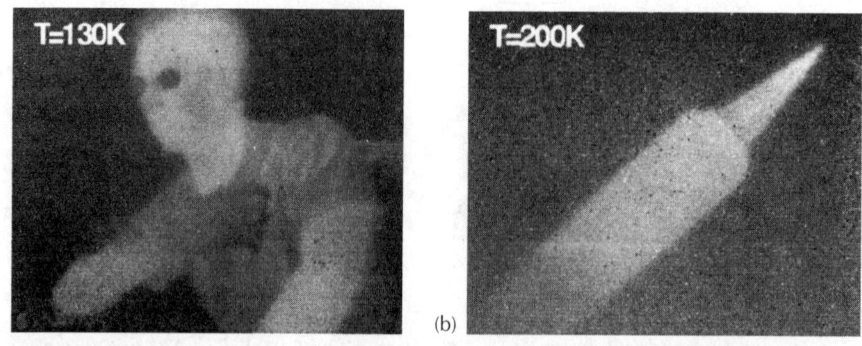

Figure 22.24   Images taken with a 320 × 256 pixel QDIP FPA camera at (a) 130 and (b) 200 K. The maximum operating temperature of the array is 200 K. After [12].

on these results, it can be concluded that the development of QD-based FPAs is feasible, even though the present QDIP FPAs show lower overall performance as compared to MCT and QWIP FPAs. A summary of QDIP FPA specifications is given in Table 22.3. Major improvements in QDIP FPAs include: high responsivity and low dark current at high temperatures, improved growth quality, and uniformity in detector material for large wafer sizes and device processing.

Table 22.3   QDIP FPA specifications. Peak wavelength is given in the response range column, when the range is not stated clearly.

Ref.	Array size	Cell pitch	Response range @ T	NEΔT @ $f\#$	Pixel operability
[10]	320×256	$30\,\mu m$	3–$5\,\mu m$ @ $80\,K$	$100\,mK$ @ 1	$>99\%$
			8–$12\,\mu m$ @ $80\,K$	$100\,mK$ @ 2	
[11]	640×512	$25\,\mu m$	$8.1\,\mu m$ @ $60\,K$	$40\,mK$ @ 2	$>99\%$
[12]	320×256	$30\,\mu m$	$4\,\mu m$ @ $120\,K$	$344\,mK$ @ 2	99%

22.7   Conclusion

The electronic transitions between energy states in the QD and QW give rise to multi-colour response in DWELL detector structures. Response peaks experimentally obtained from each detector correspond to the energy spacings calculated by theoretical models. By changing the well width and the size of the QD, detectors can be designed to operate at different wavelength regions depending on the application. The operating wavelength can be selected by varying the applied bias. Advantages of DWELL detectors over typical n-type QWIPs include their ability to operate under normal incidence and at high temperatures particularly in the VLWIR/FIR region. In a different approach, T-QDIPs designed for room temperature operation and terahertz detection were reported. As evident from the results, T-QDIPs exhibit lower dark current, and higher operating temperatures compared to typical QDIPs, which were made possible by the incorporation of double barriers into the structure. A $17\,\mu m$ T-QDIP that can operate at room temperature, and a 6 terahertz detector operating at $150\,K$, were demonstrated. T-QDIPs are photon detectors, which are inherently fast and do not use slower thermalization processes, which are the primary detection mechanism in thermal detectors such as bolometer or TGS. In comparison with the Si BIB detectors, the main advantage will be in the increase in operating temperature to $77\,K$ or higher, which will greatly reduce the cooling requirements. This would be an important feature for space-based applications where moving from $20\,K$ to higher temperatures is a major design advantage. Furthermore, T-QDIP structures open up a wide range of possible modifications and designing options to aid in wavelength tuning and performance optimization. Promising results have been observed for QD-based FPAs, even though the overall performance is low compared to MCT and QWIP FPAs.

Acknowledgements

This work is supported in part by the National Science Foundation under Grant ECCS: 0620688. The authors acknowledge the contributions of Prof. P. Bhattacharya and his group at the University of Michigan and Prof. S. Krishna and his group at the University of New Mexico, and Ms Laura Byrum for her technical assistance in preparing this article.

References

1. A. Rogalski, *Prog. Quant. Electron.* **27**, 59 (2003).
2. J. Urayama, T.B. Norris, J. Singh, and P. Bhattacharya, *Phys. Rev. Lett.* **86**, 4930 (2001).

3. E. Kim, A. Madhukar, Z. Ye, and J.C. Campbell, *Appl. Phys. Lett.* **84**, 3277 (2004).

4. B. Aslan, H.C. Liu, M. Korkusinski, S.J. Cheng, and P. Hawrylak, *Appl. Phys. Lett.* **82**, 639 (2003).

5. B. Kochman, A.D. Stiff-Roberts, S. Chakrabarti, J.D. Phillips, S. Krishna, J. Singh, and P. Bhattacharya, *IEEE. J. Quant. Electron.* **39**, 459–467, (2003).

6. H.C. Liu, M. Gao, J. McCaffrey, Z.R. Wasilewski, and S. Fafard, *Appl. Phys. Lett.* **78**, 79–81, (2001).

7. L. Jiang, S.S. Li, N.-T. Yeh, J.-I. Chyi, C.E. Ross, and K.S. Jones, *Appl. Phys. Lett.* **82**, 1986–1988, (2003).

8. S. Raghavan, P. Rotella, A. Stintz, B. Fuchs, S. Krishna, C. Morath, D.A. Cardimona, and S.W. Kennerly, *Appl. Phys. Lett.* **81**, 1369–1371, (2002).

9. S. Krishna, G. von Winckel, S. Raghavan, A. Stintz, G. Ariyawansa, S.G. Matsik, and A.G.U. Perera, *Appl. Phys. Lett.* **83**, 2745 (2003).

10. S. Krishna, D. Forman, S. Annamalai, P. Dowd, P. Varangis, T. Tumolillo, Jr, A. Gray, J. Zilko, K. Sun, M. Liu, J. Campbell, and D. Carothers, *Appl. Phys. Lett.* **86**, 193501 (2005).

11. S.D. Gunapala, S.V. Bandara, C.J. Hill, D.Z. Ting, J.K. Liu, S.B. Rafol, E.R. Blazejewski, J.M. Mumolo, S.A. Keo, S. Krishna, Y.-C. Chang, and C.A. Shott, *Infrared Physics & Technology* **50**, 149 (2007).

12. S. Tsao, H. Lim, W. Zhang, and M. Razeghi, *Appl. Phys. Lett.* **90**, 201109 (2007).

13. V. Ryzhii, *Semicond. Sci. Technol.* **11**, 759 (1996).

14. R. Leon, G.M. Swift, B. Magness, W.A. Taylor, Y.S. Tang, K.L. Wang, P. Dowd, and Y.-H. Zhang, *Appl. Phys. Lett.* **76**, 2074 (2000).

15. B. Aslan, H.C. Liu, M. Korkusinski, S.-J. Cheng, and P. Hawrylak, *Appl. Phys. Lett.* **82**, 630 (2002).

16. D. Pal, L. Chen, and E. Towe, *Appl. Phys. Lett.* **82**, 4634 (2003).

17. A. Sergeev, V. Mitin, and M. Stroscio, *Physica. B* **316**, 369 (2002).

18. Z. Ye, J.C. Campbell, Z. Chen, E.T. Kim, and A. Mad-hukar, *J. Appl. Phys.* **92**, 7462 (2002).

19. J. Phillips, K. Kamath, and P. Bhattacharya, *Appl. Phys. Lett.* **72**, 2020 (1998).

20. S. Maimon, E. Finkman, and G. Bahir, *Appl. Phys. Lett.* **73**, 2003 (1998).

21. D. Pan, E. Towe, and S. Kennerly, *Appl. Phys. Lett.* **73**, 1937 (1998).

22. H. Lim, S. Tsao, W. Zhang, and M. Razeghi, *Appl. Phys. Lett.* **90**, 131112 (2007).

23. M.R. Matthews, R.J. Steed, M.D. Frogley, C.C. Phillips, R.S. Attaluri, and S. Krishna, *Appl. Phys. Lett.* **90**, 103519 (2007).

24. C.-H. Lin, C.-Y. Yu, C.-Y. Peng, W.S. Ho, and C.W. Liu, *J. Appl. Phys.* **101**, 033117 (2007).

25. S. Tong, H.-J. Kim, and K.L. Wang, *Appl. Phys. Lett.* **87**, 081104 (2005).

26. D. Cha, M. Ogawa, C. Chen, S. Kim, J. Lee, K.L. Wang, J. Wang, and T.P. Russell, *J. Cryst. Growth* **833**, 301–302 (2007).

27. A. Vardi, N. Akopian, G. Bahir, L. Doyennette, M. Tchernycheva, L. Nevou, F.H. Julien, F. Guillot, and E. Monroy, *Appl. Phys. Lett.* **88**, 143101 (2006).

28. A. Babinski, G. Ortner, S. Raymond, M. Potemski, M. Bayer, W. Sheng, P. Hawrylak, Z. Wasilewski, S. Fafard, and A. Forchel, *Phys. Rev. B* **74**, 075310 (2006).

29. R.J.A. Hill, A. Patané, P.C. Main, L. Eaves, B. Gustafson, M. Henini, S. Tarucha, and D.G. Austing, *Appl. Phys. Lett.* **79**, 3275 (2001).

30. D. Loss and D.P. DiVincenzo, *Phys. Rev. A* **57**, 120 (1998).

31. A. Imamoglu, D.D. Awschalom, G. Burkard, D.P. DiVincenzo, D. Loss, M. Sherwin, and A. Small, *Phys. Rev. Lett.* **83**, 4204 (1999).

32. M.N. Leuenberger, M.E. Flatt, and D.D. Awschalom, *Phys. Rev. Lett.* **94**, 107401 (2005).

33. V.V. Dobrovitski, J.M. Taylor, and M.D. Lukin, *Phys. Rev. B* **73**, 245318 (2006).

34. D.A. Fay, A.M. Waxman, M. Aguilar, D.B. Ireland, J.P. Racamato, W.D. Ross, W.W. Streilein, and M.I. Braun, *IEEE Proceedings of the Third International Conference on Information Fusion 2000* **1** (2000).

35. D.A. Fay, P. Ilardi, N. Sheldon, D. Grau, R. Biehl, and A.M. Waxman, *Proc. SPIE* **5802**, 154–165 (2005).

36. A. Goldberg, P.N. Uppal, and M. Winn, *Infrared Physics & Technology* **44**, 427 (2003).

37. D.B. Law, E.M. Carapezza, C.J. Csanadi, G.D. Edwards, T.M. Hintz, and R.M. Tong, *SPIE* **2938**, 288 (1997).

38. P.M. Alsing, D.A. Cardimona, D.H. Huang, T. Apostolova, W.R. Glass, and C.D. Castillo, *Infrared Physics & Technology* **50**, 89 (2007).

39. F. Neele, *Proc. SPIE* **5787**, 134 (2005).

40. C.J. Chen, K.K. Choi, W.H. Chang, and D.C. Tsui, *Appl. Phys. Lett.* **72**, 7 (1998).

41. F. Durante, P. Alves, G. Karunasiri, N. Hanson, M. Byloos, H.C. Liu, A. Bezinger, and M. Buchanan, *Infrared Physics & Technology* **50**, 182 (2007).

42. G. Ariyawansa, A.G.U. Perera, G.S. Raghavan, G. von Winckel, A. Stintz, and S. Krishna, *IEEE Photon. Technol. Lett.* **17**, 1064–1066 (2005).

43. P. Bhattacharya, X.H. Su, S. Chakrabarti, G. Ariyawansa, and A.G.U. Perera, *Appl. Phys. Lett.* **86**, 191106 (2005).

44. H.C. Liu, P.H. Wilson, M. Lamm, A.G. Steele, Z.R. Wasilewski, J. Li, M. Buchanan, and J.G. Simmonsa, *Appl. Phys. Lett.* **64**, 475 (1994).

45. M.Z. Tidrow, K.K. Choi, A.J. DeAnni, W.H. Chang, and S.P. Svensson, *Appl. Phys. Lett.* **67**, 1800 (1995).

46. H.C. Liu, C.Y. Song, A. Shen, M. Gao, Z.R. Wasilewski, and M. Buchanan, *Appl. Phys. Lett.* **77**, 2437 (2000).

47. E. Dupont, M. Gao, Z. Wasilewski, and H.C. Liu, *Appl. Phys. Lett.* **78**, 2067 (2001).

48. J. Li, K.K. Choi, and D.C. Tsui, *Appl. Phys. Lett.* **86**, 211114 (2005).

49. A. Majumdar, K.K. Choi, J.L. Reno, and D.C. Tsui, *Appl. Phys. Lett.* **86**, 261110 (2005).

50. S.V. Bandara, S.D. Gunapala *et al.*, *Appl. Phys. Lett.* **86**, 151104 (2005).

51. M.P. Touse, G. Karunasiri, K.R. Lantz, H. Li, and T. Mei, *Appl. Phys. Lett.* **86**, 093501-1 (2005).

52. H. Schneider, T. Maier, J. Fleissner, M. Walther, P. Koidl, G. Weimann, W. Cabanski, M. Finck, P. Menger, W. Rode, and J. Ziegler, *Infrared Physics & Technology* **50**, 53 (2007).

53. R.D. Rajavel, D.M. Jamba, J.E. Jensen, O.K. Wu, J.A. Wilson, J.L. Johnson, E.A. Patten, K. Kasai, P.M. Goetz, and S.M. Johnson, *J. Electron. Mat.* **27**, 747751 (1998).

54. J.P. Zanatta, P. Ferret, R. Loyer, G. Petroz, S. Cremer, J.P. Chamonal, P. Bouchut, A. Million, and G. Destefanis, *Proc. SPIE* **4130**, 441 (2000).

55. W.E. Tennant, M. Thomas, L.J. Kozlowski, W.V. McLevige, D.D. Edwall, M. Zandian, K. Spariosu, G. Hildebrandt, V. Gil, P. Ely, M. Muzilla, A. Stoltz, and J.H. Dinan, *J. Electron. Mat.* **30**, 590594 (2001).

56. R. Breiter, W.A. Cabanski, K.-H. Mauk, W. Rode, and J. Ziegler, *Proc. SPIE* **4369**, 579–587 (2001).

57. A.I. D'Souza, L.C. Dawson, D.J. Berger, A.D. Markum, J.Bajaj, W.E. Tennant, J.M. Arias, L.J. Kozlowski, K. Vural, and P.S. Wijewarnasuriya, *Proceedings of SPIE* 3498, 192–202 (1998).

58. A.K. Dutta, N.K. Dhar, P.S. Wijewarnasuriya, M. Saif Islam, *Proc. SPIE* **6008**, 600812 (2005).

59. G. Ariyawansa, M.B.M. Rinzan, D.G. Esaev, S.G. Matsik, A.G.U. Perera, H.C. Liu, B.N. Zvonkov, and V.I. Gavrilenko, *Appl. Phys. Lett.* **86**, 143510–143513 (2005).

60. G. Ariyawansa, M.B.M. Rinzan, S.G. Matsik, G. Hastings, A.G.U. Perera, H.C. Liu, M. Buchanan, G.I. Sproule, V.I. Gavrilenko, and V.P. Kuznetsov, *Appl. Phys. Lett.* **89**, 061112 (2006).

61. G. Ariyawansa, M.B.M. Rinzan, M. Alevli, M. Strassburg, N. Dietz, A.G.U. Perera, S.G. Matsik, A. Asghar, I.T. Ferguson, H. Luo, A. Bezinger, and H.C. Liu, *Appl. Phys. Lett.* **89**, 091113 (2006).

62. A. Amtout, S. Raghavan, P. Rotella, G. von Winckel, A. Stintz, and S. Krishna, *J. Appl. Phys.* **96**, 3782–3786, (2004).

63. S. Raghavan, D. Forman, P. Hill, N.R. Weisse-Bernstein, G. von Winckel, P. Rotella, S. Krishna, S.W. Kennerly, and J.W. Little, *Appl. Phys. Lett.* **96**, 1036 (2004).

64. X. Han, J. Li, J. Wu, G. Cong, X. Liu, Q. Zhu, and Z. Wang, *J. Appl. Phys.* **98**, 053703 (2005).

65. R.S. Attaluri, S. Annamalai, K.T. Posani, A. Stintz, and S. Krishna, *Appl. Phys. Lett.* **99**, 083105 (2006).

66. S. Krishna, S. Raghavan, G. von Winckel, P. Rotella, A. Stintz, D. Le, C. Morath, and S.W. Kennerly, *Appl. Phys. Lett.* **82**, 2574 (2003).

67. B.F. Levine, *J. Appl. Phys.* **74**, R1 (1993).

68. D.M.-T. Kuo and Y.-C. Chang, *Phys. Rev. B* **67**, 035313-1 (2003).

69. V. Apalkov, *J. Appl. Phys.* **100**, 076101 (2006).

70. H. Jiang and J. Singh, *Phys. Rev. B* **56**, 4696 (1998).

71. A.G.U. Perera, W.Z. Shen, S.G. Matsik, H.C. Liu, M. Buchanan, and W.J. Schaff, *Appl. Phys. Lett.* **72**, 1596–1598 (1998).

72. B. Aslan, H.C. Liu, J.A. Gupta, Z.R. Wasiewski, G.C. Aers, S. Raymond, and M. Buchanan, *Appl. Phys. Lett.* **88**, 043103 (2006).

73. X.H. Su, J. Yang, P. Bhattacharya, G. Ariyawansa, and A.G.U. Perera, *Appl. Phys. Lett.* **89**, 031117 (2006).

74. X. Su, S. Chakrabarti, P. Bhattacharya, G. Ariyawansa, and A.G.U. Perera, *IEEE J. Quantum Electron.* **41**, 974 (2005).

75. S. Chakrabarti, A.D. Stiff-Roberts, X. Su, P. Bhattacharya, G. Ariyawansa, and A.G.U. Perera, *J. Phys. D. Appl. Phys.* **38**, 2135–2141 (2005).

76. D.G. Esaev, S.G. Matsik, M.B.M. Rinzan, A.G.U. Perera, H.C. Liu, and M. Buchanan, *J. Appl. Phys.* **93**, 1879 (2003).

77. M.S. Unlu and S. Strite, *J. Appl. Phys.* **78**, 607 (1995).

78. R.S. Attaluri, J. Shao, K.T. Posani, S.J. Lee, J.S. Brown, A. Stintz, and S. Krishna, *J. Vac. Sci. Technol. B* **25**(4), 1186 (2007).

79. F. Pulizzi, D. Walker, A. Patan, L. Eaves, M. Henini, D. Granados, J.M. Garcia, V.V. Rudenkov, P.C.M. Christianen, J.C. Maan, P. Offermans P.M. Koenraad, and G. Hill, *Phys. Rev. B* **72**, 085309 (2005).

80. E. Anemogiannis, N. Glytsis, and T.K. Gaylord, *IEEE. J. Quant. Electron.* **3**, 742 (1997).

81. K.K. Choi, B.F. Levine, R.J. Malik, J. Walker, and C.G. Bethea, *Phys. Rev. B* **35**, 4172–4175 (1987).

82. K.K. Choi, B.F. Levine, C.G. Bethea, J. Walker, and R.J. Malik, *Phys. Rev. Lett.* **59**, 4172 (1987).

83. A. Carbone and P. Mazzetti, *J. Appl. Phys.* **80**(3), 1559 (1996).

84. R. Rehm, H. Schneider, C. Schöonbein, and M. Walther, *Physica E* **7**, 124 (2000).

85. Y. Paltiel, N. Snapi, A. Zussman, and G. Jung, *Appl. Phys. Lett.* **87**, 231103 (2005).

86. A.D. Stiff, S. Krishna, P. Bhattacharya, and S.W. Kennerly, *IEEE J. Quantum Electron.* **37**, 1412 (2001).

87. Y. Haddab, A.P. Friedrich, and R.S. Popovic, *Solid-State Electron.* **43**, 413–416 (1999).

88. R. Rupani, S. Ghosh, X. Su, and P. Bhattacharya, *Microelectron. J.* **39**, 307 (2008).

89. S. Schulz, A. Schramm, C. Heyn, and W. Hansen, *Phys. Rev. B* **74**, 033311 (2006).

90. R. Wetzler, A. Wacker, E. Schöll, C.M.A. Kapteyn, R. Heitz, and D. Bimberg, *Appl. Phys. Lett.* **77**, 1671 (2000).

91. K.H. Schmidt, C. Bock, M. Versen, U. Kunze, D. Reuter, and A.D. Wieck, *Appl. Phys. Lett.* **95**, 5715 (2004).

92. S. Anand, N. Carlsson, M.-E. Pistol, L. Samuelson, and W. Seifert, *Appl. Phys. Lett.* **95**, 3016–3018 (1995).

93. C.M.A. Kapteyn, F. Heinrichsdorff, O. Stier, R. Heitz, M. Grundmann, N.D. Za-kharov, and D. Bimberg, *Phys. Rev. B* **60**, 14265–14268 (1999).

94. A.G.U. Perera, W.Z. Shen, M. Ershov, H.C. Liu, M. Buchanan, and W.J. Schaff, *Appl. Phys. Lett.* **74**, 3167 (1999).

95. M. Ershov, H.C. Liu, L. Li, M. Buchanan, Z.R. Wasilewski, and A.K. Jonscher, *IEEE Trans. Electron. Devices* **45**, 2196 (1998).

96. E.E. Haller, *Infrared Phys.* **35** (1994).

97. M. Suzuki and M. Tonouchi, *Appl. Phys. Lett.* **86**, 163504–3 (2005).

98. H.C. Liu, C.Y. Song, A.J. Spring Thorpe, and J.C. Cao, *Appl. Phys. Lett.* **84**, 4068 (2004).

99. D.G. Esaev, M.B.M. Rinzan, S.G. Matsik, A.G.U. Perera, H.C. Liu, B.N. Zhonkov, V.I. Gavrilenko, and A.A. Belyanin, *J. Appl. Phys.* **95**, 512 (2004).

100. M.B.M. Rinzan, A.G.U. Perera, S.G. Matsik, H.C. Liu, Z.R. Wasilewski, and M. Buchanan, *Appl. Phys. Lett.* **86**, 071112 (2005).

101. H. Luo, H.C. Liu, C.Y. Song, and Z.R. Wasilewski, *Appl. Phys. Lett.* **86**, 231103 (2005).

102. K.T. Posani, V. Tripathi, S. Annamalai, N.R. Weisse-Bernstein, S. Krishna, R. Perahia, O. Crisafulli, and O.J. Painter, *Appl. Phys. Lett.* **88**, 151104 (2006).

103. D. Pal and E. Towe, *Appl. Phys. Lett.* **88**, 153109 (2006).

104. E.-T. Kim and A. Madhukar, *Appl. Phys. Lett.* **84**, 3277 (2004).

105. P.V.V. Jayaweera, A.G.U. Perera, and K. Tennakone, *Appl. Phys. Lett.* **91**, 063114 (2007).

106. G. Konstantatos, I. Howard, A. Fischer, S. Hoogland, J. Clifford, E. Klem, L. Levina, and E.H. Sargent, *Nature Lett.* **442**, 180 (2006).

107. D. Pal and E. Towe, *Appl. Phys. Lett.* **100**, 084322 (2006).

CHAPTER 23

Optically Driven Schemes for Quantum Computation Based on Self-assembled Quantum dots

I. D'Amico[1] and B.W. Lovett[2]

[1]*Department of Physics, University of York, York YO10 5DD, United Kingdom*
[2]*Department of Materials, University of Oxford, OX1 3PH, United Kingdom*

23.1 Introduction and fundamentals of quantum information processing

Starting in the late 1990s, an increasing research effort has been devoted to the idea and practical implementation of the "quantum computer" [1]. While before, the quantum computer was only a theoretical speculation put forward by Feynman and Deutsch [1], in the late 1990s micro- and nanotechnology started to develop further, boosted by practical necessities such as the need for miniaturization and the speed-up of "classical" computers.[1] This improvement follows Moore's law, which states that the number of transistors per unit area on integrated circuits doubles-approximately every 18 months. Miniaturization and the related improvement of experimental techniques have made dynamical control of a few particle quantum systems a near reality, and with it ideas that the quantum computer could be transformed into actual devices.

Computation is a physical process and information has to be stored in a physical system and manipulated using physical interactions. The smallest logical unit in classical computers is the bit, which corresponds to a physical system characterized by two distinct states, labelled "0" and "1". We can think of them, for example, as the state of a capacitor (charged or discharged), or as a voltage pulse below or above a prefixed threshold. In order to speed up calculations, increase storage capacity and miniaturize computers, we can reduce the size of the physical system representing the bit, but what can we do once such a system is reduced to a single electron? Moore's law, which has ruled for the last 43 years, predicts this will happen in about ten years. Quantum computation provides one possible way forward, since it has been shown theoretically that such devices can substantially speed up the solution of problems such as factorization of large numbers and database search, and allow for the simulation of complex many-particle quantum systems.

The logical unit of a quantum computer is called the "qubit", or quantum bit. The qubit is a two-level *quantum* system, with states called $|0\rangle$ and $|1\rangle$. Qubits can be in principle realized by any two-level quantum system; for example, the horizontal or vertical polarization of a photon or the up and down states of an electronic spin. The general state of a qubit is $a_1|0\rangle + a_2|1\rangle$, a_1 and a_2 being complex numbers. In contrast with the classical bit, the general state of a qubit is not either 0 or 1, but a *superposition* of the two. This superposition property provides quantum computation with natural parallelism and it allows the possibility of processing exponentially more information than a classical computer of corresponding size. One qubit, as a superposition of two states, is represented by 2^1 complex numbers (a_1 and a_2). A general two-qubit state is written $a_1|00\rangle + a_2|01\rangle + a_3|10\rangle + a_4|11\rangle$; it is a superposition of 2^2 states, and is represented by 2^2 numbers. It is straightforward to extend this argument to N qubits: a general state is represented by 2^N numbers and by careful design of algorithms and careful control of the system's *quantum dynamics*, operations can be performed on each of these numbers. This property is known as *quantum parallelism* and provides a quantum computer with a significant advantage over its classical counterpart in certain situations.

23.1.1   Basic requirements for implementing quantum computation

A quantum computer is a physical system in which we can identify one or more collections of well-defined qubits. We refer to each of these collections as a *quantum register*. As an example, an array of well-defined self-assembled quantum dots (QDs), as depicted in Fig. 23.1, is a potential quantum register. In order to perform actual computation, we must be able to address the qubits in the quantum register and perform specific logic gates. In analogy to classical computation, to perform universal quantum computation we must be able to drive a minimum, universal, set of logic gates. A typical choice is composed of the set of *all possible* single-qubit gates and one non-trivial (or entangling) two-qubit gate, e.g. a controlled-NOT (CNOT) or a controlled-PHASE (CPHASE) gate. These are *quantum gates* which modify, in a controlled way, the state of one or more two-level systems.

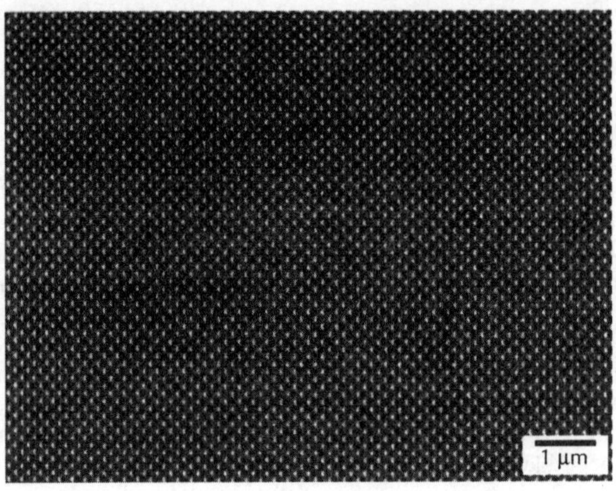

Figure 23.1   AFM image of a 2D regular array of InAs/GaAs self-assembled QDs. (From [11], Fig. 12b, courtesy of Oliver G. Schmidt and with kind permission of Springer Science and Business media.)

Single-qubit gates are represented by the set of unitary operators $\hat{U}$ which perform an arbitrary operation on the general single-qubit state $\Psi = a\,|0\rangle + b\,|1\rangle$. They can be decomposed into the 2×2 Pauli matrices $\{\sigma_0,\ \sigma_x,\ \sigma_y,\ \sigma_z\}$. Two-qubit gates are represented by specific unitary 4×4 matrices, and examples will be given later in the chapter. Each *quantum algorithm* corresponds to a well-defined sequence of quantum gates. Finally, the results of the quantum computation must be *read out*, i.e. a measuring apparatus must be connected to the quantum system and must extract information from it. The readout may be performed in different ways, depending on the properties of the hardware and implementation scheme chosen. For example, a Stern–Gerlach-type device may be used for deflecting spin-based qubits [2] or quantum point contacts [3] may detect the variation of current due to the spatial displacement of charge qubits [4, 5]. The readout can be performed on the entire qubit register or on only selected qubits, depending on the overall computer architecture and available resources.

A general quantum computation scheme may be summarized as follows: (i) preparation of the initial state (input), (ii) controlled quantum evolution of the system according to the sequence of gates forming the quantum algorithm (actual quantum computation), and (iii) measurement of the final state (readout).

This overall prescription is straightforward, but there are several considerations to be made when designing a quantum computer. We highlight three here, as follows:

- Decoherence. Nature does not provide perfectly isolated two-level systems. The real systems interact in a non-predictable way with the surrounding environment and its fluctuations. In solid-state hardware these may constitute, for example, the crystal phonon modes or the presence of spurious charges. Additionally the qubit is often physically embodied in two of the many levels of a quantum system. Unwanted transitions to the additional levels may happen. All of this spoils the quantum computation, by arbitrarily modifying the state of the systems on which the quantum gates are performed. In fact, the situation is often even worse than this: the environment entangles itself with the quantum computer in such a way as to destroy its quantum coherence and the unitarity of its quantum evolution. This phenomenon is known as *decoherence* and it underpins the inherent fragility of quantum computation, when thought of as a sequence of *controlled* quantum evolutions of an ensemble of *selected* quantum levels.

- Scalability. To perform non-trivial calculations, a quantum computer should contain a minimum number of qubits. When considering quantum algorithms that provide an exponential speed-up over classical counterparts [6], the actual number can be as small as a few tens of qubits. However, building, addressing and controlling the quantum dynamics of several two-level quantum systems is very challenging. The possibility of doing so strongly depends on the chosen hardware, qubit encoding or gating scheme.

- Error correction. As for classical computers, it is important to be able to detect and correct errors during the computation itself. Various schemes have been proposed to achieve this [1], and usually each scheme detects and corrects specific classes of errors. One of the main difficulties in designing error correction codes for quantum computation is that the collapse of the wave-function following measurement destroys the quantum information stored. This implies that it is not possible to resort to straightforward classical codes to detect (and correct) faulty computation. In general, error correction codes require redundant information to be encoded and propagated during the computation. This translates into an increased demand of resources, such as number of qubits and number and type of gates to be performed.

Since there are many criteria that must be satisfied by a quantum computer, a lot of different physical implementations have been proposed, each with its own advantages and problems. No fundamental physical obstacles exist, but from the technological point of view, building a quantum computer is extremely demanding. Some example forerunner qubit embodiments are the polarization of single photons, nuclear spins in organic molecules, phosphorus impurities in silicon, charge carriers in superconductors and hyperfine levels of ions trapped by electromagnetic fields [1]. A very promising possibility is solid-state implementations, especially schemes based on semiconductor devices. Such approaches allow for the exploitation of techniques already developed for producing micro- and nanostructures, and of the detailed work that has already been done on their electronic properties. Using sophisticated optoelectronics methods, it has also been demonstrated that excitations in semiconductors can be generated and their dynamics driven using ultra-fast laser pulses. As we will see, this allows accurate generation and control of quantum states that can be used for quantum computation. Moreover, these dynamics can be driven on sub-picosecond timescales, which are much faster than the related decoherence times. Finally, the use of semiconductor-based hardware should facilitate the connection of quantum registers and processors to traditional electronics circuits, which is certain to be an essential component of any complete device.

23.2 Semiconductor self-assembled quantum dots as hardware

In this chapter we will discuss designs which use semiconductor self-assembled QDs as hardware in some detail [7]. QDs can form when two (or more) types of semiconductor form a heterostructure. As an example, consider a GaAs/InAs structure grown by molecular beam epitaxy (MBE).

In the first stage of processing, Ga and As atoms are evaporated slowly on to a GaAs substrate such that a perfectly flat surface results. Second, Ga is replaced by In in the growth process, so that InAs grows instead of the larger band gapped GaAs. Since the lattice constant of these two materials is different, the InAs layers clump together into small three-dimensional structures that minimize the combined energy of the surface, volume and strain in the structure. This small structure, which is typically 10–50 nm in the in-plane direction, and 2–10 nm high, forms the quantum dot. Subsequent capping by more GaAs ensures electron and hole confinement in all three dimensions. Figure 23.2 is a schematic drawing comparing the bulk energy level structure of a direct gap semiconductor with that of the same semiconductor in quantum dot form. Confinement leads to discretization of both electron and hole states – and for this reason QDs are sometimes called "artificial atoms" [7]. If another layer of InAs is grown after capping with GaAs, more QDs form. These lie preferentially above the first layer due to the effect of strain propagation through the capping layer, such that quantum dot molecules formed by stacked QDs can be made with a controlled spacing between the constituents [8]. It is also important to emphasize that the spin quantum numbers of each carrier type are preserved from the bulk structure (see Fig. 23.2).

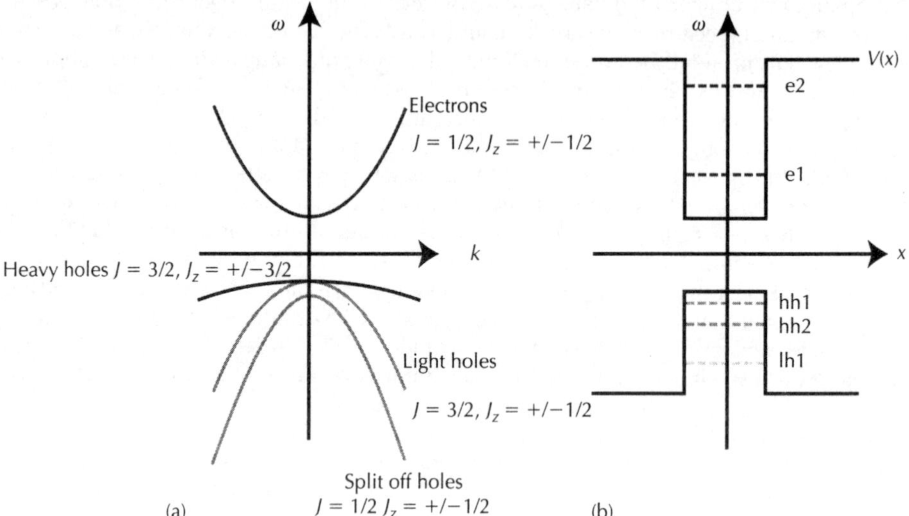

Figure 23.2   (a) Dispersion relation for a typical, direct gap, bulk semiconductor. ω is the energy and k the wavevector close to the centre of the Brillouin zone. Only the bands closest to the Fermi level are shown. (b) When a confining potential is applied, the bands split into discrete energy levels, but the spin quantum numbers are preserved. "hh" denotes heavy hole, "lh" light hole, and "e" electron.

Recent experimental developments have shown excellent control over the density, size, shape and position of QDs. Regular two- and three-dimensional arrays have been fabricated [9–11] (see, e.g., Fig. 23.1), as well as homogeneous in-plane QD chains [12]. With the use of laser pulses of specific length and frequency, selected interband transitions can be addressed and confined excitons can be controlled coherently [13–20]. Moreover, by altering the dimensions and shape of a QD, it is possible to engineer its electronic structure, and hence tune its excitation spectrum. The optical activity of QDs and photons make them ideal candidates for *hybrid* schemes in which the computation is carried out using solid-state qubits (e.g. excitons or electronic spins), while long distance communication between different processors in the computer or different computers can be carried out by transferring the quantum information to photons. All of these properties make QD-based structures very promising candidate quantum computers.

The following are the most obvious possible qubit embodiments in QDs:

- The *presence or absence* of an *exciton* (or bound state of an electron and hole).
- The *spin* of an *exciton*.

- The *spin* of an excess *electron*, added to the conduction band by doping, gating or similar means.
- The *spin* of an excess *hole*, added to the valence band by doping, gating or similar means.

The first and third items on this list are substantially more popular choices than the others, so we shall focus here on these two. Though there are several unifying features of the two representations, it turns out that the details of how to build a quantum computer using each have quite distinct features. We will therefore discuss each individually.

23.3   Exciton representation

23.3.1   Single-qubit manipulation

Consider first a single quantum dot representing a single qubit. In order to manipulate this qubit, we use a *pulsed laser*. The state of a single qubit can be represented on the surface of a sphere (the "Bloch sphere"), with the poles of the sphere representing the $|0\rangle$ and $|1\rangle$, and the azimuthal angle representing the relative phase of these basis states [1]. Rotations on the Bloch sphere can be performed using a laser that is resonant with the exciton creation energy in the dot – and by varying the amplitude, phase and duration of a laser pulse an *arbitrary* single-qubit gate can be executed.

We can see this rather easily. Consider the Hamiltonian of an exciton qubit interacting with an oscillating field in the spin representation:

$$\mathcal{H} = \epsilon\sigma_z + \Omega\sigma_x\{\exp[i(\omega_l t + \phi)] + H.c.\}. \tag{23.1}$$

The σ_x and σ_z are the Pauli matrices, ϵ is the energy of the optical gap (including the exciton binding energy), Ω is the exciton–laser coupling (also known as the Rabi frequency), and ω_l is the laser frequency (which for resonance is set to ϵ).

We now make two transformations. First, we recast the Hamiltonian in a frame rotating with the laser frequency (this is an exact transformation); second, we ignore fast oscillating exponential terms (this is the rotating wave approximation or RWA). This yields a time independent Hamiltonian:

$$\mathcal{H} = \nu\sigma_z + \Omega(\cos(\phi)\sigma_x + \sin(\phi)\sigma_y) \tag{23.2}$$

with $\nu = \epsilon - \omega_l$ the detuning of the laser from resonance. Setting $\nu = 0$ causes the eigenstates of the Hamiltonian to lie in the equatorial plane of the Bloch sphere depicted in Fig. 23.3, with the azimuthal angle of the two diametrically opposite eigenstates being ϕ. Since the natural evolution of the system is to rotate around the line connecting the two eigenstates on the Bloch sphere, *any* general single-qubit rotation can be achieved by varying ϕ and the duration of the laser pulse.

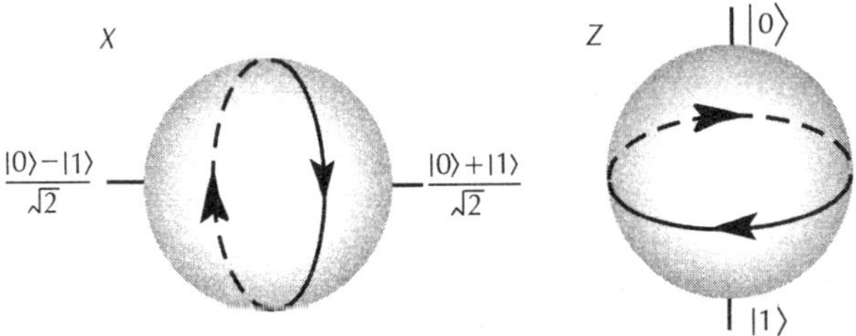

Figure 23.3   *Single-qubit rotations on the Bloch sphere. Left: The eigenstates are $(|0\rangle + |1\rangle)/\sqrt{2}$, so the state precesses around the x-axis. Right: The eigenstates are $|0\rangle$ and $|1\rangle$ so the state precesses around the z-axis. In order to effect an arbitrary single-qubit gate, two rotation axes are needed, together with control of the time of rotation.*

23.3.2   Two-qubit manipulation

In order to achieve entanglement between a pair of exciton qubits, there must exist some kind of controllable interaction between them. Two principal types of natural interaction have been proposed in this context: the *biexcitonic shift* and *Foerster energy transfer*. Each of these will now be described, and control procedures will be introduced.

23.3.2.1   Biexcitonic shift

For the schemes we are considering, we are interested in exciting *at most* one exciton in each QD. In particular, we will consider heavy-hole to conduction band ground state transitions in III–V semiconductor nanostructures, and drive them by resonant circularly polarized laser pulses. In this way the spin of the exciton is well defined and in particular each laser pulse can couple to one and only one ground state exciton in each quantum dot. Owing to the Coulomb interaction, two excitons in nearby QDs may couple and the signature of this coupling is the so-called *biexcitonic shift* [22, 23]. In the following we will consider two *vertically stacked* QDs, QD_a and QD_b (see Fig. 23.4). The biexcitonic shift $\Delta\epsilon$ is between the optical response corresponding to the ground-state excitonic transition in QD_b in the presence of an exciton in QD_a compared with the same

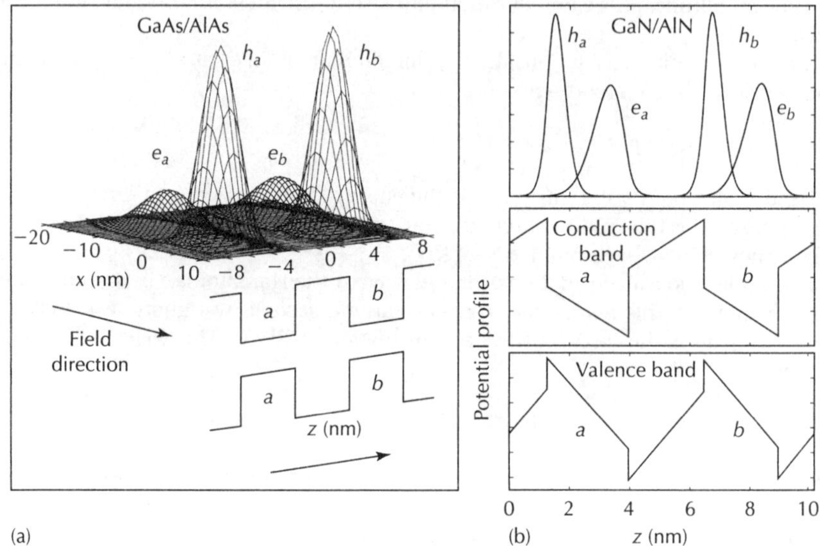

Figure 23.4   *Electron and hole particle distributions for the biexcitonic ground state in coupled GaAs/AlAs (a) and GaN/AlN (b) QD structures [28]. The distributions are calculated from direct diagonalization of the many-body Hamiltonian. The corresponding valence and conduction bands are also sketched.*

transition in QD_b with no exciton in neighbouring QD_a. $\Delta\epsilon$ is dipole–dipole to lowest non-zero order in the dot centre–centre separation R, so takes the form:

$$|\Delta\epsilon| \propto \frac{d^2}{R^3} \tag{23.3}$$

with d the length of the excitonic electron–hole dipole. If we consider a stack of QDs with a typical distance between QD centres of 100 Å, interactions beyond nearest neighbours can be neglected. We will consider structures in which single particle tunnelling between QDs is suppressed [21], so that the biexcitonic shift stems from direct Coulomb interaction.

We will concentrate here on two different heterostructures, GaAs/AlAs [22, 23] (with GaAs-based QDs) and GaN/AlN [24, 25] (with GaN-based QDs). These two structures have very different characteristics. GaAs/AlAs has a zinc-blende structure characterized by a face-centred cubic cell. This structure presents no spontaneous crystal polarization, and the natural dipole associated with a ground-state exciton is very small, so that the corresponding biexcitonic shift would be too small to be resolved by sub-picosecond laser pulses. In order to increase this interaction a *static, in-plane* electric field E may be applied [22, 23]. An estimate of the corresponding dipole length is given by:

$$d = eE\left(\frac{1}{m_e\omega_e^2} + \frac{1}{m_h\omega_h^2}\right) \tag{23.4}$$

where $m_{e(h)}$ is the electronic (hole) mass and $\omega_{e(h)}$ is the characteristic frequency of the electronic (hole) in-plane confining potential (assumed quadratic). In reality the QD confining potential has a finite height, so a strong enough applied electric field will ionize the exciton.

III–V nitride compounds, on the other hand, present a wurzite-type crystal structure, based on a hexagonal unit cell, which is compatible with spontaneous bulk polarization. The polarization is accumulated at the interfaces of GaN/AlGaN heterostructures. In QDs this combines with a strong strain-induced piezoelectric potential, and results in a strong built-in electric field, of the order of a few MV/cm [26, 27]. This is oriented along the growth direction and has opposite sign inside and outside the dot: this intrinsic field enhances the intrinsic dipoles of any exciton and at the same time confines the excitons inside the QDs.

In Fig. 23.4 the particle distributions corresponding to the biexcitonic ground states in coupled GaAs/AlAs (left) and GaN/AlN (right) QD structures are plotted. We consider parallel spin excitons only, i.e. excited by laser pulses having the same circular polarization. In GaAs/AlAs, the presence of an external electric field aligns the dipoles in the in-plane direction (which we label x for simplicity). Excitons in nearby QDs have a positive Coulomb interaction, i.e. the biexcitonic transition will be blue shifted with respect to the corresponding excitonic transition [22, 23]. The band structure in GaN/AlN includes the built-in electric field – which has the advantage that ionization does not occur, and device design is simplified. As the figure shows, this time the dipoles are stacked in the growth direction so that their interaction energy is negative, corresponding to a red shifted biexcitonic transition. In both structures, the exciton–exciton coupling can be tuned. In GaAs-based structures this is done by engineering the size and aspect ratio of the QDs, and by modifying the value of the applied field. In GaN-based QDs, varying the structure alters also the strength of the built-in field. Our theoretical studies have shown that in both materials, and for experimentally reasonable parameters, it is possible to achieve biexcitonic shifts of the order of a few meV [23, 25]. These can be resolved by sub-picosecond laser pulses. The two-qubit gating time is proportional to the inverse of the biexcitonic shift, and it is therefore typically also sub-picosecond.

In order to obtain a large biexcitonic shift without losing other important properties of the system, it is important to tune carefully the system parameters. Several parameters enter the problem and compete with each other. We will first consider GaAs-based structures [23]. To have a well-defined QD with associated low tunnelling probability between QDs, the structures need to be in the strong confinement regime. This also ensures well-separated energy states. From Eqs 23.3 and 23.4, we see that the biexcitonic shift is roughly proportional to the square of the applied field, but inversely proportional to the energy associated with the in-plane parabolic confinement: a large external field will increase the biexcitonic shift while a too strong confinement will suppress it. On the other hand, in order to obtain optical response the oscillator strength corresponding to the relevant transitions must be significantly different from zero – and a large electric field can drastically reduce it by diminishing the overlap between electron and hole wave functions. Additional constraints related to the confinement potential are that the transitions of interest in the absorption spectrum are well defined and isolated. The relevant parameter space that would satisfy all the constraints at the same time has been explored in [23] by the use of a precise analytical model and the results show that a relatively large region of experimentally accessible parameters is available for biexcitonic shifts larger than 3–4 meV, with applied fields of the order of 70 kV/cm.

A similar analysis has been undertaken for GaN-based structures [25]. We reiterate that in this case the biexcitonic shift is negative. The dipoles are in the growth direction, so in this configuration the in-plane confinement plays a minor role; the important parameters in determining the excitonic dipole strength are the QD height and the barrier width. These parameters also determine the strength of the internal field to a good approximation [24]. The biexcitonic shift decreases with increasing interdot barrier and QD height, but again there is the problem of obtaining a relatively large interaction, while maintaining the coupling to the laser field at the same time. To optimize both quantities, a figure of merit has been proposed [25], given by the product of the biexcitonic shift and the logarithm of the oscillator strength. The optimized window of parameters corresponds to a QD height of about 2.5–3 nm [25], well within experimental reach.

Figure 23.5 shows the optical response for a GaAs-based [23] (left panel) and GaN-based [24] (right panel) two-QD structure. Spectra are calculated by exact diagonalization of the corresponding many-body Hamiltonian [23]. The figure compares the calculated excitonic (solid line) and biexcitonic (dashed line) spectra. Peak "a" corresponds to the creation of an exciton in the first dot, peak "b" to the creation of an exciton in the neighbouring dot – and corresponding exciton–biexciton transitions are also seen (in the calculations of these exciton–biexciton lines, the first exciton is always created in QD$_a$). As expected, the ground-state biexcitonic transition is blue shifted for GaAs and red shifted for the GaN-based structure. Owing to the strong built-in electric field, the difference in energy between the two lowest excitonic transitions is one order of magnitude larger in GaN-based structures than in the GaAs ones.

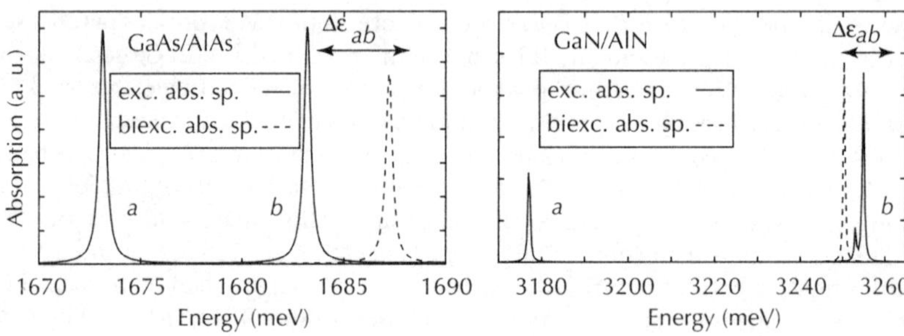

Figure 23.5   Excitonic (solid line) and biexcitonic (dashed line) absorption spectra for a GaAs-based (left panel) and GaN-based (right panel) two QD structure [28]. An in-plane field E = 75 kV/cm has been applied to the GaAs-based structure.

We now describe how to perform a quantum gate in this system. Let us consider an encoding in which $|0\rangle_l$, corresponds to the absence of ground-state exciton in QD$_l$ and $|1\rangle_l$ to the presence of ground-state exciton in QD$_l$. The whole computational space is now spanned by the basis $|n\rangle = \otimes_l |n\rangle_l$, with $n_l = 0$, 1. By introducing the operator $\hat{n}_l = |1\rangle_l \langle 1|_l$, the effective Hamiltonian of the system can be written as:

$$H = \sum_l \epsilon_l \hat{n}_l + \frac{1}{2} \sum_{ll'} \Delta\epsilon_{ll'} \hat{n}_l \hat{n}_{l'}$$

(23.5)

where ϵ_l is the energy of the ground-state exciton in QD$_l$ and $\Delta\epsilon_{ll'}$ is the biexcitonic shift between excitons in QD$_l$ and QD$_{l'}$. Equation 23.5 was first proposed in [29] and has the same structure of the effective Hamiltonian used in NMR-based quantum computation. According to this Hamiltonian, a conditional two-qubit operation, the CNOT gate, can be achieved by a laser π-pulse exploiting the energy renormalization of the ground state excitonic transition due to the biexcitonic shift. The corresponding conditional dynamics is schematically shown in Fig. 23.6, for the case of two QDs with an exciton already present in QD$_a$. In general a π-rotation of the target qubit $|n\rangle_b$ is achieved by a laser π-pulse centred at the energy $\epsilon_b + \Delta\epsilon_{ab}$ *if and only if* the state of the control qubit is $|1\rangle_a$. Since $\Delta\epsilon_{ab} \gtrsim 3$ meV, this CNOT gate can be performed on sub-picosecond timescales. In order to see how this gate can create entanglement, consider the factorizable state

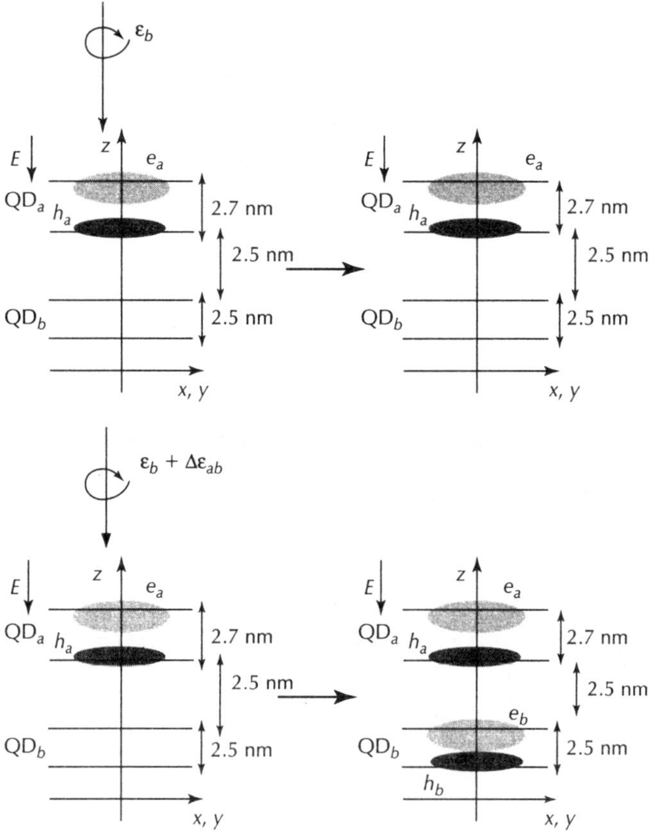

Figure 23.6 *Conditional dynamics of excitons for the case in which a ground state exciton is present in* QD$_a$. *We consider a suitable GaN-based structure. Due to the biexcitonic shift* $\Delta\epsilon_{ab}$ *between* QD$_a$ *and* QD$_b$, *a ground state exciton is excited in* QD$_b$ *if and only if the laser* π-*pulse is resonant with the renormalized transition energy* $\epsilon_b + \Delta\epsilon_{ab}$ *of* QD$_b$.

$(\alpha \, |0\rangle_a + \beta \, |1\rangle_a) \otimes |0\rangle_b$. If the CNOT is performed on it, we obtain $\alpha \, |0\rangle_a \otimes |0\rangle_b + \beta \, |1\rangle_a \otimes |1\rangle_b$, which is maximally entangled [23].

23.3.2.2   Foerster coupling

The Foerster interaction (which is also known as resonant energy transfer or dynamic dipole–dipole coupling) was first investigated in the context of sensitized luminescence of solids [30, 31], but since then has been studied in many contexts including photosynthesis in biological systems [32] and, of course, quantum information processing [33, 34].

The process of resonant energy transfer of excitons is depicted schematically in Fig. 23.7. The easiest way to think of the process is as follows. Initially, an exciton is located on one of two adjacent quantum dots. It then recombines, emitting a photon, but this photon is quickly reabsorbed by the second dot, and an exciton forms there. In fact, the photon is only ever a virtual particle, so is not "actually" emitted. It can equivalently by thought of as an interaction between the transition dipoles of the two excitons and is thus sometimes called a dynamic dipole–dipole coupling.

In order to calculate the size of the coupling, an *envelope function approximation* is used to describe the electron and hole wavefunctions:

$$\psi_p(\mathbf{r}) = \phi_p(\mathbf{r})U_p(\mathbf{r}) \tag{23.6}$$

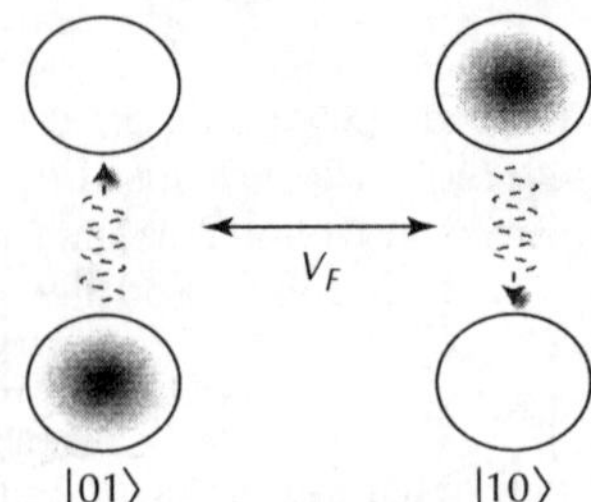

Figure 23.7   Drawing depicting Foerster coupling between two quantum dots. Excitons are shown in grey. If the dots are resonant, an exciton on one dot will recombine, emitting a virtual photon that creates an exciton on the second dot.

where $\phi_p(\mathbf{r})$ is an envelope function describing the changing wavefunction amplitude of confined states for particle type p over the dot region, and $U_p(\mathbf{r})$ is the Bloch function which has the periodicity of the atomic lattice. The matrix element describing the coupling is given by [33]:

$$V_F = \frac{e^2}{4\pi\epsilon_0\epsilon_r R^3} O_I O_{II} \left[|\langle \mathbf{r}_a \rangle|^2 - \frac{3}{R^2} (\langle \mathbf{r}_a \rangle \cdot \mathbf{R})^2 \right] \tag{23.7}$$

where the term $\langle \mathbf{r}_a \rangle$ represents the atomic position operator expectation value:

$$\langle \mathbf{r}_a \rangle = \int_{cell} U_e(\mathbf{r}) \mathbf{r} U_h(\mathbf{r}) d\mathbf{r} \tag{23.8}$$

which is the same for both dots and

$$O_i = \int_{space} \phi_e^i(\mathbf{r}) \phi_h^i(\mathbf{r}) d\mathbf{r} \tag{23.9}$$

is the overlap of electron and hole envelope functions on the appropriate dot i. Varying the choice of quantum dot size and materials can significantly affect the size of the overlap integral, allowing the grower to tailor the size of this interaction. The distance between dots R also plays an important part in determining the magnitude of the interaction, which varies as $1/R^3$. A typical spacing is around 5 nm [35], and for other typical parameters, V_F is of order 1 meV. This energy sets a timescale for the transfer of an exciton on resonance of around 1 ps.

In order to understand how we can use Foerster transfer to perform an entangling gate, let us consider the Hamiltonian for an exciton on either the right dot (state $|01\rangle$) or the left dot (state $|10\rangle$), coupled by the Foerster matrix element V_F:

$$\mathcal{H} = \epsilon |10\rangle\langle 10| + (\epsilon + \delta)|01\rangle\langle 01| + V_F(|01\rangle\langle 10| + H.c.) \tag{23.10}$$

where ϵ is the exciton energy for the left dot, and δ is the difference between the exciton creation energies of the dots. $\delta = 0$ represents the resonant condition, and in this case the eigenstates of the system are $2^{-1/2}(|10\rangle \pm |01\rangle)$. On the other hand, if $\delta \gg V_F$, the eigenstates are $|01\rangle$ and $|10\rangle$. These two situations can again be pictured in the Bloch sphere representation, and this is shown in Fig. 23.8.

If we could move between these two different regimes, a procedure for creating an entangled state would be as follows:

- Cool the double dot to its ground state ($|00\rangle$).
- In the $\delta \gg V_F$ regime, apply a π pulse to the left dot, and so create the $|10\rangle$ state.
- Move *non-adiabatically* to the $\delta = 0$ regime, and allow the system to evolve naturally. This natural evolution is precession around the axis connecting the eigenstates (which now lie in the equator of the Bloch sphere); the state is rotated towards the equator.

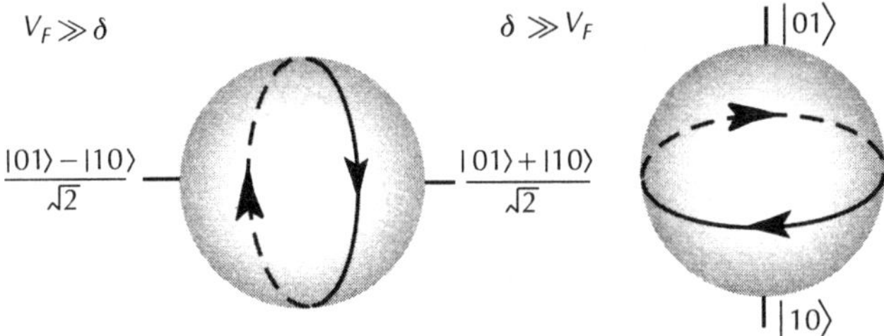

Figure 23.8   Bloch sphere picture of the $|10\rangle$, $|01\rangle$ subspace in different parameter regimes. By moving between one regime and the other quickly, an entangling gate is effected. See text for details.

- Move back to the $\delta \gg V_F$ regime once the state of the system is in the equatorial plane of the Bloch sphere. This state is maximally entangled and remains so.

Further, in a completely analogous way, a general entangling gate can be performed.

However, we must be able to move between the two regimes to do this. One possible method could be to apply an electric field to both dots, which as we have seen shifts the hole in one direction and the electron in the other. This reduces the overlap integral between them, and therefore the Foerster coupling. However, this can only be done slowly in practise, and the degree of modulation is typically only a few per cent (see [34]). It is more practical to tune the energies of the excitons to vary δ – and thus change the *effectiveness* of the Foerster coupling. This can be done using the *AC Stark effect* [36] when a detuned laser is applied to the excitonic transition, which rather than altering the state of the system changes its energy. The exciton creation energy is Stark shifted by an amount:

$$\Delta \epsilon = \frac{\Omega^2}{4\nu} \tag{23.11}$$

where Ω is the exciton–laser coupling and ν is the laser detuning. Each of the two dots have non-identical Ω and ν and so shift by different amounts, even when illuminated by the same laser. By applying fast laser pulses it is possible to tune two dots in and out of resonance on the sub-picosecond timescale.

Figure 23.9 shows a simulation of quantum dynamics during a gate. We have plotted the "entanglement of formation" as a function of time; this takes a value of unity for a maximally entangled state, and zero for a completely unentangled one.[2] The plot shows that a maximally entangled state can indeed be created using the recipe outlined above.

23.4   Spin representation

We now turn to self-assembled quantum dots in which an extra electron has been added through either doping or the application of an external potential. Again, we look at the two requirements for quantum computing: arbitrary single-qubit manipulation and an entangling two-qubit gate.

23.4.1   Single-qubit gates

The simplest method for performing arbitrary manipulation of a single electron spin proceeds along similar lines to the manipulation of a single-exciton discussed in section 23.3.1. Typically, the spin would be placed in a magnetic field in order to induce a Zeeman splitting of the magnetic sublevels. This splitting is typically on the micro-electron-volt level, and so the radiation needed to resonantly excite the spin is a few GHz. The coupling is also much weaker than before, since it proceeds via the magnetic (and not the electric) transition dipole. This means that the gates take

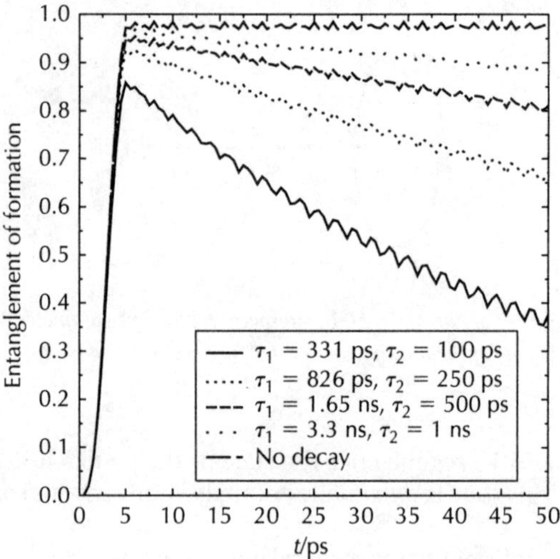

Figure 23.9   Entanglement of formation as a function of time for a Foerster coupled double dot system under the action of a laser [50]. The laser is pulsed on for 5 ns, and the AC Stark effect causes the dots to come on to resonance, thus effecting an entangling operation. The entanglement is degraded when exciton decay is simulated, the decay being characterized by time τ_i for dot i. More details of the simulation are given in [50].

much longer to perform than the optical gates discussed earlier. However, it turns out that it is possible instead to use optical radiation to rotate a spin, and as we shall now show, this can be much quicker than using microwaves.

In order to understand how this works, we must first introduce the concept of *Pauli blocking*. We show the basic principles of operation in Fig. 23.10. The qubit has a spin of either $+\frac{1}{2}$ or $-\frac{1}{2}$, and we label these as qubit state $|1\rangle$ and $|0\rangle$, respectively. The valence states are full, with the highest lying electrons (lowest lying holes) having spin $\pm\frac{3}{2}$. If a laser is applied with σ^+ polarization, tuned to the transition energy between lowest hole and lowest electron, it can only be absorbed if the QD system changes its angular momentum by $+1$. The only possible way this can happen is if an electron is promoted from the $-\frac{3}{2}$ state to the $-\frac{1}{2}$ state. Owing to the Pauli exclusion principle, such a transition is *not possible* if the system exists in qubit state $|0\rangle$. On the other hand, $|1\rangle$ will happily couple to a three-particle entity called a trion (two electrons plus one hole).

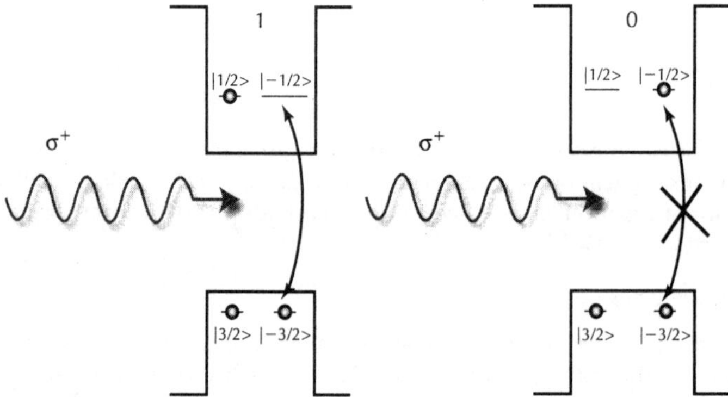

Figure 23.10   Schematic diagram showing the Pauli-blocking effect. The left-hand drawing shows the excess electron in the spin-up state (qubit $|1\rangle$), the right shows the electron spin down (qubit $|0\rangle$). Under illumination with σ^+ light, a trion state ($|X\rangle$) couples to $|1\rangle$, but not $|0\rangle$.

Under a "2π pulse" of σ^+ polarized laser radiation, the trion is created and then destroyed again, resulting in a PHASE gate:

$$|0\rangle \rightarrow |0\rangle$$
$$|1\rangle \rightarrow -|1\rangle. \tag{23.12}$$

A general phase gate:

$$|0\rangle \rightarrow |0\rangle$$
$$|1\rangle \rightarrow e^{i\phi}|1\rangle \tag{23.13}$$

can be performed using the same method, by either (a) introducing a detuning of the laser or (b) introducing a phase difference between the pulse creating the exciton and the pulse destroying it again. More details of these two approaches can be found in [38].

The phase gate provides us with an arbitrary rotation about the z-axis of the Bloch sphere (shown on the right-hand side of Fig. 23.3), but this is not enough for universal quantum computing. We need to be able to do *any* rotation – and for this we require one more possible axis of rotation to be available to us.

This second axis of rotation presents us with more difficulty than the first. This is because any other axis of rotation (apart from the z-axis) must necessarily involve some transfer of population between $|1\rangle$ and $|0\rangle$ – and the tricks we have played so far with a circularly polarized laser exciting a single exciton will no longer work. Instead, we must perform at least two different optical transitions, in order to couple *both* spin levels to one or more higher lying trion states. In this way a *Raman*-type transition can be realized, which allows population transfer. Let us refer again to Fig. 23.10. We know that our laser can couple $|1\rangle$ and $|X\rangle$, but it is also clear that $|0\rangle$ and $|X\rangle$ cannot be coupled, no matter what the laser polarization is. This is because such a coupling would involve an angular momentum change of two units. We must therefore take an alternative approach.

Light holes have a spin projection of $\pm\frac{1}{2}$, and therefore light hole excitons have angular momentum of 1 or 0 – and these are exactly what we need to perform a Raman-type manipulation of our spin. Either spin qubit state ($|0\rangle$ or $|1\rangle$) can be excited to light hole trion state (two electrons and a light hole, which has total angular momentum of $\pm\frac{1}{2}$). This is because the angular momentum change in such a transition is always either 0 or 1. Such transitions are straightforwardly achieved by using laser light of different linear polarizations. Figure 23.11 shows more detail.

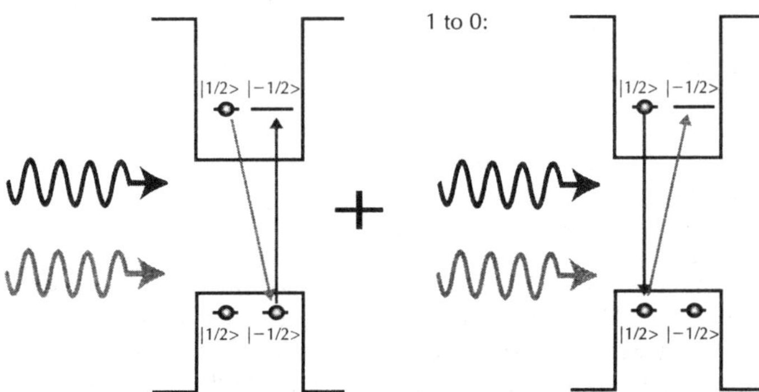

Figure 23.11   Schematic diagram showing how two linearly polarized lasers can effect an optical Raman-type transition between an "up" electron spin (qubit state $|1\rangle$) and a "down" electron spin (qubit state $|0\rangle$). Note that two pathways occur in parallel for this transition, and both involve the creation of a light-hole trion state. The reverse operation ($|0\rangle$ to $|1\rangle$) also occurs, so this scheme forms a Pauli X gate for a single spin qubit. More details can be found in [38].

23.4.2   Two-qubit gates

The most obvious method of performing a two-qubit gate on a pair of spins would probably be to control some sort of spin–spin interaction (for example, a Heisenberg exchange-type interaction). Indeed, this sort of interaction can be controlled by gating a potential barrier between spin qubits to tune wavefunction overlap (see Loss and Divincenzo's well-known paper, [39]). This method is not as effective in self-assembled dots for two main reasons. First, the degree of wavefunction overlap tends to be quite small since barrier regions are quite wide. Second, it is very hard to fabricate electrodes between these sorts of dots. Hence, we turn again to optical methods for entangling spin qubits, which, as we shall see shortly, also has the advantage of fast gate operation times (on the order of a few picoseconds).

Optical methods for spin entanglement again rely on the Pauli blocking effect and so selectively create trion states from the spin qubits. It is the interaction of the trions that indirectly couples the spins.

Consider again the macromolecule formed by the stacked dots QD_a and QD_b and consider two spin qubits driven by σ^+ laser pulses. These selectively create trion states $|X\rangle_{a(b)}$ from the qubit $|1\rangle_{a(b)}$ only. The Hamiltonian of the system is:

$$H(t) = \sum_{i=a,b} \omega_i |X\rangle_i \langle X| + \Delta\epsilon_{ab} |XX\rangle_{ab}\langle XX| + V_F(|1X\rangle\langle X1| + H.c.)$$
$$+ \sum_{i=a,b} \Omega_i \cos\omega_{li}t(|1\rangle)_i\langle X| + H.c.) \tag{23.14}$$

where $H.c.$ denotes the Hermitian conjugate, $\omega_{a(b)}$ is the trion creation energy for $QD_{a(b)}$, $\Omega_{a(b)}$ is the time-dependent coupling between laser and $QD_{a(b)}$, and $\omega_{la(b)}$ is the frequency of the laser addressing $QD_{a(b)}$. Two types of trion coupling have been included: the interdot Foerster coupling strength V_F, and the biexcitonic energy shift due to the exciton–exciton dipole interaction $\Delta\epsilon_{ab}$[23].

One possible way of performing an entangling gate is to exploit $\Delta\epsilon_{ab}$. We assume that the energy difference between states $|0\rangle$ and $|1\rangle$ is negligible on the exciton energy scales and that the two dots are different, so that Foerster interaction is inhibited [23]. A CPHASE gate can be implemented by first applying a resonant exciting laser π pulse to each dot, which creates an exciton if the dot spin state is $|1\rangle$. Now, after a time Δt, another π pulse is applied to each dot, which destroys any excitons that were created by the first pulse. A phase $\theta \propto \Delta t\Delta\epsilon_{ab}$ will be accumulated *if and only if* the system is initially in state $|11\rangle$ [40]. This scheme can be adjusted to include imperfect QD structures which present a relevant admixture of heavy-and light-hole states. This is done by using chirped laser pulses [40].

Another possibility is to use the Foerster interaction, which as we know is most effective for resonant excitons. We therefore set $\omega_a = \omega_b$, and this time we only need to use a single laser (of frequency ω_l) to excite excitons. We assume the laser couples equally strongly to each dot, with strength Ω. Making these extra assumptions allows us to recast the Hamiltonian, Eq. 23.14, in the following form:

$$H(t) = \omega_a(|X\rangle\langle X| \otimes \hat{I} + \hat{I} \otimes |X\rangle\langle X|) + \Delta_{\varepsilon_{ab}} |XX\rangle\langle XX|$$
$$+ V_F(|1X\rangle\langle X1| + H.c.)$$
$$+ \Omega\cos\omega_l t\,(|1\rangle\langle X| \otimes \hat{I} + \hat{I} \otimes |1\rangle\langle X| + H.c.) \tag{23.15}$$

This new Hamiltonian decouples into four separate subspaces each containing one of the four computation basis states. The eigenstate energy level structure of these four subspaces is depicted in Fig. 23.12 for the situation in which the laser is turned off. $|00\rangle$ is alone since it cannot couple to the laser at all. $|10\rangle$ and $|01\rangle$ couple to $|X0\rangle$ and $|0X\rangle$ respectively. $|11\rangle$ couples to the symmetric and antisymmetric combinations of $|X1\rangle$ and $|1X\rangle$, which are split by the Foerster interaction. The highest energy eigenstate is $|XX\rangle$, which lies at an energy $2\omega_a + \Delta\epsilon_{ab}$ above $|11\rangle$.

Having established the rather complicated energy level structure of the system, it is now rather simple to see how to perform an entangling gate. The laser is tuned to the energy of the $|11\rangle$ to $(|1X\rangle + |X1\rangle)/\sqrt{2}$ transition and a 2π pulse is executed. This means $|11\rangle \rightarrow -|11\rangle$. The other three

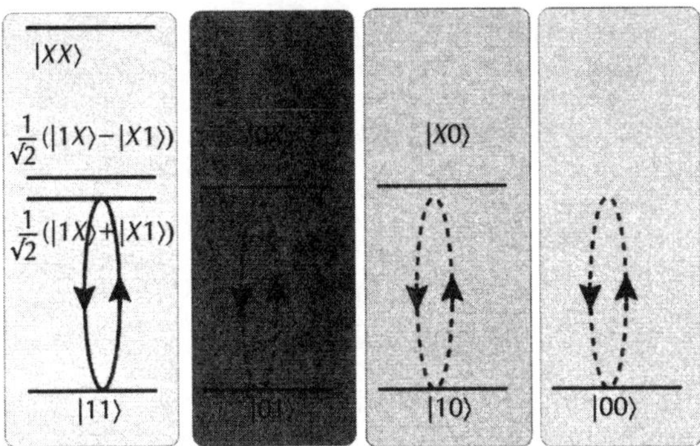

Figure 23.12   Schematic diagram showing the energy level structure of the two quantum dot system discussed in section 23.4.2. The Hamiltonian decouples into four separate subspaces, illustrated by different panels here. A pulsed laser is tuned to the $|11⟩ \leftrightarrow (|1X⟩+|X1⟩)/\sqrt{2}$ transition (shown by the solid line). It has no effect in the other subspace since it is either off-resonance or does not couple at all (shown by the dashed lines).

computation states are *unaffected* by the pulse, since they are either uncoupled (for $|00⟩$) or off-resonance (for $|10⟩$ and $|01⟩$). Hence, the action of this pulse is to effect a CPHASE gate, which is of course entangling and universal for quantum computing.

23.5   General quantum computer architecture and related quantum devices

23.5.1   Quantum dot-based quantum bus

Most of the QD-based quantum computation schemes propose the use of vertically stacked dots [7]. However, experiments show that the maximum length of such arrays is of the order of only ten QDs [8]. To implement non-trivial computation, we need arrays (or array networks) of at least a few tens of qubits, and possibly many more. Therefore, an important problem is how to communicate between different vertically stacked arrays of QDs.

The communication between distant vertical arrays of QDs can be done using quantum buses made by *in-plane* chains of quantum dots [41, 42]. Recent experimental progress has shown that it is possible to grow in-plane chains of quantum dots, whose density and QD size can be independently controlled [12]. Information could be transferred along this structure by, for example, a sequence of SWAP gates acting on excitonic or spin qubits – but this would require a high degree of external control over the quantum dynamics. A different approach is to implement information transfer across the in-plane arrays (IPAs) by exploiting the *natural dynamics* of the system [41, 42]. The key idea is that if we inject an excitation at one end of the chain, the natural quantum evolution will transfer such an excitation to the opposite end in a known time, which is determined by the chain characteristics (number of quantum dots, strength and sequence of QD–QD couplings). This concept is exemplified in Fig. 23.13. Recently, there have been interesting developments in the study of the natural dynamics of spin chains – see, for example, [43, 44]. We can translate some of these ideas into the IPA system: the use of natural dynamics should simplify the problem of transferring information across our QD-based quantum computer.

State-of-the-art in-plane QDs can be made monodisperse up to a few per cent. Let us assume that the information is stored in exciton qubits: $|0⟩_i (|1⟩_i)$ is the absence (presence) of a single ground-state exciton in QD_i. The Hamiltonian for an IPA of N QDs, in which we consider at most a single exciton in the chain, is given by:

$$H = \sum_i^N E_i |1⟩_i \langle 1|_i + \sum_{i=1}^{N-1} (V_{Fi,i+1}|1⟩_i \langle 0|_i \otimes |0⟩_{i+1} \langle 1|_{i+1} + H.c.) \tag{23.16}$$

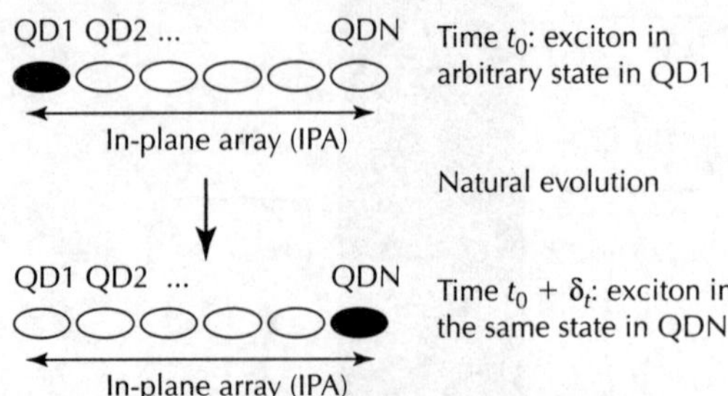

Figure 23.13   At time t_0 an excitation is injected at one end of the IPA. After the known time δt the same excitation has been transferred to the other end of the IPA by the natural quantum evolution of the system.

where $V_{Fi,i+1}$ is the Foerster coupling [45, 46] between nearby quantum dots discussed in section 23.3.2. Single-particle tunnelling is again neglected. This Hamiltonian is isomorphic to the Hamiltonian of a spin chain. We will focus on *relatively short chains*, with $N \lesssim 10$, corresponding to lengths of a few hundred nanometres: our purpose is to communicate between arrays of stacked QDs, whose relative distance is greater than the diameter of the laser spot used to address the single QDs.

From the general theory of spin chains it is known that, by applying a specific modulation to the sequence of spin–spin coupling, it is possible to perform perfect transfer of a quantum state across the chain [43]. When this scheme is considered for QD IPAs, the implementation of this strict coupling sequence can in principle be done by growth or by tuning the interdot couplings *a posteriori*, by applying in-plane electric fields [41, 42]. Our analysis though shows that both methods are highly impractical and can lead to different systematic errors [41, 42]. We will therefore consider IPAs of identical QDs, i.e. $V_{Fi,i+1} = V$ and $E_i = E$, a relatively easier requirement to satisfy. We want to address the questions (i) if transfer of a state along the chain with a high degree of accuracy is still possible by exploiting the natural dynamics of the chain and (ii) how the results are affected by imperfections. We consider IPAs with $N = 1, \ldots 9$, and simulate the evolution of the system after the first QD has been populated.

Figure 23.14 shows the time evolution of a five QD chain, following state input to one end of the chain. It presents the average fidelity of state transmission to the other end when the initial state input is varied over the Bloch sphere. As can be seen, the time evolution *as a whole* presents a periodic behaviour (within the numerical error), centred on the peak labelled with "C". Within this period many peaks are present, some of which correspond to a fidelity very close to 1. This is a typical pattern for all the spin chains analysed, and in particular for $N \leq 9$, the very first peak (labelled "A" in the figure) always corresponds to a fidelity $\langle F \rangle_{Bl} \gtrsim 95\%$ [41, 42]. The time t_A at which resonance "A" occurs and the corresponding fidelity $\langle F \rangle_{Bl}^A$ present a *weak* linear dependence with the length N of the chain (with the time getting longer for longer chains and the fidelity getting smaller). t_A remains below 2 ps for $N = 9$ [41, 42].

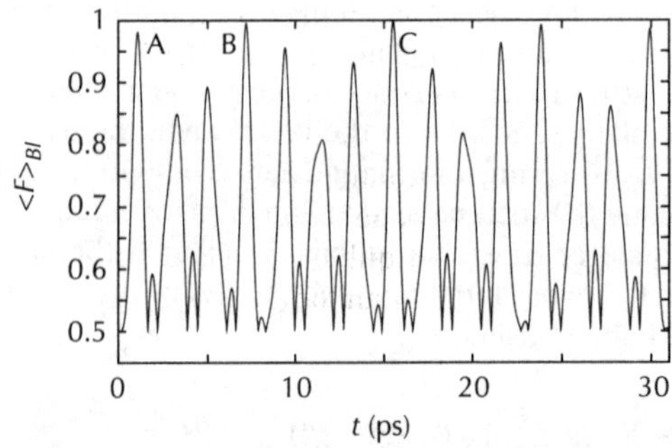

Figure 23.14 $\langle F \rangle_{Bl}$ as a function of time for $N = 5$ and $V_{Fi,i+1} = 1\,meV$.

In all cases analysed, at least one of the subsequent peaks (e.g. B and C in Fig. 23.14, for $N = 5$) corresponds to an even higher fidelity of transmission – and in some cases it could be worth waiting for these higher peaks.

In the case of imperfect chains, the exciton energies E_i might have a random distribution. Calculations have been performed in this case, assuming that the variation is bounded, and the corresponding fidelity has been averaged over an appropriate number of such configurations. Peak A is more robust to this random variation, i.e. peaks B and C tolerate a lesser amount of noise in the distribution of the energies [41, 42].

We underline that for comparable parameters, the IPA can transfer information faster than a microcavity, the typical gating time for a microcavity-mediated two-qubit gate being greater than 100 ps [47]. The use of IPA is a promising alternative to microcavity for medium-distance communication in QD-based quantum computation schemes.

23.5.2   Control arrays

For practical implementations, it is necessary to have the ability to switch the information transfer across a quantum bus on and off. This can be achieved by the use of a "control array" (CA) [42, 48]. A control array is a second IPA stacked on (or under) the IPA used as a quantum bus (see Fig. 23.15). For experimentally reasonable parameters, it is possible to optically generate a single exciton in each QD of the CA by the use of a single laser pulse (or multiple, but in-phase, laser spots) [48]. In this scheme the ground-state excitonic transition in the CA and in the quantum bus must be different by at least a few meV, so that the CA can be energy selectively addressed with respect to the quantum bus. This happens naturally in the stacking process itself, as vertically stacked QDs tend to be different in size. The ground-state excitonic transitions in the inner QDs of the bus can then be altered, as we now discuss.

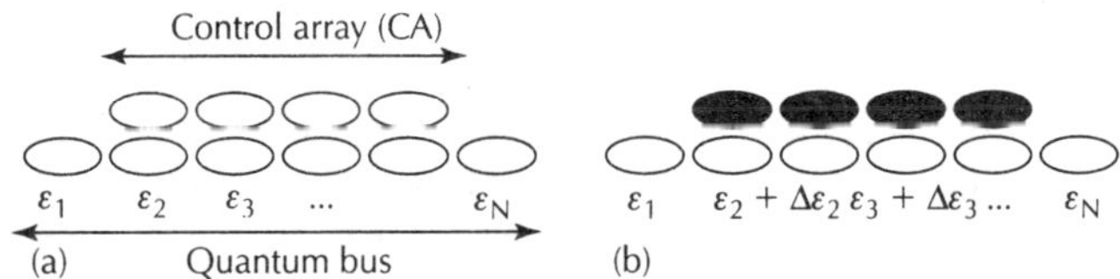

Figure 23.15   Panel (a): *No excitons in the control array, inner QDs in the quantum bus are on resonance.* Panel (b): *One exciton in each QD of the control array, inner QDs in the quantum bus are off resonance due to the related biexcitonic shift.*

Consider the scheme illustrated in Fig. 23.15. In panel (a) all ground-state excitonic transitions are on resonance, and the information transfer can occur by Foerster coupling. In panel (b) one exciton has been optically generated in each QD of the CA. The biexcitonic coupling $\Delta\epsilon_i$ between each of these excitons and the computationally relevant transition in the corresponding QD of the quantum bus detunes the inner QDs with respect to the QDs at the beginning and end of the bus chain. This detuning inhibits the Foerster coupling and hence the exciton transfer across the chain. Our calculations show that for $\Delta\epsilon_i/V_{\mathrm{Fi,i+1}} \geq 10$, the overlap between the evolution of the state of the quantum bus and its initial state (one exciton in QD1) is greater than 96% at all times [48]. We underline that the transfer time across a bus of nine QDs with a Foerster coupling as low as 0.2 meV is still less than 10 ps, so fast transfer times, together with good transfer blocking, are possible in the same device.

23.5.3   Quantum dot-based quantum computer architecture

In this section we discuss our own proposal for a quantum computer architecture which exploits vertically stacked arrays (VSA) as quantum registers, and IPAs and CAs as switchable quantum buses [41, 42]. The proposed hardware is composed of VSAs connected by IPA buses (see Fig. 23.16). Each VSA can be individually discriminated by a laser spot spatially, for example by

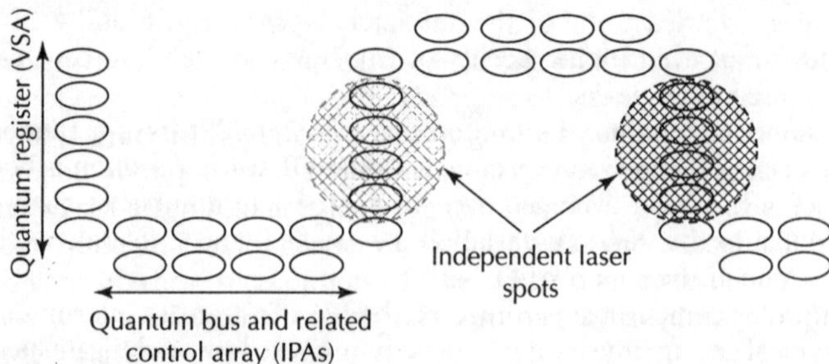

Figure 23.16 Quantum dot-based quantum computer architecture. It exploits the different characteristics of vertical and in-plane arrays of quantum dots. The first are used as quantum registers, the second as both quantum buses and to control the dynamics in the quantum buses. In this design different quantum registers can be resolved spatially by different laser spots.

near-field techniques. QDs in different VSAs may be resonant with the same laser frequency. This allows each VSA in principle to be formed by the same QD sequence. The combined spatial and energy-selective addressability of this architecture strongly enhances the hardware flexibility with respect to schemes that use energy-selective addressing only.

This need for spatial separation sets a minimum length of the order of 100 nanometres for the interconnecting IPAs. Interconnecting IPAs can both transfer qubits between different quantum registers and entangle qubits in QDs belonging to different VSAs. For example, a pair of entangled qubits in one VSA can be made using, e.g., the method described in section 23.3.2, and then one of the qubits can be transferred across the IPA to a different VSA. If the computational degrees of freedom are encoded in electronic spins in the VSA, the swapping of information between spin and excitonic qubits will be necessary before and after the transfer. This can be done using optical means and on a picosecond timescale, by exploiting the Pauli-blocking effect, as described in [48].

By using the proposed architecture, gating between qubits in different VSAs can also be performed, overcoming the problem that the physical register (the VSA) can at most be formed by 10–15 qubits (the QDs). This design allows compact implementation of a quantum computer: in principle in fact the hardware can be confined to a single slab. This allows easier lateral access (e.g. for wiring) and manipulation of each QD, which may be used to apply electric fields to specific dots in order to tune properties like biexcitonic shift or Foerster coupling. This architecture is scalable, since the same basic structure (VSA plus related IPA), and the slab structure as a whole, can be repeated at will.

23.5.4 Entanglement distributor

The results highlighted in previous sections show the feasibility of quantum communication between separated quantum registers, based on quantum buses made from chains of in-plain QDs and utilizing the natural dynamics of these systems. We will now discuss how the same hardware elements and properties can be used to design an entanglement distributor, which provides a resource of distributed entangled states on demand [48].

Entanglement is a flexible resource. Our device can be utilized in a variety of approaches to quantum processing, for example for teleportation of quantum states, to enable distributed quantum gates, or to build distributed cluster states [49]. The purpose of the device is to generate maximally entangled qubits on demand and to distribute them to *spatially separate* regions. Its structure is sketched in Fig. 23.17. It is formed by two coupled dots, QDA and QDB, where the entangled exciton pairs are produced. QDA and QDB are connected to the quantum buses BAA and BAB, respectively. The transfer of exciton along the buses is controlled by the control arrays CAA and CAB, as indicated. The buses connect QDA and QDB to QDC and QDD, respectively. These are located at the ends of the stacked quantum registers QRA and QRB. Quantum computation can be performed in QRA and QRB according to any of the all-optical methods previously

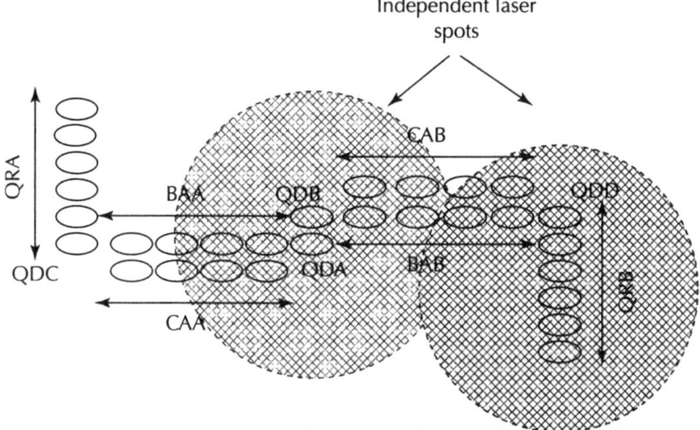

Figure 23.17   Sketch of the entanglement distributor, from [48].

described. These quantum registers and the dot pair (QDA, QDB) can be addressed by different laser spots.

The entanglement distribution process is as follows. (i) The quantum buses BAA and BAB are set off-resonance. This is achieved through the creation of an exciton in each control array (CAA and CAB) QD. The biexcitonic shift between these excitons and the excitonic transitions of the related QDs in BAA and BAB inhibits Foerster coupling between QDA(B) and BAA(B). (ii) The entangled pair is created in QDA and QDB by using the method described in section 23.3.2. IIIB based on the biexcitonic shift. (iii) The quantum buses BAA and BAB are set on resonance through the annihilation of the excitons in control arrays CAA and CAB. (iv) The entangled excitons propagate down the buses in opposite directions. Notice that each quantum bus is connected to one of the central QDs only. (v) When the excitons reach QDC and QDD, the quantum buses BAA and BAB are again set off resonance using the same method described above. (vi) The entangled excitons are now trapped in QDC and QDD and the entanglement can be delivered to the registers QRA and QRB.

Even if excitonic qubits are used to produce and deliver entanglement, actual quantum computation in the quantum arrays QRA and QRB can be based on longer-lived spin qubits. By means of the Pauli-blocking effect, the excitonic qubit can be swapped into a spin qubit. This is done by all-optical means [48]. This swapping can be essential to create a long-lived entangled resource.

The fidelity of our distribution scheme depends on several factors. By considering experimentally relevant parameters, and taking into account the correct ratio between the various energy scales (such as the ratio between biexcitonic shift and Foerster coupling discussed in section 23.5.2), we estimate that the total time for one entanglement distribution operation is no more than 20 ps, whereas the exciton decay time is about 1 ns. Our estimate of the overall fidelity of distribution is 91% [48].

One of the very important features of our entanglement distributor is that it integrates with, and thus enhances, many of the current QD-based quantum proposals. Given the impressive progress with QD structures (see e.g. [9–12]), we hope that some experimental group will pick up the challenge to build this device which constitutes a route to scalable solid-state quantum computation.

23.6   Conclusions

We have presented an overview of all-optical quantum computation schemes based on self-assembled quantum dots. We have discussed the reasons why we think these structures represent a good hardware choice for quantum devices and how an overall, scalable architecture for a quantum computer could be constructed starting from a few basic elements. This architecture exploits the natural dynamics of spin chains and in this way allows us to reduce the amount of gating – and related error sources – needed to implement information transfer across the various elements of the quantum computer. Due to the vast improvement in the last few years of growth, characterization and optical control techniques [13–20], and the creation of complex and regular quantum dot structures [9–12], we are confident that some of the devices proposed in this chapter form a feasible experimental challenge in the very near future.

Notes

1. In this chapter we will refer to existing computers as "classical computers"
2. The entanglement of formation measures the number of Bell states required to create the state of interest and for a two-qubit state it is given by: $E_F(\rho) = h((1 + \sqrt{1 - \tau})/2)$, where $h(x) = -x \log_2(x) - (1 - x)\log_2(1 - x)$ is the Shannon entropy function. τ is the "tangle" or "concurrence" squared: $\tau = C^2 = [\max\{\lambda_1 - \lambda_2 - \lambda_3 - \lambda_4, 0\}]^2$. The λs are the square roots of the eigenvalues, in decreasing order, of the matrix $\rho\tilde{\rho} = \rho\,\sigma_y^A \otimes \sigma_y^B\rho^*\sigma_y^A \otimes \sigma_y^B$, where ρ^* denotes the complex conjugation of ρ in the computational basis $100\rangle\,|10\rangle\,|11\rangle$ [37].

References

1. M.A. Nielsen and I.L. Chuang. *Quantum Computation and Quantum Information*, (CUP, Cambridge, 2000).
2. R. Ionicioiu and I. D'Amico, *Phys. Rev. B* **67**, R041307 (2003).
3. H. van Houten and C.W.J. Beenakker, *Physics Today* **49**, 22 (1996).
4. E. Pazy, I. D'Amico, P. Zanardi, and F. Rossi, *Phys. Rev. B* **64**, 195320 (2001).
5. J.M. Elzerman, R. Hanson, J.S. Greidanus, L.H. Willems van Beveren, S. De Franceschi, L.M.K. Vandersypen, S. Tarucha, and L.P. Kouwenhoven, *Phys. Rev. B* **67**, R161308 (2003).
6. P.W. Shor, *Proceedings of the 35th Annual Symposium on the Foundations of Computer Science* (IEEE Computer Society, Santa Fe, Los Alamitos, CA) 124 (1994).
7. P.M. Petroff, A. Lorke, and A. Imamoglu, *Physics Today* **54**, 46 (2001).
8. G.S. Solomon, J.A. Trezza, A.F. Marshall, and J.S. Harris, *Phys. Rev. Lett.* **76**, 952 (1996).
9. S. Kiravittaya, H. Heidemeyer, and O.G. Schmidt, *Physica E* **23**, 253 (2004).
10. S. Kiravittaya and O.G. Schmidt, *Appl. Phys. Lett.* **86**, 206101 (2005).
11. S. Kiravittaya, R. Songmuang, A. Rastelli, H. Heidemeyer, and O.G. Schmidt, *Nanoscale Res. Lett.* **1**, 1 (2006).
12. Z.M. Wang, K. Holmes, Y.I. Mazur, and G.J. Salamo, *Appl. Phys. Lett.* **84**, 1931 (2004).
13. T.H. Stievater, X. Li, D.G. Steel, D. Gammon, D.S. Katzer, D. Park, C. Piermarocchi, and L.J. Sham, *Phys. Rev. Lett.* **87**, 133603 (2001).
14. T. Guenther, C. Lienau, T. Elsaesser, M. Glanemann, V.M. Axt, T. Kuhn, S. Eshlaghi, and A.D. Wieck, *Phys. Rev. Lett.* **89**, 057401 (2002).
15. A. Zrenner, E. Beham, S. Stufle, F. Findeis, M. Bichler, and G. Abstreiter, *Nature* **418**, 612 (2002).
16. X. Li, Y. Wu, D.G. Steel, D. Gammon, T.H. Stievater, D.S. Katzer, D. Park, C. Piermarocchi, and L.J. Sham, *Science* **301**, 809 (2003).
17. T. Unold, K. Mueller, C. Lienau, T. Elsaesser, and A.D. Wieck, *Phys. Rev. Lett.* **94**, 137404 (2005).
18. A.F. Jarjour, A.M. Green, T.J. Parke, R.A. Taylor, R.A. Oliver, G.A.D. Briggs, M.J. Kappers, C.J. Humphreys, R.W. Martin, and I.M. Watson, *Physica E* **32**, 119 (2006).
19. A.J. Ramsay, R.S. Kolodka, F. Bello, P.W. Fry, W.K. Ng, A. Tahraoui, H.Y. Liu, M. Hopkinson, D.M. Whittaker, A.M. Fox, and M.S. Skolnick, *Phys. Rev. B* **75**, 113302 (2007).
20. R.S. Kolodka, A.J. Ramsay, J. Skiba-Szymanska, P.W. Fry, H.Y. Liu, A.M. Fox, and M.S. Skolnick, *Phys. Rev. B* **75**, 193306 (2007).
21. S. De Rinaldis, I. D'Amico, and F. Rossi, *Phys. Rev. B* **69**, 235316 (2004).

22. E. Biolatti, R.C. Iotti, P. Zanardi, and F. Rossi, *Phys. Rev. Lett.* **85**, 5647 (2000).

23. E. Biolatti, I. D'Amico, P. Zanardi, and F. Rossi, *Phys. Rev. B* **65**, 075306 (2002).

24. S. De Rinaldis, I. D'Amico, E. Biolatti, R. Rinaldi, R. Cingolani, and F. Rossi, *Phys. Rev. B* **65**, R081309 (2002).

25. S. De Rinaldis, I. D'Amico, and F. Rossi, *Appl. Phys. Lett.* **81**, 4236 (2002).

26. F. Widmann, J. Simon, B. Daudin, G. Feuillet, J.L. Rouvire, N.T. Pelekanos, and G. Fishman, *Phys. Rev. B* **58**, R15989 (1998).

27. R.D. Andreev and E.O. O'Reilly, *Phys. Rev. B* **62**, 15851 (2000).

28. I. D'Amico, S. De Rinaldis, P. Zanardi, and F. Rossi, *Phys. Stat. Sol. (b)* **233**, 377 (2002).

29. S. Lloyd, *Science* **261**, 1569 (1993).

30. T. Förster, *Ann. Phys. (Leipzig)* **2**, 55 (1948).

31. D.L. Dexter, *J. Chem. Phys.* **21**, 836 (1953).

32. X. Hu *et al.*, *Quarterly Rev. of Biophysics* **35**, 1 (2002).

33. B.W. Lovett, J.H. Reina, A. Nazir, and G.A.D. Briggs, *Phys. Rev. B* **68**, 205319 (2003).

34. A. Nazir, B.W. Lovett, S.D. Barrett, J.H. Reina, and G.A.D. Briggs, *Phys. Rev. B* **71**, 045334 (2005).

35. M. Bayer, P. Hawrylak, K. Hinzer, S. Fafard, M. Korkusinski, Z.R. Wailewski, O. Stern, and A. Forchel, *Science* **291**, 451 (2001).

36. A. Nazir, B.W. Lovett, S.D. Barrett, T.P. Spiller, and G.A.D. Briggs, *Phys. Rev. Lett.* **93**, 150502 (2004).

37. W.J. Munro, D.F.V. James, A.G. White, and P.G. Kwiat, *Phys. Rev. A* **64**, 030302 (2001).

38. B.W. Lovett, *New J. Phys.* **8**, 69 (2006).

39. D. Loss and D.P. DiVincenzo, *Phys. Rev. A* **57**, 120 (1998).

40. E. Pazy, E. Biolatti, T. Calarco, I. D'Amico, P. Zanardi, F. Rossi, and P. Zoller, *Europhys. Lett.* **62**, 175 (2003).

41. I. D'Amico, *Microelectron. J.* **37**, 1440 (2006).

42. I. D'Amico, Relatively short spin chains as building blocks for an all-quantum dot quantum computer architecture, in: *Semiconductor Research Trends (2007) (edited by* K.G. Sachs, Nova Science Publishers).

43. M. Christandl, N. Datta, A.C. Dorlas, A. Ekert, A. Kay, and A. Landahl, *Phys. Rev. A* **71**, 032312 (2005).

44. S. Bose, *Phys. Rev. Lett.* **91**, 207901 (2003).

45. R.S. Knox and H.V. Amerogen, *J. Phys. Chem. B* **2002**, 5289 (2002).

46. S.A. Crooker *et al.*, , *Phys. Rev. Lett.* **89**, 186802 (2002).

47. M. Feng, I. D'Amico, P. Zanardi, and F. Rossi, *Phys. Rev. A* **67**, 014306 (2003).

48. T.P. Spiller, I. D'Amico, and B.W. Lovett, *New J. Phys.* **9**, 20 (2007).

49. R. Raussendorf and H.J. Briegel, *Phys. Rev. Lett.* **76**, 722 (1996).

50. A. Nazir, B.W. Lovett, and G.A.D. Briggs, *Phys. Rev. A* **70**, 052301 (2004).

CHAPTER 24

Quantum Optics with Single CdSe/ZnS Colloidal Nanocrystals

L. Coolen,[1] X. Brokmann,[2] and J.P. Hermier[1,3]

[1]Laboratoire Kastler Brossel, Ecole normale supérieure, Université Pierre et Marie Curie Paris 6,
CNRS, 24 rue Lhomond 75231 Paris Cedex 05, France;
[2]Institut d'Electronique de Microélectronique et de Nanotechnologies,
avenue Poincaré, 59652 Villeneuve d'Ascq, France;
[3]Groupe d'Etudes de la Matière Condensée, UMR 8635 (CNRS), Université de Versailles
Saint-Quentin-en-Yvelines 45, avenue des Etats-Unis 78035 Versailles, Cedex.

24.1   Introduction

CdSe colloidal semiconductor nanocrystals illustrate the spectacular progresses made during the last few years in the synthesis of nanostructures. They are small crystalline objects containing between 1000 and 100 000 atoms. Their properties are dominated by quantum confinement effects and depend crucially on their size. Using processes first developed by three pioneer groups directed by P. Alivisatos (Berkeley), M. Bawendi (MIT), and P. Guyot-Sionnest (Chicago), their size can be tuned from 2 to 10 nm by adjusting the temperature, the concentration of chemical precursors or the duration of the reaction. The size dispersion of the nanocrystals is lower than 5%. Controlling the size during synthesis the wavelength in the visible spectrum can be tuned over several hundred of nanometres. Nanocrystals can be used at room temperature and present high photo-stability and high quantum efficiency. All these properties make them very attractive solid-state light sources, which have already been used in a wide range of applications such as the realization of optoelectronic devices as thin film LEDs or for biological labelling.

At the single molecule level, CdSe colloidal nanocrystals exhibit very remarkable properties. The fluorescence is known to display fluorescence intermittency resulting in the alternation of dark and bright periods and showing exotic statistical properties. Single-photon emission has also been demonstrated using the fluorescence of individual nanocrsytals. In the field of quantum information processing, single photons could be used for optical quantum computation or quantum cryptography. Single-photon emission has been observed for other solid-state sources such as organic molecules in a matrix, (epitaxially grown) "self-assembled" quantum dots and nitrogen vacancy centres or NE8 defects in diamond. Compared to these sources, colloidal nanocrystals exhibit specific phenomena such as strong interactions between charge carriers known as Auger processes.

The aim of this chapter is to discuss the distinctive optical properties of colloidal quantum dots from the perspective of realizing quantum optics experiments. Nanocrystals with a CdSe core show the best optical properties in terms of photostability, radiative quantum efficiency and photostability. We focus our discussion here on these nanocrystals. However, we also expose the perspectives offered by other materials such as PbSe, PbS, InP or CdTe. In section 24.2, we present the theoretical aspects on quantum optics used here for the characterization of a single-photon source. Then we summarize the basic optical properties of CdSe colloidal quantum dots. Due to quantum confinement, the fluorescence of individual nanocrystals is similar to a two-level system in terms of single-photon generation. Section 24.4 discusses the characteristics of this single-photon source. Since colloidal quantum dots are solid states structures, they are not isolated and interact strongly with their surrounding environment. The consequences on the fluorescence of CdSe nanocrystals are reported in section 24.5. We detail the properties of fluorescence intermittency and more generally show that emitter/environment interactions induce a dynamics in many fluorescence characteristics. More specifically, in section 24.6, we analyse the diffusion of the wavelength emission in terms of time coherence and coalescence of single photons. The two final sections gather preliminary results concerning new quantum optics experiments carried-out on CdSe colloidal quantum dots. In section 24.7, we discuss the possibility of generating entangled photon pairs using biexcitonic emission. Section 24.8 presents the most recent implementations of cavity quantum electrodynamics concepts applied to control the emission of CdSe nanocrystals. Finally, we conclude on the perspectives opened by colloidal quantum dots in terms of realizing new quantum optics experiments.

24.2 Theoretical background on single-photon sources

24.2.1 Introduction: quantum key distribution and single photons

Among the fields in which single-photon sources could be useful, quantum cryptography is the closest to practical realization, and some commercial products are already available [1].

The aim of quantum cryptography is to permit the secure distribution of a secret key between a sender (conventionally named Alice) and a receiver (Bob). The idea of quantum key distribution (QKD) was first proposed by S. Wiesner in the 1970s [2] and by C.H. Bennett and G. Brassard in 1984 [3]. It relies on the fundamental rule of quantum physics that any measurement perturbs a system. This allows Alice and Bob to detect the presence of an eavesdropper (Eve), over whatever technical means Eve may have. For a more precise description of QKD protocols, one can read the review by Gisin *et al.* [4].

Photons, being quickly and easily transmitted over long distances, are the most efficient way to encode the secret key. If each bit is coded on a single photon and the transmission and detection are perfect, eavesdroppers are always detected. In a real system, however, Eve can perform a few attacks which will be mistaken for transmission losses or detector dark counts. There is thus, for a given detector efficiency and dark count rate, a maximum amount of losses above which QKD is not secure (as discussed in [5]). Losses being a function of distance, this corresponds to a maximum distance over which QKD can be securely performed.

The first realization of QKD based on single photons was demonstrated in 1992, over a 30 cm distance in free space [6]. Since then, a great effort has been made on the transmission and detection devices, motivated by the perspective of creating a QKD network involving telecom-fibre networks and satellites [7–9]. Transmission distances up to 100 km of telecom fibre [10, 11], 1 km in free space at nighttime [8] and 75 m in free space under bright daylight conditions [7] can be achieved.

In these experiments, single photons were approximately obtained by attenuating laser pulses. These can be described by a coherent state with low average photon number: $\langle N \rangle \ll 1$. The probability of finding N photons in such a coherent state follows the Poisson distribution:

$$p(N) = e^{-\langle N \rangle} \frac{\langle N \rangle^N}{N!}. \tag{24.1}$$

Thus there is a non-zero probability (around $\langle N \rangle^2/2$) of having more than one photon coding a bit. Eve can then in principle select these bits, keep one photon in order to detect its state and send the others to Bob without being noticed. To prevent such a "photon number splitting attack", the ratio of multiphoton- to one-photon-qubits $\langle N \rangle/2$ can be made arbitrarily small by choosing a large pulse attenuation ($\langle N \rangle \to 0$), but then most pulses are empty ($p(0) \sim 1$) and Bob measures mostly dark counts. Eventually, this leads to a reduced secure QKD distance [5].

For this reason, as well as for further fundamental-optics and quantum-computing realizations, various "true" single-photon sources have been investigated: trapped atoms and ions [13–15], molecular impurities in a crystal [16–19] or in a polymer matrix [20], some nitrogen-vacancy defects in diamond called "colour centres" [21–23], semiconductor self-assembled quantum dots [24–26], and the semiconductor colloidal nanocrystals [27, 28, 80] which are the subject of this chapter.

In 2002, QKD with true single photons was realized, using either a colour centre [29] or a quantum dot in a micropost cavity [30]. By simulating a variable transmission distance with an attenuator, Waks *et al.* demonstrated experimentally that longer secure-transmission distances could be achieved with a true single-photon source [30].

24.2.2   Statistical properties of a light beam

Let us now set the basic elements of theory which are involved in single-photon characterization. Demonstrations and further developments on this matter can be found in quantum optics textbooks such as [31].

24.2.2.1   Quantized electromagnetic field

A mode of the electromagnetic field is defined by its wave vector $\vec{k}$, its angular frequency $\omega_{\vec{k}}$ and its polarization $\vec{\varepsilon}$ perpendicular to $\vec{k}$. The operator which creates a photon in this mode is written $\hat{a}^{\dagger}_{\vec{k},\vec{\varepsilon}}$ ($\hat{a}_{\vec{k},\vec{\varepsilon}}$ being the operator which annihilates a photon in this mode).

In the Heisenberg picture, the electric field in a cavity of volume V writes, as a function of time t and position $\vec{r}$:

$$\hat{\vec{E}}(\vec{r},t) = \hat{\vec{E}}^{+}(\vec{r},t) + \hat{\vec{E}}^{-}(\vec{r},t) \tag{24.2}$$

where the positive-frequency component $\hat{\vec{E}}^{+}$ decomposes into the different mode annihilations:

$$\hat{\vec{E}}^{+}(\vec{r},t) = i\sum_{\vec{k},\vec{\varepsilon}} \sqrt{\frac{\hbar\omega_{\vec{k}}}{2\varepsilon_0 V}} e^{i(\vec{k}\cdot\vec{r}-\omega_{\vec{k}}t)}\hat{a}^{\dagger}_{\vec{k},\vec{\varepsilon}}\vec{\varepsilon} \tag{24.3}$$

and the negative-frequency component $\hat{\vec{E}}^{-}$ is its Hermitian conjugate – a sum of creation operators.

For the following, we assume the field operator to be a scalar (meaning that the field is polarized).

24.2.2.2   Photodetection

Photodetectors in the optical range, such as photomultipliers or photodiodes, rely on a quantum process in which a bound electron is excited by the incoming light to a continuum. The free electron gives rise to a photocurrent $\hat{I}$ which is measured by electronic means. The quantum analysis of this problem, developped by Glauber in [32], shows that the mean photocurrent is:

$$\langle \hat{I}(t) \rangle = \alpha \langle \hat{E}^{-}(t)\hat{E}^{+}(t) \rangle \tag{24.4}$$

where the field is considered at the position of the detector* and α depends on the detector. If one assumes a photodetector with 100% quantum efficiency, one can define the photocurrent operator:

$$\hat{I}(t) = \alpha\hat{E}^{-}(t)\hat{E}^{+}(t) \tag{24.5}$$

which enables us to consider not only mean values but also photocurrent quantum fluctuations.

---

*One should actually sum over the whole photodetector area: let us neglect this spatial aspect.

Let us for clarity discuss here the case of a single-mode field. The photodetection signal $\hat{I}$ is proportional to the photon number $\hat{a}^{\dagger}\hat{a} = \hat{N}$. If the different photons are statistically independent, the photodetections are distributed following Poisson statistics, so that the variance of the measured N is $\Delta N = \sqrt{\langle N \rangle}$. Coherent states (defined as eigenstates of $\hat{a}$) exhibit such a Poissonian distribution. Usual laser beams, as previously mentionned, can be described by coherent states, so that their photon number fluctuates as $\sqrt{\langle N \rangle}$. This noise, called "shot noise", constitutes the lowest limit for the light fluctuations of classical sources like thermal or discharge lamps.

Sub-Poissonian instensity fluctuations can be obtained for states called "intensity squeezed states" at the cost of increased phase fluctuations. If one can produce a number state such as the single-photon state $| 1 \rangle$, N will be defined exactly ($\Delta N = 0$) but the phase will be random.

States showing sub-Poissonian intensity fluctuations can be used to enhance the sensitivity of intensity measurements. For example, if one wants to measure a weak absorption signal, it is better to send a single photon: the absence of photon detection at the output will correspond to one absorption event [33].

24.2.2.3 Second-order coherence

In order to characterize single photons and other sub-Poissonian behaviours, one usually measures the second-order** correlation function $g^{(2)}(\tau)$. When we restrict ourselves to a stationary state of light, the $g^{(2)}$ function is symmetrical and defined for $\tau > 0$ by:

$$g^{(2)}(\tau) = \frac{\langle \hat{E}^-(t)\hat{E}^-(t+\tau)\hat{E}^+(t+\tau)\hat{E}^+(t) \rangle}{\langle \hat{E}^-(t)\hat{E}^+(t) \rangle^2}. \tag{24.6}$$

Equation 24.6 corresponds to the quantum expression of the classical normalized intensity correlation $\langle I(t)I(t+\tau)\rangle/\langle I(t)\rangle^2$.

Long-time correlations are simply measured by use of a photodiode, either by measuring $I(t)$ and calculating $g^{(2)}(\tau)$, or, in the photon-counting regime, by plotting the histogram of the delays between detected photons.

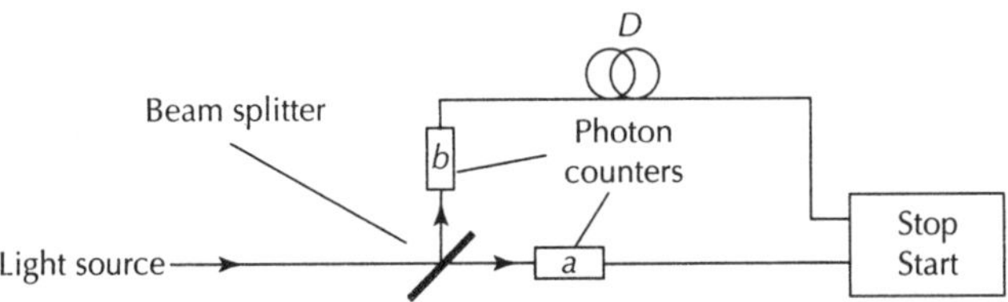

Figure 24.1   Hanbury-Brown and Twiss set-up (a 50/50 non-polarizing beam splitter and two detectors), with a "start–stop" analyser and a delay D between the two photodetectors.

This method is not appropriate for short-time correlations in the photon-counting regime, since the avalanche photodiodes used for single-photon detection exhibit a period of blindness ("dead time") of a few tens of nanoseconds after each detection. This difficulty is overcome by use of the set-up proposed by Hanbury-Brown and Twiss [34], which measures the cross-correlations between the intensities detected at the two outputs of a 50/50 beam splitter (Fig. 24.1):

$$g_{ab}(\tau) = \frac{\langle \hat{E}_a^-(t)\hat{E}_b^-(t+\tau)\hat{E}_b^+(t+\tau)\hat{E}_a^+(t) \rangle}{\langle \hat{E}_a^-(t)\hat{E}_a^+(t) \rangle \langle \hat{E}_b^-(t)\hat{E}_b^+(t) \rangle} \tag{24.7}$$

**The first-order correlation function is defined by $g^{(1)}(\tau) = \langle \hat{E}^-(t+\tau)\hat{E}^+(t)\rangle/\langle \hat{E}^-(t)\hat{E}^+(t)\rangle$ and is the Fourier transform of the emission spectrum (Wiener–Khintchine theorem). It is measured by interference experiments, such as the Fourier spectroscopy protocol described in Fig. 24.19.

In a quantum description, the fields at the outputs are given by the same relations as in a classical one[***]:

$$\hat{E}_a = \frac{1}{\sqrt{2}} \hat{E}$$
$$\hat{E}_b = -\frac{1}{\sqrt{2}} \hat{E}$$

(24.8)

so that the measured normalized cross-correlation g_{ab} is equal to the second-order correlation $g^{(2)}$ of the incident beam.

The electronic treatment of the data is then done by a "start–stop" time-interval analyser. Since the analyser cannot work on strictly zero times, a delay D is introduced so that $g^{(2)}(0) = g_{ab}(D)$ can be measured.

Let us mention that a start-stop system plots the distribution of the delays between *consecutive* photons. This is not strictly equal to the distribution of delays between *all* photon pairs which would yield $g^{(2)}$, but is approximately equal to it for delays shorter than the inverse of the photon detection rate, which is typically 10 to $100\,\mu s$ (see [35], p. 720).

Let us now consider, from a theoretical point of view, the second-order correlations of a state of light. For a coherent state, one easily finds:

$$g^{(2)}(\tau) = 1.$$

(24.9)

Thus detection of one photon tells us nothing about the probability to detect the next one. This enforces the Poissonian image of coherent states as a beam of independent photons.

Using a classical description of the electromagnetic field, Cauchy's inequality $2\langle I(t_1)\rangle$ $\langle I(t_2)\rangle \leqslant \langle I(t_1)\rangle^2 + \langle I(t_2)\rangle^2$ leads to $g^{(2)}(0) \geqslant 1$. Since $\langle (I(t_1) + I(t_2))\rangle^2 \geqslant 0$, one can also show that $g^{(2)}(0) \geqslant g^{(2)}(\tau)$. These equations indicate that photons emitted by a classical source are more likely to be detected together ("bunched"). These two inequalities become equalities in the case of a coherent state: for intensity correlations, as for intensity fluctuations, the coherent state sets the limit between super-Poissonian behaviour and non-classical sub-Poissonian behaviour.

We now turn to the number states $|n\rangle$, corresponding to exactly n photons in a certain mode. We have already shown how these states exhibit sub-Poissonian behaviour in terms of fluctuations. Here we find:

$$g^{(2)}(0) = \frac{n-1}{n} < 1.$$

(24.10)

This effect can be understood just by considering photons as particles. Since the total number of photons is known, detection of one of them makes it less likely to detect some others: the emission is called "antibunched". In particular, for the single-photon case $|1\rangle$, there is no photon left to detect after the first one and $g^{(2)}(0) = 0$. Measuring this complete antibunching is the standard way to evidence single-photon sources. Partially antibunched emission was first observed in 1977 from a beam of sodium atoms [12]. Complete antibunching was later demonstrated from a single trapped ion [13] and then from other sources such as molecules and semiconductor nanostructures.

The variation of $g^{(2)}$ corresponding to different kinds of light emitters is plotted on Fig. 24.2.

24.2.2.4 Two-level system

In a first approximation, single-photon sources are usually modelled by a two-level system: a fundamental state $|f\rangle$ and an excited state $|e\rangle$. One can easily understand that, because of the finite duration of its excitation–emission cycle, this system cannot emit two photons at the same time.

[***]The vacuum field impinging on the other side of the photodiode is omitted here as it has no effect on the calculated correlations.

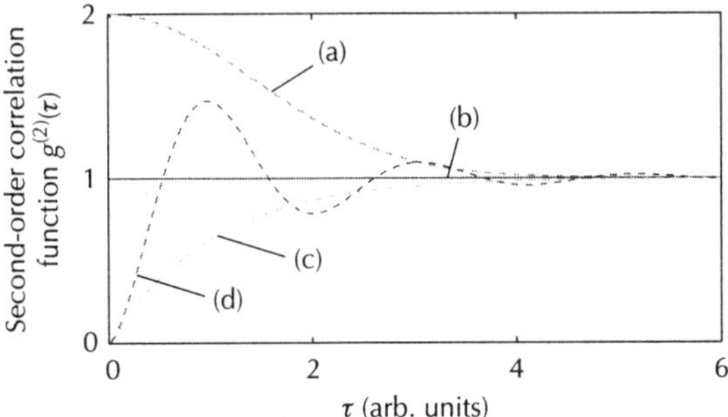

Figure 24.2   *Second-order correlation* $g^{(2)}$ *function in the cases of (a) a classical lamp with a collision broadening of 1, (b) a coherent laser beam, or a single-photon source of lifetime* $T_1 = 1$, *(c) pumped off-resonance at a rate 0.1, or (d) excited at resonance with a Rabi frequency 3 (arbitrary units).*

This is confirmed by basic quantum optics considerations. It can be shown that the emitted field, far from the source, is proportional, in the source domain, to the dipole operator. More precisely, detection (annihilation) of a photon (term $\hat{E}^+$) corresponds to a relaxation of the system (term $\mathscr{\hat{S}}^- = |f\rangle\langle e|$) – and in the same way $\hat{E}^-$ is associated to $\mathscr{\hat{S}}^+ = |e\rangle\langle f|$. The effect on expectation values is, for instance:

$$\langle \hat{E}(t_1)^- \hat{E}^+(t_2)\rangle = w^2 \langle \mathscr{\hat{S}}^+(t_1).\mathscr{\hat{S}}^-(t_2)\rangle \tag{24.11}$$

where the coefficient w depends on the source and the collection efficiency, and similar relations hold for higher-order correlations. Given that $\mathscr{\hat{S}}^+\mathscr{\hat{S}}^+ = 0$, it is easy to show that the emission from a two-level system is completely antibunched ($g^{(2)}(0) = 0$).

If one considers the emission from $\mathscr{N}$ independent dipoles, it can be shown that $g^{(2)}(0) = (\mathscr{N} - 1)/\mathscr{N}$, so that the emitted light is bunched for any $\mathscr{N} \geqslant 2$ number of dipoles. With a large number of dipoles we find the classical case: $g^{(2)}(0) = 2$. In order to obtain a single-photon source, it is thus required to collect the fluorescence of a single emitter.

In order to calculate the theoretical $g^{(2)}(\tau)$ function and compare it with experimental results, one must introduce a more specific model. Spontaneous emission has to be described in a density-matrix formalism.* The elements of the density matrix $\hat{\rho}$ are defined as:

$$\rho_{fe} = \langle \mathscr{\hat{S}}^+\rangle$$
$$\rho_{ef} = \langle \mathscr{\hat{S}}^-\rangle$$
$$\rho_{ee} = 1 - \rho_{ff} = \frac{\langle \mathscr{\hat{S}}_z\rangle + 1}{2} \tag{24.12}$$

where $\mathscr{\hat{S}}_z = |e\rangle\langle e| - |f\rangle\langle f|$ is the population inversion operator, the diagonal components ρ_{ff} and ρ_{ee} are the populations of levels $|f\rangle$ and $|e\rangle$, and the off-diagonal components ρ_{ef} and ρ_{fe} are the coherences of the dipole.

We first consider the case of an off-resonance excitation, which is the most common in practice (as photoluminescence cannot be spectrally separated from scattered excitation if excitation is resonant). By defining $\hbar\omega_0$, $\Gamma_1 = 1/T_1$, $\Gamma_2 = 1/T_2$ and r being, respectively, the energy between $|e\rangle$ and $|f\rangle$, the excited-state decay rate, the emitting-dipole decoherence rate and the pumping rate (Fig. 24.3), one writes:

$$\dot{\rho}_{ff} = \Gamma_1\rho_{ee} - r\rho_{ff}; \qquad \rho_{ef} = \rho_{fe}^*$$
$$\dot{\rho}_{fe} = (i\omega_0 - \Gamma_2)\rho_{fe}; \qquad \rho_{ff} + \rho_{ee} = 1. \tag{24.13}$$

---

*Or in the more recent "quantum jumps" formalism [36].

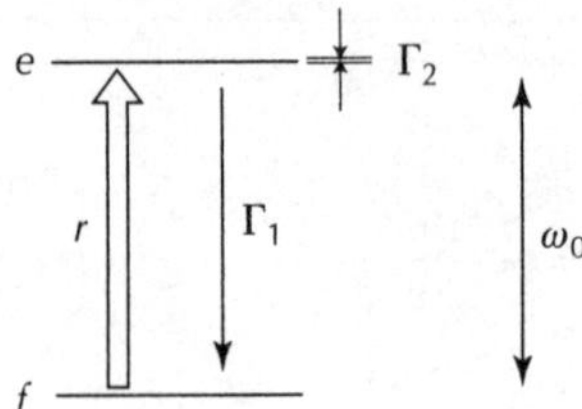

Figure 24.3   Model for a system with two levels $|f\rangle$ and $|e\rangle$ excited off-resonance: $\hbar\omega_0$, Γ_1, Γ_2 and r are, respectively, the energy between $|e\rangle$ and $|f\rangle$, the excited-state decay rate, the emitting-dipole coherence and the pumping rate.

Quantum correlations for an operator $\hat{\mathscr{O}}(t)$ are calculated by use of the quantum regression theorem, which states that [37]:

$$\text{if, for} \quad \tau > 0, \langle \hat{\mathscr{O}}(t+\tau)\rangle = \sum_k f_k(\tau)\langle \hat{\mathscr{O}}_k(t)\rangle,$$
$$\text{then} \quad \langle \hat{\mathscr{O}}_i(t)\hat{\mathscr{O}}(t+\tau)\hat{\mathscr{O}}_j(t)\rangle = \sum_k f_k(\tau)\langle (\hat{\mathscr{O}}_i\hat{\mathscr{O}}_k\hat{\mathscr{O}}_j)(t)\rangle.$$

Solving Eqs 24.13 leads to:

$$\langle \hat{\mathscr{S}}^+(t+\tau)\rangle = e^{(i\omega_0-\Gamma_2)\tau}\langle \hat{\mathscr{S}}^+(t)\rangle$$

$$\left\langle \left(\frac{\langle \hat{\mathscr{S}}_z\rangle+1}{2}\right)(t+\tau)\right\rangle = \frac{r}{r+\Gamma_1}(1-e^{-(\Gamma_1+r)\tau})+e^{-(\Gamma_1+r)\tau}\left\langle \left(\frac{\langle \hat{\mathscr{S}}_z\rangle+1}{2}\right)(t)\right\rangle \tag{24.14}$$

and, by direct application of the quantum regression theorem, one obtains:

$$g^{(1)}(\tau) = e^{(i\omega_0-\Gamma_2)|\tau|} \tag{24.15}$$

which corresponds to a Lorentzian emission spectrum of width $2\Gamma_2$, and:

$$g^{(2)}(\tau) = 1 - e^{-(\Gamma_1+r)|\tau|} \tag{24.16}$$

which indicates a complete antibunching dip of width $1/(\Gamma_1 + r)$.

As an experimental example, Fig. 24.4 shows the measured intensity correlations for a colloidal nanocrystal. They clearly agree with the present theoretical description. The excitation being performed well below saturation ($r \ll \Gamma_1$), a value $T_1 = 20$ ns can be measured, in accordance with direct lifetime measurements. A 95% antibunching is obtained. The non-zero value of $g^{(2)}(0)$ is due to the background noise, which was not included in our model. When the

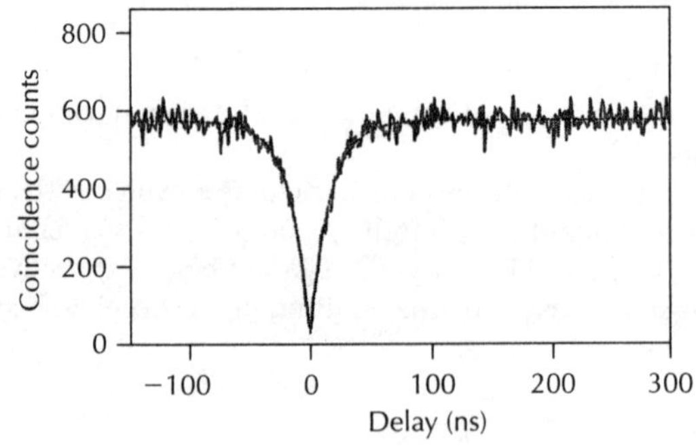

Figure 24.4   Second-order correlation function of the emission of a CdSe/ZnS colloidal nanocrystal of diameter 3.5 nm, excited well below saturation, and measured by use of a Hanbury-Brown and Twiss set-up with start–stop detection and a delay $D = 150$ ns, from [28] (authorization for reproduction asked). The solid curve is a fit by an exponential function $a[1 - b \exp(-|\tau|/\tau_0)]$, with $a = 577.5$, $b = 0.95$ and $\tau_0 = 20.1$ ns.

background is correctly substracted, one finds an antibunching greater than 99%. Other single-photon sources were evidenced in the same way.

24.2.2.5 Single photons on demand

Single-photon emission has been evidenced for a few sources by antibunching experiments. However, when the source is under continuous excitation (as on Fig. 24.4), single photons are separated by an interval which cannot be controlled. A better way to proceed is to excite the source by a pulsed laser, with a pulse duration t_{pulse} much shorter than T_1 (so that only one emission occurs during one pulse), and intervals between pulses T_{pulse} much longer than T_1 (so that two photons corresponding to consecutive excitation pulses are well separated in time) (see Fig. 24.5). The delay between two photons is then a non-zero multiple of T_{pulse}: $nT_{pulse} \pm 2T_1$, as evidenced by intensity-correlation measurements [38].

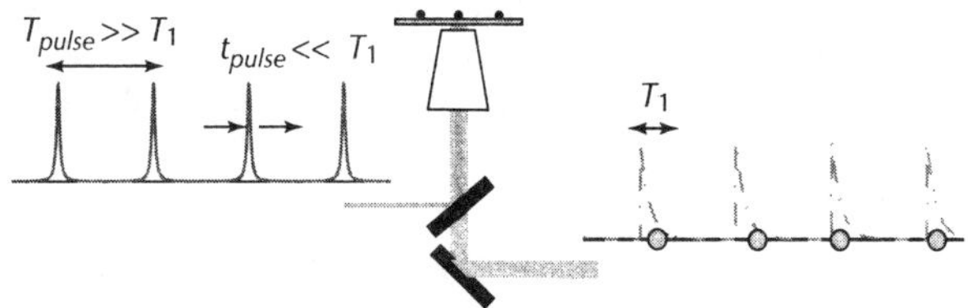

Figure 24.5   Realization of an on-demand single-photon source by pulsed excitation of a nanocrystal, with a pulse duration t_{pulse} shorter than the excited-state lifetime T_1, and an interval between two pulses T_{pulse} longer than T_1.

Although photon antibunching is sufficient for quantum key distribution, it would be preferable to generate *on demand* single photons: one photon detected for each excitation pulse (Fig. 24.5). As explained in the beginning of this section, only a certain amount of losses are acceptable to perform secure QKD, which results in a limited distribution distance. If only some of the excitations lead to single photon detection, this is equivalent to losses and reduces the secure-QKD distance. Realization of on-demand single-photon sources is one of the challenges in single-photon studies. It requires:

- A pulse intensity sufficient to saturate the transition.
- A near-unity quantum yield (defined as $\Gamma_{1,r}/(\Gamma_{1,r} + \Gamma_{1,nr})$, $\Gamma_{1,r}$ and $\Gamma_{1,nr}$ being, respectively, the radiative and non-radiative components of the decay rate Γ_1).
- 100% collection and detection efficiency.

We now outline the optical properties of colloidal nanocrystals (typically, 3.5 nm diameter CdSe/ZnS nanocrystals). Then, we will compare them to other single-photon sources, in terms of on-demand emission and other considerations.

24.3   Optical properties of a nanocrystal

24.3.1 Colloidal semiconductor nanocrystals

The chemical synthesis of quasi-spherical nanocrystals with monodisperse nanometre size is delicate as it requires the control of the kinetics of the nanocrystal growth in order to avoid fast agglomeration. One possibility is to perform the reaction in a vitrous medium at high temperature. This method was used in the Middle Ages to fabricate coloured glasses, and more recently by companies like Corning to make optical filters. The first studies on quantum confinement by A.I. Ekimov in Leningrad in the early 1980s considered such samples [39, 40]. At the same time, the group of L.E. Brus in Bell Labs synthesized semiconductor nanocrystals in solution, by a careful control of the reagents and stabilizing agents, of the temperature and duration of the reaction, and the same confinement effect was evidenced [41]. This last method appeared to be the most promising as nanocrystals in solution could be more easily handled (see [51] for more information on nanocrystal history and synthesis).

Synthesis techniques are ideal for CdSe nanocrystals, using typically spherical 3.5 nm nanocrystals of hexagonal (wurtzite) crystalline structure, emitting at 585 nm at room temperature. A size dispersion as low as 5% is obtained. In order to passivate the unbound electrons on the surface and improve the quantum yield, a shell of one to two monolayers of another semiconductor is added. This other semiconductor must have a larger band gap, so that the charges remain confined in the core. Usually, one uses ZnS; such nanocrystals are labelled "core-shell CdSe/ZnS"[42, 43].

For the moment, we will only describe the optical properties that are related to the confinement of the charge carriers in a small volume – that is, the nanocrystal can be considered as a sphere. In section 24.5, we will describe phenomena that are caused by the passivating shell and the local environment of the nanocrystal. Two of these phenomena are:

- "Blinking": the nanocrystal switches randomly between an "on" fluorescent and an "off" non-fluorescent state.
- "Spectral diffusion": the emission wavelength changes randomly in time.

24.3.2   Elements of theoretical description

Electrons and holes in a nanometre-sized semiconductor are described either (i) by means of quantum chemical calculations (pseudo-potential method, tight binding method, etc.) – it is then easier to include a complex shape of the nanocrystal [44], surface reconstruction [45], addition of shells or ligands [46] or surface defects [47] – or (ii) in the framework of the envelope function [48], which is used here as it gives a more physical picture.

The formalism of the envelope function relies on the assumption that the nanocrystal diameter $2R$, although small, is still much larger than the crystalline lattice parameter (0.43 nm for CdSe), so that the notion of valence and conduction bands, respectively containing holes and electrons, is still relevant. The wavefunction Ψ of a charge is of the form:

$$\Psi(\vec{r}) = \Phi(\vec{r})u_\mu(\vec{r}) \tag{24.17}$$

The function u_μ has the same period as the crystalline lattice and reflects the local behaviour of the charge. It is the same function as in the bulk material, with μ indicating to which (conduction or valence) band it corresponds. Φ is an envelope function reflecting the charge distribution over the nanocrystal. If one assumes that the energy bands are parabolic, Φ is the wavefunction of a free particle (electron or hole) with effective mass $m^*_{e/h}$. The further assumption that this charge is confined in the nanocrystal by an infinite potential leads to the easy boundary condition: $\Phi(r = R) = 0$ (more elaborate models have been developed in [49] and [50]). Thus the result of "quantum confinement" is that only a few discrete energy levels compatible with $\Phi(R = 0)$ are accessible (Fig. 24.6).

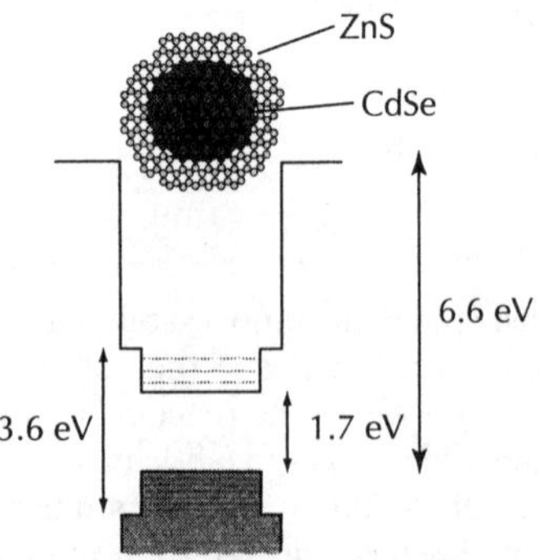

Figure 24.6   Schematic of a CdSe/ZnS nanocrystal, and the corresponding three-dimensional potential well. The dotted lines symbolize the discrete set of accessible energy levels.

As confinement and Coulombian effects scale, respectively, as $1/R^2$ and $1/R_B R$, R being the nanocrystal radius and R_B the exciton Bohr radius [51], the former effect is more important than the latter for nanocrystals smaller than R_B (5.6 nm for CdSe). In describing this "strong confinement" regime, the electron and hole are first considered separately, and Coulomb interaction is added later as a perturbation.*

For the electron, the problem is then reduced to the well-known free particle in an infinite well, which is treated in [52]. A solution in spherical coordinates is of the type $\mathscr{R}_l(k_{nl}r)\mathscr{H}_{l_z}(\theta,\phi)$, where the radial component $\mathscr{R}_l$ is expressed in terms of Bessel functions, $\mathscr{H}$ is a harmonic function, and n,l and l_z are the quantum numbers as defined for an atom. By analogy with hydrogen states, these states are labelled $n\mathscr{L}_e$, where $\mathscr{L}$ is S, P, D... when $l = 0$, 1, 2.... The discrete accessible values of k_{nl} are given by $\mathscr{R}_l(k_{nl}R) = 0$, and are thus proportional to $1/R$. The confinement energy $\hbar^2 k_{nl}^2/2m_e^*$ then varies as $1/R^2$.

For the hole, the problem is more complex as it involves the degenerate structure of the valence band [48]. A complete treatment can be found in [49]. Let us just mention that the correct quantum number is then the sum of the hole spin and the envelope momentum. It is 3/2 for the lowest hole state (labelled $1S_{3/2}$).

One can then consider the electron–hole wavefunction as the product of an electron and a hole wavefunction. The effect of the direct component of the Coulomb interaction (the smaller exchange component will be discussed later) is simply to shift the energy levels. For instance, the first $1S_{3/2}1S_e$ level is stabilized at $1.8e^2/4\pi\varepsilon_0\varepsilon_r R$ (around 150 meV in CdSe for $R = 1.7$ nm).

24.3.3 Absorption spectrum

The first decade of experiments on nanocrystals was mostly dedicated to evidencing these discrete features in the ensemble absorption spectra or photoluminescence-excitation (PLE) spectra [40, 41] (Fig. 24.7), and comparing the measured spectra with theoretical predictions. Reduction of the inhomogeneous broadening by synthesis of better samples and use of size-selective pump–probe techniques has revealed up to ten absorption lines, motivating the development of precise theoretical models including the complex valence band structure [49, 50, 53].

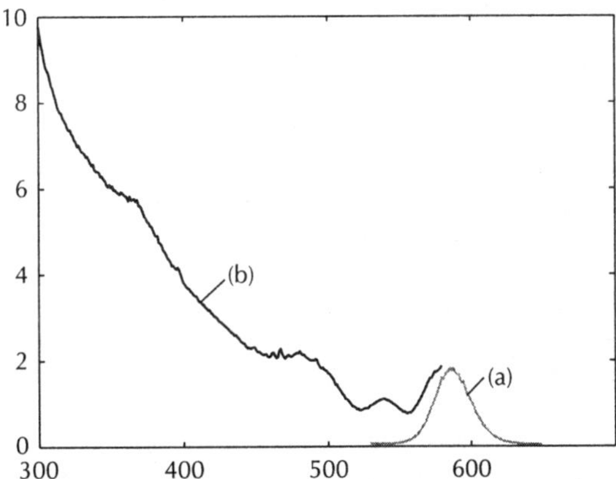

Figure 24.7   (a) Photoluminescence spectrum (with excitation at 300 nm) and (b) PLE spectrum (with photoluminescence detected at 585 nm) of a solution of 3.5 nm diameter CdSe/ZnS nanocrystals in toluene.

More than 1.2 eV above the lowest $1S_{3/2}1S_e$ level, spectrum discretization cannot be observed. The levels tend to merge into a continuum because (i) at higher energies, the dispersion relation is no longer parabolic and becomes concave, and (ii) higher levels are more degenerate, and

---

*This does not mean that it vanishes for $R \rightarrow 0$; on the contrary, Coulomb interactions are enhanced by the confinement, as it brings the charges closer to one another.

their degeneracy is lifted by perturbative effects. At these excitation energies, the absorption cross-section is proportional to the nanocrystal volume [54], and can be described in the framework of the Mie theory [55]: the nanocrystal behaves like a macroscopic sphere.

For a single nanocrystal, direct acquisition of the absorption spectrum has never been reported. For a diffraction-limited spot, the excitation area is around $1\,\mu m^2$, while the absorption cross-section is $0.1\,nm^2$ [54]. The necessary 10^7 precision could be achieved by acquiring long enough, but the acquisition time is limited as the nanocrystal blinks from "on" to statistically longer "off" states (see section 24.5).

The absorption spectrum has been measured indirectly by photothermal absorption spectroscopy, a technique that was first developed to detect non-fluorescing particles such as metallic nanocrystals. The principle is to measure, by use of a probe beam, the changes in the refractive index of the surrounding matrix caused by pump absorption-induced heating. The results were similar to those of ensemble measurements [56].

The PLE of a single nanocrystal is very delicate due to the presence of spectral diffusion. The only PLE data available to date used a constant switching between the scanning laser and a reference laser. They show the lowest $1S_{3/2}1S_e$ level separated by a $50\,meV$ "minigap" from the other levels, which form a dense quasicontinuum [57].

To sum up, because of quantum confinement, nanocrystals exhibit a discrete energy spectrum. Nanocrystals are often referred to as "artificial atoms". At this point, one might infer that this quantum-confinement discretization causes antibunched emission. In fact, other mechanisms, also related to the small size of the particles, play a major role in nanocrystal single-photon emission.

24.3.4   Exciton and multiexciton relaxation mechanisms

In a nanometre-sized structure where both charges and phonons have discrete energy spectra, a charge should *a priori* not interact with a phonon, as there is *a priori* no final state with the right energy. This "phonon bottleneck" effect [58] was expected to drastically slow down both intraband relaxation and decoherence mechanisms. It should be even more important for structures of a few nanometres like colloidal nanocrystals, as the separation between levels $1P_e$ and $1S_e$ is of a few hundred meV, while phonon energy is at most $25\,meV$.

Femtosecond transient absorption experiments showed the opposite: intraband phonon-assisted relaxation (thermalization) occurs within $500\,fs$ [59] and is faster in smaller nanocrystals [60]. The explanation is that:

- Holes relax as quickly as they do in the bulk [61]since they have a large density of states, because (i) the valence band is degenerate and (ii) the separation between states varies as the inverse of the effective mass, and holes are heavier than electrons.
- The electrons transfer their energy to the holes by a Coulombian mechanism called the Auger effect [62]. The Auger effect is enhanced by charge confinement, since it depends on the distance between the charges. Fig. 24.8a illustrates how an excited electron–hole pair decays to its lower energy state before radiative recombination.

Intraband relaxation to the lowest excited state $1S_{3/2}1S_e$ is thus much faster than radiative in-terband relaxation ($T_1 \sim 20$–$2000\,ns$), so that fluorescence emission occurs only by recombination of the exciton from its $1S_{3/2}1S_e$ state, and photoluminescence (PL) spectra show a single peak (Fig. 24.7).

Thus, in terms of emission, a nanocrystal behaves the same as a two-level system: a fundamental (no exciton) and an excited state (the $1S_{3/2}1S_e$ exciton). Excitation to a higher level followed by intraband relaxation is equivalent to the incoherent pumping which was included in the model of section 24.2. According to this model, such a system should emit single photons.

However, there is an important point which was not included in that model: it is possible, by intense excitation, to create more than one electron–hole pair in a single nanocrystal. A biexciton peak appears in absorption spectra of nanocrystal ensembles, shifted from the exciton peak by the biexciton binding energy – around $30\,meV$ [63–65]. With single nanocrystals of radius greater than 5 nm, as with self-assembled quantum dots, cascades of emissions from bi- and triexciton states have been evidenced by time-resolved spectroscopy [66]. For smaller nanocrystals, the reduced distance between the charges enhances non-radiative multiexciton-to-exciton Auger recombination, where the recombination energy is transferred to the remaining exciton instead of photons (Fig. 24.8b). Multiexciton recombination occurs within tens of ps [67, 68], faster than the radiative recombination.

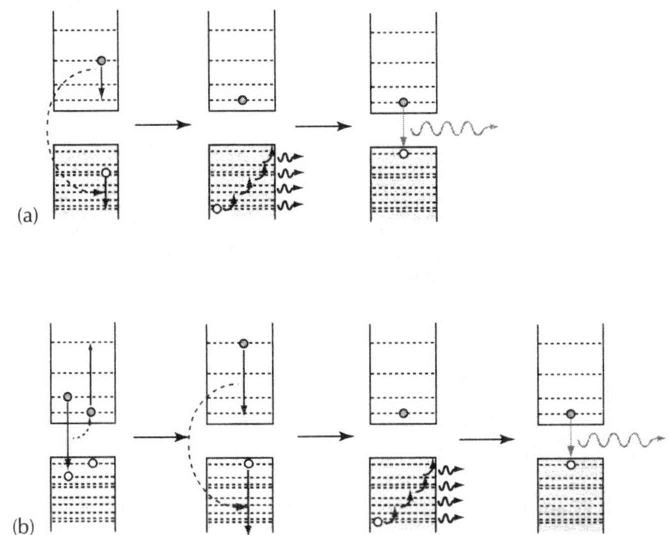

Figure 24.8   Role of Auger energy transfer in the recombination of (a) an exciton and (b) a biexciton.

For small nanocrystals, although multiexcitons can be created, there is no cascade of photons, and nanocrystal fluorescence shows complete antibunching. The possibility of observing multi-excitonic emission by nanocrystals with large radii will be analysed in detail in section 24.7.

24.3.5   Fine structure of the emission level

In the confinement model described above, the emitting level $1S_{3/2}1S_e$, having electron spin $1/2$ and hole momentum $3/2$, is eight-fold degenerate. This degeneracy can be lifted by three pertur-bative effects: the small gap between the light- and heavy-hole bands due to the hexagonal (not cubic) crystalline structure, the slightly prolate shape of the nanocrystal [42], and the exchange term in the electron–hole Coulombian interaction, which is enhanced by confinement. The first two terms just lift the degeneracy, both along the crystalline c-axis (labelled z) as the nanocrystal elongation follows the c-axis (for reasons related to the growth mechanism). The latter term couples the electron and the hole, so that the "right" quantum number is now the total momentum $\vec{F}$.

A complete mathematical treatment can be found in [70]; the result for a CdSe/ZnS nanocrystal of diameter 3.5 nm is shown in Fig. 24.9a. The main result is that the lowest sub-level, $|F_z = \pm 2\rangle$ (two-fold degenerate), cannot emit light through a linear process as it exhibits a $\pm 2\hbar$ momentum difference with the fundamental level. Emission arises mostly from two sublevels of z-momentum $\pm\hbar$, which are around 10 meV higher than the "dark exciton" state $|\pm 2\rangle$. After excitation, the system relaxes rapidly to the dark exciton state; it is later excited to the bright state by the phonon bath, and emits a photon. This process is faster at higher temperature, which explains the observed dependence of the excited-state lifetime T_1 as a function of temperature, from 20 ns at 300 K to 300 ns at 10 K and 1 μs below 2 K [72, 73] (Fig. 24.9b). At 2 K, the bright level is not thermally populated at all, and there is only a weak phonon-assisted fluorescence from the dark state [74].

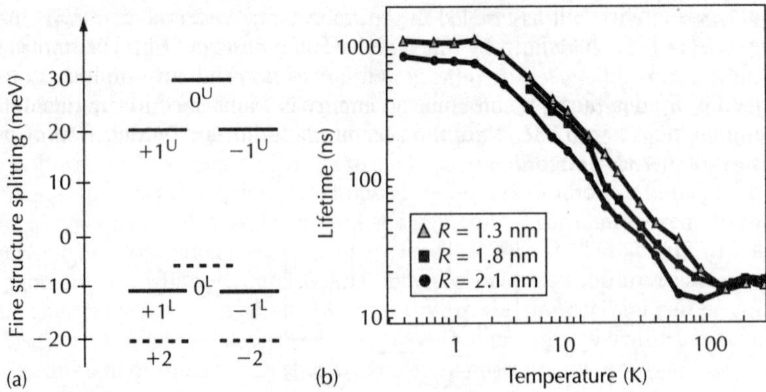

Figure 24.9   (a) Fine structure of the emitting $1S_{3/2}1S_e$ *level of a 1.8 nm radius CdSe nanocrystal, as calculated from [70] and [71], with the levels labelled by the z-component of the total electron–hole momentum* $\bar{F}$. *(b) Emission lifetime of CdSe/ZnS nanocrystals of various radii R, measured as a function of temperature (reproduced from [72], authorization for reproduction asked).*

At room temperature, such an emission from two degenerate levels $|+1\rangle$ and $|-1\rangle$ shows two original characteristics:

- It is an incoherent sum of two σ^+ and σ^- emissions.
- It is oriented according to a peanut-shaped diagram directed along the c-axis.

Both aspects have been used to confirm that the emission of a nanocrystal does not correspond to a linear dipole but to a two-dimensional degenerate dipole.*

On the one hand, the polarizations successively measured on 170 nanocrystals showed a distribution which was not compatible with a randomly oriented one-dimensional dipole [76]; polarization measurements were used to determine the orientation of a nanocrystal relative to the sample plane [77].

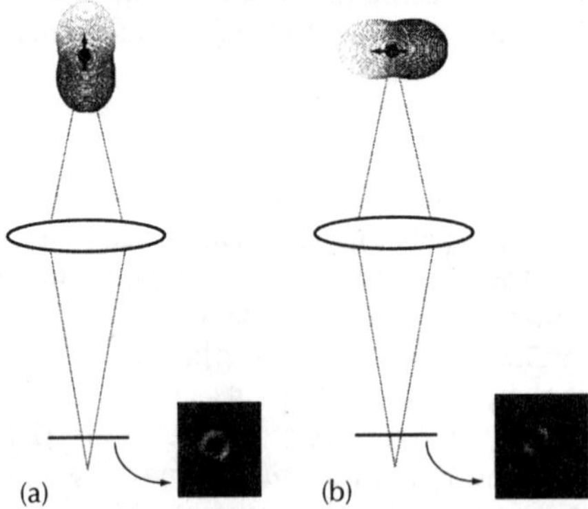

Figure 24.10   Principle and experimental result of a defocused-microscopy orientation measurement for a CdSe/ ZnS nanocrystal with c-axis (a) perpendicular and (b) parallel to the sample plane [79]. The experiment consists in measuring an image of the nanocrystal with a CCD camera at one micron from the focal plane.

On the other hand, emission diagrams and thus nanocrystal orientation were evaluated by a defocused-microscopy technique, as explained in Fig. 24.10: a nanocrystal was imaged on a slightly (1 μm) defocused CCD camera; for a nanocrystal with c-axis perpendicular to the sample

---

*At cryogenic temperature (<4 K), a splitting of 1–2 meV has recently been reported, and attributed to a non-perfect cylindrical symmetry along the c-axis [75]. We will discuss the influence of this splitting in section 24.7.

plane, the defocused image was an isotropic ring; for a nanocrystal with c-axis parallel to the sample plane, it was a ring with lobes oriented along the c-axis [78]. By varying the distance of the nanocrystal to an interface (following the procedure detailed in Fig. 24.11), one generates a change in the lifetime T_1 which depends on the nanocrystal orientation. Comparing this change and the diagram observed by defocused microscopy, it could be proved, on a single nanocrystal, that the dipole had to be in two dimensions [79].

24.4 Nanocrystals as single-photon sources

Now that the fundamental optical properties of colloidal nanocrystals have been explained, we can consider how well they qualify as single-photon sources, and how they compare to other sources.

24.4.1   Photon antibunching

A first requirement is of course to emit single photons. As described before, antibunched emission from colloidal nanocrystals was reported in 2000–2001 [24, 28, 69], and after subtraction of the background noise a $g^{(2)}(0)$ value smaller than 0.01 was found. Single-photon emission was exhibited by other sources: atoms, ions, molecules, colour centres, self-assembled quantum dots, etc.

It is worth mentioning that in fact self-assembled quantum dots do not emit single photons. For these dots, which are larger than nanocrystals, the Auger effect is not fast enough to suppress radiative multiexciton recombination, so that single photons can only be obtained by spectrally selecting the one-exciton line. This is possible only at cryogenic temperature as the exciton and biexciton lines are broadened and merge as the temperature increases. Thus, in order to obtain single photons from a quantum dot, one needs to work at cryogenic temperature and to add a spectrometer before detection. Colloidal nanocrystals, like terrylene molecules and colour centres, directly emit single photons at room temperature, which makes them quite easy to use. Moreover, working at room temperature allows the use of a high-NA immersion objective, leading to simple and efficient light collection.

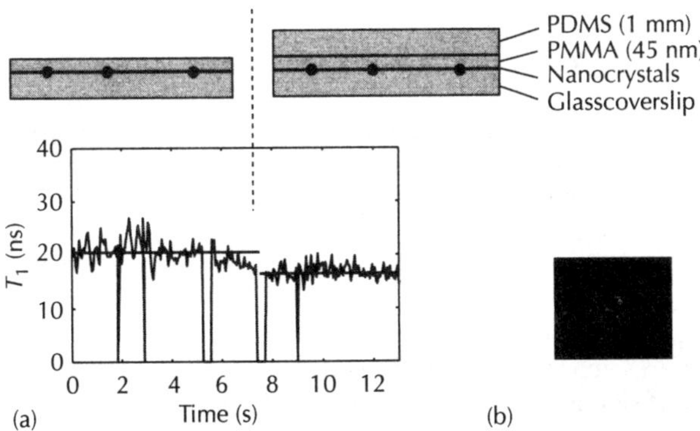

Figure 24.11   An application of the method demonstrated in [79] to measure the quantum yield of a single emitter. One considers a sample with nanocrystals spin coated on a glass coverslip, and covered by a polymethylmetacrylate (PMMA) 45 nm layer. PMMA has the same refractive index as glass, so that each nanocrystal is in a 1.5 index medium, at 45 nm from an interface with a medium of index 1 (air). At $t = 7$ s, a droplet of polydimethylsyloxane (PDMS) is deposited. PDMS is a liquid polymer of the same 1.5 index, so that after droplet deposition the nanocrystal is in a homogeneous 1.5 index medium. (a) A measure of the lifetime change $\beta = T_{1,after}/T_{1,before}$ after deposition of the droplet (here: $\beta = 0.81$). (b) A measure of the nanocrystal orientation (here: horizontal). Knowing the orientation, the radiative decay rate $\Gamma_{1,r}$ can be calculated as a function of the interface distance, and the $\Gamma_{1,r}$ change $1/\alpha = \Gamma_{1,r,after}/\Gamma_{1,r,before}$ is obtained. The quantum yield is then $(\beta - 1)/(\alpha - 1)$ (here: 0.99).

24.4.2   Single-photon emission on demand

Let us now examine the possibility of emitting single photons on demand. This requires absorption saturation, a quantum yield equal to 1, and 100% collection.

The broad absorption continuum renders nanocrystal excitation extremely easy, with any source below 530 nm. An absorption cross-section of around $0.1\,nm^2$ was measured by ensemble absorption measurement [54] or single nanocrystal saturation [80], smaller than self-assembled quantum dots ($10\,nm^2$ derived from [25]) but larger than coloured centres ($0.002\,nm^2$ derived from [81]) and terrylene molecules ($0.004\,nm^2$ [19]). Ninety five per cent saturation of the nanocrystal emission can be achieved with a standard pulsed laser diode with 5 pJ energy per pulse [38].

The fluorescence quantum yield $\Gamma_{1,r}/(\Gamma_{1,r} + \Gamma_{1,nr})$ is routinely measured by cuvette techniques, which compare the light intensities that are absorbed and emitted by a solution of emitters. This is not appropriate for nanocrystals as it gives a value (typically 40%) which is averaged on all nanocrystals, while some of them may be "off" because of the blinking. An original method has been developed for the measurement of single-emitter quantum efficiency [79]. It measures the excited-state lifetime $1/(\Gamma_{1,r} + \Gamma_{1,nr})$ consecutively with a dielectric interface close to and far from the nanocrystal, and uses theoretical calculations on the effect of a nearby interface on $\Gamma_{1,r}$ to determine $\Gamma_{1,r}$ and $\Gamma_{1,nr}$ (Fig. 24.11). An almost perfect quantum yield (95 to 100%) is derived.

A general consequence of refraction laws is that light remains mostly in the medium of higher refractive index. Colour centres and quantum dots are embedded in a medium of index higher than the objective components ($\sim$1.5), so that only a small fraction of their fluorescence is extracted and collected (Fig. 24.12a).

For a nanocrystal or a fluorescent molecule at room temperature, the standard procedure is to use an immersion objective. Nanocrystals are usually spin coated on a glass coverslip of index 1.5 and protected from oxidation by a polymer layer (usually polymethylmethacrylate – PMMA) of the same index 1.5. An objective is approached from under the coverslip, and separated from it by a droplet of immersion oil, of the same 1.5 refractive index (Fig. 24.12b). By use of a high numerical-aperture objective (1.4), most of the fluorescence is then collected: a model including the two-dimensional-degenerate emission dipole finds that around 85% of the emission is directed into the glass, and that approximately 72% of the emission is collected by the objective [38].

The main obstacle to single-photon emission on demand lies in fact in the transmission and detection step. Each optical component introduces 5 to 20% losses, and the photodiodes have at best 60% detection efficiency. With the set-up of [38], at 95% saturation and using an immersion objective with 1.4 numerical aperture and 60% efficiency photodiodes, around 8% of the pump pulses lead to photon detection, so that the transmission losses of this set-up are estimated to be 81%.

In comparison, the rate of photon detection is typically 10^{-4} per pump pulse for a quantum dot [25], 0.05 per pulse for a terrylene molecule [19], 0.02 for a trapped atom [82], and 10^{-4} for a coloured centre (derived from [21]).

For self-assembled quantum dots and colour centres, methods using cavities have been explored in order to improve the light extraction efficiency. Epitaxially grown dots have been inserted in a micropillar cavity in order to redirect the emission, yielding a photon detection rate of 0.002/pulse [26] (Fig. 24.12c). Colour centres in diamond nanocrystals have been considered: as these diamond structures are smaller than the wavelength, the low collection of light due to the high refractive index of diamond is suppressed (Fig. 24.12d). A rate of 0.013 photon detection per excitation pulse is obtained [23].

24.4.3   Other characterisitics of a single-photon source

Some other properties can be taken into account to characterize a single-photon source. A key parameter is the emitter stability. After a characteristic time of a few seconds, most fluorescent molecules "photobleach": they irreversibly become non-fluorescent. This problem is quite important for terrylene molecules, although a small fraction of them fluoresce up to 30 mn [20].

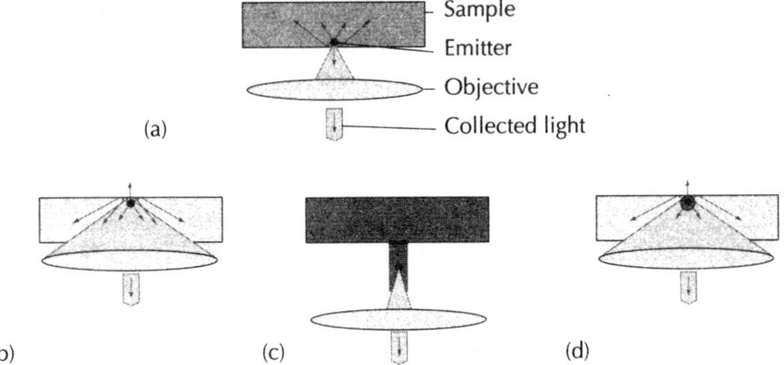

Figure 24.12 (a) The extraction problem for an emitter in (or at the surface of) a high-index sample: most of the emission remains in the sample. Three solutions: (b) the emitter is on a glass coverslip and an immersion objective is used; (c) the emitter is in a micro-cavity; (d) the emitter is in a nanocrystal.

Nanocrystals, like self-assembled quantum dots and colour centres, do not photobleach. However, as previously mentioned, they blink from "on" to "off" states. This change is reversible, but the duration of the "off" states exhibits very peculiar statistics and is on average almost infinite (see section 24.5). The explanation and eventual elimination of this "blinking" behaviour is one of the main challenges in the development of nanocrystal-based single-photon sources.

The emission wavelength could also become a crucial parameter for practical implementation of quantum cryptography protocols. In contrast with molecules and colour centres, zero-dimensional semiconductor structures offer the ability to tune the emission wavelength by adjusting their size. In particular, the size of colloidal nanocrystals is determined, with less than 5% dispersion, by the synthesis parameters. Using CdSe or CdTe cores, their emission wavelength can be tuned in the whole visible range. Commercial nanocrystals can be purchased from companies like Invitrogen or Evident Technologies.

Research on nanocrystals has focused up to now on CdSe and other II–VI materials emitting at visible wavelengths, because photodetectors and many other optics work better in this range. Silicon avalanche photodiodes exhibit very good performance (such as an efficiency of up to 60%) only between 600 and 900 nm. For large-scale applications in quantum cryptography, however, it would be better to use the existing network of telecom optical fibres, functioning between 1350 and 1550 nm. In this range, commercial avalanche photodiodes made of InGaAs layers on an InP substrate are available [1], but only have 10% efficiency [83]. Better performances have been obtained by up-converting the $1.55 \mu m$ photons to 700 nm photons and detecting them with an Si photodiode; this system achieved a 45% efficiency [84].

Synthesis of nanocrystals emitting in the near-infrared region is currently under investigation, IV–VI semiconductors like PbS or PbSe being the most promising materials. Such nanocrystals would allow transmission through telecommunication fibres or coupling to state-of-the-art photonic crystals (see section 24.8). In terms of applications of nanocrystals as labels in biology, the near-infrared constitutes a "window of biological transparency" in which it would be interesting to work. On a more theoretical level, PbS and PbSe have very large exciton Bohr radius (20 and 46 nm, respectively) so that confinement effects should be important on both electron and hole. The fluorescence properties of PbS nanocrystals are quite unsatisfying for the moment (30% quantum yield, strong photobleaching), and should be improved by progresses in synthesis reactions and addition of passivating shells [85]. Some groups are also starting experiments to investigate the synthesis of colloidal III–V quantum dots such as InP.

Another parameter is the excited-state lifetime T_1. Since the interval between two excitation pulses must be larger than T_1, the rate of single-photon emission is ultimately limited to approximately $1/T_1$. Nanocrystals have a T_1 lifetime of around 20 ns at room temperature. They can be brighter than trapped atoms and ions ($T_1 \sim 100$–200 ns), but not as bright as terrylene molecules ($T_1 \sim 4$ ns), coloured centres ($T_1 \sim 11$ ns) or self-assembled dots ($T_1 \sim 1$ ns). A solution to decrease T_1 could be to couple the nanocrystal with a cavity; this aspect will be developped in section 24.8.

In most quantum cryptography procedures, the information bit is encoded in the photon polarization by means of an electro-optical modulator which shifts the photon from its initial polarization to the desired polarization. This requires that the initial photon polarization is known: since nanocrystal emission is degenerate, one has to add a polarizer at the output, which implies the loss of half of the signal. Improvements could also be obtained by coupling to a single cavity mode of specific polarization.

24.5 Beyond the isolated emitter – QD/environment interactions

24.5.1   Introduction

Besides atomic-like properties such as photon antibunching and discrete energy levels, QDs also display a range of dynamical effects the "isolated artificial atom" picture cannot account for. Indeed, QDs significantly interact with their local environment due to their high surface-to-volume ratio,* and dynamical effects in their surroundings directly translate into random fluctuations of QD properties.

Environmentally induced fluctuations first appeared on the scene of single QD photoluminescence spectroscopy in the form of fluorescence intermittency and spectral diffusion [86, 87]. As will be shown, these effects never escaped researchers curiosity because of the surprising properties they exhibit (non-ergodicity, aging) and the window they opened on condensed matter dynamics at the nanoscale.

First considered as specific to a small number of pathological emitters embedded in glassy media or relaxing crystals, environmentally induced fluctuations have since been reported in a variety of systems (silicon nanocrystals [88], single molecules [89], fluorescing polymers [90], light harvesting complexes [91], green fluorescent proteins [92], carbon nanotubes [93]) and, incidentally now appear as the norm rather than an exception.

Compared to other blinking emitters, QDs have the distinct property of combining size tunability and intense brightness with a high immunity against photobleaching. These characteristics naturally make them ideal for measurements over long durations in order to accumulate statistics on their fluctuations, and better understand their interactions with the outside world in a variety of optical, chemical and temperature environments.

In the following, we introduce environmentally induced effects on single QDs through the phenomenon of fluorescence intermittency, i.e. any isolated QD randomly blinks between a bright "on" state and a dark "off" state, where the QD is virtually non-radiative (Fig. 24.13). Existing data are presented and their significance discussed. We explain how single QD blinking handles the trapping/detrapping dynamics of charges around a given QD, and highlight some counterintuitive effects blinking has on ensemble quantities. Further examples are provided to illustrate how, quite generally, charge motion and changes in the environment of a single QD can be tracked in the fluctuations of its optical properties (excited-state lifetime, phonon coupling and emission wavelength). Finally, current challenges and perspectives in the elucidation of environmentally induced fluctuations are presented.

24.5.2   Quantum dot fluorescence intermittency

Among the properties of single-QD fluorescence, fluorescence intermittency is probably the most striking one, seen simply by looking through a high numerical aperture microscope at the fluorescence of a dilute sample of QDs dispersed on a glass substrate. Blinking is a surprisingly ubiquitous phenomenon, routinely observed on any QD sample, and so far largely indifferent to any form of control [94].

24.5.2.1   On states, off states

Recorded over long durations with an avalanche photodiode or a CCD camera, the intensity time trace of each individual nanocrystal forms a binary "random telegraph" signal (Fig. 24.13).

---

*Twenty five per cent of the core atoms lie on the core/shell surface for a spherical 2 nm CdSe QD.

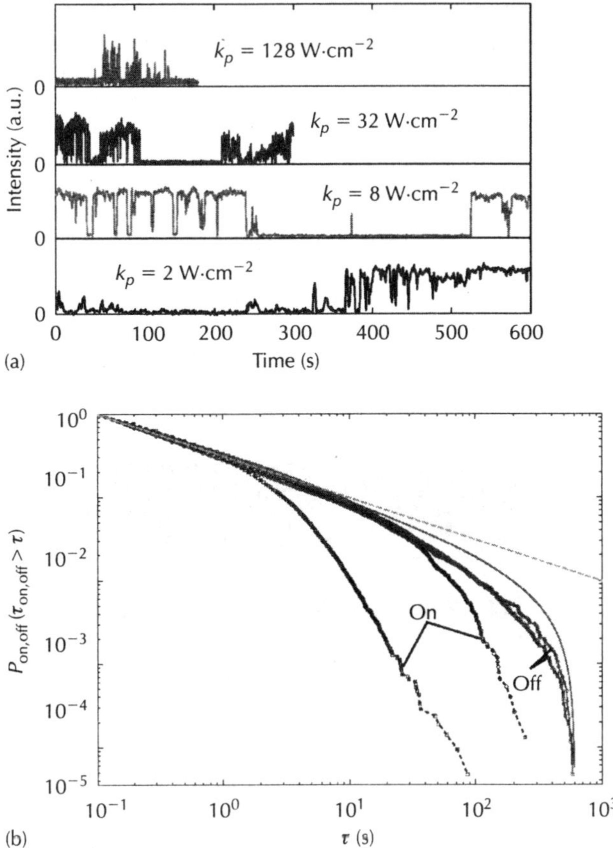

Figure 24.13   *Fluorescence intermittency of single CdSe nanocrystals under continuous laser excitation (Ar$^+$ – 488 nm) at room temperature [95]. (a) Fluorescence intensity trajectories as recorded on a CCD camera (10 ms integration time) at various laser excitation intensities k_p. The higher the excitation power, the shorter the on events. (b) Log–log plot of the on and off state statistics of an ensemble of 106 QDs at low (red, $k_p = 50\,W\cdot cm^{-2}$) and high excitation powers (blue, $k_p = 150\,W\cdot cm^{-2}$). The short timescale behaviour follows a power-law decay with $\mu = 1/2$ (dotted line). The truncation effects seen on the off states at longer timescales are mostly due to the finite observation window $\Theta = 600\,s$ of the experiment (solid line). The finite value of the acquisition time Θ, however, cannot account for the truncation effects seen with the on states – these being truncated at earlier timescales as their statistics are intrinsically truncated, truncation occuring at shorter timescales at higher excitation powers.*

Hidden in previous ensemble-averaged measurements, such intermittent fluorescence traces come as an unforeseen effect raising questions on the microscopic nature of their on and off states, and the underlying mechanism at the origin of their switching dynamics.

Nirmal *et al.* first suggested that the QDs' intrinsic radiative recombination observed in the on state would switch off, due to efficient Auger recombinations, as soon as a charge carrier escapes from the core, leaving the QD ionized – emission finally resuming when an incoming charge restores the QD neutrality [86].

This turned into a reality with three sets of experiments: (i) Klimov first demonstrated an experiment showing that Auger recombinations in a charged QD are much faster than radiative recombination [67], indicating that charging a QD with a single electron is enough to turn off its fluorescence, i.e. to put it in a off state; (ii) data in electrostatic force microscopy from the group of Louis Brus later confirmed this scenario, indicating that the charge of isolated QDs under continuous excitation fluctuates with time [96]; and (iii) plasmon enhancement of the dark-state fluorescence revealed that dark states are red shifted by ~ 25 meV compared to the bright states, in good agreement with the fact that a dark QD possess an excess charge $+e$ [97].

24.5.2.2 Environmentally induced fluctuations

Studied as an isolated phenomenon, blinking simply appears as a binary random telegraph process. Experiments correlating blinking events with the measurement of other QD properties (e.g. its excited-state lifetime, emission wavelength or lineshape), however, revealed a more complex pattern, as the ionization/neutralization dynamics underlying QD fluorescence intermittency is accompanied by charge reorganization in or around the QD. The QD therefore finds itself surrounded by a time-varying electric field, abruptly changing each time a blinking event happens.

As a result, QD on/off blinking dynamics fully correlates with sudden temporal fluctuations of many of its fluorescence properties such as spontaneous Stark shifts in its fluorescence spectra, as expected for QDs experiencing strong, time-varying internal fields under the influence of electrons relaxing in their vicinity [98, 99] (Fig. 24.14a). Similarly, time-resolved spectroscopy and photoluminescence decay measurements on single QDs reveal that blinking also correlates with spontaneous changes in the excited-state lifetime (Fig. 24.14b) or in the exciton–phonon coupling strength [100].

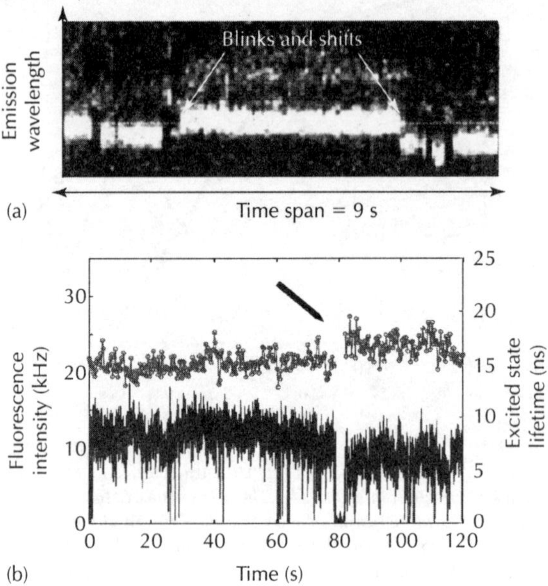

Figure 24.14 Examples of blinking-induced fluctuations in the properties of single QDs. (a) Blinking-induced spontaneous spectral jumps, reproduced from [99]. (b) Blinking-induced fluctuations of a QD excited-state fluorescence lifetime – the excited-state lifetime appears to have changed during the off state shown by the arrow [95].

QD fluorescence intermittency and its related spontaneous sudden emission properties fluctuations therefore appear as a pace clock of the QD dynamics. The remainder of this section focuses on this QD clock, its tempo, and the mechanisms accounting for its peculiar motion.

24.5.2.3 Blinking kinetics: measurement and basic properties

Numerous strategies have been developed to study the temporal dynamics of fluorescence intermittency each representing a different compromise between high temporal resolution – to resolve short on and off events and high sensitivity – meaning long integration times in order to detect QD fluorescence under low excitation.

24.5.2.3.1 Fluorescence intensity time traces

In the vast majority of experiments, the determination of single QD blinking statistics consists in recording the fluorescence intensity of many individual QDs. Due to the binary nature of the switching process, the time trace of each QD is considered as a sequence of on and off times $\{\tau_{on}^{(1)}, \tau_{off}^{(1)}, \tau_{on}^{(2)}, \tau_{off}^{(2)}, ..., \tau_{on}^{(n)}, \tau_{off}^{(n)}\}$ from which their corresponding statistical (cumulative*) distributions $P_{on}(\tau)$ and $P_{off}(\tau)$ can be determined (Fig. 24.13b). First experimental observation by Nirmal *et al.* in 1996 revealed that unlike most atomic-like single emitters, CdSe QDs have on and off state durations following non-exponential statistics [86]. Further studies by Kuno *et al.* confirmed this observation, and showed that off states are actually power-law distributed, meaning that their distribution can be written as:

$$P_{off}(\tau) \propto 1/\tau^{\mu_{off}}.$$

The exponent μ_{off} usually peaks around a value of $1/2$ for any QD sample. The corresponding power-law behaviour is observed over a broad range of timescales extending from 10^{-4}s to tens of minutes in both cryogenic (10 K) and room temperature conditions, independently of the power of the continuous optical excitation and of the QD' capping layer [101].

The statistics of on states, on the contrary, strongly depend on the experimental conditions (capping layer thickness, temperature, excitation power, *cf.* Fig. 24.13a,b). For capped QDs under vanishing excitation, on states also have a statistical cumulative distribution $P_{on}(\tau)$ following a power-law decay:

$$P_{on}(\tau) \propto 1/\tau^{\mu_{on}}$$

with $\mu_{on} \simeq \mu_{off} = 1/2$. Shimizu *et al.* later showed that raising either the excitation intensity or the QDs' temperature, or thinning their cap, results in the appearance of a truncation of the power-law statistics at shorter and shorter timescales [102]. As will appear below, these seemingly uncomplicated experimental facts are actually enough to make QD blinking a subtle and fascinating puzzle, as much for its exotic statistical properties as for anyone attempting to determine its microscopic origin.

24.5.2.3.2   Intensity autocorrelation measurements

While fluorescent time traces suffer from low time resolution,** several strategies have been developed to investigate the blinking kinetics at a faster timescale. Intensity correlation experiments, for example, consist in recording correlations in the detection times of photons and offer the same time resolution as photon antibunching experiments, i.e. down to the excited state lifetime of the emitter. Although the approach itself does not allow the determination of the on and off event statistics, it provides some insight on the short timescale intensity fluctuations [103], and shows, indeed, that blinking occurs down to timescales as short as microseconds [28, 104].

Variations of this approach have been reported, which consist either in a direct measurement of the Mandel parameter Q on single QDs [105], or in experiments performed in the frequency domain, measuring the Fourier transform of the intensity correlation function (i.e. the intensity noise power spectra) of *ensembles* of QDs [106].

24.5.2.3.3   Intensity decay measurements

As mentionned above, the on and off waiting time distributions $P_{on}(\tau)$ and $P_{off}(\tau)$ coincide in the limit of low excitation intensities and low temperatures, but differ otherwise. Assuming the sample excitation starts at time $t = 0$, the fraction of QDs in the on state at later times directly depends on the detailed shape of the distribution $P_{on}(\tau)$ and $P_{off}(\tau)$. As a result, the overall time

---

*The cumulative distribution $P(x)$ of a random variable X is the probability of observing the variable X with a value $X < x$. $P(x)$ is directly related to its histogram $h(x)$ of the random variable X by $P(x) = \int_x^\infty h(x)dx$. The cumulative distribution $P(x)$ must be preferred to the histogram when studying experimental data. Indeed, the histogram of a random variable is always more noisy than its corresponding cumulative distribution, as (i) the binning process suppresses all information on the exact ordering of the random variables belonging to a particular bin and (ii) the histogram $h(x)$ is always dependent of the set of bins chosen for its calculation. Cumulative distributions, in comparison, do not suffer these limitations, as their calculation does not involve any bins.

**This time resolution rarely exceeds milliseconds due to the integration time needed to collect enough photons and determine the state (on/off) of the QD at any given time.

evolution of the intensity $I(t)$ of an ensemble of QDs reflects the competition occurring between the on and off state under the influence of their different waiting time distributions. This approach, first used in [107] (*cf.* Fig. 24.15c) and later championed by Chung and Bawendi [108–110], appears as an efficient strategy for studying the long time (>2 hours) behaviour in the on and off states' statistical distributions and determine the dependence of their respective cut-offs against temperature and excitation intensity on large QD samples.

24.5.3 Blinking kinetics

Exploration into QD blinking kinetics has a two-fold purpose; it aims both at a broader understanding of the *statistical* properties of samples formed from large number of emitters undergoing power-law blinking statistics, and at the *physical* elucidation of the microscopic mechanisms producing power-law blinking statistics. In the following, we first review some basic properties of power-law statistics, and later investigate their influence on the temporal dynamics of collections of QDs.

24.5.3.1 Power-law blinking statistics – broad distributions

The power-law distributed waiting times are at the core of peculiar emission properties of these nanoscale emitters. To see this, let us first consider the simple situation corresponding to a QD observed at low temperature (4 K) under very low excitation power. As explained above, in this case, both its on-and off-state blinking statistics have a pure power-law behaviour with $\mu_{on} = \mu_{off} = \mu = 1/2$ over all accessible timescales, i.e. from the integration time τ_0 of the photodetector ($\tau_0 \sim 1$ ms) to total duration Θ over which blinking is recorded (up to a few hours).

The important point is that for distributions $P_{on,off}(\tau)$ with a power exponent $\mu_{on,off} = \mu$, all moments of order higher than μ diverge [111]. For QDs, we have $\mu < 1$. Hence, in the low-temperature/low-excitation regime, the average duration of the on and off states becomes formally infinite, e.g. implying that even if a number of commutations of the order of $N = (T/\tau_0)^{1/\mu} = 10^6$ occurred during the course of the QD fluorescence trajectory, time averages computed by integrating a single QD trajectory over the whole acquisition time Θ will not converge to any definite value. This situation completely contrasts with exponential blinking kinetics, for which time averages are defined within a relative accuracy of the order of $1/\sqrt{N} \sim 10^{-3}$ after $N = 10^6$ commutations! We see that compared to standard emitters with exponential blinking statistics, QDs undergoing fluorescence intermittency converge extraordinarily slowly (if ever*) towards their statistical equilibrium.

24.5.3.2 Power-law statistics in single molecule spectroscopy

Appearing in fields ranging from complex system physics (chaos [112], earthquakes [113] and sandpiles [114]) to random walks [115], anomalous diffusion and transport problems [116], broad power-law distributions – also called Lévy statistics – appear to be generally associated to unusual statistical properties and dynamics, and must be handled with special care as common intuition mostly derives from experiences involving regular distributions, i.e. with a variance or a mean value [111]. Statistical physics, for example, is largely based on the ergodic principle, i.e. on the hypothesis that ensemble and time averages of an observable coincide [117]. This hypothesis therefore implicitly assumes that systems possess a finite correlation time. How does it apply in systems driven by broad distributions of times, and therefore lacking any finite typical timescale over which fluctuations could be integrated to yield a time average? What are the dynamical properties of systems with arbitrary long correlation times?

Single-QD spectroscopy indeed provides unprecedented opportunities to investigate such questions. In condensed matter systems, answers are significantly complicated as observations proceed from macroscopic measurements leaving the detailed elementary processes unresolved due to the

---

*In practice, the long-time cut-off observed in the on-and off-state waiting time distributions however, ensures convergence of time averages to a well-defined value [110]. This asymptotic "standard" behaviour however, does not change our conclusion on the anomalously long durations required for convergence of the blinking process to reach its statistical steady state.

experimental impossibility to monitor them individually. Isolated single-QD microscopy experiments, on the contrary, can easily follow the fluorescence trajectory of each emitter within a collection of QDs, and directly investigate the relation between time- and ensemble-averaged quantities in systems driven by broad power-law distributions, and address the ergodic hypothesis, usually promoted to the status of *principle* given the practical impossibility to check its validity.

24.5.3.3 *QD blinking: a non-stationary dynamics driven by rare events*

The main tool to further grasp the properties of broad distributions is the generalized central limit theorem. This theorem, formulated by Pierre Lévy in 1937, states that the sum of N independent variables, instead of scaling as N, scales instead as $N^{1/\mu}$ when these variables follow broad power distributions of index $\mu < 1$ [118]. Following this theoretical result, the total time $\theta(N)$ spent in the off state after N switching events:

$$\theta(N) = \sum_{i=0}^{i=N} \tau_{\text{off}}^{(i)}$$

is expected to scale as $N^{1/\mu_{\text{off}}} = N^2$ for independent off-state durations following a power-law distribution with $\mu_{\text{off}} = 1/2$ [111]. As a result, as time passes, larger and larger off-events occur, and $\theta(N)$ is expected to grow faster than N. Figure 24.15a, showing $\theta(N)$ as measured on a CdSe QD by simply recording its blinking intensity over a long durations $\Theta = 10\,\text{min}$, fully supports this analysis. The sum $\theta(N)$ is dominated by a few long events of the order of $\theta(N)$ itself. The consequence of this unusual scaling property of $\theta(N)$ is that, as time grows, QDs are increasingly found "stuck" in longer and longer off states, and so the probability to observe a switch-on event decreases with time (Fig. 24.15b). Due to the occurrence of Lévy statistics, time translation is broken in the blinking process, QDs showing an aging effect of purely statistical origin ("statistical aging") even if the QD sample does not experience any progressive physical degradation (e.g. photobleaching) upon excitation.

This effect has practical importance at non-vanishing excitation power, when truncation effects appear at long timescales in the on-states distribution. In this case, the truncation defines an upper boundary for the on-states duration, while larger and larger unbounded off states continue to appear in the QD time traces, accounting for a progressive darkening in the fluorescence of the whole QD sample (Fig. 24.15c).

Knowledge concerning the blinking properties of QDs is therefore essential in the fluorescence behaviour of an ensemble of QDs. Where one might have erroneously concluded that the observation of photodarkening shows that QDs undergo significant irreversible photobleaching, we see that photodarkening on a QD sample can also originate as a manifestation of non-stationary behaviour, with the sample completely recovering its initial brightness when laser excitation is interrupted (Fig. 24.15c) [107, 108].

24.5.3.4 *Semiconductor QDs as non-ergodic systems*

Further analysis also reveals that, due to the occurrence of Lévy statistics, time averages and ensemble averages on CdSe QDs are no longer defined. This was evidenced by comparing the fraction Φ of QDs in the on state within a given sample, and the fraction of time spent in the on state by each QD individually. Under the ergodic assumption, these two quantities should coincide, and provide the probability for a QD to be in its on state. However, QDs reveal a completely different pattern (Fig. 24.16). While the ensemble-averaged fraction of QDs remains constant over time, time averages are found to fluctuate from QD to QD, even after integration times as large as 10 minutes [107]. Due to blinking, observing the fluorescence intermittency of a single QD does not provide any information on its ensemble-averaged properties.

Reported and discussed here on a particular observable (the total intensity of the QD sample), blinking-induced non-ergodic behaviour could conceivably also affect many other QD observables. Indeed, as fluorescence intermittency correlates in time with incessant charge-reorganization events randomly altering most electronic QD properties (as discussed above, see section 24.5.2.2), the broad statistics encountered in fluorescence intermittency might also appear in the fluctuating

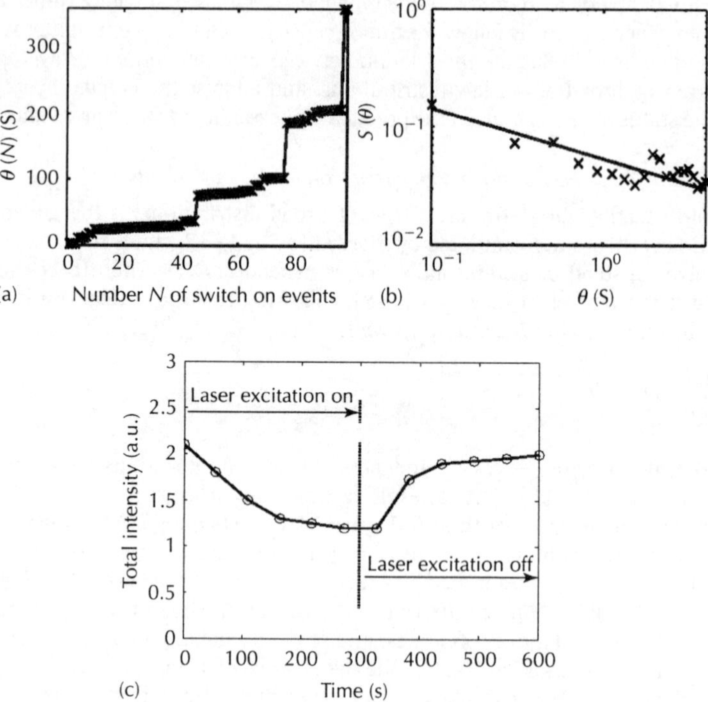

Figure 24.15 *Statistical aging in the fluorescence of single CdSe QDs. (a) Evolution of the total time θ (N) spent in the off state as a function of N, the number of switching events experienced by the QD since the beginning of its observation. The sum θ (N) is dominated by a few large events. (b) Probability density s(θ) to see a QD switching on after having spent a total time θ in the off state, as measured on an ensemble of 215 QDs (×). The "switch on" probability density s(θ) decreases in time according to a power-law decay θ⁻ᵅ with* $\alpha = \mu_{off} = 1/2$, *as expected from the theory (solide line) [107, 111]. As a result, the total intensity of the sample undergoes a progressive (yet reversible) decay (c) [95].*

dynamics of other QD properties, and yield, for example, non-trivial statistical aging effects in their ensemble fluorescence spectrum [119].

24.5.4   Origin of quantum dot blinking kinetics

Is the transition between on and off states a discontinuous quantum jump? Or does it occur in a continuous fashion which has not been temporally resolved yet [121]? Does blinking also happen at very short timescales? As questions concerning the properties of the switching process still abound, models attempting to account for available data QD blinking phenomena appeared, and continue to flourish.

24.5.4.1   Blinking models: an overview

The first class of models consists of a QD interacting with one or several traps [122]. This model, reminiscent of two-level systems coupled to a dark, triplet state, constitutes a simple mechanism for QD fluorescence intermittency. However, a single dot interacting with a single static trap only yields *exponential* on and off waiting time distributions, with rates given by the ionization rate and neutralization rate, respectively.

24.5.4.1.1   Distributed traps models

More complex models were then developed, attempting to account for the strongly non-exponential, broad power-law decay observed on QDs and their seemingly universal $\mu = 1/2$ exponent. As reviewed by Kuno [123], such models mainly explored two directions, namely the possibility for the QD (1) to be coupled not to one, but to an ensemble of traps with widely

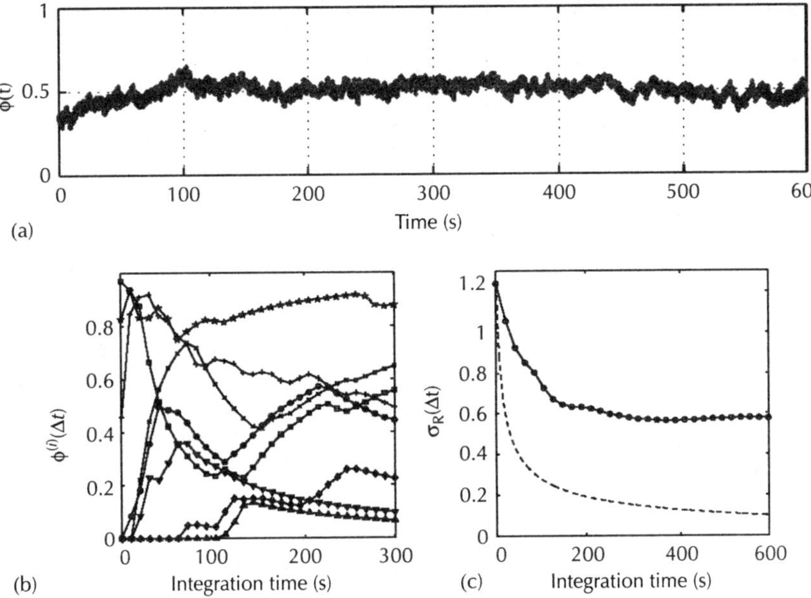

Figure 24.16 *Non-ergodic behaviour of QDs in a sample of 231 emitters under low excitation power. (a) Fraction of QDs found in the on state at time t within the sample. The constant ensemble-averaged value of 1/2 for $\Phi(t)$ is in agreement with the fact the blinking process is symetric (i.e. $\mu_{on} = \mu_{off}$) at low excitation power [111, 120]. (b) Fraction of time spent $\Phi^{(i)}(\Delta t)$ in the on state over an integration time Δt for eight different QDs. These time averages widely fluctuate from QD to QD, even in the limit of long integration times (here $\Delta t = 5\,min$). For an ergodic sample, the relative dispersion $\sigma_r(\Delta t)$ of $\Phi^{(i)}(\Delta t)$ – i.e. the standard deviation of the distribution $\Phi^{(i)}(\Delta t)$ over the set of QDs, divided by its mean value – would decay to zero over long integration times (solid line (c)). QDs, on the contrary, exhibit non–vanishing time-average fluctuations showing that time averages do not converge to any definite value, indicative of the non-ergodic character of their fluorescence properties.*

distibuted neutralization rates [124], or (2) to have a slowly varying parameter (such as the QD shell width) resulting in large fluctuations of its ionization rate [101]. Scenario (1) so far fails to account for the fact that multi-exponential decay is observed for *both* on and off waiting times, since the escape rate of a charge in a QD coupled to many traps is simply given by the sum of the individual escape rate to each trap, yielding an overall single-exponential escape probability distribution. Finally, models following scenario (2) predict correlations in the duration between successive on and off states at timescales shorter than the correlation time of the slowly varying parameter. These correlations so far failed to materialize to experimenters [101]. Importantly, these models do not provide any verified explanation for the $\mu = 1/2$ exponent observed across a variety of colloidal semiconductor QDs (CdTe, CdSe, CdS), no matter their size, temperature or excitation intensity [102].

24.5.4.2.1   First-passage models

A striking fact is that power-law statistics with the particular exponent value of $\mu = 1/2$ appear not only in QDs blinking statistics, but also – and more fundamentally – in the field of random walks theory. Indeed, the time needed for a random walker to cross the point where he started his motion (also called the first-passage time) is power-law distributed, with an exponent $\mu = 1/2$. This observation forms the basis of a third class of models, elaborating on possible mechanisms to turn what might first have appeared as an incidental coincidence into an essential and guiding pattern.

Shimizu [102] first proposed a model along this idea for a QD interacting with a single trap. The trap is assumed to perform a random walk in the (one-dimensional) energy space under the influence of fluctuations in its nanoscale environment, the QD state (on/off) subsequently switching each time an electron is exchanged between the QD and the trap, i.e. when crossing occurs between the trap energy and the excited-state energy. Despite its formal simplicity, this model has the elegance to provide a simple and direct explanation for the occurrence of power-laws with a $\mu = 1/2$ exponent.

Despite having QDs interacting with a *single* trap of *infinite* lifetime, and performing an *unbounded* random walk in the energy space seems quite unrealistic, this approach paved the way for elegant and powerful models later developed along the same "first-passage time" concept.

Inspired by diffusion-controlled electron transfer effects, Tang and Marcus replaced the (unbounded) random walk in the energy space by the diffusive walk of a reaction coordinate in a free energy harmonic (confining) potential, i.e. around a stability point of the QD [125]. Within this frame, spectral diffusion and fluorescence intermittency become the manifestation of the diffusive walk of QD reaction coordinates in the harmonic potentials of the on and off state under the influence of the slow relaxation of the QD environment.

Major achievements of their model are its unified description of both blinking and spectral diffusion – two effects heretofore treated in a disconnected fashion – and an unparalleled ability to account under rather broad assumptions for the many properties of spectral diffusion and blinking as seen in a variety of experiments (e.g. dependence of blinking and spectral diffusion against temperature and excitation intensities, significance and consistent description of the long-time cut-offs in the on and off waiting time distributions).

24.5.5   Conclusion

Seen as a pitfall leading to unwanted line broadening and instability both in spectroscopy [126] and in future optoelectronic QD-based devices (nanoprobes in biology [127], single-photon sources for quantum information processing [128, 129]), interactions of QDs with their environment, whenever properly measured and understood, also turn each of these emitters into tiny probes reporting information on local electric fields, electronic dynamics and optical properties in their surroundings [98].

From a general standpoint, blinking in QDs has so far significantly stimulated the development of new strategies and techniques to further investigate its bright and dark states' waiting time distributions and provide a clearer picture of the QD environment dynamical properties.

24.6   Time coherence of the single photons emitted by an individual nanocrystal

24.6.1   Monomode photons and quantum computing

Apart from blinking, another important effect of the local environment of the nanocrystal appears in the fluorescence spectrum. In the perspective of single-photon emission for quantum cryptography, the details of the emission spectrum are of little importance. For further quantum computing realizations, however, these aspects should be investigated much more closely.

Knill *et al.* [130] have shown that quantum logic gates can be realized with only linear optical components, provided that the photons are indistinguishable. This relies on a mechanism called "coalescence", which is the interference of two indistinguishable photons. Let us consider two such photons A and B impinging each on one side of a 50/50 non-polarizing beam splitter (Fig. 24.17). The output wavefunction is $(|\alpha\rangle - |\beta\rangle)_A \otimes (|\alpha\rangle + |\beta\rangle)_B$. This is equal to $|\alpha\rangle_A \otimes |\alpha\rangle_B + |\beta\rangle_A \otimes |\beta\rangle_B$ since $|\alpha\rangle_A \otimes |\beta\rangle_B = |\beta\rangle_A \otimes |\alpha\rangle_B$ as A and B are bosons. Thus the two photons "coalesce": they are detected at the same output.

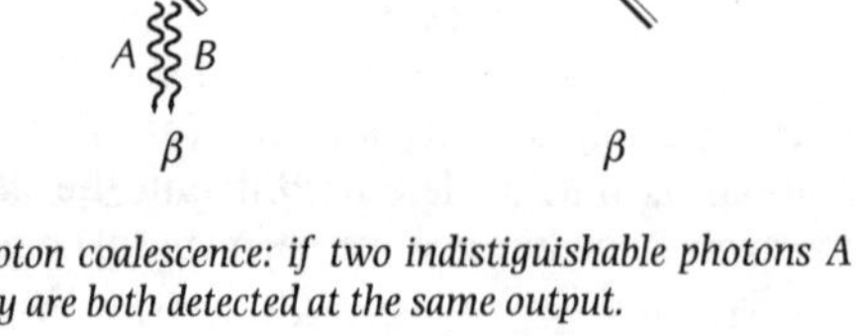

Figure 24.17   Principle of photon coalescence: if two indistiguishable photons A and B arrive on two sides of a non-polarizing beam splitter, they are both detected at the same output.

To be indistinguishable, two photons must be (i) each in a single mode and (ii) both in the same mode (in terms of wavelength and polarization). The first condition, which requires the absence of any dephasing during an emission, is characterized by $\Gamma_2 = \Gamma_1/2$, where Γ_1 and Γ_2 are the respective decay rates of the excited-state population and of the emitting dipole coherence. These conditions are fulfilled for photons emitted by a trapped atom, as evidenced by a coalescence experiment performed in 2004 [131]. This might not be the case for other sources, as interactions with the surrounding solid matrix might perturb the emission, following two limit regimes:

- Slow and large fluctuations, like fluctuations of the local electric field caused by charge movements (see section 24.5), create a spectral diffusion of the emission. If spectral diffusion occurred between the emissions of the two photons, they do not match condition (ii).
- Fast and small fluctuations, like collisions with impurities or with the phonon bath, create random dephasings during each emission. On average, these dephasings decrease the dipole coherence at a rate Γ_{deph} which adds to the rate $\Gamma_1/2$ related to emission mechanisms. The decoherence rate $\Gamma_2 = \Gamma_1/2 + \Gamma_{deph}$ is thus superior to the "transform-limited" value $\Gamma_1/2$, which characterizes mono-mode emission. This corresponds to a perturbed emission spectrally broadened by the term Γ_{deph} so that more than one mode have to be taken into account to describe the emitted photon state, and condition (i) is not fulfilled. It can also be understood as an individual photon emission undergoing a specific sequence of dephasings, which makes each emitted photon distinguishable from the others.

For some molecules at low temperature in a crystalline matrix, a transform-limited linewidth has been measured [133]. Some self-assembled dots coupled to a cavity also exhibited indistinguishable photons emission [134, 135], evidenced by indirect coalescence measurements using a Hong–Ou–Mandel set-up [136]. For coloured centres, interaction with phonons creates large broad sidebands. The zero-phonon line – the one which might be transform limited – represents only 3% of the emission spectrum and its study would not be relevant [81]. As for nanocrystals, their emission spectrum is not easy to characterize, as is detailed below.

24.6.2 Spectroscopy of a single nanocrystal

Therefore, spectroscopy of an emitter is an important aspect of its characterization, because it provides information about the influence of the local environment, and because spectral broadening and spectral diffusion are key parameters for quantum-computing applications. In order to distinguish between the linewidth $2\Gamma_2$ and the spectral-diffusion broadening during τ, which we label $2\Gamma\tau$, one must be able to perform fast spectroscopic measurements.

Time resolution is a very general problem in single-emitter experiments, as signals of typically 10 000 counts per second are detected. If one wants to plot the intensity, with a signal-to-noise ratio of 5, one needs 25 counts per time bin, so that the time resolution is limited to typically 2.5 ms. This restricts, for instance, the possibility to study very fast blinking, as discussed in section 24.5.

In the same way, if one wants to plot consecutive spectra, each spectrum requiring around 1000 counts, the resolution is limited to 100 ms. If spectral diffusion occurs faster than 100 ms, it broadens the measured spectra: following our choice of notations, the measured linewidth is $2(\Gamma_2 + \Gamma_{100ms})$. Indeed, for single nanocrystals, the linewidth measured at 10 K shows a strong dependence on the acquisition time [132], which is well accounted for by a diffusion-in-a-potential model [125]. Values of a few meV are measured – 120 μeV at best (Fig. 24.18).

Better time resolutions are achieved on ensembles of emitters. For nanocrystals, spectral diffusion over a few tens of μeV has been observed during tens of microseconds by spectral hole burning, and a linewidth of 6 μeV has been reported [137]. Three-pulse photon echo experiments have also been performed [138]. Although designed to suppress inhomogeneous broadening, these methods still consider ensemble averages and are less satisfying. Moreover, they probe the absorption spectrum, which, because of fast intraband relaxation, is substantially different from the emission spectrum. The same problem exists with the first method investigated for single-molecule time-resolved spectroscopy, which calculated the correlations of consecutive two-photon excitation spectra [139] – and the resolution was limited to a few milliseconds, which would not be sufficient for nanocrystals.

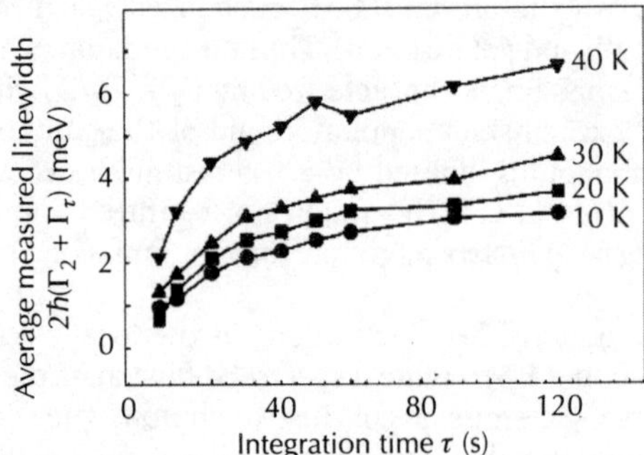

Figure 24.18   *Measured emission linewidth $2\hbar(\Gamma_2 + \Gamma_\tau)$ as a function of the spectrum integration time τ, at various temperatures (reproduced from [132] authorization for reproduction asked). Each dot is an average over several tens of CdSe/ZnS nanocrystals with 2.8 nm radius.*

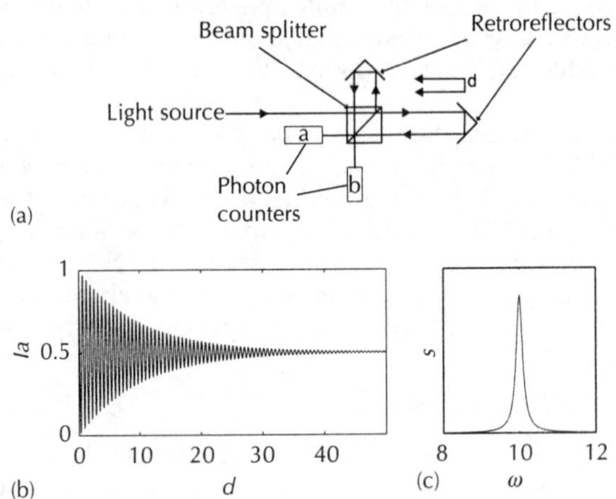

Figure 24.19   *Standard Fourier spectroscopy. (a) Michelson interferometer with a temporal path difference d and two photon-counting detectors. (b) Intensity on detector a as a function of d and (c) corresponding emission spectrum $s(\omega)$, for a two-level emitter of frequency $\omega_0 = 10$ and coherence time $T_2 = 10$ (arbitrary units). One can show that $I_a(d) = I(1 + \mathcal{R}(g^{(1)}(d)))/2$, and the Wiener–Khintchine theorem states that $g^{(1)}(d)$ is the Fourier transform of $s(\omega)$.*

24.6.3   *Photon-correlation Fourier spectroscopy*

This problem brought about the proposal of a new method, called photon-correlation Fourier spectroscopy (PCFS) [140]. Let us outline this method briefly (precise calculations can be found in [140] and [141]), by considering a classical dipole emitter with a simple spectral diffusion where spectral jumps during τ follow a Lorentzian distribution of width $4\Gamma\tau$.

Standard Fourier spectroscopy, using the scanning Michelson interferometer described in Fig. 24.19, measures the intensity I_a on photodiode a as a function of the path difference $d_t = \eta t$. If the time resolution of the intensity measurement (say, 2.5 ms) is longer than the correlation time of spectral diffusion (which is the case for nanocrystals), the measured interferometer is broadened by spectral diffusion:

$$I_a(t) = \frac{I}{2}\left(1 + e^{-(\Gamma_2 + \Gamma_{10ms})d_t}\cos(\omega_0 d_t)\right). \tag{24.18}$$

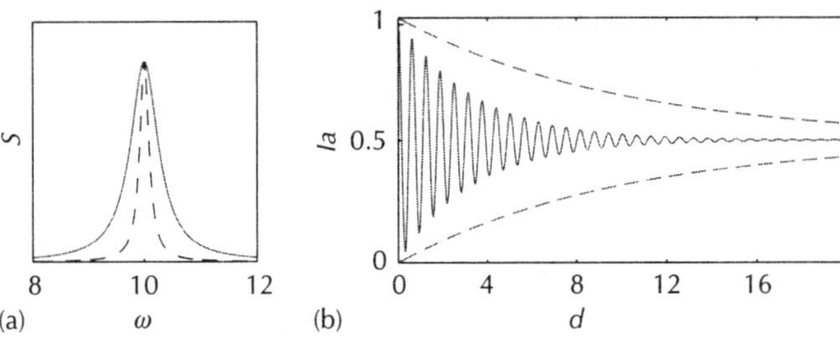

Figure 24.20 *Influence of spectral diffusion on standard (a) spectroscopy (spectrum S(ω)) and (b) Fourier spectroscopy (interferogram I_a(d), where d is the path difference), in the case of a two-level emitter of frequency 10, linewidth 0.2 and spectral-diffusion broadening 0.6 (in arbitrary units), where spectral diffusion is assumed to be faster than the spectrum or intensity measurement. The dotted line indicates the envelope of the case without spectral diffusion.*

The principle of PCFS is, during the same interferometer scan, to measure the intensity cross-correlation:

$$g_{ab}(t,\tau) = \frac{\langle I_a(t)I_b(t+\tau)\rangle_{\delta t}}{\langle I_a(t)\rangle_{\delta t}\langle I_b(t)\rangle_{\delta t}} \qquad (24.19)$$

for consecutive times t (and corresponding path differences d_t). If the averaging time δt is chosen so that one scans a large but not too large number of fringes (that is, $\eta\omega_0\delta t \gg 1$ and $\eta\Gamma_2\delta t \ll 1$):

$$g_{ab}(t,\tau) = 1 - \frac{1}{2}e^{-2(\Gamma_2+\Gamma_\tau)d_t}\cos\omega_0\eta\tau \qquad (24.20)$$

and $\cos\omega_0\eta\tau \simeq 1$ for short τ, the dependence of $g_{ab}(t,\tau)$ on d_t thus yields the decoherence rate Γ_2 and the spectral diffusion Γ_τ.

Like studies of fast blinking through bunching in the $g^{(2)}$ function [28], PCFS takes advantage of the very general fact that intensity correlations can be measured with a much better time resolution than intensity itself, provided that the acquisition time δt is long enough. More explicitly, since the number of photon pairs counted per ([t, $t + \delta t$], [τ, $\tau + \delta\tau$]) bin is of the order of $I_a I_b \delta t \delta\tau$ and should be at least 25 for a reasonable 5/1 signal-to-noise ratio, $\delta\tau$, the precision on τ is of the order of $25/I_a I_b \delta t$, and can be taken arbitrarily short by setting a sufficiently long δt integration time.

PCFS has been performed on single nanocrystals at 10 K [142]. The duration of the experiment was limited to 10 minutes by both nanocrystal blinking and slow position drift of the nanocrystal into the cryostat, resulting in a $\delta\tau = 20\,\mu$s resolution on correlation measurements. The measured $g_{ab}(t,\tau)$ showed the expected decrease with increasing d_t, with the exponential dependence assumed above. The fitted linewidth $\Gamma_2 + \Gamma_\tau$ is plotted in Fig. 24.21: a spectral diffusion of a few μeV over around 200 μs is observed, and an upper limit of 6 μeV is measured for the linewidth $2\hbar\Gamma_2$, corresponding to a coherence time $T_2 = 200$ ps.

Although longer than indicated by previous spectroscopic studies, this coherence time is much shorter than the excited-state lifetime ($T_1 \sim 200$ ns at 10K), and the limit of mono-mode photons ($T_2 = 2T_1$) seems hardly achievable. However, if one considers, for instance, the same Michelson interferometer, with a fixed $d \gg T_1$ path difference, the normalized cross-correlation will be for $\tau < d/2$ (here without spectral diffusion):

$$g_{ab}(\tau) = 1 - \frac{1}{2}e^{-(\Gamma_1+r)\tau} - \frac{1}{2}e^{-2\Gamma_2\tau} \qquad (24.21)$$

so that $g_{ab}(\tau \ll T_2) = 0$, which can be interpreted as follows: when considered on a timescale shorter than the characteristic time between two dephasings, two photons behave as if they were indistinguishable and coalesce. Thus two-photon interference effects could be observed, even with non-mono-mode photons, provided that a sufficiently good resolution on τ can be achieved.

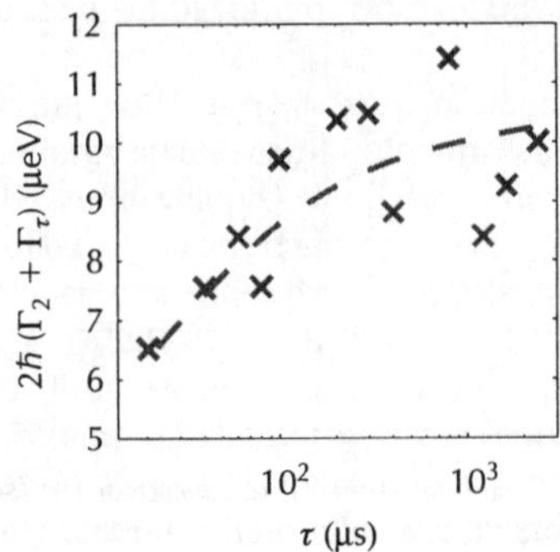

Figure 24.21   *Effective-linewidth-during-*τ *$2\hbar(\Gamma_2 + \Gamma_\tau)$ measured on a single CdSe/ZnS colloidal nanocrystal by photon-correlation Fourier spectroscopy [142].*

It might allow two-photon interferences experiments, by use of a detection system following an experiment reported in the atomic physics domain [131].

24.7   Multiexcitonic emission of colloidal quantum dots

24.7.1   *Auger processes in colloidal quantum dots*

Under pulsed and continuous excitation, colloidal CdSe quantum dots appear to be perfect single photon emitters at room and cryogenic temperature (see section 24.4). The fluorescence of these solid-state sources is then similar to that of a single two-level system such as an atom. This property reinforces the image of "artificial atoms" suggested by their atomic-like spectrum and are commonly used to describe semiconductor quantum dots. As already discussed, compared to individual molecules, ions or atoms, quantum dots display a crucial difference since various electrons can be excited simultaneously by absorption of the optical excitation. Due to the Auger effect, these multi-electron–hole pairs mostly decay non-radiatively, the energy being transferred to charge carriers.

Multiexciton states in semiconductor quantum dots have been studied in detail first for a fundamental understanding of many-body interactions in nanometre-sized confined volume which is much smaller than the volume occupied by an excition in the bulk semiconductor. They are also important for the realization of practical optoelectronic devices including quantum dots and gain applications such as lasers [143].

Auger processes are much more efficient in nanometre-sized structures than in the corresponding bulk materials. In the bulk material, the excitons' maximum density is equal to one exciton per excitonic volume which is of the order of the Bohr radius exciton cubed. Above this concentration, electron and holes build up a plasma. Resulting screening effects reduce Coulomb interactions and efficiency of the Auger processes. On the contrary, strong carrier confinement prevents dynamic screening and Coulomb interactions are enhanced. Moreover, the very small size of nanocrystals implies that there is no translation symmetry and hence no carrier momentum conservation. Auger processes forbidden in bulk semiconductor become very efficient in quantum dots.

The high efficiency of Auger effects has been reported in various experimental and theoretical works. For CdSe nanocrystals with radii ranging between 1.2 and 3.6 nm, the Auger process occurs in less than 100 ps, a time much faster than the radiative lifetime which is of the order of 20 ns [67]. The photoluminescence signal mostly comes from the recombination of single excitations generated directly by optical excitation or by Auger recombination of multi electron–hole pairs.

The dynamics of multiexcitonic states emission was first examined with very fast time resolved methods in a transient regime. Since the recombination of multiexcitonic states is mostly non-radiative, these experiments were first carried out on nanocrystal ensembles. Multiparticle Auger rate photoluminescence measurements exhibit emission bands from multi-career excited states in nanocrystals. Achermann *et al.* obtained two additional bands [144]. The first is blue shifted compared to the single excitation peak, while the other one is red shifted. The authors assigned the the first one to a negatively charged biexciton and the second one to a neutral biexciton. However, the exact nature of the state at the origin of the light remains controversial. More recently, an alternative pumping scheme was proposed [145]. It consists in an excitation pulse of long duration compared to the Auger rate, the absorption rate being kept smaller than the relaxation rate to the band edge. It enables the quantum dots to be pumped to a given number of electron hole pairs controlled by the intensity pump without passing through a large number of excitons. The authors obtained the emission spectra of higher excitonic bands and their order of appearance which enabled the nature of the electronic state involved in each line to be determined.

The size dependence of the Auger rate was investigated for quantum dots with radii R ranging between 1 and 4 nanometres. The relaxation time is found to be proportional to R^3 for two, three and four electron–hole pair states [67]. The absence of multiexcitonic emission for high pumping power is specific to colloidal quantum dots, because they can be very small. Indeed, epitaxially grown quantum dots exhibit multiexcitonic emission unless low pumping is used. As already mentionnend, from the point of view of single photon generation, this property is a clear advantage since epitaxially grown quantum dots require the use of spectral filtering in order to isolate the single-exciton emission leading to optical losses.

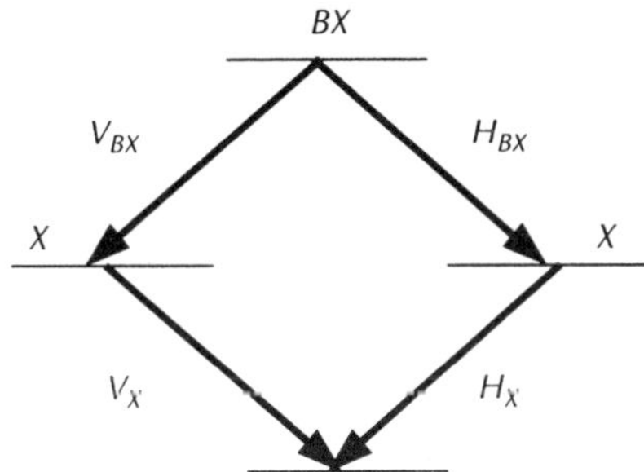

Figure 24.22   Schematic description of the biexcition – exciton cascade in quantum dots.

24.7.2   Entangled photon pair generation by solid-state sources

However, in the field of optical quantum information protocols, biexcitonic emission of an epitaxially grown quantum dot can be used to generate polarization entangled photon pairs, following the scheme of Fig. 24.22 [146, 147]. The two-exciton space of a quantum dot consists in a ground state, two single exciton states and the biexciton state. The biexciton decays through two intermediate exciton states. The proposal that the biexciton radiative cascade can provide a source of polarization entangled photon pairs is based on the fact that the two decay paths involve different polarization schemes. It also requires the indistinguishability of the two paths concerning parameters other than polarization. In an ideal situation, the polarization of the BX photon is entangled with the one of the X photon, yielding the state:

$$\frac{1}{\sqrt{2}}(|\,H_{BX}H_X\rangle + |\,V_{BX}V_X\rangle) \tag{24.22}$$

where H (respectively V) means horizontally (respectively vertically) polarized. Until recently, various experiments showed a splitting of the intermediate states. This splitting comes from the in-plane anisotropy of the structure of epitaxially grown quantum dots which induces strain and elongation. In this case, only classically correlated photons can be obtained. In 2006, two groups following different experimental schemes succeeded in getting photon pairs to satisfy entanglement conditions. Akopian *et al.* applied spectral filtering to erase the energy "which path" information [147]. The spectra of biexcitonic and monoexcitonic emission are composed of two lines corresponding to the horizonal and vertical polarization. The detuning between the two lines being lower that the linewidth of each line, a monochromator with a spectral resolution lower than the width of the spectral common part of lines project the photons in the same energy state whatever their polarization may be. Stevenson *et al.* were the first to carefully grow epitaxially quantum dots in order to suppress the splitting due the anisotropy of the structure [148, 149]. Alternatively, they applied a magnetic field to tune to zero the splitting between monoexcitonic states.

24.7.3 *Entangled photon pair generation by CdSe quantum dots*

In the case of colloidal nanocrystals, the efficiency of Auger effects must be reduced to detect multiexcitonic emission at the single-molecule level. Fisher *et al.* used nanocrystals with a large radius greater than 5 nm [66]. According to the volume proportionality of Auger lifetime, the biexcitonic lifetime becomes of the order of 1 ns. This long lifetime enables the observation of biexcitonic (and even triexcitonic) emission with a quantum efficiency of the order of 10%. At room temperature, the authors demonstrate time correlations between the various emissions.

These results could be the first step towards the generation of entangled photons. As for epitaxially grown quantum dots, the next step consists in getting the degeneracy of the two states corresponding to bright monoexcitons. In the first approximation, the two levels can be considered as degenerated (see section 24.3). However, Furis *et al.* recently reported results concerning polarization resolved resonant photoluminescence spectroscopy of these states [75]. The measured intrinsic fine structure exhibits a splitting between the bright excitons which ranges from 1.1 to 2.0 meV depending on the size of the quantum dot. This splitting is much greater than the corresponding splitting measured for CdSe epitaxially grown quantum dots (of the order of 100 μeV) and comes from the very small size of colloidal quantum dots which enhances anistropic exchange terms. This clear drawback in the perspective of getting bright exciton degeneracy may be balanced by the great spectral diffusion observed in colloidal quantum dots (typically several meV at room temperature) which may enable the spectral filtering technique used for epi-taxially grown quantum dots.

Finally, radiative multiexcitonic emission could also be obtained by using type II colloidal semiconductor quantum dots [150]. Recently synthetized, CdTe/CdSe, core–shell nanocrystals are an example of such a type II structure. When compared to standard CdSe quantum dots (denoted as type I structures), the crucial difference relies on the localization of the charges. Due to the relative position of CdTe and CdSe conduction and valence bands, the hole is confined to the CdTe core while the electron is confined to the CdSe shell when its thickness is greater than 1 nm. As a result, the carrier wave function overlap is reduced, and the Auger lifetime, like the radiative lifetime, is increased. Moreover, the Auger lifetime no longer scales as the volume of the nanocrystal and reaches values as high as 1 ns. The strong reduction of the efficiency of Auger transitions, even if it is at the cost of the reduction of the oscillator strength of the radiative transition, may lead to an increase in the quantum efficiency of radiative multiexcitonic recombinations. As a result, this kind of structure, which opens new possibilities of quantum engineering, could be a promising alternative to generate entangled photon pairs.

24.8 Controlling quantum dot emission with photonic structures

24.8.1 *Cavity quantum electodynamics*

Single photon emission by semiconductor nanocrystals comes from the discrete structure of the quantum dot carrier energy levels. Their spontaneous emission behaviour, which corresponds to the interaction between the emitter and the electromagnetic field, depends also on their optical environment – more precisely on the density of photonic states. This property is at the heart of cavity quantum electrodynamics (CQED) experiments which consist in controlling the radiative characteristics of an emitter using the modification of the light mode characteristics induced by a cavity. The first experiments were carried out in the microwave domain for which the size of the needed cavities is relatively large. Remarkable technological improvements obtained during the last 20 years offered the possibilty of realizing CQED experiments with semiconductor devices and optical frequencies. Using epitaxially grown quantum dots and three-dimensional cavities exhibiting large quality factors and small volume modes, the modification of spontaneous emission rate and the strong coupling regime between one electromagnetic mode and an individual emitter have been observed. In this section, we detail the experiments involving optical cavities and colloidal quantum dots already realized. We first summarize the crucial characteristics of the various optical cavities distinguishing dieletric structures and metallic structures.

24.8.2   Weak coupling regime

In 1946, E.M. Purcell published the paper which is the pioneer work in the field of CQED [151]. In this very short paper, he suggests that the radiative rate of an emitter can be modified using a cavity, the modification depending on the quality factor Q and the effective volume of the cavity mode V. Since then various theoretical works have dealt with the interaction between an emitter and an optical mode of a cavity. Two regimes can be destinguished. The first one is the weak coupling regime for which the spectral linewidth of the electromagnetic mode is much larger than the one of the emitter. In this case, the energy excited emitter is irreversibly transferred from the emitter to the continuum formed by the electromagnetic states. The modification of the radiative rate can be found using Fermi's golden rule. Hence, it depends also on the efficiency of the coupling between the emitter and the field. For example, to increase the radiative rate, the emitter must be located at a maximum of the electric field. The radiative rate is then multiplied by a factor usually called the Purcell factor and noted F_p which is given by:

$$F_p = \frac{3}{4\pi^2} \frac{Q\lambda^3}{V_{eff}}$$

(24.23)

where λ is the wavelength emission, Q, is the quality factor of the cavity and V_{eff} the effective volume of the electromagnetic mode. Equation 24.23 shows that high Purcell factors are achieved with cavities showing high quality factors and low mode-volumes.

The second regime, called strong coupling regime, corresponds to the case of a two-level system coupled to the mode of a cavity characterized by a high quality factor Q. The Rabi oscillation period which characterizes the energy exchange between the field and the emitter is much lower than the lifetime of the photon in the cavity. This regime is at the heart of fundamental physics experiments demonstrating the quantum non-locality or enabling the study of decoherence.

The conditions required by the strong coupling regime are the most difficult to obtain experimentally and has not been observed yet for colloidal quantum dots. We now examine the different optical cavities which can be used to get an F_p factor as large as possible.

24.8.3   Dielectric cavities

The choice of the optical resonator is determined by various factors. The ideal cavity exhibits a strong quality factor and a low mode volume, enables an easy collection of the emitted photons and is easy to fabricate.

Three types of dielectric cavities are mainly used in solid-state CQED experiments. Microdisks (or microtores) are produced by electron beam lithography and several wet etching steps. They confine light at their edges in whispering gallery modes exhibiting high quality factors. Their main drawback lies in the difficulty of collecting the light which corresponds to evanescent waves located in the plane of the disk.

Micropillars are obtained by etching planar microcavities made of two Bragg mirrors. Light confinement is due to two different effects. Vertically, it comes from the Bragg mirrors. Laterally, it is due to total internal reflection at the air/semiconductor interface. Quantum dots can be grown by epitaxy between the mirrors during the fabrication of the microcavity. Inserting colloidal quantum dots is more difficult since the temperature of epitaxy is often too high to preserve the high photoluminescence of nanocrystals.

Photonic crystals are periodic dielectric structures with a periodicity of the order of the emission wavelength. This gives rise to a particular dispersion of photons. The photon density of states can be tailored, the light localized, the propagation strongly directed and the radiative emission rate strongly modified. From a general point of view, photonic crystals are usually produced by electron beam lithography and dry etching. For technical reasons, high quality factors have not yet been achieved in three-dimensional structures. The best structures are the two-dimensional photonic crystals etched in a membrane, which ensures the confinement in the direction perpendicular to the plane corresponding to the structure. The size of the pattern depends strongly on the wavelength. Since very low-dimensional structures are more difficult to produce, devices for infrared wavelength about $1.5\,\mu m$ exhibit higher quality factors than the ones for visible light. Quality factors as high as 10^6 have been obtained [152]. Very promising for telecommunication applications, such structures cannot be used to control the emission of colloidal quantum dots which emit visible light.

Until now, various alternative solutions have been found to fabricate photonic crystals suited to modify colloidal quantum dots emission. Barth *et al.* prepared three-dimensional photonic crystals by sedimentation of polystyrene beads [153]. After inserting colloidal quantum dots, they performed defocused imaging to investigate the angular dependance of nanocrystal emission at the single molecule level. They demonstrated the anisotropy of photonic stop bands. This experiment is a proof of principle of a new application of colloidal nanocrystals as nanometric probes of confined electromagnetic fields in nanophotonic structures. As the quantum dot emission wavelength can be tuned over a wide range, this method could be applied to many nanostructures.

In order to control the radiative lifetime of nanocrystals, Lohdahl *et al.* performed an experiment with titania inverse opals which consists in face-centred cubic structures of air spheres in a titania material [154]. The photonic crystal was made by inverting template-assisted self-assembled structures. A modified lifetime was observed on quantum dots ensemble over a large frequency bandwidth. Angle-resolved spectral measurements also proved the effect on the directionality of the emission. By changing the lattice parameter, they succeeded in inhibiting (by a factor of 1.6) as well as accelerating (by a factor of 1.3) the radiative rate of CdSe nanocrystals.

The last structure was obtained by nanoimprint lithography in a composite polymer incorporating the colloidal quantum dots. Reboud *et al.* observed an enhancement of the light collection efficiency by a factor of 2 compared to an unpatterned sample [155].

24.8.4   Metallic photonic cavities

The confinement of light in dielectric structures is limited to dimensions of the order of the wavelength. The corresponding micrometre-scale of photonic components such as optical fibres has strongly limited the integration of photonic devices in electronic chips, which have reached nanometre size. Recently, devices exhibiting surface plasmons have been proposed to realize circuits with nanoscale features enabled to carry optical signals. Such plasmonic chips may solve the size compatibity problem between electronics and photonics.

Surface plasmons are an extension of the plasmon physics (for a more precise description of surface plasmons, one can read the review by H. Raether [156]). The resolution of Maxwell equations at a metal/dielectric interface shows that optical waves can propagate along the surface of the metal. The electromagnetic field decays exponentially in the direction perpendicular to the interface and the dispersion curve of surface plasmons lies at the right of the light cone. Hence, they do not propagate in this direction and, then, are called non-radiative modes. However, the corrugation of the surface or a grating structure can couple surface plasmons and propagating fields. In this case, strong light emission can be obtained. Such processes are at the origin of phenomena such as surface enhanced Raman scattering (SERS) of second harmonic generation on a metallic surface.

In 2002, the group of M. Bawendi was the first to study the effect of a rough metal film on the fluorescence of colloidal CdSe nanocrystals [97]. The gold film exhibits peaks ranging from

10 to 50 nm measured by an atomic force microscope. Using far-field microscopy and time-resolved single-photon detection, they observed a complete conversion of the emission to linear. They observed a reduction factor of the exciton lifetime larger than 10^3. At the same time, the detected efficiency is multiplied by 5. Hence, radiative and non-radiative decay rates are strongly modified. The usually dominant Auger rate is no longer so efficient. This enables the fluorescence of charged quantum dots to be detected, which are usually not fluorescent (section 24.5). The simultaneous acceleration of radiative and non-radiative decay rates illustrates the possibilities offered by the use of metallic structures as well as their drawback. Absorption and emission of the nanocrystal are increased by the coupling of the nanocrystal with the metal plasmons. At the same time, non-radiative decay chanels are enhanced by energy transfer from the quantum dot to the metal. As a result, the quantum yield of the nanocrystal is reduced while the fluorescence intensity is increased.

After the use of a rough metallic surface, Zhang *et al.* reported results concerning the modification of nanocrystal fluorescence obtained with a two-dimensional textured Ag grating [157]. The periodicity of the metallic surface is on the scale of the quantum dot emission wavelength. Then, the energy of the quantum dot can excite surface plasmons which can scatter off the metal through the coupling by the grating of the surface plasmons and propagating modes. Using an angular-resolved detection scheme, the authors demonstrated the redirection of the emission as well as a modification of its polarization. Time-resolved photoluminescence experiments showed the enhancement of non-radiative and radiative decay rates. They are due to the modification of the electromagnetic mode's density directly linked to the mode dispersion of the corrugated surface plamons. The decrease in Auger processes efficiency results in the reduction of blinking. Since the analysis is carried out on ensembles, the exact contribution of blinking reduction and quantum yield reduction cannot be quantitatively measured but the results confirm qualitatively the observations made on rough metallic surfaces.

Another scheme used to enchance colloidal quantum dot fluorescence consists in a gold nanopocket aperture [158]. The nanocrystals are partially encapsulated in a smooth Au film with an aperture lower than the emission wavelength. The enhancement depends strongly on the plasmons' characteristics which can be controlled through the gold film thickness. It is maximum for a 50 nm layer. The increase of the detection intensity is attributed to the enhancement of localized field intensity at the sharp edge of the nanopocket aperture, the high density of states due to surface plasmons and the subwavelength nanopocket aperture size which couples surface plasmons to propagating photons.

24.8.5 Perspectives

The results detailed in this section illustrate the potential applications of the coupling between colloidal quantum dots and metallic or dielectric nanophotonic structures. Very promising for the realization of optoelectronic devices or biomedical image probing, such structures could also be very useful in the field of quantum information. First, an efficient single-photon source emitting at room temperature which could be used for the implementation of quantum cryptography protocols could be developed. Light confinement inside the photonic cavities will allow control of the polarization and of the radiative pattern of the nanocrystal fluorescence. Due to the Purcell effect, the radiative emission rate will be increased improving the rate of this single photon source. The originality of the new devices will lie in the realization of photonic structures working at visible wavelengths, which requires very small patterns. Moreover, the great flexibility and ease of utilization even at the single molecule level should be used to control the position of the nanocrystals on a surface at nanometre-sized scale to optimize the coupling between the emitter and the mode of the cavity. At the single nanoobject level, this phase of CQED experiments carried on semiconductor structures remains a technical challenge.

Finally, let us mention that the colloidal nature of nanocrystals allows the addition of functions by chemical reactions. This possibility is widely used in cellular imaging, where nanocrystals are attached to biomolecules in order to be used as markers, but has been little explored in single-emitter quantum optics. It might, however, offer ways to deposit nanocrystals on specific sites of a photonic structure.

24.9   Conclusions

Colloidal quantum dots are often called artificial atoms since the three-dimensional confinement results in a discrete, atomic-like energy spectrum. Since the observation of photon antibunching, these chemically synthesized quantum dots have emerged as very promising candidates for investigating quantum optics using solid-state sources.

We first analysed in detail the generation of single photons by colloidal quantum dots. The fluorescence of an individual nanocrystal shows a perfect antibunching. This is due to the efficiency of Auger processes in small quantum dots which renders the fluorescence of a nanocrystal similar to the one of a two-level system. Auger recombination enables one to work at room temperature, which makes the collection both simple and very efficient. The main drawback of colloidal quantum dots is the blinking of their fluorescence. This phenomenon exhibits exotic characteristics which bring to light the dynamics of charge exchange between the nanocrystal and its environment. From a general point of view, this intermittency demonstrates the influence of the close environment of the nanocrystal on its fluorescence properties: the environment induces fluctuations on many optical properties such as the radiative lifetime or the emission wavelength.

The spectral diffusion has been first studied through its correlation with the fluorescence in-termittency. More recently, using a method called photon-correlation Fourier spectroscopy, it has been possible to investigate the dynamics of the spectral diffusion at a timescale of around $10\,\mu s$ and to characterize the time coherence of the single photons emitted by the nanocrystals. From the perspective of quantum computing, this characteristic is linked to the indistinguishability of photons and to the possibility of observing photon coalescence. By resolving the spectral diffusion dynamics at short timescale, it has been proved that the time coherence of the zero phonon line is about $200\,ps$, a value much longer than the one obtained by standard spectroscopy. Applying schemes used in atomic physics, and fast time-resolved detection, two photon interferences could be achieved.

The influence of the environment on the time coherence of the emission of a solid-state source makes tricky the observation of the coalescence of photons emitted by such devices. Like strong coupling or spontaneous emission rate inhibition, quantum optics phenomena have often been observed first on atomic physics systems and after on solid-state devices. However, beyond the compatibility with modern electronics, solid-state physics offers an original scheme for the generation of particular quantum states of light. The recent observation of entangled photon pairs obtained by the radiative desexcitation of biexcitons in epitaxially grown quantum dots shows that such sources could be an alternative solution to the generation of polarization entanglement by parametric down-conversion. In colloidal quantum dots, the Auger effect is more efficient than in epitaxial quantum dots because of the very small size of the structure. It even prevents any biexciton radiative recombination in quantum dots with a low radius. However, we reported and analysed recent results showing that such a radiative cascade could be obtained for the largest quantum dots or for quantum dots with alternative core–shell structures.

Finally, we discussed results concerning the control of colloidal quantum dots emission using photonic structures following concepts of CQED. Using metallic or dielectric devices, many groups have demonstrated the possibility of modifying the nanocrystal emission in terms of radiation pattern or radiative lifetime. Compared to epitaxially grown structures, colloidal quantum dots provide the opportunity to test new methods to control the position of emitters in photonic structures. Moreover, the realization of such structures working at visible wavelengths represent a technological challenge since they require very small patterns.

References

1. www.idquantique.com.
2. S. Wiesner, Conjugate coding, *SIGACT News* **15**, 78 (1983).
3. C.H. Bennett and G. Brassard, in: *Proceedings of the IEEE International Conference on Computers, Systems and Signal Processing*, Bangalore, India (IEEE, New York, pp. 175–179, (1984).

4. N. Gisin, G. Ribordy, W. Tittel, and H. Zbinden, Quantum cryptography, *Rev. Mod. Phys.* **74**, 145 (2002).

5. G. Brassard, N. Lutkenhaus, T. Mor, and B.C. Sanders, Limitations on practical quantum cryptography, *Phys. Rev. Lett.* **85**, 1330 (2000).

6. C.H. Bennett, F. Bessette, G. Brassard, L. Salvail, and J. Smolin, Experimental quantum cryptography, *J. Cryptol.* **5**, 3 (1992).

7. B.C. Jacobs and J.D. Franson, Quantum cryptography in free space, *Opt. Lett.* **21**, 1854 (1996).

8. W.T. Buttler, R.J. Hughes, P.G. Kwiat, S.K. Lamoreaux, G.G. Luther, G.L. Morgan, J.E. Nordholt, C.G. Peterson, and C.M. Simmons, Practical free-space quantum key distribution over 1 km, *Phys. Rev. Lett.* **81**, 3283 (1998).

9. E.-L. Miao, Z.-F. Han, T. Zhang, and G.-C. Guo, The feasibility of geostationary satellite-to-ground quantum key distribution, *Phys. Lett. A* **361**, 29 (2007).

10. A. Muller, J. Breguet, and N. Gisin, Experimental demonstration of quantum cryptography using polarized photons in optical fiber over more than 1 km, *Europhys. Lett.* **23**, 383 (1993).

11. E. Diamanti, H. Takesue, C. Langrock, M.M. Fejer, and Y. Yamamoto, 100 km differential phase shift quantum key distribution experiment with low jitter up-conversion detectors, *Opt. Express* **14**, 13073 (2006).

12. H.J. Kimble, M. Dagenais, and L. Mandel, Photon antibunching in resonance fluorescence, *Phys. Lett. Rev.* **39**, 691 (1977).

13. F. Diedrich and H. Walther, Nonclassical radiation of a single stored ion, *Phys. Rev. Lett.* **58**, 203 (1987).

14. A. Kuhn, M. Hennrich, and G. Rempe, Deterministic single-photon source for distributed quantum networking, *Phys. Rev. Lett.* **89**, 067901 (2002).

15. J. McKeever, A. Boca, A.D. Boozer, R. Miller, J.R. Buck, A. Kuzmich, and H.J. Kimble, Deterministic generation of single photons from one atom trapped in a cavity, *Science* **303**, 1992 (2004).

16. T. Basché, W.E. Moerner, M. Orrit, and H. Talon, Photon antibunching in the fluorescence of a single dye molecule trapped in a solid, *Phys. Rev. Lett.* **69**, 1516 (1992).

17. C. Brunel, B. Lounis, P. Tamarat, and M. Orrit, Triggered source of single photons based on controlled single molecule fluorescence, *Phys. Rev. Lett.* **83**, 2722 (1999).

18. L. Fleury, J.-M. Segura, G. Zumofen, B. Hecht, and U.P. Wild, Nonclassical photon statistics in single-molecule fluorescence at room temperature, *Phys. Rev. Lett.* **84**, 1148 (2000).

19. B. Lounis and W.E. Moerner, Single photons on demand from a single molecule at room temperature, *Nature* **407**, 491 (2000).

20. F. Treussart, A. Clouqueur, C. Grossman, and J.-F. Roch, Photon antibunching in the fluorescence of a single dye molecule embedded in a thin polymer film, *Opt. Lett.* **26**, 1504 (2001).

21. R. Brouri, A. Beveratos, J.-P. Poizat, and P. Grangier, Photon antibunching in the fluorescence of individual color centers in diamond, *Opt. Lett.* **25**, 1294 (2000).

22. C. Kurtsiefer, S. Mayer, P. Zarda, and H. Weinfurter, Stable solid-state source of single photons, *Phys. Rev. Lett.* **85**, 290 (2000).

23. A. Beveratos, R. Brouri, T. Gacoin, J.-P. Poizat, and P. Grangier, Nonclassical radiation from diamond nanocrystals, *Phys. Rev. A* **64**, 061802(R) (2001).

24. P. Michler, A. Imamoglu, M.D. Mason, P.J. Carson, G.F. Strouse, and S.K. Buratto, Quantum correlation among photons from a single quantum dot at room temperature, *Nature* **406**, 968 (2000).

25. C. Santori, M. Pelton, G. Solomon, Y. Dale, and Y. Yamamoto, Triggered single photons from a quantum dot, *Phys. Rev. Lett.* **86**, 1502 (2001).

26. M. Pelton, C. Santori, J. Vuckovic, B. Zhang, G.S. Solomon, J. Plant, and Y. Yamamoto, Efficient source of single photons: a single quantum dot in a micropost microcavity, *Phys. Rev. Lett.* **89**, 233602 (2002).

27. P. Michler, A. Kiraz, C. Becher, W.V. Schoenfeld, P.M. Petroff, L. Zhang, E. Hu, and A. Imamoglu, A quantum dot single-photon turnstile device, *Science* **290**, 2282 (2000).

28. G. Messin, J.-P. Hermier, E. Giacobino, P. Desbiolles, and M. Dahan, Bunching and anti-bunching in the fluorescence of semiconductor nanocrystals, *Opt. Lett.* **26**, 1891 (2001).

29. A. Beveratos, R. Brouri, T. Gacoin, A. Villing, J.-P. Poizat, and P. Grangier, Single photon quantum cryptography, *Phys. Rev. Lett.* **89**, 187901 (2002).

30. E. Waks, K. Inoue, C. Santori, D. Fattal, J. Vuckovic, G.S. Solomon, and Y. Yamamoto, Quantum cryptography with a photon turnstile, *Nature* **420**, 762 (2002).

31. R. Loudon. *The Quantum Theory of Light*, (Oxford University Press, Oxford, 1973).

32. R. Glauber, in: *Quantum Optics and Electronics*, edited by C. De Witt, A. Blandin, and , C. Cohen-Tannoudji. (Gordon and Breach, New York, 1965).

33. B. Lounis and M. Orrit, Single-photon sources, *Rep. Prog. Phys.* **68**, 1129 (2005).

34. R. Hanbury-Brown and R.Q. Twiss, Correlation between photons in two coherent beams of light, *Nature* **177**, 27 (1956).

35. L. Mandel and E. Wolf. *Optical Coherence and Quantum Optics*, (Cambridge University Press, Cambridge, 1995).

36. M.B. Plenio and P.L. Knight, *Rev. Mod. Phys.* **70**, 101 (1998).

37. M. Lax, Formal theory of quantum fluctuations from a driven state, *Phys. Rev.* **129**, 2342 (1963).

38. X. Brokmann, E. Giacobino, M. Dahan, and J.-P. Hermier, Highly efficient triggered emission of single photons by colloidal CdSe/ZnS nanocrystals, *Appl. Phys. Lett.* **85**, 712 (2004).

39. A.I. Ekimov, A.A. Onushchenko, and V.A. Tsekhomskii, *Fiz. Khim. Stekla.* **6**, 511 (1980).

40. A.I. Ekimov and A.A. Onushchenko, Size quantization of the electron energy spectrum in a microscopic semiconductor crystal, *Pis'ma Zh. Eksp. Teor. Fiz.* **40**, 337 (1984); *JEPT Lett.* **40**, 1136 (1984).

41. R. Rossetti, R. Hull, J.M. Gibson, and L.E. Brus, Excited electronic states and optical spectra of ZnS and CdS crystallites in the ≈15 to 50 Å size range: evolution from molecular to bulk semiconducting properties, *J. Chem. Phys.* **82**, 552 (1985).

42. B.O. Dabbousi, J. Rodriguez-Viejo, F.V. Mikulec, J.R. Heine, H. Mattoussi, R. Ober, K.F. Jensen, and M.G. Bawendi (CdSe)ZnS core-shell quantum dots: synthesis and characterization of a size series of highly luminescent nanocrystallites, *J. Phys. Chem. B* **101**, 9463 (1997).

43. M.A. Hines and P. Guyot-Sionnest, Synthesis and characterization of strongly luminescing ZnS-capped CdSe nanocrystals, *J. Phys. Chem.* **100**, 468 (1996).

44. J. Li and L.-W. Wang, Shape effects on electronic states of nanocrystals, *Nano Lett.* **3**, 1357 (2003).

45. S. Pokrant and K.B. Whaley, Tight-binding studies of surface effects on electronic structure of CdSe nanocrystals: the role of organic ligands, surface reconstruction, and inorganic capping shells, *Eur. Phys. J. D* **6**, 255 (1999).

46. L.-W. Wang and A. Zunger, Pseudopotential calculations of nanoscale CdSe quantum dots, *Phys. Rev. B* **53**, 9579 (1996).

47. M. Califano, A. Franceschetti, and A. Zunger, Temperature dependence of excitonic radiative decay in CdSe quantum dots: the role of surface hole traps, *Nano Lett.* **5**, 2360 (2005).

48. G. Bastard. *Wave Mechanics Applied to Semiconductor Heterostructures*, (Editions de Physique, 1988).

49. A.I. Ekimov, F. Hache, M.C. Schanne-Klein, D. Ricard, C. Flytzanis, I.A. Kudryavtsev, T.V. Yazeva, A.V. Rodina, and Al.L. Efros, Absorption and intensity-dependent photolumines-cence measurements on CdSe quantum dots: assignment of the first electronic transitions, *J. Opt. Soc. Am. B* **10**, 100 (1993).

50. D.J. Norris and M.G. Bawendi, Measurement and assignment of the size-dependent optical spectrum in CdSe quantum dots, *Phys. Rev. B* **53**, 16338 (1996).

51. S.V. Gaponenko. *Optical Properties of Semiconductor Nanocrystals*, (Cambridge University Press, Cambridge, 1998).

52. A.S. Davydov. *Quantum Mechanics*, (Pergamon Press, Oxford, 1965).

53. Al.L. Efros and A.V. Rodina, Confined excitons, trions and biexcitons in semiconductor microcrystals, *Solid State Com.* **72**, 645 (1989).

54. C.A. Leatherdale, W.-K. Woo, F.V. Mikulec, and M.G. Bawendi, On the absorption cross section of CdSe nanocrystal quantum dot, *J. Phys. Chem. B* **106**, 7619 (2002).

55. H.C. van de Hulst. *Light Scattering by Small Particles*, (Dover Publications, New York, 1981).

56. S. Berciaud, L. Cognet, and B. Lounis, Photothermal absorption spectroscopy of individual semiconductor nanocrystals, *Nano Lett.* **5**, 2160 (2005).

57. H. Htoon, P.J. Cox, and V.I. Klimov, Structure of excited-state transitions of individual semiconductor nanocrystals probed by photoluminescence excitation spectroscopy, *Phys. Rev. Lett.* **93**, 187402 (2004).

58. U. Bockelmann and G. Bastard, Phonon scattering and energy relaxation in two-, one-, and zero-dimensional electron gases, *Phys. Rev. B* **42**, 8947 (1990).

59. U. Woggon, H. Giessen, F. Gindele, O. Wind, B. Fluegel, and N. Peyghambarian, Ultrafast energy relaxation in quantum dots, *Phys. Rev. B* **54**, 17681 (1996).

60. V.I. Klimov, D.W. McBranch, C.A. Leatherdale, and M.G. Bawendi, Electron and hole relaxtion pathways in semiconductor quantum dots, *Phys. Rev. B* **60**, 13740 (1999).

61. S. Xu, A.A. Mikhailovsky, J.A. Hollingsworth, and V.I. Klimov, Hole intraband relaxation in strongly confined quantum dots: revisiting the "phonon bottleneck" problem, *Phys. Rev. B* **65**, 045319 (2002).

62. E. Hendry, M. Koeberg, F. Wang, H. Zhang, C. de Mello Donega, D. Vanmaekelbergh, and M. Bonn, Direct observation of electron-to-hole energy transfer in CdSe quantum dots, *Phys. Rev. Lett.* **96**, 057408 (2006).

63. Y.Z. Hu, S.W. Koch, M. Lindberg, N. Peyghambarian, E.L. Pollock, and F.F. Abraham, Biexcitons in semiconductor quantum dots, *Phys. Rev. Lett.* **64**, 1805 (1990).

64. K.I. Kang, A.D. Kepner, S.V. Gaponenko, S.W. Koch, Y.Z. Hu, and N. Peyghambarian, Confinement-enhanced biexciton binding energy in semiconductor quantum dots, *Phys. Rev. B* **48**, 15449 (1993).

65. U. Woggon, S. Gaponenko, A. Uhrig, W. Langbein, and C. Klingshirn, Homogeneous linewidth and relaxation of excited hole states in II–VI quantum dots, *Adv. Mat. Opt. Electron.* **3**, 141 (1994).

66. B. Fisher, J.-M. Caruge, D. Zehnder, and M. Bawendi, Room-temperature ordered photon emission from multiexciton states in single CdSe core-shell nanocrystals, *Phys. Rev. Lett.* **94**, 087403 (2005).

67. V.I. Klimov, A.A. Mikhailovsky, D.W. McBranch, C.A. Leatherdale, and M.G. Bawendi, Quantization of multiparticle Auger rates in semiconductor quantum dots, *Science* **287**, 1011 (2000).

68. C. Bonati, M.B. Mohamed, D. Tonti, G. Zgrablic, S. Haacke, F. van Mourik, and M. Chergui, Spectral and dynamical characterization of multiexcitons in colloidal CdSe semiconductor quantum dots, *Phys. Rev. B* **71**, 205317 (2005).

69. B. Lounis, H.A. Bechtel, D. Gerion, P. Alivisatos, and W.E. Moerner, *Chem. Phys. Lett.* **329**, 399 (2000).

70. Al.L. Efros, M. Rosen, M. Kuno, M. Nirmal, D.J. Norris, and M.G. Bawendi, Band-edge exciton in quantum dots of semiconductors with a degenerate valence band: dark and bright exciton states, *Phys. Rev. B* **54**, 4843 (1996).

71. M. Dib, Structure électronique au voisinage de la bande interdite des nanocristaux de CdSe et CdS, PhD thesis, Université Paris VII (1999).

72. S.A. Crooker, T. Barrick, J.A. Hollingsworth, and V.I. Klimov, Multiple temperature regimes of radiative decay in CdSe nanocrystal quantum dots: intrinsic limits to the dark-exciton lifetime, *Appl. Phys. Lett.* **82**, 2793 (2003).

73. O. Labeau, P. Tamarat, and B. Lounis, Temperature dependence of the luminescence lifetime of single CdSe/ZnS quantum dots, *Phys. Rev. Lett.* **90**, 257404 (2003).

74. M. Nirmal, D.J. Norris, M. Kuno, M.G. Bawendi, Al.L. Efros, and M. Rosen, Observation of the "dark exciton" in CdSe quantum dots, *Phys. Rev. Lett.* **75**, 3728 (1995).

75. M. Furis, H. Htoon, M.A. Petruska, V.I. Klimov, T. Barrick, and S.A. Crooker, Bright-exciton fine structure and anisotropic exchange in CdSe nanocrystal quantum dots, *Phys. Rev. B* **73**, 241313 (2006).

76. S.A. Empedocles, R. Neuhauser, and M.G. Bawendi, Three-dimensional orientation measurements of symmetric single chromophores using polarization microscopy, *Nature* **399**, 126 (1999).

77. I. Chung, K.T. Shimizu, and M.G. Bawendi, Room temperature measurements of the 3D orientation of single CdSe quantum dots using polarization microscopy, *Proceedings of the National Academy of Sciences* **100**, 405 (2003).

78. X. Brokmann, M.-V. Ehrensperger, J.-P. Hermier, A. Triller, and M. Dahan, Orientational imaging and tracking of single CdSe nanocrystals by defocused microscopy, *Chem. Phys. Lett.* **406**, 210 (2005).

79. X. Brokmann, L. Coolen, M. Dahan, and J.-P. Hermier, Measurement of the radiative and nonradiative decay rates of single CdSe nanocrystals through a controlled modification of their spontaneous emission, *Phys. Rev. Lett.* **93**, 107403 (2004).

80. B. Lounis, H.A. Bechtel, D. Gerion, P. Alivisatos, and W.E. Moerner, Photon antibunching in single CdSe/ZnS quantum dot fluorescence, *Chem. Phys. Lett.* **329**, 399 (2000).

81. A. Beveratos, PhD thesis, Orsay, France (2002).

82. J. McKeever, A. Boca, A.D. Boozer, R. Miller, J.R. Buck, A. Kuzmich, and H.J. Kimble, Deterministic generation of single photons from one atom trapped in a cavity, *Science* **303**, 1992 (2004).

83. G. Ribordy, N. Gisin, O. Guinnard, D. Stucki, M. Wegmuller, and H. Zbinden, Photon-counting at telecom wavelengths with commercial InAsGa/InP avalanche photodiodes: current performance, *J. Mod. Opt.* **51**, 1381 (2004).

84. E. Diamanti, H. Takesue, T. Honjo, K. Inoue, and Y. Yamamoto, Performance of various quantum-key-distribution systems using 1.55-μm up-conversion single-photon detectors,, *Phys. Rev. A* **72**, 052311 (2005).

85. J.J. Peterson and T.D. Krauss, Fluorescence spectroscopy of single lead sulfide quantum dots, *Nano Lett.* **6**, 510 (2006).

86. M. Nirmal, B.O. Dabbousi, M.G. Bawendi, J.J. Macklin, J.K. Trautman, T.D. Harris, and L.E. Brusn, Fluorescence intermittency in single cadmium selenide nanocrystals, *Nature* **383**, 802 (1996).

87. S.A. Empedocles, D.J. Norris, and M.G. Bawendi, Photoluminescence spectroscopy of single CdSe nanocrystallite quantum dots, *Phys. Rev. Lett.* **77**, 3873 (1996).

88. F. Cichos, J. Martin, and C. Von Borczyskowski, Characterizing the non-stationary blinking of silicon nanocrystal, *J. Lum.* **107**, 160 (2004).

89. F. Kulzer, S. Kummer, R. Matzke, C. Brauchle, and T. Basche, Single-molecule optical switching of terrylene in p-terphenyl, *Nature* **387**, 688 (1997).

90. D.A. VandenBout, W.T. Yip, D.H. Hu, D.K. Fu, T.M. Swager, and P.F. Barbara, Discrete intensity jumps and intramolecular electronic energy transfer in the spectroscopy of single conjugated polymer molecules, *Science* **277**, 1074 (1997).

91. M.A. Bopp, Y.W. Jia, L.Q. Li, R.J. Cogdell, and R.M. Hochstrasser, Fluorescence and pho-tobleaching dynamics of single light-harvesting complexes, *Proc. Natl. Acad. Sci.* **94**, 10630 (1997).

92. R.M. Dickson, A.B. Cubitt, R.Y. Tsien, and W.E. Moerner, Blinking and switching behavior of single green fluorescent protein molecules, *Nature* **388**, 355 (1997).

93. H. Htoon, M.J. O'Connell, P.J. Cox, S.K. Doorn, and V.I. Klimov, Low temperature emission spectra of individual single-walled carbon nanotubes: multiplicity of subspecies within single-species nanotube ensembles, *Phys. Rev. Lett.* **93**, 27401 (2004).

94. S. Hohng and T. Ha, Near-complete suppression of quantum dot blinking in ambient conditions, *J. Am. Chem. Soc.* **126**, 1324 (2004).

95. X. Brokmann, Fluorescence properties of single CdSe nanocrystals, PhD thesis, Paris (2004).

96. O. Cherniavskaya, L. Chen, and L. Brus, Imaging the photoionization of individual CdSe/CdS core–shell nanocrystals on n- and p-type silicon substrates with thin oxides, *J. Phys. Chem. B* **108**, 4946 (2004).

97. K.T. Shimizu, W.K. Woo, B.R. Fisher, H.J. Eisler, and M.G. Bawendi, Surface-enhanced emission from single semiconductor nanocrystals, *Phys. Rev. Lett.* **89**, 117401 (2002).

98. S.A. Empedocles and M.G. Bawendi, Quantum-confined stark effect in single CdSe nanocrystallite quantum dots, *Science* **278**, 2114 (1997).

99. R.G. Neuhauser, K.T. Shimizu, W.K. Woo, S.A. Empedocles, and M.G. Bawendi, Correlation between fluorescence intermittency and spectral diffusion in single semiconductor quantum dots, *Phys. Rev. Lett.* **85**, 3301 (2000).

100. B.R. Fisher, H.-J. Eisler, N.E. Stott, and M.G. Bawendi, Emission intensity dependence single-exponential behavior in single colloidal quantum dot fluorescence lifetimes, *J. Phys. Chem. B* **108**, 143 (2004).

101. M. Kuno, D.P. Fromm, H.F. Hamann, A. Gallagher, and D.J. Nesbitt, On/Off fluorescence intermittency of single semiconductor quantum dots, *J. Chem. Phys.* **115**, 1028 (2001).

102. K.T. Shimizu, R.G. Neuhauser, C.A. Leatherdale, S.A. Empedocles, W.K. Woo, and M.G. Bawendi, Blinking statistics in single semiconductor nanocrystal quantum dots, *Phys. Rev. B* **63**, 205316 (2001).

103. R. Verberk and M. Orrit, Photon statistics in the fluorescence of single molecules and nanocrystals: correlation functions versus distributions of on- and off-times, *J. Chem. Phys.* **119**, 2214 (2003).

104. R. Verberk, A.M. van Oijen, and M. Orrit, Simple model for the power-law blinking of single semiconductor nanocrystals, *Phys. Rev. B* **66**, 233202 (2002).

105. G. Margolin, V. Protasenko, M. Kuno, and E. Barkai, Photon counting statistics for blinking CdSe–ZnS quantum dots: a Lévy walk process, *J. Phys. Chem. B* **110**, 19053 (2006).

106. M. Pelton, D.G. Grier, and P. Guyot-Sionnest, Characterizing quantum-dot blinking using noise power spectra, *Appl. Phys. Lett.* **85**, 819 (2004).

107. X. Brokmann, J.-P. Hermier, G. Messin, P. Desbiolles, J.-P. Bouchaud, and M. Dahan, "Statistical aging and non-ergodicity in the fluorescence of single nanocrystals", *Phys. Rev. Lett.* **90**, 120601 (2003).

108. I. Chung and M. Bawendi, Relationship between single quantum-dot intermittency and fluorescence intensity decays from collections of dots, *Phys. Rev. B* **70**, 165304 (2004).

109. I. Chung, J.B. Witkoskie, J.S. Cao, and M. Bawendi, Description of the fluorescence intensity time trace of collections of CdSe nanocrystal quantum dots based on single quantum dot fluorescence blinking statistics, *Phys. Rev. B* **73**, 011106 (2006).

110. I. Chung, J.B. Witkoskie, J.P. Zimmer, J. Cao, and M.G. Bawendi, Extracting the number of quantum dots in a microenvironment from ensemble fluorescence intensity fluctuations, *Phys. Rev. B* **75**, 045311 (2007).

111. F. Bardou, J.-P. Bouchaud, A. Aspect, and C. Cohen-Tannoudji. *Lévy Statistics and Laser Cooling: How Rare Events Bring Atoms at Rest*, (Cambridge University Press 2001).

112. G.M. Zaslavsky, Chaotic dynamics and the origin of statistical laws, *Physics Today* **52**, 39 (1999).

113. P. Bak and C. Tang, Earthquakes as a self organized critical phenomenon, *J. Geophys. Res.-Solid* **94**, 15635 (1989).

114. M. Boguna and A. Corral, Long-tailed trapping times and lévy flights in a self-organized critical granular system, *Phys. Rev. Lett.* **78**, 4950 (1997).

115. J. Klafter, M.F. Schlesinger, and G. Zumofen, Beyond brownian motion, *Physics Today* **49**, 33 (1996).

116. M. Leadbeater, V.I. Falko, and C.J. Lambert, Lévy flights in quantum transport in quasibal-listic wires, *Phys. Rev. Lett.* **81**, 1274 (1998).

117. W. Feller. *An Introduction to Probability and its Applications*, (Wiley, New York, 1968).

118. P. Lévy. *Théorie de l'addition des variables Aléatoires* (Gauthiers-Villars, Paris, 1937, 1954).

119. E. Lutz Power-law tail distributions and nonergodicity, *Phys. Rev. Lett.* **93**, 190602 (2004).

120. C. Godrèche and J.M. Luck, Statistics of the occupation time of renewal processes, *J. Stat. Phys.* **104**, 489 (2001).

121. K. Zhang, H.Y. Chang, A.H. Fu, A.P. Alivisatos, and H. Yang, Continuous distribution of emission states from single CdSe/ZnS quantum dots, *Nano Lett.* **6**, 843 (2006).

122. A.L. Efros and M. Rosen, Random telegraph signal in the photoluminescence intensity of a single quantum dot, *Phys. Rev. Lett.* **78**, 1110 (1997).

123. M. Kuno, Modelling distributed kinetics in isolated semiconductor quantum dots, *Phys. Rev. B* **67**, 125304 (2004).

124. R. Verberk, A.M. Van Oijen, and M. Orrit, Simple model for the power-law blinking of single semiconductor nanocrystals, *Phys. Rev. B* **66**, 233202 (2002).

125. J. Tang and R.A. Marcus, Mechanisms of fluorescence blinking in semiconductor nanocrystal quantum dots, *J. Chem. Phys.* **123**, 054704 (2005).

126. S. Empedocles and M. Bawendi, Spectroscopy of single CdSe nanocrystallites, *Acc. Chem. Res.* **32**, 389 (1999).

127. M. Dahan, S. Lévi, C. Luccardini, P. Rostaing, B. Riveau, and A. Triller, Diffusion dynamics of glycine receptors revealed by single quantum dot tracking, *Science* **302**, 442 (2003).

128. S. Coe, W.K. Woo, J.S. Steckel, M. Bawendi, and V. Bulovic, Electroluminescence from single monolayers of nanocrystals in molecular organic devices, *Nature* **420**, 800 (2002).

129. O. Labeau, P. Tamarat, and B. Lounis, High resolution resonant photoluminescence excitation of CdSe/ZnS nanocrystals at low temperatures, *Appl. Phys. Lett.* **88**, 223110 (2006).

130. E. Knill, R. Laflamme, and G.J. Milburn, A scheme for efficient quantum computation with linear optics, *Nature* **409**, 46 (2001).

131. T. Legero, T. Wilk, M. Hennrich, G. Rempe, and A. Kuhn, Quantum beat of two single photons, *Phys. Rev. Lett.* **93**, 070503 (2004).

132. S.A. Empedocles and M.G. Bawendi, Influence of spectral diffusion on the line shapes of single CdSe nanocrystallite quantum dots, *J. Phys. Chem. B* **103**, 1826 (1999).

133. A. Kiraz, M. Ehrl, Th. Hellerer, Ö.E. Müstecaplioglu, C. Bräuchle, and A. Zumbusch, Indistinguishable photons from a single molecule, *Phys. Rev. Lett.* **94**, 223602 (2005).

134. B. Zhang, G.S. Solomon, M. Pelton, J. Plant, C. Santori, J. Vuckovic, and Y. Yamamoto, Fabrication of InAs quantum dots in AlAs/GaAs DBR pillar microcavities for single photon sources, *J. Appl. Phys.* **97**, 073507 (2005).

135. S. Laurent, S. Varoutsis, L. Le Gratiet, A. Lemaître, I. Sagnes, F. Raineri, A. Levenson, I. Robert-Philip, and I. Abram, Indistinguishable single photons from a single-quantum dot in a two-dimensional photonic crystal cavity, *Appl. Phys. Lett.* **87**, 163107 (2005).

136. C.K. Hong, Z.Y. Ou, and L. Mandel, Measurement of subpicosecond time intervals between two photons by interference, *Phys. Rev. Lett.* **59**, 2044 (1987).

137. P. Palinginis, S. Tavenner, M. Lonergan, and H. Wang, Spectral hole burning and zero phonon linewidth in semiconductor nanocrystals, *Phys. Rev. B* **67**, 201307 (2003).

138. D.M. Mittleman, R.W. Schoenlein, J.J. Shiang, V.L. Colvin, A.P. Alivisatos, and C.V. Shank, Quantum size dependence of femtosecond electronic dephasing and vibrational dynamics in CdSe nanocrystals, *Phys. Rev. B* **49**, 14435 (1994).

139. T. Plakhotnik and D. Walser, Time resolved single molecule spectroscopy, *Phys. Rev. Lett.* **80**, 4064 (1998).

140. X. Brokmann, M. Bawendi, L. Coolen and, and J.-P. Hermier, Photon-correlation Fourier spectroscopy, *Opt. Exp.* **14**, 6333 (2006).

141. L. Coolen, X. Brokmann, and J.-P. Hermier, Modeling coherence measurements on a spectrally-diffusing single-photon emitter, submitted to *Phys. Rev. A.*

142. L. Coolen, X. Brokmann, P. Spinicelli, and J.-P. Hermier, to be submitted to *Phys. Rev. Lett.*

143. V.I. Klimov, A.A. Mikhailovsky, S. Xu, A. Malko, J.A. Hollingsworth, C.A. Leatherdale, H.-J. Eisler, and M.G. Bawendi, Optical gain and stimulated emission in nanocrystal quantum dots, *Science* **290**, 314 (2000).

144. M. Achermann, J.A. Hollingsworth, and V.I. Klimov, Multiexcitons confined within a subexcitonic volume: spectroscopic and dynamical signatures of neutral and charged biexcitons in ultrasmall semiconductor nanocrystals, *Phys. Rev. B* **68**, 245302 (2003).

145. D. Oron, M. Kazes, I. Shweky, and U. Banin, Multiexciton spectroscopy of semiconductor nanocrystals under quasi-continuous-wave optical pumping, *Phys. Rev. B* **74**, 115333 (2006).

146. O. Benson, C. Santori, M. Pelton, and Y. Yamamoto, Regulated and entangled photons from a single quantum dot, *Phys. Rev. Lett.* **84**, 2513 (2000).

147. N. Akopian, N.H. Lindner, E. Poem, Y. Berlatzky, J. Avron, and D. Gershoni, Entangled Photon pairs from semiconductor quantum dots, *Phys. Rev. Lett.* **96**, 130501 (2006).

148. R.M. Stevenson, R.J. Young, P. Atkinson, K. Cooper, D.A. Ritchie, and A.J. Shields, A semiconductor source of triggered entangled photon pairs, *Nature* **439**, 179 (2006).

149. R.J. Young, R.M. Stevenson, P. Atkinson, K. Cooper, D.A. Ritchie, and A.J. Shields, Improved fidelity of triggered entangled photons from single quantum dots, *New. J. Phys.* **8**, 29 (2006).

150. D. Oron, M. Kazes, and U. Banin, Multiexcitons in type-II colloidal semiconductor quantum dots, *Phys. Rev. B* **75**, 035330 (2007).

151. E.M. Purcell, *Phys. Rev.* **69**, 681 (1946).

152. Y. Akahane, T. Asano, B.-S. Song, and S. Noda, High-Q photonic nanocavity in a two-dimensional photonic crystal, *Nature* **425**, 944 (2003).

153. M. Barth, R. Schuster, A. Gruber, and F. Cichos, Imaging single quantum dots in three-dimensional photonic crystals, *Phys. Rev. Lett.* **96**, 243902 (2006).

154. P. Lohdahl, A. Floris van Driel, I.S. Nikolaev, A. Irman, K. Overgaag, D. Vanmaekelbergh, and W.L. Vos, Controlling the dynamics of spontaneous emission from quantum dots by photonic crystals, *Nature* **430**, 654 (2004).

155. V. Reboud, N. Kehagias, C.M. Sotomayor Torres, M. Zelsmann, M. Striccoli, M.L. Curri, A. Agostiano, M. Tamborra, M. Fink, F. Reuther, and G. Gruetzner, Spontaneous emission control of colloidal nanocrystals using nanoimprinted photonic crystals, *Appl. Phys. Lett.* **90**, 011115 (2007).

156. H. Raether. in: *Surface Plasmons on Smooth and Rough Surfaces and on Gratings* (Springer Verlag, Berlin, 1988).

157. J. Zhang, Y.-H. Ye, X. Wang, P. Rochon, and M. Xiao, Coupling between semiconductor quantum dots and two-dimensional surface plasmons, *Phys. Rev. B* **72**, 201306R (2005).

158. G.L. Liu, J. Kim, and L.P. Lee, Fluorescence enhancement of quantum dots enclosed in Au nanopockets with subwavelength aperture, *Appl. Phys. Lett.* **89**, 163107 (2006).

CHAPTER 25

PbSe Core, PbSe/PbS and PbSe/PbSe$_x$S$_{1-x}$ Core–Shell Nanocrystal Quantum Dots: Properties and Applications

E. Lifshitz,[1] A. Kigel,[1] M. Brumer,[1] L. Etgar, V. Kloper, M. Bashouti,[1] A. Sashchiuk,[1] R. Tennenbaum,[2] M. Sirota,[3] E. Galun,[3] Z. Burshtein,[4] A.Q. Le Quang,[5] I. Ledoux-Rak,[5] and J. Zyss[5]

[1]Shulich Faculty of Chemistry, Solid State Institute and the Russell Berrie Nanotechnology Institute, Technion, Haifa 32000, Israel;
[2]Department of Chemical Engineering, Russell Berrie Nanotechnology Institute, Technion, Haifa 32000, Israel;
[3]ElOp Electro Optics Industries Ltd, PO Box 1165 Rehovot 76111, Israel;
[4]Department of Materials Engineering, Ben-Gurion University of the Negev, PO Box 653, Beer-Sheva 84105, Israel;
[5]Laboratoire de Photonique Quantique et Moléculaire, Institut d Alembert, Ecole Normale Supérieure de Cachan, 61 avenue du Président Wilson 94235 Cachan, France

25.1 Introduction

Colloidal PbSe nanocrystal quantum dots (NQDs) have currently received special interest due to their unique electronic and optical properties [1]. The bulk PbSe semiconductor has a rock-salt crystal structure, and a narrow direct band gap (0.28 eV at 300 K) with both valence and conduction band maxima being four-fold degenerate at the L-point of the Brillouin zone. Also, the PbSe semiconductor possesses a high dielectric constant ($\varepsilon_\infty = 18.0$) and a relatively large effective Bohr radius ($a_{B(PbSe)} = 46$ nm). Previous $\mathbf{k} \cdot \mathbf{p}$ calculations of the electronic structure of the PbSe NQDs suggested an ordinary effective mass, such as S, P, and D states, with similar low masses for the electron and hole, and consequently a mirror-like density of states between the valence and conduction bands [2–4]. Recent use of the tight binding model [5] and the atomistic pseudopotential [6] method have taken into consideration the strong anisotropy of the bulk L-valleys with transverse and longitudinal effective masses [4], and valley-to-valley coupling: both issues split the bulk degenerate L-states, leading to a densely spaced hole manifold.

Interband optical studies of colloidal PbSe NQDs exhibit well-defined band edge excitonic transitions tuned between 0.8 and 4 µm [7–10]. Photoluminescence (PL) lifetime measurements and transient absorption measurements revealed an intraband relaxation process on the order of ps, and an interband recombination emission on the order of hundreds of ns [8–10]. Recently, amplified spontaneous emission from PbSe NQDs was demonstrated [9] with gain parameters similar to those observed in CdSe NQDs [11]. Furthermore, the impact ionization process in PbSe NQDs, obtained upon photoexcitation with $h\nu > 3E_g$ and its competition with Auger recombination have been discussed [3, 10]. The dependence of the transient optical absorption on the power of a pumping laser was measured by Klimov et al. [9], suggesting the creation of multiple excitons (up to eight electron–hole pairs), occupying the four valleys at the L-point of the

Brillouin zone of PbSe NQDs, with an anticipated benefit in gain devices. Thus, the aforementioned investigations indicate a widespread interest in the unique physical properties of PbSe NQDs, and the feasibility of using the PbSe NQDs in telecommunications [12], near-infrared (IR) lasers [13], and as biological markers [14, 15].

Various colloidal syntheses have been developed in the last couple of years, producing PbSe NQDs with size monodispersity (<5% size distribution), uniform shape, and high crystallinity. Murray *et al.* [16] and Colvin *et al.* [17] synthesized spherical core PbSe NQDs, soluble in organic or water solutions, with narrow size distributions and band gap tuning at the near-IR spectral regime. Lifshitz *et al.* [18] reported a colloidal procedure for the preparation of spherical PbSe/PbS core–shell NQDs, with an average size ranging between 2.5 and 7.0 nm, using tributylphonsphine/tri-octylphosphine (TBP/TOP) surfactants. Lifshitz *et al.* [19, 20] also reported recently the formation of unique PbSe/PbSe$_x$S$_{1-x}$ core-alloyed–shell NQDs with tunable composition of the alloyed shell (using oleic acid (OA) and TOP surfactants), showing exceptionally high-luminescence quantum efficiency (QE). Synthesis of core–inorganic shell NQDs is usually carried out to increase the air and luminescence stability; various synthesis methods have recently been reported for PbSe/PbS core–shell NQDs [21, 22]. Lifshitz *et al.* [23] used alkyldiamine or ethylene glycol as a coordinating molecular template, which led to the formation of PbSe wires (20 nm × 1 μm), rods (20 nm × 100 nm), ribbons (60 nm × 0.5 μm), stars (500 nm) and cubes (100 nm). Furthermore, Lifshitz *et al.* [24] reported the formation of one-dimensional assemblies composed of spherical PbSe NQDs, showing high conductivity properties. The conductivity properties of PbSe NQDs' arrays were examined recently at low temperatures by Wehrenberg *et al.* [25] and at high temperatures by Drndic *et al.* [26].

This chapter discusses a few alternative synthetic routes for the synthesis of chemical stable spherical and colloidal PbSe NQDs, synthesized as PbSe cores, PbSe/PbS core–shell structures, and PbSe/PbSe$_x$S$_{1-x}$ core-alloyed–shell structures (section 25.2). A thorough investigation of the structural and optical properties of the indicated NQDs is given below (section 25.3 and 25.4), suggesting the formation of NQDs with QE up to 65%, relatively narrow emission bands, a peculiar Stokes-shift behaviour and an excited-state lifetime ranging between 70 ns and 900 ns, depending on the pumping power intensity, composition and size of the NQDs. The discussed NQDs were used as passive Q-switches in an eye-safe laser of Er:glass (section 25.5), acting as "fast" saturable absorbers with a relatively large ground-state cross-section of absorption. In addition, the gain properties of the discussed NQDs were examined, showing an amplified spontaneous emission (ASE) under conditions that are suitable for technological devices, such as optical pumping by a continuous diode laser under room temperature conditions (section 25.6). The attachment of the PbSe NQDs to metallic particles (Fe$_2$O$_3$) is discussed in section 25.7, proposing the use of the hybrid structures as biological transport and tagging agents. The conductivity properties of PbSe NQDs self-assembled solids, annealed at 200°C, were examined, showing an ohmic behaviour at the measured voltages (up to 30 volts), which is governed by a variable range-hopping transport mechanism (section 25.8).

25.2   Synthesis, chemical stability, and structural characterization of PbSe NQDs, PbSe/PbS core–shell NQDs and PbSe/PbS$_x$eS$_{1-x}$ core-alloyed–shell NQDs

25.2.1   *Synthesis of PbSe NQDs cores, covered with organic surfactants (alternative I)*

The synthesis of core PbSe NQDs followed a modified procedure to that given by Murray *et al.* [16], following the procedure given in [19] and including the preceding stages: (1) 0.71 gr of lead(II) acetate trihydrate [Pb-ac] (Pb[CH$_3$COO]$_2$·3H$_2$O, GR, Merck) were dissolved in a solution composed of 2 mL diphenyl ether [PhEt] (C$_6$H$_5$OC$_6$H$_5$, 99%, Aldrich), 1.5 mL OA (CH$_3$(CH$_2$)$_7$ CHCH(CH$_2$)$_7$COOH, 99.8%, Aldrich) and 8 mL TOP ((C$_8$H$_{17}$)$_3$P, Tech, Aldrich), under standard inert conditions in the glove box, and were inserted into a three-neck flask (flask I); (2) 10 mL of PhEt were inserted into a three-neck flask (flask II) under the inert conditions of a glove box; (3) both flasks were taken out of the glove box, placed on a Schlenk line and heated under a vacuum to 100–120°C for an hour; (4) flask I was cooled to 45°C, while flask II was heated to 180−210°C, both under a fledging of an argon gas; (5) 0.155 gr of selenium powder (Se, 99.995%, Aldrich) was dissolved in 2.0 mL TOP, forming a TOP:Se solution, under standard inert conditions of a glove box. Then, 1.7 mL of this solution was injected into flask I on the Schlenk line; (6) the content of flask I, containing the reaction precursors, was injected rapidly into the PhEt solution in flask II, reducing its temperature to 100−130°C, leading to the formation of PbSe NQDs within the first 15 min of the reaction. The described procedure produced nearly mono-dispersed NQDs with <8% size distribution, with average size between 3 and 9 nm, controlled by the temperature and by the time duration of the reaction.

25.2.2   Synthesis of PbSe NQDs cores, covered with organic surfactants (alternative II)

In alternative II, the PbSe core NQDs were synthesized according to a procedure given by Colvin *et al.* [17]. A mixture of 0.892 g of PbO (4.00 mmol), 2.825 g of OA (10.00 mmol), and technological grade 1-octadecene (ODE) (the total weight was 16 gr) was initially heated to 150°C and after the mixer was turned colourless, it was further heated to 180°C. Then, 6.4 gr of Se-TOP solution (containing 0.64 gr of selenium, 8.00 mmol) was quickly injected into the PbO hot solution. The temperature of the reaction mixture was allowed to cool to 150°C for the growth of the PbSe NQDs. All steps in the reactions were carried out under argon.

25.2.3   Synthesis of PbSe/PbS core–shell NQDs by a two-injection process

The preparation of PbSe/PbS core–shell NQDs by a two-injection process begins with formation of core PbSe NQDs and their isolation from the initial reaction solution, according to the procedure in section 25.2.1. Those core NQDs were re-dissolved in chloroform solution, forming a solution of 50 mg/mL weight concentration. A quantity of 1.4 mL of TOP was then added to the NQDs solution, while the chloroform molecules were removed by distillation under vacuum and heating at 60°C. In parallel, 0.2 gr of a Pb precursor, Pb-ac, was dissolved in a mixture of 2 mL PhEt, 1.5 mL of OA, and 8 mL of TOP, heated to 120°C for an hour, and then cooled to 45°C. Also, 0.03–0.10 gr of sulphur (S, 99.99+%, Aldrich) was dissolved in 0.3 mL of TOP and was premixed with a PbSe core NQDs in a TOP solution. This mixture was injected into the Pb-ac solution. All reagents were then injected into a PhEt mother solution and kept on a Schlenk line at 180°C, causing a reduction in temperature of the mother solution to 120°C. The indicated chemical portions caused the precipitation of 1–3 monolayers (MLs) of PbS shell over the PbSe core surface within the first 15 min of the reaction.

25.2.4   Synthesis of PbSe/PbSe$_x$S$_{1-x}$ core-alloyed–shell NQDs by a single-injection process

The preparation of PbSe/PbSe$_x$S$_{1-x}$ core-alloyed–shell structures was nearly identical to that of the core PbSe NQDs, described in section 25.2.1 using a single injection of the precursors into a single round flask. However, step (5) was altered by the use of an alternative chalcogen precursor stock solution. A stock solution of Se and S was prepared by mixing 0.15 gr Se dissolved in 1.4 mL TOP, with 0.03–0.10 gr S dissolved in 0.3 mL TOP. The amount of S in the new stock solution corresponded to a stoichiometric amount of 1–2 ML of the PbS compound. Thus, the mole ratio of the precursors Pb:Se:S ranged from 1:1:0.5 to 1:1:1.3.

Aliquots were drawn periodically from the mother solutions described in sections 25.2.1–25.2.4. A quenching process to room temperature terminated the NQDs' growth. They were isolated from the aliquots solution by the addition of methanol, and by centrifugation. The isolated NQDs were further purified by dissolving them in chloroform, followed by filtering several times through a 0.02 micron membrane. The purified NQDs were examined by structural analyses, absorption and PL spectroscopy.

25.2.5   PbSe core and core–shell NQDs, capped with water-soluble ligands

The organic capping of the PbSe core and core–shell NQDs could be replaced with strong polar groups such as 2-aminoethanethiol ([HS(CH$_2$)$_2$NH$_3$)], AET) ligands using a procedure described elsewhere [27], converting the NQDs into water-soluble species. To create water-soluble PbSe NQDs with a positively charged capping, 100 μL of an organically capped PbSe NQD solution was dissolved in 5 mL of chloroform. Subsequently, 100 μL of a 0.5 M methanol solution of AET was added. The relatively stronger affinity of the thiol group originating from AET ligands to the PbSe surface, with respect to that of the OA carboxyl group, led to a ligand exchange. Thus, the exterior surfaces of the PbSe NQDs exchanged aliphatic terminating groups (the CH_3 end-groups of the OA) with the amine groups of the AET ligands. This immediately caused flocculation of the PbSe NQDs in chloroform. Subsequently, 5 mL of water was added to the suspension, resulting in a separation of the mixer into two phases (water above chloroform). Upon further shaking, the flocculated NQDs dissolved into the water phase and formed a clear suspension.

25.2.6   Storage conditions and encapsulation of the NQDs in a polymer film

The colloidal NQDs were embedded in a polymer film or dissolved in a glassy solution (2,2,4,4,6,8,8,-heptamethyl-nonane) for the optical measurements. The polymer was prepared by mixing PbSe NQDs in chloroform solution with poly-methyl-methacrylate (PMMA) ($[-CH_2C(CH_3)(CO_2CH_3)-]_n$, analytical grade, Aldrich) polymer solution. The resultant mixture was spread on a quartz substrate and dried over 24 h to a uniform film.

The organically capped PbSe NQDs were stored either in a hexane solution or as a dry powder in air or in nitrogen ambient. The stability of these NQDs was examined over a period of time by recording the absorption spectra (see section 25.4) and following the consistency of the low exciton's energy. Plots of this exciton energy versus time suggest that the exciton energy in the core samples is blue shifted by ~400 meV over a period of days for the NQDs kept in a chloroform solution. Such a blue shift, however, occurs over months for the samples kept as dry powders. On the other hand, the energy shift is smaller for the PbSe/PbS core–shell samples, and even nearly disappeared for the NQDs coated with three PbS shell MLs. It is presumed that the exciton energy blue shift is due to oxidation of the surface, and a decrease of the effective size of the core. Obviously, the penetration of oxygen through the PbS shell is reduced as the shell becomes thicker. Furthermore, storage of the NQDs in a nitrogen environment nearly eliminates any spectral drift over a period of months, even extending to two or three years.

25.2.7   Synthesis of organically capped and water-soluble γ-Fe_2O_3 magnetic nanoparticles

The synthesis of γ-Fe_2O_3 nanoparticles (NPs) which were used for the preparation of PbSe NQD's γ-Fe_2O_3 NPs' conjugated structure, covered with organic surfactant followed a modified procedure to that documented by Held *et al.* [28]. A quantity of 0.2 mL of $Fe(CO)_5$ (1.52 mmol) was added to a mixture containing 10 mL of octyl ether and 1.28 g of OA (4.56 mmol) at 100°C. The resulting mixture was heated to 300°C and kept at that temperature for 1 h. During this time, the initial orange colour of the solution gradually changed into a black colour. The resulting black solution was cooled to room temperature and was centrifuged. The precipitate was removed, the remaining supernatant solution was then collected, and ethanol was added to produce reversible flocculation of the NPs. The solution was centrifuged again, and the precipitate was saved and subsequently suspended in hexane to produce a suspension of 6 nm iron NPs. Exposing this suspension to air for several days resulted in the complete oxidation of the iron NPs into dispersed 5 nm γ-Fe_2O_3 NPs.

The mentioned γ-Fe_2O_3 NPs were transferred into a water-soluble solution, using the procedure documented by Rotello *et al.* [29]. A quantity of 10 mg of iron oxide NPs were dissolved in 2 mL of hexane and 100 mg of polyhedral silsesquioxane (PSS) hydrate octakis were dissolved in 2 mL of water. These solutions were mixed; the solute weight ratio was 10:1. The structure of PSS hydrate octakis ($C_{32}H_{96}N_8O_{20}Si_8 \cdot xH_2O$) molecules is an octahedral featuring eight siloxy groups, therefore the resulting water-soluble γ-Fe_2O_3 NPs were capped by a negative charge. This bilayer system stayed under an inert environment for 2–3 min, and was then stirred rapidly for 24 h during which the NPs were transferred to the aqueous phase. The solution phases were carefully separated and the water-soluble iron oxide NPs were run through a 0.22 μm filter.

25.3 Structural characterization of PbSe core, PbSe/PbS core–shell and PbSe/PbS$_x$eS$_{1-x}$ core-alloyed–Shell NQDs

The morphology and crystallography of the colloidal NQDs were examined by transmission electron microscopy (TEM), high-resolution TEM (HR-TEM) and selected-area electron diffraction (SAED). High spatial frequency noise of the SAED image was filtered out using the built-in software of the measuring system, based on the Fourier transform of the original image. The TEM specimens were prepared by injecting small liquid droplets of the solution on a copper grid (300 mesh) coated with amorphous carbon film and then dried at room temperature.

Figure 25.1a shows an HR-TEM image of PbSe/PbS core–shell NQD with an overall diameter of 6.0 nm and a core diameter of 4.8 nm. The HR-TEM image of PbSe/PbSe$_x$S$_{1-x}$ core-alloyed–shell NQD, with an overall diameter of 6.0 nm, is presented in Fig. 25.1b. These images reveal the formation of spherical NQDs with well-resolved crystal planes, without distinct boundaries at the core–shell interface due to the close matching ($\sim$3%) of the PbSe and PbS crystallographic parameters. The increase in the core–shell NQDs' size by 1.2 nm as compared to the preliminary core NQDs is consistent with a PbS or PbSe$_x$S$_{1-x}$ shell thickness of 1 ML. Figure 25.1c provides a fast Fourier transform image, corresponding to the measured SAED of the sample shown in Fig. 25.1a, revealing a cubic single crystal, with lattice spacing of 6.12 Å indicating the existence of rock salt structure of space group Fm$\overline{3}$m. Figure 25.1d presents a TEM image of PbSe/PbSe$_x$S$_{1-x}$ NQDs that are 6.7 nm in diameter, capped with organic ligands. The ensemble of NQDs in the image exhibited a diameter distribution of 5%, self-organized into a hexagonal array.

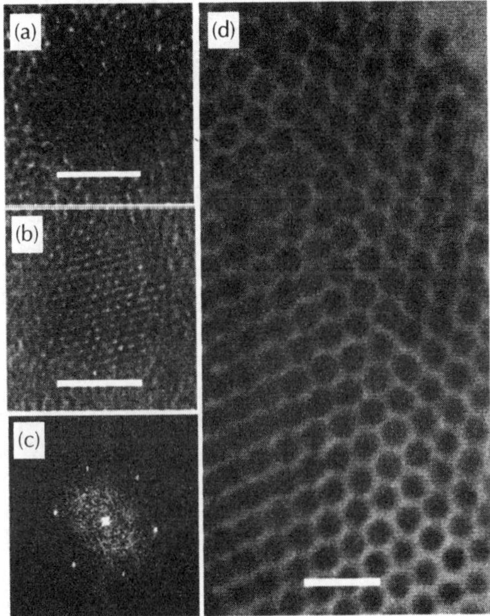

Figure 25.1 (a) HR-TEM image of a PbSe/PbS (4.8 nm/1.2 nm) core–shell NQD (bar scale = 5 nm). (b) HR-TEM image of a PbSe/PbSe$_x$S$_{1-x}$ core-alloyed–shell NQD (x = 0.5, bar scale = 5 nm). (c) Fast Fourier transform image of a measured SAED of the sample in A. (d) TEM image of 6.7 nm PbSe/PbSe$_x$S$_{1-x}$ core-alloyed–shell NQDs self-organized on a TEM grid (bar scale = 20 nm).

A representative energy-dispersive analysis of an X-ray (EDAX) of the PbSe/PbSe$_x$S$_{1-x}$ NQDs, prepared by the procedure given in section 25.2.4, is shown in Fig. 25.2. The Pb, Se and S percentages of aliquots, drawn from the reaction solution after 5 min and after 15 min are compared in Table 25.1. Comparison of the atomic percentages of the elements given in the table and the equivalent molecular formulas of the PbSe/PbSe$_x$S$_{1-x}$ NQDs suggests that a single injection of

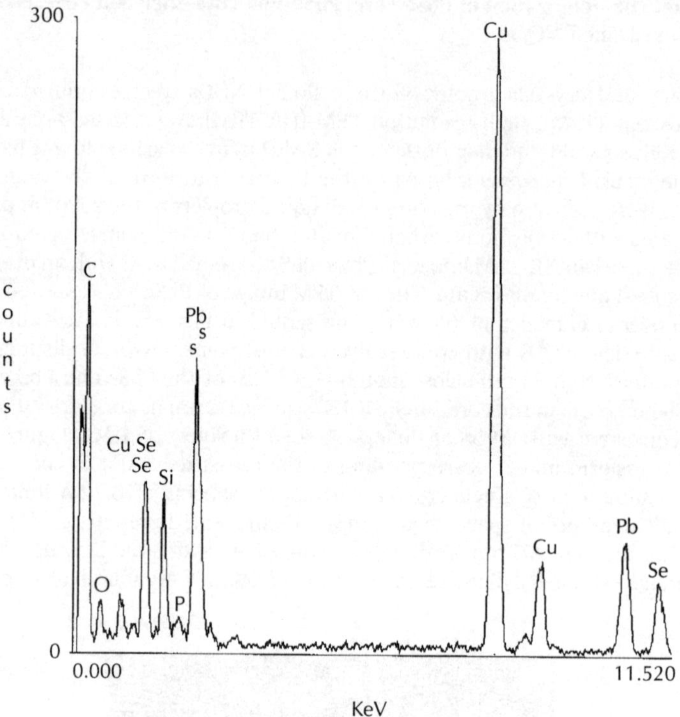

Figure 25.2   EDAX spectra of PbSe/PbSe$_x$S$_{1-x}$ core-alloyed–shell NQDs, prepared with initial Pb:Se:S = 1: 1:0.5 and extricated from the solution after 15 min.

Table 25.1 Atomic percentages of Pb, Se and S constituents in PbSe/PbSe$_x$S$_{1-x}$ NQDs, as derived from the EDAX measurements

Aliquots (elapsing time)	Initial Pb:Se:S molar ratio	Pb (%)	Se (%)	S (%)	NC molecular formula
5 minutes	1:1:0.5	51	49	–	PbSe
15 minutes	1:1:0.5	52.1	41.8	6.1	PbSe/PbSe$_{0.7}$S$_{0.3}$
15 minutes	1:1:1.3	46.7	22.6	30.7	PbSe/PbSe$_{0.1}$S$_{0.9}$

Pb, Se and S constituents starts with embryonic nucleation of PbSe cores, followed by the formation of PbSe$_x$S$_{1-x}$ alloyed shells with variable sulphur contents.

Figure 25.3 compares the TEM images of PbSe core NQDs, with an NQD average diameter of 6.0 nm, capped either with OA (a), or with AET (b). Comparison between the two images suggests that the process has not altered the size, shape and disparity of the NQDs before and after the ligand exchange. The water-soluble AET-capped NQDs exhibited a high chemical stability under a nitrogen atmosphere for an extended period of time (over months), without any indication of aggregation or photo-degradation. The electron diffraction pattern of the PbSe core NQD sample in an aqueous medium (insets in Fig. 25.3a and b) confirms that the NQDs have a perfect cubic close-packed (ccp/fcc) crystal structure with excellent agreement with the cubic (rock salt) structure of bulk PbSe and of the NQDs capped with organic ligands. The FTIR spectra (given in [27]) of the AET-capped NQDs in the aqueous solution exhibited two distinct bands, centred at 3469.3 and 3346.7 cm^{-1}, corresponding, respectively, to the asymmetric and symmetric stretching

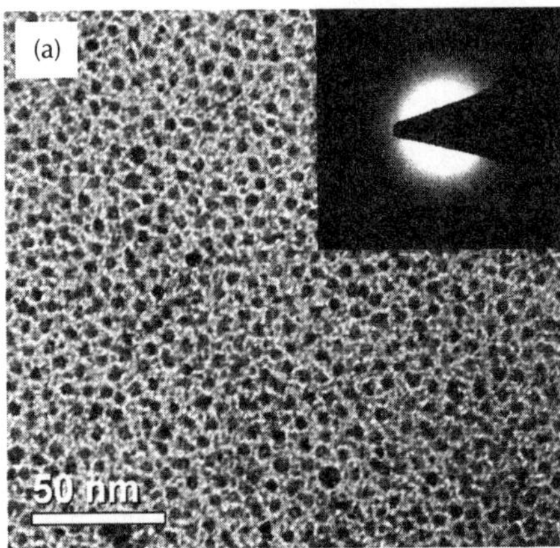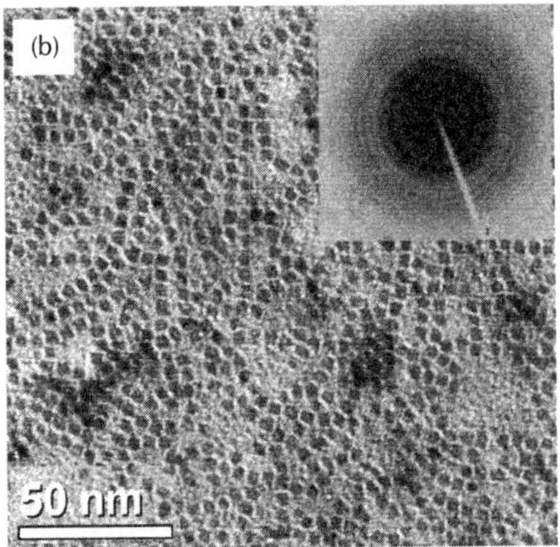

Figure 25.3   TEM images of PbSe NQDs. (a) Capped with OA/TOP ligands in an organic medium. (b) Capped with the AET ligand in an aqueous medium. The insets in panels (a) and (b) represent the corresponding electron diffraction patterns.

vibrations of terminal N–H bonds, suggesting the existence of free amine groups at the exterior side of the NQDs.

25.4   Optical properties of PbSe NQDs, PbSe/PbS core–shell NQDs and PbSe/PbS$_x$eS$_{1-x}$ core-alloyed–shell NQDs

Absorbance spectra were recorded using a UV-vis-near-IR spectrometer. PL spectra were obtained by exciting the samples with energy either above the band gap, using a Ti:sapphire laser (operating at 690–840 nm), or tuned with the band gap, using a near-IR laser diode (operating at 1463–1577 nm). The resonance excitation allowed the recording of FLN spectra. A monochromator and liquid nitrogen-cooled InSb detector detected the emission processes. All measurements were carried out at room temperature.

The PL QE was measured using an integrating sphere technique, described by de Mello *et al.* [30]. A solution of NQDs was placed inside an integrating sphere and excited by a monochromatic light. Luminescence spectra were detected by a fibre-coupled spectrometer equipped with a liquid nitrogen-cooled Ge photo detector. The entire system response was normalized against a calibrated detector, and care was taken to ensure that the sample absorption was more than 20%. The lifetime measurements were performed at room temperature using a pulsed (10 Hz) 1.064 μm ND-YAG laser, with 4.3 ns pulsewidth. For the detection we used an Acton monochromator conjugated with a Hammamtsu InGaAs PMT. We recorded the spectra using a scope, synchronized with the laser system and averaged over a given number laser pulses.

The room temperature absorption spectra of PbSe NQD core samples, with average NQD diameter varying from 2.8 nm to 8 nm, are shown in Fig. 25.4, showing a strong blue shift of the absorption edge with a decrease of the NQDs' size. The top four curves correspond to the samples prepared by a synthetic alternative given in section 25.2.2 producing high-quality NQDs with an absorption edge around 2 micron. The other absorption curves in Fig. 25.4 correspond to NQDs that were prepared by the synthesis described in section 25.2.1. It should be mentioned that the absorbance and PL spectra of the PbSe NQDs, stabilized by the AET water-soluble ligands, were identical to the corresponding NQDs capped with organic ligands, before the ligand exchange. These observations were discussed in details in [27], leading to a conclusion that the ligand exchange at the NQDs' surfaces did not alter the size and the quality of the NQDs.

Comparison between the absorption (thin lines) and the PL (bold lines) spectra of PbSe NQDs, with various diameters and capped with organic ligands, is shown in Fig. 25.5a. A plot of the 1S

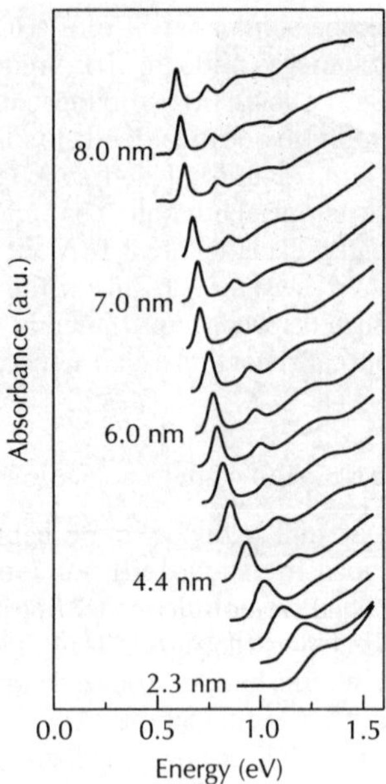

Figure 25.4   Absorption spectra of PbSe NQD core samples, with average NQD diameters varying from 2.8 nm to 8 nm. The top four curves correspond to the samples prepared by a synthetic alternative given in section 25.2.2. The other absorption curves correspond to NQDs that were prepared by the synthesis described in section 25.2.1.

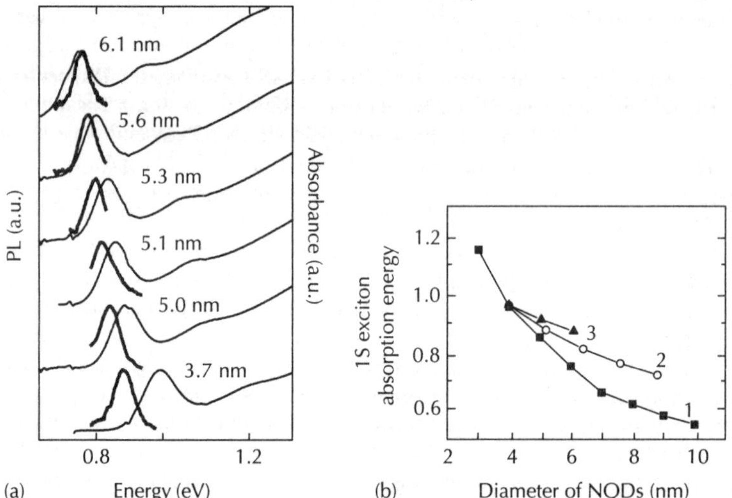

Figure 25.5   (a) Absorption (thin lines) and PL (bold lines) spectra of PbSe NQDs with various average diameters. (b) Plot (1) of the 1S exciton absorption energy versus the diameter for PbSe core NQDs, (2) of PbSe/PbS core–shell NQDs with a PbSe core diameter of 4.2 nm and n-MLs (n = 1–4) of PbS shell, and (3) of PbSe/PbSe$_x$S$_{1-x}$ core-alloyed–shell NQDs with a PbSe core diameter of 4.2 nm and an increased sulphur percentage (1 < x < 0.5).

exciton absorption energy versus NQD size is plotted in Fig. 25.5b (line 1). The PL curves shown in Fig. 25.5a exhibit an energy Stokes shift with respect to the lowest exciton absorption band, which is reduced gradually from ~100 meV for NQDs with a 4.0 nm diameter, to an anti-Stokes shift of −10 meV for NQDs with a diameter of 6.1 nm.

A rock salt PbSe semiconductor exhibits direct band transitions at the L-point of the Brillouin zone, with four-fold degeneracy. As mentioned in the Introduction, currently there are three main treatments of the electronic configuration of PbSe NQDs, based either on an effective mass approximation [2–4], tight binding model [5] or atomistic pseudo-potential method [6]. According to the effective mass approximation, ignoring the L-point lift of degeneracy, the three observed absorption bands, quoted according to an increasing energy order, correspond to the allowed $1S_h$–$1S_e$, the forbidden $1S_h$–$1P_e$ or $1P_h$–$1S_e$ and the allowed $1P_h$–$1P_e$ transitions. According to the tight binding model and to the pseudo-potential method the second absorption band corresponds to the $1P_h$–$1P_e$ allowed transition, about the L–Γ line of the Brillouin zone, while the third absorption band corresponds to the allowed $1D_h$–$1D_e$ transition. Recent, resonance-tunnelling spectroscopy measurements carried out by Liljeroth *et al.* [31] showed an agreement with the tight binding model.

The room temperature absorption (thin lines) spectra and the PL (bold lines) spectra of the PbSe/PbS core–shell NQDs, with a 4.9 nm PbSe core, and n-MLs of PbS shell (n = 0, 1, 2, 3), are presented in Fig. 25.6a. These spectra exhibit a systematic red shift of the absorption as well as the emission bands, with the increase of the shell thickness, when compared with the corresponding core sample. A plot of the 1S exciton absorption energy versus the overall diameter of the PbSe/PbS core–shell NQD is shown in Fig. 25.5b (line 2). It is interesting to note that the full width at half maximum (FWHM) of the PbSe/PbS core–shell NQDs 1S emission bands ($\sim$50 meV) is smaller than that of the PbSe corresponding core samples ($\sim$65 meV). In addition, the PL band of PbSe/PbS core–shell NQDs with one PbS ML is red shifted with respect to the first absorption band by 30 meV (Stokes shift), while the PL band of the NQDs sample with three PbS MLs is blue shifted with respect to the first absorption band by $-$10 meV (anti-Stokes shift).

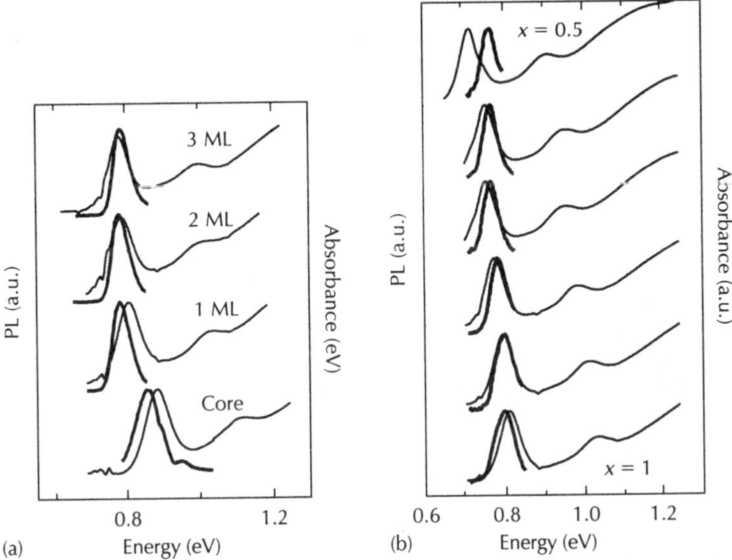

Figure 25.6   (a) Representative absorption (thin lines) and PL (bold lines) spectra of PbSe/PbS core–shell NQDs with a PbSe core diameter of 4.9 nm and an increased number of PbS shell MLs, as indicated in the figure. (b) Absorption (thin lines) and PL (bold lines) spectra of PbSe/PbSe$_x$S$_{1-x}$ core-alloyed shell NQDs with x varying from x = 1 to x = 0.5.

The absorption and PL spectra of the PbSe/PbSe$_x$S$_{1-x}$ core-alloyed–shell NQDs prepared with various stoichiometric compositions are shown in Fig. 25.6b (thin and bold lines, respectively). The PL spectra of the PbSe/PbSe$_x$S$_{1-x}$ NQDs exhibit a peculiar behaviour similar to those discussed in the preceding paragraph, showing a large (up to $-$50 meV) anti-Stokes shift with respect to the first absorption band for a shell composition of PbSe$_x$S$_{1-x}$ with x = 0.5. Also, the PL QE of the PbSe/PbSe$_x$S$_{1-x}$ (x = 0.5) core-alloyed–shell samples was measured to be 65%, higher than the corresponding PL QE of PbSe core and PbSe/PbS core–shell samples, which were measured between 40% and 50%. Again, there is a red shift of the absorption spectra with the increase of the sulphur percentage in the shell composition, as shown in Fig. 25.5b (line 3).

The absorption and emission spectra shown so far suggest that a minor anti-Stokes shift exists already in the core samples. However, this peculiar behaviour is more pronounced in the core-alloyed–shell samples (Fig. 25.6b). Moreover, it should be remembered that the entire spectrum of the core–shell samples is red shifted in comparison with that of the corresponding core NQDs. The red shift of the absorbance and PL bands of the PbSe/PbS core–shell samples and the PbSe/$PbSe_xS_{1-x}$ core-alloyed–shell samples (top curves in Fig. 25.6), with respect to that of a reference core PbSe sample (bottom curves in Fig. 25.6) can be explained by an anomalous sequence of the valence and conduction band edge energies, when $E_c(PbS) > E_c(PbSe) > E_v(PbS) \geq E_v(PbSe)$, as reported in bulk IV–VI compounds [32]. Assuming that such an anomalous behaviour is retained in the nano-size, a spread of the hole wavefunction over $E_V(PbSe)$ and $E_v(PbS)$ states might occur, leading to a red shift of the optical transitions with respect to the optical transitions of the core. Furthermore, the existence of an alloyed shell leads to tunability of the energy difference $\Delta E = E_v(PbSe)–E_v(PbSe_xS_{1-x})$, depending on the shell composition (x) and on the shell thickness. It should be mentioned that expansion of a wavefunction over the core–shell interface was also observed in CdSe/ZnS and InP/ZnS core–shell NQDs, with an exceptionally large potential barrier induced by the ZnS shell [33–37]. Also, Wise *et al.* [38] estimated recently a theoretical model, describing the electron–hole wavefunctions in PbSe/PbS core–shell NQDs. They proposed the existence of a type I NQD with a spreading of both electron and hole wavefunctions over the entire NQDs to samples with a core diameter of 3 nm and a shell thickness of 1.7 nm. When the shell thickness becomes larger than 2.5 nm, then a separation of the electron–hole wavefunctions becomes more pronounced, suggesting a type II configuration. In addition, when the core size becomes larger than 6.0 nm, the wavefunction overlap is reduced. The agreements of those calculations with the experimental data are currently being investigated. In general, the larger the separation of the electron–hole wavefunction (approaching type II configuration) the longer the recombination lifetime would be, in agreement with our preliminary results indicated above.

The red shift of the PL bands with respect to their corresponding 1S exciton absorption energy is also an issue of major concern. As mentioned above, the PbSe electronic structure consists of valence and conduction band edges at the L-point of the Brillouin zone with a four-fold degeneracy, which is doubled by the spin states. Therefore, the interband transitions may involve 8×8 possible transitions (involving the bright and dark excitons). Indeed, Delerue *et al.* [5] and Zunger *et al.* [6] recently discussed a breaking of the degeneracy due to occurrence of intervalley interactions as well as the existence of strong anisotropy of the band edge states with different values for the longitudinal (m||) and transverse (m⊥) effective masses. Thus, the PbSe NQD excitation may occur into a bright state (1 out of the 64 possibilities) and emission will take place from a lower energy bright or dark state. In addition, the energy manifold of exciton states varies with the change of the NQDs' diameter, leading to a gradual change in the Stokes shift.

The appearance of an anti-Stokes shift is a matter of debate. As mentioned above, it does exist in the PbSe core samples (-10 meV) and it is substantially more pronounced in the PbSe/PbS core–shell and PbSe/$PbSe_xS_{1-x}$ core-alloyed–shell structures (up to -50 meV). Careful observation of Figs. 25.5a and 25.6a and b reveals that the anti-Stokes shift becomes pronounced around 0.77 eV, a point were the energy shift of the emission band versus the NQDs' size nearly riches a plateau, while that of the absorption 1S exciton still continues. Currently, we presume that the peculiar appearance of the anti-Stokes shift comes about due to the fact that the smaller NQDs, emitting at the blue side of the band, are having larger emission oscillator strength, as proposed by Brus *et al.* [39] and Efros *et al.* [40]. Even though their volume fraction is not necessarily larger, an enhancement in the luminescence intensity at the high energy side of the emission band appears, artificially shifting the emission Gaussian envelope further to the blue. This is mainly pronounced in samples in which the exchange interaction is relatively small and the Stokes shift is reduced (upper curves as the indicated figures).

Figure 25.7 shows representative FLN spectra of PbSe core (A) and PbSe/PbS core–shell (B) NQD samples. These curves are compared with the non-resonance PL spectra (lower curves in

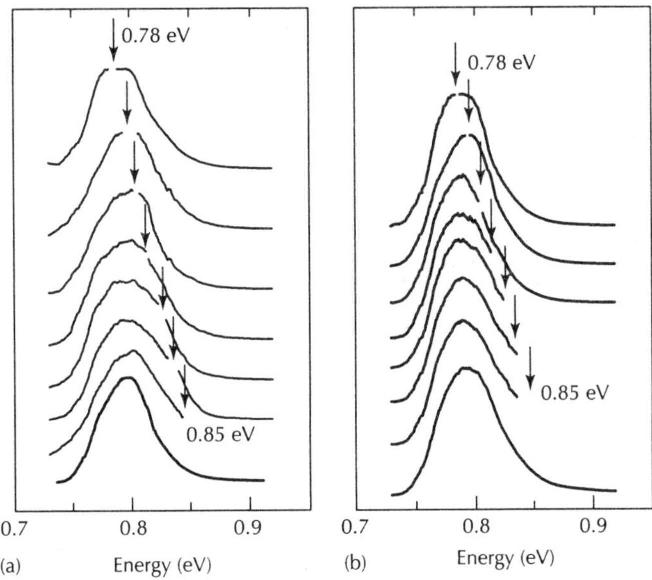

Figure 25.7 *Representative FLN spectra of (a) PbSe and (b) PbSe/PbS core–shell NQDs. The bottom curves in parts (a) and (b) correspond to the non-resonantly excited PL spectra, while all the other spectra were excited in resonance with pumping energies labelled by the arrows.*

Fig. 25.7a and b). The FLN spectra were recorded under resonance excitation at energies within the excitonic band (shown by the arrows in the figure), while the emission intensity was monitored both above and below this excitation energy. The comparison of the FLN curves with those of the non-resonance PL spectra indicates a negligible reduction of the emission band FWHM, suggesting a minor contribution of an inhomogeneous broadening due to the size distribution, while the major contribution to the line width is due to homogeneous effects on the linewidth. It should be mentioned that recent single-dot measurements of PbS NQDs, reported by Krauss *et al.* [41] also showed a relatively small PL line narrowing. According to the finding discussed here, the homogeneous broadening could be due to a lifting of the degeneracy of the band edge states, which according to Delerue [5] is due to a valence band edge energy distribution of ~40 meV, for PbSe NQDs with a similar size to our studied NQDs. Thus, an FWHM of the emission bands of PbSe and PbSe/PbS core–shell NQDs of about 50 meV can be very well associated with the energy distribution of the band edge states.

The emission lifetime of the 1S exciton, τ_f, was measured at room temperature by embedding NQDs in PMMA polymer matrix. Representative examples of the decay curve of a set of PbSe core NQDs with various average diameters are shown in Fig. 25.8a by the grassy coloured lines. The decay curves were best fitted to a double exponent function, $I_{(PL)} = A_1 \exp(-t/\tau_1) + A_2 \exp(-t/\tau_2)$ (showed by the bold lines in the figure), and the best fitted parameters of the curves shown in Fig. 25.8a are given in Table 25.2. It is seen from the table that the fast decay component, τ_1, varies, between 30 and 80 ns, while the slow component, τ_2, ranges between 200 and 900 ns, and both components become smaller with the increase of the NQDs' diameter. The existence of two components may be associated with the occurrence of recombination processes from dark and bright states among the 64 possible components of the exciton manifold. Thus, the dark states emission, with a longer lifetime, is dominated in the smallest NQDs (larger A_2 and longer τ_2), while the contribution of the bright state emission becomes more significant at the larger NQDs. It should be noted that previous reports of τ_f of core PbSe NQDs, given by Krauss *et al.* [7] and Sionnest *et al.* [8] proposed values of ~200–800 ns.

The PL decay curves of PbSe/PbS NQDs with a core diameter of 4.2 nm, with different shell thickness, are shown in Fig. 25.8b. Here the results conceal that the lifetime stays nearly the same upon the addition of 1–2 ML of PbS shell; however, it expands with the further increase of

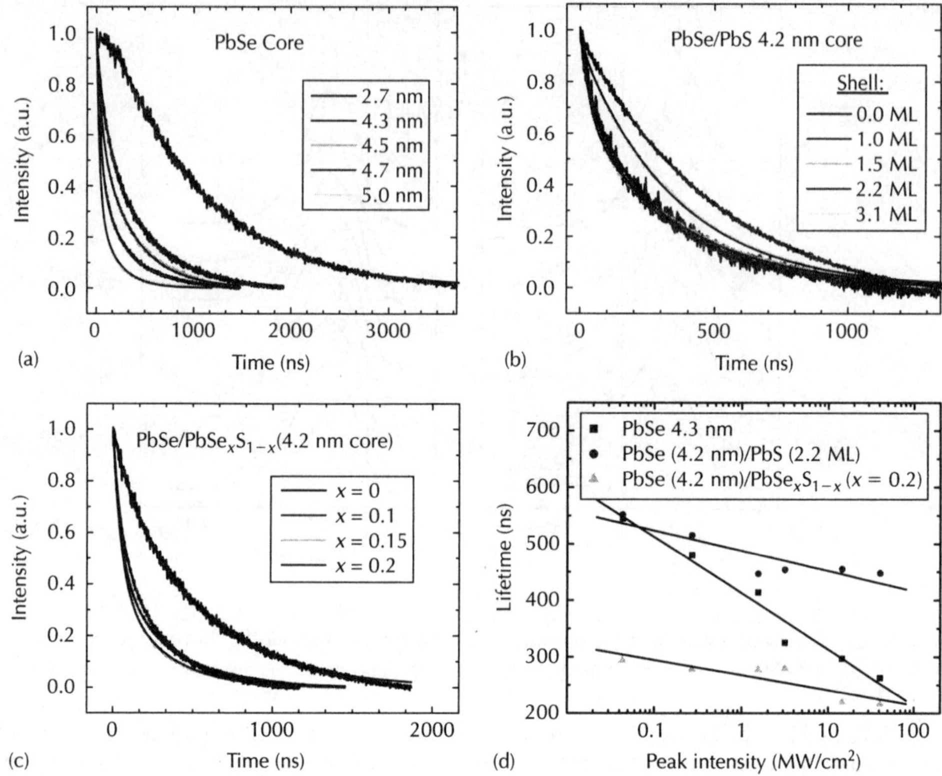

Figure 25.8   Luminescence decay of: PbSe NQDs with different diameters (a); PbSe/PbS core–shell NQDs with 4.2 nm PbSe cores and different shell thicknesses (b); PbSe/PbSe$_x$S$_{1-x}$ core-alloyed–shell with different shell compositions (c), as encoded by the colours. Each of the experimental decays (grassy coloured lines) is fitted to a second-order exponent (solid black lines). Emission lifetime as a function versus the excitation peak intensity of 4.3 nm PbSe NQDs (black), PbSe/PbS core–shell NQDs with 4.2 nm PbSe core and 2.2 ML of PbS shell (red), and PbSe/PbSe$_x$S$_{1-x}$ core-alloyed–shell NQDs with 4.2 nm PbSe core and PbSe$_{0.2}$S$_{0.8}$ shell (green). The black lines are sketched only to guide the eye (d). NQDs were embedded in PMMA matrix.

Table 25.2   Biexponential fitting of the data in Fig. 25.8a

PbSe NQD diameter (nm)	τ_1 (ns)	A_1	τ_2 (ns)	A_2
2.7	–	–	997.6	1.105
4.3	81.7	0.269	440.6	0.701
4.5	54.1	0.356	357.4	0.617
4.7	43.2	0.574	308.6	0.440
5.0	31.2	0.640	192.9	0.342

the shell thickness, although there is a maximal increase in the lifetime for 2 ML, which is slightly reduced in a sample with 3 ML. The typical lifetime of the core–shell samples vary between 300 and 520 ns. The lengthening of the lifetime in the PbSe/PbS samples may be associated with a spreading of a carrier's wavefunction (more likely that of the hole) into the shell, reducing the overlap between the electron and hole functions. Fig. 25.8c represents preliminary results, associated with the decay curves of PbSe/PbSe$_x$S$_{1-x}$ NQDs, with similar core diameter, but with different Se:S ratio (as labelled in the figure). The investigation reveals a reduction of the lifetime, upon the increase of the sulphur molar ratio in the shell and of the NQDs' overall diameter, ranging between 490 ns (for 4.2 nm NQDs with $x = 0$) to 250 ns (for 4.9 nm NQDs with $x = 0.2$). This can be correlated with the behaviour of the core PbSe structures, suggesting a spreading of the wavefunctions into the shell, leading to the reduction of the recombination rate. The reliance of the emission lifetime on the excitation peak intensity for three kinds of NQD structures is shown in Fig. 25.8d. This dependency seems to be stronger for the PbSe core NQDs, and weaker for the PbSe/PbSe core–shell NQDs and the PbSe/PbSe$_x$S$_{1-x}$ core-alloyed–shell NQDs, with dimensions and compositions as indicated in the figure. The decay curves shown here should be further investigated in the near future, clarifying the distribution of the wavefunction between the core and the shell, and the origin of the short and long component of the decay processes.

25.5 Passive Q-switching, using PbSe NQDs, PbSe/PbS core–shell NQDs and PbSe/PbSe$_x$S$_{1-x}$ core-alloyed–shell NQDs

The use of a saturable absorber inside a laser cavity may act as a passive Q-switching device, contributing to the output beam quality and simplifying the laser system design. A few materials have already been recognized as saturable absorbers operating in the near-IR regime. The systems described in past studies [42–47] provided Q-switching in well-defined wavelength ranges, i.e. $\sim$1.34, 1.44 and 1.54 µm, with typical ground-state absorption cross-sections of $\sigma_{gs} = 0.8$–5.7×10^{-19} cm^2. The PbSe NQDs can function as Q-switches over wide spectral regions, and as will be shown below, they exhibit exceptionally large σ_{gs}.

The present work describes an attempt to use the PbSe NQDs, PbSe/PbS core–shell NQDs and PbSe/PbSe$_x$S$_{1-x}$ core-alloyed–shell NQDs as passive Q-switches in Er:glass lasers operating at 1.54 µm. This laser is of special interest, due to its potential applications in light detection and ranging, surgery, and telecommunications. This stems from the fact that its emission wavelength coincides with the so-called eye-safe spectral region, corresponding also to one of the atmospheric transmission windows, and to the third transmission window of silica fibres.

The performance of these NQDs as passive Q-switches was initially investigated by characterization of their behaviour as saturable absorbers. This was done by following the change in optical transmission versus the power of the pumping laser. For that purpose, we used 1.537 µm signal output pulses from a 1.064 µm pumped KTiOPO$_4$ optical parametric oscillation laser. The output (signal) beam, with energy up to 150 mJ/pulse and typical 10 ns duration (FWHM), was vertically polarized with a near-TEM$_{00}$ transverse energy distribution. The laser pulse fluency (energy density) at the investigated sample was varied by moving the sample along the propagation axis of the laser beam, which was focused by a focusing lens. The transverse beam energy distribution at each point was measured using the knife-edge scanning method.

Figure 25.9 shows representative examples of the 1.54 µm transmission through an NQD solution versus the pumping pulse intensity (upper scale). In the case of PbSe NQDs (top curve), the transmission starts at approximately $T_0 = 62\%$ in the low power range, rises with increasing pulse power, and saturates at approximately $T_{max} = 77\%$ in the high power range. This general behaviour was also observed when the light was transmitted through the PbSe/PbS core–shell and PbSe/PbSe$_x$S$_{1-x}$ core-alloyed–shell NQDs samples (lower curves) as indicated in the figure. The saturation occurs due to depletion of the ground state above a given laser fluency. The fact that the transmission never reaches 100% is explained by the existence of excited-state absorption.

The 1S exciton singlets τ_f are of the order of 70–1000 ns (Fig. 25.8), which are considerably longer than the pump laser duration of $\sim$10 ns. Thus, initially we assumed that PbSe NQDs may act as "slow" saturable absorbers, with respect to the pumping laser. It should, however, be remembered that τ_f was estimated from the PL spectrum observed under low-power illumination, ensuring that there was never more than a single exciton created in a single NQD. Then, the ground-state absorption coefficient, $\sigma_{gs} = -(1/NL)\ln T_0$, where N is the NQD volume density and L is the sample thickness, could be calculated to be $\sigma_{gs} \cong 10^{-15} - 10^{-16}$ cm^2. This preliminary assumption, however, yielded a serious internal inconsistency. The calculated saturation fluency, $J_S \equiv h\nu/\sigma_{gs}$, is of the order of several hundred µJ/cm^2, in contradiction with the experimental value of $\sim$1 J/cm^2 (Fig. 25.9). This inconsistency led to a second thought that actually under absorption

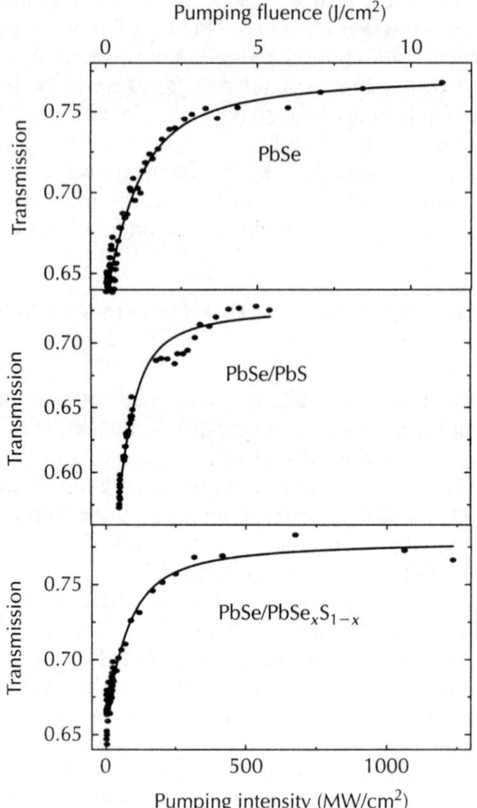

Figure 25.9   Transmission versus laser intensity of 5.6 nm PbSe NQDs (top), 5.8 nm PbSe/PbS 1.5 ML core–shell NQDs (middle), and 6.0 nm PbSe/PbSe$_x$S$_{1−x}$ core-alloyed–shell NQDs (lower), all suspended in chloroform. The dots represent the experimental data. Solid curves correspond to the simulation, utilizing Eq. 25.1.

saturation conditions, one generates multiple excitons in a single NQD with an effective lifetime τ_{eff} that is substantially shorter than the pumping laser pulse ($\sim$10 ns). Thus, the NQDs under those conditions would not behave as a "slow" absorber, but rather as a "fast" saturable absorber. On the basis of this understanding, the following theoretical expressions previously developed by Burshtein *et al.* [48] for transmission of a fast saturable absorber, including excited-state absorption σ_{es}, could be considered. This theory provides a closed equation for the transmission $T \equiv I_{tr}/I_0$, where I_{tr} and I_0 are the transmitted and incident pulse power densities, respectively:

$$I_0 = \frac{S[1 - (T_0/T)^{1/D}]}{(T_0/T)^{1/D} - T} \tag{25.1}$$

where $S \equiv 1/(\tau_{eff} \cdot \sigma_{gs})$ and $D \equiv (\sigma_{gs} - \sigma_{es})/\sigma_{es}$. A best fit of the data in Fig. 25.6 to Eq. 25.1 yields the values of σ_{gs} and τ_{eff}, as given in Table 25.3, suggesting σ_{gs} of the order of 10^{-16}–10^{-15} cm^2, and τ_{eff} of the order of several ps. The obtained σ_{es} values were approximately 40% of the obtained σ_{gs} values. Delerue [5] recently calculated a homogeneous width of approximately 30–80 meV (depending on the NQDs' size) of the 1S exciton band in PbSe caused by splitting among its four degenerate states. Our measured absorption curves exhibit widths of approximately 50 meV (Figs 25.5 and 25.6). Due to the closeness between these values, further consideration of inhomogeneous broadening can be neglected. The saturation power density defined as $I_S \equiv h\nu \cdot S = h\nu/(\tau_{eff} \cdot \sigma_{gs})$ is then calculated as $I_S = 45 - 137$ MW/cm^2, as given in Table 25.3.

Table 25.3   Cross-section of absorption, σ_{gs}, effective lifetimes, τ_{eff}, and saturation fluence, I_S, for several representative NQD samples. The NQDs' diameters D and concentrations N are indicated

Composition	D (nm)	N_0 (10^{15} NQDs/mL)	I_S (MW/cm^2)	σ_{gs} (10^{-16} cm^2)	τ_{eff} (ps)
PbSe*	5.6	2.9	137.0	2.05	4.8
PbSe	5.4	3.1	139.7	1.25	7.4
PbSe/PbS 1.5ML*	5.8	1.5	45.8	4.7	6.0
PbSe/PbSe$_{0.5}$S$_{0.5}$*	6.0	1.3	45.2	2.6	11.0
PbSe/PbSe$_{0.5}$S$_{0.5}$	6.0	2.2	117.0	2.4	4.6

Klimov *et al.* [10] and Nozik *et al.* [3] recently showed that coexistence of two excitons in PbSe NQDs recombine on a timescale of 20–50 ps, while Klimov *et al.* [49] showed a radiative decay time of a few ps for a multiple exciton emission. As discussed above, every NQD can accommodate up to 64 excitons, and thus multiple exciton interactions set in under saturation, which apparently renders their effective recombination time very short. The simulation of the transmission saturation experiments (Table 25.3, Fig. 25.9) revealed a radiative lifetime of a few ps, opposing the creation of multiple excitons ($n > 2$). The Q-switch performance is determined by the product of σ_{gs} and τ_{eff}. Thus, PbSe NQDs act as "fast" saturable absorbers, suggesting a poor quality of the Q-switching devices. This fact, however, is compensated for by the large σ_{gs} values of these materials.

The utilization of the NQDs' saturable absorber samples as Q-switching components in an Er: glass laser was examined by inserting them in a chloroform solution of 1.54 μm antireflective (AR)-coated quartz cuvette or embedding them in a PMMA polymer that was placed between two AR-coated glass windows. Then, the glass sandwich or cuvette was placed inside the Er:glass laser resonator. The laser itself consisted of a 3.0 mm diameter, 4.0 cm long laser rod, hosted in a diffused reflector cavity equipped with a linear Xe flashtube. The flashtube was powered by a current pulse of ~700 μs in duration. The distance between the mirrors was approximately 7 cm. The pulse temporal behaviour was measured using a fast p-i-n InGaAs photodiode, with a response time <175 ps. Figure 25.10 provides an oscilloscope trace of the Er:glass laser output under Q-switching conditions, operating at 1.54 μm. The duration of each pulse was between 40 and 50 ns, with pulse energy between 0.8 mJ and 3.5 mJ and pulse power between 1.6×10^4 W and 8.75×10^4 W. The Q-switch based on the PbSe/PbSe$_x$S$_{1-x}$ core-alloyed–shell NQDs always showed the better performance in terms of the pulsewidth and output power, and was by far more robust upon illumination, over long periods of time. It should be commented that the baseline at long times in the trace shown in Fig. 25.10 does not retain its zero value due to an experimental artefact, associated with the parasitic capacitance and inductance, caused by the prolonged impulse attenuation.

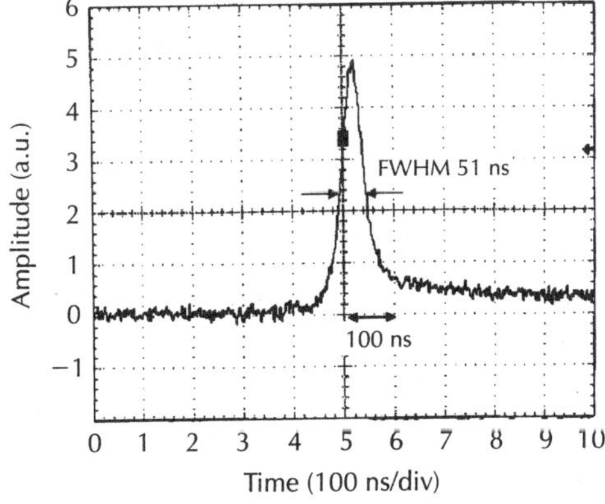

Figure 25.10   Oscilloscope trace of a single output pulse of an Er:glass laser (plot of intensity versus time) using PbSe NQD colloidal solution as a Q-switch.

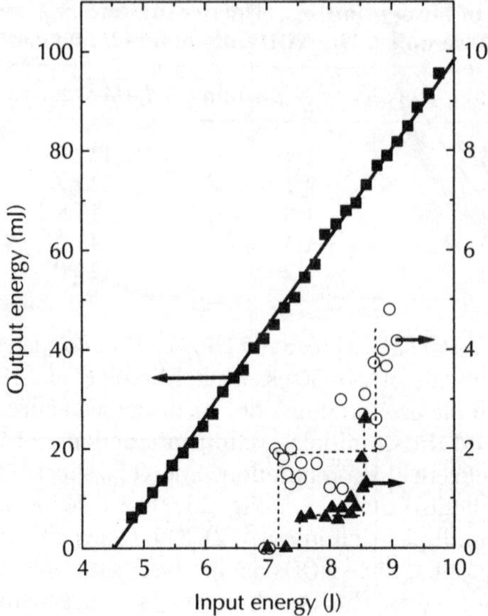

Figure 25.11   Output versus input pump energies of an Er:glass laser under free-running and under passive Q-switching conditions. Solid squares, free-running; open circles, PbSe/PbS NQDs in PMMA; solid triangles, PbSe/ PbS NQDs in chloroform. Scale for the Q-switched output is on the right.

Figure 25.11 presents the laser output versus input pump energy of the free-running and passively Q-switched laser. The free-running laser exhibits threshold energy of 4.5 J, and a slope efficiency of 1.8%. Pulse durations were approximately 500 μs. The two passively Q-switched cases (using a PbSe/PbS 6.0 nm diameter NQDs in the polymer, and PbSe/PbS same diameter NQDs in chloroform) exhibit increased thresholds of 5.8 J and 7.3 J, respectively, and slope efficiencies of 0.17% and ~0.2%, respectively, and pulse durations of approximately 50 ns (Fig. 25.10). Both Q-switched curves reveal two steep steps, each step raiser reflecting the appearance of a new pulse. Similar steps were observed in the passive Q-switching of Nd:YAG lasers.

Table 25.4   Saturable absorber Q-switch parameters used in Er:glass lasers

Q-switch	Relaxation time (ns)	σ_{gs} (cm^2)	σ_{es}/σ_{gs} (cm^2)	Ref.
Co:LaMgAl$_{11}$O$_{10}$	220	1.2×10^{-19}	~0.07	[50]
Co:MgAl$_2$O$_4$	350	3.5×10^{-19}	~0.03	[51]
Co:ZnSe	2.9×10^5	$5–5.7 \times 10^{-19}$	0.05–0.22	[52, 53]
Cr:ZnSe	8×10^3	2.7×10^{-19}	~0.075	[52]
U:CaF$_2$	4.7×10^3	$7–9 \times 10^{-19}$	~0.4	[53, 54]
V:YAG (at 1.34 μm)	20–30	7.2×10^{-18}	0.1	[56, 55]
PbSe NQDs	$\sim 10^{-3}$	$\sim 4 \times 10^{-16}$	~0.5	Present study

In Table 25.4, we compare the spectroscopic characteristics of some other saturable absorbers studied for the near-1.54 μm spectral region [50–56] to those of the PbSe/PbS core–shell NQDs of our present study. The latter differ by two prominent basic features: the effective lifetime (relaxation time) τ_{eff} is many orders of magnitude shorter, and the ground-state absorption cross-section σ_{gs} is many orders of magnitude larger. Thus, the other materials (except for the semiconductor saturable absorber mirror case) can be regarded as *slow* absorbers (effective lifetime longer than the typical Q-switched laser output pulsewidth of ~50 ns). The PbSe/PbS core–shell NQDs, on the other hand, are definitely fast absorbers under all circumstances. The property that renders them suitable for Q-switching of the Er:glass laser is their huge ground-state absorption cross-section, which allows the saturation of their absorption at reasonable pulse intensities. While the PbSe/PbS core–shell NQDs seem to be equivalent to other alternatives for Er:glass laser passive Q-switching, they seem to have a higher chemical stability, and they can be tuned to operate in different spectral regimes. In addition, these NQDs indicate potential superiority for laser mode locking in the near-IR spectral region. This is an aspect we intend to investigate in future studies.

25.6   PbSe NQDs used in a gain device and integrated into microcavities

As indicated above, the PbSe NQDs exhibit tunability in the near-IR spectral regime, and when coated with an inorganic shell, they show a chemical robustness over years and stability under intense illumination, as well as relatively large σ_{gs} values (10^{-16} to 10^{-15} cm^2) as observed in section 25.5. These properties make them potentially useful as near-IR lasers. Here we experimentally observed the ASE properties of PbSe NQDs at room temperature.

PbSe NQDs were embedded in a PMMA polymer matrix, as discussed in section 25.2 and [19]. The nominal mass concentration of the NQDs in the polymer was only 5%. A chloroform solution of a PMMA NQD is stirred for 30 min, filtered through a 0.2 μm filter to remove impurities, and then spin coated at 2000 rpm during 20 s on an Si-wafer oxidized with a 7.0 μm thick silica layer, then annealed at 90°C for 2 h to stabilize the films. The optical gain and the ASE properties were performed with variable pump intensity on a 4.2 μm thick NQD-doped PMMA film with a variable stripe length, using a continuous wave (CW) laser diode at 980 nm as the pump source. The pump beam is concentrated with a cylindrical lens, and a narrow stripe of uniform intensity is selected using a 300 μm wide slit made of an aluminium foil, directly deposited on the film surface. The irradiated length was adjusted with the edge of a movable foil, between 100 and 1000 μm.

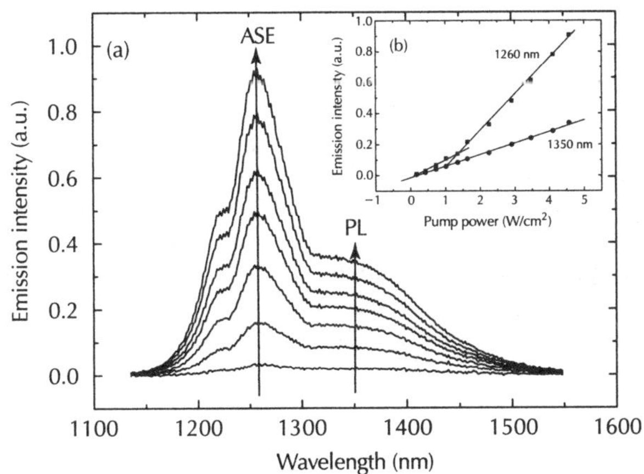

Figure 25.12   ASE spectra of PbSe NQDs embedded in a PMMA film with pump power ranging from 0.42 W/cm^2 (lower curve) to 4.56 W/cm^2 (upper curve). The initial mass concentration of the NQDs is 5%, but this value might be significantly reduced by the use of a millipore 0.2 μm filter during film deposition. The inset represent plot of ASE at 1260 nm (squares) and spontaneous luminescence at 1350 nm (circles) emissions as a function of pump intensity at 980 nm. Length of the excited zone l = 0.5 mm.

Figure 25.12 shows representative emission spectra of an NQD-doped PMMA film recorded at various pumping intensities. An ASE signal appeared progressively centred around 1255 nm, distinct from the spontaneous emission signal centred on 1350 nm. The ASE band is clearly narrower than the spontaneous emission band, but its width is independent of the pumping power, suggesting the absence of inhomogeneous broadening due to the NQDs' size distribution. The ASE spectrum features obtained for these NQDs are very similar to those observed in [9], which were recorded at 80 K and excited with a ps laser at 100 Hz repetition rate. The current ASE curves, however, were recorded at room temperature, utilizing a CW pumping of a standard laser diode, thus qualifying PbSe NQDs for practical applications in the domain of integrated amplifiers for optical communications. More likely, due to the Stokes shift between the absorption and emission, and the relatively low concentration of NQDs within the transparent films, reabsorption processes by neighbouring NQDs is mainly avoided. In addition, it is possible to assume that the emission process occurs, after an efficient excitation (with relatively large cross-section for absorption), from bright and dark states, resembling a three-level system. In such a case, the probability for emission in an ensemble of NQDs at a certain time should be larger than the probability for absorption, leading to a population inversion situation and to a stimulated emission, even under CW excitation.

The radical difference between the spontaneous emission and the ASE signals is emphasized again by a plot of the respective emission intensities versus the laser pumping power (inset of Fig. 25.12). The spontaneous emission intensity linearly increases with the pumping power, whereas the ASE intensity behaves as a spontaneous emission at low pumping powers, but sharply increases above a threshold of 1.4 W/cm^2. The appearance of a breaking point in the luminescence intensity versus laser power curve suggests the emergence of an amplifying emission regime above a given pump threshold.

Moreover, the gain coefficient, g, can be determined according to the following relation, valid in one-dimensional approximation and far from saturation:

$$I(\lambda)_{ASE} = \frac{A(\lambda)I_p}{g(\lambda)} \cdot (\exp(g(\lambda) \cdot l) - 1) \tag{25.2}$$

where $I(\lambda)_{ASE}$ is the ASE output intensity, l is the length of the pumped stripe, the term $A(\lambda)I_p$ corresponds to a spontaneous emission that is proportional to excitation intensity, I_p while g is the net gain. The net gain was estimated by fitting the experimental data from Fig. 25.13a to Eq. 25.2. The deviation of the fit at larger l is due to the fact that the model described by Eq. 25.2 do not consider gain saturation. However, the plots of Fig. 25.13a reveal significantly high g values, of 2.63 cm^{-1} and 6.68 cm^{-1} (when pumped with different intensities, as indicated in the figure), in comparison with other ASE systems operating in the near IR spectral regime, such as erbium-doped PMMA polymer films and waveguides. This exceptionally high gain coefficient may be attributed to the large absorption cross-section mentioned previously (approximately 10^{-16} cm^{-2}), substantially higher than that of an erbium complex (less than 10^{-20} cm^{-1} in the 980 nm region).

Finally, the directionality of the ASE signal was examined. A plot of the signal intensity as a function of the collection angle is shown in Fig. 25.13b, which clearly shows that the ASE signal

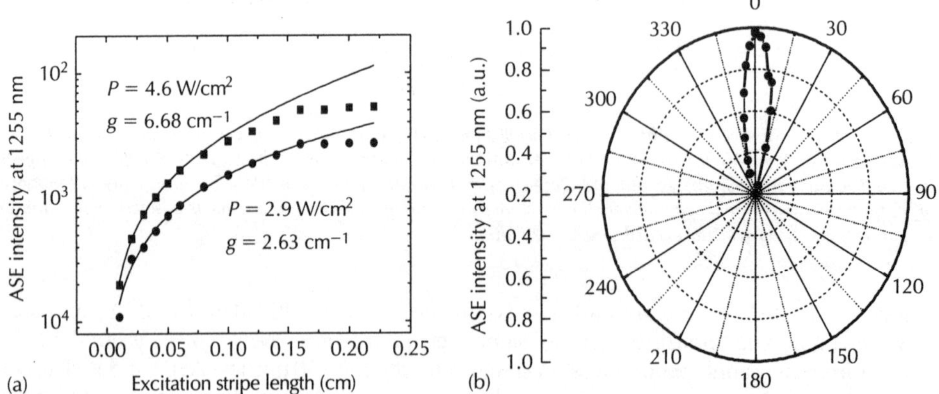

Figure 25.13   (a) Plot of ASE intensity at 1255 nm as a function of l with a pump power of 2.9 W/cm^2 (circles) and of 4.6 W/cm^2 (squares). (b) Plot of ASE emission at 1255 nm for different values of collection angle.

is strongly collimated, a signature of the emitted beam spatial coherence due to the amplification process.

25.7   Coupling of PbSe NQDs with γ-Fe$_2$O$_3$ magnetic nanoparticles, for a future biological application

The present section discusses the preparation and characterization of PbSe NQDs' γ-Fe$_2$O$_3$ NPs' conjugated structure. The main challenge in the utilization of the NQDs' constituents as fluorescent probes and NPs as magnetically controlled transport platforms in biological applications is the concern regarding their compatibility with the biological environment, i.e. whether they have the ability to maintain their dispersion and their properties in aqueous media. Previous efforts to prepare the semiconductor NQDs and the magnetic NPs in aqueous media were met with serious obstacles associated with chemical stability and particle dispersion. This work concentrates on a novel, alternative procedure, based on the initial preparation of the NQDs and NPs by traditional synthesis and capping methods with organic stabilizers, followed by a carefully tailored exchange of the surfactants with selective water-soluble ligands. Furthermore, given the hierarchical design approach taken in this work, these water-soluble ligands (e.g. 2-aminoethanethiol (AET) and PSS hydrate octakis) were chosen to include specific functional groups that would enable the subsequent chemical bonding between the NQDs and NPs in aqueous solutions, as discussed in section 25.2.7. The coupling process leading to the formation of the conjugate structures of γ-Fe$_2$O$_3$ nanoparticles and PbSe quantum dots was straightforward. It involved a simple mixing of equal volumes of the water-based solutions of the AET-capped PbSe quantum dots with a water-based solution of the PSS-capped γ-Fe$_2$O$_3$ nanoparticles, as shown schematically in Fig. 25.14. No clouding of the solution was observed during the mixing process and the resulting solution remained clear.

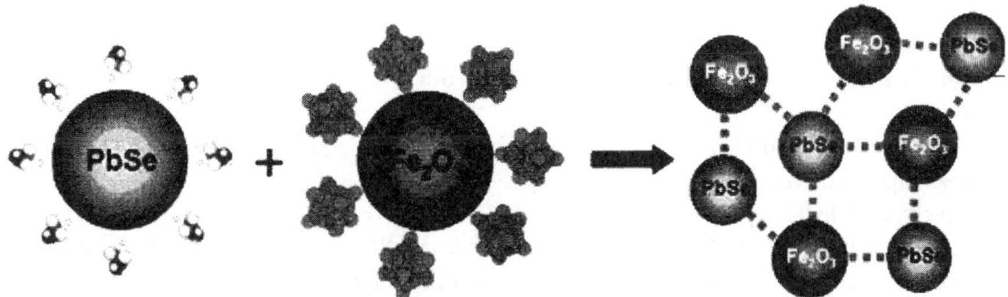

Figure 25.14   Coupling between γ-Fe$_2$O$_3$ NPs and PbSe NQDs forming the conjugated hierarchical structures.

The formation of the water-soluble PbSe NQDs functionalized with amine groups, and the γ-Fe$_2$O$_3$ NPs functionalized with silicone oxide ionic groups, generated the building blocks for the subsequent hierarchical coupling reaction resulting in the formation of the structurally conjugated fluorescing and magnetic nanoparticles. Figure 25.15a shows a TEM image of the conjugated structure ensembles of nanoparticles, while Fig. 25.15b exhibits an HR-TEM image of a small domain of the ensembles. This high-resolution image reveals the existence of particles with different lattice fringes (marked by the arrows in the figure) and a common repeating distance of about 1.3 nm, which is in agreement with the sum of the AET and PSS molecular lengths, forming a hydrogen bond between them.

The electron diffraction pattern of the conjugated structure, shown in Fig. 25.15c, is composed of the diffraction bands associated with both the magnetic γ-Fe$_2$O$_3$ NPs and the semiconductor PbSe NQDs. The assignment of the diffraction rings are labelled in Fig. 25.15c, proposing the existence of the behaviour of γ-Fe$_2$O$_3$ NPs' lattice [57, 58] and the characteristic PbSe NQDs with cubic closed-packed (ccp/fcc) crystal structure.

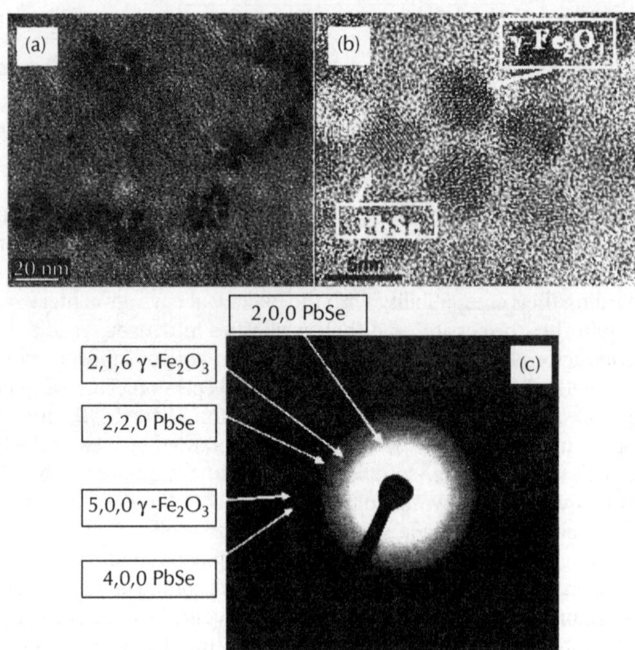

Figure 25.15   (a) TEM image of the magnetic γ-Fe₂O₃ NPs' and semiconductor PbSe NQDs' conjugated ensemble. (b) HR-TEM image of a small domain of the conjugated ensemble shown in panel (a) Panel (b) reveals the existence of two types of particles, the γ-Fe₂O₃ NPs and PbSe NQDs, with different crystallographic fringes. (c) Electron diffraction pattern of the conjugated structure comprised of γ-Fe₂O₃ NPs and PbSe NQDs.

25.8   Electrical properties of PbSe NQDs' ordered solids

The NQDs of PbSe core and the PbSe/PbS core–shell structures discussed above exhibit a relatively small band gap, predicting electrical conductivity and rendering them as suitable building blocks in nanoelectronic devices. Here we describe the investigation of the conductivity properties of PbSe NQDs, forming an ordered array at the temperature range between 100°C and 200°C.

The NQDs discussed earlier in the chapter, i.e. PbSe NQDs capped with OA and TOP, were spread over a TEM grid by drying a drop of the NQDs dispersed in a chloroform solution. Controlled evaporation of the drop induced a self-organization of the NQDs into an ordered array. Fig. 25.1d exhibited a hexagonal arrangement of a single monolayer of PbSe core NQDs with 6.7 nm in diameter with a 5% size distribution and 2 nm interdot spacing. Such spacing corresponds to an intimate contact with the OA, each of which has a length of 1 nm. This NQD arrangement has an extremely high resistivity (1 TΩ) due to the insulating organic capping.

Instead, the as-deposited NQD array was heated under reducing conditions, using a forming gas (95% Ar and 5% H_2) to evaporate partially or completely the organic capping in the absence of oxygen and eventually leading to the reduction of the interdot spacing. For example, heating to 150°C reduced the interdot spacing to 1 nm, while heating to 300°C for the same length of time sintered the NQDs into one another. Thus, the annealing temperature and time was crucial for the formation of closely packed arrays. Fig. 25.16a represents a two-dimensional arrangement of PbSe NQDs (with average diameter of 5 nm) annealed at 150°C for 15 min, showing nearly a direct contact between NQDs' cores. As will be discussed below, the annealing process greatly enhanced the conductivity in the NQDs array. Also, the absorption spectra of the annealed solids are red shifted by 20–60 meV, depending on the relative reduction of the interdot spacing (not shown). The red shift can be attributed to the change of the dielectric environment surrounding the NQDs.

The electrical conductivity of the PbSe NQD array was measured by depositing it on a highly doped Si wafer coated with 500 nm thick SiO_2. The source and drain were made of Ti/Au and were spaced 200 μm apart. The leads were patterned on the wafer after deposition of the NQDs. The

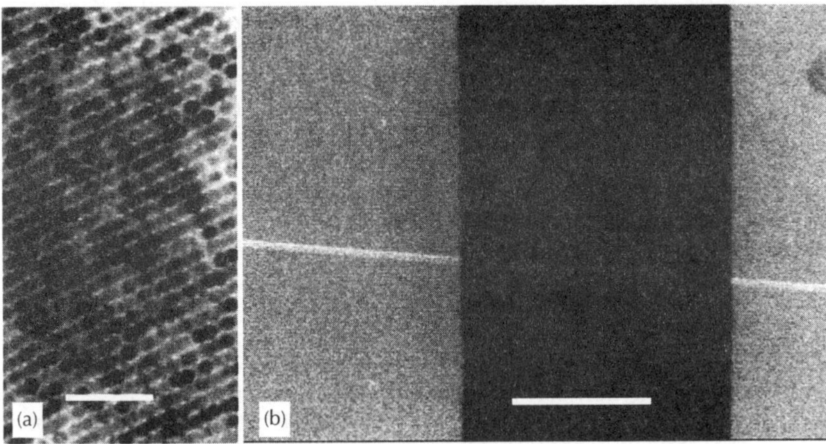

Figure 25.16 (a) TEM image of self-organized 5 nm PbSe NQDs annealed at 150°C for 15 min (scale bar = 60 nm). (b) SEM image of a 5 nm NQD solid, deposited on an SiO$_2$ layer, placed between Ti/Au source and drain leads (scale bar = 100 μm).

NQDs themselves were injected into poly ethylene oxide (PEO) polymer solution and were deposited on the silicon wafer by an electro-deposition method. Then polymer was removed by plasma of nitrogen, while the remaining NQDs were annealed according to the procedure discussed above. Fig. 25.16b shows a representative scanning electron microscopy (SEM) image of a 5.0 nm diameter PbSe NQD array, between the source and drain leads. The size of the NQD array was 200 μm in length, 20 μm in width and approximately 4–5 MLs in thickness, containing ~2.0 · 10^9 NQDs.

Figure 25.17 represents current–voltage (I–V) curves of PbSe NQD solids annealed at 200°C and measured at three different temperatures, as indicated in the figure. It should be mentioned that the current increased by one to two orders of magnitude when the annealing temperature is increased (still below 300°C). It is seen from the figure that the I–V curves have linear behaviour besides a slight asymmetry at negative bias and there is no threshold voltage V_{th}.

Charge transport in NQD arrays is controlled by the following parameters: (1) the NQDs preferably should have a close diameter, permitting energy resonance among their valence and

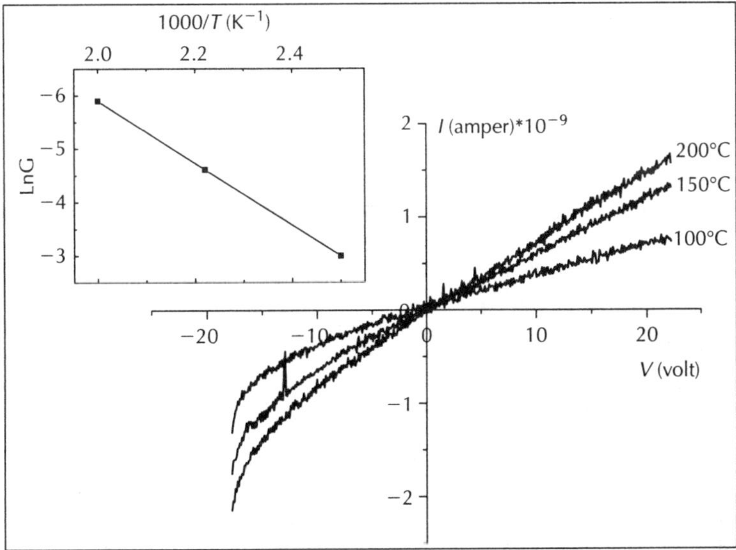

Figure 25.17 Current–voltage (I–V) curve of PbSe NQD solids annealed at 200°C and measured at three different temperatures, as indicated in the figure. Inset: Plot of the conductivity versus the measured temperature.

conduction band states; (2) exchange coupling between NQDs is given by $\beta \sim \exp[-\kappa(d + \delta)]$, where d is the NQDs' diameter, δ is the interdot spacing, and κ^{-1} describes the length scale of the wavefunction leakage outside the NQD; (3) Coulomb charging energy of the NQD array is given by $E_c = e^2/2C_0$. This parameter is influenced by the surrounding medium, when $C_0 = 2\pi\varepsilon_m\varepsilon_0 r$ ($\varepsilon_m = 2.0$ is the dielectric constant of the surrounding medium, while $\varepsilon_0 = 24.0$ is the dielectric constant of the NQDs with radius r). Thus, the relatively large dielectric constant of the PbSe NQDs and evaporation of the organic ligands would lead to a small E_c value of ~ 0.1 eV. Also, the large Bohr radius of electrons and holes in PbSe (each ~ 46 nm) and the absence of the organic ligands permits the wavefunctions to spill to outside the volume of the NQD, facilitating exchange interactions. Also, when δ decreases, the β value increases. Thus, the selected PbSe NQDs, with the large dielectric constant, allow smaller E_c and larger β values, compared to the situation found in CdSe NQD solids [59], allowing a better conductivity through the NQD array.

The transport mechanism of NQD arrays may follow either a variable range-hopping conduction, or Coulomb blockage transport [26]. The lack of threshold voltage in Fig. 25.17 eliminates the existence of a Coulomb blockage process, and thus suggests the occurrence of a hopping mechanism. Across the range of the measured temperature, the hopping process should comply with an Arrhenius-like behaviour of the conductivity $G = G_0 \exp(-E_a/k_BT)$, where E_a is the activation for charge transport. Thus a plot of the conductivity ($G = I/V$) versus the measured temperature is shown in the inset of Fig. 25.17. From the slope of the plot we found that $E_a \sim 0.5$ eV. Also the activation energy is given as $E_a = E_g/2 + \Delta E$, where E_g is the band gap energy, while ΔE is the maximum charging energy that is close to the value of $E_c \sim 0.1$ eV mentioned in the preceding paragraph. Considering a band gap of 0.8 eV for NQDs with 5 nm diameter (Fig. 25.4, line 7), the activation energy for hopping between nearest-neighbour NQDs is evaluated to be ~ 0.5 eV.

As mentioned in the Introduction, the conductivity properties of PbSe NQD arrays were examined recently by Wehrenberg *et al.* [25] utilizing a spectro-chemical method at the temperature range between $-269°C$ and $-50°C$. This study excluded the occurrence of thermal activated conductivity process at the studied temperature range. However, Drndic *et al.* [26] investigated the conductivity properties of PbSe NQD arrays at similar temperatures to those used in the present work (between $100°C$ and $200°C$). Although Drndic *et al.* also proposed a thermal activation process with Arrhenius behaviour, the present study revealed a conductivity which is a thousand times larger than that published in [26]. This suggests the existence of high-quality NQDs and their impact-packed arrangement. In addition, it shines light on the efficiency in removing the organic ligands and protection of the surface sites by hydrogenation, using forming gas at elevated temperatures. Hence, the results presented in this section could have important implications for the proposed applications of PbSe NQDs in optoelectronic devices.

25.9   Summary

This chapter showed the preparation of air-stable PbSe/PbS core–shell and PbSe/PbSe$_x$S$_{1-x}$ core-alloyed–shell NQDs. The structural characterization showed a continuous growth of the crystallographic plans without a distinct boundary at the core–shell interface due to a close crystallographic matching. However, the absorption and emission spectra of the core–shell structures were red shifted with respect to the relevant core samples, due to a wavefunction spilling into the low barrier shell. Furthermore, the emission spectra are shifted with respect to their absorption bands, merging from a Stokes shift into an anti-Stokes shift behaviour, depending on the NQDs and shell thickness and composition. In any event, the core–shell structures showed a unique chemical stability in air over periods of months and years, tolerance under intense illumination and a large quantum efficiency of 65%. The typical excited-state lifetime of PbSe NQDs, PbSe/PbS core–shell and PbSe/PbSe$_x$S$_{1-x}$ core-alloyed–shell NQDs is tens to hundreds of ns, depending on the pumping power intensity, the shell thickness and the composition.

The described core–shell structures and the corresponding PbSe core NQDs were used as passive Q-switches in eye-safe laser Er:glass, acting as saturable absorbers. The absorber saturation investigations revealed a relatively large ground-state cross-section of absorption ($\sigma_{gs} = \sim 10^{-15}$ cm^2) and a behaviour of a "fast" absorber with an effective lifetime ($\tau_{eff} \sim 4.0$ ps) is proposed. The

τ_{eff} was associated with the formation of multiple excitons at the measured pumping power. The product of σ_{gs} and τ_{eff} enables a sufficient Q-switching performance and tunability in the near-IR spectral regime. The ASE properties of PbSe NQDs were examined under continuous illumination by a diode laser at room temperature, suitable for standard device conditions. The results revealed a relatively large gain parameter ($g = 2.63\,\text{cm}^{-1}$ and $6.68\,\text{cm}^{-1}$). The conductivity properties of a PbSe NQD self-assembled solid showed an ohmic behaviour at the measured voltages (up to 30 V), which is governed by a variable range-hopping charge transport mechanism.

Acknowledgments

The authors wish to express their gratitude to the German Israel Foundation (GIF) contract no. #156/03-12.6, to the German-Israel Program (DIP) project no. #D 3.2, to the Ministry of Industry, Trade and Labor, Magneton #1000052, and to the Russell Berrie foundation for the contribution of the excellent infrastructure and to the donation of Matilda and Gariel Barnett for the establishment of the Nanocrystalline Laboratory. The authors thank Prof. N. Tessler and Dr O. Solomesh for performing the PL QE measurements and to Lilac Amirav for performing the TEM and HR-TEM measurements. Dr A. Sashchiuk and M. Brumer both express their deep gratitude to the Ministry of Absorption of the State of Israel for the KAMEA Fellowship, and to the Israel Ministry of Science and Education, Eshkol #867.

References

1. I. Kang and F.W. Wise, *J. Opt. Soc. Am. B* **14**, 1632 (1997).
2. F.W. Wise, *Acc. Chem. Res*, **33**, 773 (2000).
3. R.J. Ellingson, M.C. Beard, J.C. Johnson, P. Yu, O.I. Micic, A.J. Nozik, A. Shabaev, and A.L. Efros, *Nano Lett.* **5**, 865 (2005).
4. A.D. Andreev and A.A. Lipovskii, *Phys. Rev. B* **59**, 15402 (1999).
5. G. Allan and C. Delerue, *Phys. Rev. B* **70**, 24531-1, (2004).
6. J.M. An, A. Franceschetti, S.V. Dudiy, and A. Zunger, *Nano Lett.* **6**, 2728 (2006).
7. J.M. Harbold, H. Du, C. Chen, T.D. Krauss, K. Cho, C. Murray, R. Krishnan, and F.W. Wise, *Phys. Rev. B* **72**, 195312 (2005).
8. B.L. Wehrenberg, C. Wang, and P. Guyot-Sionnest, *J. Phys. Chem. B* **106**, 10634 (2002).
9. R.D. Schaller, M.A. Petruska, and V.I. Klimov, *J. Phys. Chem. B* **107**, 13765 (2003).
10. R.D. Schaller and V.I. Klimov, *Phys. Rev. Lett.* **92**, 186601-1 (2004).
11. V.I. Klimov, A.A. Mikhailovsky, S. Xu, A. Malko, J.A. Hollingsworth, C.A. Leatherdale, H.J. Eisler, and M.G. Bawendi, *Science,* **290**, 314 (2000).
12. V.L. Colvin, M.C. Schlamp, and A.P. Alivisatos, *Nature,* **370**, 354 (1994).
13. M. Sirota, E. Galun, V. Krupkin, A. Glushko, A. Kigel, M. Brumer, A. Sashchiuk, L. Amirav, and E. Lifshitz, *SPIE,* **9**, 5510 (2004).
14. M. Dahan, T. Laurence, F. Pinaud, D.S. Chemla, A.P. Alivisatos, M. Suer, and S. Weiss, *Phys. Rev. B* **6320**, 5309 (2001).
15. C.J. Murphy, E.B. Brauns, and L. Gearheart, in: Advances in Microcrystalline and Nanocrystalline Semiconductors, *Symposium. Mater. Res. Soc.* Pittsburgh, PA, USA 1997, p. 597.
16. C.B. Murray, S. Shouheng, W. Gaschler, H. Doyle, T.A. Betley, and C.R. Kagan, *IBM J. Res. & Dev.* **45**, 47 (2001).
17. W.W. Yu, J.C. Falkner, B.S. Shih, and V.L. Colvin, *Chem. Mater.* **16**, 3318 (2004).
18. A. Sashchiuk, L. Langof, R. Chaim, and E. Lifshitz, *J. Cryst. Growth,* **240**, 431 (2002).
19. M. Brumer, A. Kigel, L. Amirav, A. Sashchiuk, O. Solomesch, N. Tessler, and E. Lifshitz, *Adv. Funct. Mater.* **15**, 1111 (2005).
20. E. Lifshitz, M. Brumer, A. Kigel, A. Sashchiuk, M. Bashouti, M. Sirota, E. Galun, Z. Burshtein, A.Q. Le Quang, I. Ledoux-Rak, and J. Zyss, *J. Phys. Chem. B* **110**(50), 25356 (2006).
21. J. Xu, D. Cui, T. Zhu, G. Paradee, Z. Liang, Q. Wang, S. Xu, and A.Y. Wang, *Nanotechnology,* **17**, 5428 (2006).
22. J.W. Stouwdam, J. Shan, F.C.J.M. Van Veggel, A.G. Pattantyus-Abraham, J.F. Young, and M. Raudsepp, *J. Phys.Chem. C,* **111**, 1086 (2007).

23. E. Lifshitz, M. Bashouti, V. Kloper, A. Kigel, M.S. Eisen, and S. Berger, *Nano Lett.* **3**, 857 (2003).

24. A. Sashchiuk, L. Amirav, M. Bashouti, M. Krueger, U. Sivan, and E. Lifshitz, *Nano Lett.* **4**, 159 (2003).

25. B.L. Wehrenberg, D. Yu, J. Ma, and P. Guyot-Sionnest, *J. Phys. Chem. B* **109**, 20192 (2005).

26. H.E. Romero and M. Drndic, *Phys. Rev. Lett.* **95**, 156801 (2005).

27. L. Etgar, E. Lifshitz, and R. Tannenbaum, *J. Phys. Chem. C*, **111**, 6238 (2007).

28. E.E. Larken, S.G. Grancharov, S. O'Brien, T.J. Deming, G.D. Stucky, C.B. Murray, and G.A. Held, *Nano Lett.* **3**, 1489 (2003).

29. B.L. Frankamp, N.O. Fischer, R. Hong, S. Srivastava, and V.M. Rotello, *Chem. Mater.* **18**, 956 (2006).

30. J.C. de Mello, H.F. Wittmann, and R.H. Friend, *Adv. Mater.* **9**, 230 (1997).

31. P. Liljeroth, P.A.Z. van Emmichoven, S.G. Hickey, H. Weller, B. Grandidier, G. Allan, and D. Vanmaekelbergh, *Phys. Rev. Lett.* **95**, 0868011 (2005).

32. S.H. Wei and A. Zunger, *Phys. Rev. B* **55**, 13605 (1997).

33. B.O. Dabbousi, J. Rodriguez-Viejo, F.W. Mikulec, J.R. Heine, H. Mattoussi, R. Ober, K.F. Jensen, and M.G. Bawendi, *J. Phys. Chem.* **101**, 9463 (1997).

34. J.S. Steckel, J.P. Zimmer, S. Coe-Sullivan, N.E. Stott, V. Bulovic, and M.G. Bawendi, *Angew. Chem. Int. Ed.* **43**, 2154 (2004).

35. R.G. Neuhaouser, K.T. Shimizu, W.K. Woo, S.A. Empedocles, and M.G. Bawendi, *Phys. Rev. Lett.* **85**, 3301 (2000).

36. O.I. Micic, B.B. Smith, and A.J. Nozik, *J. Phys. Chem. B* **104**, 12149 (2000).

37. O. Millo, D. Katz, Y.W. Cao, and U. Banin, *Phys. Stat. Sol. B* **224**, 271 (2001).

38. A.C. Bartnik, F.W. Wise, A. Kigel, and E. Lifshitz, *Phys. Rev. B.* (2007). in press.

39. R. Rossetti, J.L. Ellison, J.M. Gibson, and L.E. Brus, *J. Chem. Phys.* **80**, 4464 (1984).

40. Al.L. Efros, M. Rosen, M. Kuno, M. Nirmal, D.J. Norris, and M. Bawendi, *Phys. Rev. B* **54**, 4843 (1996).

41. J.J. Peterson and T.D. Krauss, *Nano. Lett.* **6**, 510 (2006).

42. S.A. Zolotovskaya, K.V. Yumashev, N.V. Kuleshov, and A.V. Sandulenko, *Appl. Opt.* **44**, 1704 (2005).

43. R.D. Stultz, M.B. Camargo, and M. Birnbaum, *J. Appl. Phys.* **78**, 2959 (1995).

44. M.B. Camargo, R.D. Stultz, M. Birnbaum, and M. Kokta, *Opt. Lett.* **20**, 339 (1995).

45. K.V. Yumashev, I.A. Denisov, N.N. Posnov, V.P. Mikhailov, R. Moncorge, D. Vivien, B. Ferrand, and Y.J. Guyot, *Opt. Soc. Am. B* **16**, 2189 (1999).

46. K.V. Yumashev, I.A. Denisov, N.N. Posnov, P.V. Prokoshin, and V.P. Mikhailov, *Appl. Phys. B* **70**, 179 (2000).

47. T.Y. Tsai and M. Birnbaum, *J. Appl. Phys.* **87**, 25 (2000).

48. Z. Burshtein, P. Blau, Y. Kalisky, Y. Shimony, and M.R. Kokta, *IEEE J. Quantum Electron.* **34**, 292 (1998).

49. R.D. Schaller and V.I. Klimov, *Phys. Rev. Lett.* **96**, 097402 (2006).

50. K.V. Yumashev, I.A. Denisov, N.N. Posnov, V.P. Mikhailov, R. Moncorge, D. Vivien, B. Ferrand, and Y. Guyot, *J. Opt. Soc. Am. B* **16**, 2189 (1999).

51. K.V. Yumashev, I.A. Denisov, N.N. Posnov, P.V. Prokoshin, and V.P. Mikhailov, *Appl. Phys. B* **70**, 179 (2000).

52. A.V. Podlipensky, V.G. Shcherbitsky, N.V. Kuleshov, V.P. Mikhailov, V.I. Levchenko, and V.N. Yakimovich, *Opt. Lett.* **24**, 960 (1999).

53. Z. Burshtein, Y. Shimony, R. Feldman, V. Krupkin, A. Glushko, and E. Galun, *Opt. Mater.* **15**, 285 (2001).

54. R.D. Stultz, M.B. Carmargo, M. Birnbaum, and M. Kokta. *Advanced Solid-State Lasers*, vol. 24 of OSA proceeding series (Optical Society of America, Washington, DC, 1995). pp. 460–464.

55. A.M. Malyarevich, I.A. Denisov, K.V. Yumashev, V.P. Mikhailov, R.S. Conroy, and B.D. Sinclair, *Appl. Phys. B* **67**, 555 (1998).

56. A.V. Podlipensky, K.V. Yumashev, N.V. Kuleshov, H.M. Kretschmann, and G. Huber, *Appl. Phys. B* **76**, 245 (2003).

57. R.M. Cornell, and U. Schwertmann. *The Iron Oxides: Structures, Properties, Reactions, Occurrence and Uses*, 1st ed. (Wiley-VCH Verlag GmbH & Co, Weinheim, Germany, 1996).

58. G. Brown, ed. *X-Ray Identification and Crystal Structures of Clay* (The Mineralogical Society of America, Chantilly, VA, USA, 1961).

59. M. Drndic, M.V. Jarosz, N.Y. Morgan, M.A. Kastner, and M.G. Bawendi, *J. App. Phys.* **92**, 7498 (2002).

CHAPTER 26

Semiconductor Quantum Dots for Biological Applications

Beate S. Santos,[1,3] Patrícia M.A. Farias,[2,3] and Adriana Fontes[2,3]

[1]*Departamento de Ciências Farmacêuticas – CCS – Universidade Federal de Pernambuco-Recife – PE, Brazil, CEP: 50670-901;*
[2]*Departamento de Biofísica e Radiobilogia – CCB – Universidade Federal de Pernambuco-Recife – PE, Brazil, CEP: 50670-901;*
[3]*Research group on Nanostructures and Biological Interfaces (NIB)*

26.1 Introduction

Nanophotonics, the fusion of biophotonics and nanotechnology, is an emerging multidisciplinary field that describes nanoscale optical science and technology. It deals with the interaction of light with matter in the nanometre-size scale. Nanomaterials constitute a major area of nanophotonics. Materials can be scaled down for many orders of magnitude, from macroscopic- to nanoscopic-size scale. For many semiconductor materials, the nanometric-size region confers huge alterations in some physical–chemical properties, if compared with the properties of the same material in the macroscopic-size scale. By manipulating size and structure of nanometrical semiconductor particles, one can, for instance, perform changes in their optical properties. These unique characteristics of semiconductor nanoparticles occur due to the fact that they are in the quantum confinement regime. These nanoparticles are called quantum dots (QDs).

QDs are semiconductor three-dimensional nanoparticles, with typical dimensions ranging from nanometres to tens of nanometres. A QD is often described as an artificial atom because the electron is dimensionally confined just like in a real atom and similarly it shows only discrete energy levels. The energy levels of QDs can be probed by optical spectroscopy techniques as well as in atoms. In other words, it is possible to excite these semiconductor nanocrystals, generate optical signals and use their optical properties. For example, it is possible to use QDs to generate fluorescence. On the other hand, in contrast to atoms, the energy spectrum of a QD can be engineered by controlling size, shape and strength of the confinement potential. QDs of the same material, but with different sizes, can emit light of different colours. As energy is related to wavelength (or colour), this means that the optical properties of the particle can be finely tuned depending on its size. Thus, particles can be produced in order to emit or absorb specific wavelengths, merely by controlling their size. The larger the dot, the more towards to the red end (lower energy) of the spectrum is its corresponding light emission. The smaller the dot, the more towards to the blue end (higher energy) it is. The light emission is directly related to the energy levels of the QD. Quantitatively speaking, the band gap energy that determines the energy of the fluorescent light is inversely proportional to the square of the size of the quantum dot. Larger quantum dots have more energy levels which are more closely spaced.

The ability to change optical properties by tuning the size of the quantum dot is advantageous for many applications. Basically, for this reason, quantum dots have found applications for solar cells [1], diodes and lasers [2], detectors [3], switches [4], telecommunication [5] and also as fluorescent biological labels [6, 7]. Although the first works using quantum dots of CdS and CdSe in telecommunications arose in the 1980s, the first biological applications of quantum dots as fluorescent labels for biological application was first reported in literature in 1998 simultaneously by Bruchez *et al.* [8] and Chan *et al.* [9]. Both groups used CdSe quantum dots coated with silica and mercapto-acetic acid layers, respectively. These layers constitute the *functionalization shell*, which was labelled by covalent coupling of specific biomolecules to the QDs' surface. This reaction step is called *bioconjugation*, resulting in QD bioconjugates, which are ready to interact with biological systems. Subsequently, many authors reported the labelling of whole cells and tissue sections, using several different experimental procedures for modification of the QDs' surface [10–12]. Since the first results reported in 1998, the use of semiconductor quantum dots for biological purposes has increased exponentially. Currently, fluorescent semiconductor quantum dots became a promising class of materials in the labelling of biological systems, as well as in the diagnostics of pathologies, such as different kinds of cancer [13].

Fluorescent semiconductor quantum dots have been used mainly for bioimaging, using conventional or confocal fluorescence microscopy to study cellular biology and also for optical diagnostics. However, they can also be used in any experiment that is conventionally performed with fluorescent dyes, such as: flow cytometry, fluorescence resonance energy transfer (FRET) and fluorescence lifetime measurements (FLIM), as well as in two-photon/multiphoton microscopy [14, 15]. Biological applications of fluorescent semiconductor quantum dots will be further detailed.

One of the fundamental goals in biology is the comprehension of the complex spatio-temporal interplay of biomolecules from the cellular to the integrative level. To study these interactions, researchers commonly use fluorescent labelling for both *in vivo*/*in vitro* cellular or tissular imaging and assay detection. Fluorescence provides a unique method for the investigation of basic physical properties of biological structures. The high sensitivity of fluorescence, combined with the advances in measurement techniques, permits detection of ultra-small quantities of specific compounds present in biological systems. In modern biological analysis, many kinds of fluorescent organic dyes are used. However, with each passing year, more flexibility is being required of these dyes, and the traditional dyes are simply unable to meet the necessary standards at times. The intrinsic photophysical properties of organic and genetically encoded fluorophores, which generally have broad absorption/emission profiles and low photobleaching thresholds, have limited their effectiveness in long-term imaging and "multiplexing" (simultaneous detection of multiple signals) without complex instrumentation and processing. To this end, quantum dots have quickly filled in the role and showed that their unique properties could overcome these issues and therefore be superior to traditional organic dyes on several counts.

Compared to organic fluorophores, the major advantages of interest to biologists offered by fluorescent semiconductor quantum dots are the following:

(a) Narrower emission bands compared to organic dyes, symmetric luminescence spectra (full width at half-maximum $\sim$25–40 nm) spanning the UV to near-infrared (as exemplified in Fig. 26.1) and large effective Stokes shifts. Thus, the complication in simultaneous quantitative multiplexing detection, posed by cross-talks between different detection channels from spectral overlap, is significantly reduced.

(b) Longer emission lifetimes (hundreds of nanoseconds) compared to organic fluorophores, thus allowing, for example, one to utilize time-gated detection to suppress autofluorescence, which has a considerably shorter lifetime.

(c) Higher brightness (due to the high-fluorescence quantum yields) as well as photostability (in general, no photodegradation is observed for a time interval of many hours) which leads to a high resistance to photobleaching (1000 smaller than in organic dyes) and exceptional resistance to photo- and chemical degradation (Fig. 26.2). As a consequence, single QDs can easily be detected in living cells, and their localization can be monitored over minutes to days.

(d) Compared with molecular dyes, two properties in particular stand out: the unparalleled ability to tune fluorescent emission spectra as a function of the nanoparticle's size and the broad excitation spectra, which allow excitation of mixed quantum dot populations at a

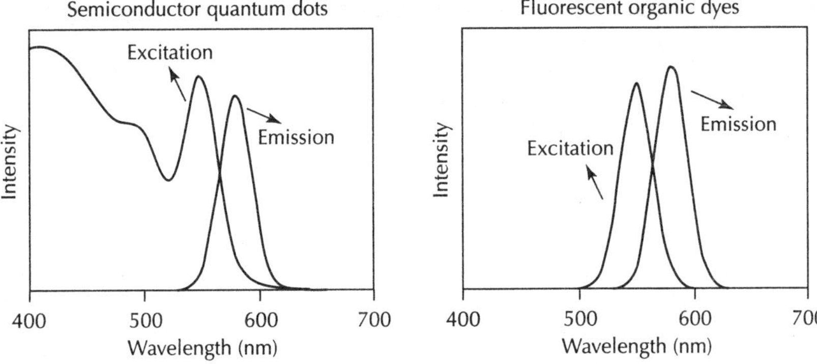

Figure 26.1   *Comparative excitation and emission spectra of quantum dots and fluorescent organic dyes.*

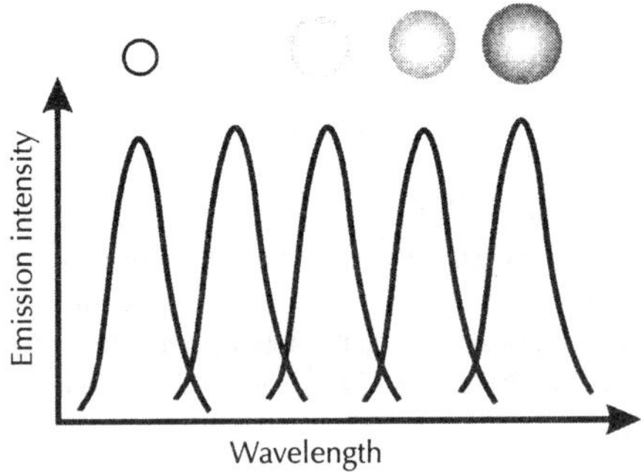

Figure 26.2   *Fluorescent emission as a function of size of the quantum dots. The larger the dot, the more towards to the red end of the spectrum is its corresponding light emission.*

single wavelength far removed (>100 nm) from their respective emissions bands (refer to Fig. 26.1).

(e) Larger two-photon cross-sections allowing, for example, *in vivo* imaging at greater depths [16].

Quantum dots used in biological applications are exclusively obtained by colloidal synthesis methodologies, which result in colloidal suspensions, in which the solid nanocrystals are dispersed into a liquid phase. Several energetic parameters have to be taken into account in order to control the growth of the particles, as well as to maintain the stability, and the conditions to make them feasible for biological labelling. The major problems and challenges in the use of quantum dots in biological systems are:

(a) Surface-induced quenching of emission efficiency due to the high surface area of the nanocrystals. This requires surface passivation by encapsulation or by using core–shell techniques, as will be discussed in the next sections.

(b) To be able to prepare quantum dots of different sizes, in order to have distinct emission spectra and to be able to control their size dispersion, a parameter closely related to the width of the emission spectra.

(c) Dispersibility of the quantum dots in biological media. The nanocrystals need to be biologically compatible, either lipid soluble or mainly water soluble. Some researchers ascribe the use of silica layer encapsulation [8], others use synthetic routes in aqueous medium which produce quantum dots with hydrophilic surfaces [13, 17].

(d) The colloidal dot suspensions also need to be isotonic and in physiological pH media to be compatible with the intra-cellular/tissular medium.

(e) Labelling and functionalizing the quantum dots to target specific cells and other biological systems. This step consists in a great challenge for this area.

(f) The quantum dots in general, bearing from 4000 to 6000 atoms, are much bigger than the usual dyes, so they have to be internalized in the biological systems using specific ways.

All these boundary conditions make the application of semiconductor quantum dots for biological labelling purposes an emerging and important field which demands a lot of effort towards interdisciplinary expertise, and promises to have a potential impact for novel nanomedical applications on disease diagnosis, therapy and prevention, which can lead to fundamental changes in health care now and in the future.

In the next sections there will be presented and discussed a short review of the experimental procedures which yield water-soluble fluorescent quantum dots. There will also be presented some applications of these QDs as biological labels as well as potential tools for cancer diagnostics.

26.2   Creating a fluorescent biolabel out of semiconductor nanocrystals

For an optimal performance in biological imaging (and competitive to the commercial organic fluorophores), semiconductor quantum dots are being developed in order to optimize their luminescent, surface and chemical stability properties. These conditions result in a very complex multilayered chemical assembly where the nanocrystal core determines its emission colour, the passivation shell determines its brightness and photostability and the organic capping determines its stability and functionality.

A schematic representation of what this "hybrid bio/organic/inorganic nanostructured assembly" looks like is shown in Fig. 26.3.

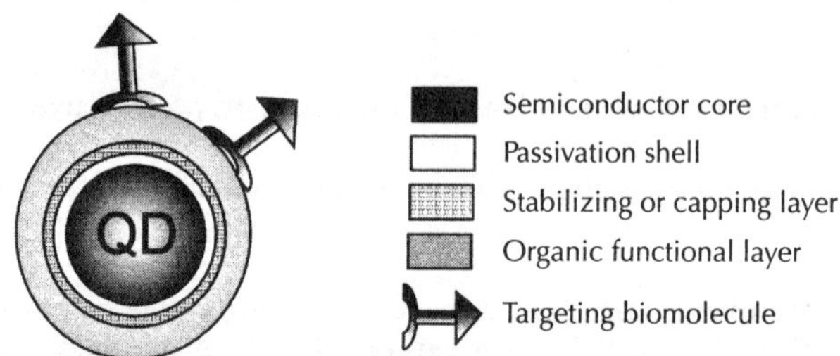

Figure 26.3   Schematic representation of a semiconductor quantum dot adapted for biolabelling purposes. Each part of this hybrid bio/organic/inorganic assembly is designed to optimize its functionality.

Nanostructured semiconductor systems received increased attention in the early 1980s and are still being optimized to render good quality nanoparticles. Two independent research fields brought to existence stable semiconductor nanocrystals (<10 nm). It was expected that lowering the size of semiconductor particles would increase their oxidation/reduction performance in catalysis and, on the other hand, it was also expected that semiconductor materials in the nanometre size range would show quantum confinement properties. In order to test this hypothesis, researchers such as the groups of A. Henglein, M. Grätzel and L. Brus adapted methodologies based on colloidal chemistry developed in the 19th century. They produced large quantities of selected semiconductor materials, mainly CdS and ZnS which, aside from a large surface area, also possessed unique optical and electrical properties. The theoretical model of quantum confinement regime was successfully applied in the description of their behaviour. Brus, Efros and

Ekimov published the first papers describing three-dimensional size quantization effects using effective mass approximations [18, 19].

In order to control the growth, chemical nature and surface of the particles, the bottom-up technique based on colloidal chemistry was and still is the primary choice for the preparation and control of large numbers of these assemblies. After the initial success in the obtention of QDs, their photophysical processes are currently being extensively studied.

There are four basic steps in the assembly of these complex chemical biofunctional semiconductor nanoparticles and each one must be developed to optimize its performance in biological imaging:

- The semiconductor core particle growth in a solution containing its chemical precursors and stabilizing molecules.
- The passivation procedure of the core particle by introducing a capping layer of a higher semiconductor band gap. As defect structures on surface can alter the total number of photons emitted, this step is fundamental for the optimization of their fluorescence quantum yield.
- The solubilization of the particles in aqueous solution. This step is necessary for methodologies where the core–shell particles are obtained in organic solvents and must be extracted to aqueous media.
- Surface functionalization of the particles in order to direct their interaction with biomolecules or biological systems.

Synthetic routes for the obtention of colloidal II–VI nanoparticles (2–20 nm) were extensively investigated and several reviews in this field have been published [20–26].

Colloidal methods involve the controlled formation (due to its low solubility product, Kps) of a crystalline precipitate from a solution containing its ionic precursors. Just as in the crystal growth kinetics from supersaturated solutions, there are three main processes which determine the formation and aging of these particles:

(a) The nucleation stage in which the first aggregates are formed in the mother solution.
(b) The growth of these nuclei by consuming the species present in the mother solution to form primary crystalline particles.
(c) The aging of the primary particle dispersion during which changes in shape, structure, size, agglomeration and flocculation may take place.

The success observed for the colloidal methods in the obtention of nanostructured crystals relies on the ability to stop the crystal growth process immediately after nucleation begins. This is accomplished by controlling the equilibrium between the solid crystalline precipitate and the solvated metal ions in solution and this can be experimentally achieved by selecting an appropriate reaction temperature and solvent.

Another crucial point to be observed in the chemical stability of these colloidal systems is the occurrence of long-term growth processes, mainly the diffusion growth and the Ostwald ripening, which end up producing bigger agglomerates which may lead to an irreversible flocculation and phase separation. These processes can be controlled by using appropriate stabilizing conditions, mainly by rendering particles with charged surfaces and/or sterically built systems using polymers. This surface stabilization may also have an additional purpose: the nanocrystal solubilization in water media.

In the next sections the evolution of the currently known methodologies for the obtention of soluble surface passivated semiconductor nanocrystals using colloidal chemistry techniques is described. Special attention is given in section 26.2.4 to the functionalization procedures which are used to optimize the chemical association of the nanocrystal with a specific targeting biomolecule.

26.2.1   *How did people work out how to control the particles down to the nanoscale size range? The evolution of the synthetic procedures*

The first methods that successfully described II–VI semiconductor nanocrystals reported the preparation of CdS and ZnS particles in the 3–6 nm range [27–30]. This was accomplished by applying an arrested precipitation technique at room temperature involving either the injection of the metal salt into a solution of ammonium, sodium sulphide or the injection of $H_2S(g)$ in a Cd^{2+} containing solution, both bearing a suitable reaction solvent (water, methanol, 2-propanol or acetonitrile) in the presence or not of a stabilizing agent, such as polyphosphate and thioglycerol [28] or AOT (sodium bis(2-ethylhexyl) sulphosuccinate) reverse micelle systems [31, 32]. Figure 26.4 shows two representative transmission electronic micrographs of CdS nanocrystals obtained from the reaction of Cd^{2+} ions with $H_2S(g)$ in the presence of either polyphosphate chains $[NaPO_3]n$ or mercaptoacetic acid as the stabilizing agents. The Cd:S ratio was the same in both cases. The size dispersion of the polyphosphate-stabilized CdS nanocrystals is 15% while for the thiol stabilized ones it is <10%. The polyphosphate method for the preparation of CdS (and ZnS, (CdZn)S, HgS) particles is still used for the preparation of highly stable, fluorescent and biocompatible colloidal aqueous systems. On the other hand, the stabilization of the colloidal suspensions was longer for larger chains than for smaller capping molecules.

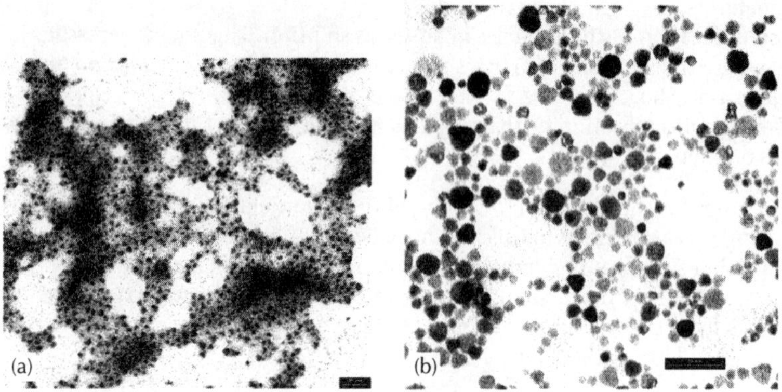

Figure 26.4   Transmission electronic microscopic images of CdS nanocrystals obtained in water, at room temperature, using $Cd(ClO_4)_2$ and H_2S (1:1) as the precursors and two different stabilizing agents: (a) mercaptoacetic acid, bar = 40 nm and (b) polyphosphate anions, bar = 50 nm.

The precipitation technique in aqueous medium was extended by Weller and his group to other semiconductor materials such as CdSe and HgTe by using different stabilizers (also referred to as capping agents) such as those containing amine and thiol groups [33–35] similar to a previously described synthesis by Nosaka *et al.* [36]. They reported the synthesis of gram-scale water redispersible stable powders showing luminescence quantum yields up to 50% and high photostability. Although all these previously obtained systems present simpler methodologies and good optical properties, they show a poor degree of crystallinity, with high concentration of surface defects and large size variations (relative standard deviation >15%) requiring size-selective precipitation procedures [37, 38] or chromatographic gel separation [39] to obtain narrower-size distributions.

The size-selective precipitation method is based on the size-dependent solubility of the nanocrystals. Adding controlled amounts of a non-solvent to the colloidal suspension will lead to the precipitation of the large nanocrystals, which can then be separated by either filtration or centrifugation. Such a procedure may be carried out consecutively to obtain a set of nanocrystal fractions with different diameters. A similar procedure may in principle be applied again to each fraction to further narrow the size distribution [37, 40].

In 1988, Steigerwald [41] reported the preparation, in micelle systems, of II–VI nanoparticles which were then coated with covalently bound phenyl groups. This first report brought up the possibility of organometallic rather than inorganic precursors. Later on Bawendi and collaborators [42] also used inverse micelle systems to prepare similar nanoparticles, which after annealing were dispersed in Lewis base solvents such as trialkylphosphines/phosphine oxides.

The 1990s showed an extensive diversification of these new methodologies driven by the potential application of these materials in non-linear optics. High-quality CdSe nanocrystals became available using high-temperature crystal formation in organometallic solvents. Knowing that a great number of divalent transition ions show strong coordination ability towards various solvent molecules such as amine and phosphine containing molecules [43], forming low-dimensional structures referred to as chalco(genido)metalates [44], researchers directed these reactions towards the obtention of II–VI nanocrystals. In 1993 Bawendi and collaborators [37] introduced the "organometallic TOP/TOPO" procedure by synthesizing highly luminescent CdSe quantum dots with nearly perfect crystal structures and narrow-size variations (relative standard deviation <5%). This route was based on the pyrolysis of organometallic reagents (e.g. dimethylcadmium – $Cd(CH_3)_2$ – and bis(trimethylsilyl)selenium) by injection into high boiling point coordinating solvents (like tri-*n*-octylphosphine oxide (TOPO) and tri-*n*-octylphosphine (TOP)). A series of organometallic routes was described aimed at the obtention of II–VI nanocrystals, mainly CdSe, but also CdS, CdTe and ZnSe [45, 46]. CdSe nanoparticles showed narrower-size dispersion, consequently a greater control on the amplitude of emission colours (ranging from violet to red).

The central point in this methodology is the control of the nucleation and growing steps of the particles performed at high temperatures ($T > 300°C$) for an extended period of time (ranging from minutes to hours, depending on the desired particle size) [47]. In this process, smaller nanocrystals are broken down, and the dissolved ions are transferred to larger crystals. The rate of this "ripening" process is dependent on both the temperature and the amount of the limiting reagent [47, 48]. Continuous injection of precursor solutions into the CdSe reaction mixture at 300°C also produces larger nanocrystals. As a matter of fact, when using different TOP:TOPO ratios for the preparation of ZnSe, Hines and Guyot-Sionnest commented that the reaction could render nanocrystals so small that they could not be isolated by standard solvent/non-solvent precipitation techniques, neither could they precipitate as large aggregates [46]. The authors suggested that these difficulties arose from TOPO binding too strongly and TOP too weakly to Zn.

This new organometallic route succeeded in obtaining homogeneous nanocrystals with a small size dispersion but the fluorescence quantum yields were still relatively low ($\sim$10%). Moreover, it was observed that the TOP/TOPO system showed better growth control for CdSe nanocrystals not larger than 4.5 nm (i.e. in the strong quantum-confinement size regime, with first exciton absorption peak <600 nm) [49].

Currently the organometallic procedures are still the most popular choice for the development of a variety of colloidal nanocrystals in non-aqueous solutions although the main chemicals used are highly toxic, expensive, pyrophoric and explosive (the systems are heated above TOPO's flash point: 350°C). To improve this methodology, some researchers tried to introduce other coordinating systems in the reaction solution, such as amines which were a logical choice for ligands of intermediate strength. Amines are slightly weaker bases than phosphine oxide and long-chain alkylamines have much higher boiling points (hexadecylamine, bp: 330°C). In addition, the less sterically hindered amine creates a larger capping density, which probably increases the surface passivation. As such, dodecylamine (DDA) and hexadecylamine (HDA) have been successfully used to cap the surface of CdSe nanocrystals [49, 50–53]. Another feature of these organometallic methodologies is the possibility of the nanocrystal's final shape control (e.g. obtaining nanorods) by altering the ratio of the surfactants (e.g. TOPO and hexylphosphonic acid, HPA) [53–55]).

Peng and Peng [49], motivated by the instability and the low reproducibility of the optical properties of the nanocrystals obtained by the organometallic route, developed a series of methodologies based on CdO, $CdCO_3$ and $Cd(acetate)_2$ in fatty acid solutions aimed at the substitution of the pyrophoric $Cd(CH_3)_2$. They argumented that this cadmium precursor actually decomposes in hot TOPO and generates insoluble metallic cadmium precipitates. Studying a different set of experimental conditions and precursors these authors suggested that the existence of any anion of a strong acid, either in the form of the cadmium precursor or as an added cadmium ligand, made it impossible to form high-quality CdSe nanocrystals in the current systems. Thiols, which bind strongly to cadmium, were found to inhibit the nucleation process. Using this procedure they reported a size range of nearly monodisperse CdSe crystals, from about 1.5 nm to above 25 nm, a much broader range than that achieved by the original organometallic method.

Peng and Peng [49] further reported that the temporal evolution of the size and size distribution of CdSe nanocrystals in fatty acid systems were quite reproducible, although the reaction rates were fast. They considered this to be due probably to the more controllable nucleation step initiated by cadmium carboxylates less active than $Cd(CH_3)_2$ used in the traditional organometallic approach. This phenomenon implies that the control of the nucleation process may be the key step towards a fully controllable synthesis. Moreover, the authors state that in practice, the fatty acid systems are not recommended to synthesize small nanocrystals because of their fast growth rates.

In the late 1990s two rather different alternative methodologies were also described for the synthesis of selenides and tellurides. In the first one the chalcogenide precursors were produced *in situ* by using reducing agents such as KBH_4, which converts Se^{2+} and Te^{2+} to Se^{2-} and Te^{2-}, respectively [56, 57]. Wang and collaborators proposed a simple solution synthesis for pure quantum dots of M chalcogenides (M = Bi, Cu, Cd, Sn, Zn; chalcogenide = S, Se) by providing the *in situ* reduction of S or Se in the presence of KBH_4 and the corresponding metal salt at room temperature in strong basic solvents. They showed that the solvent significantly influenced the quality of the final product, yielding small uniform nanoparticles (4–6 nm) in the case of ethylenediamine and a mixed metal/chalcogenide precipitate with poor crystallinity and low yield in the case of pyridine. In fact, pyridine is known to provide stable capping through the N atom, but its low boiling point suggests limitations as a growth solvent.

Recently, amine-capped PbSe nanoparticles of tunable sizes and shapes were also obtained with this method [58]. As in the previous methods, the main limitation is the difficulty to achieve narrow-size distributions and high crystallinity.

The second methodology, proposed for large-scale production, involves the application of ultrasound (formation and implosive localized hot spots induced by acoustic cavitations) on chemical reactions [59, 60]. Zhu *et al.* reported the preparation of spherical ZnSe nanoparticles of average sizes of 3, 4, and 5 nm by reacting $Zn(acetate)_2$ and selenourea in water followed by sonication with a high-intensity ultrasonic probe under inert atmosphere for a determined period of time [60]. Pb and Cu selenide nanoparticles were also obtained by using the corresponding acetates.

A recent search for "greener" and simpler procedures that could produce semiconductor nanocrystals directly in aqueous media, aimed at bioapplications, readapted the thiol stabilized CdTe synthetic methods originally reported by Rogach *et al.* [33, 34] and the one reported by Nosaka *et al.* for the synthesis of CdS [36]. By using different approaches Gaponik *et al.* [38], Zhang *et al.* [61], and Menezes *et al.* [17] prepared highly luminescent CdSe or CdTe nanoparticles directly in water. Gaponik reported the dissolution of Al_2Te_3 in an acidic solution to render the Te^{2-} precursor ions as $H_2Te(g)$, while, Zhang *et al.* [61] and Menezes *et al.* [17] used $NaBH_4$ in aqueous solution to reduce Te to Te^{2-}. This reduction process generates Na–Te–Te–Na a stable intermediate complex which will be converted to CdTe after injection of the metal precursor complexed with a thiol molecule in water under inert atmosphere. These colloidal systems render particles in the 2–6 nm range and show luminescence after a certain period of time suggesting a slow kinetic surface passivation process [17]. This observation will be discussed in the next section.

Regarding the inherent toxicity of these systems for *in vivo* applications, for example, it was recently reported by Pradhan *et al.* [62] that the synthesis of pure and doped ZnSe QDs as an alternative for CdSe nanocrystal aimed at the obtention of a less toxic labelling material which could be envisioned for a safe use in *in vivo* experiments and diagnostics.

On the other hand, to avoid the inconvenient autofluorescence observed in biological systems when these are excited in the UV blue region of the spectrum, lower band gap II–VI semiconductor QDs are also being converted into biolabels. These systems when quantized present the onset of the absorption and emission bands in the near-infrared region (700–1300 nm). As an example, Kumar and Jahkmola reported on the RNA-mediated fluorescent PbS nanoparticles as novel tools for biophotonic applications [63]. Also a recent report of DNA-directed semiconductor quantum dot synthesis described highly optically emissive PbS nanocrystals [64]. Furthermore, Hinds *et al.* [65] were able to synthesize infrared emitting PbS QDs (4 nm) stabilized with Guanine-triphosphates (GTP). The authors systematically investigated how nucleotide functionalities (base, sugar and phosphate) influenced nanoparticle growth. They proposed a set of rules for using nucleic acids as ligands in order to profit from the natural biorecognition properties of DNA and programmable templates for nanoparticle synthesis.

In summary, much work has been done in the synthesis of II–VI nanostructured semiconductor compounds in the past three decades. Still, researchers are looking for the ideal preparation methodology, aimed at particles with fewer surface defects and small size dispersion using less expensive, non-toxic, less risky and simpler experimental conditions which can be extended to a large-scale production. The need to improve their surfaces in order to increase their fluorescence quantum yields, for example, prompted a close look at the QDs surfaces. The next section comprises the processes used to overcome this problem.

26.2.2 Still some problems: quantum dots have imperfect surfaces! The passivation process

The observed fluorescence in semiconductor nanoparticles is produced upon the recombination of the charge carriers which are generated by light absorption. The first colloidally obtained nanocrystals showed a very low fluorescence quantum yield ($\varphi < 1\%$). The non-radiative processes involved in semiconductor nanocrystals are said to have the same physico-chemical nature as those observed in bulk semiconductor materials [66]. Taking into account the high number of surface atoms compared to bulk atoms, it was suggested that the main contribution for this was that the prepared colloidal particles had a lot of surface defects (shallow and deep traps) where radiationless recombination of the charge carriers occurred. A schematic representation of these defects is illustrated in Fig. 26.5.

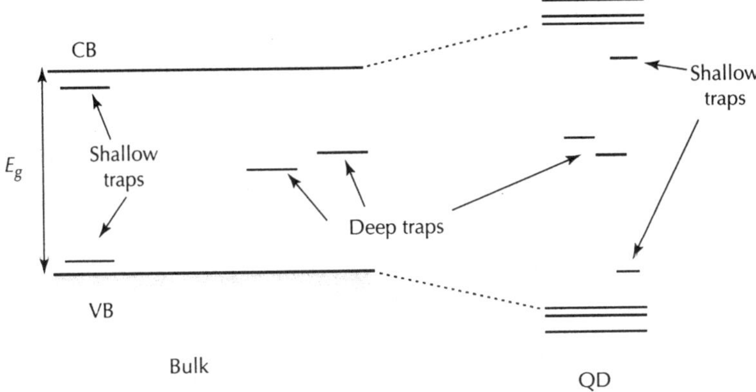

Figure 26.5 *A schematic representation of deep and shallow traps (originated from structural defects) claimed to be responsible for the radiationless processes in bulk and quantized semiconductor particles.*

Soon it was realized that if the defect sites, most probably resulting from dangling bonds, were located at the surface of the colloidal particles, there was a chance to chemically modify these sites. Several studies on the chemical influence on the spectroscopic properties of semiconductor quantum dots were reported. The fluorescence intensity (and also the fluorescence spectrum pattern) of CdS nanocrystals was shown to be drastically increased by certain surface modification procedures such as exchanging the aqueous solvent by alcohol [31], adsorbing triethylamine in low concentrations [31], and covering the surface with cadmium hydroxide or silver sulphide [28].

By coating the CdS nanocrystals with a layer of Cd(OH)$_2$ the resulting system is a core semiconductor coated by a shell of another semiconducting material possessing a higher band gap. The resulting assembly was later called a core–shell system. The increased luminescence quantum yields of the core–shell particles is explained by preventing from the photogenerated excitons spreading over the entire particle, forcing them to recombine while being spatially confined to the core. This luminescence enhancement was taken as an indication of the formation of the proposed structure due to the difficulty at that time to investigate such a thin chemical layer (just a few atomic monolayers). Figure 26.6 illustrates schematically the band gap offset created in the interface of the CdSe core ($E_g = 1.71$ eV, wurtzite) and the ZnS shell ($E_g = 3.68$ eV, zinc-blende) during the passivation process.

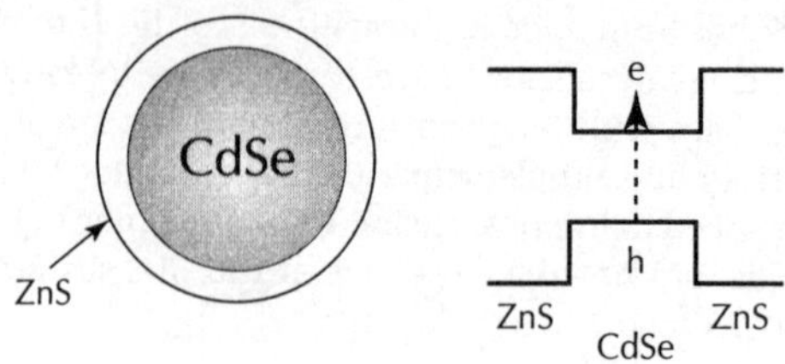

Figure 26.6   Schematic representation of the energy band gap offset between the CdSe core and the ZnS shell during the surface passivation procedure.

Using the synthetic methods described in the previous section, several core–shell semiconductor nanoassemblies have been prepared since the CdS/Cd(OH)$_2$ CdS/Ag$_2$S [67], CdS/ZnS [68], CdS/HgS [69], CdSe/ZnS [70–73], CdSe/CdS [50], CdTe/CdS [17, 38, 61]. The capping of II–VI nanocrystals with long chain organic surfactants was also utilized to passivate surface atoms, but at room temperature the luminescence quantum yield was as high as 10% with a very long fluorescence lifetime and some non-band edge luminescence [74, 75].

The deposition of a second layer onto the particle's surface represents the chemical growth of another crystalline phase. Two main problems may arise during this process: (i) formation of single particles in the colloidal suspension instead of the second layer growth and (ii) imperfect growth patterns or even alloying of the second layer due to a large or too small crystal mismatch, respectively. This last structural problem either may lead to unstable chemical assemblies presenting a large quantity of surface defects or may alter the original optical properties of the core itself.

The first difficulty is overcome by controlling the experimental parameters during the formation of the second layer. Very low concentrations of the capping precursors and a fast crystallization rate are recommended. Dabbousi *et al.* [71] reported that the growth of ZnS shell on CdSe particles was accomplished, without the precipitation of ZnS, by dropwise adding diethylzinc and hexamethyldisilathiane (as Zn and S precursors) in equimolar amounts to vigorously stirred CdSe colloidal solutions held at temperatures between 140 and 220°C, depending on the core sizes. ZnS has a wider band gap than CdSe (E_g = 3.91 eV and 1.71 eV, respectively [76, 77]) and, in fact, the authors report an enhancement of ~10 to >50% of the quantum yield of CdSe nanocrystals after this passivation procedure. The thickness of the capping layer was monitored by small angle X-ray scattering (SAXS) measurements indicating 0.65 monolayers to 5.3 monolayers. The authors also suggest, by correlating the structural and optical properties, that the ZnS layer tends to generate structural defects when more than 1.3 monolayers are grown, decreasing the luminescence quantum yield.

The second difficulty is the possible chemical alloy between the core and shell layers which would influence the core optical properties. By choosing capping layers with different crystal structures or even with same crystal structure but possessing different bond lengths one prevents this possibility. For example, the growth of wurtzite-type CdS on wurtzite CdSe nanocrystals shows a lattice mismatch of 3.9% which is small enough to allow epitaxial growth while still preventing alloying. The band gap of CdS (E_g = 2.50 eV [76, 77]) is greater than the CdSe band gap (E_g = 1.71 eV [76, 77]) promoting a quantum yield enhancement [50]. On the other hand, in CdSe/ZnS core–shell nanostructures the Zn–S bond length is about 12% larger than Cd–Se [77]. This chemical mismatch that would prevent the growth of flat heterostructures in epitaxial growth is believed to be relaxed in nanocrystals with short facets [71] explaining the chemical stability and excellent optical properties of these core–shell nanoparticles. Regarding the optical properties of the core after passivation, several reports on different core–shell systems show that there is no observable modification of the original properties of the core semiconductor nanocrystal. As an example, Fig. 26.7a shows the absorption spectra of CdS nanoparticles of $d \sim 6\pm$ nm prior and after passivation with a Cd(OH)$_2$ layer.

In this methodology the surface of the CdS nanocrystals is loaded with hydroxyl groups by increasing the pH of the suspension up to 10.5 [28, 78], followed by a dropwise addition of a low-concentration solution of Cd^{2+}. The increase of the fluorescence intensity may be monitored

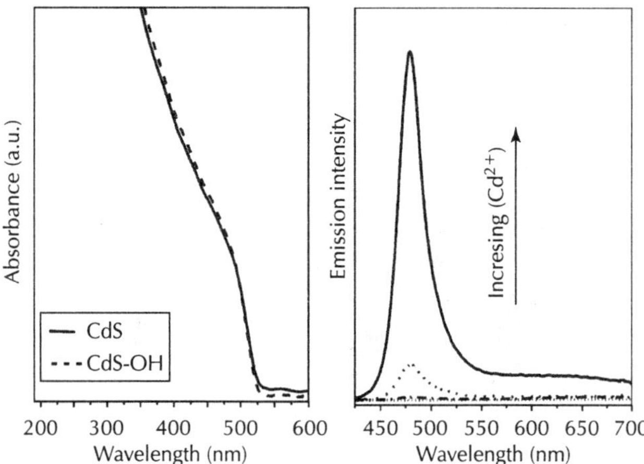

Figure 26.7   (a) Absorption spectra of a polyphosphate stabilized CdS (~6 nm) suspension prior to and after the passivation procedure. (b) Monitoring the emission intensity increase as the cadmium hydroxide layer is being deposited onto the particles's surface.

by emission spectroscopy (as observed in Fig. 26.7). The quantum yield increases to ~20% and the mechanism suggested is that the excess of OH^- will convert SH^- surface states into S^{2-}, which in turn will bind to the excess Cd^{2+} ions. The cadmium-rich surface will then associate to the polyphosphate anions present.

The passivation on the thiol-capped cubic CdTe nanocrystals prepared in aqueous solution [61, 79–81] was brought up in a different form. It was mentioned in the previous section that the photoluminescence of the colloidal suspensions of CdTe nanoparticles (2–6 nm) developed slowly after its synthesis take up to several days to become readily observable even in daylight. This suggested a slow kinetic mechanism of surface defect supression. Rockenberger *et al.* [79] studied these systems using extended X-ray absorption fine structure measurements (EXAFS) and observed that the particles possessed CdS bonds although they were larger than the bulk value for the CdS bond. Later Bouchert *et al.* [81] and Zhang *et al.* [61] investigated and reported on the nature of the surface of these particles by photoelectron spectroscopy (XPS). Zhang *et al.* suggested that the carboxyl groups of the mercapto-carboxylic acid stabilizers coordinated to the surface of the CdTe particles have a great influence on both their photoluminescence and stability by substituting some Te^{2-} ions with oxygen of the carboxyl group. They also observed CdS bonds, which they suggested result from the decomposition of the thiol groups. On the other hand, Bouchert *et al.* showed strong evidence for Cd–SR bonds near the surface and no indications for the presence of Te atoms at the surface. They suggested that this was the result of the hydrolysis of the thiol molecules which ended up incorporated onto the particle's surface, substituting Te^{2-} ions and satisfying Cd^{2+} dangling bonds. Figure 26.8 shows a structure model of the CdS layer growing on the surface of the cubic CdTe particle.

Thinking of a structural match one may observe that both the core and the shell possess cubic crystalline structures. A reasonable extension of the thiol hydrolysis process would end up creating a CdS shell, which in turn possesses a higher band gap, and would be responsible for the passivation of the CdTe surface defects.

Once the particles are optimized to show a high fluorescence quantum yield they may be turned into fluorescent biolabels by adapting their surface to be (i) water soluble and (ii) biocompatible. In the next sections further modifications of the nanocrystal's surface are discussed.

26.2.3   How to render QDs soluble (hydrophilic surface) for biological applications. Solubilization of the nanoparticles

In order to interact with biological systems there are two forms for the QDs to present themselves: solubilized in aqueous medium or encapsulated inside lipophylic carrying systems, such

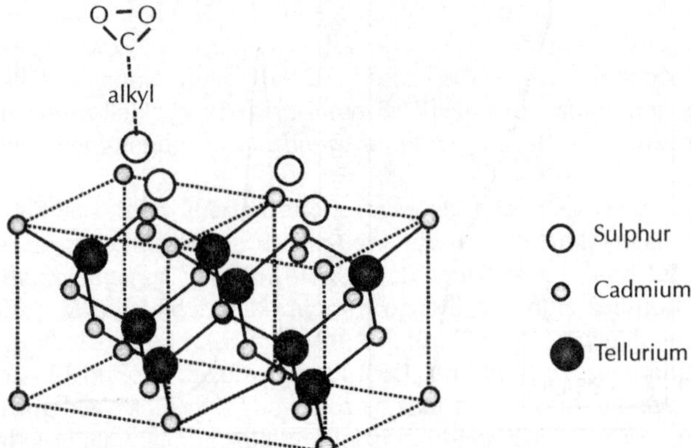

Figure 26.8   A schematic model of a cubic CdS layer growing on the surface of a cubic CdTe lattice from the hydrolysis of the thiol alkyl carboxilate capping molecules.

as micelles. Although the general properties of the nanocrystalline surface appear to be understood, the exact surface chemistry involved in such processes remains to be elucidated.

While the QDs prepared in aqueous media such as the polyphosphate-and/or thiol-capped CdS, CdSe and CdTe nanocrystals have inherent solubility, QDs obtained by the organometallic route must overcome the aqueous incompatibility problem by altering the nature of their surface. Among the different techniques developed for surface modification, the four most general methods described in the literature [82] are: the surface-exchange procedure, the silanization of the nanocrystals, the formation of a hydrophilic interface with amphiphilic molecules and the micellization of the nanocrystals. Figure 26.10 represents three of these methods.

The surface exchange of the hydrophobic surfactant molecules (e.g. TOPO) for bifunctional binding water-soluble molecules [9, 83] has been employed using a wide variety of species. These bifunctional linkers (such as HS–alkyl–COOH molecules) act in similar fashions to solubilize the phosphine-capped nanocrystals and to provide functional groups (e.g. carboxylic acid and amine) which may be used, in a posterior step, to conjugate to biomolecules using well-established protocols. This capping procedure yields gram quantities of water-soluble QDs, but the thiol ligands are not completely stable. Slow desorption of thiol–alkyl–acid molecules and the possibility of intermolecular reactions resulting in disulphides decreases the capping layer and leads to the flocculation of the particles [82]. Moreover, these particles show an overall decrease in their fluorescence quantum yields [84].

Another approach is the silanization with organosilicone molecules containing $-NH_2$ or $-SH$ functional groups [8, 84–88]. A silica/siloxane shell is formed on the surface by the induced hydrolysis of the silanol groups. Polymerizing silanol groups help to stabilize the nanocrystals against flocculation but only small amounts (milligram quantities) can be prepared per batch [24]. Also, residual silanol groups on the nanocrystal's surface often lead to precipitation and gel formation at neutral pH.

A third possibility is to apply phase-transfer methods using amphiphilic molecules that act as detergents for solubilizing the QDs coated with hydrophobic groups [17, 26]. This method has been particularly advantageous in allowing the retention of the native surfactant molecules, which appear to increase the stability and fluorescence efficiency over those samples where the native layer has been replaced with a bifunctional binding molecule. Encapsulating QDs and their initial ligands with macromolecules such as polymers or lipids can preserve the emission quantum efficiency, but generally adds a large volume to the nanocrystals, resulting in a final size that may be greater than desired. This may diminish imaging sensitivity by decreasing the number of QDs that can be attached to a target. For *in vivo* imaging, bulky nanocrystals may have limited accessibility to target systems.

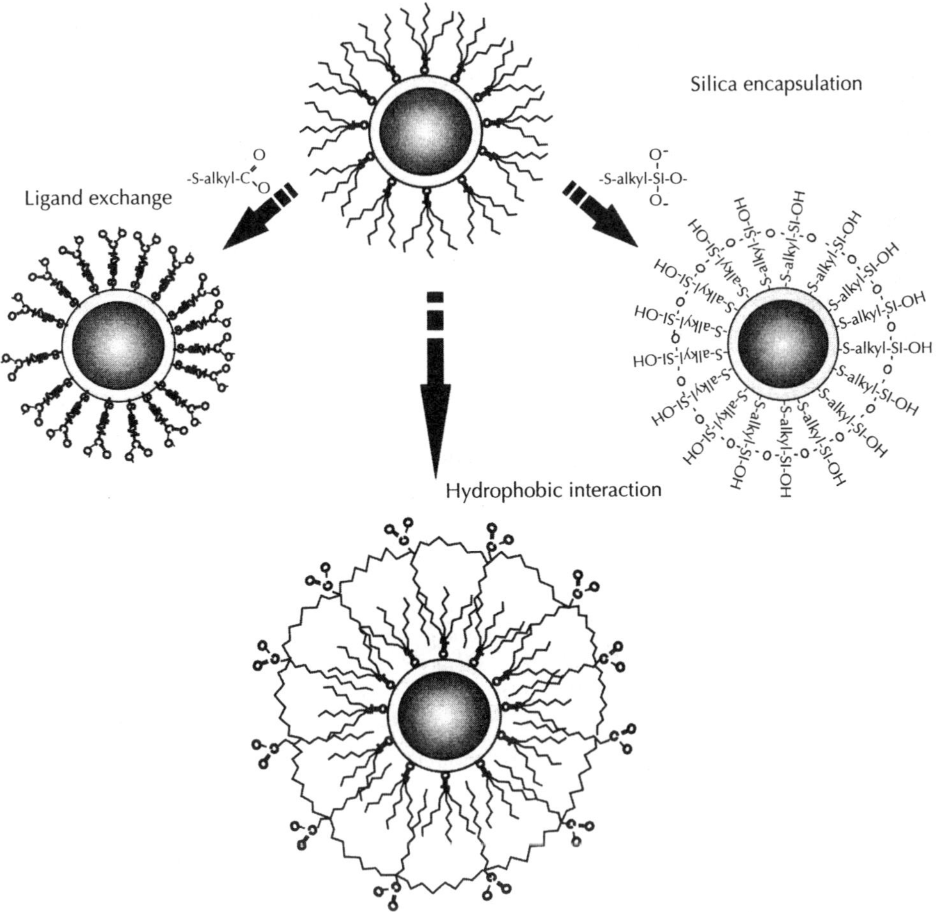

Figure 26.9   Representation of three different experimental modifications described in the literature to produce water-soluble QDs.

Two different solubilization approaches based on the micellization of the hydrophobic nanostructures were recently reported by Dubertret and collaborators and also by Korgel and Monbouquete. By encapsulating individual CdSe/CdS QDs in phospholipid block–copolymer micelles Dubertret *et al.* overcame the solubility problems and also demonstrated a longer stability of the particles in a living body [89]. Moreover, by adapting DNA to the nanocrystal micelles, these systems acted as *in vitro* fluorescent probes to hybridize to specific complementary sequences. Korgel and Monbouquete successfully used phosphatidylcholine vesicles to prepare mixed core and layered (Zn,Cd)S and (Hg,Cd)S nanocrystals [90].

26.2.4   How to target the QDs to biological systems. The functionalization step

The synthesis, passivation and stabilization steps for the obtention of core–shell fluorescent semiconductor quantum dots prepared for biological labelling are generally succeeded by a functionalization procedure. The term *functionalization* (also known as organic capping and ligand conjugation) refers to the chemical modifications of the quantum dot's surface to render biocompatible systems feasible for binding with specific biomolecules which will target the QDs to specific sites in the biological systems. This process is thermodynamically favoured due to the highly active chemical surfaces of the nanocrystals, either presenting dangling bonds at the passivation shell or active binding groups in the stabilizing species.

There are several different experimental routes for this functionalization [91–94]. The *functionalizing* agents may be small organic molecules (e.g. thiol molecules [38], nucleotides [95, 96], carbohydrates [97], polyphosphates [98]), or larger biomolecules (e.g. proteins [7, 99], biotin [100], avidin [101], polyethylene-glycol [102], streptavidin, RNA [63], DNA [103], glucose oxidase and IgG). Usually the combination of small and large molecules will promote adequate modifications on the inorganic QDs' surfaces, providing functional groups that will intermediate the QD/cell interaction [104]. For a schematic illustration of a functionalized core–shell semiconductor quantum dot, see Fig. 26.10. Figure 26.11 represents, as an example, the reaction steps involved in the obtention of fluorescent CdS/Cd(OH)$_2$ QDs, functionalized with polyethylene-glycol (PEG), a biocompatible polymer.

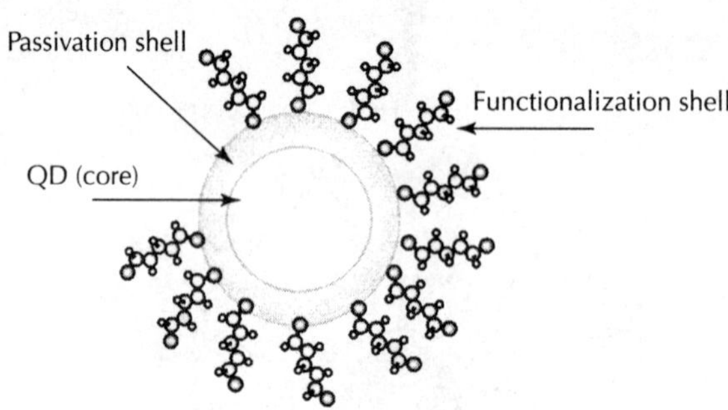

Figure 26.10   *A functionalized core–shell quantum dot is schematically represented here.*

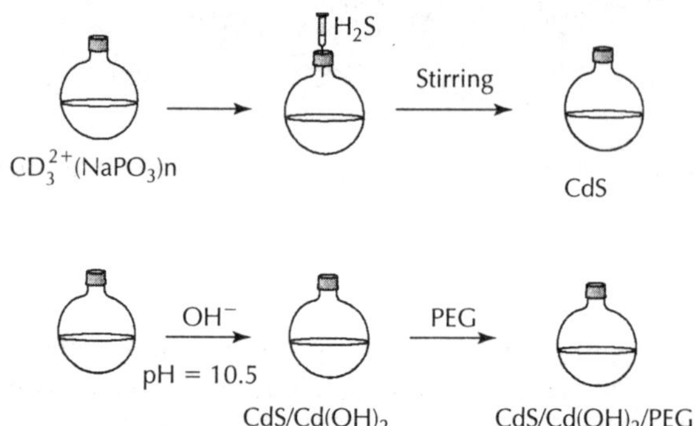

Figure 26.11   *Synthesis, surface passivation and functionalization: reaction steps involved in the obtention of fluorescent CdS/Cd(OH)$_2$ QDs, functionalized with polyethylene-glycol (PEG).*

In order to promote an adequate functionalization procedure, every single modification performed in the colloidal suspension conditions (e.g. temperature, pH, composition) has to take into account the fact that the colloidal suspension must remain stable. Another important feature which cannot be neglected is the maintenance of the QDs characteristic luminescence. Moreover, the choice of the adequate functionalizing agent is someway driven by the desired QD function/application in the bio-system.

More complex functionalization procedures may employ conjugation protocols adapted from those used in the binding of biomolecules and organic fluorophores. The functionalization of proteins (which represents the chemical nature of a great amount of biofunctional molecules, such as antibodies, antigens, enzymes, growth factor molecules, etc.) to QDs may be performed in different forms, which will depend on the protein type and the nature of the nanocrystal's capping surface. Some described methods are: (i) the condensation of amine groups of the protein to carboxyl groups of the QDs' surface by using EDC, 1-ethyl-3-(3-dimethylaminopropyl) carbodiimide as a chemical activator; (ii) direct binding to the QD surface using thiolated peptides or polyhistidine (HIS) residues; and (iii) adsorption or non-covalent self-assembly using engineered proteins [91, 105].

Optimal functionalization techniques should also be able to keep the quantum dots' fundamental properties for periods of time longer than several weeks. An example described in the literature refers to "engineered-protein" functionalized quantum dots which retain their quantum efficiency and offer longer shelf life [91]. These systems were also further functionalized with multiple functional groups without decreasing their quantum efficiency [93].

It has been shown that the optical properties of semiconductor nanocrystals are very sensitive to certain capping molecules attached to their surfaces. Some stabilizing agents or hydrophobic surfactant molecules used in the solubilization procedure decrease the overall fluorescence quantum yield and the explanation for this effect remains in the "electronic blinding capacity" of the passivation shell [61, 92, 106, 107]. This has been demonstrated in the case of mercaptoacetic acid-treated QDs where the quantum efficiency was drastically reduced [93, 94]. However, for some probing applications this feature is taken as an advantage, especially if the change in fluorescence is generated by resonance energy transfer (FRET) to target analyte molecules [91]. This opens up other applications in biosensing, where the extension of this FRET process can be controlled by the thickness of the passivation shell and the distance to the binding species.

The choice of the functionalizers will provide low or high specificity concerning the association of the functionalized QDs with biomolecules. As already mentioned in the previous section, thiol–alkyl–COOH ligands (such as mercapto-acetic acid) used as stabilizing agents in the synthesis of CdTe QDs in aqueous medium also act as functionalizing agents and provide surface carboxylate groups, making the QDs feasible to react with primary amines expressed by cell surface proteins. This kind of reaction produces relatively stable amides [107]. Glutaraldehyde (at very low concentrations), thiol-containing molecules and polyethylene-glycol can be cited as efficient functionalizing agents, in an increasing order concerning specificity in the bioconjugation process [91].

26.2.5 Bioconjugation of the QDs

The term *bioconjugation* is used to denominate the interaction between the functionalized quantum dots and the molecules present in a biological system. A more general definition includes also the conjugation of functionalized QDs with biomolecules produced by the bio-system and dispersed in a liquid medium (e.g. the conjugation of circulating antibodies in the human serum). As mentioned previously, QDs are prepared to bind chemically (covalent bonds) or physically (adsorption phenomena) to molecules present in the bio-system. The nature of the functionalizing groups will, in theory, define the nature of the interaction with the biological system. For instance, if the target biomolecule is expressed in the cell membrane then most probably a superficial labelling will be observed.

Depending on the kind of interaction observed one may identify different internalization mechanisms related to the bioconjugation process. In *specific biomolecules recognition* the QDs will bind specifically to the target molecule [94, 95]. Biotinylated antigens, for example, interact strongly with QDs functionalized with avidin or streptavidin molecules [94, 99] rendering great specificity labelling.

The QDs may be functionalized with molecules that interact specifically to species presented in *distinct cellular organelles/structures*, this is exemplified by the cell cytoskeleton microtubules labelled with CdSe/ZnS quantum dots functionalized with anti-tubulin [92]; another example is the CdS/Cd(OH)$_2$ polyphosphate stabilized QDs which were directed to high consuming energy internal sites of living leishmania parasites [98].

When the functionalizing groups are not specific, the interaction may occur indistinctively with all the similar molecules and in this case the QDs are internalized by a natural process known as endocytosis and end up bound to *internal cell compartments* [13, 109]. Endocytosis is the process whereby cells engulf vitamins and nutrients from their outside surroundings. The nanocrystals can also be artificially internalized by inserting them using micromanipulators [91]. In fixed cells the membrane pores are increased and the QDs can pass through these channels easily without any other kind of internalization.

Various covalent and non-covalent strategies employing crosslinkers [7, 13, 91–94, 109–115] have been developed for conjugating biomolecules to the QDs' surfaces. Biomolecules can be linked to the QDs via functional groups such as –COOH, –NH$_2$ or –SH, which may be present at their modified surface. These groups may be provided by functionalizing compounds such as protein A, avidin, streptavidin, glucose oxidase horseradish peroxidase, and IgG [13, 114–116]. Figure 26.12 summarizes the steps discussed up to this section, concerning the preparation and use of fluorescent semiconductor quantum dots for biological labelling purposes.

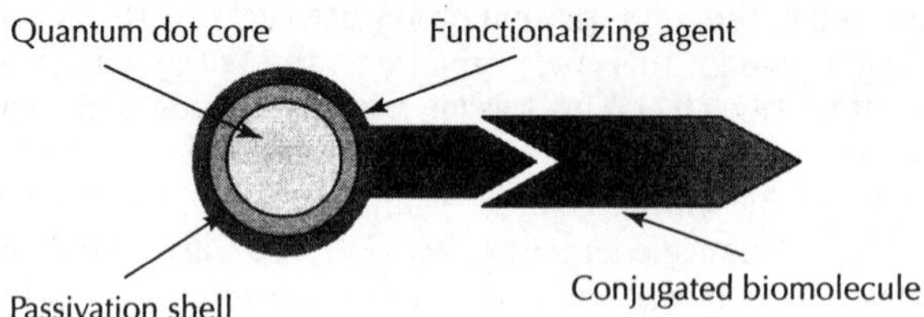

Figure 26.12   A schematic representation of a passivated, functionalized and bioconjugated quantum dot.

For living cells labelled with quantum dots, the bioconjugation step may be processed into the cell culture medium [7, 13] or even in another appropriate medium, such as saline solution. The resulting systems to be used for bioconjugation must be compatible with cellular/tissular conditions, i.e. they must be used in physiological pH, present low toxicity, be isotonic with the intracellular (or intra-tissular) medium, and the resulting QDs must remain hydrophilic and retain their characteristic fluorescence.

In the next section some applications of QDs in biolabelling procedures are presented.

26.3   Applications

Since the first reports published in 1998 [8, 94] concerning the application of quantum dots as biolabels, much progress has appeared in this field. There are several reports which indicate the fast improvement on conjugating these nanostructures to biological molecules, cells, microorganisms and tissues. It is worthwhile mentioning some representative examples:

- Multiplexed biological detection and imaging [24, 117, 118].
- Sensing trace analytes in biological samples [119].
- Tools for rapid and sensitive diagnosis of viruses [120].
- Molecular characterization in combination with optical imaging to detect the progression of precancerous lesions [121].
- Nanoparticles for targeting *in vivo* [122].
- Cervical, breast and brain cancer diagnostics [13, 109], as well as many other applications in life sciences.

Multiplexed biological detection and imaging, using the unique properties of photostability and especially the narrower emission spectra of QDs, have been applied for fluorescence *in situ* hybridization (FISH) to study molecular biology. For instance, Chan *et al.* used FISH and QDs for monitoring mRNA transcripts [107]. This work demonstrated an increased sensitivity of FISH using QDs in comparison with organic fluorophores which can facilitate the ultrasensitive simultaneous study of multiple mRNA and protein markers in tissue cultures and histological sections.

The feasibility of *in vivo* targeting by using QDs has being explored, for instance, by Akerman *et al.* using ZnS-capped CdSe nanoparticles [11]. They showed that ZnS-capped CdSe QDs coated with a lung-targeting peptide accumulate in the lungs of mice after i.v. injection. Two other kinds of peptides direct the QDs specifically to blood and lymphatic vessels in tumours. They also show that QDs with polyethylene glycol prevent non-selective accumulation of QDs in reticuloendothelial tissues. These studies encourage the synthesis of nanostructures associated with drug delivery systems.

After this brief overview of biological applications of quantum dots, this section will now focus on and discuss some results obtained by our research group: (i) determination of the antigen-A expression in living red blood cells; (ii) obtention of non-linear microspectroscopy in an optical tweezers system – application to living macrophage cells marked with quantum dots – and (iii) cancer diagnostics.

26.3.1   CdS quantum dots for red blood cells antigen – A labelling

Core–shell $CdS/Cd(OH)_2$ quantum dots obtained in aqueous medium were successfully used as efficient fluorescent labels for living human red blood cells. The aim of this investigation was to precisely determine the antigen-A expression in subgroups of group A erythrocytes.

Luminescent CdS nanoparticles were functionalized via a one-pot cross-linking glutaraldehyde procedure. These functionalized nanoparticles were conjugated to monoclonal antibody anti-A for 5 hours. Living human red blood cells, before the contact with QDs, were diluted in 0.9% saline solution, centrifuged and separated from the liquid phase. The resulting conjugate' QDs/anti-A were incubated with human erythrocytes of blood groups A^+, A_2^+ and O^+ for 30 minutes at 37°C. Prior to visualization, the samples were centrifugated for 2 minutes (3000 rpm) in saline buffer solution and washed. A schematic diagram of the conjugation of the QDs/anti-A as well as of their interaction with red cell membrane antigen is depicted in Fig. 26.13. Functionalization and conjugation steps were performed in an aqueous medium at physiological pH.

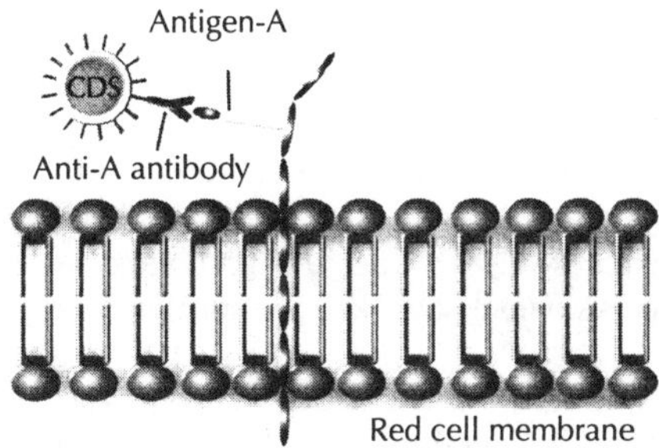

Figure 26.13   Schematic diagram for the specific conjugation QDs/anti-A interaction with red cell membrane antigen-A.

The cells conjugated with the quantum dots were characterized by confocal laser scanning microscopy. The conjugates' QDs/anti-A intensely marked group A erythrocytes, showing different intensities of luminescence for the A_2 group investigated, and did not show any luminescence for group O erythrocytes [7]. Figure 26.14 shows respectively a confocal image obtained for QDs/anti-A marking A^+ erythrocytes and for QDs/anti-A marking O^+. The lack of emission of the type O^+ erythrocytes is explained by the absence of anti-A binding, indicating the absence of antigen-A.

These results show the high potential for the use of $CdS/Cd(OH)_2$ semiconductor luminescent nanoparticles as fluorescent labels for red cells. The specificity of the QDs' conjugation in the erythrocyte cell membrane opens up the possibility of using this methodology as a quantitative

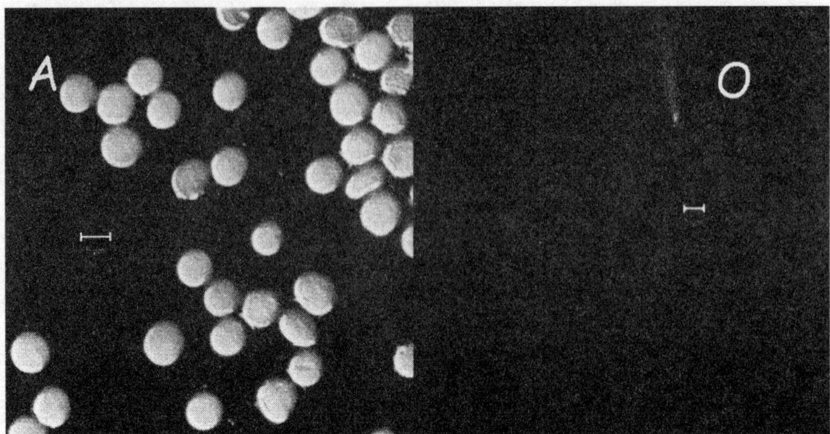

Figure 26.14   Microscopic confocal images obtained for QDs/anti-A marking living A^+ (left) and O^+ (right) erythrocytes.

tool to investigate the distribution and expression of alloantigens in red blood cells. The results obtained show the high potential presented by these new fluorescent labels for living cells.

26.3.2   Non-linear microspectroscopy in an optical tweezers system – application to cells marked with quantum dots

Optical tweezers have been used as a tool to manipulate and measure mechanical properties at the cellular level [123]. Adding linear and non-linear spectroscopic capacity to the optical tweezers would allow one to observe real time biochemical reactions of captured living cells. The dynamic, mechanical and spectroscopic information of triggered events can clarify biological events such as cell infection by parasites. Here we present a set-up consisting of an optical tweezers combined to a non-linear microspectroscopy system that was used to perform scanning microscopy and observe spectra from two-photon excited (TPE) luminescence of trapped living single cells conjugated with quantum dots (CdS and CdTe) [124, 125]. Confocal microscopy using TPE luminescence of living cells with infrared lasers has the advantages of high resolution in the vertical direction, the absence of damage caused by heating and larger light penetration revealing underneath structures [126]. Using this combined set-up it is also possible to obtain images and spectra of hyper-Rayleigh also known sometimes in the literature as second harmonic generation (SHG).

The combined system used a cw Nd:YAG laser (for trapping) and a femtosecond Ti:sapphire laser (for TPE luminescence and hyper-Rayleigh). The laser beams were focused through 100× oil immersion in a microscope. The back-scattered spectroscopic signals were collected with the 30 cm monochromator-equipped CCD. For TPE luminescence and hyper-Rayleigh a short pass colour filter was used to transmit the visible and cut the infrared. In these first experiments image scanning was performed with a mechanical translation stage.

The luminescent CdS nanoparticles were functionalized with glutaraldehyde solution while the CdTe QDs were functionalized with mercaptoacetic acid (AMA). The CdTe and also CdS were obtained using synthetic routes in aqueous medium which produce quantum dots with hydrophilic surface in a water solution [7, 17]. After testing the performance of the system, we obtained TPE luminescence spectra and also hyper-Rayleigh images and spectra for trapped and non-trapped cells and particles. Figure 26.15 shows the image from TPE luminescence of macrophages conjugated with CdS QDs.

It was shown that a system of optical tweezers combined with non-linear spectroscopy is capable of analysing trapped single particles and living cells from the visible to the infrared, and also to perform spectroscopic image reconstruction. The acquired spectra and images include

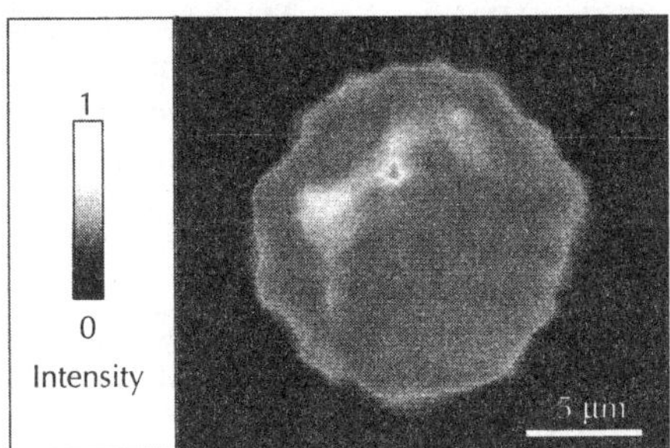

Figure 26.15   Image from TPE luminescence of a living macrophage cell conjungated with CdS/Cd(OH)$_2$ QDs.

hyper-Rayleigh and TPE luminescence of trapped and non-trapped particles. This system demonstrated that it is possible to trigger and observe chemical reactions and mechanical properties in real time of living trapped microorganisms in any neighbourhood in combination with quantum dot fluorescent markers which show several advantages over the conventional fluorophores.

26.3.3   Quantum dots as fluorescent bio-labels in cancer diagnostics

Quantum dot applications in the investigation of neoplastic processes (which may give rise to cancer) constitute a topic of interest with many questions still waiting for precise answers. In the pursuit of sensitive and quantitative methods to detect and diagnose cancer, nanotechnology has been identified as a field of great promise. Hydrophylic quantum dots at physiological pH conditions have the potential to expand conventional protocols used for cancer diagnostics, which need previous tissue/cell fixation, and extend to investigate living cellular and tissular neoplastic mechanisms in real time.

Some results concerning the application of water-soluble colloidal semiconductor quantum dots for the purpose of diagnostics in living cells are presented here. The fluorescence was used as a primary tool in order to explore and differentiate the labelling of the samples. Tissues and cells conjugated with QDs were analysed by laser scanning confocal microscopy. All the images were collected with the same acquisition parameters for comparison. In order to confirm the presence of QDs inside the cells, some of the conjugated systems were also characterized by transmission electronic microscopy (TEM). This kind of measurement complements the fluorescence analysis, as they show where the QDs were internalized in the cell. The images obtained show that the nanocrystals accumulate near the nuclear envoltorium. Figure 26.16 shows a representative

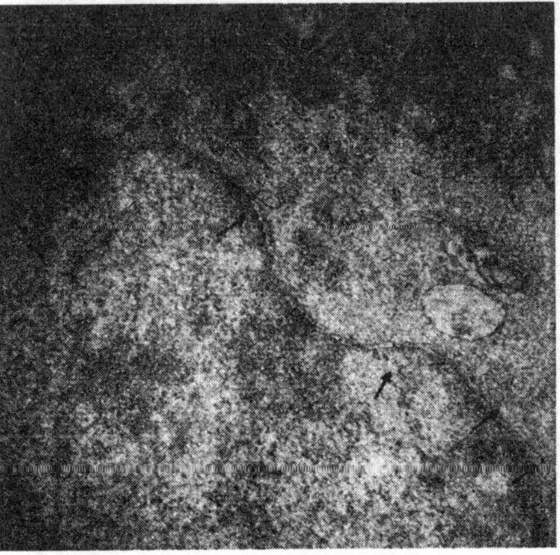

Figure 26.16   Transmission electronic microscopy image of a glioblastoma-labelled cell, in which the highest QDs concentration is nearby the nuclear envoltorium (arrows).

TEM image of core–shell CdS/Cd(OH)$_2$ quantum dots functionalized with glutaraldehyde (QD/Glut) and conjugated *in vitro* with live human glioblastoma cells. The cells showed no signs of damage after the conjugation procedure and maintained their integrity even after 5 days of incubation time, demonstrating the low toxicity of the QDs for *in vitro* studies.

In the first application, the quantum dots (CdTe/CdS or CdS/Cd(OH)$_2$ at physiological pH) were functionalized with a 0.01% glutaraldehyde solution and then incubated with living healthy and neoplastic cells (glial, glioblastoma and cervical) and tissues (breast) in culture medium [13]. For the purpose of diagnostics with living cells, the CdS/Cd(OH)$_2$ presented the best results, maintaining high levels of luminescence as well as high stability in biological media. The measurements were performed for different time intervals in order to monitor the time evolution of the interaction between the QDs and the cells. Figures 26.17 and 26.18 show confocal microscopy images and the corresponding fluorescence intensity maps for the time evolution (1–3 min incubation time) of the interaction of healthy and neoplastic glial cells. In the fluorescence intensity maps, the dark grey regions correspond to the absence of fluorescence, while the lighter but more intense grey corresponds to regions of highest fluorescence intensities, respectively.

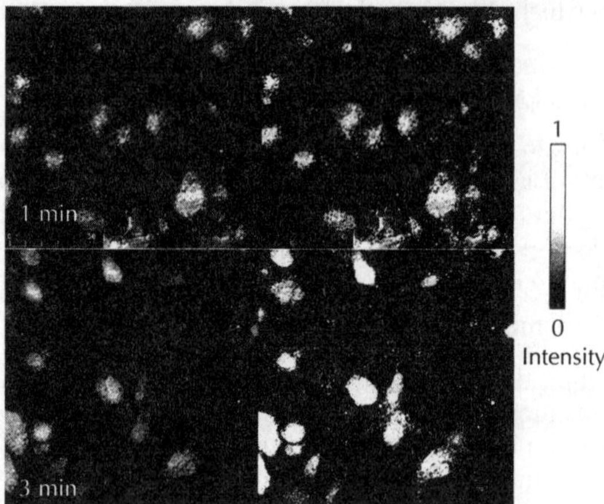

Figure 26.17   Time evolution of the fluorescence pattern of healthy glial cells incubated with QDs/Glut. Confocal microscopy images (left) and the corresponding fluorescence intensity maps (right).

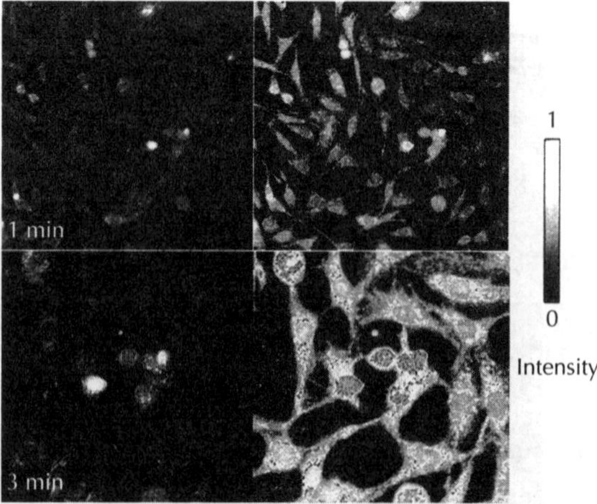

Figure 26.18   Time evolution of the fluorescence pattern of neoplastic glial (glioblastoma) cells incubated with QDs/Glut. Confocal microscopy image (left) and the corresponding fluorescence intensity maps (right).

For all kinds of cells analysed, the time evolution of the interaction clearly reveals different labelling patterns as well as different fluorescence intensities. It can also be noticed that the QD/Glut easily interacts with both healthy and neoplastic cells. As the glutaraldehyde is a homofunctional bidentade ligand, it establishes hemi-acetal interaction with the QDs' outer shell at the same time as it binds to cell proteins by Schiff's base interactions.

As a final application, we report the use of QDs/Glut conjugated to the concanavalin-A lectin (Con-A) to investigate cell alterations regarding carbohydrate profile in human mammary tissues diagnosed as fibroadenoma (a benign type of mammary tumour). The Con-A lectin is a biomolecule which binds specifically to glucose/mannose residues present in the cellular membrane. The bioconjugated particles were incubated with normal tissue sections of fibroadenoma tissue sections. The tissue sections were deparafinized, hydrated in graded alcohol and treated with a solution of Evans Blue (EB) in order to avoid autofluorescence.

One class of targeting biomolecules that is commonly used with organic fluorophores are the lectins. Lectins are structurally diverse carbohydrate-binding proteins of non-immune origin that agglutinate cells and recognize carbohydrates in oligosaccharides and glycoconjugates. They have been used in the medical and biological fields. In histochemistry, lectins with different carbohydrate specificity can provide a sensitive detection system for changes in glycosylation and carbohydrate expression that may occur during embryogenesis, growth and disease. Tumour lectinology has so far shown cytochemical and histochemical differences between normal and transformed tissues such as mammary [127, 128] and brain [129]. Quantitative and qualitative changes in glycoconjugates of cell membranes play significant roles in the development and progression of pathologies, including neoplasias. Higher or weaker and even the absence of staining patterns between normal and transformed tissues suggests a dearrangement of secretory mechanisms. The observation that, in general, the more anaplastic the cell becomes the more intense is its staining, seems to indicate that the site and nature of cell surface glycoconjugates are altered. In addition, tissue factors may influence and induce differentiation/dedifferentiation reflected in the different lectin binding patterns [128]. Figure 26.19 depicts schematically how the bioconjugate QD–lectin interacts with the carbohydrate residues of the glycoprotein on the cell membrane.

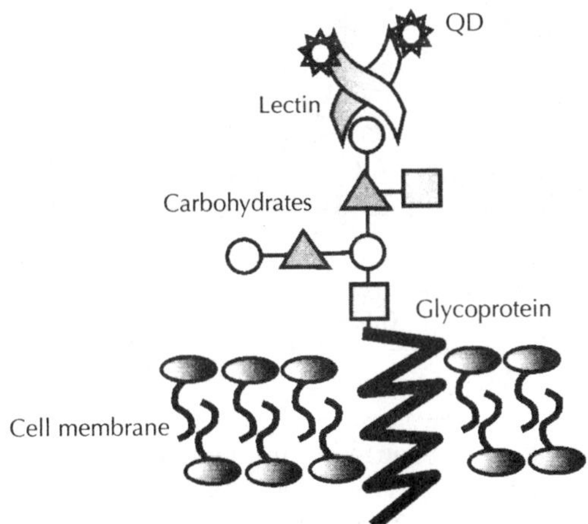

Figure 26.19   Scheme of the cell membrane carbohydrate residues labelled with QDs-lectin.

In the bioconjugation of quantum dots with concanavalin A, the pH of the suspension containing the quantum dots was decreased to 7.2 with 0.1 M HClO$_4$. QDs were functionalized with a 0.1% glutaraldehyde solution (100 μL) to 5 mL of QD suspension (QD/Glut). Finally, Con-A (Sigma, 1 mg/mL) was incubated with QD/Glut (10^{13} nanoparticles/mL) for 2 h at 25°C in 100 mM phosphate buffer pH 7.2 containing 150 mM (w/v) NaCl (reaction buffer). The specimens used were, mammary normal and transformed tissues, diagnosed as fibroadenoma, obtained from the Tissue Bank of the Sector of Pathology of the Keizo Asami Immunopathology Laboratory at the Federal University of Pernambuco, Brazil. In lectin histochemistry, sections (4 μm) of specimens were deparaffinized in xyline and hydrated in graded alcohols (100–70%). Tissue slices were treated with a 0.1% (w/v) trypsin solution for 2 min at 35°C and then incubated with a solution of Con-A–QD (50 μg/mL) and QD/Glut, prepared or not with Evans Blue solution (5 mg%).

The samples were prepared in sets described as follows: (1) NEB (normal tissue sections treated with EB); (2) NEB–Con-A–QD (normal tissue sections treated with EB and conjugated to CdS/Cd(OH)$_2$/Glut/ConA); (3) FIB–EB (fibroadenoma tissue sections treated with EB); (4) FIB–EB–QD (fibroadenoma tissues treated with EB and conjugated to CdS/Cd(OH)$_2$/Glut QDs); and (5) FIB–EB–Con-A–QD (fibroadenoma tissues treated with EB and conjugated to CdS/Cd(OH)$_2$/Glut/ConA QDs). Analysing the confocal microscopic images of the fibroadenoma and normal tissue samples in the 500–535 nm range the following observations were made: (a) FIB–EB tissues presented a very faint luminescence; (b) FIB–EB–QD tissue samples showed a slightly bright luminescence in the green region. This emission showed a homogeneous pattern which is expected for a non-specific labelling which can be the result of the occurrence of covalent attachment between glutaraldehyde and the amino acid residues, such as lysine in cell proteins, through Schiff bases; (c) in FIB–EB–Con-A–QD tissue samples all the samples where brightly luminescent and showed a specific labelling corresponding to the internal structures of the mammalian ducts; (d) norm-EB tissue samples presented no detectable luminescence; (e) NEB–Con-A–QD tissue samples showed luminescence and also showed selected structures brighter than the overall tissue sample (data not shown). Figure 26.20 presents the fluorescence pattern of the normal and fibroadenoma tissues obtained under the same acquisition parameters.

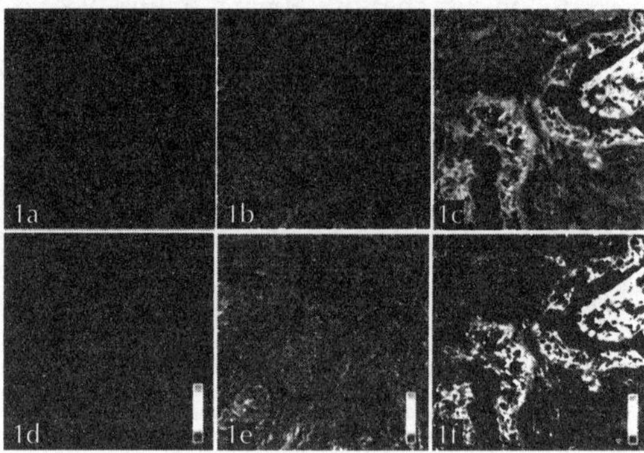

Figure 26.20   Human mammary tissue diagnosed as fibroadenoma (Fib). (1a) Fib treated with Evan's Blue solution (EB); (1b) Fib incubated with QDs diluted in EB (FIB–EB–QD); (1c) Fib incubated with Con-A–QD conjugate diluted in EB (FIB–EB–Con-A–QD). (1d), (1e) and (1f) are the intensity maps of (1a), (1b) and (1c), respectively. The bright pattern observed in (1c) is related to regions of high glucose/mannose expression in the tissue.

In summary, the fluorescence intensity of QD–Con-A stained tissues showed different patterns which reflect the carbohydrate expression of glucose/mannose in fibroadenoma when compared to the detection of the normal carbohydrate expression. The pattern of inespecific labelling of the tissues with QD-Glut is compared to the targeting driven by the Con-A lectin. These results, although preliminary, show a very promising tool in the detection of tumour differentiation in human mammary tissues.

It is important to note that in both applications, due to the differences that rise in neoplastic processes (such as metabolic rates and regimen, cell membrane permeability and fluidity), the fluorescence intensities and patterns are quite different for healthy and neoplastic cells. The neoplastic mechanisms almost always result in cancer. These results show that a simple procedure of synthesis, functionalization and incubation with healthy and neoplastic cells of quantum dots at physiological conditions may represent a potential tool for fast and precise diagnostics of different kinds of cancer.

References

1. V. Aroutiounian and S. Petrosyan, *J. Appl. Phys.* **89**, 2268–2271 (2001).
2. A.R. Kovsh *et al.*, *Elect. Lett.* **35**, 1161–1163 (1999).
3. H.C. Liu, M. Gao, J. McCaffrey, Z.R. Wasilewski, and S. Fafard, *Appl. Phys. Lett.* **78**, 79–81 (2000).
4. I. Amlani, A.O. Orlov, G. Toth, G.H. Bernstein, C.S. Lent, and G.L. Snider, *Science*, **284**, 289–291 (1999).
5. G. Jacob, L.C. Barbosa, and C.L. Cesar, *Proceedings of the SPIE*, **5734**, 124–129 (2005).
6. B. Ballou, B.C. Lagerholm, L.A. Ernst, M.P. Bruche, and A.S. Waggoner, *Bioconjug. Chem.* **15**, 79–86 (2004).
7. P.M.A. Farias, B.S. Santos, F.D. Menezes, C.L. Cesar, A. Fontes, P.R.M. Lima, M.L. Barjas-Castro, V. Castro, and R. Ferreira, *J. Microsc.* **219**, 103–108 (2005).
8. M. Bruchez, Jr., M. Morrone, P. Gin, S. Weiss, and A.P. Alivisatos, *Science*, **281**, 2013–2016 (1998).
9. W.C.W. Chan and S. Nie, *Science* **281**, 2016–2018 (1998).
10. X. Gao and S. Nie, *Trends Biotechnol.* **21**, 371–373 (2003).
11. M.E. Äkerman, W.C.W. Chan, P. Laakkonen, S.N. Bhatia, and E. Ruoslahti, *Appl. Biol. Sci.* **99**, 12617–12621 (2002).
12. L.C. Mattheakis, J.M. Dias, Y. Choi, J. Gong, M.P. Bruchez, J. Liu, and E. Wang, *Anal. Biochem.* **327**, 200–208 (2004).
13. P.M.A. Farias, B.S. Santos, F.D. Menezes, R. Ferreira, A. Fontes, H.F. Carvalho, L. Romão, V. Moura-Neto, J.C.O.F. Amaral, C.L. Cesar, R.C.B.Q. Figueiredo, and F.R.B. Lorenzato, *Phys. Stat. Sol. (c)* **11**, 4001–4008 (2006).
14. A.R. Clapp, I.L. Medintz, J.M. Mauro, B.R. Fisher, M.G. Bawendi, and H. Mattoussi, *J. Am. Chem. Soc.* **126**, 301–310 (2004).
15. D.E. Ferrara, D. Weiss, P.H. Carnell, R.P. Vito, D. Vega, X. Gao, S. Nie, and W.R. Taylor, *Am. J. Physiol. Regul. Integr. Comp. Physiol.* **290**, 114–123 (2006).
16. D.R. Larson, W.R. Zipfel, R.M. Williams, S.W. Clark, M.P. Bruchez, F.W. Wise, and W.W. Webb, *Science* **300**, 1434–1436 (2003).
17. F.D. Menezes, A.G. Brasil J., W.L. Moreira, L.C. Barbosa, C.L. Cesar, R. Ferreira, P.M.A. Farias, and B.S. Santos, *Microelectr. J.* **36**, 989–991 (2005).
18. A.l.L. Efros and A.L. Efros, *Sov. Phys. Semicond. (Engl. Transl.)* **16**, 772–775 (1982).
19. L.E. Brus, *J. Chem. Phys.* **79**, 5566–5571 (1983).
20. M.P. Pileni, *Catal. Today* **58**, 151–166 (2000).
21. T. Trindade, P. O'Brien, and N.L. Pickett, *Chem. Mater.* **13**, 3843–3858 (2001).
22. A.C.C. Esteves and T. Trindade, *Current Opinion in Solid State Mater. Sci.* **6**, 347–353 (2002).
23. X. Peng, *Chem. Eur. J.* **8**, 335–339 (2002).
24. W.C.W. Chan, D.J. Maxwell, X. Gao, R.E. Bailey, M. Han, and S. Nie, *Current Opinion in Biotechnol.* **13**, 40–46 (2002).
25. O. Masala and R. Seshadri, *Annu. Rev. Mater. Res.* **34**, 41–81 (2004).
26. G. Schmid, ed. In: *Nanoparticles. From Theory to Applications.* (VCH2004).
27. R. Rossetti, J.L. Ellison, J.M. Gibson, and L.E. Brus, *J. Chem. Phys.* **80**, 4464–4469 (1984).
28. A. Fojtik, H. Weller, U. Koch, and A. Henglein, *Ber. Bunsenges. Phys. Chem.* **88**, 969–977 (1984).
29. J.J. Ramsden and M.J. Gratzel, *Chem. Soc. Faraday Trans. 1* **80**, 919–933 (1984).
30. R. Rossetti, R. Hull, J.M. Gibson, and L.E. Brus, *J. Chem. Phys.* **82**, 552–559 (1985).
31. T. Dannhauser, M. O'Neil, K. Johansson, D. Whitten, and G. McLendon, *J. Phys. Chem.* **90**, 6076–6077 (1986).
32. A.R. Kortan, R. Hull, R.L. Opila, M.G. Bawendi, M.L. Steigerwald, P.J. Carroll, and L.E. Brus, *J. Am. Chem. Soc.* **112**, 1327–1332 (1990).
33. A.L. Rogach, A. Kornowski, M. Gao, A. Eychmüller, and H. Weller, *J. Phys. Chem. B* **103**, 3065–3072 (1999).
34. A.L. Rogach *et al.*, *Adv. Mater.* **11**, 552–555 (1999).
35. M.T. Harrison, S.V. Kershaw, M.G. Burt, A. Eychmüller, H. Weller, and A.L. Rogach, *Mater. Sci. Eng. B* **355**, 69–70 (2000).
36. Y. Nosaka, N. Ohta, T. Fukuyama, and N. Fujii, *J. Coll. Interf. Sci.* **155**, 23–29 (1993).

37. C.B. Murray, D.J. Norris, and M.G. Bawendi, *J. Am. Chem. Soc.* **115**, 8706–8715 (1993).
38. N. Gaponik, D.V. Talapin, A.L. Rogach, A. Eychmüller, and H. Weller, *Nano Letters* **02**, 803–806 (2002).
39. A. Eychmüller, A. Hässelbarth, L. Katsikas, and H. Weller, *J. Lumin.* **48–49**, 745–749 (1991).
40. A.A. Guzelian, U. Banin, A.V. Kadanich, X. Peng, and A.P. Alivisatos, *Appl. Phys. Lett.* **69**, 1432–1434 (1996).
41. M.L. Steigerwald, A.P. Alivisatos, J.M. Gibson, T.D. Harris, R. Kortan, A.J. Muller, A.M. Thayer, T.M. Duncan, D.C. Douglass, and L.E. Brus, *J. Am. Chem. Soc.* **110**, 46–50 (1988).
42. M.G. Bawendi, A.R. Kortan, M.L. Steigerwald, and L.E. Brus, *J. Chem. Phys.* **91**, 7282–7290 (1989).
43. J. Li, Z. Chen, R.J. Wang, and D.M. Proserpio, *Coordination Chem. Rev.* **190–192**, 707–735 (1999).
44. A. P. Cheetham and P. Day. eds, in: *Solid State Chemistry: Compounds*, (Clarendon Press, Oxford, 1992).
45. X. Peng, M.C. Schlamp, V.A. Kadavanich, and A.P. Alivisatos, *J. Am. Chem. Soc.* **119**, 7019–7029 (1997).
46. M.A. Hines and P. Guyot-Sionnest, *J. Phys. Chem. B* **102**, 3655–3657 (1998).
47. X. Peng, J. Wickham, and A.P. Alivisatos, *J. Am. Chem. Soc.* **120**, 5343–5344 (1998).
48. Y. Desmet, L. Deriemaeker, E. Parloo, and R. Finsy, *Langmuir* **15**, 2327–2332 (1999).
49. L. Qu, Z.A. Peng, and X. Peng, *Nano Letters* **01**, 333–337 (2001).
50. X. Peng, M.C. Schlamp, A.V. Kadavanich, and A.P. Alivisatos, *J. Am. Chem. Soc.* **119**, 7019–7029 (1997).
51. M.A. Hines and P. Guyot-Sionnest, *J. Phys. Chem. B* **102**, 3655–3660 (1998).
52. D.V. Talapin, A.L. Rogach, A. Kornowski, M. Haase, and H. Weller, *NanoLetters* **01**, 207–211 (2001).
53. P. Xiaogang, L. Yang, W. Manna, J. Wickham, E. Sher, A. Kadavanich, and A.P. Alivisatos, *Nature* **407**, 981–983 (2000).
54. L. Manna, E.C. Sher, and A.P. Alivisatos, *J. Am. Chem. Soc.* **122**, 12700–12706 (2000).
55. M.B. Mohammed, C. Burda, and M.A. El-Sayed, *Nano Letters* **01**, 589–593 (2001).
56. W. Wang, Y. Geng, P. Yan, F. Liu, Y. Xie, and Y. Qian, *J. Am. Chem. Soc.* **121**, 4062–4063 (1999).
57. W. Wang, I. Germanenko, and M.S. El-Shall, *Chem. Mater.* **14**, 3028–3033 (2002).
58. E. Lifshitz, M. Bashouti, V. Kloper, A. Kigel, M.S. Eisen, and S. Berger, *Nano Letters* **03**, 857–862 (2003).
59. J. Zhu, S.T. Aruna, Y. Koltypin, and A. Gedanken, *Chem. Mater.* **12**, 143–147 (2000).
60. J. Zhu, Y. Koltypin, and A. Gedanken, *Chem. Mater.* **12**, 73–78 (2000).
61. H. Zhang, Z. Zhou, B. Yang, and M. Gao, *J. Phys. Chem. B* **107**, 08–13 (2003).
62. N. Pradhan, D. Goorskey, J. Thessing, and X. Peng, *J. Am. Chem. Soc.* **127**, 17586–17587 (2005).
63. A. Kumar and A. Jakhmola, *Langmuir* **23**, 2915–2918 (2007).
64. L. Levina, V. Sukhovatkin, S. Musikhin, S. Cauchi, R. Nisman, D. Bazett-Jones, and E.H. Sargent, *Adv. Mater.* **17**, 1857 (2005).
65. S. Hinds, B.J. Taft, L. Levina, V. Sukhovatkin, C.J. Dooley, M.D. Roy, D.D. MacNeil, E.H. Sargent, and S.O. Kelley, *J. Am. Chem. Soc.* **128**, 64–65 (2006).
66. G. Rothenberger, J. Moser, M. Grätzel, N. Serpone, and D.K. Sharma, *J. Am. Chem. Soc.* **107**, 8054–8059 (1985).
67. L. Spanhel, H. Weller, A. Fojtik, and A. Henglein, *Ber. Bunsenges. Phys. Chem.* **91**, 88–95 (1987).
68. H.-C. Youn, S. Baral, and J.H. Fendler, *J. Phys. Chem.* **92**, 6320–6327 (1988).
69. Eychmüller, A.A. Hässelbarth, H. Weller, and J. Lumin. **53** 113 (1992).
70. M.A. Hines and P. Guyot-Sionnest, *J. Phys. Chem.* **100**, 468–471 (1996).
71. B.O. Dabbousi, J. Rodriguez-Viejo, F.V. Mikulec, J.R. Heine, H. Mattoussi, R. Ober, K.F. Jensen, and M.B. Bawendi, *J. Phys. Chem. B* **101**, 9436–9440 (1997).
72. J. RodriguezViejo, K.F. Jensen, H. Mattoussi, J. Michel, B.O. Dabbousi, and M.B. Bawendi, *Appl. Phys. Lett.* **70**, 2132–2134 (1997).
73. M. Kuno, J.K. Lee, B.O. Dabbousi, F.V. Mikulec, and M.G. Bawendi, *J. Chem. Phys.* **106**, 9869–9882 (1997).
74. W. Hoheisel, V.L. Colvin, C.S. Johnson, and A.P. Alivisatos, *J. Chem. Phys.* **101**, 8455–8460 (1994).

75. D.J. Norris, A. Sacra, C.B. Murray, and M.G. Bawendi, *Phys. Rev. Lett.* **72**, 2612–2615 (1994).
76. Landolt-Börnstein, *Numerial Data and Functional Relationships in Science and Technology* ed. by O. Madelung, Vol. III-17, *Semiconductors*, (Springer Verlag, Berlin, 1988).
77. *Semiconductors, Data in Science and Technology* (Springer-Verlag: Berlin, 1992).
78. D. Petrov, B.S. Santos, G.A.L. Pereira, and C. de M. Donegá, *J. Phys. Chem. B* **106**, 5325–5334 (2002).
79. J. Rockenberger, L. Tröger, A. Kornowski, T. Vossmeyer, A. Eychmüller, J. Feldhaus, H. Weller, and B. Bunsenges, *Phys. Chem.* **101**, 1613 (1997).
80. J. Rockenberger, L. Tröger, A. Kornowski, T. Vossmeyer, A. Eychmuller, J. Feldhaus, and H. Weller, *J. Phys. Chem. B* **101**, 2691 (1997).
81. H. Borchert, D.V. Talapin, N. Gaponik, C. McGinley, S. Adam, A. Lobo, T. Möller, and H. Weller, *J. Phys. Chem. B* **107**, 9662–9668 (2003).
82. W.W. Yu, E. Chang, R. Drezek, and V.L. Colvin, *Biochem. Biophys. Res. Com.* **348**, 781–786 (2006).
83. H. Mattoussi, *J. Am. Chem. Soc.* **122**, 12142–12150 (2000).
84. S. Kim and M.G. Bawendi, *J. Am. Chem. Soc.* **125**, 14652–14653 (2003).
85. D. Gerion, F. Pinaud, S.C. Williams, W.J. Parak, D. Zanchet, S. Weiss, and A.P. Alivisatos, *J. Phys. Chem. B.* **105**, 8861–8871 (2001).
86. A.N. Rogach, D. Nagesha, J.W. Ostrander, M. Giersig, and N.A. Kotov, *Chem. Mater.* **12**, 2676–2685 (2000).
87. W. Tan, K. Wang, H. Drake, X. He, X.J. Zhao, T. Drake, L. Wang, and R.P. Bagwe, *Med. Res. Rev.* **24**, 621–638 (2004).
88. S.T. Selvan, T.T. Tan, and J.Y. Ying, *Adv. Mater. (Weinheim, Germany)* **17**, 1620–1625 (2005).
89. B. Dubertret, P. Skourides, D.J. Norris, V. Noireaux, A.H. Brivanlou, and A. Libchaber, *Science* **298**, 1759–1762 (2002).
90. B.A. Korgel and H.G. Monbouquette, *Langmuir* **16**, 3588–3594 (2000).
91. I.L. Medintz, H.T. Uyeda, E.R. Goldman, and H. Mattoussi, *Nature Mater.* **4**, 435–446 (2005).
92. M. Bruchez, M. Moronne, P. Gin, S. Weiss, and A.P. Alivisatos, *Science* **281**, 2013–2015 (1998).
93. H. Mattoussi, J.M. Mauro, E.R. Goldman, G.P. Anderson, V.C. Sundar, F.V. Mikulec, and M.G. Bawendi, *J. Am. Chem. Soc.* **122**, 12142–12150 (2000).
94. W.C.W. Chan and S. Nie, *Science* **281**, 2016 (1998).
95. M. Green, R. Taylor, and G. Wakefield, *J. Mater. Chem.* **13**, 1859–1861 (2003).
96. M. Green, D.S.-B., J. Harries, and R. Taylor, *Chem. Commun.* 4830–4832 (2005).
97. A. Robinson, J.-M. Fang, P.-T. Chou, K.-W. Liao, R.-M. Chu, and S.-J. Lee, *Chem. Bio. Chem.* **6**, 1–8 (2005).
98. B.S. Santos, P.M.A. de Farias, F.D. Menezes, E.L. Mariano, R.C. Ferreira, S. Giorgio, M.C. Bosetto, D.C. Ayres, P.M. Lima, A. Fontes, A.A. de Thomas, and C.L. Cesar, *Proceedings of SPIE Optical Molecular Probes for Biomedical Applications* **6097**, 6–13 (2006).
99. D.S. Lidke, P. Nagy, R. Heintzmann, D.J. Arndt-Jovin, J.N. Post, H.E. Grecco, E.A. Jares-Erijman, and T.M. Jovin, *Nature Biotech.* **22**(2), 198–204 (2004).
100. B.M. Lingerfelt, H. Mattoussi, E.R. Goldman, J.M. Mauro, and G.P. Anderson, *Anal. Chem.* **75**, 4043–4049 (2003).
101. E.R. Goldman, E.D. Balighian, H. Mattoussi, M.K. Kuno, J.M. Mauro, P.T. Tran, and G.P. Anderson, *J. Am. Chem. Soc.* **124**, 6378–6382 (2002).
102. D. Geho, N. Lahar, P. Gurnani, M. Huebschman, P. Herrmann, V. Espina, A. Shi, J. Wulfkuhle, H. Garner, E. Petricoin, III, L.A. Liotta, and K.P. Rosenblatt, *Bioconjugate Chem.* **16**, 559–566 (2005).
103. D. Gerion *et al.*, *J. Am. Chem. Soc.* **124**, 7070–7074 (2002).
104. E.R. Goldman, E.D. Balighian, H. Mattoussi, M.K. Kuno, J.M. Mauro, P.T. Tran, and G.P. Anderson, *J. Am. Chem. Soc.* **124**, 6372–6378 (2002). Anonymous, *Immunochemistry Labfax*, (Oxford, UK: BIOS Scientific Publishers; 1994).
105. S.-Y. Ding, M. Jones, M.P. Tucker, J.M. Nedeljkovic, J. Wall, M.N. Simon, G. Rumbles, and M.E. Himmel, *Nano Lett.* **3**(11), 1581–1585 (2003).
106. A. Kumar and D.P.S. Negi, *J. Coll. Interf. Sci.* **238**, 310–317 (2001).
107. N. Marmé, J.-P. Knemeyer, M. Sauer, and J. Wolfrum, *Bioconjugate Chem.* **14**, 1139–1333 (2003).

108. T.W. Graham Solomons and C.B. Fryhle, *Organic Chemistry*, 8th ed. 1224 pp., (2004).

109. B.S. Santos, P.M.A. de Farias, F.D. Menezes, R.C. Ferreira, S.A. Júnior, R.C.B.Q. Figueiredo, L.B. de Carvalho Jr, and E.I.C. Beltrão, *Phys. Stat. Sol. (c)* **11**, 4017 (2006).

110. W.C.W. Chan, D.J. Maxwell, X. Gao, R.E. Bailey, M. Han M., and S. Nie, *Curr. Opin. Biotechnol.* **13**, 40–46 (2002).

111. C.M. Niemeyer, *Angew. Chem. Int. Ed. Engl.* **40**, 4128–4158 (2001).

112. Y. Chen and Z. Rosenzweig, *Anal. Chem.* **74**, 5132–5138 (2002).

113. J.K. Jaiswal, H. Mattoussi, J.M. Mauro, and S.M. Simon, *Nat. Biotechnol.* **21**, 47–51 (2003).

114. S.J. Rosenthal, I.D. Tomlinson, E.M. Adkins., S. Schroeter, S. Adams, L. Swafford, J. McBride, Y. Wang, L.J. DeFelice, and R.D. Blakely, *J. Am. Chem. Soc.* **124**, 4586–4594 (2002).

115. R.P. Bagwe., X. Zhao, and W. Tan, *J. Dispersion. Sci. Technol.* **24**, 453–464 (2003).

116. X. Wu, H. Liu, K.N. Haley, J.A. Treadway, J.P. Larson, N. Ge, F. Peale, and M.P. Bruchez, *Nature Biotechnology* **21**, 41–46 (2003).

117. P. Chan, T. Yuen, F. Ruf, J. Gonzalez-Maeso, and S.C. Sealfon, *Nucleic Acids Research* **33**, 18 (2005).

118. S. Pathak, S.-K. Choi, N. Arnheim, and M.E. Thompson, *J. Am. Chem. Soc.* **123**, 4103–4104 (2001).

119. S. Liang, D.T. Pierce, C. Amiot, and X. Zhao, *Synth. and React. Inorg., Metal-Org., Nano-Metal Chem.* **35**, 661–668 (2005).

120. E. Bentzen, D.W. Wright, and J. E Crowe, Jr., *Fut. Virology* **6**, 769–781 (2006).

121. S. Kumar and R. Richards-Kortum, *Nanomedicin* **1**, 23–30 (2006).

122. M.E. Åkerman, W.C.W. Chan, P. Laakkonen, S.N. Bhatia, and E. Ruoslahti, *Appl. Biol. Science* **99**, 12617–12621 (2002).

123. A. Ashkin and J.M. Dziedzic, *Science* **235**, 1517–1520 (1987).

124. A. Fontes, A.A. de Thomaz, W.L. Moreira, A.A.R. Neves, L.C. Barbosa, P.M.A. Farias, B.S. Santos, and C.L. Cesar, *Proceedings of the SPIE* **5700**, 273–278 (2005).

125. A. Fontes, K. Ajito, A.A. Neves, W.L. Moreira, A.A. de Thomaz, L.C. Barbosa, A.M. de Paula, and C.L. Cesar, *Phys. Rev E* **72**, 012903 (2005).

126. K. Konig, *Hyst. and Cell Biol.* **114**, 79–91 (2000).

127. E.I.C. Beltrão, M.T.S. Correia, J.F. Silva, and L.C.B.B. Coelho, *Appl. Biochem. Biotech.* **74**, 125 (1998).

128. E.I.C. Beltrão, J.F. Silva, L.C.B.B. Coelho, and L.B. Carvalho Jr, *Anais Fac. Méd. Univ. Fed. Pernambuco* **46**, 32 (2001).

129. E.I.C. Beltrão, P.L. Medeiros, O.G. Rodrigues, J.F. Silva, M.M. Valença, L.C.B.B. Coelho, and L.B. Carvalho, Jr., *Eur. J. Histochem.* **47**, 139–142 (2003).

CHAPTER 27

Quantum Dot Modification and Cytotoxicity

Kenji Yamamoto[1] and Akiyoshi Hoshino[1,2]

[1]International Clinical Research Center, Research Institute, International Medical
Center of Japan, Tokyo, Japan
[2]Department of Pharmacodynamics and Pharmacokinetics, Hospital Pharmacy,
Tokyo Medical and Dental University School of Medicine, Tokyo, Japan

27.1 Quantum dots are born to be an electronic device

A quantum dot (QD) is a semiconductor nanostructure that confines the motion of conduction
band electrons in three spatial directions, so-called zero dimension. Small QDs have 1–10 nanometre diameter with 100 to 100 000 atoms, and the confined electrons do not move in free
space, but in the semiconductor host crystal. Confinement electrons in the semiconductor host
crystal have a capacity to receive photon energy when QDs are excited by visible, ultraviolet
and infrared light. Then QDs release their excited energy by emitting fluorescence to its intrinsic
potential energy band gap that is regulated by the size of QDs. The fact that enormous advances
in the field of nanotechnology have recently been made can give conventional material an additional function based on nanoscale effects already predicted by Prof. Ryogo Kubo in 1960s [1–
3], but it has taken more than 40 years to achieve the production of nanomaterials [4, 5]. The
QDs that have specific functions based on nanometre-size effect are now widely attracting much
attention due to the intrinsic characters of nanomaterials which are enhanced by their significantly broader surface area on nanomaterials [5–7]. Nowadays, QDs are widely used in biological
study because they are attractive fluorophores for multicolour imaging due to broad absorption
and narrow emission spectra, and ave brighter and far more photostable than organic dyes. QDs
emit far brighter fluorescence owing to the high quantum yield and are applied to the electronic
assembled parts for semiconductor devices because of their photoelectrical properties such as
quantum memory effect [8]. By applying small voltages to the QDs, the flow of electrons through
the QDs can be controlled to make precise measurements of the spin. Therefore, QD technology is
a useful tool in solid-state quantum computation.

27.2 Quantum dots acquire a colourful personality with their environmental magic

The physical properties of QDs have been well investigated. Many advanced improvements on the
modification of QDs are performed to make far brighter nanocrystals. The photophysical properties are defined by the following factors: (1) the QD core itself, (2) surface modifications, and
(3) extraparticlar conditions. These three factors change the properties and behaviour of QD
particles.

QD core structure is one of the important factors. A core–shell structure with ZnS shells up to three monolayers in thickness over the core CdSe particle increased the photostability and photoluminescence quantum yields of QD particles [8–10]. In addition, CdSe cores with small amounts of some transition metal dopant such as manganese and titanium make the fluorescence intensity decrease, and fluorescent emission shifted red with the same particle size [11–14]. Thus, the QD core itself regulates the photoluminescence property.

The chemical property of molecules which cover the surface of QD particles is the other factor that defines the property of QD particles. The surface modification of QDs affect the fluorescence intensity. The fluorescence intensity and peak wavelength were changed by the surface capping molecules of the QDs; QDs with carboxyl groups had higher luminescence than the other groups (Fig. 27.1). Furthermore, the peak wavelength varied according to the surface modification. QDs containing amino groups emitted a shorter wavelength (513 nm wavelength) than the originally synthesized QD (TOPO-capped QD with 518 nm emission). In contrast, QDs with hydroxyl groups were slightly red shifted. We assumed that a longer carbon chain and the carboxyl groups of MUA (mercaptoundecanoic acid, HS-(CH2)10-COOH) may contribute to the long lifetime and high quantum yield of QD-COOH. In contrast, the decreasing fluorescence of amino-QD particles may be caused by the oxidation of QD-metal because of the leakage of electrons from QDs

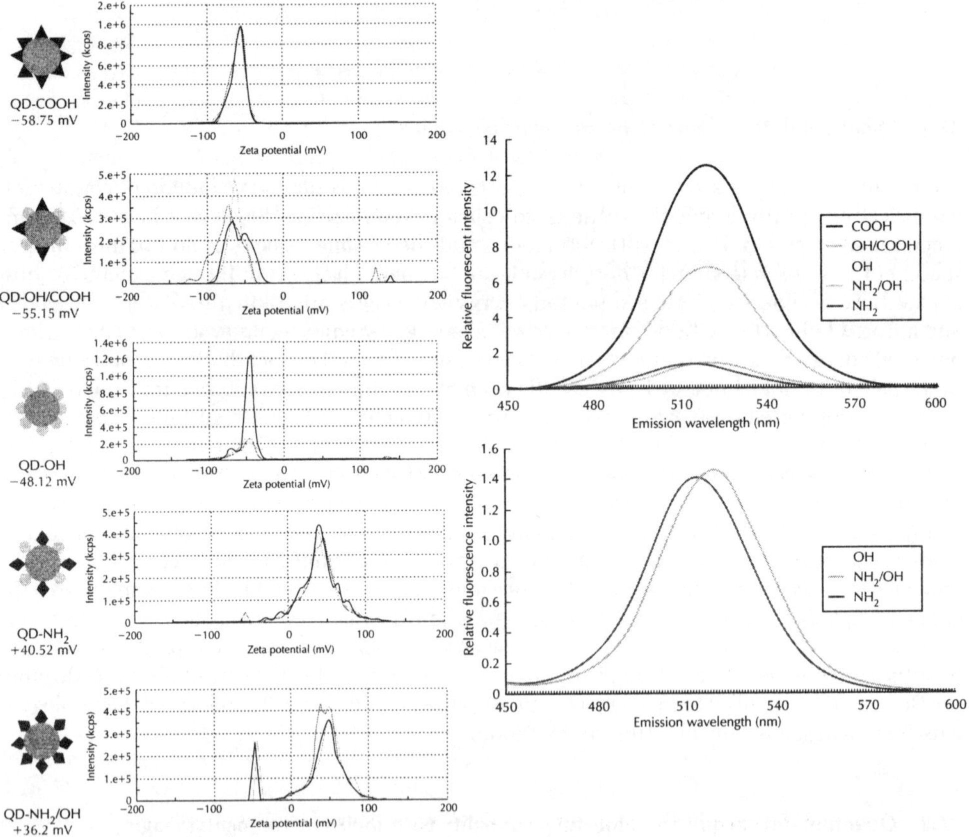

Figure 27.1 Surface ζ-potential and fluorescence intensity of QDs varied by their surface modification. (a) Cartoon (left) and ζ-potential (right) of the novel modified QDs. Surface zeta-potential of QDs is measured by electrophoresis. Each line shows the electrophoretic mobility of QDs in the stationary layers. The data are the average of 30 assays. (b) Relative fluorescence intensity and peak wavelength of QD-COOH (red); QD-COOH/-OH (orange); QD-OH (yellow); QD-NH$_2$/OH (green) and QD-NH$_2$ (blue) measured by fluorescence spectrometry. (Lower) enlarged panel of QD-OH (yellow); QD-NH$_2$/OH (green) and QD-NH$_2$ (blue) in the upper panel. The peak emission wavelengths of QD-COOH, QD-OH/COOH, QD-OH, QD-NH$_2$/OH, and QD-NH$_2$ varied at 519, 520, 526, 520, and 513 nm, respectively.

through NH_2 groups in aqueous solution. This result suggested that the luminescence intensity of QDs may increase according to the surface-covered molecule structures [15].

The third factor is the intermolecular effects between the QD particle and the extraparticular conditions surrounding it. It is especially influenced by a specific photophysical feature of QDs called blinking. Blinking is a fluctuation of photoluminescence which every QD shows intrinsically when QDs are observed under steady laser illumination such as evanescent fields [16–19]. Blinking is one of the specific characters of single fluorescent nanomaterials, and is measured by fluorospectrometry as the bright "on-phase" fraction and on-dark "off-phase" fraction events in the span of several minutes (Fig. 27.2a) [20, 21]. In addition, sudden hopping enhancement of photoluminescence intensity in a span of milliseconds is observed, and continues after transition from "on-phase" to "off-phase" of the QD (Fig. 27.2b). Another fluctuation called millisecond oscillation is a different phenomenon to conventional photo-intermittence blinking.

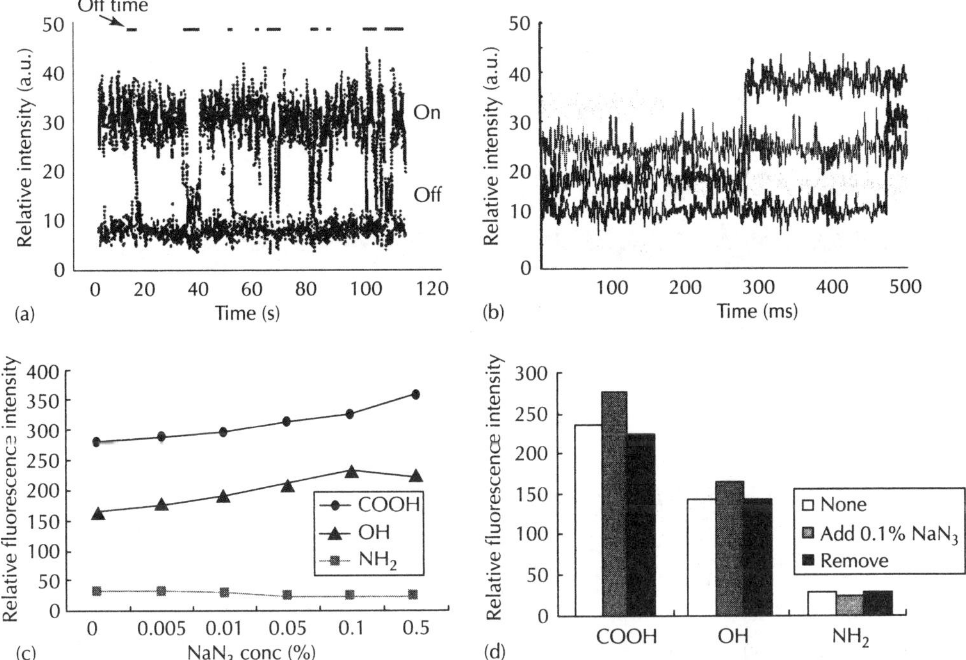

Figure 27.2   The mechanism of the millisecond oscillation is different from that of blinking. (a) A blinking change of QD-COOH fluorescence intensity in aqueous solution. Fluorescence intensity was continuously measured for 120 s with a fluorospectrometer. The red line upper graph indicates the "off" phase of blinking. (b) The millisecond oscillation of QD-COOH in aqueous solution was observed on the evanescent field with ultra-sensitive high-speed fibre-coupled CCD camera (FASTCAM® MAXI-I², Photoron Corp., Tokyo, Japan). x-axis indicates the time and y-axis indicates the relative fluorescence intensity. Red, blue, and yellow lines that indicate the emission from individual QD during the on–off switching phase (n = 3) were overlaid. Green line indicates the photoemission of the "on phase". (c) External environment on QDs varies with fluorescence intensity. NaN_3 in aqueous solution enhances fluorescence intensity of QD. 100 μM of QD-COOH (circle), QD-OH (triangle), and QD-NH₂ (square) were diluted to 0.1% NaN_3 solution. (d) NaN_3 removal by dialysis recovered the enhanced fluorescence intensity. QD solutions with 0.1% NaN_3 were dialysed by an ultrafiltration membrane. The concentration of collected solution was adjusted with an adequate volume of distilled water. Data are the mean aggregation area ± 1 SD of triplicate experiments.

The surrounding environmental molecules could affect the emission properties of QDs and blinking. The preservative chemical, sodium azide (NaN_3), in storage buffer has the ability to enhance fluorescent intensity of QDs in a dose-dependent manner (Fig. 27.2c). QD-NH₂ weakens the fluorescent intensity whereas NaN_3 enhances QD-COOH and QD-OH that were negatively charged. These data suggest that the electric charge of QDs was significantly concerned with their enhancements of fluorescent intensity by NaN_3. Removal of NaN_3 by dialysis from the solution recovered the QDs' enhanced fluorescence (Fig. 27.2c), implying that the solutes around the QD solution influenced the fluorescence intensity of QDs. These results suggest that not only the surface covered molecules but also surrounding external environmental solutes around the QD solution regulate the fluorescence intensity. To support the possibility that aqueous solvents were implicated in the millisecond oscillation, QD-COOH fixed on the surface of the cover slip are filled with some organic solvents, implying that the polarity of the solvents was also operated as the electron donor in the solution. In a non-polar organic solution such as toluene, each QD displays no detectable fluorescent oscillation (Fig. 27.3). The results suggest that both solutes and solvents that are surrounded by QDs may function as the donor/interceptor of excited electrons on nanocrystal QDs. The theory that electron transfer processes regulate the power-law distribution for the lifetime of a blinking of QDs has already been proposed. These results demonstrate that the electron condition on the core of the QD, beside the QD covered molecules, and even around QD external environments by observation of millisecond oscillation, imply that emission of QD

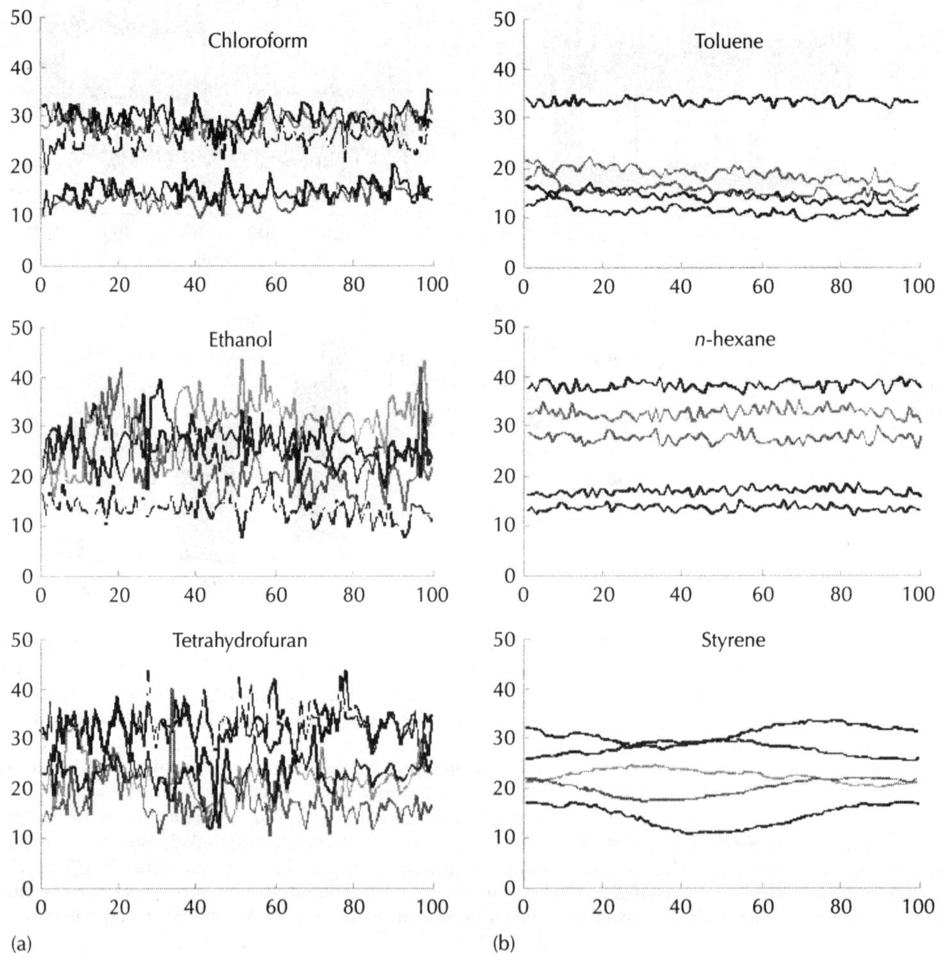

(a)   (b)

Figure 27.3   Polarity of the solvent regulates the frequency of the millisecond oscillation. The millisecond oscillation of QD-COOH in (a) polar organic solvents; chloroform, ethanol, and tetrahydrofuran, and (b) non-polar inorganic solvents; toluene, n-hexane, and styrene, were observed on the evanescent field by the total-reflective fluorescent microscopy with a CCD camera unit FASTCAM®. The blinking frequency of each monodispersed QD (n = 6) on each one millisecond is measured by one millisecond exposure. QDs x-axis indicates the time (millisecond) and y-axis indicates the fluorescence intensity. Data shown are one of three independent experiments that gave the same results.

depends on at least these factors: surface-covered substance, the solvent and solutes in solution, and the specified fluorescence activity of the QD particle.

Studies suggest that QD property depends on multiple factors derived from both the inherent physicochemical properties of QDs and environmental conditions. For example, QD properties like QD size, charge, concentration, outer coating bioactivity (capping material and functional groups), and oxidative, photolytic, and mechanical stability are factors that, collectively and individually, can determine the toxicity of QDs [15, 22]. The fact that the photoluminescence property of QDs also depends on multiple factors like environmental conditions supports the possibility that the brightness of the whole QD particle can be improved by the extraparticular modification including capping materials, functional bioactive groups, and even periparticular molecules.

27.3 Quantum dots go to the biological field, wearing a fluorescent dress

In modern biological analysis, various kinds of organic dyes are used to visualize the objective molecules. At present, many organic fluorophores have been used in various biological applications including fluorescent-labelled antibodies, molecules that are used to stain cells and cellar organs. Experiments using organic dyes are limited to short-time assays such as flow cytometry due to their lifetime of fluorescence. Those conventional dyes were less suitable for extended periods of bio-imaging observations using fluorescent and confocal microscopy because organic fluorophores tend to quench rapidly [23, 24]. Furthermore, it is sometimes difficult or impossible to record fine fluorescent images while the organic colouring probes fade in the course of adjusting the focus. Therefore, much brighter fluorophores are required.

Currently, QDs are widely used in biological and even in medical studies because QDs are attractive fluorophores for multicolour imaging due to broad absorption and narrow emission spectra, and are brighter and far more photostable than organic fluorescent dyes [5–7, 23–29].

27.4 Quantum dot toxicty: a forgotton glass slipper

Great quantities of these "artificial nanomaterials" including QDs are produced and consumed on the unfounded inference that nanomaterials are biologically and environmentally harmless, whereas only few studies have reported the toxicity of these nanomaterials [22, 30–43]. Recently, nanomaterial researchers have had to deal with nanomaterial-induced toxicity problems [15, 44–55]. Why does research on the toxicity of nanomaterials progress so slowly? The reason is attributed to the following factors: (1) some researches on nanomaterials have been performed under the premise that there is no toxicity; (2) various kinds of nanomaterials which have their own unique properties have been synthesized, and a variety of QD concentrations have been reported – the toxicity of QDs in dose dependent manner remains a problematic area for discussion; (3) sufficient nanomaterials cannot be supplied to carry out a series of toxic studies. Some studies were designed for toxicologists.

We and other nanomaterial researchers have reported that QD behaviour in the body (called ADME; absorption, distribution, metabolism, excretion) and toxicity are dependent on multiple factors derived from both individual physicochemical properties and environmental conditions [22, 40, 56–58]. The toxicity is primarily caused by the inherent physicochemical properties. In addition, some additional environmental conditions including oxidative, photolytic, and mechanical stability often affect QDs, resulting in other QD toxicities.

The QD itself has its own physicochemical properties, which exercise the toxicity dependent on the size, electric charge, concentration, and surface-coating materials including capping material and functional groups. We assessed the dose and size dependency of QD-mediated cytotoxicity at quite a high (approximately 1000-fold concentration of usual cellular application) concentration. We observed the dose-dependent (Fig. 27.4a) and size-dependent (Fig. 27.4b) cytotoxity of QDs [55]. The observed cytotoxicity is proportional to the number of QDs that are incorporated

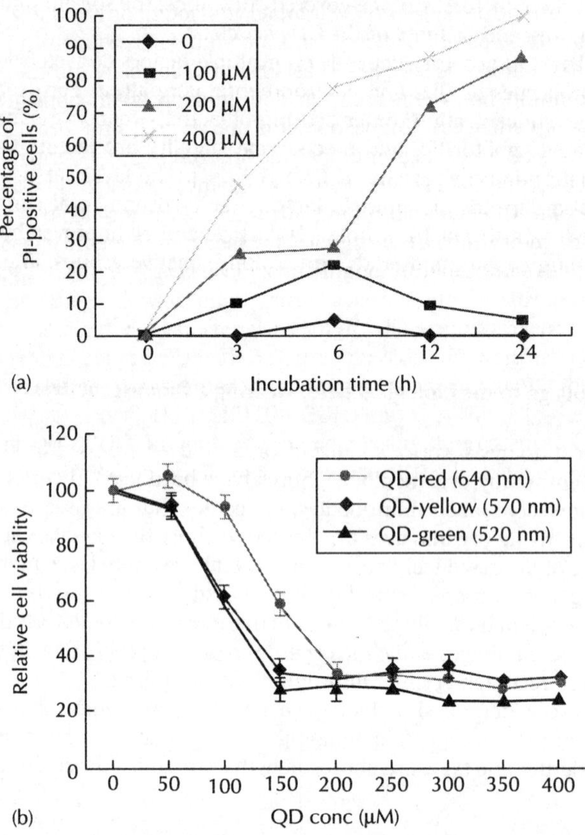

Figure 27.4   Dose and particle size-dependent QD toxicity at high concentrations. (a) The cytotoxicity of QD520 (green) solution depends on its concentration. Vero cells were plated at 1×10^5 cells and stimulated with culture media alone or with QD at the concentration of 100, 200, and 400 μM. The cells were incubated for 3, 6, 12, and 24 h at 37°C. The cells were harvested, stained for observation of dead cells with propodium iodide (PI), and measured by flow cytometric analysis. Fluorescence was measured by FACS Calibur (Beckton Dickinson). (b) Relative cell viability of different size QDs were measured with MTT assay. Vero cells were plated at 1×10^5 cells on 24-well plate and incubated for 12 h with indicated concentration of QDs. Results are representative of three separate experiments. $p < 0.05$.

into the cells. Actually, we demonstrated that the red QDs are incorporated into cells less than green ones [55]. Other groups also demonstrated that QD-induced cell death was due to chromatin condensation and the size of QDs contributed to their subcellular distribution [59]. Moreover, cadmium and selenium, two of the most widely used constituent metals in our QD core crystalline, are reported to cause acute and chronic toxicities in vertebrates when QDs were irradiated to the UV-mediated oxidative stress [60]. Furthermore, the capping materials and biomolecules which cover the QD surface also have defined the total biological behaviour of whole nanocrystal QDs including toxicity. We demonstrated that some surface-covered toxic molecules caused severe cytotoxicity at dose dependent manner whereas some surface-covered less toxic molecules caused less cytotoxicity (Fig. 27.5). This result suggested that the bioactive behaviour of QDs in biological systems is not only dependent on the nanocrystal particle itself but on the biochemical properties of surface-covered molecules [56]. These results provide evidence that some hydrophilic compounds which capped on the surface of QDs are responsible for the biological effect of QD whole molecules. This result prompted us to attach the medicine to QDs and observe whether the complex exerted its medicinal effect *in vivo*. First, we conjugated QDs with an anti-hypertension medicine, captopril [61, 62]. Then we measured the effect of QD-conjugated captopril (QD-cap)

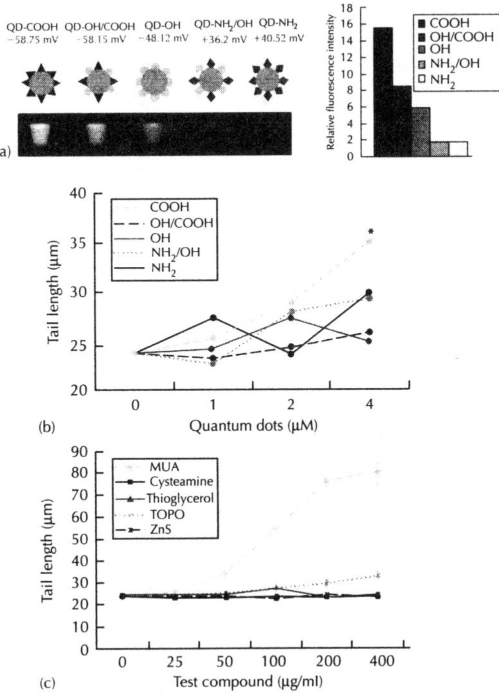

Figure 27.5 *Surface-capped molecules participate in QD-induced cell damage and lethal cytotoxicity due to the toxicity of the molecules. (a) ZnS-coated CdSe nanocrystal QDs (fluorescence wavelength: approximately 518 nm, emitted green) capped with 11-mercaptoundecanioc acid (QD-COOH), 2-aminoethanethiol (QD-NH₂), and 3-mercapto1,2-propanediol (QD-OH) by thiol exchange reactions. Cartoon images and solution photos of QDs capped with various surface molecules. The snapshots of QD aqueous solution (final concentration is 100 nM) were captured using a digital camera with 1/30 s exposure excited by a 365 nm wavelength (UV-A). Relative fluorescence intensity of each QD (shown in right) was measured by fluorescence spectrometry. (b, c) DNA-damaging effects of hydrophilic nanocrystalline QDs by comet assay. When WTK-1 cells were incubated with each sample, the fragmented DNA of apoptotic cells (called comet tail) was eliminated from the nuclei by electrophoresis. The length of comet tail stained with ethidium bromide was measured. (b) QD solution induced genotoxic cell death. Cells were treated with QD-COOH (   ), QD-NH₂ (—), QD-OH (—), QD-OH/COOH (– –), or QD-NH₂/OH (·····) for 2 h. (c) Ingredients capped with QDs also induce cell death. Cells were treated with MUA (   ), cysteamine (·····), thioglycerol (—), TOPO (—), or ZnS (– –) for 2 h. The difference between the means in the treated and control plates was compared with the Dunnett test after one-way ANOVA. *, p < 0.05. Results are representative of three separate experiments.*

in vitro, and administered it to the spontaneously hypertensive rat and assessed the effect of QD medicine *in vivo* [63]. As we expected, we synthesized the functional nanocomposite particles of QD and captopril without losing the fluorescence activity and anti-hypertensive effects (Fig. 27.6). These results suggested that the surface treatment of nanocrystals (surface-capped functional groups and biomolecules covering the surface of QDs) has defined the biological behaviour of whole nanocrystal QDs [31, 32, 64–66].

We revealed that the toxicity of QDs in biological systems is not only dependent on the nanocrystal "core particle" itself but also on the surface molecules. We observed no cytotoxicity from their ingredients or the QD core itself, suggesting that surface processing will overcome the toxicity of nanomaterials unless a core structure is broken. Surface modifications of functional molecules combined with nanomaterials can work as novel bio-nanomachines conforming to the functions designated by their surface molecules [56, 63]. On the other hand, these results suggest that inappropriate treatment and disposal of QDs may still cause environmental pollution including risks to human health under certain conditions.

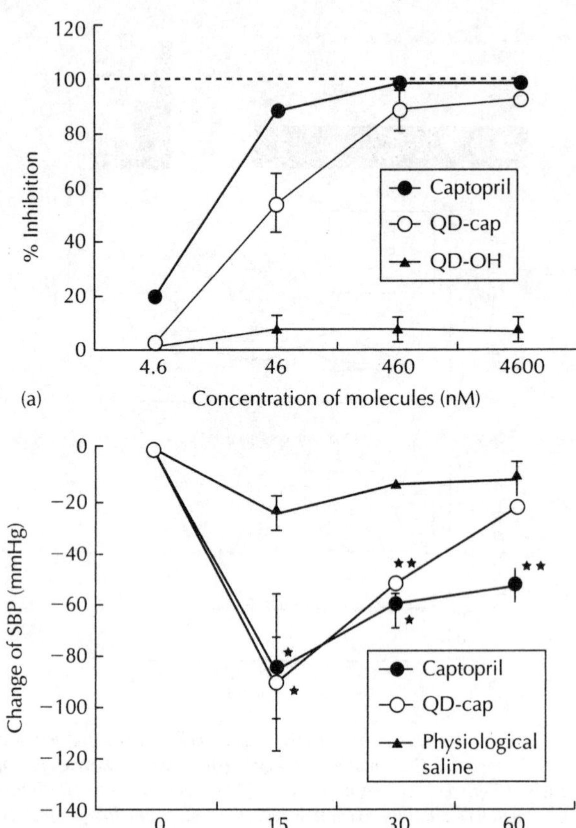

Figure 27.6 Captopril-capped QD (QD-cap) exerts its medicinal effect both in vitro *and* in vivo. *Captoprol, an anti-hypertension medicine, is directly conjugated with QDs (QD-cap). Each QD-cap particle has approximately 350 captopril molecules by peptide content assay. (a) Inhibition activity of angiotensin-1 converting enzyme (ACE, the target molecule of captopril) by QD-cap was measured by the Cushman and Cheung method. The data presented are the mean value ± standard deviation of triplicated samples. QD-cap shows inhibitory activity as well as captopril alone. (b) QD-cap decreases the blood pressure in a hypertensive rat. QD-cap (5 mg/kg) was injected i.v. into the SHR-SP spontaneous hypertensive rat via carotid artery. After injection, systolic blood pressure was measured with the tail-cuff method at an indicated time (n = 3). The results are representative of three separate experiments.*

References

1. R. Kubo, Statistical-mechanical theory of irreversible processes, *J. Phys. Soc. Jpn.* **12**, 570–586 (1957).

2. R. Kubo, Generalized cumulant expansion method, *J. Phys. Soc. Jpn.* **17**, 1100–1120 (1962).

3. R. Kubo, Stochastic Liouville equations, *J. Math. Phys.* **4**, 174–183 (1962).

4. G. Springholz, V.V. Holy, M. Pinczolits, and G. Bauer, Self-organized growth of three-dimensional quantum-dot crystals with fcc-like stacking and a tunable lattice constant, *Science* **282**, 734–737 (1998).

5. W.C. Chan *et al.*, Luminescent quantum dots for multiplexed biological detection and imaging, *Curr. Opin. Biotechnol.* **13**, 40–46 (2002).

6. W.C. Chan and S. Nie, Quantum dot bioconjugates for ultrasensitive nonisotopic detection, *Science* **281**, 2016–2018 (1998).

7. J.K. Jaiswal, H. Mattoussi, J.M. Mauro, and S.M. Simon, Long-term multiple color imaging of live cells using quantum dot bioconjugates, *Nat. Biotechnol.* **21**, 47–51 (2003).

8. S. Maenosono, C.D. Dushkin, S. Saita, and Y. Yamaguchi, Optical memory media based on excitation-time dependent luminescence from a thin film of semiconductor nanocrystals, *Jpn. J. of Appl. Phys. Part 1* **39**, 4006–4012 (2000).

8a. M.A. Hines, and P. Guyot-Sionnest, Synthesis and characterization of strongly luminescing ZnS-capped CdSe nanocrystals. *J. Phys. Chem.* **100**, 468–471 (1996).

9. X. Peng, M.C. Schlamp, A.V. Kadavanich, and A.P. Alivisatos, Epitaxial growth of highly luminescent CdSe/CdS core/shell nanocrystals with photostability and electronic accessibility, *J. Am. Chem. Soc.* **119**, 7019–7029 (1997).

10. B.O. Dabbousi *et al.*, (CdSe)ZnS core-shell quantum dots: synthesis and characterization of a size series of highly luminescent nanocrystallites, *J. Phys. Chem.* **101**, 9463–9475 (1997).

11. H. Yang, P.H. Holloway, and S. Santra, Water-soluble silica-overcoated CdS:Mn/ZnS semiconductor quantum dots, *J. Chem. Phys.* **121**, 7421–7426 (2004).

12. S. Santra, H. Yang, P.H. Holloway, J.T. Stanley, and R.A. Mericle, Synthesis of water-dispersible fluorescent, radio-opaque, and paramagnetic CdS:Mn/ZnS quantum dots: a multifunctional probe for bioimaging, *J. Am. Chem. Soc.* **127**, 1656–1657 (2005).

13. H. Yang, P.H. Holloway, G. Cunningham, and K.S. Schanze, CdS:Mn nanocrystals passivated by ZnS: synthesis and luminescent properties, *J. Chem. Phys.* **121**, 10233–10240 (2004).

14. S. Sapra, A. Prakash, A. Ghangrekar, N. Periasamy, and D.D. Sarma, Emission properties of manganese-doped ZnS nanocrystals, *J. Phys. Chem. B* **109**, 1663–1668 (2005).

15. A. Hoshino *et al.*, Physicochemical properties and cellular toxicity of nanocrystal quantum dots depend on their surface modification, *Nano Lett.* **4**, 2163–2169 (2004).

16. R.G. Neuhauser, K.T. Shimizu, W.K. Woo, S.A. Empedocles, and M.G. Bawendy, Correlation between fluorescence intermittency and spectral diffusion in single semiconductor quantum dots, *Phys. Rev. Lett.* **85**, 3301–3304 (2000).

17. N. Eiha, S. Maenosono, K. Hanaki, K. Yamamoto, and Y. Yamaguchi, Collective fluorescence oscillation in a water dispersion of colloidal quantum dots, *Jpn. J. Appl. Phys.* **42**, 310–313 (2003).

18. T. Uematsu, J. Kimura, and Y. Yamaguchi, The reversible photoluminescence enhancement of a CdSe/ZnS nanocrystal thin film, *Nanotechnology* **15**, 822–827 (2004).

19. J. Kimura, T. Uematsu, S. Maenosono, and Y. Yamaguchi, Photoinduced fluorescence enhancement in CdSe/ZnS quantum dot submonolayers sandwiched between insulating layers: influence of dot proximity, *J. Phys. Chem. B* **108**, 13258–13264 (2004).

20. J. Tang and R.A. Marcus, Mechanisms of fluorescence blinking in semiconductor nanocrystal quantum dots, *J. Chem. Phys.* **123**, 054704 (2005).

21. J. Yao, D.R. Larson, H.D. Vishwasrao, W.R. Zipfel, and W.W. Webb, Blinking and nonradiant dark fraction of water-soluble quantum dots in aqueous solution, *Proc. Natl. Acad. Sci.* **102**, 14284–14289 (2005).

22. R. Hardman, A toxicologic review of quantum dots: toxicity depends on physicochemical and environmental factors, *Environ. Health Perspect.* **114**, 165–172 (2006).

23. M.E. Akerman, W.C. Chan, P. Laakkonen, S.N. Bhatia, and E. Ruoslahti, Nanocrystal targeting in vivo, *Proc. Natl. Acad. Sci. USA*, **99**, 12617–12621 (2002).

24. M. Bruchez, M. Moronne, P. Gin, S. Weiss, and A.P. Alivisatos, Semiconductor nanocrystals as fluorescent biological labels, *Science* **281**, 2013–2016 (1998).

25. A. Hoshino, K. Hanaki, K. Suzuki, and K. Yamamoto, Applications of T-lymphoma labelled with fluorescent quantum dots to cell tracing markers in mouse body, *Biochem. Biophys. Res. Commun.* **314**, 46–53 (2004).

26. X. Gao and S. Nie, Quantum dot-encoded beads, *Methods Mol. Biol.* **303**, 61–71 (2005).

27. H. Xu *et al.*, Multiplexed SNP genotyping using the Qbead system: a quantum dot-encoded microsphere-based assay, *Nucleic Acids Res.* **31**, e43 (2003).

28. B. Dubertret *et al.*, In vivo imaging of quantum dots encapsulated in phospholipid micelles, *Science* **298**, 1759–1762 (2002).

29. K. Hanaki *et al.*, Semiconductor quantum dot/albumin complex is a long-life and highly photostable endosome marker, *Biochem. Biophys. Res. Commun.* **302**, 496–501 (2003).

30. P.J. Borm *et al.*, The potential risks of nanomaterials: a review carried out for ECETOC, *Part Fibre Toxicol.* **3**, 11 (2006).

31. J.Y. Bottero, J. Rose, and M.R. Wiesner, Nanotechnologies: tools for sustainability in a new wave of water treatment processes, *Integr. Environ. Assess. Manag.* **2**, 391–395 (2006).

32. M.C. Garnett and P. Kallinteri, Nanomedicines and nanotoxicology: some physiological principles, *Occup. Med. (Lond.)* **56**, 307–311 (2006).

33. K.A. Guzman, M.R. Taylor, and J.F. Banfield, Environmental risks of nanotechnology: National Nanotechnology Initiative funding, 2000–2004, *Environ. Sci. Technol.* **40**, 1401–1407 (2006).

34. S. Lanone and J. Boczkowski, Biomedical applications and potential health risks of nanomaterials: molecular mechanisms, *Curr. Mol. Med.* **6**, 651–663 (2006).

35. C.C. Lawson *et al.*, Workgroup report: implementing a national occupational reproductive research agenda – decade one and beyond, *Environ. Health Perspect.* **114**, 435–441 (2006).

36. M.N. Moore, Do nanoparticles present ecotoxicological risks for the health of the aquatic environment? *Environ. Int.* **32**, 967–976 (2006).

37. D.V. Talapin and C.B. Murray, PbSe nanocrystal solids for n- and p-channel thin film field-effect transistors, *Science* **310**, 86–89 (2005).

38. A. Nel, T. Xia, L. Madler, and N. Li, Toxic potential of materials at the nanolevel, *Science* **311**, 622–627 (2006).

39. G. Oberdorster, E. Oberdorster, and J. Oberdorster, Nanotoxicology: an emerging discipline evolving from studies of ultrafine particles, *Environ. Health. Perspect.* **113**, 823–839 (2005).

40. C. Robichaud, D. Tanzil, U. Weilenmann, and M. Wiesner, Relative risk analysis of several manufactured nanomaterials: an insurance industry context, *Environ. Sci. Technol.* **39**, 8985–8994 (2005).

41. R.F. Service, Nanotechnology. Calls rise for more research on toxicology of nanomaterials, *Science* **310**, 1609 (2005).

42. M.R. Wiesner, Responsible development of nanotechnologies for water and wastewater treatment, *Water Sci. Technol.* **53**, 45–51 (2006).

43. J.M. Worle-Knirsch, K. Pulskamp, and H.F. Krug, Oops they did it again! Carbon nanotubes hoax scientists in viability assays, *Nano Lett.* **6**, 1261–1268 (2006).

44. L.K. Adams, D.Y. Lyon, and P.J. Alvarez, Comparative eco-toxicity of nanoscale $TiO(2)$, $SiO(2)$, and ZnO water suspensions, *Water Res.* **40**, 3527–3532 (2006).

45. L. Braydich-Stolle, S. Hussain, J.J. Schlager, and M.C. Hofmann, In vitro cytotoxicity of nanoparticles in mammalian germline stem cells, *Toxicol. Sci.* **88**, 412–419 (2005).

46. J.D. Fortner *et al.*, C60 in water: nanocrystal formation and microbial response, *Environ. Sci. Technol.* **39**, 4307–4316 (2005).

47. J.S. Kim *et al.*, Toxicity and tissue distribution of magnetic nanoparticles in mice, *Toxicol. Sci.* **89**, 338–347 (2006).

48. A. Magrez *et al.*, Cellular toxicity of carbon-based nanomaterials, *Nano Lett.* **6**, 1121–1125 (2006).

49. C.M. Sayes *et al.*, Nano-C60 cytotoxicity is due to lipid peroxidation, *Biomaterials* **26**, 7587–7595 (2005).

50. C.M. Sayes *et al.*, Correlating nanoscale titania structure with toxicity: a cytotoxicity and inflammatory response study with human dermal fibroblasts and human lung epithelial cells, *Toxicol. Sci.* **92**, 174–185 (2006).

51. F. Tian, C. Cui, H. Schwarz, G.G. Estrada, and H. Kobayashi, Cytotoxicity of single-wall carbon nanotubes on human fibroblasts, *Toxicol. In Vitro* **20**, 1202–1212 (2006).

52. H. Yamawaki and N. Iwai, Cytotoxicity of water-soluble fullerene in vascular endothelial cells, *Am. J. Physiol. Cell. Physiol.* **290**, C1495–C1502 (2006).

53. K. Yang, L. Zhu, and B. Xing, Adsorption of polycyclic aromatic hydrocarbons by carbon nanomaterials, *Environ. Sci. Technol.* **40**, 1855–1861 (2006).

54. S. Zhu, E. Oberdorster, and M.L. Haasch, Toxicity of an engineered nanoparticle (fullerene, C60) in two aquatic species, Daphnia and fathead minnow, *Mar. Environ. Res. Suppl.* S5–S9 (2006).

55. A. Shiohara, A. Hoshino, K. Hanaki, K. Suzuki, and K. Yamamoto, On the cyto-toxicity caused by quantum dots, *Microbiol. Immunol.* **48**, 669–675 (2004).

56. S. Kako *et al.*, A gallium nitride single-photon source operating at 200 K *Nat. Mater.* **5**, 887–892 (2006).

57. J.P. Ryman-Rasmussen, J.E. Riviere, and N.A. Monteiro-Riviere, Surface coatings determine cytotoxicity and irritation potential of quantum dot nanoparticles in epidermal keratinocytes, *J. Invest. Dermatol.* (2006).

58. T. Zhang *et al.*, Cellular effect of high doses of silica-coated quantum dot profiled with high throughput gene expression analysis and high content cellomics measurements, *Nano Lett.* **6**, 800–808 (2006).

59. J. Lovric *et al.*, Differences in subcellular distribution and toxicity of green and red emitting CdTe quantum dots, *J. Mol. Med.* **83**, 377–385 (2005).

60. J. Lovric, S.J. Cho, F.M. Winnik, and D. Maysinger, Unmodified cadmium telluride quantum dots induce reactive oxygen species formation leading to multiple organelle damage and cell death, *Chem. Biol.* **12**, 1227–1234 (2005).

61. R. Laffan *et al.*, Antihypertensive activity in rats for SQ 14225, an orally active inhibitor of angiotensin I-converting enzyme, *J. Pharmacol. Exp. Ther.* **204**, 281–288 (1978).

62. B. Rubin *et al.*, SQ 14225 (D-3 mercapto-methylpropanoyl-L-proline), a novel orally active inhibitor of angiotensin I-converting enzyme, *Pharmacol. Exp. Ther.* **204**, 271–280 (1978).

63. N. Manabe *et al.*, Quantum dot conjugated with medicine as the drug-tracer in vitro and in vivo, *IEEE Transactions on NanoBioscience* in press (2006).

64. U. Pison, T. Welte, M. Giersig, and D.A. Groneberg, Nanomedicine for respiratory diseases, *Eur. J. Pharmacol.* **533**, 341–350 (2006).

65. A. Bianco, K. Kostarelos, and M. Prato, Applications of carbon nanotubes in drug delivery, *Curr. Opin. Chem. Biol.* **9**, 674–679 (2005).

66. H. Lin and R.H. Datar, Medical applications of nanotechnology, *Natl. Med. J. India* **19**, 27–32 (2006).

CHAPTER 28

Colloidal Quantum Dots (QDs) in Optoelectronic Devices – Solar Cells, Photodetectors, Light-emitting Diodes

Mitra Dutta,[1,2] Ke Sun,[1] Yang Li,[1] Vaishnavi Narayanamurthy,[3] Kitt Reinhardt,[4] and Michael A. Stroscio[1,2,3]

[1]*Electrical and Computer Engineering Department, University of Illinois at Chicago (UIC), 851 S. Morgan Street, Chicago, Illinois 60607*
[2]*Physics Department, University of Illinois at Chicago, 851 S. Morgan Street, Chicago, Illinois 60607*
[3]*Bioengineering Department, University of Illinois at Chicago, 851 S. Morgan Street, Chicago, Illinois 60607*
[4]*Physics and Electronics Directorate, Air Force Office of Scientific Research, Suite 325, 875 N.Randolph Street, Arlington, Virginia 22203*

28.1 Introduction

This review focuses on the applications of ensembles of colloidal quantum dots. Although colloidal quantum dots have been studied since the pioneering work of Michael Faraday, it has been quantum dots self-assembled during growth on a two-dimensional semiconductor surface that were initially studied over the last two decades by the international semiconductor device community. Representative studies of this type include those of Bockelmann and Bastard [1], Inoshita and Sakaki [2] and Leburton *et al.* [3]. Indeed, a number of reviews have been written on the physical properties and applications of these self-assembled quantum dots; examples include the review of Bhattacharya, Ghosh, and Stiff-Roberts [4] and that of Skolnick and Mowbray [5]. Recent, state-of-the-art successes in the application of these self-assembled quantum dots include the works of Bhattacharya and Ghosh [6] on the tunnel injection $In_{0.4}Ga_{0.6}As/GaAs$ quantum dot lasers with impressive performance indicated by a 15 GHz modulation bandwidth at room temperature, Blakesley *et al.* [7] on efficient single photon detection by quantum dot resonant tunnelling diodes, and Shields *et al.* [8] on the detection of single photons using a field-effect transistor gated by a layer of quantum dots.

During the last decade the semiconductor device community has, however, focused increasingly on the study of applications of colloidal semiconductor quantum dots, the subject of this chapter. Applications of colloidal quantum dot (QD) ensembles have emerged as a result of numerous advances in quantum dot synthesis as well as a number of other advances including: advances in the use of molecular linkers/spacers; the ability to assemble quantum dots in ensembles; the ability to disperse quantum dots in conductive polymers; and modelling efforts that have predicted the advantages of using such quantum dot ensembles in device applications.

Frequently discussed advantages of using colloidal quantum dots (QDs) in optoelectronic devices such as solar cells, photodetectors, and light-emitting diodes include: the potential for multi-colour capability from a variety of available QDs; the many matrix materials potentially available; the flexibility of polymer matrix materials making it possible to fabricate flexible detecting elements that may be moulded into a variety of forms for optimum light collection on complex surfaces; the capability to incorporate QDs in conductive polymers, sol-gels, or porous films matrices; the potentially low temperature processing; and the variety of inexpensive substrates such as glass, metal sheets, plastics, etc. that are potentially suitable for devices based on QD ensembles in specific matrix materials. These advantages will be considered in this chapter as well as other possible benefits of using such QD ensembles such as the possibility of using three-dimensional (3D) arrays forming minibands that lead to enhanced transport and collection of carriers.

The great diversity of possible QD polymer systems is illustrated by Fig. 28.1 summarizing commonly used semiconductors and Fig. 28.2 providing the LUMO and HOMO values of some available conductive polymers.

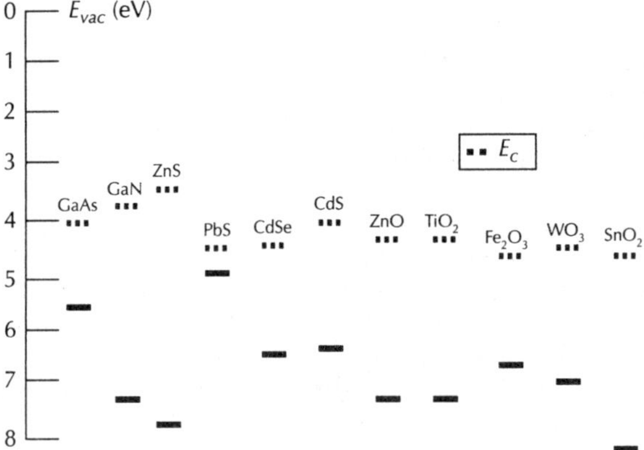

Figure 28.1 Conduction band energies, E_c, of various semiconductors are denoted with a dashed line, and the valence band energies, E_v, with a solid line. All energies are measured relative to the vacuum level in electron volts (eV).

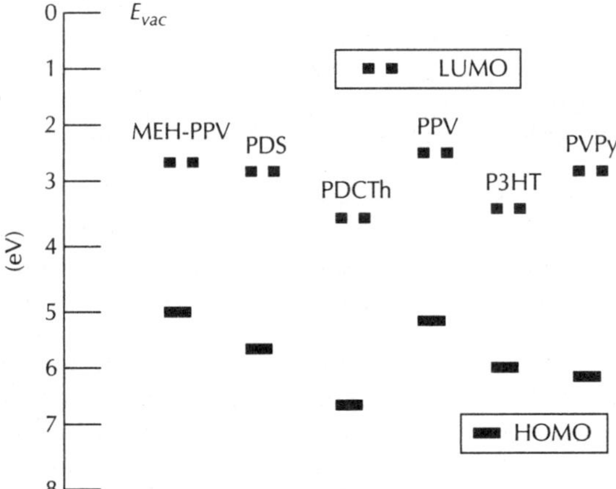

Figure 28.2 The lowest unoccupied molecular orbitals (LUMOs) and the highest occupied molecular orbitals (HOMOs) are depicted here for a selection of conducting polymers. Poly(2-methoxy-5(2-ethyl) hexoxy- phenylene-vinylene (MEH-PPV), peroxydisulphate (PDS), poly(3,4-dicyanothiophene) (PDCTh), poly-(phenylenevinylene) (PPV), poly-3-hexylthiophene (P3HT), and polyvinylpyrrolidone (PVPy) have been selected for inclusion in this summary as a result of the suitability of their LUMO and HOMO energies for device applications. The LUMO and HOMO energies are measured relative to the vacuum energy, E_{vac}, in electron volts (eV).

28.2   Advances in QD synthesis, QD array synthesis and properties of the resulting structures

Pacifico *et al.* [9] have provided a review concerning tunable three-dimensional arrays of quantum dots emphasizing synthesis and luminescence of quantum dot arrays.

The realization of electronic and optical devices based on coupled quantum dot systems depends on an understanding of the interaction of coupled quantum dot systems; see Zhao *et al.* [10], Kato *et al.* [11], Bakkers *et al.* [12], Brumer *et al.* [13], Crooker *et al.* [14], Constantine *et al.* [15] and Jaffar *et al.* [16].

Zhao *et al.* [10] have recorded the electroluminescence and photoluminescence spectra of CdSe/ZnS core–shell quantum dots functionalized with organic ligands and incorporated into multi-layered light-emitting diodes; in this study, peak energies of the electroluminescence and photoluminescence exhibited spectra shifts of only several nanometres for diluted quantum dots and revealed no dependence on the quantum dot orientation, surface ligands, or conductive polymers.

Kato *et al.* [11] have reported new latent chemically cross-linked gel electrolyte precursors for quasi-solid dye-sensitized solar cells (QDSC). In the studies of Kato *et al.* [11], the gel electrolyte precursors consisted of nano-particles and dicarboxylic acids as the latent gelators. It was observed that the viscosity of the precursor is low at first and did not increase during storage at room temperature. When the precursor was baked at 80°C, it was observed to solidify immediately. In addition, the photovoltaic performance was found to be maintained after solidification. In addition, Bakkers *et al.* [12] have described a class of assemblies in which photoexcitable quantum dots (QDs) – as an example – are linked covalently to a metal by spacer molecules of variable length; in these studies, a series of bisulphide spacer molecules – with lenghts of 0.34 nm, 0.77 nm and 1.18 nm – are used to assemble quantum dot ensembles. Bakkers *et al.* [12] point to the robust mechanical properties of the metal/spacer/QD (CdSe) as evidence for a covalent linking of the nanocrystals to the spacer molecule by S–Cd bonds.

Crooker *et al.* [14] have reported results on the dynamics of resonant energy transfer in monodisperse, mixed-size, and energy gradient (layered) assemblies of CdSe nanocrystal quantum dots. In their studies, time-resolved and spectrally resolved photoluminescence determined the energy-dependent transfer rate of excitons from smaller to larger dots. The data show a rapid – 0.7 to 1.9 ns – energy transfer directly across a large tens of meV energy gap (i.e. between dots of disparate size), and provided evidence that interdot energy transfer can approach picosecond timescales in structurally optimized systems.

Constantine *et al.* [15] demonstrated that layer-by-layer (LBL) assembly techniques may be used to fabricate an ultrathin film of polyelectrolytes. The architecture for the LBL structures of Constantine *et al.* [15] was composed of chitosan and organophosphorus hydrolase polycations along with thioglycolic acid-capped CdSe quantum dots (QDs) as the polyanion. Constantine *et al.* [15] studied the topography of the films using epifluorescence microscopy imaging. The photoluminescence property of the functionalized QDs was found to improve when sandwiched between the polycation layers. This enhancement in the optical property of QDs facilitated the monitoring of LBL growth and detection of paraoxon with high sensitivity. The presence of organophosphorus compounds was confirmed through UV-vis and emission spectroscopies. In a related work, Jaffer *et al.* [16] have discussed the modification of ZnS-capped CdSe quantum dot (QD) surfaces with polyelectrolyte coatings and subsequent layer-by-layer deposition to build hierarchical structures.

Tang *et al.* [17] and Pacifico *et al.* [18] have discussed the synthesis of quantum dot films using self-assembled monolayer (SAM) methods. Tang *et al.* [17] studied the self-assembly of CdSe/CdS core–shell quantum dots onto thiolcarboxylic acid functionalized gold surfaces by hydrogen bonding. Tang *et al.* [17] found that control of the pH during deposition facilitated the production of a high coverage photoactive surface for use in a surface-sensitized Schottky barrier photovoltaic structure. The photoresponse obtained from the QD photovoltaic device confirmed the strong electronic coupling between the QDs and the Au surface. The hydrogen bonding strategy of Tang *et al.* [17] provides a flexible and easy way to increase the sensitizer uptake on a large-area scale. Moreover, Pacifico *et al.* [18] have reported the fabrication of two-dimensional ordered arrays of quantum dots. The ordered arrays were created using nanosphere lithography to generate metallic islands with thicknesses of about 30 nanometers, and by using bifunctional alkanes to form self-assembled monolayers of a chemical linker on the metallic island; finally core–shell quantum dots are bound to these linkers to produce two-dimensional arrays.

The use of bifunctional chemical linkers, studied as a way of providing control over the spacings between adjacent quantum dots, have been discussed by Ouyang and Awschalom [19]. Also, in the application of self-assembly techniques, Stroscio *et al.* [20] have used biomolecular peptides to chemically assemble alternating layers of CdSe-ZnS and CdS on an Au substrate; Fig. 28.3 depicts a schematic of the resulting structure. Ouyang and Awschalom [19] have demonstrated coherent spin transfer between molecularly bridged quantum dots. By using femtosecond time-resolved Faraday rotation spectroscopy, Quyang and Awschalom [19] have demonstrated the transfer of spin coherence through conjugated molecular bridges spanning quantum dots of different sizes over a broad range of temperature. In particular, the room temperature spin transfer efficiency is 20%. This work demonstrates that conjugated molecules can be used not only as interconnections for the hierarchical assembly of functional networks but also as efficient spin channels. Moreover, the results of Ouyang and Awschalom [19] suggest that such molecularly bridged QD ensembles may be useful as two-spin quantum devices operating at ambient temperatures and may offer promising opportunities for future versatile molecule-based spintronic technologies.

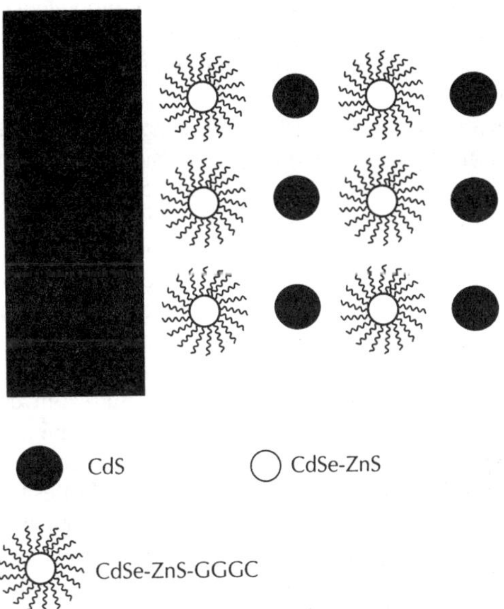

Figure 28.3   Chemically assembled colloidal quantum dots of CdS and ZnS-coated CdSe form an ensemble of dense $>10^{17}$ QDs per cm^3; a biomolecular peptide of three glycines and one cysteine, GGGC, form the linkers.

Van Embden and Mulvaney [21] have studied the competing effects of two ligands, oleic acid (OA) and bis-(2,2,4-trimethylpentyl) phosphinic acid (TMPPA), on the nucleation rate and growth of CdSe nanocrystals in octadecene; these authors found that TMPPA acts as a high boiling point "non-solvent" or "nucleating agent". Moreover, it was found that the addition of TMPPA resulted in higher initial particle yields and smaller particle diameters. On the other hand, van Embden and Mulvaney [21] found that oleic acid inhibits nucleation and results in a drastic increase in early time ripening (ETR) and, consequently, a rapid reduction in the number of particles. Van Embden and Mulvaney [21] also found that by controlling the number of nuclei formed with TMPPA and tuning the rate of ETR with oleic acid, high yields of particles could be obtained with sizes between 3 and 7 nm. Furthermore, van Embden and Mulvaney [21] observed that the absence of OA facilitated the preparation of very small nanocrystals with diameters of ~2 nm.

Williams *et al.* [22] have determined the extinction coefficients of CdTe, CdSe, and CdS nanocrystals; these authors found the extinction coefficient per mole of nanocrystals at the first exitonic absorption peak, for high-quality CdTe, CdSe, and CdS nanocrystals, to be strongly dependent on the size of the nanocrystals with a square and a cubic dependence. The measurements of Williams *et al.* [22] were carried out using either nanocrystals purified with specified purification procedures or nanocrystals prepared through controlled etching methods. Williams *et al.* [22] state that within experimental error, (a) the nature of the surface ligands, (b) the refractive index of the solvents, (c) the PL quantum yield of the nanocrystals, (d) the methods used for the synthesis of the nanocrystals, and (e) the temperature for the measurements, did not all have a detectable influence on the extinction coefficient for a given sized nanocrystal.

Burnside *et al.* [23] provided an early report on the synthesis of nanocrystalline titanium dioxide colloids using a sol-gel technique followed by growth under hydrothermal conditions at temperatures between 190 and 270°C; characterization studies by these authors revealed these nanostructures were primarily in the anatase phase and that low-temperature growth produced predominantly rod-like particles, while high-temperature growth largely produced truncated tetragonal or tetrahydral bipyramidal nanocrystallites.

Al Kuhaimi [24] has reported studies of the conduction and valence band offsets of CdS/CdTe solar cells fabricated by several different processes; the value of the conduction band offsets was found to be between 0.23 and 0.3 eV, and that of the valence band offsets was found to be between 0.67 and 0.74 eV.

In synthesis studies, Kim and Bawendi [25] have reported a new family of oligomeric phosphine ligands that electronically passivate quantum dots and form thin organic shells that facilitate potential sensor applications involving energy transfer to and from quantum dots. Moreover, Liang *et al.* [26] have fabricated ensembles of colloidal quantum dots using layer-by-layer assembly of conjugated polymers and CdSe nanoparticles; as noted by the authors, this method provides a means of preparing robust and uniform functional thin films using common organic solvents. Brumer *et al.* [13] have reported the synthesis of PbSe/PbS and PbSe/PbSe$_x$S$_{1-x}$ core–shell nanocrystals with luminescence quantum efficiencies of 45–55%.

Lazarenkova and Balandin [27] have modelled the carrier energy band structure in three-dimensional arrays of uniformly spaced semiconductor quantum dots to study the expected splitting of carrier energy levels that results in three-dimensional miniband formation; in this study, it was found, as expected, that changing the sizes of quantum dots as well as the spacings between quantum dots and the associated barrier heights facilitates the control of the electronic band structure of these artificial quantum dot crystals. Lazarenkova and Balandin [27] demonstrated that the effective-mass tensor and density-of-states for these quantum dot crystals are different from those of bulk semiconductors; these authors also demonstrated that the electronic properties of these artificial crystals are more sensitive to interdot spacings than dot geometries. By extending the techniques of Lazarenkova and Balandin [27], Balandin and Lazarenkova [28] have demonstrated a thermoelectric figure-of-merit enhancement in regimented quantum dot superlattices that can be an order of magnitude or more greater than that in bulk semiconductors depending on the Fermi energy of the system. These results provide clear evidence that the thermal and electrical, of three-dimensional quantum dot arrays may be controlled or "engineered" to tune device properties as suggested by the works of other authors including Lyanda-Geller and Leburton [29] and Dmitriev and Suris [30]. Exploiting these concepts of band engineering of three-dimensional quantum dot arrays, Vasudev *et al.* [31] and Yamanaka *et al.* [32] have modelled the transmission coefficients of carriers of TiO$_2$, CdSe and GaN quantum dots forming three-dimensional arrays in a variety of conductive-polymer matrices, for both regular and irregular interdot spacings. These results indicate that carrier transmission coefficients may be large (>50%) at selected band energies for suitably designed arrays; as an example, Fig. 28.4 depicts the electronic minibands for an array of CdSe QDs embedded in the conductive polymer PDCTh (poly(3,4-dicyanothiophene)).

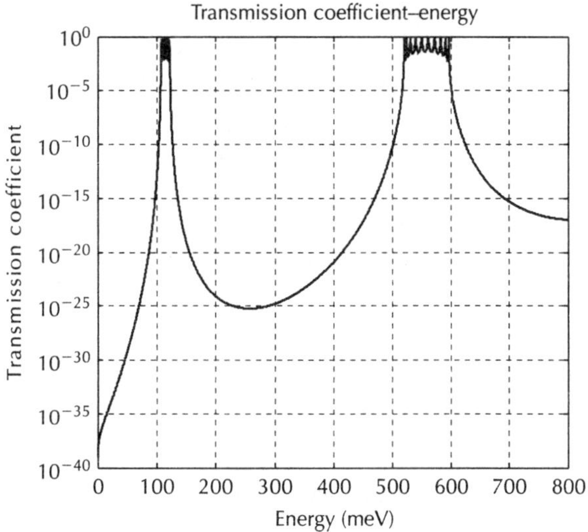

Figure 28.4   Minibands formed when 10 CdSe QDs with 3 nm diameters are placed 1 nm apart in a PDCTh matrix. The miniband energies are measured relative to the bottom of the CdSe confining potential.

28.3   Quantum dot arrays and related studies underlying photoluminescence and potential light-emitting diode applications

The integration of organic and inorganic materials on the nanoscale to form hybrid optoelectronic structures opens the way to applications in light-emitting diodes.

As an example, Coe-Sullivan *et al.* [33] have reported promising techniques for fabricating large-area ordered quantum dot monolayers via phase separation during spin casting; the areas of these layers may exceed 1 cm^2.

Coe-Sullivan *et al.* [34] have shown that a phase segregation process can be applied to the fabrication of QD-LEDs containing a wide range of CdSe particle sizes and ZnS overcoating thicknesses; in this work peak electroluminescence was tuned from 540 nm to 635 nm by varying the QD core diameter from 3.2 to 5.8 nm. Coe-Sullivan *et al.* [34] have demonstrated that for both QD-LEDs and archetypical all-organic LEDs with thin emissive layers, there is an increase in the exciton recombination region width as the drive current density is increased; these results illustrate that integration of QDs into organic LEDs has the potential to enhance the performance of thin film light emitters.

Coe-Sullivan *et al.* [35] have studied a hybrid light-emitting diode (LED) that combines the ease of processability of organic materials with the narrow-band, efficient luminescence of colloidal quantum dots (QDs); these authors have fabricated a quantum dot LED (QD-LED) that contains only a single monolayer of QDs, sandwiched between two organic thin films in order to isolate the luminescence processes from charge conduction. To fabricate these structures, Coe-Sullivan *et al.* [35] employed a method that used material phase segregation between the QD aliphatic capping groups and the aromatic organic materials; in these studies, with QDs functioning exclusively as lumophores, a 25-fold improvement in luminescence efficiency (1.6 cd A^{-1} at 2,000 cd m^{-2}) over the best previous QD-LED results was observed.

Lee *et al.* [36] have pointed out that the development of full colour emitting devices is one of the main challenges in optical displays. Lee *et al.* [36] have demonstrated nearly full colour emission using semiconductor nanocrystals, quantum dots and polymer composites. In this work, composites were fabricated by stabilizing chemically synthesized II–VI semiconductor QDs into polylaurylmethacrylate (PLMA) matrices in the presence of tri-*n*-octylphosphine (TOP). Lee *et al.* [36] found that the fluorescence of the resulting composites, optically excited by ultraviolet or blue light sources, covered the entire visible range with narrow emission profiles and high photoluminescence (PL) quantum yields; in addition, they found that mixed colours were readily produced by controlling the mixing ratio of different sized QDs. Lee *et al.* [36] noted that the specific advantages of these QDs over organic phosphors include their range of emission frequencies, their greater stability and their high density of absorbing states. In additional studies related to the photoluminescent properties of quantum dots, Anikeeva *et al.* [37] have recently reported a study of the enhancement of photoluminescence of CdSe/ZnS core–shell quantum dots by energy transfer from a phosphorescent donor; in these studies the photoluminescent dynamics was shown to be dominated by exciton diffusion within a *fac* tris(2-phenylpyridine) iridium (Ir(ppy)3) film to the QD layer. An especially promising device-enhancing technique has been considered by Lin *et al.* [38] who have employed self-organized CdS-coated PMMA (poly(methyl methacrylate)) microspheres to form a photonic crystal that provides spectrally and angularly dependent electromagnetic structural resonances; transmission and photoluminescence measurements revealed the interaction of the photonic stopband with photoluminescence from the nanocrystals.

28.4 Quantum dot arrays as potential photodetectors

Oertel *et al.* [39] have investigated photodetectors based on treated CdSe quantum dot films. The conductive polymer used in these studies was poly-3,4-ethylenedioxythiophene doped with polystyrene sulphonate (PEDOT:PSS). These authors found that under 110 mW/cm^2 illumination with light at $\lambda = 514$ nm the photocurrent to dark current ratio was 10^3 at $V = 0$ volts, and that the 3 dB frequency was ~50 kHz; the observed zero-bias external quantum efficiencies ranged from 0.08% to 0.23% in the wavelength range from $\lambda = 350$ nm to $\lambda = 575$ nm.

Kang *et al.* [40] have studied the photoluminescence (PL) characteristics of the self-assembled silica nanospheres containing coupled CdSe/ZnS core–shell quantum dots (QDs) on the surface of the sphere; these authors observed an effective band gap energy shift to lower energy via strong Förster energy transfer.

Konstantatos *et al.* [41] have reported the fabrication of solution-processed infrared photodetectors that are superior in their normalized detectivity (D^*, the Figure of merit for detector sensitivity) to the best epitaxially grown devices operating at room temperature; these devices were fabricated by overcoating a prefabricated planar electrode array with an unpatterned layer of PbS colloidal quantum dot nanocrystals.

McDonald *et al.* [42] have used a nanocomposite approach in which PbS nanocrystals tuned by the quantum size effect sensitize the conjugated polymer – poly[2-methoxy-5-(2′-ethylhexyloxy-phenylenevinylene)] (MEH-PPV) into the infrared. By varying the size of the nanocrystals during processing, McDonald *et al.* [42] were able to realize photocurrent spectra with peaks tailored to 980 nm, 1.200 μm and 1.355 μm.

28.5 Quantum dot arrays as potential solar cells

Nozik *et al.* [43] have assessed quantum dot-based solar cells and found that they have the potential to increase the maximum attainable thermodynamic conversion efficiency of solar photon conversion up to 66% exploiting hot photogenerated carriers to produce higher photovoltages or higher photocurrents.

As pointed out by Nozik [43], there is both early experimental and theoretical support that the relaxation dynamics of photogenerated carriers may be markedly affected by quantization effects in dimensionally confined semiconductors; see the works of Boudreaux *et al.* [44] and Benistry [45].

Boudreaux *et al.* [44] provided an early calculation of the energy of injected electronic minority carriers from an illuminated semiconductor into an electrolyte; these energies were predicted to be larger than predicted in earlier treatments. Phonon-bottleneck effects in quantum dots are introduced in Benistry [45]; phonon-related effects in semiconductor nanostructures have been discussed and surveyed at length by Stroscio and Dutta [46].

In early work on nanoparticle-based solar cells, Hagen *et al.* [47] have presented a three-layer concept for efficient solid-state solar cells; these devices consisted of a TiO_2 nanocrystalline layer for electron conduction, a surface-absorbed ruthenium dye complex for light absorption, and an organic triphenyldiamine layer for the transport of holes; in this work, the high charge storage capability of porous titanium dioxide layers was demonstrated by junction-recovery measurements.

Arango *et al.* [48] have investigated charge transfer in photovoltaics consisting of interpenetrating networks of conjugated polymer – poly[2-methoxy-5(2-ethyl)hexoxy-phenylene-vinylene] (MEH-PPV) – and TiO_2 nanoparticles; a two order of magnitude increase in photoconductivity and sharp saturation were observed for layered versus blended structures.

Gratzel [49] has provided an assessment and review of photoelectrochemical cells based, in part, on novel photovoltaic cells fabricated from nanocrystalline materials and conducting polymer films; structures fabricated from the components were noted to offer the prospect of inexpensive fabrication and flexibility as well as high conversion efficiencies of selected device designs. Among the classes of structures considered by Gratzel [49] are dye-sensitized solid heterojunctions, extremely thin absorber (ETA) solar cells and organic solar cells incorporating interpenetrating polymer networks. The photovoltaic devices reviewed by Gratzel [49] are based on interpenetrating mesoscopic networks and they are characterized by ultra-fast initial charge separation and much slower back-reaction that facilitates the collection of charge carriers as an electric current before recombination can occur.

O'Regan and Gratzel [50] have considered low-cost high-efficiency solar cells based on dye-sensitized colloidal TiO_2 films; the material structures considered in this work have enormous internal surface area and are attractive as a result of the need for dye-sensitized solar cells to absorb more incident light. In the studies of O'Regan and Gratzel [50] interconnected mesoporous materials were considered to allow for electronic conduction.

Kroeze *et al.* [51] have highlighted solid-state dye-sensitized solar cells employing a solid organic hole-transport material (HTM) and have pointed out that they offer a number of practical advantages over liquid-electrolyte junction devices. Kroeze *et al.* [51] have emphasized the practical importance of the control of interfacial charge transfer in the design of such devices. Kroeze *et al.* [51] have identified the factors that determine the yield of hole transfer at the dye/HTM interface and its correlation with solid-state cell performance.

Suraprapapich *et al.* [52] have studied InAs self-assembled quantum dots incorporated in an AlGaAs/GaAs heterostructure for solar cell applications and have reported potentially high efficiencies for devices based on these structures.

While the great variety of possible nanostructured solar cells enhances and expands the types of potential applications of such devices, there are some known difficulties associated with realizing efficient nanostructured solar cells. For quantum well-based solar cells, these difficulties include: (a) the localization of carriers increases the difficulty of carrier collection and has motivated carrier collection through complex means such as hot-carrier transport or miniband transport; (b) the polarization-sensitive nature of the absorption reduces absorption efficiency and necessitates the use of elaborate incident optics; and (c) the voltage drop across complex structures leads to reduced power output. For quantum dot-based nanostructured solar cells, the same difficulties arise except for that associated with the severely restrictive polarization dependence of quantum well-based devices. Nevertheless, there are ongoing efforts aimed at realizing quantum dot-based solar cells. As an example illustrating the difficulty of achieving high efficiencies, Ruangdet *et al.* [53] have studied the performance of structures based on multi-stacked high-density InAs QDs. While this system is based on self-assembled quantum dots, it does yield some insights relevant to the case of colloidal quantum dot-based devices. Ruangdet *et al.* [53] use a thin-capping-and-regrowth molecular-beam epitaxy (MBE) process to fabricate multiple layers of InAs-based QDs with areal densities of 10^{10} to 10^{12} cm^{-2}. These authors show by electrical characterization of homojunction p-n solar cells with one layer of QDs and five layers of QDs that a short-circuit current of 14.4 mA/cm^2 results for the five-layer case as compared to 9.6 mA/cm^2 for the one-layer case; however, the efficiency for the five-layer case is only 5.1%. In other studies, Marti *et al.* [54], Luque and Marti [55], and Luque and Marti [56] have investigated quantum dot-based solar cells employing intermediate bands. Luque and Marti [56] presented an analysis of a solar cell with an impurity level in the semiconductor band gap under ideal conditions. Under these ideal conditions, it was argued that an efficiency of 63.1% was possible instead of the 40.7% predicted in the Shockley and Queisser model limit, as discussed by Werner *et al.* [57]. While this analysis of an idealized system with an impurity having an energy in the semiconductor band gap predicts high efficiencies, the implementation of such a scheme through the use of quantum dots to provide the intermediate level (or band) has resulted recently [55] in the observation and conclusion that the absorption of light in a ten-layer system is low and that increasing the number of layers might lead to material defects.

28.6   Summary

This chapter has summarized a number of major trends underlying the continuing effort to realize practical optoelectronic, electronic and information-processing devices based on ensembles of quantum dots assembled in a variety of matrix materials. The great diversity of such structures has opened the possibility of numerous device applications and stimulated research underlying photoluminescent devices, light-emitting diodes, displays, photodetectors, photovoltaic devices, solar cells and novel spin-based information processing devices. It is expected that research underlying these applications will continue to thrive due to the enormous number of possible device embodiments possible with colloidal quantum dots and available matrix materials.

References

1. I. Bockelmann and G. Bastard, Phonon scattering and energy relaxation in two-, one-, and zero-dimensional electron gases, *Phys. Rev. B* **42**(14), 8947–8951 (1990).
2. T. Inoshita and H. Sakaki, Electron relaxation in a quantum dot: significance of multiphonon processes, *Phys. Rev. B* **46**(11), 7260–7263 (1992).
3. J.-P. Leburton, R.C. Fonseca, S. Nagaraja, J. Shumway, D. Ceperley, and R.M. Martin, Electronic structure and many-body effects in self-assembled quantum dots, *J. Phys. Cond. Matt.*, **11**, 5953–5967 (1999).
4. P. Bhattacharya, S. Ghosh, and A.D. Stiff-Roberts, Quantum dot optoelectronic devices, *Ann. Rev. Mat. Res.* **34**, 1–40 (2004).
5. M.S. Skolnick and D.J. Mowbray, Self-assembled semiconductor quantum dots: fundamental physics and device applications, *Annu. Rev. Mater. Res.* **34**, 181–218 (2004).
6. P.K. Bhattacharya and S. Ghosh, Tunnel injection $In_{0.4}Ga_{0.6}As$/GaAs quantum dot lasers with a 15 GHz modulation bandwidth at room temperature, *Appl. Phys. Lett.* **80**(19), 3482–3484 (2002).
7. C. Blakesley, P. See, A.J. Shields, B.E. Kardynal, P. Atkinson, I. Farrer, and D.A. Ritchie, Efficient single photon detection by quantum dot resonant tunneling diodes, *Phys. Rev. Lett.* **94**, 067401–067414 (2005).
8. A.J. Shields, M.P. O'Sullivan, I. Farrer, D.A. Ritchie, R.A. Hogg, M.L. Leadbeater, C.E. Norman, and M. Pepper, Detection of single photons using a field-effect transistor gated by a layer of quantum dots, *Appl. Phys. Lett.* **76**(25), 3673–3675 (2000).
9. J. Pacifico, J. Jasieniak, D. Gomez, and P. Mulvaney, Tunable 3D arrays of quantum dots: synthesis and luminescent properties, *Small.* **2**(2), 199–203 (2006).
10. J. Zhao, J. Zhang, C. Jiang, J. Bohnenberger, T. Basche, and A. Mews, Electroluminescence from isolated CdSe/ZnS quantum dots in multilayered light-emitting diodes, *J. Appl. Phys.* **96**, 3206–3210 (2004).
11. T. Kato, A. Okazaki, and S. Hayase, Latent gel electrolyte precursors for quasi-solid dye sensitized solar cells, *Chem. Commun.* **3**, 363–365 (2005).
12. E.P.A.M. Bakkers, A.W. Marsman, L. Jenneskens, and D. Vanmaekelbergh, Distance-dependent electron transfer in Au/spacer/Q-CdSe assemblies, *Angew. Chem.* **39**(13), 2297–2299 (2000).
13. M. Brumer, A. Kigel, L. Amirav, A. Sashchiuk, O. Solomesch, N. Tessler, and E. Lifshitz, PbSe/PbS and $PbSe/PbSe_xS_{1-x}$ core–shell nanocrystals, *Adv. Funct. Mat.* **15**, 1111–1116 (2005).

14. S.A. Crooker, J.A. Hollingsworth, S. Tretiak, and V.I. Klimov, Spectrally resolved dynamics of energy transfer in quantum-dot assemblies: toward engineered energy flows in artificial materials, *Phys. Rev. Lett.* **89**(18), 186802–186809 (2002).

15. C.A. Constantine, K.M. Gattas-Asfura, S.V. Mello, G. Crespo, V. Rastogi, T.C. Cheng, J.J. DeFrank, and R.M. Leblanc, Layer-by-layer biosensor assembly incorporating functionalized quantum dots, *Langmuir* **19**(23), 9863–9867 (2003).

16. S. Jaffar, K.T. Nam, A. Khademhosseini, J. Xing, R.S. Langer, and A.M. Belcher, Layer-by-layer surface modification and patterned electrostatic deposition of quantum dots, *Nano Lett.* **4**, 1421–1425 (2004).

17. J. Tang, H. Birkeldal, E.W. McFarland, and G.D. Stucky, Self-assembly of CdSe/CdS quantum dots by hydrogen bonding on Au surfaces for photoreception, *Chem. Commun.* **18**, 2278–2279 (2003).

18. J. Pacifico, D. Gomez, and P. Mulvaney, A simple route to tunable two-dimensional arrays of quantum dots, *Adv. Mater.* **17**(4), 415–419 (2005).

19. M. Ouyang and D.D. Awschalom, Coherent spin transfer between molecularly bridged quantum dots, *Science.* **301**, 1074–1078 (2003).

20. M.A. Stroscio, M. Dutta, D. Ramadurai, P. Shi, Y. Li, M. Vasudev, D. Alexson, B. Kohanpour, A. Sethuraman, V. Saini, A. Raichura, and J. Yang, Optical and electrical properties of colloidal quantum dots in electrolytic environments: using biomolecular links in chemically-directed assembly of quantum dot networks, *J. Comp. Electron.* **4**, 21–25 (2005).

21. J. van Embden and P. Mulvaney, Nucleation and growth of CdSe nanocrystals in a binary ligand system, *Langmuir* **21**(22), 10226–10233 (2005).

22. W.W. Yu, L. Qu, W. Guo, and X. Peng, Experimental determination of the extinction coefficient of CdTe, CdSe, and CdS nanocrystals, *Chemistry of Materials.* **15**(14), 2854–2860 (2003).

23. S.D. Burnside, V. Shklover, C. Barbé, P. Comte, F. Arendse, K. Brooks, and M. Grätze, Self-organization of TiO$_2$ nanoparticles in thin films, *Chemistry of Materials.* **10**, 2419–2425 (1998).

24. S.A. Al Kuhaimi, Conduction and valence band offsets of CdS/CdTe solar cells, *Energy.* **25**(8), 731–739 (2000).

25. S. Kim and M.G. Bawendi, Oligomeric ligands for luminescent and state nanocrystal quantum dots, *J. Amer. Chem. Soc.* **125**, 14652–14653 (2003).

26. Z. Liang, K.L. Dzienis, J. Xu, and Q. Wang, Covalent layer-by-layer assembly of conjugated polymers and CdSe nanoparticles: multilayer structure and photovoltaic properties, *Adv. Funct. Mat.* **16**, 542–548 (2006).

27. O.L. Lazarenkova and A.A. Balandin, Miniband formation in a quantum dot crystal, *J. Appl. Phys.* **89**(10), 5509–5515 (2001).

28. A.A. Balandin and O.L. Lazarenkova, Mechanism for thermoelectric figure-of-merit enhancement in regimented quantum dot superlattices, *Appl. Phys. Lett.* **82**(3), 415–417 (2003).

29. Y.B. Lyanda-Geller and J.-P. Leburton, Resonant tunnelling through arrays of nanostructures, *Semiconductor Science and Technology.* **13**, 35–42 (1998).

30. I.A. Dmitriev and R.A. Suris, Electron localization and Bloch oscillations in quantum-dot superlattices under a constant electric field, *Low-Dimensional Systems.* **35**(2), 219–226 (2001).

31. M. Vasudev, T. Yamanaka, K. Sun, Y. Li, J. Yang, D. Ramadurai, M.A. Stroscio, and M. Dutta, Colloidal quantum dots as optoelectronic elements, in: *Quantum Sensing and Nanophotonic Devices IV*, ed. by M. Razeghi and G.J. Brown, SPIE Vol. 6479, 64790I-1-12 (2007).

32. T. Yamanaka, K. Sun, Y. Li, M. Dutta, and M.A. Stroscio, Spontaneous polarizations, electrical properties, and phononic properties of GaN nanostructures and systems, in: *GaN Materials and Devices II*, ed. by H. Morkoc and C. W. Litton, SPIE Vol. 6473, 64730F-1-14 (2007).

33. S. Coe-Sullivan, J.S. Steckel, W.-R. Woo, M.G. Bawendi, and V. Bulovic, Large-area ordered quantum-dot monolayers via phase separation during spin-casting, *Adv. Funct. Mat.* **15**(7), 1117–1124 (2005).

34. S. Coe-Sullivan, W.-R. Woo, J.S. Steckel, M. Bawendi, and V. Bulovic, Tuning the performance of hybrid organic/inorganic quantum dot light-emitting devices, *Org. Electron.* **4**, 123–130 (2003).

35. S. Coe-Sullivan, W.-K. Woo, M. Bawendi, and V. Bulovic, Electroluminescence from single monolayers of nanocrystals in molecular organic devices, *Nature* **420**, 800–803 (Letter to Nature) (2002).

36. J. Lee, V.C. Sundar, J.R. Heine, M.G. Bawendi, and K.F. Jensen, Full color emission from II–VI semiconductor quantum dot-polymer composites, *Adv. Mat.* **12**(15), 1102–1105 (2000).

37. P.O. Anikeeva, C.F. Madigan, S.A. Coe-Sullivan, J.A. Steckel, M.G. Bawendi, and V. Bulovic, Photoluminescence of CdSe/ZnS core/shell quantum dots enhanced by energy transfer from a phosphorescent donor, *Chem. Phys. Lett.* **424**, 120–125 (2006).

38. Y. Lin, J. Zhang, E.H. Sargent, and E. Kumacheva, Photonic pseudo-gap-based modification of photoluminescence from CdS nanocrystal satellites around polymer microspheres in a photonic crystal, *Appl. Phys. Lett.* **81**, 3134–3136 (2002).

39. D.C. Oertel, M.G. Bawendi, A.C. Arango, and V. Bulović, Photodetectors based on treated CdSe quantum-dot films, *Appl. Phys. Lett.* **87**, 213505-1-3 (2005).

40. K.S. Kang, H.L. Ju, W.H. Han, J.H. Lee, J.G. Choi, and D.W. Boo, Photoluminescence characteristics of coupled CdSe/ZnS quantum dots on self-assembled silica nanospheres, *Appl. Phys. Lett.* **87**, 141909–141911 (2005).

41. G. Konstantatos, I. Howard, A. Fischer, S. Hoogland, J. Clifford, E. Klem, L. Levina, and E.H. Sargent, Utrasensitive solution-cast quantum dot photodetectors, *Nature* **442**, 180–183 (Letters to Editor) (2006).

42. S. McDonald, G. Konstantatos, S. Zhang, P.W. Cyr, E.J.D. Klem, L. Levina, and E.H. Sargent, Solution-processed PbS quantum dot infrared photodetectors and photovoltaics, *Nature Mat.* **4**, 138–142 (Letter to Editor) (2005).

43. A.J. Nozik, Quantum dot solar cells, *Physica E*, **14**, 115–120 (2002).

44. D.S. Boudreaux, F. Williams, and A.J. Nozik, Hot carrier injection at semiconductor–electrolyte junctions, *J. Appl. Phys.* **51**, 2158–2163 (1980).

45. H. Benistry, Reduced electron-phonon relaxation rates in quantum box systems: theoretical analysis, *Phys. Rev. B* **51**, 13281–13293 (1995).

46. M.A. Stroscio and M. Dutta. *Phonons in Nanostructures*, (Cambridge University Press, Cambridge, 2001.

47. J. Hagen, W. Schaffrath, P. Otschik, R. Fink, A. Bacher, H. Schmidt, and D. Haarer, Novel hybrid solar cells consisting of inorganic nanoparticles and an organic hole transport material, *Synth. Metals* **89**, 215–220 (1997).

48. A.C. Arango, S.A. Carter, and P.J. Brock, Charge transfer in photovoltaics consisting of interpenetrating networks of conjugated polymer and TiO2 nanoparticles, *Appl. Phys. Lett.* **74**(12), 1698–1700 (1999).

49. M. Grätzel, Photoelectrochemical cells, *Nature* **414**, 338–344 (2001).

50. B. O'Regan and M. Gratzel, A low-cost, high-efficiency solar cell based on dye-sensitized colloidal TiO$_2$ films, *Nature* **353**, 737–740 (1991).

51. E. Kroeze, N. Hirata, L. Schmidt-Mende, C. Orizu, S.D. Ogier, K. Carr, M. Gratzel, and J.R. Durrant, Parameters influencing charge separation in solid-state dye-sensitized solar cells using novel hole conductors, *Adv. Funct. Mat.* **16**, 1832–1838 (2006).

52. S. Suraprapapich, S. Thainoi, S. Kanjanachuchai, and S. Panyakeow, Quantum dot integration in heterostructure solar cells, *Solar Energy Materials & Solar Cells* **90**, 2968–2974 (2006).

53. S. Ruangdet, S. Thainoi, S. Kanajnachuchai, and S. Panyakeow, Improvement of PV performance by using multi-stacked high density InAs quantum dot molecules, *IEEE Proceedings of the 4th World Conference on Photovoltaic Energy Conversion*, 225–228, May 7–12, Waikoloa, Hawaii; 1-4244-0016-3 (2006).

54. A. Marti, L. Cuadra, and A. Luque, Partial filling of a quantum dot intermediate band for solar cells, *IEE Transactions on Electron Devices* **48**(10), 2394–2399 (2001).

55. A. Luque and A. Marti, Recent progress in intermediate band solar cells, *IEEE Proceedings of the 4th World Conference on Photovoltaic Energy Conversion*, 49–52, May 7–12, Waikoloa, Hawaii; 1-4244-0016-3 (2006).

56. A. Luque, A. Marti Increasing the efficiency of ideal solar cells by photon induced transitions at intermediate levels,, *Phys. Rev. Lett.* **78**(26), 5014–5017 (1997).

57. J.H. Werner, S. Kodolinski, and H.J. Queisser, Novel optimization principles and efficiency limits for semiconductor solar cells, *Phys. Rev. Lett.* **72**, 3851–3854 (1994)

Index